Michał Kleiber (Ed.)

Handbook of Computational Solid Mechanics

Springer
Berlin
Heidelberg
New York
Barcelona
Budapest
Hong Kong
London
Milano
Paris
Singapore
Tokyo

Michał Kleiber (Ed.)

Handbook of Computational Solid Mechanics

Survey and Comparison of Contemporary Methods

With 183 Figures

Springer

Professor Michał Kleiber
ul.Krzywickiego
02-078 Warszawa
Poland

ISBN 3-540-62862-2 Springer-Verlag Berlin Heidelberg New York

Library of Congress Cataloging-in-Publication Data applied for

Handbook of Computational Solid Mechanics: Survey and Comparison of Contemporary Methods / Ed.: Michał Kleiber. - Berlin; Heidelberg; New York; Barcelona; Budapest; Hong Kong; London; Milano; Paris; Singapore; Tokyo: Springer, 1998
ISBN 3-540-62862-2

Typesetting: Camera-ready by authors
Cover-Design: de'blik, Berlin
SPIN 10569527 62/3020-5 4 3 2 1 0 - Printed on acid -free paper

Preface

The volume scope encompasses the whole variety of issues related to the influence of modern computational techniques upon ways to formulate and solve boundary-value problems typical of contemporary solid and structural mechanics. Until quite recently a prevailing opinion was that the research area by the name of Computational Mechanics did not require any separate treatment – it was widely believed that in order to effectively solve any problem, however difficult, the work of the researcher or engineer using conventional tools of continuum mechanics had to be possibly supplemented only by the work of some independently working numerical analysts and computer programmers. Such a viewpoint is today considered by many as incorrect at least for the following reasons:

- to assure the maximum efficiency of the modelling and solution process, the capabilities offered by the available computational techniques must be accounted for already at the very beginning of the analysis,
- specific needs of solid mechanics researchers and engineers concerning reliable and well established results in the area of approximation theory, numerical analysis and programming techniques very often go far beyond all what the researchers in these disciplines can offer – the autonomous research must therefore be carried out to provide results of crucial significance for the advancement of mechanics.

Even in such a brief introduction yet another aspect of dealing in this book specifically with computational mechanics issues should certainly be mentioned. For regardless the current opinions as to the autonomy of this field the role to be played by computer assisted techniques in any engineering discipline in the future can hardly be overstated. A few decades of using the computational methods in mechanics do not suffice to fully realize their real potential – one can safely predict their ever increasing role in the years to come, though.

The way this volume is structured corresponds to specific computational mechanics methodologies rather than to classes of problems typically encountered in solid and structural mechanics. In other words, it will probably be easier for the reader to get acquainted with the current status of specific methods (finite elements, boundary elements, finite differences) than to find an answer as to which of these techniques should best be used for solving the mechanical problem on hand. There are in the book some exceptions to the above rule – Parts V and VI are devoted to structural optimization and design sensitivity problems, respectively, with the emphasis put on the formulation of the problem rather than on using any specific computational technique.

Computational mechanics is clearly a transdisciplinary research area. In particular, it can be considered as:

- a typical engineering subject with the emphasis placed on clear mechanical interpretation of the methodology,
- a branch of computational science with the efficiency of numerical algorithms and computer implementation lying at its core,
- a branch of mathematics in which approximation theory and its specific properties like convergence and accuracy play the primary role.

Being fully aware of the significance of all the above three faces of computational mechanics to its concordant further development, this book treats the subject as the engineering discipline, the remaining two aspects invoked only casually to broaden the reader's perspective. Such an approach to the subject forjudges its readership – it is researchers and graduate students at mechanical, aerospace and civil engineering departments of universities who may benefit the most from studying the book.

No computational approach can be displayed in a fully satisfactory manner without taking reference to the underlying computer implementation of the theory. In a book of this generality the inclusion of computer programs, or even their detailed descriptions, would go far beyond its intended scope. However, the reader should be aware that most of the developments described in the book have been based on extensive authors' programming experience – the reader may find the relevant information in the literature listed at the end of each Part.

The current English version of the book is based on its Polish edition published in 1995. Apart of some minor improvements and reference updating the text is a direct translation of the Polish original – a notable exception is Part III which has been entirely reworked by its author.

March 1998 Michał Kleiber

Contents

Authors:

A. Borkowski – Part V, Chap. 2
T. Burczyński – Part IV
L. Demkowicz – Part II, Chap. 2
K. Dems – Part V, Chap. 1; Part VI
J. Grabacki – Part IV
W. Gutkowski – Part V, Chaps. 3, 4
Z. Kączkowski – Part II, Chap. 5
M. Kleiber – Part I; Part II, Chaps. 1, 3, 6
J. Kruszewski – Part II, Chap. 4
Z. Mróz – Part V, Chap. 1; Part VI
J. Orkisz – Part III
Z. Waszczyszyn – Part II, Chap. 7

Part I

General Introduction

1 On Solving Problems of Mechanics by Computer Methods

Before we begin the discussion of specific ways to formulate and solve solid mechanics problems it is worthwhile to consider some very general features of computer assisted solution methodologies. This should facilitate a better understanding of the place occupied by computational mechanics in the whole area of contemporary mechanics of deformable bodies.

Computer assisted treatment of solid mechanics problems consists basically of the following steps:

- Modelling of the real system. This corresponds to an idealization of the real system on hand towards accounting for only those its features which are likely to be important in the analysis to follow. The physical model so obtained is usually expressed in terms of some PDE's.
- Discrete representation of the physical model. Because of the nature of digital computers (which handle algebraic, or discrete problems better than those of functional analysis, or continuum ones) the PDE's which characterize the physical model are typically subject to discretization. The discretization may be complete, resulting in a discrete model represented by means of some system of algebraic equations, or only partial (so-called semi-discrete) which most often leads to a system of ODE's with respect to time. The most common discretization procedures are offered by the finite element method, boundary element method and finite difference method as discussed in detail later in the book. It should be pointed out that some real systems do not require any generation of physical models since a discrete model can be constructed directly from the real system considerations.
- Numerical modelling of the discrete system. This step corresponds to the selection of numerical algorithms for solving the equations describing the discrete model. This choice may for specific applications quite significantly influence the efficiency of the solution procedure. What is generally required at this stage is the analyst's familiarity with modern numerical algorithms of service to efficient solving of large systems of linear and nonlinear algebraic equations, algebraic eigenproblems, ordinary differential equations, linear and nonlinear mathematical programming problems, etc.
- Computer programming. The computer implementation step results in a code written in one of the available computer languages (Fortran has been so far the most popular for scientific computations) – the code is

just another version of the system model constructed to fit the existing hardware.

- Generation of input data in a pre-set format.
- Carrying out computations.
- Verification of the models used and interpretation of the results.

We emphasize again that the effectiveness of the overall solution process depends very much on knowing all the essential ingredients of the process already at the stage of physical modelling. An example is provided by the necessity of making a deliberate choice as to a particular shell theory (for a 'shell-like' real problem) in a way which is a trade-off between the accuracy required by the application and the capabilities of software and hardware on hand.

The main goal of the book is the discussion of ways to construct discrete models for different physical models of deformable bodies and presentation of numerous illustrative examples. Quite often details of numerical modelling will also be given; they frequently turn out absolutely essential for computational effectiveness of the discrete models derived.

2 Basic Equations of Nonlinear Solid Mechanics

2.1 Introductory Comments

The starting point for any discussion of computational mechanics issues must clearly be a review of all the relevant equations describing the typical initial–boundary value problem of solid mechanics. There are literally tens of instructive monographs and textbooks on the subject; those listed as [3–6,8–10,12–17] in the bibliography can serve as a reliable source of information very much in spirit of the present volume.

The aim of this section is a concise review of the nonlinear mechanics equations needed later in the book. The emphasis is put on uniformity of the formulation and its usefulness in dealing with computational aspects. Discussing particular groups of the equations efforts will be made to provide the reader with transparent physical interpretations of the quantities considered and their mutual relationships. Any attempt to give a complete account of all the mathematical aspects of nonlinear continuum mechanics will be intentionally avoided. This appears fully warranted in view of the very nature of the book aimed at bringing the derivations down to the level of specific numerical applications.

For the sake of transparency of this chapter all the quantities are referred to a single, fixed Cartesian coordinate system x_k, $k = 1, 2, 3$. Summation convention applies to twice repeated indices.

2.2 Description of Strain

Let us assume that the motion of a deformable body $\mathcal{B}$ subject to some changing external load is considered in the time interval $[0, \bar{t}]$ We shall focus our attention on the part of the deformation process corresponding to changes in the body configuration C^τ, $\tau \in [0, \bar{t}]$ ranging from a fixed, typical time instant $\tau = t$ to a 'subsequent' time instant $\tau = t + \Delta t$, Fig. 2.1. (Symbols t, $\bar{t}$, ... etc. stand for fixed time instants along the time axis τ). As we shall see soon, to describe the deformation process in the interval $[t, t + \Delta t]$ (in fact, in any time interval) a fixed reference state C^r must be selected. Basically, any known real or fictitious configuration of the body can be employed to this purpose – it by no means implies that all the choices lead to comparable

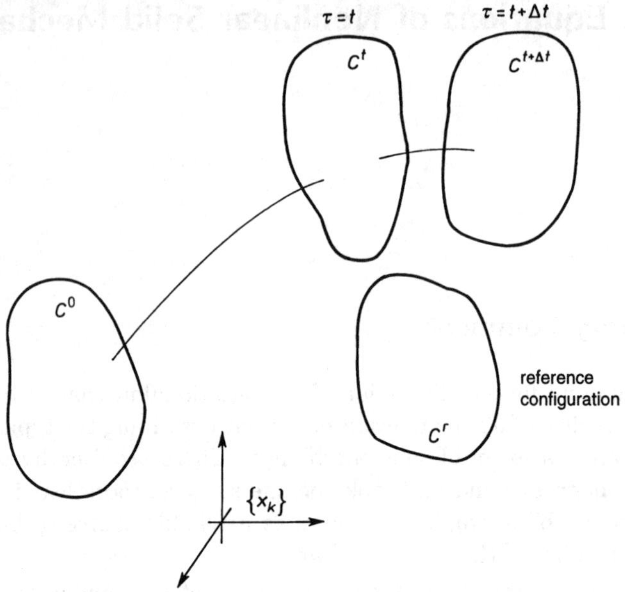

Fig. 2.1. Configurations of the body in the deformation process

computational efficiency of the corresponding solution procedures.

Because the time instant t and the time increment Δt are selected arbitrarily, the solution methodology to be worked out for the interval $[t, t+\Delta t]$ makes it possible to obtain the solution in the whole interval $[0, \bar{t}]$ just by proceeding step-wise from 0 to $\bar{t}$.

Let the symbol $\mathbf{x}^\tau = \{x_k^\tau\}$ stand for the location of the particle P in the coordinate system x_k at time τ. The motion of P may be described by means of a smooth mapping

$$\mathbf{x}^\tau = \mathbf{x}^\tau(\mathbf{x}^0, \tau), \qquad x_k^\tau = x_k^\tau(x_l^0, \tau).$$

The total displacement of P from the beginning of the deformation process ($\tau = 0$) to the current time instant $\tau = t$ is denoted by the symbol ${}^0\mathbf{u}^t = \{{}^0u_k^t\}$ or, shorter, by the symbol $\mathbf{u}^t = \{u_k^t\}$ i.e.

$$\mathbf{u}^t = \mathbf{x}^t - \mathbf{x}^0, \qquad u_k^t = x_k^t - x_k^0.$$

The displacement increment, or the displacement from $\tau = t$ to $\tau = t + \Delta t$, is denoted as $\Delta\mathbf{u}^t = \{\Delta u_k^t\}$ i.e.

$$\Delta\mathbf{u}^t = \mathbf{x}^{t+\Delta t} - \mathbf{x}^t, \qquad \Delta u_k^t = x_k^{t+\Delta t} - x_k^t.$$

Whenever no confusion is likely to arise the symbols $\mathbf{u}$ and $\Delta\mathbf{u}$ will be used to replace $\mathbf{u}^t$ and $\Delta\mathbf{u}^t$, respectively.

The gradient of deformation $C^0 \to C^\tau$ is defined as

$$ {}^0\mathbf{F}^\tau = \frac{\partial \mathbf{x}^\tau(\mathbf{x}^0, \tau)}{\partial \mathbf{x}^0}, \qquad {}^0F^\tau_{kl} = \frac{\partial x^\tau_k(x^0_l, \tau)}{\partial x^0_l}, $$

with the index '0' frequently dropped for simplicity, i.e. using the notation ${}^0\mathbf{F}^\tau = \mathbf{F}^\tau$.

Since the reference configuration C^r for the deformation process $C^t \to C^{t+\Delta t}$ is known by assumption, the mappings

$$ \mathbf{x}^r = \mathbf{x}^r(\mathbf{x}^0), \qquad \mathbf{x}^t = \mathbf{x}^t(\mathbf{x}^r) $$

deformation gradients

$$ {}^0\mathbf{F}^r = \frac{\partial x^r}{\partial \mathbf{x}^0}, \qquad {}^r\mathbf{F}^t = \frac{\partial x^t}{\partial \mathbf{x}^r} $$

and displacement vectors

$$ {}^0\mathbf{u}^r = \mathbf{x}^r - \mathbf{x}^0, \qquad {}^r\mathbf{u}^t = \mathbf{x}^t - \mathbf{x}^r $$

can all be considered known as well.

The (total) strain at $\tau = t$ corresponding to the deformation process $C^0 \to C^t$ is defined by

$$ {}^0_r\varepsilon^t = \frac{1}{2}\left[{}^r\overset{T}{\mathbf{F}^t}\ {}^r\mathbf{F}^t - \left({}^0\mathbf{F}^r\ {}^0\overset{T}{\mathbf{F}^r} \right)^{-1} \right]. \tag{2.1} $$

It is said that ${}^0_r\varepsilon^t$ describes the strain at $\tau = t$ from $\tau = 0$ to $\tau = t$ measured with respect to the real or fictitious reference state C^r. The strain ${}^0_r\varepsilon^t$ may be expressed as a function of the displacement components ${}^0\mathbf{u}^t$ and ${}^r\mathbf{u}^t$ as

$$ {}^0_r\varepsilon^t_{ij} = \frac{1}{2}\left[\frac{\partial\, {}^0u^t_i}{\partial x^r_j} + \frac{\partial\, {}^0u^t_j}{\partial x^r_i} + \frac{\partial\, {}^ru^t_k}{\partial x^r_i}\frac{\partial\, {}^0u^t_k}{\partial x^r_j} + \frac{\partial\, {}^0u^t_k}{\partial x^r_i}\frac{\partial\, {}^ru^t_k}{\partial x^r_j} - \frac{\partial\, {}^0u^t_k}{\partial x^r_i}\frac{\partial\, {}^0u^t_k}{\partial x^r_j} \right], \tag{2.2} $$

where

$$ {}^0u^t_i = {}^0u^t_i(x^r_j), \qquad {}^ru^t_i = {}^ru^t_i(x^r_j). $$

Before we illustrate by a simple example the real significance of the state C^r in the definition (2.2) let us observe that for the so-called total Lagrangian (T.L.) description for which we take $C^r = C^0$ the equations (2.1), (2.2) yield

$$ \begin{aligned} {}^0_0\varepsilon^t &= \frac{1}{2}\left[{}^0\overset{T}{\mathbf{F}^t}\ {}^0\mathbf{F}^t - \mathbf{1} \right] = \tfrac{1}{2}[\mathbf{C} - \mathbf{1}], \\ {}^0_0\varepsilon^t_{ij} &= \frac{1}{2}\left[\frac{\partial u^t_i}{\partial x^0_j} + \frac{\partial u^t_j}{\partial x^0_i} + \frac{\partial u^t_k}{\partial x^0_i}\frac{\partial u^t_k}{\partial x^0_j} \right], \end{aligned} \tag{2.3} $$

whereas for the so-called updated Lagrangian (U.L.) description for which $C^r = C^t$ we obtain

$$
\begin{aligned}
{}^0_t\boldsymbol{\varepsilon}^t &= \tfrac{1}{2}\left[\mathbf{1} - \left({}^0\overset{T}{\mathbf{F}^t}\,{}^0\mathbf{F}^t\right)^{-1}\right] = \tfrac{1}{2}[\mathbf{1} - \mathbf{B}^{-1}], \\
{}^0_t\varepsilon^t_{ij} &= \frac{1}{2}\left[\frac{\partial u^t_i}{\partial x^t_j} + \frac{\partial u^t_j}{\partial x^t_i} - \frac{\partial u^t_k}{\partial x^t_i}\frac{\partial u^t_k}{\partial x^t_j}\right].
\end{aligned}
\tag{2.4}
$$

The quantities $\mathbf{C}$ and $\mathbf{B}$ defined by

$$
\mathbf{C} = {}^0\overset{T}{\mathbf{F}^t}\,{}^0\mathbf{F}^t, \qquad \mathbf{B} = {}^0\mathbf{F}^t\,{}^0\overset{T}{\mathbf{F}^t}.
$$

are known in the continuum mechanics literature as the right and left Cauchy–Green deformation tensors, [15, 16]. It is easily recognized that ${}^0_0\boldsymbol{\varepsilon}^t$ coincides with the Green strain tensor typically denoted by $\mathbf{E}$ whereas ${}^0_t\boldsymbol{\varepsilon}^t$ corresponds to the so-called Almansi strain tensor typically denoted by $\mathbf{A}$ [6].

In order to illustrate the significance of Eq. (2.2) let us consider the configuration C^0 with two closely located points $\mathbf{x}^0$ and $\mathbf{x}^0 + \mathrm{d}\mathbf{x}^0$; the distance between them is given as $\mathrm{d}s^0 = \sqrt{\mathrm{d}\overset{T}{\mathbf{x}^0}\,\mathrm{d}\mathbf{x}^0}$. Assume that the length of the linear element in question is given in the configurations C^t and C^r as

$$
\mathrm{d}s^t = \sqrt{\mathrm{d}\overset{T}{\mathbf{x}^t}\,\mathrm{d}\mathbf{x}^t}, \qquad \mathrm{d}s^r = \sqrt{\mathrm{d}\overset{T}{\mathbf{x}^r}\,\mathrm{d}\mathbf{x}^r}.
$$

respectively. Clearly, the linear element $\mathrm{d}\mathbf{x}^0$ in C^0 becomes generally curvilinear in any other configuration C^τ – the symbol $\mathrm{d}\mathbf{x}^t$ corresponds to the chord connecting the points $\mathbf{x}^t$ and $\mathbf{x}^t + \mathrm{d}\mathbf{x}^t$ lying on the curve.

The basic property of the strain measure ${}^0_r\boldsymbol{\varepsilon}^t$ may best be seen by considering the relationship

$$
\tfrac{1}{2}[(\mathrm{d}s^t)^2 - (\mathrm{d}s^0)^2] = \mathrm{d}\overset{T}{\mathbf{x}^r}\,{}^0_r\boldsymbol{\varepsilon}^t\,\mathrm{d}\mathbf{x}^r = {}^0_r\varepsilon^t_{kl}\,\mathrm{d}x^r_k\,\mathrm{d}x^r_l\,,
$$

which in the uniaxial case of a bar subject to tension, cf. Fig. 2.2, becomes

$$
{}^0_r\varepsilon^t = \frac{1}{2}\frac{(l^t)^2 - (l^0)^2}{(l^r)^2}.
$$

A closer look at either of the two last equations confirms our earlier statement that ${}^0_r\boldsymbol{\varepsilon}^t$ represents the strain corresponding to the deformation $C^0 \to C^t$ taken with respect to the configuration C^r.

Let us further adopt the following compact notation for time increment of any function $\mathbf{a}(\mathbf{x}^r, \tau)$

$$
\Delta\mathbf{a}(\mathbf{x}^r, t) = \mathbf{a}(\mathbf{x}^r, t + \Delta t) - \mathbf{a}(\mathbf{x}^r, t).
$$

The incremental strain $\Delta\,{}^0_r\boldsymbol{\varepsilon}^t$ is defined in such a way that

$$
\Delta\left\{\tfrac{1}{2}\left[(\mathrm{d}s^t)^2 - (\mathrm{d}s^0)^2\right]\right\} = \mathrm{d}\overset{T}{\mathbf{x}^r}\,\Delta\,{}^0_r\boldsymbol{\varepsilon}^t\,\mathrm{d}\mathbf{x}^r = \Delta\,{}^0_r\varepsilon^t_{kl}\,\mathrm{d}x^r_k\,\mathrm{d}x^r_l\,, \tag{2.5}
$$

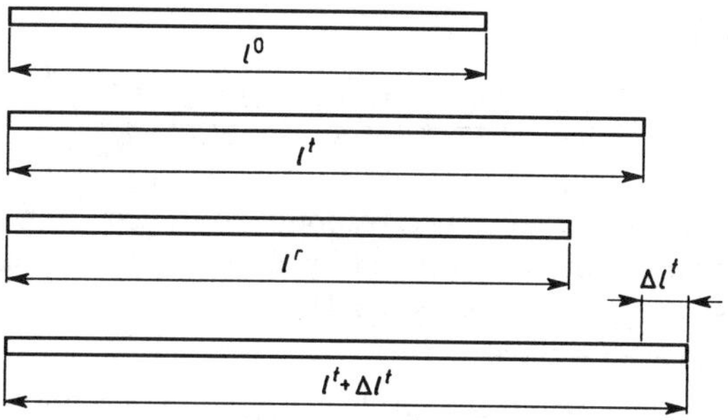

Fig. 2.2. Uniaxial tension of a bar

It can be expressed in terms of the components of the displacement vectors ${}^0\mathbf{u}^t$, ${}^r\mathbf{u}^t$ and $\Delta\mathbf{u}^t$ as, cf. [6]

$$\Delta\,{}^0_r\varepsilon^t_{ij} = \frac{1}{2}\left[\frac{\partial\,\Delta u^t_i}{\partial x^r_j} + \frac{\partial\,\Delta u^t_j}{\partial x^r_i} + \frac{\partial\,{}^r u^t_k}{\partial x^r_i}\frac{\partial\,\Delta u^t_k}{\partial x^r_j} + \frac{\partial\,\Delta u^t_k}{\partial x^r_i}\frac{\partial\,{}^r u^t_k}{\partial x^r_j} + \frac{\partial\,\Delta u^t_k}{\partial x^r_i}\frac{\partial\,\Delta u^t_k}{\partial x^r_j}\right]. \tag{2.6}$$

It is seen that the expression on the right-hand side in Eq. (2.6) is independent of C^0 which allows to replace the symbol for the incremental strain $\Delta\,{}^0_r\varepsilon^t_{ij}$ by a similar one $\Delta\,{}_r\varepsilon^t_{ij}$. We denote

$$\Delta\,{}_r\boldsymbol{\varepsilon}^t = \Delta\,{}_r\bar{\boldsymbol{\varepsilon}}^t + \Delta\,{}_r\bar{\bar{\boldsymbol{\varepsilon}}}^t,$$

where

$$\Delta\,{}_r\bar{\varepsilon}^t_{ij} = \frac{1}{2}\left[\frac{\partial\,\Delta u^t_i}{\partial x^r_j} + \frac{\partial\,\Delta u^t_j}{\partial x^r_i} + \frac{\partial\,{}^r u^t_k}{\partial x^r_i}\frac{\partial\,\Delta u^t_k}{\partial x^r_j} + \frac{\partial\,\Delta u^t_k}{\partial x^r_i}\frac{\partial\,{}^r u^t_k}{\partial x^r_j}\right] \tag{2.7}$$

is the linear function of the incremental displacement while

$$\Delta\,{}_r\bar{\bar{\varepsilon}}^t_{ij} = \frac{1}{2}\frac{\partial\,\Delta u^t_k}{\partial x^r_i}\frac{\partial\,\Delta u^t_k}{\partial x^r_j} \tag{2.8}$$

is its quadratic function. It is emphasized that $\Delta\,{}_r\bar{\bar{\boldsymbol{\varepsilon}}}^t$ is not a quadratic approximation to the incremental strain but its exact value.

For the T.L. description ($C^r = C^0$) the expression for the incremental strain (2.6) becomes

$$\Delta\,{}_0\varepsilon^t_{ij} = \frac{1}{2}\left[\frac{\partial\,\Delta u^t_i}{\partial x^0_j} + \frac{\partial\,\Delta u^t_j}{\partial x^0_i} + \frac{\partial u^t_k}{\partial x^0_i}\frac{\partial\,\Delta u^t_k}{\partial x^0_j} + \frac{\partial\,\Delta u^t_k}{\partial x^0_i}\frac{\partial u^t_k}{\partial x^0_j} + \frac{\partial\,\Delta u^t_k}{\partial x^0_i}\frac{\partial\,\Delta u^t_k}{\partial x^0_j}\right],$$

which for infinitesimal increments corresponds to the material time derivative of the Green strain $\mathbf{E}$ multiplied by Δt. For the U.L. description ($C^r = C^t$)

Eq. (2.6) becomes

$$\Delta\,{}_t\varepsilon^t_{ij} = \frac{1}{2}\left[\frac{\partial\,\Delta u^t_i}{\partial x^t_j} + \frac{\partial\,\Delta u^t_j}{\partial x^t_i} + \frac{\partial\,\Delta u^t_k}{\partial x^t_i}\frac{\partial\,\Delta u^t_k}{\partial x^t_j}\right],$$

which for infinitesimal increments corresponds to the convective time derivative of the Almansi strain $\mathbf{A}$ multiplied by Δt (and it is equal to the deformation rate tensor typically denoted by $\mathbf{D}$).

In the uniaxial case we have

$$\Delta\,{}_r\varepsilon^t = \frac{1}{2}\frac{2l^t\Delta l^t + (\Delta l^t)^2}{(l^r)^2} = \frac{l^t\Delta l^t}{(l^r)^2} + \frac{1}{2}\left(\frac{\Delta l^t}{l^r}\right)^2,$$

$$\Delta\,{}_r\bar{\varepsilon}^t = \frac{l^t\Delta l^t}{(l^r)^2}, \qquad \Delta\,{}_r\bar{\bar{\varepsilon}}^t = \frac{1}{2}\left(\frac{\Delta l^t}{l^r}\right)^2,$$

$$\Delta\,{}_t\varepsilon^t = \frac{\Delta l^t}{l^t} + \frac{1}{2}\left(\frac{\Delta l^t}{l^t}\right)^2.$$

2.3 State of Stress

Let us consider for any point P of the body $\mathcal{B}$ in the configuration C^t and any cross-sectional plane passing through P a surface element $\mathrm{d}\omega$ with the unit normal $\mathbf{n}^t$, Fig. 2.3. Denoting the force transmitted through the surface $\mathrm{d}\omega$ by $\mathrm{d}\mathbf{f}^t$ we define the stress vector $\mathbf{t}^t$ as

$$\mathbf{t}^t = \frac{\mathrm{d}\mathbf{f}^t}{\mathrm{d}\omega^t} \tag{2.9}$$

and the Cauchy stress tensor $\boldsymbol{\sigma}^t$ as

$$\mathbf{t}^t = \boldsymbol{\sigma}^t\,\mathbf{n}^t, \qquad t^t_k = \sigma^t_{kl}n^t_l\,. \tag{2.10}$$

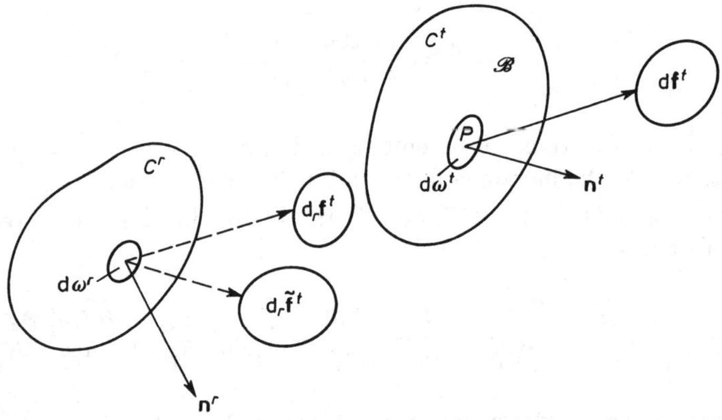

Fig. 2.3. State of stress

We shall consider next the reference configuration C^r and define two types of fictitious forces related to it:

$$\mathrm{d}\,{}_r\mathbf{f}^t = \mathrm{d}\mathbf{f}^t, \qquad \mathrm{d}\,{}_r\tilde{\mathbf{f}}^t = \frac{\partial \mathbf{x}^r}{\partial \mathbf{x}^t}\mathrm{d}\mathbf{f}^t.$$

Similarly to Eq. (2.9) two auxiliary stress vectors ${}_r\mathbf{t}^t, {}_r\tilde{\mathbf{t}}^t$ are defined as

$${}_r\mathbf{t}^t = \frac{\mathrm{d}\,{}_r\mathbf{f}^t}{\mathrm{d}\omega^t}, \qquad {}_r\tilde{\mathbf{t}}^t = \frac{\mathrm{d}\,{}_r\tilde{\mathbf{f}}^t}{\mathrm{d}\omega^t}.$$

which are used to define two stress tensors ${}_r\boldsymbol{\sigma}^t, {}_r\tilde{\boldsymbol{\sigma}}^t$ as

$${}_r\mathbf{t}^t = {}_r\boldsymbol{\sigma}^t\,\mathbf{n}^t, \qquad {}_r\tilde{\mathbf{t}}^t = {}_r\tilde{\boldsymbol{\sigma}}^t\,\mathbf{n}^t,. \tag{2.11}$$

The tensors ${}_r\boldsymbol{\sigma}^t$ and ${}_r\tilde{\boldsymbol{\sigma}}^t$ are called the first and second Piola–Kirchhoff stress tensor, respectively. They describe the state of stress at time t with reference to the configuration C^r. It can easily be seen that the second Piola–Kirchhoff stress tensor ${}_r\tilde{\boldsymbol{\sigma}}^t$ is symmetric regardless the choice of C^r. For the reference configuration which coincides with the current state (i.e., for the U.L. approach) all the three stress tensors just defined are equal, i.e.

$$\boldsymbol{\sigma}^t = {}_t\boldsymbol{\sigma}^t = {}_t\tilde{\boldsymbol{\sigma}}^t.$$

The stress increments are defined as

$$\Delta\,{}_r\boldsymbol{\sigma}^t = {}_r\boldsymbol{\sigma}^{t+\Delta t} - {}_r\boldsymbol{\sigma}^t, \qquad \Delta\,{}_r\tilde{\boldsymbol{\sigma}}^t = {}_r\tilde{\boldsymbol{\sigma}}^{t+\Delta t} - {}_r\tilde{\boldsymbol{\sigma}}^t.$$

It is important that we subtract here the quantities which can be given a mechanical interpretation of forces acting through the fixed (the same for both the time instants t and $t+\Delta t$) surface element $\mathrm{d}\omega^r$. Such a property is not shared by the Cauchy stress (for which ${}_t\boldsymbol{\sigma}^t$ and ${}_t\boldsymbol{\sigma}^{t+\Delta t}$ are measured with respect to generally different surface elements $\mathrm{d}\omega^t$ and $\mathrm{d}\omega^{t+\Delta t}$, respectively) and thus the stress ${}_t\boldsymbol{\sigma}^t$ can not be rationally employed to describe the deformation process with finite increments. Let us also observe that in spite of the relation ${}_t\boldsymbol{\sigma}^t = {}_t\tilde{\boldsymbol{\sigma}}^t$ the increments of these quantities are different

$$\Delta\,{}_t\boldsymbol{\sigma}^t = {}_t\boldsymbol{\sigma}^{t+\Delta t} - {}_t\boldsymbol{\sigma}^t \;\neq\; \Delta\,{}_t\tilde{\boldsymbol{\sigma}}^t = {}_t\tilde{\boldsymbol{\sigma}}^{t+\Delta t} - {}_t\tilde{\boldsymbol{\sigma}}^t.$$

Because of the so-called material objectivity requirement, in nonlinear continuum mechanics some generalized (objective) time derivatives of the Cauchy stress tensor are considered instead of the ordinary time derivatives. Among the infinity of the objective derivatives of $\boldsymbol{\sigma}^t$ the one frequently employed in applications is called the Truesdell derivative – we shall discuss the matter in a more detailed way in Chap. 6, Part II. Here we only note that the components of the Truesdell derivative of the Cauchy stress are equal to the components of the ordinary derivative of the second Piola–Kirchhoff stress defined on the current configuration. This makes it possible to use $\Delta\,{}_t\tilde{\boldsymbol{\sigma}}^t$ as a basic incremental stress measure which, as we shall see, is very convenient.

2.4 Equations of Motion

The incremental equations of motion can be derived by considering the dynamic equilibrium conditions at t and $t+\Delta t$. Three types of forces act upon a selected body part in the configuration C^t: the volumetric forces $\varrho^r\,\mathbf{f}^t$, the surface forces ${}_r\mathbf{t}^t$ and the inertial (d'Alembert) forces $-\varrho^r\,\left(\mathrm{d}^2\,\mathbf{x}^t/\mathrm{d}\tau^2\right)$, Fig. 2.4. The configuration C^r has been taken here as the reference state. The same configuration is taken as the reference state to define the forces acting upon the same part of the body in the configuration $C^{t+\Delta t}$ – the forces are denoted as $\varrho^r\,\mathbf{f}^{t+\Delta t}$, ${}_r\mathbf{t}^{t+\Delta t}$ and $-\varrho^r\,\left(\mathrm{d}^2\,\mathbf{x}^{t+\Delta t}/\mathrm{d}\tau^2\right)$, respectively. The dynamic equilibrium conditions at t and $t+\Delta t$ can be expressed in the following form:

$$\int_{\partial\Omega^r} {}_r\mathbf{t}^t\,\mathrm{d}(\partial\Omega) + \int_{\Omega^r} \varrho^r\,\mathbf{f}^t\,\mathrm{d}\Omega = \int_{\Omega^r} \varrho^r\,\frac{\mathrm{d}^2\,\mathbf{x}^t}{\mathrm{d}\tau^2}\,,$$

$$\int_{\partial\Omega^r} {}_r\mathbf{t}^{t+\Delta t}\,\mathrm{d}(\partial\Omega) + \int_{\Omega^r} \varrho^r\,\mathbf{f}^{t+\Delta t}\,\mathrm{d}\Omega = \int_{\Omega^r} \varrho^r\,\frac{\mathrm{d}^2\,\mathbf{x}^{t+\Delta t}}{\mathrm{d}\tau^2}\,.$$

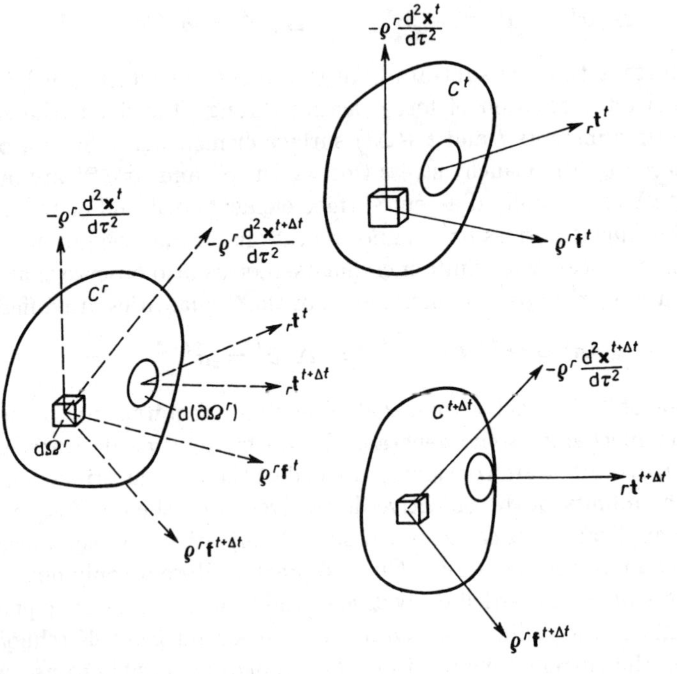

Fig. 2.4. Forces acting upon the body

By subtracting side-wise we obtain

$$\int_{\partial\Omega^r} \Delta\,{}_r\mathbf{t}^t \,\mathrm{d}(\partial\Omega) + \int_{\Omega^r} \varrho^r \Delta\mathbf{f}^t \,\mathrm{d}\Omega = \int_{\Omega^r} \varrho^r \,\frac{\mathrm{d}^2\Delta\mathbf{u}^t}{\mathrm{d}\tau^2}\,, \tag{2.12}$$

where in accordance with the earlier notation

$$\Delta\,{}_r\mathbf{t}^t = {}_r\mathbf{t}^{t+\Delta t} - {}_r\mathbf{t}^t$$

$$\Delta\mathbf{f}^t = \mathbf{f}^{t+\Delta t} - \mathbf{f}^t$$

$$\Delta\mathbf{u}^t = \mathbf{u}^{t+\Delta t} - \mathbf{u}^t = \mathbf{x}^{t+\Delta t} - \mathbf{x}^r - (\mathbf{x}^t - \mathbf{x}^r) = \mathbf{x}^{t+\Delta t} - \mathbf{x}^t .$$

By substituting Eq. (2.11) written incrementally as ($\mathbf{n}^r$ is kept constant in passing from t to $t + \Delta t$)

$$\Delta\,{}_r\mathbf{t}^t = \Delta\,{}_r\boldsymbol{\sigma}^t\,\mathbf{n}^r$$

in Eq. (2.12), and using the divergence theorem of the form

$$\int_{\partial\Omega^r} v\mathbf{n}^r \,\mathrm{d}(\partial\Omega) = \int_{\Omega^r} \mathrm{div}\, v \,\mathrm{d}\Omega$$

for any continuously differentiable function $v = v(\mathbf{x}^r)$, we arrive at

$$\int_{\Omega^r} [\Delta\,{}_r\sigma^t_{kl,l} + \varrho^r \Delta f^t_k] \,\mathrm{d}\Omega = \int_{\Omega^r} \varrho^r \,\frac{\mathrm{d}^2\Delta u^t_k}{\mathrm{d}\tau^2}\,\mathrm{d}\Omega\,.$$

Since this equation has to hold true for any part of the body, the local form of the incremental equation of motion finally results as

$$\Delta\,{}_r\sigma^t_{kl,l} + \varrho^r \Delta f^t_k = \varrho^r \,\frac{\mathrm{d}^2\Delta u^t_k}{\mathrm{d}\tau^2}\,. \tag{2.13}$$

2.5 Constitutive Equations

In this section a certain general form of the constitutive equation useful in dealing with a broad class of materials will be briefly discussed. Only mechanical effects will be accounted for. Specific forms of the general formulation of this section will be reviewed later in subsequent chapters; the material models to be specified include linear elasticity, small and large deformation elasto-plasticity with different hardening rules (including additionally some thermo-mechanical couplings neglected in this section), elasto-viscoplasticity (requiring some extensions to the general form considered in this Section), inelasticity with effects of voids nucleation, growth and coalescence, etc.

Let us assume that the mechanical properties of the material on hand may be described by means of the so-called strain rate potential. It is a scalar function $\bar{W}$ dependent on the strain rate and some parameters known as the state variables, which at the given stage of the process provides a link between stress rate and strain rate measures according to a symbolically written constitutive equation

$$\text{stress rate} = \frac{\partial \bar{W}}{\partial(\text{strain rate})} \,. \tag{2.14}$$

Selection of stress rate and strain rate measures to enter Eq. (2.14) should account for the following requirements:

- the constitutive equation should be objective under the change of observer, [1, 5, 9, 10],
- material constants in Eq. (2.14) should lend themselves to (as straightforward as possible) experimental determination,
- rates (and, after time discretization, finite time increments) of stress and strain should be conjugate in the sense of the virtual work expression – the condition satisfied by the stress increment $\Delta\,{}_r\tilde{\sigma}^t$ and strain increment $\Delta\,{}_r\varepsilon^t$, for instance, since

$$\frac{1}{\varrho^{r_1}}\,\Delta\,{}_{r_1}\tilde{\sigma}^t_{kl}\,\Delta\,{}_{r_1}\varepsilon^t_{kl} = \frac{1}{\varrho^{r_2}}\,\Delta\,{}_{r_2}\tilde{\sigma}^t_{kl}\,\Delta\,{}_{r_2}\varepsilon^t_{kl}$$

holds true for any reference configurations C^{r_1} and C^{r_2}.

By defining

$$ {}_r\dot{\varepsilon}^t_{ij} = \lim_{\Delta t \to 0} \frac{{}_r\varepsilon^{t+\Delta t}_{ij} - {}_r\varepsilon^t_{ij}}{\Delta t}\,,$$

$$ {}_r\dot{\tilde{\sigma}}^t_{ij} = \lim_{\Delta t \to 0} \frac{{}_r\tilde{\sigma}^{t+\Delta t}_{ij} - {}_r\tilde{\sigma}^t_{ij}}{\Delta t}\,,$$

and assuming the potential $\bar{W}$ to have the general form

$$\bar{W} = \tfrac{1}{2} C^r_{ijkl}\, {}_r\dot{\varepsilon}^t_{ij}\, {}_r\dot{\varepsilon}^t_{kl} \tag{2.15}$$

Eq. (2.14) becomes

$$ {}_r\dot{\tilde{\sigma}}^t_{ij} = \frac{\partial \bar{W}}{\partial\, {}_r\dot{\varepsilon}^t_{ij}} = C^r_{ijkl}\, {}_r\dot{\varepsilon}^t_{kl}\,, \tag{2.16}$$

which in finitely incremental form reads

$$ {}_r\tilde{\sigma}^{t+\Delta t}_{ij} = {}_r\tilde{\sigma}^t_{ij} + \int_t^{t+\Delta t} C^r_{ijkl}\, {}_r\dot{\varepsilon}^t_{kl}\, \mathrm{d}\tau\,. \tag{2.17}$$

Equation (2.17) can easily be transformed to the equivalent form involving quantities related to an arbitrary selected other reference configuration.

For many materials of engineering significance the tensor of material properties C^r_{ijkl} is quite a complex function of the state variables – examples will be given in Chap. 6, Part II.

For sufficiently small time steps it may sometimes be assumed that the tensor C^r_{ijkl} is constant within the time step considered – Eq. (2.17) then becomes

$$\Delta\,_r\tilde{\sigma}^t_{ij} = C^r_{ijkl}\,\Delta\,_r\varepsilon^t_{kl}\,.$$

which is called the explicitly integrated form of the rate-type constitutive equation (2.16). The problem of accuracy of the above incremental form (and, indeed, of some more sophisticated, implicitly integrated forms) with respect to the original rate-type formulation will be undertaken in Chap. 6, Part II.

2.6 Fundamental System of Equations for Nonlinear Mechanics of Deformable Bodies

In previous sections we have derived all the field equations describing the general problem of nonlinear solid mechanics in the framework of the incremental methodology. Some pretty complex notation turned out unavoidable to assure the correct interpretation of the particular symbols. Now we shall summarize the results by formulating the boundary value problem of nonlinear solid mechanics using somewhat simplified notation. In particular, we propose now to assume that the current and reference configurations C^t and C^r are fixed; thus, we may drop the indices t and r in all the equations without any risk of confusion.

The fundamental system of equations for the problem considered assumes the following form:

- equations of motion

$$\Delta\sigma_{kl,l} + \varrho\Delta f_k = \varrho\Delta\ddot{u}_k\,, \qquad \mathbf{x}\in\Omega,\ \ \tau\in[0,t], \tag{2.18}$$

- geometric equations

$$\begin{aligned}\Delta\varepsilon_{kl} &= \tfrac{1}{2}[\Delta u_{k,l} + \Delta u_{l,k} + u_{i,k}\Delta u_{i,l} + \Delta u_{i,k}u_{i,l} + \Delta u_{i,k}\Delta u_{i,l}]\\ &= \Delta\bar{\varepsilon}_{kl} + \Delta\bar{\bar{\varepsilon}}_{kl}\,, \qquad \mathbf{x}\in\Omega,\ \ \tau\in[0,t],\end{aligned} \tag{2.19}$$

- constitutive equations

$$\Delta\tilde{\sigma}_{ij} = \int_t^{t+\Delta t} C_{ijkl}\dot{\varepsilon}_{kl}\,\mathrm{d}\tau\,, \qquad \mathbf{x}\in\Omega,\ \ \tau\in[0,t], \tag{2.20}$$

- stress-type boundary conditions

$$\Delta\sigma_{kl}n_l = \Delta\hat{t}_k\,, \qquad \mathbf{x}\in\partial\Omega_\sigma,\ \ \tau\in[0,t], \tag{2.21}$$

- displacement-type boundary conditions

$$\Delta u_k = \Delta\hat{u}_k\,, \qquad \mathbf{x}\in\partial\Omega_u,\ \ \tau\in[0,t], \tag{2.22}$$

- initial conditions

$$u_k = \hat{u}^0_k\,, \qquad \dot{u}_k = \hat{\dot{u}}^0_k\,, \qquad \mathbf{x}\in\Omega,\ \ \tau = 0. \tag{2.23}$$

- relationship between the first and second incremental Piola–Kirchhoff stress tensor components

$$\Delta\sigma_{kl} = \Delta u_{k,m}\tilde{\sigma}_{ml} + F_{km}\Delta\tilde{\sigma}_{ml} + \Delta u_{k,m}\Delta\tilde{\sigma}_{ml}\,. \tag{2.24}$$

It has been assumed that the part of the boundary $\partial\Omega_\sigma$ is subject to a surface external load $\hat{t}_k$ while at the remaining part $\partial\Omega_u$ a prescribed displacement $\hat{u}_k$ is given.

The solution of the initial–boundary value problem (2.18)–(2.24) consists in finding functions $u_k(\mathbf{x},\tau)$ and, by means of Eqs. (2.19), (2.20), functions $\varepsilon_{kl}(\mathbf{x},\tau)$, $\tilde{\sigma}_{kl}(\mathbf{x},\tau)$ for $\mathbf{x} \in \Omega$, $\tau \in [0,t]$.

Different special cases of such a general formulation are considered in engineering applications. The quasi-static problem is described by the same Eqs. (2.18)–(2.24) with the inertial effects neglected by setting $\varrho\Delta\ddot{u}_k = 0$. It should be emphasized, however, that the solution to any quasi-static problem is a function of space and time variables $\mathbf{x}$ and τ, quite similarly to the problem of dynamics.

The linear problem of solid dynamics is defined by the following system of non-incremental differential equations (the moduli C_{ijkl} assumed constant):

$$\left.\begin{aligned} &\sigma_{kl,l} + \varrho f_k = \varrho\ddot{u}_k \\ &\sigma_{kl} = C_{klmn}\varepsilon_{mn} \\ &\varepsilon_{kl} = \tfrac{1}{2}(u_{k,l} + u_{l,k}) \end{aligned}\right\} \quad \mathbf{x} \in \Omega,\ \ \tau \in [0,t],$$

$$\begin{aligned} &\sigma_{kl}n_l = \hat{t}_k && \mathbf{x} \in \partial\Omega_\sigma,\ \ \tau \in [0,t], \\ &u_k = \hat{u}_k && \mathbf{x} \in \partial\Omega_u,\ \ \tau \in [0,t], \end{aligned} \tag{2.25}$$

$$\left.\begin{aligned} &u_k = \hat{u}_k^0 \\ &\dot{u}_k = \hat{\dot{u}}_k \end{aligned}\right\} \quad \mathbf{x} \in \Omega,\ \ \tau = 0.$$

For linear problems any configuration change is neglected. Consequently, only the Cauchy stress tensor needs to be considered.

By neglecting in Eqs. (2.25) the term $\varrho\ddot{u}_k$ we obtain the system of differential equations describing the problem of linear statics. The solution to it depends on the space variable $\mathbf{x} \in \Omega$ only.

In accordance with the remarks given at the beginning of this part of the book we have been so far using the indicial notation. In other words, we have been developing all the relevant equations in terms of tensor representations taken with respect to the fixed Cartesian coordinate system x_k. Quite often we have been using the term 'matrix' to emphasize that, say, the strain tensor components ε_{ij} , $i,j = 1,2,3$ may also be viewed as the 3×3 matrix $\boldsymbol{\varepsilon} = [\varepsilon_{ij}]$.

In continuum mechanics formulations aimed at computational applications another matrix notation has been standardized in the literature. In it[1], the symmetric 3×3 stress and strain matrices are replaced by some 6-dimensional

[1] 3D formulation is only examplified for the sake of brevity.

column vectors consisting of six independent components of the corresponding 3×3 matrices, respectively. The fourth-order tensor representations like C_{ijkl} (i.e. exhibiting symmetries with respect to the interchange of the first and last two indices) are accordingly replaced by some 6×6 matrices. Such a notation offers a lot of advantages as it lends itself nicely to direct computer implementation. We shall illustrate the notation below using the linear problem of statics as an example. No further comments will be provided as any (unlikely) confusion may be cleared up by taking reference to the indicial counterpart of any symbol. We may thus write:

- stress vector (a slightly redefined version of it will also be later used)

$$\boldsymbol{\sigma}_{6\times1} = \{\sigma_{11}\ \sigma_{22}\ \sigma_{33}\ \sigma_{23}\ \sigma_{13}\ \sigma_{12}\}. \tag{2.26}$$

- displacement vector

$$\mathbf{u}_{3\times1} = \{u_1\ u_2\ u_3\}. \tag{2.27}$$

- strain vector

$$\boldsymbol{\varepsilon}_{6\times1} = \{\varepsilon_{11}\ \varepsilon_{22}\ \varepsilon_{33}\ \gamma_{23}\ \gamma_{13}\ \gamma_{12}\}. \tag{2.28}$$

- geometric relations

$$\boldsymbol{\varepsilon}_{6\times1} = \mathbf{D}_{6\times3}\mathbf{u}_{3\times1}\,, \tag{2.29}$$

  where

$$\mathbf{D}_{6\times3} = \begin{bmatrix} \frac{\partial}{\partial x_1} & 0 & 0 \\ 0 & \frac{\partial}{\partial x_2} & 0 \\ 0 & 0 & \frac{\partial}{\partial x_3} \\ 0 & \frac{\partial}{\partial x_3} & \frac{\partial}{\partial x_2} \\ \frac{\partial}{\partial x_3} & 0 & \frac{\partial}{\partial x_1} \\ \frac{\partial}{\partial x_2} & \frac{\partial}{\partial x_1} & 0 \end{bmatrix}.$$

- constitutive equation (generalized Hooke's law)

$$\boldsymbol{\sigma}_{6\times1} = \mathbf{C}_{6\times6}\boldsymbol{\varepsilon}_{6\times1}\,, \tag{2.30}$$

  where

$$\mathbf{C}_{6\times6} = \frac{E}{(1+\nu)(1-2\nu)} \begin{bmatrix} 1-\nu & \nu & \nu & 0 & 0 & 0 \\ & 1-\nu & \nu & 0 & 0 & 0 \\ & & 1-\nu & 0 & 0 & 0 \\ & & & \frac{1-2\nu}{2} & 0 & 0 \\ & \text{sym.} & & & \frac{1-2\nu}{2} & 0 \\ & & & & & \frac{1-2\nu}{2} \end{bmatrix}.$$

- equilibrium equations

$$\mathbf{D}^T_{3\times6}\boldsymbol{\sigma}_{6\times1} + \varrho\,\hat{\mathbf{f}}_{3\times1} = \mathbf{0}\,. \tag{2.31}$$

- displacement-type boundary conditions

$$\mathbf{u}_{3\times1} = \hat{\mathbf{u}}_{3\times1}\,. \tag{2.32}$$

- stress-type boundary conditions

$$\mathbf{N}_{3\times 6}\boldsymbol{\sigma}_{6\times 1} = \hat{\mathbf{t}}_{3\times 1}\,, \tag{2.33}$$

where

$$\mathbf{N}_{3\times 6} = \begin{bmatrix} n_1 & 0 & 0 & 0 & n_3 & n_2 \\ 0 & n_2 & 0 & n_3 & 0 & n_1 \\ 0 & 0 & n_3 & n_2 & n_1 & 0 \end{bmatrix}, \qquad \mathbf{n}_{3\times 1} = \{n_1\; n_2\; n_3\}\,.$$

2.7 Variational Formulation

So far, we have only considered the so-called local formulation for problems of nonlinear solid mechanics. An exact analytical solution to most practical problems described in such a manner can hardly be expected. This is one of the reasons why an alternative formulation further referred to as the variational approach has been extensively investigated in the literature and is briefly discussed below.

The importance of variational statements of physical laws goes far beyond their use as simply an alternative to the local formulation. In fact, the variational (or weak) forms of the laws of solid mechanics are often considered to be the only natural and rigorously correct way to think about them. However, aside from this basic observation it is sufficient for us to note that the use of variational statements makes it possible to concentrate in a single functional all of the intrinsic features of the problem on hand: the governing field equations as well as the boundary and initial conditions.

The variational formulation has several definite advantages over the classical (local) formulation. Firstly, the functional which is subjected to variation usually has a definite physical meaning and is invariant under coordinate transformation. Consequently, once the variational principle has been formulated in one coordinate system, the governing equations expressed in another coordinate system can be obtained by writing the invariant quantity in the new coordinates and then applying the variational procedure. Secondly, the variational principles sometimes lead to estimates for lower and upper bounds of the exact solution. More generally, the variational methods often provide an approximate formulation of the problem which yields a solution compatible with the degree of approximation assumed. In this class of problems the construction of various finite element models is included; such models are now widely considered as the most effective tools for the numerical analysis of solid and structures. Finally, a very important method of investigating stability and uniqueness problems via the variational principles should be mentioned in this context.

Ways to formulate and use different variational principles have been widely considered in the literature, cf. [2, 6, 11], for instance. Only a very limited discussion of the problem is offered below with the objective of setting up the background for further specific considerations related to techniques typical of computational mechanics.

Let us begin by considering the linear problem of statics of a deformable body in the form, cf. Eq. (2.25)

$$\left.\begin{aligned} &\sigma_{kl,l} + \varrho f_k = 0 \\ &\sigma_{kl} = C_{klmn}\varepsilon_{mn} \\ &\varepsilon_{kl} = \frac{1}{2}(u_{k,l} + u_{l,k}) \end{aligned}\right\} \quad \mathbf{x} \in \Omega, \tag{2.34}$$
$$\sigma_{kl}n_l = \hat{t}_k \qquad \mathbf{x} \in \partial\Omega_\sigma\,,$$
$$u_k = \hat{u}_k \qquad \mathbf{x} \in \partial\Omega_u\,,$$

The potential energy for the system on hand is defined as the functional

$$J_P[u_i] = \int_\Omega \left(\tfrac{1}{2}C_{ijkl}\varepsilon_{ij}\varepsilon_{kl} - \varrho\hat{f}_i u_i\right) \mathrm{d}\Omega - \int_{\partial\Omega_\sigma} \hat{t}_i u_i \,\mathrm{d}(\partial\Omega)\,, \tag{2.35}$$

which is a sum of the internal elastic strain energy

$$W = \int_\Omega \tfrac{1}{2}C_{ijkl}\varepsilon_{ij}\varepsilon_{kl}\,\mathrm{d}\Omega \tag{2.36}$$

and the potential energy of the external volumetric and surface loads

$$L = -\int_\Omega \varrho\hat{f}_i u_i \,\mathrm{d}\Omega - \int_{\partial\Omega_\sigma} \hat{t}_i u_i \,\mathrm{d}(\partial\Omega)\,. \tag{2.37}$$

In Eq. (2.35) it is assumed that the strain ε_{ij} is the given function of the displacement u_i, cf. Eq. $(2.34)_3$, so that the functional may be treated as dependent on $u_i(\mathbf{x})$ only.

The principle of minimum total potential energy states that among all the geometrically possible configurations which a body can take up (i.e. of all sufficiently smooth displacement fields $u_i(\mathbf{x})$ satisfying the kinematic boundary conditions) the true one, corresponding to internal and external equilibrium between stresses and applied loads (i.e. satisfying Eqs. $(2.34)_{1,2}$) is identified by a minimum value of the total potential energy.

Thus, for the true displacement field $u_i(\mathbf{x})$ there holds

$$\delta J = 0, \qquad \delta^2 J \geq 0$$

where $\delta^2 J = 0$ only if $\delta\varepsilon_{ij} = \frac{1}{2}(\delta u_{i,j} + \delta u_{j,i}) = 0$, i.e. when the body is subjected to a rigid motion.

We note that in the above variational statement the kinematic boundary conditions are taken into account a priori, i.e. they are satisfied by limiting the kinematic admissible displacement fields to those satisfying the kinematic boundary conditions $(2.34)_5$. However, the formulation may be generalized by introducing in it the conditions $(2.34)_5$ in an entirely explicit manner. This will allow to consider a more general class of displacements not necessarily satisfying the conditions $(2.34)_5$. To this aim we define the functional

$$J_P^*[u_i,\sigma_{ij}] = \int_\Omega \left(\tfrac{1}{2}C_{ijkl}\varepsilon_{ij}\varepsilon_{kl} - \varrho\hat{f}_i u_i\right) \mathrm{d}\Omega - \int_{\partial\Omega_\sigma} \hat{t}_i u_i \,\mathrm{d}(\partial\Omega) + \int_{\partial\Omega_u} \sigma_{kl} n_l (\hat{u}_k - u_k)\,\mathrm{d}(\partial\Omega)\,, \tag{2.38}$$

in which clearly the compatibility (geometric) equation $(2.34)_3$ is again assumed to hold. It is easy to check, cf. [6], that vanishing of the first variation of the functional (2.38), i.e. $\delta J_P^* = 0$, implies vanishing of Eqs. $(2.34)_{1,2}$ and, additionally, Eq. $(2.34)_5$. It should be emphasized that in this case the second variation will not generally preserve its sign so that the computed displacement field implies the value of J_P^* which may only be stationary and not necessarily minimum.

A generalization of the minimum potential energy principle which accounts for the inertial terms is known as the Hamilton principle. It states that among all sufficiently smooth displacement fields $u_i(\mathbf{x},\tau)$ satisfying the kinematic boundary conditions $(2.34)_5$ and the prescribed conditions at the limits $\tau = t_{(1)}$ and $\tau = t_{(2)}$ the actual displacement field renders the functional

$$J_1[u_i] = \int_{t_{(1)}}^{t_{(2)}} (T - J_P)\,\mathrm{d}\tau\,, \tag{2.39}$$

stationary. The symbol T stands for the kinetic energy of the body

$$T = \tfrac{1}{2}\int_\Omega \varrho\dot{u}_i\dot{u}_i \,\mathrm{d}\Omega \tag{2.40}$$

while J_P^* is again assumed to be expressed in terms of the displacement field $u_i(\mathbf{x},\tau)$. The Hamilton principle states in other words that the actual displacement field assures vanishing of the first variation of the functional (2.39).

For the nonlinear quasi-static problem described by Eqs. (2.18)–(2.24) with $\varrho\Delta\ddot{u}_k = 0$ the functional of the incremental potential energy assumes at the incremental step $t \to t + \Delta t$ the form

$$\check{J}_P[\Delta u_i] = \int_\Omega \left(\tfrac{1}{2}C_{ijkl}\Delta\varepsilon_{ij}\Delta\varepsilon_{kl} + \tfrac{1}{2}\tilde{\sigma}_{ij}\Delta\bar{\bar{\varepsilon}}_{ij} - \varrho\Delta\hat{f}_i\Delta u_i\right)\mathrm{d}\Omega - \int_{\partial\Omega_\sigma} \Delta\hat{t}_i\Delta u_i\,\mathrm{d}(\partial\Omega)\,, \tag{2.41}$$

where $\Delta\varepsilon_{ij}$ and $\Delta\bar{\bar{\varepsilon}}_{ij}$ are understood to be expressed in terms of u_i by Eqs. (2.19) while the constitutive equation (2.20) is taken for simplicity[2] in the explicitly integrated form

$$\Delta\tilde{\sigma}_{ij} = {}^tC_{ijkl}\Delta\bar{\varepsilon}_{kl}$$

[2] A complete discussion of the problem can be found in [6], for instance.

with the constitutive moduli corresponding to the beginning of the step and the linearized incremental strain.

It is straightforward to show that the first variation of the functional (2.41) vanishes for the true solution Δu_i – the only subsidiary condition imposed on Δu_i being the kinematic boundary condition (2.22).

The interested reader is referred to [6] for a thorough account of issues related to the formulation and use of different variational principles in computational solid mechanics.

In order to illustrate the use of the matrix notation introduced in Sect. 2.6 we shall rewrite the functionals (2.35), (2.38) and (2.41) as follows

$$
\begin{aligned}
J_P[\mathbf{u}] &= \int_\Omega \left[\tfrac{1}{2}(\mathbf{Du})^T\mathbf{C}\,(\mathbf{Du}) - \varrho\hat{\mathbf{f}}^T\mathbf{u}\right] \mathrm{d}\Omega - \int_{\partial\Omega_\sigma} \hat{\mathbf{t}}^T\mathbf{u}\,\mathrm{d}(\partial\Omega)\,,\\
J_P^*[\mathbf{u},\boldsymbol{\sigma}] &= \int_\Omega \left[\tfrac{1}{2}(\mathbf{Du})^T\mathbf{C}\,(\mathbf{Du}) - \varrho\hat{\mathbf{f}}^T\mathbf{u}\right] \mathrm{d}\Omega - \int_{\partial\Omega_\sigma} \hat{\mathbf{t}}^T\mathbf{u}\,\mathrm{d}(\partial\Omega)\\
&\quad + \int_{\partial\Omega_u} (\mathbf{N}\boldsymbol{\sigma})^T(\hat{\mathbf{u}}-\mathbf{u})\,\mathrm{d}(\partial\Omega)\,, \qquad (2.42)\\
\check{J}_P[\Delta\mathbf{u}] &= \int_\Omega \left[\tfrac{1}{2}(\mathbf{D}(\mathbf{u})\Delta\mathbf{u})^T\mathbf{C}\,(\mathbf{D}(\mathbf{u})\Delta\mathbf{u}) + \tfrac{1}{2}\tilde{\boldsymbol{\sigma}}^T\Delta\bar{\bar{\boldsymbol{\varepsilon}}} - \varrho\Delta\hat{\mathbf{f}}^T\Delta\mathbf{u}\right] \mathrm{d}\Omega\\
&\quad - \int_{\partial\Omega_\sigma} \Delta\hat{\mathbf{t}}^T\Delta\mathbf{u}\,\mathrm{d}(\partial\Omega)\,.
\end{aligned}
$$

The variational notation discussed so far is typical of engineering mechanics. Mathematicians normally use a slightly different formalism which for further use in Chap. 2, Part II, will be briefly introduced below.

Let K be a set of kinematically admissible displacements as defined before. Let $\mathbf{u} \in K$ be the true displacement which renders the total potential energy of the system minimum. Consider a virtual displacement $\mathbf{v} \in V$ where V is the space of sufficiently smooth displacements satisfying the homogeneous boundary conditions on $\partial\Omega_u$. Let us calculate the system potential energy for the displacement $\mathbf{u}+\varepsilon\mathbf{v}$, which for small ε is close to the true displacement $\mathbf{u}$:

$$
\begin{aligned}
J_P[\mathbf{u}+\varepsilon\mathbf{v}] &= \int_\Omega \left\{\tfrac{1}{2}[\mathbf{D}(\mathbf{u}+\varepsilon\mathbf{v})]^T\mathbf{C}\,\mathbf{D}(\mathbf{u}+\varepsilon\mathbf{v}) - \varrho\hat{\mathbf{f}}^T(\mathbf{u}+\varepsilon\mathbf{v})\right\} \mathrm{d}\Omega\\
&\quad - \int_{\partial\Omega_\sigma} \hat{\mathbf{t}}^T(\mathbf{u}+\varepsilon\mathbf{v})\,\mathrm{d}(\partial\Omega)\,.
\end{aligned}
$$

For a fixed $\mathbf{v} \in \mathbf{V}$ we denote

$$J_P[\mathbf{u}+\varepsilon\mathbf{v}] = H(\varepsilon).$$

Since the total potential energy of the system J_P attains its minimum at $\mathbf{u}$ the function $H(\varepsilon)$ has its minimum at $\varepsilon = 0$ regardless the form of $\mathbf{v} \in V$. The derivative of H with respect to ε calculated at $\varepsilon = 0$ must therefore vanish. We obtain the equation

$$\int_{\Omega} (\mathbf{Dv})^T \mathbf{C} (\mathbf{Du}) \, d\Omega = \int_{\Omega} \mathbf{v}^T \varrho \hat{\mathbf{f}} \, d\Omega + \int_{\partial\Omega_\sigma} \mathbf{v}^T \hat{\mathbf{t}} \, d(\partial\Omega) \quad \text{for every } \mathbf{v} \in V \tag{2.43}$$

which is the variational equilibrium equation for the body on hand. Defining the bilinear strain energy form

$$b(\mathbf{u}, \mathbf{v}) = \int_{\Omega} (\mathbf{Dv})^T \mathbf{C} (\mathbf{Du}) \, d\Omega$$

and the linear load functional

$$l(\mathbf{v}) = \int_{\Omega} \mathbf{v}^t \varrho \hat{\mathbf{f}} \, d\Omega + \int_{\partial\Omega_\sigma} \mathbf{v}^T \hat{\mathbf{t}} \, d(\partial\Omega) ,$$

Eq. (2.43) becomes

$$b(\mathbf{u}, \mathbf{v}) = l(\mathbf{v}) \qquad \text{for every } \mathbf{v} \in V . \tag{2.44}$$

Assuming the strain energy to be non-negative

$$b(\mathbf{u}, \mathbf{u}) > 0 \qquad \text{for every } \mathbf{u} \in K, \ \mathbf{u} \neq 0,$$

the use of the Lax–Milgram theorem allows to conclude that, cf. Chap. 2, Part II, Eq. (2.44) has a unique solution $\mathbf{u} \in K$. Equation(2.44) is clearly the statement of the virtual work for linear elastic body.

2.8 Heat Conduction

The three-dimensional nonlinear problem of nonstationary heat transfer in the anisotropic body is described by the equation

$$\varrho c \frac{\partial T}{\partial \tau} = \frac{\partial}{\partial x_i} \left(\lambda_{ik} \frac{\partial T}{\partial x_k} \right) + \dot{g} , \tag{2.45}$$

in which T is the temperature, $\dot{g}$ is the internal heat generation rate per unit volume, $\boldsymbol{\lambda} = [\lambda_{ik}]$ is the matrix of conductivity coefficients generally dependent on temperature, ϱ is the material density and c is the material specific heat which can also be temperature dependent.

The boundary conditions are given as follows:

$$\begin{aligned} T &= \hat{T} , \qquad \mathbf{x} \in \partial\Omega_T , \\ \lambda_{ij} n_j \frac{\partial T}{\partial n_i} &= \hat{q} , \quad \mathbf{x} \in \partial\Omega_q , \end{aligned} \tag{2.46}$$

where $\partial\Omega_T$ and $\partial\Omega_q$ are the boundary parts with prescribed temperature $\hat{T}$ and heat flux $\hat{q}$, respectively, while $\mathbf{n}$ is the unit outward-drawn vector normal to the boundary surface.

By appropriately modifying the heat flux expression, Eq. (2.46) may be specified to include the convection and radiation boundary conditions, [7].

The condition imposed on the initial temperature distribution has the form

$$T = \hat{T}_0\,, \qquad \mathbf{x} \in \Omega,\ \tau = 0. \tag{2.47}$$

Equation (2.45) may be presented in the matrix form as

$$\varrho c\,\frac{\partial T}{\partial \tau} = \boldsymbol{\nabla}^T[\boldsymbol{\lambda}\boldsymbol{\nabla}T] + \dot{g}\,, \tag{2.48}$$

where

$$\boldsymbol{\nabla} = \{\frac{\partial}{\partial x_1}\ \frac{\partial}{\partial x_2}\ \frac{\partial}{\partial x_3}\}\,, \qquad \boldsymbol{\lambda} = \begin{bmatrix} \lambda_{11} & \lambda_{12} & \lambda_{13} \\ \lambda_{21} & \lambda_{22} & \lambda_{23} \\ \lambda_{31} & \lambda_{32} & \lambda_{33} \end{bmatrix}.$$

For the thermally orthotropic material (with axes of orthotropy coinciding with the coordinate axes) Eq. (2.46) becomes

$$\varrho c\,\frac{\partial T}{\partial \tau} = \frac{\partial}{\partial x_1}\left(\lambda_1\,\frac{\partial T}{\partial x_1}\right) + \frac{\partial}{\partial x_2}\left(\lambda_2\,\frac{\partial T}{\partial x_2}\right) + \frac{\partial}{\partial x_3}\left(\lambda_3\,\frac{\partial T}{\partial x_3}\right) + \dot{g}\,, \tag{2.49}$$

which for the thermally isotropic material is further reduced to

$$\boldsymbol{\nabla}^T[\lambda\boldsymbol{\nabla}T] + \dot{g} = \varrho c\,\frac{\partial T}{\partial \tau}\,, \qquad \mathbf{x} \in \Omega,\ \tau \in [0, \bar{t}]. \tag{2.50}$$

For the thermal conductivity independent of the spatial coordinates and temperature the last equation becomes

$$\nabla^2 T + \frac{\dot{g}}{\lambda} = \frac{1}{\alpha}\frac{\partial T}{\partial \tau}\,, \qquad \mathbf{x} \in \Omega,\ \tau \in [0, \bar{t}], \tag{2.51}$$

where $\alpha = \dfrac{\lambda}{\varrho c}$, $\nabla^2 = \boldsymbol{\nabla}^T\boldsymbol{\nabla}$. In the stationary case in which the temperature is independent of time, Eq. (2.51) reduces to

$$\nabla^2 T + \frac{\dot{g}}{\lambda} = 0\,, \qquad \mathbf{x} \in \Omega\,. \tag{2.52}$$

3 On Approximate Solving Systems of Differential Equations

It results form the previous discussion that addressing realistic problems of solid mechanics comes down in practice to computer assisted seeking for approximate solutions to certain nonlinear initial–boundary value problems. Some very general comments concerning the nature of such approximate solution procedures are given in this chapter.

In order to find an approximate solution to a system of differential equations we normally expand the unknown functions in terms of a certain number of so-called trial functions. To illustrate such a procedure let us consider a system of n linear partial differential equations of the form

$$A_{ij}(\mathbf{x})u_j(\mathbf{x}) + B_i(\mathbf{x}) = 0, \qquad \mathbf{x} \in \Omega,\ i,j = 1,2,\dots,n \tag{3.1}$$

together with an appropriate set of boundary conditions of the form

$$\bar{A}_{rj}(\mathbf{x})u_j(\mathbf{x}) + \bar{B}_r(\mathbf{x}) = 0, \qquad \mathbf{x} \in \partial\Omega,\ r = 1,2,\dots,n'. \tag{3.2}$$

According to the convention adopted in this book summation in Eqs. (3.1), (3.2) holds with respect to the index j. The symbols A_{ij} and $\bar{A}_{rj}$ stand for matrices whose entries are linear differential operators; the unknown functions $u_i(\mathbf{x})$ are assumed to be smooth enough to assure the existence of all the derivatives required.

Let the functions $u_i(\mathbf{x})$ be expressed as

$$u_i(\mathbf{x}) \cong \tilde{u}_i(\mathbf{x}) = \varphi_{i\alpha}q_\alpha\,, \qquad \text{summation over } \alpha = 1,2,\dots,N, \tag{3.3}$$

where $\varphi_{11}, \varphi_{12}, \dots, \varphi_{nN}$ are a set of linearly independent known functions not necessarily satisfying the boundary conditions (3.2) while $q_1, q_2, \dots, q_N$ are a set of unknown parameters to be determined. Substituting Eq. (3.3) in Eqs. (3.1), (3.2) results in

$$\begin{aligned} A_{ij}\tilde{u}_j + B_i &= R_i, \qquad && \mathbf{x} \in \Omega\,,\\ \bar{A}_{rj}\tilde{u}_j + \bar{B}_r &= \bar{R}_r, \qquad && \mathbf{x} \in \partial\Omega\,, \end{aligned} \tag{3.4}$$

with the residual vectors $R_i, \bar{R}_r$ usually different from zero on account of only an approximate character of the assumed solution (3.3). In order to derive equations from which to solve for the parameters $q_1, q_2, \dots, q_N$ let us form the weighted sum of the residual functions R_i, $\bar{R}_r$ by writing

$$\int_{\Omega} \psi_{i\alpha}(\mathbf{x}) R_i(\mathbf{x})\,\mathrm{d}\Omega + \int_{\partial\Omega} \bar{\psi}_{r\alpha}(\mathbf{x}) \bar{R}_r(\mathbf{x})\,\mathrm{d}(\partial\Omega) = 0, \tag{3.5}$$

$$\begin{aligned} \alpha &= 1,2,\ldots,N, \\ i &= 1,2,\ldots,n \text{ (summation)}, \\ r &= 1,2,\ldots,n' \text{ (summation)}. \end{aligned}$$

It is intuitively obvious that if Eq. (3.5) is satisfied for a large enough number N of the weighting functions $\psi_{i\alpha}, \bar{\psi}_{r\alpha}$ then the expansion $\tilde{u}_i$ in Eq. (3.3) should be a 'good' approximation to the unknown solution u_i. By substituting Eq. (3.3) into Eq. (3.4) and the so computed residual vectors into Eq. (3.5) we arrive at

$$\left(\int_{\Omega} \psi_{i\alpha} A_{ij} \varphi_{j\beta}\,\mathrm{d}\Omega\right) q_\beta + \left(\int_{\partial\Omega} \bar{\psi}_{r\alpha} \bar{A}_{rj} \varphi_{j\beta}\,\mathrm{d}(\partial\Omega)\right) q_\beta$$
$$+ \int_{\Omega} \psi_{i\alpha} B_i\,\mathrm{d}\Omega + \int_{\partial\Omega} \bar{\psi}_{r\alpha} \bar{B}_r\,\mathrm{d}(\partial\Omega) = 0. \tag{3.6}$$

This equation can be rewritten in a more compact form as

$$K_{\alpha\beta} q_\beta = Q_\alpha \qquad \alpha = 1,2,\ldots,N, \tag{3.7}$$

or

$$\mathbf{K}_{N\times N}\mathbf{q}_{N\times 1} = \mathbf{Q}_{N\times 1},$$

where

$$\begin{aligned} K_{\alpha\beta} &= \int_{\Omega} \psi_{i\alpha} A_{ij} \varphi_{j\beta}\,\mathrm{d}\Omega + \int_{\partial\Omega} \bar{\psi}_{r\alpha} \bar{A}_{rj} \varphi_{j\beta}\,\mathrm{d}(\partial\Omega), \\ Q_\alpha &= -\int_{\Omega} \psi_{i\alpha} B_i\,\mathrm{d}\Omega - \int_{\partial\Omega} \bar{\psi}_{r\alpha} \bar{B}_r\,\mathrm{d}(\partial\Omega) \end{aligned} \tag{3.8}$$

In case the matrix $\mathbf{K}$ is non-singular the system of the linear algebraic equations (3.7) may be solved for the parameters $q_1, q_2, \ldots, q_N$ which in turn, after using Eq. (3.3), provides the approximate solution to the boundary value problem (3.1), (3.2).

The method just described is known as the weighted residual method. Its performance depends clearly on the choice of the trial and weighting functions $\varphi_{i\alpha}$ and $\psi_{i\alpha}$, $\bar{\psi}_{r\alpha}$, respectively.

Three situations can be rationally distinguished when looking closer at the selection of the trial functions $\varphi_{i\alpha}$:

- $\varphi_{i\alpha}$ are such that neither Eqs. (3.1) nor Eqs. (3.2) are automatically (i.e. regardless of the values of $q_1, q_2, \ldots, q_N$) satisfied; the consequence is that both R_i and $\bar{R}_r$ do not vanish.
- $\varphi_{i\alpha}$ are such that only Eqs. (3.1) are satisfied for any values of $q_1, q_2, \ldots, q_N$; we have then $R_i = 0$, $\bar{R}_r \neq 0$ and the weighted residual method generated by this choice of $\varphi_{i\alpha}$'s is known as the boundary

method. The most widely used version of this approach will be presented in Part IV of the book.

- $\varphi_{i\alpha}$ are such that only Eqs. (3.2) are satisfied resulting in $R_i \neq 0$, $\bar{R}_r = 0$

In the last case it turns out useful to consider a set of functions $\chi_i(\mathbf{x})$ such that

$$\bar{A}_{rj}\chi_j + \bar{B}_r = 0, \qquad \mathbf{x} \in \partial\Omega. \tag{3.9}$$

By selecting the trial functions $\varphi_{i\alpha}$ in such a way that

$$\bar{A}_{rj}\varphi_{j\alpha} = 0, \qquad r = 1, 2, \ldots, n'; \; \alpha = 1, 2, \ldots, N, \tag{3.10}$$

we may present the solution in the form

$$u_i(\mathbf{x}) \cong \tilde{u}_i(\mathbf{x}) = \chi_i(\mathbf{x}) + \varphi_{i\alpha}(\mathbf{x})\, q_\alpha\,. \tag{3.11}$$

which assures that the boundary conditions are satisfied regardless of the values $q_1, q_2, \ldots, q_N$. We have then

$$R_i = A_{ij}(\chi_j + \varphi_{j\alpha}q_\alpha) + B_i\,, \qquad \bar{R}_r = 0,$$

and, by using Eq. (3.5)

$$\left(\int_\Omega \psi_{i\alpha}A_{ij}\varphi_{j\beta}\,\mathrm{d}\Omega\right)q_\beta + \int_\Omega \psi_{i\alpha}(A_{ij}\chi_j + B_i)\,\mathrm{d}\Omega = 0. \tag{3.12}$$

By using the definitions (3.8) we finally arrive at

$$K_{\alpha\beta}q_\beta = Q'_\alpha\,, \qquad \alpha = 1, 2, \ldots, N, \tag{3.13}$$

in which

$$Q'_\alpha = -\int_\Omega \psi_{i\alpha}(A_{ij}\chi_j + B_i)\,\mathrm{d}\Omega$$

is a known vector.

The most popular choice of the functions $\psi_{i\alpha}$ is that

$$\psi_{i\alpha} = \varphi_{i\alpha}, \qquad i = 1, 2, \ldots, n; \; \alpha = 1, 2, \ldots, N,$$

The weighted residual method in known in such a case as the Galerkin method in which

$$\begin{aligned} K_{\alpha\beta} &= \int_\Omega \varphi_{i\alpha}A_{ij}\varphi_{j\beta}\,\mathrm{d}\Omega\,, \\ Q_\alpha &= -\int_\Omega \varphi_{i\alpha}(A_{ij}\chi_j + B_i)\,\mathrm{d}\Omega\,. \end{aligned} \tag{3.14}$$

The next method to be briefly reviewed is the so-called Rayleigh–Ritz method. Let us consider again the boundary value problem (3.1), (3.2) and assume that some functions $v_i(\mathbf{x})$, $i = 1, 2, \ldots, n$, satisfy the homogeneous boundary conditions in the form

$$\bar{A}_{rj}v_j = 0, \qquad \mathbf{x} \in \partial\Omega. \tag{3.15}$$

Let us recall that an operator A_{ij} is called symmetric (self-adjoint) in the region Ω with respect to the set of functions satisfying (3.15) if for any $v_i^1(\mathbf{x})$, $v_i^2(\mathbf{x})$ belonging to the set there holds

$$\int_\Omega v_i^1 A_{ij} v_j^2 \,\mathrm{d}\Omega = \int_\Omega v_i^2 A_{ij} v_j^1 \,\mathrm{d}\Omega. \tag{3.16}$$

The symmetric operator A_{ij} is called positive definite in Ω with respect to the above set of functions if for any v_i from this set there holds

$$\int_\Omega v_i A_{ij} v_j \,\mathrm{d}\Omega \geq 0,$$

and the integral equals zero if and only if $v_i(\mathbf{x}) \equiv 0$, $\mathbf{x} \in \Omega$.

Let us assume now that the operator A_{ij} in Eqs. (3.1), (3.2) is linear and symmetric. By taking reference to the previously introduced functions χ_i, $\varphi_{i\alpha}$, $i = 1, 2, \ldots, n$; $\alpha = 1, 2, \ldots, N$, we adopt again the expansion for the solution vector in the form

$$u_i(\mathbf{x}) \cong \tilde{u}_i(\mathbf{x}) = \chi_i(\mathbf{x}) + \varphi_{i\alpha}(\mathbf{x}) q_\alpha . \tag{3.17}$$

Thus, we satisfy the boundary conditions (3.2) regardless of the values $q_1, q_2, \ldots, q_N$. The Rayleigh–Ritz method is based on considering a functional

$$\Pi[\tilde{u}_i] = \int_\Omega \left\{ (\tilde{u}_i - \chi_i) \left[\tfrac{1}{2} A_{ij} (\tilde{u}_j - \chi_j) + A_{ij}\chi_j + B_i \right] \right\} \mathrm{d}\Omega , \tag{3.18}$$

which is to be made stationary by appropriately selecting the parameters $q_1, q_2, \ldots, q_N$. Let us note that

$$\delta\Pi = \int_\Omega \left[A_{ij}(\tilde{u}_j - \chi_j) + A_{ij}\chi_j + B_i \right] \delta\tilde{u}_i \,\mathrm{d}\Omega = \int_\Omega (A_{ij}\tilde{u}_j + B_i)\, \delta\tilde{u}_i \,\mathrm{d}\Omega$$

so that the sufficient condition for the vanishing of the functional first variation is that the functions $\tilde{u}_i$ satisfy the original differential equation system (3.1).

Using the expansion (3.17), Eq. (3.18) becomes

$$\Pi[\tilde{u}_i] = \frac{q_\alpha q_\beta}{2} \int_\Omega \varphi_{i\alpha} A_{ij} \varphi_{j\beta} \,\mathrm{d}\Omega + q_\alpha \int_\Omega \varphi_{i\alpha} (A_{ij}\chi_j + B_i) \,\mathrm{d}\Omega . \tag{3.19}$$

The stationarity conditions

$$\frac{\partial \Pi}{\partial q_1} = 0, \qquad \frac{\partial \Pi}{\partial q_2} = 0, \qquad \ldots, \qquad \frac{\partial \Pi}{\partial q_N} = 0,$$

generate the following system of N linear algebraic equations for the unknown $q_1, q_2, \ldots, q_N$:

$$q_\alpha \int_\Omega \varphi_{i\alpha} A_{ij} \varphi_{j\beta} \,\mathrm{d}\Omega = - \int_\Omega \varphi_{i\alpha} (A_{ij}\chi_j + B_i) \,\mathrm{d}\Omega ,$$

which coincides with the previously derived system of equations (3.12). However, unlike the previous case, the operator A_{ij} has now to be symmetric. This is in fact a general property linking the Galerkin and Rayleigh–Ritz methods: both techniques generate the same solutions provided the operator A_{ij} is symmetric (which assures the existence of the standard variational formulation for the original boundary value problem).

We have so far been dealing with linear problems. In order to see how the discussion can be generalized to the case of nonlinear systems let us consider the following system of nonlinear differential equations:

$$A_{ij}(\mathbf{x}; u_k)u_j(\mathbf{x}) + B_i(\mathbf{x}) = 0, \qquad \mathbf{x} \in \Omega, \tag{3.20}$$

together with appropriate boundary conditions of the form

$$\bar{A}_{rj}(\mathbf{x}; u_k)u_j(\mathbf{x}) + \bar{B}_r(\mathbf{x}) = 0, \qquad \mathbf{x} \in \partial\Omega. \tag{3.21}$$

The operator matrices are now assumed to depend on the solution $u_i(\mathbf{x})$, $i = 1, 2, \ldots, n$. By employing the weighted method residual method in much the same fashion as before we may obtain the following system of N nonlinear algebraic equations

$$K_{\alpha\beta}(q_\gamma)q_\beta = Q_\alpha\,, \tag{3.22}$$

in which

$$K_{\alpha\beta}(q_\gamma) = \int_\Omega \psi_{i\alpha}[A_{ij}(\varphi_{k\gamma}q_\gamma)]\varphi_{j\beta}\,\mathrm{d}\Omega + \int_{\partial\Omega} \bar{\psi}_{r\alpha}[\bar{A}_{rj}(\varphi_{k\gamma}q_\gamma)]\varphi_{j\beta}\,\mathrm{d}(\partial\Omega),$$

whereas the vector Q_α is given in Eq. (3.8). The solution to the nonlinear system (3.22) can be sought by using anyone of many existing algorithms. Conceptually the simplest is the so-called direct iteration method which is based on:

1. assuming the zero-th estimate for the solution vector $q_\alpha = q_\alpha^{(0)}$, $\alpha = 1, 2, \ldots, N$,
2. finding the zero-th estimate for the secant matrix
$$K_{\alpha\beta}^{(0)} = K_{\alpha\beta}(q_\gamma^{(0)}).$$
3. solving for the first solution estimate from ($K_{\alpha\beta}^{(0)}$ must be nonsingular)
$$q_\alpha^{(1)} = \overset{-1}{K_{\alpha\beta}^{(0)}}\, Q_\beta\,.$$
4. using the so obtained solution estimate to compute the first estimate to the secant matrix
$$K_{\alpha\beta}^{(1)} = K_{\alpha\beta}(q_\gamma^{(1)}).$$
5. continuing the iteration until the difference between two successive solution estimates is sufficiently small in the vector norm adopted, i.e.
$$\|q^{(i)} - q^{(i-1)}\| < \varepsilon.$$

Another very widely used method is called the Newton–Raphson iteration which at a typical iteration is based on:

1. expanding the unknown i-th residual in the Taylor series and retaining only the first order term as

$$R_\alpha^{(i)} = R_\alpha^{(i-1)} + \left.\frac{\partial R_\alpha}{\partial q_\beta}\right|_{(i-1)} \delta q_\beta^{(i)}$$

   where

$$R_\alpha^{(i-1)} = Q_\alpha - K_{\alpha\gamma}(\mathbf{q}^{(i-1)})q_\gamma^{(i-1)}$$

   is the residual from the previous iteration while $\delta q_\alpha^{(i)}$ is the iterative correction sought,
2. computing $\delta q_\alpha^{(i)}$ by making $R_\alpha^{(i)}$ vanish, i.e.

$${}^{(T)}K_{\alpha\beta}^{(i-1)}\,\delta q_\beta^{(i)} = R_\alpha^{(i-1)}$$

   where

$${}^{(T)}K_{\alpha\beta} = -\left(K_{\alpha\beta} + \frac{\partial K_{\alpha\gamma}}{\partial q_\beta}\,q_\gamma\right) \tag{3.23}$$

   is the so-called tangent system matrix,
3. updating the solution according to the rule

$$\mathbf{q}^{(i)} = \mathbf{q}^{(i-1)} + \delta\mathbf{q}^{(i)}\,,$$

4. continuing the iteration until the difference between two successive estimates is sufficiently small.

Still another method, used as a rule in the so-called path-dependent problems of mechanics, can be best discussed by considering the right-hand side and solution vectors $\mathbf{Q}$ and $\mathbf{q}$ as functions of time, or time-like parameter τ in the form

$$Q_\alpha = Q_\alpha(\tau), \qquad q_\alpha = q_\alpha(\tau). \tag{3.24}$$

By differentiating Eq. (3.17) with respect to τ we arrive at

$$\left[K_{\alpha\beta}(q_\gamma) + \frac{\partial K_{\alpha\gamma}}{\partial q_\beta}q_\gamma\right]\dot{q}_\beta = \dot{Q}_\alpha\,,$$

or, more compactly, cf. Eq. (3.23), at

$${}^{(T)}K_{\alpha\beta}(q_\gamma)\dot{q}_\beta = \dot{Q}_\alpha\,, \tag{3.25}$$

with the dot indicating the derivative with respect to τ. Eq. (3.25) is a system of ordinary differential equations which, when complemented with the initial condition

$$q_\alpha(0) = \hat{q}_\alpha\,,$$

may be regarded as equivalent to the original algebraic problem (3.22).

Many algorithms exist in the literature for solving sets of ODE's of the type (3.25). In the area of mechanics only so-called one-step time integration algorithms are used which is due to typically very large dimensions of the tangent matrix ${}^{(T)}\mathbf{K}$ (N typically of the order 10^3 or more).

After replacing the derivatives in Eq. (3.25) by the corresponding finite difference expressions the class of algorithms of our interest can be somewhat symbolically presented as

$$ {}^{(T)}\mathbf{K}(\mathbf{q} + \xi \Delta\mathbf{q})\Delta\mathbf{q} = \Delta\mathbf{Q} \,. \tag{3.26} $$

The resulting system of N nonlinear algebraic equations should be solved for N components of the solution vector $\Delta\mathbf{q}$ at each time station considered in the problem. For $\xi = 0$ the algorithms generated by Eq. (3.26) are referred to as explicit while the assumption $\xi \neq 0$ leads to the class of algorithms known as implicit. It is noted that the explicit algorithms are non-iterative (but typically less accurate) whereas the implicit ones call for iteration at each time station. The use of different time-stepping algorithms will be discussed in more detail at many places throughout the book.

References

1. Eringen AC (1962) Nonlinear theory of continuous media. McGraw–Hill, New York
2. Finlayson BA (1972) The method of weighted residuals and variational Principles. Academic Press, New York
3. Fung Y–C (1965) Foundations of solid mechanics. Prentice Hall, Englewood Cliffs
4. Green AE, Adkins JE (1980) Large elastic deformations and nonlinear continuum mechanics. Clarendon Press, Oxford
5. Jaunzemis W (1967) Continuum mechanics. Macmillan, New York
6. Kleiber M (1989) Incremental finite element modelling in non-linear solid mechanics. PWN–Ellis Horwood
7. Kleiber M, Antúnez H, Hien TD, Kowalczyk P (1997) Parameter sensitivity in nonlinear mechanics. Wiley, Chichester
8. Kleiber M, Woźniak C (1990) Nonlinear mechanics of structures. Kluwer–PWN, Dordrecht–Warsaw
9. Lai WM, Rubin D, Krempl E (1979) Introduction to continuum mechanics. Pergamon Press, Oxford
10. Malvern LE (1969) Introduction to the mechanics of continuous media. Prentice Hall, Englewood Cliffs
11. Mikhlin A (1964) Variational methods in mathematical physics. Macmillan, New York
12. Perzyna P (1966) Theory of viscoplasticity (in Polish). Polish Scientific Publishers PWN, Warsaw
13. Prager W (1961) Introduction to mechanics of continua. Ginn & Co., New York
14. Sedow LJ (1965) Introduction to the mechanics of continuous media. Addison–Wesley, New York
15. Truesdell C, Noll W (1965) The non-linear field theories of mechanics, vol III/3 of Handbuch der Physik. Springer Verlag, Berlin
16. Truesdell C, Toupin RA (1960) The classical field theories, vol III/1 of Handbuch der Physik. Springer Verlag, Berlin
17. Woźniak C (1969) Fundations of dynamics of deformable bodies (in Polish). Polish Scientific Publishers PWN, Warsaw

Part II

Finite Element Method

1 Introduction

We have already indicated that an efficient way to approximately solve the boundary value problems typical of solid mechanics applications can be based on using the Galerkin or Rayleigh–Ritz methods. However, a serious and common difficulty in using these methods in computational practice is the necessity to select a set of trial functions approximating the solution well enough in the whole domain considered. For in accordance with what we have discussed so far, these functions must satisfy the problem boundary conditions at the same time being able to properly represent the geometry of the domain and different other characteristics of the problem. For problems with complex shape and anticipated non-uniform character of the solution the task of generating such global trial functions becomes increasingly difficult until, for most situations of practical interest, it becomes quite impossible to handle.

With the development of powerful computer techniques it has become possible to construct the approximating functions in a localized manner. Let us imagine a solid divided into a number of finite sized and conveniently shaped sub-regions or elements. The distribution of the required quantities is regarded as relatively simple within these elements but by using a sufficient number of them an acceptable representation of the overall situation is obtained. The basic premise here is that each element is small and so the variation of the test functions over its entire area will be small as well. To improve the approximation one can either divide the domain into smaller elements or use more degrees of freedom per element, or do both things. On the other hand, the repetitive character of the element shapes as well as of the approximating functions makes it possible to automate the computing process thus making it particularly suitable to the use of computer techniques.

The version of the Galerkin (or Rayleigh–Ritz, if one only can construct the appropriate variational functional) method based on the above methodology is called the finite element method (FEM). The very basic features of FEM will be illustrated below by taking reference to the general boundary value problem of linear 3D elasticity. The Rayleigh–Ritz method will be employed by adapting the potential energy functional as the basis for seeking the approximate solution.

Suppose that a given domain Ω is divided into a finite number, say E, of open disjoint subregions – finite elements Ω_e , $e = 1, 2, \ldots, E$; $\Omega_e \cap \Omega_f = \emptyset$ for $e \neq f$. The boundary of the element e is denoted by $\partial\Omega_e$, the boundary common to two adjacent elements e and f by $\partial\Omega_{ef} = \partial\Omega_e \cap \partial\Omega_f$ while the part of the boundary of the element e which also belongs to the boundary of the whole region $\partial\Omega$ by $\partial\Omega_{\bar{e}}$, $\partial\Omega_{\bar{e}} = \partial\Omega \cap \partial\Omega_e$.

For every finite element e we introduce a local (i.e. defined for this particular element) Cartesian[1] coordinate system $x_k^{(e)}$, $e = 1, 2, \dots, E$. We assume that the transformation of the vector components from the global coordinate system x_k to the local one $x_k^{(e)}$ is performed by means of the transformation matrix $\mathbf{A}^{(e)}$ so that for any vector $\mathbf{a}$ we have $a_k^{(e)} = A_{kl}^{(e)} a_l$, etc.

In order to obtain the fundamental set of equations for the so-called conforming displacement model of FEM we shall use the principle of minimum potential energy. The potential energy for the assembly of elements may be expressed as, cf. Eqs. (2.3), Part I

$$J_P[\mathbf{u}] = \sum_{e=1}^{E} \left[\int_{\Omega_e} \left(\tfrac{1}{2} \boldsymbol{\varepsilon}^T \mathbf{C} \boldsymbol{\varepsilon} - \varrho \hat{\mathbf{f}}^T \mathbf{u} \right) \mathrm{d}\Omega - \int_{\partial\Omega_{\bar{e}}^{\sigma}} \hat{\mathbf{t}}^T \mathbf{u} \, \mathrm{d}(\partial\Omega) \right] \tag{1.1}$$

where the strain $\boldsymbol{\varepsilon}$ is treated as the function of the displacement $\mathbf{u}$ according to $\boldsymbol{\varepsilon} = \mathbf{D}\mathbf{u}$ and it is assumed that $\mathbf{u}$ satisfies the kinematic boundary conditions (2.34), Part I, for $\mathbf{x} \in \partial\Omega_u$. The symbol $\partial\Omega_{\bar{e}}^{\sigma}$ stands here for that part of the e-th element boundary on which the traction boundary conditions are prescribed The formulation (1.1) is equivalent to the expression (2.35), Part I , provided the functions $\mathbf{u}(\mathbf{x})$ are continuous in the whole domain Ω and differentiable within each subdomain Ω_e , $e = 1, 2, \dots, E$; differentiability is not required at the interelemental boundaries $\partial\Omega_{ef}$, $e, f = 1, 2, \dots, E$; $e \neq f$.

Generally, in order to assure convergence of the approximate solution to the exact one with the mesh becoming more and more dense, in the variational problem described by a functional involving the derivatives of the unknown functions up to the m-th order, one has to fulfil the following requirements:

1. compatibility conditions – at the interelemental boundaries the unknown functions must be differentiable up to the order $(m-1)$,
2. completeness conditions – within the elements the unknown functions must have continuous derivatives up to the order m.

Assume now that the components $u_k(\mathbf{x})$ can be approximated within each element Ω_e by means of the so-called shape functions $\varphi_k(\mathbf{x})$ as

$$u_k(\mathbf{x}) = \varphi_{k\,1\times N_e}(\mathbf{x})\, \mathbf{q}^{(e)}_{N_e\times 1} , \qquad \mathbf{u}_{3\times 1}(\mathbf{x}) = \varphi_{3\times N_e}(\mathbf{x})\, \mathbf{q}^{(e)}_{N_e\times 1} , \tag{1.2}$$

where the $3\times N_e$ matrix φ has the form

$$\varphi = \{\varphi_1\, \varphi_2\, \varphi_3\} = \begin{bmatrix} \varphi_{11} & \varphi_{12} & \cdots & \varphi_{1\,N_e} \\ \varphi_{21} & \varphi_{22} & \cdots & \varphi_{2\,N_e} \\ \varphi_{31} & \varphi_{32} & \cdots & \varphi_{3\,N_e} \end{bmatrix} . \tag{1.3}$$

[1] As in Chap. 2, Part I, only Cartesian coordinate systems are employed for simplicity of the presentation.

The row vectors $\varphi_k(\mathbf{x}) = [\varphi_{k1}\ \varphi_{k2}\ \cdots\ \varphi_{k\,N_e}]$ and column vectors $\mathbf{q}^{(e)} = \{q_1^{(e)}\ q_2^{(e)}\ \ldots\ q_{N_e}^{(e)}\}$ represent the element shape functions and the element generalized coordinates (degrees of freedom), respectively while N_e stands for the number of the degrees of freedom in the e-th finite element on hand.

By using the strain vector definition (2.29), Part I, and the relationship (1.2) we arrive at

$$\varepsilon_{6\times1}(\mathbf{x}) = \mathbf{B}_{6\times N_e}(\mathbf{x})\,\mathbf{q}^{(e)}_{N_e\times1}\,, \tag{1.4}$$

where

$$\mathbf{B}_{6\times N_e}(\mathbf{x}) = \mathbf{D}_{6\times3}\,\varphi_{3\times N_e}(\mathbf{x})\,. \tag{1.5}$$

By Eq. (1.5), the e-th element contribution to the total potential energy (1.1) can be presented as

$$J_P^{(e)}[\mathbf{q}^{(e)}] = \tfrac{1}{2}\overset{T}{\mathbf{q}}{}^{(e)}\,\mathbf{k}^{(e)}\,\mathbf{q}^{(e)} - \overset{T}{\mathbf{Q}}{}^{(e)}\,\mathbf{q}^{(e)}\,, \tag{1.6}$$

where

$$\begin{aligned}\mathbf{k}^{(e)}_{N_e\times N_e} &= \int_{\Omega_e} \mathbf{B}^T_{N_e\times6}\,\mathbf{C}_{6\times6}\,\mathbf{B}_{6\times N_e}\,\mathrm{d}\Omega,\\ \mathbf{Q}^{(e)}_{N_e\times1} &= \int_{\Omega_e} \varphi^T_{N_e\times3}\,\varrho\,\hat{\mathbf{f}}_{3\times1}\,\mathrm{d}\Omega + \int_{\partial\Omega_e^\sigma} \varphi^T_{N_e\times3}\,\hat{\mathbf{t}}_{3\times1}\,\mathrm{d}(\partial\Omega).\end{aligned} \tag{1.7}$$

The matrix $\mathbf{k}^{(e)}$ is called the elastic stiffness matrix for the element considered while the vector $\mathbf{Q}^{(e)}$ is the generalized nodal force vector acting upon the element through its nodes. We emphasize that the components of the vectors $\mathbf{q}^{(e)}$ and $\mathbf{Q}^{(e)}$ and the matrix $\mathbf{k}^{(e)}$ are taken with respect to the local coordinate system $x_k^{(e)}$, $k = 1, 2, 3$.

Suppose now that in the discretized system consisting of all the individual elements linked together at nodes a global numeration of the nodes has been carried out. Assigning to each node its number of degrees of freedom we may in this way establish the numbering of all the degrees of freedom in the system on hand. Let the column vector of so ordered generalized displacement components taken with respect to the global coordinates be denoted as

$$\mathbf{q}_{N\times1} = \{q_\alpha\} = \{q_1\ q_2\ \ldots\ q_N\}\,, \tag{1.8}$$

where N is the total number of the degrees of freedom in the system.

Assume further for simplicity that the external load acts upon the structure through the nodes only – otherwise the load has to be reduced to the statically equivalent nodal forces. The column vector of the external nodal load components taken with respect to the global coordinates is denoted as

$$\mathbf{Q}_{N\times1} = \{Q_\alpha\} = \{Q_1\ Q_2\ \ldots\ Q_N\}\,. \tag{1.9}$$

The components of $\mathbf{Q}$ are ordered in exactly the same way as those of $\mathbf{q}$ – this makes it possible to express the work of external forces as $\frac{1}{2}\mathbf{Q}^T\mathbf{q}$.

Because the components of all the elemental quantities are referred to the local (for each element) coordinate system $x_k^{(e)}$ in order to carry out the assembly process we have first to transform all the elemental matrices and vectors from $x_k^{(e)}$ to the global system x_k. Denoting by $\tilde{\mathbf{q}}^{(e)}$ the vector of the generalized displacement components for the typical e-th element taken with respect to the global system and by ${}^{I}\mathbf{a}^{(e)}$ the $N_e \times N_e$ matrix which transforms $\tilde{\mathbf{q}}^{(e)}$ into $\mathbf{q}^{(e)}$ we have

$$\mathbf{q}^{(e)}_{N_e\times 1} = {}^{I}\mathbf{a}^{(e)}_{N_e\times N_e}\,\tilde{\mathbf{q}}^{(e)}_{N_e\times 1}\,. \tag{1.10}$$

The transformation matrix ${}^{I}\mathbf{a}^{(e)}$ consists of diagonally located submatrices of transformation from the coordinates x_k to the coordinates $x_k^{(e)}$ for vectors representing generalized displacements of particular element nodes. For a general 3D problem we thus have, for instance

$${}^{I}\mathbf{a}^{(e)} = \begin{bmatrix} \mathbf{A}^{(e)}_{3\times 3} & & & \\ & \mathbf{A}^{(e)}_{3\times 3} & & \\ & & \ddots & \\ & & & \mathbf{A}^{(e)}_{3\times 3} \end{bmatrix}_{N_e\times N_e}\,, \tag{1.11}$$

since

$$\begin{bmatrix} {}^{(1)}q_1^{(e)} \\ {}^{(1)}q_2^{(e)} \\ {}^{(1)}q_3^{(e)} \end{bmatrix} = \mathbf{A}^{(e)}_{3\times 3} \begin{bmatrix} {}^{(1)}q^{(e)}_{1(e)} \\ {}^{(1)}q^{(e)}_{2(e)} \\ {}^{(1)}q^{(e)}_{3(e)} \end{bmatrix}, \qquad \text{etc.} \tag{1.12}$$

The matrix $\mathbf{A}^{(e)}$ consists of direction cosines of coordinate axes x_k with respect to the axes of coordinate set $x_k^{(e)}$, respectively. Using Eq. (1.10) the element potential energy (1.6) becomes

$$J_P^{(e)}[\mathbf{q}^{(e)}] = \tfrac{1}{2}\,\overset{T}{\tilde{\mathbf{q}}}{}^{(e)}\, {}^{I}\overset{T}{\mathbf{a}}{}^{(e)}\, \mathbf{k}^{(e)}\, {}^{I}\mathbf{a}^{(e)}\, \tilde{\mathbf{q}}^{(e)} - \overset{T}{\tilde{\mathbf{Q}}}{}^{(e)}\, \tilde{\mathbf{q}}^{(e)}\,, \tag{1.13}$$

in which we have used the condition

$${}^{I}\overset{T}{\mathbf{a}}{}^{(e)}\, {}^{I}\mathbf{a}^{(e)} = \mathbf{I}_{N_e}\,.$$

Equation (1.13) is written in a more compact form as

$$J_P^{(e)}[\mathbf{q}^{(e)}] = \tfrac{1}{2}\,\overset{T}{\tilde{\mathbf{q}}}{}^{(e)}\, \tilde{\mathbf{k}}^{(e)}\, \tilde{\mathbf{q}}^{(e)} - \overset{T}{\tilde{\mathbf{Q}}}{}^{(e)}\, \tilde{\mathbf{q}}^{(e)}\,, \tag{1.14}$$

We emphasize that the dimension of the elemental stiffness matrix

$$\tilde{\mathbf{k}}^{(e)} = {}^{I}\overset{T}{\mathbf{a}}{}^{(e)}\, \mathbf{k}^{(e)}\, {}^{I}\mathbf{a}^{(e)} \tag{1.15}$$

and of the elemental load and displacement vectors $\tilde{\mathbf{Q}}^{(e)}$, $\tilde{\mathbf{q}}^{(e)}$ remain unchanged, i.e. $N_e\times N_e$, $N_e\times 1$, $N_e\times 1$, respectively.

The next step in the process of assembling the elements is to expand each elemental matrix $\tilde{\mathbf{k}}^{(e)}$ and each elemental vectors $\tilde{\mathbf{Q}}^{(e)}$, $\tilde{\mathbf{q}}^{(e)}$ to the dimensions $N\times N$, $N\times 1$, $N\times 1$, respectively, in the way which corresponds to the global numbering of the degrees of freedom. Thus, for the e-th element generalized displacement components expressed in the global coordinate system, denoted as $\tilde{q}_1^{(e)}, \tilde{q}_2^{(e)}, \ldots, \tilde{q}_{N_e}^{(e)}$ and having respectively the numbers $\alpha_1, \alpha_2, \ldots, \alpha_{N_e}$ within the global numbering scheme, the expanded vectors $\bar{\mathbf{Q}}^{(e)}$, $\bar{\mathbf{q}}^{(e)}$ and matrix $\mathbf{K}^{(e)}$ take the form

$$\begin{aligned}\bar{\mathbf{q}}^{(e)}_{N\times 1} &= \{\, 0 \ldots 0 \ \tilde{q}_1^{(e)} \ 0 \ldots 0 \ \tilde{q}_2^{(e)} \ 0 \ldots\ldots 0 \ \tilde{q}_{N_e}^{(e)} \ 0 \ldots 0 \,\} \\ &= \{\, 0 \ldots 0 \ q_{\alpha_1} \ 0 \ldots 0 \ q_{\alpha_2} \ 0 \ldots\ldots 0 \ q_{\alpha_{N_e}} \ 0 \ldots 0 \,\}\,,\end{aligned} \tag{1.16}$$

$$\begin{aligned}\bar{\mathbf{Q}}^{(e)}_{N\times 1} &= \{\, 0 \ldots 0 \ \tilde{Q}_1^{(e)} \ 0 \ldots 0 \ \tilde{Q}_2^{(e)} \ 0 \ldots\ldots 0 \ \tilde{Q}_{N_e}^{(e)} \ 0 \ldots 0 \,\} \\ &= \{\, 0 \ldots 0 \ Q_{\alpha_1} \ 0 \ldots 0 \ Q_{\alpha_2} \ 0 \ldots\ldots 0 \ Q_{\alpha_{N_e}} \ 0 \ldots 0 \,\}\,,\end{aligned} \tag{1.17}$$

$$\mathbf{K}^{(e)}_{N\times N} = \begin{array}{c} \begin{matrix} & \text{column} & \text{column} & & \text{column} & \\ & \alpha_1 & \alpha_2 & & \alpha_{N_e} & \end{matrix} \\ \left[\begin{matrix} & \vdots & & \vdots & & \vdots & \\ \cdots & \tilde{k}_{11}^{(e)} & \cdots & \tilde{k}_{12}^{(e)} & \cdots\cdots\cdots & \tilde{k}_{1\,N_e}^{(e)} & \cdots \\ & \vdots & & \vdots & & \vdots & \\ \cdots & \tilde{k}_{12}^{(e)} & \cdots & \tilde{k}_{22}^{(e)} & \cdots\cdots\cdots & \tilde{k}_{2\,N_e}^{(e)} & \cdots \\ & \vdots & & \vdots & & \vdots & \\ \cdots & \tilde{k}_{1\,N_e}^{(e)} & \cdots & \tilde{k}_{2\,N_e}^{(e)} & \cdots\cdots\cdots & \tilde{k}_{N_e\,N_e}^{(e)} & \cdots \\ & \vdots & & \vdots & & \vdots & \end{matrix}\right] \end{array} \begin{matrix} \\ \text{row } \alpha_1 \\ \\ \text{row } \alpha_2 \\ \\ \text{row } \alpha_{N_e} \\ \end{matrix} \tag{1.18}$$

We stress once again that apart from the components of $\mathbf{K}^{(e)}$ and $\bar{\mathbf{q}}^{(e)}$, $\bar{\mathbf{Q}}^{(e)}$ which are equal to the corresponding entries of the matrix $\tilde{\mathbf{k}}^{(e)}$ and vectors $\tilde{\mathbf{q}}^{(e)}$, $\tilde{\mathbf{Q}}^{(e)}$, the remaining entries in Eqs. (1.16)–(1.18) are zeros. The above algebraic transformation may formally be described by defining a Boolean matrix ${}^{II}\mathbf{a}^{(e)}$ such that

$$\tilde{\mathbf{q}}^{(e)}_{N_e\times 1} = {}^{II}\mathbf{a}^{(e)}_{N_e\times N}\, \mathbf{q}_{N\times 1}\,, \tag{1.19}$$

where

$${}^{II}\mathbf{a}^{(e)}_{N_e\times N} = \begin{array}{c} \begin{matrix} \qquad\quad \alpha_1 & \qquad \alpha_2 & \quad \cdots & \qquad \alpha_{N_e} \\ \qquad\quad \vdots & \qquad \vdots & & \qquad \vdots \end{matrix} \\ \begin{bmatrix} 0 \cdots 0\,1\,0 \cdots\cdots\cdots\cdots & \cdots & \cdots\cdots\cdots\cdots\, 0 \\ 0 \cdots\cdots\cdots\cdots 0\,1\,0 \cdots & \cdots & \cdots\cdots\cdots\cdots\, 0 \\ \vdots \qquad\quad \vdots \qquad\quad \vdots & & \vdots \qquad \vdots \\ 0 \cdots\cdots\cdots\cdots\cdots\cdots & \cdots & \cdots 0\,1\,0 \cdots 0 \end{bmatrix} \end{array} .$$

It is an elementary task to check that

$$\begin{aligned}\tilde{\mathbf{q}}^{(e)}_{N_e\times 1} &= {}^{II}\mathbf{a}^{(e)}_{N_e\times N}\, \bar{\mathbf{q}}^{(e)}_{N\times 1}\,, \\ \tilde{\mathbf{Q}}^{(e)}_{N_e\times 1} &= {}^{II}\mathbf{a}^{(e)}_{N_e\times N}\, \bar{\mathbf{Q}}^{(e)}_{N\times 1}\,.\end{aligned} \tag{1.20}$$

which, when used in Eq. (1.14) leads to the expression for the element potential energy in the form

$$J_P^{(e)}[\mathbf{q}] = \tfrac{1}{2}\overset{T}{\mathbf{q}}\, {}^{II}\overset{T}{\mathbf{a}}{}^{(e)}\, \tilde{\mathbf{k}}^{(e)}\, {}^{II}\mathbf{a}^{(e)}\, \mathbf{q} - \overset{T}{\tilde{\mathbf{Q}}}{}^{(e)}\, \mathbf{q}\,. \tag{1.21}$$

In deriving Eq. (1.21) the condition was used that

$$
{}^{II}\overset{T}{\mathbf{a}}{}^{(e)}\, {}^{II}\mathbf{a}^{(e)} =
\begin{array}{c}
\begin{array}{ccc} \alpha_1 & \alpha_2 & \alpha_{N_e} \end{array} \\
\left[\begin{array}{ccccccc}
 & \vdots & & \vdots & & \vdots & \\
\cdots\cdots & 1 & \cdots\cdots & 1 & \cdots\cdots\cdots\cdots & 1 & \cdots\cdots \\
 & \vdots & & \vdots & & \vdots & \\
\cdots\cdots & 1 & \cdots\cdots & 1 & \cdots\cdots\cdots\cdots & 1 & \cdots\cdots \\
 & \vdots & & \vdots & & \vdots & \\
\cdots\cdots & 1 & \cdots\cdots & 1 & \cdots\cdots\cdots\cdots & 1 & \cdots\cdots \\
 & \vdots & & \vdots & & \vdots &
\end{array}\right]
\end{array}
\begin{array}{c} \\ \alpha_1 \\ \\ \alpha_2 \\ \\ \alpha_{N_e} \\ \end{array}
$$

Equation (1.21) is written in a more compact way as

$$J_P^{(e)}[\mathbf{q}] = \tfrac{1}{2}\overset{T}{\mathbf{q}}_{1\times N}\, \mathbf{K}^{(e)}_{N\times N}\, \mathbf{q}_{N\times 1} - \overset{T}{\mathbf{Q}}{}^{(e)}_{1\times N}\, \mathbf{q}_{N\times 1}\,. \tag{1.22}$$

Let us observe that the transformation of the element potential energy from the form (1.6) to the form (1.22) may be concisely represented by using the transformation matrix defined as

$$\mathbf{a}^{(e)}_{N_e\times N} = {}^{I}\mathbf{a}^{(e)}_{N_e\times N_e}\; {}^{II}\mathbf{a}^{(e)}_{N_e\times N} \tag{1.23}$$

which embraces both the transformation $x_k^{(e)} \to x_k$ and the procedure of identifying the local and global degrees of freedom.

Having represented the e-th element potential energy in the form (1.22) we may now easily calculate the potential energy for the whole system. The functional J_P is given simply as the sum of its elemental contributions so that

$$J_P[\mathbf{q}] = \sum_{e=1}^{E} J_P^{(e)}[\mathbf{q}] = \tfrac{1}{2}\mathbf{q}^T\mathbf{K}\mathbf{q} - \mathbf{Q}^T\mathbf{q}\,, \tag{1.24}$$

where

$$\mathbf{K}_{N\times N} = \sum_{e=1}^{E}\mathbf{K}^{(e)}_{N\times N}\,, \qquad \mathbf{Q}_{N\times 1} = \sum_{e=1}^{E}\mathbf{Q}^{(e)}_{N\times 1}\,. \tag{1.25}$$

In accordance with the previous notation the vector $\mathbf{Q}$ represents the external load acting upon the nodes of the system. To justify this interpretation we note that since the expression $-\sum_{e=1}^{E}\mathbf{Q}^{(e)}$ groups all the resultant vectors of forces acting upon the particular nodes from their surrounding elements,

Eq. (1.25) is nothing else but the equilibrium equations for all the nodes in the system. To maintain equilibrium we thus need to enforce the condition

$$-\sum_{e=1}^{E} \mathbf{Q}^{(e)} + \mathbf{Q} = 0\,,$$

which is just another form of Eq. (1.25).

The stationarity condition for the functional of potential energy

$$\frac{\partial J_P}{\partial \mathbf{q}} \delta \mathbf{q} = 0 \tag{1.26}$$

leads, due to the independence of each particular variation in $\delta \mathbf{q} = \{\delta q_1, \delta q_2, \ldots, \delta q_N\}$, to

$$\mathbf{K} \mathbf{q} = \mathbf{Q}\,. \tag{1.27}$$

The relationship (1.27) is referred to as the fundamental algebraic equation describing the linear problem of statics of the deformable body within the framework of the finite element displacement model.

We emphasize that the variational formulation here has been based displacement functions $\mathbf{u}$ satisfying the kinematic boundary conditions. The consequence of this fact is that Eq. (1.27) may be solved provided the appropriate boundary conditions have been imposed on the generalized nodal displacements $q_1, q_2, \ldots, q_N$.

We summarize the findings so far by enumerating the steps typical of the displacement-based linear FEM approach:

1. domain discretization, selection of element types and numbering of the nodes,
2. evaluation of the element stiffness matrices,
3. assembly of the global stiffness matrix and load vector,
4. imposing kinematic constraints,
5. solution of the fundamental system of equations,
6. calculation of secondary variables (strains, stresses),
7. assessment of the solution accuracy.

The FEM equations of statics can be generalized to incorporate inertial and damping effects by taking reference to the Hamilton principle discussed in Sect. 2.6 of Part I. To this aim we assume the expansion for the displacement field $\mathbf{u}(\mathbf{x}, \tau)$ in the form

$$\begin{aligned} \mathbf{u}_{3\times1}(\mathbf{x}, \tau) &= \varphi_{3\times N_e}(\mathbf{x})\, \mathbf{q}^{(e)}_{N_e\times1}(\tau) \\ &= \varphi_{3\times N_e}(\mathbf{x})\, \mathbf{a}^{(e)}_{N_e\times N}\, \mathbf{q}_{N\times1}(\tau) = \mathbf{\Phi}_{3\times N}(\mathbf{x})\, \mathbf{q}_{N\times1}(\tau)\,, \end{aligned} \tag{1.28}$$

different from the one employed so far by explicitly indicating the dependence of the generalized coordinates on time. The strain vector becomes

$$\begin{aligned}\boldsymbol{\varepsilon}_{6\times 1}(\mathbf{x},\tau) &= \mathbf{B}_{6\times N_e}(\mathbf{x})\,\mathbf{q}^{(e)}_{N_e\times 1}(\tau)\\ &= \mathbf{B}_{6\times N_e}(\mathbf{x})\,\mathbf{a}^{(e)}_{N_e\times N}\,\mathbf{q}_{N\times 1}(\tau) = \tilde{\mathbf{B}}_{6\times N}(\mathbf{x})\,\mathbf{q}_{N\times 1}(\tau)\,. \end{aligned} \tag{1.29}$$

By using Eqs. (1.28), (1.29) in the Hamilton principle (2.39), Part I, we arrive at

$$\delta\int_{t_{(1)}}^{t_{(2)}}\left[\frac{1}{2}\sum_{e=1}^{E}\dot{\mathbf{q}}^{T(e)}\,\mathbf{m}^{(e)}\,\dot{\mathbf{q}}^{(e)} - \frac{1}{2}\sum_{e=1}^{E}\mathbf{q}^{T(e)}\,\mathbf{k}^{(e)}\,\mathbf{q}^{(e)} + \sum_{e=1}^{E}\mathbf{Q}^{T(e)}\,\mathbf{q}^{(e)}\right]\mathrm{d}\tau = 0\,, \tag{1.30}$$

or, after assembling the elements, at

$$\delta\int_{t_{(1)}}^{t_{(2)}}\left[\tfrac{1}{2}\dot{\mathbf{q}}^T\,\mathbf{M}\,\dot{\mathbf{q}} - \tfrac{1}{2}\mathbf{q}^T\,\mathbf{K}\,\mathbf{q} + \mathbf{Q}^T\,\mathbf{q}\right]\mathrm{d}\tau = 0\,. \tag{1.31}$$

The element and global mass matrices are here defined as

$$\begin{aligned}\mathbf{m}^{(e)}_{N_e\times N_e} &= \int_{\Omega_e}\varrho(\mathbf{x})\,\mathbf{B}^T_{N_e\times 6}(\mathbf{x})\,\mathbf{B}_{6\times N_e}(\mathbf{x})\,\mathrm{d}\Omega\,,\\ \mathbf{M}_{N\times N} &= \int_{\Omega}\varrho(\mathbf{x})\,\tilde{\mathbf{B}}^T_{N\times 6}(\mathbf{x})\,\tilde{\mathbf{B}}_{6\times N}(\mathbf{x})\,\mathrm{d}\Omega\,,\end{aligned} \tag{1.32}$$

the stiffness matrices are given by the expressions already discussed, i.e.

$$\begin{aligned}\mathbf{k}^{(e)}_{N_e\times N_e} &= \int_{\Omega_e}\mathbf{B}^T_{N_e\times 6}(\mathbf{x})\,\mathbf{C}_{6\times 6}(\mathbf{x})\,\mathbf{B}_{6\times N_e}(\mathbf{x})\,\mathrm{d}\Omega\,,\\ \mathbf{K}_{N\times N} &= \int_{\Omega}\tilde{\mathbf{B}}^T_{N\times 6}(\mathbf{x})\,\mathbf{C}_{6\times 6}(\mathbf{x})\,\tilde{\mathbf{B}}_{6\times N}(\mathbf{x})\,\mathrm{d}\Omega\,.\end{aligned} \tag{1.33}$$

and the load vector depends now on time, $\mathbf{Q} = \mathbf{Q}(\mathbf{x},\tau)$. By taking the variation and integrating by parts with respect to time Eq. (1.31) becomes

$$\dot{\mathbf{q}}^T\,\mathbf{M}\,\delta\mathbf{q}\,\Big|_{t_{(1)}}^{t_{(2)}} - \int_{t_{(1)}}^{t_{(2)}}(\ddot{\mathbf{q}}^T\,\mathbf{M} + \mathbf{q}^T\,\mathbf{K} - \mathbf{Q}^T)\,\delta\mathbf{q}\,\mathrm{d}\tau = 0\,.$$

By observing that

$$\delta\mathbf{q}(t_{(1)}) = 0\,, \qquad \delta\mathbf{q}(t_{(2)}) = 0\,,$$

we obtain the final form of the stationarity condition for the functional in the Hamilton principle as

$$\mathbf{M}\ddot{\mathbf{q}} + \mathbf{K}\mathbf{q} = \mathbf{Q}\,. \tag{1.34}$$

This is the equation of motion for the discretized system on hand. The mass matrix defined by Eqs. (1.32) is called consistent (with the stiffness matrix, as both have been derived using the same shape functions). The consistent mass matrix for the element is in general fully populated, and the global mass matrix is bounded. In computational practice another mass matrix is frequently employed based on the inertial model of the system in the form of masses concentrated at nodes. Such an elemental mass matrix, referred to in the literature as inconsistent, has a diagonal build-up which implies the same population pattern in the global mass matrix for the whole system.

The equation of motion (1.34) is usually augmented with the term $\mathbf{C}\dot{\mathbf{q}}$ which represents damping properties of the system, thus yielding

$$\mathbf{M}\ddot{\mathbf{q}} + \mathbf{C}\dot{\mathbf{q}} + \mathbf{K}\mathbf{q} = \mathbf{Q}\,. \tag{1.35}$$

The damping matrix $\mathbf{C}$ is usually postulated as the so-called proportional damping matrix which has the form

$$\mathbf{C} = \alpha_1 \mathbf{M} + \alpha_2 \mathbf{K}\,, \tag{1.36}$$

with the coefficients α_1, α_2 taken from modal considerations.

Equation (1.34) has to be considered together with appropriate initial conditions of the general form

$$\mathbf{q}(0) = \hat{\mathbf{q}}\,, \qquad \dot{\mathbf{q}}(0) = \hat{\dot{\mathbf{q}}}\,. \tag{1.37}$$

Only a very rudimentary discussion of the FEM fundamentals has been given in this section – a much more detailed information regarding further applications of the general FEM methodology to problems of solid mechanics will be given in subsequent chapters.

2 Selected Topics from the Mathematical Theory of Finite Elements

2.1 Variational Formulation

The mathematical foundations of Finite Elements are based on the modern theory of partial differential equations (PDE's) which in turn is rooted in Functional Analysis, another fundamental branch of mathematics concerned with function spaces (sets of functions with a specific algebraic structure). As every branch of mathematics, the Functional Analysis uses its own precise language to describe the objects of interest. Understanding the language does not reduce only to acquiring a new vocabulary and developing the corresponding intuition, but requires a systematic mathematical study on the subject [34, 127]. For that reason lectures on the mathematical theory of FEM are usually preceded by elements of real analysis including foundations of modern mathematics, linear algebra in infinite dimensional spaces, Lebesgue integration, topology and metric spaces, all of these subjects leading eventually to the actual topics of Functional Analysis – the theory of Banach and Hilbert spaces [126].

It is obvious that the scope of this chapter precludes such a depth of presentation, forcing us to focus on a specific selection of topics which may be presented to a general engineering audience in a comparatively simple way. We hope that the "story-telling" fashion of this chapter will not scare away readers with more advanced mathematical background and, at the same time, will help the "practitioners" of the FEM to add some precision to their every day struggle.

As a framework for our discussions, we choose the equations of two-dimensional elastostatics represented in terms of the unknown displacement vector $\mathbf{u} = (u_i), i = 1, 2$ (Lame's equations).

Find $\mathbf{u}(\mathbf{x}), \mathbf{x} \in \Omega$, such that

$$
\begin{aligned}
-\sigma_{kl,l} &= f_k && \text{in } \Omega, \\
u_k &= 0 && \text{on } \partial\Omega_u\,, \\
\sigma_{kl} n_l &= t_k && \text{on } \partial\Omega_\sigma\,,
\end{aligned}
\tag{2.1}
$$

where

$$\sigma_{kl} = C_{klmn}\epsilon_{mn} \tag{2.2}$$

and

$$\epsilon_{mn} = \frac{1}{2}(u_{m,n} + u_{n,m}). \tag{2.3}$$

Strains ϵ_{mn} and stresses σ_{kl} must not be understood as new unknowns but simply as expressions (functions of the unknown displacement $\mathbf{u}$) facilitating a concise (and physically clear) statement of the differential equation $(2.1)_1$ and traction boundary conditions $(2.1)_3$.

In order to simplify the presentation, the kinematic boundary conditions $(2.1)_2$ have been assumed in the homogeneous form, i.e. with the zero right-hand side [comp. $(2.34)_5$, Part I]. In (2.1), f_k stand for prescribed volumetric forces ($f_k = \rho\hat{f}_k$), and $t_k(=\hat{t}_k)$ denote the prescribed tractions on part $\partial\Omega_\sigma$ of the boundary.

We assume that the tensor of elasticities C_{klmn} satisfies the classical symmetry conditions:

$$C_{klmn} = C_{klnm}, \quad C_{klmn} = C_{lkmn}, \quad C_{klmn} = C_{mnkl}, \tag{2.4}$$

for $k, l, m, n = 1, 2$, as well as the *ellipticity condition*:

$$C_{klmn}\xi_{kl}\xi_{mn} \geq \alpha\xi_{kl}\xi_{kl}, \tag{2.5}$$

for every symmetric tensor $\xi_{kl} = \xi_{lk}$, whereas $\alpha > 0$.

It is easy to check that both conditions are satisfied in particular by the Hooke law, where:

$$C_{klmn} = \lambda\delta_{kl}\delta_{mn} + 2G\delta_{km}\delta_{ln} \tag{2.6}$$

with λ and G denoting the Lame constants.

The boundary-value problem (2.1) is classified as the elliptic *self-adjoint problem*. The definition of the self-adjointness reduces to the fact that the problem admits an equivalent formulation in terms of minimizing the potential energy functional (comp. Chap. 2, Part I). One of the consequences of the minimum energy theorem is the fact that the solution $\mathbf{u}$ satisfies the *Principle of Virtual Work*, also known as the *variational* or *weak* formulation of the problem (comp. (2.43), Part I):

Find $\mathbf{u}(\mathbf{x}), \mathbf{u} = 0$ on $\partial\Omega_u$, such that:

$$\int_\Omega \sigma_{kl}(\mathbf{u})\epsilon_{kl}(\mathbf{v})\, d\Omega = \int_\Omega f_k v_k\, d\Omega + \int_{\partial\Omega_\sigma} t_k v_k\, d(\partial\Omega), \tag{2.7}$$

for every (*virtual displacement*) $\mathbf{v}$, such that $\mathbf{v} = \mathbf{0}$ on $\partial\Omega_u$.

The collection of all functions $\mathbf{u}(\mathbf{x})$ satisfying the homogeneous kinematic boundary condition on $\partial\Omega_u$ is identified as the *space of kinematically admissible displacements*. The word *space* refers here to the mathematical notion of the *vector space* understood as a collection of objects identified as *vectors* [in our case, these are the displacements fields $\mathbf{u}(\mathbf{x})$] in which two operations have been introduced: vector addition and multiplication by a number. For function spaces these operations are defined as follows:

$$(\mathbf{u}+\mathbf{v})(\mathbf{x}) \stackrel{\text{def}}{=} \mathbf{u}(\mathbf{x})+\mathbf{v}(\mathbf{x}),$$
$$(\alpha\mathbf{u})(\mathbf{x}) \stackrel{\text{def}}{=} \alpha\mathbf{u}(\mathbf{x}) \tag{2.8}$$

The sum of two displacement fields is, therefore, understood as the new displacement obtained through point-wise superposition of the fields. Analogously, multiplying the displacement $\mathbf{u}(\mathbf{x})$, at each point $\mathbf{x}$, by the same number α, we obtain a new displacement identified as the product $\alpha\mathbf{u}$.

Any vector space must be *closed* with respect the two operations, i.e. for instance, adding two kinematically admissible fields we must obtain a field which is also kinematically admissible. This explains why we have assumed the homogeneous kinematic boundary conditions. When the kinematic boundary conditions are not homogeneous we can speak only about the *set* of kinematically admissible displacements. This set does not have the structure of a vector space and, consequently, more advanced algebraic notions have to be used.

The vector space operations must additionally satisfy a number of axioms, e.g. they must be additive and commutative. Sometimes objects looking like "vectors" do not satisfy those axioms and, consequently, cannot be classified as vectors in the mathematical sense, i.e. cannot be described using the means of the linear algebra. The classical example of such a situation is provided by the "vectors" of finite rotation, see e.g. [126].

We shall denote the space of kinematically admissible displacements by

$$\mathbf{V} = \{\mathbf{u}(\mathbf{x}) \; : \; \mathbf{u}(\mathbf{x}) = \mathbf{0} \text{ on } \partial\Omega_u\}. \tag{2.9}$$

The right-hand side of the variational equation (2.7), interpreted as the *work of external forces* on virtual displacement $\mathbf{v}(\mathbf{x})$, is identified with a *linear functional* $l(\mathbf{v})$,

$$l(\mathbf{v}) \stackrel{\text{def}}{=} \int_\Omega f_k v_k \,\mathrm{d}\Omega + \int_{\partial\Omega_\sigma} t_k v_k \,\mathrm{d}(\partial\Omega). \tag{2.10}$$

By the *functional* we understand a function that operates itself on functions (in our case the displacement fields $\mathbf{v}(\mathbf{x})$, and takes values in $\mathbb{R}$ – the set of real numbers. A functional is said to be *linear*, if it satisfies the condition:

$$l(\alpha_1\mathbf{v}_1+\alpha_2\mathbf{v}_2) = \alpha_1 l(\mathbf{v}_1)+\alpha_2 l(\mathbf{v}_2) \tag{2.11}$$

for every pair of numbers α_i and vectors (i.e. functions treated as elements of the vector space) $\mathbf{v}_i$. An alternative name of a *linear form* means exactly the same notion.

The left-hand side of equation (2.7) depends upon two arguments $\mathbf{u}$ and $\mathbf{v}$ and is identified as a *bilinear functional* (or *bilinear form*),

$$b(\mathbf{u},\mathbf{v}) \stackrel{\text{def}}{=} \int_\Omega \sigma_{kl}(\mathbf{u})\epsilon_{kl}(\mathbf{v})\,\mathrm{d}\Omega = \int_\Omega C_{klmn}\epsilon_{mn}(\mathbf{u})\epsilon_{kl}(\mathbf{v})\,\mathrm{d}\Omega. \tag{2.12}$$

The bilinearity of functional $b(\mathbf{u}, \mathbf{v})$ indicates the linearity with respect each of the two variables, i.e.

$$\begin{aligned} b(\alpha_1 \mathbf{u}_1 + \alpha_2 \mathbf{u}_2, \mathbf{v}) &= \alpha_1 b(\mathbf{u}_1, \mathbf{v}) + \alpha_2 b(\mathbf{u}_2, \mathbf{v}), \\ b(\mathbf{u}, \alpha_1 \mathbf{v}_1 + \alpha_2 \mathbf{v}_2) &= \alpha_1 b(\mathbf{u}, \mathbf{v}_1) + \alpha_2 b(\mathbf{u}, \mathbf{v}_2). \end{aligned} \tag{2.13}$$

From the point of view of mechanics, functional $b(\mathbf{u}, \mathbf{v})$ is interpreted as the work of stresses corresponding to displacement $\mathbf{u}$ on strains corresponding to virtual displacement $\mathbf{v}$. Expression

$$\frac{1}{2} b(\mathbf{u}, \mathbf{u}) \tag{2.14}$$

(comp. (2.36) in Part I) denotes the strain energy corresponding to displacement field $\mathbf{u}$.

Summing up, we can rewrite the Principle of Virtual Work (2.7) in the concise abstract form:

Find $\mathbf{u} \in \mathbf{V}$, such that:

$$b(\mathbf{u}, \mathbf{v}) = l(\mathbf{v}), \tag{2.15}$$

for every $\mathbf{v} \in \mathbf{V}$.

Remark 1. The bilinear form (2.12) satisfies also the symmetry condition, i.e.

$$b(\mathbf{u}, \mathbf{v}) = b(\mathbf{v}, \mathbf{u}), \tag{2.16}$$

for every pair $\mathbf{u}, \mathbf{v} \in \mathbf{V}$. The symmetry condition is characteristic for the *self-adjoint* problems only. □

Remark 2. Variational formulation (2.7), or equivalently (2.15), may be derived without referring to the minimum energy principle. The main tool for the derivation is provided by a generalization of integration by parts formula, the so called *Fundamental Green Formula*,

$$\int_\Omega \frac{\partial u}{\partial x_i} v \, \mathrm{d}\Omega = -\int_\Omega u \frac{\partial v}{\partial x_i} \, \mathrm{d}\Omega + \int_{\partial\Omega} u v n_i \, \mathrm{d}(\partial\Omega), \tag{2.17}$$

that holds for every pair of sufficiently regular functions $\mathbf{u}(\mathbf{x})$ and $\mathbf{v}(\mathbf{x})$, with $\mathbf{n} = (n_i)$ denoting the outward normal unit for boundary $\partial\Omega$.

In order to derive the Principle of Virtual Work (2.7), we multiply equations (2.1) by an arbitrary virtual displacement (called also the *test function*) $\mathbf{v} \in \mathbf{V}$, integrate over domain Ω and, finally, use the Fundamental Green Formula (2.17) to "move" derivatives from $\mathbf{u}$ onto $\mathbf{v}$. Due to the kinematic boundary conditions on the virtual displacement, the resulting boundary term vanishes on $\partial\Omega_u$, and the traction boundary condition is used to replace the stress vector on $\partial\Omega_t$ with the traction forces t_i.

In that way we show that every classical solution to (2.1) is also a *weak* solution, i.e. it satisfies the Principle of Virtual Work. Under an *additional* assumption on the regularity of the weak solution, we can reverse the procedure, and show that the weak solution is also a classical one. The assumption on the regularity is very essential and we shall return to it in the next paragraphs.

Finally, we would like to emphasize that the described procedure allows to derive variational (weak) formulations (the Principle of Virtual Work) also for non self-adjoint problems, e.g. problems in mechanics involving friction. □

Remark 3. An additional comment is necessary for the case when $\partial\Omega_u = \emptyset$, i.e. the boundary-value problem with traction boundary conditions only. In order for the solution to exist, the volumetric and traction forces f_k, t_k must satisfy the global equilibrium conditions:

$$\begin{aligned} &\int_\Omega f_k \, \mathrm{d}\Omega + \int_{\partial\Omega_\sigma} t_k \, \mathrm{d}(\partial\Omega) = 0, \qquad k = 1,2, \\ &\epsilon_{ijk}\left(\int_\Omega x_j f_k \, \mathrm{d}\Omega + \int_{\partial\Omega_\sigma} x_j t_k \, \mathrm{d}(\partial\Omega)\right) = 0, \qquad i = 1, \end{aligned} \tag{2.18}$$

or, equivalently, in the Functional Analysis language,

$$l(\mathbf{v}) = 0 \tag{2.19}$$

for every infinitesimal rigid body motion $\mathbf{v}$.

Condition (2.18) is a necessary condition for the existence of solutions to problem (2.1) that can be determined then only up to an (infinitesimal) rigid body motion. The underlying mathematics becomes more complicated (the notion of a *quotient space* has to be used), and for that reason we will exclude this case from our presentation assuming that the set $\partial\Omega_u$ is always non-empty. More precisely, we need to assume that $\partial\Omega_u$ has a non-zero measure, i.e. it contains a part of the boundary with a positive length and does not reduce to a finite number of isolated points. □

2.2 Regularity of the Solution. Sobolev Spaces

One of the essential differences between the classical (2.1), and variational (2.7) formulations is the required regularity of the solution. In the classical formulation we use the second order derivatives of the displacements what, in the language of classical PDE's, translates into the assumption that solution $\mathbf{u}(\mathbf{x})$ is sought in class $\mathbf{C}^2(\Omega)$ of functions that are twice differentiable in Ω with the second order derivatives being continuous. In many practical problems, e.g. those involving irregular loads, such a regularity assumption is *not* satisfied, i.e. the corresponding solution is less regular and it must be understood in a different, more general sense. Probably the most important generalization is provided by the variational formulation that involves only *first* order derivatives of the solution. Moreover, those derivatives appear only under the integral sign which implies that the notion of the derivative can be interpreted in a global sense without the necessity of defining it point-wise. This leads to the definition of the so-called ***distributional derivative.***

We say that a function $\mathbf{w}(\mathbf{x})$ is a distributional derivative of function $\mathbf{u}(\mathbf{x})$ with respect variable x_i, if

$$\int_\Omega u(\mathbf{x}) \frac{\partial \phi}{\partial x_i} \, d\Omega = - \int_\Omega w(\mathbf{x}) \phi(\mathbf{x}) \, d\Omega, \tag{2.20}$$

for every $C^1(\Omega)$-function vanishing on boundary $\partial\Omega$. Functions $\mathbf{u}(\mathbf{x})$ and $\mathbf{w}(\mathbf{x})$ need to satisfy additional, technical assumptions to guarantee that the involved integrals are well defined. All integrals discussed in this chapter are understood in the Lebesgue sense. The Lebesgue integral is a generalization of the Riemann integral allowing to integrate less regular functions. For functions that are sufficiently regular, e.g. that are continuous everywhere except for a finite number of isolated points and curves, the two notions of the integrals are identical.

It is easy to see (comp. (2.17)) that, if a function is differentiable in the classical sense, then its derivative $\partial u / \partial x_i$ is also the derivative in the distributional sense and equation (2.20) reduces simply to the Fundamental Green Formula. In order to present a non-trivial example of a function that is differentiable in the distributional but not in the classical sense, let us partition domain Ω into a finite number of subdomains $\Omega_e, e = 1, \ldots, E$. Assume next that we are given a function u that is C^1 in the classical sense in each of the subdomains, and it is globally continuous in the whole domain Ω. We do not assume anything about the existence of derivatives at points from subdomains boundaries $\partial\Omega_e$. The C^1 regularity implies that, for each of the subdomains Ω_e,

$$\int_{\Omega_e} \frac{\partial \phi}{\partial x_i} \, d\Omega_e = - \int_{\Omega_e} \frac{\partial u}{\partial x_i} \phi \, d\Omega_e + \int_{\partial\Omega_e} u \phi n_i \, d(\partial\Omega_e). \tag{2.21}$$

Summing up over all subdomains we get:

$$\begin{aligned} \int_\Omega u \frac{\partial \phi}{\partial x_i} \, d\Omega &= \sum_{e=1}^{E} \int_{\Omega_e} u \frac{\partial \phi}{\partial x_i} \, d\Omega_e \\ &= - \sum_{e=1}^{E} \int_{\Omega_e}^{E} \frac{\partial u}{\partial x_i} \phi \, d\Omega_e + \sum_{e=1}^{E} \int_{\partial\Omega_e} u \phi n_i \, d(\partial\Omega_e). \end{aligned} \tag{2.22}$$

The second of the sums on the right-hand side vanishes because:

- $u(\mathbf{x})$ is continuous in the whole domain Ω;
- unit vectors $\mathbf{n}_e = (n_i)$, normal to boundary $\partial\Omega_e$, and $\mathbf{n}_f = (n_i)$, normal to boundary $\partial\Omega_f$ are opposite to each other on the common boundary $\partial\Omega_{ef} = \partial\Omega_e \cap \partial\Omega_f$;
- the test function $\phi(\mathbf{x})$ vanishes on boundary $\partial\Omega$.

Therefore, introducing a function $w(\mathbf{x})$ that coincides with functions $\partial u/\partial x_i$ in subdomains Ω_e and takes on arbitrary values on the subdomains boundaries, we obtain

$$\int_\Omega u \frac{\partial \phi}{\partial x_i}\, d\Omega = -\int_\Omega w\phi\, d\Omega, \tag{2.23}$$

which proves that w is a derivative of u in the distributional sense. One of the crucial points in this example is the fact that the Lebesgue integral is insensitive to values of the integrand on subsets of *measure zero* such as the union of all of the interdomains boundaries. The considered function is typical for FE discretizations where globally defined functions are constructed by taking unions (gluing together) regular functions defined on finite elements.

The notion of the distributional derivative allows us now to introduce the fundamental space of functions appropriate for specifying regularity assumptions for the variational formulation, the *Sobolev space of first order* $H^1(\Omega)$. We define

$$H^1(\Omega) \stackrel{\text{def}}{=} \left\{ u \in L^2(\Omega) \ : \ \frac{\partial u}{\partial x_i} \in L^2(\Omega),\ i = 1, 2 \right\}, \tag{2.24}$$

where the derivative is understood in the distributional sense, and $L^2(\Omega)$ denotes the space of square integrable functions, i.e. those for which the integral

$$\int_\Omega u^2(\mathbf{x})\, d\Omega \tag{2.25}$$

exists and it is finite.

Therefore, the Sobolev space of order one, $H^1(\Omega)$ contains all square integrable functions that have distributional derivatives and those derivatives are *also* square differentiable on Ω. The function used in the example above will fall into that category provided that, in each of the subdomains, both function u and its derivatives are square integrable.

As usual, the word *space* emphasizes that $H^1(\Omega)$ is a vector space. The structure of $H^1(\Omega)$ is actually much more complex, as it falls into the category of so-called *Hilbert spaces*, and being a Hilbert space, it is automatically a *Banach space* as well.

The fundamental characteristics of a Hilbert space V is the existence of the so-called *scalar* or *inner product* defined on V. A bilinear form $a(u, v)$ defined on space V is an inner product, if it is *symmetric* and *positive-definite*. By the positive definiteness we understand the fact that the scalar product of a function u with itself, $a(u, u)$ vanishes if and only if function u is identically equal zero. The symbol for a specific inner product frequently incorporates the symbol for the space,

$$(u, v)_V \quad (= a(u, v)).$$

The most important example of the Hilbert space structure is provided by the space of square-integrable functions $L^2(\Omega)$ where the inner product is defined as:

$$(u,v)_{L^2(\Omega)} \overset{\text{def}}{=} \int_\Omega u(\mathbf{x})v(\mathbf{x})\,d\Omega. \tag{2.26}$$

In the case of the Sobolev space $H^1(\Omega)$ the corresponding inner product takes into account also the derivatives and it is defined as follows:

$$(u,v)_{H^1(\Omega)} \overset{\text{def}}{=} \int_\Omega \left(u(\mathbf{x})v(\mathbf{x}) + \sum_i \frac{\partial u}{\partial x_i}(\mathbf{x})\frac{\partial v}{\partial x_i}(\mathbf{x}) \right) d\Omega. \tag{2.27}$$

We shall leave for the reader to verify that both functions (2.26) and (2.27) satisfy the three axioms defining an inner product. In particular, it follows from the *Schwarz inequality*:

$$\int_\Omega u(\mathbf{x})v(\mathbf{x})\,d\Omega \le \left(\int_\Omega u^2(\mathbf{x})\,d\Omega \right)^{\frac{1}{2}} \left(\int_\Omega v^2(\mathbf{x})\,d\Omega \right)^{\frac{1}{2}} \tag{2.28}$$

and the assumption that we have restricted ourselves to the square-integrable functions only (including the derivatives), that all integrals are finite.

The name *scalar product* refers to the fact that the simplest example of the abstract inner product structure is provided by the space of free vectors (on a plane or in a space) and the classical notion of the scalar product.

Each inner product induces the corresponding *norm* of a vector defined as:

$$\|u\|_V \overset{\text{def}}{=} [(u,u)_V]^{\frac{1}{2}}\,. \tag{2.29}$$

By a *norm* we understand a functional defined on a vector space that takes on non-negative values and it satisfies three fundamental properties:

- it is positive-definite, i.e.

$$\|u\|_V = 0 \quad \text{implies} \quad u = 0; \tag{2.30}$$

- it is *homogeneous*, i.e.

$$\|\lambda u\|_V = |\lambda|\,\|u\|_V, \tag{2.31}$$

  for every real number λ, whereas $|\lambda|$ stands for the absolute value of λ;
- it satisfies the triangle inequality:

$$\|u+v\|_V \le \|u\|_V + \|v\|_V\,. \tag{2.32}$$

It is easy to verify that functional (2.29) satisfies the three axioms for a norm. In other words, every inner product induces a norm. Not every norm, however, must be introduced through an inner product. There are many norms that do not have corresponding inner products. Intuitively speaking, the notion of the norm is intended to generalize the notion of the *magnitude* of a vector in the classical geometry. If functions are identified as vectors in the abstract sense, like in the case of spaces $L^2(\Omega)$ and $H^1(\Omega)$, a norm is used to measure the *magnitude* of the function letting us know how much the function differs from the zero function.

A vector space with a norm introduced in it that is additionally *complete*[1], is called a *Banach space*, if additionally the norm has been derived from a scalar product, we talk about a *Hilbert space*. Thus, every Hilbert space is a Banach space but not conversely.

Having discussed the regularity assumptions, we are ready now to define precisely the *space of kinematically admissible displacements* [comp. (2.9)],

$$\mathbf{V} \stackrel{\text{def}}{=} \{\mathbf{u} = (u_i) \;:\; u_i \in H^1(\Omega), u_i(\mathbf{x}) = 0 \text{ on } \partial\Omega_u\}. \tag{2.33}$$

In order to make sure that integrals (2.19) and (2.12) are finite, we may assume for instance that all data given in the problem, i.e. elasticities C_{ijkl}, volume forces f_k and tractions t_k, are bounded from above.

The space of kinematically admissible displacements is itself a Hilbert space with the product induced by the scalar product from space $H^1(\Omega)$,

$$(\mathbf{u}, \mathbf{v})_V \stackrel{\text{def}}{=} \sum_{k=1}^{2} (u_k, v_k)_{H^1(\Omega)} \,. \tag{2.34}$$

Remark 4. We are still quite far from being fully precise from the mathematical point of view. The incorporation of the kinematic boundary condition in definition (2.33) is based on the famous Lions' *Trace Theorem* that we have not explained. We also have not made any assumptions on the regularity of the domain Ω. These and other missing technical details should not, however, prevent understanding the main ideas presented in this chapter. □

2.3 Existence, Uniqueness and Regularity Results

An ill-posed problem may have no solution. That does not prevent an *ad hoc* formulated approximate problem to have no solution as well. To the contrary, most of the time, the computer will return *some* numbers and, with a bit of luck, we may be even able to interprete them from the mechanical point of view. The trouble begins when we refine the mesh and increase the number of degrees of freedom. The corresponding solutions will diverge, for that simple reason that there is "nothing" to converge to!

In practice, most of the boundary-value problems investigated in practice are far from being fully understood mathematically. An increasing complexity of problems being considered most often offers little time and chance for a full mathematical analysis. A compromise is made by investigating the so-called *model problems* that, on one side, share at least main features with the actual problems being solved and, on the other side, are simple enough to allow for a precise mathematical analysis. The linear elasticity is clas-

[1] See [126]. We shall not need the notion of completeness directly in this presentation.

sified as one of such model problems leading to more complex problems of elasto-plasticity, visco-elasto-plasticity etc.

From the point of view of the convergence analysis, therefore, the question on existence and uniqueness of solutions to the elasticity problem is crucially important. The answer is provided by one of the most fundamental results of Functional Analysis.

Theorem 2.1. (Lax–Milgram Lemma) *Let* $\mathbf{V}$ *be a Hilbert space with inner product* $(\mathbf{u},\mathbf{v})_{\mathbf{V}}$ *and the corresponding norm* $\|\mathbf{u}\|_{\mathbf{V}}$. *Assume that the bilinear form* $b(\mathbf{u},\mathbf{v})$ *and linear form* $l(\mathbf{v})$ *from the abstract variational problem (2.15) satisfy the following assumptions:*

- $l(\mathbf{v})$ *is* continuous *on* $\mathbf{V}$, *i.e. there exists a constant* $C > 0$ *such that:*

$$|l(\mathbf{v})| \leq C\|\mathbf{v}\|_{\mathbf{V}} \quad \textit{for every } \mathbf{v} \in \mathbf{V}; \tag{2.35}$$

- $b(\mathbf{u},\mathbf{v})$ *is* continuous *on* $\mathbf{V}$, *i.e. there exists a constant* $M > 0$ *such that:*

$$|b(\mathbf{u},\mathbf{v})| \leq M\|\mathbf{u}\|_{\mathbf{V}}\|\mathbf{v}\|_{\mathbf{V}} \quad \textit{for all } \mathbf{u},\mathbf{v} \in \mathbf{V}; \tag{2.36}$$

- $b(\mathbf{u},\mathbf{v})$ *is* coercive *on* $\mathbf{V}$, *i.e. there exists a constant* $\alpha > 0$ *such that:*

$$\alpha\|\mathbf{u}\|_{\mathbf{V}}^2 \leq b(\mathbf{u},\mathbf{u}) \quad \textit{for every } \mathbf{u} \in \mathbf{V}. \tag{2.37}$$

There exists then a unique *solution* $\mathbf{u}$ *to the abstract variational boundary-value problem (2.15).* □

Referring with the proof to [126], we shall focus here on interpreting the assumptions of the theorem. First of all, we would like to attract the reader's attention to the *coercivity condition* (2.37). This is a rather strong assumption. It implies the *positive-definiteness* of form $b(\mathbf{u},\mathbf{v})$ which in turn is sufficient to prove uniqueness. Indeed, if $\mathbf{u}_1$ and $\mathbf{u}_2$ were two solutions to (2.15), i.e.

$$b(\mathbf{u}_i,\mathbf{v}) = l(\mathbf{v}), \quad i = 1,2, \tag{2.38}$$

for every $\mathbf{v} \in \mathbf{V}$, then, by subtracting the two equations from each other, we get:

$$b(\mathbf{u}_1 - \mathbf{u}_2,\mathbf{v}) = 0 \quad \text{for every } \mathbf{v} \in \mathbf{V}. \tag{2.39}$$

Substituting $\mathbf{v} = \mathbf{u}_1 - \mathbf{u}_2$, we obtain that

$$b(\mathbf{u}_1 - \mathbf{u}_2,\mathbf{u}_1 - \mathbf{u}_2) = 0. \tag{2.40}$$

But positive definiteness of form $b(\mathbf{u},\mathbf{v})$ implies that $\mathbf{u}_1 - \mathbf{u}_2 = \mathbf{0}$, or equivalently, $\mathbf{u}_1 = \mathbf{u}_2$. The positive definiteness, therefore, implies the uniqueness. We mention, however, that it is not sufficient to accomplish the existence result.

In order to prove the existence and uniqueness of solution to the elasticity problem we need now to verify the assumptions of the Lax–Milgram lemma. The first two of them are a simple consequence of the definition of

Sobolev space, Schwarz inequality, and the regularity assumptions on functions C_{ijkl}, f_k, t_k (the trace theorem mentioned in Remark 2.4 is needed once more to show continuity of the linear form). The proof of the coercivity result, based on the *Korn's inequality* is much more advanced [46].

Constant C depends on the magnitude of forces f_k and t_k (the bigger the forces, the bigger is the constant C), whereas constant M depends upon the elasticities C_{ijkl} (again, it grows with the stiffness of the material). The coercivity constant α is related to the smallest resonant frequency of the body and it depends on the size od domain and boundary conditions as well. Intuitively speaking, the more is the boundary fixed, the bigger is the constant α. In particular, for a given domain Ω, the largest value of α corresponds to the pure kinematic boundary conditions, i.e. when the body is fixed along its whole boundary.

A separate issue, fundamental for estimating the anticipated convergence rates of the FEM discussed next, is that of the regularity of the solution. At this point, we know only that the solution sits in a rather "large" space $H^1(\Omega)$. Whether we can estimate the regularity more precisely, it will depend upon three factors:

- regularity of the domain Ω,
- regularity of the coefficients in the operator (elasticities C_{ijkl}),
- regularity of the load (forces f_k, t_k).

Customary, the regularity of the solution is measured using Sobolev spaces of higher order $H^m(\Omega), m = 2, 3, \ldots$. These spaces are defined analogously to space $H^1(\Omega)$, except that derivatives of the first order are now replaced with derivatives of order m. It is also possible to introduce a more sophisticated notion of Sobolev spaces of *fractional order*.

The increasingly demanding conditions on higher order derivatives imply that the Sobolev spaces may be ordered into a "decreasing" sequence,

$$H^1(\Omega) \not\supseteq H^2(\Omega) \not\supseteq H^3(\Omega) \not\supseteq \ldots \quad (2.41)$$

Without getting into technical details we mention for example that, if domain Ω is bounded and *convex*, its boundary consists of smooth curves (vertices allowed), elasticities C_{ijkl} are C^1–functions, and loads f_k, t_k are of class L^2, then the solution $\mathbf{u}(\mathbf{x})$ that *a-priori* sits in space $H^1(\Omega)$, is actually an element of the "smaller" space $H^2(\Omega)$.

2.4 Fundamental Notions of the Mathematical Theory of Finite Elements

We begin with the formal definition of a *finite element* [34]. More precisely, by a *Lagrange finite element* we will understand a triple consisting of

- the actual geometrical figure Ω_e occupied by the element,

- the *element space of shape functions* P_e (a set of specific functions, most often polynomials or functions "close" to polynomials),
- a collection of *nodes* $\mathbf{a}_1^e, \ldots, \mathbf{a}_{N_e}^e$.

By the nodes we understand here specific points from domain Ω^e selected in some special way. We shall assume that the set of nodes is P_e-*unisolvent*, i.e. that *for every sequence of numbers* $q_1^e, \ldots, q_{N^e}^e$, *there exists a unique function* $p(x_1, x_2)$ *from space* P_e *such that*

$$p(\mathbf{a}_i) = q_i^e, \quad i = 1, \ldots, N_e \,. \tag{2.42}$$

In particular, selecting $q_i^e = \delta_{ij}$, we obtain the *shape function* ϕ_j^e corresponding to node $\mathbf{a}_j^e$, i.e. such a function from space P_e that vanishes at all nodes, except for node $\mathbf{a}_j^e$ where it takes the value 1. Values of the functions at the nodes are also known as the *degrees of freedom* of the element; for scalar-valued functions, the number of element degrees of freedom coincides simply with the number of nodes.

Remark 5. In a more general, abstract definition of the finite element, degrees of freedom are viewed as arbitrary linear functionals, defined not only through function values but their derivatives (Hermite finite elements) or integrals over parts or the whole of the element. Through an appropriate definition of degrees of freedom, it is also possible to introduce the so-called *hierarchical shape functions*, a notion fundamental for the p-version of the FEM where the convergence is achieved not by refining the elements (version h) but rather by increasing the order of approximation. Those more advanced notions remain beyond the scope of this presentation. The definition presented above is a special case of a general definition given by Ciarlet [34]. □

For each function $u(\mathbf{x})$ continuous over element Ω^e (more precisely, over the closure of it, including the element boundary), there exists a unique function $u_I^e(\mathbf{x})$ from the space of element shape functions P_e that assumes at nodes $\mathbf{a}_i^e$ values identical with those of function $u(\mathbf{x})$. This function is given by the formula:

$$u_I^e(\mathbf{x}) = \sum_{i=1}^{N_e} u(\mathbf{a}_i^e)\phi_i^e(\mathbf{x}) \,. \tag{2.43}$$

Indeed, as a linear combination of the element shape functions, the function belongs to space P_e, and it follows from the definition of shape functions $\phi_i^e(\mathbf{x})$ that:

$$u_I^e(\mathbf{a}_j^e) = \sum_{i=1}^{N_e} u(\mathbf{a}_i^e)\phi_i^e(\mathbf{a}_j^e) = \sum_{i=1}^{N_e} u(\mathbf{a}_i^e)\delta_{ij} = u(\mathbf{a}_j^e) \,. \tag{2.44}$$

We shall identify the function $u_I^e(\mathbf{x})$ as the P_e-*interpolant* of function $u(\mathbf{x})$. It is easy to see that, for a function $u(\mathbf{x})$ coming from the very family P_e, its

P_e-interpolant coincides with the original function, i.e.

$$u(\mathbf{x}) = u_I^e(\mathbf{x}) = \sum_{i=1}^{N_e} u(\mathbf{a}_i^e)\phi_i^e(\mathbf{x}). \tag{2.45}$$

This allows to interprete the element space of shape functions as the collection of linear combinations of the nodal shape functions $\phi_i^e(\mathbf{x})$. Mathematically speaking, functions $\phi_i^e(\mathbf{x})$ provide a basis for space P_e, whereas the degrees of freedom $u(\mathbf{a}_i^e)$ form the corresponding dual basis.

We shall proceed now with the simplest examples of Lagrange finite elements.

A triangular element of order 1 (linear triangle). Let Ω_e be a triangle with vertices $\mathbf{a}_1, \mathbf{a}_2, \mathbf{a}_3$. Assuming for space P_e the space of polynomials of order 1, and identifying the nodes with the vertices, we obtain the simplest (and oldest) example of a finite element known as the *Courant triangle.* The unisolvence condition follows from the fact that every linear polynomial is uniquely determined by its values at three non-colinear points (comp. Fig. 2.1).

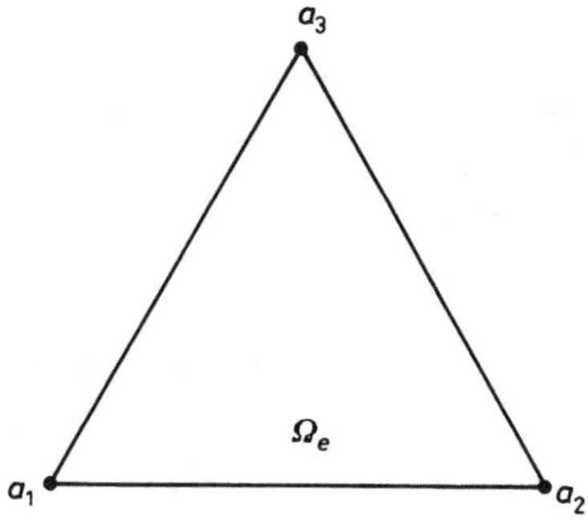

Fig. 2.1. Courant's triangle

Triangular element of order 2 (quadratic triangle). This time we identify space P_e with polynomials of order 2, $P_e = P^2$ and add three more nodes $\mathbf{a}_4, \mathbf{a}_5, \mathbf{a}_6$, coinciding with mid-points of the element edges (see Fig. 2.2). The unisolvence follows from the fact that every polynomial of degree 2 is uniquely determined by its values at the nodes $\mathbf{a}_1, \ldots, \mathbf{a}_6$. The concept of the triangular element can be generalized to polynomials of arbitrary degree p.

Rectangular element of order 1. Identifying the element Ω_e with a rectangle with vertices $\mathbf{a}_1, \ldots, \mathbf{a}_4$ (see Fig. 2.3), we select for the element space P_e the space of bilinear shape functions Q^1. More generally, by space Q^p, we understand the collection of all polynomials of order p, but *with*

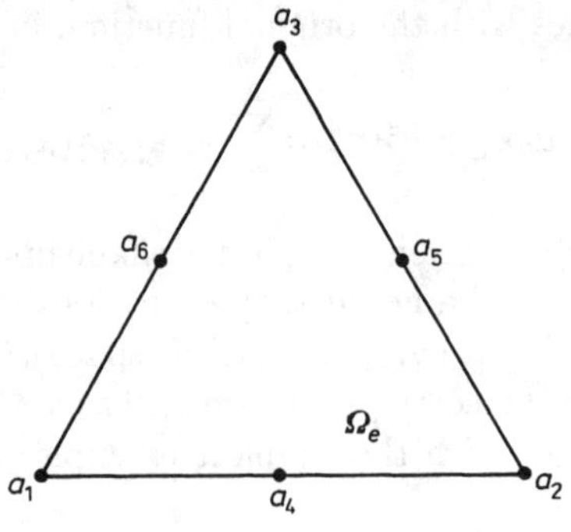

Fig. 2.2. Triangular element of order 2

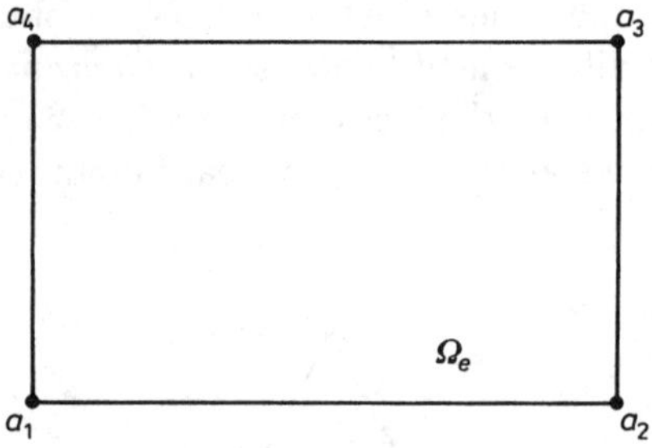

Fig. 2.3. Rectangular element of order 1 Q^1

respect each of the variables separately. Consequently, space Q^1 is generated by monomials

$$1,\ x_1\ ,\ x_2\ ,\ x_1x_2\ . \tag{2.46}$$

In particular, the last monomial is of order 1 with respect to each of the two variables x_1 and x_2, but as a polynomial of *two* variables is of order 2. Adding additionally the monomials:

$$x_1^2\ ,\ x_2^2\ ,\ x_1^2x_2\ ,\ x_1x_2^2\ ,\ x_1^2x_2^2\ , \tag{2.47}$$

we obtain a basis for space Q^2 etc. Finally, identifying the nodes as the rectangle vertices $\mathbf{a}_1, \mathbf{a}_2, \mathbf{a}_3, \mathbf{a}_4$, we complete the definition of the rectangular element of order 1. As before, the unisolvence condition follows from the fact that any Q^1-polynomial is uniquely determined through its values at the vertices.

Rectangular element of order 2. Assuming $P_e = Q^2$ and adding five more nodes (compare with the number of added monomials, (2.47)), including four mid-points of the element sides and the middle point of the rectangle, we end up with the definition of the quadratic rectangle (comp. Fig. 2.4). Similarly

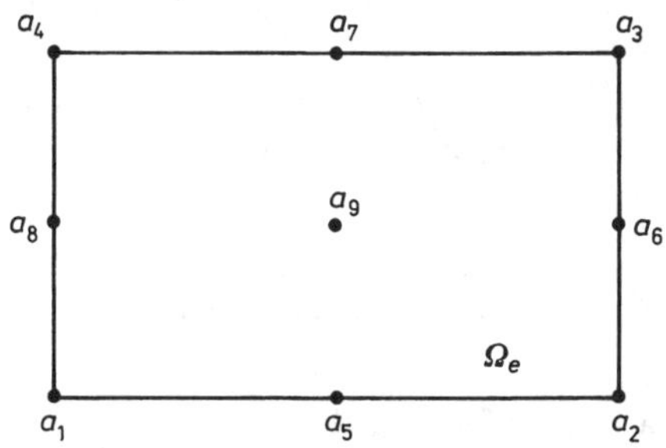

Fig. 2.4. Rectangular element of order 2 Q^2

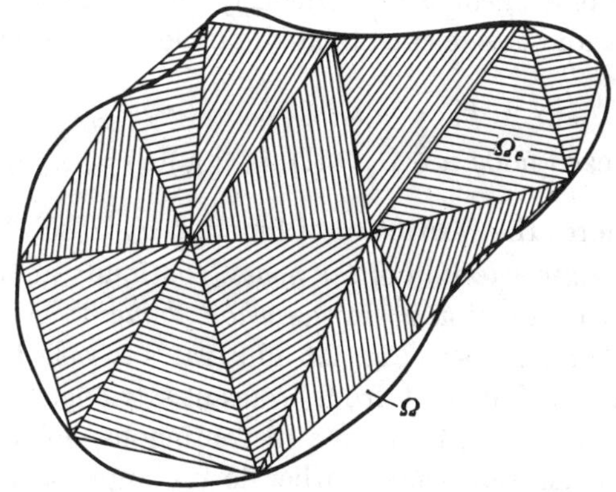

Fig. 2.5. Approximation of a domain Ω with finite elements. Notice the difference between the original domain Ω and its (shaded) approximation Ω_h

as for triangles, the definition of rectangular elements can be extended to polynomials Q^p of arbitrary order p.

The element nodes and the corresponding element shape functions allow us to define the interpolant $u_I(\mathbf{x})$, i.e. to approximate a given function $u(\mathbf{x})$ over each of the elements Ω_e separately. In order to define a *global approximation*, over the whole domain Ω, we follow now the procedure described in Chap. 1. We start by noticing that, in the case of domains with complex geometry and elements with straight sides, it is, in general, impossible to partition the domain Ω precisely into finite elements. Rather, we have to introduce an additional domain Ω_h being the union of all finite elements, that only approximates the actual domain Ω (comp. Fig. 2.5). In what follows, just to simplify the presentation, we will restrict ourselves to polygonal domains which can be partitioned into elements exactly, i.e. $\Omega = \Omega_h$.

We are ready now to introduce a counterpart of the element space of shape functions on the global level – the *finite element space* X_h. The space X_h

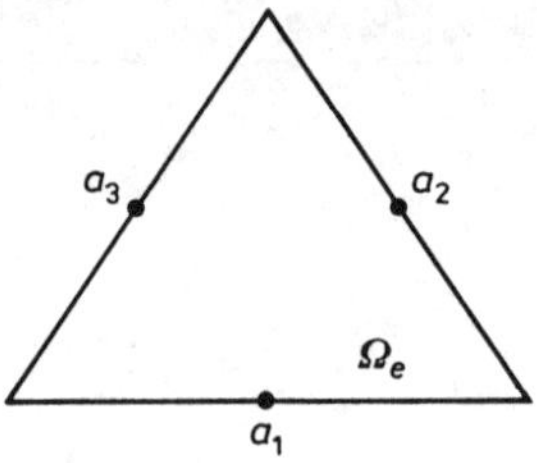

Fig. 2.6. Example of a non-conforming element

consists of all functions that, over each element Ω_e, belong to the element space P_e and that are continuous at the nodes. In other words, if a node $\mathbf{a}$ belongs to a number of elements simultaneously, then the corresponding values of shape functions for those elements at the node are identical. Formally:

$$\begin{aligned} X_h = \{u_h(\mathbf{x}) \ : \ u_h|_{\Omega_e} \in P_e,\ e = 1, \ldots, E, \\ \text{and, for each node } \mathbf{a} \in \Omega_e \cap \Omega_f, (u_h|_{\Omega_e})(\mathbf{a}) = (u_h|_{\Omega_f})(\mathbf{a})\} \end{aligned} \tag{2.48}$$

Symbol $u_h|_{\Omega_f}$ denotes the *restriction* of function u_h to subdomain Ω_f.

With an *appropriate selection of the nodes*, continuity at the nodes implies a global continuity of function u_h in the *whole domain.* This need not to be true in general. For instance, Fig. 2.6 illustrates a well-defined triangular element of first order (different from Courant's triangle!) which, despite the continuity at the nodes, leads to a discontinuous global approximation. However, all the elements presented earlier do guarantee the global continuity. We speak then about the *conforming elements.* For the conforming elements we have

$$X_h \subset H^1(\Omega), \tag{2.49}$$

i.e., all functions from the finite element space are elements of the Sobolev space $H^1(\Omega)$ in which the exact solution is being sought.

Given a particular mesh, we introduce now a global denumeration of element nodes $\mathbf{a}_i, i = 1, \ldots, N$, for all elements in the mesh, and define the *global degrees of freedom* understood as values of $u(\mathbf{x})$ at nodes $\mathbf{a}_i$, where $u(\mathbf{x})$ is an arbitrary continuous function. It follows from the definition of finite elements and the finite element space X_h that the global degrees of freedom are X_h-unisolvent, i.e., *for an arbitrary sequence of numbers* $q_1, \ldots, q_N$, *there exists a unique function* $v_h(\mathbf{x}) \in X_h$ *such that*

$$v_h(\mathbf{a}_i) = q_i\,. \tag{2.50}$$

In particular, for $q_i = \delta_{ij}$, we obtain a basis function $e_j(\mathbf{x})$ corresponding to node $\mathbf{a}_j$, a global counterpart of the element shape function. From the geometrical point of view, the basis function $e_j(\mathbf{x})$ is obtained by "gluing" together corresponding element shape functions for all elements adjacent to

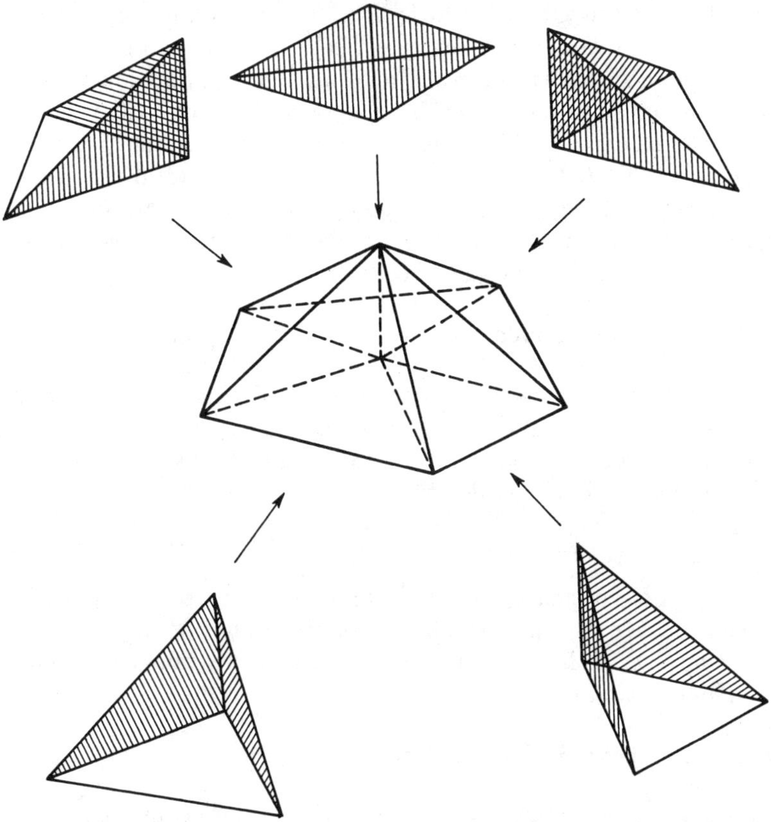

Fig. 2.7. "Gluing" element shape functions into a global basis function

the node[2], see Fig. 2.7.

A function $u_I(\mathbf{x})$ from the finite element space X_h coinciding at the nodes $\mathbf{a}_i, i = 1, \ldots, N$, with a given continuous function $u(\mathbf{x})$, is called the X_h-*interpolant of function* $u(\mathbf{x})$. Analogously to the element level, it can be expressed as a linear combination of the corresponding basis functions:

$$u_I(\mathbf{x}) = \sum_{i=1}^{N} u(\mathbf{a}_i) e_i(\mathbf{x}). \tag{2.51}$$

The finite element space X_h may, similarly to the element considerations, may be identified with a collection of all linear combinations of basis functions $e_i(\mathbf{x})$, or in another words, the basis functions $e_i(\mathbf{x})$ form a basis in vector space X_h.

[2] More precisely, $e_j(\mathbf{x})$ is equal to the *union* of the element shape functions, for the adjacent elements, and the zero function elsewhere.

All notions discussed so far have been concerned with the approximation of a single, scalar-valued function $u(\mathbf{x})$. In order to approximate the vector-valued displacement function $\mathbf{u}(\mathbf{x})$, we simply apply the described approximation procedure to each of the components of vector $\mathbf{u}(\mathbf{x})$. This leads to vector-valued degrees of freedom $\mathbf{q}_i^e$ for an element Ω_e, and vector-valued global degrees of freedom $\mathbf{q}_i$ at the global level. The total number of degrees of freedom doubles and it is equal to twice the total number of the Lagrange nodes (comp. Chap. 1, Part I). The scalar finite element space X_h gets replaced with its vector-valued counterpart $\mathbf{X}_h$.

Kinematic Boundary Conditions

Taking into the account the kinematic boundary conditions is relatively simple. One needs to eliminate from the finite element space $\mathbf{X}_h$ those displacement fields that violate the kinematics conditions. This leads to the *discrete space of kinematically admissible fields*:

$$\mathbf{V}_h = \{\mathbf{v}_h \in \mathbf{X}_h \;:\; \mathbf{v}_h(\mathbf{a}_i) = \mathbf{0} \text{ for all nodes } \mathbf{a}_i \in \partial\Omega_u\}. \tag{2.52}$$

For appropriately designed meshes (endpoints of connected components of $\partial\Omega_u$ must coincide with nodes), and conforming elements, the discrete space $\mathbf{V}_h$ is a subspace of the continuous space $\mathbf{V}$, and we again speak about the *internal approximation.*

Interpolation Theory in Sobolev Spaces

The difference $e_I(\mathbf{x})$ between a function $u(\mathbf{x})$ and its X_h-interpolant $u_I(\mathbf{x})$:

$$e_I(\mathbf{x}) \overset{\text{def}}{=} u(\mathbf{x}) - u_I(\mathbf{x}), \tag{2.53}$$

is called the *FE interpolation error.* The interpolation error is thus a function which, accordingly to its construction, vanishes at the nodes. We shall measure its magnitude using *Sobolev seminorms,* a notion related to the Sobolev norms discussed in paragraph 2.2. The *Sobolev seminorm of order k* of function $u(\mathbf{x})$, $k = 0, 1, 2, \ldots$, is defined as follows,

$$|u|_{k,\Omega} = \left[\sum_{\alpha_1+\alpha_2=k} \int_\Omega \left(\frac{\partial^{\alpha_1+\alpha_2} u(\mathbf{x})}{\partial x_2^{\alpha_1} \partial x_2^{\alpha_2}} \right)^2 \mathrm{d}\mathbf{x} \right]^{\frac{1}{2}}, \tag{2.54}$$

for $k > 0$. For $k = 0$, the notions of L^2-norm, Sobolev norm, and the Sobolev seminorm, coincide with each other.

Contrary to the Sobolev norm, the seminorm takes into account only the behavior of derivatives of the highest order. From the axiomatic point of view, seminorms are slightly "worse" than norms. They still satisfy conditions (2.31) and (2.32), but they need not satisfy condition (2.30). If the Sobolev seminorm of order k is zero, it follows only that the derivatives of order k

vanish and, consequently, the function must be a polynomial of order $k-1$ but not necessary a zero function.

It is easy to notice that by summing up all seminorms up to (including) order k, we recover the norm of order k. More precisely,

$$\|u\|^2_{H^k(\Omega)} = |u|^2_{0,\Omega} + \ldots + |u|^2_{k,\Omega} \,. \tag{2.55}$$

We shall introduce now the notion of a *regular family of affine FE meshes*. The geometry of each element Ω_e will be characterized with two parameters: *external diameter* $h(\Omega_e)$ and *internal diameter* $\rho(\Omega_e)$. By the external diameter we understand simply the maximum linear dimension of the element:

$$h(\Omega_e) = \max_{\mathbf{x},\mathbf{y}\in\Omega_e} \sqrt{(x_1-y_1)^2 + (x_2-y_2)^2}, \tag{2.56}$$

whereas the internal diameter is identified as the diameter of the maximal circle that may be inscribed into the element. The ratio of the external and the internal diameters reflects a deformation of the element. In particular for "flat" elements, degenerating to a segment of line, the ratio will approach infinity.

The elements discussed earlier are usually defined first on special, reference domains like the right unit triangle and the unit square, and identified as the *master elements.* An element defined on an arbitrary triangle or a parallelogram is called *affine element* if it can be identified as the image of one of the master elements through an *affine map.* Recall that by an affine map we understand a composition of a linear map and a translation. In particular, straight lines and parallel lines are invariant under affine maps but for instance orthogonal lines are not. Consequently, an affine finite element must have straight edges and, for the rectangular master element, the opposite edges must be parallel to each other. Let us emphasize that, when talking about the image of the master element, we understand that the affine element nodes coincide precisely with the images of the master element nodes. That implies for instance that the Lagrange nodes for the affine elements are uniformly distributed along the element sides, etc. A mesh consisting of affine elements only, is said to be an *affine FE mesh.*

Finally, a family of affine FE meshes is said to be *regular*, if there exists a positive constant $C > 0$ such that:

$$\frac{h(\Omega_e)}{\rho(\Omega_e)} < C, \tag{2.57}$$

for each element Ω_e belonging to any of the meshes from the family. An example of a regular family can be obtained by starting with an arbitrary mesh consisting of triangles or rectangles and then performing uniform refinements dividing each of the elements in the mesh into four congruent triangles or rectangles, comp. Fig. 2.8. Intuitively speaking, the regularity of a family of the FE meshes means that, during the refinement process, the elements must not degenerate into segments of lines.

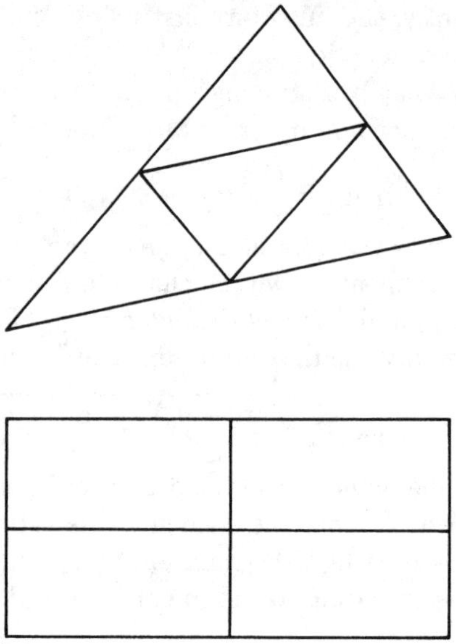

Fig. 2.8. Mesh refinements into congruent elements

We conclude this section with a fundamental result from the interpolation theory in Sobolev spaces.

Theorem 2.2. *Suppose, we are given a regular family of affine FE meshes or order k, i.e., for each element Ω_e, the element space of shape functions P_e contains all polynomials of order k.*

There exists then a constant $C > 0$, depending upon k, the kind of elements being used, domain Ω, but independent of function $u(\mathbf{x})$ such that:

$$|u - u_I|_{m,\Omega} \leq Ch^{k+1-m}|u|_{k+1,\Omega}, \quad m = 0, 1, \ldots, k, \tag{2.58}$$

for every function $u(\mathbf{x})$ from Sobolev space $H^{k+1}(\Omega)$, where h stands for the maximum element size in the mesh, i.e.

$$h = \max_{e=1,\ldots,E} h_e \,. \qquad \square$$

We say that the interpolation error, measured in the seminorm of order m, has the rate of convergence $k+1-m$. The rate of convergence, therefore, depends upon:

- order of approximation k,
- regularity of function $u(\mathbf{x})$ ($u(\mathbf{x})$ must be from the space $H^{k+1}(\Omega)$),
- m – the order of the seminorm in which the error is measured.

2.5 Convergence Analysis for Conforming Elements. Cea's Lemma

Let us recall the main assumptions made so far. We are considering *conforming elements* of order k, the domain Ω is polyhedral and, therefore, it can be partitioned exactly into the finite elements. The approximate counterpart of problem (2.15) is now formulated as follows.

Find $\mathbf{u}_h \in \mathbf{V}_h$ such that:

$$b(\mathbf{u}_h, \mathbf{v}_h) = l(\mathbf{v}_h), \tag{2.59}$$

for every $\mathbf{v}_h \in \mathbf{V}_h$.

Recall that both the approximate solution $\mathbf{u}_h$ and the test function $\mathbf{v}_h$ (approximate virtual displacement) are linear combinations of the global basis functions:

$$\mathbf{u}_h(\mathbf{x}) = \sum_i \mathbf{u}_h^i e_i(\mathbf{x}), \quad \mathbf{v}_h = \sum_j \mathbf{v}_h^j e_j(\mathbf{x}), \tag{2.60}$$

where vector $\mathbf{u}_h^i = (u_{ih}^i, u_{2h}^i)^T$ is equal to the displacement at node $\mathbf{a}_i$.

For a more practical interpretation of the approximate problem (2.59), it is convenient to represent both the bilinear form $b(\mathbf{u}_h, \mathbf{v}_h)$ and linear form $l(\mathbf{v}_h)$ as sums of corresponding bilinear and linear forms acting on particular components of the displacement fields. We have:

$$\begin{aligned} b(\mathbf{u}_h, \mathbf{v}_h) &= b_{11}(u_1, v_1) + b_{12}(u_1, v_2) + b_{21}(u_2, v_1) + b_{22}(u_2, v_2) \\ l(\mathbf{v}_h) &= l_1(v_1) + l_2(v_2). \end{aligned} \tag{2.61}$$

The formula for each of the component bilinear forms $b_{kl}(u_k, v_l), k, l = 1, 2$ and linear forms $l_k(v_k), k = 1, 2$ can easily be deduced from the Principle of Virtual Work. We have for instance:

$$\begin{aligned} l_1(v_1) &= \int_\Omega f_1 v_1 \, d\Omega + \int_{\partial\Omega} t_1 v_1 \, d(\partial\Omega), \\ b_{11}(u_1, v_1) &= \int_\Omega C_{klmn} \frac{1}{2}(u_{1m,n} + u_{1n,m}) \frac{1}{2}(v_{1k,l} + v_{1l,k}) \, d\Omega. \end{aligned} \tag{2.62}$$

Substituting (2.60) into (2.59) and making use of linearity with respect to both test function $\mathbf{v}$ and solution $\mathbf{u}$, we end up with the following system of $2N$ linear equations to be solved for unknown nodal displacements $\mathbf{u}_i^h = (u_{1h}^i, u_{2h}^i)$ (for simplicity we drop mesh symbol h from the notation),

$$\begin{aligned} \sum_i u_1^i b_{11}(e_i, e_j) + \sum_i u_2^i b_{21}(e_i, e_j) &= l_1(e_j), \\ \sum_i u_1^i b_{12}(e_i, e_j) + \sum_i u_2^i b_{22}(e_i, e_j) &= l_2(e_j), \end{aligned} \tag{2.63}$$

for every j. First of the equations in system (2.63) corresponds to choosing for the test function $\mathbf{v} = (e_j, 0)^T$, the second to $\mathbf{v} = (0, e_j)^T$.

The elements of the global stiffness matrix $b_{kl}(e_i, e_j)$ and the load vector $l_k(e_j)$ are evaluated by splitting the integrals over domain Ω or boundary $\partial\Omega_\sigma$ into sums of integrals over finite elements Ω_e or portions of their boundaries. This corresponds to the usual FE assembling procedure.

We emphasize that, introducing the approximate problem (2.59), *we have not* taken into account the numerical integration. All integrals present in the formulation are assumed to be integrated exactly. The effect of numerical integration will be discussed in next sections.

One of the immediate consequences of the *internal approximation*, i.e. the inclusion $\mathbf{V}_h \subset \mathbf{V}$, is the existence and uniqueness of the approximate solution $\mathbf{u}_h$. One needs only to verify again the assumptions of the Lax–Milgram lemma. The continuity conditions (2.35) and (2.36) are satisfied with the *same constants* as on the continuous level. This follows from the fact that both conditions are satisfied *for every* $\mathbf{v} \in \mathbf{V}$ and therefore also for every $\mathbf{v}_h \in \mathbf{V}_h$, as $\mathbf{V}_h \subset \mathbf{V}$. In the same way, the coercivity condition is satisfied with the same constant α (as a matter of fact, the discrete coercivity constant may be larger as a result of a smaller space). Concluding, the discrete problem is well posed; there exists a unique discrete solution.

The second fundamental consequence of the internal approximation is the so-called *orthogonality condition.* As equation (2.15) is satisfied for every $\mathbf{v} \in \mathbf{V}$, and $\mathbf{V}_h$ is contained in $\mathbf{V}$, we have in particular:

$$b(\mathbf{u}, \mathbf{v}_h) = l(\mathbf{v}_h), \quad \text{for every } \mathbf{v}_h \in \mathbf{V}_h\,. \tag{2.64}$$

Subtracting (2.9) from (2.64) and taking advantage of linearity of $b(\mathbf{u}, \mathbf{v})$ with respect to the first argument, we obtain:

$$b(\mathbf{u} - \mathbf{u}_h, \mathbf{v}_h) = 0, \quad \text{for every } \mathbf{v}_h \in \mathbf{V}_h\,. \tag{2.65}$$

The difference $\mathbf{e}_h \stackrel{\text{def}}{=} \mathbf{u} - \mathbf{u}_h$ is known as the *approximation error* (not to be confused with the interpolation error), and condition (2.65) is known as the *orthogonality condition.* The name follows from the observation that, for self-adjoint problems [form $b(u,v)$ is then symmetric], form $b(u,v)$ satisfies the axioms for a scalar product and can indeed be viewed as an equivalent inner product in the space of kinematically admissible functions $\mathbf{V}$. Condition (2.65) then means that the approximation error $\mathbf{e}_h = \mathbf{u} - \mathbf{u}_h$ is orthogonal (in the sense of form $b(\mathbf{u}, \mathbf{v})$) to subspace $\mathbf{V}_h$.

The following result is fundamental for proving convergence of the FEM.

Theorem 2.3. (Cea's lemma) *Let* $\mathbf{u}$ *denote the solution of the abstract boundary-value problem (2.15), and let* $\mathbf{u}_h$ *be the corresponding solution of the approximate problem (2.59), whereas* $\mathbf{V}_h \subset \mathbf{V}$ *(internal approximation). Under the assumptions of the Lax–Milgram lemma, we have*

$$\|\mathbf{u} - \mathbf{u}_h\|_\mathbf{V} \le \frac{M}{\alpha} \|\mathbf{u} - \mathbf{v}_h|_\mathbf{V} \quad \textit{for every } \mathbf{v}_h \in \mathbf{V}_h \tag{2.66}$$

(see the Lax–Milgram lemma for the explanation of constants M and α).
Proof. Substituting $\mathbf{u} := \mathbf{u} - \mathbf{u}_h$ in the coercivity condition (2.37), we get:

$$\alpha \|\mathbf{u} - \mathbf{u}_h\|_{\mathbf{V}}^2 \leq b(\mathbf{u} - \mathbf{u}_h, \mathbf{u} - \mathbf{u}_h). \tag{2.67}$$

By adding and subtracting term $b(\mathbf{u}-\mathbf{u}_h, \mathbf{v}_h)$, we convert the right-hand side to:

$$b(\mathbf{u} - \mathbf{u}_h, \mathbf{u} - \mathbf{v}_h) + b(\mathbf{u} - \mathbf{u}_h, \mathbf{v}_h - \mathbf{u}_h), \tag{2.68}$$

where, by the orthogonality condition (2.65), the second of the integrals must vanish (since $\mathbf{v}_h - \mathbf{u}_h$ is an element of $\mathbf{V}_h$).

Finally, using the continuity of the bilinear form, we obtain:

$$\alpha \|\mathbf{u} - \mathbf{u}_h\|_{\mathbf{V}}^2 \leq M \|\mathbf{u} - \mathbf{u}_h\|_{\mathbf{V}} \|\mathbf{u} - \mathbf{v}_h\|_{\mathbf{V}}, \quad \text{for every } \mathbf{v}_h \in \mathbf{V}_h. \tag{2.69}$$

Now, if $\|\mathbf{u} - \mathbf{u}_h\|_{\mathbf{V}} = 0$ (i.e. $\mathbf{u} = \mathbf{u}_h$) then inequality (2.66) is trivially satisfied, otherwise dividing both sides of (2.69) by $\alpha\|\mathbf{u} - \mathbf{u}_h\|_{\mathbf{V}}$ gives the required result. □

Since inequality (2.66) is satisfied *for every* $\mathbf{v}_h \in \mathbf{V}_h$, as a special choice, we may select for $\mathbf{v}_h$ function $\mathbf{u}_I$ – the $\mathbf{V}_h$-interpolant of the exact solution $\mathbf{u}$. We then obtain:

$$\|\mathbf{u} - \mathbf{u}_h\|_{\mathbf{V}} \leq \frac{M}{\alpha} \|\mathbf{u} - \mathbf{u}_I\|_{\mathbf{V}}. \tag{2.70}$$

Thus, one of the immediate consequences of the Cea's lemma is the fact that, for internal approximations, *the approximation error is always bounded by the interpolation error.* By forcing then the interpolation error to decrease, we *automatically* decrease the approximation error as well. This, very strong *stability condition,* is typical for conforming finite elements applied to coercive problems.

Finally, by combining the Cea lemma with the interpolation error estimate, we obtain the following standard error estimate for conforming elements.

Theorem 2.4. *Let us assume that the assumptions of both Cea's lemma and the interpolation error estimate hold. Let k denote the order of approximation, and let exact solution $\mathbf{u}$ be sufficiently regular to be in space $H^{l+1}(\Omega), l \geq 1$. The following estimate holds:*

$$\|\mathbf{u} - \mathbf{u}_h\|_{\mathbf{H}^1(\Omega)} \leq C h^s |\mathbf{u}|_{s+1,\Omega}, \tag{2.71}$$

where $s = \min(k, l)$, and constant C is independent of solution $\mathbf{u}$ and mesh size h.
Proof. We have, by the Cea lemma,

$$\|\mathbf{u} - \mathbf{u}_h\|_{\mathbf{H}^1(\Omega)} \leq \frac{M}{\alpha} \|\mathbf{u} - \mathbf{u}_I\|_{\mathbf{H}^1(\Omega)}. \tag{2.72}$$

But from (2.55) and (2.56) follows that

$$\begin{aligned} \|\mathbf{u} - \mathbf{u}_I\|_{\mathbf{H}^1(\Omega)}^2 &\leq C_1^2 h^{2(s+1)} |\mathbf{u}|_{s+1,\Omega}^2 + C_2^2 h^{2s} |\mathbf{u}|_{s+1,\Omega}^2 \\ &= (C_1^2 h^2 + C_2^2) h^{2s} |\mathbf{u}|_{s+1,\Omega}^2. \end{aligned} \tag{2.73}$$

This leads directly to the final estimate. □

Refining a finite element mesh (recall the assumptions on the regular families of finite elements), we force the mesh size h to approach zero and, according to estimate (2.71), the finite element approximation error $\mathbf{e}_h = \mathbf{u} - \mathbf{u}_h$, measured in the Sobolev norm of order one, converges to zero as well. We say shortly that the *FE solution* $\mathbf{u}_h$ *converges to the exact solution* $\mathbf{u}$. The exponent s describes the *rate of convergence* of the method. The higher the order of convergence, the faster the right-hand side of estimate (2.70) and the finite element error will converge to zero.

Please notice the main factors affecting the rate of convergence:

- order of approximation k,
- "order" l of regularity of the solution.

In particular, for singular solutions to the elasticity equations (not to be confused with solutions with high gradients!), the order of regularity l may be very low (close to one). The use of higher order elements then, intended to accelerate the order of convergence, is rather limited as the order of convergence will be dominated by the low regularity of the solution.

2.6 Convergence in Norm L^2. The Aubin–Nitsche Argument

When discussing the interpolation error estimate, we noticed that the error, when measured in the lower order norm, converges to zero faster, i.e. at a higher rate. Intuitively speaking, by selecting to work with less demanding norm, we obtain a better convergence rate. It turns out that, under appropriate conditions, a similar effect takes place for the approximation error as well.

We begin by introducing the *adjoint problem.* The easiest way to arrive at the adjoint problem is to consider equation (2.15) and interchange the role of the arguments in the bilinear form, assuming that now $\mathbf{v}$ is a solution, and $\mathbf{u}$ will be the test function. We shall also assume that the linear functional (the "load" in the adjoint problem) contains no boundary integrals and, therefore, can be reduced to the L^2 inner product. We arrive at the problem:

Find $\mathbf{v_g} \in \mathbf{V}$ such that:

$$b(\mathbf{u}, \mathbf{v_g}) = (\mathbf{g}, \mathbf{u})_{\mathbf{L}^2(\Omega)}, \quad \text{for every } \mathbf{u} \in \mathbf{V}. \tag{2.74}$$

For the elasticity problem, the bilinear form $b(\mathbf{u}, \mathbf{v})$ is *symmetric* and the adjoint problem will coincide with the original one. It explains why the equations of elasticity fall into the category of *self-adjoint* problems. In general, however, form $b(\mathbf{u}, \mathbf{v})$ may be nonsymmetric, and the adjoint and original problems will be different. The classical form of the adjoint problem, i.e. the corresponding differential equations and boundary conditions may be deduced from the variational formulation using the fundamental Green formula.

Function $\mathbf{g}(\mathbf{x})$ in (2.74) is vector-valued, with each of its components in $L^2(\Omega)$. The term on the right-hand side of (2.74):

$$(\mathbf{g}, \mathbf{u})_{\mathbf{L}^2(\Omega)} \stackrel{\text{def}}{=} (g_1, u_1)_{L^2(\Omega)} + (g_2, u_2)_{L^2(\Omega)} = \int_\Omega g_1 u_1 \, d\Omega + \int_\Omega g_2 u_2 \, d\Omega, \tag{2.75}$$

defines the inner product in the Hilbert space $\mathbf{L}^2(\Omega)$ of vector-valued functions with square integrable components.

By the Lax–Milgram lemma, adjoint problem (2.74) has a unique solution $\mathbf{v_g}$. Index $\mathbf{g}$ emphasizes the dependence of the solution upon the right-hand side of (2.74) defined by function $\mathbf{g}$.

Let $\mathbf{u}$ denote one more time the exact solution, and let $\mathbf{u}_h$ by the FE approximate solution of (2.59). The difference $\mathbf{u} - \mathbf{u}_h$ sits in space $\mathbf{V}$ which implies that

$$b(\mathbf{u} - \mathbf{u}_h, \mathbf{v_g}) = (\mathbf{g}, \mathbf{u} - \mathbf{u}_h) \tag{2.76}$$

Taking into account orthogonality condition (2.65), we get:

$$(\mathbf{g}, \mathbf{u} - \mathbf{u}_h) = b(\mathbf{u} - \mathbf{u}_h, \mathbf{v_g}) - b(\mathbf{u} - \mathbf{u}_h, \mathbf{v}_h) = b(\mathbf{u} - \mathbf{u}_h, \mathbf{v_g} - \mathbf{v}_h), \tag{2.77}$$

for every $\mathbf{v}_h \in \mathbf{V}_h$.

Taking advantage of the continuity of form $b(\mathbf{u}, \mathbf{v})$, we get finally

$$(\mathbf{g}, \mathbf{u} - \mathbf{u}_h) \le M \|\mathbf{u} - \mathbf{u}_h\|_{\mathbf{H}^1(\Omega)} \|\mathbf{v_g} - \mathbf{v}_h\|_{\mathbf{H}^1(\Omega)} . \tag{2.78}$$

Employing for $\mathbf{v}_h$ on the right-hand side of (2.65) an *approximate solution to the adjoint problem* and using Theorem 2.4, we obtain:

$$\|\mathbf{v_g} - \mathbf{v}_h\|_{\mathbf{H}^1(\Omega)} \le Ch |\mathbf{v_g}|_{2,\Omega} , \tag{2.79}$$

where constant $C > 0$, as usual, is independent of mesh size h and function $\mathbf{v_g}$.

One can show that, under *appropriate regularity assumptions*, second order derivatives of the solution $\mathbf{v_g}$ to the adjoint problem can be estimated by the "load" function $\mathbf{g}(\mathbf{x})$, i.e.

$$|\mathbf{v_g}|_{2,\Omega} \le C |\mathbf{g}|_{\mathbf{L}^2(\Omega)} \tag{2.80}$$

with some other constant $C > 0$, independent of function $\mathbf{g}(\mathbf{x})$. The norm in $\mathbf{L}^2(\Omega)$ is induced by the inner product (2.75), i.e.

$$\|\mathbf{g}\|^2_{\mathbf{L}^2(\Omega)} = \|g_1\|^2_{L^2(\Omega)} + \|g_2\|^2_{L^2(\Omega)} . \tag{2.81}$$

Taking into account formulas (2.78),(2.79),(2.80) and the estimate of the approximation error in norm $\mathbf{H}^1(\Omega)$, we obtain:

$$(\mathbf{g}, \mathbf{u} - \mathbf{u}_h) \le Ch^s |\mathbf{u}|_{s+1,\Omega} h |\mathbf{g}|_{\mathbf{L}^2(\Omega)} , \tag{2.82}$$

still for an arbitrary function $\mathbf{g} \in \mathbf{L}^2(\Omega)$. Taking $\mathbf{g} = \mathbf{u} - \mathbf{u}_h$ and diving both sides of the estimate by $\|\mathbf{g}\|_{\mathbf{L}^2(\Omega)}$, we obtain finally:

$$\|\mathbf{u} - \mathbf{u}_h\|_{\mathbf{L}^2(\Omega)} \le Ch^{s+1} |\mathbf{u}|_{s+1,\Omega} . \tag{2.83}$$

The use of the adjoint equation in deriving estimate (2.83), measuring the approximation error in the $\mathbf{L}^2(\Omega)$-norm, is known as the Aubin–Nitsche argument [34].

We emphasize that the improvement of the convergence rate in the L^2-norm is related to the regularity of the adjoint problem. For the self-adjoint problems (including elasticity) and standard regularity assumptions guaranteeing existence of solution in $H^2(\Omega)$, these assumptions are satisfied automatically.

2.7
Numerical Integration. Strang's First Lemma

In practice, all integrals defining bilinear and linear forms are integrated numerically using the standard formula:

$$\int_{\Omega_e} \phi(\mathbf{x})\,\mathrm{d}\mathbf{x} \approx \sum_{l=1}^{L} \phi(\mathbf{b}_l) w_l \,. \tag{2.84}$$

Points $\mathbf{b}_l$, $l = 1, \ldots, L$, are known as *integration points* and w_l, $l = 1, \ldots, L$, are the corresponding *integration weights*. The most important rules of the *numerical integration* procedure are the *Gauss rules* where points $\mathbf{b}_l$ and weights w_l are selected in such a way as to yield exact values for polynomials of order p with p as big as possible.

For problems with piecewise (over each element) constant elasticities and loads, affine elements, and the round-off error neglected, one can always implement a Gaussian integration procedure with sufficiently high order to obtain exact values for all involved integrals. Consequently, the numerical integration will produce no extra error.

The situation is different if the coefficients of the differential operator (in our case, the elasticities), or the right-hand side of the differential equation or Neumann boundary conditions, are arbitrary functions, and the evaluation of the stiffness and load matrices does not reduce to integrating polynomials only. Another typical situation leading to the integration error corresponds to the use of *curvilinear elements* (not discussed here) for which the element shape functions are no longer polynomials and the integration cannot be done in practice exactly, even for piecewise constant coefficients and right-hand sides.

Let us also mention that, in practice, by an appropriate change of variables, all integrals are integrated over master elements only. For an affine element, the Jacobian of the corresponding transformation is constant, and the change of variables does not imply a need for an increased order of integration.

From the convergence analysis point of view, we need to consider a new abstract approximate problem:

Find $\mathbf{u}_h \in \mathbf{V}_h$ such that,

$$b_h(\mathbf{u}_h, \mathbf{v}_h) = l_h(\mathbf{v}_h), \quad \text{for every } \mathbf{v}_h \in \mathbf{V}_h \,. \tag{2.85}$$

The additional indices h accompanying the symbols for the bilinear and linear forms, emphasize that the corresponding values for these forms are *mesh dependent*, the fact corresponding to the use of the numerical quadrature. More precisely, the new forms are defined as follows:

$$\begin{aligned} b_h(\mathbf{u}_h, \mathbf{v}_h) &= \sum_{e=1}^{E} (\text{num}) \int_{\Omega_e} C_{klmn}\epsilon_{kl}(\mathbf{u}_h)\epsilon_{mn}(\mathbf{v}_h)\, \mathrm{d}\Omega_e\,, \\ l_h(\mathbf{v}) &= \sum_{e=1}^{E} (\text{num}) \int_{\Omega_e} \mathbf{f}\mathbf{v}_h\, \mathrm{d}\Omega_e \\ &\quad + (\text{num}) \int_{\partial\Omega_e \cap \partial\Omega_\sigma} \mathbf{t}\mathbf{v}_h\, \mathrm{d}(\partial\Omega_e), \end{aligned} \tag{2.86}$$

where all integrals are evaluated numerically, i.e. for instance:

$$(\text{num}) \int_{\Omega_e} \mathbf{f}\mathbf{v}_h\, \mathrm{d}\Omega_e \stackrel{\text{def}}{=} \sum_{l=1}^{L_e} \mathbf{f}(\mathbf{b}_l)\mathbf{v}_h(\mathbf{b}_l) w_l\,. \tag{2.87}$$

This time, therefore, we are dealing not only with a *family of discrete approximate spaces* $\mathbf{V}_h$ but also a *a family of approximate bilinear forms* $b_h(\mathbf{u}_h, \mathbf{v}_h)$ and *family of approximate linear forms* $l_h(\mathbf{v}_h)$.

We will say that a family of discrete bilinear forms $b_h(\mathbf{u}_h, \mathbf{v}_h)$ is *uniformly coercive* if there exists a coercivity constant $\alpha > 0$, independent of h, such that:

$$\alpha \|\mathbf{u}_h\|_{\mathbf{V}}^2 \le b_h(\mathbf{u}_h, \mathbf{u}_h) \quad \text{for all } \mathbf{u}_h \in \mathbf{V}_h\,. \tag{2.88}$$

Notice that, when the definition of b_h reduces to that of form b, condition (2.88) is an immediate consequence of the coercivity and the internal approximation.

Finally, assuming that forms b_h and l_h are continuous[3], we can immediately conclude that the approximate solution $\mathbf{u}_h$ exists and it is unique (Lax–Milgram lemma).

Let then $\mathbf{u}$ be the exact solution and let $\mathbf{u}_h$ denote the corresponding approximate solution. The convergence analysis is based on the following result.

Theorem 2.5. (Strang's first lemma) *Assume that the assumptions of the Lax–Milgram lemma hold and the family of approximate bilinear forms $b_h(\mathbf{u}_h, \mathbf{v}_h)$ is uniformly coercive, i.e. it satisfies (2.88). The following estimate holds:*

$$\|\mathbf{u} - \mathbf{u}_h\|_{\mathbf{V}} \le C\{\|\mathbf{u} - \mathbf{v}_h\|_{\mathbf{V}} + e_b(\mathbf{v}_h) + e_l\}, \quad \textit{for all } \mathbf{v}_h \in \mathbf{V}_h\,, \tag{2.89}$$

[3] For *finite*-dimensional spaces linearity implies continuity and this assumption is trivially satisfied.

where constant C is independent of $\mathbf{u}$ *and* h, *and*

$$
\begin{aligned}
e_b(\mathbf{v}_h) &= \max_{\mathbf{w}_h \in \mathbf{V}_h} \frac{|b(\mathbf{v}_h, \mathbf{w}_h) - b_h(\mathbf{v}_h, \mathbf{w}_h)|}{\|\mathbf{w}_h\|_{\mathbf{V}}}, \\
e_l &= \max_{\mathbf{w}_h \in \mathbf{V}_h} \frac{|l(\mathbf{w}_h) - l_h(\mathbf{w}_h)|}{\|\mathbf{w}_h\|_{\mathbf{V}}}.
\end{aligned}
\tag{2.90}
$$

Proof. Making use of the uniform coercivity, we get:

$$
\begin{aligned}
\alpha\|\mathbf{u}_h - \mathbf{v}_h\|_{\mathbf{V}}^2 &\le b_h(\mathbf{u}_h - \mathbf{v}_h, \mathbf{u}_h - \mathbf{v}_h) \\
&= b(\mathbf{u} - \mathbf{v}_h, \mathbf{u}_h - \mathbf{v}_h) + \{b(\mathbf{v}_h, \mathbf{u}_h - \mathbf{v}_h) - b_h(\mathbf{v}_h, \mathbf{u}_h - \mathbf{v}_h)\} \\
&\quad + \{l_h(\mathbf{u}_h - \mathbf{v}_h) - l(\mathbf{u}_h - \mathbf{v}_h)\}.
\end{aligned}
\tag{2.91}
$$

Taking advantage of continuity of $b(\mathbf{u}, \mathbf{v})$ and dividing both sides by $\alpha\|\mathbf{u}_h - \mathbf{v}_h\|_{\mathbf{V}}$, we obtain:

$$
\begin{aligned}
\|\mathbf{u}_h - \mathbf{v}_h\|_{\mathbf{V}} \le \frac{M}{\alpha}\|\mathbf{u} - \mathbf{v}_h\|_{\mathbf{V}} &+ \frac{1}{\alpha}\frac{|b(\mathbf{v}_h, \mathbf{u}_h - \mathbf{v}_h) - b_h(\mathbf{v}_h, \mathbf{u}_h - \mathbf{v}_h)|}{\|\mathbf{u}_h - \mathbf{v}_h\|_{\mathbf{V}}} \\
&+ \frac{1}{\alpha}\frac{|l_h(\mathbf{u}_h - \mathbf{v}_h) - l(\mathbf{u}_h - \mathbf{v}_h)|}{\|\mathbf{u}_h - \mathbf{v}_h\|_{\mathbf{V}}}.
\end{aligned}
\tag{2.92}
$$

Introducing the appropriate suprema and applying the triangle inequality:

$$
\|\mathbf{u} - \mathbf{u}_h| \le \|\mathbf{u} - \mathbf{v}_h\| + \|\mathbf{u}_h - \mathbf{v}_h\|, \tag{2.93}
$$

we finish the proof. □

Notice that, substituting for $\mathbf{v}_h$ interpolant $\mathbf{u}_I$, we obtain on the right-hand side, similarly as in the Cea lemma, the interpolation error. This time, however, we have the two additional terms on the right-hand side of the estimate, corresponding to the quadrature error for the stiffness matrix and the load vector.

Based on the Strang first lemma we may attempt now to specify sufficient conditions with which, when satisfied, the numerical integration would not alter the convergence rates for the approximate problem with exact integration. It is sufficient to guarantee that the *consistency errors* in the estimate above, would converge to zero with the *same rate* as the interpolation error. Let us comment that such conditions usually hold *asymptotically* in h, i.e. for h sufficiently small. The question "how small" is related to regularity assumptions and usually remains without answer.

As an example, we present without proof the following result for the case of pure kinematic boundary conditions ($\partial\Omega_\sigma = \emptyset$).

Theorem 2.6. (Ciarlet) *Assume that we are dealing with a regular family of affine elements of order k and that the quadrature rule is exact for polynomials of order $2k-2$. Assume additionally that*

- *exact solution* $\mathbf{u} \in \mathbf{H}^{k+1}(\Omega)$,
- *the coefficients of the operator, i.e. elasticities $C_{klmn}(\mathbf{x})$ are of class C^k,*

- *the right-hand side, i.e. volume forces* $\mathbf{f}(\mathbf{x})$ *are from space* $\mathbf{H}^k(\Omega)$.

The following estimate then holds:

$$\|\mathbf{u} - \mathbf{u}_h\|_{\mathbf{H}^1(\Omega)} \le Ch^k \left(|\mathbf{u}|_{k+1,\Omega} + \sum_{k,l,m,n} \|C_{klmn}\|_{k,\infty,\Omega} + \|\mathbf{f}\|_{k,\Omega} \right) , \quad (2.94)$$

where $\|C_{klmn}\|_{k,\infty,\Omega}$ *denote the maximum values of the* k*-th order derivative of functions* $C_{klmn}(\mathbf{x})$ *in domain* Ω. □

We shall comment on the regularity assumptions yet in the concluding section.

2.8 Nonconforming Elements. Strang's Second Lemma

The linear element with nodes placed at the mid-points of its edges, discussed earlier in Sect. 2.5, is the classical example of a *non-conforming* approximation. As a result of a *wrong* location of the nodes, continuity at the nodes does not imply the global continuity of approximate solution $\mathbf{u}_h$. That does not necessary imply that the corresponding FE method will not converge, to the contrary, the elements of this type may have additional attractive properties (e.g. stability for mixed approximations). The fundamental difficulty, though, in proving the convergence, lies in the fact that the approximation ceases to be *internal*, i.e. discrete spaces $\mathbf{V}_h$ are no longer subspaces of $\mathbf{V}$ (a similar effect comes into play for finite difference methods).

We begin by introducing a *mesh-dependent* norm[4]:

$$\|\mathbf{v}_h\|_h = \left(\sum_{e=1}^{E} |\mathbf{v}_h|^2_{1,\Omega_e} \right)^{\frac{1}{2}} . \quad (2.95)$$

In a similar way, we define the discrete bilinear forms $b_h(\mathbf{u}_h, \mathbf{v}_h)$ and linear forms $l_h(\mathbf{v}_h)$[5]. Again, the family of the bilinear forms b_h is said to be *uniformly continuous* if there exists a constant $\alpha > 0$, independent of h such that:

$$\alpha \|\mathbf{u}_h\|_h^2 \le b_h(\mathbf{u}_h, \mathbf{u}_h). \quad (2.96)$$

The difference between conditions (2.96) and (2.88) lies in the use of the mesh-dependent norm in (2.96). Also, the continuity condition for the discrete

[4] In fact, the quantity defined below is *a priori* only a *seminorm*. Once a specific nonconforming element is assumed, one proceeds by showing that this seminorm is, in fact, a norm.

[5] The crucial point here is that, due to the nonconformity of the approximation, the sum of the integrals over elements cannot be longer interpreted as the integral over the whole domain.

bilinear forms has now to be rewritten in a slightly different form:

$$|b_h(\mathbf{u},\mathbf{v})| \leq M\|\mathbf{u}\|_h\|\mathbf{v}\|_h, \quad \text{for every } \mathbf{u},\mathbf{v} \in \mathbf{V}_h,\mathbf{V}. \tag{2.97}$$

Recalling that the discrete linear forms are automatically continuous, conditions (2.96) and (2.97) one more time imply that the discrete solution $\mathbf{u}_h$ exists and it is unique.

The following result provides a foundation for the convergence analysis.

Theorem 2.7. (Strang's second lemma) *Let $b_h(\mathbf{u}_h,\mathbf{v}_h)$ be a family of bilinear forms satisfying conditions (2.96) and (2.97). There exists then a constant $C > 0$, independent of h such that the following estimate holds,*

$$\|\mathbf{u}-\mathbf{u}_h\|_h \leq C\left(\|\mathbf{u}-\mathbf{v}_h\|_h + \max_{\mathbf{w}_h\in\mathbf{V}_h}\frac{|b_h(\mathbf{u}_h,\mathbf{w}_h)-l(\mathbf{w}_h)|}{\|\mathbf{w}_h\|_h}\right), \tag{2.98}$$

for every $\mathbf{v}_h \in \mathbf{V}_h$.

Proof. Taking advantage of the uniform coercivity, we get:

$$\begin{aligned}\alpha\|\mathbf{u}_h-\mathbf{v}_h\|_h^2 &\leq b_h(\mathbf{u}_h-\mathbf{v}_h,\mathbf{u}_h-\mathbf{v}_h)\\ &= b(\mathbf{u}-\mathbf{v}_h,\mathbf{u}_h-\mathbf{v}_h) + \{l(\mathbf{u}_h-\mathbf{v}_h) - b_h(\mathbf{u},\mathbf{u}_h-\mathbf{v}_h)\}.\end{aligned} \tag{2.99}$$

Making use of the continuity of b_h, and dividing both sides by $\alpha\|\mathbf{u}_h-\mathbf{v}_h\|_h$, we obtain:

$$\|\mathbf{u}_h-\mathbf{v}_h\|_h \leq \frac{M}{\alpha}\|\mathbf{u}-\mathbf{v}_h\|_h + \frac{1}{\alpha}\frac{l(\mathbf{u}_h-\mathbf{v}_h)-b_h(\mathbf{u},\mathbf{u}_h-\mathbf{v}_h)}{\|\mathbf{u}_h-\mathbf{v}_h\|_h}. \tag{2.100}$$

Taking maximum with respect to $\mathbf{v}_h$ in the second term, and using the triangle inequality, we get the final result. □

As before, the first term on the right-hand side of estimate (2.98) may be reduced to the interpolation error. The second one corresponds to the "crime" of using non-conforming elements. Its estimate involves the analysis of a particular element and it is related to the satisfaction of the so-called *patch test*. For an example of such a estimate we refer to [34].

Finally, let us note one more time that, for conforming elements, the second term in the estimate is identically zero, and the estimate reduces to Cea's lemma.

2.9 Steady State Vibrations as an Example of a Non-coercive Problem

We shall replace now equations $(2.1)_1$ with more general equations of steady-state vibrations:

$$-\rho\omega^2 u_k - \sigma_{kl,l} = f_k \quad \text{in } \Omega. \tag{2.101}$$

Here $\rho(\mathbf{x})$ is the density of the body and ω denotes the frequency of vibrations. For the static case $\omega = 0$ and equations (2.101) reduce to (2.1).

Equations (2.101) will be accompanied by the same kinematic and traction boundary conditions as for the original, static problem.

The equations of steady-state vibrations are obtained from the transient equations of motion:

$$\rho \frac{\partial^2 u_k}{\partial t^2} - \sigma_{kl,l} = f_k \tag{2.102}$$

by assuming time-periodic loadings:

$$f_k(\mathbf{x}, t) = \Re(e^{i\omega t} f_k(\mathbf{x})), \quad t_k(\mathbf{x}, t) = \Re(e^{i\omega t} t_k(\mathbf{x})) \tag{2.103}$$

and seeking the transient solution in the same form:

$$u_k(\mathbf{x}, t) = \Re(e^{i\omega t} u_k(\mathbf{x})) \tag{2.104}$$

The *phasors* $\mathbf{f}(\mathbf{x}), \mathbf{t}(\mathbf{x}), \mathbf{u}(\mathbf{x})$ are complex-valued functions. For the elastodynamics equations, however, the equations for the real and the imaginary parts of the unknown phasor $u_k(\mathbf{x})$ decouple from each other, and we can return to the real-valued interpretation with the understanding that f_k and g_k denote the real or imaginary parts of the actual phasors and u_k are the corresponding real or imaginary components of the displacement vector.

Alternatively, equations (2.101) can be obtained by Fourier-transforming equations (2.102) and interpreting $\mathbf{f}, \mathbf{t}$ and $\mathbf{u}$ as Fourier transforms of the original quantities. In any case, once the solution to (2.101) is known, the inverse Fourier transform or the Fourier series can be used to calculate the solution to the transient equations of elastodynamics.

Following the same steps as for the static problem, we arrive at the variational formulation:

Find $\mathbf{u}(\mathbf{x}), \mathbf{u} = 0$ on $\partial\Omega_u$, such that:

$$\int_\Omega \sigma_{kl}(\mathbf{u})\epsilon_{kl}(\mathbf{v}) \mathrm{d}\Omega - \omega^2 \int_\Omega \rho u_k v_k \mathrm{d}\Omega = \int_\Omega f_k v_k \mathrm{d}\Omega + \int_{\partial\Omega_\sigma} t_k v_k \mathrm{d}(\partial\Omega), \tag{2.105}$$

for every test function $\mathbf{v}$, such that $\mathbf{v} = \mathbf{0}$ on $\partial\Omega_u$.

The left-hand side of (2.105) can again be interpreted as a bilinear form that is not longer, however, coercive or positive-definite. More precisely, we have:

$$b(\mathbf{u}, \mathbf{v}) = a(\mathbf{u}, \mathbf{v}) - \omega^2 (\mathbf{u}, \mathbf{v})_\rho \tag{2.106}$$

where the contribution $a(\mathbf{u}, \mathbf{v})$ corresponds to the original bilinear form for the static case, and $(\mathbf{u}, \mathbf{v})_\rho$ denotes the inertia term interpreted as a *weighted* L^2*-inner product:*

$$(\mathbf{u}, \mathbf{v})_\rho = \int_\Omega \rho(\mathbf{x}) u_k(\mathbf{x}) v_k(\mathbf{x}) \, \mathrm{d}\Omega. \tag{2.107}$$

Thus, we again obtain the abstract variational boundary-value problem (2.15) and the corresponding FE approximation (2.59), except that a new theory has to be developed to handle the non-coercive bilinear form.

The key to understanding of the convergence mechanism for the steady-state vibrations comes from the related eigenvalue problem:

Find $\mathbf{u} \in \mathbf{V}$ and $\omega \geq 0$ such that

$$a(\mathbf{u},\mathbf{v}) = \omega^2(\mathbf{u},\mathbf{v})_\rho \tag{2.108}$$

for every test function $\mathbf{v} \in \mathbf{V}$.

It can be shown (see e.g. [126]) that problem (2.108) admits an infinite sequence of positive eigenvalues:

$$0 < \omega_1^2 \leq \omega_2^2 \leq \ldots \leq \omega_k^2 \leq \ldots \tag{2.109}$$

with $\omega_k \to \infty$ as $k \to \infty$. The corresponding eigenvectors $\mathbf{e}_i(\mathbf{x})$ are orthogonal in the sense of both $a(\mathbf{u},\mathbf{v})$ and $(\mathbf{u},\mathbf{v})_\rho$ forms:

$$a(\mathbf{e}_i,\mathbf{e}_j) = (\mathbf{e}_i,\mathbf{e}_j)_\rho = 0 \quad \text{for } i \neq j \tag{2.110}$$

and form an orthonormal basis[6] in $\mathbf{L}^2(\Omega)$. In particular, fields $\mathbf{u},\mathbf{v} \in \mathbf{V}$ can be expanded into the infinite series:

$$\mathbf{u} = \sum_{n=1}^{\infty} u_n \mathbf{e}_n\,, \qquad \mathbf{v} = \sum_{n=1}^{\infty} v_n \mathbf{e}_n\,, \tag{2.111}$$

where $u_n = (\mathbf{u},\mathbf{e}_n)_\rho\,,\ v_n = (\mathbf{v},\mathbf{e}_n)_\rho$ are known as the *spectral or modal components* of vectors $\mathbf{u}$ and $\mathbf{v}$.

Introducing the FE approximation of the eigenvalue problem:

Find $\mathbf{u}_h \in \mathbf{V}_h$ and $\omega_h \geq 0$ such that

$$a(\mathbf{u}_h,\mathbf{v}_h) = \omega_h^2(\mathbf{u}_h,\mathbf{v}_h)_\rho\,, \tag{2.112}$$

for every test function $\mathbf{v}_h \in \mathbf{V}_h$, we record identical properties for the approximate eigenpairs $(\omega_{nh}^2, \mathbf{e}_{nh})$ except that, of course, the number of the eigenpairs equals now the dimension of the FE space, and the infinite series are replaced with finite sums.

It is well known that the first eigenvalue can be obtained by minimizing the Rayleigh quotient:

$$\omega_1^2 = \min_{\substack{\mathbf{u}\in\mathbf{V}\\ \mathbf{u}\neq\mathbf{0}}} \frac{a(\mathbf{u},\mathbf{u})}{(\mathbf{u},\mathbf{u})_\rho} = \min_{\substack{\mathbf{u}\in\mathbf{V}\\ \|\mathbf{u}\|_\rho=1}} a(\mathbf{u},\mathbf{u}) \tag{2.113}$$

with an identical representation for the first approximate eigenvalue:

$$\omega_{1h}^2 = \min_{\substack{\mathbf{u}_h\in\mathbf{V}_h\\ \mathbf{u}_h\neq\mathbf{0}}} \frac{a(\mathbf{u}_h,\mathbf{u}_h)}{(\mathbf{u}_h,\mathbf{u}_h)_\rho} = \min_{\substack{\mathbf{u}_h\in\mathbf{V}_h\\ \|\mathbf{u}_h\|_\rho=1}} a(\mathbf{u}_h,\mathbf{u}_h)\,. \tag{2.114}$$

[6] Normalized with respect to the weighted L^2-product, $(\mathbf{e}_i,\mathbf{e}_i)_\rho = 1$.

For internal approximations $\mathbf{V}_h \subset \mathbf{V}$, and the minimum in (2.114) is taken over a smaller set that in (2.113). Consequently, it must be:

$$\omega_1 \le \omega_{1h}\,, \tag{2.115}$$

An analogous result for the next eigenvalues is derived by the same reasoning from the following minimum-maximum principle [6] that holds for all eigenvalues,

$$\omega_n^2 = \min_{\substack{\mathbf{V}_n \subset \mathbf{V} \\ \dim \mathbf{V}_n = n}} \max_{\mathbf{u} \in \mathbf{V}_n} \frac{a(\mathbf{u},\mathbf{u})}{(\mathbf{u},\mathbf{u})_\rho}\,. \tag{2.116}$$

We have thus

$$\omega_n \le \omega_{nh} \tag{2.117}$$

for every $n = 1, 2, \ldots$.

As $h \to 0$, the approximate eigenpairs $(\omega_{nh}^2, \mathbf{e}_{nh})$ converge to the exact eigenpairs $(\omega_n^2, \mathbf{u}_n)$. For a comprehensive review on the corresponding error estimates we refer to [6]. It follows from the arguments used above that for a sequence of *nested meshes*, i.e. for $\mathbf{V}_{h_1} \subset \mathbf{V}_{h_2}$ with $h_1 < h_2$, the corresponding approximate eigenvalues ω_{nh}^2 will converge *monotonically* from the right to the exact ones.

We are ready now to analyze the problem of steady-state vibrations. Existence and uniqueness of solution follows from the following generalization of the Lax–Milgram lemma.

Theorem 2.8. (Babuška) *Let* $\mathbf{V}$ *be a Hilbert space with inner product* $(\mathbf{u},\mathbf{v})_\mathbf{V}$ *and the corresponding norm* $\|\mathbf{u}\|_\mathbf{V}$. *Assume that the bilinear form* $b(\mathbf{u},\mathbf{v})$ *and linear form* $l(\mathbf{v})$ *from the abstract variational problem (2.15) satisfy the following assumptions:*

- $l(\mathbf{v})$ *is* continuous *on* $\mathbf{V}$,
- $b(\mathbf{u},\mathbf{v})$ *is* continuous *on* $\mathbf{V}$,

the following stability condition holds:

$$\gamma \stackrel{\text{def}}{=} \inf_{\substack{\mathbf{u} \in \mathbf{V} \\ \|\mathbf{u}\|_\mathbf{V}=1}} \sup_{\substack{\mathbf{v} \in \mathbf{V} \\ \|\mathbf{v}\|_\mathbf{V}=1}} b(\mathbf{u},\mathbf{v}) > 0\,. \tag{2.118}$$

There exists then a unique *solution* $\mathbf{u}$ *to the abstract variational boundary-value problem (2.15).* □

Remark 6. For a coercive form $b(\mathbf{u},\mathbf{v})$ (condition (2.37)), we have for $\|\mathbf{u}\|_\mathbf{V} = 1$,

$$\sup_{\|\mathbf{v}\|=1} b(\mathbf{u},\mathbf{v}) \ge b(\mathbf{u},\mathbf{u}) \ge \alpha \|\mathbf{u}\|_\mathbf{V}^2 = \alpha, \tag{2.119}$$

and, consequently, $\gamma \ge \alpha$. □

We shall verify now the inf–sup stability condition (2.118) for our problem. In the calculations to follow, it is convenient to replace the original norm in $\mathbf{V}$

with an equivalent[7] energy norm:

$$\|\mathbf{u}\|_{\mathbf{V}}^2 = a(\mathbf{u}, \mathbf{u}). \tag{2.120}$$

Using the spectral decomposition (2.111), we can now conveniently express the bilinear form $b(\mathbf{u}, \mathbf{v})$ in terms of spectral components of $\mathbf{u}$ and $\mathbf{v}$,

$$\begin{aligned} b(\mathbf{u}, \mathbf{v}) &= b\left(\sum_{n=1}^{\infty} u_n \mathbf{e}_n, \sum_{m=1}^{\infty} v_m \mathbf{e}_m\right) \\ &= \sum_{n=1}^{\infty}\sum_{m=1}^{\infty} u_n v_m \{a(\mathbf{e}_n, \mathbf{e}_m) - \omega^2 (\mathbf{e}_n, \mathbf{e}_m)_\rho\} \\ &= \sum_{n=1}^{\infty} u_n v_n (\omega_n^2 - \omega^2) \end{aligned} \tag{2.121}$$

A direct, elementary calculation based on the Lagrange multipliers gives:

$$\gamma = \min_{n=1,2,\ldots} \frac{|\omega_n^2 - \omega^2|}{\omega_n^2}. \tag{2.122}$$

Thus, as long as the forcing frequency ω *does not coincide* with any of the resonant frequencies ω_n, the steady-state vibrations problem has a unique solution.

The convergence analysis is made possible by the following fundamental result of Babuška.

Theorem 2.9. (Babuška) *Assume that a discrete counterpart of the inf–sup stability condition (2.118) holds:*

$$\gamma_h \stackrel{\text{def}}{=} \inf_{\substack{\mathbf{u}_h \in \mathbf{V}_h \\ \|\mathbf{u}_h\|_{\mathbf{V}}=1}} \sup_{\substack{\mathbf{v}_h \in \mathbf{V}_h \\ \|\mathbf{v}_h\|_{\mathbf{V}}=1}} b(\mathbf{u}_h, \mathbf{v}_h) > 0. \tag{2.123}$$

The following estimate then holds:

$$\|\mathbf{u} - \mathbf{u}_h\|_{\mathbf{V}} \le (1 + M\gamma_h^{-1}) \|\mathbf{u} - \mathbf{w}_h\|_{\mathbf{V}}, \tag{2.124}$$

for every $\mathbf{w}_h \in \mathbf{V}_h$*, with* M *denoting the continuity constant for form* $b(\mathbf{u}, \mathbf{v})$.

[7] Norms $\|\mathbf{v}\|_1$ and $\|\mathbf{v}\|_2$ are equivalent if there exist constants $C_1, C_2 > 0$ such that:

$$\|\mathbf{v}\|_1 \le C_1 \|\mathbf{v}\|_2, \quad \|\mathbf{v}\|_2 \le C_2 \|\mathbf{v}\|_1,$$

for every vector $\mathbf{v}$. Replacing a norm with an equivalent one in error estimates changes constants in the estimates but not the convergence rates.

Proof. Making use of (2.123) and orthogonality condition (2.65), we get:

$$\begin{aligned}
\gamma \|\mathbf{u}_h - \mathbf{w}_h\| &\leq \sup_{\mathbf{v}_h \neq \mathbf{0}} \frac{b(\mathbf{u}_h - \mathbf{w}_h, \mathbf{v}_h)}{\|\mathbf{v}_h\|} \\
&= \sup_{\mathbf{v}_h \neq \mathbf{0}} \frac{b(\mathbf{u} - \mathbf{u}_h, \mathbf{v}_h) + b(\mathbf{u}_h - \mathbf{w}_h, \mathbf{v}_h)}{\|\mathbf{v}_h\|} \\
&= \sup_{\mathbf{v}_h \neq \mathbf{0}} \frac{b(\mathbf{u} - \mathbf{w}_h, \mathbf{v}_h)}{\|\mathbf{v}_h\|} \\
&\leq \sup_{\mathbf{v}_h \neq \mathbf{0}} \frac{M\|\mathbf{u} - \mathbf{w}_h\| \|\mathbf{v}_h\|}{\|\mathbf{v}_h\|} \\
&= M\|\mathbf{u} - \mathbf{w}_h\| \,.
\end{aligned} \tag{2.125}$$

Consequently,

$$\begin{aligned}
\|\mathbf{u} - \mathbf{u}_h\| &\leq \|\mathbf{u} - \mathbf{w}_h\| + \|\mathbf{u}_h - \mathbf{w}_h\| \\
&\leq (1 + M\gamma_h^{-1})\|\mathbf{u} - \mathbf{w}_h\|.
\end{aligned} \tag{2.126}$$

□

The critical issue thus is the behavior of the discrete stability constant γ_h. If we can guarantee that γ_h stays bounded away from zero, i.e. there exists a positive constant γ_0 such that

$$\gamma_h \geq \gamma_0 > 0, \tag{2.127}$$

then

$$\|\mathbf{u} - \mathbf{u}_h\|_{\mathbf{V}} \leq (1 + M\gamma_0^{-1})\|\mathbf{u} - \mathbf{w}_h\|_{\mathbf{V}} \,, \tag{2.128}$$

for every $\mathbf{w}_h \in \mathbf{V}_h$. The whole analysis can then be reduced once more to the interpolation error estimates. We say frequently that the *discrete stability* and *approximability* (i.e. the fact that the interpolation error converges to zero) implies convergence. Note that, for the coercive problems, the discrete stability follows directly from the continuous stability and in that sense it is "granted for free".

The same, direct, evaluation as for the continuous problem yields the formula for the discrete inf–sup constant:

$$\gamma_h = \min_{n=1,\ldots,N_h} \frac{|\omega_{nh}^2 - \omega^2|}{\omega_{nh}^2} \,. \tag{2.129}$$

Now comes the main point. As $h \to 0$, the approximate eigenvalues converge to the exact ones monotonically from the right, i.e. the whole approximate spectrum gets shifted to the left. During the shifting, a particular approximate eigenvalue may cross the forcing frequency ω or even hit it. The discrete inf–sup constant γ_h may then decrease, or may even be equal zero! Once, however, all discrete eigenvalues reach the *right* side of the forcing frequency (i.e. all exact eigenvalues and their approximate counterparts are on the same side of the forcing frequency), the discrete inf–sup constant γ_h will start converging monotonically to the exact inf–sup constant γ. We say that the problem is *asymptotically stable.* In other words, there exists a threshold value h_0 such that for $h \leq h_0$ (meshes fine enough), the discrete stability constant is bounded away from zero. Consequently, (2.128) holds and the standard error estimates follow.

Physically, reaching the *region of asymptotic stability* is related to the sufficient resolution of the eigenvalues neighboring the forcing frequency. Obviously, the distance between frequency ω and the nearest resonant frequency ω_n is crucial. For more related information on the subject we refer to [43].

2.10 Conclusions

2.10.1 Relaxing Regularity Assumptions

At first glance, the emphasis placed on a possible low regularity of the exact solution, when deriving the variational formulation, seems to be in contradiction with comparatively high regularity assumptions used either in Theorems 2.2, 2.4 and 2.6, or the considerations leading to the L^2-error estimates. The big point with all those assumptions is that they can be replaced with analogous conditions over subdomains of domain Ω, provided the finite element meshes are constructed in such a way that the interelement boundaries overlap with the interfaces between the subdomains. In practice, the irregularities happen across interfaces between subdomains corresponding to different material constants or loads, and the condition reduces to an appropriate mesh generation.

This possibility of treating the problems with interfaces is perhaps one of the most significant differences between finite elements and finite difference methods[8].

2.10.2 Summary

The presented "crash course" on the mathematical theory of finite elements is neither complete nor up to date. The main goal of the presentation has been to introduce the engineering reader into the subject with the hope that adding some precision to the FE analysis may enhance the understanding of the every day computations. The classical results presented here were obtained in sixties and early seventies. They do not include more advanced topics such as the analysis of elements with curved geometry (e.g. isoparametric elements), mixed and hybrid elements, the p-version of the FEM, not to

[8] At least those that are based on the local formulation.

mention such subjects like adaptivity, *a-posteriori* error estimation or mathematical postprocessing. Finally, we have not touched even the subject of applying finite elements to parabolic and hyperbolic problems, or problems that are non-linear.

For an introduction to necessary mathematical tools we refer to [126], and for more comprehensive presentations on mathematical theory of finite elements to [34, 127, 161].

3 Fundamentals of nonlinear analysis

3.1
Solid Mechanics Problems

Let us consider a deformation process of a deformable solid under an external load changing in the time interval $[0, \bar{t}]$. In accordance with the concept of the incremental description presented in Chap. 2, Part I, we shall now consider just a part of the process extending from the time instant $\tau = t$ to the 'next' time instant $\tau = t + \Delta t$, τ being the running time coordinate. We further assume that the process is in the state of dynamic equilibrium at t which means that at the beginning of the step the equilibrium conditions (including the d'Alambert forces) have been enforced at the end of the previous step in the computational process. Considering only three dimensional problems the system of differential equations describing the situation on hand within the updated Lagrangian description is recalled as follows:

- equations of motion

$$\Delta\sigma_{kl,l} + \varrho\Delta\hat{f}_k = \varrho\Delta\ddot{u}_k\,, \qquad x \in \Omega\,, \quad \tau \in [0, \bar{t}]\,, \tag{3.1}$$

- geometric relations

$$\Delta\varepsilon_{kl} = \frac{1}{2}(\Delta u_{k,l} + \Delta u_{l,k})\,, \qquad x \in \Omega\,, \quad \tau \in [0, \bar{t}]\,, \tag{3.2}$$

- constitutive equations

$$\Delta\tilde{\sigma}_{kl} = \int_t^{t+\Delta t} C_{klmn}\dot{\varepsilon}_{mn}\,\mathrm{d}\tau\,, \qquad x \in \Omega\,, \quad \tau \in [0, \bar{t}]\,, \tag{3.3}$$

- traction boundary conditions

$$\Delta\sigma_{kl}n_l = \Delta\hat{t}_k\,, \qquad x \in \partial\Omega_\sigma\,, \quad \tau \in [0, \bar{t}]\,, \tag{3.4}$$

- displacement boundary conditions

$$\Delta u_k = \Delta\hat{u}_k\,, \qquad x \in \partial\Omega_u\,, \quad \tau \in [0, \bar{t}]\,. \tag{3.5}$$

Here, $\Delta\sigma_{kl}$ and $\Delta\tilde{\sigma}_{kl}$ are the incremental first and second Piola–Kirchhoff stress tensors based on the configuration C^t, respectively, ϱ is the mass density in C^t, $\varrho\Delta\hat{f}_k$ the vector of body forces, n_k the unit vector normal to the boundary $\partial\Omega_\sigma$, $\Delta\hat{t}_k$ the vector of (configuration-independent) tractions on $\partial\Omega_\sigma$, and $\Delta\hat{u}_k$ the vector of known incremental displacements on $\partial\Omega_u$. The above equations should be augmented by the relation between the incremental components of both the Piola–Kirchhoff stresses given by

$$\Delta\sigma_{kl} = \Delta\tilde{\sigma}_{kl} + \Delta u_{k,m}\tilde{\sigma}_{ml} + \Delta u_{k,m}\Delta\tilde{\sigma}_{ml}\,, \tag{3.6}$$

cf. Eq. (2.24), Part I. In the limiting case of $\Delta t \to 0$, Eqs. (3.1)–(3.5) reduce to the rate form

$$\left.\begin{aligned} &\dot{\sigma}_{kl,l} + \varrho\dot{\hat{f}}_k = \varrho\dddot{u}_k\,,\\ &\dot{\varepsilon}_{kl} = \tfrac{1}{2}(\dot{u}_{k,l} + \dot{u}_{l,k})\,,\\ &\dot{\tilde{\sigma}}_{kl} = \int_t^{t+\Delta t} C_{klmn}\dot{\varepsilon}_{mn}\,\mathrm{d}\tau\,, \end{aligned}\right\}\quad x\in\Omega\,,\ \ \tau\in[0,\bar{t}]\,,$$
$$\begin{aligned} &\dot{\sigma}_{kl}n_l = \dot{\hat{t}}_k\,, && x\in\partial\Omega_\sigma\,,\ \ \tau\in[0,\bar{t}]\,,\\ &\dot{u}_k = \dot{\hat{u}}_k\,, && x\in\partial\Omega_u\,,\ \ \tau\in[0,\bar{t}]\,. \end{aligned} \tag{3.7}$$

Equation $(3.7)_3$ represents a broad class of the so-called rate-independent materials including the linear elasticity and the flow theory of elasto-plasticity as the important special cases. By eliminating in Eq. (3.7) the rates of stress and strain and using the relation

$$\dot{\sigma}_{kl} = \dot{\tilde{\sigma}}_{kl} + \dot{u}_{k,m}\tilde{\sigma}_{kl} \tag{3.8}$$

we arrive at

$$\left[C_{ijkl}\dot{u}_{k,l} + \tilde{\sigma}_{mj}\dot{u}_{i,m}\right]_{,j} + \varrho\dot{\hat{f}}_i - \varrho\dddot{u}_i = 0\,. \tag{3.9}$$

Equations (3.9) form a system of partial differential equations with $\dot{u}_i$ as the unknown function. The system has to be solved in the domain Ω together with the appropriate boundary conditions of the form

$$\begin{aligned} \dot{u}_k &= \dot{\hat{u}}_k\,, && x\in\partial\Omega_u\,,\ \ \tau\in[0,\bar{t}]\,,\\ \left[C_{ijkl}\dot{u}_{k,l} + \tilde{\sigma}_{mj}\dot{u}_{i,m}\right]n_j &= \dot{\hat{t}}_i\,, && x\in\partial\Omega_\sigma\,,\ \ \tau\in[0,\bar{t}]\,, \end{aligned} \tag{3.10}$$

in which the right-hand sides are considered given, and the appropriate initial conditions imposed on the solution $\dot{u}_i(\tau)$ at time $\tau = t$.

In view of essential difficulties with the analytical procedures for solving the system (3.9), (3.10), which, moreover, has to be typically solved at each subsequent time instant with a different array of the constitutive coefficients and in a gradually changing domain, the use of numerical techniques becomes unavoidable. To derive the equations of the finite element method appropriate for the situation on hand we first recall the displacement expansion typical of the method, cf. (1.2), (1.28)

$$u_i(x,\tau) = \Phi_{i\alpha}(x)\,q_\alpha(\tau)\,, \qquad \alpha = 1,2,\ldots,N, \tag{3.11}$$

in which $\Phi_{i\alpha}(x)$ are known shape functions, q_α – generalized nodal displacements and N – the global number of the degrees of freedom reduced by explicitly accounting for the displacement-type boundary conditions on the boundary $\partial\Omega_u$. Similarly as in Chap. 1, Eq. (3.11) is considered in the context of the body discretized into finite elements with the functions $\Phi_{i\alpha}$ assumed localized, i.e. each is different from zero only in the elements adjacent to the node at which the α-th generalized coordinate is defined.

By substituting Eq. (3.11) in Eqs. (3.9) and (3.10) we obtain

$$\begin{aligned} &[C_{ijkl}\Phi_{k\alpha,l} + \Phi_{i\alpha,m}\tilde{\sigma}_{mj}]_{,j}\,\dot{q}_\alpha \\ &\qquad - \varrho\Phi_{i\alpha}\,\dddot{q}_\alpha + \varrho\dot{\hat{f}}_i = R_i\,, \qquad x \in \Omega, \qquad \tau \in [0,\bar{t}]\,, \\ &- [C_{ijkl}\Phi_{k\alpha,l} + \Phi_{i\alpha,m}\tilde{\sigma}_{mj}]\,n_j\dot{q}_\alpha + \dot{\hat{t}}_i = \bar{R}_i\,, \qquad x \in \partial\Omega_\sigma\,, \quad \tau \in [0,\bar{t}]\,, \end{aligned} \tag{3.12}$$

with the residual vectors R_i, $\bar{R}_i$ generally different from zero on account of only approximate nature of the assumed displacement solution (3.11). In order to explicitly derive the FEM equations we shall use the Galerkin method based on the minimization of the residual forces by means of the condition, cf. Eq. (3.5), Part I,

$$\int_\Omega R_i\Phi_{i\alpha}\,\mathrm{d}\Omega + \int_{\partial\Omega_\sigma} \bar{R}_i\Phi_{i\alpha}\,\mathrm{d}(\partial\Omega) = 0\,, \qquad \alpha = 1, 2, \ldots, N, \tag{3.13}$$

with the weighting functions assumed the same as the shape functions in the displacement expansion. The minimization of the residuals comes down to the evaluation of the N generalized coordinates $\dot{q}_\alpha$ using N conditions of the type (3.13). It should be emphasized that in the self-adjoint problems we deal now with, the minimization (3.13) is optimal in every respect.

By substituting Eq. (3.12) in Eq. (3.13) and using the Gauss–Ostrogradski theorem we arrive at

$$\begin{aligned} &\int_\Omega \left\{ [C_{ijkl}\Phi_{k\alpha,l} + \Phi_{i\alpha,m}\tilde{\sigma}_{mj}]_{,j}\,\dot{q}_\alpha - \varrho\Phi_{i\alpha}\,\dddot{q}_\alpha + \varrho\dot{\hat{f}}_i \right\} \Phi_{i\beta}\,\mathrm{d}\Omega \\ &\quad + \int_{\partial\Omega_\sigma} \left\{ - [C_{ijkl}\Phi_{k\alpha,l} + \Phi_{i\alpha,m}\tilde{\sigma}_{mj}]\,n_j\dot{q}_\alpha + \dot{\hat{t}}_i \right\} \Phi_{i\beta}\,\mathrm{d}(\partial\Omega) = 0\,, \\ &\int_\Omega [C_{ijkl}\Phi_{k\alpha,l}\Phi_{i\beta,j} + \tilde{\sigma}_{mj}\Phi_{i\alpha,m}\Phi_{i\beta,j}]_{,j}\,\dot{q}_\alpha\,\mathrm{d}\Omega \\ &\quad + \int_{\partial\Omega_\sigma} [C_{ijkl}\Phi_{k\alpha,l}\Phi_{i\beta} + \tilde{\sigma}_{mj}\Phi_{i\alpha,m}\Phi_{i\beta}]\,n_j\dot{q}_\alpha\,\mathrm{d}(\partial\Omega) \\ &\quad - \int_\Omega \left[\varrho\Phi_{i\alpha}\,\dddot{q}_\alpha - \varrho\dot{\hat{f}}_i \right] \Phi_{i\beta}\,\mathrm{d}\Omega \\ &\quad + \int_{\partial\Omega_\sigma} \left\{ - [C_{ijkl}\Phi_{k\alpha,l} + \Phi_{i\alpha,m}\tilde{\sigma}_{mj}]\,n_j\dot{q}_\alpha + \dot{\hat{t}}_i \right\} \Phi_{i\beta}\,\mathrm{d}(\partial\Omega) = 0\,. \end{aligned} \tag{3.14}$$

Using the notation

$$
\begin{aligned}
K^{(con)}_{\alpha\beta} &= \int_\Omega C_{ijkl}\Phi_{k\alpha,l}\Phi_{i\beta,j}\,\mathrm{d}\Omega\,,\\
K^{(\sigma)}_{\alpha\beta} &= \int_\Omega \tilde{\sigma}_{mj}\Phi_{i\alpha,m}\Phi_{i\beta,j}\,\mathrm{d}\Omega\,,\\
K_{\alpha\beta} &= K^{(con)}_{\alpha\beta} + K^{(\sigma)}_{\alpha\beta} \qquad (3.15)\\
M_{\alpha\beta} &= \int_\Omega \varrho\Phi_{i\alpha}\Phi_{i\beta}\,\mathrm{d}\Omega\,,\\
\dot{Q}_\alpha &= \int_\Omega \varrho\dot{\hat{f}}_i\Phi_{i\alpha}\,\mathrm{d}\Omega + \int_{\partial\Omega_\sigma}\dot{\hat{t}}_i\Phi_{i\alpha}\,\mathrm{d}(\partial\Omega)\,,
\end{aligned}
$$

taking the advantage of the symmetry of the matrices $K^{(con)}_{\alpha\beta}$, $K^{(\sigma)}_{\alpha\beta}$, $M_{\alpha\beta}$ and observing that some of the surface integrals cancel out, Eq. (3.14) is reduced to

$$
M_{\alpha\beta}\stackrel{\dots}{q}_\beta + K_{\alpha\beta}\dot{q}_\beta = \dot{Q}_\alpha\,. \qquad (3.16)
$$

The system (3.16) is the fundamental system of equations describing the spatially (or semi)-discretized problem of nonlinear dynamics. By complementing the above equation with the term describing the so-called structural damping we obtain

$$
M_{\alpha\beta}\stackrel{\dots}{q}_\beta + C_{\alpha\beta}\ddot{q}_\beta + K_{\alpha\beta}\dot{q}_\beta = \dot{Q}_\alpha\,, \qquad \tau \in [0,\bar{t}]. \qquad (3.17)
$$

The above system of the second order ordinary differential equations with respect to the velocity vector $\dot{q}_\alpha$ has to be solved together with the following initial conditions:

$$
\begin{aligned}
\dot{q}_\alpha(t) &= \dot{\hat{q}}_\alpha\,,\\
\ddot{q}_\alpha(t) &= -\overset{-1}{M}_{\alpha\beta}\left[C_{\beta\gamma}\dot{\hat{q}}_\gamma(t) + K_{\beta\gamma}\hat{q}_\gamma(t)\right]\,,
\end{aligned} \qquad (3.18)
$$

where $\dot{\hat{q}}_\alpha$, $\hat{q}_\alpha$ are known from the analysis at the time step proceeding the step currently discussed.

The order of the system (3.17), by one higher than normally expected in problems of dynamics, results from our wish to explicitly introduce the tangent stiffness matrix $K_{\alpha\beta}$ in the formulation of the nonlinear problem on hand. Before transforming Eq. (3.17) to a more standard form we note that in the case of constant matrices $M_{\alpha\beta}$, $K_{\alpha\beta}$ and $C_{\alpha\beta}$ Eqs. (3.17) may clearly be integrated to yield

$$
M_{\alpha\beta}\ddot{q}_\beta + C_{\alpha\beta}\dot{q}_\beta + K_{\alpha\beta}q_\beta = Q_\alpha\,. \qquad (3.19)
$$

The dynamic problems with geometric nonlinearities (large displacements, for instance) may be described by the second order ODE's provided the so-called secant, rather than tangent, matrix is employed. We then have

$$
M_{\alpha\beta}\ddot{q}_\beta + C_{\alpha\beta}\dot{q}_\beta + \hat{K}_{\alpha\beta}(q_\gamma)\,q_\beta = Q_\alpha\,, \qquad (3.20)
$$

where

$$K_{\alpha\beta} = \frac{\partial \hat{K}_{\alpha\gamma}}{\partial q_\beta} q_\gamma + \hat{K}_{\alpha\beta}\,, \tag{3.21}$$

which directly results from differentiating Eq. (3.20) with respect to time and comparing it with Eq. (3.17). The second order equation applicable to any type of nonlinearity may be presented in the form

$$M_{\alpha\beta}\ddot{q}_\beta + C_{\alpha\beta}\dot{q}_\beta + F_\alpha = Q_\alpha\,, \tag{3.22}$$

where F_α is the vector of the internal generalized nodal forces acting upon the subsequent nodes from the elements surrounding them. The vector is computed from the current stresses according to

$$F_\alpha = \int_\Omega \tilde{\sigma}_{ij}\,\tfrac{1}{2}(\Phi_{i\alpha,j} + \Phi_{j\alpha,i})\,\mathrm{d}\Omega\,. \tag{3.23}$$

It is obvious that by neglecting the inertial and damping effects the above equations reduce to those describing the corresponding problems of statics. From Eq. (3.17) we thus have, for instance

$$K_{\alpha\beta}\,\dot{q}_\beta = \dot{Q}_\alpha\,, \tag{3.24}$$

whereas the reduced equation (3.22)

$$F_\alpha = Q_\alpha \tag{3.25}$$

expresses directly the equilibrium condition of all the forces acting upon the successive system nodes.

Equations (3.17) may be essentially solved by any numerical technique known in the area of ordinary differential equations. However, many specific features of the computational mechanics formulations have influence the way in which the solution algorithms are normally discussed in the relevant literature. Only a few very general remarks on the subject will be given below – the interested reader should consult the available literature [14, 87, 183].

Replacing the first time derivatives of the functions $\mathbf{q}(\tau)$ and $\mathbf{Q}(\tau)$ by the finite difference expressions

$$\dot{\mathbf{q}}(\tau) = \frac{\Delta\mathbf{q}}{\Delta\tau}\,, \qquad \dot{\mathbf{Q}}(\tau) = \frac{\Delta\mathbf{Q}}{\Delta\tau}\,, \tag{3.26}$$

and using it in Eq. (3.17) results in

$$\mathbf{M}\,\Delta\ddot{\mathbf{q}} + \mathbf{C}\,\Delta\dot{\mathbf{q}} + \mathbf{K}\,\Delta\mathbf{q} = \Delta\mathbf{Q}\,. \tag{3.27}$$

By adopting, similarly as before, the notation

$$\Delta\mathbf{q} = \mathbf{q}^{t+\Delta t} - \mathbf{q}^t\,, \qquad \Delta\mathbf{Q} = \mathbf{Q}^{t+\Delta t} - \mathbf{Q}^t\,,$$

where $\mathbf{q}^t = \mathbf{q}(t)$ etc., we arrive at

$$\mathbf{M}\,\ddot{\mathbf{q}}^{t+\Delta t} + \mathbf{C}\,\dot{\mathbf{q}}^{t+\Delta t} + \mathbf{K}(\mathbf{q})\,\Delta\mathbf{q} = \mathbf{Q}^{t+\Delta t} - \mathbf{Q}^t + \mathbf{M}\,\ddot{\mathbf{q}}^t + \mathbf{C}\,\dot{\mathbf{q}}^t\,. \tag{3.28}$$

We intentionally leave unspecified the time instant at which to compute the matrix $\mathbf{K}$. Also, for compactness of the general discussion in this chapter, only $\mathbf{q}$ is indicated as the argument in the stiffness matrix – this seems sufficient for the discussion of the basic structure of the formulation but will require essential extensions later on when dealing with inelastic problems (when stresses and internal variables will appear as the arguments in the constitutive matrix and consequently in the stiffness matrix as well).

By noting that the dynamic equilibrium condition at time t may be written as

$$\mathbf{M}\,\ddot{\mathbf{q}}^t + \mathbf{C}\,\dot{\mathbf{q}}^t + \mathbf{F}^t = \mathbf{Q}^t\,, \tag{3.29}$$

Eq. (3.28) may be replaced by

$$\mathbf{M}\,\ddot{\mathbf{q}}^{t+\Delta t} + \mathbf{C}\,\dot{\mathbf{q}}^{t+\Delta t} + \mathbf{K}(\mathbf{q})\,\Delta\mathbf{q} = \mathbf{Q}^{t+\Delta t} - \mathbf{F}^t\,. \tag{3.30}$$

The vector $\mathbf{F}^t$ may clearly be regarded as known at the time step $[t, t+\Delta t]$. The form (3.30) is widely accepted as the starting point to discuss the non-linear dynamics numerical procedures in more detail. Many theoretical and practical aspects of such procedures will be discussed later in the book. A few additional comments below will serve the purpose of just introducing some useful terminology.

Let us consider Eq. (3.29) and assume in it the so-called central difference scheme for finite difference approximation of derivatives, i.e.

$$\begin{aligned} \ddot{\mathbf{q}}^t &= \frac{1}{(\Delta t)^2}[\mathbf{q}^{t-\Delta t} - 2\mathbf{q}^t + \mathbf{q}^{t+\Delta t}]\,, \\ \dot{\mathbf{q}}^t &= \frac{1}{2\Delta t}[-\mathbf{q}^{t-\Delta t} + \mathbf{q}^{t+\Delta t}]\,. \end{aligned} \tag{3.31}$$

This leads directly to

$$\begin{aligned} \left[\frac{1}{(\Delta t)^2}\mathbf{M} + \frac{1}{2\Delta t}\mathbf{C}\right]\mathbf{q}^{t+\Delta t} &= \mathbf{Q}^t - \mathbf{F}^t + \frac{2}{(\Delta t)^2}\mathbf{M}\,\mathbf{q}^t \\ &\quad - \left[\frac{1}{(\Delta t)^2}\mathbf{M} - \frac{1}{2\Delta t}\mathbf{C}\right]\mathbf{q}^{t-\Delta t}. \end{aligned} \tag{3.32}$$

The so obtained system of algebraic equations is linear in the unknown vector $\mathbf{q}^{t+\Delta t}$ which can easily be solved for provided the vectors $\mathbf{q}^t$, $\mathbf{q}^{t-\Delta t}$, $\mathbf{Q}^t$ and $\mathbf{F}^t$ are known. There is another aspect of the system (3.32) which should be pointed out, though. For the vast experience in solving different practical problems of mechanics suggests that it is very often warranted to adopt the diagonal approximation for the mass and damping matrices. In such a case the system (3.32) becomes extremely attractive since its form makes it possible to solve for each component of the vector $\mathbf{q}^{t+\Delta t}$ independently – and no matrix inversion is needed at all. This practically means that all the algebraic operations can be carried out at the element level enormously reducing the computer storage required by the solution algorithm.

The above approach, known as the central difference method, is an example of a class of the direct integration algorithms called the explicit methods. A typical drawback of the explicit methods is the lack of the so-called unconditional stability property. This imposes severe constraints on the length of the time step Δt which must be smaller than a certain critical value not known in advance and dependent on the properties of the whole system.

In order to very briefly introduce the second basic class of the integration algorithms, known as the implicit methods, let us consider the condition

$$\mathbf{M}\,\ddot{\mathbf{q}}^{t+\Delta t} + \mathbf{C}\,\dot{\mathbf{q}}^{t+\Delta t} + \mathbf{K}^{t+\Delta t}\Delta\mathbf{q} = \mathbf{Q}^{t+\Delta t} - \mathbf{F}^{t}\,, \tag{3.33}$$

where

$$\mathbf{K}^{t+\Delta t} = \mathbf{K}(\mathbf{q}^{t+\Delta t}) = \mathbf{K}[\mathbf{q}(t+\Delta t)]\,.$$

Let the acceleration and velocity be expressed in the following finite difference form:

$$\begin{aligned} \ddot{\mathbf{q}}^{t+\Delta t} &= \frac{1}{\alpha(\Delta t)^2}\left[\mathbf{q}^{t+\Delta t} - \mathbf{q}^{t} - \dot{\mathbf{q}}^{t}\Delta t - (\Delta t)^2\left(\tfrac{1}{2}-\alpha\right)\ddot{\mathbf{q}}^{t}\right], \\ \dot{\mathbf{q}}^{t+\Delta t} &= \frac{\delta}{\alpha\Delta t}(\mathbf{q}^{t+\Delta t} - \mathbf{q}^{t}) + \left(1-\frac{\delta}{\alpha}\right)\dot{\mathbf{q}}^{t} + \Delta t\left(1-\frac{\delta}{2\alpha}\right)\ddot{\mathbf{q}}^{t}\,. \end{aligned} \tag{3.34}$$

The integration method based on the above approximation is called the Newmark method. In it, the parameters α and δ can be adjusted to assure the best properties of the algorithm under given circumstances – they are usually taken as $\delta = 0.50$ and $\alpha = 0.25$.

By substituting Eqs. (3.34) in Eq. (3.33) we obtain

$$\begin{aligned} &\left[\mathbf{K}^{t+\Delta t} + \frac{1}{(\alpha\Delta t)^2}\mathbf{M} + \frac{\delta}{\alpha\Delta t}\mathbf{C}\right]\Delta\mathbf{q} \\ &\quad = \mathbf{Q}^{t+\Delta t} + \mathbf{M}\left[\frac{1}{\alpha\Delta t}\dot{\mathbf{q}}^{t} + \left(\frac{1}{2\alpha}-1\right)\ddot{\mathbf{q}}^{t}\right] \\ &\quad\quad + \mathbf{C}\left[\left(\frac{\delta}{\alpha}-1\right)\dot{\mathbf{q}}^{t} + \frac{\Delta t}{2}\left(\frac{\delta}{\alpha}-2\right)\ddot{\mathbf{q}}^{t}\right] - \mathbf{F}^{t}\,, \end{aligned} \tag{3.35}$$

or, shortly

$$\mathbf{K}^{t+\Delta t}_{(ef)}\,\Delta\mathbf{q} = \Delta\mathbf{Q}_{(ef)}\,. \tag{3.36}$$

where the definitions of the effective stiffness matrix $\mathbf{K}^{t+\Delta t}_{(ef)}$ and effective right-hand side vector $\Delta\mathbf{Q}_{(ef)}$ result directly by comparing Eq. (3.35) and Eq. (3.36). The last equation may be written as

$$\mathbf{K}_{(ef)}(\mathbf{q}^{t} + \Delta\mathbf{q})\,\Delta\mathbf{q} = \Delta\mathbf{Q}_{(ef)}\,. \tag{3.37}$$

We first observe that this time the matrix governing the system contains the stiffness matrix $\mathbf{K}$ – this makes it effectively impossible to decouple the system. Moreover, the matrix $\mathbf{K}^{t+\Delta t}$ is unknown at the time step on hand as it depends itself on the unknown vector $\mathbf{q}^{t+\Delta t}$. This feature of the system calls for iteration and may significantly influence the overall cost of the calculation. However, the increased computational effort at each step is typically compensated by the possibility of adopting much larger time step lengths – a property resulting from much better stability characteristics of the implicit methods.

We also note that by neglecting the inertial and damping effects the implicit integration methods become directly applicable for solving nonlinear static problems – it is not so with the explicit methods as seen from Eq. (3.32).

It should be clear from the discussion in Chap. 2, Part I, and those of the present chapter that the equations of FEM may be derived using any configuration as the reference state for the incremental computations. So far in this chapter we have been considering in this capacity only the current configuration C^t. To briefly illustrate a more general approach the semi-discretized problem will now be derived using any (but fixed) reference state C^r. For simplicity, we shall confine ourselves to the quasi-static problem and employ the variational procedure typical of the Rayleigh–Ritz methodology.

Let us consider the variational functional J_P written for the collection of finite elements as, cf. Eq. (2.41), Part I,

$$J_P[\Delta u_i] = \sum_{e=1}^{E} \left\{ \int_{\Omega_e} \left[\tfrac{1}{2} \mathbf{C}_{klmn} \Delta\varepsilon_{kl} \Delta\varepsilon_{mn} + \tfrac{1}{2} \tilde{\sigma}_{ij} \Delta u_{k,i} \Delta u_{k,j} - \right.\right.$$
$$\left.\left. - \varrho \Delta \hat{f}_k \Delta u_k \right] \mathrm{d}\Omega - \int_{\partial\Omega_e^\sigma} \Delta\hat{t}_k \Delta u_k \,\mathrm{d}(\partial\Omega) \right\} . \quad (3.38)$$

For simplicity, the constitutive moduli are assumed constant at the given step – the way to accomodate a more general situation will be given a detailed treatment in Chap. 6. Under the same smoothness conditions as before we adopt the expansion for the displacement vector in the form

$$\Delta u_k(\mathbf{x}, \tau) = \varphi_{k\xi}(\mathbf{x}) \, \Delta q_\xi^{(e)}(\tau), \qquad k = 1, 2, 3; \;\; \xi = 1, 2, \ldots, N_e\,, \quad (3.39)$$

or, more compactly

$$\Delta\mathbf{u}(\mathbf{x}, \tau) = \varphi_{3\times N_e}(\mathbf{x}) \, \Delta\mathbf{q}^{(e)}_{N_e \times 1} = \mathbf{\Phi}_{3\times N}(\mathbf{x}) \, \Delta\mathbf{q}_{N\times 1} \,.$$

By using the definitions of the incremental strains, Eqs. (2.7), (2.8), Part I, and Eq. (3.39) we arrive at

$$\begin{aligned} \Delta\bar{\varepsilon}_{kl}(\mathbf{x}, \tau) &= \left[\bar{B}^{(1)}_{kl\xi} + \bar{B}^{(2)}_{kl\xi} \right] \Delta q_\xi^{(e)} = \bar{B}_{kl\xi} \Delta q_\xi^{(e)} \,, \\ \Delta\bar{\bar{\varepsilon}}_{kl}(\mathbf{x}, \tau) &= \bar{\bar{B}}_{kl\xi\zeta} \Delta q_\xi^{(e)} \Delta q_\zeta^{(e)} \,. \end{aligned} \quad (3.40)$$

where

$$\begin{aligned} \bar{B}^{(1)}_{kl\xi} &= \tfrac{1}{2} (\varphi_{k\xi,l} + \varphi_{l\xi,k}) \,, \\ \bar{B}^{(2)}_{kl\xi} &= \tfrac{1}{2} (\varphi_{i\xi,k} \varphi_{i\zeta,l} + \varphi_{i\xi,l} \varphi_{i\zeta,k}) q_\zeta^{(e)} \,, \\ \bar{\bar{B}}_{kl\xi\zeta} &= \tfrac{1}{2} \varphi_{i\xi,k} \varphi_{i\zeta,l} \,. \end{aligned}$$

Substituting Eqs. (3.40) in the functional (3.38) and neglecting the higher order terms[1], the contribution of each finite element to the system potential energy results as

$$J_P^{(e)}[\Delta q_\xi^{(e)}] = \tfrac{1}{2} k_{\xi\zeta} \Delta q_\xi^{(e)} \Delta q_\zeta^{(e)} \tag{3.41}$$

where

$$k_{\xi\zeta} = \int_{\Omega_e} \left(C_{klmn} \bar{B}_{kl\xi} \bar{B}_{mn\zeta} + \tilde{\sigma}_{kl} \bar{\bar{B}}_{kl\xi\zeta} \right) \mathrm{d}\Omega \,. \tag{3.42}$$

The vector $\Delta\mathbf{Q}^{(e)}$ is the incremental generalized force vector representing forces acting upon the e-th element through its nodes. The elemental stiffness matrix $\mathbf{k}$ may be decomposed into three parts as follows

$$k_{\xi\zeta} = k_{\xi\zeta}^{(con)} + k_{\xi\zeta}^{(\sigma)} + k_{\xi\zeta}^{(u)} \,, \tag{3.43}$$

where

$$\begin{aligned} k_{\xi\zeta}^{(con)} &= \int_{\Omega_e} C_{klmn} \bar{B}_{kl\xi}^{(1)} \bar{B}_{mn\zeta}^{(1)} \, \mathrm{d}\Omega \,, \\ k_{\xi\zeta}^{(\sigma)} &= \int_{\Omega_e} \tilde{\sigma}_{kl} \bar{\bar{B}}_{kl\xi\zeta} \, \mathrm{d}\Omega \,, \\ k_{\xi\zeta}^{(u)} &= \int_{\Omega_e} C_{klmn} \left(\bar{B}_{kl\xi}^{(1)} \bar{B}_{mn\zeta}^{(2)} + \bar{B}_{kl\xi}^{(2)} \bar{B}_{mn\zeta}^{(1)} + \bar{B}_{kl\xi}^{(2)} \bar{B}_{mn\zeta}^{(2)} \right) \mathrm{d}\Omega \,. \end{aligned} \tag{3.44}$$

are the constitutive, initial stress and initial displacement stiffness matrices, respectively.

By using the above relationships one can easily obtain the functional J_P for the whole discretized system. The procedure is identical to that presented in Chap. 1 and is thus omitted here. The conditions of stationarity for J_P with respect to the unknown vector $\Delta\mathbf{q}_{N\times 1}$ lead to the incremental displacement-based equilibrium condition for the finite element system in the form

$$\left[\mathbf{K}^{(con)} + \mathbf{K}^{(\sigma)} + \mathbf{K}^{(u)} \right] \Delta\mathbf{q} = \Delta\mathbf{Q} \,. \tag{3.45}$$

in which the system (or global) stiffness matrices $\mathbf{K}^{(con)}$, $\mathbf{K}^{(\sigma)}$ and $\mathbf{K}^{(u)}$ correspond to the element matrices $\mathbf{k}^{(con)}$, $\mathbf{k}^{(\sigma)}$ and $\mathbf{k}^{(u)}$, and bear the same names. According to the formalism employed while deriving Eq. (3.45), they can be used with any reference configuration.

In large displacement problems for the Hooke materials it is often convenient to employ the stationary Lagrangian description using the total (as opposed to incremental) quantities. Let the constitutive equation describing this class of materials has the form

$$\tilde{\sigma}_{kl} = C_{klmn} \varepsilon_{mn} \,, \tag{3.46}$$

[1] A more general derivation which includes the higher order terms can be found in [87].

with the material property tensor C_{klmn} assumed constant in the whole range of straining. The total strain is defined as, cf. Eq. (2.3), Part I,

$$\varepsilon_{kl} = \tfrac{1}{2}(u_{k,l} + u_{l,k} + u_{i,k}u_{i,l}) \,. \tag{3.47}$$

The stationarity condition for the potential energy of the finite element system is written as

$$\delta J_P = \sum_{e=1}^{E} \left[\int_{\Omega_e} \left(C_{klmn}\varepsilon_{kl}\delta\varepsilon_{mn} - \varrho \hat{f}_k \delta u_k \right) \mathrm{d}\Omega + \int_{\partial\Omega_e^\sigma} \hat{t}_k \delta u_k \,\mathrm{d}(\partial\Omega) \right] = 0 \,. \tag{3.48}$$

The finite element approximation is adopted in the form, cf. Eqs. (1.28), (3.11),

$$u_k(\mathbf{x}) = \varphi_{k\xi}(\mathbf{x}) q_\xi^{(e)} = \varphi_{k\xi}(\mathbf{x}) a_{\xi\alpha} q_\alpha = \Phi_{k\alpha} q_\alpha \,,$$

which allows to rewrite the strain as

$$\varepsilon_{kl} = \tfrac{1}{2}(\Phi_{k\alpha,l} + \Phi_{l\alpha,k})\, q_\alpha + \tfrac{1}{2}\Phi_{i\alpha,k}\Phi_{i\beta,l}\, q_\alpha\, q_\beta \,,$$

or, more compactly

$$\varepsilon_{kl} = [B'_{kl\alpha} + B''_{kl\alpha}(q_\beta)]\; q_\alpha \,, \tag{3.49}$$

where

$$B'_{kl\alpha} = \tfrac{1}{2}(\Phi_{k\alpha,l} + \Phi_{l\alpha,k}) \,, \qquad B''_{kl\alpha} = \tfrac{1}{2}\Phi_{i\alpha,k}\Phi_{i\beta,l}\, q_\beta \,.$$

Note that since the function $B''_{kl\alpha}$ is linear and homogeneous with respect to q_β, then

$$B''_{kl\alpha}(q_\beta)\, \breve{q}_\alpha = B''_{kl\alpha}(\breve{q}_\beta)\, q_\alpha$$

for any vectors $\mathbf{q}_{N\times 1}$, $\breve{\mathbf{q}}_{N\times 1}$. Using this property, the strain variation as required in Eq. (3.48) becomes

$$\delta\varepsilon_{kl} = [B'_{kl\alpha} + B''_{kl\alpha}(q_\beta)]\; \delta q_\alpha + B''_{kl\alpha}(\delta q_\beta)\, q_\alpha = [B'_{kl\alpha} + 2B''_{kl\alpha}(q_\beta)]\; \delta q_\alpha \,,$$

which, when used in (3.48), yields

$$\int_\Omega C_{klmn}[B'_{kl\beta} + B''_{kl\beta}(q_\gamma)]\,[B'_{mn\alpha} + 2B''_{mn\alpha}(q_\gamma)]\, q_\beta\, \mathrm{d}\Omega = Q_\alpha \tag{3.50}$$

in which, cf. Eq. $(3.15)_5$

$$Q_\alpha = \int_\Omega \varrho \hat{f}_k \Phi_{k\alpha}\, \mathrm{d}\Omega + \int_{\partial\Omega_\sigma} \hat{t}_k \Phi_{k\alpha}\, \mathrm{d}(\partial\Omega) \,.$$

It results from Eq. (3.50) that

$$\left\{ \int_\Omega C_{klmn} B'_{kl\beta} B'_{mn\alpha}\, \mathrm{d}\Omega + \int_\Omega C_{klmn} B''_{kl\beta}(q_\gamma) B'_{mn\alpha}\, \mathrm{d}\Omega + \right.$$
$$\left. + 2 \int_\Omega C_{klmn} \left[B'_{kl\beta} + B''_{kl\beta}(q_\gamma) \right] B''_{mn\alpha}(q_\gamma)\, \mathrm{d}\Omega \right\} q_\beta = Q_\alpha \,.$$

The last integral on the right-hand side in this equation may be written as

$$2\int_\Omega \tilde{\sigma}_{kl} B''_{kl\alpha}(q_\gamma)\,\mathrm{d}\Omega = \int_\Omega \tilde{\sigma}_{kl}\Phi_{i\alpha,k}\Phi_{i\beta,l}q_\beta\,\mathrm{d}\Omega\,.$$

By introducing the notation

$$\begin{aligned}\overset{*}{K}{}^{(s)}_{\alpha\beta} &= \int_\Omega C_{klmn}B'_{kl\beta}B'_{mn\alpha}\,\mathrm{d}\Omega\,,\\ \overset{*}{K}{}^{(\sigma)}_{\alpha\beta} &= \int_\Omega \tilde{\sigma}_{kl}\Phi_{i\alpha,k}\Phi_{i\beta,l}q_\beta\,\mathrm{d}\Omega\,,\\ \overset{*}{K}{}^{(u)}_{\alpha\beta} &= \int_\Omega C_{klmn}B''_{kl\beta}B'_{mn\alpha}\,\mathrm{d}\Omega\,.\end{aligned} \tag{3.51}$$

we arrive at

$$\left[\overset{*}{\mathbf{K}}{}^{(s)} + \overset{*}{\mathbf{K}}{}^{(\sigma)} + \overset{*}{\mathbf{K}}{}^{(u)}\right]\mathbf{q} = \mathbf{Q}\,, \tag{3.52}$$

or, more compactly

$$\overset{*}{\mathbf{K}}(\mathbf{q})\,\mathbf{q} = \mathbf{Q}\,. \tag{3.53}$$

Equation (3.52) is the counterpart of the previously derived incremental equation (3.45). Using it, one may attempt to solve any problem in the area of large displacements of linear elastic structures. The symbol '*' used in the definitions of the stiffness matrices in Eq. (3.51) is meant to indicate the differences between the total and incremental description. In the non-incremental and incremental description the total stiffness matrices may be given an interpretation of the global secant and the current tangent, respectively – in the latter case the tangent should rather be thought of as the incremental secant unless the increments are 'truly infinitesimal'. The stiffness matrices $\overset{*}{\mathbf{K}}{}^{(s)}$, $\overset{*}{\mathbf{K}}{}^{(\sigma)}$ and $\overset{*}{\mathbf{K}}{}^{(u)}$ are called elastic, initial stress and initial displacement, respectively.

By performing on Eq. (3.52) the finite difference operation and neglecting the higher order terms we easily arrive at the incremental equation of the form

$$\left[\mathbf{K}^{(s)} + \mathbf{K}^{(\sigma)} + \mathbf{K}^{(u)}\right]\Delta\mathbf{q} = \Delta\mathbf{Q}\,,$$

The following relations are observed:

$$\begin{aligned}K^{(s)}_{\alpha\beta} &= \overset{*}{K}{}^{(s)}_{\alpha\beta}\,, \qquad K^{(\sigma)}_{\alpha\beta} = \overset{*}{K}{}^{(\sigma)}_{\alpha\beta}\,,\\ K^{(u)}_{\alpha\beta} &= 2\left(\overset{*}{K}{}^{(u)}_{\alpha\beta} + \overset{*}{K}{}^{(u)}_{\beta\alpha}\right) + 4\int_\Omega C_{klmn}B''_{kl\alpha}B''_{mn\beta}\,\mathrm{d}\Omega\,.\end{aligned} \tag{3.54}$$

3.2 Heat Conduction Problem

By using virtually the same reasoning as in the previous section the equations typical of the finite element method can also be derived for the heat transfer problem described by Eq. (2.45), Part I. For simplicity of notation the following thermally nonhomogeneous and orthotropic problem is considered, cf. Eq. (2.49), Part I:

$$\frac{\partial}{\partial x_1}\left(\lambda_1 \frac{\partial \vartheta}{\partial x_1}\right) + \frac{\partial}{\partial x_2}\left(\lambda_2 \frac{\partial \vartheta}{\partial x_2}\right) + \frac{\partial}{\partial x_3}\left(\lambda_3 \frac{\partial \vartheta}{\partial x_3}\right) + \dot{g} = \varrho c \frac{\partial \vartheta}{\partial \tau}, \quad (3.55)$$

in which $\vartheta = T - T_0$ while T_0 is a constant reference temperature in the whole domain Ω. The boundary conditions are taken in the form

$$\begin{aligned} \vartheta &= \hat{\vartheta}, \qquad && \mathbf{x} \in \partial\Omega_T, \\ \lambda_1 n_1 \frac{\partial \vartheta}{\partial x_1} + \lambda_2 n_2 \frac{\partial \vartheta}{\partial x_2} + \lambda_3 n_3 \frac{\partial \vartheta}{\partial x_3} &= \hat{q}, && \mathbf{x} \in \partial\Omega_q, \end{aligned} \quad (3.56)$$

In order to obtain the equations of FEM we adopt the expansion, cf. Eq. (3.11)

$$\vartheta(\mathbf{x}, \tau) = \boldsymbol{\Phi}_{1\times M}(\mathbf{x})\, \boldsymbol{\theta}_{M\times 1}(\tau), \quad (3.57)$$

where $\boldsymbol{\theta}$ is the vector of the nodal temperatures in the whole domain while $\boldsymbol{\Phi}$ – the vector of shape functions assumed to identically (i.e. regardless the values of $\boldsymbol{\theta}$) satisfy the boundary conditions $(3.56)_1$. Substituting Eq. (3.57) in Eqs. (3.55), (3.56) leads to

$$\begin{aligned} &\left[\frac{\partial}{\partial x_1}\left(\lambda_1 \frac{\partial \boldsymbol{\Phi}}{\partial x_1}\right) + \frac{\partial}{\partial x_2}\left(\lambda_2 \frac{\partial \boldsymbol{\Phi}}{\partial x_2}\right) + \frac{\partial}{\partial x_3}\left(\lambda_3 \frac{\partial \boldsymbol{\Phi}}{\partial x_3}\right)\right] \boldsymbol{\theta} \\ &\qquad\qquad + \dot{g} - \varrho c\, \boldsymbol{\Phi} \dot{\boldsymbol{\theta}} = R_1, \quad \mathbf{x} \in \Omega, \quad (3.58) \\ &-\left[\lambda_1 n_1 \frac{\partial \boldsymbol{\Phi}}{\partial x_1} + \lambda_2 n_2 \frac{\partial \boldsymbol{\Phi}}{\partial x_2} + \lambda_3 n_3 \frac{\partial \boldsymbol{\Phi}}{\partial x_3}\right] \boldsymbol{\theta} + \hat{q} = R_2, \quad \mathbf{x} \in \partial\Omega_q, \end{aligned}$$

in which the residuals R_1, R_2 are generally different from zero on account of only approximate character of the postulated solution form (3.57). Using the Galerkin method the residuals are minimized by solving the following system of M equations:

$$\int_\Omega \boldsymbol{\Phi}^T_{M\times 1} R_1 \, d\Omega + \int_{\partial\Omega_q} \boldsymbol{\Phi}^T_{M\times 1} R_2 \, d(\partial\Omega) = 0. \quad (3.59)$$

By noting that

$$\begin{aligned} &\int_\Omega \boldsymbol{\Phi}^T \frac{\partial}{\partial x_1}\left(\lambda_1 \frac{\partial \boldsymbol{\Phi}}{\partial x_1}\right) d\Omega \\ &\quad = \int_\Omega \frac{\partial}{\partial x_1}\left[\left(\varphi^T \frac{\partial \boldsymbol{\Phi}}{\partial x_1}\right) \lambda_1\right] d\Omega - \int_\Omega \lambda_1 \frac{\partial \boldsymbol{\Phi}^T}{\partial x_1} \frac{\partial \boldsymbol{\Phi}}{\partial x_1} \, d\Omega \\ &\quad = \int_{\partial\Omega_q} \varphi^T \frac{\partial \boldsymbol{\Phi}}{\partial x_1} \lambda_1 n_1 \, d(\partial\Omega) - \int_\Omega \lambda_1 \frac{\partial \boldsymbol{\Phi}^T}{\partial x_1} \frac{\partial \boldsymbol{\Phi}}{\partial x_1} \, d\Omega, \text{ etc.,} \end{aligned}$$

and

$$\int_{\partial\Omega_q} \boldsymbol{\Phi}^T \left[\lambda_1 n_1 \frac{\partial\boldsymbol{\Phi}}{\partial x_1} + \lambda_2 n_2 \frac{\partial\boldsymbol{\Phi}}{\partial x_2} + \lambda_3 n_3 \frac{\partial\boldsymbol{\Phi}}{\partial x_3}\right] \mathrm{d}(\partial\Omega) = \int_{\partial\Omega_q} \boldsymbol{\Phi}^T \hat{q}\,\mathrm{d}(\partial\Omega) - \int_{\partial\Omega_q} R_2\,\mathrm{d}(\partial\Omega),$$

we arrive at

$$-\left\{\int_\Omega \left[\lambda_1 \frac{\partial\boldsymbol{\Phi}^T}{\partial x_1}\frac{\partial\boldsymbol{\Phi}}{\partial x_1} + \lambda_2 \frac{\partial\boldsymbol{\Phi}^T}{\partial x_2}\frac{\partial\boldsymbol{\Phi}}{\partial x_2} + \lambda_3 \frac{\partial\boldsymbol{\Phi}^T}{\partial x_3}\frac{\partial\boldsymbol{\Phi}}{\partial x_3}\right]\mathrm{d}\Omega\right\}\boldsymbol{\theta} + \int_\Omega \boldsymbol{\Phi}^T \dot{g}\,\mathrm{d}\Omega - \left[\int_\Omega \varrho c\,\boldsymbol{\Phi}^T\boldsymbol{\Phi}\,\mathrm{d}\Omega\right]\dot{\boldsymbol{\theta}} + \int_{\partial\Omega_q} \boldsymbol{\Phi}^T \hat{q}\,\mathrm{d}(\partial\Omega) = 0\,. \tag{3.60}$$

The last equation may be written in a more compact form as

$$\mathbf{C}^{(\vartheta)}_{M\times M}\,\dot{\boldsymbol{\theta}}_{M\times 1} + \mathbf{K}^{(\vartheta)}_{M\times M}\,\boldsymbol{\theta}_{M\times 1} = \mathbf{P}_{M\times 1}\,, \tag{3.61}$$

where the capacitance matrix $\mathbf{C}^{(\vartheta)}_{M\times M}$, conductance matrix $\mathbf{K}^{(\vartheta)}_{M\times M}$ and the vector $\mathbf{P}_{M\times 1}$ are defined as

$$\begin{aligned}
\mathbf{C}^{(\vartheta)}_{M\times M} &= \int_\Omega \varrho c\,\boldsymbol{\Phi}^T_{M\times 1}\boldsymbol{\Phi}_{1\times M}\,\mathrm{d}\Omega\,,\\
\mathbf{K}^{(\vartheta)}_{M\times M} &= \int_\Omega \left[\lambda_1 \frac{\partial\boldsymbol{\Phi}^T_{M\times 1}}{\partial x_1}\frac{\partial\boldsymbol{\Phi}_{1\times M}}{\partial x_1} + \lambda_2 \frac{\partial\boldsymbol{\Phi}^T_{M\times 1}}{\partial x_2}\frac{\partial\boldsymbol{\Phi}_{1\times M}}{\partial x_2}\right.\\
&\qquad \left. + \lambda_3 \frac{\partial\boldsymbol{\Phi}^T_{M\times 1}}{\partial x_3}\frac{\partial\boldsymbol{\Phi}_{1\times M}}{\partial x_3}\right]\mathrm{d}\Omega\,,\\
\mathbf{P}_{M\times 1} &= \int_\Omega \boldsymbol{\Phi}^T_{M\times 1}\dot{g}\,\mathrm{d}\Omega + \int_{\partial\Omega_q} \boldsymbol{\Phi}^T_{M\times 1}\,\hat{q}\,\mathrm{d}(\partial\Omega)\,.
\end{aligned} \tag{3.62}$$

The system (3.61) is a system of M ordinary differential equations with (generally) variable coefficients, which are to be solved for M unknown functions $\theta_\alpha(\tau)$, $\alpha = 1, 2, \ldots, M$.

The system (3.61) may be solved by replacing the time derivative of the vector $\boldsymbol{\theta}$ by an appropriate finite difference scheme. We may use, for instance

$$\boldsymbol{\theta}_{t+\alpha\Delta t} = (1-\alpha)\boldsymbol{\theta}_t + \alpha\boldsymbol{\theta}_{t+\Delta t}\,, \quad \dot{\boldsymbol{\theta}}_{t+\alpha\Delta t} = \frac{1}{\Delta t}(\boldsymbol{\theta}_{t+\Delta t} - \boldsymbol{\theta}_t)\,, \qquad \alpha \in [0,1]\,,$$

which substituted in Eq. (3.61) specified at time $t + \alpha\Delta t$ gives

$$\left[\frac{1}{\Delta t}\mathbf{C}^{(\vartheta)}_{t+\alpha\Delta t} + \alpha\mathbf{K}^{(\vartheta)}_{t+\alpha\Delta t}\right]\boldsymbol{\theta}_{t+\Delta t} = \mathbf{P}_{t+\alpha\Delta t} + \left[\frac{1}{\Delta t}\mathbf{C}^{(\vartheta)}_{t+\alpha\Delta t} + (1-\alpha)\mathbf{K}^{(\vartheta)}_{t+\alpha\Delta t}\right]\boldsymbol{\theta}_t\,. \tag{3.63}$$

This equation may be solved for the unknown vector $\boldsymbol{\theta}_{t+\Delta t}$ by using any available iterative scheme in which $\boldsymbol{\theta}_t$ should be assumed known while the matrices $\mathbf{C}^{(\vartheta)}_{t+\alpha\Delta t}$ and $\mathbf{K}^{(\vartheta)}_{t+\alpha\Delta t}$ approximated in an iterative manner. The direct iteration leads to the following system of M linear algebraic equations, for instance

$$\left[\frac{1}{\Delta t}\mathbf{C}^{(\vartheta)(i-1)}_{t+\alpha\Delta t} + \alpha\mathbf{K}^{(\vartheta)(i-1)}_{t+\alpha\Delta t}\right]\boldsymbol{\theta}^{(i)}_{t+\Delta t}$$

$$= \mathbf{P}_{t+\alpha\Delta t} + \left[\frac{1}{\Delta t}\mathbf{C}^{(\vartheta)(i-1)}_{t+\alpha\Delta t} + (1-\alpha)\mathbf{K}^{(\vartheta)(i-1)}_{t+\alpha\Delta t}\right]\boldsymbol{\theta}_t\,, \qquad (3.64)$$

$$i = 1, 2, \ldots$$

4 Problems of Dynamics

This chapter deals only with deterministic vibrations. Stability of motion will not be discussed generally and only simple problems of dynamic stability will be described. The limited size of the chapter does not allow us to discuss problems connected with accuracy of calculations and numerical stability of results.

When discussing equations of motion and methods of forming their characteristic matrices the rigid finite element method will be described because of its simplicity in applications to vibration analysis and its entirely Polish origin [94, 95].

In order to fully understand this chapter the reader is referred to books on vibrations and waves, see [73], for instance.

4.1 Classification of Computer Methods Used for Analysis of Dynamic Problems

After discretization of real systems, most problems of applied dynamics can be described by N differential equations which can be written in the following matrix form (compare (1.35))

$$\mathbf{M}\,\ddot{\mathbf{q}} + \mathbf{C}\,\dot{\mathbf{q}} + \mathbf{K}\,\mathbf{q} = \mathbf{Q} \tag{4.1}$$

where $\mathbf{M}$ is the $N \times N$ mass matrix, $\mathbf{C}$ is the $N \times N$ damping matrix, $\mathbf{K}$ is the $N \times N$ stiffness matrix, $\mathbf{q}$ is the vector of generalized displacements of N components, $\mathbf{Q}$ is the given vector of generalized forces (non-potential and non-dissipative) which is called the force input, N is the number of degrees of freedom of the system analysed.

If, apart from the force input, there is a kinematic input to the system (for example, displacements of some elements of the system are given) then differential equations of motion can also be presented in the form (4.1). To this end we release the system from constraints by replacing them with forces called constraint reactions.

If w constraints are eliminated then the system will have

$$\tilde{N} = N + w \tag{4.2}$$

degrees of freedom and will take the following form

$$\tilde{\mathbf{M}}\,\ddot{\tilde{\mathbf{q}}} + \tilde{\mathbf{C}}\,\dot{\tilde{\mathbf{q}}} + \tilde{\mathbf{K}}\,\tilde{\mathbf{q}} = \tilde{\mathbf{Q}} \tag{4.3}$$

where $\tilde{\mathbf{M}}$, $\tilde{\mathbf{C}}$, $\tilde{\mathbf{K}}$ are characteristic matrices of the released system,

$$\tilde{\mathbf{q}} = \{\mathbf{q}, \mathbf{z}\} \tag{4.4}$$

is the vector of generalized displacements of the system released from constraints consisting of $\tilde{N}$ components, $\mathbf{z}$ is the vector of given generalized displacements with w components which describes a kinematic input,

$$\tilde{\mathbf{Q}} = \{\mathbf{Q}, \mathbf{R}\} \tag{4.5}$$

is the vector of non-potential and non-dissipative generalized forces acting on the released system and is called below the vector of generalized forces (with $\tilde{N}$ components), $\mathbf{R}$ is the vector of constraint reactions with w components.

If constraints are static the respective elements of the vector $\mathbf{z}$ are constant and in particular cases equal to zero.

Having inserted (4.4) and (4.5) into (4.3) and divided matrices $\tilde{\mathbf{M}}$, $\tilde{\mathbf{C}}$, $\tilde{\mathbf{K}}$ into blocks according to the size of blocks of vectors $\tilde{\mathbf{q}}$ and $\tilde{\mathbf{Q}}$ the following is obtained

$$\begin{bmatrix} \mathbf{M} & \mathbf{M}'' \\ \mathbf{M}''^{T} & \mathbf{M}' \end{bmatrix} \begin{bmatrix} \ddot{\mathbf{q}} \\ \ddot{\mathbf{z}} \end{bmatrix} + \begin{bmatrix} \mathbf{C} & \mathbf{C}'' \\ \mathbf{C}''^{T} & \mathbf{C}' \end{bmatrix} \begin{bmatrix} \dot{\mathbf{q}} \\ \dot{\mathbf{z}} \end{bmatrix} + \begin{bmatrix} \mathbf{K} & \mathbf{K}'' \\ \mathbf{K}''^{T} & \mathbf{K}' \end{bmatrix} \begin{bmatrix} \mathbf{q} \\ \mathbf{z} \end{bmatrix} = \begin{bmatrix} \mathbf{Q} \\ \mathbf{R} \end{bmatrix}, \tag{4.6}$$

where $\mathbf{M}'$, $\mathbf{C}'$, $\mathbf{K}'$ are blocks of $w \times w$ elements of characteristic matrices of the released system, $\mathbf{M}''$, $\mathbf{C}''$, $\mathbf{K}''$ are rectangular blocks of $N \times w$ characteristic matrices of connections between that part of the model without constraints and that with.

Having multiplied blocks in (4.6) the following two matrix equations are obtained after simple transformations

$$\mathbf{M}\,\ddot{\mathbf{q}} + \mathbf{C}\,\dot{\mathbf{q}} + \mathbf{K}\,\mathbf{q} = \mathbf{Q} - \mathbf{M}''\,\ddot{\mathbf{z}} - \mathbf{C}''\,\dot{\mathbf{z}} - \mathbf{K}''\,\mathbf{z}\,, \tag{4.7}$$

$$\mathbf{R} = \mathbf{M}''^{T}\,\ddot{\mathbf{q}} + \mathbf{M}'\,\ddot{\mathbf{z}} + \mathbf{C}''^{T}\,\dot{\mathbf{q}} + \mathbf{C}'\,\dot{\mathbf{z}} + \mathbf{K}''^{T}\mathbf{q} + \mathbf{K}'\,\mathbf{z}\,. \tag{4.8}$$

For a given kinematic input the right side of the equation (4.7) is a known function of time. Using the following notation

$$\mathbf{b} = \mathbf{Q} - \mathbf{M}''\,\ddot{\mathbf{z}} - \mathbf{C}''\,\dot{\mathbf{z}} - \mathbf{K}''\,\mathbf{z} \tag{4.9}$$

Eq. (4.7) takes the form

$$\mathbf{M}\,\ddot{\mathbf{q}} + \mathbf{C}\,\dot{\mathbf{q}} + \mathbf{K}\,\mathbf{q} = \mathbf{b} \tag{4.10}$$

which is mathematically identical with the form of Eq. (4.1). Vector $\mathbf{b}$ which includes both kinematic and force inputs is called a substitute input. Thus in further theoretical considerations there is no need to separate vibrations caused kinetically or by force.

After solution of Eq. (4.10), that is, after calculation of the vector of generalized coordinates, we can calculate the vector of constraint reactions from Eq. (4.8).

For linear systems characteristic matrices $\mathbf{M}$, $\mathbf{C}$, $\mathbf{K}$ are constant or time dependent but they do not depend on vector $\mathbf{q}$ or its derivatives. Vibrations described by such equations are called linear. When matrices $\mathbf{M} = \mathbf{M}(t)$, $\mathbf{C} = \mathbf{C}(t)$, $\mathbf{K} = \mathbf{K}(t)$ are explicit functions of time vibrations are called parametric. Moreover, if characteristic matrices are functions of displacements and/or velocities, $\mathbf{M} = \mathbf{M}(\dot{\mathbf{q}}, \mathbf{q}, t)$, $\mathbf{C} = \mathbf{C}(\dot{\mathbf{q}}, \mathbf{q}, t)$, $\mathbf{K} = \mathbf{K}(\dot{\mathbf{q}}, \mathbf{q}, t)$, the systems are called non-linear and vibrations described by them are also called non-linear.

Most cases of non-linear vibrations can be described using the following equations

$$\mathbf{M}(\dot{\mathbf{q}}, \mathbf{q}, t)\,\ddot{\mathbf{q}} + \mathbf{C}(\dot{\mathbf{q}}, \mathbf{q}, t)\,\dot{\mathbf{q}} + \mathbf{K}(\dot{\mathbf{q}}, \mathbf{q}, t)\,\mathbf{q} = \mathbf{b}(\dot{\mathbf{q}}, \mathbf{q}, t)\,. \tag{4.11}$$

In computer-oriented formulation of dynamics we propose a slightly different classification of problems in comparison to that used in traditional course books. In using computer methods it is more important which method is used for the solution of equations of motion than the type of equations being solved. For example the "step by step" integration method can be well used in analysis of both linear and non-linear vibrations, in analysis of forced as well as free vibrations.

Table 4.1. shows methods used most often in computer analysis of vibrations.

4.2 Formulation of Equations of Motion Using Rigid Finite Element Method

In previous chapters we have presented methods of formulation of equations of motion and their characteristic matrices $\mathbf{M}$, $\mathbf{C}$, $\mathbf{K}$ using the finite element method. In this chapter we will describe how to formulate those matrices for systems being discretized by the rigid finite element method (RFEM).

The RFEM has been created for vibration analysis of ship structures (torsional, longitudinal and bending vibrations of main engine shafting, vibrations of devices on spring washers, bending vibrations of hulls [94, 95]). On the basis of the method a computer system HESAS–PC for analysis of vibrations of mechanical systems has been worked out. The RFEM has also been applied to vibration analysis of machine tools, pipelines when liquid flow has been taken into account and systems with changing configuration as well as in analysis of dynamic stability.

The RFEM cannot compete with the method of flexible finite elements in exactness of calculations for two and three-dimensional problems. However, because it is simple and easy to learn without deep theoretical knowledge, it has been applied in analysis of beam-like systems and, especially, systems with bodies which can be treated as non-deformable. The RFEM can also be easily connected with the finite element method and rigid elements can be treated as one of many types of finite elements.

A physical model used in the rigid finite element method consists of u non-deformable bodies (RFEM), each of which has 6 degrees of freedom.

Table 4.1. Methods used most often in computer analysis of vibrations

No.	Name of the method of solution	Initial equation of motion	Form of equation for direct solution	Examples of applications
1	Standard eigenvalue problem	$\mathbf{M}\ddot{\mathbf{q}} + \mathbf{K}\mathbf{q} = \mathbf{0}$ $\mathbf{M}$ – diagonal, $\mathbf{K}$ – symmetrical	$(\mathbf{U} - \mathbf{E}\omega^2)\mathbf{x}^0 = \mathbf{0}$ $\mathbf{U} = \mathbf{M}^{-\frac{1}{2}}\mathbf{K}\mathbf{M}^{-\frac{1}{2}}$ $\mathbf{q}^0 = \mathbf{M}^{-\frac{1}{2}}\mathbf{x}^0$	frequencies and forms of free vibrations of undamped linear systems with diagonal inertial matrices
2	General eigenvalue problem	$\mathbf{M}\ddot{\mathbf{q}} + \mathbf{K}\mathbf{q} = \mathbf{0}$ $\mathbf{M}$ – symmetrical, $\mathbf{K}$ – symmetrical	$(\mathbf{K} - \mathbf{M}\omega^2)\mathbf{q}^0 = \mathbf{0}$	frequencies and forms of free vibrations of undamped linear systems with symmetrical (non-diagonal) inertial matrices
3	Complex eigenvalue problem	$\mathbf{M}\ddot{\mathbf{q}} + \mathbf{C}\dot{\mathbf{q}} + \mathbf{K}\mathbf{q} = \mathbf{0}$ $\mathbf{M}$, $\mathbf{C}$, $\mathbf{K}$ – symmetrical	$(\mathbf{M}\lambda^2 + \mathbf{C}\lambda + \mathbf{K})\,\mathbf{a}^0 = \mathbf{0}$, $\mathbf{q} = \mathbf{a}^0 \mathrm{e}^{\lambda t}$ $\mathbf{a}^0$, λ – complex variables	frequencies and forms of free vibrations of linear systems with viscous damping
4	Non-symmetrical complex eigenvalue problem	$\mathbf{M}\ddot{\mathbf{q}} + \mathbf{C}\dot{\mathbf{q}} + \mathbf{K}\mathbf{q} = \mathbf{0}$ $\mathbf{C}$, $\mathbf{K}$ – non-symmetrical	$(\mathbf{M}\lambda^2 + \mathbf{C}\lambda + \mathbf{K})\,\mathbf{a}^0 = \mathbf{0}$, $\mathbf{q} = \mathbf{a}^0 \mathrm{e}^{\lambda t}$ $\mathbf{a}^0$, λ – complex variables	frequencies and forms of free vibrations of linear systems with viscous damping and gyroscope effect considered
5	Fourier transformations and solutions of sets of algebraic equations in domain of complex numbers	$\mathbf{M}\ddot{\mathbf{q}} + \mathbf{C}\dot{\mathbf{q}} + \mathbf{K}\mathbf{q} = \mathrm{col}\left(\mathbf{b}_i^0 \sin(\omega t + \varphi_i)\right)$	$[(\mathbf{K} - \mathbf{M}\omega^2) + \mathbf{C}\mathrm{j}\omega]\, q(\mathrm{j}\omega)$ $= \mathbf{b}(\mathrm{j}\omega)$ $\mathrm{j} = \sqrt{-1}$	stable harmonically forced vibrations of linear systems with viscous damping (frequency courses of amplitudes and phases)

6	Fourier transformations with harmonic analysis of input and solutions of sets of algebraic equations in domain of complex numbers; superposition method	$\mathbf{M}\,\ddot{\mathbf{q}} + \mathbf{C}\,\dot{\mathbf{q}} + \mathbf{K}\,\mathbf{q} = \mathbf{b}(t)$, where a) $\mathbf{b}(t) = \mathbf{b}(t+T)$ – periodical input expanded into Fourier series b) $\mathbf{b}(t) = \sum_{\nu=0}^{k} \mathrm{col}\left(\mathbf{b}^0_{\nu i} \sin(\omega_\nu t + \varphi_{\nu i})\right)$ – pseudo-periodical input c) $\mathbf{b}(t)$ – non-periodical input approximately assumed as periodical with very long period	$[(\mathbf{K} - \mathbf{M}\omega_\nu^2) + \mathbf{C}\mathrm{j}\omega_\nu]\,\mathbf{q}_\nu(\mathrm{j}\omega_\nu) = \mathbf{b}_\nu(\mathrm{j}\omega_\nu)$, $\nu = 0, 1, 2, \ldots, k$	a) stable periodically and pseudo-periodically forced vibrations of linear systems with viscous damping b) vibrations forced non-periodically (approximate method). Frequency courses of individual harmonic components and time courses
7	Integration "step by step" of equations of second order	for $\mathbf{q}(0) = \mathbf{s}_0$ and $\dot{\mathbf{q}}(0) = \mathbf{u}_0$ a) $\mathbf{M}\,\ddot{\mathbf{q}} + \mathbf{C}\,\dot{\mathbf{q}} + \mathbf{K}\,\mathbf{q} = \mathbf{b}(t)$ b) $\mathbf{M}(\dot{\mathbf{q}},\mathbf{q},t)\,\ddot{\mathbf{q}} + \mathbf{C}(\dot{\mathbf{q}},\mathbf{q},t)\,\dot{\mathbf{q}} + \mathbf{K}(\dot{\mathbf{q}},\mathbf{q},t)\,\mathbf{q} = \mathbf{0}$ c) $\mathbf{M}(\dot{\mathbf{q}},\mathbf{q},t)\,\ddot{\mathbf{q}} + \mathbf{C}(\dot{\mathbf{q}},\mathbf{q},t)\,\dot{\mathbf{q}} + \mathbf{K}(\dot{\mathbf{q}},\mathbf{q},t)\,\mathbf{q} = \mathbf{b}(\dot{\mathbf{q}},\mathbf{q},t)$	as initial equations	unstable states: a) vibrations of linear system forced with optional input b) free vibrations of non-linear systems c) vibrations of non-linear systems forced with optional input
8	Integration "step by step" of state equations (of the first order)	$\dot{\mathbf{y}} = \mathbf{S}\,\mathbf{y} + \mathbf{U}\,\mathbf{b}(t)$ $\mathbf{S} = \begin{bmatrix} -\mathbf{M}_E^{-1}\mathbf{C} \\ -\mathbf{M}_0^{-1}\mathbf{K} \end{bmatrix}$, $\mathbf{U} = \begin{bmatrix} \mathbf{M}^{-1} \\ \mathbf{0} \end{bmatrix}$, $\mathbf{y} = \begin{bmatrix} \dot{\mathbf{q}} \\ \mathbf{q} \end{bmatrix}$	as initial equation	as in point 7
9	Method of harmonic balance	a) $\mathbf{M}(t)\,\ddot{\mathbf{q}} + \mathbf{C}(t)\,\dot{\mathbf{q}} + \mathbf{K}(t)\,\mathbf{q} = \mathbf{b}^0 \sin\omega t$, b) $\mathbf{M}(\dot{\mathbf{q}},\mathbf{q},t)\,\ddot{\mathbf{q}} + \mathbf{C}(\dot{\mathbf{q}},\mathbf{q},t)\,\dot{\mathbf{q}} + \mathbf{K}(\dot{\mathbf{q}},\mathbf{q},t)\,\mathbf{q} = \mathbf{b}(\dot{\mathbf{q}},\mathbf{q},t)$		dynamic stability

Those bodies are connected to each other and with the base by v weightless elements with spring-damping features. Those elements are called spring-damping elements (SDE) and are denoted in figures by the symbol -⊗-. We assume that features of SDEs comply with the Kelvin–Voight rheological model, that is that elastic forces transmitted by SDEs are proportional to deformations and damping forces are proportional to deformation velocities.

In analysis of small vibrations, usually deflections of mass centres of bodies along axes x_{r1}, x_{r2}, x_{r3}, $(r = 1, 2, \ldots, u)$ which coincide with central, main inertial axes of bodies in a state of equilibrium as well as rotations about those axes, are usually taken as generalized coordinates. Those displacements create the vector

$$\mathbf{q} = \{\mathbf{q}_1, \mathbf{q}_2, \ldots, \mathbf{q}_u\} \tag{4.12}$$

in which

$$\mathbf{q}_r = \{q_{r1}, q_{r2}, \ldots, q_{r6}\} \tag{4.13}$$

is the vector of displacements of body r. The first three displacements q_{r1}, q_{r2}, q_{r3} are translations of the mass centre along x_{r1}, x_{r2}, x_{r3} and the remaining (q_{r4}, q_{r5}, q_{r6}) are rotations about those axes (Fig. 4.1). The coordinate systems thus assumed are local coordinate systems.

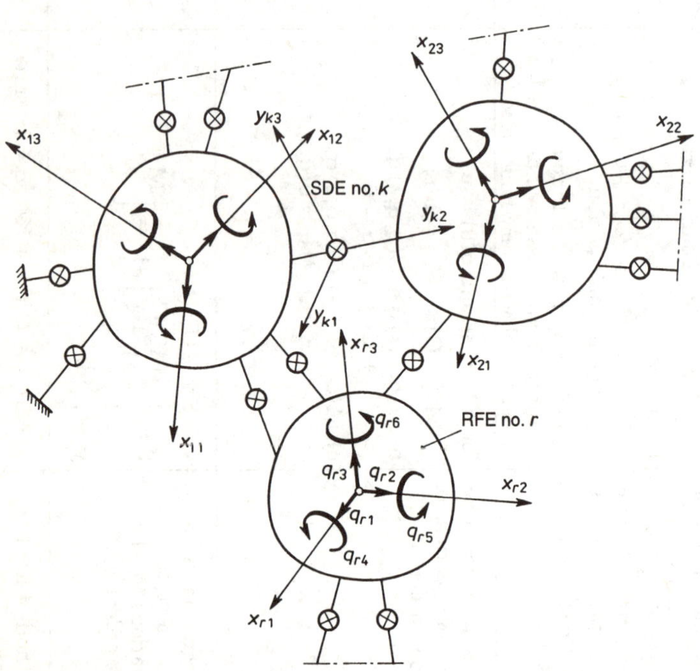

Fig. 4.1. System of rigid finite elements (RFE) connected by spring-damping elements (SDE)

Generalized forces acting on the system are presented similarly and they form the vector

$$\mathbf{Q} = \{\mathbf{Q}_1, \mathbf{Q}_2, \ldots, \mathbf{Q}_u\} \tag{4.14}$$

in which

$$\mathbf{Q}_r = \{Q_{r1}, Q_{r2}, \ldots, Q_{r6}\} \tag{4.15}$$

is the vector of generalized forces acting on body r. The first three components (Q_{r1}, Q_{r2}, Q_{r3}) of that vector are forces acting along axes x_{r1}, x_{r2}, x_{r3}, the remaining (Q_{r4}, Q_{r5}, Q_{r6}) are pairs of forces with respect to those axes. If forces and pairs of forces do not act consistently with the axes they must be reduced to those directions using, for example, the principle of virtual work.

Matrices $\mathbf{M}$, $\mathbf{C}$, $\mathbf{K}$ are derived using the Lagrange equations of the second order. Detailed derivations can be found in [95] and here we present only final formulae.

The inertial matrix $\mathbf{M}$ has the following form

$$\mathbf{M} = \mathrm{diag}\,(\mathbf{M}_r), \qquad r = 1, 2, \ldots, u, \tag{4.16}$$

where

$$\mathbf{M}_r = \mathrm{diag}\,(m_r, m_r, m_r, I_{r1}, I_{r2}, I_{r3}) \tag{4.17}$$

is the inertial matrix of the body r, m_r is a mass of body r, and I_{r1}, I_{r2}, I_{r3} are the main central moments of inertia of the body r with respect to axes x_{r1}, x_{r2}, x_{r3}, respectively.

If local coordinate systems are not the main central inertial axes the mass matrix is not diagonal but has the following form

$$\mathbf{M}_r = \begin{bmatrix} m_r & 0 & 0 & 0 & S_{r3} & -S_{r2} \\ & m_r & 0 & -S_{r3} & 0 & S_{r1} \\ & 0 & m_r & S_{r2} & -S_{r1} & 0 \\ & & & I_{r11} & -I_{r12} & -I_{r13} \\ & \text{sym.} & & & I_{r22} & -I_{r23} \\ & & & & & I_{r33} \end{bmatrix} \tag{4.18}$$

where

$m_r = \int_V \rho \, \mathrm{d}V$ – mass of body r;

$S_{r\alpha} = \int_V \rho x_\alpha \, \mathrm{d}V, \quad \alpha = 1, 2, 3,$ – mass static moments;

$I_{r\alpha\beta} = \int_V \rho x_\alpha x_\beta \, \mathrm{d}V, \quad \alpha, \beta = 1, 2, 3,$ – mass moments of inertia.

The stiffness matrix takes the form

$$\mathbf{K} = \begin{bmatrix} \mathbf{K}_{11} & \mathbf{K}_{12} & \ldots & \mathbf{K}_{1u} \\ & \mathbf{K}_{22} & \ldots & \mathbf{K}_{2u} \\ & & \ddots & \vdots \\ & \text{sym.} & & \mathbf{K}_{uu} \end{bmatrix} \tag{4.19}$$

where

$$\mathbf{K}_{rr} = \sum_{\alpha=1}^{i_r} \mathbf{S}_{r\alpha}^T \boldsymbol{\theta}_{r\alpha}^T \mathbf{K}_\alpha \boldsymbol{\theta}_{r\alpha} \mathbf{S}_{r\alpha}\,, \tag{4.20}$$

$$\mathbf{K}_{rp} = -\sum_{\alpha=1}^{i_{rp}} \mathbf{S}_{r\alpha}^T \boldsymbol{\theta}_{r\alpha}^T \mathbf{K}_\alpha \boldsymbol{\theta}_{p\alpha} \mathbf{S}_{p\alpha}\,, \qquad r < p, \tag{4.21}$$

$$\mathbf{S}_{r\alpha} = \begin{bmatrix} 1 & 0 & 0 & 0 & s_{r\alpha 3} & -s_{r\alpha 2} \\ 0 & 1 & 0 & -s_{r\alpha 3} & 0 & s_{r\alpha 1} \\ 0 & 0 & 1 & s_{r\alpha 2} & -s_{r\alpha 1} & 0 \\ 0 & 0 & 0 & 1 & 0 & 0 \\ 0 & 0 & 0 & 0 & 1 & 0 \\ 0 & 0 & 0 & 0 & 0 & 1 \end{bmatrix}, \tag{4.22}$$

$s_{r\alpha 1}$, $s_{r\alpha 2}$, $s_{r\alpha 3}$ are coordinates of a fixing point of SDE no. α to body no. r with respect to the system of axes x_{r1}, x_{r2}, x_{r3},

$$\boldsymbol{\theta}_{r\alpha} = \begin{bmatrix} \cos\theta_{r\alpha 11} & \cos\theta_{r\alpha 12} & \cos\theta_{r\alpha 13} & 0 & 0 & 0 \\ \cos\theta_{r\alpha 21} & \cos\theta_{r\alpha 22} & \cos\theta_{r\alpha 23} & 0 & 0 & 0 \\ \cos\theta_{r\alpha 31} & \cos\theta_{r\alpha 32} & \cos\theta_{r\alpha 33} & 0 & 0 & 0 \\ 0 & 0 & 0 & \cos\theta_{r\alpha 11} & \cos\theta_{r\alpha 12} & \cos\theta_{r\alpha 13} \\ 0 & 0 & 0 & \cos\theta_{r\alpha 21} & \cos\theta_{r\alpha 22} & \cos\theta_{r\alpha 23} \\ 0 & 0 & 0 & \cos\theta_{r\alpha 31} & \cos\theta_{r\alpha 32} & \cos\theta_{r\alpha 33} \end{bmatrix}, \tag{4.23}$$

$\theta_{r\alpha ij}$ is the angle between the i main axis of deformation[1] of the α SDE and the j main central inertial axis of body r $(i, j = 1, 2, 3)$,

$$\mathbf{K}_\alpha = \mathrm{diag}\,(k_{\alpha 1}, k_{\alpha 2}, \ldots, k_{\alpha 6}), \tag{4.24}$$

$k_{r\alpha 1}$, $k_{r\alpha 2}$, $k_{r\alpha 3}$ are coefficients of translational stiffness ($\mathrm{N\,m^{-1}}$), $k_{r\alpha 4}$, $k_{r\alpha 5}$, $k_{r\alpha 6}$ are coefficients of rotational stiffness ($\mathrm{N\,m\,rad^{-1}}$), i_r is the number of the SDE connected to body r, and i_{rp} is the number of the SDE which connect body r to body p.

There is a mnemonic rule which makes it easier to form matrix $\mathbf{K}$. It can be seen that blocks $\mathbf{K}_{rr}$ on the diagonal of matrix $\mathbf{K}$ include only parameters which describe features of body r and of elements connected to it (Fig. 4.2). Thus, block $\mathbf{K}_{rr}$ "responds" with body r. Blocks $\mathbf{K}_{rp}$ $(r \neq p)$ include only parameters of the SDE connecting body r to body p and parameters of those bodies. Thus, block $\mathbf{K}_{rp}$ "responds" to the connection between elements r

[1] The main axes of deformation of the SDE are characterized by the following: if a force acts on an element along one of those axes the element deforms only in the direction of the acting force and if a pair of forces acts on an element about one of those axes then the element deforms rotationally only in the direction of the acting pair of forces.

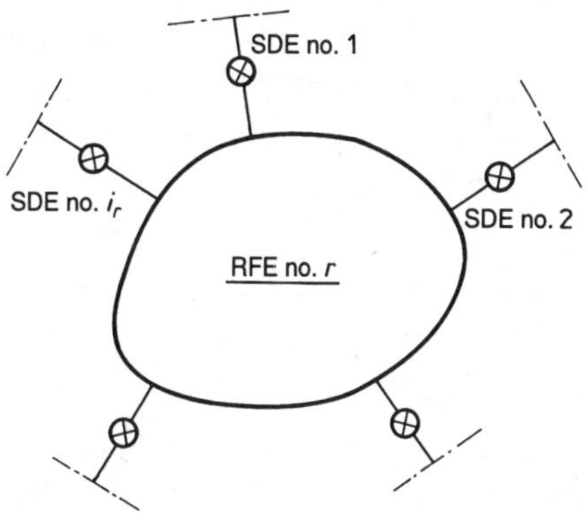

Fig. 4.2. Rigid finite element r with spring-damping elements

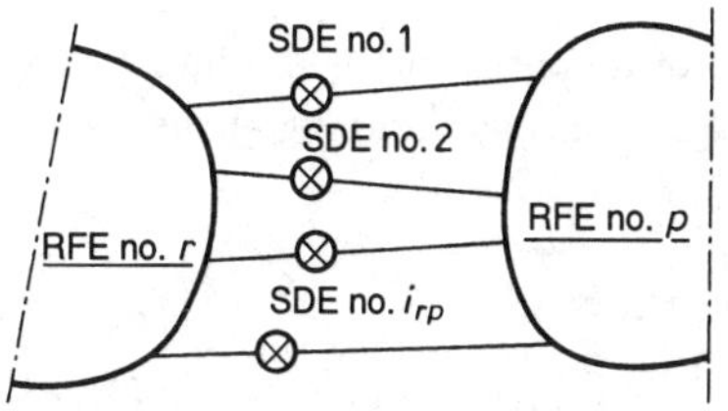

Fig. 4.3. Spring-damping elements connecting rigid finite elements r and p

and p (Fig. 4.3). Of course, if body r is not connected with body p then $\mathbf{K}_{rp} = \mathbf{0}$.

For systems in which the main central axes of inertia of all bodies and the main deformation axes of SDEs are parallel, blocks $\boldsymbol{\theta}_{r\alpha}$ become unit matrices and formulae (4.20) and (4.21) simplify to the following form

$$\mathbf{K}_{rr} = \sum_{\alpha=1}^{i_r} \mathbf{S}_{r\alpha}^T \mathbf{K}_\alpha \mathbf{S}_{r\alpha} , \tag{4.25}$$

$$\mathbf{K}_{rp} = -\sum_{\alpha=1}^{i_{rp}} \mathbf{S}_{r\alpha}^T \mathbf{K}_\alpha \mathbf{S}_{p\alpha} . \tag{4.26}$$

For planar systems blocks $\mathbf{K}_{rp}$, $r, p = 1, 2, \ldots, u$ are of 3×3 dimension and

$$\mathbf{S}_{r\alpha} = \begin{bmatrix} 1 & 0 & -s_{r\alpha 2} \\ 0 & 1 & s_{r\alpha 1} \\ 0 & 0 & 1 \end{bmatrix} , \qquad \boldsymbol{\theta}_r = \begin{bmatrix} \cos\theta_{r\alpha} & -\sin\theta_{r\alpha} & 0 \\ \sin\theta_{r\alpha} & \cos\theta_{r\alpha} & 0 \\ 0 & 0 & 1 \end{bmatrix} , \tag{4.27}$$

where $\theta_{r\alpha}$ is the angle between the first main axis of deformation of SDE no. α and the first main central axis of inertia of RFE no. r.

The damping matrix in the case of viscous damping has a form similar to

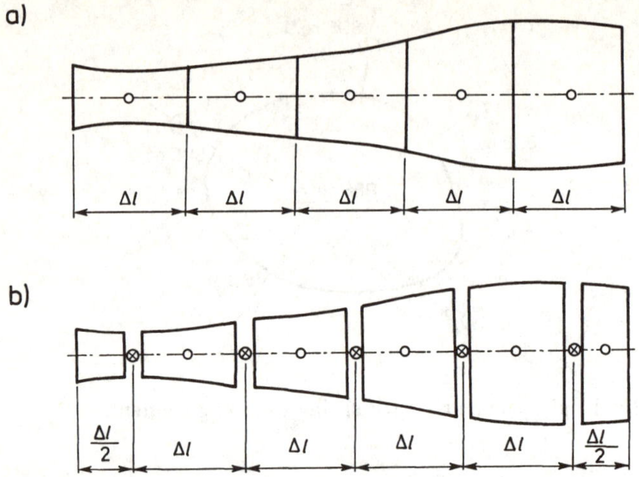

Fig. 4.4. Discretization of a continuous beam into two steps: **a)** initial division, **b)** final division by rigid and spring–damping elements

that of the stiffness matrix

$$\mathbf{C} = \begin{bmatrix} \mathbf{C}_{11} & \mathbf{C}_{12} & \dots & \mathbf{C}_{1u} \\ & \mathbf{C}_{22} & \dots & \mathbf{C}_{2u} \\ & \text{sym.} & \ddots & \vdots \\ & & & \mathbf{C}_{uu} \end{bmatrix} \tag{4.28}$$

and blocks $\mathbf{C}_{rp}$, $r, p = 1, 2, \dots, u$ are calculated using formulae (4.20) and (4.21) in which blocks $\mathbf{K}_\alpha$ defined by (4.24) are replaced by the following blocks of damping coefficients

$$\mathbf{C}_\alpha = \operatorname{diag}(c_{\alpha 1}, c_{\alpha 2}, \dots, c_{\alpha 6}), \tag{4.29}$$

where $c_{\alpha 1}$, $c_{\alpha 2}$, $c_{\alpha 3}$ are coefficients of translational damping (N s m^{-1}), and $c_{\alpha 4}$, $c_{\alpha 5}$, $c_{\alpha 6}$ are coefficients of rotational damping (N m s rad^{-1}).

In order to model a continuous beam by the RFEM a two step discretization is used. First (Fig. 4.4a), the beam is initially divided into elements of length Δl_i. We assume that each element is prismatic and we concentrate its spring-damping features in the middle of each element, obtaining a massless SDE. Then (Fig. 4.4b) we assume that segments of a beam between the points in which SDE have been placed are rigid bodies with mass m_r and mass moments of inertia I_{r1}, I_{r2}, I_{r3}, $r = 1, 2, \dots, u$. Those bodies are rigid finite elements.

By comparing deformations of each segment of the beam with deformations of the SDEs and using elementary knowledge of material resistance, the formulae defining stiffness coefficients of the SDEs can be obtained:

- for stretching

$$k_1 = \frac{EA}{\Delta l}, \tag{4.30}$$

- for shearing

$$k_2 = \frac{1}{\frac{GA}{\kappa \Delta l} + \frac{12EJ_2}{\Delta l^3}}, \qquad k_3 = \frac{1}{\frac{GA}{\kappa \Delta l} + \frac{12EJ_3}{\Delta l^3}}, \tag{4.31}$$

- for torsion (only for circular cross-section)

$$k_4 = \frac{GJ_\circ}{\Delta l}, \tag{4.32}$$

- for bending

$$k_5 = \frac{EJ_2}{\Delta l}, \qquad k_6 = \frac{EJ_3}{\Delta l}, \tag{4.33}$$

where E is the Young modulus, G is the shear modulus, A is the area of the beam cross-section, $J_\circ$ is the geometrical, polar moment of inertia of the cross-section (only for a circular cross-section); J_2, J_3 are the second moments of the cross-section with respect to neutral bending axes x_2, x_3; k is the shape coefficient of the cross-section, which takes into account the non-uniform pattern of shearing stress.

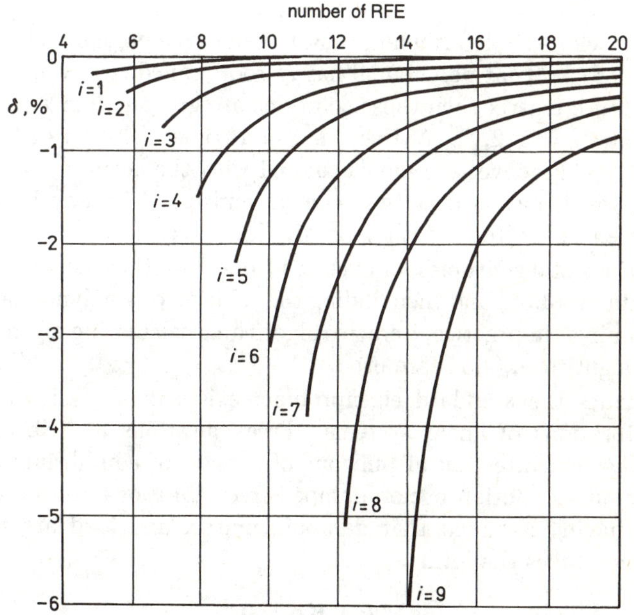

Fig. 4.5. Comparison of results of free frequencies of a prismatic free beam divided into different number of RFEs with analytical ones obtained for continuous Timoshenko beam; i – number of frequency

For non-symmetrical and non-circular cross-sections, coefficient k_4 is calculated according to certain formulae in strength of materials. Such a discrete model of a continuous beam corresponds to the continuous model of the Timoshenko beam because it takes into account the influence of shear forces.

Figure 4.5 shows the comparison of results of free frequencies of a prismatic free beam which has been divided into a different number of RFEs with those of analytical solution for the continuous Timoshenko beam. It can be seen that for 20 RFEs the error for first eigenfrequencies is less than 1%.

4.3 Eigenvalue Problem

The standard problem of natural vibrations is the problem of solution of a homogeneous linear system of algebraic equations in the form

$$(\mathbf{A} - \mathbf{E}\lambda)\mathbf{x} = \mathbf{0} \tag{4.34}$$

in which $\mathbf{A}$ is a $N \times N$ matrix, $\mathbf{E}$ is a $N \times N$ unit matrix.

As it is known from algebra, Eq. (4.34) has a non-trivial solution (not equal to 0) when

$$\det(\mathbf{A} - \mathbf{E}\lambda) = 0. \tag{4.35}$$

Having developed a determinant (4.35) we obtain a polynomial of N degree with respect to λ. That polynomial has N roots called eigenvalues. If $\mathbf{A}$ is a positive definite matrix then those solutions are greater than zero. For each eigenvalue λ_i, $i = 1, 2, \ldots, N$ there is a non-trivial solution in the form of eigenvector $\mathbf{x}_i$. Eigenvectors are calculated with the accuracy of a constant multiplier, which means that they can be multiplied by any constant and they still are eigenvectors i.e. they satisfy Eq. (4.34).

Calculations of eigenvalues and vectors from definition, i.e. expanding out the determinant (4.35) and then finding the solution of a polynomial, are very time-consuming. In practice, iterative algorithms calculating eigenvalues and vectors straight from (4.34) are used.

In dynamics the standard eigenproblem exists in analysis of free non-damped vibrations of linear systems. Free vibrations are vibrations of a system on which, after initial pull out of a state of equilibrium, no forces act. Equations of motion of non-damped free vibrations are obtained from Eq. (4.1), having assumed that damping matrix and load are zero. The equation then takes the form

$$\mathbf{M}\ddot{\mathbf{q}} + \mathbf{K}\mathbf{q} = \mathbf{0}\,. \tag{4.36}$$

Assuming the solution in the form

$$\mathbf{q} = \mathbf{q}^0 \sin(\omega_0 t) \tag{4.37}$$

or more general

$$\mathbf{q} = \mathbf{q}^0 \, \mathrm{e}^{\mathrm{j}\omega_0 t} \tag{4.38}$$

where $\mathrm{j} = \sqrt{-1}$, $\mathbf{q}^0 = \{q_1^0, q_2^0, \ldots, q_N^0\}$ is the vector of displacement amplitudes (of free vibrations) called the vector of vibration form, ω_0 is an eigenfrequency ($\mathrm{rad\,s^{-1}}$), t is time, the system of algebraic linear equations of the following form is obtained

$$(\mathbf{K} - \mathbf{M}\omega_0^2)\,\mathbf{q}^0 = \mathbf{0}\,. \tag{4.39}$$

The mass matrix is decomposed into a multiplication of two triangular matrices

$$\mathbf{M} = \mathbf{L}^T\,\mathbf{L} \tag{4.40}$$

where $\mathbf{L}$ is the upper triangular matrix.

Having substituted (4.40) into (4.39) and after some simple transformations Eq. (4.39) takes the form

$$(\mathbf{A} - \mathbf{E}\omega_0^2)\mathbf{x} = \mathbf{0} \tag{4.41}$$

where

$$\mathbf{A} = \mathbf{L}^{-T}\,\mathbf{K}\,\mathbf{L}^{-1}\,, \tag{4.42}$$

$$\mathbf{x}^0 = \mathbf{L}\,\mathbf{q}^0\,. \tag{4.43}$$

Equation (4.41) has the same form as (4.34). The eigenvalues of matrix $\mathbf{A}$ are then squares of circular frequencies of free vibrations

$$\lambda_i = \omega_{0i}^2\,, \qquad i = 1, 2, \ldots, N, \tag{4.44}$$

and vectors of forms of vibrations are calculated using eigenvectors of matrix $\mathbf{A}$ from (4.43) obtaining the following

$$\mathbf{q}^0 = \mathbf{L}^{-1}\,\mathbf{x}^0\,. \tag{4.45}$$

Calculations of matrix $\mathbf{L}$ are considerably simpler when the mass matrix is diagonal

$$\mathbf{M} = \mathrm{diag}\,(M_i), \qquad i = 1, 2, \ldots, N. \tag{4.46}$$

Then

$$\mathbf{L} = \mathbf{M}^{\frac{1}{2}} = \mathrm{diag}\left(\sqrt{M_i}\right), \tag{4.47}$$

and

$$\mathbf{q}^0 = \left\{\frac{x_i^0}{\sqrt{M_i}}\right\}, \qquad i = 1, 2, \ldots, N. \tag{4.48}$$

Since eigenvectors $\mathbf{x}^0$ and thus vectors of forms of vibrations $\mathbf{q}^0$ are calculated with the accuracy to a multiplier, we usually normalize them in a certain way. Most often one of the four following methods is used:

Method 1: a selected coefficient, for example k, of the vector of vibration forms was equal to 1. If $\bar{\mathbf{q}}^0$ is a non-normalized vector of vibration forms then normalized vector is calculated as follows

$$\mathbf{q}^0 = \bar{\mathbf{q}}^0 \frac{1}{\bar{q}_k^0} . \tag{4.49}$$

Method 2: the maximum absolute value of the component of the vector of vibration forms was equal to 1. The normalized vector is then calculated according to the following formula

$$\mathbf{q}^0 = \bar{\mathbf{q}}^0 \frac{\operatorname{sign} \bar{q}_\alpha^0}{\bar{q}_\alpha^0} \tag{4.50}$$

where α is the number of the maximum coefficient of a non-normalized vector.

Method 3 (orthonormalization): vectors of vibration forms are multiplied by such a number that the sum of squares of their coefficients equals 1. The normalized vector is then calculated according to the formula

$$\mathbf{q}^0 = \bar{\mathbf{q}}^0 \frac{1}{\sqrt{\bar{\mathbf{q}}^{0T} \bar{\mathbf{q}}^0}} = \bar{\mathbf{q}}^0 \frac{1}{\sqrt{\sum_{i=1}^N (\bar{q}_i^0)^2}} . \tag{4.51}$$

Method 4 (M-orthonormalization): vectors are normalized according to the formula

$$\mathbf{q}^0 = \bar{\mathbf{q}}^0 \frac{1}{\sqrt{\bar{\mathbf{q}}^{0T} \mathbf{M} \bar{\mathbf{q}}^0}} = \bar{\mathbf{q}}^0 \frac{1}{\sqrt{\sum_{i=1}^N \sum_{j=1}^N (M_{ij} \bar{q}_i^0 \bar{q}_j^0)}} . \tag{4.52}$$

Formula (4.52) for the diagonal mass matrix takes the form

$$\mathbf{q}^0 = \bar{\mathbf{q}}^0 \frac{1}{\sqrt{\sum_{i=1}^N [M_i (\bar{q}_i^0)^2]}} . \tag{4.53}$$

Figures 4.6, 4.7 and 4.8 present example results of calculations of frequencies and forms of vibrations for three systems modelled using finite elements.

The standard eigenvalue problem is also used in dynamics in order to calculate the principal moments and axes of inertia of solids. If we know the mass matrix for a solid with respect to axes x_1, x_2, x_3 (Fig. 4.9)

$$\mathbf{J} = \begin{bmatrix} I_{11} & -I_{12} & -I_{13} \\ & I_{22} & -I_{23} \\ \text{sym.} & & I_{33} \end{bmatrix} \tag{4.54}$$

where

$$I_{ij} = \int_V \rho x_i x_j \, \mathrm{d}V, \qquad i,j = 1,2,3,$$

then

$$(\mathbf{J} - \mathbf{E}\lambda)\mathbf{x} = \mathbf{0} . \tag{4.55}$$

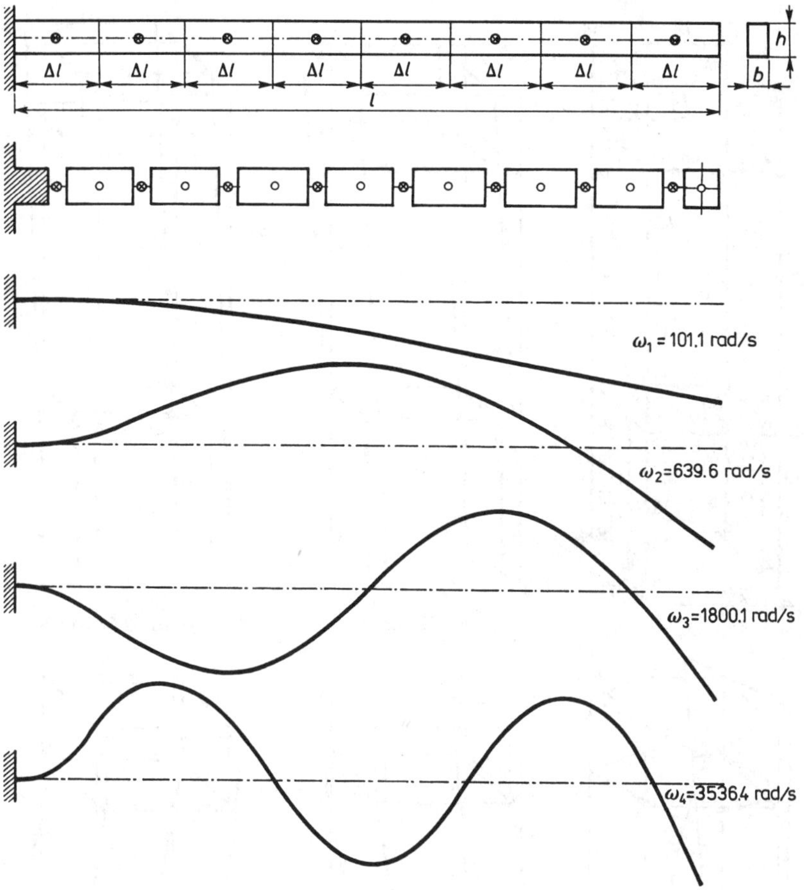

Fig. 4.6. Cantilever beam, its division into rigid finite elements and four first frequencies and forms of free vibrations in the plane of figure. Data: $l = 1000$ mm, $b = 12$ mm, $h = 20$ mm, $E = 2.06 \times 10^5$ MPa, $\nu = 0.3$, $m = 2$ kg, $\kappa = 1.2$

The eigenvalues of matrix **J** are the principal moments of inertia

$$J_i = \lambda_i\,, \qquad i = 1, 2, 3, \tag{4.56}$$

and the respective eigenvectors after normalization (method 3) are directional cosines defining the position of the principal axes of inertia

$$\cos\alpha_{ij} = x_{ij}\,, \qquad i, j = 1, 2, 3, \tag{4.57}$$

where i is the number of eigenvalue, j is the number of the component of eigenvector i.

In practice also the general eigenvalue problem is used in calculations. It is

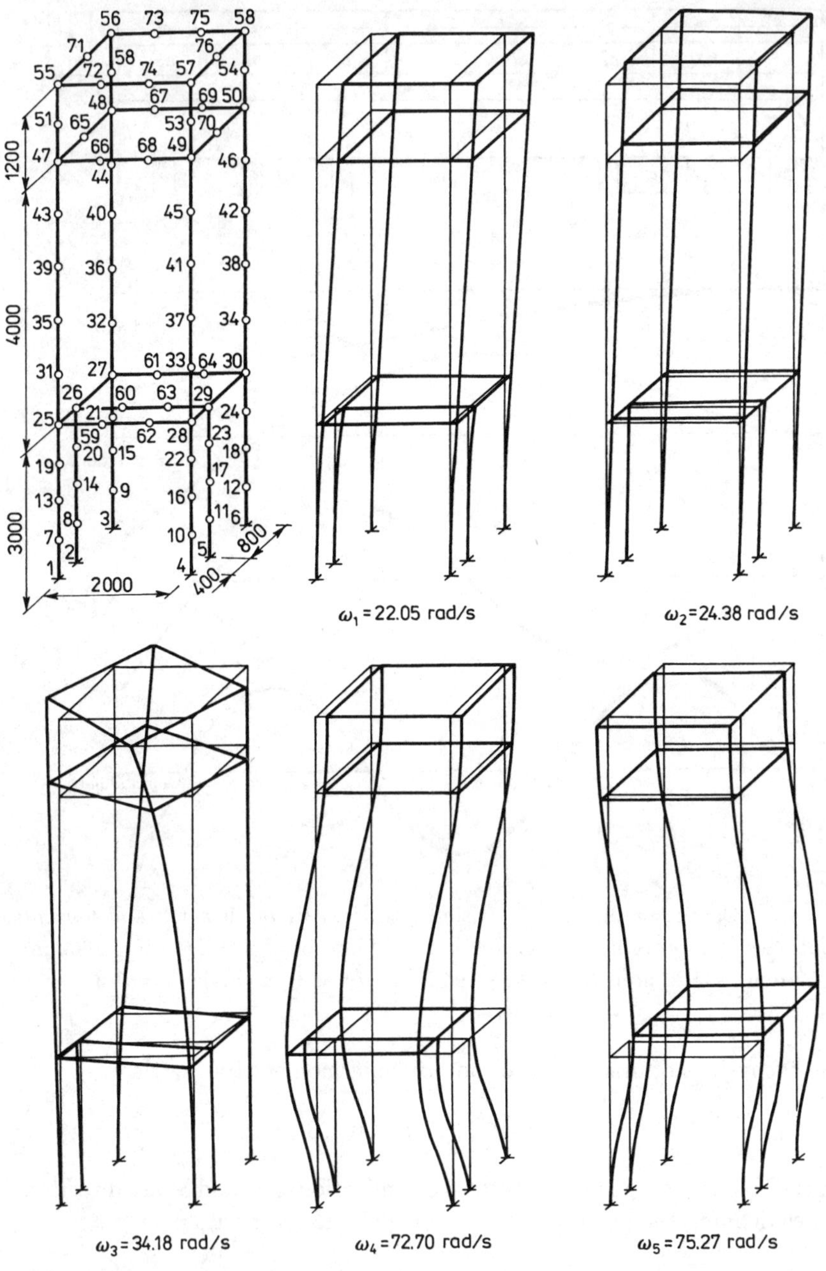

Fig. 4.7. Spatial frame divided into 75 finite beam-like elements and 66 nodes and also first five frequencies and forms of free vibrations. Data: rod cross-section – square $a \times a = 20 \times 20$ mm, $E = 2.06 \times 10^5$ MPa, $\nu = 0.3$, $\rho = 0.78 \times 10^4$ kg m^{-3}, $\kappa = 1.2$

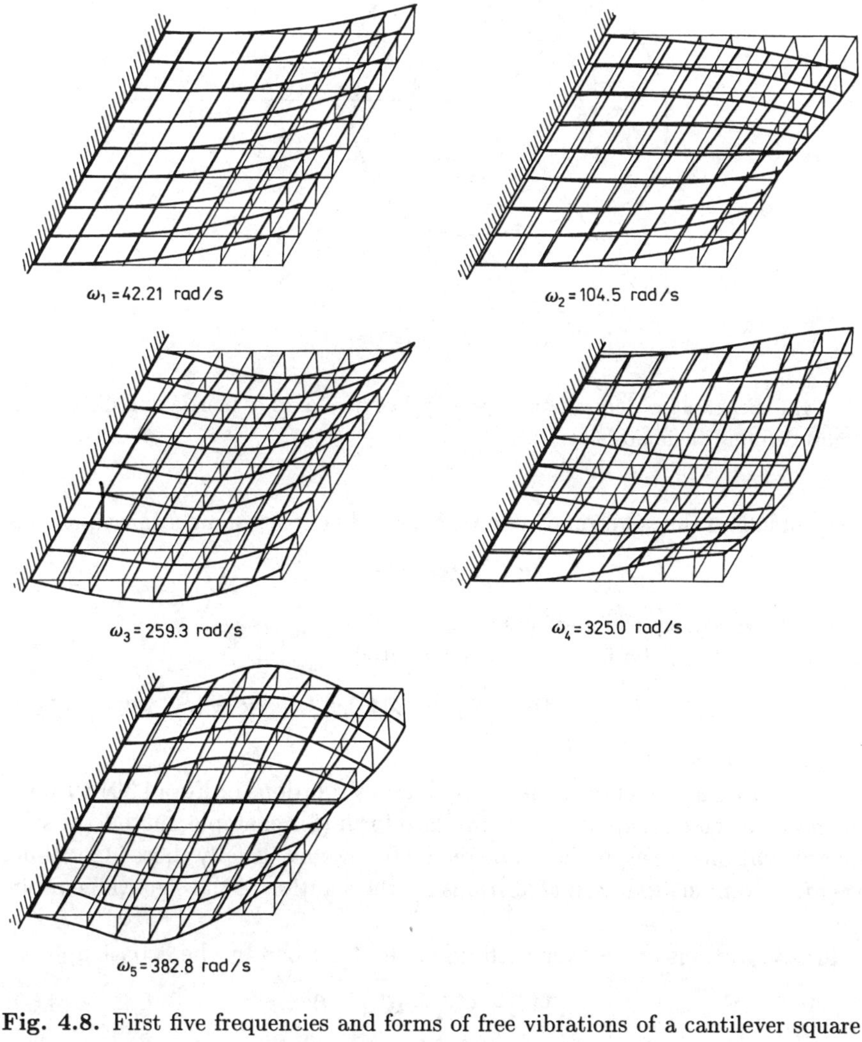

Fig. 4.8. First five frequencies and forms of free vibrations of a cantilever square plate divided into 64 rectangular finite elements. Data: $a \times a \times h = 1200 \times 1200 \times 15.1$ mm, $E = 1.47 \times 10^5$ MPa, $\nu = 0.3$, $\rho = 10^4\ \mathrm{kg\,m^{-3}}$

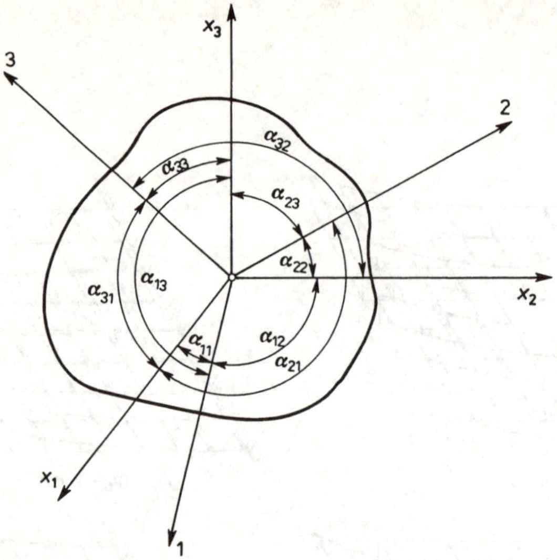

Fig. 4.9. Scheme for denotation of principal central axes of inertia and directional angles defining their position

the problem of the solution of a homogeneous linear set of algebraic equations

$$(\mathbf{A} - \mathbf{B}\lambda)\mathbf{x} = \mathbf{0} \tag{4.58}$$

where **A**, **B** are square matrices.

If in Eq. (4.58) the following is substituted

$$\mathbf{A} = \mathbf{K}, \qquad \mathbf{B} = \mathbf{M}, \qquad \lambda = \omega_0^2, \qquad \mathbf{x} = \mathbf{q}^0, \tag{4.59}$$

we obtain Eq. (4.39).

There are many iterative methods of direct solutions of Eq. (4.39) without the need to reformulate it into a standard form (for example, method of subspace iterations). Yet, it should be remembered that usually those algorithms are more complicated than algorithms for the solution of the eigenvalue problem.

Free vibrations of systems with damping described by the equation

$$\mathbf{M}\,\ddot{\mathbf{q}} + \mathbf{C}\,\dot{\mathbf{q}} + \mathbf{K}\,\mathbf{q} = \mathbf{0} \tag{4.60}$$

can be solved using the eigenvalue problem in the domain of complex numbers. To this end Eq. (4.60) is presented in state coordinates using the following

$$\mathbf{y} = \{\dot{\mathbf{q}}, \mathbf{q}\}\,. \tag{4.61}$$

Equation of motion (4.60) in state coordinates takes the form

$$\dot{\mathbf{y}} + \mathbf{V}\,\mathbf{y} = \mathbf{0} \tag{4.62}$$

in which

$$\mathbf{V} = \begin{bmatrix} -\mathbf{M}^{-1}\mathbf{C} & -\mathbf{M}^{-1}\mathbf{K} \\ \mathbf{E} & \mathbf{0} \end{bmatrix}. \tag{4.63}$$

We forecast the solution in the form

$$\mathbf{y} = \mathbf{y}^0 \, \mathrm{e}^{\lambda t} \tag{4.64}$$

where $\mathbf{y}^0$ is a complex vector and λ is a complex number.

Substituting (4.64) into (4.62), after some transformation, the following homogeneous set of $2N$ algebraic equations in the domain of complex numbers is obtained

$$(\mathbf{V} - \mathbf{E}\lambda)\,\mathbf{y}^0 = \mathbf{0}. \tag{4.65}$$

Equations (4.65) have non-trivial solutions only when

$$\det(\mathbf{V} - \mathbf{E}\lambda) = 0\,. \tag{4.66}$$

Thus, we obtain complex conjugate eigenvalues

$$\lambda_i = \mu_i + \mathrm{j}\omega_{*i}\,, \qquad \bar{\lambda}_i = \mu_i - \mathrm{j}\omega_{*i}\,, \qquad i = 1, 2, \ldots, N, \tag{4.67}$$

in which $\mathrm{j} = \sqrt{-1}$ is a complex unit, ω_{*i} is an i-th free frequency of a system with damping, μ_i is a quantity defining speed of disappearance of vibrations.

For complex conjugate eigenvalues there are corresponding complex conjugate vectors of forms of vibrations

$$\mathbf{a}_i^0 = \mathbf{r}_i + \mathrm{j}\mathbf{p}_i\,, \qquad \bar{\mathbf{a}}_i^0 = \mathbf{r}_i - \mathrm{j}\mathbf{p}_i\,, \qquad i = 1, 2, \ldots, N. \tag{4.68}$$

The k-th component of the vector of modules of the eigenvector is calculated from the following

$$q_{ik}^0 = \sqrt{r_{ik}^2 + p_{ik}^2}\,, \qquad i, k = 1, 2, \ldots, N, \tag{4.69}$$

and its amplitude from the following

$$\varphi_{ik} = \arctan \frac{p_{ik}}{r_{ik}}\,, \qquad i, k = 1, 2, \ldots, N. \tag{4.70}$$

In practice, this method is not used very often because calculation of eigenvalues and eigenvectors of complex eigenvalue problem with non-symmetrical matrix $\mathbf{V}$ with $2N \times 2N$ components is very time-consuming. That is why usually free vibrations with damping are calculated either using the step-by-step iteration method (compare Sect. 4.5) or the modal method (compare Sect. 4.6). Further information about this can be found in books [18, 23, 92, 96, 97, 139].

4.4 Fourier Transformation

The Fourier transformation is used in the analysis of linear systems when substitute inputs $\mathbf{b}(t)$ [cf. (4.9)] are periodic or pseudoperiodic signals. For periodic inputs

$$\mathbf{b}(t) = \mathbf{b}(t+T) \tag{4.71}$$

in which T is a vibration period, harmonic analysis of input $\mathbf{b}(t)$ is carried out. The limited number of terms of the Fourier series is taken into account and substitute inputs (4.71) are presented as follows

$$\mathbf{b}(t) \approx \sum_{\nu=0}^{k} \mathbf{b}_\nu(t) = \left\{ \sum_{\nu=0}^{k} b_{\nu i}^0 \sin(\omega_\nu t + \varphi_{\nu i}) \right\}, \qquad i = 1, 2, \ldots, N. \tag{4.72}$$

where $b_{\nu i}^0$, $\varphi_{\nu i}$ are amplitude and phase respectively of ν-th harmonic of i-th component of the substitute input vector, $\omega_\nu = 2\pi\nu/T$ is the circular frequency of ν-th harmonic component, k is the number of harmonic components taken into account. Many algorithms of harmonic analysis are known. The most popular and the fastest uses the so-called *Fast Fourier Transformation* (FFT) [129].

For pseudo-periodical inputs, which means that inputs are not periodical but composed of harmonic functions, formula (4.72) is used immediately. If the system is linear the principle of superposition is used and calculations are carried out separately for each harmonic component and results are then summed. Then following k equations are solved

$$\mathbf{M}\,\ddot{\mathbf{q}}_\nu + \mathbf{C}\,\dot{\mathbf{q}}_\nu + \mathbf{K}\,\mathbf{q}_\nu = \left\{ b_{\nu i}^0 \sin(\omega_\nu t + \varphi_{\nu i}) \right\} \qquad \begin{array}{l} i = 1, 2, \ldots, N, \\ \nu = 0, 1, 2, \ldots, k. \end{array} \tag{4.73}$$

For steady conditions after the Fourier transformation we obtain

$$\left[(\mathbf{K} - \mathbf{M}\omega_\nu^2) + \mathbf{C}\mathrm{j}\omega_\nu \right] \mathbf{q}_\nu(\mathrm{j}\omega_\nu) = \mathbf{b}_\nu(\mathrm{j}\omega_\nu) \tag{4.74}$$

where $\mathbf{q}_\nu(\mathrm{j}\omega_\nu) = \{q_{\nu i}(\mathrm{j}\omega_\nu)\}$, $i = 1, 2, \ldots, N$ is the complex form of the vector of displacements caused by ν-th harmonic component of substitute input, $\mathbf{b}_\nu(\mathrm{j}\omega_\nu) = \{b_{\nu i}(\mathrm{j}\omega_\nu)\}$, $i = 1, 2, \ldots, N$ is the complex form of the vector of n-th harmonic component of substitute input.

Since $\mathbf{b}_\nu(t)$ is a harmonic function

$$\mathbf{b}_\nu(t) = \left\{ \mathbf{b}_{\nu i}^0 \sin(\omega_\nu t + \varphi_{\nu i}) \right\} \tag{4.75}$$

then its complex form (Fourier transformation) for steady conditions is calculated (see Fig. 4.10) from the relation

$$b_{\nu i}(\mathrm{j}\omega_\nu) = b_{\nu i}^0 \cos\varphi_{\nu i} + \mathrm{j}\, b_{\nu i}^0 \sin\varphi_{\nu i}\,. \tag{4.76}$$

Equation (4.74) is a linear algebraic equation in the domain of complex numbers which can be solved without difficulties using one of the known methods, for example the Gauss elimination method. Of course, the number of equations (4.74) solved has to be the same as the number of harmonic components of input taken into account. As a result of solution of Eqs. (4.74) the complex forms of displacement vectors are obtained

$$\mathbf{q}(\mathrm{j}\omega_\nu) = \{q_{\nu i}(\mathrm{j}\omega_\nu)\}\,, \qquad i = 1, 2, \ldots, N. \tag{4.77}$$

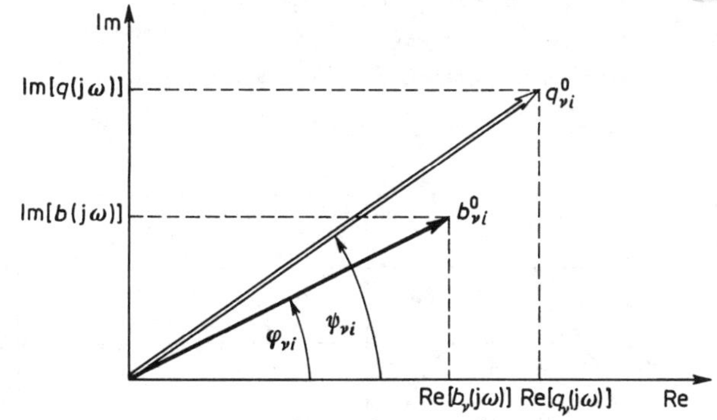

Fig. 4.10. Geometrical interpretation of complex components of displacement and substitute input vectors

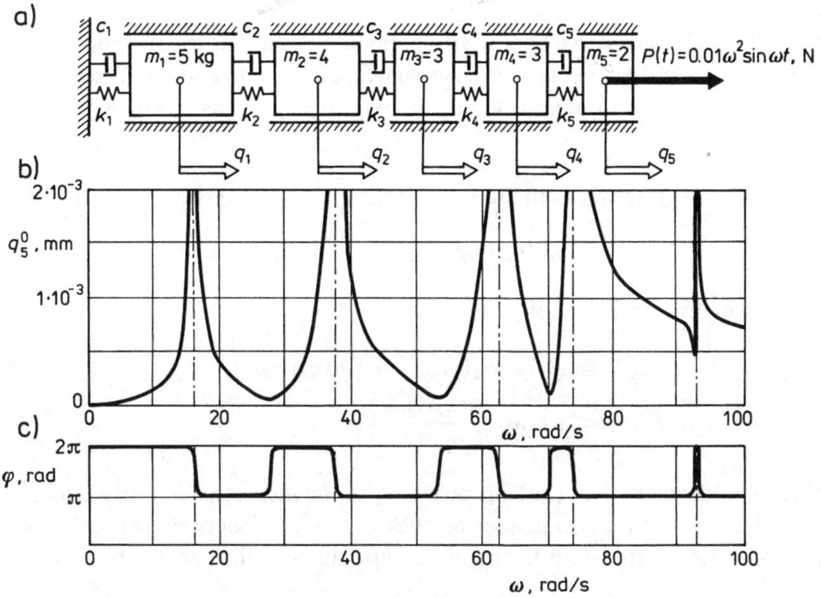

Fig. 4.11. **a)** System of five bodies with harmonic force input with amplitude proportional to the square of input frequency; **b)** amplitude characteristics; **c)** phase characteristics. Data: $k_1 = k_2 = k_3 = 10\,000$ N m^{-1}, $k_4 = k_5 = 5\,000$ N m^{-1}, $c_i = 0.001 k_i$ N s m^{-1}, $i = 1, \ldots, 5$

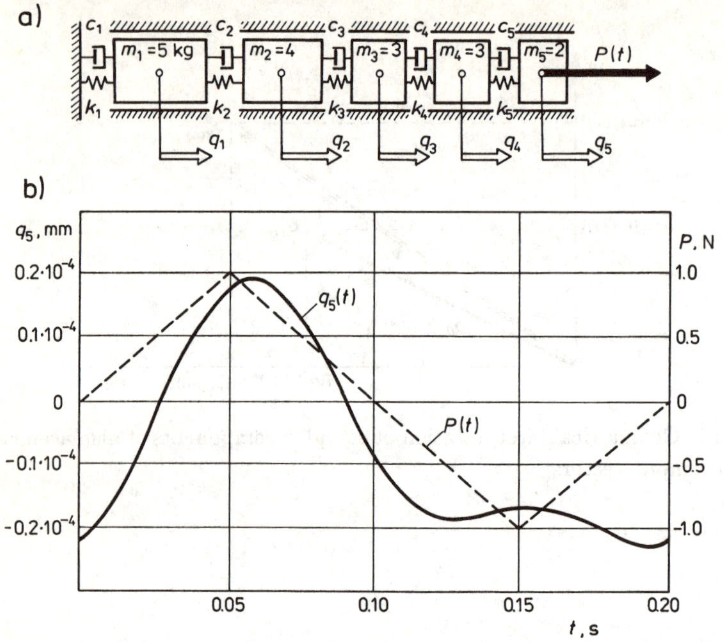

Fig. 4.12. **a)** System of five bodies with periodic force input; **b)** time course of the input force (resolved into 9 harmonic components) and time course of displacements of the body No. 5. Data: stiffness and damping coefficients as in Fig. 4.11

Generalized displacements in time are harmonic functions in the form

$$q_{\nu i}(t) = q^0_{\nu i} \sin(\omega_\nu t + \psi_{\nu i}) \tag{4.78}$$

where, according to Fig. 4.10,

$$q^0_{\nu i} = \sqrt{[\mathrm{Re}\, q_{\nu i}(\mathrm{j}\omega_\nu)]^2 + [\mathrm{Im}\, q_{\nu i}(\mathrm{j}\omega_\nu)]^2}\,, \tag{4.79}$$

$$\psi_{\nu i} = \arctan \frac{\mathrm{Im}\, q_{\nu i}(\mathrm{j}\omega_\nu)}{\mathrm{Re}\, q_{\nu i}(\mathrm{j}\omega_\nu)}\,. \tag{4.80}$$

Global displacement from all harmonic components considered is generally not a harmonic but a periodical or pseudo-periodical function. It is calculated as a geometrical sum of all harmonic components of displacements

$$q_i = \sum_{\nu=0}^{k} q^0_{\nu i} \sin(\omega_\nu t + \psi_{\nu i})\,. \tag{4.81}$$

Figure 4.11 shows an example of calculation results of steady harmonic vibrations with periodic input and Fig. 4.12 those vibrations with periodic "triangular" input signal of a system with five degrees of freedom.

4.5 Numerical Integration

Numerical integration, often called step-by-step integration, is the most universal and most often used method of computer analysis of dynamic problems in the time domain. Results in the form of time courses are simulations of real courses. Using this method, analysis of free vibrations as well as forced vibrations can be carried out and, moreover, inputs can be optional. Characteristic matrices $\mathbf{M}$, $\mathbf{C}$, $\mathbf{K}$ as well as substitute inputs $\mathbf{b}(t)$ can be functions of time, displacements and velocities of displacements. Moreover, not only coefficients of matrices $\mathbf{M}$, $\mathbf{C}$, $\mathbf{K}$ can change in time but also the structure of those matrices can be changeable. Thus, the method can be used in analysis of non-linear vibrations. However, we also must remember about some disadvantages of the method, for example a long calculation time and the possibility of unstable solutions for some non-linear systems.

Methods of numerical integration can be divided into two groups.

In the first group there are methods which integrate directly differential equations of the second order in the forms (4.10) or (4.11). The simplest and

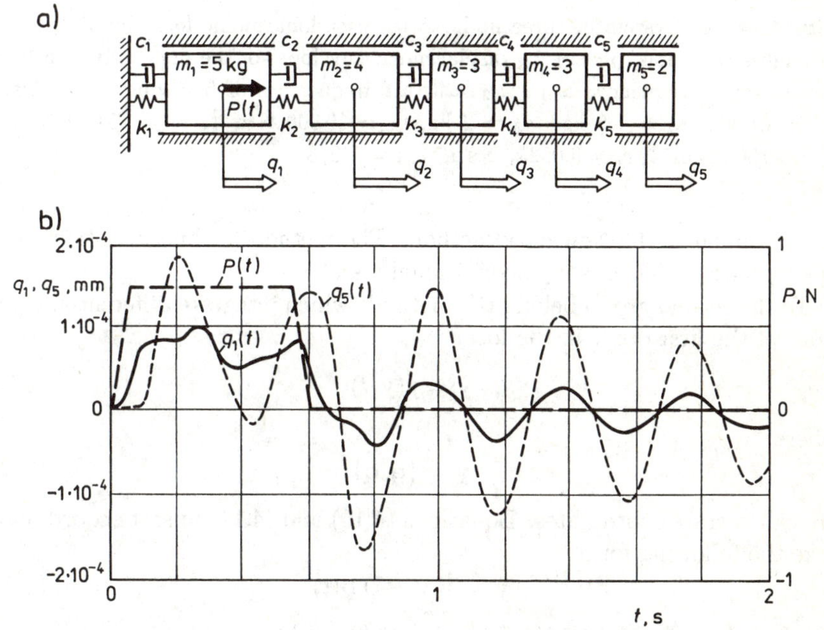

Fig. 4.13. a) System of five bodies with non-periodical force input of trapezoid course imposed on body No. 1, **b)** time courses of input force as well as displacements of bodies No. 1 and 5. Data: stiffness coefficients as in Fig. 4.11, damping coefficients $c_i = 0.004k_i$ $\mathrm{N\,s\,m^{-1}}$, $i = 1, \ldots, 5$

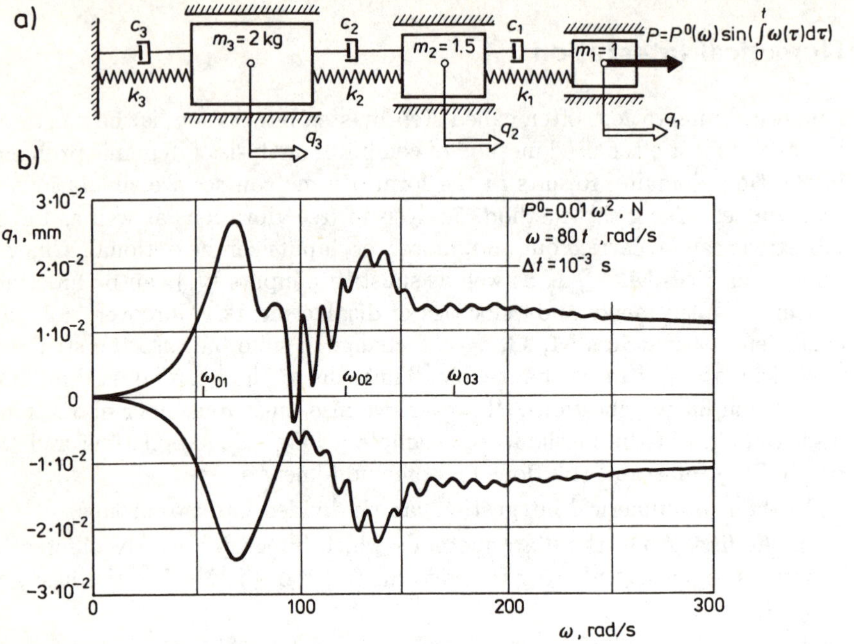

Fig. 4.14. a) System of three bodies with pseudoharmonic force input (passing through a resonance phase), **b)** displacement envelopes of the first body as a function of input frequency. ω_{01} , ω_{02} , ω_{03} are frequencies of free vibrations. Data: $m_1 = 61$ kg, $m_2 = 1.5$ kg, $m_3 = 2$ kg, $k_1 = 10\,000$ N m^{-1}, $k_2 = 15\,000$ N m^{-1}, $k_3 = 20\,000$ N m^{-1}, $c_i = 0.002 k_i$ N s m^{-1}, $i = 1, 2, 3$

most popular is the Newmark method. The method for linear systems with appropriate coefficients is unconditionally stable.

To the second group belong the methods which integrate differential equations of the first order in the form

$$\dot{\mathbf{y}} = f(\mathbf{y}, t) \tag{4.82}$$

where

$$\mathbf{y} = \{\dot{\mathbf{q}}, \mathbf{q}\} \tag{4.83}$$

are called state coordinates. Equations (4.10) and (4.11) in state coordinates take the following form

$$\dot{\mathbf{y}} = \mathbf{V}\,\mathbf{y} + \mathbf{U}\,\mathbf{b}(t) \tag{4.84}$$

in which $\mathbf{V}$ is defined by the formula (4.63), and

$$\mathbf{U} = \begin{bmatrix} \mathbf{M}^{-1} \\ \mathbf{0} \end{bmatrix} . \tag{4.85}$$

If matrix $\mathbf{M}$ is diagonal then coefficients of matrices $\mathbf{V}$ and $\mathbf{U}$ are very

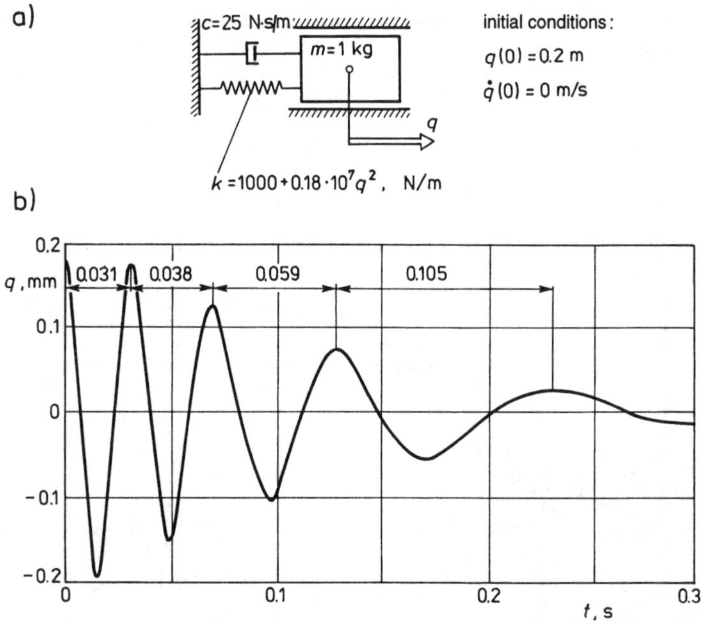

Fig. 4.15. **a)** Non-linear system with one degree of freedom, **b)** displacement course as a function of time (free vibrations)

easy to calculate. In order to use this method larger computer memory is required because matrix $\mathbf{V}$ has $2N \times 2N$ dimensions and is non-symmetrical.

The most popular method of the second group is the Runge–Kutta method of the fourth order.

In order to illustrate the possibilities of using numerical integration method we will present some examples.

Figure 4.13 shows unstable vibrations of the first and fifth body of the system of five bodies which are caused by a non-periodical trapezoid signal with zero initial conditions.

Figure 4.14 shows unstable vibrations of the first body of a system of three solids when passing through three phases of resonance caused by pseudoharmonic force with zero initial conditions. There can be seen a "wavy motion" at the maximal displacements which is caused by superposition of free vibrations occasioned by a sudden increase of amplitude and forced vibrations. This figure also shows an occurrence of maximal displacements for input frequencies larger than frequencies of free vibrations (denoted by ω_{01} , ω_{02} , ω_{03} in Fig. 4.14b).

Figure 4.15 shows free vibrations of a non-linear system with one degree of freedom. Since, in this example, stiffness of springs increases with the square of displacements, deflections (spring deformation) are larger as time intervals called "periods" are shorter.

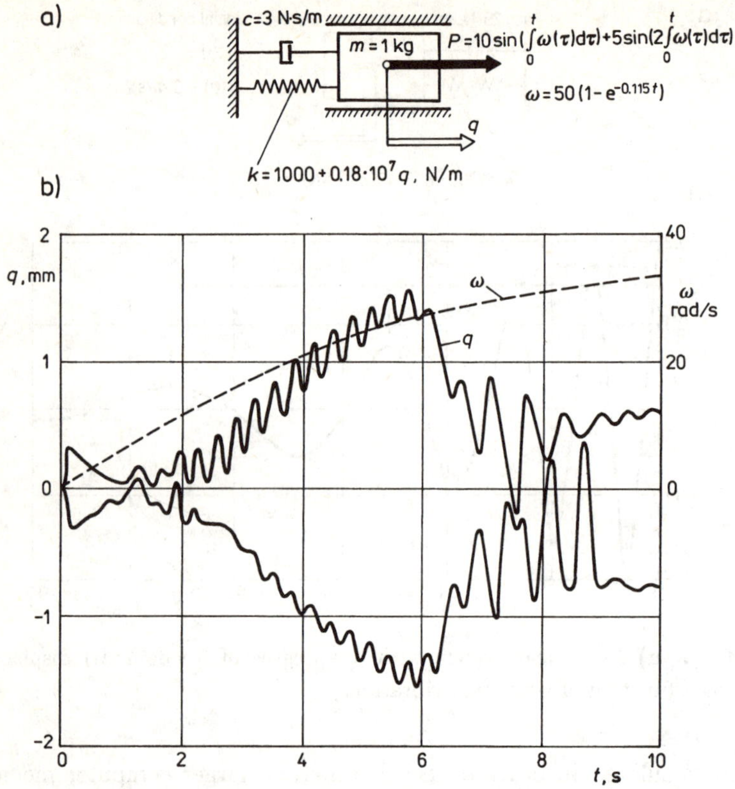

Fig. 4.16. **a)** Non-linear system with one degree of freedom with two pseudoharmonic input forces, **b)** displacement envelope as a function of time and exponential course of input frequency

Figure 4.16 presents unstable vibrations of a non-linear system going through a resonance phase caused by two pseudoharmonic input forces with different periods.

4.6 The Modal Method

The modal method involves choosing such coordinates (called modal or principal coordinates) so that the set of equations (4.10) is uncoupled. This means that as a result N equations are obtained and each has only one unknown function. To this end we use the following transformation

$$\mathbf{q} = \mathbf{\Phi}\,\mathbf{q}^* \tag{4.86}$$

where $\mathbf{q}$ is the vector of generalized coordinates, $\mathbf{q}^*$ is the vector of displacements in modal coordinates, $\mathbf{\Phi}$ is the transformation matrix.

Let us assume that the set of equations considered is a linear system with proportional damping which means that the damping matrix is a linear combination of inertial and stiffness matrices

$$\mathbf{C} = \alpha \mathbf{M} + \beta \mathbf{K} \tag{4.87}$$

where α and β are constants.

For such a system the transformation matrix uncoupling the set of equations (4.10) is a matrix whose columns are M-orthonormalized [cf. (4.52) and (4.53)] vectors of free vibration forms calculated without consideration of damping

$$\mathbf{\Phi} = \left[\mathbf{q}_1^0, \mathbf{q}_2^0, \ldots, \mathbf{q}_N^0\right]_{N \times N} . \tag{4.88}$$

Having used transformation (4.86) the following is obtained

$$\mathbf{M}^* \ddot{\mathbf{q}}^* + \mathbf{C}^* \dot{\mathbf{q}}^* + \mathbf{K}^* \mathbf{q}^* = \mathbf{b}^*(t) \tag{4.89}$$

where

$$\mathbf{M}^* = \mathbf{\Phi}^T \mathbf{M} \mathbf{\Phi} = \mathbf{E} \tag{4.90}$$

is a modal mass matrix,

$$\mathbf{K}^* = \mathbf{\Phi}^T \mathbf{K} \mathbf{\Phi} = \mathbf{\Omega}^2 = \operatorname{diag}(\omega_{0j}^2), \qquad j = 1, 2, \ldots, N, \tag{4.91}$$

is a modal stiffness matrix,

$$\mathbf{C}^* = \mathbf{\Phi}^T \mathbf{C} \mathbf{\Phi} = \alpha \mathbf{E} + \beta \mathbf{\Omega}^2 \tag{4.92}$$

is a modal damping matrix, and

$$\mathbf{b}^*(t) = \mathbf{\Phi}^T \mathbf{b}(t) \tag{4.93}$$

is a modal substitute input.

Since modal matrices are diagonal we obtain N independent equations of motion in the form

$$\ddot{q}_j^* + (\alpha + \beta \omega_{0j}^2) \dot{q}_j^* + \omega_{0j}^2 q_j^* = b_j^*(t), \qquad j = 1, 2, \ldots, N. \tag{4.94}$$

Having multiplied matrices in (4.93) we obtain

$$b_j^*(t) = \sum_{i=1}^{N} q_{ji}^0 b_i(t) . \tag{4.95}$$

Initial conditions $\mathbf{q}^*(0)$ and $\dot{\mathbf{q}}^*(0)$ in modal coordinates generally case are calculated from the relations

$$\mathbf{q}^*(0) = \mathbf{\Phi}^T \mathbf{M} \mathbf{q}(0) , \tag{4.96}$$

$$\dot{\mathbf{q}}^*(0) = \mathbf{\Phi}^T \mathbf{M} \dot{\mathbf{q}}(0) . \tag{4.97}$$

Having calculated all modal displacements, displacements in generalized coordinates are calculated from (4.86) obtaining the following

$$q_i(t) = \sum_{j=1}^{N} q_{ji}^0 q_j^*(t), \qquad i = 1, 2, \ldots, N. \tag{4.98}$$

The modal method is considerably faster than methods in which coupled equations are solved and gives the same solution (does not introduce simplifications).

Time of calculations can be shortened even further if only some (the most essential) vibration forms are taken into account. Results are then approximate, but – if we know that the influence of vibration forms omitted is small – we can get good accuracy in practical terms. However, such approach requires great experience from someone who carries out calculations.

If $k < N$ forms of vibrations are taken into account the transformation matrix $\mathbf{\Phi}$ becomes rectangular of $N \times k$ dimensions and matrix $\mathbf{\Omega}^2$ is a $k \times k$ square matrix. All formulae given above are valid. Only the number of components in the sum (4.98) changes and, of course, k (not N) uncoupled equations of motion (4.94) are obtained.

The modal method can also be used in the case of nonproportional damping when (4.87) is not fulfilled. This, however, requires calculation of complex forms of damped free vibrations which complicates calculation algorithms.

Further information about forced vibrations can be found in [18, 73, 95, 96, 97].

4.7 Vibration Analysis of Systems with Changing Configuration[2]

In this chapter we are concerned with vibrations of systems which consist of deformable subsystems. Subsystems are in motion because of generalized forces or kinematic inputs. We assume that displacements of subsystems can be large and deformations are small. Because of large displacements the relative position of each subsystem changes and thus so does the configuration of the whole system.

The concept of a subsystem is treated very widely. A subsystem can be a simple structural element (for example, a link of a slider crank mechanism) but also the whole complex unit (for example, a vehicle as a mass-spring-damping system). During motion subsystems are in contact because there is a possibility of connections (constraints) between subsystems and between subsystems and the base. This is assured by all kinds of kinematic pairs.

The general form of a system with changing configuration is presented in Fig. 4.17 and some real examples in Fig. 4.18.

[2] Author: Edmund Wittbrodt

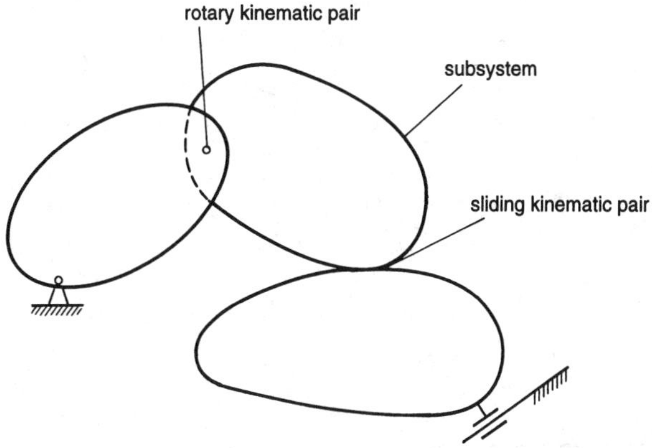

Fig. 4.17. General form of a system with changing configuration

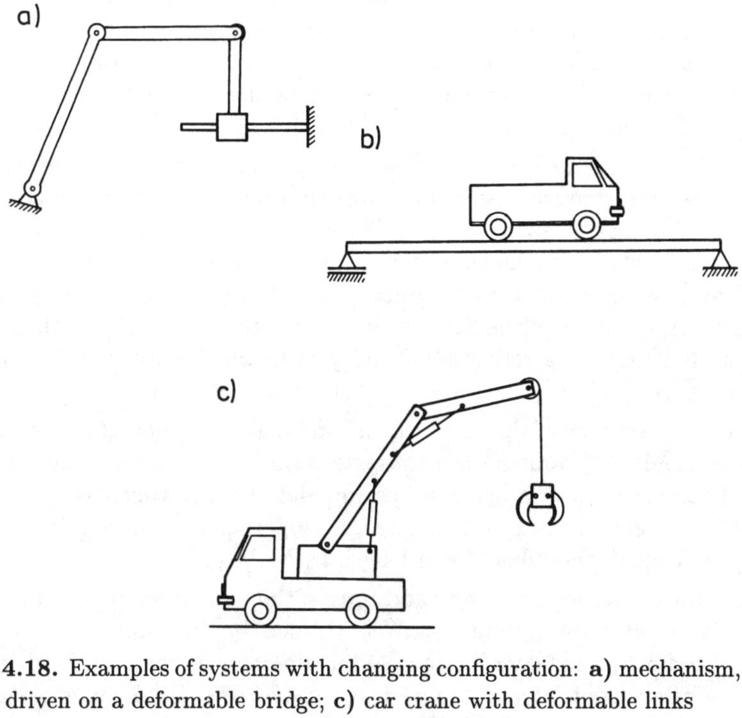

Fig. 4.18. Examples of systems with changing configuration: **a)** mechanism, **b)** vehicle driven on a deformable bridge; **c)** car crane with deformable links

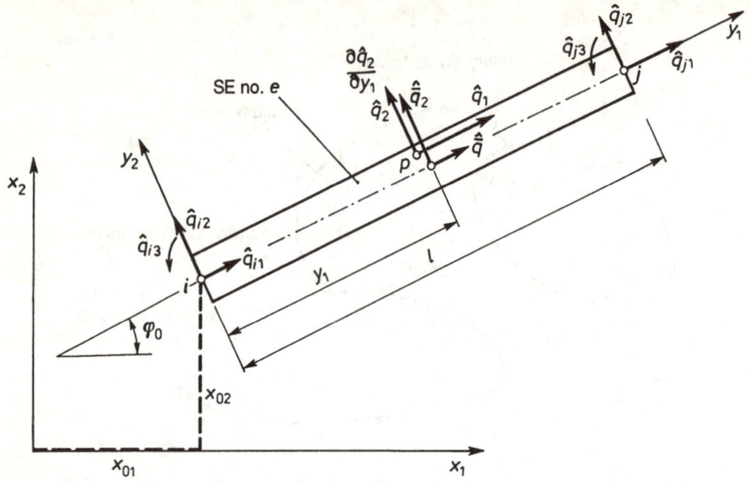

Fig. 4.19. Subsystem modelled using a deformable finite beam-like element and axes of coordinate systems: inertial system x_1, x_2 and mobile system y_1, y_2

Considerations are limited to planar systems, i.e. those in which movements of each subsystem take place in parallel planes. Interactions between subsystems occur in the form of constraints which are rotary or sliding kinematic pairs and the pairs are ideal (without friction). Those constraints are treated as bilateral constraints.

In order to define deformation of subsystems, and thus the whole system, subsystems are discretized using the finite element method. Depending on the physical nature of elements we can use flexible elements (continuous medium) or rigid finite elements connected by means of massless spring-damping elements (systems in which rigid fragments of large mass and massless flexible elements can be distinguished). The rigid finite element method can also be successfully used for discretization of subsystems which are continuous media (cf. Sect. 4.2).

Let us consider one of the cases which can be used for modelling, the slider crank mechanism. To model the subsystem (link) we will use a deformable finite element (FE). It is a beam-like binodal element which is a segment of prismatic beam with a cross-section A, superficial moment of inertia J, density ρ, Young's modulus E and length l (Fig. 4.19).

The position of the subsystem, and thus of the finite element, with respect to inertial coordinate system x_1, x_2 is defined by translation coordinates x_{01}, x_{02} and rotational coordinate φ_o while its deformation is defined by the relative displacements of nodes i and j: $\hat{q}_{i1}, \hat{q}_{i2}, \hat{q}_{i3}, \hat{q}_{j1}, \hat{q}_{j2}, \hat{q}_{j3}$ in non-inertial local system y_1, y_2 (Fig. 4.19). Thus, the vector of generalized coordinates equal to the vector of relative generalized coordinates of the deformable finite element is written in the form

$$\hat{\mathbf{q}}_c = \hat{\mathbf{q}}_e = \{\hat{q}_{i1}, \hat{q}_{i2}, \hat{q}_{i3}, \hat{q}_{j1}, \hat{q}_{j2}, \hat{q}_{j3}\}. \tag{4.99}$$

Point of coordinate y_1 which lies on neutral axis of the element undergoes the following displacements

$$\hat{\bar{\mathbf{q}}}(y_1) = \{\hat{\bar{q}}_1, \hat{\bar{q}}_2\}, \tag{4.100}$$

where $\hat{\bar{q}}_1$, $\hat{\bar{q}}_2$ are displacements along directions y_1 and y_2. We make the vector of displacements dependent on the vector of generalized coordinates

$$\hat{\bar{\mathbf{q}}}(y_1) = \bar{\mathbf{N}}(y_1)\,\hat{\mathbf{q}}_e \tag{4.101}$$

where

$$\bar{\mathbf{N}}(y_1) = \begin{bmatrix} (1-\eta) & 0 & 0 & \eta & 0 & 0 \\ 0 & (1-3\eta^2+2\eta^3) & l\eta(1-2\eta+\eta^2) & 0 & \eta^2(3-2\eta) & l\eta^2(\eta-1) \end{bmatrix} \tag{4.102}$$

is a matrix of shape functions of an element which approximates displacements of points lying on the neutral axis of the element and $\eta = y_1/l$. Point P (Fig. 4.19) lying off the neutral axis undergoes the following displacements

$$\hat{\mathbf{q}}(y_1, y_2) = \{\hat{q}_1, \hat{q}_2\} = \hat{\bar{\mathbf{q}}} - y_2 \begin{bmatrix} \frac{\partial \hat{\bar{\mathbf{q}}}_2}{\partial y_1} \\ 0 \end{bmatrix} \tag{4.103}$$

which, having taken into account (4.101), can be written in the form

$$\hat{\mathbf{q}}(y_1, y_2) = \mathbf{N}(y_1, y_2)\,\hat{\mathbf{q}}_e \tag{4.104}$$

in which

$$\mathbf{N}(y_1, y_2) = \bar{\mathbf{N}}(y_1) - y_2 \bar{\mathbf{N}}^*(y_1) \tag{4.105}$$

and

$$\bar{\mathbf{N}}^*(y_1) = \begin{bmatrix} 0 & -6\frac{\eta}{l}(1-\eta) & 1-4\eta+3\eta^2 & 0 & 6\frac{\eta}{l}(1-\eta) & \eta(3\eta-2) \\ 0 & 0 & 0 & 0 & 0 & 0 \end{bmatrix}. \tag{4.106}$$

For a beam-like element matrix $\mathbf{B}_e$ is obtained from (1.5), while the stiffness matrix coefficients is a scalar [96]

$$\mathbf{D}_e = E\,. \tag{4.107}$$

Relation (4.103) can also be presented in a simplified way [180] in the form of two equations

$$\begin{aligned} \hat{q}_1 &= [\mathbf{H}_1(y_1)\,\mathbf{R}_1 - y_2\mathbf{H}_0(y_1)\,\mathbf{R}_0]\,\hat{\mathbf{q}}_e\,, \\ \hat{q}_2 &= \mathbf{H}_0(y_1)\,\mathbf{R}_0\,\hat{\mathbf{q}}_e\,, \end{aligned} \tag{4.108}$$

in which $\mathbf{H}_1$ and $\mathbf{H}_0$ contain non-zero elements of the matrix of shape function (4.102) and have the form

$$\begin{aligned} \mathbf{H}_1(y_1) &= [(1-\eta),\ \eta]\,, \\ \mathbf{H}_0(y_1) &= [(1-3\eta^2+2\eta^3),\ l\eta(1-2\eta+\eta^2),\ \eta^2(3-2\eta),\ l\eta^2(\eta-1)]\,, \end{aligned} \tag{4.109}$$

while matrices $\mathbf{R}_1$ and $\mathbf{R}_0$, used for "choosing" adequate elements from the vector of generalized coordinates of FE, have the form

$$\mathbf{R}_1 = \begin{bmatrix} 1 & 0 & 0 & 0 & 0 & 0 \\ 0 & 0 & 0 & 1 & 0 & 0 \end{bmatrix}, \qquad \mathbf{R}_2 = \begin{bmatrix} 0 & 1 & 0 & 0 & 0 & 0 \\ 0 & 0 & 1 & 0 & 0 & 0 \\ 0 & 0 & 0 & 0 & 1 & 0 \\ 0 & 0 & 0 & 0 & 0 & 1 \end{bmatrix}. \tag{4.110}$$

Using such notation the matrix of shape function of the element is written in the form

$$\bar{\mathbf{N}}(y_1, y_2) = \begin{bmatrix} \mathbf{H}_1\mathbf{R}_1 - y_2\mathbf{H}_0'\mathbf{R}_0 \\ \mathbf{H}_0\mathbf{R}_0 \end{bmatrix} \tag{4.111}$$

where $(\ldots)'$ means derivative with respect to y_1.

Having taken into account (4.111), the matrix $\mathbf{B}_e$ of the element is obtained in the form

$$\mathbf{B}_e = \begin{bmatrix} \frac{\partial}{\partial y_1} & 0 \\ 0 & \frac{\partial}{\partial y_2} \\ \frac{\partial}{\partial y_2} & \frac{\partial}{\partial y_1} \end{bmatrix} \begin{bmatrix} \mathbf{H}_1\mathbf{R}_1 - y_2\mathbf{H}_0'\mathbf{R}_0 \\ \mathbf{H}_0\mathbf{R}_0 \end{bmatrix} = \begin{bmatrix} \mathbf{H}_1'\mathbf{R}_1 - y_2\mathbf{H}_0''\mathbf{R}_0 \\ 0 \\ 0 \end{bmatrix}. \tag{4.112}$$

Differential equations of motion of the subsystem are obtained on the basis of the Lagrange equations, taking into account that the local system is mobile. Having omitted damping the final form of the equations of motion is as follows

$$\hat{\mathbf{A}}_c\,\ddot{\hat{\mathbf{q}}}_c + \hat{\mathbf{B}}_c\,\dot{\hat{\mathbf{q}}}_c + \hat{\mathbf{C}}_c\,\hat{\mathbf{q}}_c = \hat{\mathbf{f}}_c - \hat{\mathbf{f}}_0 \tag{4.113}$$

where

$$\hat{\mathbf{A}}_c = \int_0^l \rho\left[A(\mathbf{R}_0^T\,\mathbf{H}_0^T\,\mathbf{H}_0\,\mathbf{R}_0 + \mathbf{R}_1^T\,\mathbf{H}_1^T\,\mathbf{H}_1\,\mathbf{R}_1) + J\,\mathbf{R}_0^T\,\mathbf{H}_0'^T\,\mathbf{H}_0'\,\mathbf{R}_0\right] \mathrm{d}y_1\,, \tag{4.114}$$

$$\hat{\mathbf{B}}_c = 2\dot{\varphi}_0 \int_0^l \rho\,A(\mathbf{R}_0^T\,\mathbf{H}_0^T\,\mathbf{H}_1\,\mathbf{R}_1 - \mathbf{R}_1^T\,\mathbf{H}_1^T\,\mathbf{H}_0\,\mathbf{R}_0)\,\mathrm{d}y_1\,, \tag{4.115}$$

$$\begin{aligned} \hat{\mathbf{C}}_c = \int_0^l \{\rho A\,[&\ddot{\varphi}_0(\mathbf{R}_0^T\,\mathbf{H}_0^T\,\mathbf{H}_1\,\mathbf{R}_1 - \mathbf{R}_1^T\,\mathbf{H}_1^T\,\mathbf{H}_0\,\mathbf{R}_0) \\ &- \dot{\varphi}_0^2(\mathbf{R}_0^T\,\mathbf{H}_0^T\,\mathbf{H}_0\,\mathbf{R}_0 + \mathbf{R}_1^T\,\mathbf{H}_1^T\,\mathbf{H}_1\,\mathbf{R}_1)] \\ &- \rho J\dot{\varphi}_0^2\,\mathbf{R}_0^T\,\mathbf{H}_0'^T\,\mathbf{H}_0'\,\mathbf{R}_0 \\ &+ EJ\,\mathbf{R}_0^T\,\mathbf{H}_0''^T\,\mathbf{H}_0''\,\mathbf{R}_0 + EA\,\mathbf{R}_1^T\,\mathbf{H}_1'^T\,\mathbf{H}_1'\,\mathbf{R}_1\}\,\mathrm{d}y_1\,, \end{aligned} \tag{4.116}$$

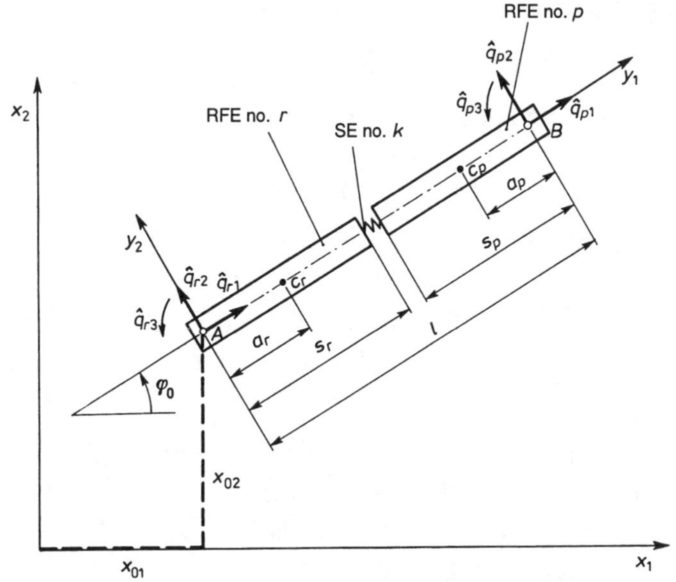

Fig. 4.20. Subsystem modelled using 2 RFEs and SE as well as the axes of coordinate systems: inertial system x_1, x_2 and mobile system y_1, y_2

and the vector of additional loads from Eq. (4.113) has the form

$$\begin{aligned}\hat{\mathbf{f}}_0 = \int_0^l \rho\, \{A\, [(\ddot{x}_{01} \cos\varphi_0 + \ddot{x}_{02} \sin\varphi_0 - y_1 \dot{\varphi}_0^2)\, \mathbf{R}_1^T\, \mathbf{H}_1^T \\ + (-\ddot{x}_{01} \sin\varphi_0 + \ddot{x}_{02} \cos\varphi_0 - y_1 \ddot{\varphi}_0)\, \mathbf{R}_0^T\, \mathbf{H}_0^T] \\ + J\, \ddot{\varphi}_0\, \mathbf{R}_0^T\, \mathbf{H}_0'^{\,T}\}\, \mathrm{d}y_1 \,. \qquad (4.117)\end{aligned}$$

Let us now consider the case in which the subsystem is modelled by two rigid finite elements connected by a spring element (SE) (Fig. 4.20). The rigid finite elements are prismatic rods. The local system is assumed in such a way that its origin lies in point A and axis y_1 coincides with the axes of rods. The local systems of elements r and p have the same directions as the directions of axes of the local coordinate system of the subsystem. The origins of those systems are placed at points A and B, which are the ends of RFEs.

In this case the vector of the generalized coordinates of the subsystem has the form

$$\hat{\mathbf{q}}_e = \{\hat{q}_{r1}, \hat{q}_{r2}, \hat{q}_{r3}, \hat{q}_{p1}, \hat{q}_{p2}, \hat{q}_{p3}\} \qquad (4.118)$$

where $\hat{q}_{r1}$, $\hat{q}_{r2}$, $\hat{q}_{r3}$ are relative displacements of RFE r; $\hat{q}_{p1}$, $\hat{q}_{p2}$, $\hat{q}_{p3}$ are relative displacements of RFE p (Fig. 4.20), and matrices from (4.113) have the following form

$$\hat{\mathbf{A}}_c = \begin{bmatrix} m_r & 0 & 0 & & & \\ & m_r & m_r a_r & & \mathbf{0} & \\ & & I_r & & & \\ & & & m_p & 0 & 0 \\ & \text{sym.} & & & m_p & m_p a_p \\ & & & & & I_p \end{bmatrix}, \tag{4.119}$$

$$\hat{\mathbf{B}}_c = \dot{\varphi}_0 \begin{bmatrix} 0 & -2m_r & -m_r a_r & & & \\ 2m_r & 0 & 0 & & \mathbf{0} & \\ m_r a_r & 0 & 0 & & & \\ & & & 0 & -2m_p & -m_p a_p \\ & \mathbf{0} & & 2m_p & 0 & 0 \\ & & & m_p a_p & 0 & 0 \end{bmatrix}, \tag{4.120}$$

$$\hat{\mathbf{C}}_c = \begin{bmatrix} -m_r\dot{\varphi}_0^2 & -m_r\ddot{\varphi}_0 & 0 & & & \\ -m_r\ddot{\varphi}_0 & -m_r\dot{\varphi}_0^2 & 0 & & \mathbf{0} & \\ m_r a_r\ddot{\varphi}_0 & 0 & 0 & & & \\ & & & -m_p\dot{\varphi}_0^2 & -m_p\ddot{\varphi}_0 & 0 \\ & \mathbf{0} & & m_p\ddot{\varphi}_0 & -m_p\dot{\varphi}_0^2 & 0 \\ & & & m_p a_p\ddot{\varphi}_0 & 0 & 0 \end{bmatrix}$$

$$+ \begin{bmatrix} c_1 & 0 & 0 & -c_1 & 0 & 0 \\ & c_2 & c_2 s_r & 0 & -c_2 & -c_2 s_p \\ & & c_3 + c_2 s_r^2 & 0 & -c_2 s_r & -(c_3 + c_2 s_r s_p) \\ & & & c_1 & 0 & 0 \\ & \text{sym.} & & & c_2 & c_2 s_p \\ & & & & & c_3 + c_2 s_p^2 \end{bmatrix}, \tag{4.121}$$

$$\hat{\mathbf{f}}_0 = \begin{bmatrix} m_r(\ddot{x}_{01}\cos\varphi_0 + \ddot{x}_{02}\sin\varphi_0 - a_r\dot{\varphi}_0^2) \\ m_r(-\ddot{x}_{01}\sin\varphi_0 + \ddot{x}_{02}\cos\varphi_0 + a_r\ddot{\varphi}_0) \\ m_r a_r[-\ddot{x}_{01}\sin\varphi_0 + \ddot{x}_{02}\cos\varphi_0 \\ \qquad - \dot{\varphi}_0(\ddot{x}_{01}\cos\varphi_0 + \ddot{x}_{02}\sin\varphi_0)] + I_r\ddot{\varphi}_0 \\ m_p(\ddot{x}_{01}\cos\varphi_0 + \ddot{x}_{02}\sin\varphi_0 - (l + a_p)\dot{\varphi}_0^2) \\ m_p(-\ddot{x}_{01}\sin\varphi_0 + \ddot{x}_{02}\cos\varphi_0 + (l + a_p)\ddot{\varphi}_0) \\ m_p a_p[-\ddot{x}_{01}\sin\varphi_0 + \ddot{x}_{02}\cos\varphi_0 \\ \qquad - \dot{\varphi}_0(\ddot{x}_{01}\cos\varphi_0 + \ddot{x}_{02}\sin\varphi_0) + l\ddot{\varphi}_0] + I_p\ddot{\varphi}_0 \end{bmatrix}, \tag{4.122}$$

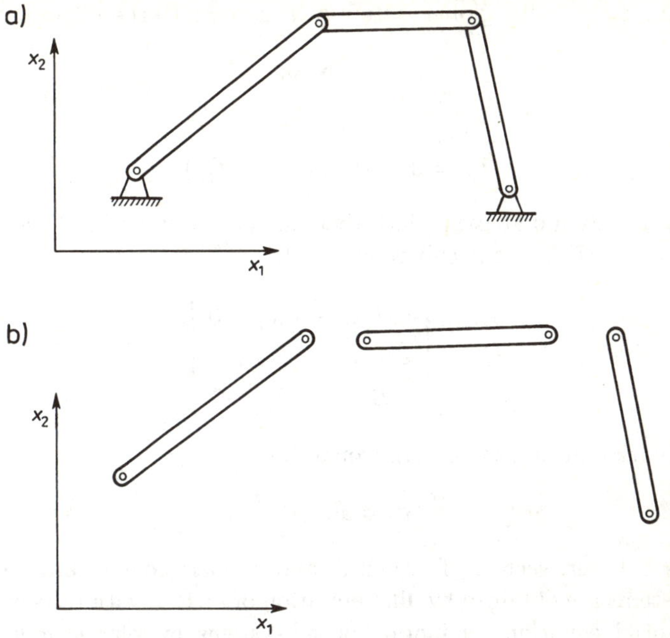

Fig. 4.21. Models of calculation system: **a)** with constraints given by rotary kinematic pairs, **b)** of free subsystem

where m_r, m_p, I_r, I_p are masses and mass moments of inertia of RFEs; a_r, a_p are coordinates of mass centres of RFEs; s_r, s_p are coordinates of fixing points of SE to RFEs r and p; c_1, c_2, c_3 are stiffness coefficients of SE.

Equations of motion of the calculation model are derived in two stages. In the first stage we obtain equations of motion of the free system in which constraints imposed on the system are omitted (Fig. 4.21a), and in the second stage we obtain equations of motion of the system with constraints (Fig. 4.21b).

In order to obtain equations of motion of the free model, we introduce vectors of generalized coordinates and forces of the free model

$$\mathbf{q}_s = \{\mathbf{q}_1, \mathbf{q}_2, \ldots, \mathbf{q}_c, \ldots, \mathbf{q}_u\} \tag{4.123}$$

$$\mathbf{f}_s = \{\mathbf{f}_1, \mathbf{f}_2, \ldots, \mathbf{f}_c, \ldots, \mathbf{f}_u\} \tag{4.124}$$

where $\mathbf{q}_c$ is the vector of generalized coordinates of the subsystem c, and $\mathbf{f}_c$ is the vector of generalized forces of the subsystem c.

Vectors $\mathbf{q}_s$ and $\mathbf{f}_s$ have as many coefficients as the sum of degrees of freedom of all subsystems belonging to the system analysed. Components $\mathbf{q}_c$ and $\mathbf{f}_c$ of (4.123) and (4.124) are defined in the same directions for different subsystems. This requires transformation of the generalized coordinate vector of the subsystem in the mobile system $\hat{\mathbf{q}}_c$ to the generalized coordinate vector of the subsystem in the global system $\mathbf{q}_c$ according to the relation

$$\mathbf{q}_c = \mathbf{R}_c\,\hat{\mathbf{q}}_c \tag{4.125}$$

where

$$\mathbf{R}_c = \operatorname{diag}(\mathbf{Q}_c, \mathbf{Q}_c, \ldots, \mathbf{Q}_c) \tag{4.126}$$

is a transformation matrix in which there are as many blocks $\mathbf{Q}_c$ as there are constraints or RFEs in the subsystem, and

$$\mathbf{Q}_c = \begin{bmatrix} \cos\varphi_0 & -\sin\varphi_0 & 0 \\ \sin\varphi_0 & \cos\varphi_0 & 0 \\ 0 & 0 & 1 \end{bmatrix}. \tag{4.127}$$

Transformation matrix $\mathbf{R}_c$ is orthogonal, thus

$$\hat{\mathbf{q}}_c = \mathbf{R}_c^T\,\mathbf{q}_c\,. \tag{4.128}$$

Relations between vectors of velocities and accelerations in both local and global systems are obtained by differentiation of (4.128) with respect to time.

Differential equations of motion of subsystems in relative mobile coordinates are transformed to the global system by substituting (4.128) into (4.113) and then premultiplying by $\mathbf{R}_c$, and the following is obtained

$$\mathbf{A}_c\ddot{\mathbf{q}}_c + \mathbf{B}_c\dot{\mathbf{q}}_c + \mathbf{C}_c\mathbf{q}_c = \mathbf{f}_c \tag{4.129}$$

where

$$\begin{aligned} \mathbf{A}_c &= \mathbf{R}_c\,\hat{\mathbf{A}}_c\,\mathbf{R}_c^T\,, \\ \mathbf{B}_c &= \mathbf{R}_c\,\hat{\mathbf{B}}_c\,\mathbf{R}_c^T + 2\dot{\varphi}_0\,\mathbf{R}_c\,\hat{\mathbf{A}}_c\,\mathbf{R}_c'^T\,, \\ \mathbf{C}_c &= \mathbf{R}_c\,\hat{\mathbf{C}}_c\,\mathbf{R}_c^T + \dot{\varphi}_0\,\mathbf{R}_c\,\hat{\mathbf{B}}_c\,\mathbf{R}_c^T + \dot{\varphi}_0^2\,\mathbf{R}_c\,\hat{\mathbf{A}}_c\,\mathbf{R}_c''^T + \ddot{\varphi}_0\,\mathbf{R}_c\,\hat{\mathbf{A}}_c\,\mathbf{R}_c'^T\,, \\ \mathbf{f}_c &= \mathbf{R}_c\,\hat{\mathbf{f}}_c\,, \end{aligned} \tag{4.130}$$

while

$$\mathbf{R}_c' = \frac{\mathrm{d}\mathbf{R}_c}{\mathrm{d}\varphi_0}\,.$$

In order to obtain equations of motion of the system with constraints components of the vector of generalized coordinates of the free system are divided into three groups:

1. independent coordinates,
2. dependent coordinates,
3. given coordinates.

Independent coordinates are those which describe independent positions of joints or rigid finite elements.

Dependent coordinates are those whose values depend on previously defined independent coordinates. Those relationships are the results of imposing constraints on the system.

Given coordinates are those whose values are known. Most often those are coordinates joined to the base and then their values are zero.

Thus, the vector of generalized coordinates of the free system (4.123) can be arranged as follows

$$\mathbf{q}_s = \{\mathbf{q}_\mathrm{i}, \mathbf{q}_\mathrm{d}, \mathbf{q}_\mathrm{g}\} \tag{4.131}$$

where $\mathbf{q}_\mathrm{i}$, $\mathbf{q}_\mathrm{d}$, $\mathbf{q}_\mathrm{g}$ are vectors of independent, dependent and given coordinates, respectively.

Equation of motion of all subsystems of a free system after adequate transposition of rows and columns in matrices of the system (also necessary as when separating unknown and given motion [96]) take the form

$$\begin{bmatrix} \mathbf{A}_\mathrm{i} & \mathbf{A}_\mathrm{id} & \mathbf{A}_\mathrm{ig} \\ \mathbf{A}_\mathrm{di} & \mathbf{A}_\mathrm{d} & \mathbf{A}_\mathrm{dg} \\ \mathbf{A}_\mathrm{gi} & \mathbf{A}_\mathrm{gd} & \mathbf{A}_\mathrm{g} \end{bmatrix} \begin{bmatrix} \ddot{\mathbf{q}}_\mathrm{i} \\ \ddot{\mathbf{q}}_\mathrm{d} \\ \ddot{\mathbf{q}}_\mathrm{g} \end{bmatrix} + \begin{bmatrix} \mathbf{B}_\mathrm{i} & \mathbf{B}_\mathrm{id} & \mathbf{B}_\mathrm{ig} \\ \mathbf{B}_\mathrm{di} & \mathbf{B}_\mathrm{d} & \mathbf{B}_\mathrm{dg} \\ \mathbf{B}_\mathrm{gi} & \mathbf{B}_\mathrm{gd} & \mathbf{B}_\mathrm{g} \end{bmatrix} \begin{bmatrix} \dot{\mathbf{q}}_\mathrm{i} \\ \dot{\mathbf{q}}_\mathrm{d} \\ \dot{\mathbf{q}}_\mathrm{g} \end{bmatrix} + \begin{bmatrix} \mathbf{C}_\mathrm{i} & \mathbf{C}_\mathrm{id} & \mathbf{C}_\mathrm{ig} \\ \mathbf{C}_\mathrm{di} & \mathbf{C}_\mathrm{d} & \mathbf{C}_\mathrm{dg} \\ \mathbf{C}_\mathrm{gi} & \mathbf{C}_\mathrm{gd} & \mathbf{C}_\mathrm{g} \end{bmatrix} \begin{bmatrix} \mathbf{q}_\mathrm{i} \\ \mathbf{q}_\mathrm{d} \\ \mathbf{q}_\mathrm{g} \end{bmatrix} = \begin{bmatrix} \mathbf{f}_\mathrm{i} \\ \mathbf{f}_\mathrm{d} \\ \mathbf{f}_\mathrm{g} \end{bmatrix} \tag{4.132}$$

where

$$\begin{matrix} \mathbf{A}_\mathrm{i} & \mathbf{B}_\mathrm{i} & \mathbf{C}_\mathrm{i} \\ \mathbf{A}_\mathrm{d} & \mathbf{B}_\mathrm{d} & \mathbf{C}_\mathrm{d} \\ \mathbf{A}_\mathrm{g} & \mathbf{B}_\mathrm{g} & \mathbf{C}_\mathrm{g} \end{matrix}$$

are blocks corresponding to independent, dependent and given coordinates, respectively, while

$$\begin{matrix} \mathbf{A}_\mathrm{id} & \mathbf{A}_\mathrm{di} & \mathbf{A}_\mathrm{ig} & \mathbf{A}_\mathrm{gi} & \mathbf{A}_\mathrm{dg} & \mathbf{A}_\mathrm{gd} \\ \mathbf{B}_\mathrm{id} & \mathbf{B}_\mathrm{di} & \mathbf{B}_\mathrm{ig} & \mathbf{B}_\mathrm{gi} & \mathbf{B}_\mathrm{dg} & \mathbf{B}_\mathrm{gd} \\ \mathbf{C}_\mathrm{id} & \mathbf{C}_\mathrm{di} & \mathbf{C}_\mathrm{ig} & \mathbf{C}_\mathrm{gi} & \mathbf{C}_\mathrm{dg} & \mathbf{C}_\mathrm{gd} \end{matrix}$$

are blocks of coupling between those coordinates.

Then, we present Eq. (4.132) in the form of two sets of equations, the first of which,

$$\begin{bmatrix} \mathbf{A}_\mathrm{i} & \mathbf{A}_\mathrm{id} \\ \mathbf{A}_\mathrm{di} & \mathbf{A}_\mathrm{d} \end{bmatrix} \begin{bmatrix} \ddot{\mathbf{q}}_\mathrm{i} \\ \ddot{\mathbf{q}}_\mathrm{d} \end{bmatrix} + \begin{bmatrix} \mathbf{B}_\mathrm{i} & \mathbf{B}_\mathrm{id} \\ \mathbf{B}_\mathrm{di} & \mathbf{B}_\mathrm{d} \end{bmatrix} \begin{bmatrix} \dot{\mathbf{q}}_\mathrm{i} \\ \dot{\mathbf{q}}_\mathrm{d} \end{bmatrix} + \begin{bmatrix} \mathbf{C}_\mathrm{i} & \mathbf{C}_\mathrm{id} \\ \mathbf{C}_\mathrm{di} & \mathbf{C}_\mathrm{d} \end{bmatrix} \begin{bmatrix} \mathbf{q}_\mathrm{i} \\ \mathbf{q}_\mathrm{d} \end{bmatrix} = \begin{bmatrix} \mathbf{f}_\mathrm{i} \\ \mathbf{f}_\mathrm{d} \end{bmatrix} - \left(\begin{bmatrix} \mathbf{A}_\mathrm{ig} \\ \mathbf{A}_\mathrm{dg} \end{bmatrix} \ddot{\mathbf{q}}_\mathrm{g} + \begin{bmatrix} \mathbf{B}_\mathrm{ig} \\ \mathbf{B}_\mathrm{dg} \end{bmatrix} \dot{\mathbf{q}}_\mathrm{g} + \begin{bmatrix} \mathbf{C}_\mathrm{ig} \\ \mathbf{C}_\mathrm{dg} \end{bmatrix} \mathbf{q}_\mathrm{g} \right), \tag{4.133}$$

is used for calculation of unknown generalized coordinates $\mathbf{q}_i$ and $\mathbf{q}_d$ and the second one

$$\mathbf{f}_g = [\,\mathbf{A}_{gi}\;\mathbf{A}_{gd}\;\mathbf{A}_g\,]\begin{bmatrix}\ddot{\mathbf{q}}_i\\ \ddot{\mathbf{q}}_d\\ \ddot{\mathbf{q}}_g\end{bmatrix} + [\,\mathbf{B}_{gi}\;\mathbf{B}_{gd}\;\mathbf{B}_g\,]\begin{bmatrix}\dot{\mathbf{q}}_i\\ \dot{\mathbf{q}}_d\\ \dot{\mathbf{q}}_g\end{bmatrix} + [\,\mathbf{C}_{gi}\;\mathbf{C}_{gd}\;\mathbf{C}_g\,]\begin{bmatrix}\mathbf{q}_i\\ \mathbf{q}_d\\ \mathbf{q}_g\end{bmatrix} \tag{4.134}$$

is used for definition of the vector of generalized forces $\mathbf{f}_g$ which are usually reaction forces of the base.

Constraints in the system cause an occurrence of relationships between dependent $\mathbf{q}_d$ and independent coordinates $\mathbf{q}_i$ and also additional unknown reaction forces $\mathbf{r}_d$ and $\mathbf{r}_i$ depending on type of coordinates. Having taken into account reaction vectors in (4.133), the following is obtained

$$\begin{bmatrix}\mathbf{A}_i & \mathbf{A}_{id}\\ \mathbf{A}_{di} & \mathbf{A}_d\end{bmatrix}\begin{bmatrix}\ddot{\mathbf{q}}_i\\ \ddot{\mathbf{q}}_d\end{bmatrix} + \begin{bmatrix}\mathbf{B}_i & \mathbf{B}_{id}\\ \mathbf{B}_{di} & \mathbf{B}_d\end{bmatrix}\begin{bmatrix}\dot{\mathbf{q}}_i\\ \dot{\mathbf{q}}_d\end{bmatrix} + \begin{bmatrix}\mathbf{C}_i & \mathbf{C}_{id}\\ \mathbf{C}_{di} & \mathbf{C}_d\end{bmatrix}\begin{bmatrix}\mathbf{q}_i\\ \mathbf{q}_d\end{bmatrix}$$
$$= \begin{bmatrix}\mathbf{f}_i\\ \mathbf{f}_d\end{bmatrix} + \begin{bmatrix}\mathbf{r}_i\\ \mathbf{r}_d\end{bmatrix} - \left(\begin{bmatrix}\mathbf{A}_{ig}\\ \mathbf{A}_{dg}\end{bmatrix}\ddot{\mathbf{q}}_g + \begin{bmatrix}\mathbf{B}_{ig}\\ \mathbf{B}_{dg}\end{bmatrix}\dot{\mathbf{q}}_g + \begin{bmatrix}\mathbf{C}_{ig}\\ \mathbf{C}_{dg}\end{bmatrix}\mathbf{q}_g\right). \tag{4.135}$$

Relationships between dependent and independent coordinates are assumed in the form of the following constraint equation [180]

$$[\,\mathbf{W}_i\;\mathbf{W}_d\,]\begin{bmatrix}\mathbf{q}_i\\ \mathbf{q}_d\end{bmatrix} = 0 \tag{4.136}$$

or

$$\mathbf{q}_d = \mathbf{W}\,\mathbf{q}_i \tag{4.137}$$

where

$$\mathbf{W} = -\mathbf{W}_d^{-1}\,\mathbf{W}_i$$

is a constraint matrix whose coefficients depend on the kind of kinematic pair (Fig. 4.22).

Relationships between derivatives of dependent and independent coordinates are obtained by differentiation of (4.137) with respect to time.

In order to eliminate dependent coordinates Eq. (4.135) is written in the form of two sets of equations. Then, Eq. (4.137) is taken into account and, having premultiplied the second equation by $\mathbf{W}^T$, both equations are added side by side. Finally, we obtain the equation of motion of the model with constraints in the form

$$\mathbf{A}^*\,\ddot{\mathbf{q}}_i + \mathbf{B}^*\,\dot{\mathbf{q}}_i + \mathbf{C}^*\,\mathbf{q}_i = \mathbf{f}^* \tag{4.138}$$

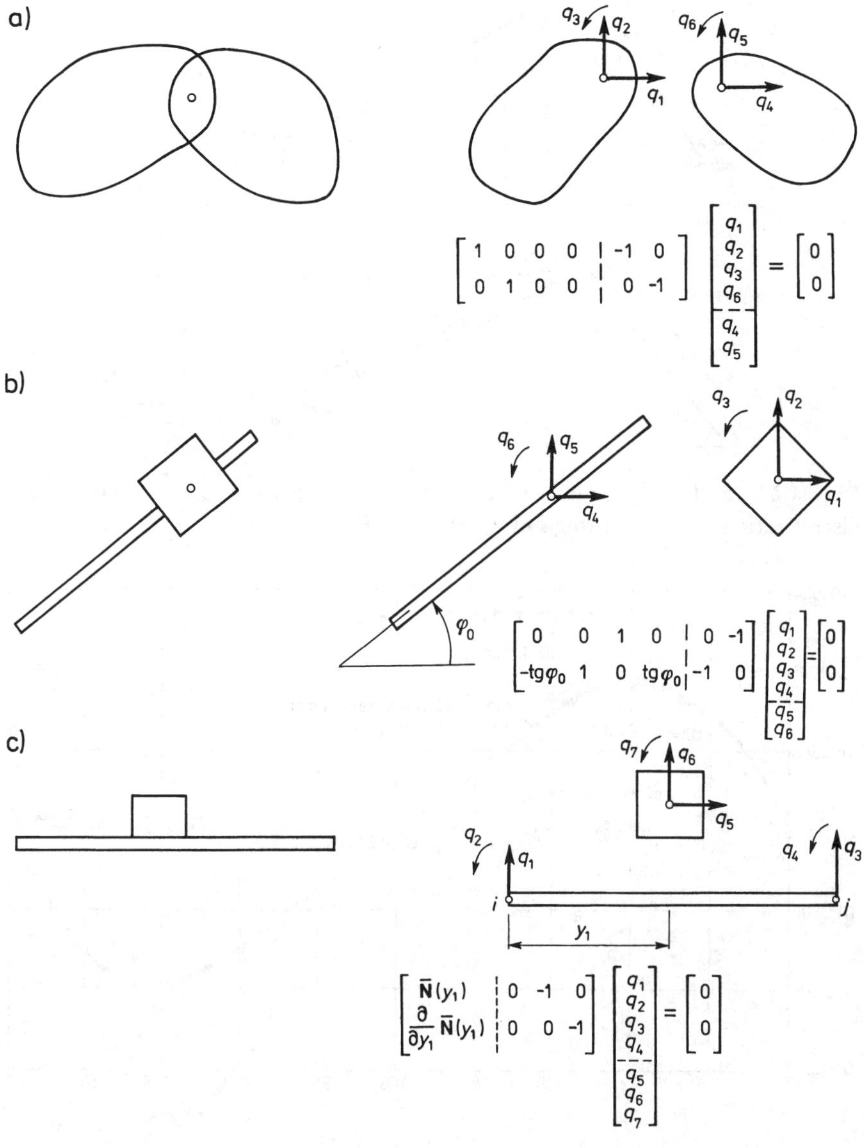

Fig. 4.22. Example constraints and corresponding constraint equations: **a)** rotary kinematic pair, **b)** sliding kinematic pair composed of 2 RFEs, **c)** sliding kinematic pair composed of RFE moving along beam-like finite element

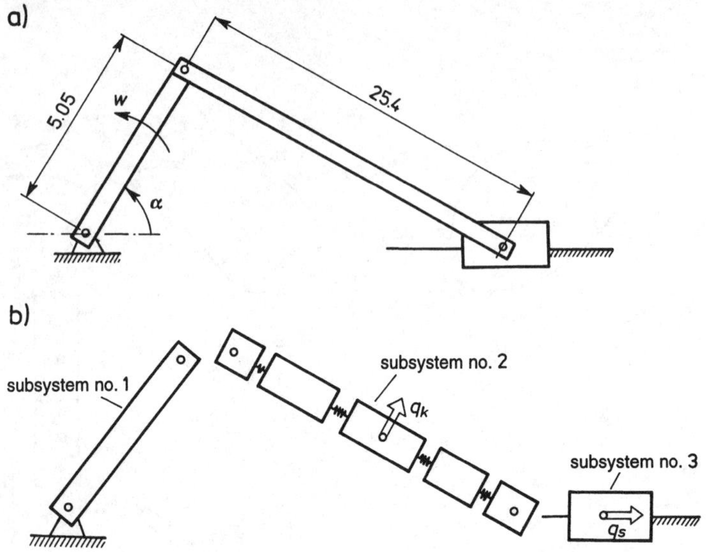

Fig. 4.23. Slider crank mechanism: **a)** scheme, **b)** division into subsystems and discretization of a connecting link by the 5 RFEs

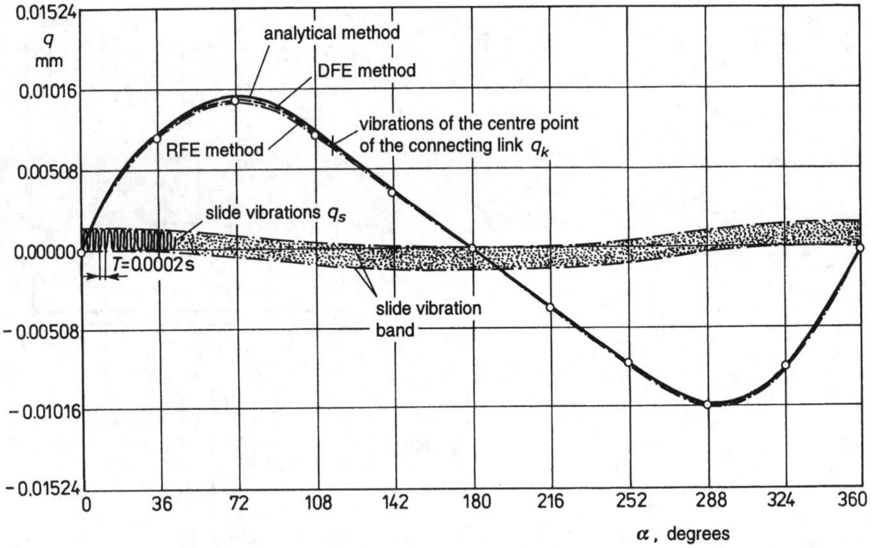

Fig. 4.24. Relative displacements of the centre point of the connecting link q_k obtained using: analytical method – full line, finite elements (3 DFEs) – broken line, rigid finite elements (5RFEs) – dotted and broken line, and relative displacements of the slide q_s obtained using the RFEM. Data: $r = 50.5$ mm, $l = 254$ mm, $b \times h = 6.35 \times 25.4$ mm, $\rho = 7.9 \times 10^{-5}$ kg cm^{-3}, $E = 2.1 \times 10^6$ N cm^{-2}, $G = 0.85 \times 10^6$ N cm^{-2}, $\omega = 10$ rad s^{-1}

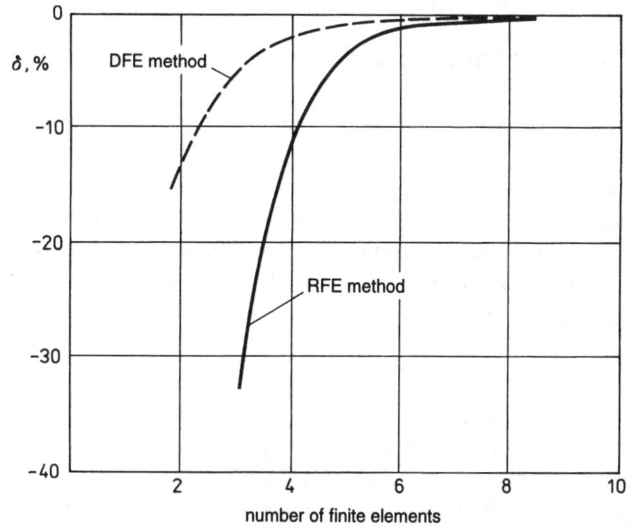

Fig. 4.25. Error $\delta(\%)$ of relative displacement q_k (Fig. 4.24) of the centre point of the connecting link for RFE and DFE methods in comparison to analytical solution with respect to density of discretization. Data: as in Fig. 4.24

where

$$\mathbf{A}^* = \mathbf{A}_\mathrm{i} + \mathbf{A}_\mathrm{id}\,\mathbf{W} + \mathbf{W}^T\,\mathbf{A}_\mathrm{di} + \mathbf{W}^T\,\mathbf{A}_\mathrm{d}\,\mathbf{W}\,, \tag{4.139}$$

$$\begin{aligned}\mathbf{B}^* &= \mathbf{B}_\mathrm{i} + \mathbf{B}_\mathrm{id}\,\mathbf{W} + \mathbf{W}^T\,\mathbf{B}_\mathrm{di} + \mathbf{W}^T\,\mathbf{B}_\mathrm{d}\,\mathbf{W}\\ &\quad + 2\mathbf{A}_\mathrm{id}\,\dot{\mathbf{W}} + 2\mathbf{W}^T\,\mathbf{A}_\mathrm{d}\,\dot{\mathbf{W}}\,,\end{aligned} \tag{4.140}$$

$$\begin{aligned}\mathbf{C}^* &= \mathbf{C}_\mathrm{i} + \mathbf{C}_\mathrm{id}\,\mathbf{W} + \mathbf{W}^T\,\mathbf{C}_\mathrm{di} + \mathbf{W}^T\,\mathbf{C}_\mathrm{d}\,\mathbf{W}\\ &\quad + \mathbf{B}_\mathrm{id}\,\dot{\mathbf{W}} + \mathbf{W}^T\,\mathbf{B}_\mathrm{d}\,\dot{\mathbf{W}} + \mathbf{A}_\mathrm{id}\,\ddot{\mathbf{W}} + \mathbf{W}^T\,\mathbf{A}_\mathrm{d}\,\ddot{\mathbf{W}}\,,\end{aligned} \tag{4.141}$$

$$\begin{aligned}\mathbf{f}^* &= \mathbf{f}_\mathrm{i} - (\mathbf{A}_\mathrm{ig}\,\ddot{\mathbf{q}}_\mathrm{g} + \mathbf{B}_\mathrm{ig}\,\dot{\mathbf{q}}_\mathrm{g} + \mathbf{C}_\mathrm{ig}\,\mathbf{q}_\mathrm{g})\\ &\quad + \mathbf{W}^T\,[\mathbf{f}_\mathrm{d} - (\mathbf{A}_\mathrm{dg}\,\ddot{\mathbf{q}}_\mathrm{g} + \mathbf{B}_\mathrm{dg}\,\dot{\mathbf{q}}_\mathrm{g} + \mathbf{C}_\mathrm{dg}\,\mathbf{q}_\mathrm{g})]\,.\end{aligned} \tag{4.142}$$

A slider crank mechanism is an example of a system with changing configuration (Fig. 4.23). The system consists of 3 subsystems: crank, connecting link and slider and it is assumed that the connecting link is flexible.

Figure 4.24 shows results of calculations of mechanism vibration with zero initial conditions and constant rotary speed of the crank. The results have been obtained using the finite element method (the connecting link has been discretized with 3 beam-like elements), the rigid finite element method (the connecting link has been discretized by 5 RFEs), and also the analytical method (method of harmonic series [7]). Error dependency of both FEM and RFEM compared with the analytical solution is presented in Fig. 4.25.

4.8 Investigation of Dynamic Stability of a Linear System[3]

In order to define stability, first we will define the concept of initial configuration of a system. Under this concept we understand the state of a system when there are not any external loads acting on the system and free motion of the system has been damped. Initial configuration of a rod loaded with axial force is the undeformed form of the rod. Dynamic stability of structures includes cases when the system considered is loaded with forces changeable in time. In this point we will consider dynamic stability only when loads are periodical.

There are different criteria assumed when investigating stability which help to classify states of the systems considered. Usually, analysis of solution of equations of motion is carried out using the method of quality research. Under the concept of quality analysis we understand investigations of general characteristics of functions such as: extremum, inflexion points, limits, asymptotes. Further we use stability criteria based on the Lyapunov concept [24].

Stability of a system depends both on its physical and geometrical parameters as well as on external loads. Change in external load can cause loss of stability. Thus, for an axially compressed rod, the loss of stability is caused by a force equal to the critical force. If the external load changes periodically, then for certain frequencies of input the system can be unstable, which means that vibrations with increasing amplitude occur.

Dynamic stability refers to selected forms of system vibrations. The same external load can for some forms of vibrations cause stable vibrations and for other forms non-stable vibrations. We will explain this problem using the example of a beam (Fig. 4.26a). It can be shown that changeable axial force $P(t)$ excites non-stable bending vibrations. For a certain frequency of the continuous load $p(t)$ acting on a ring (Fig. 4.26b) there can be loss of stability. Then bending vibrations with increasing amplitude occur. Similarly, for certain frequencies of forces loading the frame (Fig. 4.26c) there is loss of dynamic stability. Then bending vibrations of posts and spandrel beam with increasing amplitude occur. Force loading an arch (Fig. 4.26d) can cause loss of dynamic stability, and it can be shown that only non-symmetrical forms of bending vibrations are non-stable. Non-stable vibrations on the surface of a plate can also occur when a plate is loaded with forces acting on its middle layer. In the case of loss of dynamic stability of a system increase of vibration amplitude occur usually at frequencies of loading different from frequencies of free vibrations of the system, and critical frequencies of vibrations also depend on loading of the system.

Critical loading of structures loaded dynamically differs so considerably

[3] Author: Wiesław Ostachowicz

from static critical loading that there is no sense in using methods of investigations of static stability for investigations of dynamic stability. For some frequencies of loading, external loads with amplitude close to zero can cause loss of dynamic stability even though the value of static critical load is much larger. Or, on the contrary, momentary loading of the system can be considerably larger than the value of static critical loading and the system is dynamically stable.

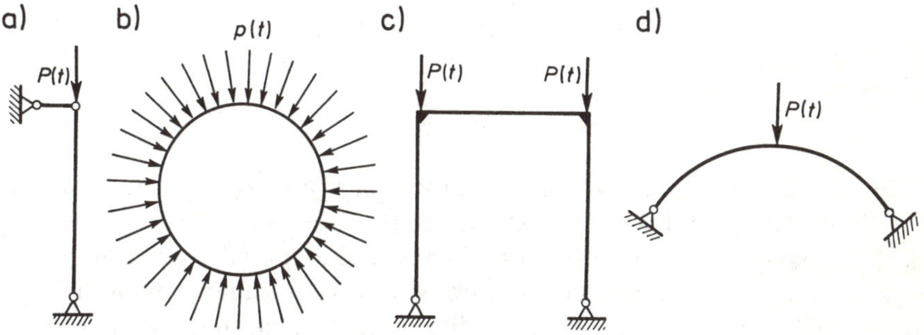

Fig. 4.26. Connecting rods systems

The problem of dynamic stability is described by N linear differential equations of the second order

$$\mathbf{M}\,\ddot{\mathbf{q}} + \mathbf{C}\,\dot{\mathbf{q}} + [\mathbf{K} - \mathbf{K}_G(t)]\,\mathbf{q} = \mathbf{0} \tag{4.143}$$

in which $\mathbf{M}$ is the mass matrix, $\mathbf{C}$ is the damping matrix, $\mathbf{K}$ is the stiffness matrix, $\mathbf{K}_G(t)$ is the geometrical stiffness matrix$\mathbf{q}$ is the vector of generalized coordinates, N is a number of degrees of freedom of the system considered.

In Eq. (4.143) loading occurs as a parameter on the left side. If the loading changes in time vibrations caused by it are called parametrical.

We can meet parametrical vibrations in many machines and technical devices. One of the examples is main engine shaft which is loaded with an axial force created by a propeller. The force is a sum of constant part and periodically changeable part. Another example is a shaft of non-symmetrical cross-section. Parametrical vibrations are also created in parts of the engine and piston compressor such as crankshafts and rigging screws of the block.

If external loading can be written as a sum of constant and changeable parts, the matrix of geometrical stiffness is written in the following form

$$\mathbf{K}_G(t) = \mathbf{K}_G^S + \mathbf{K}_G^Z(t)\,, \tag{4.144}$$

where $\mathbf{K}_G^S$ is the matrix of geometrical stiffness which contains constant parts of external loading, $\mathbf{K}_G^Z(t)$ is the matrix of geometrical stiffness containing the changeable part of the external loading. The method calculating of those matrices is presented in [96].

If external loading changes periodically, then matrix $\mathbf{K}_G^Z(t)$ is presented in the form

$$\mathbf{K}_G^Z(t) = f(t)\mathbf{\Pi} \tag{4.145}$$

in which $\mathbf{\Pi}$ is the matrix of parametrical loading and $f(t)$ are periodical functions.

Having substituted (4.144) into (1.143) and taken into account (4.145) we obtain the following:

$$\mathbf{M}\,\ddot{\mathbf{q}} + \mathbf{C}\,\dot{\mathbf{q}} + [\hat{\mathbf{K}} - f(t)\mathbf{\Pi}]\,\mathbf{q} = \mathbf{0} \tag{4.146}$$

where

$$\hat{\mathbf{K}} = \mathbf{K} - \mathbf{K}_G^S\,.$$

In the case of loss of dynamic stability which appears as an increase of amplitude of non-stable vibrations, we speak about occurrence of parametrical resonance. Initially, there is an increase of amplitude of vibrations by exponential function and then periodical vibrations with large amplitude are established. Analysis of dynamic stability of the system, the motion of which is described by Eqs. (4.146), aims to define the boundaries of the parametrical resonance area in the domain of loading and frequency of input.

The theory of differential equations with changeable coefficients holds that solutions of Eqs. (4.146) increase without limit in time around frequencies defined by relationships

$$\omega \approx \frac{\omega_i}{k} \tag{4.147}$$

and

$$\omega \approx \left|\frac{\omega_i \pm \omega_j}{2k}\right|, \qquad i \neq j \tag{4.148}$$

where i, j, k are natural numbers and ω_i, ω_j are circular frequencies of free vibrations of the system considered.

If the system vibrates near frequency ω defined by (4.147) then we speak about periodical resonance. Vibrations near frequency ω defined by (4.148) cause combined resonance.

Methods of analysis of Eqs. (4.146) are divided into two basic groups. Methods of numerical integration of differential equations, which are called direct methods (see 4.5), belong to the first group. The most widespread methods of the second group are the small parameter methods and especially the perturbation method, the averaging method, the method based on the Krylow–Bogolyubow–Mitropolski method as well as the method of harmonic balance. This latter enables a full analysis of combined resonance to be carried out.

For periodical resonance there are two independent solutions not equal to zero of Eqs. (4.146). We assume the solution of those equations in the form of the Fourier series

$$\mathbf{q}(t) = \sum_{k=1,3,5,\ldots}^{\infty} (\mathbf{a}_k \sin k\omega t + \mathbf{b}_k \cos k\omega t) \tag{4.149}$$

and

$$\mathbf{q}(t) = \sum_{k=0,2,4,\ldots}^{\infty} (\mathbf{a}_k \sin k\omega t + \mathbf{b}_k \cos k\omega t) \tag{4.150}$$

where $\mathbf{a}_k$, $\mathbf{b}_k$ are vectors of N components independent of time, 2ω is frequency of input. Having substituted (4.149) and (4.150) into (4.146) and equated coefficients at the same trigonometric functions with zero we obtain an infinite system of algebraic equations. Below we will consider the procedure when function $f(t)$ from (4.146) is a harmonic function of the form

$$f(t) = \cos 2\omega t\,. \tag{4.151}$$

Generally, function $f(t)$ is presented in the form of the Fourier series.

Having assumed input in the form of a harmonic function (4.151), Eqs. (4.146) take the form

$$\mathbf{M}\,\ddot{\mathbf{q}} + \mathbf{C}\,\dot{\mathbf{q}} + (\hat{\mathbf{K}} - \mathbf{\Pi}\cos 2\omega t)\,\mathbf{q} = \mathbf{0}\,. \tag{4.152}$$

Having substituted (4.149) into (4.152) and equated coefficients at the same trigonometric functions with zero we obtain a homogeneous linear equation with respect to $\mathbf{a}_k$ and $\mathbf{b}_k$

$$\begin{bmatrix} & \vdots & \vdots & \vdots & \vdots & \\ \cdots & (\hat{\mathbf{K}}-9\omega^2\mathbf{M}) & -\frac{1}{2}\mathbf{\Pi} & \mathbf{0} & -6\omega\mathbf{C} & \cdots \\ \cdots & -\frac{1}{2}\mathbf{\Pi} & (\hat{\mathbf{K}}+\frac{1}{2}\mathbf{\Pi}-\omega^2\mathbf{M}) & -2\omega\mathbf{C} & \mathbf{0} & \cdots \\ \cdots & \mathbf{0} & 2\omega\mathbf{C} & (\hat{\mathbf{K}}-\frac{1}{2}\mathbf{\Pi}-\omega^2\mathbf{M}) & -\frac{1}{2}\mathbf{\Pi} & \cdots \\ \cdots & 6\omega\mathbf{C} & \mathbf{0} & -\frac{1}{2}\mathbf{\Pi} & (\hat{\mathbf{K}}-9\omega^2\mathbf{M}) & \cdots \\ & \vdots & \vdots & \vdots & \vdots & \end{bmatrix} \begin{bmatrix} \vdots \\ \mathbf{a}_3 \\ \mathbf{a}_1 \\ \mathbf{b}_1 \\ \mathbf{b}_3 \\ \vdots \end{bmatrix} = \begin{bmatrix} \vdots \\ \mathbf{0} \\ \mathbf{0} \\ \mathbf{0} \\ \mathbf{0} \\ \vdots \end{bmatrix}. \tag{4.153}$$

Having substituted (4.150) into (4.152) and proceeding as before we obtain

$$\begin{bmatrix} & \vdots & \vdots & \vdots & \vdots & \vdots & \\ \cdots & (\hat{\mathbf{K}}-16\omega^2\mathbf{M}) & -\frac{1}{2}\mathbf{\Pi} & \mathbf{0} & \mathbf{0} & -16\omega\mathbf{C} & \cdots \\ \cdots & -\frac{1}{2}\mathbf{\Pi} & (\hat{\mathbf{K}}-4\omega^2\mathbf{M}) & \mathbf{0} & -4\omega\mathbf{C} & \mathbf{0} & \cdots \\ \cdots & \mathbf{0} & \mathbf{0} & \hat{\mathbf{K}} & -\mathbf{\Pi} & \mathbf{0} & \cdots \\ \cdots & \mathbf{0} & 4\omega\mathbf{C} & -\frac{1}{2}\mathbf{\Pi} & (\hat{\mathbf{K}}-4\omega^2\mathbf{M}) & -\frac{1}{2}\mathbf{\Pi} & \cdots \\ \cdots & -16\omega\mathbf{C} & \mathbf{0} & \mathbf{0} & -\frac{1}{2}\mathbf{\Pi} & (\hat{\mathbf{K}}-16\omega^2\mathbf{M}) & \cdots \\ & \vdots & \vdots & \vdots & \vdots & \vdots & \end{bmatrix} \begin{bmatrix} \vdots \\ \mathbf{a}_4 \\ \mathbf{a}_2 \\ \mathbf{b}_0 \\ \mathbf{b}_2 \\ \mathbf{b}_4 \\ \vdots \end{bmatrix} = \begin{bmatrix} \vdots \\ \mathbf{0} \\ \mathbf{0} \\ \mathbf{0} \\ \mathbf{0} \\ \mathbf{0} \\ \vdots \end{bmatrix} . \qquad (4.154)$$

Matrices and column matrices in Eqs. (4.153) and (4.154) have infinite dimensions. Non-trivial solutions of those systems are obtained if we equate determinants of infinite matrices from (4.153) and (4.154) with zero.

We calculate frequencies ω from those determinants. In practice, roots of those determinants are defined on the basis of determinants with a finite number of rows and columns. Blocks of those determinants have dimensions $N \times N$. Determinants with finite dimensions have kN rows and columns, where k is a natural number. It can be shown that with increase of k we obtain greater exactness of calculations of ω, and moreover while increasing k we calculate periodical resonance of the superior harmonics.

Calculations can be simplified by omitting damping of a system (when damping in the system is not large). In a system without damping periodical resonance occurs at a wide range of loading; critical loading then equals zero.

Moreover, in the above systems, we obtain a larger interval of critical frequencies of periodical resonance, and thus in a real system the range of resonance frequencies is smaller than calculated values. In damped systems the loss of dynamic stability results, starting from a certain non-zero value of loading. In such a situation real critical loading is larger than the calculated one. Matrix determinants from (4.153) and (4.154) equations after omitting damping take the form

$$\det \begin{bmatrix} (\hat{\mathbf{K}}+\frac{1}{2}\mathbf{\Pi}-\omega^2\mathbf{M}) & -\frac{1}{2}\mathbf{\Pi} & \mathbf{0} & \cdots \\ -\frac{1}{2}\mathbf{\Pi} & (\hat{\mathbf{K}}-9\omega^2\mathbf{M}) & -\frac{1}{2}\mathbf{\Pi} & \cdots \\ \mathbf{0} & -\frac{1}{2}\mathbf{\Pi} & (\hat{\mathbf{K}}-25\omega^2\mathbf{M}) & \cdots \\ \vdots & \vdots & \vdots & \end{bmatrix} = 0 , \qquad (4.155)$$

$$\det\begin{bmatrix} \hat{\mathbf{K}} & -\mathbf{\Pi} & \mathbf{0} & \cdots \\ -\frac{1}{2}\mathbf{\Pi} & (\hat{\mathbf{K}}-4\omega^2\mathbf{M}) & -\frac{1}{2}\mathbf{\Pi} & \cdots \\ \mathbf{0} & -\frac{1}{2}\mathbf{\Pi} & (\hat{\mathbf{K}}-16\omega^2\mathbf{M}) & \cdots \\ \vdots & \vdots & \vdots & \end{bmatrix} = 0\,. \quad (4.156)$$

Calculation of resonance frequencies from (4.155) leads to solution of the eigenvalue problem

$$\det(\tilde{\mathbf{K}} - \omega^2\tilde{\mathbf{M}}) = 0 \quad (4.157)$$

where

$$\tilde{\mathbf{K}} = \begin{bmatrix} (\hat{\mathbf{K}}+\frac{1}{2}\mathbf{\Pi}) & -\frac{1}{2}\mathbf{\Pi} & \mathbf{0} & \cdots \\ -\frac{1}{2}\mathbf{\Pi} & \hat{\mathbf{K}} & -\frac{1}{2}\mathbf{\Pi} & \cdots \\ \mathbf{0} & -\frac{1}{2}\mathbf{\Pi} & \hat{\mathbf{K}} & \cdots \\ \vdots & \vdots & \vdots & \end{bmatrix}, \qquad \tilde{\mathbf{M}} = \begin{bmatrix} \mathbf{M} & \mathbf{0} & \mathbf{0} & \cdots \\ \mathbf{0} & 9\mathbf{M} & \mathbf{0} & \cdots \\ \mathbf{0} & \mathbf{0} & 25\mathbf{M} & \cdots \\ \vdots & \vdots & \vdots & \end{bmatrix}.$$

In the case of Eq. (4.156) the problem is more complicated. In this case the iterative method can be used in order to solve the general eigenvalue problem. The work mentioned presents some other methods of calculations of resonance frequencies for periodical resonance.

The method of harmonic balance enables a full analysis of combined resonance (Table 4.1). For resonance of the first order, i.e. for $k = 1$, the solution of the system of equations is assumed to take the form

$$\mathbf{q}(t) = \mathbf{a}_i \sin\theta_i t + \mathbf{b}_i \cos\theta_i t + \mathbf{a}_j \sin\theta_j t + \mathbf{b}_j \cos\theta_j t \quad (4.158)$$

in which

$$\theta_i \approx \omega_i\,, \qquad \theta_j \approx \omega_j\,, \quad (4.159)$$

$\mathbf{a}_i, \mathbf{b}_i, \mathbf{a}_j, \mathbf{b}_j$ are column matrices of dimension N. Elements of those matrices are constant. For resonance of the first order, relationship (4.148) takes the form

$$2\omega \approx |\omega_i \pm \omega_j|\,, \quad (4.160)$$

thus, having taken into account (4.159), critical frequencies are defined by

$$2\omega = |\theta_i \pm \theta_j|\,. \quad (4.161)$$

Having substituted (4.158) into (4.146) and equated coefficients at the same trigonometric functions with zero for harmonic input of frequency (4.161), we obtain a system of algebraic equations in the form

$$\begin{bmatrix} (\hat{\mathbf{K}}-\theta_i^2\mathbf{M}) & \frac{1}{2}\mathbf{\Pi} & -\theta_i\mathbf{C} & \mathbf{0} \\ \frac{1}{2}\mathbf{\Pi} & (\hat{\mathbf{K}}-\theta_j^2\mathbf{M}) & \mathbf{0} & -\theta_j\mathbf{C} \\ \theta_i\mathbf{C} & \mathbf{0} & (\hat{\mathbf{K}}-\theta_i^2\mathbf{M}) & -\frac{1}{2}\mathbf{\Pi} \\ \mathbf{0} & \theta_j\mathbf{C} & -\frac{1}{2}\mathbf{\Pi} & (\hat{\mathbf{K}}-\theta_j^2\mathbf{M}) \end{bmatrix} \begin{bmatrix} \mathbf{a}_i \\ \mathbf{a}_j \\ \mathbf{b}_i \\ \mathbf{b}_j \end{bmatrix} = \begin{bmatrix} \mathbf{0} \\ \mathbf{0} \\ \mathbf{0} \\ \mathbf{0} \end{bmatrix}. \quad (4.162)$$

Relationship (4.162) is equivalent to a system of $4N$ linear homogenous algebraic equations with respect to vectors $\mathbf{a}_i$, $\mathbf{b}_i$, $\mathbf{a}_j$, $\mathbf{b}_j$. Non-trivial solution of

Eqs. (4.162) exists when the characteristic determinant of the matrix equals zero

$$\det\begin{bmatrix} (\hat{\mathbf{K}}-\theta_i^2\mathbf{M}) & \frac{1}{2}\mathbf{\Pi} & -\theta_i\mathbf{C} & \mathbf{0} \\ \frac{1}{2}\mathbf{\Pi} & (\hat{\mathbf{K}}-\theta_j^2\mathbf{M}) & \mathbf{0} & -\theta_j\mathbf{C} \\ \theta_i\mathbf{C} & \mathbf{0} & (\hat{\mathbf{K}}-\theta_i^2\mathbf{M}) & -\frac{1}{2}\mathbf{\Pi} \\ \mathbf{0} & \theta_j\mathbf{C} & -\frac{1}{2}\mathbf{\Pi} & (\hat{\mathbf{K}}-\theta_j^2\mathbf{M}) \end{bmatrix} = 0. \qquad (4.163)$$

Equation (4.163) includes two optional quantities θ_i and θ_j. Another condition is necessary which, together with the condition of setting the determinant to zero, will enable us to calculate θ_i and θ_j at the same time. This condition is that all sub-determinants of $4N-1$ order, which have been obtained in expansion of the principal determinant (4.163) with respect to elements different to zero, should be equated to zero. In some problems it is enough that one of those sub-determinants equals zero because the other sub-determinants are then also zero. Yet, in general, calculation of frequencies θ_i and θ_j for systems with a large number of degrees of freedom requires considerable calculation time. Iterative procedures of calculations of polynomial roots used then give approximate results.

When damping of the system is small it can be omitted and relationship (4.162) is written in the form of two independent sets of equations

$$\begin{bmatrix} (\hat{\mathbf{K}}-\theta_i^2\mathbf{M}) & \frac{1}{2}\mathbf{\Pi} \\ \frac{1}{2}\mathbf{\Pi} & (\hat{\mathbf{K}}-\theta_j^2\mathbf{M}) \end{bmatrix} \begin{bmatrix} \mathbf{a}_i \\ \mathbf{a}_j \end{bmatrix} = \begin{bmatrix} \mathbf{0} \\ \mathbf{0} \end{bmatrix}, \qquad (4.164)$$

$$\begin{bmatrix} (\hat{\mathbf{K}}-\theta_i^2\mathbf{M}) & -\frac{1}{2}\mathbf{\Pi} \\ -\frac{1}{2}\mathbf{\Pi} & (\hat{\mathbf{K}}-\theta_j^2\mathbf{M}) \end{bmatrix} \begin{bmatrix} \mathbf{b}_i \\ \mathbf{b}_j \end{bmatrix} = \begin{bmatrix} \mathbf{0} \\ \mathbf{0} \end{bmatrix}. \qquad (4.165)$$

Equations (4.164) and (4.165) include two unknowns θ_i and θ_j. It is necessary to state additional conditions which enable us to define θ_i and θ_j or the sum or subtraction of those quantities. For a supported prismatic beam (Fig. 4.27) boundaries of areas of periodical and combined resonance have been calculated (Fig. 4.28). To this end the beam has been divided into 6 equal finite elements and vibrations of the beam in one plane have been considered. Damping has been omitted. Critical frequencies of periodical resonance have been calculated from Eq. (4.155), having assumed the deter-

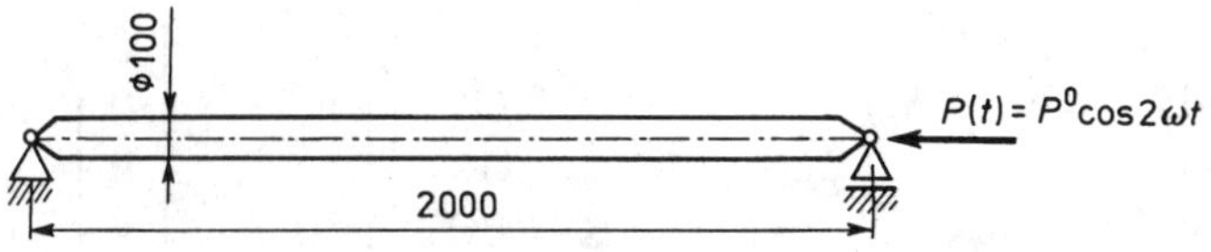

Fig. 4.27. A beam excited with an axial force changeable in time. Data: $\rho = 0.783 \times 10^4$ kg m^{-3}, $E = 2.1 \times 10^5$ MPa, $G = 0.85 \times 10^5$ MPa

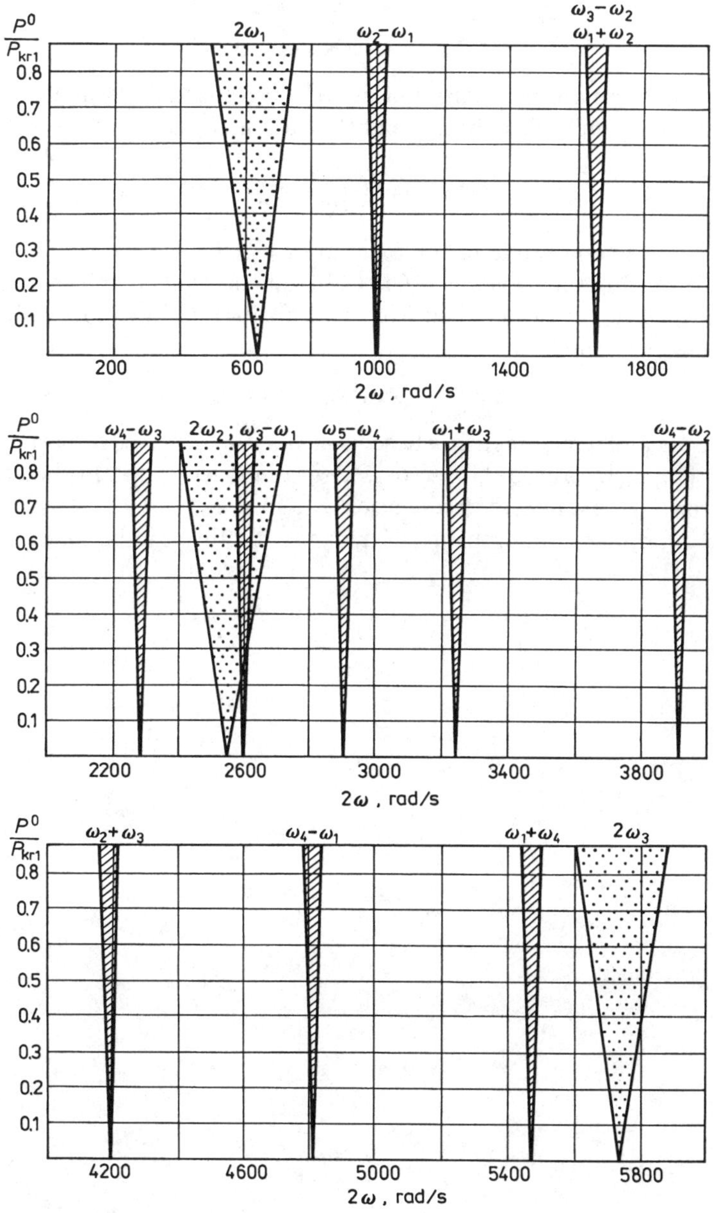

Fig. 4.28. Areas of periodical resonance (dotted) and combined resonance (oblique lines) of the beam (Fig. 4.27) divided into 6 finite elements. The system is treated as planar

minant for a planar system in the form

$$\begin{bmatrix} (\hat{\mathbf{K}} + \frac{1}{2}\boldsymbol{\Pi} - \omega^2\mathbf{M}) & -\frac{1}{2}\boldsymbol{\Pi} \\ -\frac{1}{2}\boldsymbol{\Pi} & (\hat{\mathbf{K}} - 9\omega^2\mathbf{M}) \end{bmatrix} = 0\,. \tag{4.166}$$

In Fig. 4.28 there are three areas of periodical resonance. Graphs are made as a function of input frequencies and the parameter

$$\alpha_p = \frac{P^0}{P_{\mathrm{kr1}}} \tag{4.167}$$

where P^0 is the amplitude of the axial force and P_{kr1} is the lowest value of critical force (for static loading). Critical frequencies of combined resonance (Fig. 4.28) have been calculated from Eqs. (4.164) and (4.165), which have been brought to the form

$$[\boldsymbol{\Lambda}_k(\beta) - (2\omega)^2\mathbf{E}]\,\mathbf{w}(\beta) = \mathbf{0}\,, \tag{4.168}$$

where $\boldsymbol{\Lambda}_k$ is a diagonal matrix whose elements are functions of parameter β, $\mathbf{w}(\beta) = \{\mathbf{b}_i, \mathbf{a}_i, \mathbf{b}_j, \mathbf{a}_j\}$ is the vector whose components are also functions of parameter β, β is a ratio of frequencies ω_j and ω_i .

More information on problems of dynamic stability can be found in [96].

5 Space–Time Element Method

5.1 Introductory Remarks and General Relations

The Space–Time Element Method (STEM) is a numerical procedure allowing us to solve the problems of dynamic analysis of structures and continuous media subjected to transient loadings or kinematic actuations. It is also suitable for the calculation of non-stationary thermal fields occuring in arbitrarily shaped regions. In STEM we are going to treat time t as the fourth dimension complementary to the Cartesian co-ordinates x_i. We assume further that all four co-ordinate axes in the space–time are mutually orthogonal.

Instead of three-dimensional physical bodies we are going to consider in the sequel certain four-dimensional space–time objects. Taking a cross-section of such an object with the hyperplane $t = t^s$ one obtains the configuration of the physical body at the instant t^s. We assume each 4D-object to be made of anisotropic material inextensible in the direction of the t-axis. This allows us to remain within the frame of Newtonian mechanics where displacements parallel to the time-axis can not occur. In Sect. 5.6 devoted to the problems of thermal flux this assumption will be abandoned.

Let the vector

$$\mathbf{u}_{4\times 1}(\mathbf{x}, t) = \{u_1\; u_2\; u_3\; 0\}. \tag{5.1}$$

describe the displacements relative to the known initial configuration inside the region Ω occupied by the 4D-object. Note that the last component of that vector is always equal to zero. For the sake of brevity we shall use in the sequel the definition (2.27) from Part I of the present book instead of (5.1).

The single column matrix of strains defined by Eq. (2.28) in Part I must be expanded now since strain rates in the space–time appear as the shear strains of 4D-object:

$$\begin{aligned} \boldsymbol{\varepsilon}_{9\times 1} &= \{\varepsilon_{11}\; \varepsilon_{22}\; \varepsilon_{33}\; \gamma_{23}\; \gamma_{31}\; \gamma_{12}\; \gamma_{t1}\; \gamma_{t2}\; \gamma_{t3}\} \\ &= \mathbf{D}_{9\times 3}\, \mathbf{u}_{3\times 1}. \end{aligned} \tag{5.2}$$

$$\mathbf{D}_{9\times 3} = \begin{bmatrix} \partial/\partial x_1 & 0 & 0 \\ 0 & \partial/\partial x_2 & 0 \\ 0 & 0 & \partial/\partial x_3 \\ 0 & \partial/\partial x_3 & \partial/\partial x_2 \\ \partial/\partial x_3 & 0 & \partial/\partial x_1 \\ \partial/\partial x_2 & \partial/\partial x_1 & 0 \\ \partial/\partial t & 0 & 0 \\ 0 & \partial/\partial t & 0 \\ 0 & 0 & \partial/\partial t \end{bmatrix} . \tag{5.3}$$

Similarly, the single column matrix of stresses defined by Eq. (2.26) in Part I should be expanded in the space–time by 3 additional components. These new "stresses" are the internal impulses[1] taken with negative sign at the cross-sections perpendicular to the time axis

$$\boldsymbol{\sigma}_{9\times 1} = \{\sigma_{11}\ \sigma_{22}\ \sigma_{33}\ \tau_{23}\ \tau_{31}\ \tau_{12}\ \tau_{t1}\ \tau_{t2}\ \tau_{t3}\} = \mathbf{D}_{9\times 3}\,\mathbf{u}_{3\times 1}. \tag{5.4}$$

In the expanded elasticity matrix

$$\mathbf{C}_{9\times 9} = \begin{bmatrix} \mathbf{C}'_{3\times 3} & \mathbf{0}_{3\times 3} & \mathbf{0}_{3\times 3} \\ & \mathbf{C}''_{3\times 3} & \mathbf{0}_{3\times 3} \\ \text{sym.} & & -\varrho\,\mathbf{I}_{3\times 3} \end{bmatrix}, \tag{5.5}$$

there would be

$$\mathbf{C}'_{3\times 3} = \frac{E}{(1+\nu)(1-2\nu)} \begin{bmatrix} 1-\nu & \nu & \nu \\ \nu & 1-\nu & \nu \\ \nu & \nu & 1-\nu \end{bmatrix}, \quad \mathbf{C}''_{3\times 3} = G\,\mathbf{I}_{3\times 3}. \tag{5.6}$$

under the assumption of linearly elastic material defined by Eq. (2.30) in Part I. Since we intend to consider also the dynamics of viscoelastic bodies governed by the Kelvin–Voigt model (compare, e.g., [74, 124, 122, 137]), matrices (5.6) must be modified. Our aim is to introduce the deviators of stress and strain states into the constitutive equations. Taking

$$\sigma = \frac{1}{3}\sigma_{kk}, \quad \varepsilon = \frac{1}{3}\varepsilon_{kk} \tag{5.7}$$

together with the definition (2.25) from Part I, we can express such deviators as

$$s_{kl} = \sigma_{kl} - \sigma\,\delta_{kl}, \quad e_{kl} = \varepsilon_{kl} - \varepsilon\,\delta_{kl}, \tag{5.8}$$

where δ_{kl} is the Kronecker symbol.

---

[1] We call the product of the mass density and the velocity of motion an *internal impulse* in order to distinguish it from the product of the force and the duration of its action. The latter is called an *external impulse* or simply an *impulse*.

It is usual to assume in the theory of viscoelasticity that the mean normal stress σ depends linearly upon the dilatation 3ε:

$$\sigma = 3K\varepsilon. \tag{5.9}$$

The constant

$$K = \frac{1}{3}\frac{E}{1-2\nu} \tag{5.10}$$

appearing in this equation is the compressibility modulus of the material. On the other hand, the relation

$$s_{kl} = 2\left(G + \eta\frac{\partial}{\partial t}\right) e_{kl}. \tag{5.11}$$

between the deviators of stress and strain follows from the Kelvin–Voigt model. Here G and η are the shear modulus and the Newtonian viscosity coefficient, respectively.

Using the notation from Eqs. (2.26), (2.28), Part I, we convert relations (5.7), (5.8) into the following matrix form:

$$\boldsymbol{\sigma}_{6\times 1} = \mathbf{A}_{6\times 7}\,\mathbf{s}_{7\times 1}, \quad \mathbf{e}_{7\times 1} = \mathbf{B}_{7\times 6}\,\boldsymbol{\varepsilon}_{6\times 1}. \tag{5.12}$$

Here

$$\begin{aligned} \mathbf{s}_{7\times 1} &= \{s\; s_{11}\; s_{22}\; s_{33}\; s_{23}\; s_{31}\; s_{12}\}, \\ \mathbf{e}_{7\times 1} &= \{e\; e_{11}\; e_{22}\; e_{33}\; e_{23}\; e_{31}\; e_{12}\}, \end{aligned} \tag{5.13}$$

$$\mathbf{A}_{6\times 7} = \begin{bmatrix} 1 & & \\ 1 & \mathbf{I}_{3\times 3} & \mathbf{0}_{3\times 3} \\ 1 & & \\ \mathbf{0}_{3\times 4} & & \mathbf{I}_{3\times 3} \end{bmatrix}, \quad \mathbf{B}_{7\times 6} = \begin{bmatrix} \frac{1}{3} & \frac{1}{3} & \frac{1}{3} & \\ \frac{2}{3} & -\frac{1}{3} & -\frac{1}{3} & \\ -\frac{1}{3} & \frac{2}{3} & -\frac{1}{3} & \mathbf{0}_{4\times 3} \\ -\frac{1}{3} & -\frac{1}{3} & \frac{2}{3} & \\ & \mathbf{0}_{3\times 3} & & \frac{1}{2}\mathbf{I}_{3\times 3} \end{bmatrix}. \tag{5.14}$$

Taking into account Eqs. (5.9), (5.11), we can write the constitutive relations as

$$\boldsymbol{\sigma}_{6\times 1} = \mathbf{A}_{6\times 7}\begin{bmatrix} 3K & \mathbf{0}_{1\times 6} \\ \mathbf{0}_{6\times 1} & 2\mathbf{I}_{6\times 6}(G + \eta\frac{\partial}{\partial t}) \end{bmatrix}\mathbf{B}_{7\times 6}\,\boldsymbol{\varepsilon}_{6\times 1} = \mathbf{C}_{6\times 6}\,\boldsymbol{\varepsilon}_{6\times 1}. \tag{5.15}$$

After relevant matrix multiplications were performed, the following matrices are obtained for viscoelastic body:

$$\mathbf{C}_{6\times 6} = \begin{bmatrix} \mathbf{C}'_{3\times 3} & \mathbf{0}_{3\times 3} \\ \mathbf{0}_{3\times 3} & \mathbf{C}''_{3\times 3} \end{bmatrix}, \tag{5.16}$$

$$\mathbf{C}'_{3\times3} = \begin{bmatrix} K+\frac{4}{3}(G+\eta\frac{\partial}{\partial t}) & K-\frac{2}{3}(G+\eta\frac{\partial}{\partial t}) & K-\frac{2}{3}(G+\eta\frac{\partial}{\partial t}) \\ & K+\frac{4}{3}(G+\eta\frac{\partial}{\partial t}) & K-\frac{2}{3}(G+\eta\frac{\partial}{\partial t}) \\ \text{sym.} & & K+\frac{4}{3}(G+\eta\frac{\partial}{\partial t}) \end{bmatrix}, \tag{5.17}$$

$$\mathbf{C}''_{3\times3} = \mathbf{I}_{3\times3}\left(G+\eta\frac{\partial}{\partial t}\right). \tag{5.18}$$

It is easy to check that for $\eta = 0$ matrix (5.16) coincides with that of Eq. (2.30) in Part I.

The equation of dynamic equilibrium (2.25), Part I, i.e. the equation of equilibrium of impulses, takes under the assumed notation the following form [compare Eq. (2.31), Part I]:

$$\mathbf{D}^{\mathrm{T}}_{3\times9}\,\boldsymbol{\sigma}_{9\times1} + \mathbf{U}_{3\times1} = \mathbf{0}_{3\times1}. \tag{5.19}$$

We replaced the term $\varrho\hat{\mathbf{f}}_{3\times1}$ used in Part I by the shorter symbol $\mathbf{U}_{3\times1}$.

The kinematic boundary conditions defined by Eq. (2.32) in Part I remain valid. On the other hand, the static boundary conditions given by Eq. (2.33) in Part I must be slightly modified. We have now

$$\mathbf{N}_{3\times9}\,\boldsymbol{\sigma}_{9\times1} = \hat{\mathbf{t}}_{3\times1}, \tag{5.20}$$

where

$$\mathbf{N}_{3\times9} = \begin{bmatrix} n_1 & 0 & 0 & 0 & n_3 & n_2 & n_t & 0 & 0 \\ 0 & n_2 & 0 & n_3 & 0 & n_1 & 0 & n_t & 0 \\ 0 & 0 & n_3 & n_2 & n_1 & 0 & 0 & 0 & n_t \end{bmatrix}. \tag{5.21}$$

Since the space–time is not a metric space, Eqs. (5.20), (5.21) should be considered only symbolically in general. It would be difficult to express the directional cosines as the quotients of the co-ordinates that have different dimensions. However, when we consider the hyperplanes bounding the 4D-object that are either parallel or perpendicular to the time axis, then 2 geometric relations remain valid. One of them is given by Eq. (2.33), Part I, whereas the other reads

$$n_t\,\{\tau_{t1}\ \tau_{t2}\ \tau_{t3}\}_{3\times1} = \hat{\mathbf{t}}_{3\times1}. \tag{5.22}$$

Here n_t takes the value 1 when the normal to the plane t = const points along the positive direction of the t-axis and it takes the value -1 when that normal points in the opposite direction. The components of the vector $\hat{\mathbf{t}}_{3\times1}$ coincide with the components of impulses: the initial one ($n_t = -1$) and the final one ($n_t = 1$) related to the corresponding cross-sections that define the position of the 4D-body in time.

Since it is difficult to imagine the 4D space–time and it is impossible to represent it graphically, let us illustrate the introduced notions on simple examples.

We begin with the case of a concentrated mass m pushed along the axis x_1 by the force $U_1(t)$. In the 3D physical space such a mass reduces to a

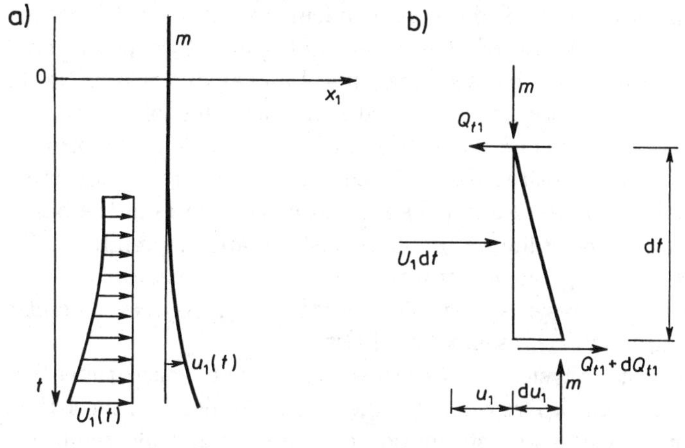

Fig. 5.1. a) Projection of line of life of concentrated mass on plane x_1, t; **b)** infinitesimal segment of line of life.

point but its image in the 4D space–time is a line. Cutting this line by the hyperplane $t = t^s$ we obtain the position of mass at the instant t^s. As long as the mass does not move that line, called in astronomy the line of life, remains straight and parallel to the time axis.

The external force acting upon the mass makes its line of life curvilinear and the slope of such a curve measured against the time axis gives the velocity of the mass. Figure 5.1a shows the projection of the line of life segment on the plane x_1, t. It follows from the 2nd law of Newton that

$$U_1(t)\,\mathrm{d}t = \frac{\mathrm{d}}{\mathrm{d}t}\left(m\,\frac{\mathrm{d}u_1}{\mathrm{d}t}\right)\,\mathrm{d}t. \tag{5.23}$$

Substituting into this equation a new variable

$$Q_{t1} = -\,m\,\frac{\mathrm{d}u_1}{\mathrm{d}t}, \tag{5.24}$$

that represents the internal impulse of mass taken with negative sign, we obtain the new form of Eq. (5.23):

$$\frac{\mathrm{d}Q_{t1}}{\mathrm{d}t} + U_1 = 0. \tag{5.25}$$

Equations (5.24), (5.25) can be interpreted as the equilibrium equations for the vectors acting on the element dt of the line of life of the mass m (Fig. 5.1b). In particular, Eq. (5.24) can be understood as the equilibrium equation for the moments, if we treat m as a pair of vectors acting on the distance du_1. There appears a doubt at once, whether we can treat mass as a vector quantity.

As long as the line of life remains uncut, we must treat mass as a scalar. Similarly, the axial force of a bar is a scalar because it is a product of 3 scalars: Young's modulus E, cross-sectional area A and axial strain ε. Neither a force acting inside a bar nor a mass related to an intact line of life are vector quantities. After a bar has been cut its axial force appears in the form of 2 vectors attached to the opposite sides of the cut. These vectors are not internal forces anymore: they are external loads acting on the 2 parts of the bar.

Cutting the line of life we simultaneously create and annihilate the matter. These acts can be given vector interpretation: the mass vector pointing along the time axis represents creation, whereas the mass vector pointing in the opposite direction represents annihilation.

In relativistic physics one uses the metric form of space–time. This is done by taking $ct = x_4$, where c denotes the velocity of light, as the 4-th dimension instead of t. Such a transformation of space–time would require to modify the loads acting on the element dx_4. In particular, the mass vectors would be replaced by the well known products mc^2 representing energy released by the annihilation of matter or required for its creation.

In order to maintain clear physical meaning of the considered phenomena we will not use the above mentioned metric space–time. Moreover, it seems to be very inconvenient to replace a single second by the line segment having length of 300 000 km.

It has been demonstrated already that a mass concentrated in a dimensionless point is represented in the space–time by a line. Similarly, linear structural schemes (bars, beams, arches, cables, frames, trusses, etc.) are converted into 2D-space–time–objects and 2D-schemes (discs, plates or shells) are represented in the space–time as 3D-objects. Following the same pattern, we can convert linear or planar problems of continuum mechanics into 2D or 3D-problems in the space–time.

Let us consider an example of a body unbounded in the direction x_3 and vibrating in the plane x_1, x_2. This is the planar problem of the plane strain state and its image in the space–time has 3 dimensions. Let us cut an infinitesimal element $dx_1 \times dx_2 \times dx_3 \times dt$ out of that image. Figure 5.2 shows the projection of such an element on the hyperplane $x_3 = \text{const}$. We can not show the segment dx_3 but the magnitudes of all vectors were multiplied by this length.

In order to keep the picture readable, we show only the shear strain γ_{t1}, i.e. the component of displacement velocity parallel to the axis x_1, and the components of the stress tensor parallel to that axis. All vectors acting on the element remain in equilibrium when the following equation holds:

$$\left(\frac{\partial\sigma_{11}}{\partial x_1} + \frac{\partial\tau_{21}}{\partial x_2} + \frac{\partial\tau_{t1}}{\partial t} + U_1\right) \mathrm{d}x_1\,\mathrm{d}x_2\,\mathrm{d}x_3\,\mathrm{d}t = 0. \tag{5.26}$$

Here,

$$\tau_{t1} = -\varrho\,\frac{\partial u_1}{\partial t} \tag{5.27}$$

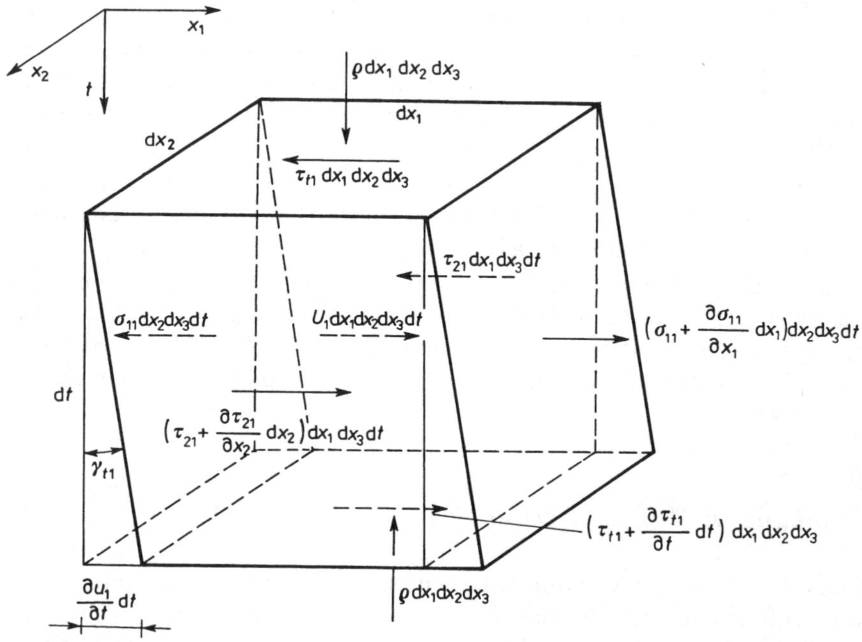

Fig. 5.2. Projection of infinitesimal element of 4D-object on hyperplane $x_3 = \text{const}$.

according to Eqs. (5.4), (5.5).

Note that the internal impulse τ_{t1} appears in the cross-section perpendicular to the time axis. The relevant projection of 4D-object is the 3D-cube having dimensions $dx_1 \times dx_2 \times dx_3$. The resultant internal impulse (with negative direction) for such a cube, i.e. $\tau_{t1}\, dx_1\, dx_2\, dx_3$ is depicted on the upper side of the element shown in Fig. 5.2. Similarly, the cross-section by the hyperplane $x_1 = \text{const}$ gives the cube $dx_2 \times dx_3 \times dt$. Such a cube is acted by the stress σ_{11} whose resultant is the impulse $\sigma_{11}\, dx_2\, dx_3\, dt$.

Assuming that the masses of such cubes $\varrho\, dx_1\, dx_2\, dx_3$ act as opposite vectors on the sides perpendicular to the time axis and shifted by $\frac{\partial u_1}{\partial t} dt$, we can interpret Eq. (5.27) as the moment equilibrium equation for the plane x_1, t.

It is worth to point out an additional consequence of introduction of time as the 4th dimension. There is no motion in the space–time. Objects immersed into that space remain immobile but undergo deformations: the displacement velocities are interpreted as additional components of the strain tensor (compare Fig. 5.2). Similarly, the equations representing Newton's 2nd law of motion take the form of equilibrium equations for stresses since impulses were replaced by additional components of the stress tensor. As a result, each problem of dynamics can be treated almost as a problem of statics. The only difference appears in the boundary conditions. In statics we usually deal with bodies that are bounded in physical space and boundary conditions are given on the entire external surface of the body. Objects in the space–time are unbounded in the direction of t-axis and the initial conditions have to be specified on the hyperplane $t = 0$ that corresponds to the onset of observation.

The above interpretation of the space–time and its resulting dynamics were given in [75, 76]. In [178] it was assumed that the space–time–object can undergo displacement parallel to the time axis.

It should be stressed that though finite elements in space and time were discussed in papers [4, 3, 183], these elements were not identical with the space–time objects defined above. Only in [125] the space–time has been treated as an object divided into finite elements but this idea was not further developed.

5.2 Principle of Virtual Action

In Sect. 2.7, Part I variational principles were formulated by introducing certain functional and defining the conditions for its minimum. Such a procedure leads in dynamics to the Hamilton principle. However, this principle is based upon the assumption given prior to Eq. (2.39), Part I that restricts very severely its application. Namely, the variations of displacements must vanish at the end points of the observed time interval. This excludes the derivation of the initial conditions for the motion from Hamilton's principle. Attempts to generalize that principle were made, e.g., in [57, 60, 142]. The Hamilton principle was considered also in [8, 13, 25, 128].

It is possible to omit the formulation of functional and to introduce a new principle in analogous way as the principle of virtual work was derived in statics (compare, e.g., [123]). Such a derivation is based upon the equilibrium equation for impulses (5.19). Let a certain space–time object occupying the 4D-region Ω bounded by the surface $\partial\Omega$ be subject to virtual increments of displacements

$$\delta \mathbf{u}_{3\times 1}(\mathbf{x},\, t) = \{\delta u_1\ \delta u_2\ \delta u_3\}. \tag{5.28}$$

These increments are small when compared with the dimensions of the physical body and can be described by functions of at least C1 continuity. We denote by $\delta\hat{\mathbf{u}}$ the values of virtual displacements on the boundary $\partial\Omega$.

Let us multiply each equation (5.19) by the corresponding component of the vector $\delta\mathbf{u}$ and let us integrate the sum of such products over the entire 4D-object:

$$\int_\Omega \delta\mathbf{u}^{\mathrm{T}}_{1\times 3}\,(\mathbf{D}^{\mathrm{T}}_{3\times 9}\,\boldsymbol{\sigma}_{9\times 1} + \mathbf{U}_{3\times 1})\,\mathrm{d}\Omega = 0. \tag{5.29}$$

Integrating by parts we can use the Gauss–Ostrogradski relation that after simple derivations leads to the following result:

$$\int_{\partial\Omega} \delta\hat{\mathbf{u}}_{1\times 3}^{\mathrm{T}}\, \hat{\mathbf{t}}_{3\times 1}\, \mathrm{d}(\partial\Omega) + \int_{\Omega} \delta\mathbf{u}_{1\times 3}^{\mathrm{T}}\, \mathbf{U}_{3\times 1}\, \mathrm{d}\Omega = \int_{\Omega} \delta\boldsymbol{\varepsilon}_{1\times 9}^{\mathrm{T}}\, \boldsymbol{\sigma}_{9\times 1}\, \mathrm{d}\Omega. \tag{5.30}$$

Here

$$\delta\boldsymbol{\varepsilon}_{9\times 1} = \mathbf{D}_{9\times 3}\, \delta\mathbf{u}_{3\times 1}. \tag{5.31}$$

Let us analyse the first integral in Eq. (5.30). We know that the stresses $\hat{\mathbf{t}}$ expressed by $\mathrm{N\,m^{-2}}$ act on the segments of $\partial\Omega$ parallel to the time axis and having the dimension $\mathrm{m^2\,s}$. The product of such 2 variables has the dimension of impulse N s. On the other hand, the segments of $\partial\Omega$ perpendicular to the time axis have the dimension $\mathrm{m^3}$ and are influenced by the densities of impulses with the dimension $\mathrm{kg\,m^{-2}s^{-1}}$. Their product has also the dimension of impulse: $\mathrm{kg\,m\,s^{-1}} = \mathrm{N\,s}$. Multiplying such impulses by the virtual displacements $\delta\mathbf{u}$ we obtain the quantity having dimension of the work multiplied by time: $\mathrm{N\,m\,s} = \mathrm{J\,s}$.

We shall call a quantity expressed in terms of J s an internal or external action[2]. Then Eq. (5.30) corresponds to the following principle of virtual action:

Principle of virtual action. *The external action developed by the generalized forces acting on the hypersurface bounding the space–time object together with the mass forces acting inside it is equal to the internal action accumulated in that object.*

Such a principle is formally analogous to the principle of virtual work (compare [160], p. 86). It was introduced in less general form in [75, 76]. The principle of virtual action forms the base of the Space–Time Element Method (STEM).

5.3 Rectangular Space–Time Elements

We call rectangular element a space–time element having edges either parallel or perpendicular to the time axis. Cutting such an element by hyperplanes perpendicular to the time axis we obtain 3D-objects that can have nothing in common with rectangle. Figure 5.3 shows 2 examples of rectangular space–time elements. The cross-section of the first one perpendicular to the time axis is a line segment (Fig. 5.3a), whereas similar cross-section of the second object is a triangular segment of shell (Fig. 5.3b).

Note that at the ends of time interval $t = t^a$ and $t = t^p$ the shape of the object is identical. Let N_e be the number of degrees of freedom (DOF's) of the nodes situated in one of those hyperplanes. Then the total number of DOF's for the element is $2N_e$.

[2] These names replace the terms *four-work, four-energy* used by the author in his earlier publications [75, 76], and *time-work, time-energy* used by Bajer [12].

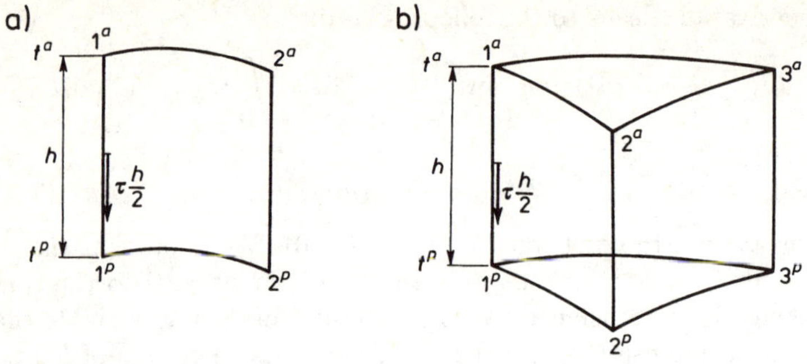

Fig. 5.3. Rectangular space–time elements

Let the field of displacements be defined inside the space–time element by the nodal displacements multiplied by the shape functions (compare Eqs. (1.2), (1.3), Part II):

$$\mathbf{u}_{3\times1}(\mathbf{x},\, t) = \varphi_{3\times2N_e}(\mathbf{x},\, t)\, \mathbf{q}^{(e)}_{2N_e\times1}. \tag{5.32}$$

In the case of rectangular space–time element the matrix of shape functions can be expressed by the following tensor product[3]

$$\varphi_{3\times2N_e}(\mathbf{x},\, t) = \varphi^t_{1\times2}(t) \otimes \varphi^x_{3\times N_e}(\mathbf{x}). \tag{5.33}$$

We assume that the first component of the product (5.33) has the following form:

$$\varphi^t_{1\times2}(t) = \begin{bmatrix} \dfrac{1-\tau}{2} & \dfrac{1+\tau}{2} \end{bmatrix}, \quad \tau = \frac{t - t^a - h/2}{h/2}, \quad \tau \in (-1,\, 1). \tag{5.34}$$

The form of the second component can be arbitrary. It depends upon the shape of the element in 3D-space (compare Fig. 5.3).

It follows from Eqs. (5.2), (5.3) that

$$\varepsilon_{9\times1} = \mathbf{B}_{9\times2N_e}(\mathbf{x},\, t)\, \mathbf{q}^{(e)}_{2N_e\times1}, \tag{5.35}$$

where

$$\mathbf{B}_{9\times2N_e} = \mathbf{D}_{9\times3}\, \varphi_{3\times2N_e} = \begin{bmatrix} \varphi^t_{1\times2} \otimes \mathbf{B}^x_{6\times N_e} \\ \mathbf{B}^t_{1\times2} \otimes \varphi^x_{3\times N_e} \end{bmatrix}. \tag{5.36}$$

[3] The tensor product is defined as:

$$\mathbf{A}_{i\times j} \otimes \mathbf{B}_{k\times l} = \mathbf{C}_{ik\times jl} = \begin{bmatrix} A_{11}\,\mathbf{B} & A_{12}\,\mathbf{B} & \dots & A_{1j}\,\mathbf{B} \\ A_{21}\,\mathbf{B} & A_{22}\,\mathbf{B} & \dots & A_{2j}\,\mathbf{B} \\ \vdots & \vdots & \vdots & \vdots \\ A_{i1}\,\mathbf{B} & A_{i2}\,\mathbf{B} & \dots & A_{ij}\,\mathbf{B} \end{bmatrix}$$

We introduced here the following notation:

$$\mathbf{B}^x_{6\times N_e} = \mathbf{D}_{6\times 3}\,\varphi^x_{3\times N_e}, \quad \mathbf{B}^t_{1\times 2} = \frac{\mathrm{d}}{\mathrm{d}t}\varphi^t_{1\times 2} = \begin{bmatrix} -\frac{1}{h} & \frac{1}{h} \end{bmatrix}. \tag{5.37}$$

Symbol $\mathbf{D}_{6\times 3}$ has been explained after Eq. (2.29), Part I.

It follows further from Eqs. (5.4), (5.5), (5.17) and (5.18) that

$$\boldsymbol{\sigma}_{9\times 1} = \mathbf{C}_{9\times 9}\,\mathbf{B}_{9\times 2N_e}\,\mathbf{q}^{(e)}_{2N_e\times 1}. \tag{5.38}$$

The product of the 2 first matrices can be written as:

$$\begin{aligned}\mathbf{C}_{9\times 9}\,\mathbf{B}_{9\times 2N_e} &= \begin{bmatrix} \mathbf{C}^E_{6\times 6} + \mathbf{C}^\eta_{6\times 6}\frac{\partial}{\partial t} & \mathbf{0}_{6\times 3} \\ \mathbf{0}_{3\times 6} & -\varrho\,\mathbf{I}_{3\times 3} \end{bmatrix} \begin{bmatrix} \varphi^t_{1\times 2}\otimes \mathbf{B}^x_{6\times N_e} \\ \mathbf{B}^t_{1\times 2}\otimes \varphi^x_{3\times N_e} \end{bmatrix} \\ &= \begin{bmatrix} \varphi^t_{1\times 2}\otimes(\mathbf{C}^E_{6\times 6}\,\mathbf{B}^x_{6\times N_e}) + \mathbf{B}^t_{1\times 2}\otimes(\mathbf{C}^\eta_{6\times 6}\mathbf{B}^x_{6\times N_e}) \\ -\mathbf{B}^t_{1\times 2}\otimes\varrho\,\varphi^x_{3\times N_e} \end{bmatrix},\end{aligned} \tag{5.39}$$

where new symbols have the following meaning:

$$\mathbf{C}^E_{6\times 6} = \begin{bmatrix} K+\frac{4}{3}G & K-\frac{2}{3}G & K-\frac{2}{3}G & \\ & K+\frac{4}{3}G & K-\frac{2}{3}G & \mathbf{0}_{3\times 3} \\ & & K+\frac{4}{3}G & \\ \text{sym.} & & & G\,\mathbf{I}_{3\times 3} \end{bmatrix}, \tag{5.40}$$

$$\mathbf{C}^\eta_{6\times 6} = \frac{\eta}{3}\begin{bmatrix} 4 & -2 & -2 & \\ & 4 & -2 & \mathbf{0}_{3\times 3} \\ & & 4 & \\ \text{sym.} & & & 3\,\mathbf{I}_{3\times 3} \end{bmatrix}. \tag{5.41}$$

In order to take advantage from the principle of virtual action, we must assume, similarly as it is done in the FEM, that the first integral in Eq. (5.30) can be replaced by the sum of products of the nodal impulses $\hat{\mathbf{Q}}^{(e)}_{2N_e\times 1}$ equivalent to the impulses distributed over the surface $\partial\Omega_e$ and the virtual nodal displacements $\delta\mathbf{q}^{\mathrm{T}(e)}_{1\times 2N_e}$. Hence, we have

$$\begin{aligned}&\delta\mathbf{q}^{\mathrm{T}(e)}_{1\times 2N_e}\,\hat{\mathbf{Q}}^{(e)}_{2N_e\times 1} + \int_{\Omega_e}\delta\mathbf{q}^{\mathrm{T}(e)}_{1\times 2N_e}\,\varphi^{\mathrm{T}}_{2N_e\times 3}\,\mathbf{U}_{3\times 1}\,\mathrm{d}\Omega_e \\ &= \int_{\Omega_e}\delta\mathbf{q}^{\mathrm{T}(e)}_{1\times 2N_e}\,\mathbf{B}^{\mathrm{T}}_{2N_e\times 9}\,\mathbf{C}_{9\times 9}\,\mathbf{B}_{9\times 2N_e}\,\mathbf{q}^{(e)}_{2N_e\times 1}\,\mathrm{d}\Omega_e.\end{aligned} \tag{5.42}$$

Since virtual nodal displacements entering the vector $\delta\mathbf{q}^{(e)}$ can assume arbitrary values, the scalar equation (5.42) leads to $2N_e$ equations that relate the vector of nodal impulses $\hat{\mathbf{Q}}^{(e)}$ with the vector of nodal displacements $\mathbf{q}^{(e)}$ and with the loading $\mathbf{U}$ acting over the space–time element:

$$\hat{\mathbf{Q}}^{(e)}_{2N_e\times 1} = \mathbf{K}^{(e)}_{2N_e\times 2N_e}\,\mathbf{q}^{(e)}_{2N_e\times 1} - \mathbf{Q}^{(e)}_{2N_e\times 1}. \tag{5.43}$$

Here the stiffness matrix can be expressed in the following way taking into account Eqs. (5.36) to (5.41):

$$\begin{aligned}\mathbf{K}^{(e)}_{2N_e\times 2N_e} &= \int_{\Omega_e} \mathbf{B}^{\mathrm{T}}_{2N_e\times 9}\,\mathbf{C}_{9\times 9}\,\mathbf{B}_{9\times 2N_e}\,\mathrm{d}\Omega_e = \\ &= \int_{-1}^{1} \overset{\mathrm{T}t}{\boldsymbol{\varphi}}_{2\times 1}\,\boldsymbol{\varphi}^{t}_{1\times 2}\,\frac{h}{2}\,\mathrm{d}\tau \otimes \int_{V_e} \overset{\mathrm{T}}{\mathbf{B}}{}^{x}_{N_e\times 6}\,\mathbf{C}^{E}_{6\times 6}\,\mathbf{B}^{x}_{6\times N_e}\,\mathrm{d}V_e + \\ &+ \int_{-1}^{1} \overset{\mathrm{T}t}{\boldsymbol{\varphi}}_{2\times 1}\,\mathbf{B}^{t}_{1\times 2}\,\frac{h}{2}\,\mathrm{d}\tau \otimes \int_{V_e} \overset{\mathrm{T}}{\mathbf{B}}{}^{x}_{N_e\times 6}\,\mathbf{C}^{\eta}_{6\times 6}\,\mathbf{B}^{x}_{6\times N_e}\,\mathrm{d}V_e - \\ &- \int_{-1}^{1} \overset{\mathrm{T}}{\mathbf{B}}{}^{t}_{2\times 1}\,\mathbf{B}^{t}_{1\times 2}\,\frac{h}{2}\,\mathrm{d}\tau \otimes \int_{V_e} \overset{\mathrm{T}x}{\boldsymbol{\varphi}}_{N_e\times 3}\,\varrho\,\boldsymbol{\varphi}^{x}_{3\times N_e}\,\mathrm{d}V_e, \end{aligned} \tag{5.44}$$

where V_e is the volume of the body in the 3D-space. The following relations were used when deriving the above result:

$$(\mathbf{A}_{i\times j} \otimes \mathbf{B}_{k\times l})^{\mathrm{T}} = \mathbf{A}^{\mathrm{T}}_{j\times i} \otimes \mathbf{B}^{\mathrm{T}}_{l\times k}, \tag{5.45}$$

$$(\mathbf{A}_{i\times j} \otimes \mathbf{B}_{k\times l})(\mathbf{C}_{j\times i} \otimes \mathbf{D}_{l\times k}) = (\mathbf{A}_{i\times j}\mathbf{C}_{j\times i}) \otimes (\mathbf{B}_{k\times l}\mathbf{D}_{l\times k}) = \mathbf{E}_{ik\times ik}. \tag{5.46}$$

Introducing

$$\mathbf{J}_{2\times 2} = \int_{-1}^{1} \overset{\mathrm{T}t}{\boldsymbol{\varphi}}_{2\times 1}\,\boldsymbol{\varphi}^{t}_{1\times 2}\,\frac{h}{2}\,\mathrm{d}\tau = \frac{h}{6}\begin{bmatrix} 2 & 1 \\ 1 & 2 \end{bmatrix}, \tag{5.47}$$

$$\mathbf{J}'_{2\times 2} = \int_{-1}^{1} \overset{\mathrm{T}t}{\boldsymbol{\varphi}}_{2\times 1}\,\mathbf{B}^{t}_{1\times 2}\,\frac{h}{2}\,\mathrm{d}\tau = \frac{1}{6}\begin{bmatrix} -1 & 1 \\ -1 & 1 \end{bmatrix}, \tag{5.48}$$

$$\mathbf{J}''_{2\times 2} = \int_{-1}^{1} \overset{\mathrm{T}}{\mathbf{B}}{}^{t}_{2\times 1}\,\mathbf{B}^{t}_{1\times 2}\,\frac{h}{2}\,\mathrm{d}\tau = \frac{1}{h}\begin{bmatrix} 1 & -1 \\ -1 & 1 \end{bmatrix}, \tag{5.49}$$

$$\mathbf{S}^{(e)}_{N_e\times N_e} = \int_{V_e} \overset{\mathrm{T}}{\mathbf{B}}{}^{x}_{N_e\times 6}\,\mathbf{C}^{E}_{6\times 6}\,\mathbf{B}^{x}_{6\times N_e}\,\mathrm{d}V_e, \tag{5.50}$$

$$\mathbf{W}^{(e)}_{N_e\times N_e} = \int_{V_e} \overset{\mathrm{T}}{\mathbf{B}}{}^{x}_{N_e\times 6}\,\mathbf{C}^{\eta}_{6\times 6}\,\mathbf{B}^{x}_{6\times N_e}\,\mathrm{d}V_e, \tag{5.51}$$

$$\mathbf{M}^{(e)}_{N_e\times N_e} = \int_{V_e} \overset{\mathrm{T}x}{\boldsymbol{\varphi}}_{N_e\times 3}\,\varrho\,\boldsymbol{\varphi}^{x}_{3\times N_e}\,\mathrm{d}V_e, \tag{5.52}$$

we can write the stiffness matrix of the rectangular space–time element as

$$\mathbf{K}^{(e)}_{2N_e\times 2N_e} = \mathbf{J}_{2\times 2} \otimes \mathbf{S}^{(e)}_{N_e\times N_e} + \mathbf{J}'_{2\times 2} \otimes \mathbf{W}^{(e)}_{N_e\times N_e} - \mathbf{J}''_{2\times 2} \otimes \mathbf{M}^{(e)}_{N_e\times N_e}. \tag{5.53}$$

A transition from the local reference frame of the element e to the global one requires to transform the stiffness matrix

$$\begin{aligned}\tilde{\mathbf{K}}^{(e)}_{2N_e\times 2N_e} &= (\mathbf{I}_{2\times 2} \otimes {}^{I}\overset{\mathrm{T}}{\mathbf{a}}{}^{(e)}_{N_e\times N_e})\,\mathbf{K}^{(e)}_{2N_e\times 2N_e}\,(\mathbf{I}_{2\times 2} \otimes {}^{I}\mathbf{a}^{(e)}_{N_e\times N_e}) \\ &= \mathbf{J}_{2\times 2} \otimes \tilde{\mathbf{S}}_{N_e\times N_e} + \mathbf{J}'_{2\times 2} \otimes \tilde{\mathbf{W}}_{N_e\times N_e} - \mathbf{J}''_{2\times 2} \otimes \tilde{\mathbf{M}}_{N_e\times N_e}. \end{aligned} \tag{5.54}$$

in accordance with Eq. (1.15). The following notation was introduced here:

$$\begin{aligned}\tilde{\mathbf{S}}^{(e)} &= {}^{I}\tilde{\mathbf{a}}^{\mathrm{T}(e)}\,\mathbf{S}^{(e)}\,{}^{I}\mathbf{a}^{(e)},\\ \tilde{\mathbf{W}}^{(e)} &= {}^{I}\tilde{\mathbf{a}}^{\mathrm{T}(e)}\,\mathbf{W}^{(e)}\,{}^{I}\mathbf{a}^{(e)},\\ \tilde{\mathbf{M}}^{(e)} &= {}^{I}\tilde{\mathbf{a}}^{\mathrm{T}(e)}\,\mathbf{M}^{(e)}\,{}^{I}\mathbf{a}^{(e)}.\end{aligned} \tag{5.55}$$

The last term in Eq. (5.43) represents the nodal impulses

$$\mathbf{Q}^{(e)}_{2N_e\times 1} = \int_{\Omega_e} \varphi^{\mathrm{T}}_{2N_e\times 3}\,\mathbf{U}_{3\times 1}\,\mathrm{d}\Omega_e. \tag{5.56}$$

equivalent to the internal impulses of inertia forces distributed over the space–time element. If some additional impulses distributed over a surface bounding the volume of the element or situated inside it are present, then such impulses should be added to the nodal values. In the global co-ordinate system the nodal impulses read

$$\tilde{\mathbf{Q}}^{(e)}_{2N_e\times 1} = \left(\mathbf{I}_{2\times 2}\otimes {}^{I}\tilde{\mathbf{a}}^{\mathrm{T}(e)}_{N_e\times N_e}\right)\mathbf{Q}^{(e)}_{2N_e\times 1}. \tag{5.57}$$

The independent on time matrices (5.55) can be assembled, in the way described in Part I, into the global matrices $\mathbf{S}_{N\times N}$, $\mathbf{W}_{N\times N}$, $\mathbf{M}_{N\times N}$ related to the 3D-body with N degrees of freedom. Hence, the assembled stiffness matrix of the space–time element E bounded by the hyperplanes $t = t^a$ and $t = t^p$ has the following structure:

$$\begin{aligned}\mathbf{K}^{(E)}_{2N\times 2N} &= \mathbf{J}_{2\times 2}\otimes \mathbf{S}_{N\times N} + \mathbf{J}'_{2\times 2}\otimes \mathbf{W}_{N\times N} - \mathbf{J}''_{2\times 2}\otimes \mathbf{M}_{N\times N} =\\ &= \begin{bmatrix}\mathbf{A}_{N\times N} & \mathbf{B}_{N\times N}\\ \mathbf{C}_{N\times N} & \mathbf{D}_{N\times N}\end{bmatrix}.\end{aligned} \tag{5.58}$$

Here

$$\begin{aligned}\left.\begin{matrix}\mathbf{A}_{N\times N}\\ \mathbf{D}_{N\times N}\end{matrix}\right\} &= \frac{h}{3}\mathbf{S}_{N\times N} \mp \frac{1}{2}\mathbf{W}_{N\times N} - \frac{1}{h}\mathbf{M}_{N\times N},\\ \left.\begin{matrix}\mathbf{B}_{N\times N}\\ \mathbf{C}_{N\times N}\end{matrix}\right\} &= \frac{h}{6}\mathbf{S}_{N\times N} \pm \frac{1}{2}\mathbf{W}_{N\times N} + \frac{1}{h}\mathbf{M}_{N\times N}.\end{aligned} \tag{5.59}$$

are the new matrices introduced for brevity.

We group all nodal displacements that appear in the hyperplane $t = t^s$ separating two adjacent assembled elements into the vector $\mathbf{q}^s_{N\times 1}$. The given nodal impulses (5.57) can be assembled into the vector of external impulses $\mathbf{Q}^s_{N\times 1}$. That allows us to write the equilibrium equation for nodes situated on the hyperplane $t = t^s$ as

$$\mathbf{C}^s_{N\times N}\,\mathbf{q}^{s-1}_{N\times 1} + (\mathbf{D}^s_{N\times N} + \mathbf{A}^{s+1}_{N\times N})\,\mathbf{q}^s_{N\times 1} + \mathbf{B}^{s+1}_{N\times N}\,\mathbf{q}^{s+1}_{N\times 1} = \mathbf{Q}^s_{N\times 1}. \tag{5.60}$$

We introduced the superscript s for matrices $\mathbf{A}, \mathbf{B}, \mathbf{C}, \mathbf{D}$ in order to allow time increments $h^s = t^s - t^{s-1}$ to differ along the time axis. In the case of $h^s = h = \text{const}$ such a superscript can be omitted what we do in the sequel.

The matrix relation

$$\mathbf{B}_{N\times N}\,\mathbf{q}^{1}_{N\times 1} = \mathbf{Q}^{0}_{N\times 1} - \mathbf{A}_{N\times N}\,\mathbf{q}^{0}_{N\times 1}, \tag{5.61}$$

expresses the equilibrium condition for the impulses acting at the initial instant $t^0 = 0$ $(s = 0)$. All quantities that are given by the initial value conditions of the problem were placed at the right hand side. The formula

$$\mathbf{q}^{1}_{N\times 1} = \mathbf{B}^{-1}_{N\times N}(\mathbf{Q}^{0}_{N\times 1} - \mathbf{A}_{N\times N}\,\mathbf{q}^{0}_{N\times 1}). \tag{5.62}$$

gives us means to calculate N unknown components of the vector $\mathbf{q}^1$. In order to obtain nodal displacements in the subsequent time steps, we use the recurrent relation

$$\mathbf{q}^{s+1}_{N\times 1} = \mathbf{B}^{-1}_{N\times N}(\mathbf{Q}^{s}_{N\times 1} - \mathbf{H}_{N\times N}\,\mathbf{q}^{s}_{N\times 1} - \mathbf{C}_{N\times N}\,\mathbf{q}^{s-1}_{N\times 1}), \tag{5.63}$$

where

$$\mathbf{H}_{N\times N} = \mathbf{A}_{N\times N} + \mathbf{D}_{N\times N} = \frac{2h}{3}\mathbf{S}_{N\times N} - \frac{2}{h}\mathbf{M}_{N\times N}. \tag{5.64}$$

Thus, using almost standard FEM-procedures, we come to the algebraic equations of the type (5.60) or to the recurrent formula (5.63). All other methods reduce problems of non-stationary dynamics to sets of ordinary differential equations (1.35) that must be further algebraized. The STEM method excludes such an intermediate phase leading directly to the set of algebraic equations suitable for solving by computer.

Introducing matrices (5.59) into the set of equations (5.60), we arrive after simple derivation at the following difference scheme:

$$\mathbf{M}\,\frac{\mathbf{q}^{s-1} - 2\mathbf{q}^{s} + \mathbf{q}^{s+1}}{h^2} + \mathbf{W}\,\frac{\mathbf{q}^{s+1} - \mathbf{q}^{s-1}}{2h} + \mathbf{S}\,\frac{\mathbf{q}^{s-1} + 4\mathbf{q}^{s} + \mathbf{q}^{s+1}}{6} = \frac{\mathbf{Q}^{s}}{h}. \tag{5.65}$$

Going to the limit $h \to 0$ we obtain the equation

$$\mathbf{M}\ddot{\mathbf{q}} + \mathbf{W}\dot{\mathbf{q}} + \mathbf{S}\,\mathbf{q} = \mathbf{P}. \tag{5.66}$$

that coincides up to the notation with Eq. (1.35).

In [110, 183] more general difference scheme has been developed. Given our notation and given Eqs. (3.24) that scheme reads

$$\begin{aligned}\mathbf{M}\,\frac{\mathbf{q}^{s-1} - 2\mathbf{q}^{s} + \mathbf{q}^{s+1}}{h^2} + \mathbf{W}\,\frac{\delta\,\mathbf{q}^{s+1} + (1-2\delta)\,\mathbf{q}^{s} - (1-\delta)\,\mathbf{q}^{s-1}}{h} + \\ +\mathbf{S}\left(\frac{1-2\delta+2\alpha}{2}\,\mathbf{q}^{s-1} + \frac{1+2\delta-4\alpha}{2}\,\mathbf{q}^{s} + \alpha\,\mathbf{q}^{s+1}\right) = \frac{\mathbf{Q}^{s}}{h}.\end{aligned} \tag{5.67}$$

It is easy to check that the substitution

$$\delta = \frac{1}{2}, \quad \alpha = \frac{1}{6} \tag{5.68}$$

makes Eqs. (5.65) and (5.67) identical.

The process governed by the recurrent relation (5.67) would be numerically stable if the constants δ and α satisfy 3 inequalities:

$$\alpha \geq \frac{1}{4}\left(\frac{1}{2}+\delta\right)^2, \quad \delta \geq \frac{1}{2}, \quad \frac{1}{2}-\delta+\alpha \geq 0. \tag{5.69}$$

In our case the first inequality is not fulfilled. Hence, the recurrent process is conditionally stable and the stability condition reads

$$h \leq \frac{2}{\sqrt{\left(\frac{1}{2}+\delta\right)^2 - 4\alpha}} \frac{1}{\omega_{\max}}. \tag{5.70}$$

Here ω_{max} is the maximum eigenfrequency of the assembled structure of N degrees of freedom. Substituting the values (5.68) we obtain

$$h \leq \frac{\sqrt{12}}{\omega_{\max}} = \frac{\sqrt{3}}{\pi} T_{\min} = 0.551\, T_{\min}, \quad T_{\min} = \frac{2\pi}{\omega_{\max}}. \tag{5.71}$$

Thus, in order to establish the maximum time dimension h of the space–time element under which the recurrent procedure remains stable one has to find the maximum eigenfrequency of the discretized structure which is a lot of work.

The maximum value of h can be found different way. The authors of the papers [72, 100] have shown that the positive definiteness of the matrix

$$\check{\mathbf{S}}_{N\times N} = \mathbf{B}_{N\times N} + \mathbf{C}_{N\times N} - \mathbf{H}_{N\times N} \tag{5.72}$$

is sufficient for the stability of calculations. Symbolically this condition can be written as

$$\check{\mathbf{S}}_{N\times N} > 0. \tag{5.73}$$

It is easy to show that this condition is satisfied when the stiffness matrices of individual elements are positive definite:

$$\check{\mathbf{S}}^{(e)}_{N_e\times N_e} = \mathbf{B}^{(e)}_{N_e\times N_e} + \mathbf{C}^{(e)}_{N_e\times N_e} - \mathbf{H}^{(e)}_{N_e\times N_e} > 0. \tag{5.74}$$

Obviously, the condition (5.74) is more restrictive than (5.73) and leads to the smaller dimension h of space–time elements.

Numerical stability and accuracy of solutions were considered in many papers, e.g. in [98, 41, 42].

The version of STEM presented above is based upon the paper [77].

The conditional stability of the recurrent procedure can be viewed as a certain disadvantage of STEM. This disadvantage can be diminished or entirely eliminated by certain ways of performing calculations. Let us discuss shortly some of them.

It is a common practice in the FEM to divide large structures into super-elements (compare [14]). By eliminating displacements of nodes situated inside super-elements one reduces the vector of unknowns to displacements of nodes situated on the boundary of such super-elements.

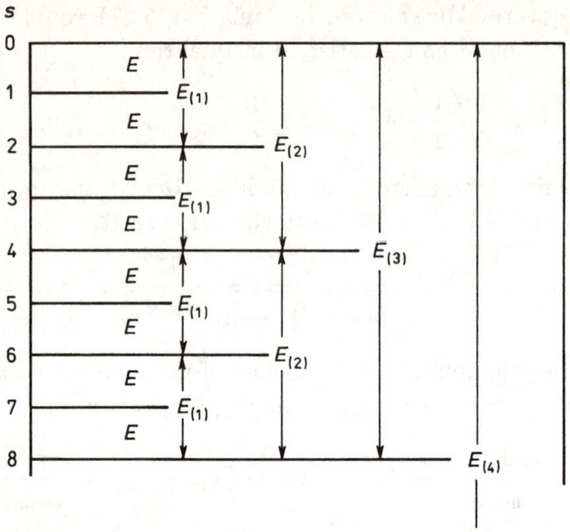

Fig. 5.4. Space–time super-elements

Space–time objects representing in the cross-sections $t = t^s$ a physical body in the subsequent time instances can be divided into space–time super-elements. Each of them consists of two assembled elements E mentioned previously. Let us assume that the super-elements $E_{(1)}$ are bounded by the hyperplanes of even indices s (Fig. 5.4). The time dimension of such a super-element is equal to the double dimension of the assembled element E. The vectors of displacements $\mathbf{q}^s$ occurring at the instances t^s of odd s must be eliminated. This can be done according to the following scheme:

$$\begin{array}{r|c|} 1 & \\ -\mathbf{BH}^{-1} & -\mathbf{CH}^{-1} \\ & 1 \\ & -\mathbf{BH}^{-1} \\ & \end{array} \begin{bmatrix} \mathbf{A} & \mathbf{B} & & & \\ \mathbf{C} & \mathbf{H} & \mathbf{B} & & \\ & \mathbf{C} & \mathbf{H} & \mathbf{B} & \\ & & \mathbf{C} & \mathbf{H} & \mathbf{B} \\ & & \dots & \dots & \dots \end{bmatrix} \begin{bmatrix} \mathbf{q}^0 \\ \mathbf{q}^1 \\ \mathbf{q}^2 \\ \mathbf{q}^3 \\ \dots \end{bmatrix} = \begin{bmatrix} \mathbf{Q}^0 \\ \mathbf{Q}^1 \\ \mathbf{Q}^2 \\ \mathbf{Q}^3 \\ \dots \end{bmatrix} . \tag{5.75}$$

The dimensions of the matrices have been omitted because all square matrices in this formula have the dimensions $N \times N$ and all vectors have N components.

On the left side we wrote the factors by which the relevant equations must be multiplied in order to eliminate $\mathbf{q}^s$ of odd indices from the resulting set of equations. Hence, the subset $s = 0$ of the equations [compare (5.61)] is added to the subset $s = 1$ multiplied by $-\mathbf{BH}^{-1}$. The result is the subset $s = 0$ after the first elimination

$$\mathbf{A}_{(1)}\, \mathbf{q}^0 + \mathbf{B}_{(1)}\, \mathbf{q}^2 = \mathbf{Q}^0_{(1)}, \tag{5.76}$$

where the following matrices were introduced:

$$\mathbf{A}_{(1)} = \mathbf{A} - \mathbf{B}\mathbf{H}^{-1}\mathbf{C}, \quad \mathbf{B}_{(1)} = -\mathbf{B}\mathbf{H}^{-1}\mathbf{B}, \quad \mathbf{Q}^0_{(1)} = \mathbf{Q}^0 - \mathbf{B}\mathbf{H}^{-1}\mathbf{Q}^1. \tag{5.77}$$

Similarly, each subset of even $s = 2r(r = 1, 2, 3, \ldots)$ is summed with the adjacent subsets multiplied by either $-\mathbf{C}\mathbf{H}^{-1}$ or $-\mathbf{B}\mathbf{H}^{-1}$. This operation gives the s-th subset after the first elimination

$$\mathbf{C}_{(1)}\,\mathbf{q}^{2(r-1)} + \mathbf{H}_{(1)}\,\mathbf{q}^{2r} + \mathbf{B}_{(1)}\,\mathbf{q}^{2(r+1)} = \mathbf{Q}^{2r}_{(1)}, \tag{5.78}$$

where

$$\mathbf{C}_{(1)} = -\mathbf{C}\,\mathbf{H}^{-1}\mathbf{C}, \qquad \mathbf{H}_{(1)} = \mathbf{H} - \mathbf{C}\,\mathbf{H}^{-1}\mathbf{B} - \mathbf{B}\,\mathbf{H}^{-1}\mathbf{C}, \tag{5.79}$$

$$\mathbf{Q}^{2r}_{(1)} = \mathbf{Q}^{2r} - \mathbf{C}\,\mathbf{H}^{-1}\mathbf{Q}^{2r-1} - \mathbf{B}\,\mathbf{H}^{-1}\mathbf{Q}^{2r+1}. \tag{5.80}$$

In order to accelerate even more the recurrent calculation of displacements, we combine 2 adjacent super-elements $E_{(1)}$ after the first reduction into a super-element $E_{(2)}$ containing 4 assembled elements. After the second reduction step the displacements at the instances $s = 2, 6, 10, \ldots$ disappear from the set of equations and the reduced set is similar to (5.76) and (5.78).

After m such steps each super-element $E_{(m)}$ consists of 2^m assembled elements E. Introducing the matrices

$$\begin{aligned}
\mathbf{A}_{(m)} &= \mathbf{A}_{(m-1)} - \mathbf{B}_{(m-1)}\mathbf{H}^{-1}_{(m-1)}\mathbf{C}_{(m-1)},\\
\mathbf{B}_{(m)} &= -\mathbf{B}_{(m-1)}\mathbf{H}^{-1}_{(m-1)}\mathbf{B}_{(m-1)},\\
\mathbf{C}_{(m)} &= -\mathbf{C}_{(m-1)}\mathbf{H}^{-1}_{(m-1)}\mathbf{C}_{(m-1)}\\
\mathbf{H}_{(m)} &= \mathbf{H}_{(m-1)} - \mathbf{C}_{(m-1)}\mathbf{H}^{-1}_{(m-1)}\mathbf{B}_{(m-1)} - \mathbf{B}_{(m-1)}\mathbf{H}^{-1}_{(m-1)}\mathbf{C}_{(m-1)},
\end{aligned} \tag{5.81}$$

$$\mathbf{Q}^{2^m r}_{(m)} = \mathbf{Q}^{2^m r}_{(m-1)} - \mathbf{C}_{(m-1)}\mathbf{H}^{-1}_{(m-1)}\mathbf{Q}^{2^m(r-1/2)}_{(m-1)} - \mathbf{B}_{(m-1)}\mathbf{H}^{-1}_{(m-1)}\mathbf{Q}^{2^m(r+1/2)}_{(m-1)}, \tag{5.82}$$

we obtain the set of equations

$$\begin{aligned}
\mathbf{A}_{(m)}\,\mathbf{q}^0 + \mathbf{B}_{(m)}\,\mathbf{q}^{2^m} &= \mathbf{Q}^0_{(m)},\\
\mathbf{C}_{(m)}\,\mathbf{q}^{2^m(r-1)} + \mathbf{H}_{(m)}\,\mathbf{q}^{2^m r} + \mathbf{B}_{(m)}\,\mathbf{q}^{2^m(r+1)} &= \mathbf{Q}^{2^m r}_{(m)} \quad (r = 1,\, 2,\, 3, \ldots).
\end{aligned} \tag{5.83}$$

Thus, each reduction doubles the time dimension of the super-element. It should be noted that this transformation has no influence on the accuracy of solution. The paper [70] includes a detailed presentation of the above procedure, together with the discussion of its efficiency.

A different proposal on the increase of the time dimension h of the rectangular space–time elements can be found in [133]. Namely, the applicability of the so-called Serendipian finite elements known in the FEM (compare [183]) has been investigated. Additional nodes on the edges parallel to the time axis were introduced. Figure 5.5 shows 2 Serendipian elements of this type based on a triangle.

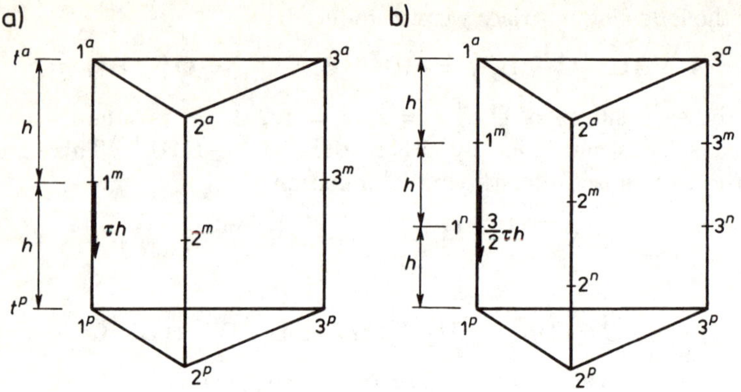

Fig. 5.5. Serendipian space–time elements

Applying the procedure described above to the element shown in Fig. 5.5a [compare Eqs. (5.32) and (5.33)], we obtain

$$\mathbf{u}_{3\times1}(\mathbf{x},t) = \boldsymbol{\varphi}_{3\times3N_e}(\mathbf{x},t)\,\mathbf{q}^{(e)}_{3N_e\times1}, \tag{5.84}$$

where

$$\boldsymbol{\varphi}_{3\times3N_e}(\mathbf{x},t) = \boldsymbol{\varphi}^t_{1\times3}(t)\otimes\boldsymbol{\varphi}^x_{3\times N_e}(\mathbf{x}). \tag{5.85}$$

We assume the following shape matrix dependent on time:

$$\boldsymbol{\varphi}^t_{1\times3} = \left[\begin{array}{ccc} \frac{\tau^2-\tau}{2} & 1-\tau^2 & \frac{\tau^2+\tau}{2}\end{array}\right], \qquad \tau = \frac{t-t^a-h}{h}, \qquad \tau\in(-1,\,1). \tag{5.86}$$

Then we calculate according to Eqs. (5.37) and (5.47) to (5.49):

$$\mathbf{B}^t_{1\times3}(t) = \frac{\mathrm{d}}{\mathrm{d}t}\boldsymbol{\varphi}^t_{1\times3} = \frac{1}{h}\left[\begin{array}{ccc}\tau-\frac{1}{2} & -2\tau & \tau+\frac{1}{2}\end{array}\right], \tag{5.87}$$

$$\mathbf{J}_{3\times3} = \int_{-1}^{1}\overset{\mathrm{T}}{\boldsymbol{\varphi}}{}^t_{3\times1}\,\boldsymbol{\varphi}^t_{1\times3}\,h\,\mathrm{d}\tau = \frac{h}{15}\begin{bmatrix} 4 & 2 & -1\\ 2 & 16 & 2\\ -1 & 2 & 4\end{bmatrix}, \tag{5.88}$$

$$\mathbf{J}'_{3\times3} = \int_{-1}^{1}\overset{\mathrm{T}}{\boldsymbol{\varphi}}{}^t_{3\times1}\,\mathbf{B}^t_{1\times3}\,h\,\mathrm{d}\tau = \frac{1}{6}\begin{bmatrix} -3 & 4 & -1\\ -4 & 0 & 4\\ 1 & -4 & 3\end{bmatrix}, \tag{5.89}$$

$$\mathbf{J}''_{3\times3} = \int_{-1}^{1}\overset{\mathrm{T}}{\mathbf{B}}{}^t_{3\times1}\,\mathbf{B}^t_{1\times3}\,h\,\mathrm{d}\tau = \frac{1}{6h}\begin{bmatrix} 7 & -8 & 1\\ -8 & 16 & -8\\ 1 & -8 & 7\end{bmatrix}. \tag{5.90}$$

and we obtain the stiffness matrix of Serendipian space–time element as

$$\mathbf{K}^{(e)}_{3N_e\times3N_e} = \mathbf{J}_{3\times3}\otimes\mathbf{S}^{(e)}_{N_e\times N_e} + \mathbf{J}'_{3\times3}\otimes\mathbf{W}^{(e)}_{N_e\times N_e} + \mathbf{J}''_{3\times3}\otimes\mathbf{M}^{(e)}_{N_e\times N_e}. \tag{5.91}$$

The stiffness matrix of the assembled element E is similar in its structure to that of Eq. (5.58):

$$\mathbf{K}^{(E)}_{3N\times 3N} = \begin{bmatrix} \mathbf{A}_{N\times N} & \mathbf{F}_{N\times N} & \mathbf{B}_{N\times N} \\ \mathbf{G}_{N\times N} & \mathbf{E}_{N\times N} & \mathbf{F}_{N\times N} \\ \mathbf{C}_{N\times N} & \mathbf{G}_{N\times N} & \mathbf{D}_{N\times N} \end{bmatrix}, \tag{5.92}$$

where

$$\begin{aligned} \left.\begin{matrix}\mathbf{A}_{N\times N}\\ \mathbf{D}_{N\times N}\end{matrix}\right\} &= \frac{4h}{15}\mathbf{S}_{N\times N} \mp \frac{1}{2}\mathbf{W}_{N\times N} - \frac{7}{6h}\mathbf{M}_{N\times N}, \\ \left.\begin{matrix}\mathbf{B}_{N\times N}\\ \mathbf{C}_{N\times N}\end{matrix}\right\} &= -\frac{h}{15}\mathbf{S}_{N\times N} \mp \frac{1}{6}\mathbf{W}_{N\times N} - \frac{1}{6h}\mathbf{M}_{N\times N}, \\ \left.\begin{matrix}\mathbf{F}_{N\times N}\\ \mathbf{G}_{N\times N}\end{matrix}\right\} &= \frac{2h}{15}\mathbf{S}_{N\times N} \pm \frac{2}{3}\mathbf{W}_{N\times N} + \frac{4}{3h}\mathbf{M}_{N\times N}, \\ \mathbf{E}_{N\times N} &= \frac{16}{15}h\,\mathbf{S}_{N\times N} - \frac{8}{3h}\mathbf{M}_{N\times N}. \end{aligned} \tag{5.93}$$

Hence, the global set of equations can be written as

$$\left[\begin{array}{c|cc|cc|cc|c} \mathbf{A} & \mathbf{F} & \mathbf{B} & & & & & \\ \mathbf{G} & \mathbf{E} & \mathbf{F} & & & & & \\ \hline \mathbf{C} & \mathbf{G} & \mathbf{H} & \mathbf{F} & \mathbf{B} & & & \\ & & \mathbf{G} & \mathbf{E} & \mathbf{F} & & & \\ \hline & & \mathbf{C} & \mathbf{G} & \mathbf{H} & \mathbf{F} & \mathbf{B} & \\ & & & & \mathbf{G} & \mathbf{E} & \mathbf{F} & \\ \hline & & & & & \cdots & \cdots & \cdots \end{array}\right] \left[\begin{array}{c} \mathbf{q}^0 \\ \hline \mathbf{q}^1 \\ \mathbf{q}^2 \\ \hline \mathbf{q}^3 \\ \mathbf{q}^4 \\ \hline \mathbf{q}^5 \\ \mathbf{q}^6 \\ \hline \cdots \end{array}\right] = \left[\begin{array}{c} \mathbf{Q}^0 \\ \mathbf{Q}^1 \\ \hline \mathbf{Q}^2 \\ \mathbf{Q}^3 \\ \hline \mathbf{Q}^4 \\ \mathbf{Q}^5 \\ \hline \cdots \end{array}\right], \tag{5.94}$$

where

$$\mathbf{H} = \mathbf{D} + \mathbf{A}. \tag{5.95}$$

Introducing

$$\begin{aligned} \tilde{\mathbf{q}}^s_{2N\times 1} &= \begin{bmatrix} \mathbf{q}^{2s-1}_{N\times 1} \\ \mathbf{q}^{2s}_{N\times 1} \end{bmatrix}, \quad s = 1, 2, 3, \ldots, \\ \tilde{\mathbf{Q}}^s_{2N\times 1} &= \begin{bmatrix} \mathbf{Q}^{2s}_{N\times 1} \\ \mathbf{Q}^{2s+1}_{N\times 1} \end{bmatrix}, \quad s = 0, 1, 2, \ldots, \end{aligned} \tag{5.96}$$

$$\begin{aligned} \tilde{\mathbf{C}}_{2N\times 2N} &= \begin{bmatrix} \mathbf{0}_{N\times N} & \mathbf{C}_{N\times N} \\ \mathbf{0}_{N\times N} & \mathbf{0}_{N\times N} \end{bmatrix}, \qquad \tilde{\mathbf{H}}_{2N\times 2N} = \begin{bmatrix} \mathbf{G}_{N\times N} & \mathbf{H}_{N\times N} \\ \mathbf{0}_{N\times N} & \mathbf{G}_{N\times N} \end{bmatrix}, \\ \tilde{\mathbf{B}}_{2N\times 2N} &= \begin{bmatrix} \mathbf{F}_{N\times N} & \mathbf{B}_{N\times N} \\ \mathbf{E}_{N\times N} & \mathbf{F}_{N\times N} \end{bmatrix}, \end{aligned} \tag{5.97}$$

we can divide the set (5.94) into the following subsets:

$$\tilde{\mathbf{B}}\tilde{\mathbf{q}}^{1} = \tilde{\mathbf{Q}}^{0} - \begin{bmatrix} \mathbf{A} \\ \mathbf{G} \end{bmatrix} \mathbf{q}^{0}, \qquad \tilde{\mathbf{H}}\tilde{\mathbf{q}}^{1} + \tilde{\mathbf{B}}\tilde{\mathbf{q}}^{2} = \tilde{\mathbf{Q}}^{1} - \begin{bmatrix} \mathbf{C} \\ \mathbf{0} \end{bmatrix} \mathbf{q}^{0},$$
$$\tilde{\mathbf{C}}\tilde{\mathbf{q}}^{s-1} + \tilde{\mathbf{H}}\tilde{\mathbf{q}}^{s} + \tilde{\mathbf{B}}\tilde{\mathbf{q}}^{s+1} = \tilde{\mathbf{Q}}^{s}, \qquad s = 2, 3, 4, \ldots. \tag{5.98}$$

These subsets contains twice as many unknown components of the vector $\tilde{\mathbf{q}}^{s+1}$ as we would obtain for the elements shown in Fig. 5.3.

Eliminating the unknowns appearing in the intermediate nodes we arrive at the set of equations

$$\check{\mathbf{A}}\mathbf{q}^{0} + \check{\mathbf{B}}\mathbf{q}^{2} = \check{\mathbf{Q}}^{0}, \quad \check{\mathbf{C}}\mathbf{q}^{2s-2} + \check{\mathbf{H}}\mathbf{q}^{2s} + \check{\mathbf{B}}\mathbf{q}^{2s+2} = \check{\mathbf{Q}}^{2s}, \quad s = 1, 2, 3, \ldots, \tag{5.99}$$

where

$$\check{\mathbf{A}}_{N\times N} = \mathbf{A} - \mathbf{F}\mathbf{E}^{-1}\mathbf{G}, \qquad \check{\mathbf{B}}_{N\times N} = \mathbf{B} - \mathbf{F}\mathbf{E}^{-1}\mathbf{F},$$
$$\check{\mathbf{C}}_{N\times N} = \mathbf{C} - \mathbf{G}\mathbf{E}^{-1}\mathbf{G}, \qquad \check{\mathbf{H}}_{N\times N} = \mathbf{H} - \mathbf{G}\mathbf{E}^{-1}\mathbf{F}, -\mathbf{F}\mathbf{E}^{-1}\mathbf{G}, \tag{5.100}$$

$$\check{\mathbf{Q}}^{2s}_{N\times N} = \mathbf{Q}^{2s} - \mathbf{G}\mathbf{E}^{-1}\mathbf{Q}^{2s-1} - \mathbf{F}\mathbf{E}^{-1}\mathbf{Q}^{2s+1}. \tag{5.101}$$

Each of the above subsets contains the same number of unknowns as encountered when the elements of the first type were used.

In order the above recurrent scheme to be stable, the following 2 matrices must be positive definite:

$$\tilde{\mathbf{B}} + \tilde{\mathbf{C}} - \tilde{\mathbf{H}} = \begin{bmatrix} \mathbf{F}-\mathbf{G} & \mathbf{B}+\mathbf{C}-\mathbf{H} \\ \mathbf{E} & \mathbf{F}-\mathbf{G} \end{bmatrix} > 0,$$
$$\tilde{\mathbf{B}} + \tilde{\mathbf{C}} + \tilde{\mathbf{H}} = \begin{bmatrix} \mathbf{F}+\mathbf{G} & \mathbf{B}+\mathbf{C}+\mathbf{H} \\ \mathbf{E} & \mathbf{F}+\mathbf{G} \end{bmatrix} > 0. \tag{5.102}$$

Neglecting damping we obtain such stability constraints on the time dimension:

$$h^2 \le \frac{5}{2\omega^2}, \qquad \frac{3}{\omega^2} \le h^2 \le \frac{15}{\omega^2_{\max}}. \tag{5.103}$$

Taking the largest value $h^2 = 15/\omega^2_{max}$, we must additionally check whether none of the lower eigenfrequencies falls into the interval $6^{-1/2}\omega_{max}$ to $5^{-1/2}\omega_{max}$. When this additional constraint is satisfied, the time dimension $2h$ of the Serendipian element is 2.236 times longer than the same dimension of the element of the first type.

It is worth noting that dumping reduces the gap between stable regions. Under sufficiently large dumping this gap disappears.

Similar procedure holds for Serendipian elements with more than one intermediate node. For the element shown in Fig. 5.5b the stable region is described by the following inequalities (damping neglected):

$$h^2 \leq \frac{1.097}{\omega^2}, \qquad \frac{10}{9\omega^2} \leq h^2 \leq \frac{14}{3\omega^2}, \qquad \frac{20}{3\omega^2} \leq \frac{18.903}{\omega_{\max}^2}. \tag{5.104}$$

This means that taking $h^2 = 18.903/\omega_{max}^2$ we must check whether each of the lower eigenfrequencies falls into one of the following intervals:

$$\omega \leq 0.2409\omega_{\max}, \qquad 0.2424\omega_{\max} \leq \omega \leq 0.4969\omega_{\max}, \qquad \omega \geq 0.5939\omega_{\max}. \tag{5.105}$$

Provided this condition is satisfied, we can use Serendipian elements that have the time dimension $3h$ 3.765 times larger than the same dimension of the elements of the first type.

Therefore, the time dimension of the Serendipian elements is not substantially greater but their accuracy is usually higher.

Finally let us consider shortly applying different shape functions for virtual increments than the shape functions representing real displacements. Since virtual displacement increments can be arbitrary, there is no reason to restrict this freedom to the nodal values. The approach presented here has been proposed and analysed in [71].

Hence, let us assume that real and virtual displacements are described inside the rectangular space–time element as follows:

$$\begin{aligned} \mathbf{u}_{3\times1}(\mathbf{x}, t) &= [\varphi^t_{1\times2}(t) \otimes \varphi^x_{3\times N_e}(\mathbf{x})]\, \mathbf{q}^{(e)}_{2N_e\times1}, \\ \delta\mathbf{u}_{3\times1}(\mathbf{x}, t) &= [\psi^t_{1\times2}(t) \otimes \varphi^x_{3\times N_e}(\mathbf{x})]\, \delta\mathbf{q}^{(e)}_{2N_e\times1}, \end{aligned} \tag{5.106}$$

The shape matrix φ^t is given by Eq. (5.34) and the matrix of weight functions ψ^t has the following entries:

$$\psi^t_{1\times2} = \left[\frac{1-\tau}{2} + 5\beta\frac{\tau-\tau^3}{2} \quad \frac{1+\tau}{2} - 5\beta\frac{\tau-\tau^3}{2} \right]. \tag{5.107}$$

Performing the same operations as shown before and taking into account relations [compare Eqs. (5.47) to (5.49)]

$$\begin{aligned} \int_{-1}^{1} \psi^{t\,\mathrm{T}}_{2\times1}\, \varphi^t_{1\times2} \frac{h}{2}\, \mathrm{d}\tau &= \mathbf{J}_{2\times2} - \frac{\beta}{6} h^2 \mathbf{J}''_{2\times2}, \\ \int_{-1}^{1} \psi^{t\,\mathrm{T}}_{2\times1}\, \mathbf{B}^t_{1\times2} \frac{h}{2}\, \mathrm{d}\tau &= \mathbf{J}'_{2\times2}, \\ \int_{-1}^{1} \left(\frac{\mathrm{d}}{\mathrm{d}t}\psi^t\right)^{\mathrm{T}}_{2\times1} \mathbf{B}^t_{1\times2} \frac{h}{2}\, \mathrm{d}\tau &= \mathbf{J}''_{2\times2}, \end{aligned} \tag{5.108}$$

we obtain similar stiffness matrix of the space–time element as the one given in Eq. (5.33):

$$\mathbf{K}^{(e)}_{2N_e\times 2N_e} = \left(\mathbf{J}_{2\times 2} - \frac{\beta}{6}h^2\mathbf{J}''_{2\times 2}\right)\otimes \mathbf{S}^{(e)}_{N_e\times N_e} + \mathbf{J}'_{2\times 2}\otimes \mathbf{W}^{(e)}_{N_e\times N_e} + \mathbf{J}''_{2\times 2}\otimes \mathbf{M}^{(e)}_{N_e\times N_e}. \tag{5.109}$$

After transition to the global frame of reference [compare Eqs. (5.54), (5.55)] and after matrices $\mathbf{S}^{(e)}, \mathbf{W}^{(e)}, \mathbf{M}^{(e)}$ were assembled, we obtain the stiffness matrix of the assembled element E given by Eq. (5.58) where

$$\beta = 6\alpha - 1, \tag{5.110}$$

and

$$\begin{aligned} \left.\begin{matrix}\mathbf{A}_{N\times N}\\ \mathbf{D}_{N\times N}\end{matrix}\right\} &= h\left(\frac{1}{2}-\alpha\right)\mathbf{S}_{N\times N} \mp \frac{1}{2}\mathbf{W}_{N\times N} - \frac{1}{h}\mathbf{M}_{N\times N},\\ \left.\begin{matrix}\mathbf{B}_{N\times N}\\ \mathbf{C}_{N\times N}\end{matrix}\right\} &= h\alpha\mathbf{S}_{N\times N} \pm \frac{1}{2}\mathbf{W}_{N\times N} + \frac{1}{h}\mathbf{M}_{N\times N}. \end{aligned} \tag{5.111}$$

The difference scheme (5.65) takes now the form

$$\mathbf{M}\frac{\mathbf{q}^{s-1} - 2\mathbf{q}^s + \mathbf{q}^{s+1}}{h^2} + \mathbf{W}\frac{\mathbf{q}^{s+1} - \mathbf{q}^{s-1}}{2h} + \mathbf{S}\left[\alpha\mathbf{q}^{s-1} + (1-2\alpha)\mathbf{q}^s + \alpha\mathbf{q}^{s+1}\right] = \frac{1}{h}\mathbf{Q}^s. \tag{5.112}$$

It easy to notice that this result coincides with the Newmark scheme (5.67) under $\delta = 1/2$. Moreover, assuming $\alpha = 1/6$ we would obtain the relation (5.65).

The condition of numerical stability for the recurrent process (5.112) reads

$$h^2 \le \frac{12}{(1-2\beta)\,\omega^2_{\max}}. \tag{5.113}$$

It is seen that for $\beta = 1/2$ the time dimension of the element can grow arbitrary and the recurrent process becomes unconditionally stable. Numerical experiments revealed, however, that such elements lead to big errors in amplitudes and frequencies. It is better, therefore, to apply the conditionally stable recurrence under $\beta = 0$.

Speaking in the present Section about the specific version of the STEM, we considered an isotropic viscoelastic 3D-solid. The same approach remains valid for dynamics of such 1D structures as bars and cables or 2D-structures like plates, membranes and shells. Some differences appear only in the notation for co-ordinates and functions describing displacements, strains and stresses.

Let us recall a part of such a notation for a prismatic bar having solid cross-section (the most frequently used structural element). We shall partially use the notation introduced in [83].

First, for practical reasons we introduce the co-ordinates x, y, z instead of $x_i, (i = 1, 2, 3)$. We assume that the x-axis of the local reference frame coincides with the axis of the bar, whereas y and z are the main central axes of the inertia for the cross-section.

Further we replace the displacement vector $\mathbf{u}_{3\times1}(\mathbf{x},t)$ given by Eq. (2.27), Part I and the relevant vector of inertia forces $\mathbf{U}_{3\times1}(\mathbf{x},t)$ [compare (5.19)] by the vector describing the displacements of the bar (the axial displacements and the rotations of the cross-section)

$$\mathbf{f}_{6\times1}(x,\, t) = \{u,\, v,\, w,\, \varphi,\, \psi,\, \vartheta\}, \tag{5.114}$$

and the vector of loads acting on the unit length of the bar

$$\mathbf{F}_{6\times1}(x,\, t) = \{U,\, V,\, W,\, \Phi,\, \Psi,\, \Theta\}. \tag{5.115}$$

Similarly, we replace matrices $\boldsymbol{\varepsilon}_{9\times1}$ – Eq. (5.2), $\mathbf{D}_{9\times3}$ – Eq. (5.3), $\boldsymbol{\sigma}_{9\times1}$ – Eq. (5.4) and $\mathbf{C}_{9\times9}$ – Eq. (5.5) by

$$\mathbf{s}_{12\times1}(x,\, t) = \{\varepsilon,\, \beta_y,\, \beta_z,\, \kappa_x,\, \kappa_y,\, \kappa_z,\, \gamma_{tx},\, \gamma_{ty},\, \gamma_{tz},\, \kappa_{tx},\, \kappa_{ty},\, \kappa_{tz}\}, \tag{5.116}$$

$$\mathbf{D}_{12\times6} = \left[\begin{array}{cc} \mathbf{I}_{3\times3}\frac{\partial}{\partial x} & \mathbf{e}_{3\times3} \\ \mathbf{0}_{3\times3} & \mathbf{I}_{3\times3}\frac{\partial}{\partial x} \\ \hline \multicolumn{2}{c}{\mathbf{I}_{6\times6}\frac{\partial}{\partial t}} \end{array}\right], \qquad \mathbf{e}_{3\times3} = \begin{bmatrix} 0 & 0 & 0 \\ 0 & 0 & -1 \\ 0 & 1 & 0 \end{bmatrix}. \tag{5.117}$$

$$\mathbf{S}_{12\times1}(x,\, t) = \{N,\, T_y,\, T_z,\, M_x,\, M_y,\, M_z,\, T_{tx},\, T_{ty},\, T_{tz},\, M_{tx},\, M_{ty},\, M_{tz}\}, \tag{5.118}$$

$$\begin{aligned}\mathbf{C}_{12\times12} = \operatorname{diag}[&E^* A_x,\, G^* A_y,\, G^* A_z,\, G^* J_x,\, E^* J_y,\, E^* J_z, \\ &-\mu,\, -\mu,\, -\mu,\, -I_x,\, -I_y,\, -I_z],\end{aligned} \tag{5.119}$$

where ε means the axial strain, β_y, β_z are the mean angles of shear, κ_x denotes the axial warping of the bar, κ_y, κ_z are its curvatures, γ_{tx}, γ_{ty}, γ_{tz} are linear velocities, κ_{tx}, κ_{ty}, κ_{tz} – angular velocities, N is the axial force, T_y, T_z – shear forces, M_x – the twisting moment, M_y, M_z – the bending moments, T_{tx}, T_{ty}, T_{tz} – linear impulses with reversed sign , M_{tx}, M_{ty}, M_{tz} – angular impulses with reversed sign , A_x – actual area of the cross-section, $A_y = A_x/k_y, A_z = A_x/k_z$ – reduced cross-sectional area for shear (compare Eq. (4.32), [83]),

$$k_y = \frac{A_x}{J_y^2}\int_{A_x}\frac{S_y^2}{b_y^2}\mathrm{d}A_x, \qquad k_z = \frac{A_x}{J_z^2}\int_{A_x}\frac{S_z^2}{b_z^2}\mathrm{d}A_x \tag{5.120}$$

are the shape factors for the cross-section with J_x standing for torsional rigidity (compare Sect. 2.2, [83]),

$$J_y = \int_{A_x} z^2\,\mathrm{d}A_x\,, \qquad J_z = \int_{A_x} y^2\,\mathrm{d}A_x \tag{5.121}$$

are the moments of inertia of the cross-section,

$$E^* = E\left(1 + t^*\frac{\partial}{\partial t}\right), \qquad G^* = G\left(1 + t^*\frac{\partial}{\partial t}\right) \tag{5.122}$$

– the generalized modulae of elasticity, $t^* = \frac{\eta}{G}$ – the so-called retardation time,

$$\mu = \int_A \varrho \, \mathrm{d}A \tag{5.123}$$

– the mass attached to the unit length of the bar, and

$$I_y = \int_A \varrho \, z^2 \, \mathrm{d}A, \qquad I_z = \int_A \varrho \, y^2 \, \mathrm{d}A, \qquad I_x = I_y + I_z \tag{5.124}$$

– the mass moments of inertia of the cross-section.

In Eqs. (5.123) and (5.124) one has to take into account the cross-sectional area A_x of the bar itself together with the same area of the parts of structure rigidly attached to that bar. Hence, integration is performed over the total cross-sectional area $A \geq A_x$.

Rectangular space–time elements, build according to the above mentioned principles, were applied for solving various test problems. An exact agreement with analytical solutions was obtained for the longitudinal vibrations of the bar subjected to the axial impact [76] or for the bar falling on the rigid obstacle [85]. For other problems numerical solutions were approximate. In particular, in [69] the dynamic response of the industrial chimney to the displacements of its foundation has been calculated, vibrations of a string caused by the force moving at constant velocity were considered in [75] and in [27] vibrations of a car driven over a rough terrain were found.

Viscoelastic space–time elements and some geometrically non-linear problems are treated in the papers [136, 137].

5.4 Non-rectangular Space–Time Elements

In the previous Section we presented the general method for solving non-stationary problems in dynamics of structures. This method can be applied to any 3D-, 2D- or 1D-solid. Now we are going to deal with skeletal structures only but we intend to generalize the way in which space–time objects are digitized.

Figure 5.6a shows an example of a planar space–time object divided into triangular elements often used in the FEM. When a static problem is solved by the FEM, the density of mesh is increased at those places where the concentration of stresses is anticipated. That improves the accuracy of solution. Similarly, using the STEM it is reasonable to take more dense mesh in the vicinity of points where external impulses are applied. Hence, if a concentrated load acts during a short time (the thick line segment at Fig. 5.6a) at a certain point of the structure, then we anticipate large gradients of displacements and stresses and we cover this region with refined mesh. It is important to keep the number of equations of equilibrium for nodal impulses equal to the number of unknown nodal displacements. Due to the given ini-

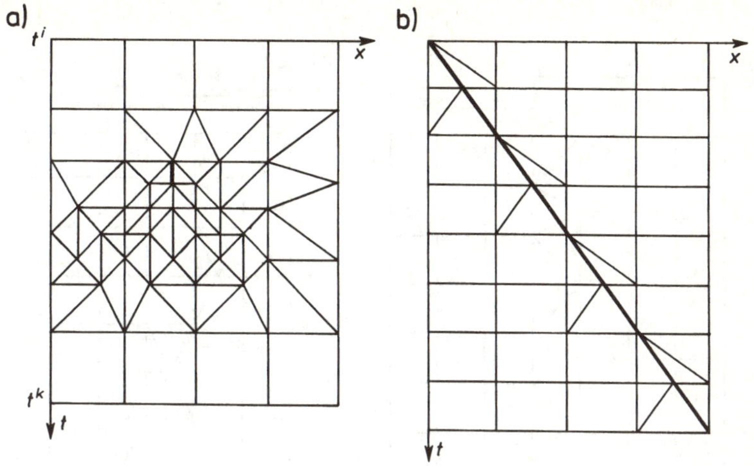

Fig. 5.6. Increasing mesh density near: **a)** fixed point load, **b)** moving load

tial values, the unknowns include the displacements of all nodes except the nodes situated at the line $t = t^i$ corresponding to the onset of the observed process. At our disposal are the equations of equilibrium for all nodes except the nodes laying on the line $t = t^k$ that marks the end of observation. In order to maintain the balance between the number of equations and the number of unknowns, one has to keep equal number of nodes in the cross-sections $t = t^i$ and $t = t^k$. Some authors seem to be unaware of that, as follows from a figure presented in [62].

If one considers a concentrated force moving along the beam or along the string, then it is convenient to replace rectangular elements along the straight line (Fig. 5.6b) that represents the position of such a load in the space–time by triangular elements. Then the force acts all the time along the edges of elements. For the loading moving along the string such a digitization results in the exact solution [79]. As mentioned before, the rectangular mesh gives an approximate result for this problem [75].

It may happen that using the rectangular mesh it is preferable to adopt different time dimensions in different regions. A beam partially loaded by heavy mass is such a case (Fig. 5.7a). However, the boundary of inertial load need not be necessarily parallel to the time axis. Such a boundary is skew when a column of heavy vehicles moves along the bridge (Fig. 5.7b). Then triangular elements are again useful in the transition zone.

Let us consider a triangular space–time element having nodes arbitrary positioned on the plane xt (Fig. 5.8). We describe the displacements through their nodal values in the known way

$$\mathbf{f}_{6\times1}(x,\, t) = \boldsymbol{\varphi}_{6\times18}(x,\, t)\, \mathbf{q}^{(e)}_{18\times1}, \tag{5.125}$$

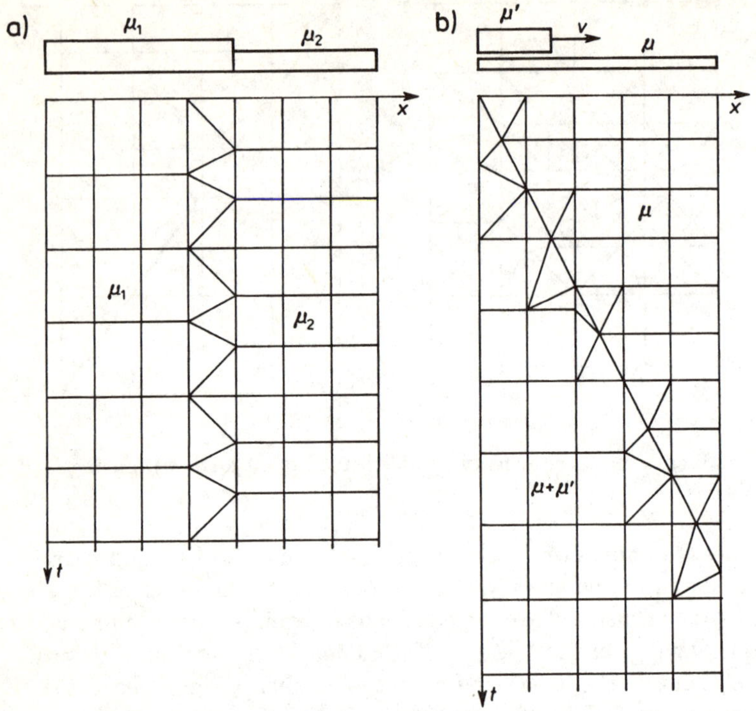

Fig. 5.7. Examples of triangular space–time elements used to connect regions of different mesh size

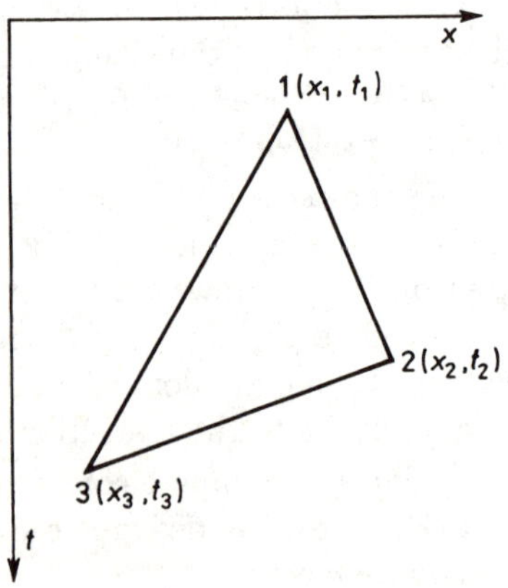

Fig. 5.8. Triangular space–time element

where

$$\varphi_{6\times 18}(x,\,t) = [\varphi_1\;\varphi_2\;\varphi_3] \otimes \mathbf{I}_{6\times 6},$$

$$\varphi_i = \frac{1}{2\Omega}\det\begin{bmatrix} 1 & x & t \\ 1 & x_{i+1} & t_{i+1} \\ 1 & x_{i+2} & t_{i+2} \end{bmatrix}, \tag{5.126}$$

describe the shape matrix and the vector of nodal displacements has the following components:

$$\mathbf{q}^{(e)}_{18\times 1} = \{\mathbf{q}_1,\,\mathbf{q}_2,\,\mathbf{q}_3\}, \qquad \mathbf{q}_i = \{u_i,\,v_i,\,w_i,\,\varphi_i,\,\psi_i,\,\vartheta_i\}. \tag{5.127}$$

The symbol Ω denotes space–time area of the triangle:

$$\Omega = \frac{1}{2}\det\begin{bmatrix} 1 & x_1 & t_1 \\ 1 & x_2 & t_2 \\ 1 & x_3 & t_3 \end{bmatrix}, \tag{5.128}$$

and the quantities x_4, x_5, t_4, t_5 that can appear in Eq. (5.126) should be replaced by x_1, x_2, t_1, t_2.

Using Eqs. (5.116) to (5.119) we calculate the strain matrix

$$\mathbf{B}_{12\times 18} = \mathbf{D}_{12\times 6}\,\varphi_{6\times 18} = [\mathbf{B}_1\;\mathbf{B}_2\;\mathbf{B}_3], \tag{5.129}$$

$$\mathbf{B}_i = \mathbf{D}\,\varphi_i \mathbf{I}_{6\times 6} = \left[\begin{array}{cc} -\frac{\Delta t_i}{2\Omega}\mathbf{I}_{3\times 3} & \varphi_i \mathbf{e}_{3\times 3} \\ \mathbf{0}_{3\times 3} & -\frac{\Delta t_i}{2\Omega}\mathbf{I}_{3\times 3} \\ \hline \multicolumn{2}{c}{\frac{\Delta x_i}{2\Omega}\mathbf{I}_{6\times 6}} \end{array}\right], \qquad \begin{aligned} \Delta t_i &= t_{i+2} - t_{i+1}, \\ \Delta x_i &= x_{i+2} - x_{i+1}. \end{aligned} \tag{5.130}$$

The stiffness matrix of the considered element can be written as

$$\mathbf{K}^{(e)}_{18\times 18} = \int_\Omega \mathbf{B}^{\mathrm{T}}_{18\times 12}\,\mathbf{C}_{12\times 12}\,\mathbf{B}_{12\times 18}\,\mathrm{d}\Omega = \begin{bmatrix} \mathbf{K}_{11} & \mathbf{K}_{12} & \mathbf{K}_{13} \\ \mathbf{K}_{21} & \mathbf{K}_{22} & \mathbf{K}_{23} \\ \mathbf{K}_{31} & \mathbf{K}_{32} & \mathbf{K}_{33} \end{bmatrix}, \tag{5.131}$$

where

$$\begin{aligned} \mathbf{K}_{ik} &= \int_\Omega \mathbf{B}_i^{\mathrm{T}}\mathbf{C}\,\mathbf{B}_k \mathrm{d}\Omega \\ &= \frac{\Delta t_i\,\Delta t_k}{4\Omega}\mathrm{diag}\,[EA_x\; GA_y\; GA_z\; GJ_x\; EJ_y\; EJ_z] \\ &\quad - \frac{\Delta x_i\,\Delta x_k}{4\Omega}\mathrm{diag}\,[\,\mu \quad \mu \quad \mu \quad I_x \quad I_y \quad I_z\,] + \end{aligned}$$

$$
+\begin{bmatrix}
0 & 0 & 0 & 0 & 0 & 0 \\
0 & 0 & 0 & 0 & 0 & \frac{GA_y\Delta t_i}{6}+\frac{\eta A_y\Delta t_i\Delta x_k}{4\Omega} \\
0 & 0 & 0 & 0 & -\frac{GA_z\Delta t_i}{6}-\frac{\eta A_z\Delta t_i\Delta x_k}{4\Omega} & 0 \\
0 & 0 & 0 & 0 & 0 & 0 \\
0 & 0 & -\frac{GA_z\Delta t_k}{6} & 0 & \frac{GA_z\Omega}{12}(1+\delta_{ik})+\frac{\eta A_z\Delta x_k}{6} & 0 \\
0 & \frac{GA_y\Delta t_k}{6} & 0 & 0 & 0 & \frac{GA_y\Omega}{12}(1+\delta_{ik})+\frac{\eta A_y\Delta x_k}{6}
\end{bmatrix}, \tag{5.132}
$$

and δ_{ik} is the Kronecker's symbol. Note that the linearity of shape functions precludes us from modelling exactly the viscous damping. The latter appears only in the shear forces dependent on shear strains as well as on their velocities. This deficiency could be eliminated by adopting, e.g., triangular elements with 6 nodes.

Let us note finally that in [9, 12] trapezoidal elements were used for solving the beam that rests unilaterally on the elastic foundation. The originality of this idea lies in the fact that elements change their shape: the skew edges adapt themselves to the moving area of the contact between the deformed beam and the foundation.

5.5 Uncoupled Equilibrium Equations for Impulses

A possibility to uncouple equations in the STEM has been pointed out by Oden in [125]. This is achieved by subdividing each rectangular space–time element into 2 triangular elements. Fig. 5.9 shows several regular meshes of this type. It is easy to notice that writing the nodal equilibrium equations for impulses in the specific sequence indicated under the figure we obtain a single new unknown in each equation.

This proposal remained unused for a long time until Witkowski [178, 179] recalled it. The advantages of this version of STEM were demonstrated on several examples. Especially the contributions [10, 11] should be mentioned, where the Oden's idea was generalized for 2D and 3D-problems and the relevant computer code was developed.

The disadvantage of the above mentioned STEM-version is that it introduces anisotropy of the space–time object (Fig. 5.9a) or requires to consider separately 2 subsets of nodes. In each of them the equations of equilibrium must be written in different way (Figs. 5.9b,c,d). Hence, we proceed now to the presentation of alternative method free from such deficiencies.

Let us subdivide the space–time object related to an arbitrary skeletal

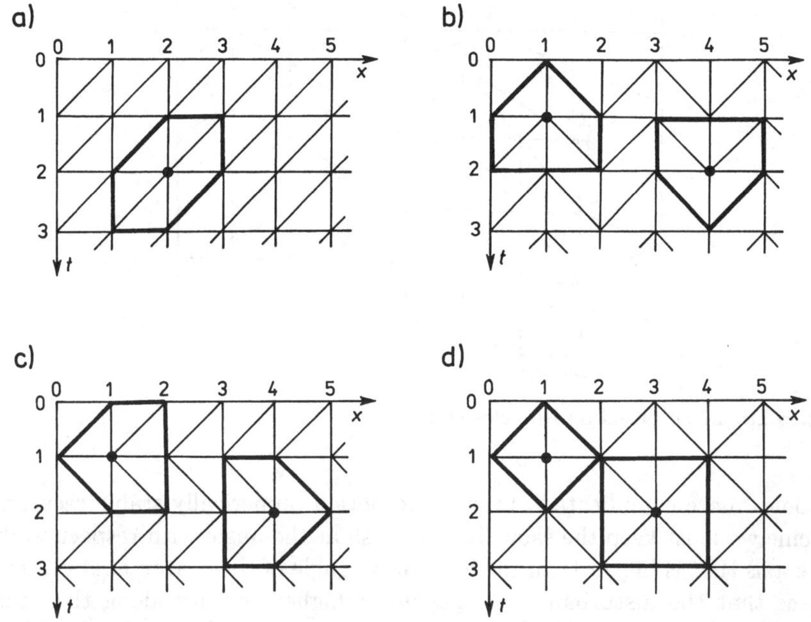

Fig. 5.9. Examples of regular meshes based on rectangular triangles

structure into the isosceles triangles as shown in Fig. 5.10. Each node of such a mesh has 2 indices: i related to the spatial position and s related to the time instant. Under the enumeration scheme given in Fig. 5.10 the sum of those indices is always even. This simply means that the nodes with odd indices i appear in odd instances s and the nodes with even i-indices correspond to even s-instances. Only in the cross-section $t = 0$, i.e. at the instance $s = 0$ all indices i are present. The initial conditions can be formulated in such a way that the cross-section $t = 0$ becomes obsolete.

The equilibrium equations of nodes situated in the cross-section $s = 0$ and having odd i-indices allow us to compute the displacements $\mathbf{q}_i^1$. Considering the nodes situated in the same cross-section $s = 0$ but having even i-indices we can use a single equilibrium equation for each node and find the displacements $\mathbf{q}_i^2$. Further we exploit a single equation for each subsequent node (i, s) containing a single new unknown $\mathbf{q}_i^{s+2}$. A single equation means here a set of equations written for each degree of freedom of a particular node. A single unknown means a vector containing all displacements of that node.

This version of the STEM is also conditionally stable. For wave propagation problems the skew mesh lines should coincide with the characteristics (compare [84]). Numerical experiments have shown that such a mesh leads to the most accurate, for some problems even exact, results.

For equations of motion that allow for the infinite velocity of propagation of disturbances (e.g. for a beam vibrating in bending mode) the problem

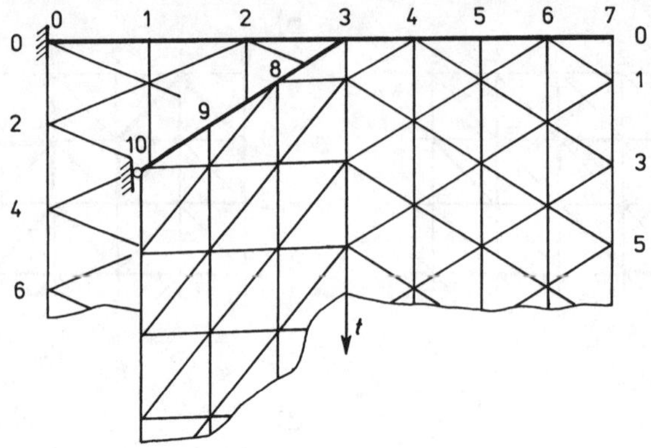

Fig. 5.10. Mesh based on isosceles triangles

becomes more complicated. In order to obtain numerically stable recurrent scheme we must keep the skew lines of mesh at the angle with respect to the time axis that is bigger than the inclination angle of the characteristics. This means that the disturbances propagate at higher velocity along the beam than the sound waves. However that velocity remains finite. This result of the STEM is in better agreement with the fundamental principles of physics than the exact solutions of differential equations that lead to the velocity of propagation exceeding the velocity of light. In [81] we used the above mentioned space–time elements for solving the vibrations of elastic cable network under falling mass

The problem of longitudinal vibrations of the cable of shaft hoist belongs to the more interesting practical problems solved by the STEM. The cage attached to a long cable undergoes very large vertical oscillations after being halted deep under the earth's surface. In order to diminish them one has to apply an adequate control to the hoisting drum. This problem has been formulated in [151] and an attempt has been made to solve it analytically. A numerical solution by means of the triangular space–time elements proposed by Oden has been given in [178]. Now we present an application of the isosceles triangular elements to the same problem.

Let us consider a cable of mass μ [kg m^{-1}] with a concentrated mass m [kg] attached to its end. Let the point G corresponding to the level of the axis of elevator's wheel (Fig. 5.11a) be situated at $z = 0$ and the concentrated mass be at $z = l_0$. At the time instant $t = 0$ the wheel begins to rotate which brings gradually into motion the points of the hanging cable. These points having initial position z undergo the displacement $w(z, t)$. On the other hand, we treat the displacement of the concentrated mass

$$w(l_0,\, t) = f(t). \tag{5.133}$$

as being given.

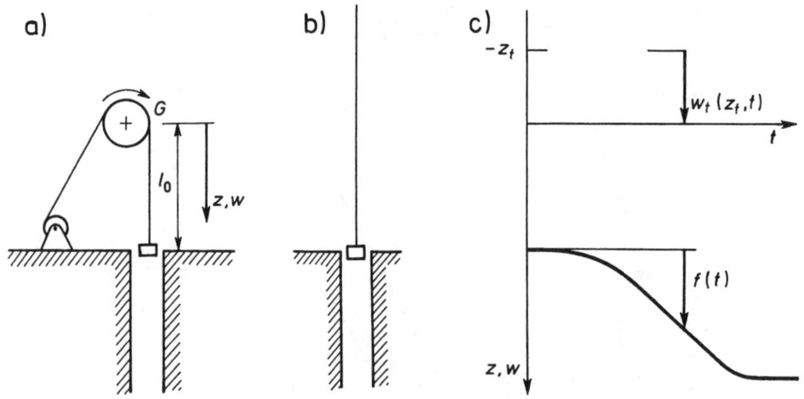

Fig. 5.11. **a)** Scheme of shaft hoist, **b)** concentrated mass on infinitely long cable, **c)** space–time interpretation of motion of mass–cable system

Hence, we deal with a peculiar problem of dynamic analysis: all initial conditions are given as usual but additionally all boundary conditions are prescribed at the lower end of the cable. On the contrary, at its upper end neither the velocity of cable nor the force acting in it is known.

It is convenient to replace such a problem by an equivalent one where the concentrated mass is attached to the infinitely long cable (Fig. 5.11b). The unknown function of displacement $w(z,t)$ must fulfil in the interval $0 < z < l_0$ the initial conditions

$$w(z,\,0) = 0, \qquad \left.\frac{\partial w}{\partial t}\right|_{t=0} = 0 \tag{5.134}$$

and the boundary condition (5.133). To simplify the problem we neglect the displacements caused by the self-weight of the cable and the mass. In other words, we treat $w(z,t)$ as dynamic increments of static displacements.

In order to establish which point of the cable defined by the co-ordinate $z_t < 0$ will cross at the instant t the level of the point G and will enter the considered interval (Fig. 5.11c), we must solve the equation

$$z_t + w(z_t,\,t) = 0. \tag{5.135}$$

Only the part of cable situated in the interval

$$z_t = -w(z_t,\,t) < z < l_0, \tag{5.136}$$

will undergo free longitudinal vibrations. The rest of the cable (its upper part) will be constrained by the wheel or drum.

We subdivide the considered space–time object into the triangular elements (Fig. 5.12) and we build the stiffness matrices of such elements according to Eq. (5.132). In this particular case we have the longitudinal vibrations

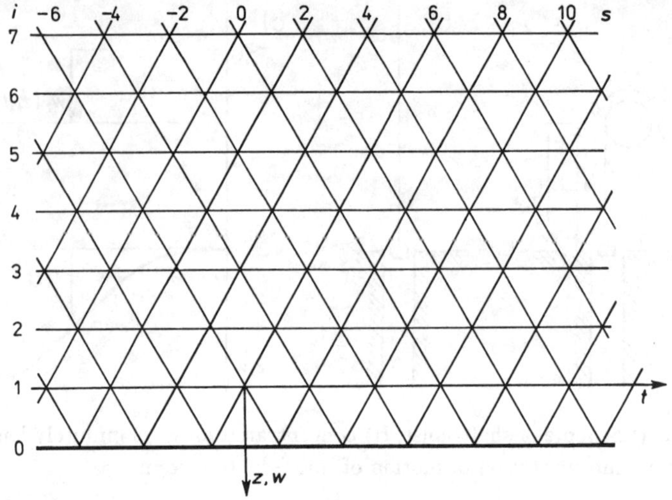

Fig. 5.12. Triangular space–time mesh

only. Hence, the stiffness matrix reads

$$\mathbf{K}^{(e)}_{3\times 3} = \frac{EAh}{4a}\begin{bmatrix} 1 & -2 & 1 \\ -2 & 4 & -2 \\ 1 & -2 & 1 \end{bmatrix} - \frac{\mu a}{4h}\begin{bmatrix} 1 & 0 & -1 \\ 0 & 0 & 0 \\ -1 & 0 & 1 \end{bmatrix}. \tag{5.137}$$

The assumption that the skew edges of elements coincide with the characteristics amounts to the assumption that the ratio a/h of the dimensions is equal to the velocity of longitudinal wave in the cable:

$$\frac{a}{h} = c_E = \sqrt{\frac{EA}{\mu}}. \tag{5.138}$$

Under this assumption the stiffness matrix becomes

$$\mathbf{K}^{(e)}_{3\times 3} = \frac{EA}{2c_E}\begin{bmatrix} 0 & -1 & 1 \\ -1 & 2 & -1 \\ 1 & -1 & 0 \end{bmatrix}. \tag{5.139}$$

Additionally we must define the stiffness matrix for the 1D-element representing a segment of the line of life of the concentrated mass (Fig. 5.13). Taking the vector of nodal displacements and the shape matrix as

$$\mathbf{q}^{(m)}_{2\times 1} = \{w_a,\, w_p\}, \qquad \boldsymbol{\varphi}^{(t)}_{1\times 2} = \left[\frac{1-\tau}{2} \quad \frac{1+\tau}{2}\right], \tag{5.140}$$

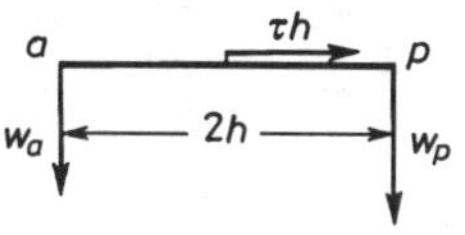

Fig. 5.13. One-dimensional space–time element

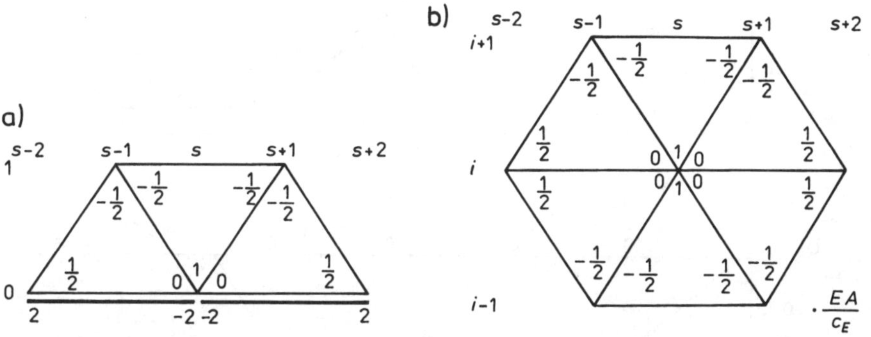

Fig. 5.14. Space–time elements adjacent to: **a)** node situated at the mass level, **b)** arbitrary node (i, s)

we obtain

$$\mathbf{B}_{1\times 2} = \frac{\mathrm{d}}{\mathrm{d}t}\varphi = \frac{1}{2h}[-1 \;\; 1],$$
$$\mathbf{K}^{(m)}_{2\times 2} = -\int_{-1}^{1} \mathbf{B}^{\mathrm{T}}\, m\, \mathbf{B}\, h\, \mathrm{d}\tau = -\frac{m}{2h}\begin{bmatrix} 1 & -1 \\ -1 & 1 \end{bmatrix}. \tag{5.141}$$

It was assumed in the numerical example that

$$m = 4a\mu = 4aEA/c_E^2, \tag{5.142}$$

This gives

$$\mathbf{K}^{(m)}_{2\times 2} = \frac{2EA}{c_E}\begin{bmatrix} -1 & 1 \\ 1 & -1 \end{bmatrix}. \tag{5.143}$$

After stiffness matrices were established, we can build the equilibrium equations for the nodes. Due to the omitted self-weight, these equations are homogeneous. For the nodes situated at the level of the concentrated mass (Fig. 5.14a) such equations read

$$2,5\,(w_0^{s-2} + w_0^{s+2}) - 3w_0^s - w_1^{s-1} - w_1^{s+1} = 0, \tag{5.144}$$

whereas for the rest of nodes (Fig. 5.14b) they have the form

$$w_i^{s-2} + w_i^{s+2} + 2w_i^s - w_{i-1}^{s-1} - w_{i-1}^{s+1} - w_{i+1}^{s-1} - w_{i+1}^{s+1} = 0. \tag{5.145}$$

Contrary to the problems considered previously, the unknowns are now the displacements w_{i+1}^{s+1} instead of w_i^{s+2}, since all boundary values are given at

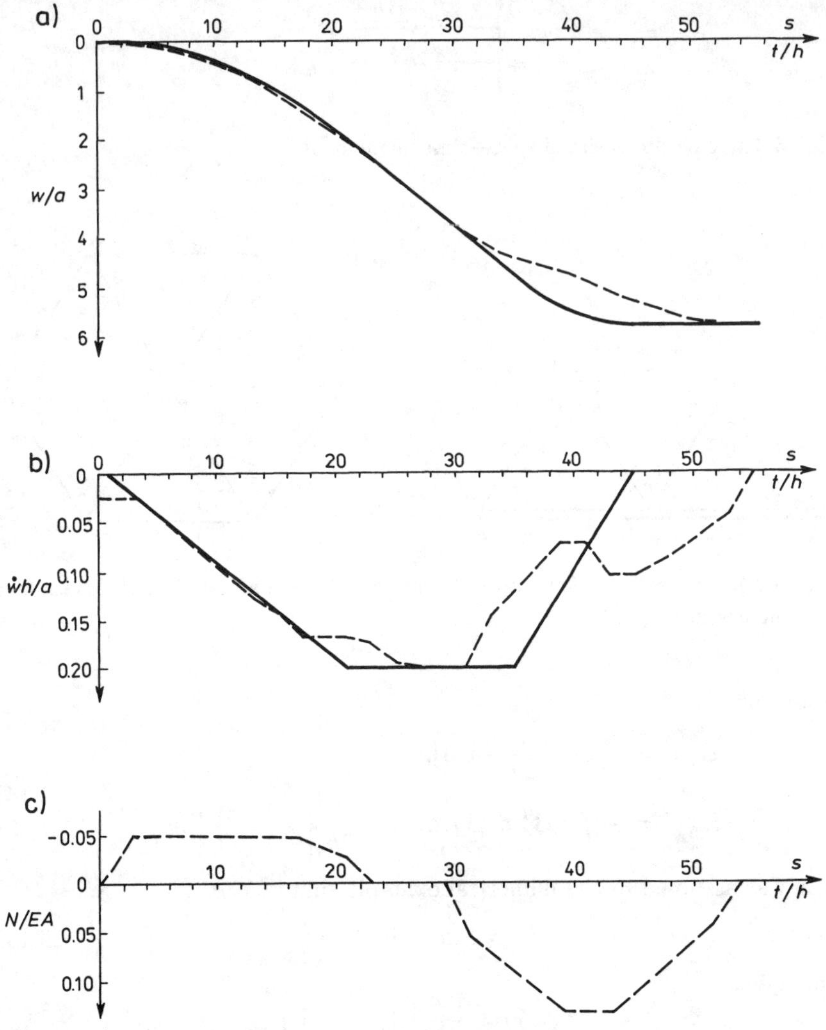

Fig. 5.15. Graphs of: **a)** displacements, **b)** velocities, **c)** force increments at the level of mass (solid line) and at the level of wheel (dotted line)

the line $i = 0$. Hence, the recurrent formulae assume the form

$$\begin{aligned} w_1^{s+1} &= 2,5\,(w_0^{s-2} + w_0^{s+2}) - 3w_0^s - w_1^{s-1}, \\ w_{i+1}^{s+1} &= w_i^{s-2} + w_i^{s+2} + 2w_i^s - w_{i-1}^{s-1} - w_{i-1}^{s+1} - w_{i+1}^{s-1}. \end{aligned} \tag{5.146}$$

Assuming $l_0 = a = 100$ m and taking the required for of the life line (5.133) for the concentrated mass as shown in Fig. 5.15a by a solid line, we performed the calculations according to the algorithm given in Fig. 5.16. The obtained values of displacements are given in Table 5.1. It's last row contains the

Table 5.1. Nodal displacements of shaft hoist

i \ s	-6		−4		−2		0		2		4		6		8		10		12		14
7	0		5		12		21		32		45		60		77		93		113		137
6		0		5		12		21		32		45		60		74		92		114	
5	0		0		5		12		21		32		45		57		73		93		117
4		0		0		5		12		21		32		42		56		74		96	
3	0		0		0		5		12		21		29		41		57		77		101
2		0		0		0		5		12		18		28		42		60		82	
1	0		0		0		0		5		9		17		29		45		65		89
0		0		0		0		0		2		8		18		32		50		72	
w_t^s	0		0		0		0		5		10		18		31		48		69		95

i \ s		16		18		20		22		24		26		28		30		32		34
7		160		185		212		241		272		305		340		370		396		418
6	140		165		192		221		252		285		320		360		390		416	
5		145		172		201		232		265		300		340		380		410		436
4	122		152		181		212		245		280		320		360		400		430	
3		129		161		192		225		260		300		340		380		420		450
2	108		138		172		205		240		280		320		360		400		440	
1		117		149		185		220		260		300		340		380		420		460
0	98		128		162		200		240		280		320		360		400		440	
w_t^s		124		158		192		226		261		300		340		380		409		433

i \ s		36		38		40		42		44		46		48		50		52		54
7		436		450		470		490		516		538		556		570		580		580
6	438		456		470		490		516		538		556		570		580		580	
5		458		476		490		516		538		556		570		580		580		580
4	456		478		496		516		538		556		570		580		580		580	
3		476		498		522		538		556		570		580		580		580		580
2	470		496		524		544		556		570		580		580		580		580	
1		490		522		546		562		570		580		580		580		580		580
0	480		516		544		564		576		580		580		580		580		580	
w_t^s		452		467		482		503		524		543		559		571		580		580

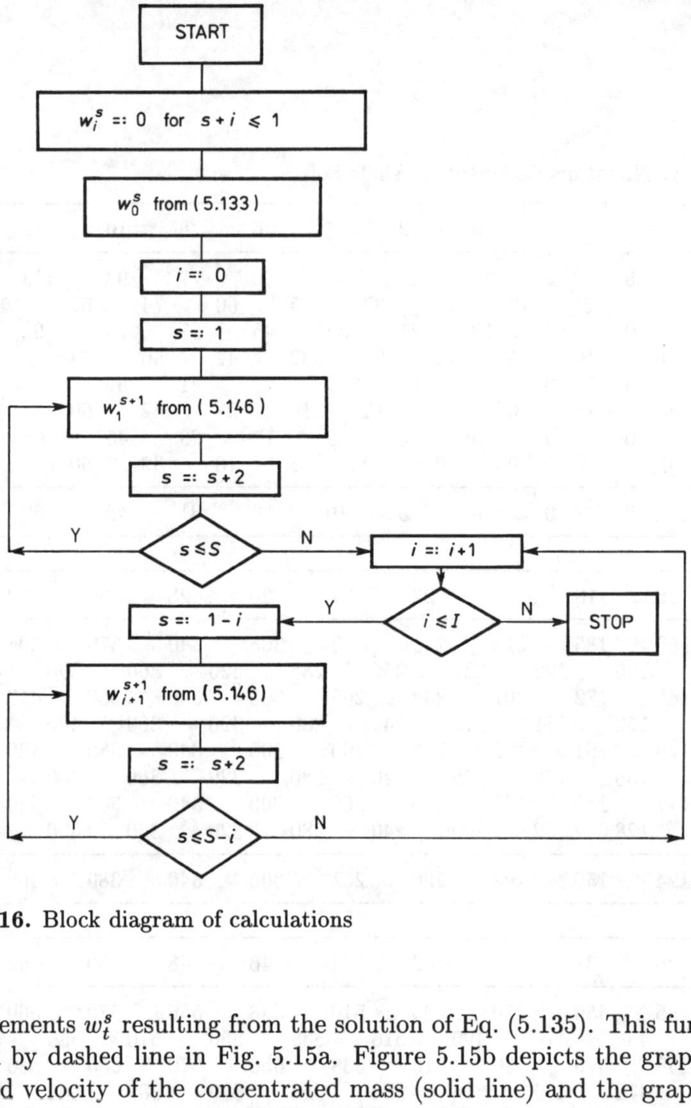

Fig. 5.16. Block diagram of calculations

displacements w_i^s resulting from the solution of Eq. (5.135). This function is plotted by dashed line in Fig. 5.15a. Figure 5.15b depicts the graph of the required velocity of the concentrated mass (solid line) and the graph of the tangential velocity of the wheel (dashed line). The dynamic force increments in the cable at the level of wheel axis are given in Fig. 5.15c. The mechanical parameters $(EA, a, \mu, m,$ etc.) were taken in this example on purpose far from technical reality. Such a choice allowed us to demonstrate in details the process of calculations as well as to shown qualitatively the influence of control of the velocity of the wheel on the vibrations of the concentrated mass.

Finally, let us note that the idea of isosceles triangular elements can be easily extended for 2D-problems. Figure 5.17 shows a segment of 2D-body subdivided into triangles. The vertices of those triangles fall into 3 distinct

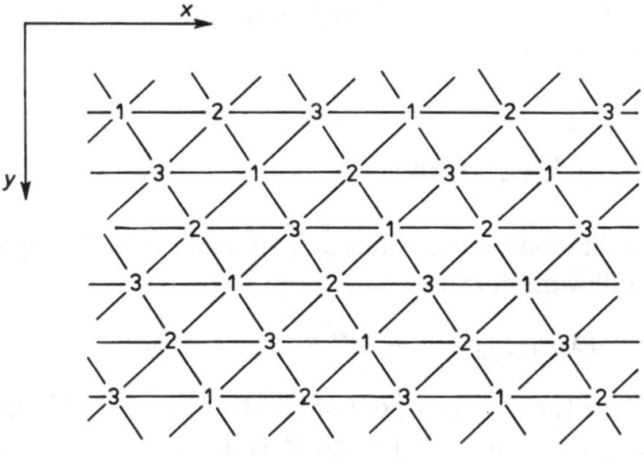

Fig. 5.17. Segment of 2D-body divided into triangles

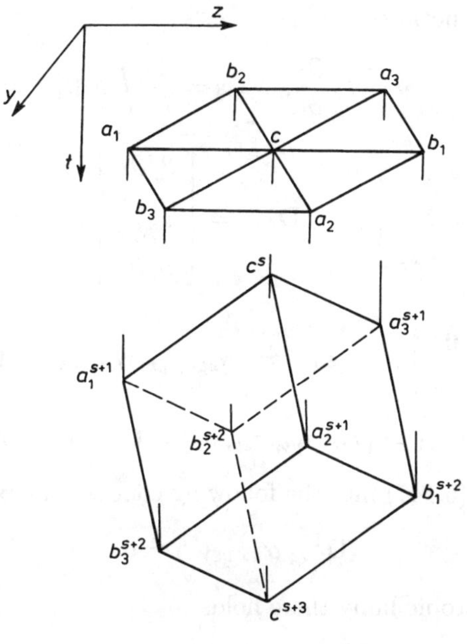

Fig. 5.18. Compound of 6 tetrahedral space–time elements building a parallelepiped

sets $z = 1, 2, 3$ and a member of each set is adjacent to members of other sets only. In the space–time the lines of life of the vertices belonging to the set z contain the nodes related to the time instances $(3s + z)h$. Hence, a 3D-object in the space–time is subdivided into tetrahedrons. Six adjacent tetrahedrons build a parallelepiped that can be brought by a proper linear transformation of the co-ordinates into a cube (Fig. 5.18). About 1/3 of all nodes falls into the hyperplane $t = sh$. Each displacement of such a node can be found from a single uncoupled equation.

5.6 Non-stationary Heat Flow

Let us take the differential equations of heat conduction [compare Eq. (2.45), Part I] in the following form:

$$(\lambda_{ik}\,\vartheta_{,k})_{,i} - c_\varepsilon\,\dot{\vartheta} + S = 0; \qquad i,\, k = 1,\, 2,\, 3. \tag{5.147}$$

Here $\vartheta = T - T_0$ is the increment of temperature measured from the reference temperature T_0 [the latter can be the so-called temperature of the natural state (compare [124, 122])], $c_\varepsilon = c\varrho$ is the specific heat of the volume of material, λ_{ik} are the heat conduction coefficients for anisotropic body and S is the capacity of the volume heat sources.

Introducing the notation

$$\vartheta = T_0\,\frac{\partial s}{\partial t}, \qquad s = \frac{1}{T_0}\int \vartheta\,\mathrm{d}t, \tag{5.148}$$

$$\boldsymbol{\lambda}_{3\times 3} = \begin{bmatrix} \lambda_{xx} & \lambda_{xy} & \lambda_{xz} \\ \lambda_{yx} & \lambda_{yy} & \lambda_{yz} \\ \lambda_{zx} & \lambda_{zy} & \lambda_{zz} \end{bmatrix}, \qquad \mathbf{D}_{4\times 1} = \begin{bmatrix} \partial/\partial x \\ \partial/\partial y \\ \partial/\partial z \\ \partial/\partial t \end{bmatrix}, \qquad \mathbf{D}'_{4\times 1} = \begin{bmatrix} \partial/\partial x \\ \partial/\partial y \\ \partial/\partial z \\ 1 \end{bmatrix}, \tag{5.149}$$

$$\mathbf{C}_{4\times 4} = \begin{bmatrix} \boldsymbol{\lambda} & \mathbf{0} \\ \mathbf{0} & -c_\varepsilon \end{bmatrix}, \qquad \boldsymbol{\varepsilon}_{4\times 1} = \{\gamma_{xt},\, \gamma_{yt},\, \gamma_{zt},\, \varepsilon_t\} = \mathbf{D}_{4\times 1}\, s, \tag{5.150}$$

$$\boldsymbol{\sigma}_{4\times 1} = \{\tau_{xt},\, \tau_{yt},\, \tau_{zt},\, \sigma_t\} = \mathbf{C}_{4\times 4}\,\mathbf{D}'_{4\times 1}\,\vartheta \tag{5.151}$$

we transform Eq. (5.147) into the following concise matrix form:

$$\mathbf{D}^{\mathrm{T}}_{1\times 4}\,\boldsymbol{\sigma}_{4\times 1} + S = 0. \tag{5.152}$$

Obviously, for isotropic body there holds

$$\boldsymbol{\lambda}_{3\times 3} = \lambda\,\mathbf{I}_{3\times 3}. \tag{5.153}$$

The function $s(\mathbf{x}, t)$ has the dimension of time and can be interpreted as the 4th component of the displacement vector parallel to the time axis. Such an interpretation of s allows us to apply the STEM for the problems of non-stationary heat flow.

In order to derive the equation expressing the virtual action principle (compare [78]) we proceed in a similar way as described previously in Sect. 5.2. Namely, we multiply Eq. (5.152) by the virtual increment of the displacement s and we integrate the result over the 4D-volume of the relevant space–time object. Using the Gauss–Ostrogradski relations, subject to constraints mentioned after Eq. (5.21), we obtain:

$$\int_{\partial\Omega} \delta s\, \mathbf{N}_{1\times 4}\, \boldsymbol{\sigma}_{4\times 1} \mathrm{d}(\partial\Omega) + \int_{\Omega} \delta s\, S\, \mathrm{d}\Omega = \int_{\Omega} \delta\boldsymbol{\varepsilon}^{\mathrm{T}}_{1\times 4}\, \boldsymbol{\sigma}_{4\times 1}\, \mathrm{d}\Omega. \tag{5.154}$$

where

$$\mathbf{N}_{1\times 4} = \begin{bmatrix} n_x & n_y & n_z & n_t \end{bmatrix}. \tag{5.155}$$

We assume as usually that the displacement s inside the space–time element can be expressed by its nodal values and the shape functions:

$$s(\mathbf{x},\, t) = \boldsymbol{\varphi}_{1\times N_e}(\mathbf{x},\, t)\, \mathbf{s}^{(e)}_{N_e\times 1}. \tag{5.156}$$

The temperature field $\vartheta(\mathbf{x},\, t)$ inside the element defined by Eq. (5.148) depends upon a smaller number of parameters ($n_e < N_e$) than the function $s(\mathbf{x}, t)$:

$$\vartheta(\mathbf{x},\, t) = T_0\, \frac{\partial}{\partial t} \boldsymbol{\varphi}_{1\times N_e}\, \mathbf{s}^{(e)}_{N_e\times 1} = \boldsymbol{\varphi}'_{1\times n_e}\, \boldsymbol{\vartheta}^{(e)}_{n_e\times 1}. \tag{5.157}$$

Using Eqs. (5.150), (5.151) we find

$$\boldsymbol{\varepsilon}_{4\times 1} = \mathbf{D}_{4\times 1}\, \boldsymbol{\varphi}_{1\times N_e}\, \mathbf{s}^{(e)}_{N_e\times 1} = \mathbf{B}_{4\times N_e}\, \mathbf{s}^{(e)}_{N_e\times 1}, \tag{5.158}$$

$$\boldsymbol{\sigma}_{4\times 1} = \mathbf{C}_{4\times 4}\, \mathbf{D}'_{4\times 1}\, \boldsymbol{\varphi}'_{1\times n_e}\, \boldsymbol{\vartheta}^{(e)}_{n_e\times 1} = \mathbf{C}_{4\times 4}\, \mathbf{B}'_{4\times n_e}\, \boldsymbol{\vartheta}^{(e)}_{n_e\times 1}, \tag{5.159}$$

where

$$\mathbf{B}_{4\times N_e} = \mathbf{D}_{4\times 1}\, \boldsymbol{\varphi}_{1\times N_e}, \qquad \mathbf{B}'_{4\times n_e} = \mathbf{D}'_{4\times 1}\, \boldsymbol{\varphi}'_{1\times n_e}. \tag{5.160}$$

We replace the heat impulses distributed over the hypersurface $\partial\Omega$ of the space–time element by impulses $\hat{\mathbf{S}}^{(e)}_{N_e\times 1}$ concentrated in its nodes. This allows us to write the equation of virtual action (5.154) in the form

$$\delta\mathbf{s}^{\mathrm{T}(e)}_{1\times N_e}\, \hat{\mathbf{S}}^{(e)}_{N_e\times 1} + \int_{\Omega} \delta\mathbf{s}^{\mathrm{T}(e)}_{1\times N_e}\, \boldsymbol{\varphi}^{\mathrm{T}}_{N_e\times 1}\, S\, \mathrm{d}\Omega$$
$$= \int_{\Omega} \delta\mathbf{s}^{\mathrm{T}(e)}_{1\times N_e}\, \mathbf{B}^{\mathrm{T}}_{N_e\times 4}\, \mathbf{C}_{4\times 4}\, \mathbf{B}'_{4\times n_e}\, \boldsymbol{\vartheta}^{(e)}_{n_e\times 1} \mathrm{d}\Omega. \tag{5.161}$$

Hence, the following relation holds

$$\hat{\mathbf{S}}^{(e)}_{N_e\times 1} = \mathbf{K}^{(e)}_{N_e\times n_e}\, \boldsymbol{\vartheta}^{(e)}_{n_e\times 1} - \mathbf{S}^{(e)}_{N_e\times 1}, \tag{5.162}$$

where

$$\mathbf{K}^{(e)}_{N_e\times n_e} = \int_{\Omega} \mathbf{B}^{\mathrm{T}}_{N_e\times 4}\, \mathbf{C}_{4\times 4}\, \mathbf{B}'_{4\times n_e}\, \mathrm{d}\Omega, \qquad \mathbf{S}^{(e)}_{N_e\times 1} = \int_{\Omega} \boldsymbol{\varphi}^{\mathrm{T}}_{N_e\times 1}\, S\, \mathrm{d}\Omega. \tag{5.163}$$

It is seen that the heat stiffness matrix $\mathbf{K}^{(e)}_{N_e\times n_e}$ of the space–time element has less columns than rows.

In the special case of rectangular space–time elements (Fig. 5.3) the vector of nodal displacements $\mathbf{s}^{(e)}_{2N_e\times1}$ can be split into 2 sub-vectors (compare [80])

$$\mathbf{s}^{(e)}_{2N_e\times1} = \begin{bmatrix} \mathbf{s}^{(a)}_{N_e\times1} \\ \mathbf{s}^{(p)}_{N_e\times1} \end{bmatrix}, \tag{5.164}$$

containing the displacements of nodes situated on the hyperplanes $t = t^a$ and $t = t^p$. Assuming the shape functions [compare Eq. (5.33)]

$$\varphi_{1\times2N_e}(\mathbf{x}, t) = \varphi^t_{1\times2}(t) \otimes \varphi^x_{1\times N_e}(\mathbf{x}), \tag{5.165}$$

where the matrix $\varphi^t_{1\times2}$ is defined by Eq. (5.34) and using Eq. (5.148) we obtain

$$\vartheta(\mathbf{x}, t) = T_0 \left(\frac{1}{h} [-1 \;\; 1] \otimes \varphi^x_{1\times N_e} \right) \mathbf{s}^{(e)}_{2N_e\times1} = \varphi'_{1\times N_e}\, \boldsymbol{\vartheta}^{(e)}_{N_e\times1}. \tag{5.166}$$

Here

$$\varphi'_{1\times N_e} = \varphi^x_{1\times N_e}(\mathbf{x}), \qquad \boldsymbol{\vartheta}^{(e)}_{N_e\times1} = \frac{T_0}{h}(\mathbf{s}^{(p)}_{N_e\times1} - \mathbf{s}^{(a)}_{N_e\times1}). \tag{5.167}$$

Hence, the temperature field inside the element does not depend on t and the components of the matrix $\boldsymbol{\vartheta}^{(e)}_{N_e\times1}$ are the temperatures at the edges parallel to the time axis.

It follows from Eqs. (5.160) that

$$\mathbf{B}_{4\times2N_e} = \mathbf{D}_{4\times1}\, \varphi_{1\times2N_e} = \begin{bmatrix} \varphi^t_{1\times2} \otimes \mathbf{B}^x_{3\times N_e} \\ \mathbf{B}^t_{1\times2} \otimes \varphi^x_{1\times N_e} \end{bmatrix}, \qquad \mathbf{B}'_{4\times N_e} = \begin{bmatrix} \mathbf{B}^x_{3\times N_e} \\ \varphi^x_{1\times N_e} \end{bmatrix}, \tag{5.168}$$

$$\mathbf{B}^x_{3\times N_e} = \begin{bmatrix} \partial/\partial x \\ \partial/\partial y \\ \partial/\partial z \end{bmatrix} \varphi^x_{1\times N_e}, \qquad \mathbf{B}^t_{1\times2} = \frac{\mathrm{d}}{\mathrm{d}t}\varphi^t_{1\times2} = \frac{1}{h}[-1 \;\; 1]. \tag{5.169}$$

The heat stiffness matrix of the rectangular space–time element can be found from Eq. (5.163):

$$\mathbf{K}^{(e)}_{2N_e\times N_e} = \begin{bmatrix} \mathbf{A}^{(e)}_{N_e\times N_e} \\ \mathbf{P}^{(e)}_{N_e\times N_e} \end{bmatrix}, \tag{5.170}$$

$$\left.\begin{matrix} \mathbf{A}^{(e)}_{N_e\times N_e} \\ \mathbf{P}^{(e)}_{N_e\times N_e} \end{matrix}\right\} = \frac{h}{2} \int_V \overset{\mathrm{T}}{\mathbf{B}}{}^x_{N_e\times3}\, \boldsymbol{\lambda}_{3\times3}\, \mathbf{B}^x_{3\times N_e}\, \mathrm{d}V \pm \int_V \overset{\mathrm{T}}{\varphi}{}^x_{N_e\times1}\, c_\varepsilon\, \varphi^x_{1\times N_e}\, \mathrm{d}V. \tag{5.171}$$

The stiffness matrix of the assembled element E can be found in similar way [compare Eq. (5.58)]:

$$\mathbf{K}^{(E)}_{2N\times N} = \begin{bmatrix} \mathbf{A}_{N\times N} \\ \mathbf{P}_{N\times N} \end{bmatrix}, \tag{5.172}$$

and the set of equilibrium equations for the heat impulses applied to the nodes situated in the hyperplane $t = t^m$ assumes the form:

$$\mathbf{P}_{N\times N}\,\boldsymbol{\vartheta}^{m}_{N\times 1} + \mathbf{A}_{N\times N}\,\boldsymbol{\vartheta}^{m+1}_{N\times 1} = \mathbf{S}^{m}_{N\times 1}. \tag{5.173}$$

Here the entries of matrix $\boldsymbol{\vartheta}^{m}_{N\times 1}$ are the temperatures at each node of the 3D-body constant in the time interval $t^{m-1} < t < t^m$.

Equation (5.173) leads to the simple recurrent formula

$$\boldsymbol{\vartheta}^{m+1}_{N\times 1} = \mathbf{G}_{N\times N}\,\boldsymbol{\vartheta}^{m}_{N\times 1} + \tilde{\boldsymbol{\vartheta}}^{m}_{N\times 1}, \tag{5.174}$$

where

$$\mathbf{G}_{N\times N} = -\mathbf{A}^{-1}_{N\times N}\,\mathbf{P}_{N\times N}, \qquad \tilde{\boldsymbol{\vartheta}}^{m}_{N\times 1} = \mathbf{A}^{-1}_{N\times N}\,\mathbf{S}^{m}_{N\times 1}. \tag{5.175}$$

In the case of constant heat sources ($\mathbf{S}^{m}_{N\times 1} = \mathbf{S}_{N\times 1}$, $\tilde{\boldsymbol{\vartheta}}^{m}_{N\times 1} = \tilde{\boldsymbol{\vartheta}}_{N\times 1}$) the step of recurrence can be significantly increased

$$\boldsymbol{\vartheta}^{m+n}_{N\times 1} = \mathbf{G}^{n}_{N\times N}\,\boldsymbol{\vartheta}^{m}_{N\times 1} + \sum_{k=0}^{n-1} \mathbf{G}^{k}_{N\times N}\,\tilde{\boldsymbol{\vartheta}}_{N\times 1}. \tag{5.176}$$

The symbol $\mathbf{G}^{k}_{N\times N}$ denotes the k-th power of the matrix $\mathbf{G}_{N\times N}$.

The assumption $n = 2^p$ leads to further reduction of computing time, since then (compare [184]):

$$\mathbf{G}^{n}_{N\times N} = \mathbf{G}_{N\times N}\prod_{k=0}^{p-1}\mathbf{G}^{(2^k)}_{N\times N}, \qquad \sum_{k=0}^{n-1}\mathbf{G}^{k}_{N\times N} = \prod_{k=0}^{p-1}(\mathbf{I}_{N\times N} + \mathbf{G}^{(2^k)}_{N\times N}). \tag{5.177}$$

In the case of unidirectional heat flow, e.g. the flow across the plane wall, we can use triangular elements (compare Fig. 5.10) obtaining uncoupled equations. The presence of mixed derivatives of the shape function in the heat stiffness matrix excludes elements with 3 nodes. Linear shape functions for such elements would lead to the stiffness matrix independent on λ, whereas non-linear shape functions [82] require non-conforming elements. Hence, we describe the wall consisting of layers of different thermal parameters λ_i, c_i and different thicknesses a_i by the 6-node elements having initial shape of convex or concave pentagons (Fig. 5.19). If all layers would have the same thickness ($a_i = a$), then we would have isosceles triangles instead of pentagons.

Figure 5.20a shows one of pentagonal space–time elements related to the local reference frame. A piecewise-linear transformation of co-ordinates

$$\xi = \frac{x}{2a_\alpha}\,\mathrm{H}(a_\alpha - x) + \frac{a_\beta - a_\alpha + x}{2a_\beta}\,\mathrm{H}(x - a_\alpha), \qquad \tau = \frac{t}{2h}, \tag{5.178}$$

brings that element into the isosceles triangle shown in Fig. 5.20b.

We assume the following shape matrix:

$$\boldsymbol{\varphi}_{1\times 6} = \Big[\ (\xi+\tau)\tfrac{\xi+\tau-1}{2} \quad 2\xi\,(1-\xi-\tau) \quad (1-\xi)^2-\tau^2 \quad \xi\,(2\xi-1) \quad 2\xi\,(1-\xi+\tau) \quad (\xi-\tau)\tfrac{\xi-\tau-1}{2}\ \Big]. \tag{5.179}$$

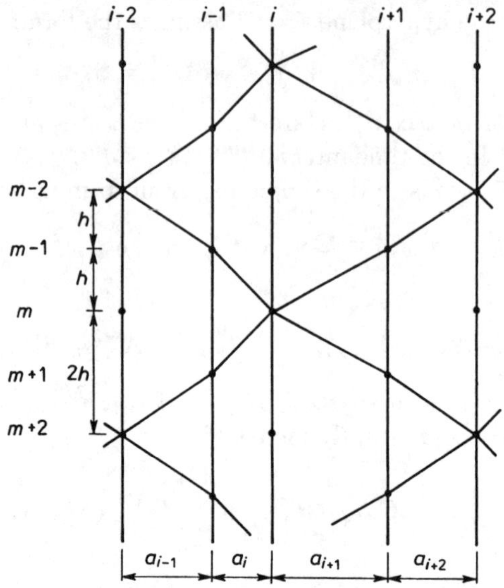

Fig. 5.19. Subdivision of non-homogeneous space–time object into pentagonal 6-node elements

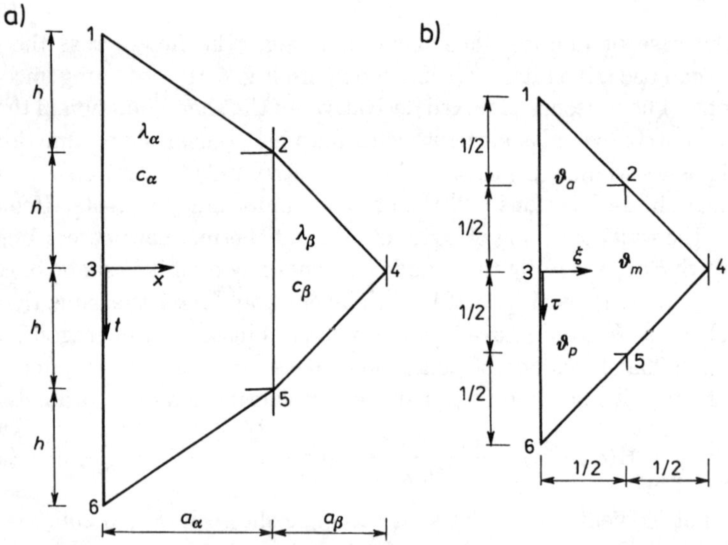

Fig. 5.20. Space–time elements with 6 nodes: **a)** pentagonal, **b)** triangular (after non-smooth transformation of co-ordinates)

According to Eq. (5.175) the temperature field inside the element is given by the product

$$\begin{aligned}\vartheta(\xi,\tau) &= \frac{T_0}{2h}\left[\,-\tfrac{1}{2}+\xi+\tau \quad -2\xi \quad -2\tau \quad 0 \quad 2\xi \quad \tfrac{1}{2}-\xi+\tau\,\right]\mathbf{s}^{(e)}_{6\times 1}\\ &= \boldsymbol{\varphi}'_{1\times 3}\,\boldsymbol{\vartheta}^{(e)}_{3\times 1}.\end{aligned} \tag{5.180}$$

Here

$$\begin{aligned}\boldsymbol{\varphi}'_{1\times 3} &= \left[\,\tfrac{1}{2}-\xi-\tau \quad 2\xi \quad \tfrac{1}{2}-\xi+\tau\,\right],\\ \boldsymbol{\vartheta}^{(e)}_{3\times 1} &= \frac{T_0}{2h}\left\{\, s_3-s_1,\ s_5-s_2,\ s_6-s_3\,\right\} = \left\{\,\vartheta_a,\ \vartheta_m,\ \vartheta_p\,\right\}.\end{aligned} \tag{5.181}$$

Further we find the following matrices:

$$\begin{aligned}\mathbf{B}_{2\times 6} &= \begin{bmatrix}\partial/\partial x\\ \partial/\partial t\end{bmatrix}\boldsymbol{\varphi}_{1\times 6} = \begin{bmatrix}\frac{1}{2a_\alpha}\mathrm{H}(\frac{1}{2}-\xi)+\frac{1}{2a_\beta}\mathrm{H}(\xi-\frac{1}{2}) & 0\\ 0 & \frac{1}{2h}\end{bmatrix}\\ &\times\begin{bmatrix}\xi{+}\tau{-}\frac{1}{2} & 2{-}2\tau{-}4\xi & -2{+}2\xi & 4\xi{-}1 & 2{+}2\tau{-}4\xi & \xi{-}\tau{-}\frac{1}{2}\\ \xi{+}\tau{-}\frac{1}{2} & -2\xi & -2\tau & 0 & 2\xi & -\xi{+}\tau{+}\frac{1}{2}\end{bmatrix},\end{aligned} \tag{5.182}$$

$$\begin{aligned}\mathbf{B}'_{2\times 3} &= \begin{bmatrix}\partial/\partial x\\ 1\end{bmatrix}\boldsymbol{\varphi}'_{1\times 3}\\ &= \begin{bmatrix}\frac{1}{2a_\alpha}\mathrm{H}(\frac{1}{2}{-}\xi)+\frac{1}{2a_\beta}\mathrm{H}(\xi{-}\frac{1}{2}) & 0\\ 0 & 1\end{bmatrix}\begin{bmatrix}-1 & 2 & -1\\ \frac{1}{2}{-}\xi{-}\tau & 2\xi & \frac{1}{2}{-}\xi{+}\tau\end{bmatrix},\end{aligned} \tag{5.183}$$

$$\mathbf{C}_{2\times 2} = \begin{bmatrix}\lambda_\alpha\,\mathrm{H}(\frac{1}{2}-\xi)+\lambda_\beta\,\mathrm{H}(\xi-\frac{1}{2}) & 0\\ 0 & c_\alpha\,\mathrm{H}(\frac{1}{2}-\xi)+c_\beta\,\mathrm{H}(\xi-\frac{1}{2})\end{bmatrix}. \tag{5.184}$$

After the relevant integration [compare Eq. (5.163)] the heat stiffness matrix assumes the form:

$$\mathbf{K}^{(e)}_{6\times 3} = \int_\Omega \mathbf{B}^{\mathrm{T}}_{6\times 2}\,\mathbf{C}_{2\times 2}\,\mathbf{B}'_{2\times 3}\,\mathrm{d}\Omega = \begin{bmatrix}K_{1a} & K_{1m} & K_{1p}\\ K_{2a} & K_{2m} & K_{2p}\\ K_{3a} & K_{3m} & K_{3p}\\ K_{4a} & K_{4m} & K_{4p}\\ K_{5a} & K_{5m} & K_{5p}\\ K_{6a} & K_{6m} & K_{6p}\end{bmatrix}. \tag{5.185}$$

Denoting

$$L_\alpha = \frac{\lambda_\alpha h}{24a_\alpha},\qquad L_\beta = \frac{\lambda_\beta h}{24a_\beta},\qquad C_\alpha = \frac{c_\alpha a_\alpha}{24},\qquad C_\beta = \frac{c_\beta a_\beta}{24} \tag{5.186}$$

we obtain the following components of the matrix (5.185):

$$
\begin{aligned}
\left.\begin{matrix} K_{1a} \\ K_{6p} \end{matrix}\right\} &= 5L_\alpha - L_\beta \pm (11C_\alpha + C_\beta), \\
\left.\begin{matrix} K_{1m} \\ K_{6m} \end{matrix}\right\} &= -10L_\alpha + 2L_\beta \pm 3(C_\alpha - C_\beta), \\
\left.\begin{matrix} K_{1p} \\ K_{6a} \end{matrix}\right\} &= 5L_\alpha - L_\beta \mp 4C_\alpha, \\
\left.\begin{matrix} K_{2a} = K_{2p} \\ K_{5m} = K_{5p} \end{matrix}\right\} &= -20L_\alpha + 4L_\beta \pm 3(C_\alpha - C_\beta), \\
\left.\begin{matrix} K_{2m} \\ K_{5m} \end{matrix}\right\} &= 40L_\alpha - 8L_\beta \pm (10C_\alpha + 22C_\beta), \\
\left.\begin{matrix} K_{3a} \\ K_{3p} \end{matrix}\right\} &= 28L_\alpha + 4L_\beta \mp (15C_\alpha + C_\beta), \\
\left.\begin{matrix} K_{4a} \\ K_{4p} \end{matrix}\right\} &= 2L_\alpha - 10L_\beta, \\
K_{3m} &= -56L_\alpha - 8L_\beta, \\
K_{4m} &= -4L_\alpha + 20L_\beta.
\end{aligned}
\tag{5.187}
$$

In particular, for an isotropic body we obtain:

$$
a_\alpha = a_\beta = a, \qquad \lambda_\alpha = \lambda_\beta = \lambda, \qquad c_\alpha = c_\beta = c_\varepsilon, \tag{5.188}
$$

$$
L_\alpha = L_\beta = \frac{\lambda h}{24a} = \kappa \frac{c_\varepsilon a}{24}, \qquad C_\alpha = C_\beta = \frac{c_\varepsilon a}{24}, \qquad \kappa = \frac{\lambda h}{c_\varepsilon a^2}, \tag{5.189}
$$

with the heat stiffness matrix (5.185) reduced to

$$
\mathbf{K}^{(e)}_{6\times 3} = \frac{c_\varepsilon a}{6} \begin{bmatrix} 3+\kappa & -2\kappa & -1+\kappa \\ -4\kappa & 8+8\kappa & -4\kappa \\ -4+8\kappa & -16\kappa & 4+8\kappa \\ -2\kappa & 4\kappa & -2\kappa \\ -4\kappa & -8+8\kappa & -4\kappa \\ 1+\kappa & -2\kappa & -3+\kappa \end{bmatrix}. \tag{5.190}
$$

Equilibrium conditions for the heat impulses acting on the nodes (i, m), $(i-1, m+1)$, $(i+1, m+1)$ and $(i, m+2)$ (compare Fig. 5.19) lead to the set of 4 equations with the temperatures $\vartheta_i^{m+1}, \vartheta_{i-1}^{m+2}, \vartheta_{i+1}^{m+2}, \vartheta_i^{m+3}$. Hence, independently from the number of layers into which the wall has been divided, we obtain uncoupled sets of equations. Each of them contains 4 unknowns. Obviously, these equations become modified near the boundaries.

The solution can be even further simplified by introducing a non-stationary space–time element mesh. Let us build 3 other families of space–time elements on the same nodes as used by the mesh shown in Fig. 5.19. We assume that the numbers (i, m) defining the position of all nodes in the space–time confirm to the condition $(-1)^{i+m} = 1$. However, 6 elements meet only at each fourth node (compare Fig. 5.19). Therefore, we have 4 sets of nodes in which the vertices of triangular elements can be located. The indices of nodes belonging to each set fulfil the conditions given in Table 5.2.

Table 5.2. Conditions fulfilled by indices of nodes belonging to different sets

	$(-1)^i$	$(-1)^{(i+m)/2}$
a	−1	−1
b	−1	1
c	1	−1
d	1	1

If we want to find the temperature, for example, ϑ_i^{m+1}, we construct 2 space–time elements having vertices in the nodes $(i, m-2), (i-2, m), (i+2, m), (i, m+2)$. The equilibrium condition for heat impulses acting at the node (i, m) gives a single equation with one unknown:

$$(K_{3a}^{(l)} + K_{3a}^{(r)})\,\vartheta_i^{m-1} + K_{3m}^{(l)}\,\vartheta_{i-1}^{m} + K_{3m}^{(r)}\,\vartheta_{i+1}^{m} + (K_{3p}^{(l)} + K_{3p}^{(r)})\,\vartheta_i^{m+1} = S_i^m . \tag{5.191}$$

In that equation the entries of matrix (5.185) are related either to the left (l) or to the right (r) element.

In order to find the temperature ϑ_{i-1}^{m+2} we consider the equilibrium of heat impulses in the node $(i-1, m+1)$ situated between 2 elements with the nodes $(i-1, m-1), (i-3, m+1), (i+1, m+1), (i-1, m+3)$. The resulting equation with a single unknown is similar to Eq. (5.191). This way we can compute the values of temperature in the subsequent nodes.

The procedure described above can be generalized for the case of 2D heat flow. Let us consider a planar region subjected to the temperature field and related to the co-ordinate system x, y. We assume that the region is non-homogeneous and his thermal properties in each of the rectangles of the mesh are described by the parameters $\lambda_{i,j}$ and $c_{i,j}$ (Fig. 5.21).

The main space–time element is the 9-node octahedral (Fig. 5.22a). It's projection on the x, y plane can be a convex or concave tetragon (Fig. 5.21), as a special case this projection can be also a triangle. When described by the non-dimensional co-ordinates

$$\begin{aligned} \xi &= \frac{x}{2a_\alpha}\,\mathrm{H}(a_\alpha - x) + \frac{a_\beta - a_\alpha + x}{2a_\beta}\,\mathrm{H}(x - a_\alpha), \\ \eta &= \frac{y}{2b_\alpha}\,\mathrm{H}(b_\alpha - y) + \frac{b_\gamma - b_\alpha + y}{2b_\gamma}\,\mathrm{H}(y - b_\alpha), \qquad \tau = \frac{t}{2h}, \end{aligned} \tag{5.192}$$

this element takes the shape of tetrahedral shown in Fig. 5.22b.

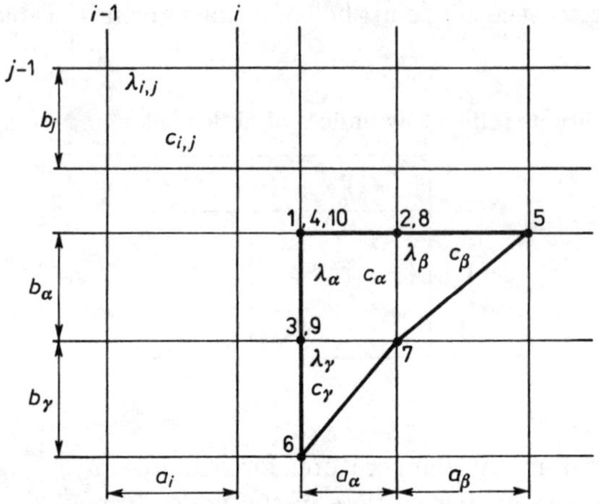

Fig. 5.21. Thermally non-homogeneous 2D-region

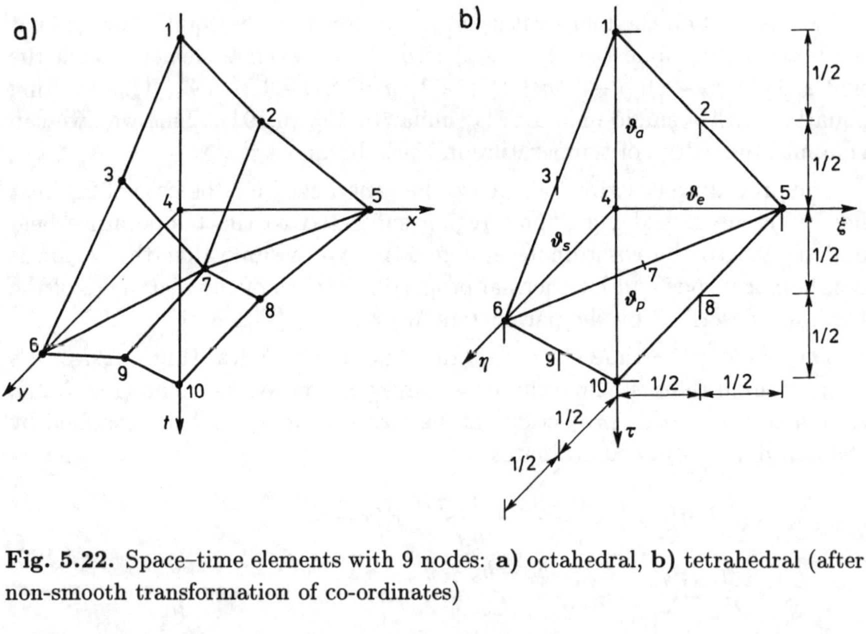

Fig. 5.22. Space–time elements with 9 nodes: **a)** octahedral, **b)** tetrahedral (after non-smooth transformation of co-ordinates)

Assuming the following shape matrix:

$$\begin{aligned}\varphi_{1\times 10} = [\ & \tfrac{1-\xi-\eta-\tau}{2}(-\xi-\eta-\tau) \quad 2\xi(1-\xi-\eta-\tau) \quad 2\eta(1-\xi-\eta-\tau) \\ & (1-\xi-\eta)^2-\tau^2 \quad \xi(2\xi-1) \quad \eta(2\eta-1) \quad 4\xi\eta \\ & 2\xi(1-\xi-\eta+\tau) \quad 2\eta(1-\xi-\eta+\tau) \quad \tfrac{1-\xi-\eta+\tau}{2}(-\xi-\eta+\tau)\ ],\end{aligned} \tag{5.193}$$

we could perform the derivations described by Eqs. (5.157) to (5.163) and obtain the stiffness matrix $\mathbf{K}^{(e)}_{10\times 4}$. However, using non-stationary space–time element mesh we can calculate only the fourth row of the stiffness matrix [compare Eq. (5.191)]. Introducing the notation

$$\begin{aligned}C_\alpha &= \tfrac{c_\alpha a_\alpha b_\alpha}{120}, & C_\beta &= \tfrac{c_\beta a_\beta b_\alpha}{120}, & C_\gamma &= \tfrac{c_\gamma a_\alpha b_\gamma}{120}, \\ L_{\alpha x} &= \tfrac{\lambda_\alpha h}{120}\tfrac{b_\alpha}{a_\alpha}, & L_{\beta x} &= \tfrac{\lambda_\beta h}{120}\tfrac{b_\alpha}{a_\beta}, & L_{\gamma x} &= \tfrac{\lambda_\gamma h}{120}\tfrac{b_\gamma}{a_\alpha}, \\ L_{\alpha y} &= \tfrac{\lambda_\alpha h}{120}\tfrac{a_\alpha}{b_\alpha}, & L_{\beta y} &= \tfrac{\lambda_\beta h}{120}\tfrac{a_\beta}{b_\alpha}, & L_{\gamma y} &= \tfrac{\lambda_\gamma h}{120}\tfrac{a_\alpha}{b_\gamma},\end{aligned} \tag{5.194}$$

we obtain the following components of that row:

$$\begin{aligned}\left.\begin{matrix}K_{4a}\\ K_{4p}\end{matrix}\right\} &= \mp(30C_\alpha + C_\beta + C_\gamma) + 70(L_{\alpha x} + L_{\alpha y}) + 5(L_{\beta x} + L_{\beta y} + L_{\gamma x} + L_{\gamma y}), \\ K_{4e} &= -140L_{\alpha x} - 10L_{\beta x} - 10L_{\gamma x}, \\ K_{4s} &= -140L_{\alpha y} - 10L_{\beta y} - 10L_{\gamma y}.\end{aligned} \tag{5.195}$$

In the particular case of an isotropic body we have $a = b$ and taking κ according to Eq. (5.189), we write the fourth row of the stiffness matrix as

$$\mathbf{K}_4 = \frac{4}{15} c_\varepsilon a^2 \left[\ -1+5\kappa \quad -5\kappa \quad -5\kappa \quad 1+5\kappa\ \right]. \tag{5.196}$$

Then we assemble 4 tetrahedral elements (Fig. 5.22b) into the octahedral element shown in Fig. 5.23 and we combine the matrices of type (5.196) into the single-row heat stiffness matrix for the internal node 4 (i, j, m). This brings us to the equilibrium equation of heat impulses acting at this node:

$$A\,\vartheta^{m-1}_{i,j} + E\,\vartheta^m_{i+1,j} + S\,\vartheta^m_{i,j+1} + W\,\vartheta^m_{i-1,j} + N\,\vartheta^m_{i,j-1} + P\,\vartheta^{m+1}_{i,j} = S^m_{i,j}. \tag{5.197}$$

The coefficients $A, E, \ldots$ of this equation are composed of the entries of matrices (5.195) or (5.196). In the latter case we obtain

$$\left.\begin{matrix}A\\ P\end{matrix}\right\} = \frac{16}{15}\, c_\varepsilon\, a^2(\mp 1 + 5\kappa), \qquad E = S = W = N = -\frac{8}{3}\, c_\varepsilon\, a^2\, \kappa. \tag{5.198}$$

Based upon Eq. (5.197) under the substitution (5.198) and the assumption $\kappa = 1/5$ we end up with the following recurrent expression:

$$\vartheta^{m+1}_{i,j} = \frac{1}{4}(\vartheta^m_{i,j-1} + \vartheta^m_{i-1,j} + \vartheta^m_{i+1,j} + \vartheta^m_{i,j+1}) + \tilde{\vartheta}^{m+1}_{i,j}, \qquad \tilde{\vartheta}^{m+1}_{i,j} = \frac{15}{16}\frac{S^m_{i,j}}{c_\varepsilon\, a^2}. \tag{5.199}$$

Let us use this formula for solving the following example. We are looking for the time-dependent field of temperature in a planar square region divided

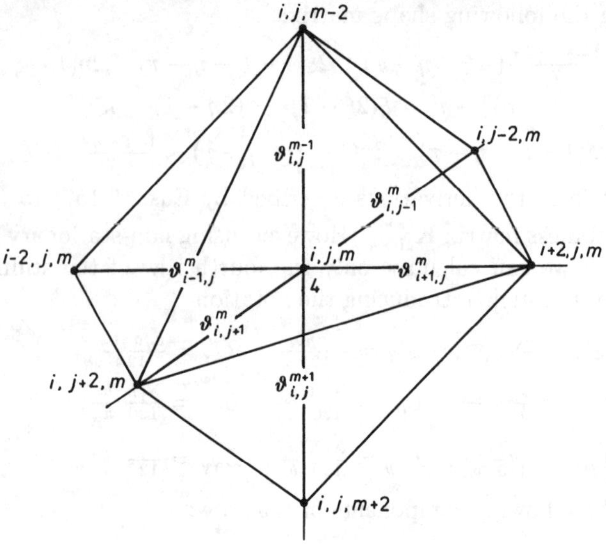

Fig. 5.23. Compound of 4 tetrahedral space–time elements building regular octahedral

into 100 elements of dimensions $a \times a$. Up to the instant $t = 0$ the entire region remained under temperature $T_0(\vartheta(x, y, 0) = 0)$ and in that instant a parabolic distribution of temperature appears on the boundary of the region:

$$\begin{aligned} &\vartheta(x,\,0,\,t) = \vartheta(x,\,l,\,t) = 400\frac{x}{l}\,(1 - \frac{x}{l}), \\ &\vartheta(0,\,y,\,t) = \vartheta(l,\,y,\,t) = -400\frac{y}{l}\,(1 - \frac{y}{l}), \qquad l = 10a. \end{aligned} \tag{5.200}$$

Taking advantage of the symmetry of the problem with respect to the axes $x = l/2, y = l/2$ and the skew-symmetry with respect to the both diagonals of the square, we will consider only 1/8 of the region (Fig. 5.24). There are 10 nodes in that part of the region and in each of them we want to find the increment of temperature. It follows from the boundary conditions (5.200) that

$$\begin{aligned} &\vartheta^m_{i,i} = 0, \qquad \vartheta^m_{6,j} = \vartheta^m_{4,j}, \qquad \vartheta^m_{1,0} = 36, \qquad \vartheta^m_{2,0} = 64, \\ &\qquad \vartheta^m_{3,0} = 84, \qquad \vartheta^m_{4,0} = 96, \qquad \vartheta^m_{5,0} = 100. \end{aligned} \tag{5.201}$$

We assume that the indices of nodes fulfil the condition

$$(-1)^{i+j+m} = 1. \tag{5.202}$$

Hence, all nodes can be divided into 2 subsets. One of them groups the nodes with indices satisfying the condition

$$(-1)^{i+j} = (-1)^m = -1, \tag{5.203}$$

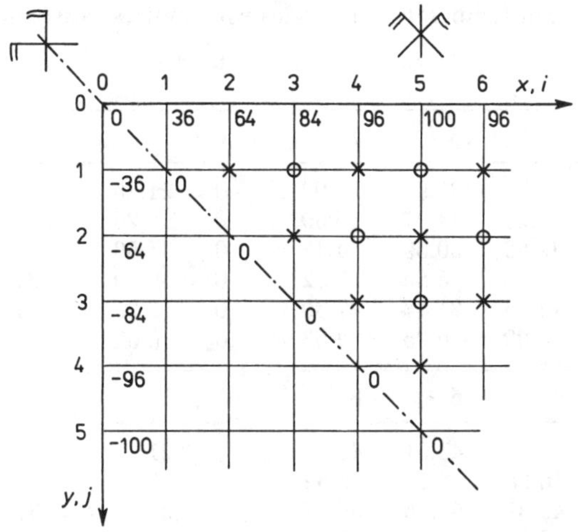

Fig. 5.24. 2D-region bounded by 2 axes of symmetry

The nodes with indices satisfying

$$(-1)^{i+j} = (-1)^m = 1. \tag{5.204}$$

build the other set. In Fig. 5.24 the nodes belonging to different set are labelled by circles and crosses.

The increments of temperature in the nodes belonging to each set are grouped in the relevant matrix:

$$\begin{aligned} {}^o\boldsymbol{\vartheta}^m_{6\times1} &= \{\vartheta^m_{2,1},\ \vartheta^m_{3,2},\ \vartheta^m_{4,1},\ \vartheta^m_{4,3},\ \vartheta^m_{5,2},\ \vartheta^m_{5,4}\} \quad && (-1)^m = -1, \\ {}^e\boldsymbol{\vartheta}^m_{4\times1} &= \{\vartheta^m_{3,1},\ \vartheta^m_{4,2},\ \vartheta^m_{5,1},\ \vartheta^m_{5,3}\}, && (-1)^m = 1. \end{aligned} \tag{5.205}$$

Using Eq. (5.199) and introducing the matrices

$$\mathbf{G}^{oe}_{6\times4} = \frac{1}{4}\begin{bmatrix} 1 & 0 & 0 & 0 \\ 1 & 1 & 0 & 0 \\ 1 & 1 & 1 & 0 \\ 0 & 1 & 0 & 1 \\ 0 & 2 & 1 & 1 \\ 0 & 0 & 0 & 1 \end{bmatrix}, \qquad \mathbf{G}^{eo}_{4\times6} = \frac{1}{4}\begin{bmatrix} 1 & 1 & 1 & 0 & 0 & 0 \\ 0 & 1 & 1 & 1 & 1 & 0 \\ 0 & 0 & 2 & 0 & 1 & 0 \\ 0 & 0 & 0 & 2 & 1 & 1 \end{bmatrix}, \tag{5.206}$$

$${}^o\tilde{\boldsymbol{\vartheta}}_{6\times1} = \frac{1}{4}\{64,\ 0,\ 96,\ 0,\ 0,\ 0\}, \qquad {}^e\tilde{\boldsymbol{\vartheta}}_{4\times1} = \frac{1}{4}\{84,\ 0,\ 100,\ 0\}, \tag{5.207}$$

we arrive at the following relations between the matrices (5.205):

$$\begin{aligned} {}^o\boldsymbol{\vartheta}^{m+1}_{6\times1} &= \mathbf{G}^{oe}_{6\times4}\ {}^e\boldsymbol{\vartheta}^m_{4\times1} + {}^o\tilde{\boldsymbol{\vartheta}}_{6\times1}, \quad && (-1)^m = 1, \\ {}^e\boldsymbol{\vartheta}^{m+1}_{4\times1} &= \mathbf{G}^{eo}_{4\times6}\ {}^o\boldsymbol{\vartheta}^m_{6\times1} + {}^e\tilde{\boldsymbol{\vartheta}}_{4\times1}, && (-1)^m = -1. \end{aligned} \tag{5.208}$$

Table 5.3. Values of temperature in nodes with indices satisfying conditions

a) $(-1)^{i+j+m} = 1$

i,j \ m	1	3	5	7
2, 1	16	23.75	25.97	26.90
3, 2	0	9.25	14.07	16.59
4, 1	24	42.50	50.39	54.48
4, 3	0	1.50	5.05	7.72
5, 2	0	12.25	21.47	27.30
5, 4	0	0.00	0.95	2.03

i,j \ m	0	2	4	6	8
3, 1	0	31	39.88	43.61	45.49
4, 2	0	6	16.38	22.75	26.52
5, 1	0	37	49.31	55.56	59.07
5, 3	0	0	3.81	8.13	11.19

b) $(-1)^{i+j+m} = -1$

i,j \ m	0	2	4	6	8
2, 1	0	21.25	25.13	26.52	27.17
3, 2	0	5.25	12.06	15.51	17.39
4, 1	0	35.50	47.14	52.74	55.78
4, 3	0	0.00	3.33	6.52	8.68
5, 2	0	6.25	17.34	24.74	29.29
5, 4	0	0.00	0.39	1.53	2.46

i,j \ m	1	3	5	7
3, 1	21	36.50	42.08	44.69
4, 2	0	11.75	19.97	24.88
5, 1	25	44.31	52.91	57.56
5, 3	0	1.56	6.10	9.83

These formulae allow us to compute ${}^{o}\boldsymbol{\vartheta}^1_{6\times1}$, ${}^{o}\boldsymbol{\vartheta}^2_{6\times1}$, ${}^{o}\boldsymbol{\vartheta}^3_{6\times1}$, etc. The obtained values are given in Table 5.3a.

In order to refine the mesh, we must change the assumption (5.202) taking $(-1)^{i+j+m} = -1$. Then the temperatures ${}^{o}\boldsymbol{\vartheta}^m$ will appear in the even instants ($(-1)^m = 1$) and the temperatures ${}^{e}\boldsymbol{\vartheta}^m$ will appear in the odd instances. Table 5.3b contains the results.

If we want to increase the step of integration in time, then the matrices ${}^{e}\boldsymbol{\vartheta}^m$ should be eliminated from Eq. (5.208). This gives

$$ {}^{e}\boldsymbol{\vartheta}^{m+2}_{4\times1} = \mathbf{G}_{4\times4}\, {}^{e}\boldsymbol{\vartheta}^{m}_{4\times1} + {}^{e}\breve{\boldsymbol{\vartheta}}_{4\times1}, \tag{5.209} $$

where

$$ \mathbf{G}_{4\times4} = \mathbf{G}^{eo}_{4\times6}\,\mathbf{G}^{oe}_{6\times4} = \frac{1}{16}\begin{bmatrix} 3 & 2 & 1 & 0 \\ 2 & 5 & 2 & 2 \\ 2 & 4 & 3 & 1 \\ 0 & 4 & 1 & 4 \end{bmatrix}, \tag{5.210} $$

$$ {}^{e}\breve{\boldsymbol{\vartheta}}_{4\times1} = {}^{e}\tilde{\boldsymbol{\vartheta}}_{4\times1} + \mathbf{G}^{eo}_{4\times6}\, {}^{o}\tilde{\boldsymbol{\vartheta}}_{6\times1} = \{31,\ 6,\ 37,\ 0\}. \tag{5.211} $$

The operation (5.209) performed n times gives [compare Eq. (5.176)]:

$$ {}^{e}\boldsymbol{\vartheta}^{m+2n}_{4\times1} = \mathbf{G}^{n}_{4\times4}\, {}^{e}\boldsymbol{\vartheta}^{m}_{4\times1} + \sum_{k=1}^{n-1} \mathbf{G}^{k}_{4\times4}\, {}^{e}\breve{\boldsymbol{\vartheta}}_{4\times1}, \tag{5.212} $$

or

$$ {}^{e}\boldsymbol{\vartheta}^{m+2^{p+1}}_{4\times1} = \mathbf{G}_{4\times4}\left[\prod_{k=0}^{p-1} \mathbf{G}^{(2^k)}_{4\times4}\right] {}^{e}\boldsymbol{\vartheta}^{m}_{4\times1} + \left[\prod_{k=0}^{p-1} (\mathbf{I}_{4\times4} + \mathbf{G}^{(2^k)}_{4\times4})\right] {}^{e}\breve{\boldsymbol{\vartheta}}_{4\times1}. \tag{5.213} $$

in the case $n = 2^p, p \geq 1$ [compare Eq. (5.177)].

Table 5.4. Values of temperature obtained at increasing steps of recurrence

i,j \ m	0	2	4	8	16	32(∞)
3, 1	0	31	39.88	45.49	47.70	48
4, 2	0	6	16.38	26.52	31.36	32
5, 1	0	37	49.31	59.07	63.39	64
5, 3	0	0	3.81	11.19	15.39	16

Table 5.4 contains the entries of matrix ${}^{e}\boldsymbol{\vartheta}^{m}$ calculated according to this formula. It is seen that already after 4 time steps, each of them being twice longer than the previous one, we obtain the values of temperature that would appear in the region under stationary heat flow, i.e. after infinitely long time. The values of temperature in the remaining nodes follow from Eq. (5.208) :

$$ {}^{o}\boldsymbol{\vartheta}^{\infty}_{6\times 1} = \{28,\ 20,\ 60,\ 12,\ 36,\ 16\} $$

Solutions of many other test problems solved by means of the STEM coincide practically with analytical solutions given, e.g., in the monograph [68].

6 Plasticity Problems

6.1 Forms of Constitutive Equations

6.1.1 Introductory Comments

In Sect. 6.1 a class of the constitutive equations typical of the inelastic material models will be discussed. The considerations will deliberately be limited to only some representative material models. However, the scope of the formulations discussed will be broad enough to discuss in quite general terms issues involved in developing effective numerical schemes needed for solving the corresponding boundary-value problems of solid mechanics.

The small thermo-elastic-plastic deformation problems will be taken up first in this chapter with our attention focussed on rate-independent materials with isotropic strain-hardening, Sect. 6.1.2. A generalization of the theory to cover a model of the mixed isotropic–kinematic hardening will be given in Sect. 6.1.3, followed by the discussion of ways to incorporate rate-dependence effects typical of creep and visco-plasticity, Sect. 6.1.4. A further generalization of the material model will be demonstrated in Sect. 6.1.5 in which a simple continuum theory of material damage will be presented. Section 6.1.6 is devoted to modifications necessary in the case the deformations analysed cannot be assumed small, while Sect. 6.1.7 contains a brief exposition of the theory of rigid-plasticity and rigid-visco-plasticity – a highly useful approach to problems in which elastic deformations are negligibly small in comparison to the permanent ones.

In the whole of Sect. 6.1 only the three-dimensional formulations will be discussed. The constitutive models to be presented are believed to be directly applicable to the description of many metals – other materials, such as soils and rock, are out of the scope of this presentation.

6.1.2 Thermo-Elasto-Plasticity with Isotropic Hardening

Confining ourselves to the kinematically linearized theory our starting assumption concerns the additivity of the three strain contributions

$$\boldsymbol{\varepsilon} = \boldsymbol{\varepsilon}^{(\mathrm{e})} + \boldsymbol{\varepsilon}^{(\mathrm{p})} + \boldsymbol{\varepsilon}^{(\vartheta)}, \tag{6.1}$$

where $\boldsymbol{\varepsilon}^{(e)}$, $\boldsymbol{\varepsilon}^{(p)}$ and $\boldsymbol{\varepsilon}^{(\vartheta)}$ stand for the elastic, plastic and thermal part of the total strain, respectively. The elastic strain is related to the stress by means of the Hooke's law in the form

$$\boldsymbol{\sigma} = \mathbf{C}\boldsymbol{\varepsilon}^{(e)}. \tag{6.2}$$

For reasons which will become clear later the definitions of the strain and stress vectors is adopted as, cf. Eqs. (2.26), (2.28), Part I,

$$\begin{aligned} \boldsymbol{\varepsilon} &= \left\{\varepsilon_{11}\ \varepsilon_{22}\ \varepsilon_{33}\ \sqrt{2}\varepsilon_{23}\ \sqrt{2}\varepsilon_{13}\ \sqrt{2}\varepsilon_{12}\right\} \\ &= \left\{\varepsilon_{11}\ \varepsilon_{22}\ \varepsilon_{33}\ \frac{1}{\sqrt{2}}\gamma_{23}\ \frac{1}{\sqrt{2}}\gamma_{13}\ \frac{1}{\sqrt{2}}\gamma_{12}\right\}, \\ \boldsymbol{\sigma} &= \left\{\sigma_{11}\ \sigma_{22}\ \sigma_{33}\ \sqrt{2}\sigma_{23}\ \sqrt{2}\sigma_{13}\ \sqrt{2}\sigma_{12}\right\}, \end{aligned} \tag{6.3}$$

where

$$\varepsilon_{12} = \frac{1}{2}\left(\frac{\partial u_1}{\partial x_2} + \frac{\partial u_2}{\partial x_1}\right), \qquad \gamma_{12} = \frac{\partial u_1}{\partial x_2} + \frac{\partial u_2}{\partial x_1}, \qquad \text{etc.}$$

For elastically isotropic materials the constitutive matrix $\mathbf{C}$ may be presented as, cf. Eq. (2.30), Part I,

$$\mathbf{C}_{6\times6} = \frac{E}{(1+\nu)(1-2\nu)} \begin{bmatrix} 1-\nu & \nu & \nu & 0 & 0 & 0 \\ & 1-\nu & \nu & 0 & 0 & 0 \\ & & 1-\nu & 0 & 0 & 0 \\ & & & 1-2\nu & 0 & 0 \\ & \text{sym.} & & & 1-2\nu & 0 \\ & & & & & 1-2\nu \end{bmatrix} \tag{6.4}$$

or, equivalently, as

$$\begin{aligned} \mathbf{C}_{6\times6} &= \begin{bmatrix} K+\frac{4}{3}G & K-\frac{2}{3}G & K-\frac{2}{3}G & 0 & 0 & 0 \\ & K+\frac{4}{3}G & K-\frac{2}{3}G & 0 & 0 & 0 \\ & & K+\frac{4}{3}G & 0 & 0 & 0 \\ & & & 2G & 0 & 0 \\ & \text{sym.} & & & 2G & 0 \\ & & & & & 2G \end{bmatrix} \\ &= K\,\mathbf{1}\mathbf{1}^T + 2G\left[\mathbf{I}_6 - \tfrac{1}{3}\mathbf{1}\mathbf{1}^T\right], \end{aligned} \tag{6.5}$$

with

$$K = \frac{E}{3(1-2\nu)}, \qquad G = \frac{E}{2(1+\nu)}, \tag{6.6}$$

and

$$\mathbf{1}_{6\times 1} = \{1\,1\,1\,0\,0\,0\}\,, \qquad \mathbf{I}_6 = \begin{bmatrix} 1 & 0 & 0 & 0 & 0 & 0 \\ & 1 & 0 & 0 & 0 & 0 \\ & & 1 & 0 & 0 & 0 \\ & & & 1 & 0 & 0 \\ & \text{sym.} & & & 1 & 0 \\ & & & & & 1 \end{bmatrix} . \tag{6.7}$$

We note that

$$\mathbf{1}\mathbf{1}^T = \begin{bmatrix} 1 & 1 & 1 & 0 & 0 & 0 \\ & 1 & 1 & 0 & 0 & 0 \\ & & 1 & 0 & 0 & 0 \\ & & & 0 & 0 & 0 \\ & \text{sym.} & & & 0 & 0 \\ & & & & & 0 \end{bmatrix} .$$

Let the thermal part of the strain be dependent on temperature in an experimentally established manner, say

$$\boldsymbol{\varepsilon}^{(\vartheta)} = \boldsymbol{\varepsilon}^{(\vartheta)}(T)\,, \tag{6.8}$$

By differentiating Eq. (6.8) with respect to time we obtain

$$\dot{\boldsymbol{\varepsilon}}^{(\vartheta)} = \frac{\mathrm{d}\boldsymbol{\varepsilon}^{(\vartheta)}}{\mathrm{d}T}\dot{T} = \mathbf{a}(T)\,\dot{T}\,, \tag{6.9}$$

where $\mathbf{a}(T) = (\mathrm{d}\boldsymbol{\varepsilon}^{(\vartheta)}/\mathrm{d}T)$ may clearly be treated as a known function of the temperature. Equation (6.8) is typically assumed in the form

$$\boldsymbol{\varepsilon}^{(\vartheta)} = \alpha(T - T_0)\mathbf{1} = \alpha\vartheta\,\mathbf{1}\,, \tag{6.10}$$

in which α is the thermal expansion coefficient (generally temperature-dependent), $\vartheta = T - T_0$, and T_0 is a fixed reference temperature. The change of the thermal strain with the change of the temperature is computed as

$$\dot{\boldsymbol{\varepsilon}}^{(\vartheta)} = \overset{*}{\alpha}(T)\,\dot{T}\,\mathbf{1} = \overset{*}{\alpha}(T)\,\dot{\vartheta}\,\mathbf{1}\,, \tag{6.11}$$

in which

$$\overset{*}{\alpha} = \frac{\mathrm{d}\alpha(T)}{\mathrm{d}T}(T - T_0) + \alpha(T) = \frac{\mathrm{d}\alpha(T)}{\mathrm{d}T}\,\vartheta + \alpha(T)\,.$$

For $\alpha = \text{const}$ we clearly have $\alpha = \overset{*}{\alpha}$.

In order to describe the non-viscous plastic part of the material response we first postulate the form of the yield condition as

$$f(\boldsymbol{\sigma}, \sigma_y) = F(\boldsymbol{\sigma}) - \sigma_y(\bar{\varepsilon}^{(\mathrm{p})}, T) = \sqrt{\tfrac{3}{2}\,{}^{\mathrm{D}}\boldsymbol{\sigma}^T\,{}^{\mathrm{D}}\boldsymbol{\sigma}} - \sigma_y = \bar{\sigma} - \sigma_y = 0\,, \tag{6.12}$$

where σ_y is a material function which characterizes the yield limit in the uniaxial experiment, $\bar{\varepsilon}^{(\mathrm{p})}$ is the effective plastic strain computed by the time integration of its rate defined as

$$\dot{\bar{\varepsilon}}^{(\mathrm{p})} = \sqrt{\tfrac{2}{3}\,\dot{\boldsymbol{\varepsilon}}^{(\mathrm{p})T}\,\dot{\boldsymbol{\varepsilon}}^{(\mathrm{p})}} = \sqrt{\tfrac{2}{3}}\;\|\dot{\boldsymbol{\varepsilon}}^{(\mathrm{p})}\|\,, \tag{6.13}$$

and

$$\bar{\sigma} = \sqrt{\tfrac{3}{2}\, {}^D\boldsymbol{\sigma}^T\, {}^D\boldsymbol{\sigma}} = \sqrt{\tfrac{3}{2}}\, \|{}^D\boldsymbol{\sigma}\| \tag{6.14}$$

is the effective Huber–Mises stress. The symbol $\boldsymbol{\mu}_D$ is defined as

$$\boldsymbol{\mu}_D = \frac{1}{3}\begin{bmatrix} 2 & -1 & -1 & 0 & 0 & 0 \\ & 2 & -1 & 0 & 0 & 0 \\ & & 2 & 0 & 0 & 0 \\ & & & 3 & 0 & 0 \\ & \text{sym.} & & & 3 & 0 \\ & & & & & 3 \end{bmatrix}$$

so that the deviatoric stress may be presented in the form

$${}^D\boldsymbol{\sigma} = \boldsymbol{\mu}_{D6\times6}\,\boldsymbol{\sigma} = \left(\mathbf{I}_6 - \tfrac{1}{3}\mathbf{1}\mathbf{1}^T\right)\boldsymbol{\sigma}\,. \tag{6.15}$$

The flow rule associated with the yield condition (6.12) has the form

$$\dot{\boldsymbol{\varepsilon}}^{(\mathrm{p})} = \dot{\lambda}\frac{\partial f}{\partial \boldsymbol{\sigma}^T} = \dot{\lambda}\sqrt{\tfrac{3}{2}}\,\mathbf{n}\,, \tag{6.16}$$

in which

$$\mathbf{n} = \sqrt{\frac{3}{2}}\frac{{}^D\boldsymbol{\sigma}}{\sigma_y} = \frac{{}^D\boldsymbol{\sigma}}{\|{}^D\boldsymbol{\sigma}\|}$$

is the unit vector normal to the yield surface,

$$\|{}^D\boldsymbol{\sigma}\| = \sqrt{\frac{3}{2}}\sigma_y\,,$$

cf. Eqs. (6.12), (6.14), while $\dot{\lambda}$ is a certain scalar function. By substituting (6.16) in the previously defined expression for the effective plastic strain rate (6.13), we arrive at

$$\dot{\lambda} = \dot{\bar{\varepsilon}}^{(\mathrm{p})} \tag{6.17}$$

To assure that any stress point remains on the yield surface during plastic flow the so-called consistency condition is adopted as

$$\dot{f} = \frac{\partial f}{\partial \boldsymbol{\sigma}}\dot{\boldsymbol{\sigma}} - \left.\frac{\partial \sigma_y}{\partial \bar{\varepsilon}^{(\mathrm{p})}}\right|_{T=\mathrm{const.}} \dot{\bar{\varepsilon}}^{(\mathrm{p})} - \left.\frac{\partial \sigma_y}{\partial T}\right|_{\bar{\varepsilon}^{(\mathrm{p})}=\mathrm{const.}} \dot{T} = 0\,. \tag{6.18}$$

Defining the isotropic hardening parameter

$$\zeta = \left.\frac{\partial \sigma_y}{\partial \bar{\varepsilon}^{(\mathrm{p})}}\right|_{T=\mathrm{const.}} \tag{6.19}$$

and using the relation

$$\frac{\partial f}{\partial \boldsymbol{\sigma}^T} = \frac{3}{2}\frac{{}^D\boldsymbol{\sigma}}{\sigma_y}\,,$$

Eq. (6.18) is reduced to

$$\dot{\bar{\sigma}} - \zeta\dot{\bar{\varepsilon}}^{(\mathrm{p})} - \frac{\partial \sigma_y}{\partial T}\dot{T} = 0 \tag{6.20}$$

which may also be written as

$$\zeta = \frac{\mathrm{d}\bar{\sigma}}{\mathrm{d}\bar{\varepsilon}^{(\mathrm{p})}} - \frac{\partial \sigma_y}{\partial T} \frac{1}{\frac{\mathrm{d}\bar{\varepsilon}^{(\mathrm{p})}}{\mathrm{d}T}} . \tag{6.21}$$

Defining an auxiliary hardening modulus

$$\zeta^* = \frac{\mathrm{d}\bar{\sigma}}{\mathrm{d}\bar{\varepsilon}^{(\mathrm{p})}} \tag{6.22}$$

Eq. (6.21) becomes

$$\zeta^* = \zeta + \frac{\partial \sigma_y}{\partial T} \frac{1}{\frac{\mathrm{d}\bar{\varepsilon}^{(\mathrm{p})}}{\mathrm{d}T}} . \tag{6.23}$$

It is sometimes more convenient to use the hardening moduli defined by

$$h = \tfrac{2}{3}\zeta , \qquad h^* = \tfrac{2}{3}\zeta^* , \tag{6.24}$$

which makes it possible to present the flow rule (6.16) in the form

$$\dot{\boldsymbol{\varepsilon}}^{(\mathrm{p})} = \frac{1}{h^*} (\mathbf{n}^T \dot{\boldsymbol{\sigma}})\mathbf{n} , \tag{6.25}$$

or, by means of Eq. (6.23), in the form

$$\dot{\boldsymbol{\varepsilon}}^{(\mathrm{p})} = \frac{1}{h} \mathbf{n} \left[\mathbf{n}^T \dot{\boldsymbol{\sigma}} - \sqrt{\frac{2}{3}} \frac{\partial \sigma_y}{\partial T} \dot{T} \right] . \tag{6.26}$$

The first term on the right-hand side in this equation represents the plastic strain increase under the isothermal conditions ($\dot{T} = 0$), while the second term describes the thermal effects.

By calculating the strain $\boldsymbol{\varepsilon}^{(\mathrm{e})}$ from Eq. (6.1), substituting it in Eq. (6.2) and differentiating the result with respect to time we obtain

$$\dot{\boldsymbol{\sigma}} = \mathbf{C} \left(\dot{\boldsymbol{\varepsilon}} - \dot{\boldsymbol{\varepsilon}}^{(\mathrm{p})} - \overset{*}{\alpha} \dot{T} \mathbf{1} \right) + \overset{*}{\boldsymbol{\sigma}} \tag{6.27}$$

with

$$\overset{*}{\boldsymbol{\sigma}} = \dot{\mathbf{C}} \left[\boldsymbol{\varepsilon} - \boldsymbol{\varepsilon}^{(\mathrm{p})} - \alpha (T - T_0) \mathbf{1} \right] . \tag{6.28}$$

We have assumed here that the elastic moduli may depend on temperature. By substituting now the flow rule (6.26) in Eq. (6.27) and solving the resulting equation for $\dot{\boldsymbol{\sigma}}$ we arrive at the constitutive law of the form

$$\dot{\boldsymbol{\sigma}} = \left[\mathbf{C} - \frac{(\mathbf{C}\mathbf{n})(\mathbf{n}^T \mathbf{C})}{h + \mathbf{n}^T \mathbf{C} \mathbf{n}} \right] (\dot{\boldsymbol{\varepsilon}} - \overset{*}{\alpha} \dot{\vartheta} \mathbf{1}) + \overset{*}{\boldsymbol{\sigma}} + \overset{**}{\boldsymbol{\sigma}} = \mathbf{C}^{(\mathrm{e-p})} (\dot{\boldsymbol{\varepsilon}} - \overset{*}{\alpha} \dot{\vartheta} \mathbf{1}) + \overset{*}{\boldsymbol{\sigma}} + \overset{**}{\boldsymbol{\sigma}} , \tag{6.29}$$

where

$$\overset{**}{\boldsymbol{\sigma}} = \mathbf{C}\mathbf{n} \left[\frac{-(\mathbf{n}^T \mathbf{C} \mathbf{n}) \sqrt{\frac{2}{3}} \frac{1}{h} \frac{\partial \sigma_y}{\partial T} \dot{T} - \overset{*}{\boldsymbol{\sigma}}{}^T \mathbf{n}}{h + \mathbf{n}^T \mathbf{C} \mathbf{n}} + \sqrt{\frac{2}{3}} \frac{1}{h} \frac{\partial \sigma_y}{\partial T} \dot{T} \right] , \tag{6.30}$$

while the vector $\overset{*}{\boldsymbol{\sigma}}$ is given by Eq. (6.28). We note that since

$$\dot{\mathbf{C}} = \frac{\mathrm{d}\mathbf{C}}{\mathrm{d}\vartheta}\,\dot{\vartheta}\,, \tag{6.31}$$

we have

$$\overset{*}{\boldsymbol{\sigma}} = \frac{\mathrm{d}\mathbf{C}}{\mathrm{d}\vartheta}\left[\boldsymbol{\varepsilon} - \boldsymbol{\varepsilon}^{(\mathrm{p})} - \alpha\vartheta\mathbf{1}\right]\dot{\vartheta} = \mathbf{a}\,\dot{\vartheta} \tag{6.32}$$

and

$$\overset{**}{\boldsymbol{\sigma}} = \mathbf{C}\mathbf{n}\left[\frac{\sqrt{\frac{2}{3}}\,\frac{\partial\sigma_y}{\partial\vartheta} - \left(\boldsymbol{\varepsilon} - \boldsymbol{\varepsilon}^{(\mathrm{p})} - \alpha\vartheta\mathbf{1}\right)^T\frac{\mathrm{d}\mathbf{C}}{\mathrm{d}\vartheta}\,\mathbf{n}}{h + \mathbf{n}^T\mathbf{C}\mathbf{n}}\right]\dot{\vartheta} = \mathbf{b}\dot{\vartheta}\,. \tag{6.33}$$

Using the explicit form of the elastic moduli matrix the final form of the constitutive law for the class of materials on hand is obtained as

$$\dot{\boldsymbol{\sigma}} = \left[K\,\mathbf{1}\mathbf{1}^T + 2G\left(\mathbf{I}_6 - \tfrac{1}{3}\,\mathbf{1}\mathbf{1}^T\right) - 2G\,\frac{\mathbf{n}\mathbf{n}^T}{1 + \frac{h}{2G}}\right]\dot{\boldsymbol{\varepsilon}} + \mathbf{c}\dot{\vartheta}\,, \tag{6.34}$$

where

$$\begin{aligned}\mathbf{c} = -3K\,\overset{*}{\alpha}\mathbf{1} &+ \left[\frac{\mathrm{d}K}{\mathrm{d}\vartheta}\,\mathbf{1}\mathbf{1}^T + 2\,\frac{\mathrm{d}G}{\mathrm{d}\vartheta}\left(\mathbf{I}_6 - \tfrac{1}{3}\,\mathbf{1}\mathbf{1}^T\right)\right]\left(\boldsymbol{\varepsilon} - \boldsymbol{\varepsilon}^{(\mathrm{p})} - \alpha\vartheta\mathbf{1}\right)\\ &+ 2G\,\frac{\sqrt{\frac{2}{3}}\,\frac{\partial\sigma_y}{\partial\vartheta} - 2\left(\boldsymbol{\varepsilon} - \boldsymbol{\varepsilon}^{(\mathrm{p})} - \alpha\vartheta\mathbf{1}\right)^T\frac{\mathrm{d}\mathbf{C}}{\mathrm{d}\vartheta}\,\mathbf{n}}{h + 2G}\,.\end{aligned} \tag{6.35}$$

By using the notation

$$\gamma = \frac{1}{1 + \frac{h}{2G}} = \frac{1}{1 + \frac{\zeta}{3G}}\,, \qquad \hat{\gamma} = 2G\gamma\,, \tag{6.36}$$

Eq. (6.34) may be transformed to the form

$$\begin{aligned}\dot{\boldsymbol{\sigma}} &= \left[\mathbf{C} - \hat{\gamma}\mathbf{n}\mathbf{n}^T\right]\left(\dot{\boldsymbol{\varepsilon}} - \overset{*}{\alpha}\dot{\vartheta}\mathbf{1}\right) + \overset{*}{\boldsymbol{\sigma}} + \overset{**}{\boldsymbol{\sigma}}\\ &= \left[\left(K - \tfrac{2}{3}G\right)\mathbf{1}\mathbf{1}^T + 2G\left(\mathbf{I}_6 - \gamma\mathbf{n}\mathbf{n}^T\right)\right]\left(\dot{\boldsymbol{\varepsilon}} - \overset{*}{\alpha}\dot{\vartheta}\mathbf{1}\right) + \overset{*}{\boldsymbol{\sigma}} + \overset{**}{\boldsymbol{\sigma}}\\ &= \left[K\,\mathbf{1}\mathbf{1}^T + 2G\left(\mathbf{I}_6 - \tfrac{1}{3}\,\mathbf{1}\mathbf{1}^T - \gamma\mathbf{n}\mathbf{n}^T\right)\right]\left(\dot{\boldsymbol{\varepsilon}} - \overset{*}{\alpha}\dot{\vartheta}\mathbf{1}\right) + \overset{*}{\boldsymbol{\sigma}} + \overset{**}{\boldsymbol{\sigma}}\,.\end{aligned} \tag{6.37}$$

In accordance with the basic idea of the elastic-plastic response modelling the constitutive equation (6.34) is valid only for the (local) loading processes – the loading/unloading conditions will be specified slightly later.

With no thermal effects included Eq. (6.34) is reduced to

$$\begin{aligned}\dot{\boldsymbol{\sigma}} &= \left[\mathbf{C} - \hat{\gamma}\mathbf{n}\mathbf{n}^T\right]\dot{\boldsymbol{\varepsilon}}\\ &= \left[\left(K - \tfrac{2}{3}G\right)\mathbf{1}\mathbf{1}^T + 2G\left(\mathbf{I}_6 - \gamma\mathbf{n}\mathbf{n}^T\right)\right]\dot{\boldsymbol{\varepsilon}}\\ &= \left[K\,\mathbf{1}\mathbf{1}^T + 2G\left(\mathbf{I}_6 - \tfrac{1}{3}\,\mathbf{1}\mathbf{1}^T - \gamma\mathbf{n}\mathbf{n}^T\right)\right]\dot{\boldsymbol{\varepsilon}}\,,\end{aligned} \tag{6.38}$$

since in such case $\overset{*}{\boldsymbol{\sigma}} = \overset{**}{\boldsymbol{\sigma}} = 0$. By finally defining the tangent constitutive matrix $\mathbf{C}_T = \mathbf{C}^{(e-p)}$ as

$$\mathbf{C}_T = \mathbf{C} - \hat{\gamma}\mathbf{n}\mathbf{n}^T \tag{6.39}$$

we arrive at

$$\dot{\boldsymbol{\sigma}} = \mathbf{C}_T\, \dot{\boldsymbol{\varepsilon}}\,. \tag{6.40}$$

For convenience of the reader the constitutive equations for different classes of materials discussed so far are collected below in both the matrix and index notation.

- Linear elasticity

$$\begin{aligned} \operatorname{tr}\boldsymbol{\sigma} = \mathbf{1}^T\boldsymbol{\sigma} = 3K\operatorname{tr}\boldsymbol{\varepsilon}^{(e)} = 3K\,\mathbf{1}^T\boldsymbol{\varepsilon}^{(e)}\,; \qquad & \sigma_{kk} = 3K\varepsilon_{kk}^{(e)}\,, \\ {}^D\boldsymbol{\sigma} = \boldsymbol{\sigma} - \tfrac{1}{3}(\operatorname{tr}\boldsymbol{\sigma})\mathbf{1} = 2G\left(\mathbf{I}_6 - \tfrac{1}{3}\mathbf{1}\mathbf{1}^T\right) = 2G\,{}^D\boldsymbol{\varepsilon}^{(e)}\,; \quad & {}^D\sigma_{ij} = 2G\,{}^D\varepsilon_{ij}^{(e)}\,. \end{aligned} \tag{6.41}$$

- Linear thermoelasticity

$$\begin{aligned} \operatorname{tr}\boldsymbol{\sigma} = 3K\left[\operatorname{tr}\boldsymbol{\varepsilon}^{(e)} - 3\alpha\vartheta\right]\,; \qquad & \sigma_{kk} = 3K\left[\varepsilon_{kk}^{(e)} - 3\alpha\vartheta\right]\,, \\ {}^D\boldsymbol{\sigma} = 2G\,{}^D\boldsymbol{\varepsilon}^{(e)}\,; \qquad & {}^D\sigma_{ij} = 2G\,{}^D\varepsilon_{ij}^{(e)}\,. \end{aligned} \tag{6.42}$$

- Thermoelasticity with temperature dependent elastic moduli (rate formulation)

$$\begin{aligned} \operatorname{tr}\dot{\boldsymbol{\sigma}} &= 3K\left[\operatorname{tr}\dot{\boldsymbol{\varepsilon}}^{(e)} - 3\overset{*}{\alpha}\dot{\vartheta} - \operatorname{tr}\boldsymbol{\sigma}\left(\frac{1}{3K}\right)^{\cdot}\right]\,; \\ \dot{\sigma}_{kk} &= 3K\left[\dot{\varepsilon}_{kk}^{(e)} - 3\overset{*}{\alpha}\dot{\vartheta} - \sigma_{kk}\left(\frac{1}{3K}\right)^{\cdot}\right]\,, \\ {}^D\dot{\boldsymbol{\sigma}} &= 2G\,{}^D\dot{\boldsymbol{\varepsilon}}^{(e)} - G\,{}^D\boldsymbol{\sigma}\left(\frac{1}{G}\right)^{\cdot} = 2G\left[{}^D\dot{\boldsymbol{\varepsilon}}^{(e)} - {}^D\boldsymbol{\sigma}\left(\frac{1}{2G}\right)^{\cdot}\right]\,; \\ {}^D\dot{\sigma}_{ij} &= 2G\left[{}^D\dot{\varepsilon}_{ij}^{(e)} - {}^D\sigma_{ij}\left(\frac{1}{2G}\right)^{\cdot}\right]\,. \end{aligned} \tag{6.43}$$

- Thermo-elasto-plasticity

$\operatorname{tr}\dot{\boldsymbol{\sigma}} =$ as above,

$$\begin{aligned} {}^D\dot{\boldsymbol{\sigma}} &= 2G\left[\mathbf{I}_6 - \frac{\mathbf{n}\mathbf{n}^T}{1+\frac{h}{2G}}\right]{}^D\dot{\boldsymbol{\varepsilon}} + \mathbf{n}\,\frac{\sqrt{\frac{2}{3}}\,\frac{\partial\sigma_y}{\partial\vartheta}\,\dot{\vartheta}}{1+\frac{h}{2G}} - \frac{\left(\frac{1}{G}\right)^{\cdot}\,h\,{}^D\boldsymbol{\sigma}}{2\left(1+\frac{h}{2G}\right)}\,; \\ {}^D\dot{\sigma}_{ij} &= 2G\left[\delta_{ik}\delta_{jl} - \frac{n_{ij}n_{kl}}{1+\frac{h}{2G}}\right]{}^D\dot{\varepsilon}_{kl} + n_{ij}\,\frac{\sqrt{\frac{2}{3}}\,\frac{\partial\sigma_y}{\partial\vartheta}\,\dot{\vartheta}}{1+\frac{h}{2G}} - \frac{\left(\frac{1}{G}\right)^{\cdot}\,h\,{}^D\sigma_{ij}}{2\left(1+\frac{h}{2G}\right)}\,. \end{aligned} \tag{6.44}$$

6.1.3 Mixed Isotropic–Kinematic Hardening

We shall show in this section how the isotropic hardening model may be formally generalized to describe a more general hardening called the mixed isotropic–kinematic hardening. The yield condition is adopted in the form

$$f(\boldsymbol{\sigma},\boldsymbol{\alpha},\kappa) = F(\boldsymbol{\sigma},\boldsymbol{\alpha}) - \kappa(\bar{\varepsilon}^{(p)},T) = \sqrt{\tfrac{3}{2}\,{}^{D}\bar{\boldsymbol{\sigma}}^{T}\,{}^{D}\bar{\boldsymbol{\sigma}}} - \kappa = \bar{\sigma} - \kappa = 0\,, \quad (6.45)$$

where the vector $\boldsymbol{\alpha}$ (so-called back stress) represents the displaced center of the current yield surface in the stress space. The symbol ${}^{D}\bar{\boldsymbol{\sigma}}$ stands for the deviator of the new stress defined as

$$\bar{\boldsymbol{\sigma}} = \boldsymbol{\sigma} - \boldsymbol{\alpha}\,, \quad (6.46)$$

the effective stress is to be now understood as

$$\bar{\sigma} = \sqrt{\tfrac{3}{2}\,{}^{D}\bar{\boldsymbol{\sigma}}^{T}\,{}^{D}\bar{\boldsymbol{\sigma}}}\,, \quad (6.47)$$

and

$$\kappa = (1-\beta)\sigma_y^0 + \beta\sigma_y\,, \qquad \beta \in [0,1]\,,$$

is the 'radius' of the current yield surface, σ_y^0 being its initial value. The parameter β represents the amount of isotropic hardening in the total hardening of the material – the purely isotropic hardening corresponds to $\beta = 1$ while the purely kinematic hardening results when $\beta = 0$. The parameter β is assumed here as a constant; a more general formulation with β dependent on the process history may also be considered. We emphasize that the model on hand has a computational significance as it may be used to work out numerical procedures encompassing a broad class of hardening effects. On the other hand, due to the lack of experimental data regarding the possible values of the parameter β, few applications have so far been based on the model considered in this section.

To describe the translation of the yield surface the so-called Ziegler's modification to the Prager's kinematic hardening rule is adopted which reads

$$\dot{\boldsymbol{\alpha}} = \dot{\mu}(\boldsymbol{\sigma} - \boldsymbol{\alpha}) \quad (6.48)$$

where $\dot{\mu}$ is a scalar parameter to be determined from the consistency condition. Employing this condition we are able to obtain the constitutive moduli of the form (6.34) with $\boldsymbol{\sigma}$ replaced by $\boldsymbol{\sigma} - \boldsymbol{\alpha}$ and the evolution laws for the functions $\boldsymbol{\alpha}$ and κ as

$$\begin{aligned} \dot{\boldsymbol{\alpha}} &= (1-\beta)h^{*}\dot{\boldsymbol{\varepsilon}}^{(p)} = \tfrac{2}{3}(1-\beta)\zeta^{*}\dot{\boldsymbol{\varepsilon}}^{(p)}\,, \\ \dot{\kappa} &= \tfrac{3}{2}\beta h^{*}\dot{\bar{\varepsilon}}^{(p)} = \beta\zeta^{*}\dot{\bar{\varepsilon}}^{(p)}\,. \end{aligned} \quad (6.49)$$

The loading/unloading conditions may be taken as

- elastic process

$$\bar{\sigma} < \kappa \quad \text{or} \quad \left(\bar{\sigma} = \kappa \quad \text{and} \quad \mathbf{n}^T \dot{\boldsymbol{\sigma}} - \sqrt{\frac{2}{3}} \frac{\partial \kappa}{\partial T} \dot{T} \leq 0 \right) , \tag{6.50}$$

- plastic process

$$\bar{\sigma} = \kappa \quad \text{and} \quad \mathbf{n}^T \dot{\boldsymbol{\sigma}} - \sqrt{\frac{2}{3}} \frac{\partial \kappa}{\partial T} \dot{T} > 0 . \tag{6.51}$$

However, as indicated by many authors, this form is numerically inconvenient. To propose an alternative, we define the so-called trial (or elastic) stress rate as

$$\dot{\boldsymbol{\sigma}}^{(\mathrm{e})} = \mathbf{C} \left(\dot{\boldsymbol{\varepsilon}} - \overset{*}{\alpha} \dot{T} \, \mathbf{1} \right) \tag{6.52}$$

and note the following relations to hold true identically:

$$\mathbf{C} \left(\dot{\boldsymbol{\varepsilon}} - \overset{*}{\alpha} \dot{T} \, \mathbf{1} \right) = \dot{\boldsymbol{\sigma}} + \frac{1}{h} (\mathbf{C}\mathbf{n})(\mathbf{n}^T \dot{\boldsymbol{\sigma}}) - (\mathbf{C}\mathbf{n}) \sqrt{\frac{2}{3}} \frac{1}{h} \frac{\partial \kappa}{\partial T} \dot{T} , \tag{6.53}$$

$$\mathbf{n}^T \dot{\boldsymbol{\sigma}} = \mathbf{n}^T \dot{\boldsymbol{\sigma}}^{(\mathrm{e})} \frac{h}{h + \mathbf{n}^T \mathbf{C} \mathbf{n}} + \frac{\mathbf{n}^T \mathbf{C} \mathbf{n}}{h + \mathbf{n}^T \mathbf{C} \mathbf{n}} \sqrt{\frac{2}{3}} \frac{\partial \kappa}{\partial T} \dot{T} . \tag{6.54}$$

The loading/unloading conditions are redefined to read

- elastic process

$$\bar{\sigma} < \kappa \quad \text{or} \quad \left(\bar{\sigma} = \kappa \quad \text{and} \quad \mathbf{n}^T \dot{\boldsymbol{\sigma}}^{(\mathrm{e})} - \sqrt{\frac{2}{3}} \frac{\partial \kappa}{\partial T} \dot{T} \leq 0 \right) , \tag{6.55}$$

- plastic process

$$\bar{\sigma} = \kappa \quad \text{and} \quad \mathbf{n}^T \dot{\boldsymbol{\sigma}}^{(\mathrm{e})} - \sqrt{\frac{2}{3}} \frac{\partial \kappa}{\partial T} \dot{T} > 0 . \tag{6.56}$$

We note that this form of the loading/unloading condition corresponds to the so-called strain-space formulation of plasticity and applies to any hardening of the material (hardening, perfect plasticity, softening).

The fundamental system of equations valid for the class of materials considered consists of:

- constitutive equation (6.34) (and Eqs. (6.35), (6.36)) with the argument $\boldsymbol{\sigma}$ replaced by $\boldsymbol{\sigma} - \boldsymbol{\alpha}$,
- evolution equations (6.49),
- loading/unloading condition (6.55), (6.56).

It should be noted that for $h^* < 0$, $\beta > 0$ a stopping criterion needs to be introduced to ensure that the radius of the yield surface does not become negative; this is particularly important for isothermally softening materials, i.e. for $\zeta < 0$.

For isothermal processes we obtain

$$
\begin{aligned}
\dot{\boldsymbol{\sigma}} &= \left[\left(K - \tfrac{2}{3}G\right)\mathbf{1}\mathbf{1}^T + 2G\left(\mathbf{I}_6 - \gamma\mathbf{n}\mathbf{n}^T\right)\right]\dot{\boldsymbol{\varepsilon}} \\
\mathbf{n} &= \sqrt{\frac{3}{2}}\,\frac{{}^D\boldsymbol{\sigma} - \boldsymbol{\alpha}}{\sigma_y} \\
\dot{\boldsymbol{\alpha}} &= (1-\beta)h\,\dot{\boldsymbol{\varepsilon}}^{(p)} \\
\dot{\kappa} &= \tfrac{3}{2}\beta h^* \dot{\bar{\varepsilon}}^{(p)}
\end{aligned}
\tag{6.57}
$$

- elastic process

$$
\bar{\sigma} < \kappa \qquad \text{or} \qquad \left(\bar{\sigma} = \kappa \qquad \text{and} \qquad \mathbf{n}^T\dot{\boldsymbol{\sigma}}^{(e)} \leq 0\right), \tag{6.58}
$$

- plastic process

$$
\bar{\sigma} = \kappa \qquad \text{and} \qquad \mathbf{n}^T\dot{\boldsymbol{\sigma}}^{(e)} > 0\,. \tag{6.59}
$$

Another form of the above equations is sometimes used in applications. To show it we denote

$$
\dot{\boldsymbol{\alpha}} = \tfrac{2}{3}H'_{(\alpha)}\dot{\boldsymbol{\varepsilon}}^{(p)}\,, \qquad \dot{\kappa} = H'\dot{\boldsymbol{\varepsilon}}^{(p)}\,, \tag{6.60}
$$

considering the moduli

$$
H'_{(\alpha)} = (1-\beta)\zeta\,, \qquad H' = \beta\,\frac{\partial\kappa}{\partial\bar{\varepsilon}^{(p)}} = \beta\zeta\,, \tag{6.61}
$$

as arbitrary, independent of each other functions of the effective plastic strain $\bar{\varepsilon}^{(p)}$, $H'_{(\alpha)} = H'_{(\alpha)}(\bar{\varepsilon}^{(p)})$, $H' = H'(\bar{\varepsilon}^{(p)})$. These moduli are treated as the material functions to be determined experimentally for each material and can be employed instead of the functions ζ and β. It can be verified that

$$
\zeta = H' + H'_{(\alpha)}\,, \tag{6.62}
$$

so that also, cf. Eq. (6.24)

$$
h = \tfrac{2}{3}\,H' + \tfrac{2}{3}\,H'_{(\alpha)} \tag{6.63}
$$

and, cf. Eq. (6.36)

$$
\gamma = \frac{1}{1 + \frac{H' + H'_{(\alpha)}}{3G}}\,.
$$

The elastic-plastic constitutive law assumes the form

$$
\dot{\boldsymbol{\sigma}} = \left[\left(K - \tfrac{2}{3}G\right)\mathbf{1}\mathbf{1}^T + 2G\left(\mathbf{I}_6 - \frac{\mathbf{n}\mathbf{n}^T}{1 + \frac{H' + H'_{(\alpha)}}{3G}}\right)\right]\dot{\boldsymbol{\varepsilon}}\,. \tag{6.64}
$$

The isotropic hardening is very often expressed as a linear relationship of the form

$$
\sigma_y(\bar{\varepsilon}^{(p)}) = \sigma_0 + a\bar{\varepsilon}^{(p)}\,, \tag{6.65}
$$

in which

$$H' = \frac{\partial \sigma_y}{\partial \bar{\varepsilon}^{(p)}} = a = \text{const},$$

The exponential form

$$\sigma_y(\bar{\varepsilon}^{(p)}) = \sigma_0 + (\sigma_\infty - \sigma_0)\left[1 - e^{-\bar{\gamma}\bar{\varepsilon}^{(p)}}\right], \tag{6.66}$$

with

$$H' = \bar{\gamma}(\sigma_\infty - \sigma_0)\, e^{-\bar{\gamma}\bar{\varepsilon}^{(p)}},$$

a, σ_∞, σ_0 and $\bar{\gamma}$ being the material constants such that $\sigma_\infty > \sigma_0 > 0$, $\bar{\gamma} > 0$, has also been frequently used in the literature.

The so-called generalized power law has the form

$$\sigma_y(\bar{\varepsilon}^{(p)}) = \sigma_0 \frac{\left(1 + \frac{\bar{\varepsilon}^{(p)}}{\varepsilon_0}\right)^{N(\bar{\varepsilon}^{(p)})}}{1 + \left(\frac{\bar{\varepsilon}^{(p)}}{\varepsilon_1}\right)^2}, \tag{6.67}$$

in which $\varepsilon_0 = \frac{\sigma_0}{E}$ while $N(\bar{\varepsilon}^{(p)})$ and ε_1 are a function and a constant characteristic of the material under consideration. Putting $\varepsilon_1 \to \infty$, Eq. (6.67) assumes the more conventional power form.

To describe the kinematic hardening we usually take

$$H'_{(\alpha)}(\bar{\varepsilon}^{(p)}) = \text{const}.$$

The hardening curve for the thermo-elastic-plastic material is frequently postulated as, cf. Eq. (6.66)

$$\sigma_y(\bar{\varepsilon}^{(p)}, T) = \left\{\sigma_0 + (\sigma_\infty - \sigma_0)\left[1 - e^{-\bar{\gamma}\bar{\varepsilon}^{(p)}}\right]\right\}\left[1 - (\bar{\gamma}_\vartheta T)^{n_1}\right]^{n_2} \tag{6.68}$$

where $\bar{\gamma}_\vartheta$, n_1, n_2 are the material constants.

6.1.4 Creep and Visco-Plasticity

In order to account in the constitutive equations for the rate-dependency effects, we replace Eq. (6.1) by

$$\boldsymbol{\varepsilon} = \boldsymbol{\varepsilon}^{(e)} + \boldsymbol{\varepsilon}^{(p)} + \boldsymbol{\varepsilon}^{(\vartheta)} + \boldsymbol{\varepsilon}^{(c)}, \tag{6.69}$$

with $\boldsymbol{\varepsilon}^{(c)}$ standing for the creep strain, or by

$$\boldsymbol{\varepsilon} = \boldsymbol{\varepsilon}^{(e)} + \boldsymbol{\varepsilon}^{(vp)} + \boldsymbol{\varepsilon}^{(\vartheta)}, \tag{6.70}$$

if the visco-plastic properties result in a strain contribution $\boldsymbol{\varepsilon}^{(vp)}$, assumed generally unseparable into its two viscous and plastic ingredients. Generally, the creep effects make themselves visible in solids subject to long-lasting loads, particularly in elevated temperatures, while visco-plastic behaviour is typical of solids deforming at high speeds (impact loading problems, for instance).

Let us first observe that the reasoning similar to that presented in Sect. 6.1.3 leads in the case of creep to the rate-type constitutive equation of the form

$$\dot{\boldsymbol{\sigma}} = \left[K\,\mathbf{1}\mathbf{1}^T + 2G\left(\mathbf{I}_6 - \tfrac{1}{3}\mathbf{1}\mathbf{1}^T\right) - 2G\gamma\mathbf{n}\mathbf{n}^T\right]\left(\dot{\boldsymbol{\varepsilon}} - \dot{\boldsymbol{\varepsilon}}^{(c)} - \mathring{\alpha}\dot{T}\,\mathbf{1}\right) + \overset{*}{\boldsymbol{\sigma}} + \overset{**}{\boldsymbol{\sigma}}, \tag{6.71}$$

with the functions $\overset{*}{\boldsymbol{\sigma}}$ and $\overset{**}{\boldsymbol{\sigma}}$ (different from zero for the non-isothermal processes only) given by

$$\begin{aligned} \overset{*}{\boldsymbol{\sigma}} &= \left[\dot{K}\,\mathbf{1}\mathbf{1}^T + 2\dot{G}\left(\mathbf{I}_6 - \tfrac{1}{3}\mathbf{1}\mathbf{1}^T\right)\right](\boldsymbol{\varepsilon} - \boldsymbol{\varepsilon}^{(p)} - \boldsymbol{\varepsilon}^{(c)} - \alpha\vartheta\mathbf{1})\,, \\ \overset{**}{\boldsymbol{\sigma}} &= \mathbf{n}\gamma\left(\sqrt{\frac{2}{3}}\,\frac{\partial\sigma_y}{\partial\vartheta}\,\dot{\vartheta} - \overset{*}{\boldsymbol{\sigma}}^T\mathbf{n}\right). \end{aligned} \tag{6.72}$$

regardless of the specific flow rule adopted for the creep strain description. The local loading/unloading conditions may be taken in the same form as before.

Assume for instance that the uniaxial creep strain may be described by an expression of the form (thermal effects omitted)

$$\varepsilon^{(c)} = A\sigma^n\tau^k \tag{6.73}$$

in which a, n and k are material constants and τ denotes time. The rate of the creep strain then results as

$$\dot{\varepsilon}^{(c)} = kA\sigma^n\tau^{k-1} = \tilde{G}(\sigma,\tau)\,, \tag{6.74}$$

or, which amounts to the same

$$\dot{\varepsilon}^{(c)} = kA^{\frac{1}{k}}\sigma^{\frac{n}{k}}[\varepsilon^{(c)}]^{1-\frac{1}{k}} = \bar{G}(\sigma,\varepsilon^{(c)})\,. \tag{6.75}$$

At a fixed stress value σ the functions $\tilde{G}$ and $\bar{G}$ describe the strain rate as a function of time and the current creep strain, respectively. We note that Eqs. (6.74), (6.75) are basically applicable only to problems characterized by a given, fixed stress level σ – using them to calculate creep effects under different, varying stress histories will generally lead to different values of the creep strain. We shall be using below creep law function without specifying its precise built-up.

Let us note, moreover, that even if using the function $\bar{G}(\sigma,\varepsilon^{(c)})$ may look quite natural in describing processes involving creep, Eq. (6.75) may (at least in theory) be solved for σ to yield

$$\sigma = \bar{H}(\varepsilon^{(c)},\dot{\varepsilon}^{(c)})\,. \tag{6.76}$$

For the description of the visco-plastic effects we may take the flow rule in the form

$$\dot{\varepsilon}^{(vp)} = \dot{\varepsilon}_0\left[\frac{\sigma}{g(\varepsilon^{(vp)})}\right]^{\frac{1}{m}}\,, \tag{6.77}$$

for instance, where m is a material constant, $\dot{\varepsilon}_0$ is a reference strain rate while $g(\varepsilon^{(vp)})$ represents stress generated experimentally in the uniaxial tension test under a given strain rate $\dot{\varepsilon} = \dot{\varepsilon}_0$. It is frequently assumed that, cf. Eq. (6.67),

$$g(\varepsilon^{(vp)}) = \sigma_0 \frac{\left(1 + \frac{\varepsilon^{(vp)}}{\varepsilon_0}\right)^N}{1 + \left(\frac{\varepsilon^{(vp)}}{\varepsilon_1}\right)^2}, \tag{6.78}$$

where σ_0 is the initial static yield limit, N is a material constant, $\varepsilon_0 = \frac{\sigma_0}{E}$ while ε_1 is a reference strain often for simplicity taken as an infinitely large number. For this material we have

$$\dot{\varepsilon}^{(vp)} = \bar{G}(\sigma, \varepsilon^{(vp)}) = \dot{\varepsilon}_0 \left[\frac{\sigma}{g(\varepsilon^{(vp)})}\right]^{\frac{1}{m}} = \dot{\varepsilon}_0 \left[\frac{\sigma}{\sigma_0} \frac{1 + \left(\frac{\varepsilon^{(vp)}}{\varepsilon_1}\right)^2}{\left(1 + \frac{\varepsilon^{(vp)}}{\varepsilon_0}\right)^N}\right]^{\frac{1}{m}} \tag{6.79}$$

and

$$\sigma = H(\varepsilon^{(vp)}, \dot{\varepsilon}^{(vp)}) = \left(\frac{\dot{\varepsilon}^{(vp)}}{\dot{\varepsilon}_0}\right)^m \sigma_0 \frac{\left(1 + \frac{\varepsilon^{(vp)}}{\varepsilon_0}\right)^N}{1 + \left(\frac{\varepsilon^{(vp)}}{\varepsilon_1}\right)^2}, \tag{6.80}$$

The above uniaxial considerations can be formally generalized to apply to more complex stress states by assuming the Huber–Mises potential for computing the multiaxial strain rates, replacing the uniaxial stress σ by the effective stress defined as $\bar{\sigma} = \sqrt{\frac{3}{2}\, {}^D\boldsymbol{\sigma}^T\, {}^D\boldsymbol{\sigma}}$ and the uniaxial strain rate $\dot{\varepsilon}^{(vp)}$ (or $\dot{\varepsilon}^{(c)}$) by the effective visco-plastic (or creep) strain rate $\dot{\bar{\varepsilon}}^{(vp)} = \sqrt{\frac{2}{3}\, \dot{\boldsymbol{\varepsilon}}^{(vp)T}\, \dot{\boldsymbol{\varepsilon}}^{(vp)}}$ Consequently, the multidimensional counterpart to Eqs. (6.75), (6.79) becomes

$$\dot{\boldsymbol{\varepsilon}}^{(vp)} = \dot{\bar{\varepsilon}}^{(vp)} \frac{\partial f}{\partial \boldsymbol{\sigma}^T} = \dot{\bar{\varepsilon}}^{(vp)} \sqrt{\frac{3}{2}}\, \mathbf{n} = \dot{\bar{\varepsilon}}^{(vp)} \frac{3}{2} \frac{{}^D\boldsymbol{\sigma}}{\bar{\sigma}} = \frac{3}{2} \bar{G}(\bar{\sigma}, \varepsilon^{(vp)}) \frac{{}^D\boldsymbol{\sigma}}{\bar{\sigma}}, \tag{6.81}$$

in which, as before, $\mathbf{n}$ is the unit vector normal to the potential surface, which effectively determines the direction of the inelastic strain rate vector.

In the case of creep it is common to adopt in Eq. (6.81) the function $\bar{G}$ in the form of the so-called Norton law

$$\bar{G}(\sigma, \cdot) = \dot{\varepsilon}_c \left(\frac{\sigma}{\sigma_c}\right)^p, \tag{6.82}$$

with $\dot{\varepsilon}_c$, σ_c being material constants. This function is a special case of the creep law (6.75). For the multidimensional case we then have

$$\dot{\boldsymbol{\varepsilon}}_{(c)} = \frac{3}{2} \dot{\varepsilon}_c \left(\frac{\bar{\sigma}}{\sigma_c}\right)^p \frac{{}^D\boldsymbol{\sigma}}{\bar{\sigma}}. \tag{6.83}$$

It has been widely documented experimentally that under dynamic load conditions the yield limit σ_y is a function of both the total inelastic effective strain $\bar{\varepsilon}^{(\mathrm{vp})}$ and its rate $\dot{\bar{\varepsilon}}^{(\mathrm{vp})}$. Equation (6.76) may then be seen as a definition of the 'dynamic' yield limit σ_y^d in terms of $\bar{\varepsilon}^{(\mathrm{vp})}$ and $\dot{\bar{\varepsilon}}^{(\mathrm{vp})}$:

$$\sigma_y^d = H(\bar{\varepsilon}^{(\mathrm{vp})}, \dot{\bar{\varepsilon}}^{(\mathrm{vp})}) . \tag{6.84}$$

The visco-plastic constitutive equation of the form

$$\dot{\boldsymbol{\sigma}} = \mathbf{C}(\dot{\boldsymbol{\varepsilon}} - \dot{\boldsymbol{\varepsilon}}^{(\mathrm{vp})}) = \mathbf{C}\dot{\boldsymbol{\varepsilon}} - \mathbf{C}\dot{\boldsymbol{\varepsilon}}^{(\mathrm{vp})} ,$$

does not lend itself too easily to the description in terms of the tangent constitutive matrix. In the finite element context the direct use of the above equation would result in the stiffness matrix based on the elastic constitutive matrix $\mathbf{C}$, the inelastic effects taken care of by some initial load terms entering the right-hand side vector in the finite element system. In problems with extensive plastic straining this would results in very poor stability properties of the iterative solution. The following discussion is meant to at least partially circumvent this problem.

We shall be separately treating below the two cases corresponding to the constitutive description using the functions $\sigma_y^d = H(\bar{\varepsilon}^{(\mathrm{vp})}, \dot{\bar{\varepsilon}}^{(\mathrm{vp})})$ and $\dot{\bar{\varepsilon}}^{(\mathrm{vp})} = \bar{G}(\bar{\sigma}, \bar{\varepsilon}^{(\mathrm{vp})})$, respectively. We shall assume the Huber–Mises yield condition

$$f = \sqrt{\tfrac{3}{2}\, {}^{D}\bar{\boldsymbol{\sigma}}^T\, {}^{D}\bar{\boldsymbol{\sigma}}} - \sigma_y^d = 0 . \tag{6.85}$$

to be at the same time the potential for computing the inelastic strain rate.

A. The constitutive function given in the form of $\sigma_y^d = H(\bar{\varepsilon}^{(\mathrm{vp})}, \dot{\bar{\varepsilon}}^{(\mathrm{vp})})$.

The consistency condition $\Delta f = 0$ which assures that the stress point remains on the yield surface during the inelastic straining implies that

$$\sqrt{\tfrac{3}{2}}\, \mathbf{n}^T (\Delta^{D}\boldsymbol{\sigma} - \Delta\boldsymbol{\alpha}) - \Delta\sigma_y^d = 0 . \tag{6.86}$$

Therefore

$$\begin{aligned} \Delta^{D}\boldsymbol{\sigma} &= 2G(\Delta^{D}\boldsymbol{\varepsilon} - \Delta\boldsymbol{\varepsilon}^{(\mathrm{vp})}) , \\ \Delta\boldsymbol{\alpha} &= \tfrac{2}{3}\, H'_{(\alpha)} \Delta\boldsymbol{\varepsilon}^{(\mathrm{vp})}, \\ \Delta\boldsymbol{\varepsilon}^{(\mathrm{vp})} &= \sqrt{\tfrac{3}{2}}\, \Delta\bar{\varepsilon}^{(\mathrm{vp})} \mathbf{n} . \end{aligned} \tag{6.87}$$

The increment of the function σ_y^d may be expressed for sufficiently small increments of its arguments as

$$\Delta\sigma_y^d = \frac{\partial H}{\partial \bar{\varepsilon}^{(\mathrm{vp})}} \Delta\bar{\varepsilon}^{(\mathrm{vp})} + \frac{\partial H}{\partial \dot{\bar{\varepsilon}}^{(\mathrm{vp})}} \Delta\dot{\bar{\varepsilon}}^{(\mathrm{vp})} = H' \Delta\bar{\varepsilon}^{(\mathrm{vp})} + H'_1 \Delta\dot{\bar{\varepsilon}}^{(\mathrm{vp})} . \tag{6.88}$$

The increment of $\dot{\bar{\varepsilon}}^{(\mathrm{vp})}$ is written as

$$\Delta\dot{\bar{\varepsilon}}^{(\mathrm{vp})} = \dot{\bar{\varepsilon}}^{(\mathrm{vp})\, t+\Delta t} - \dot{\bar{\varepsilon}}^{(\mathrm{vp})\, t} , \tag{6.89}$$

where in accordance with the notation introduced in Sect. 2.1

$$\dot{\bar{\varepsilon}}^{(\mathrm{vp})\, t} = \dot{\bar{\varepsilon}}^{(\mathrm{vp})}(t) , \quad \text{etc.}$$

The linear approximation for $\Delta\bar{\varepsilon}^{(\mathrm{vp})}$ in the time interval $[t, t+\Delta t]$

$$\Delta\bar{\varepsilon}^{(\mathrm{vp})} = \dot{\bar{\varepsilon}}^{(\mathrm{vp})\,t+\Delta t}\xi\Delta t + \dot{\bar{\varepsilon}}^{(\mathrm{vp})\,t}(1-\xi)\Delta t\,, \qquad 0 < \xi \leq 1\,, \tag{6.90}$$

results in

$$\dot{\bar{\varepsilon}}^{(\mathrm{vp})\,t+\Delta t} = \frac{\Delta\bar{\varepsilon}^{(\mathrm{vp})} - \dot{\bar{\varepsilon}}^{(\mathrm{vp})\,t}(1-\xi)\Delta t}{\xi\Delta t}\,. \tag{6.91}$$

From Eqs. (6.89), (6.91) there results the relation

$$\Delta\dot{\bar{\varepsilon}}^{(\mathrm{vp})} = \frac{\Delta\bar{\varepsilon}^{(\mathrm{vp})} - \dot{\bar{\varepsilon}}^{(\mathrm{vp})\,t}\Delta t}{\xi\Delta t}\,. \tag{6.92}$$

By using Eq. (6.92) in Eq. (6.88) we may express $\Delta\sigma_y^d$ as a function of $\Delta\bar{\varepsilon}^{(\mathrm{vp})}$ and a known value $\dot{\bar{\varepsilon}}^{(\mathrm{vp})\,t}$. The so obtained equation together with Eq. (6.87) is then substituted in Eq. (6.86) to yield

$$\Delta\bar{\varepsilon}^{(\mathrm{vp})} = \frac{2G\,\mathbf{n}^T\Delta\boldsymbol{\varepsilon} + \frac{2}{3}H_1'\frac{\dot{\bar{\varepsilon}}^{(\mathrm{vp})\,t}}{\xi}}{\sqrt{\frac{3}{2}}\left(2G + \frac{2}{3}H'_{(\alpha)}\right) + \left(H' + \frac{H_1'}{\xi\Delta t}\right)\sqrt{\frac{2}{3}}}\,. \tag{6.93}$$

By substituting Eq. (6.93) in Eq. (6.87) and Eq. (6.87) in Eq. (6.87) we arrive at the final form of the incremental visco-plastic constitutive equation as

$$\Delta\boldsymbol{\sigma} = \left[\left(K - \tfrac{2}{3}G\right)\mathbf{1}\mathbf{1}^T + 2G\left(\mathbf{I}_6 - \overset{*}{\gamma}\mathbf{n}\mathbf{n}^T\right)\right]\Delta\boldsymbol{\varepsilon} - \hat{\boldsymbol{\sigma}}\,, \tag{6.94}$$

where

$$\overset{*}{\gamma} = \frac{1}{1 + \frac{H' + H'_{(\alpha)} + \frac{H_1'}{\xi\Delta t}}{3G}}\,, \qquad \hat{\boldsymbol{\sigma}} = 2G\overset{*}{\gamma}\sqrt{\frac{2}{3}}\frac{H_1'}{\xi\Delta t}\mathbf{n}\,. \tag{6.95}$$

It may easily be observed that for $\sigma_y^d = \sigma_y(\bar{\varepsilon}^{(\mathrm{vp})})$ (i.e. for $H_1' = 0$), which corresponds to the inviscid plasticity, the above relationships reduce to those derived earlier:

$$\overset{*}{\gamma} = \frac{1}{1 + \frac{H' + H'_{(\alpha)}}{3G}}\,, \qquad \hat{\boldsymbol{\sigma}} = 0\,,$$

$$\Delta\boldsymbol{\sigma} = \left[\left(K - \tfrac{2}{3}G\right)\mathbf{1}\mathbf{1}^T + 2G\left(\mathbf{I}_6 - \gamma\,\mathbf{n}\mathbf{n}^T\right)\right]\Delta\boldsymbol{\varepsilon}\,,$$

$$\Delta\bar{\varepsilon}^{(\mathrm{vp})} = \frac{\mathbf{n}^T\Delta\boldsymbol{\varepsilon}}{\left(1 + \frac{H' + H'_{(\alpha)}}{3G}\right)\sqrt{\frac{3}{2}}} = \sqrt{\frac{2}{3}}\frac{2G\mathbf{n}^T\Delta\boldsymbol{\varepsilon}}{2G + \frac{2}{3}(H' + H'_{(\alpha)})}$$

$$= \sqrt{\frac{2}{3}}\frac{\mathbf{n}^T\Delta\boldsymbol{\sigma}^{(\mathrm{e})}}{2G + h} = \sqrt{\frac{2}{3}}\frac{\mathbf{n}^T\Delta\boldsymbol{\sigma}}{h}\,,$$

$$\Delta\boldsymbol{\varepsilon}^{(\mathrm{vp})} = \sqrt{\frac{3}{2}}\Delta\bar{\varepsilon}^{(\mathrm{vp})}\mathbf{n} = \frac{1}{h}(\mathbf{n}^T\Delta\boldsymbol{\sigma})\,\mathbf{n} = \frac{1}{h+2G}(\mathbf{n}^T\Delta\boldsymbol{\sigma}^{(\mathrm{e})})\,\mathbf{n}\,.$$

It is sometimes convenient numerically to replace Eq. (6.88) by the relation

$$\Delta\sigma_y^d = H'\Delta\bar{\varepsilon}^{(\mathrm{vp})} + \sigma_y^d\left(\cdot, \dot{\bar{\varepsilon}}^{(\mathrm{vp})\,t+\Delta t}\right) - \sigma_y^d\left(\cdot, \dot{\bar{\varepsilon}}^{(\mathrm{vp})\,t}\right)\,. \tag{6.96}$$

Substituting Eq. (6.87) in Eq. (6.87) and Eq. (6.96) in Eq. (6.86) leads to

$$\Delta\bar{\varepsilon}^{(\mathrm{vp})} = \frac{\mathbf{n}^T\Delta\boldsymbol{\varepsilon} - \sqrt{\frac{2}{3}}(\sigma_y^{d,t+\Delta t} - \sigma_y^{d,t})}{1+\frac{H'+H'_{(\alpha)}}{3G}}\,\mathbf{n} \tag{6.97}$$

and

$$\Delta\boldsymbol{\sigma} = \left[\left(K - \tfrac{2}{3}G\right)\mathbf{1}\mathbf{1}^T + 2G\left(\mathbf{I}_6 - \gamma\,\mathbf{n}\mathbf{n}^T\right)\right]\Delta\boldsymbol{\varepsilon} + \frac{\sqrt{\frac{2}{3}}(\sigma_y^{d,t+\Delta t} - \sigma_y^{d,t})\,\mathbf{n}}{1+\frac{H'+H'_{(\alpha)}}{3G}}\,, \tag{6.98}$$

in which the coefficient γ is given, as in the case of inviscid plasticity, by Eq. (6.36).

B. The constitutive function given in the form of $\dot{\bar{\varepsilon}}^{(\mathrm{vp})} = \bar{G}(\bar{\sigma}, \bar{\varepsilon}^{(\mathrm{vp})})$.

Similarly as before we assume

$$\Delta\bar{\varepsilon}^{(\mathrm{vp})} = \dot{\bar{\varepsilon}}^{(\mathrm{vp})\,t+\Delta t}\xi\Delta t + \dot{\bar{\varepsilon}}^{(\mathrm{vp})\,t}(1-\xi)\Delta t\,, \qquad 0 < \xi \le 1\,. \tag{6.99}$$

The increment of $\dot{\bar{\varepsilon}}^{(\mathrm{vp})}$ may for sufficiently small increments of $\Delta\bar{\varepsilon}^{(\mathrm{vp})}$ and $\Delta\bar{\sigma}$ be written as

$$\Delta\dot{\bar{\varepsilon}}^{(\mathrm{vp})} = \dot{\bar{\varepsilon}}^{(\mathrm{vp})\,t+\Delta t} - \dot{\bar{\varepsilon}}^{(\mathrm{vp})\,t} = \frac{\partial\bar{G}}{\partial\bar{\varepsilon}^{(\mathrm{vp})}}\Delta\bar{\varepsilon}^{(\mathrm{vp})} + \frac{\partial\bar{G}}{\partial\bar{\sigma}}\Delta\bar{\sigma} = \bar{G}'_1\,\Delta\bar{\varepsilon}^{(\mathrm{vp})} + \bar{G}'_2\,\Delta\bar{\sigma}\,. \tag{6.100}$$

By eliminating $\dot{\bar{\varepsilon}}^{(\mathrm{vp})\,t+\Delta t}$ from Eqs. (6.99), (6.100) we obtain

$$\Delta\bar{\varepsilon}^{(\mathrm{vp})} = \bar{G}^t\Delta t + \xi\Delta t\left(\bar{G}'_1\,\Delta\bar{\varepsilon}^{(\mathrm{vp})} + \bar{G}'_2\,\Delta\bar{\sigma}\right)\,, \tag{6.101}$$

where $\bar{G}^t$ stands for the value of the function $\bar{G}$ at the time t, i.e. $\bar{G}^t = \bar{G}(\bar{\varepsilon}^{(\mathrm{vp}),t}, \bar{\sigma}^t)$.

The consistency condition implies

$$\Delta\bar{\sigma} = \sqrt{\tfrac{3}{2}}\,\mathbf{n}^T(\Delta^D\boldsymbol{\sigma} - \Delta\boldsymbol{\alpha}) = \sqrt{\tfrac{3}{2}}\left[2G\,\mathbf{n}^T\Delta\boldsymbol{\varepsilon} - \sqrt{\tfrac{3}{2}}\left(2G + \tfrac{2}{3}H'_{(\alpha)}\right)\Delta\bar{\varepsilon}^{(\mathrm{vp})}\right]\,, \tag{6.102}$$

in which we have used Eqs. (6.87). From the last two equation there results

$$\Delta\bar{\varepsilon}^{(\mathrm{vp})} = \Delta t\left[\frac{\bar{G}^t + \sqrt{\frac{3}{2}}\,2G\bar{G}'_2\xi\,\mathbf{n}^T\Delta\boldsymbol{\varepsilon}}{1 + \bar{G}'_2\xi\Delta t(3G + H'_{(\alpha)}) - \bar{G}'_1\xi\Delta t}\right]\,, \tag{6.103}$$

which when used in Eqs. (6.87) allows to write the flow rule in the form

$$\Delta\boldsymbol{\varepsilon}^{(\mathrm{vp})} = \mathbf{n}\Delta t\left[\frac{\sqrt{\frac{3}{2}}\bar{G}^t + 3G\bar{G}'_2\xi\,\mathbf{n}^T\Delta\boldsymbol{\varepsilon}}{1 + \bar{G}'_2\xi\Delta t(3G + H'_{(\alpha)}) - \bar{G}'_1\xi\Delta t}\right]\,. \tag{6.104}$$

Knowing $\Delta\boldsymbol{\varepsilon}^{(\mathrm{vp})}$ the stress increment is computed from Eq. (6.87) (the deviatoric part) and from the elastic constitutive law (6.41) (the spherical part).

Let us observe that for $\xi = 0$ Eqs. (6.103), (6.104) reduce to the already known equations (6.16), (6.17), i.e.

$$\Delta\bar{\varepsilon}^{(\mathrm{vp})} = \Delta t\, \bar{G}^t\,, \qquad \Delta\boldsymbol{\varepsilon}^{(\mathrm{vp})} = \Delta t \sqrt{\tfrac{3}{2}}\, \bar{G}^t \mathbf{n}\,.$$

Denoting

$$\tilde{h} = 3G - \frac{\bar{G}_1'}{\bar{G}_2'}\,, \qquad \beta = \xi \Delta t\, \tilde{h} \bar{G}_2' = \xi \Delta t\, 3G\bar{G}_2' - \xi \Delta t\, \bar{G}_1'\,, \tag{6.105}$$

and neglecting for simplicity the kinematic hardening ($H'_{(\alpha)} = 0$), Eq. (6.103) allows to obtain the relationship

$$\Delta\bar{\varepsilon}^{(\mathrm{vp})} = \Delta t \frac{\bar{G}^t}{1+\beta} + \frac{2G\sqrt{\frac{3}{2}}}{\tilde{h}} \frac{\beta}{1+\beta} \mathbf{n}^T \Delta\boldsymbol{\varepsilon}\,. \tag{6.106}$$

By dividing the both sides in this equation by Δt, taking the limit at $\Delta t \to 0$ and substituting the result in

$$\dot{\boldsymbol{\sigma}} = \mathbf{C}\dot{\boldsymbol{\varepsilon}} - \mathbf{C}\sqrt{\tfrac{3}{2}}\, \mathbf{n}\dot{\bar{\varepsilon}}^{(\mathrm{vp})} \tag{6.107}$$

we arrive at

$$\dot{\boldsymbol{\sigma}} = \mathbf{C}_T\, \dot{\boldsymbol{\varepsilon}} - \frac{2\sqrt{\frac{3}{2}}\, G\bar{G}^t}{1+\beta} \mathbf{n}\,, \tag{6.108}$$

in which the (symmetric) algorithmic tangent constitutive stiffness matrix $\mathbf{C}_T$ is given by

$$\mathbf{C}_T = \mathbf{C} - \frac{6G^2}{\tilde{h}} \frac{\beta}{1+\beta} \mathbf{n}\mathbf{n}^T\,. \tag{6.109}$$

Assuming the explicit form for the function $\bar{G}$ as, cf. Eq. (6.79)

$$\dot{\bar{\varepsilon}}^{(\mathrm{vp})} = \bar{G}(\bar{\varepsilon}^{(\mathrm{vp})}, \bar{\sigma}) = \dot{\varepsilon}_0 \left[\frac{\bar{\sigma}}{g(\bar{\varepsilon}^{(\mathrm{vp})})} \right]^{\frac{1}{m}}\,, \tag{6.110}$$

the modulus $\tilde{h}$ is obtained as

$$\tilde{h} = 3G + \left[\frac{\dot{\bar{\varepsilon}}^{(\mathrm{vp})}}{\dot{\varepsilon}_0} \right]^m \frac{\mathrm{d}g}{\mathrm{d}\bar{\varepsilon}^{(\mathrm{vp})}}\,, \tag{6.111}$$

while the coefficient β becomes

$$\beta = \frac{\xi \Delta t\, \tilde{h} \dot{\bar{\varepsilon}}^{(\mathrm{vp})\, t}}{m\bar{\sigma}}\,. \tag{6.112}$$

Note that for $\Delta t \to 0$ there holds $\beta \to 0$, while the matrix $\mathbf{C}_T$ then reduces to the elastic constitutive matrix $\mathbf{C}$, cf. (6.109). In the limiting case of the inviscid material we have $m \to 0$ (which at $\xi \Delta t \neq 0$ implies that $\beta \to \infty$) while $\tilde{h} \to 3G + \zeta$, cf. Eq. (6.36). Equation (6.108) assumes the form

$$\dot{\boldsymbol{\sigma}} = \mathbf{C}_T\, \dot{\boldsymbol{\varepsilon}} \tag{6.113}$$

where

$$\mathbf{C}_T = \mathbf{C} - \frac{6G^2}{3G + \frac{3}{2}h}\,\mathbf{n}\mathbf{n}^T = \mathbf{C} - \hat{\gamma}\,\mathbf{n}\mathbf{n}^T\,, \tag{6.114}$$

which fully coincides with Eq. (6.39) derived for the elastic-plastic material in a different way.

In the frequently employed power creep law we assume $g(\bar{\varepsilon}^{(\mathrm{vp})}) = 0$ (i.e. also $\mathrm{d}g/\mathrm{d}\bar{\varepsilon}^{(\mathrm{vp})} = 0$). In the constitutive equation (6.108) we then take $\tilde{h} = 3G$, cf. Eq. (6.111).

Before moving on to consider still another class of the constitutive equations we note that in the literature there exists a different approach to modelling the visco-plastic material behaviour. For it is sometimes more convenient to represent the dynamic properties of the material by giving explicitly only the static yield limit $\sigma_y = \sigma_y(\bar{\varepsilon}^{(\mathrm{vp})})$, but modifying at the same time the flow rule incorporating a new material parameter ω to represent the material viscosity. The yield condition in such a formulation ceases to be a real constraint imposed on the state of stress, in other words, stresses for which

$$f = F(\boldsymbol{\sigma}) - \sigma_y = \bar{\sigma} - \sigma_y > 0\,. \tag{6.115}$$

are admissible. For the class of the constitutive laws on hand we assume the flow rule to have the form

$$\dot{\boldsymbol{\varepsilon}}^{(\mathrm{vp})} = \omega\left\langle \Phi\left(\frac{F - \sigma_y}{\sigma_y}\right)\right\rangle \frac{\partial f}{\partial \boldsymbol{\sigma}^T}\,, \tag{6.116}$$

where the symbol $\langle\Phi(\cdot)\rangle$ is defined as

$$\left\langle \Phi\left(\frac{F - \sigma_y}{\sigma_y}\right)\right\rangle = \begin{cases} \Phi\left(\frac{F-\sigma_y}{\sigma_y}\right) & \text{for } F - \sigma_y \geq 0, \\ 0 & \text{for } F - \sigma_y < 0, \end{cases} \tag{6.117}$$

ω is the material viscosity parameter and Φ is assumed as, for instance

$$\Phi = \left(\frac{F - \sigma_y}{\sigma_y}\right)^n\,, \tag{6.118}$$

or

$$\Phi = e^{m\left(\frac{F-\sigma_y}{\sigma_y}\right)} - 1\,, \tag{6.119}$$

m and n being the material constants. Confining ourselves to the case (6.118) with $n = 1$ and using the Huber–Mises potential for computing the inelastic strain rate we obtain

$$\dot{\boldsymbol{\varepsilon}}^{(\mathrm{vp})} = \frac{3}{2}\,\omega\,\frac{\bar{\sigma} - \sigma_y}{\sigma_y}\,\frac{{}^{D}\boldsymbol{\sigma}}{\bar{\sigma}} \qquad \text{for } \bar{\sigma} - \sigma_y \geq 0\,. \tag{6.120}$$

We observe that the above law (as well as its more general form (6.96)) is consistent with the previously postulated expression of the form

$$\dot{\boldsymbol{\varepsilon}}^{(\mathrm{vp})} = \frac{3}{2}\,\frac{G}{\sigma_y}\,{}^{D}\boldsymbol{\sigma}\,, \tag{6.121}$$

in which

$$G = G(\bar{\sigma}, \bar{\varepsilon}^{(\mathrm{vp})}) = \omega\,\frac{\bar{\sigma} - \sigma_y(\bar{\varepsilon}^{(\mathrm{vp})})}{\bar{\sigma}}\,. \tag{6.122}$$

Moreover, there holds

$$\dot{\bar{\varepsilon}}^{(\mathrm{vp})} = \sqrt{\frac{2}{3}\,\dot{\boldsymbol{\varepsilon}}^{(\mathrm{vp})T}\,\dot{\boldsymbol{\varepsilon}}^{(\mathrm{vp})}} = \sqrt{\omega^2\,\frac{3}{2}\,\frac{{}^{D}\boldsymbol{\sigma}^{T}\,{}^{D}\boldsymbol{\sigma}}{\bar{\sigma}^2}\left(\frac{\bar{\sigma} - \sigma_y}{\sigma_y}\right)^2} = \omega\,\frac{\bar{\sigma} - \sigma_y}{\sigma_y}\,, \tag{6.123}$$

so that

$$\bar{\sigma} - \sigma_y\left(\frac{\dot{\bar{\varepsilon}}^{(\mathrm{vp})}}{\omega} + 1\right) = 0\,. \tag{6.124}$$

Equation (6.124) may be regarded as the 'viscous' counterpart of the yield condition (6.12). It is seen that for $\omega \to \infty$ Eq. (6.124) coincides with Eq. (6.12). For finite values of ω Eq. (6.124) implies that the current (deviatoric) stress point ${}^{D}\boldsymbol{\sigma}$ may lie outside the yield surface (6.12) – the higher the effective strain rate $\dot{\bar{\varepsilon}}^{(\mathrm{vp})}$ and lower the value of viscosity ω, the greater the excess stress.

Denoting, similarly as before, the dynamic yield limit by the symbol σ_y^d, we have

$$\sigma_y^d = H(\bar{\varepsilon}^{(\mathrm{vp})}, \dot{\bar{\varepsilon}}^{(\mathrm{vp})}) = \sigma_y(\bar{\varepsilon}^{(\mathrm{vp})})\left(\frac{\dot{\bar{\varepsilon}}^{(\mathrm{vp})}}{\omega} + 1\right) \tag{6.125}$$

(σ_y is the static yield limit). By using Eq. (6.124) in Eq. (6.120) to eliminate $\bar{\sigma}$ we obtain

$$\dot{\boldsymbol{\varepsilon}}^{(\mathrm{vp})} = \frac{3}{2}\,\frac{\dot{\bar{\varepsilon}}^{(\mathrm{vp})}}{\sigma_y\left(\frac{\dot{\bar{\varepsilon}}^{(\mathrm{vp})}}{\omega} + 1\right)}\,{}^{D}\boldsymbol{\sigma}\,. \tag{6.126}$$

Remembering that $\mathbf{n} = \sqrt{\frac{3}{2}}\,\frac{{}^{D}\boldsymbol{\sigma}}{\sigma_y}$ for the inviscid plasticity ($\omega \to \infty$) Eq. (6.126) implies

$$\dot{\boldsymbol{\varepsilon}}^{(\mathrm{vp})} = \sqrt{\frac{3}{2}}\,\dot{\bar{\varepsilon}}^{(\mathrm{vp})}\mathbf{n} \tag{6.127}$$

which has been derived before in a different way, cf. Eq. (6.16). It is seen that for high values of viscosity ω the viscous strain increment in the elastic-visco-plastic material will be close, for the same stress, to the plastic strain increment in the corresponding elastic-plastic material.

6.1.5
Elasto-Plasticity with Damage

Only one rather specific example of the inelastic material with damage will be described in this section to illustrate the possible complications in the description. No thermal effects will be included.

Let us rewrite the flow rule (6.25) and the tangent constitutive matrix (6.39) as, cf. Eq. (6.24)

$$\dot{\boldsymbol{\varepsilon}}^{(\mathrm{p})} = \frac{1}{\zeta}(\mathbf{s}^T\,\dot{\boldsymbol{\sigma}})\mathbf{s}\,, \qquad \mathbf{s} = \sqrt{\frac{3}{2}}\,\mathbf{n} \tag{6.128}$$

$$\mathbf{C}_T = \mathbf{C} - \frac{E}{1+\nu} \frac{\mathbf{s}\,\mathbf{s}^T}{\frac{3}{2} + \frac{1+\nu}{E}\zeta} . \tag{6.129}$$

Note that the numerator in the second term on the right-hand side of the last equation involving the expression $\mathbf{s}\,\mathbf{s}^T$ assures the symmetry of $\mathbf{C}_T$ (provided the elastic constitutive matrix $\mathbf{C}$ is symmetric).

A natural generalization of the above formulation appears to be abandoning the normality of the plastic strain vector to the yield surface. Such a modified flow rule, known as non-associated, has the general form

$$\dot{\varepsilon}^{(\mathrm{p})} = \frac{1}{\zeta} ({}^{(2)}\mathbf{s}^T \dot{\sigma})\, {}^{(1)}\mathbf{s} , \tag{6.130}$$

in which ${}^{(2)}\mathbf{s}$ is, as before, the vector normal to the current yield surface $f = 0$ while ${}^{(1)}\mathbf{s}$ is the vector normal to the surface of a certain plastic potential, generally different form the yield surface. It appears that some materials may be better described by the non-associated flow rule (6.130) than by the associated flow rule (6.128) – as examples we may mention metals exhibiting some microdamage effects and some rocks and soils.

A very simple model useful in modelling ductile materials with microscopic voids will be briefly described below. The model, proposed in [56], is based on the following assumptions:

- damage effects are represented by means of spherical voids,
- the continuum description applies with the only parameter representing the damage effects being the so-called porosity ξ defined as the limit of the ratio of the void volume to the material volume encompassing the given point, taken at the material volume tending to zero.
- the matrix material exhibits plastic incompressibility and may be described by the Huber–Mises elasto-plasticity (involving the associated plastic flow rule).

Due to the changing void volumes the resulting continuum is compressible – in other words, the plastic dilatancy is the apparent macroscopic effect of the void growth. Moreover, it turns out that the inclusion of plastic dilatancy and pressure sensitivity of yield leads in a natural way to the relations describing non-associated plasticity – the material model on hand may thus be regarded as a special case of the class defined by the flow rule (6.130).

The plastic potential is postulated in the following way:

$$\tilde{f}(\boldsymbol{\sigma}, \sigma_M, \xi) = \frac{3}{2} \frac{{}^{D}\boldsymbol{\sigma}^T\, {}^{D}\boldsymbol{\sigma}}{\sigma_M^2} + 2\xi \cosh\left(\frac{3\sigma}{2\sigma_M}\right) - (1+\xi^2) = 0 , \tag{6.131}$$

in which $\boldsymbol{\sigma}$ is the macroscopic Cauchy stress, ${}^{D}\boldsymbol{\sigma}$ its deviator, ξ is the material porosity and σ_M is the effective plastic strain in the matrix material while $\sigma = \frac{1}{3}\mathbf{1}^T\boldsymbol{\sigma}$ is the averaged normal stress. Only isotropic hardening will be considered for simplicity. Note that for $\xi = 0$ (no voids) Eq. (6.131) becomes identical with the Huber–Mises yield condition. Furthermore, the vector ${}^{(1)}\mathbf{s}$ which is normal to the plastic potential surface (6.131) will no longer be deviatoric which implies in turn that the plastic strain rate vector will loose this property either.

The yield surface is in this model generally different from the plastic potential surface. However, there is no need to specify the former surface analytically – it will serve sufficiently well our purpose to merely define the current normal to the yield surface corresponding to the stress point lying at the time instant considered at both the surfaces intersecting each other.

The evolution equation for the porosity parameter ξ is assumed in the form

$$\dot{\xi} = (1-\xi)\,\mathbf{1}^T\dot{\boldsymbol{\varepsilon}}^{(\mathrm{p})} + A(\dot{\sigma}_M + \dot{\sigma}) + B\dot{\varepsilon}_M^{(\mathrm{p})}\,, \tag{6.132}$$

in which the first term represents the growth of the existing voids while the remaining two terms describe the void nucleation effects – the maximum normal stress and the effective plastic strain-controlled, respectively. The parameter A (or B) is different from zero only if the value of $\sigma_M + \sigma$ (or $\varepsilon_M^{(\mathrm{p})}$) exceeds at a given step its maximum value assumed so far in the course of the analysis. It will further be assumed that

$$B = 0\,, \qquad A = \frac{\hat{K}}{\sigma_M}\,, \tag{6.133}$$

where $\hat{K}$ is a material constant.

Using the definition of the plastic potential

$$\dot{\boldsymbol{\varepsilon}}^{(\mathrm{p})} = \dot{\lambda}\,\frac{\partial \tilde{f}}{\partial \boldsymbol{\sigma}^T}\,, \tag{6.134}$$

additivity of the elastic and plastic strain rate contributions

$$\dot{\boldsymbol{\varepsilon}} = \dot{\boldsymbol{\varepsilon}}^{(\mathrm{e})} + \dot{\boldsymbol{\varepsilon}}^{(\mathrm{p})}\,, \tag{6.135}$$

the generalized Hooke's law

$$\dot{\boldsymbol{\sigma}} = \mathbf{C}\dot{\boldsymbol{\varepsilon}}^{(\mathrm{e})} \tag{6.136}$$

and Eqs. (6.130)–(6.132), one can obtain after some tedious calculations the constitutive equation

$$\dot{\boldsymbol{\sigma}} = \mathbf{C}_T\,\dot{\boldsymbol{\varepsilon}}^{(\mathrm{e})} \tag{6.137}$$

and the flow rule

$$\dot{\boldsymbol{\varepsilon}}^{(\mathrm{p})} = \frac{1}{\bar{\zeta}}({}^{(2)}\mathbf{s}^T\,\dot{\boldsymbol{\sigma}})\,{}^{(1)}\mathbf{s}\,, \tag{6.138}$$

where

$$\mathbf{C}_T = \mathbf{C} - \frac{E}{1+\nu}\,\frac{\left(\dfrac{3}{2}\dfrac{{}^D\boldsymbol{\sigma}}{\sigma_M} + \beta\dfrac{1+\nu}{1-2\nu}\mathbf{1}\right)\left(\dfrac{3}{2}\dfrac{{}^D\boldsymbol{\sigma}}{\sigma_M} + \mu\dfrac{1+\nu}{1-2\nu}\mathbf{1}\right)^T}{\dfrac{3}{2}\dfrac{\bar{\sigma}^2}{\sigma_M^2} + \dfrac{1+\nu}{E}\bar{\zeta} + 3\beta\mu\dfrac{1+\nu}{1-2\nu}} \tag{6.139}$$

and

$$
{}^{(1)}\mathbf{s} = \frac{3}{2}\frac{{}^{D}\boldsymbol{\sigma}}{\sigma_M} + \beta\mathbf{1}\,, \qquad {}^{(2)}\mathbf{s} = \frac{3}{2}\frac{{}^{D}\boldsymbol{\sigma}}{\sigma_M} + \mu\mathbf{1}\,,
$$

$$
\beta = \xi \sinh\left(\frac{3\sigma}{2\sigma_M}\right), \qquad \mu = \beta + \frac{\hat{K}\left[\cosh\left(\frac{3\sigma}{2\sigma_M}\right) - \xi\right]}{3}\,,
$$

$$
\bar{\zeta} = \zeta\, \frac{\left[\omega + \xi\frac{3\sigma}{2\sigma_M}\sinh\left(\frac{3\sigma}{2\sigma_M}\right)\right]^2}{1-\xi} - \frac{\left[\cosh\left(\frac{3\sigma}{2\sigma_M}\right) - \xi\right]\sigma_M}{2} \tag{6.140}
$$

$$
\times \left[3\xi(1-\xi)\sinh\left(\frac{3\sigma}{2\sigma_M}\right) + \frac{2\hat{K}}{\sigma_M}\,\xi\,\frac{\omega + \xi\frac{3\sigma}{2\sigma_M}\sinh\left(\frac{3\sigma}{2\sigma_M}\right)}{1-\xi}\right],
$$

$$
\omega = \frac{\bar{\sigma}^2}{\sigma_M^2} = 1 + \xi^2 - 2\xi\cosh\left(\frac{3\sigma}{2\sigma_M}\right), \qquad \bar{\sigma} = \left(\frac{3}{2}\,{}^{D}\boldsymbol{\sigma}^T\,{}^{D}\boldsymbol{\sigma}\right)^{\frac{1}{2}}.
$$

The local loading condition reads

$$
\frac{1}{\bar{\zeta}}\left(\frac{3}{2}\frac{{}^{D}\boldsymbol{\sigma}}{\sigma_M} + \mu\mathbf{1}\right)^T \dot{\boldsymbol{\sigma}} \geq 0\,. \tag{6.141}
$$

We note that ${}^{(1)}\mathbf{s} \neq {}^{(2)}\mathbf{s}$ (i.e. $\beta \neq \mu$) which implies non-symmetry of the constitutive matrix C_T in Eq. (6.139). However, by limiting the validity of the model to the description of the void growth (i.e. neglecting the nucleation of any new deffects) we obtain $\beta = \mu$ and the matrix C_T becomes symmetric again.

To fully specify the model we need the elastic constants E and ν, the initial yield limit σ_M^0 in the matrix material, the initial porosity ξ_0 and the material function describing the changes of the matrix yield limit with the changes of the effective plastic strain in the matrix material $\sigma_M = \sigma_M(\varepsilon_M^{(\mathrm{p})})$.

The above constitutive model (frequently with minor modifications and improvements) has gained significant popularity among the researchers. Some of the reason for the success are:

- the structure of the model resembles that of classical plasticity,
- the material degradation effects are represented in a simple way,
- the existence of some experimental evidence that void-containing materials do not obey the normality rule and thus the conviction that the model may correctly, at least in the qualitative sense, describe ductile materials with microvoids.

6.1.6 Large Elastic-Plastic Deformations

There are many possibilities to formally generalize the constitutive equations discussed so far so that after the appropriate parameter identification they become applicable to the description of large deformation processes as well.

Such a generalization of the theory proceeds typically by substituting in the rate form of the constitutive law (such as Eq. (6.37)) a so-called objective stress rate for $\dot{\sigma}_{ij}$ and replacing $\dot{\varepsilon}_{ij}$ by the rate of deformation tensor

$$d_{ij} = \frac{1}{2}(v_{i,j} + v_{j,i}) , \tag{6.142}$$

where $v_i = v_i(\mathbf{x}, \tau)$ is the spatial velocity field, and taking due account of the configuration changes. Owing to the lack of reliable experimental data, such an approach is frequently accepted as sufficiently accurate for computational purposes.

A constitutive model based on the above methodology and the updated Lagrangian approach will be briefly described below. Because the indicial tensor notation turns out more convenient in this case, we shall begin by recalling the elastic-plastic constitutive law in the form

$$\dot{\sigma}_{ij} = C^{(\text{e-p})}_{ijkl} d_{kl} . \tag{6.143}$$

As the rate of stress which satisfies the required objectivity we take a rate of the so-called Kirchhoff stress tensor, which is defined for any reference configuration C^o as

$$\tau^{ij} = \det \left[\frac{\partial \overset{(\tau)}{\mathbf{x}}}{\partial \overset{(o)}{\mathbf{x}}} \right] \sigma^{ij} . \tag{6.144}$$

Clearly, for compressible materials the Kirchhoff stress coincides with the true (Cauchy) stress. The rate of $\dot{\tau}^{ij}$ is adopted in the form called the Jaumann rate denoted by $\overset{\circ}{\tau}{}^{ij}$ with the tensor components understood as expressed in the convective coordinates coinciding with Cartesian coordinates at time $\tau = t$. Thus, we obtain

$$\overset{\circ}{\tau}{}^{ij} = C^{(\text{e-p})\,ijkl} \, d_{kl} . \tag{6.145}$$

The components $\overset{\circ}{\tau}{}^{ij}$ can be expressed in terms of the conventional time derivative of the second Piola–Kirchhoff stress based on the current configuration denoted by $\dot{\tilde{\sigma}}^{ij}$, cf. Eq. (2.20), resulting in the final form of the postulated large deformation constitutive law as

$$\dot{\tilde{\sigma}}_{ij} = \tilde{C}^{(\text{e-p})}_{ijkl} \, \dot{\varepsilon}_{kl} \tag{6.146}$$

with

$$\tilde{C}^{(\text{e-p})}_{ijkl} = C^{(\text{e-p})\,ijkl} - \frac{1}{2} \left(\tau^{ik}\delta^{jl} + \tau^{jk}\delta^{il} + \tau^{il}\delta^{jk} + \tau^{jl}\delta^{ik} \right) \tag{6.147}$$

and $\dot{\varepsilon}_{ij}$ coinciding with d_{ij} in the updated Lagrangian description adopted here. A more detailed discussion of the issues involved in defining different stress rate may be found in [87], for instance.

A crucial property of the above equation is that it preserves the symmetry under the change of indices $i, j \leftrightarrow k, l$ which allows for using the concept of a strain rate potential as, cf. Eq. (2.16), Part I,

$$\bar{W} = \frac{1}{2} \tilde{C}^{(\text{e-p})}_{ijkl} \, \dot{\varepsilon}_{ij} \dot{\varepsilon}_{kl} , \tag{6.148}$$

so that

$$\dot{\tilde{\sigma}}_{ij} = \frac{\partial \bar{W}}{\partial \dot{\varepsilon}_{ij}} = \tilde{C}^{(\mathrm{e-p})}_{ijkl}\, \dot{\varepsilon}_{kl}\,. \tag{6.149}$$

Replacing the stress rate by the corresponding finite increment results in

$$\Delta\tilde{\sigma}_{ij} = \int_t^{t+\Delta t} \tilde{C}^{(\mathrm{e-p})}_{ijkl}\, \dot{\varepsilon}_{kl}\, \mathrm{d}\tau\,, \tag{6.150}$$

cf. Eq. (2.17), Part I. Equation (6.150) can be taken as the basis for the numerical analysis for a broad class of inelastic materials. Updating the incremental quantities at the end of the step by using the rule

$$(\ldots)^{t+\Delta t} = (\ldots)^t + \Delta(\ldots)$$

the analysis at current step is completed by transforming the components of the so computed second Piola–Kirchhoff stress tensor ${}_t\tilde{\sigma}^{t+\Delta t}_{ij}$ based on the configuration C^t (beginning of the step) to the corresponding components of this tensor referred to the new reference state $C^{t+\Delta t}$. The transformation is carried out by using the equation

$${}_{t+\Delta t}\tilde{\sigma}^{t+\Delta t}_{kl} \cong (1 - \Delta u_{i,i})(\delta_{km} + \Delta u_{k,m})\, {}_t\tilde{\sigma}^{t+\Delta t}_{mn}\, (\delta_{ln} + \Delta u_{l,n})\,, \tag{6.151}$$

since, cf. [87],

$$\tilde{\boldsymbol{\sigma}}^{t+\Delta t} = {}_{t+\Delta t}\boldsymbol{\sigma}^{t+\Delta t} = \frac{\varrho^{t+\Delta t}}{\varrho^t}\, {}^t\mathbf{F}^{t+\Delta t}\, {}_t\tilde{\boldsymbol{\sigma}}^{t+\Delta t}\, {}^t\mathbf{F}^{T\,t+\Delta t}\,.$$

6.1.7 Rigid-Plasticity and Rigid Visco-Plasticity

In many metal forming processes the plastic (or visco-plastic) strains are so much bigger than the elastic ones that neglecting any elastic effects appears entirely warranted. As a result of such an approximation the material becomes virtually a non-Newtonian fluid.

The visco-plastic flow rule is adopted in the form, cf. Eqs. (6.118), (6.120)

$$\dot{\boldsymbol{\varepsilon}}^{(\mathrm{vp})} = \omega \left(\frac{\bar{\sigma} - \sigma_y}{\sigma_y}\right)^n \frac{3}{2}\frac{{}^D\boldsymbol{\sigma}}{\bar{\sigma}}\,, \tag{6.152}$$

where ω stands for the material viscosity which is changing in the process while n is an additional material parameter. The following relations hold true, cf. Eqs. (6.118), (6.123), (6.124)

$$\dot{\bar{\varepsilon}}^{(\mathrm{vp})} = \sqrt{\tfrac{2}{3}\dot{\boldsymbol{\varepsilon}}^{(\mathrm{vp})T}\, \dot{\boldsymbol{\varepsilon}}^{(\mathrm{vp})}} = \omega^*(\bar{\sigma} - \sigma_y)^n\,, \tag{6.153}$$

$$\bar{\sigma} = \sigma_y + \left(\frac{\dot{\bar{\varepsilon}}^{(\mathrm{vp})}}{\omega^*}\right)^{\frac{1}{n}}\,, \tag{6.154}$$

where

$$\omega^* = \frac{\omega}{\sigma_y^n}\,. \tag{6.155}$$

Using Eqs. (6.153) and (6.154) the flow rule (6.152) may be written as

$$\dot{\boldsymbol{\varepsilon}}^{(\mathrm{vp})} = \frac{1}{2\mu^*}(\boldsymbol{\sigma} - \sigma\mathbf{1}) = \frac{1}{2\mu^*}\,{}^{D}\boldsymbol{\sigma}\,, \tag{6.156}$$

where

$$\sigma = \tfrac{1}{3}\,\mathbf{1}^T\boldsymbol{\sigma} = \tfrac{1}{3}(\sigma_{11} + \sigma_{22} + \sigma_{33})\,,$$

while a new function representing the instantaneous visco-plastic properties of the material is defined as

$$\mu^* = \frac{\bar{\sigma}}{3\dot{\bar{\varepsilon}}^{(\mathrm{vp})}} = \frac{\sigma_y + \left(\frac{\dot{\bar{\varepsilon}}^{(\mathrm{vp})}}{\omega^*}\right)^{\frac{1}{n}}}{3\dot{\bar{\varepsilon}}^{(\mathrm{vp})}}\,. \tag{6.157}$$

For the inviscid material we have $\omega^* \to \infty$, cf. Eq. (6.127) and

$$\mu^* = \frac{\sigma_y}{3\dot{\bar{\varepsilon}}^{(\mathrm{vp})}}\,. \tag{6.158}$$

Equation (6.156) with the function μ^* given by Eq. (6.158) is the flow rule for the rigid-plastic material.

The complete system of equations for the quasi-static flow of the rigid-visco-plastic material has the form

$$\begin{aligned} \sigma_{ij,j} &= 0\,, & \mathbf{D}^T\boldsymbol{\sigma} &= 0\,,\\ \dot{\varepsilon}_{ij}^{(\mathrm{vp})} &= \frac{1}{2\mu^*}(\sigma_{ij} - \sigma\delta_{ij})\,, & \dot{\boldsymbol{\varepsilon}}^{(\mathrm{vp})} &= \frac{1}{2\mu^*}(\boldsymbol{\sigma} - \sigma\mathbf{1})\,,\\ \dot{\varepsilon}_{ij}^{(\mathrm{vp})} &= \tfrac{1}{2}(v_{i,j} + v_{j,i})\,, & \dot{\boldsymbol{\varepsilon}}^{(\mathrm{vp})} &= \tfrac{1}{2}\left[\boldsymbol{\nabla}\mathbf{v} + (\boldsymbol{\nabla}\mathbf{v})^T\right]\,, \end{aligned} \tag{6.159}$$

where $\mathbf{v}$ is the velocity vector. At this point we would like to recall the system of equations describing the problem of linear incompressible elasticity:

$$\begin{aligned} \sigma_{ij,j} &= 0\,, & \mathbf{D}^T\boldsymbol{\sigma} &= 0\,,\\ \varepsilon_{ij}^{(\mathrm{e})} &= \frac{1}{2G}(\sigma_{ij} - \sigma\delta_{ij})\,, & \boldsymbol{\varepsilon}^{(\mathrm{e})} &= \frac{1}{2G}(\boldsymbol{\sigma} - \sigma\mathbf{1})\,,\\ \varepsilon_{ij}^{(\mathrm{e})} &= \tfrac{1}{2}(u_{i,j} + u_{j,i})\,, & \boldsymbol{\varepsilon}^{(\mathrm{e})} &= \tfrac{1}{2}\left[\boldsymbol{\nabla}\mathbf{u} + (\boldsymbol{\nabla}\mathbf{u})^T\right]\,, \end{aligned} \tag{6.160}$$

in which G is the shear modulus. It is now straightforward to ascertain that both the above systems of equations exhibit the complete formal analogy of their structure provided the following quantities are taken as corresponding to each other:

$$\begin{aligned} \mathbf{v} &\longleftrightarrow \mathbf{u}\\ \dot{\boldsymbol{\varepsilon}}^{(\mathrm{vp})} &\longleftrightarrow \boldsymbol{\varepsilon}^{(\mathrm{e})}\\ \mu^* = \frac{\sigma_y + \left(\frac{\dot{\bar{\varepsilon}}^{(\mathrm{vp})}}{\omega^*}\right)^{\frac{1}{n}}}{3\dot{\bar{\varepsilon}}^{(\mathrm{vp})}} &\longleftrightarrow G = \mathrm{const.} \end{aligned} \tag{6.161}$$

The above analogy has a great computational significance since it allows to treat advanced plastic flow problems using computer software developed for linear elasticity. To do so, one simply has to allow the elastic shear modulus G

to be a given function of the current state variables and to interpret the displacements coming out of the calculations as the instantaneous velocities.

We have so far established the analogy between the rigid-visco-plastic flow of the no-damage material and the deformations of the incompressible linear elastic material. In the light of the discussion presented in Sect. 6.1.5 we may expect a similar analogy to hold for the advanced rigid-visco-plastic flow of the material with voids and the deformations of linear compressible elasticity. To see this in more precise terms let us assume the visco-plastic material properties to be described by the relationship, cf. Eq. (6.77)

$$\dot{\bar{\varepsilon}}_M^{(\mathrm{vp})} = \dot{\varepsilon}_0 \left[\frac{\sigma_M}{g(\bar{\varepsilon}_M^{(\mathrm{vp})})} \right]^{\frac{1}{m}} , \tag{6.162}$$

in which m is a material constant, $\dot{\varepsilon}_0$ a reference effective strain velocity and the subscript 'M' refers to the matrix material. For the macroscopic stresses $\boldsymbol{\sigma}$ at the end of the step and the current value of the porosity ξ the dynamic yield limit for the matrix material σ_M may be found iteratively from the relation, cf. Eq. (6.131)

$$\sigma_M = \sqrt{\frac{\frac{3}{2}\, {}^{D}\boldsymbol{\sigma}^T\, {}^{D}\boldsymbol{\sigma}}{\omega}} , \tag{6.163}$$

in which

$$\omega = 1 - 2\xi \cosh\left(\frac{3\sigma}{2\sigma_M} \right) + \xi^2 . \tag{6.164}$$

The function g represents the yield limit in the material subject to tension at the strain rate $\dot{\bar{\varepsilon}}_M^{(\mathrm{vp})} = \dot{\varepsilon}_0$. This function may be taken as, cf. Eq. (6.78)

$$g(\bar{\varepsilon}_M^{(\mathrm{vp})}) = \sigma_0 \left(1 + \frac{\bar{\varepsilon}_M^{(\mathrm{vp})}}{\varepsilon_0} \right)^N , \qquad \varepsilon_0 = \frac{\sigma_0}{E} . \tag{6.165}$$

We note that the model does not call for an explicit inclusion of the elastic unloading of the material as for the small values of the parameter m considered here the plastic strain rates are negligible for any σ_M smaller than $g(\bar{\varepsilon}_M^{(\mathrm{vp})})$.

By taking the visco-plastic strain rate potential in the form (6.131) we have, cf. Eq. (6.133)

$$\dot{\boldsymbol{\varepsilon}}^{(\mathrm{vp})} = \dot{\lambda} \frac{\partial \tilde{f}}{\partial \boldsymbol{\sigma}^T} . \tag{6.166}$$

The function $\dot{\lambda}$ is determined by equating the macroscopic plastic energy rate and the matrix dissipation as

$$\boldsymbol{\sigma}^T \dot{\boldsymbol{\varepsilon}}^{(\mathrm{vp})} = (1-\xi)\sigma_M \dot{\bar{\varepsilon}}_M^{(\mathrm{vp})} , \tag{6.167}$$

which results in

$$\dot{\lambda} = (1-\xi)\sigma_M \dot{\bar{\varepsilon}}_M^{(\mathrm{vp})} \frac{1}{2\left(\omega + \frac{\beta}{\sigma_M} 3\sigma\right)} , \tag{6.168}$$

where

$$\beta = \frac{\xi \sinh\left(\frac{3\sigma}{2\sigma_M}\right)}{2} .$$

Equations (6.166), (6.168) imply that

$$\dot{\boldsymbol{\varepsilon}}^{(\mathrm{vp})} = \frac{3(1-\xi)\dot{\bar{\varepsilon}}_M^{(\mathrm{vp})}}{2(\omega\sigma_M + 3\beta\sigma)} \left(\boldsymbol{\sigma} - \frac{3\sigma - 2\beta\sigma_M}{3\sigma}\,\sigma\,\mathbf{1}\right) . \tag{6.169}$$

We neglect elasticity in the porous material on hand and recall the constitutive equation for the linear compressible elastic material as

$$\boldsymbol{\varepsilon}^{(\mathrm{e})} = \frac{1}{2G}\left(\boldsymbol{\sigma} - \frac{3\nu}{1+\nu}\sigma\mathbf{1}\right) . \tag{6.170}$$

By rewriting Eq. (6.169) as

$$\dot{\boldsymbol{\varepsilon}}^{(\mathrm{vp})} = \frac{1}{2\mu^*}\left(\boldsymbol{\sigma} - \frac{3\nu^*}{1+\nu^*}\sigma\mathbf{1}\right) , \tag{6.171}$$

the analogy between Eqs. (6.170) and (6.171) is readily established. Here, the additional notation

$$\mu^* = \frac{\omega\sigma_M + 3\beta\sigma}{3(1-\xi)\dot{\bar{\varepsilon}}_M^{(\mathrm{vp})}} , \qquad \nu^* = \frac{3\sigma - 2\beta\sigma_M}{2(3\sigma - \beta\sigma_M)} . \tag{6.172}$$

has been introduced.

Let us note that for the no-damage material, i.e. when $\xi=0$, $\beta=0$, $\omega=1$, $\sigma_M=\bar{\sigma}$, $\varepsilon_M^{(\mathrm{vp})}=\varepsilon^{(\mathrm{vp})}$, the moduli become

$$\mu^* = \frac{\bar{\sigma}}{3\dot{\bar{\varepsilon}}^{(\mathrm{vp})}} , \qquad \nu^* = \frac{1}{2} , \tag{6.173}$$

exactly coinciding with the ones derived earlier, cf. Eqs. (6.157). By substituting Eqs. (6.162), (6.165) in Eq. (6.172) we arrive at

$$\mu^* = \frac{(\omega\sigma_M + 3\beta\sigma)\sigma_0^{1/m}\left(1 + \frac{\bar{\varepsilon}_M^{(\mathrm{vp})}}{\varepsilon_0}\right)^{\frac{N}{m}}}{3(1-\xi)\sigma_M^{1/m}} , \qquad \nu^* = \frac{3\sigma - 2\beta\sigma_M}{2(3\sigma - \beta\sigma_M)} . \tag{6.174}$$

The material functions μ^*, ν^* depend on the current values of stresses, effective plastic strain and porosity but are independent of their rates. Similarly as before we conclude that the inelastic flow of the voided material may be analysed by using the computer programs developed for the linear elastic materials provided the appropriate correspondence of the variables is observed.

The formal analogy of the equations describing the two physically different deformation processes excludes by its nature any possibility of directly including elastic effects in the visco-plastic problem. However, the elastic effects can be accounted for in an approximate way by adopting the following reasoning. First, we rewrite Eq. (6.171) in the inverse form as

$$\boldsymbol{\sigma} = \mathbf{C}^* \, \dot{\boldsymbol{\varepsilon}}^{(\mathrm{vp})} \,, \tag{6.175}$$

where $\mathbf{C}^*$ is the standard elasticity matrix in which the Lame constants have been replaced by the functions μ^*, ν^*, respectively. Using the additivity of the elastic and visco-plastic strain rates

$$\dot{\boldsymbol{\varepsilon}} = \dot{\boldsymbol{\varepsilon}}^{(\mathrm{vp})} + \dot{\boldsymbol{\varepsilon}}^{(\mathrm{e})} \,,$$

Eq. (6.175) implies

$$\boldsymbol{\sigma} = \mathbf{C}^*(\dot{\boldsymbol{\varepsilon}} - \dot{\boldsymbol{\varepsilon}}^{(\mathrm{e})}) = \mathbf{C}^*(\boldsymbol{\sigma}, \ldots)\, \dot{\boldsymbol{\varepsilon}} - \mathbf{C}^* \, \dot{\boldsymbol{\varepsilon}}^{(\mathrm{e})} = \mathbf{C}^* \, \dot{\boldsymbol{\varepsilon}} - \boldsymbol{\sigma}^* \,. \tag{6.176}$$

Assuming at the beginning of the step that the ‘initial’ stresses $\boldsymbol{\sigma}^*$ (or the elastic strain rates $\dot{\boldsymbol{\varepsilon}}^{(\mathrm{e})}$) are known the problem is reduced to the iterative solution of the FEM equations in the form

$$\mathbf{K}^* \dot{\mathbf{q}} = \mathbf{Q} + \mathbf{Q}^* \,, \tag{6.177}$$

where $\mathbf{K}^*$ is the elastic stiffness matrix with the appropriately replaced material constants while $\mathbf{Q}^*$ is the initial load vector corresponding to the initial stresses $\boldsymbol{\sigma}^*$.

6.2 Solving Plasticity Problems by FEM

6.2.1 Boundary-Value Problem of Plasticity

The fundamental system of equations describing the boundary-value problem for large deformation theory of thermo-elastic-plastic bodies in the updated Lagrangian description has the form[1], cf. Eqs. (2.18)–(2.24), Part I,

$$\begin{aligned}
&\dot{\sigma}_{ij,j} + \varrho \dot{f}_k = 0 \,, && \mathbf{D}^T \boldsymbol{\sigma} + \varrho \dot{\mathbf{f}} = \mathbf{0} \,, \\
&\dot{\varepsilon}_{ij} = \tfrac{1}{2}(\dot{u}_{i,j} + \dot{u}_{j,i}) \,, && \dot{\boldsymbol{\varepsilon}} = \mathbf{D}\, \dot{\mathbf{u}} \,, \\
&\dot{\tilde{\sigma}}_{ij} = C^{(\mathrm{e-p})}_{ijkl}\, \dot{\varepsilon}_{kl} + c_{ij} \dot{\vartheta} \,, && \dot{\tilde{\boldsymbol{\sigma}}} = \mathbf{C}^{(\mathrm{e-p})}\, \dot{\boldsymbol{\varepsilon}} + \mathbf{c} \dot{\vartheta} \,, \\
&\dot{\sigma}_{ij} = \dot{u}_{i,k} \tilde{\sigma} + \dot{\tilde{\sigma}}_{ij} \,, && \dot{\boldsymbol{\sigma}} = \boldsymbol{\nabla} \dot{\mathbf{u}}\, \tilde{\boldsymbol{\sigma}} + \dot{\tilde{\boldsymbol{\sigma}}} \,, \\
&\dot{u}_i = \hat{\dot{u}}_i \,, && \dot{\mathbf{u}} = \hat{\dot{\mathbf{u}}} \,, \\
&\dot{\sigma}_{ij} n_j = \hat{\dot{t}}_i \,, && \mathbf{N} \dot{\boldsymbol{\sigma}} = \hat{\dot{\mathbf{t}}} \,.
\end{aligned} \tag{6.178}$$

[1] Only quasi-static processes for materials without damage are considered for simplicity.

Eliminating the stress and strain rates the system may be expressed in terms of the velocities only as

$$\begin{gathered}
\left[C^{(e-p)}_{ijkl}\,\dot{u}_{k,l} + c_{ij}\dot{\vartheta} + \tilde{\sigma}_{mj}\dot{u}_{i,m}\right]_{,j} + \varrho\dot{f}_i = 0, \\
\left[C^{(e-p)}_{ijkl}\,\dot{u}_{k,l} + c_{ij}\dot{\vartheta} + \tilde{\sigma}_{mj}\dot{u}_{i,m}\right] n_j = \hat{\dot{t}}_i\,, \\
\dot{u}_i = \hat{\dot{u}}_i\,, \\
\boldsymbol{\nabla}\left[\mathbf{C}^{(e-p)}\mathbf{D}\dot{\mathbf{u}} + \mathbf{c}\dot{\vartheta} + \boldsymbol{\nabla}\dot{\mathbf{u}}\tilde{\boldsymbol{\sigma}}\right] + \varrho\dot{\mathbf{f}} = \mathbf{0}, \\
\mathbf{N}\left[\mathbf{C}^{(e-p)}\mathbf{D}\dot{\mathbf{u}} + \mathbf{c}\dot{\vartheta} + \boldsymbol{\nabla}\dot{\mathbf{u}}\tilde{\boldsymbol{\sigma}}\right] = \hat{\dot{\mathbf{t}}}\,, \\
\dot{\mathbf{u}} = \hat{\dot{\mathbf{u}}}\,,
\end{gathered} \tag{6.179}$$

where

$$\boldsymbol{\nabla}(\ldots) = [(\ldots)_{,i}]\,.$$

In case of local unloading the elastic constitutive matrix $\mathbf{C}$ replaces in Eq. (6.179) the elastic-plastic matrix $\mathbf{C}^{(e-p)}$.

The system of equations (6.178), and the constitutive equation (6.178) in particular, may be employed provided at the time instant $\tau = t$ considered (and each point in the space domain) both the temperature $\vartheta = \vartheta(\mathbf{x}, t)$ and its rate $\dot{\vartheta} = \dot{\vartheta}(\mathbf{x}, t)$ are known. They may be either given a priori or obtained at $\tau = t$ as the solution to the heat transfer equation of the form, cf. Eq. (2.45), Part I,

$$\boldsymbol{\nabla}(\lambda\boldsymbol{\nabla}^T T) + \dot{g} = \varrho c\,\frac{\partial T}{\partial \tau}\,, \qquad \vartheta = T - T_0\,. \tag{6.180}$$

The above equation is assumed here to be independent of any mechanical variable resulting in what is called the uncoupled formulation of the thermo-mechanical problem. In it, one can solve at each time step the thermal boundary-value problem and then use the solution for solving the mechanical part of the formulation.

The isothermal theory results from Eqs. (6.178) by setting $\vartheta = 0$, $\dot{\vartheta} = 0$ and neglecting any dependence of the mechanical variables on temperature.

The small deformation problems may be analysed by identifying the stress rate tensors $\dot{\sigma}_{ij}$ and $\dot{\tilde{\sigma}}_{ij}$ and neglecting the configuration changes.

It is known that the mechanical variables may quite often influence the distribution of temperature. Among many possible factors which may underlie this coupling just one will be incorporated below into the structure of the problem equations. To this aim we observe that the term $\dot{g}$ in Eq. (6.180) represents the rate of internal heat generation within the material. An assumption widely adopted in the case of the advanced plastic flow is that significant part of the plastic work is converted to heat resulting in the relationship

$$\dot{g} = \chi\,\boldsymbol{\sigma}^T\,\dot{\boldsymbol{\varepsilon}}^{(p)}\,, \tag{6.181}$$

in which χ is an experimental parameter having for many metals the value in the range of 0.85 to 0.95.

Let us further note that the following sequence of relations holds true, cf. Eqs. (6.27)–(6.35):

$$\dot{\sigma} = \mathbf{C}^{(e-p)}\,\dot{\varepsilon} + \mathbf{c}\dot{\vartheta} = (\mathbf{C} - \mathbf{C}^{(p)})\dot{\varepsilon} + \mathbf{c}\dot{\vartheta}\,,$$

$$\dot{\sigma} = \mathbf{C}\dot{\varepsilon}^{(e)} + \dot{\mathbf{C}}\varepsilon^{(e)} = \mathbf{C}\dot{\varepsilon}^{(e)} + \mathbf{a}\dot{\vartheta}\,,$$

$$(\mathbf{C} - \mathbf{C}^{(p)})\dot{\varepsilon} + \mathbf{c}\dot{\vartheta} = \mathbf{C}\dot{\varepsilon}^{(e)} + \dot{\mathbf{C}}\varepsilon^{(e)} = \mathbf{C}\dot{\varepsilon}^{(e)} + \mathbf{a}\dot{\vartheta}\,,$$

$$\mathbf{C}\dot{\varepsilon}^{(p)} = \mathbf{C}^{(p)}\dot{\varepsilon} - \mathbf{c}\dot{\vartheta} - \mathbf{C}\dot{\varepsilon}^{(\vartheta)} + \mathbf{a}\dot{\vartheta} = \mathbf{C}^{(p)}\dot{\varepsilon} - (\mathbf{c} + \alpha\mathbf{C}^{(p)}\!:\!\mathbf{1} - \mathbf{a})\,\dot{\vartheta}\,,$$

$$\dot{\varepsilon}^{(p)} = \overset{-1}{\mathbf{C}}[\mathbf{C}^{(p)}\dot{\varepsilon} - (\mathbf{c} + \alpha\mathbf{C}^{(p)}\!:\!\mathbf{1} - \mathbf{a})\,\dot{\vartheta}]\,, \tag{6.182}$$

$$\dot{g} = \chi\boldsymbol{\sigma}^T\,\dot{\varepsilon}^{(p)} = \mathbf{s}_1\dot{\varepsilon} + s_2\dot{\vartheta}\,,$$

$$\mathbf{s}_1 = \chi\boldsymbol{\sigma}^T\mathbf{n}\,\frac{\mathbf{n}^T\mathbf{C}}{h + \mathbf{n}^T\mathbf{C}\mathbf{n}}\,,\qquad s_2 = \chi\boldsymbol{\sigma}^T\,\overset{-1}{\mathbf{C}}\mathbf{c}'\,,\qquad \mathbf{c}' = \alpha\mathbf{C}^{(p)}\!:\!\mathbf{1} + \mathbf{b}\,,$$

$$\nabla(\lambda\nabla\vartheta) - (c\varrho - s_2)\dot{\vartheta} + \mathbf{s}_1\dot{\varepsilon} = 0\,,$$

$$\nabla(\lambda\nabla\vartheta) - (c\varrho - s_2)\dot{\vartheta} + \mathbf{s}_1\mathbf{D}\dot{\mathbf{u}} = 0\,.$$

The so derived the heat transfer equation contains explicitly terms dependent on the current values of the velocities, which implies that the thermal problem can no longer be solved independently of the corresponding mechanical problem. Thus, the formulation becomes fully coupled and requires in principle the simultaneous solution of the problem with respect to both the velocities and temperature. The solution algorithms for the problem on hand will be discussed in the subsequent section.

6.2.2 Finite Element Equations

The FEM equations for the nonlinear dynamics problem of solid mechanics have been derived in Sect. 3.1 in the form, cf. Eq. (3.33)

$$\mathbf{M}\ddot{\mathbf{q}}^{t+\Delta t} + \mathbf{C}\dot{\mathbf{q}}^{t+\Delta t} + [\mathbf{K}^{(con)} + \mathbf{K}^{(\sigma)}]\Delta\mathbf{q} = \mathbf{Q}^{t+\Delta t} - \mathbf{F}^t\,. \tag{6.183}$$

in which the constitutive stiffness matrix $\mathbf{K}^{(con)} = \mathbf{K}^{(e-p)}$ for the elastic-plastic material assumes the form, cf. Eq. (3.15)

$$K^{(e-p)}_{\alpha\beta} = \int_\Omega C^{(e-p)}_{ijkl}\,\Phi_{k\alpha,l}\Phi_{i\beta,j}\,\mathrm{d}\Omega\,, \tag{6.184}$$

or, in the matrix notation

$$\mathbf{K}^{(e-p)} = \int_\Omega \mathbf{B}^T\mathbf{C}^{(e-p)}\mathbf{B}\,\mathrm{d}\Omega\,.$$

Let us assume that a finite difference scheme has been employed to replace the time derivatives in Eq. (6.183) by the corresponding finite difference expressions – an example of such an approach has been given in Sect. 3.1, Eqs. (3.23)–(3.27). The resulting system of algebraic equations may be presented as

$$\mathbf{K}_{(ef)}^{t+\Delta t}\Delta\mathbf{q} = \Delta\mathbf{Q}_{(ef)}^{t+\Delta t}, \tag{6.185}$$

in which $\mathbf{K}_{(ef)}$ is the so-called effective stiffness matrix characteristic of the finite difference scheme used and $\Delta\mathbf{Q}_{(ef)}$ is the effective load vector. Clearly, in the case of statics the effective stiffness matrix and load vector become the standard stiffness matrix and load vector, respectively. Equation (6.185) serves the purpose of computing the unknown vector $\Delta\mathbf{q}$ which it turn makes it possible to determine the total displacement vector at the end of the step as

$$\mathbf{q}^{t+\Delta t} = \mathbf{q}^t + \Delta\mathbf{q}. \tag{6.186}$$

Since the matrix $\mathbf{K}_{(ef)}^{t+\Delta t}$ depends generally on the solution $\mathbf{q}^{t+\Delta t}$ (and values of some other functions at $t+\Delta t$, such as stresses or hardening functions), the solution of Eq. (6.185) typically proceeds in an iterative fashion. They will be analysed in some detail in Chap. 7 – here, we only remark that a typical algorithm of this type is based on the expression

$$\mathbf{K}_{(i-1)}^{t+\Delta t}\,\delta\mathbf{q}_{(i)} = \delta\mathbf{Q}_{(i-1)}^{t+\Delta t}, \tag{6.187}$$

in which $\delta\mathbf{q}_{(i)}$, $i = 1,2,\ldots$ are the subsequent corrections to the incremental solution at the given step such that

$$\Delta\mathbf{q}_{(i)} = \Delta\mathbf{q}_{(i-1)} + \delta\mathbf{q}_{(i)}, \qquad \mathbf{q}_{(i)}^{t+\Delta t} = \mathbf{q}_{(i-1)}^{t+\Delta t} + \Delta\mathbf{q}_{(i)}. \tag{6.188}$$

$\mathbf{K}_{(i-1)}^{t+\Delta t}$ is the effective stiffness matrix formed using the solution after the $(i-1)$-th iteration, while $\delta\mathbf{Q}_{(i-1)}^{t+\Delta t}$ is the vector of the nonequilibrated dynamic nodal forces. For better transparency the superscript indicating the effective character of the stiffness matrix and load vector has been omitted. Usually it is taken that

$$\mathbf{q}_{(0)}^{t+\Delta t} = \mathbf{q}^t, \qquad \mathbf{K}_{(0)}^{t+\Delta t} = \mathbf{K}_{(ef)}^t, \qquad \delta\mathbf{Q}_{(0)}^{t+\Delta t} = \Delta\mathbf{Q}_{(ef)}^{t+\Delta t}. \tag{6.189}$$

It should be quite obvious by now that in order to determine the value of the matrix $\mathbf{K}_{(i-1)}^{t+\Delta t}$ representing inelastic properties of the underlying material one has to determine first the stresses $\boldsymbol{\sigma}_{(i-1)}^{t+\Delta t}$. The way to do so is crucial for accuracy of the overall integration procedure and will be discussed later on.

In the case of the non-associated plasticity which generates the non-symmetric stiffness matrix, the solution procedure requires some additional comments. In the quasi-static case we have

$$[\mathbf{K}^{(e-p)} + \mathbf{K}^{(\sigma)}]\dot{\mathbf{q}} = \dot{\mathbf{Q}}, \tag{6.190}$$

where, cf. Eqs. (1.7), (6.39),

$$\mathbf{K}^{(e-p)} = \int_\Omega \mathbf{B}^T[\mathbf{C} - \mathbf{C}^{(p)}]\mathbf{B}\,d\Omega = \mathbf{K} - \mathbf{K}^{(p)}, \tag{6.191}$$

the matrix $\mathbf{C}^{(p)}$ and, consequently, the matrix $\mathbf{K}^{(p)}$ being generally non-symmetric

$$\dot{\mathbf{C}}^{(p)} = \frac{E}{1+\nu}\,\frac{\left(\dfrac{3}{2}\dfrac{{}^D\boldsymbol{\sigma}}{\sigma_M} + \beta\dfrac{1+\nu}{1-2\nu}\mathbf{1}\right)\left(\dfrac{3}{2}\dfrac{{}^D\boldsymbol{\sigma}}{\sigma_M} + \mu\dfrac{1+\nu}{1-2\nu}\mathbf{1}\right)^T}{\dfrac{3}{2}\dfrac{\bar{\sigma}^2}{\sigma_M^2} + \dfrac{1+\nu}{E}\bar{\zeta} + 3\beta\mu\dfrac{1+\nu}{1-2\nu}}\,. \tag{6.192}$$

To facilitate the solution algorithm we may decompose the constitutive stiffness matrix into the symmetric and antisymmetric parts

$$\mathbf{C}^{(p)} = \mathbf{C}^{(p)}_{\text{sym}} + \mathbf{C}^{(p)}_{\text{asym}} \tag{6.193}$$

whereby Eq. (6.190) becomes

$$[\mathbf{K} - \mathbf{K}^{(p)}_{\text{sym}} + \mathbf{K}^{(\sigma)}]\dot{\mathbf{q}} = \dot{\mathbf{Q}} + \mathbf{K}^{(p)}_{\text{asym}}\dot{\mathbf{q}}, \tag{6.194}$$

with

$$\mathbf{K}^{(p)}_{\text{sym}} = \int_\Omega \mathbf{B}^T\mathbf{C}^{(p)}_{\text{sym}}\mathbf{B}\,d\Omega, \qquad \mathbf{K}^{(p)}_{\text{asym}} = \int_\Omega \mathbf{B}^T\mathbf{C}^{(p)}_{\text{asym}}\mathbf{B}\,d\Omega. \tag{6.195}$$

The matrix $[\mathbf{K} - \mathbf{K}^{(p)}_{\text{sym}} + \mathbf{K}^{(\sigma)}]$ is now symmetric while the term $\mathbf{K}^{(p)}_{\text{asym}}\dot{\mathbf{q}}$ may be accounted for at the stage of the iterative procedure. The resulting convergence properties of the algorithm will likely be worsened only slightly which is due to the generally rather small non-symmetry effects – a feature typical of the problems on hand.

By limiting ourselves to the quasi-static situations we shall now briefly present the algorithm for solving the coupled thermo-mechanical problem. The spatially discretized thermal problem is given by, cf. Eq. (3.61)

$$\mathbf{C}^{(\vartheta)}_{M\times M}\,\dot{\boldsymbol{\theta}}_{M\times 1} + \mathbf{K}^{(\vartheta)}_{M\times M}\,\boldsymbol{\theta}_{M\times 1} = \mathbf{P}_{M\times 1}, \tag{6.196}$$

in which the vector $\mathbf{P}$ has the form

$$\mathbf{P}_{M\times 1} = \int_\Omega \overset{(\vartheta)}{\boldsymbol{\Phi}}{}^T_{M\times 1}\dot{g}\,d\Omega + \int_{\partial\Omega_q} \overset{(\vartheta)}{\boldsymbol{\Phi}}{}^T_{M\times 1}\hat{q}\,d(\partial\Omega), \tag{6.197}$$

with the superscript 'ϑ' indicating that the temperature interpolation may generally be given in terms of shape functions different than those adopted in the mechanical problem. Using Eq. (6.85) and the relation $\dot{\varepsilon} = \mathbf{B}\dot{\mathbf{q}}$ Eq. (6.196) is transformed to the form

$${}^{(TT)}\mathbf{K}\dot{\boldsymbol{\theta}} + {}^{(TM)}\mathbf{K}\dot{\mathbf{q}} = {}^{(T)}\dot{\mathbf{Q}}, \tag{6.198}$$

in which

$$
\begin{aligned}
{}^{(TT)}\mathbf{K}_{M\times M} &= \int_{\Omega}(\varrho c - s_2)\overset{(\vartheta)}{\mathbf{\Phi}}{}^{T}_{M\times 1}\overset{(\vartheta)}{\mathbf{\Phi}}_{1\times M}\,\mathrm{d}\Omega = \mathbf{C}^{(\vartheta)}_{M\times M} - \int_{\Omega} s_2\,\overset{(\vartheta)}{\mathbf{\Phi}}{}^{T}\overset{(\vartheta)}{\mathbf{\Phi}}\,\mathrm{d}\Omega\\
{}^{(TM)}\mathbf{K}_{M\times N} &= \int_{\Omega}\overset{(\vartheta)}{\mathbf{\Phi}}{}^{T}_{M\times 1}\,\mathbf{s}_{1\,1\times 6}\,\mathbf{B}_{6\times N}\,\mathrm{d}\Omega\\
{}^{(T)}\dot{\mathbf{Q}}_{M\times 1} &= \mathbf{K}^{(\vartheta)}_{M\times M}\boldsymbol{\theta}_{M\times 1} + \int_{\partial\Omega_q}\overset{(\vartheta)}{\mathbf{\Phi}}{}^{T}_{M\times 1}\,\hat{q}\,\mathrm{d}(\partial\Omega)\,.
\end{aligned}
$$

The mechanical problem with the thermal effects included is presented as

$$
{}^{(MM)}\mathbf{K}\dot{\mathbf{q}} + {}^{(MT)}\mathbf{K}\dot{\boldsymbol{\theta}} = {}^{(M)}\dot{\mathbf{Q}}\,, \tag{6.199}
$$

in which

$$
\begin{aligned}
{}^{(MM)}\mathbf{K}_{N\times N} &= \mathbf{K}^{(\mathrm{e-p})}_{N\times N}\,,\\
{}^{(MT)}\mathbf{K}_{N\times M} &= \int_{\Omega}\mathbf{B}^{T}_{N\times 6}\,\mathbf{c}_{6\times 1}\overset{(\vartheta)}{\mathbf{\Phi}}_{1\times M}\,\mathrm{d}\Omega\,,\\
{}^{(M)}\dot{\mathbf{Q}}_{N\times 1} &= \dot{\mathbf{Q}}_{N\times 1}\,.
\end{aligned}
$$

The coupled system of equations (6.198), (6.199) may be compactly written as

$$
\begin{bmatrix} {}^{(MM)}\mathbf{K} & {}^{(MT)}\mathbf{K}\\ {}^{(TM)}\mathbf{K} & {}^{(TT)}\mathbf{K}\end{bmatrix}\begin{bmatrix}\dot{\mathbf{q}}\\ \dot{\boldsymbol{\theta}}\end{bmatrix} = \begin{bmatrix}{}^{(M)}\dot{\mathbf{Q}}\\ {}^{(T)}\dot{\mathbf{Q}}\end{bmatrix} \tag{6.200}
$$

in which the submatrices ${}^{(MT)}\mathbf{K}$ and ${}^{(TM)}\mathbf{K}$ are the terms coupling the mechanical and thermal problems.

Very many algorithms exist in the literature for solving systems of equations of the above type. Perhaps the simplest conceptually is the direct approach in which all the equations in the system (6.200) are solved simultaneously by using any integration procedure which is applicable to large systems of ordinary differential equations. In case the coupling terms are small a frequently employed algorithm is based on solving the transformed system

$$
\begin{bmatrix} {}^{(MM)}\mathbf{K} & \mathbf{0}\\ \mathbf{0} & {}^{(TT)}\mathbf{K}\end{bmatrix}\begin{bmatrix}\dot{\mathbf{q}}\\ \dot{\boldsymbol{\theta}}\end{bmatrix} = \begin{bmatrix}{}^{(M)}\dot{\mathbf{Q}}\\ {}^{(T)}\dot{\mathbf{Q}}\end{bmatrix} - \begin{bmatrix} \mathbf{0} & {}^{(MT)}\mathbf{K}\\ {}^{(TM)}\mathbf{K} & \mathbf{0}\end{bmatrix}\begin{bmatrix}\dot{\mathbf{q}}\\ \dot{\boldsymbol{\theta}}\end{bmatrix}\,. \tag{6.201}
$$

Assuming that the right-hand side vectors are known the overall problem splits into two subproblems for which many well established procedures exist. Then, the right-hand side vectors are updated and a new solution is obtained. The procedure is continued until no significant changes in the solution result from the subsequent iterations. Clearly, this iterative process may be performed together with the iteration anyhow required by the implicit integration of both the equation systems.

6.2.3 Time Integration of the Constitutive Equations

From the computational point of view the problem of the constitutive equation time integration is strain-driven – the stress history at each point in the body is determined on the basis of the strain history at this point. The so-called return mapping algorithms are currently considered as the most effective way to perform such an integration, [159]. The basic idea of the algorithm may be presented in the following terms. The stress value $\boldsymbol{\sigma}^t$ obtained as a result of the convergent iterative procedure at time $\tau = t$ is used at the i-th iteration of the step $t \to t+\Delta t$ to compute a certain new, so-called trial stress value $\hat{\boldsymbol{\sigma}}_{(i)}^{t+\Delta t}$. For $\hat{\boldsymbol{\sigma}}_{(i)}^{t+\Delta t}$ being outside of the yield surface $f(\boldsymbol{\sigma}) = 0$ the real stress at the iteration i is then defined as a projection of $\hat{\boldsymbol{\sigma}}_{(i)}^{t+\Delta t}$ onto the surface. For convex surfaces the problem is thus reduced to the classical problem of looking for the minimum distance of a point to the given surface. Note also that each iteration at the step $t \to t+\Delta t$ takes reference to the stress $\boldsymbol{\sigma}^t$ which has been obtained as a result of the convergent iterative process (the stress is therefore physically admissible) and not to the last available stress $\boldsymbol{\sigma}_{(i-1)}^{t+\Delta t}$.

In case of the Huber–Mises yield condition the above algorithm becomes very simple and bears the name of the radial return algorithm. The version of the algorithm presented below strictly applies to the case of stepwise linear hardening ($h = \text{const}$ for $\tau \in [t, t+\Delta t]$) – the algorithm has also been used successfully for an approximate solving of more complex problems. An algorithm applicable to any hardening will be presented later.

The radial return algorithm discussed so far is summarized as follows:

1. Compute the i-th approximation to the incremental strain using the displacement correction $\delta\mathbf{q}_{(i)}$ obtained by solving Eq. (6.187), i.e.

$$\Delta\boldsymbol{\varepsilon}_{(i)} = \mathbf{B}\left(\Delta\mathbf{q}_{(i-1)} + \delta\mathbf{q}_{(i)}\right) . \tag{6.202}$$

2. Compute the i-th approximation to the trial stress increment as

$$\Delta\hat{\boldsymbol{\sigma}}_{(i)} = \mathbf{C}\left(\Delta\boldsymbol{\varepsilon}_{(i)} - \overset{*}{\alpha}{}^{t+\Delta t}\Delta\vartheta\,\mathbf{1}\right) . \tag{6.203}$$

3. Compute the i-th trial stresses as

$$\hat{\boldsymbol{\sigma}}_{(i)}^{t+\Delta t} = \boldsymbol{\sigma}^t + \Delta\hat{\boldsymbol{\sigma}}_{(i)} , \qquad \hat{\hat{\boldsymbol{\sigma}}}{}_{(i)}^{t+\Delta t} = \hat{\boldsymbol{\sigma}}_{(i)}^{t+\Delta t} - \boldsymbol{\alpha}^t . \tag{6.204}$$

4. Compute the i-th deviatoric trial stress

$$^{D}\hat{\hat{\boldsymbol{\sigma}}}{}_{(i)}^{t+\Delta t} = \hat{\hat{\boldsymbol{\sigma}}}{}_{(i)}^{t+\Delta t} - \tfrac{1}{3}\mathbf{1}^T\hat{\hat{\boldsymbol{\sigma}}}{}_{(i)}^{t+\Delta t} . \tag{6.205}$$

5. Compute the square of $^{D}\hat{\hat{\boldsymbol{\sigma}}}{}_{(i)}^{t+\Delta t}$,

$$A = {}^{D}\hat{\hat{\boldsymbol{\sigma}}}{}_{(i)}^{t+\Delta t\,T}\;{}^{D}\hat{\hat{\boldsymbol{\sigma}}}{}_{(i)}^{t+\Delta t} . \tag{6.206}$$

6. Check the yield condition

$$\text{if}\ \ \tfrac{3}{2}A \le (\sigma_y^t)^2\ \ \text{then}\ \ \begin{aligned} \boldsymbol{\sigma}_{(i)}^{t+\Delta t} &= \hat{\boldsymbol{\sigma}}_{(i)}^{t+\Delta t}, \\ \sigma_{y\,(i)}^{t+\Delta t} &= \sigma_y^t, \\ \boldsymbol{\alpha}_{(i)}^{t+\Delta t} &= \boldsymbol{\alpha}^t, \\ &\text{go back to step 1}\quad \text{(elastic process);} \end{aligned} \tag{6.207}$$

$$\text{if}\ \ \tfrac{3}{2}A > (\sigma_y^t)^2\ \ \text{then}\ \ \text{proceed to step 7}\quad \text{(plastic process).}$$

7. Compute the i-th approximation to the yield surface normal vector

$$\mathbf{n}_{(i)} = \frac{{}^D\hat{\bar{\boldsymbol{\sigma}}}_{(i)}^{t+\Delta t}}{\sqrt{A}}\,. \tag{6.208}$$

8. Compute the i-th effective incremental plastic strain

$$\Delta\bar{\varepsilon}_{(i)}^{(\mathrm{p})} = \frac{1}{\frac{1}{3}\,\mathbf{n}_{(i)}^T\,\mathbf{C}\,\mathbf{n}_{(i)} + \frac{3}{2}\,h}\left(\sqrt{\tfrac{3}{2}A} - \sigma_y^t\right)\,. \tag{6.209}$$

9. Update the values of particular functions

$$\begin{aligned} \boldsymbol{\sigma}_{(i)}^{t+\Delta t} &= \hat{\boldsymbol{\sigma}}_{(i)}^{t+\Delta t} - \mathbf{C}\,\mathbf{n}_{(i)}\Delta\bar{\varepsilon}_{(i)}^{(\mathrm{p})}\sqrt{\tfrac{3}{2}}\,, \\ \sigma_{y\,(i)}^{t+\Delta t} &= \sigma_y^t + \beta\zeta\,\Delta\bar{\varepsilon}_{(i)}^{(\mathrm{p})}\,, \\ \boldsymbol{\alpha}_{(i)}^{t+\Delta t} &= \boldsymbol{\alpha}^t + \tfrac{2}{3}(1-\beta)\zeta\,\Delta\bar{\varepsilon}_{(i)}^{(\mathrm{p})}\,\mathbf{n}_{(i)}\,, \\ \bar{\boldsymbol{\sigma}}_{(i)}^{t+\Delta t} &= \boldsymbol{\sigma}_{(i)}^{t+\Delta t} - \boldsymbol{\alpha}_{(i)}^{t+\Delta t}\,, \\ \boldsymbol{\varepsilon}_{(i)}^{(\mathrm{p})\,t+\Delta t} &= \boldsymbol{\varepsilon}^{(\mathrm{p})\,t} + \Delta\bar{\varepsilon}_{(i)}^{(\mathrm{p})}\,\mathbf{n}_{(i)}\,, \\ \bar{\varepsilon}_{(i)}^{(\mathrm{p})\,t+\Delta t}\,\mathbf{n}_{(i)} &= \bar{\varepsilon}^{(\mathrm{p})\,t} + \Delta\bar{\varepsilon}_{(i)}^{(\mathrm{p})}\,. \end{aligned} \tag{6.210}$$

The above algorithm may be slightly improved in terms of its accuracy by replacing the steps 2 and 3 by the following operations

$$\begin{aligned} \boldsymbol{\varepsilon}_{(i)}^{t+\Delta t} &= \boldsymbol{\varepsilon}^t + \Delta\boldsymbol{\varepsilon}(i)\,, \\ \boldsymbol{\varepsilon}_{(i)}^{(\mathrm{e})\,t+\Delta t} &= \boldsymbol{\varepsilon}_{(i)}^{t+\Delta t} - \boldsymbol{\varepsilon}^{(\mathrm{p})\,t} - \boldsymbol{\varepsilon}^{(\vartheta)\,t}\,, \\ \hat{\boldsymbol{\sigma}}_{(i)}^{t+\Delta t} &= \frac{\partial W(\boldsymbol{\varepsilon}_{(i)}^{(\mathrm{e})\,t+\Delta t})}{\partial\boldsymbol{\varepsilon}_{(i)}^{(\mathrm{e})\,t+\Delta t}}\,, \\ \hat{\bar{\boldsymbol{\sigma}}}_{(i)}^{t+\Delta t} &= \hat{\boldsymbol{\sigma}}_{(i)}^{t+\Delta t} - \boldsymbol{\alpha}^t\,, \end{aligned} \tag{6.211}$$

in which $W(\ldots)$ is the total elastic strain energy. The modification turns out particularly helpful in large deformation problems for which the use of the original step 2 may generate errors related to the so-called incremental objectivity violation.

We have already pointed out that the above algorithm is based on the assumption of the constant hardening modulus h (and the parameter β) within the time step analysed. Considering now for simplicity only elastically isotropic, isothermal elastic-plastic deformations we shall now very briefly describe the algorithm valid for any nonlinear hardening function. In other words, we consider Eq. (6.64) recalled here for convenience

$$\begin{aligned}\dot{\boldsymbol{\sigma}} &= \left[\left(K - \tfrac{2}{3}\right)\mathbf{1}\mathbf{1}^T + 2G\left(\mathbf{I} - \frac{\mathbf{n}\,\mathbf{n}^T}{1 + \frac{\sigma_y' + H_{(\alpha)}'}{3G}}\right)\right]\dot{\boldsymbol{\varepsilon}} \\ &= \left[K\,\mathbf{1}\mathbf{1}^T + 2G\left(I - \frac{1}{3}\mathbf{1}\mathbf{1}^T\right) - 2G\gamma\,\mathbf{n}\,\mathbf{n}^T\right]\dot{\boldsymbol{\varepsilon}}\,. \end{aligned} \tag{6.212}$$

Also, the following relations are valid:

$$\begin{aligned} \mathbf{n} &= \frac{{}^D\bar{\boldsymbol{\sigma}}}{\|\,{}^D\bar{\boldsymbol{\sigma}}\,\|}\,, \\ \dot{\bar{\varepsilon}}^{(\mathrm{p})} &= \sqrt{\tfrac{2}{3}}\,\|\dot{\boldsymbol{\varepsilon}}^{(\mathrm{p})}\|\,, \\ \dot{\boldsymbol{\alpha}} &= \tfrac{2}{3}\,H_{(\alpha)}'(\bar{\varepsilon}^{(\mathrm{p})})\,\dot{\boldsymbol{\varepsilon}}^{(\mathrm{p})}\,, \\ {}^D\bar{\boldsymbol{\sigma}}_{(i)}^{t+\Delta t} &= {}^D\boldsymbol{\sigma}_{(i)}^{t+\Delta t} - \boldsymbol{\alpha}_{(i)}^{t+\Delta t}\,, \\ {}^D\boldsymbol{\sigma}_{(i)}^{t+\Delta t} &= {}^D\hat{\boldsymbol{\sigma}}_{(i)}^{t+\Delta t} - 2G\sqrt{\tfrac{3}{2}}\Delta\bar{\varepsilon}_{(i)}^{(\mathrm{p})}\,\mathbf{n}_{(i)}\,, \\ {}^D\hat{\boldsymbol{\sigma}}_{(i)}^{t+\Delta t} &= {}^D\boldsymbol{\sigma}^t + 2G\Delta\boldsymbol{\varepsilon}_{(i)}\,. \end{aligned} \tag{6.213}$$

In order to assure the consistency conditions at the step end it is necessary to find the location of the yield surface at $t + \Delta t$ which in turn requires the values of the hardening parameter and the function $\boldsymbol{\alpha}$ at that time instant. We have

$$\begin{aligned} \bar{\varepsilon}_{(i)}^{(\mathrm{p})\,t+\Delta t} &= \bar{\varepsilon}^{(\mathrm{p})\,t} + \Delta\bar{\varepsilon}_{(i)}^{(\mathrm{p})}\,, \\ \boldsymbol{\alpha}_{(i)}^{t+\Delta t} &= \boldsymbol{\alpha}^t + \frac{2}{3}\,H_{(\alpha)}'\left(\bar{\varepsilon}_{(i)}^{(\mathrm{p})\,t+\frac{\Delta t}{2}}\right)\Delta\bar{\varepsilon}_{(i)}^{(\mathrm{p})}\sqrt{\frac{3}{2}}\,\mathbf{n}_{(i)} \\ &= \boldsymbol{\alpha}^t + \sqrt{\frac{2}{3}}\,\frac{H_{(\alpha)}\left(\bar{\varepsilon}_{(i)}^{(\mathrm{p})\,t+\Delta t}\right) - H_{(\alpha)}(\bar{\varepsilon}^{(\mathrm{p})\,t})}{\Delta\bar{\varepsilon}_{(i)}^{(\mathrm{p})}}\,\Delta\bar{\varepsilon}_{(i)}^{(\mathrm{p})}\,\mathbf{n}_{(i)} \\ &= \boldsymbol{\alpha}^t + \sqrt{\tfrac{2}{3}}\,\Delta H_{(\alpha)\,(i)}\,\mathbf{n}_{(i)}\,, \end{aligned} \tag{6.214}$$

where

$$\Delta H_{(\alpha)\,(i)} = H_{(\alpha)}\left(\bar{\varepsilon}_{(i)}^{(\mathrm{p})\,t+\Delta t}\right) - H_{(\alpha)}\left(\bar{\varepsilon}^{(\mathrm{p})\,t}\right).$$

From Eqs. (6.213) and (6.214) there results

$$\begin{aligned} {}^{D}\bar{\boldsymbol{\sigma}}_{(i)}^{t+\Delta t} &= {}^{D}\boldsymbol{\sigma}_{(i)}^{t+\Delta t} - \boldsymbol{\alpha}_{(i)}^{t+\Delta t} \\ &= {}^{D}\boldsymbol{\sigma}_{(i)}^{t+\Delta t} - \boldsymbol{\alpha}^{t} - \sqrt{\tfrac{2}{3}}\,\Delta H_{(\alpha)\,(i)}\,\mathbf{n}_{(i)} \\ &= {}^{D}\hat{\boldsymbol{\sigma}}_{(i)}^{t+\Delta t} - \boldsymbol{\alpha}^{t} - \left[2G\sqrt{\tfrac{3}{2}}\,\Delta\bar{\varepsilon}_{(i)}^{(\mathrm{p})} + \sqrt{\tfrac{2}{3}}\,\Delta H_{(\alpha)\,(i)}\right]\mathbf{n}_{(i)}\,. \end{aligned} \tag{6.215}$$

Using the notation

$${}^{D}\hat{\boldsymbol{\sigma}}_{(i)}^{t+\Delta t} - \boldsymbol{\alpha}^{t} = {}^{D}\hat{\bar{\boldsymbol{\sigma}}}_{(i)}^{t+\Delta t} \tag{6.216}$$

we obtain

$${}^{D}\bar{\boldsymbol{\sigma}}_{(i)}^{t+\Delta t} = {}^{D}\hat{\bar{\boldsymbol{\sigma}}}_{(i)}^{t+\Delta t} - \left[2G\sqrt{\tfrac{3}{2}}\,\Delta\bar{\varepsilon}_{(i)}^{(\mathrm{p})} + \sqrt{\tfrac{2}{3}}\,\Delta H_{(\alpha)\,(i)}\right]\mathbf{n}_{(i)}\,. \tag{6.217}$$

Since

$${}^{D}\bar{\boldsymbol{\sigma}} = \|\,{}^{D}\bar{\boldsymbol{\sigma}}\,\|\,\mathbf{n}_{(i)}\,,$$

the last relationship implies that

$$\mathbf{n}_{(i)} = \frac{{}^{D}\hat{\bar{\boldsymbol{\sigma}}}_{(i)}^{t+\Delta t}}{\|\,{}^{D}\hat{\bar{\boldsymbol{\sigma}}}_{(i)}^{t+\Delta t}\,\|}\,.$$

Equation (6.215) is then used to obtain

$$\|\,{}^{D}\bar{\boldsymbol{\sigma}}_{(i)}^{t+\Delta t}\,\|\,\mathbf{n}_{(i)} = \left\{\|\,{}^{D}\hat{\bar{\boldsymbol{\sigma}}}_{(i)}^{t+\Delta t}\,\| - \left[2G\sqrt{\tfrac{3}{2}}\,\Delta\bar{\varepsilon}_{(i)}^{(\mathrm{p})} + \sqrt{\tfrac{2}{3}}\,\Delta H_{(\alpha)\,(i)}\right]\right\}\mathbf{n}_{(i)}\,, \tag{6.218}$$

so that also

$$\begin{aligned} e(\Delta\bar{\varepsilon}_{(i)}^{(\mathrm{p})/k}) = {} & -\sqrt{\tfrac{2}{3}}\sigma_y\left(\bar{\varepsilon}_{(i)}^{(\mathrm{p})\,t+\Delta t}\right) + \|\,{}^{D}\hat{\bar{\boldsymbol{\sigma}}}_{(i)}^{t+\Delta t}\,\| \\ & - \left[2G\sqrt{\tfrac{3}{2}}\,\Delta\bar{\varepsilon}_{(i)}^{(\mathrm{p})} + \sqrt{\tfrac{2}{3}}\,\Delta H_{(\alpha)\,(i)}\right] = 0\,. \end{aligned} \tag{6.219}$$

Since $\bar{\varepsilon}_{(i)}^{(\mathrm{p})\,t+\Delta t} = \bar{\varepsilon}_{(i)}^{(\mathrm{p})\,t} + \Delta\bar{\varepsilon}_{(i)}^{(\mathrm{p})}$, Eq. (6.219) is a generally nonlinear equation for the scalar $\Delta\bar{\varepsilon}_{(i)}^{(\mathrm{p})}$. the functions $\sigma_y(\cdot)$ and $\Delta H_{(\alpha)}(\cdot)$ being given.

The integration algorithm thus becomes the following:

1.–7. – as before.

8. Compute $\Delta\bar{\varepsilon}_{(i)}^{(\mathrm{p})}$ from the iterative procedure

   (a) $\bar{\varepsilon}_{(i)}^{(\mathrm{p})\,t+\Delta t/k} = \bar{\varepsilon}_{(i)}^{(\mathrm{p})\,t+\Delta t/k-1} + \Delta\bar{\varepsilon}_{(i)}^{(\mathrm{p})/k}\,, \qquad k = 1, 2, \ldots,$

   (b) $\mathrm{De}(\Delta\bar{\varepsilon}_{(i)}^{(\mathrm{p})/k}) = 2G\left[1 + \frac{\sigma_y' + H_{(\alpha)}'}{3G}\right]^{(k)}\,,$

   (c) $\Delta\bar{\varepsilon}_{(i)}^{(\mathrm{p})/k+1} = \Delta\bar{\varepsilon}_{(i)}^{(\mathrm{p})/k} - \frac{e(\Delta\bar{\varepsilon}_{(i)}^{(\mathrm{p})/k})}{\mathrm{De}(\Delta\bar{\varepsilon}_{(i)}^{(\mathrm{p})/k})}\,,$

   if $|e(\Delta\bar{\varepsilon}_{(i)}^{(\mathrm{p})/k})| > \eta$ go back to (a),

   if $|e(\Delta\bar{\varepsilon}_{(i)}^{(\mathrm{p})/k})| \leq \eta$ proceed to 9.

9. as before.

6.2.4 The Consistent (Algorithmic) Tangent Matrix

The inelastic materials models have so far been characterized by means of some 'tangent' constitutive matrices, i.e. the matrices which occur in the rate-type constitutive laws. From the viewpoint of computational accuracy it is crucial to realize that the finite element algorithms involve finite time increments rather than the corresponding rates. On the other hand, we have already indicated that the solution of the incremental problem hinges heavily upon the use of the real tangent coefficient matrix. As a consequence of both the above aspects we may suspect that using the constitutive tangent matrix (i.e. the one characteristics of the rate formulation) for solving the time discretized problem may not be optimal computationally. Therefore, finding a real tangent matrix for the incremental problem, which will certainly incorporate some features of the tangent constitutive matrix on the one hand and some features of the time discretization scheme on the other, appears as a fundamental problem for the analysis we have in mind. Such a matrix is known as the consistent, or algorithmic, tangent matrix. The form of the matrix for the elastic-plastic case will be discussed below.

By analysing the radial return procedure given in the previous section we may conclude that we deal in fact with the nonlinear problem of the following type:

$$\begin{aligned} \text{Given:} \quad & \Delta\boldsymbol{\varepsilon} = \boldsymbol{\varepsilon}^{t+\Delta t} - \boldsymbol{\varepsilon}^{t} \\ \text{Find:} \quad & \boldsymbol{\sigma}^{t+\Delta t} = \boldsymbol{\sigma}^{t+\Delta t}\big(\boldsymbol{\sigma}^{t}, \bar{\varepsilon}^{(\mathrm{p})\,t}, \boldsymbol{\varepsilon}^{t}, \boldsymbol{\varepsilon}^{t+\Delta t} - \boldsymbol{\varepsilon}^{t}\big) \;, \end{aligned} \tag{6.220}$$

whereas the quantities defined at $\tau = t$ i.e. $\boldsymbol{\sigma}^t$, $\bar{\varepsilon}^{(\mathrm{p})\,t}$ and $\boldsymbol{\varepsilon}^t$ are known while $\boldsymbol{\sigma}^{t+\Delta t}$ and $\boldsymbol{\varepsilon}^{t+\Delta t}$ are to be found. For the value of $\Delta\boldsymbol{\varepsilon}$ computed from the incremental problem Eq. (6.212) serves the purpose of finding $\boldsymbol{\sigma}^{t+\Delta t}$.

Neglecting the inertial and large deformation effects which are not essential in our search for the consistent tangent matrix the weak (variational) formulation of the body equilibrium has the form, cf. Eq. (2.43), Part I,

$$\begin{aligned} G(\mathbf{u};\mathbf{v}) = & \int_{\Omega^t} \overset{T}{\boldsymbol{\sigma}}{}^{t+\Delta t}\left(\boldsymbol{\sigma}^{t}, \bar{\varepsilon}^{(\mathrm{p})\,t}, \boldsymbol{\varepsilon}^{t}, \mathbf{D}\mathbf{u}^{t+\Delta t} - \boldsymbol{\varepsilon}^{t}\right) \mathbf{D}\mathbf{v}\,\mathrm{d}\Omega \\ & - \int_{\Omega^t} \varrho\,\mathbf{f}^T \mathbf{v}\,\mathrm{d}\Omega - \int_{\partial\Omega^t_\sigma} \mathbf{t}^T \mathbf{v}\,\mathrm{d}(\partial\Omega) = 0 \end{aligned} \tag{6.221}$$

for each $\mathbf{v}$ satisfying the homogeneous kinematic boundary conditions on $\partial\Omega^t_u$. The iterative Newton–Raphson method, cf. Chap. 7, consists in solving the linearized equation system of the form

$$G\left(\mathbf{u}^{t+\Delta t}_{(i)};\mathbf{v}\right) + \mathcal{D}G\left(\mathbf{u}^{t+\Delta t}_{(i)};\mathbf{v}\right)\delta\mathbf{u}_{(i)} = 0 \tag{6.222}$$

with respect to the vector $\delta\mathbf{u}_{(i)}$, i.e. solving the system

$$\mathcal{D}G\left(\mathbf{u}^{t+\Delta t}_{(i)};\mathbf{v}\right)\delta\mathbf{u}_{(i)} = -G\left(\mathbf{u}^{t+\Delta t}_{(i)};\mathbf{v}\right) \tag{6.223}$$

for $i = 1, 2, \ldots$ until the residuum $G(\mathbf{u}_{(i)}^{t+\Delta t}; \mathbf{v})$ becomes small enough in a given norm. The symbol $\mathcal{D}G$ stands here for

$$\begin{aligned}\mathcal{D}G\Big(\mathbf{u}_{(i)}^{t+\Delta t}; \mathbf{v}\Big) &= \int_{\Omega^t} (\mathbf{D}\mathbf{v})^T \frac{\partial \boldsymbol{\sigma}^{t+\Delta t}}{\partial \mathbf{u}_{(i)}^{t+\Delta t}} \,\mathrm{d}\Omega \\ &= \int_{\Omega^t} (\mathbf{D}\mathbf{v})^T \frac{\partial \boldsymbol{\sigma}^{t+\Delta t}}{\partial \boldsymbol{\varepsilon}_{(i)}^{t+\Delta t}} \frac{\partial \boldsymbol{\varepsilon}_{(i)}^{t+\Delta t}}{\partial \mathbf{u}_{(i)}^{t+\Delta t}} \,\mathrm{d}\Omega \\ &= \int_{\Omega^t} (\mathbf{D}\mathbf{v})^T \, \mathbf{C}_{(i)}^{t+\Delta t} \, \mathbf{D} \,\mathrm{d}\Omega \,,\end{aligned} \tag{6.224}$$

so that Eq. (6.223) assumes the form

$$\int_{\Omega^t} (\mathbf{D}\mathbf{v})^T \, \mathbf{C}_{(i)}^{t+\Delta t} \, \mathbf{D} \, \delta\mathbf{u}_{(i)} \,\mathrm{d}\Omega = -G\Big(\mathbf{u}_{(i)}^{t+\Delta t}; \mathbf{v}\Big) \,. \tag{6.225}$$

Our goal now is to determine the matrix $\mathbf{C}_{(i)}^{t+\Delta t}$ in a way which is consistent with the radial return algorithm presented in Sect. 6.2.3. The matrix is defined as

$$\mathbf{C}^{t+\Delta t} = \left.\frac{\partial \boldsymbol{\sigma}^{t+\Delta t}(\boldsymbol{\sigma}^t, \bar{\varepsilon}^{(\mathrm{p})\,t}, \boldsymbol{\varepsilon}^t, \boldsymbol{\varepsilon} - \boldsymbol{\varepsilon}^t)}{\partial \boldsymbol{\varepsilon}}\right|_{\boldsymbol{\varepsilon} = \boldsymbol{\varepsilon}^{t+\Delta t}} , \tag{6.226}$$

whereas in the radial return case we have

$$\boldsymbol{\sigma}^{t+\Delta t} = \boldsymbol{\alpha}^{t+\Delta t} + \sqrt{\tfrac{2}{3}}\, \sigma_y\big(\bar{\varepsilon}^{(\mathrm{p})\,t+\Delta t}\big)\, \mathbf{n}^{t+\Delta t} + K\, \mathbf{1}(\mathbf{1}^T \boldsymbol{\varepsilon}^{t+\Delta t}) \,. \tag{6.227}$$

Dropping for simplicity the indices we observe that

$$\begin{aligned}&\frac{\partial \mathbf{n}}{\partial\, {}^D\!\boldsymbol{\sigma}} = \frac{1}{\|{}^D\!\boldsymbol{\sigma}\|} [\mathbf{I} - \mathbf{n}\,\mathbf{n}^T] \,, && \mathbf{n} = \frac{{}^D\!\boldsymbol{\sigma}}{\|{}^D\!\boldsymbol{\sigma}\|} \,, \\ &\frac{\partial(\mathbf{1}^T \boldsymbol{\varepsilon})}{\partial \boldsymbol{\varepsilon}} = \mathbf{1}^T \,, && \frac{\partial\left[(\mathbf{I} - \tfrac{1}{3}\mathbf{1}\mathbf{1}^T)\,\boldsymbol{\varepsilon}\right]}{\partial \boldsymbol{\varepsilon}} = \mathbf{I} - \tfrac{1}{3}\mathbf{1}\mathbf{1}^T \,.\end{aligned} \tag{6.228}$$

which allows to write the matrix $\mathbf{C}^{t+\Delta t}$ as

$$\begin{aligned}\mathbf{C}^{t+\Delta t} &= K\, \mathbf{1}\mathbf{1}^T + \frac{\partial \boldsymbol{\alpha}^{t+\Delta t}}{\partial \boldsymbol{\varepsilon}^{t+\Delta t}} + \sqrt{\frac{2}{3}}\, \mathbf{n}^{t+\Delta t} \frac{\partial \sigma_y^{t+\Delta t}}{\partial \boldsymbol{\varepsilon}^{t+\Delta t}} \\ &\quad + 2G \frac{\sqrt{\tfrac{2}{3}}\sigma_y^{t+\Delta t}}{\|{}^D\hat{\boldsymbol{\sigma}}^{t+\Delta t}\|} \left[\mathbf{I} - \frac{1}{3}\mathbf{1}\mathbf{1}^T - \mathbf{n}^{t+\Delta t}\, \mathbf{n}^{t+\Delta t\; T}\right] .\end{aligned} \tag{6.229}$$

Furthermore, we have

$$\mathbf{n}^{t+\Delta t}\frac{\partial \sigma_y^{t+\Delta t}}{\partial \boldsymbol{\varepsilon}^{t+\Delta t}} = \sqrt{\frac{2}{3}}\,\frac{\sigma_y'^{\,t+\Delta t}}{1+\frac{\sigma_y'^{\,t+\Delta t}+H_{(\alpha)}'^{\,t+\Delta t}}{3G}}\,\mathbf{n}^{t+\Delta t}\,\mathbf{n}^{t+\Delta t\,T}\,,$$

$$\frac{\partial \boldsymbol{\alpha}^{t+\Delta t}}{\partial \boldsymbol{\varepsilon}^{t+\Delta t}} = \sqrt{\frac{2}{3}}\,\frac{\Delta H_{(\alpha)}}{\|{}^{D}\hat{\boldsymbol{\sigma}}^{t+\Delta t}\|}\left[\mathbf{I}-\frac{1}{3}\mathbf{1}\mathbf{1}^T-\mathbf{n}^{t+\Delta t}\,\mathbf{n}^{t+\Delta t\,T}\right] + \frac{2}{3}\,\frac{H_{(\alpha)}'^{\,t+\Delta t}}{1+\frac{\sigma_y'^{\,t+\Delta t}+H_{(\alpha)}'^{\,t+\Delta t}}{3G}}\,\mathbf{n}^{t+\Delta t}\,\mathbf{n}^{t+\Delta t\,T}\,, \tag{6.230}$$

By substituting the above in Eq. (6.229) we arrive at the final form of the tangent matrix consistent with the radial return algorithm in case of the elastic-plastic material with the mixed isotropic-kinematic hardening as

$$\mathbf{C}^{t+\Delta t} = K\,\mathbf{1}\mathbf{1}^T + 2G\beta\left[\mathbf{I}-\tfrac{1}{3}\,\mathbf{1}\mathbf{1}^T\right] - 2G\tilde{\gamma}\,\mathbf{n}^{t+\Delta t}\,\mathbf{n}^{t+\Delta t\,T}\,, \tag{6.231}$$

in which

$$\beta = \sqrt{\frac{2}{3}}\,\frac{\sigma_y^{t+\Delta t}+\Delta H_{(\alpha)}}{\|{}^{D}\hat{\boldsymbol{\sigma}}^{t+\Delta t}\|} \le 1\,, \qquad \tilde{\gamma} = \frac{1}{1+\frac{\sigma_y'^{\,t+\Delta t}+H_{(\alpha)}'^{\,t+\Delta t}}{3G}} - (1-\beta)\,. \tag{6.232}$$

It is interesting to compare the matrix just obtained against the constitutive tangent matrix typical of the rate problem. We observe that the time integration algorithm made the shear modulus G be reduced by means of a coefficient β given in Eq. $(6.232)_1$. Also, the multiplier $2G\gamma$ in Eq. (6.36) has been replaced by $2G\tilde{\gamma}$ with the coefficient $\tilde{\gamma}$ given in Eq. $(6.232)_2$; there holds $\gamma-1 \le \tilde{\gamma} \le \gamma$. Moreover, since $\hat{\boldsymbol{\sigma}}^{t+\Delta t}$ may lie far outside of the current yield surface, the coefficient β may be much smaller than 1. It is thus seen that, particularly for large steps, the consistent matrix may be greatly different from the matrix given in Eq. (6.39). In other words, the use in the iterative procedure of the latter matrix may essentially worsen the convergence properties typical of the Newton algorithm with the correct tangent matrix (i.e. the consistent one).

In terms of the terminology widely employed in the numerical analysis of differential equations we may say that the algorithm for the time integration of the constitutive equations presented in this chapter belongs to the class of one-step implicit methods of the predictor-corrector type. It is implicit because it attempts to satisfy the equation at the end of the step, the one-step property refers to using just the beginning-of-the-step values and no other values from other time instants, the predictor is the trial stress and the corrector is obtained as a result of the stress projection upon the current yield surface.

It has to be also realized that an attempt to satisfy the equation at the end of the step (with the consistency condition satisfied, i.e. the stress point lying on the yield surface) in not necessarily optimal in view of the accuracy of the algorithm. In the computational practice more general algorithms based on the so-called generalized mid-point rules (of which the above approach is an example) are also used. Such algorithms attempt to satisfy the equation at a certain time instant $\tau = t + \alpha\Delta t$, $\alpha \in [0,1]$, within the time step considered, together with the satisfaction of the consistency condition at the same moment. The numerical experiences so far tend to indicate that for small steps the value of $\alpha = \frac{1}{2}$ leads to the best accuracy while for larger steps the better results may be obtained for greater values of α (up to $\alpha = 1$).

6.3 Computational Illustrations

A good advanced test of the quality of any specific constitutive model is using it for solving a complex boundary-value problem for which some experimental data exist. A typical test of this kind is the analysis of a plastic instability phenomenon known as the axisymmetric necking problem. Generally, two groups of methods exist for the analysis of instability phenomena which can be classified as the bifurcational approach and the direct integration approach, respectively. Under the idealized loading conditions, specimen geometry and material homogeneity instability of the equilibrium mode of deformation is usually associated with bifurcation from the fundamental equilibrium path into a secondary equilibrium path. The Hill's theory of uniqueness and stability in elastic-plastic solids provides an effective analytical tool to deal with such situations. However, unavoidable deviations from the perfect conditions often result in significant reductions of the critical loads or strains at which failure occurs. Therefore, it is important to know how sensitive is the specimen behaviour to different imperfection patterns. Such imperfect problems form a starting point for the direct integration approach to the instability analysis. According to this concept the nonlinear differential equations describing the behaviour of the imperfect specimen are integrated in a step-by-step manner leading to the overall force-displacement characteristics often exhibiting some singular points which signal the possible loss of stability. The method is relatively costly and requires repetitive computations for different imperfections, supplies however a very detailed information on the specimen behaviour. In particular, for problems with non-uniform pre-critical response and those requiring post-critical behaviour evaluation the direct integration methods seem to be superior to those based on the bifurcation analysis.

Let us consider the axisymmetric problem defined in Fig. 6.1 – the analysis is limited to such deformation patterns which make it possible to employ in the computations one quadrant of the mesh only. The cylindrical bar is subject to the uniaxial tension enforced by an axial displacement rate prescribed at the ends of the bar. The geometric data for the problem are given in Fig. 6.1; the material properties are: $E = 207$ GPa, $\nu = 0.3$, $\sigma_0 =$

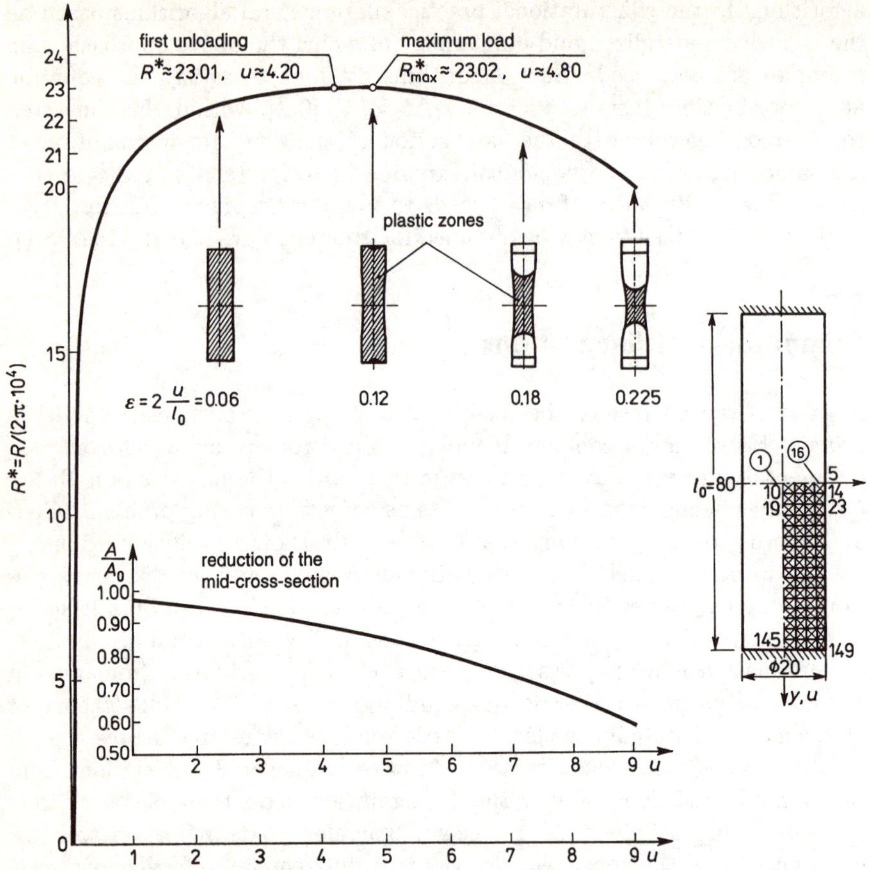

Fig. 6.1. Necking of a cylindrical specimen with built-in ends – J_2 flow theory

196 MPa, isotropic hardening with the Ramberg–Osgood hardening curve of the form

$$\varepsilon^{(\mathrm{p})} = \frac{1.1\sigma_0}{8E}\left[\left(\frac{\sigma_M}{1.1\sigma_0}\right)^m - \left(\frac{1}{1.1}\right)^m\right], \qquad m = 8.$$

In each calculation to be discussed the overall elongation of 0.225 times the initial specimen length is imposed in 300 equal steps of the prescribed end displacements so that finally the engineering strain of $\varepsilon_{\max} = 300 \times \frac{0.03}{40} = 0.225$ is achieved. The boundary-value problem is posed in two ways: in one the ends are assumed to be cemented to rigid grips, while in the other they remain shear free. The bifurcation from the state of uniform tensile stress occurs with the latter end-condition only; nevertheless, we start with a short discussion of the results obtained for the bar with built-in ends.

The specimen is first analysed within the framework of the so-called

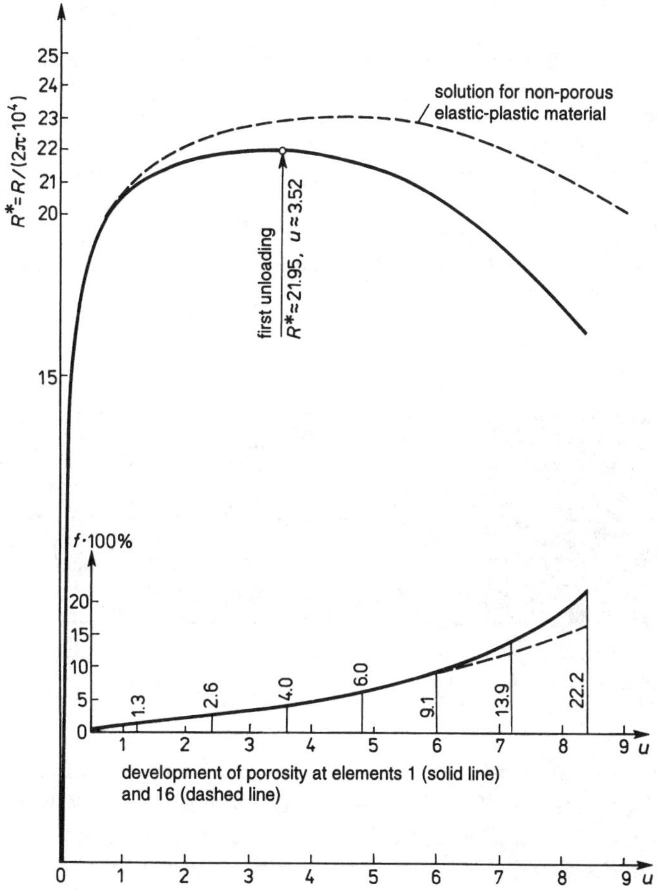

Fig. 6.2. Necking of a cylindrical specimen with built-in ends – Gurson's theory for $\hat{K} = 0.001$, $\xi_0 = 0.00$

J_2–theory of elastic-plastic materials. The curve of the resultant reaction R^* versus the prescribed displacement u, the development of unloading zones with the increasing deformation and the gradual reduction of the cross-sectional area at neck are shown in Fig. 6.1. It is observed that the whole specimen goes plastic before it starts to unload. Furthermore, the unloading starts at distinctly smaller strains than those accompanying the maximum load point. As expected, no sudden change in the slope of the cross-section reduction curve is observed.

The calculations are next repeated for the porous material assuming the theory of Sect. 6.1.5 to hold. The results, including the development of porosity in the neck at the element 1 (near the axis of the bar) and at the element 16 (near the lateral surface of the bar) are shown in Fig. 6.2.

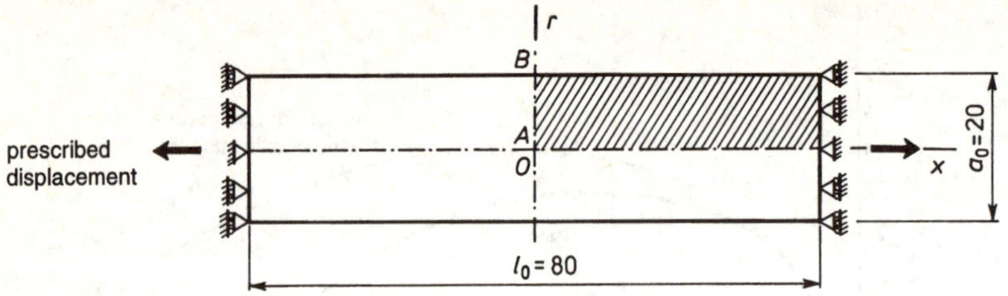

Fig. 6.3. Geometry and displacement boundary conditions for axisymmetric specimen

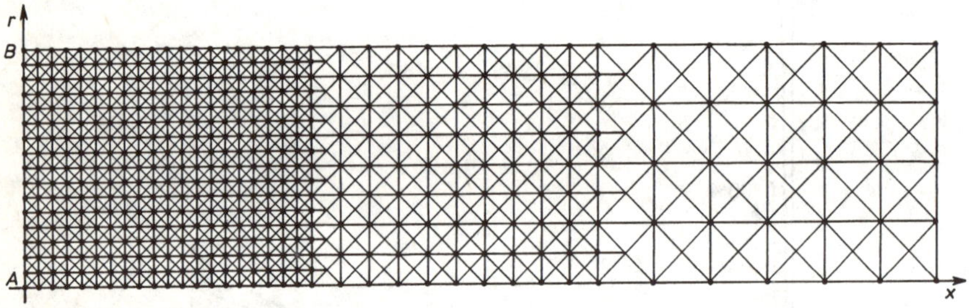

Fig. 6.4. Finite element mesh

A much more interesting discussion can be carried out for the bar with the shear-free ends, Fig. 6.3. The discretization mesh is shown in Fig. 6.4; its non-uniform character is essential for the correct capturing of the localization effects to be discussed. The numerical results obtained for the no-voids case are shown in Fig. 6.5; the curve I refers to the ideal specimen while the curve II describes the specimen with an initial imperfection in the form of the bar mid-cross-section reduced by 0.5%. Both the curves virtually coincide up to a point B at which they start to branch off in a distinct way. The first unloading takes place at the point B in the vicinity of the symmetry axis near the bar ends. The point B is clearly identified as the primary bifurcation point; the results confirmed by the precise bifurcational analysis in [86]. At about the same deformation the area of the neck starts to decrease much more rapidly than that of the ideal specimen. The continuing unloading process is qualitatively similar to that presented earlier in Fig. 6.1 but this time it develops much faster.

A similar, but more complex still is the deformation pattern typical of the material whose model accounts for the voids effects, Fig. 6.6. In it, the curve I refers for comparison to the previous analysis and coincides with the curve I of Fig. 6.5. The curve II corresponds to the initially ideal specimen (the subsequent material fracture is included in the model, though) and it is thus taken as the fundamental solution for the case on hand. The

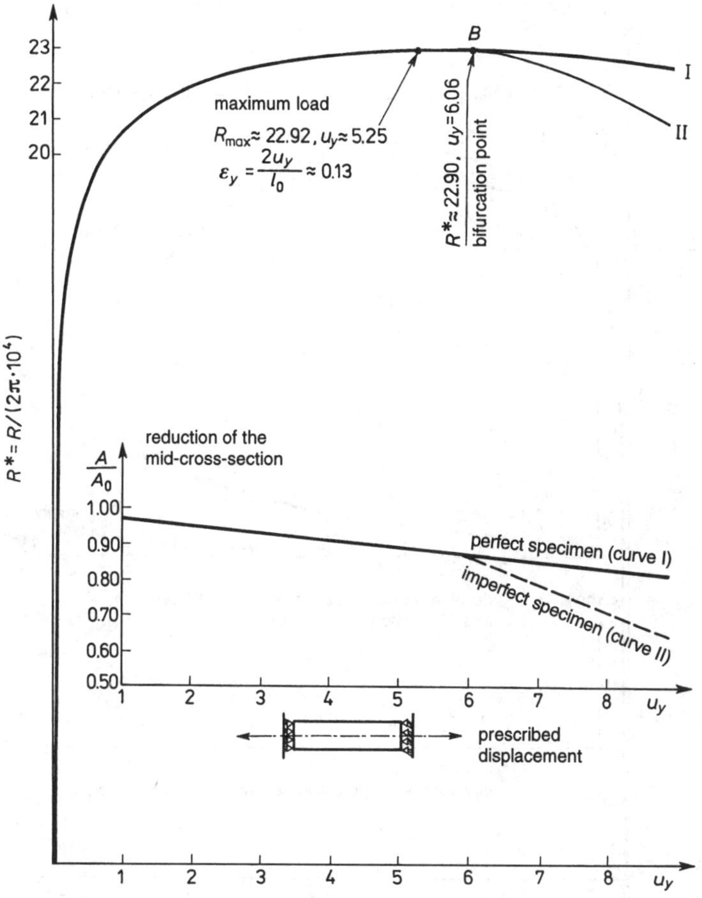

Fig. 6.5. Necking of a cylindrical specimen with simply supported (shear-free) ends – J_2 flow theory

curve III is obtained for the specimen with an initial imperfection in the form of the initial porosity of in the elements adjacent to the mid-cross-section of the bar. The point B corresponds to the first unloading in the imperfect bar and, at the same time, to the beginning of the accelerated mid-cross-section reduction in it, Fig. 6.7. The porosity development at two selected points A and B for both the ideal and imperfect specimen are shown in Fig. 6.8.

A more advanced analysis carried out in [86] took additionally into account the phenomenon of voids coalescence all the way up to the creation of the macroscopic crack. Assuming that the coalescence starts when the porosity reaches the value of $\xi = 0.20$ the paper [86] presented the approximate analysis in the whole range of the zig-zag crack formation up to the

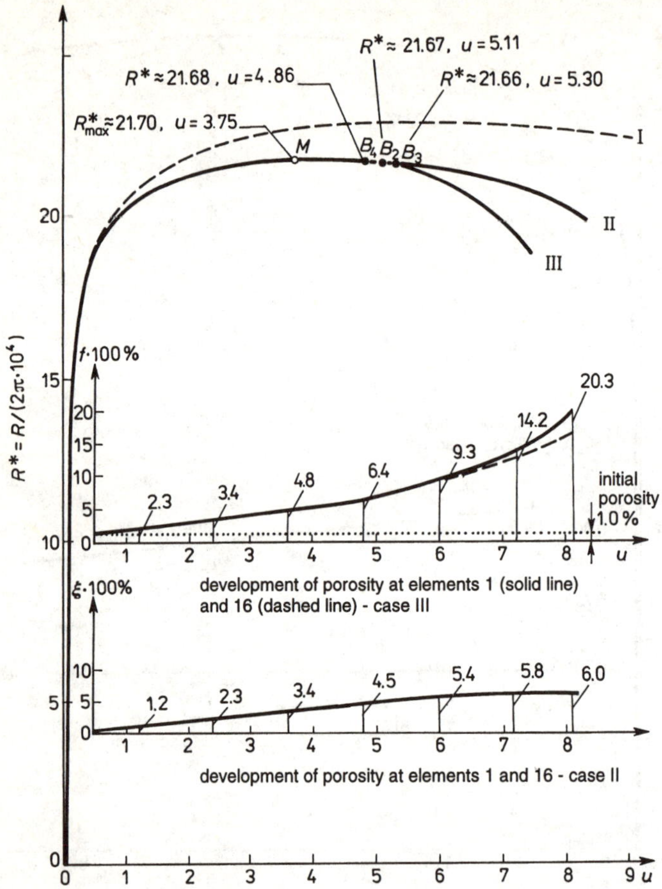

Fig. 6.6. Necking of a cylindrical specimen with simply supported (shear-free) ends – Gurson's theory

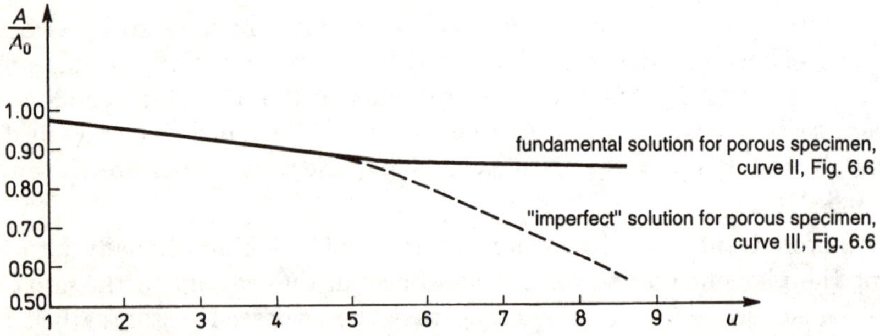

Fig. 6.7. Reduction of the mid-cross-section for the imperfect specimen

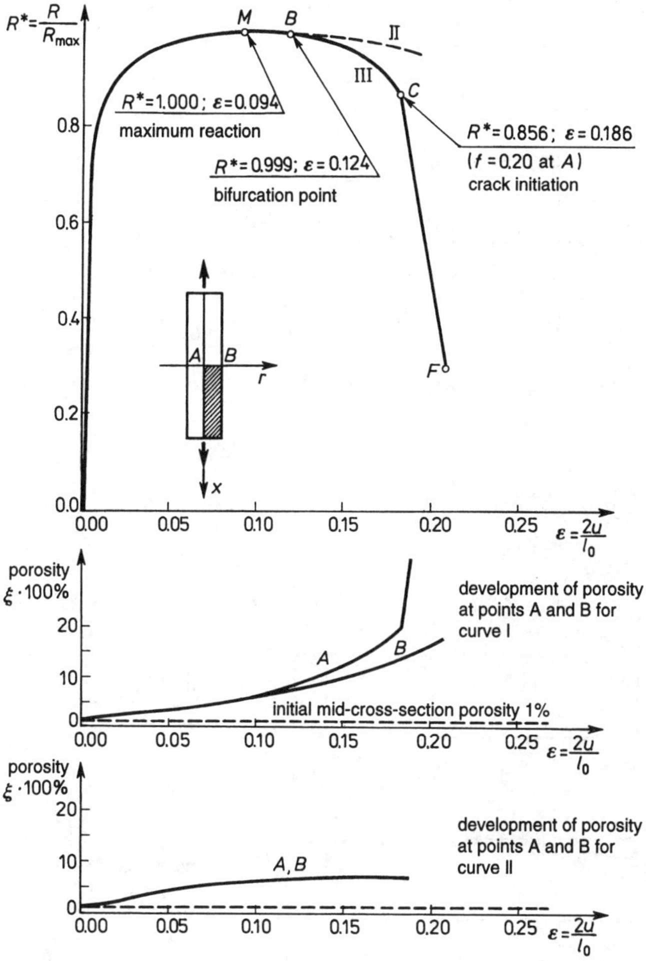

Fig. 6.8. Necking of a cylindrical specimen with simply supported (shear-free) ends including effects of porosity development

entire loss of the specimen load carrying capacity in the typical cup–cone fracture mode observed experimentally. The development of such a process is presented in Fig. 6.8. An interesting observation is the characteristic arrest of further necking effects at the point C of the cracking initiation (Fig. 6.9). The process of the crack growth is shown in Fig. 6.10.

A further interesting discussion can be carried on by analysing a similar configuration with the additional thermal effect included – the theory presented in Sec. 6.1.2 can be used to this purpose. The problem considered is shown in Fig. 6.11 – note slightly different geometric and material parameters adopted in this case, cf. [181]. We also note that in spite of the shear-free

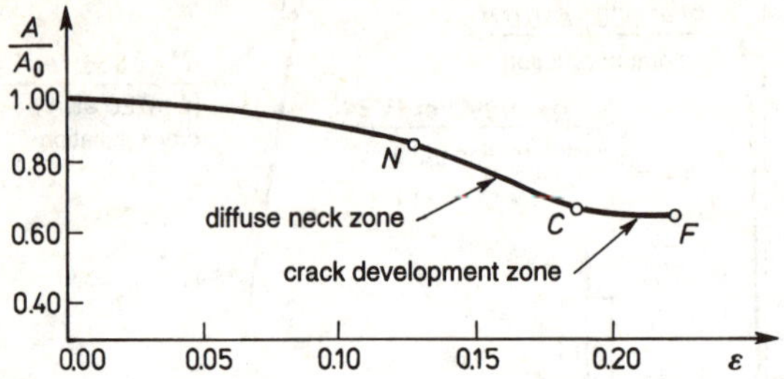

Fig. 6.9. Reduction of the mid-cross-section including effects of void coalescence

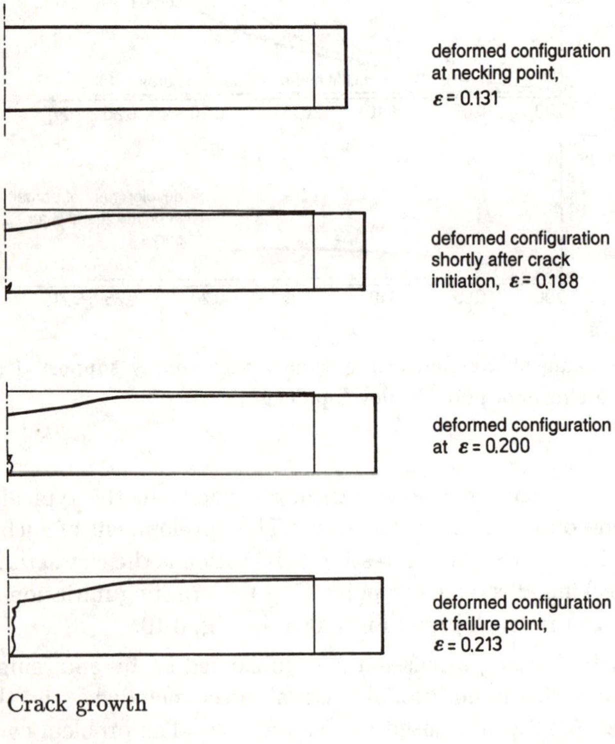

Fig. 6.10. Crack growth

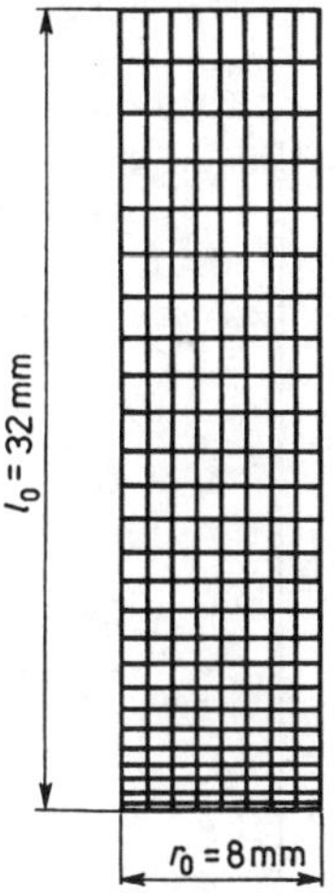

Mechanical properties	Thermal properties
$K = 164.206$ GPa	$\rho_R = 7.8 \cdot 10^{-6}\,\mathrm{kg\,mm^{-3}}$
$G = 80.194$ GPa	$c = 460\,\mathrm{J\,kg^{-1}\,K^{-1}}$
$Y_0 = 0.45$ GPa	$k = 0.045\,\mathrm{J\,s^{-1}\,K^{-1}\,mm^{-1}}$
$Y_\infty = 0.715$ GPa	$\alpha_t = 1.2 \cdot 10^{-5}$
$\delta = 16.93$	$H_\vartheta = 0.002$
$H_t = 0.1292$	
$\chi = 0.9$	

Fig. 6.11. Necking of a cylindrical specimen – thermomechanical problem. Finite element mesh, geometric and material data

boundary conditions considered now, the present analysis will be qualitatively more similar to the previous case with the fixed specimen ends rather than to the simply supported one. In particular, no bifurcation point will now be observed. The reason for this behaviour is the effect of temperature – due to the thermal boundary conditions the distribution of the temperature is non-uniform from the very beginning of the process thus generating a non-uniform deformation. This implies in turn that there is no fundamental homogeneous state which is clearly a prerequisite for the branching point to exists at all.

The above comment is valid in case we perform the fully coupled thermomechanical analysis of the bar – as discussed in [181], two limiting processes in the form of the purely isothermal and purely adiabatic responses do admit the bifurcation points since their pre-critical behaviour is homogeneous. In fact, the isothermal response provides an upper bound to the thermo-mechanical load–deflection curve whereas the adiabatic response bounds this curve from below.

The three curves corresponding to the different material thermo-mechanical behaviour are displayed in Fig. 6.12 – the isothermal process and the characteristic points M (maximum force) and B (bifurcation) being in full correspondence with the previous analysis. The bifurcation point is also identified on the curve representing the purely adiabatic behaviour. Figure 6.13 illustrates the temperature distribution in time for the adiabatic and coupled thermo-mechanical behaviour at two selected points P and R while Fig. 6.14 demonstrates the temperature distribution along the specimen axis at different time instants for the latter process.

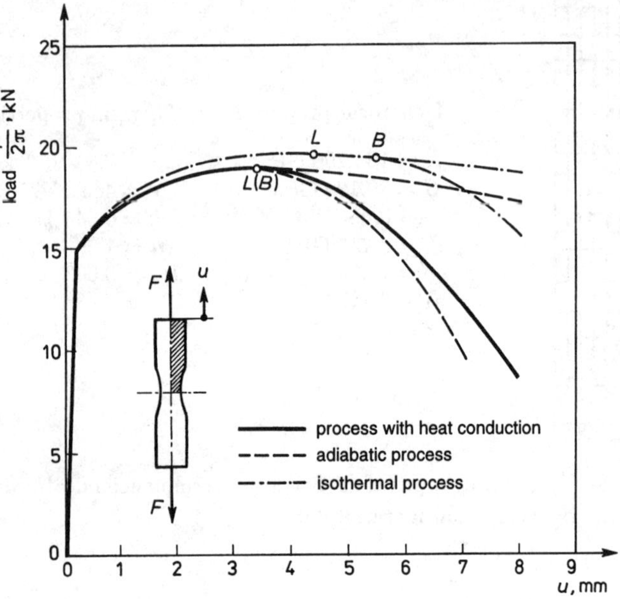

Fig. 6.12. Load–deflection curves for the necking specimen

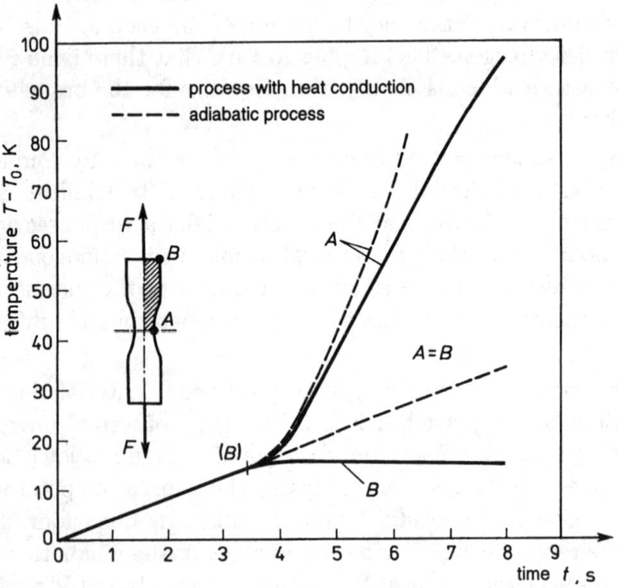

Fig. 6.13. Temperature increase in two points of the necking specimen

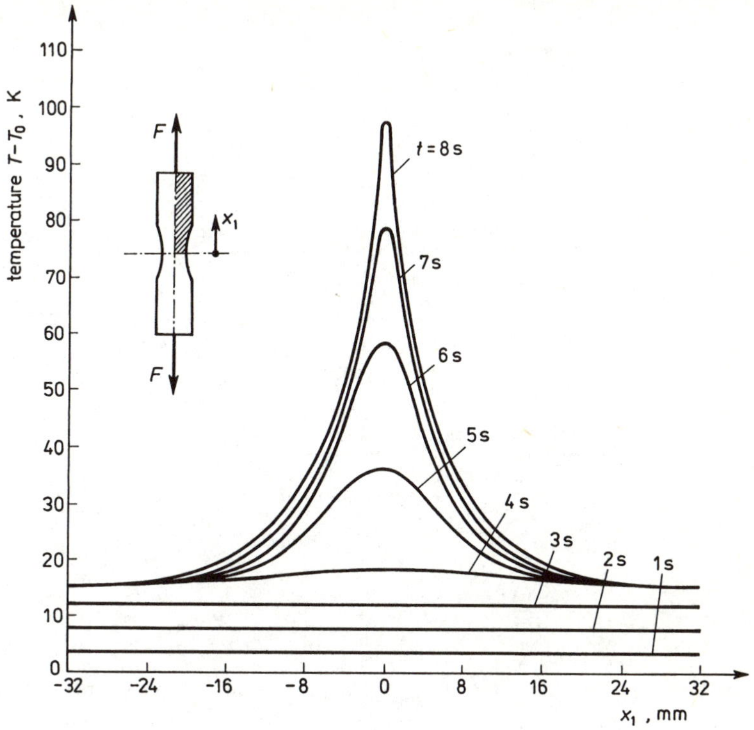

Fig. 6.14. Temperature distribution along the necking specimen

A still more general case was discussed in [88], where an approximate model was used to perform the coupled thermo-mechanical analysis for the specimen modelled as elastic-plastic with voids effects included.

Fig. [illegible] Temperature distribution along the necking specimens

[illegible] nonisothermal case was discussed [illegible], which [illegible] approximate [illegible] model was used to perform the coupled thermo-mechanical analysis for the [illegible] model [illegible] plastic [illegible] [illegible].

7 Stability Problems and Methods of Analysis of Nonlinear FEM Equations

7.1 Remarks on Stability Analysis of Mechanical Systems Equilibrium States

7.1.1 Loads in Stability Analysis of Mechanical Systems

Stability of equilibrium is a special case of stable motion of mechanical systems. The term 'static stability' can be used if inertia forces and kinetic energy are ignored. The definition of stability is very wide and it enables introduction of a number of additional completions, cf. [55]. In what follows only the basic definitions which are associated with stable equilibrium states of structures discretized by means of FEM are mentioned.

Stability is considered with respect to the *parameters of generalized loads* which are the components of vector $\boldsymbol{\lambda}^{\tau}$:

$$\boldsymbol{\lambda}^{\tau}_{M+1} = \{\lambda_1, \ldots, \lambda_M\}^{\tau}, \tag{7.1}$$

where τ is a monotonically increasing parameter, called *conventional time* (quasi-time). In stationary problems (called statics for short) the choice of τ parameter is arbitrary and it usually corresponds to computational aspects. Like in the previous chapters the superscript τ is omitted whenever it introduces any doubt.

In general, the loads can also depend on coordinates of the material point of a deformable body (material coordinates ξ^i should be distinguished from spatial coordinates $x_k(\xi^i)$). The load direction can be spatially fixed or dependent on displacements $u_k(\xi^i)$.

In what follows the load field, corresponding to quasi-time τ and points of a deformable body, is:

$$\mathbf{p}^{\tau} = \mathbf{p}(\boldsymbol{\lambda}^{\tau}, \mathbf{u}^{\tau}, \boldsymbol{\xi}). \tag{7.2}$$

Such a precise description of loads is not needed in many problems of statics and dynamics but it must not be simplified in stability problems. Let us go back to the example of buckling of a perfect column under the action of concentrated load P applied to the free edge Fig. 7.1. In case of spatially fixed load (Euler's problem) the critical load is $P_{\mathrm{cr}} = (\pi^2/4)EI/L^2$. If the follower load is applied (Beck's problem), then the critical value is $P_{\mathrm{cr}} = 20.05EI/L^2$,

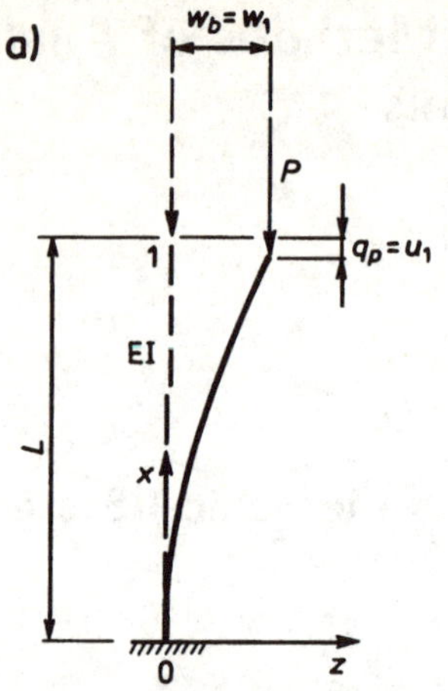

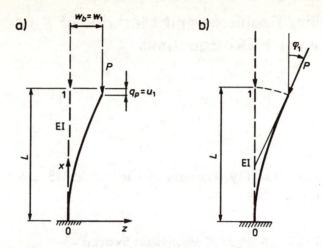

Fig. 7.1. Buckling of perfect column under: **a**) spatially fixed load, **b**) follower load

i.e. the follower buckling load is over eight times higher than in the case of spatially fixed direction of buckling load action, cf. [185].

For stationary problems of FEM the generalized nodal loads are:

$$\mathbf{Q} = \mathbf{Q}(\boldsymbol{\lambda}, \mathbf{q})\,. \tag{7.3}$$

The loads (7.3) are independent of velocities. Such loads can be *potential* (conservative) or *nonpotential* (nonconservative). It depends on the specification of the function (7.3) and the boundary conditions of FE system.

In stability problems the *single parameter load* is usually considered:

$$\mathbf{Q}^{\tau} = \lambda^{\tau}\mathbf{Q}^{*}(\mathbf{q})\,, \tag{7.4a}$$

where: λ^{τ} – load parameter, $\mathbf{Q}^{*}$ – reference load vector. The load (7.4a) is a special case of *multiple parameter loads*:

$$\mathbf{Q}^{\tau}_{N\times 1} = \mathbf{Q}^{*}_{N\times M}\boldsymbol{\lambda}^{\tau}_{M\times 1}\,, \tag{7.4b}$$

where: N – number of degrees of freedom (DOF) of the system, M – number of independent load parameters.

7.1.2 Equilibrium Paths and Buckling of Structures

In order to simplify the following considerations let us assume a single parameter load and displacements measured from the initial equilibrium state, i.e. $\mathbf{q} = \mathbf{0}$ for $\lambda = 0$. The relations $\mathbf{q}$–λ, fulfilling the equilibrium conditions, define the so-called *equilibrium paths in load-configuration space* $\mathbb{R}^{N+1}$ of the

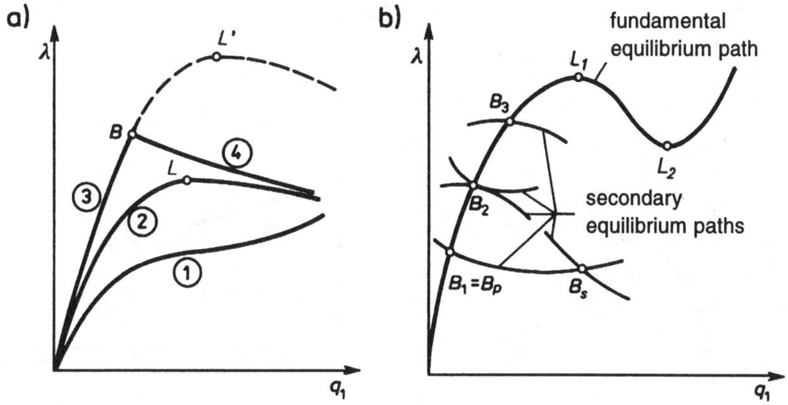

Fig. 7.2. Equilibrium paths on the plane (λ, q_1): **a)** with single primary bifurcation point **b)** with multiple and secondary bifurcation points

position vector:

$$\tilde{\mathbf{q}}_{(N+1)\times 1} = \{\mathbf{q}_{N+1}, \lambda\}\,. \tag{7.5}$$

The equilibrium path that fulfills the initial condition $\mathbf{q}(\lambda = 0) = \mathbf{0}$ is called a *fundamental equilibrium path* (primary equilibrium path).

For practical reasons scalar functions $\lambda(q_\beta)$ are considered, where: q_β – is a selected displacement. The choice of displacement q_β is usually suggested for physical or computational reasons – mainly to have a unique relation corresponding to the function $\lambda(q_\beta)$.

In Fig. 7.2 there are shown three fundamental equilibrium paths for a mechanical system of different geometrical parameters. The path (1) is defined by a unique one-to-one relation $\lambda - q_1$. The *limit point L* corresponds to the maximum value $\lambda_L = \lambda_{\max}$ at the equilibrium path (2). The fundamental path (3) intersects the *secondary equilibrium path* (4) at the *bifurcation point* B before the limit point L is reached.

The points L and B at equilibrium paths are called *critical points*. The limit critical point L is associated with *local* or *global load carrying capacity*. In the case of local limit point the *snap-through* phenomenon takes place and the system jumps to a stable equilibrium state. Large displacements and dynamic effects usually accompany the snap-through of the system. The global load carrying capacity can lead to an unbounded increase of deformation, e.g. instability corresponding to development of a 'neck' in tension specimens.

Bifurcation critical points are associated with the formation of new, qualitatively different equilibrium states, e.g. the straight column in Fig. 7.1 undergoes bending in the postcritical state. This means that after buckling certain constraints become active and the corresponding displacements in-

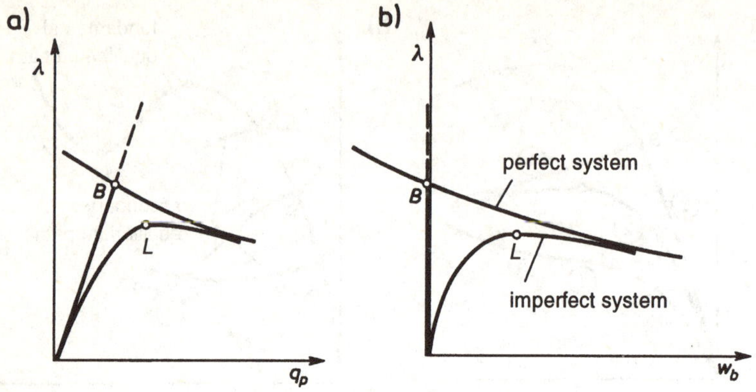

Fig. 7.3. Equilibrium paths for: **a)** prebuckling displacement q_p, **b)** leading buckling displacement w_b

crease significantly. The respective degrees of freedom can be called *buckling DOFs* and they correspond to $\mathbf{q}_b = \{q_{b1}, \ldots, q_{bn}\}$, where $n < N$.

Buckling, associated with bifurcation of equilibrium states, occurs in so-called *perfect structures.* In order to formulate such models additional assumptions have to be adopted. The assumptions concern the geometry of the perfect structure, boundary conditions, material characteristics, distribution and application of loads. Due to these assumptions it is possible to construct a model with nonactive buckling DOFs in the prebuckling state (fundamental equilibrium state). In Fig. 7.3 the equilibrium paths are shown for a linear model in the prebuckling state. The buckling is associated with a leading displacement w_b. It corresponds to a secondary equilibrium path. In the example shown in Fig. 7.1a the prebuckling representative displacement $q_p = u_1$ after buckling is completed with the amplitude of buckling form $w_b = w_1$.

Closer to real structures are systems with *imperfections* in which buckling DOFs (all or only selected components of vector q_b) are active in the prebuckling state. This changes the type of critical points – instead of the bifurcation point B the limit point L occurs in an imperfect structure, cf. Fig. 7.3b.

The analysis of perfect structures can often be simpler than the analysis of imperfect structures. The influence of perturbation of parameters on the behaviour of perfect structures is considered in the frame of *sensitivity analysis.* It is the basis for the evaluation of the influence of defined imperfections on decreasing or increasing the critical loads of perfect structures.

Engineering applications usually need computations of the *first critical points* at the fundamental equilibrium path. These points are also called *primary* critical points (point B_p in Fig. 7.2b) in order to distinguish them from the secondary critical points at postbifurcation, secondary equilibrium

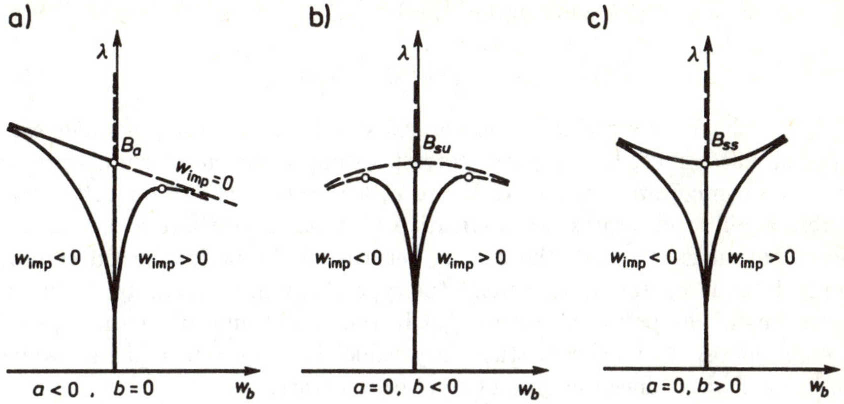

Fig. 7.4. Types of bifurcation points: **a**) asymmetric, **b**) symmetric stable, **c**) symmetric unstable

paths (points B_s in Fig. 7.2b). The bifurcation point is *single* if it is associated with only one secondary path (points B_1 and B_3 in Fig. 7.2b).

Three types of primary, single bifurcation points are shown in Fig. 7.4. The names of bifurcation points: a) *asymmetric*, b) *symmetric–unstable*, c) *symmetric–stable* result from the shape and type of postbifurcation equilibrium paths. The systems of two first bifurcation points B_a and B_su respectively are sensitive to imperfections (to the imperfections corresponding to the buckling form to be more exact) but the structures can react differently to the sign of imperfection amplitude w_imp.

Therefore, the analysis of perfect structures in the vicinity of primary critical points can give essential information about buckling and initial postbuckling behaviour of structures. This aspect of stability analysis was discussed by W. T. Koiter in his fundamental book [90] which originated the *asymptotic analysis* of structural stability. According to this approach the load parameter λ can be calculated at the vicinity of a critical point by means of the power series

$$\lambda/\lambda_\mathrm{b} = 1 + a w_\mathrm{b} + b w_\mathrm{b}^2 + \cdots . \tag{7.6}$$

The case associated with values $a \neq 0$, $b > 0$ corresponds to the asymmetric bifurcation point and for $a = 0$, $b \neq 0$ symmetric bifurcation point occurs. In the case $a < 0$ and imperfection $w_\mathrm{imp} > 0$ the sensitivity of structure can be evaluated by the following simple formula:

$$\lambda_L/\lambda_\mathrm{b} \approx 1 - 2(-\alpha a w_\mathrm{imp})^{1/2} , \tag{7.7}$$

where: α – parameter associated with the 'shape' of the imperfection form. Similarly, for the symmetric–unstable bifurcation point, i.e. for $a = 0$, $b < 0$, the following sensitivity evaluation can be used:

$$\lambda_L/\lambda_b \approx 1 - 3\,(-b/4)^{1/3}\,(\alpha w_{imp})^{2/3}\,. \tag{7.8}$$

In the theory of stability of mechanical systems a definition of *stable equilibrium states* is used. The paths corresponding to the equilibrium states in the load-configuration space $(\mathbf{q}, \lambda)$ are called *stable equilibrium paths.* The stable equilibrium states are insensitive to small perturbation, i.e. after a perturbation is removed, the system returns to the nonperturbed equilibrium state. In conservative systems the type of equilibrium can be evaluated by means of the potential energy J_p. In the equilibrium of the mechanical system energy J_p reaches a stationary value, i.e. $\delta J_p = 0$, and the second variation $\delta^2 J_p$ defines the type of equilibrium state:

$$\delta J_p = 0 \text{ and } \delta^2 J_p = \begin{cases} > 0 \text{ stable equilibrium state} \\ = 0 \text{ transitional (critical) equilibrium} \\ < 0 \text{ unstable equilibrium} \end{cases} \tag{7.9}$$

The above definitions are in agreement with the *Lagrange–Dirichlet theorem:*

Theorem 7.1. *If the potential energy of a mechanical system has a minimum value in the equilibrium state then the equilibrium is stable.*

This is one of the most general theorems of the theory of stability and after additional assumptions concerning the type of minimum (the minimum should be isolated) the Lagrange–Dirichlet theorem is the necessary and sufficient condition for the stable equilibrium of mechanical systems (cf. [47, 58]).

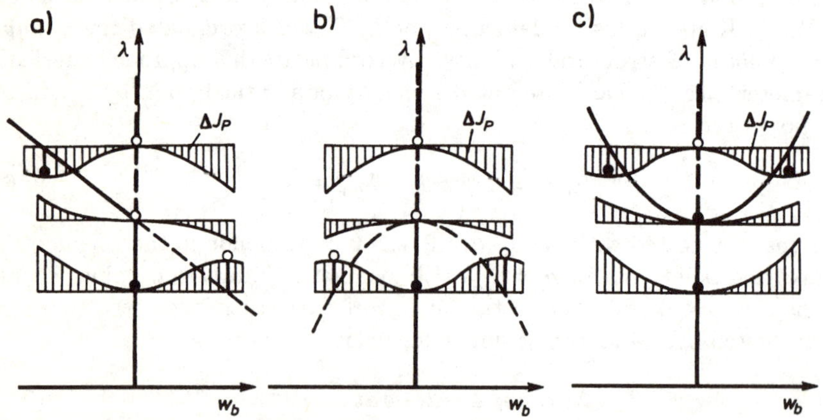

Fig. 7.5. Increments of potential energy ΔJ_p for structures of different primary bifurcation points, fundamental and postbifurcation paths.

The increment of potential energy ΔJ_P, measured from the fundamental equilibrium state enables us to evaluate the stability of equilibrium. In Fig. 7.5, taken from [162], the graphics of function $\Delta J_{\mathrm{p}}(w_{\mathrm{b}})$ are shown. Black points correspond to the stable equilibrium, white points are associated with unstable (local maxima) or critical points. The stable equilibrium paths are shown as continuous lines, unstable paths are drawn as broken lines.

7.1.3
A More General Stability Analysis of Structures

The analysis of structural stability is usually limited to the computation of the first (primary) critical points of perfect mechanical systems. In many cases it is enough to apply *linear stability equations* and analyse nonstandard *eigenvalue problems.* If the fundamental equilibrium state is nonlinear, then the stability equations are linearized with respect to the displacement increments $\Delta\mathbf{q}$.

The analysis of bifurcation states of perfect structures is satisfactory only in cases of stable symmetric critical points since they correspond to safe evaluations from the viewpoint of imperfect structures. Such a situation takes place in columns, framed structures and plates.

But there are many structures (arches, many thin shells) which are sensitive to imperfections. That means that the *nonlinear stability analysis* has to be applied to compute limit critical points.

The critical equilibrium state is usually analysed on the basis of necessary condition (7.9):

$$\delta^2 J_{\mathrm{p}} \equiv \frac{\partial^2 J_{\mathrm{p}}}{\partial q_\alpha \partial q_\beta} \delta q_\alpha \delta q_\beta = 0 \quad \text{for} \quad \alpha, \beta = 1, 2, \ldots, N\,. \tag{7.10}$$

In case of FE approximation we assume that the vector of displacement increments $\Delta\mathbf{q}$ is colinear with the vector of variations $\delta\mathbf{q}$ and the condition (7.10) is written as the zero value of the following quadratic form:

$$\Delta\mathbf{q}^{\mathrm{T}}\mathbf{K}_T\Delta\mathbf{q} = 0\,, \tag{7.11}$$

where: $\mathbf{K}_T$ – tangent stiffness matrix, which is used in the incremental equation (3.46), linearized with respect to $\Delta\mathbf{q}$:

$$\mathbf{K}_T\Delta\mathbf{q} = \Delta\lambda\,\mathbf{Q}^*\,. \tag{7.12}$$

It should be added that the tangent matrix formulated in (3.46) corresponds only to spatially fixed loads. When the loads are dependent on displacements, the matrix $\mathbf{K}_T(\mathbf{q},\lambda)$ has to be formulated more generally. It will be discussed in the following sections. The condition (7.11) implies the so-called *static criterion of instability* which is associated with the singularity of tangent matrix $\mathbf{K}_T$. This criterion is:

- The necessary condition of instability of the conservative mechanical systems is the zero value of the *stability determinant*

$$D \equiv \det \mathbf{K}_T(\mathbf{q}, \lambda) = 0 \,. \tag{7.13}$$

The condition (7.13) is only the necessary condition since there are cases which require discussion of higher variations of potential energy. This condition is of a global character and it cannot give an insight into local forms of instabilities (instead of (7.13) the complete Sylvester theorems should be used).

The condition (7.13) is valid for both bifurcation and limit critical points. In these points additional relations can be formed on the basis of Cramer's formula applied to (7.12):

$$D \Delta \mathbf{q} = \mathbf{d}\, \Delta\lambda \tag{7.14}$$

where the components of vector **d** are:

$$d_\alpha = \frac{\partial D}{\partial K_{\beta\alpha}} Q^*_\beta \quad \text{for} \quad \alpha, \beta = 1, \ldots, N \,. \tag{7.15}$$

The zero value of the stability determinant D implies the zero value of the right-hand side of (7.14). This can occur if:

$$\begin{aligned} &1)\ \mathbf{d} \neq \mathbf{0}, \quad \Delta\lambda = 0 \ \text{– limit point,} \\ &2)\ \mathbf{d} = \mathbf{0}, \quad \Delta\lambda \neq 0 \ \text{– bifurcation point.} \end{aligned} \tag{7.16}$$

The static criterion of instability, associated with the bifurcation of equilibrium, corresponds to *divergence* of stability. This means that the critical states are defined by the equations $\omega_j(\lambda) = 0$, where: ω_j – eigenfrequencies of oscillation motion of the mechanical system with the eigenforms $\delta\mathbf{q}_j \sin \omega_j t$. The first and single bifurcation point is associated with the lowest and single eigenfrequency ω_1 computed at an arbitrary (e.g. unit) mass matrix **M** of the conservative system.

The case is different for nonconservative systems. The loss of stability can occur by *flatter*. According to the kinetic criterion of instability in a narrower sense, (cf. [185]). The flatter takes place if two neighbouring frequencies become equal to each other, i.e. $\omega_j(\lambda) = \omega_{j+1}(\lambda)$. In that case the oscillation is of form $\delta\mathbf{q}_j t \sin \omega_j t$ with infinitely growing amplitudes $\delta\mathbf{q}_j t$.

The *dynamic criteria of instability* mentioned above are more general than the static criterion. On the other hand the dynamic criteria are more difficult for computations. This is a reason why for conservative systems the static criterion is recommended. This criterion can be applied also for an approximate analysis of some nonconservative systems [185].

7.1.4
Instability under Multiple Parameter Loads

So far only single parameter loads have been discussed. In the case of *multiple parameter loads* of the form (7.4a) solutions of the equation :

$$\mathbf{K}_T \Delta \mathbf{q} = \mathbf{Q}^* \Delta \boldsymbol{\lambda} \tag{7.17}$$

give *equilibrium surfaces* in the load-configuration space $\mathbb{R}^{N+M}$ of the position vector $\tilde{\mathbf{q}} = \{\mathbf{q}_{N+1}, \boldsymbol{\lambda}\}$. The primary critical points form the *critical zone* at the fundamental equilibrium surface. The projection of these zones on the load subspace gives the *stability boundary* which bound the domains of stable loads. The stability boundary can also be a boundary of existence of equilibrium if it corresponds to limit points at the fundamental surface.

As an illustration the equilibrium surfaces and equilibrium paths in the space $(q_1, \lambda_1, \lambda_2)$ are shown in Fig. 7.6. The critical zone can be formed either as a geometric locus of limit points at the fundamental equilibrium space Fig. 7.6a or as an interaction line of the fundamental and secondary equilibrium surfaces Fig. 7.6b. As a result of projection of a smooth critical zone on the load plane (λ_1, λ_2) a singular point can occur at the stability boundary. A classification of points at the critical zones and critical surfaces was formulated in [63].

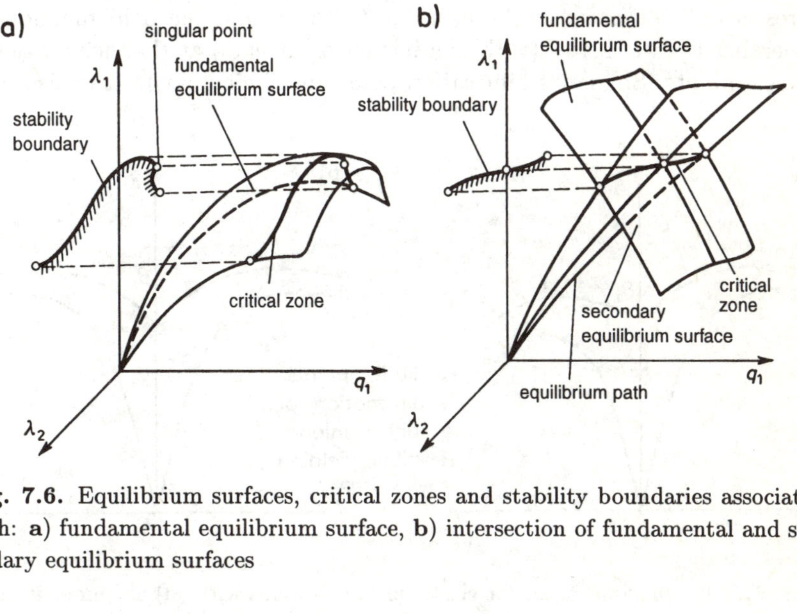

Fig. 7.6. Equilibrium surfaces, critical zones and stability boundaries associated with: **a)** fundamental equilibrium surface, **b)** intersection of fundamental and secondary equilibrium surfaces

7.1.5
Instability of Elastic-Plastic Systems

The analysis of elastic-plastic mechanical systems is much more difficult than the analysis of elastic systems. Papers by Engesser, Jasinsky and von Kármán now historical (cf. [131]), tried to extend the Euler approach to the analysis of elastic-plastic columns. The crucial paper by Shanley [156] introduced a new approach of increasing load at the bifurcation of equilibrium. The basic papers by Hill confirmed this approach in two- and three-dimensional stress states. Those papers set off the role of the development of passive processes (local unloading) in stability of elastic-plastic solids. An extensive review of papers on those problems can be found in [30, 65, 155].

The asymptotic analysis of stability of elastic-plastic structures was developed mainly by Hutchinson, Needleman and Tvergaard. In Fig. 7.7 there are shown equilibrium paths corresponding to asymmetric or symmetric bifurcation points. In [65] the following evaluation was given for the branch $w_\mathrm{b} \geq 0$:

$$\lambda/\lambda_B = 1 + a_1 w_\mathrm{b} + a_2 w_\mathrm{b}^{1+\beta} + \cdots , \tag{7.18}$$

where the parameters $a_1 > 0$, $a_2 < 0$ are used and parameter $0 < \beta < 1$ depends on the distribution of the local unloading zone. The formula (7.18) corresponds to (7.6) and it enables us to produce the most important phenomena associated with postbifurcation elastic-plastic behaviour, i.e.: i) initial increase of load, ii) reaching of limit state, iii) influence of local unloading.

The asymptotic analysis of postcritical behaviour of elastic-plastic structures is difficult because of the necessity to expand the field functions in generalized power series (with non-integer exponents) at the vicinity of two critical points (Shanley's bifurcation point and limit point). It is also most

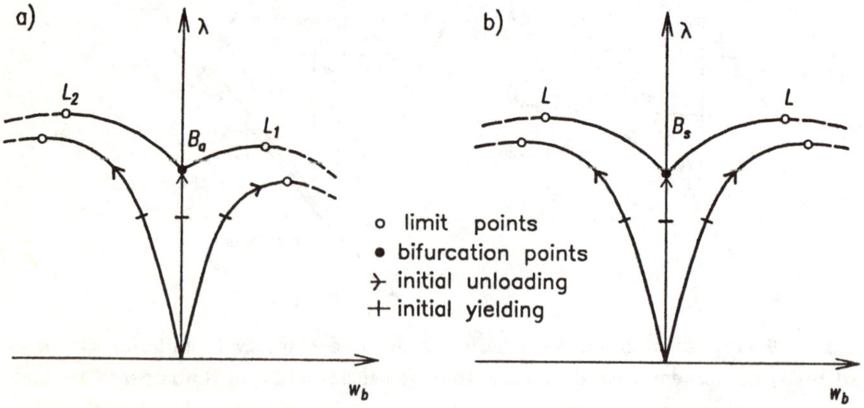

Fig. 7.7. Equilibrium paths for elastic-plastic systems with: **a**) asymmetric bifurcation point, **b**) symmetric bifurcation point

difficult to take into consideration the development of local unloading zones. A substantial assumption is to neglect the unloading zones and consider nonlinear elastic material, e.g. hypoelastic material. However, such an assumption can lead to significant errors (cf. the sensitivity analysis of a structure to imperfections in [165]).

The problems discussed above are reflected in the shapes of equilibrium paths and a possibility of fulfilling the static criterion of stability (7.13). The transition from elastic to plastic deformations and the beginning of local unloadings can cause a jump-type change of structural stiffness. This can mean that despite decreasing the conventional time step $\Delta\tau$, in the vicinity of $\tau_{\rm cr}$ the stability determinant can only change its sign, i.e. $D(\tau_{\rm cr}) \cdot D(\tau_{\rm cr} + \Delta\tau) < 0$ for $\Delta\tau \to 0$.

The problems mentioned above and those associated with the stability analysis of elastic-plastic systems have not been fully explained. This is why the elastic-plastic analysis should be carried out with great care and results ought to be treated with stronger criticism than in case of elastic analysis.

7.1.6 Further Remarks

In what follows the FEM equations for the stability analysis of mechanical systems are formulated. The methods of analysis of nonlinear algebraic equations are discussed. Moreover, attention is focused on the computation of limit points. Extensive literature is devoted to the stability analysis of mechanical systems, cf. lists of books and review papers in [52, 185]. The problems discussed in this chapter are more extensively discussed in books [26, 63, 163], state-of-the-art papers [28, 30, 65] and proceedings of advanced courses, e.g. [166, 168]. The problems of FEM applications to the stability of structures are also considered in [170].

7.2 FEM Equations for Structural Stability

The stability equations are formulated for the analysis of critical loads (load parameters) and forms of critical equilibrium. The primary critical points are at fundamental equilibrium paths (surfaces) which can be nonlinear in general. That is why *nonlinear FE equations*, usually in an *incremental form* have to be used. Such equations were formulated in Chap. 3. In what follows those equations are generalized on cases of loads dependent on displacements.

The analysis of perfect structures can give satisfactory evaluation under the assumption of linear behaviour up to when the bifurcation point is reached. The analysis is then performed by means of *linear equations* of *initial* or *linearized buckling equations.*

Some structures (e.g. many types of shells) need a nonlinear prebuckling analysis. The bifurcation points at the fundamental equilibrium paths are computed by means of equations of *adjacent equilibrium states adjacent.*

In both cases mentioned above the computation for bifurcation states is carried out applying the *analysis of eigenproblems*, by means of initial or linearized buckling equations. Those equations are formulated on the basis of general, nonlinear FE incremental equations.

The critical points can be easily completed if the FE equations are completed with additional equations (constraint equations, equations of critical equilibrium). The set of *extended equations* is especially useful for the computation of limit critical points. In the case of two parameter loads the extended equations are applied to the tracing of stability boundary.

7.2.1 Generalisation of FE Incremental Equations

The nonlinear EF equations were formulated in Chap. 3 on the basis of stationarity of the potential energy of a mechanical system and under the assumption of spatially fixed action of loads $\hat{\mathbf{f}}$ and $\hat{\mathbf{t}}$. In a more general case of displacement dependent loads the system can be nonpotential (nonconservative). That is why the *principle of virtual work* is applied in what follows. The additional terms to FE equations given in Chap. 3 are found from the virtual work of loads.

The principle of virtual work can be formulated with respect to increments of strains and displacements:

$$\int_{\Omega} (\tilde{\boldsymbol{\sigma}} + \Delta\tilde{\boldsymbol{\sigma}})^{\mathrm{T}} \Delta\varepsilon \, \mathrm{d}\Omega = \int_{\Omega} \rho(\hat{\mathbf{f}} + \Delta\hat{\mathbf{f}})\delta\Delta\mathbf{u} \, \mathrm{d}\Omega + \int_{\partial\Omega^{\sigma}} (\hat{\mathbf{t}} + \Delta\hat{\mathbf{t}})\delta\Delta\mathbf{u} \, \mathrm{d}(\partial\Omega) + \int_{\partial\Omega^{\sigma}} (\mathbf{p} + \Delta\mathbf{p})\delta\Delta\mathbf{u} \, \mathrm{d}(\partial\Omega) \,, \qquad (7.19)$$

where the appropriate integrals correspond to the virtual work of internal forces and external loads:

$$\delta_{\mathrm{r}} W_{\mathrm{int}}^{t+\Delta t} = \delta_{\mathrm{r}} W_{(f)}^{t+\Delta t} + \delta_{\mathrm{r}} W_{(t)}^{t+\Delta t} + \delta_{\mathrm{r}} W_{(p)}^{t+\Delta t} \qquad (7.19\mathrm{a})$$

In (7.19) and (7.20) the following vectors of loads are considered: $\rho\hat{\mathbf{f}}$ – mass-type forces, $\hat{\mathbf{t}}$ conservative surface loads, $\mathbf{p}$ – displacement dependent surface loads. The integrals in (7.19) are written using the abbreviated notation (cf. Chap. 3):

$$\tilde{\boldsymbol{\sigma}} + \Delta\tilde{\boldsymbol{\sigma}} \equiv {}_{\mathrm{r}}\tilde{\boldsymbol{\sigma}}^{\tau} + \Delta\,{}_{\mathrm{r}}\tilde{\boldsymbol{\sigma}}^{\tau}, \quad \Omega \equiv \Omega^{\mathrm{r}} \quad \text{etc.} \qquad (7.20)$$

All the quantities used in (7.19) are, of course, consistent with respect to the reference configuration C^{r}.

Let us consider more closely the case of surface load $\mathbf{p}$ acting perpendicularly to the deformed body surface. This load is called a *follower load.* It is assumed that a single parameter load is distributed according to the function

${}_0f = f(\varphi^i)$ associated with the initial configuration C^0:

$$\mathbf{p}^\tau = \lambda^{(\tau)}\, {}_0f\mathbf{n}^\tau \tag{7.21}$$

The load is applied to a part of surface $S^{(0)} \subset \partial_0\Omega^\sigma \to S^{(\tau)}$ with the external normal vector $\mathbf{n}^\tau(\varphi^i)$, where φ^i for $i = 1, 2$ are material coordinates of the surface $S^{(\tau)}$.

The virtual work of the follower load at the time $\tau = t + \Delta t$ is:

$$\delta W_{(p)}^{t+\Delta t} = \lambda^{t+\Delta t} \int_{S^{(t+\Delta t)}} {}_0f(\mathbf{n}^{t+\Delta t})^{\mathrm{T}}\delta\Delta\mathbf{u}\,\mathrm{d}S^{(t+\Delta t)}\,. \tag{7.22}$$

The components of vector $\mathbf{n}\,\mathrm{d}S$ are written according to the formula

$$n_i^{(t+\Delta t)}\,\mathrm{d}S^{(t+\Delta t)} = e_{ijk}\frac{\partial x_j^{(t+\Delta t)}}{\partial\xi^1}\frac{\partial x_k}{\partial\xi^2}\,\mathrm{d}\xi^1\,\mathrm{d}\xi^2\,, \tag{7.23}$$

where e_{ijk} is Ricci's permutation tensor.

Similarly as in (7.20), the abbreviated notation is used in what follows, i.e.

$$x_j^{(t+\Delta t)} = x_j^{(t)} + {}^t u_j^{(t+\Delta t)} \equiv x_j + \Delta u_j\,. \tag{7.24}$$

On the basis of (7.23) and (7.24) the virtual work (7.22) can be expressed by increments of the displacement vector $\Delta\mathbf{u}$. Conserving linear terms Δu_j, the virtual work $\delta W_L = \delta W_{(p)}^{(t+\Delta t)}$ is:

$$\delta W_L = (\lambda + \Delta\lambda)e_{ijk}\int_{\xi_1}\int_{\xi_2} {}_0f\frac{\partial x_j}{\partial\xi^1}\frac{\partial x_k}{\partial\xi^2}\delta\Delta u_i\,\mathrm{d}\xi^1\,\mathrm{d}\xi^2$$

$$+(\lambda + \Delta\lambda)e_{ijk}\int_{\xi_1}\int_{\xi_2} {}_0f\left(\frac{\partial\Delta u_j}{\partial\xi^1}\frac{\partial x_k}{\partial\xi^2} + \frac{\partial x_j}{\partial\xi^1}\frac{\partial u_k}{\partial\xi^2}\right)\delta u_i\,\mathrm{d}\xi^1\,\mathrm{d}\xi^2\,. \tag{7.25}$$

The displacement increments are approximated by means of shape functions (3.11):

$$\Delta u_j = \phi_{j\alpha}(\mathbf{x}^0)\Delta q_\alpha \quad \text{for } \alpha = 1, 2, \ldots, N\,, \tag{7.26}$$

which are associated with the finite element of domain Ω_e and boundary $\partial\Omega_e$.

The transformation from local to global coordinates and aggregation of finite elements were described in Chap. 2. That is why those procedures are omitted in what follows and all variables correspond to the whole domain of the finite element system $\Omega = \bigcup_e \Omega_e$.

After substitution (7.26) to (7.25) the following *vector of nodal forces* of follower surface loads is formulated:

$$Q_\alpha^{(p)} + \Delta Q_\alpha^{(p)} = (\lambda + \Delta\lambda)e_{ijk}\iint_{S(\varphi)} {}_0f\phi_{i\alpha}\frac{\partial x_j}{\partial\xi^1}\frac{\partial x_k}{\partial\xi^2}\,\mathrm{d}\xi^1\,\mathrm{d}\xi^2\,. \tag{7.27}$$

The second term of (7.25) gives the *load stiffness matrix*:

$$K_{\alpha\beta}^{(p)} = -(\lambda + \Delta\lambda)e_{ijk}\iint_{S(\xi)} {}_0f\phi_{i\alpha}\left(\frac{\partial\phi_{i\beta}}{\partial\xi_1}\frac{\partial x_k}{\partial\xi^2} + \frac{\partial\phi_{k\beta}}{\partial\xi^2}\right)\mathrm{d}\xi^1\,\mathrm{d}\xi^2\,. \tag{7.28}$$

The sign minus in (7.28) means that the forms $K^{(p)}_{\alpha\beta}\Delta q_\beta$ are transmitted to the left-hand side of the incremental equilibrium equations:

$$(K^{(\text{con})}_{\alpha\beta} + K^{(\sigma)}_{\alpha\beta} + K^{(u)}_{\alpha\beta} + K^{(p)}_{\alpha\beta})\Delta q_\beta = \Delta Q_\alpha - Q_\alpha - F_\alpha$$

$$\text{for } \alpha, \beta = 1, 2, \ldots, N\,. \tag{7.29}$$

On the right-hand side of (7.29) there are the nodal forces from loads

$$Q_\alpha = Q^{(f)}_\alpha + Q^{(t)}_\alpha + Q^{(p)}_\alpha \tag{7.30}$$

as well as the vector of internal forces

$$F_\alpha = \int_\Omega \tilde{\sigma}_{kl}(B'_{kl\alpha} + B''_{kl\alpha})\,\mathrm{d}\Omega\,, \tag{7.31}$$

where the matrices $B'_{kl\alpha}$ and $B''_{kl\alpha}$ are formulated in (3.49).

In the equilibrium state the residual forces

$$R_\alpha = Q_\alpha - F_\alpha \tag{7.32}$$

vanish, cf. (3.25). They serve to lead down the computed path to the equilibrium path solutions of incremental FE equations (7.29) in course of an iteration procedure.

In the case of stationary systems the principle of virtual work (PVW) is equivalent to the stationarity condition of potential energy. More details about the application of PVW to the formulation of FE incremental equations for spatially fixed loads can be found in many papers, cf. [17, 16] and for follower loads in [61].

Let us consider a special case of a uniform surface load, i.e. ${}_0f$ = const and $\lambda\,{}_0f = p$. In this case the second term in (7.25) can be transformed by means of Green's formula to the following form:

$$\begin{aligned}\delta W^{II}_L = \frac{1}{2}(p+\Delta p)e_{ijk}\Bigg\{&\int_{\Gamma(\varphi)}\left(\frac{\partial x_k}{\partial\xi^1}\nu_1 + \frac{\partial x_k}{\partial\xi_2}\nu_2\right)\Delta u_j\delta u_i\,\mathrm{d}\Gamma\\ &+\iint_{S(\xi)}\left[\frac{\partial x_k}{\partial\xi^2}\left(\frac{\partial\Delta u_i}{\partial\xi_1}\delta\Delta u_i + \frac{\partial\delta u_j}{\partial\xi^1}\Delta u_j\right)\right.\\ &\left.+\frac{\partial x_j}{\partial\xi^1}\left(\frac{\partial\Delta u_k}{\partial\xi_2}\delta\Delta u_i + \frac{\partial\delta u_k}{\partial\xi^2}\Delta u_i\right)\right]\mathrm{d}\xi^1\,\mathrm{d}\xi^2\Bigg\}\,,\end{aligned} \tag{7.33}$$

where: ν_i – components of the vector normal to the boundary. Substitution of approximation (7.26) to (7.33) leads to the load stiffness matrix $K^{(p)}_{\alpha\beta}$. The matrix $K^{(p)}_{\alpha\beta}$ is composed of two parts. The part associated with integration over the surface $S(\xi)$ is symmetric and the part corresponding to integration along $\Gamma(\xi)$ is antisymmetric because of factor $e_{ijk}\Delta u_j\delta u_i$ in (7.33).

The above result is illustrated on the case of thin shells under the assumption of small strains and moderate rotations. Instead of transformation of displacements to the surface coordinates associated with vectors (${}_0\mathbf{e}_i$, ${}_0\mathbf{n}$), let us write the formula (7.23) in the following form:

$$\mathbf{n}\,\mathrm{d}S = (\varphi_i\,{}_0\mathbf{e}^i + {}_0\mathbf{n})(1 + \varepsilon_k^k)\,\mathrm{d}\xi^1\,\mathrm{d}\xi^2\,, \tag{7.34}$$

where the left-hand subscript 0 means that the vectors ${}_0\mathbf{e}^i$ and ${}_0\mathbf{n}$ are associated with the initial configuration C^0. All the variables without the subscript 0 are associated with the current configuration C^τ. The components of the vector of small rotations and tensor of small strain are:

$$\begin{aligned} \varphi_i &= -{}_0b_{ij}u^j - w_{,i}\,, \\ \varepsilon_{ij} &= \tfrac{1}{2}(u_{i|j} + u_{j|i}) - {}_0b_{ij}w\,, \end{aligned} \tag{7.35}$$

where: ${}_0b_{ij}$ – tensor of curvature changes, $()_{i|j}$ – covariant differentials, $i, j, k = 1, 2, 3$ – indices for surface variables. The variation of displacement vector

$$\delta\mathbf{u} = \delta u^i{}_0\mathbf{e}_i + \delta w\;{}_0\mathbf{n} \tag{7.36}$$

is used in the following scalar function:

$$\begin{aligned} \mathbf{n}^{\mathrm{T}}\delta\mathbf{u}\,\mathrm{d}S &= \left[\varphi_i(1+\varepsilon_k^k)\delta u^i + (1+\varepsilon_k^k)\delta w\right]\,\mathrm{d}\xi^1\,\mathrm{d}\xi^2 \\ &\approx \left[-({}_0b_{ij}u^j + w_{,i})\delta u^i + (1 + u^k|_k - {}_0b_k^k w\right]\,\mathrm{d}\xi^1\,\mathrm{d}\xi^2\,. \end{aligned} \tag{7.37}$$

In the approximate relation (7.37) there are conserved only the terms linear with respect to displacements u_i, w and *gradient* $w_{,i}$.

After Green's formula is applied

$$\iint_{S(\varphi)} w_{,i}\delta u^i\,\mathrm{d}\xi^1\,\mathrm{d}\xi^2 = -\iint_{S(\varphi)} w\delta u^k|_k\,\mathrm{d}\xi^1\,\mathrm{d}\xi^2 + \int_{\Gamma(\varphi)} w\delta u^k\nu_k\,\mathrm{d}\Gamma\,, \tag{7.38}$$

and (7.37) is used in PVW the virtual work δW_L can be expressed as:

$$\delta W_L = -\delta\Pi_{\mathrm{p}} + \delta W_\Gamma\,, \tag{7.39}$$

where $\delta\Pi_{\mathrm{p}}$ is the variation of load functional

$$\delta\Pi_{\mathrm{p}} = -p\iint_{S(\xi)} \left\{(1 -{}_0\,b_k^k w)\delta w - \left[{}_0b_{ij}u^j\delta u^i - (w\delta u^k|_k + u^k|_k\delta w)\right]\right\}\,\mathrm{d}\xi^1\mathrm{d}\xi^2 \tag{7.40}$$

and δW_Γ is the virtual work of boundary loads

$$\delta W_\Gamma = -p\int_\Gamma w\delta u^k\nu_k\,\mathrm{d}\Gamma\,. \tag{7.41}$$

The above manipulations confirm the conclusion corresponding to (7.33). The symmetric part of the load stiffness matrix is

$$K_{\alpha\beta}^{(p)S} = \frac{\partial\Pi_{\mathrm{p}}}{\partial q_\alpha\partial q_\beta}\,, \tag{7.42}$$

where Π_p is the *load potential*

$$\Pi_\mathrm{p} = -\iint_{S(\xi)} \left[w(1 + u^k|_k) - \tfrac{1}{2}({}_0b^k_k w^2 + {}_0b_{ij}u^i u^j)\right] \mathrm{d}\xi^1\, \mathrm{d}\xi^2 \,, \tag{7.43}$$

The asymmetric part corresponds to (7.41). The structure of the asymmetric part is clear with respect to the incremental vector $\Delta q = \{\Delta \mathbf{q}_u, \Delta \mathbf{q}_w\}$:

$$\mathbf{K}^{(p)\Gamma} = \begin{bmatrix} \mathbf{0} & \mathbf{K}_\Gamma \\ \mathbf{0} & \mathbf{0} \end{bmatrix} , \tag{7.44}$$

where:

$$\mathbf{K}_\Gamma = -p \int_\Gamma (\phi_u \nu)^\mathrm{T} \phi_{,w}\, \mathrm{d}\Gamma \,. \tag{7.45}$$

The integral (7.45) vanishes if one of the following conditions takes place: 1) there is no boundary load, i.e. $p_\Gamma = 0$; 2) normal displacement vanishes, i.e. $w_\Gamma = 0$; 3) there is zero value of in-midsurface displacements $u^k_\Gamma \nu_k = 0$. That means that the uniform follower load $p_\Gamma \neq 0$ is of conservative type if the boundary has special kinematics boundary conditions (simple supporting or there are constraints against movement in midsurface). In that case the load stiffness matrix is symmetric $\mathbf{K}^\mathrm{p} \equiv \mathbf{K}^{(p)S}$ since $\mathbf{K}_\Gamma = \mathbf{0}$.

General considerations on the existence of the load potential for follower-type loads were given in [29]. The paper [153], written from the FEM point of view, discussed special cases of conservative and nonconservative loads.

7.2.2 Eigenproblems in the Buckling Analysis

Critical points can be analysed by means of appropriately formulated eigenequations, as mentioned in Sect. 7.1 and at the beginning of Sect. 7.2. Let us consider a conservative system in equilibrium state, i.e. Eqs. (7.12) are satisfied. According to the static criterion (7.13) the matrix $\mathbf{K}_T$ is singular in the critical state. A more convenient analysis can be done if matrix $\mathbf{K}_T$ is transformed to the diagonal form $\mathbf{K}_\kappa$, corresponding to the standard eigenproblem:

$$(\mathbf{K}_\kappa - \kappa_\alpha \mathbf{I})\mathbf{W}_\alpha = \mathbf{0} \,. \tag{7.46}$$

In conservative systems the stiffness matrix is symmetric, i.e. $\mathbf{K}_T = \mathbf{K}_T^\mathrm{T}$. This leads to real eigenvalues which can be arranged according to the increasing eigenvalues

$$\kappa_1 \leq \kappa_2 \cdots \leq \kappa_N \,. \tag{7.47}$$

The diagonal matrix $\mathbf{K}_\kappa$ and the matrix of eigenvectors $\mathbf{W}$, corresponding to (7.46) are:

$$\mathbf{K}_\kappa = \begin{bmatrix} \kappa_1 & 0 & \dots & 0 \\ 0 & \kappa_2 & \dots & 0 \\ \dots & \dots & \dots & \dots \\ 0 & 0 & \dots & \kappa_N \end{bmatrix}, \quad \mathbf{W} = [\mathbf{w_1} \cdots \mathbf{w_N}] . \tag{7.48}$$

The equilibrium equation (7.12) can be transformed to the space of eigenvectors $\mathbf{w}_\alpha$:

$$\mathbf{K}_\kappa \Delta \mathbf{q}_\kappa = \Delta\lambda\, \mathbf{Q}^*_\kappa , \tag{7.49}$$

where the following relations occur:

$$\Delta\mathbf{q} = \mathbf{W}\Delta\mathbf{q}_\kappa, \quad \mathbf{Q}^*_\kappa = \mathbf{W}^{\mathrm{T}}\mathbf{Q}, \quad \mathbf{K}_\kappa = \mathbf{W}^{\mathrm{T}}\mathbf{K}_T\mathbf{W} , \tag{7.50}$$

and an evident relation for the determinant D:

$$D \equiv \det \mathbf{K}_\kappa = \kappa_1\kappa_2 \cdots \kappa_N . \tag{7.51}$$

During increase of the load parameter along the fundamental path the transition $\lambda \to \lambda_{\mathrm{cr}}$ is associated with $\kappa_1 \to 0$ and in case of singular primary critical point the following inequalities are valid:

$$0 < \kappa_2 \le \kappa_3 \le \cdots \le \kappa_N . \tag{7.52}$$

In this case the first equation from (7.41), corresponding to $\kappa_1 = 0$, takes the form:

$$\Delta\lambda\, \mathbf{w}_1^{\mathrm{T}}\mathbf{Q}^* = 0 . \tag{7.53}$$

Equation (7.53) is satisfied for two types of critical points:

$$\begin{aligned} &1) \quad \Delta\lambda = 0, \quad \mathbf{w}_1^{\mathrm{T}}\mathbf{Q}^* \neq 0 \quad \text{– limit point,} \\ &2) \quad \Delta\lambda \neq 0, \quad \mathbf{w}_1^{\mathrm{T}}\mathbf{Q}^* = 0 \quad \text{– bifurcation point.} \end{aligned} \tag{7.54}$$

In the case of bifurcation of equilibrium the primary eigenvector $\mathbf{w}_1^{\mathrm{T}}$ is orthogonal to the vector of reference load $\mathbf{Q}^*$. This is also valid for subsequent bifurcation points (also multiple bifurcation points) since vanishing eigenvalues $\kappa_\beta = 0$ imply $\mathbf{w}_\beta^{\mathrm{T}}\mathbf{Q}^* = 0$, according to (7.49).

In the analysis of structural stability instead of (7.46) other eigenproblems are usually formulated. Let us assume that for increment of load parameter $\Delta\lambda$ two different solutions $\Delta\mathbf{q}_1$ and $\Delta\mathbf{q}_2$ correspond which fulfil the following equilibrium equations:

$$\mathbf{K}_T\Delta\mathbf{q}_1 = \Delta\lambda\, \mathbf{Q}^*, \quad \mathbf{K}_T\Delta\mathbf{q}_2 = \Delta\lambda\, \mathbf{Q}^* . \tag{7.55}$$

The subtraction of both sides of (7.55) leads to the eigenproblem equation in the following form:

$$\mathbf{K}_T\mathbf{v} = \mathbf{0} , \tag{7.56}$$

where: $\mathbf{v} = \mathbf{q}_2 - \mathbf{q}_1$. Non-zero solution $\mathbf{v} \neq \mathbf{0}$ occurs if the static condition of instability (7.13) is satisfied, i.e. $\det \mathbf{K}_T = 0$.

The components of stiffness matrix $\mathbf{K}_T$ are as follows:

$$\mathbf{K}_T = \mathbf{K}^{\text{con}} + \mathbf{K}^{\sigma} + \mathbf{K}^{(u1)}(\mathbf{u}) + \mathbf{K}^{(u2)}(\mathbf{u},\mathbf{u}) + \mathbf{K}^{(p)}(\lambda,\mathbf{u})\,, \tag{7.57}$$

where the matrix of initial displacements $\mathbf{K}^{(u)}$ is split into two parts. The matrices $\mathbf{K}^{(u1)}$ and $\mathbf{K}^{(u2)}$ depend linearly and quadratically on the displacement vector $\mathbf{u}$ respectively, cf. $(3.44)_3$.

Let us assume that the critical state occurs within the range $[t, t+\Delta t]$ and is defined by the factor $\mu \in (0,1)$ in the eigenproblem:

$$\mathbf{K}_T(\boldsymbol{\sigma} + \mu\Delta\boldsymbol{\sigma}, \mathbf{u} + \mu\Delta\mathbf{u}, \boldsymbol{\lambda} + \mu\Delta\boldsymbol{\lambda})\mathbf{v} = \mathbf{0}\,. \tag{7.58}$$

The stiffness matrix $\mathbf{K}_T$ can be expanded in Taylor series regarding μ as a small parameter which decreases $\mu \to 0$ if $t \to t_{\text{cr}}$. If the partial derivatives of $\mathbf{K}_T$ are calculated at $\mu = 0$:

$$\mathbf{K}_{T,r} \equiv \frac{\partial^r}{\partial \mu^r}\mathbf{K}_T(\mathbf{q} + \mu\Delta\mathbf{q})|_{\mu=0}\,,$$

the following expansion is obtained:

$$\mathbf{K}_T(\mathbf{q} + \mu\Delta\mathbf{q}) = \mathbf{K}_T(\mathbf{q}) + \mathbf{K}_T(\Delta\mathbf{q})\mu + \mathbf{K}_{T,2}\frac{\mu^2}{2} + \cdots\,. \tag{7.59}$$

In this way (7.56) is transformed. If only the first terms are conserved, then the following equation is formulated:

$$\begin{aligned}&\Big\{\Big[\mathbf{K}^{\text{con}} + \mathbf{K}^{\sigma}(\boldsymbol{\sigma}) + \mathbf{K}^{u}(\mathbf{u},\mathbf{u}) + \mathbf{K}^{(p)}(\lambda,\mathbf{u})\Big]\\ &+\mu\Big[\mathbf{K}^{(\sigma)}(\Delta\boldsymbol{\sigma}) + \mathbf{K}^{(u1)}(\Delta\mathbf{u}) + \big(\mathbf{K}^{(u2)}(\Delta\mathbf{u},\mathbf{u}) + \mathbf{K}^{(u2)}(\mathbf{u},\Delta\mathbf{u})\big)\\ &+\big(\mathbf{K}^{(p)}(\Delta\lambda,\mathbf{u}) + \mathbf{K}^{(p)}(\lambda,\Delta\mathbf{u})\big)\Big]\\ &+\mu^2\Big[\mathbf{K}^{(u2)}(\Delta\mathbf{u},\Delta\mathbf{u}) + \mathbf{K}^{(p)}(\Delta\lambda,\Delta\boldsymbol{\sigma})\Big]\Big\}\mathbf{v} = \mathbf{0}\,.\end{aligned} \tag{7.60}$$

Equation (7.60) corresponds to the *quadratic eigenvalue problem*. The analysis of this problem is complex because it is necessary to leave the frame of linear algebra and a greater number of matrices is needed. That is why linear equations are usually formulated. This can be done if the increment of stiffness matrix is calculated:

$$\Delta\mathbf{K}_T = \mathbf{K}_T(\mathbf{q} + \Delta\mathbf{q}) - \mathbf{K}_T(\Delta\mathbf{q}) \tag{7.61}$$

and an approximate *linear eigenproblem* equation is in the form:

$$[\mathbf{K}_T(\mathbf{q}) + \mu\Delta\mathbf{K}_T]\,\mathbf{v} = \mathbf{0}\,. \tag{7.62}$$

Another possibility is offered by writing (7.56) in the form:

$$\left[\mathbf{K}^{(\mathrm{con})} + \mu(\mathbf{K}^{(\sigma)} + \mathbf{K}^{(u)} + \mathbf{K}^{(p)})\right] \mathbf{v} = \mathbf{0}\,, \tag{7.63}$$

which corresponds to the following computational procedure: if $t \to t_{\mathrm{cr}}$ then $\mu \to 1$.

In the analysis of engineering structures the load parameter λ is usually used as the eigenvalue. The linear reference displacement $\mathbf{q}^* = \mathbf{q}(\mathbf{Q}^*)$ is accepted for the fundamental equilibrium path. This leads to relations:

$$\sigma = \lambda\,\sigma^*, \quad \mathbf{u} = \lambda\,\mathbf{u}^*, \quad p = \lambda\,p^*\,, \tag{7.64}$$

which enable us to formulate the following nonlinear equation:

$$\begin{aligned}&\{\mathbf{K}^{(\mathrm{con})} + \lambda[\mathbf{K}^{(\sigma)}(\boldsymbol{\sigma}^*) + \mathbf{K}^{(u1)}(\mathbf{u}^*) + \mathbf{K}^{(p1)}(\mathbf{p}^*)]\\ &+\lambda^2[\mathbf{K}^{(u2)}(\mathbf{u}^*,\mathbf{u}^*) + \mathbf{K}^{(p2)}(p^*,\mathbf{u}^*)]\}\mathbf{v} = \mathbf{0}\,,\end{aligned} \tag{7.65}$$

where the matrix $\mathbf{K}^{(\sigma)}(\boldsymbol{\sigma}^*)$ is independent of $\mathbf{u}^*$ and λ, and matrix $\mathbf{K}^{(u1)}$ is linearly dependent on displacement vector $\mathbf{u}^*$.

Neglecting of the quadratic term in (7.65) leads to the *linearized buckling equation*

$$\{\mathbf{K}^{(\mathrm{con})} + \lambda[\mathbf{K}^{\sigma}(\boldsymbol{\sigma}^*) + \mathbf{K}^{(u1)}(\mathbf{u}^*) + \mathbf{K}^{(p1)}]\}\mathbf{v} = \mathbf{0}\,. \tag{7.66}$$

If additionally the matrix $\mathbf{K}^{(u1)}$, dependent on displacement $\mathbf{u}^*$, is omitted then the *initial buckling equation* is obtained:

$$\{\mathbf{K}^{(\mathrm{con})} + \lambda[\mathbf{K}^{(\sigma)}(\boldsymbol{\sigma}^*) + \mathbf{K}^{(p1)}]\}\mathbf{v} = \mathbf{0}\,. \tag{7.67a}$$

In the case of spatially fixed load Eq. (7.66) is reduced to the classical form:

$$[\mathbf{K}^{(\mathrm{con})} + \lambda\mathbf{K}^{(\sigma)}(\boldsymbol{\sigma}^*)]\mathbf{v} = \mathbf{0}\,. \tag{7.67b}$$

The above formulated equations (7.65)–(7.67) are based on the linear relations (7.64). In case of strongly nonlinear behaviour in the prebuckling state (this concerns also nonlinear constitutive relations) the incremental approach has to be applied and in the vicinity of buckling points the equations (7.60), (7.62), or (7.63) can be analysed.

All the equations discussed in this section were formulated for conservative systems for which the stiffness matrix is symmetric, i.e. $\mathbf{K}_T = \mathbf{K}_T^{\mathrm{T}}$. In nonconservative problems the matrix $\mathbf{K}_T$ is asymmetric and the formulation of eigenproblem is much more complicated, cf. [64]. Certain nonconservative problems can be well evaluated by means of appropriate symmetrized matrices, cf. [64, 102]. This approach is discussed in Sect. 7.5.

7.2.3 Extended Sets of Equations

The computation of nonlinear equilibrium paths can be easier if the *extended set of equations* is used:

$$\tilde{\mathbf{G}}(\mathbf{q},\lambda;\tau) = \begin{bmatrix} \mathbf{R}(\mathbf{q},\lambda) \\ \varphi(\mathbf{q},\lambda;\tau) \end{bmatrix} = \mathbf{0}\,, \tag{7.68}$$

where the non-incremental FEM equations

$$\mathbf{R}(\mathbf{q},\lambda) \equiv \mathbf{F}(\mathbf{q}) - \lambda \mathbf{Q}^* = \mathbf{0} \tag{7.69}$$

are completed with the *constraint equation*

$$\varphi(\mathbf{q},\lambda;\tau) = \mathbf{0}\,. \tag{7.70}$$

In the constraint equation the conventional time τ is used. It is called a *control parameter.* The simplest cases of function φ at the left-hand side of (7.70) correspond to continuation of the computational process by means of:

$$\begin{aligned} &1) \quad \varphi = \lambda - \tau && \text{– load control,} \\ &2) \quad \varphi = q_\beta - \tau && \text{– displacement control.} \end{aligned} \tag{7.71}$$

The extended set of Eqs. (7.68) was introduced to computational mechanics mainly due to Riks' papers, cf. references in [148]. Efficient algorithms can be formulated by means of incremental extended equations. This question is discussed in detail in Sect. 7.3.2.

Extended sets of equations can be formulated for the computation of critical points without tracing the whole equilibrium path. In paper [1] the extended equations

$$\tilde{\mathbf{G}}(\mathbf{q},\lambda) \equiv \begin{bmatrix} \mathbf{G}(\mathbf{q},\lambda) \\ \det \mathbf{K}_T(\mathbf{q},\lambda) \end{bmatrix} = \mathbf{0} \tag{7.72}$$

were used to the computation of bifurcation points. More general extended equations

$$\tilde{\mathbf{G}}(\mathbf{q},\mathbf{w},\lambda) \equiv \begin{bmatrix} \mathbf{G}(\mathbf{q},\lambda) \\ \mathbf{K}_T(\mathbf{q},\lambda)\mathbf{w} \\ \psi(\mathbf{w}) - 1 \end{bmatrix} = \mathbf{0} \tag{7.73}$$

explore a functional $\boldsymbol{\psi}$ which was used in [174, 176] for the analysis of bifurcation and limit points.

The combination of Eqs. (7.70) and (7.72) gives the following set of extended equations:

$$\tilde{\mathbf{G}}_c(\mathbf{q}, \mathbf{w}, \lambda) = \begin{bmatrix} \mathbf{G}_c(\mathbf{q}, \lambda) \\ \varphi(\mathbf{q}, \tau) \\ \det \mathbf{K}_T(\mathbf{q}, \tau) \end{bmatrix} = \mathbf{0}\,, \tag{7.74}$$

where in the relation

$$\mathbf{G}_c(\mathbf{q}, \lambda) = \mathbf{F}(\mathbf{q}) + \mathbf{R}_c - \lambda \mathbf{Q}^* \tag{7.75}$$

the reactions $\mathbf{R}_c$ for unilateral constraints are introduced, cf. [182].

The extended set of equations mentioned above was given for the total displacement $\mathbf{q}$ and the load parameter λ. These equations are usually transformed into incremental equations as a basis for iterative procedures. Incremental equations can also be formulated directly by means of incremental FE equations completed with an incremental constraint equation:

$$\tilde{\mathbf{G}}(\Delta\mathbf{q}, \Delta\lambda; \Delta\tau) = \begin{bmatrix} \mathbf{K}_T(\mathbf{q})\Delta\mathbf{q} - \Delta\lambda \mathbf{Q}^* - \mathbf{R}(\mathbf{q}) \\ \varphi(\Delta\mathbf{q}, \Delta\lambda; \Delta\tau) \end{bmatrix} = \mathbf{0}\,. \tag{7.76}$$

In Sect. 7.5, Example 15, the stability boundary is computed for two parameter loads $\{\lambda_1, \lambda_2\}$. An appropriate extended set of equations corresponds to Eqs. (7.74), cf. [169]:

$$\tilde{\mathbf{G}}(\Delta\mathbf{q}, \Delta\lambda_1, \Delta\lambda_2; \Delta\tau) = \begin{bmatrix} \mathbf{K}_T(\mathbf{q})\Delta\mathbf{q} - \Delta\lambda_1 \mathbf{Q}_1^* - \Delta\lambda_2 \mathbf{Q}_2^* - \mathbf{R}(\mathbf{q}) \\ \varphi(\Delta\mathbf{q}, \Delta\lambda_1, \Delta\lambda_2; \Delta\tau) \\ \det \mathbf{K}_T(\mathbf{q} + \Delta\mathbf{q}) \end{bmatrix} = \mathbf{0}\,. \tag{7.77}$$

7.3 Solution Methods for Nonlinear FEM Equations

For the solutions of FEM nonlinear algebraic equations there were formulated both general methods and special procedures well matching the specific features of structures and approximations applied. The methods can be classified in general as:

1. *incremental methods* combined with various iterative procedures,
2. iterative *non-incremental* methods,
3. different *approximate* methods based mainly on the analysis of simplified discrete models,
4. *asymptotic procedures* associated with the perturbation method,
5. *special methods* efficient for the analysis of special problems.

Incremental, step-by-step procedures are the most efficient. They can by treated as discrete continuation methods [173, 143]. In methods commonly used at each increment the zero value of residuals is tried to be achieved in an iterative way.

Non-incremental methods have rather a limited range of applications since there are difficulties with efficient iterative algorithms. The non-incremental methods cannot be applied for the solutions of history dependent problems, i.e. to the analysis of elastic-plastic problems.

In the group of approximate methods the *reduced basis methods* should be distinguished. These methods depend on searching for solutions in n – dimension basis spaces for $n \ll N$, where: N – original dimension of space for FE system, cf. [114, 119].

The asymptotic (perturbation) procedures are related rather to analytical methods and their applications are limited to the vicinity of fundamental equilibrium states or to the first terms of expansion into power series of small parameters. This approach needs multidimensional matrices. It makes the methods somewhat rather complicated and reduces the generality of perturbation methods despite the efforts to combine them with FEM, cf. [163, 53].

'Artificial springs' methods can refer to special procedures. They are related to adding fictitious components to the stiffness matrix [157] or to the method of fictitious viscotic damping [93].

Incremental methods are discussed more thoroughly. Selected methods of the reduced basis are considered as well.

The procedures for the computation of critical points and postcritical equilibrium are also discussed for cases of singular or negative defined stiffness matrices $\mathbf{K}_T$.

In the majority of methods the FEM equations are analysed adding a constraint equation to trace the equilibrium path as the function $\lambda(\mathbf{q};\tau)$.

The attention is first focussed on the load control $\tau \equiv \lambda$ applied for the solution of incremental FEM equations which can be classified as:

1. Newtonian methods, usually called *Newton–Raphson methods* (NR),
2. Quasi–Newton methods (QN),
3. Conjugate lines methods (CL),
4. Secant-Newton methods (SN).

The NR and QN methods are discussed in what follows since these methods are commonly used in nonlinear procedures for the analysis of nonlinear problems by means of FEM.

7.3.1 Incremental Methods

Let us discuss the analysis of an incremental set of FEM equations (7.29) which is written in the matrix form:

$$\mathbf{K}\Delta\mathbf{q} = \Delta\lambda\,\mathbf{Q}^* + \mathbf{R}\,, \tag{7.78}$$

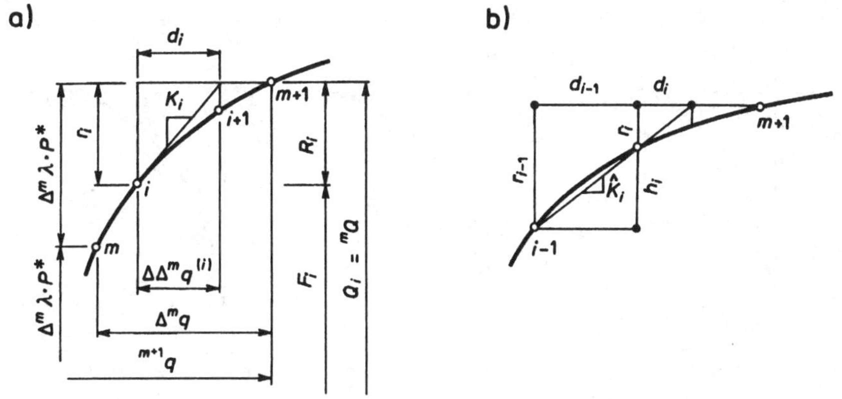

Fig. 7.8. Increments for: **a)** tangent direction, **b)** locally secant direction

where it is assumed that the stiffness matrix $\mathbf{K} \equiv \mathbf{K}_T$ is symmetric and the vector of residual forces vanishes in the equilibrium state (7.69).

The incremental procedure depends on discrete continuation of the computational process. This means that a known solution ${}^m\mathbf{q} = \mathbf{q}({}^m\tau)$ is extended to ${}^{m+1}\mathbf{q} = \mathbf{q}({}^{m+1}\tau)$ for the next value of control parameter ${}^{m+1}\tau = {}^m\tau + \Delta\,{}^m\tau$:

$$ {}^{m+1}\mathbf{q} = {}^m\mathbf{q} + \Delta\,{}^m\mathbf{q} . \tag{7.79} $$

The selection of an appropriate control parameter (conventional time) affects the efficiency of the computation of increments $\Delta\,{}^m\mathbf{q}$. Let us assume for simplicity of considerations that the load control parameter is applied, i.e. $\tau \equiv \lambda$.

Iterative methods are used to compute the increment $\Delta\,{}^m\mathbf{q}$. After a finite number of iterative steps n the following result is achieved:

$$ \Delta\,{}^m\mathbf{q} = \sum_{i=1}^{n} \Delta\Delta\,{}^m\mathbf{q}^{(i)} \equiv \sum_{i=1}^{n} \mathbf{d}_i \, , \tag{7.80} $$

if the sequence $\{\mathbf{d}_i\}$ is convergent to the limit $\mathbf{d}_n$. The number of iterations n is evaluated by the convergence condition

$$ \|d_{n+1}\| < \varepsilon_q \, , \tag{7.81} $$

where ε_q is an admissible error. An abbreviated notation is used whenever it does not lead to vagueness, e.g. the increment of residual forces can be written as (cf. Fig. 7.8a)

$$ \mathbf{r}_i \equiv {}^m\mathbf{r}^{(i)} = \alpha_i \Delta\,{}^m\lambda \mathbf{Q}^* + \mathbf{R}_i \tag{7.82} $$

and used in (7.78) written in the form

$$ \mathbf{K}_i \mathbf{d}_i = \mathbf{r}_i \, . \tag{7.83} $$

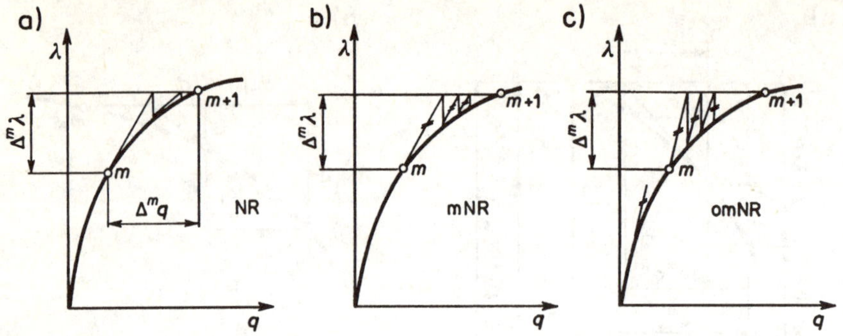

Fig. 7.9. Unidimensional interpretation of Newton procedures corresponding to: **a)** classical Newton–Raphson methods, **b)** modified RN method, **c)** method of initial stiffness.

In relation (7.82) the coefficient α_i is introduced in order to have the first approximation (predictor) and subsequent approximations (correctors):

$$\alpha_i = \begin{cases} 1 & \text{for} \quad i = 1 \quad \text{– predictor,} \\ 0 & \text{for} \quad i > 1 \quad \text{– corrector.} \end{cases} \tag{7.84}$$

At the first iteration step the residual forces are near zero $\mathbf{R}^{(1)} \approx \mathbf{0}$ since the iteration procedure should start from nearly equilibrium state which is evaluated by the condition

$$\| {}^m\mathbf{R}^{(1)} \| \equiv \|\mathbf{R}_1\| < \varepsilon_R \,. \tag{7.85}$$

If the tangent matrix $\mathbf{K}_i \equiv \mathbf{K}_T^{(i)}$ is nonsingular then the NR formula corresponds to the following relation

$$\mathbf{d}_i = \mathbf{K}_i^{-1}\mathbf{r}_i \,. \tag{7.86}$$

The specification of matrix $\mathbf{K}_i$ leads to different versions of Newton methods.

The unidimensional versions of NR methods are shown in Fig. 7.9: a) *classical method* (NR), b) *modified method* (mNR), c) *initial stiffness method* (omNR). All the Newton–Raphson methods need the computation of unbalanced residual forces $\mathbf{r}_i$. It ensures coming back to the equilibrium path $\mathbf{q}(\lambda)$ at each iteration step. In the NR procedure the matrix $\mathbf{K}_i$ is updated at every iteration. In case of (mNR) the matrix $\mathbf{K}_i \equiv {}^n\mathbf{K}_0 = \mathbf{K}({}^m\mathbf{q})$ is computed only once at the beginning of iteration point m. The method (omNR) is associated with the fixed matrix ${}^0\mathbf{K}$ which is computed at the beginning of deformation process.

It is possible to introduce further modifications to the above mentioned versions of the Newton–Raphson methods, e.g. in [134] the stiffness matrix in mNR procedure is fixed after the first iteration, i.e. $\mathbf{K}_i \equiv {}^m\mathbf{K}_1 = \mathbf{K}({}^m\mathbf{q}+\mathbf{d}_1)$.

The efficiency of Newton method strongly depends on nonlinearity of the analysed problem. It can be roughly evaluated as a 'deviation' of the equilibrium path from its tangent at origin of λ, q_β coordinates. The selection of methods depends also on the number of degrees of freedom N, material characteristics, selection of control parameter τ and its step $\Delta\,{}^m\tau$.

The non-iterative incremental method are not worth discussing because of difficult evaluation of the errors caused by the deviation of the computed path from the equilibrium path. The procedure (omNR) is also called the method of initial loads or initial stress in order to distinguish it from the method of variable stiffnesses which corresponds to the classical NR methods [183].

In the NR method the direction of correction is given by the tangent vector $\mathbf{q}_i + \mathbf{d}_i$ – cf. Fig. 7.7a. In *Quasi-Newton methods* (QN) the direction of correction is locally secant as shown in Fig. 7.7b.

The locally secant matrix $\hat{\mathbf{K}}_i$ has to fulfil the condition – cf. Fig. 7.7b:

$$\hat{\mathbf{K}}_i\,\mathbf{d}_{i-1} = \mathbf{h}_i\,, \tag{7.87}$$

where the vector $\mathbf{h}_i$ is:

$$\mathbf{h}_i = \mathbf{r}_{i-1} - \mathbf{r}_i\,. \tag{7.88}$$

In case of nonsingular matrix $\hat{\mathbf{K}}_i$ the Quasi-Newton analogue of the NR formula (7.86) is of the form:

$$\mathbf{d}_i = \hat{\mathbf{K}}_i^{-1}\mathbf{r}_i\,. \tag{7.89}$$

After Eq. (7.87) is multiplied by $\mathbf{d}_i^{\mathrm{T}}$ and taken into account (7.89) the following secant condition is obtained:

$$\mathbf{d}_i^{\mathrm{T}}\mathbf{h}_i = \mathbf{r}_i^{\mathrm{T}}\mathbf{d}_{i-1}\,. \tag{7.90}$$

In the QN methods the matrix $\hat{\mathbf{K}}_i$ is calculated by modification of matrix $\hat{\mathbf{K}}_{i-1}$, computed at the previous iteration step. This problem was discussed in detail in the review paper [44].

A simple formula of the first order can be formulated by addition of the matrix $\hat{\mathbf{K}}_{i-1}$ to a matrix which fulfills the secant condition. Let us quote the Davidon formula:

$$\hat{\mathbf{K}}_i = \hat{\mathbf{K}}_{i-1} + \frac{(\mathbf{h}_i - \hat{\mathbf{K}}_{i-1}\mathbf{d}_{i-1})(\mathbf{h}_i - \hat{\mathbf{K}}_{i-1}\mathbf{d}_{i-1})^{\mathrm{T}}}{\mathbf{d}_{i-1}^{\mathrm{T}}(\mathbf{h}_i - \hat{\mathbf{K}}_{i-1}\mathbf{d}_{i-1})} \tag{7.91}$$

or in a simplified notation:

$$\hat{\mathbf{K}}_{\mathrm{D}} = \mathbf{K} + \frac{(\mathbf{h} - \mathbf{Kd})\,(\mathbf{h} - \mathbf{Kd})^{\mathrm{T}}}{\mathbf{d}^{\mathrm{T}}\,(\mathbf{h} - \mathbf{Kd})}\,. \tag{7.91a}$$

Looking at the Davidon formula it is clear that the locally secant matrix $\hat{\mathbf{K}}_i$ can be computed on the basis of the matrix $\hat{\mathbf{K}}_{i-1}$ (not only a locally secant matrix) and vectors $\mathbf{d}_{i-1}$, $\mathbf{h}_i = \mathbf{r}_{i-1} - \mathbf{r}_i$ taken from the iterative steps i and $i-1$. In such a way there is no need to return to the finite elements to formulate of the updated stiffness matrix.

The formula of first order can be unstable (it can occur if the vector $\mathbf{d}$ is almost orthogonal to the vector $\mathbf{h}-\mathbf{Kd}$). This is why the second order formulae are commonly used basing on the superposition of the following matrices:

$$\hat{\mathbf{K}} = \mathbf{K} + \mathbf{k}\mathbf{l}^{\mathrm{T}} + \mathbf{l}\mathbf{k}^{\mathrm{T}}\,, \tag{7.92}$$

where the vectors $\mathbf{k}, \mathbf{l}$ fulfil not only the secant condition but satisfy additional criteria discussed in [44] as well.

The most popular second order formulae are: DFP formula (corresponding to names Davidon, Fletcher and Powell) and BFGS formula (Broyden, Fletcher, Goldfarb, Shanno):

$$\hat{\mathbf{K}}_{\mathrm{DFP}} = \left(\mathbf{I} - \frac{\mathbf{h}\mathbf{d}^{\mathrm{T}}}{\mathbf{d}^{\mathrm{T}}\mathbf{h}}\right)\mathbf{K}\left(\mathbf{I} - \frac{\mathbf{d}\mathbf{h}^{\mathrm{T}}}{\mathbf{d}^{\mathrm{T}}\mathbf{h}}\right), \tag{7.93}$$

$$\hat{\mathbf{K}}_{\mathrm{BFGS}} = \mathbf{K} + \frac{\mathbf{h}\mathbf{h}^{\mathrm{T}}}{\mathbf{d}^{\mathrm{T}}\mathbf{h}} - \frac{\mathbf{K}\mathbf{d}\mathbf{d}^{\mathrm{T}}\mathbf{K}}{\mathbf{d}^{\mathrm{T}}\mathbf{K}\mathbf{d}}. \tag{7.94}$$

In view of FEM it is important to formulate directly the inverse locally secant matrices:

$$\hat{\mathbf{K}}_{\mathrm{D}}^{-1} = \mathbf{K}^{-1} + \frac{(\mathbf{d}-\mathbf{K}^{-1}\mathbf{h})(\mathbf{d}-\mathbf{K}^{-1}\mathbf{h})^{\mathrm{T}}}{\mathbf{h}^{\mathrm{T}}(\mathbf{d}-\mathbf{K}^{-1}\mathbf{h})}, \tag{7.95}$$

$$\hat{\mathbf{K}}_{\mathrm{FDP}}^{-1} = \mathbf{K}^{-1} + \frac{\mathbf{d}\mathbf{d}^{\mathrm{T}}}{\mathbf{d}^{\mathrm{T}}\mathbf{h}} - \frac{\mathbf{K}^{-1}\mathbf{h}\mathbf{h}^{\mathrm{T}}\mathbf{K}^{-1}}{\mathbf{h}^{\mathrm{T}}\mathbf{K}\mathbf{h}}, \tag{7.96}$$

$$\hat{\mathbf{K}}_{\mathrm{BFGS}}^{-1} = \left(\mathbf{I} - \frac{\mathbf{d}\mathbf{d}^{\mathrm{T}}}{\mathbf{d}^{\mathrm{T}}\mathbf{h}}\right)\mathbf{K}^{-1}\left(\mathbf{I} - \frac{\mathbf{h}\mathbf{d}^{\mathrm{T}}}{\mathbf{d}^{\mathrm{T}}\mathbf{h}}\right) + \frac{\mathbf{d}\mathbf{d}^{\mathrm{T}}}{\mathbf{d}^{\mathrm{T}}\mathbf{h}}. \tag{7.97}$$

The interest in Quasi-Newton methods increased after the vector version of matrix updating was proposed in [105]. This version for BFGS formula is based on the following formula:

$$\hat{\mathbf{K}}_i^{-1} = (\mathbf{I} + \mathbf{w}_i\mathbf{v}_i^{\mathrm{T}})\hat{\mathbf{K}}_{i-1}^{-1}(\mathbf{I} + \mathbf{v}_i\mathbf{w}_i^{\mathrm{T}})\,, \tag{7.98}$$

where the vectors v_i and w_i are:

$$\mathbf{v}_i = \mathbf{r}_{i-1}\left[1 + \left(-\frac{\mathbf{d}_{i-1}^{\mathrm{T}}\mathbf{h}_i}{\mathbf{d}_{i-1}^{\mathrm{T}}\mathbf{r}_{i-1}}\right)^{1/2}\right] - \mathbf{r}_i, \quad \mathbf{w}_i = \frac{\mathbf{d}_{i-1}}{\mathbf{d}_{i-1}^{\mathrm{T}}\mathbf{h}_i}. \tag{7.99}$$

The algorithm of BFGS starts from the predictor computation:

$$\mathbf{d}_1 = (\mathbf{I} + \mathbf{w}_1\mathbf{v}_1^{\mathrm{T}})\mathbf{K}_{\wedge}^{-1}(\mathbf{I} + \mathbf{v}_1\mathbf{w}_1^{\mathrm{T}})\mathbf{r}_1\,,$$

as a result of vector multiplication from the right-hand side of the above formula, i.e. starting from the scalar product $\mathbf{w}_1^{\mathrm{T}}\mathbf{r}_1$. The next step corresponds to repeated multiplication by vectors according to the formula:

$$\mathbf{d}_2 = (\mathbf{I} + \mathbf{w}_2\mathbf{v}_2^{\mathrm{T}})(\mathbf{I} + \mathbf{w}_1\mathbf{v}_1^{\mathrm{T}})\mathbf{K}_{\wedge}^{-1}(\mathbf{I} + \mathbf{v}_1\mathbf{w}_1^{\mathrm{T}})(\mathbf{I} + \mathbf{v}_2\mathbf{w}_2^{\mathrm{T}})\mathbf{r}_2\,.$$

The algorithm needs the factorized matrix $\mathbf{K}_\wedge = \mathbf{L}_\wedge \mathbf{D}_\wedge \mathbf{L}_\wedge^{\mathrm{T}}$ (the upper triangle matrix $\mathbf{L}_\wedge$ and the diagonal matrix $\mathbf{D}_\wedge$ to be more precise) and the subsequent vectors $\mathbf{v}_i$ and $\mathbf{w}_i$.

The algorithm described in [104, 134] is continued by the computation of the following vectors:

$$\begin{aligned} \overset{*}{\mathbf{r}}_i &= \left[\prod_{j=1}^{i} (\mathbf{I} + \mathbf{v}_j \mathbf{w}_j^{\mathrm{T}}) \right] \mathbf{r}_i , \\ \overset{*}{\mathbf{d}}_i &= \mathbf{K}_\wedge^{-1} \overset{*}{\mathbf{r}}_i , \\ \mathbf{d}_i &= \left[\prod_{j=1}^{i} (\mathbf{I} + \mathbf{w}_j \mathbf{v}_j^{\mathrm{T}}) \right] \overset{*}{\mathbf{d}}_i . \end{aligned} \tag{7.100}$$

The algorithm requires storing of an increasing number of vectors. That is why for a fixed number of iterations $i \leq 10 - 15$ the matrix $\mathbf{K}_\wedge$ is updated as in the classical NR procedure.

A similar recursive algorithm for Davidon method was given in [54].

The *methods of conjugate lines* (CL) were discussed in review paper [39]. The methods are based on a *conjugate line relationship*

$$\mathbf{d}_i^{\mathrm{T}} \mathbf{h}_i = 0 \tag{7.101}$$

and on the condition of optimal step

$$\mathbf{r}_{i+1}^{\mathrm{T}} \mathbf{d}_i = 0 . \tag{7.102}$$

Using (7.101) and (7.102) the parameter of line search β_i and parameter of step length η_i are calculated in the formula

$$\mathbf{q}_{i+1} = \mathbf{q}_i + \eta_i \mathbf{d}_i , \tag{7.103}$$

where

$$\mathbf{d}_i = \overset{*}{\mathbf{d}}_i + \beta_i \mathbf{d}_{i-1}, \quad \overset{*}{\mathbf{d}} = \mathbf{K}_a^{-1} \mathbf{r}_i . \tag{7.104}$$

The matrix $\mathbf{K}_a \neq \mathbf{I}$ is used in the so-called *scaled gradient method.* In papers by Crisfield (cf. [39]) the formulation of matrix $\mathbf{K}_a$ was discussed, e.g. the matrix $\mathbf{D}$ composed only of diagonal components of the stiffness matrix can be used as $\mathbf{K}_a$.

The CL methods are sensitive to the step length parameters. That is why Crisfield combines them with QN methods. This leads to the so-called *Secant-Newton methods* (SN). The general SN formula is:

$$\mathbf{d}_i = a_i \overset{*}{\mathbf{d}}_i + b_i \eta_{i-1} \mathbf{d}_{i-1} + c_i \overset{*}{\mathbf{d}}_{i-1} . \tag{7.105}$$

The parameters a_i, b_i, c_i, η_{i-1} are computed from the secant condition (7.90) and additional optimality conditions but without the conjugate line relationship (7.101). Therefore, the SN methods can be treated as a generalization of QN methods.

In formula (7.105) the scaled displacements are used

$$\overset{*}{\mathbf{d}}_j = \mathbf{K}_a^{-1}\mathbf{r}_j \quad \text{for} \quad j = i, i-1 \,. \tag{7.106}$$

They are calculated by means of matrix $\mathbf{K}_a$ which is not updated during the iteration process. After the formula (7.106) is substituted to the inverse BFGS formula the following coefficients can be derived, cf. [39]:

$$\begin{aligned} c_i &= -\gamma\, \mathbf{d}_{i-1}^{\mathrm{T}}\mathbf{r}_i, \qquad a_i = 1 - c_i, \\ b_i &= -c_i + \frac{\gamma}{\eta_{i-1}}\left[(\overset{*}{\mathbf{d}}_i - \overset{*}{\mathbf{d}}_{i-1})^{\mathrm{T}}\mathbf{r}_i + c_i(\overset{*}{\mathbf{d}}_i - \overset{*}{\mathbf{d}}_{i-1})^{\mathrm{T}}\mathbf{h}_i\right], \end{aligned} \tag{7.107}$$

where $\gamma = 1/(\mathbf{d}_{i-1}^{\mathrm{T}}\mathbf{h}_i)$. The parameter η_{i-1} can be computed from the previous iteration step

$$\eta_{i-1} = \gamma\, \mathbf{d}_{i-1}^{\mathrm{T}}\mathbf{r}_{i-1} \,. \tag{7.108}$$

In paper [38] the two parameter SN formula was derived, called a 'faster modified Newton–Raphson method'

$$\mathbf{d}_i = a_i\overset{*}{\mathbf{d}}_i + \bar{b}_i\eta_{i-1}\mathbf{d}_{i-1} \,. \tag{7.109}$$

This formula results from (7.105) after the scaled displacement $\overset{*}{\mathbf{d}}_{i-1}$ is assumed to be

$$\overset{*}{\mathbf{d}}_{i-1} = \eta_{i-1}\mathbf{d}_{i-1} \,. \tag{7.110}$$

The coefficient a_i is defined in (7.107) and the coefficient $\bar{b}_i$ is in the form

$$\bar{b}_i \equiv b_i + c_i = -c_i - a_i\frac{\gamma}{\eta_{i-1}}\overset{*}{\mathbf{d}}_i{}^{\mathrm{T}}\mathbf{h}_i \,. \tag{7.111}$$

The two parameter formula (7.109) is more efficient for computation since the scaled displacement $\overset{*}{\mathbf{d}}_{i-1}$ does not have to be memorized. As in Sect. 7.5, Example 3, the efficiency of the application of formula (7.105) and (7.109) is comparable if the number of iteration is taken into account.

Other three parameter formula were given in [39, 104]. In [40] one parameter formula were also discussed but their efficiency, measured in the number of iterations, is smaller than for two- or three-parameter formula (7.105) and (7.109) respectively.

From among the methods discussed above the most frequently used are the NR, mNR and BFGS methods. As shown in Sect. 7.5, Examples 1 and 2, the mNR method can be non-convergent, especially for larger iteration steps. The QN methods lead to comparable efficiency. The consistent formulation in elastoplastic problems makes the NR methods as efficient as the QN methods.

In review papers [39, 148] there were also discussed other incremental methods and appropriate references were quoted.

7.3.2 Computation of Equilibrium Paths by Incremental Methods

Different methods in the previous section were discussed under the assumption of load control. At this control the number of iteration increases and the iteration process is divergent even if the limit point is approached.

From among the different procedures only the Batoz–Dhatt algorithm [20], generalized in [138], is discussed below. Corresponding to (7.83) the displacement increments $\mathbf{d}$ are in the following form

$$\mathbf{d} = a_* \, \mathbf{d}^* + a_R \, \mathbf{d}^{(R)} \,, \tag{7.112}$$

where $\mathbf{d}^*$ and $\mathbf{d}^{(R)}$ are solutions of the equations

$$\mathbf{K}\mathbf{d}^* = \mathbf{Q}^*, \quad \mathbf{K}\mathbf{d}_i^{(R)} = \mathbf{R}_i \,. \tag{7.113}$$

If $a_* = 0$ and $a_R = 1$ is assumed then the transition to the NR method takes place. The case $a_* \neq 0$ and $a_R = 0$ corresponds to the incremental, non-iterative procedure. In the Batoz–Dhatt algorithm $a_R = 1$ and the coefficient a_* is computed from the condition

$$\Delta \tau = a_* d^*_\beta + d^{(R)}_\beta \equiv d^*_\beta \Delta\Delta\lambda + d^{(R)}_\beta \,. \tag{7.114}$$

At the first iteration step (predictor) $i = 1$ the residual forces are assumed to vanish, i.e. $\mathbf{R}_1 = \mathbf{0}$ which implies $\mathbf{d}^{(R)} = \mathbf{0}$, and for the next iterations $i > 1$ (correctors) $\Delta\, {}^m\tau = 0$ is assumed. This leads to the following formula:

$$a_* \equiv \Delta\Delta\lambda_i = \begin{cases} \dfrac{\Delta\, {}^m\tau}{\mathbf{t}^{\mathrm{T}}\mathbf{d}^*} & \text{for } i = 1 \quad \text{– predictor,} \\[2ex] -\dfrac{\mathbf{t}^{\mathrm{T}}\mathbf{d}_i^{(R)}}{\mathbf{t}^{\mathrm{T}}\mathbf{d}^*} & \text{for } i > 1 \quad \text{– corrector,} \end{cases} \tag{7.115}$$

where the *control vector*

$$\mathbf{t}^{\mathrm{T}} = [0, \ldots, 0, \underset{\beta}{1}, 0, \ldots, 0] \tag{7.116}$$

has the value 1 on the β-position. This vector serves to select the component d_β from the displacement vector $\mathbf{d}$.

The procedure given above needs a single solution of equation $(7.113)_1$ and memorizing the vector $\mathbf{d}^*$ and then the computation of vector $\mathbf{d}_i^{(R)}$ for $i > 1$ is carried out from equation $(7.113)_2$, where

$$\mathbf{R}_i = \Delta\lambda_{i-1}\mathbf{Q}^* - \mathbf{F} \quad \text{for} \quad \Delta\lambda_{i-1} = \sum_{j=1}^{i-1} \Delta\Delta\lambda_j \,. \tag{7.117}$$

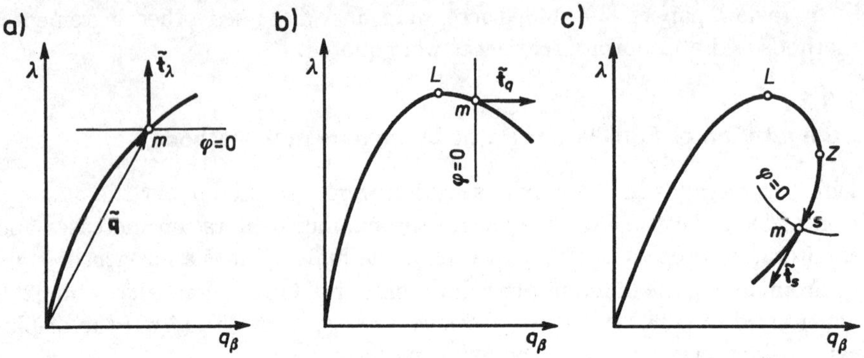

Fig. 7.10. Interpretation of the constraint equation $\varphi = 0$ and types of control: **a**) load control, **b**) displacement control, **c**) arc-length control

The formula (7.115) can be applied even in the very near vicinity of limit points. In case when the tangent stiffness matrix $\mathbf{K}$ is singular, other formulae have to be used. One possible approach is to analyse a subset of equations corresponding to nonsingular matrix $\mathbf{K}_A$ by removing the column and row of number β (cf. [167]).

There are more complicated cases for which it is difficult to point out the control parameter τ which increases monotonically during the deformation process of the analysed structure. It is possible, of course, to change the control parameter in the course of deformation process but such an approach complicates the procedure and consistency of computations. That is why a more general approach is associated with the analysis of extended equation (7.76) which can be written in the compact form:

$$\begin{cases} \bar{\mathbf{K}}\tilde{\mathbf{d}} = \mathbf{R} \\ \varphi(\tilde{\mathbf{d}}, \Delta\tau) = 0\,, \end{cases} \tag{7.118}$$

where: $\bar{\mathbf{K}}_{N\times(N+1)} = [\mathbf{K}, -\mathbf{Q}^*]$, $\tilde{\mathbf{d}}_{(N+1)\times 1} = \{\mathbf{d}, \Delta\Delta\lambda\}$.

The simple cases of constraint equations are shown in Fig. 7.9. The hyperplanes $\varphi = 0$ of the normal vectors $\tilde{\mathbf{t}}_\lambda$ or $\tilde{\mathbf{t}}_q$ correspond to the load control $\tau = \lambda$ or $\tau = q_\beta$, respectively (Fig. 7.9a, b). A more general case is shown in Fig. 7.9c. The constraint surface $\varphi = 0$ crosses the equilibrium path at the point m of the control vector $\tilde{\mathbf{t}}_s$ which is tangent to the path and normal to the constraint surface. In this case the control parameter $\tau = s$, called *arc-length parameter*, corresponds to the length s of the equilibrium path.

It is clear from Fig. 7.10 that the load control cannot be applied to the computation of limit point L. The displacement control cannot be used to compute the turning point R. That is why a more general control by means of the arc-length parameter can be recommended. Below there are discussed

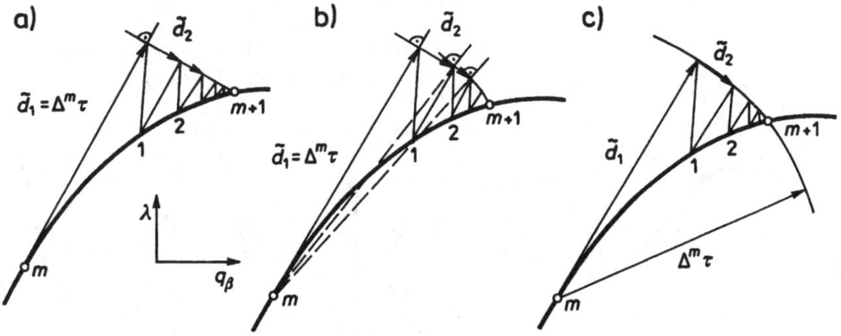

Fig. 7.11. Iterative computation of displacement increments by means of constraint equations; **a**) Riks–Wempner equation, **b**) updated RW equation, **c**) Crisfield equation

four constraint equations. The *Riks–Wempner equation*

$$ {}^m\tilde{\mathbf{t}}^{\mathrm{T}}\,\tilde{\mathbf{d}}_i = \alpha_i \Delta\,{}^m\tau\,, \tag{7.119}$$

is frequently used [144, 175]. The control vector ${}^m\tilde{\mathbf{t}}$ is tangent to the equilibrium path at point m and the factor α_i corresponds to predictor or corrector in the iteration process according to (7.84). The value $\alpha_1 = 1$ appoints the tangent direction $\tilde{\mathbf{d}}_1$ and $\alpha_i = 0$ for $i > 1$ corresponds to the direction $\tilde{\mathbf{d}}_i$ perpendicular to $\tilde{\mathbf{t}}_i$. In Fig. 7.11a the Riks–Wempner approach is shown as coupled with the modified Newton–Raphson method.

An updated modification of Riks–Wempner procedure is shown in Fig. 7.11b, cf. [140, 167]. In this case the relationship used for the corrector is

$$\tilde{\mathbf{t}}^{\mathrm{T}}\,\tilde{\mathbf{d}} = 0 \quad \text{for} \quad i > 1\,, \tag{7.120}$$

where the control vector is locally secant (it passes by points m and i):

$$\tilde{\mathbf{t}}_{i-1} = \frac{\Delta\tilde{\mathbf{q}}_{i-1}}{|\Delta\tilde{\mathbf{q}}_{i-1}|}, \quad \Delta\tilde{\mathbf{q}}_{i-1} \equiv \{\Delta\mathbf{q}_{i-1}, \Delta\lambda_{i-1}\} = \sum_{j=1}^{i-1}\tilde{\mathbf{d}}_j\,. \tag{7.121}$$

The modification discussed above gives results close to those by the *Crisfield equation* [38, 167] which corresponds to the spherical surface in the load-displacement space (Fig. 7.11c):

$$\Delta\mathbf{q}_i^{\mathrm{T}}\Delta\mathbf{q}_i + (\Delta\lambda_i)^2 = (\Delta\,{}^m\tau)^2\,. \tag{7.122}$$

Equation (7.122) is used for the computation of the load parameter increment $\Delta\Delta\lambda_i$. At the first iteration step $i = 1$ the solution $\mathbf{d}^*$ of $(7.113)_1$ can be used in order to predict the first approximation:

$$(\Delta\Delta\lambda_1)^2 = \frac{(\Delta\,{}^m\tau)^2}{1 + \overset{*}{\mathbf{d}}{}^{\mathrm{T}}\overset{*}{\mathbf{d}}}\,. \tag{7.123}$$

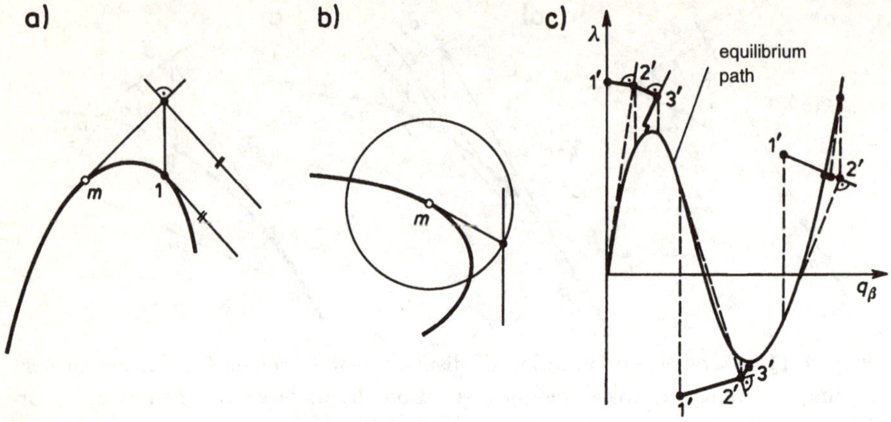

Fig. 7.12. Divergence of: **a**) Riks–Wempner method and **b**) Crisfield method and convergence of **c**) Fried's method

The following quadratic equation is applied to compute the increments $\Delta\Delta\lambda_i$ for $i \geq 2$:

$$(\Delta\Delta\lambda_i)^2 + 2\Delta\lambda_{i-1}\Delta\Delta\lambda_i + \mathbf{d}_i^{\mathrm{T}}\mathbf{d}_i + 2\mathbf{d}_i^{\mathrm{T}}\Delta\mathbf{q}_{i-1} = 0\,, \tag{7.124}$$

where the displacement increments $\mathbf{d}_i = \Delta\Delta q_i$ are computed from $(7.113)_2$ for residual forces (7.117).

Both Riks–Wempner and Crisfield methods can be divergent as shown in Figs. 7.12a, b. This can be overcome by means of the *orthogonal trajection method* proposed by Fried [51]. With respect to the corrector the following linear constraint equation is used:

$$\Delta\Delta\lambda = \mathbf{a}^{\mathrm{T}}\mathbf{d}\,. \tag{7.125a}$$

Substitution of $\mathbf{a} = -\mathbf{K}^{-1}\mathbf{Q}^* \equiv -\mathbf{d}^*$ and $\mathbf{d} = \mathbf{K}^{-1}(\mathbf{R} + \mathbf{Q}^*\Delta\Delta\lambda)$ leads to the formula:

$$\Delta\Delta\lambda_i = -\frac{\overset{*}{\mathbf{d}}{}^{\mathrm{T}}\mathbf{d}_i^{(R)}}{1 + \overset{*}{\mathbf{d}}{}^{\mathrm{T}}\mathbf{d}} \quad \text{for} \quad i \geq 2\,, \tag{7.125b}$$

The quadratic constraint equation was formulated in [15] on the basis of increment of load work $\Delta\,{}^m\tau \equiv \Delta\,{}^mW_L$:

$$({}^m\lambda^{(i-1)} + \frac{1}{2}\Delta\Delta\lambda_i)\overset{*}{\mathbf{Q}}{}^{\mathrm{T}}(\Delta\Delta\lambda_i\,\overset{*}{\mathbf{d}} + \mathbf{d}_i^{(R)}) = \alpha\Delta_i\,{}^mW_L\,. \tag{7.126}$$

The constraint equations can be formulated similarly assuming the current stiffness parameter [21] or on the basis of a hypereliptic surface and increment of deformation energy [130].

The methods discussed above can be classified as two-step methods. They depend on a separate solution of the FEM method and the constraint equation at each iteration step. It is possible to apply a one-step method by means of extended set of equations (7.118) cf. [48, 148, 167].

In [167] the one-step method was called the computation in the load-displacement space $\mathbb{R}^{N+1}$. The FEM equation $(7.118)_1$ is completed with the constraint equation (7.119). The extended set of equations can be written in a shortened form:

$$\tilde{\mathbf{K}}\tilde{\mathbf{d}} = \tilde{\mathbf{R}}\,, \tag{7.127}$$

where the matrix $\tilde{\mathbf{K}}_{(N+1)\times(N+1)}$ is formulated by adding the column $-\mathbf{Q}^*$ and the row $\tilde{\mathbf{t}}$ to the stiffness matrix $\mathbf{K}$. The extended vector of residual forces is $\tilde{\mathbf{R}}_{(n+1)\times 1} = \{\mathbf{R}, \alpha_i \Delta\,{}^m\tau\}$.

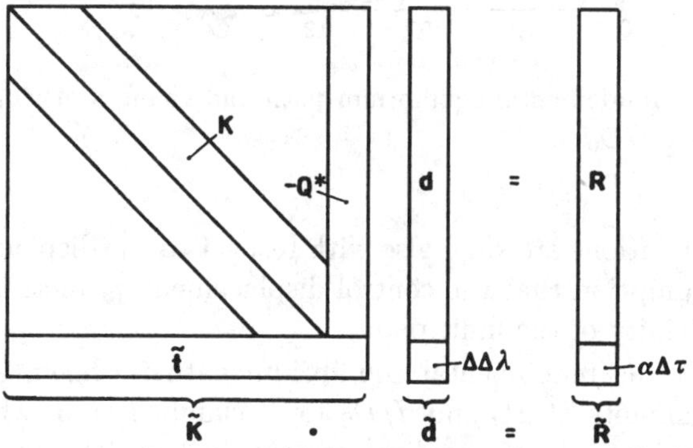

Fig. 7.13. Structure of extended equation (7.127)

In Fig. 7.13 the structure of extended Eq. (7.127) is shown. The extended matrix $\tilde{\mathbf{K}}$ is asymmetric but its structure makes possible an easy modification of standard solvers of the set of linear equations.

The following specification of the control parameter and control vectors gives:

$$\begin{aligned}
&1)\quad \tau \equiv \lambda, \quad \tilde{\mathbf{t}}_\lambda = \{0,\ldots,0,1\} && \text{– load control,}\\
&2)\quad \tau \equiv q_\beta, \quad \tilde{\mathbf{t}}_q = \{0,\ldots,0,\underset{\beta}{1},0,\ldots,0\} && \text{– displacement control,}\\
&3)\quad \tau \equiv s, \quad \tilde{\mathbf{t}}_s = \{\Delta\tilde{\mathbf{q}}/|\Delta\tilde{\mathbf{q}}|\} && \text{– arc-length control.}
\end{aligned} \tag{7.128}$$

If the arc-length parameter is used as the control parameter $\tau \equiv s$ then at the limit point the extended matrix $\tilde{\mathbf{K}}$ is nonsingular but it is singular at the bifurcation point . This feature of matrix $\tilde{\mathbf{K}}$ leads to an efficient method of distinguishing types of critical points (cf. [145, 146, 167]):

$$D \equiv \det \mathbf{K}_T = 0 \;\text{ and }\; J \equiv \det \tilde{\mathbf{K}} \begin{cases} \neq 0 & \text{– limit point,}\\ = 0 & \text{– bifurcation point.} \end{cases} \tag{7.129}$$

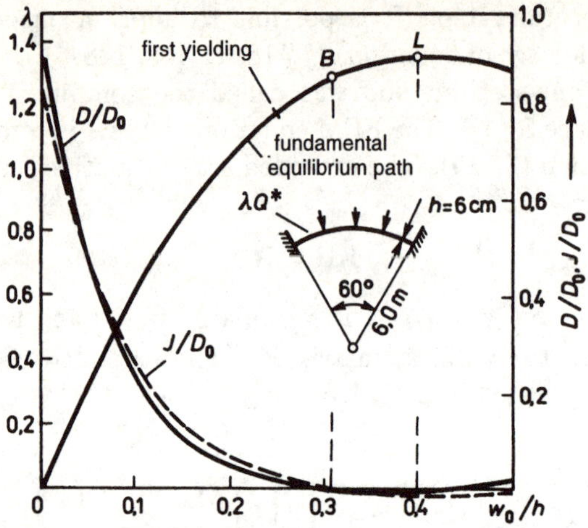

Fig. 7.14. The fundamental equilibrium path and changes of values of determinants D/D_0 and J/D_0

The above relations are valid also with respect to the displacement control under the assumption that the control displacement q_β increases monotonically in the vicinity of the limit point.

In Fig. 7.14 the fundamental equilibrium path is shown as well as the related determinants D/D_0 and J/D_0 for a clambed arch, where D_0 is the value of determinant $\det \mathbf{K}_0$ at displacement $w_0 = 0$. The computation was carried out in [36] for elastoplastic material and reference load vector $Q^* = 10$ kN/m. The displacement w_0 in the middle of arc length was adopted as the control parameter $\tau = w_0$ which made it possible to evaluate the position of critical points B and L according to (7.129).

In case of the arc-length control parameter $\tau \equiv s$ the convergence of iterations can be accelerated by scaling the components in the constraint equation:

$$\Delta\bar{\mathbf{q}} = \mathbf{v}\Delta\mathbf{q}, \quad \Delta\bar{\lambda} = \beta\Delta\lambda, \tag{7.130}$$

where $\mathbf{v}$ is a diagonal matrix. Adopting $\beta = 0$ the efficiency of iteration process can be improved for flexible structures and $v_{\mathrm{rot}} = 1$, $v_{\mathrm{trans}} = 0$ is good for structures with dominating rotational degrees of freedom. In general other scalings are possible (e.g. $v_\alpha = {}^m q_\alpha$) but it is rather difficult to give appropriate recommendations since they strongly depend on the features of analysed structures and deformation process. These problems were discussed more thoroughly in [154] .

As shown in Fig. 7.12 the selection of step length ${}^m\tau$ can have essential influence on the convergence of the iteration procedure. The choice of optimal values of $\Delta\, {}^m\tau$ is not easy since it depends on the structural parameters, material parameters, adopted control as well as development of nonlinearities.

Usually the following formula is used

$$\Delta\,{}^{m+1}\tau = \left(\frac{I_a}{I_m}\right)^{\alpha} \Delta\,{}^{m}\tau\,, \tag{7.131}$$

where: I_a – assumed number of iterations, I_m – number of iterations from the previous step $\Delta^m\tau$. The parameter $\alpha \in [0,1]$ is fixed on the basis of numerical experiments, for instance in [140] the value $\alpha = 1/2$ is recommended.

The length of load control parameter can depend on the value of current stiffness parameter

$$\Delta\,{}^{m+1}\lambda = \left(\frac{\Delta S_a}{|\Delta S|}\right)^{\beta} \Delta\,{}^{m}\lambda\,, \tag{7.132}$$

where the current stiffness parameter is defined by the following formula [21]:

$$\Delta\,{}^{m}S = \left(\frac{\Delta\,{}^{m}\lambda}{\Delta^{1}\lambda}\right)^{2} \frac{\left(\Delta\mathbf{q}^{\mathrm{T}}\mathbf{K}\Delta\mathbf{q}\right)_1}{\left(\Delta\mathbf{q}^{\mathrm{T}}\mathbf{K}\Delta\mathbf{q}\right)_m} \tag{7.133}$$

and $\beta \in [1,2]$.

A more complicated criterion for the choice of load parameter length was proposed in [152], where in the formula

$$\Delta\,{}^{m+1}\lambda = f\,({}^{m}\delta)\,\Delta\,{}^{m}\lambda\,. \tag{7.134}$$

the function f depends on the parameter ${}^{m}\delta$ of contracting mapping

$${}^{m}\delta = \max_{i}\left(\|d_{i+1}\|/\|d_i\|\right)_m \tag{7.135}$$

In material nonlinear problems the choice of optimal step is much more complicated since it can depend also on local effects, for instance the change of yielding parameters should be taken into account.

In the case of an elastoplastic material the solution depends on the length of control parameter despite convergent computational procedures. It refers especially to the cases of deformation process with local unloadings (passive yielding). Good results are obtained if the so-called ***consistent approach*** is applied at the point level [141].

The questions mentioned above are partially discussed in Sect. 7.5, Example 3, where the displacement control and arc-length control are compared to each other and different lengths of control parameters are used.

The solution depends on the length of control parameter also in the case of nonconservative loads. The stiffness matrix is in general non-symmetric, which complicates the algorithms, needs storing more extended data and elongates the computational time. That is why there are different ways for *symmetrizing the stiffness matrix* [50, 153].

So only fundamental equilibrium paths have been considered. The computation of secondary paths initiated by bifurcation points needs first the computation of these points. The criterion (7.129) is useful in evaluating the range of isolation and then the eigenvalue problem (7.60), (7.62) or (7.63)

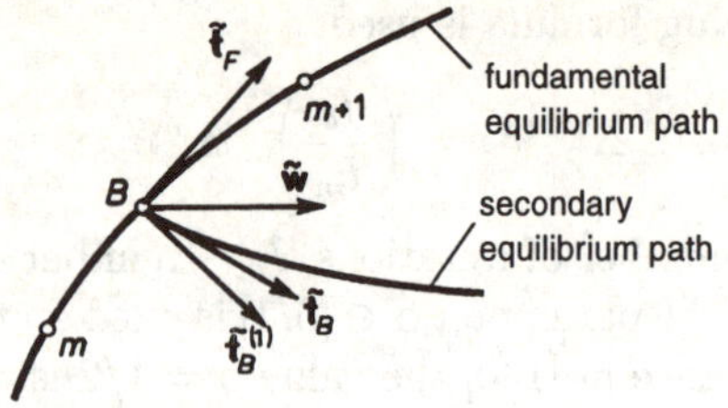

Fig. 7.15. Vectors at the bifurcation point B

can be analysed. Essential information can be obtained on the basis of the eigenvalue equation (7.46).

In the case of a single bifurcation point the vector tangent to the secondary equilibrium path $\tilde{\mathbf{t}}_B$ can be calculated as a linear combination of the vector tangent to the path $\tilde{\mathbf{t}}_F = \{\mathbf{t}_F, t_F^{N+1}\}$ and eigenvector $\tilde{\mathbf{w}} = \{\mathbf{w}, 0\}$, cf. Fig. 7.15:

$$\tilde{\mathbf{t}}_\beta = \alpha\left(\tilde{\mathbf{w}} + \eta\,\tilde{\mathbf{t}}_F\right) , \tag{7.136}$$

where $\alpha = |\tilde{\mathbf{w}} + \eta\,\tilde{\mathbf{t}}_F|^{-1} = \left(1 + 2\eta\,\tilde{\mathbf{w}}^{\mathrm{T}}\tilde{\mathbf{t}}_F + \eta^2\right)^{-1/2}$. The coefficient η is defined by the following formula (cf. [145, 146]):

$$\eta = \frac{K_{\alpha\beta\gamma} w_\alpha w_\beta w_\gamma}{2\left(K_{\alpha\beta\gamma} w_\alpha w_\beta t_{F\gamma} + K^\lambda_{\alpha\beta} w_\alpha w_\beta t_F^{N+1}\right)} , \tag{7.137}$$

where: $K_{\alpha\beta\gamma} \equiv \partial K_{\alpha\beta}/\partial q_\gamma$, $K^\lambda_{\alpha\beta} \equiv \partial K_{\alpha\beta}/\partial\lambda$. The application of formula (7.137) to further computations is much more difficult since the derivatives of the tangent stiffness matrix have to be computed. That is why an approximated value of the parameter η is commonly used.

As an initial vector $\mathbf{t}_B^{(1)}$ a vector placed in the plane $\left(\tilde{\mathbf{t}}_F, \tilde{\mathbf{w}}\right)$ can be adopted. The vector orthogonal to $\tilde{\mathbf{t}}_F$, cf. [145], can be used, as shown in Fig. 7.15. It corresponds to the orthogonality of pre- and post-bifurcation forms which lead to [22]:

$$\eta = -\mathbf{w}^{\mathrm{T}}\mathbf{t}_F . \tag{7.138}$$

In case of symmetric bifurcation points $\eta = 0$ and the vector $\tilde{\mathbf{t}}_B$ tangent to secondary equilibrium path is:

$$\tilde{\mathbf{t}}_B = \{\mathbf{w}, 0\} \equiv \tilde{\mathbf{w}} \tag{7.139}$$

The eigenvector $\tilde{\mathbf{w}}$ can be adopted as a predictor for the asymmetric bifurcation points since it usually well approximates the vector $\mathbf{t}_B$ in the displacement space $\mathbb{R}_N$, cf. [167].

The approximation of the predictor $\tilde{\mathbf{t}}_B^{(1)}$ as a vector non-collinear with the vector $\tilde{\mathbf{t}}_F$ makes it difficult to return to the fundamental path in the course of iterations. That is why the stiffness matrix is modified to the form

$$\mathbf{K}^{(i)} = \mathbf{K}_T\left(\tilde{\mathbf{q}}_F + \Delta\tilde{\mathbf{q}}_B^{(i)}\right) \quad \text{where} \quad \Delta\tilde{\mathbf{q}}_B^{(1)} = \beta\tilde{\mathbf{t}}_B^{(1)}\Delta\,{}^B\tau \quad \text{for} \quad \beta < 1\,. \tag{7.140}$$

The continuation of the iteration process agrees with the incremental methods discussed earlier in this section.

More refined methods for tracing the post-bifurcation paths were formulated in [147].

The procedures for the computation of bifurcation points and secondary equilibrium paths discussed above are illustrated in Sect. 7.5, Example 6, where large displacements of planar portal frame are analysed.

7.3.3 Methods of Reduced Basis

The analysis of structures by means of FEM can need a great number of degrees of freedom N. In the case of nonlinear problems it causes increasing the number of operations and a longer computational time. One of the possibilities to overcome such difficulties is to formulate a model with a smaller number of DOF, i.e. for $n < N$.

Let us assume that the vector of nodal displacements $\mathbf{q}$ is formulated as the following mapping

$$\mathbf{q}_{N\times 1} = \mathbf{\Gamma}_{N\times n}\boldsymbol{\psi}_{n\times 1}\,, \tag{7.141}$$

where the matrix $\mathbf{\Gamma}$ is composed of the basis vectors $\mathbf{g}_i$:

$$\mathbf{\Gamma}_{N\times n} = [\mathbf{g}_1 \ldots \mathbf{g}_n] \quad \text{for} \quad \mathbf{g}_i \in \mathbb{R}^N\,, \tag{7.142}$$

and the coefficients ψ_j are called *generalized degrees of freedom.* The vectors $\mathbf{g}_i$ are assumed to be linearly independent and the matrix $\mathbf{\Gamma}_{N\times n}$ is called the *matrix of reduced basis.*

Similarly to (7.141) the incremental relation is assumed

$$\Delta\mathbf{q} = \mathbf{\Gamma}\Delta\boldsymbol{\psi}\,. \tag{7.143}$$

After (7.143) is substituted to the incremental Eqs. (7.78) and then multiplied the left-hand side by $\mathbf{\Gamma}^{\mathrm{T}}$ the following incremental equation is formulated:

$$\mathbf{K}\Delta\boldsymbol{\psi} = \Delta\lambda\,\mathbf{Q}_R^* + \mathbf{R}_R\,, \tag{7.144}$$

where:

$$\mathbf{K}_{n\times n} = \mathbf{\Gamma}^{\mathrm{T}}\mathbf{K}_T\mathbf{\Gamma}, \quad \mathbf{Q}_R^* = \mathbf{\Gamma}^{\mathrm{T}}\mathbf{Q}^*, \quad \mathbf{R}_R = \mathbf{\Gamma}^{\mathrm{T}}\mathbf{R}\,. \tag{7.145}$$

The transformation associated with (7.145) is on the global level, i.e. it refers to the equations of assembled FE systems. Eqs. (7.144) are analysed by means of procedures discussed in the previous section. Due to relation (7.143) the return to original model can be carried out. After the vector of displacement increments $\Delta\mathbf{q}$ is computed also the vector of residual forces $\mathbf{R}$

can be computed. It needs the computation on the level of integration points and separated finite elements.

In papers by Noor and Peters (cf. e.g. [116]) the error corresponding to the application of the reduced basis is evaluated by the following formula:

$$e = \frac{1}{\lambda N}\sqrt{\frac{\mathbf{R}_R^{\mathrm{T}}\mathbf{R}_R}{\mathbf{Q}_R^{*T}\mathbf{Q}_R^{*}}}\,. \tag{7.146}$$

When the error is greater than the admissible error $e > e_a$ a modification of the basis vectors should be performed and their number n can be increased.

The main problem is the selection of the vectors of reduced basis $\mathbf{g}_i$. Besides the linear independence the generation of vectors $\mathbf{g}_i$ should be simple and efficient (low number of operations and data transmissions). The number of basis vectors should be very small, i.e. $n \ll N$, automatically selected and the approximation by them should be good enough to compute long parts of the equilibrium path without modifications of the reduced basis. Moreover the vectors $\mathbf{g}_i$ should enable us to compute the limit and turning points. The solutions by the reduced basis method are called approximate solutions as opposed to 'exact' solutions which result from the non-reduced, original FE model.

Satisfaction of all the criteria mentioned above, is practically impossible. The existing methods of the basis vectors selection can be classified in four groups: 1) eigenvectors associated with different formulation of eigenproblems, 2) displacement vectors as solution of (7.127) and their derivatives, 3) displacement vectors, the predictor and correctors of the FEM nonlinear set of equations, 4) other basis of orthogonal vectors.

The *method of modal superposition* can be mentioned as a commonly used method of reduced basis in the dynamics of deformable mechanical systems. In nonlinear problems the local modal superposition is used [111]. This approach corresponds to small harmonic vibrations imposed on large static deflections. The analysis of small changes is carried out by means of eigenvalues and eigenvectors of the equation:

$$\left[\mathbf{K}_T\left({}^m\mathbf{q}\right) - \omega_j^2\mathbf{M}\right]\mathbf{w}_j = \mathbf{0}\,. \tag{7.147}$$

An analogue to the approximation (7.143) is the vector (cf. [66, 111]):

$$\Delta\mathbf{q} = \mathbf{W}^{(n)}\left({}^m\mathbf{q}\right)\mathbf{z}\,, \tag{7.148}$$

where the reduced matrix is composed of n eigenvectors of Eq. (7.147):

$$\mathbf{W}^{(n)}\left({}^m\mathbf{q}\right) \equiv \mathbf{W}_{N\times n} = \left[\mathbf{w}_1\left({}^m\mathbf{q}\right), \ldots, \mathbf{w}_n\left({}^m\mathbf{q}\right)\right]\,. \tag{7.149}$$

In nonlinear problems the vectors $\mathbf{w}_j$ depend on current displacements ${}^m\mathbf{q}$ and if nonlinearities are significant the reduced basis has to be frequently updated. The same situation takes place if instead of (7.147) the stability equation (7.46) is used. The approximation can be improved if besides the basis vectors also their derivatives are used [67]:

$$\mathbf{\Gamma} = \left[\mathbf{w}_j, \frac{\partial \mathbf{w}_j}{\partial z_k} + \frac{\partial \mathbf{w}_k}{\partial z_j}, \dots \right] . \tag{7.150}$$

In statics the reduced basis can be formulated by means of eigenvectors $\mathbf{v}_j$ associated with eigenproblems (7.66) or (7.67). In papers [89, 108, 109, 177] the matrix of reduced basis

$$\mathbf{X}^{(n)} \equiv \mathbf{X}_{N\times n} = [\mathbf{v}_1, \dots, \mathbf{v}_n] \tag{7.151}$$

was composed of the initial buckling vectors $\mathbf{v}_j$. This approach restricts applications to small or moderate nonlinearities (at this approach limit points cannot be computed).

A different approach was developed in many papers by Noor and Peters [113]–[121]. The basis vectors correspond to the solution of FE equations $\bar{\mathbf{q}} \equiv {}^m\bar{\mathbf{q}}$ and subsequent derivatives $\partial^r \bar{\mathbf{q}}/\partial \eta^r$ for $r = 1, \dots, R < n - 1$ are used:

$$\mathbf{\Gamma} = \left[\bar{\mathbf{q}}, \frac{\partial \bar{\mathbf{q}}}{\partial \eta}, \dots, \frac{\partial^R \bar{\mathbf{q}}}{\partial \eta^R} \right] , \tag{7.152}$$

where bars over symbols denote scaling and η is a parameter of the equilibrium path. This parameter is selected similarly to the choice of control parameter τ.

In the quoted papers by Noor and Peters the non-incremental formulation of FEM equations was used:

$$\mathbf{K}^{(\mathrm{con})}\mathbf{q} + \mathbf{N}(\mathbf{q}) = \lambda\, \mathbf{q}^* , \tag{7.153}$$

where the vector of internal forces $\mathbf{N}$ depends on the displacement vector $\mathbf{q}$, i.e. $\mathbf{N} = \left[\overset{*}{\mathbf{K}}{}^{(\sigma)}(\mathbf{q}) + \overset{*}{\mathbf{K}}{}^{(u)}(\mathbf{q}) \right] \mathbf{q}$.

One of the main disadvantages of the reduced basis (7.152) is the tendency of the basis vectors to be linearly dependent if their number increases. In order to evaluate this effect the Gram matrix is formulated

$$\mathbf{G} = \mathbf{\Gamma}^{\mathrm{T}}\mathbf{\Gamma} , \tag{7.154}$$

and then the condition number is computed

$$\gamma = \omega_{\max}(\mathbf{G})/\omega_{\min}(\mathbf{G}) . \tag{7.155}$$

In case of $\gamma = 1$ the basis is orthonormal and $\gamma \to \infty$ if collinear vectors appear.

In papers by Noor and Peters a recursion type algorithm was applied to the computation of reduced basis using a procedure associated with the perturbation method [53]. From among many analysed problems in Examples, Sect. 7.5, there are discussed results corresponding to large displacement and computation of limit points for an arch and cylindrical shell.

The efficiency of the approach, measured by the number of modifications of the reduced basis vectors, increases if mixed finite elements are used [115]. From among various applications the computation of secondary equilibrium path [117, 118, 121] and the stability analysis for two parameters loads [120] should be mentioned.

An especially simple approach to the selection of basis vectors was proposed in [31, 32]. The vectors are accepted as the nodal displacement vectors from the previous steps $m-1$ and m as well as the predictor and correctors from the mNR method:

$$\mathbf{\Gamma} = \left[{}^{m-1}\mathbf{q},\ {}^{m}\mathbf{q}, \mathbf{d}_1, \mathbf{d}_2, \ldots \right] . \tag{7.156}$$

The efficiency of the approach can be increased by normalization and orthogonalization of these vectors by means of the Gram–Schmidt algorithm.

In [32] direct and indirect reductions of FE system were considered. The direct reduction is based on the explicit formulation of FE equations which can be written in the following non-incremental form corresponding to (7.153):

$$F_\alpha \equiv K^{(\mathrm{con})}_{\alpha\beta}\, q_\beta + N^1_{\alpha\beta\gamma}\, q_\beta\, q_\gamma + N^2_{\alpha\beta\gamma\delta}\, q_\beta\, q_\gamma\, q_\delta = \lambda\, \mathbf{Q}^*_\alpha \tag{7.157}$$

for $\alpha, \beta, \gamma, \delta = 1, \ldots, N$. Due to explicit form (7.157) the computation of residual forces of the original system $R_\alpha = \lambda Q^*_\alpha - F_\alpha$ does not need returning to the elements and the transformation $(7.145)_3$ can be performed on the level of FE assembly.

The direct reduction, associated with equations (7.157), is possible only for linear elastic material. In the case of materials of nonlinear characteristics the matrices $\mathbf{N}^{(1)}$ and $\mathbf{N}^{(2)}$ are not digital matrices and the indirect reduction requires an analysis on the level of individual finite elements.

The direct reduction is much more efficient than the indirect reduction and can give considerable reduction of computational time. This is shown in Sect. 7.5, Example 9 where for a spatial truss at comparatively low number of DOF, i.e. for $N = 111$, the computational time is half of the time corresponding to indirect reduction or the analysis of original system.

The efficiency of indirect reduction will presumably decrease for a higher number of DOF. Besides the paper [32], there is lack of comparison of the efficiency of various approaches and remarks about the possibility of significant reduction of the execution time (cf. [113]) should be accepted with great criticism. This refers especially to those methods which base not only on the indirect reduction but also generated eigenvectors as the basis vectors. Such conclusions were formulated in [150] where a good approximation was obtained by means of mixed bases composed of the eigenvectors of the stiffness matrix $\mathbf{K}_T$ and basis (7.156) associated with the NR method. It is worth mentioning that the idea of reduced bases was introduced to the global Rayleigh–Ritz global approximation [2]. A development of this approach to automatic generation of orthogonal functions was given in [66].

In this section only those methods of reduced bases which are associated with the transformation (7.143) have been discussed. It enables us to have 'a contact' between the original and reduced systems in order to evaluate the errors of approximation.

Other approaches are associated with local-global approximations [37, 107] and especially they are associated with the strip method [33, 158].

7.4 Initial Buckling and Evaluation of Solution of Nonlinear and Nonconservative Problems

In many engineering problems the primary critical points can be calculated with satisfactory accuracy applying initial buckling equation (7.67). Such solutions can be a basis for approximate solutions of nonlinear problems.

The initial computation is associated with linear solution

$$\mathbf{K}^{(\mathrm{con})}\mathbf{q}^* = \mathbf{Q}^* \to \boldsymbol{\sigma}^* = \boldsymbol{\sigma}(\mathbf{q}^*)\,. \tag{7.158}$$

Then eigenvalues and eigenvectors are computed λ_α, $\mathbf{v}_\alpha$ for $\alpha = 1, \ldots, n \ll N$ using Eqs. (7.13) and (7.67). The computed values are combined as matrices:

$$\boldsymbol{\Lambda}^{(n)} \equiv \boldsymbol{\Lambda}_{n\times n} = \begin{bmatrix} \lambda_1 & 0 & \ldots & 0 \\ 0 & \lambda_2 & \ldots & 0 \\ \ldots & \ldots & \ldots & \ldots \\ 0 & 0 & \ldots & \lambda_n \end{bmatrix}, \quad \mathbf{X}^{(n)}_{N\times n} = [\mathbf{v}_1, \ldots, \mathbf{v}_n]\,, \tag{7.159}$$

which enable us to write (7.67b) in the matrix form:

$$\mathbf{K}^{(\mathrm{con})}\mathbf{X}^{(n)} = -\mathbf{K}^{(\sigma)}(\boldsymbol{\sigma}^*)\mathbf{X}^{(n)}\boldsymbol{\Lambda}^{(n)}. \tag{7.160}$$

It should be mentioned that for practical reasons only $n \ll N$ eigenvalues and eigenvectors are computed. For the sequence $\lambda_\alpha \ll \lambda_{\alpha+1}$ the number n defines 'essential' eigenvalues, corresponding to condition $|\lambda_1/\lambda_{n+1}| \ll 1$.

Matrices (7.159) satisfy the orthogonality conditions (index n is omitted):

$$\mathbf{X}^{\mathrm{T}}\mathbf{K}^{(\mathrm{con})}\mathbf{X} = \boldsymbol{\Lambda}, \quad \mathbf{X}^{\mathrm{T}}\mathbf{K}^{(\sigma)}\mathbf{X} = -\mathbf{I}\,. \tag{7.161}$$

The eigenvectors can be normalized in such a way as to fulfil the condition

$$\bar{\mathbf{v}}^{\mathrm{T}}_\alpha\mathbf{K}^{(\mathrm{con})}\bar{\mathbf{v}}_\beta = \delta_{\alpha\beta} \quad \text{for } \alpha, \beta = 1, \ldots, n \ll N\,. \tag{7.162}$$

Equations (7.67) and (7.160) correspond to generalized (nonstandard) eigenproblems. The majority of algorithms of linear algebra were formulated the analyse of standard eigenproblems according to (7.46). That is why the eigenproblem equation is transformed to the following matrix form:

$$\mathbf{AY} = \mathbf{Y}\boldsymbol{\Omega}\,, \tag{7.163}$$

where

$$\mathbf{A} = \mathbf{U}^{-\mathrm{T}}\mathbf{K}^{(\sigma)}\mathbf{U}^{-1}, \quad \mathbf{Y} = \mathbf{U}\mathbf{X} \quad \text{or} \quad \mathbf{X} = \mathbf{U}^{-1}\mathbf{Y}, \\ \boldsymbol{\Omega} \equiv \lceil \omega_1 \ldots \omega_n \rfloor = \boldsymbol{\Lambda}^{-1}. \tag{7.164}$$

In the above relations the upper triangular matrix $\mathbf{U}$ is a result of factorization of the matrix $\mathbf{K}^{(\mathrm{con})}$:

$$\mathbf{K}^{(\mathrm{con})} = \mathbf{U}^{\mathrm{T}}\mathbf{U}, \tag{7.165}$$

and the diagonal matrix $\boldsymbol{\Omega}$ is composed of roots $\omega_1 \geq \omega_2 \geq \cdots \geq \omega_n$ of the equation

$$\det|\mathbf{A} - \omega\mathbf{I}| = 0. \tag{7.166}$$

The orthogonality condition (7.162) takes the form:

$$\mathbf{Y}^{\mathrm{T}}\mathbf{Y} = \mathbf{I}. \tag{7.167}$$

In the case of linear deformation in the prebuckling state the eigenvectors $\mathbf{v}_\alpha$ are orthogonal to the displacement vector $\mathbf{q}^*$. In the nonlinear problems the vector $\mathbf{q}^*$ can be split into two parts:

$$\mathbf{q}^* = \overset{\parallel}{\mathbf{q}} + \overset{\perp}{\mathbf{q}}, \tag{7.168}$$

where the vector $\overset{\parallel}{\mathbf{q}}$ is 'parallel' to eigenvectors in the sense of linear combination of vectors $\mathbf{v}$:

$$\overset{\parallel}{\mathbf{q}}_{N\times 1} = \mathbf{X}_{N\times n}\,\mathbf{a}_{n\times 1}. \tag{7.169}$$

The 'orthogonal' vector $\overset{\perp}{\mathbf{q}}$ satisfies the following equation:

$$\mathbf{X}^{\mathrm{T}}\mathbf{K}^{(\mathrm{con})}\overset{\perp}{\mathbf{q}} = 0. \tag{7.170}$$

After the relations (7.169), (7.170) and (7.161) are taken into account the vector of coefficients $\mathbf{a}$ is:

$$\mathbf{a} = \boldsymbol{\Lambda}^{-1}\mathbf{X}^{\mathrm{T}}\mathbf{K}^{(\mathrm{con})}\mathbf{q}^* \equiv \boldsymbol{\Lambda}^{-1}\mathbf{X}^{\mathrm{T}}\mathbf{Q}^*, \tag{7.171}$$

or corresponding to (7.162):

$$a_\alpha = \bar{\mathbf{v}}_\alpha^{\mathrm{T}}\mathbf{K}^{(\mathrm{con})}\mathbf{q}^* \quad \text{for} \quad \alpha = 1, \ldots, n. \tag{7.172}$$

The coefficients a_α can be used for the evaluation of an inclination from the linearity of the prebuckling state according to the following inequality:

$$\frac{a_\alpha}{\sqrt{W}} < \varepsilon_\alpha \quad \text{where} \quad W = \overset{*}{\mathbf{Q}}{}^{\mathrm{T}}\mathbf{q}^*. \tag{7.173}$$

The evaluation $\varepsilon \ll 1$ was computed on the basis of eigenvector spectrum in course of the iterative eigenanalysis [108, 109].

A slightly different approach was suggested in [177]. From the scalar product $\overset{*}{\mathbf{q}}{}^T\mathbf{Q}^* = \overset{*}{\mathbf{q}}{}^T\mathbf{K}^{(\mathrm{con})}\overset{*}{\mathbf{q}}$ the following relation results if (7.168) and (7.169) are taken into account:

$$\overset{\shortparallel}{\mathbf{q}}{}^{\mathrm{T}}\mathbf{K}^{(\mathrm{con})}\overset{\shortparallel}{\mathbf{q}} + \overset{\perp}{\mathbf{q}}{}^{\mathrm{T}}\mathbf{K}^{(\mathrm{con})}\overset{\perp}{\mathbf{q}} = \overset{*}{\mathbf{q}}{}^T\mathbf{Q}^* \,.$$

This leads to the relation

$$\overset{\shortparallel}{\zeta} + \overset{\perp}{\zeta} = 1 \tag{7.174}$$

of the following dimensionless variables:

$$\overset{\shortparallel}{\zeta} = \frac{1}{W}\mathbf{a}^{\mathrm{T}}\mathbf{\Lambda}\,\mathbf{a}, \quad \overset{\perp}{\zeta} = \frac{1}{W}\overset{\perp}{\mathbf{q}}{}^{\mathrm{T}}\mathbf{K}^{(\mathrm{con})}\overset{\perp}{\mathbf{q}} \tag{7.175}$$

where: $W = \overset{*}{\mathbf{q}}{}^T\mathbf{Q}^*$.

The global variable $\overset{\shortparallel}{\zeta}$ can be used for the evaluation of nonlinearity of the prebuckling state. In case

$$\overset{\shortparallel}{\zeta} < \varepsilon_\zeta \tag{7.176}$$

the system is *slightly nonlinear*. The value $\varepsilon_\zeta = 0.01$ was computed in [177] for numerous examples of structural systems.

Another evaluation of nonlinearities corresponds to the computation of buckling loads for the initial buckling and linearized buckling, λ_I and λ_L, respectively. The matrix of initial displacements $\mathbf{K}^{(u1)}(\mathbf{u}^*)$, taken into account in (7.66), makes the structural system more flexible, which leads to the inequality

$$\lambda_I \geq \lambda_L \,. \tag{7.177}$$

The approach to linear systems is reflected by transition $\lambda_L \to \lambda_I$ thus the level of nonlinearity can be evaluated by the following parameter [177]:

$$\beta = \frac{\lambda_I - \lambda_L}{\lambda_L} 100\% \,. \tag{7.178}$$

Numerous computed examples have shown that for $\varepsilon_\zeta \approx 0.01$ there is $\beta \leq 2\%$.

From among the numerical examples discussed in [177] only the analysis of *William's frame* is presented in Sect. 7.5, Example 10. In this example the values of parameters $\overset{\shortparallel}{\zeta}$ and β are given for the initial and linearized buckling analysis, respectively.

The initial nonlinear deformation can be computed by means of the modal superposition method. This method can be treated as a special case of the method of reduced basis.

Let us assume the approximate Eq. (3.52), corresponding to (7.67b):

$$\left[\mathbf{K}^{(\mathrm{con})} + \lambda\mathbf{K}^{(\sigma)}(\sigma^*)\right]\mathbf{q} = \lambda\,\mathbf{Q}^* \,. \tag{7.179}$$

The matrices $\mathbf{K}^{(\mathrm{con})}$ and $\mathbf{K}^{(\sigma)}(\boldsymbol{\sigma}^*)$ are formulated on the basis of (7.161) assuming $n = N$:

$$\mathbf{K}^{(\mathrm{con})} = \mathbf{X}^{-\mathrm{T}}\mathbf{\Lambda}\,\mathbf{X}^{-1}, \quad \mathbf{K}^{(\sigma)}(\boldsymbol{\sigma}^*) = -\mathbf{X}^{-\mathrm{T}}\mathbf{X} \,. \tag{7.180}$$

After substitution (7.180) to (7.179) the displacement vector $\mathbf{q}(\lambda)$ can be expressed by the following formula:

$$\mathbf{q}(\lambda) = \mathbf{X}\left[\frac{1}{\lambda}\mathbf{\Lambda} - \mathbf{I}\right]^{-1} \mathbf{X}^{\mathrm{T}}\mathbf{Q}^{*} \equiv \mathbf{X}\left[\frac{1}{\lambda}\mathbf{\Lambda} - \mathbf{I}\right]^{-1} \mathbf{a}\,. \tag{7.181}$$

Formula (7.181) is valid also for $n < N$, for the reduced basis $\mathbf{\Gamma} \equiv \mathbf{X}^{(n)}\left[\lambda^{-1}\mathbf{I} - \mathbf{\Lambda}^{(n)^{-1}}\right]^{-1}$ and at the vector of generalized displacements $\boldsymbol{\psi} \equiv \mathbf{a}$. The appropriate formula, suitable for computation takes the form [89]:

$$\mathbf{q}(\lambda) = \sum_{\alpha=1}^{n} \frac{\lambda}{\lambda_\alpha - \lambda}\mathbf{v}_\alpha \mathbf{v}_\alpha^{\mathrm{T}}\mathbf{Q}^{*}\,. \tag{7.181a}$$

It is clear that the formula (7.181a) is valid for $\lambda \neq \lambda_\alpha$. In practical applications the formula is used for $\lambda < \lambda_1$.

The next Example 11 in Sect. 7.5 corresponds to a portal planar frame. The perfect frame, perturbed by the horizontal force, can be analysed with reasonable accuracy by means, for instance of four eigenvectors only, i.e. for $n = 4$ if the structure is weakly nonlinear.

In the case of structures with a free edge the influence of non-symmetric part in the load matrix $\mathbf{K}^{(p)}$ can be evaluated by an approximate approach [102, 103]. This can be done by the computation of critical values of load parameters for two symmetric matrices:

$$\det\left|\mathbf{K}^{(\mathrm{con})} + \lambda^{(0)}\left[\mathbf{K}^{(\sigma)}(\boldsymbol{\sigma}^{*}) + \mathbf{K}^{(p1)S}(p^{*})\right]\right| = 0, \tag{7.182a}$$

$$\det\left|\mathbf{K}^{(\mathrm{con})} + \lambda^{(1)}\left[\mathbf{K}^{(\sigma)}(\boldsymbol{\sigma}^{*}) + \mathbf{K}^{(p1)S}(p^{*}) + \mathbf{K}^{(p)\Gamma S}(p^{*}_{\Gamma})\right]\right| = 0, \tag{7.182b}$$

where the matrices defined in (7.42) and (7.44) are used. The tangent matrix of initial loads $\mathbf{K}^{(p1)S}$ is formulated from (7.42) taking into account only the part dependent on reference load p^{*}. The matrix of non-potential load, defined by (7.44), is symmetrized:

$$\mathbf{K}^{(p)\Gamma S} = \frac{1}{2}\left[(\mathbf{K}^{(p)\Gamma})^{\mathrm{T}} + \mathbf{K}^{(p)\Gamma}\right]. \tag{7.183}$$

The influence of matrix $\mathbf{K}^{(p)\Gamma}$ is small if the following condition is satisfied:

$$\mu \equiv \left|\frac{\lambda^{(1)} - \lambda^{(0)}}{\lambda^{(0)}}\right| \ll 1. \tag{7.184}$$

It was proved in [103] that the value $\lambda^{(1)}$ can be treated as the first approximation $\lambda^{(1)} = \lambda^{(0)} + \Delta\lambda$, where:

$$\Delta\lambda \equiv \lambda^{(1)} - \lambda^{(0)} = \lambda^{(0)}\mathbf{v}_0^{T}\mathbf{K}^{(p)\Gamma}\mathbf{v}^{0}\,, \tag{7.185}$$

and $\mathbf{v}_0$ is the eigenvector of (7.67a), corresponding to (7.182a).

In Sect. 7.5, Example 13, it is shown that for a cylindrical shell under external pressure the influence of non-potential effects associated with the free edge can be negligibly small.

7.5 Numerical Examples

Several numerical examples have been selected from the existing literature in order to illustrate the problems discussed in previous sections of this chapter. Only basic data are shown – more precise information can be found in the quoted references.

The numerical examples deal mainly with the stability of bar structures, then also with the stability of plates and shells.

Taking into account the illustrative aspects only less complex structures are analysed. Due to such a selection it is easier to discuss the basic aspects of the considered problems.

Example 1. Large Deflections of a Circular Plate

The aim of this example, taken from [104] is to show a comparison of the efficiency of different methods for the analysis of FE nonlinear equations. A rotationally symmetric plate is made of elastic-plastic material which obeys the Huber–Mises–Hencky yield condition and isotropic linear hardening. The

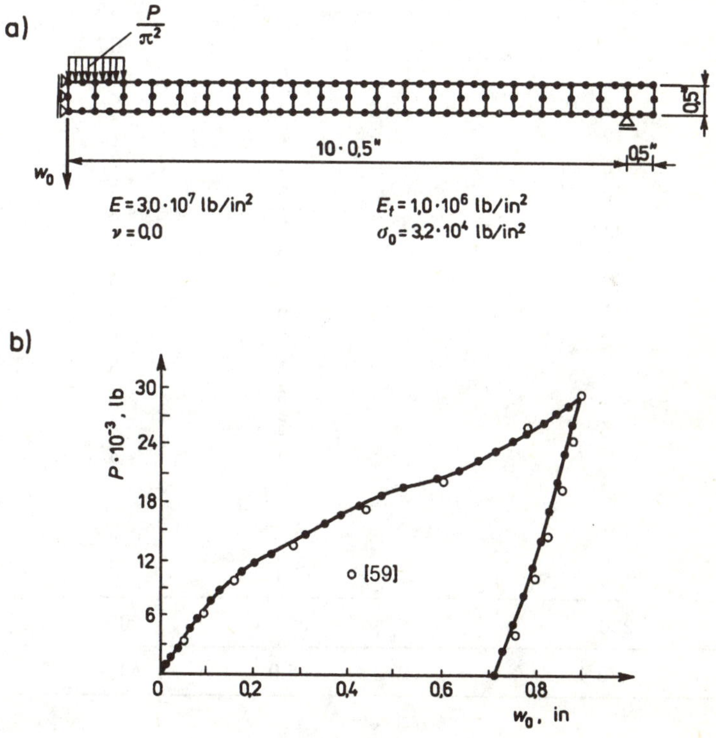

Fig. 7.16. Deflections of rotationally–symmetric loaded circular plate

Table 7.1. Number of iterations and CPU time for the plate in Fig. 7.16

Load $P \cdot 10^{-3}$ (lb)	Number of iterations in the methods				
	NR	mNR	BFDS	SN3	SN2
1.0	1	1	1	1	1
2.0	4	4	4	4	4
.					
.					
.					
8.0	4	16	7	6	7
9.0	4	12	8	8	8
10.0	5	10	7	7	7
11.0	5	7	5	5	5
.					
.					
.					
14.0	6	7	5	5	5
15.0	6	diverg.	10	12	12
16.0	6		9	9	10
17.0	7		5	6	6
18.0	6		5	5	5
19.0	7		5	5	8
20.0	12		7	7	8
21.0	16		6	7	8
22.0	10		5	6	6
23.0	11		6	6	6
24.0	7		6	6	6
25.0	9		5	5	6
26.0	5		6	6	6
27.0	8		5	5	6
28.0	4		5	5	5
29.0	7		5	5	5
30.0	5		5	5	5
27.0	4		4	4	4
.					
.					
.					
0.0	5		4	4	4
$\sum =$	226		204	209	219
CPU (s)	82.4		45.0	44.3	46.0

plate is discretized by means of 21 finite elements of 108 nodes and the number of DOFs is $N = 214$, cf. Fig. 7.16.

The load control parameter was used, i.e. $\tau \equiv \lambda$ applying 20 equal steps for loading and 10 steps for unloading. The results of computation are compared with the results of experiments [59].

The number of iterations and total CPU time is put together for the following methods: Newton–Raphson (NR), modified Newton–Raphson (mNR), Quasi-Newton (BFDS) and Secant-Newton (SN3 and SN2). The methods SN3 and SN2 correspond to the three-parameter formula (7.105) and two-parameter formula (7.109), respectively. In the last three methods the first approximation at each step was computed by means of the mNR method. In Table 7.1 there are omitted steps at which the number of iterations was not changed.

The number of iterations is comparable for different methods. The updating of stiffness matrix at each iteration leads to nearly doubled time in the NR method when it is compared with the QN methods. The mNR method is divergent because of development of the yielding zone. Geometrical nonlinearities also cause a significant increase of iterations in the NR method (16 iterations at the 21st step). The computational time is given for the CDC 7600 computer.

Example 2. Cylindrical Shell under External Pressure

A cylindrical shell called also the *Scordelis–Lo roof* is used as a bench-mark test for the validation of finite elements and algorithms for nonlinear analysis. This is associated with coupling of membrane and bending stress fields in the meridional and circumferential directions.

In Fig. 7.17a the analysed shell is shown. The shell was considered to be loaded by external, follower pressure. Because of symmetry only a quarter of the shell was computed. The mesh of 5×5 eight-node serendipity type finite elements with reduced integration was used [141]. Perfect elastic–plastic material and the Huber–Mises–Hencky yield condition was assumed. The consistent stiffness matrix was used at the point level and the layered model for computation of the generalized resultant forces was performed (in this example the Simpson quadrature formula was used for 7 layers). In the model considered the geometrical nonlinearities and load stiffness matrix were taken into account.

In Fig. 7.17a the basic data are presented and in Fig. 7.17b the fundamental equilibrium path is shown on the plane (p, u_y^A), computed in [141], where p – external pressure intensity, u_y^A – displacement along y axis at the point A of the edge.

The arc-length control parameter $\tau \equiv s$ was used. At the points 1, ... ,4 at the equilibrium path there is written the percentage of Gauss integration points which undergo yielding.

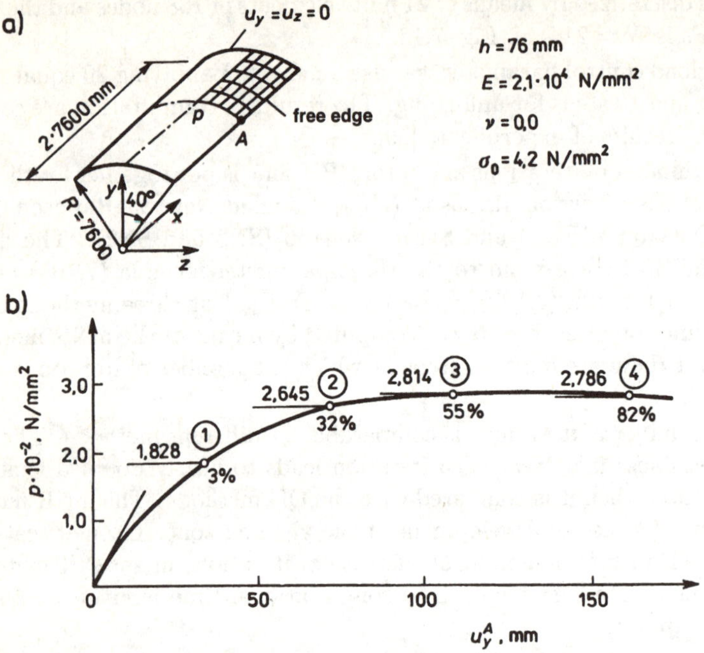

Fig. 7.17. Cylindrical panel under external pressure: **a)** data, **b)** equilibrium path

Table 7.2. Number of iterations and average CPU time for the panel shown in Fig. 7.17

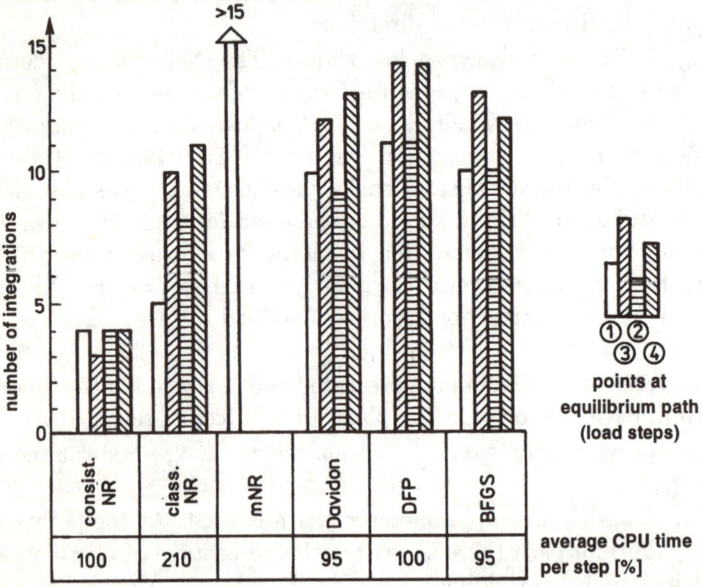

The number of iterations and average CPU time are put together in Table 7.2, referred to points 1, ... ,4 at the equilibrium path shown in Fig. 7.17b. The reference time was accepted as 100 for the NR method with consistent stiffness matrix. This time is nearly equal to the computational time for the QN methods, similarly as in Example 1.

Example 3. Cylindrical Shallow Shell under Concentrated Force

An elastic, shallow cylindrical shell, shown in Fig. 7.18a, has free curvilinear edges and simply supported meridional edges. Assuming symmetry of deformation only a quarter of the shell is considered, similarly as in Example 2. Since the papers [149] by Sabir and Lock this shell has been considered as a bench–mark test for the validation of various finite elements and computational procedures, cf. e.g. [106].

The computation reported in [39] was carried out for the mesh of 4 × 4 rectangular finite elements. A finite element had 28 DOFs (4 corner nodes each of 5 DOF, 2 middle edge nodes of 2 DOF per one node). That leads to $N = 151$ DOF of the analysed FE system shown in Fig. 7.18a.

In Table 7.3 the number of iterations is given for the points A,B, ... ,K at the equilibrium path in Fig. 7.18a. The results were obtained on the basis of mNR method and for the SN methods. The formulae (7.105) and (7.109) correspond to the three parameter method SN3 and two parameter method

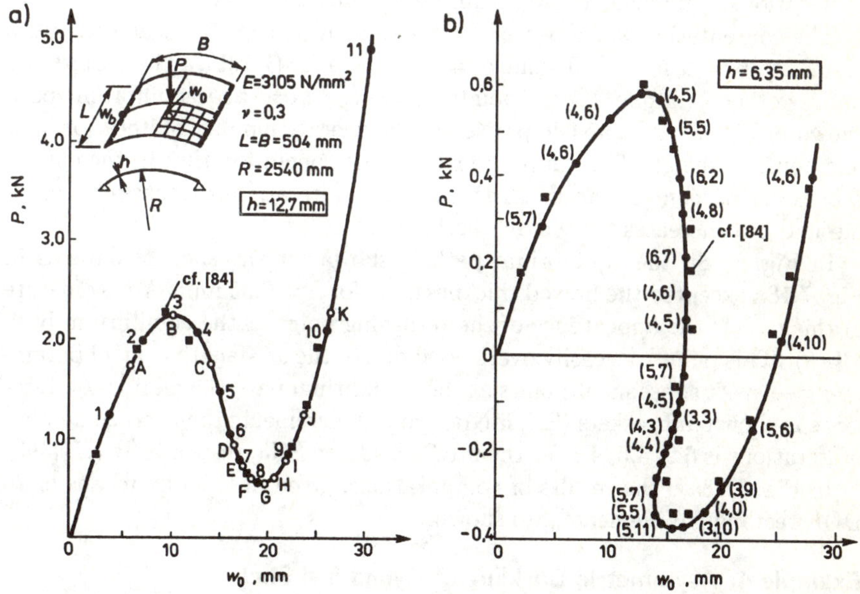

Fig. 7.18. Equilibrium paths for a shallow cylindrical shell of thickness: **a)** $h = 12.7$ mm, **b)** $h = 6.35$ mm

Table 7.3. Average number of iterations for the shell in Fig. 7.18a under displacement control $\tau \equiv w_0$

Methods of analysis	Points at equilibrium path											Average number of iterations
	A	B	C	D	E	F	G	H	I	J	K	
mNR	6	7	13	13	6	4	3	4	4	4	5	6.3
SN3-(7.105)	5	5	6	6	2	2	2	2	2	3	3	3.5
SN2-(7.109)	5	5	6	7	3	3	2	2	3	3	3	3.7

Table 7.4. Average number of iterations for the panel in Fig. 7.18a under arc-length control parameter $\tau \equiv s$

Methods of analysis	Points at equilibrium path											Average number of iterations
	1	2	3	4	5	6	7	8	9	10	11	
mNR	4	4	4	4	6	5	4	5	6	7	8	5.2
SN3	4	4	4	4	5	5	3	4	5	4	4	4.2
SN2	4	4	4	4	5	4	3	4	4	4	5	4.1

SN2, respectively. The computation was carried out under the displacement control $\tau = w_0$. As can be seen, the SN methods give a comparable number of iterations, significantly lower than for the mNR method.

The computation was repeated for the arc-length control parameter $\tau \equiv s$ assuming $\beta = 0$ in (7.130) and $\alpha = 1$, $I_z = 4$ in (7.131). The results are put together in Table 7.4 for points 1, 2, ... , 11 at the equilibrium path shown in Fig. 7.18a. The comparison of the average number of iterations per one point at the equilibrium path leads to the conclusion that in the case of the arc-length control parameter the number of iterations decreases for mNR method but increases for the SN methods.

In Fig. 7.18b the equilibrium path is shown for the shell of data as in Fig. 7.18a excepted the halved thickness, i.e. for $h = 6.35$ mm. A new feature of this case is the appearance of the returning point at the equilibrium path $P(w_0)$. This point was easily overpassed due to the arc-length control parameter $\tau \equiv s$. At subsequent points at the equilibrium path the number of iterations is shown in brackets (SN, mNR). For 29 increments the average number of iterations is 6.9, 4.5, 4.1 for the mNR, SN2 and SN3, respectively (cf. [39]).

In Fig. 7.18a,b the results of computations, given in [106] by means of 20 DOF rectangular elements, are shown.

Example 4. Asymmetric Buckling of Cylindrical Shell

The simply supported cylindrical shell shown in Fig. 7.18b has the span of 503 mm and height of 12.5 mm. Now the height is doubled, which leads to

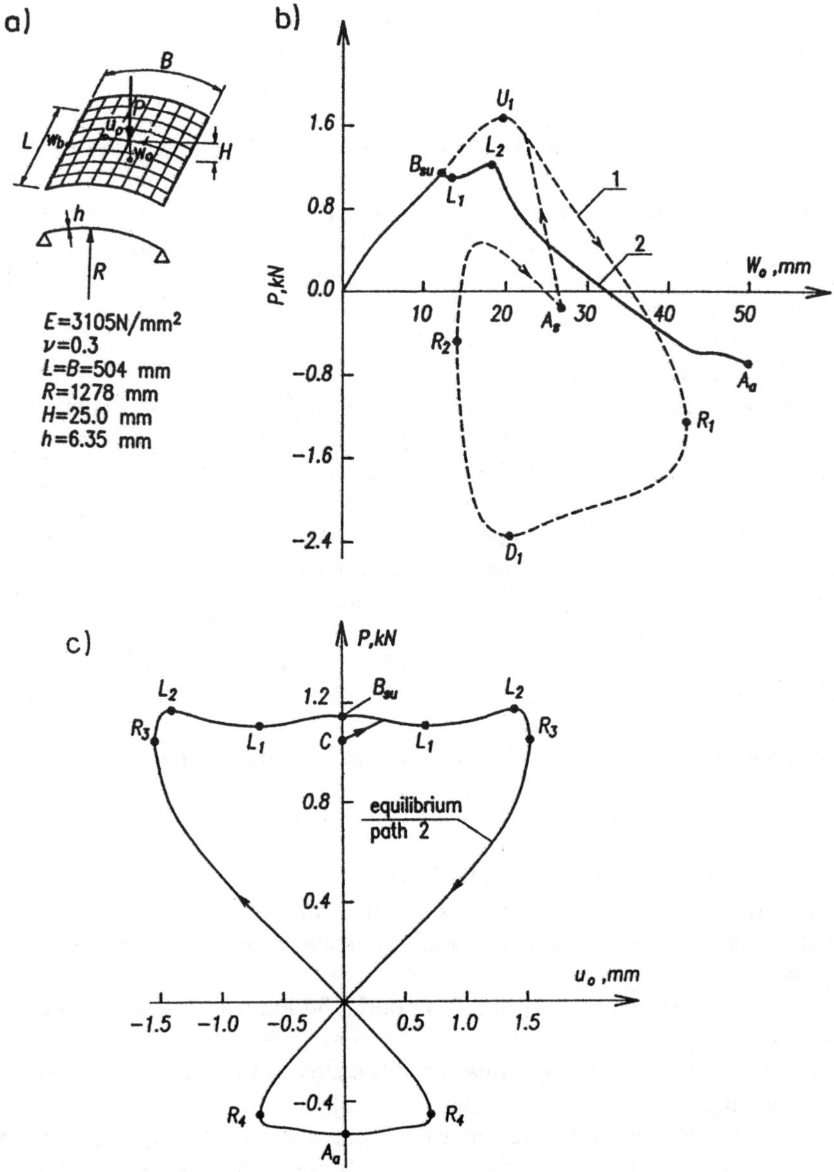

Fig. 7.19. Asymmetric buckling of a shallow cylindrical shell: **a)** data, **b)** equilibrium paths on the plane (P, w_0), **c)** equilibrium paths on the plane (P, u_0)

the shell radius $R = 1278$ mm. The thickness $h = 6.35$ mm and other data are as those for the shell in Fig. 7.19b.

For the whole shell there were applied 64 serendipity type finite elements as those described in Example 2. The arc-length control parameter was used in [170] with the step $\Delta\tau \equiv \Delta s = 1.0 - 10.0$ mm (the parameter $\beta = 0$ was assumed).

In Fig. 7.19a the equilibrium path 1 corresponds to the unforced symmetric deformation. It corresponds to adding extra constraints in the middle of the shell, which eliminated asymmetric displacements. The upper and lower limit points U_1 and D_1 occurred at the equilibrium path 1. The snap-back points R_1, R_2 were computed as well. At the point A_a a snap-through took place and then the fundamental path 1 was followed at monotonically increasing arc-length parameter.

The secondary equilibrium path 2 was computed by means of a small perturbation of the symmetric concentrated load P. At the load $P = 1.07$ kN a small load $P_u = 10^{-3} \cdot P$ was added acting in the direction of horizontal direction u_0. The asymmetric path 2 was then followed. It is visible in Fig. 7.19b,c that besides limit points L_1, L_2 also returning points R_3, R_4 occur in plane (P, u_0). The equilibrium path 2 is nonsensitive to the same of load P_u so the path is symmetric with respect to the axis P. Due to arc-length control the whole secondary equilibrium path could be computed and the bifurcation point B_{su} could be evaluated.

At small step $\Delta\tau \equiv \Delta s = 1.0$ mm the number of iterations in the NR method was on average 4, independent of which equilibrium path was traced. More details were given in [170].

Example 5. Large Displacements of a Two-member Frame

A planar two-member frame, shown in Fig. 7.20b, is frequently analysed in literature as *Lee's frame.* In paper [99] the critical value of spatially fixed load was computed, $P_{\mathrm{cr}} = 18.55$ kN.

In Fig. 7.21 there are shown results of computations made by different authors for various types of materials and loads, applying different finite elements.

At equilibrium paths λ–u and λ–w both the limit and return points are visible for displacements comparable with the frame member length. That is why exact formulae had to be used for the curvature of deformed axis and the arc-length parameter was applied, cf. [35].

In paper [35] the finite element ELEB was used. Besides standard nodal displacements u_i, w_i, φ_i additional, local degree of freedom $u_{i,\xi}$ was introduced, cf. Fig. 7.20b. The 5-points *Lobatto's quadrature formula* [139] was used in order to integrate stiffness and force functions along the element axis and along the thickness if the material was elastic-plastic, cf. Fig 7.20a. In Fig. 7.21a the results of computation are shown for elastic and elastic-plastic material, respectively.

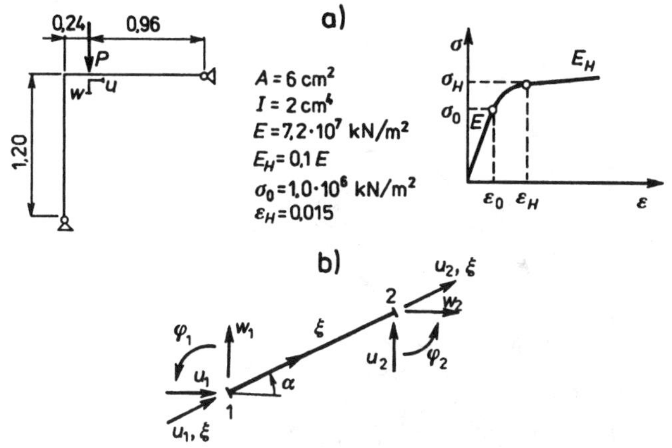

Fig. 7.20. **a**) Geometrical and material data for a two-member plane frame, **b**) Finite element ELEB used for the analysis

The equilibrium paths 1u, 1w and 2u, 2w were computed by means of 10 ELEB elements. For the first step the load control parameter $\Delta\tau \equiv \Delta p = 5$ was used and then the arc-length parameter of constant length $\Delta\tau \equiv \Delta s = 8$ was used for elastic material, respectively. At selected points at the paths 1w and 2u the number of iterations in the NR method is shown.

At the paths 1w and 1u there are marked results from [50] where 9 isoparametric finite elements, each of 12 DOF, were used.

In Fig. 7.21a the equilibrium paths are also shown for the follower force. The paths 3u, 3w were computed by means of 20 standard frame elements [5].

The equilibrium paths 4u, 4w were computed for the elastic-plastic material and the follower force using three so-called exact finite elements [171]. Matrices of these elements were computed integrating the field equations by means of finite differences [170].

The values of limit loads are higher for the follower load than for spatially fixed (dead) load. In Fig. 7.21b the subsequent forms of deformed frame are shown. It is visible that the horizontal component of the follower load tries to move the frame to the initial configuration. This can explain higher values of limit loads in comparison with dead load action.

Example 6. Buckling and Postcritical Displacement of a Portal Frame

In paper [36] a portal frame shown in Fig. 7.22a was analysed. Bifurcation load for this frame was computed in [101]. Basic data and division into ELEB type elements are also shown in Fig. 7.22a. The elastic-plastic characteristic of material is as that in Fig. 7.22a. The frame columns and beam are composed of the same American profiles W33×130.

In Fig. 7.22b,c the equilibrium paths are shown on the plane (λ, w_0), where

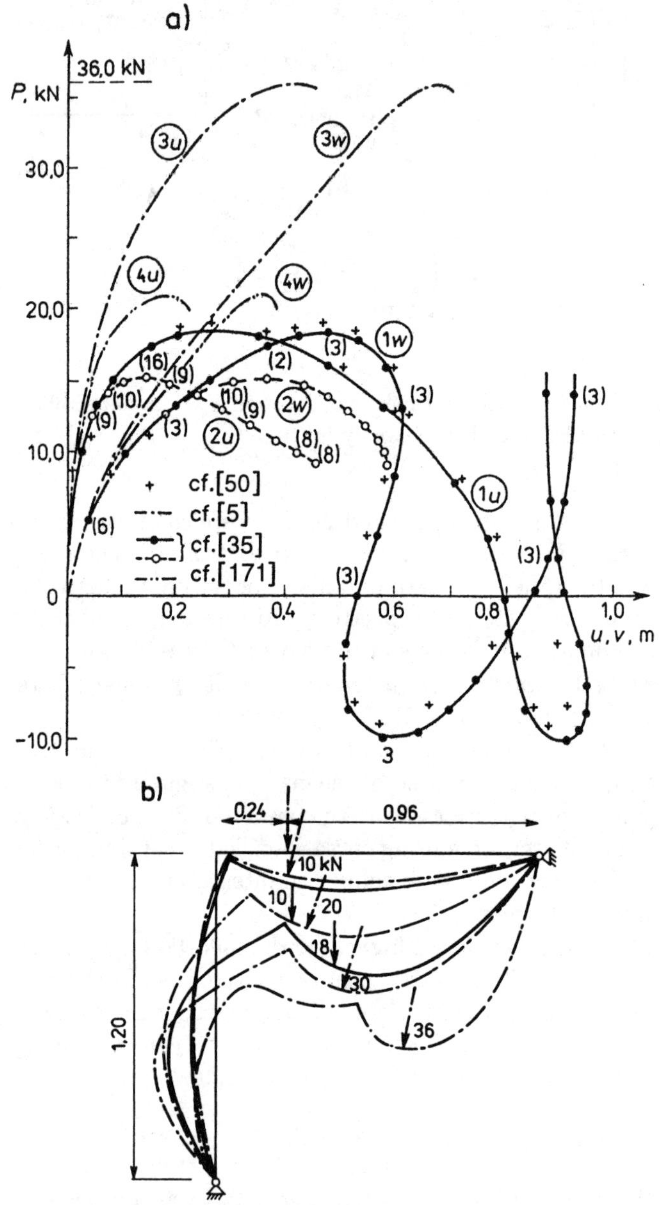

Fig. 7.21. **a**) Equlibrium paths for dead load: 1u, 1w – elastic material, 2u, 2w – elastic-plastic material; for follower load: 3u, 3w – elastic material , 4u,4w – elastic-plastic material, **b**) equilibrium forms for elastic material

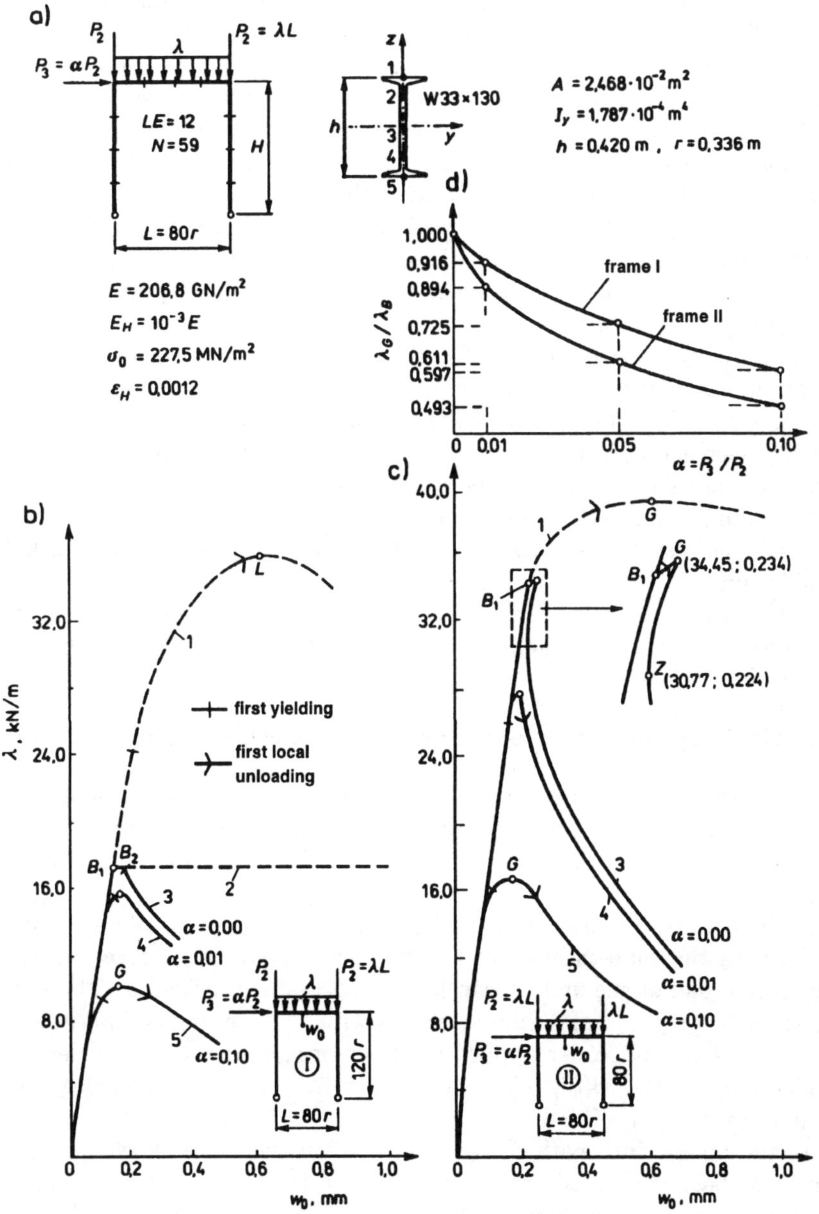

Fig. 7.22. **a)** Portal plane frame, **b)** Equilibrium paths for the frame I of $H = 120r$, **c)** Equilibrium paths for the frame II of $H = 80r$, **d)** Sensitivity graphics for imperfection $\alpha = P_3/P_2$

λ is the intensity of uniform load applied to the beam. The concentrated horizontal force $P_3 = \alpha P_2 = \alpha\lambda L$ disturbs the symmetric deformation of the frame. The force can be considered as an imperfection of parameter α.

At the fundamental equilibrium path 1 the first bifurcation point B_1 occurs. In case of higher frame I the bifurcation of perfect frame (for $\alpha = 0$) appears before the first yielding arises. Yielding develops after the frame buckling and it implies the appearance of the secondary bifurcation point B_2 at the equilibrium path 2 and the next path 3.

In case of imperfection $\alpha \neq 0$ instead of primary bifurcation point the limit points occur at the equilibrium paths 4, 5. In both frames yielding develops before the limit points L are reached and local unloading appears after the limit points are overpassed.

The primary bifurcation point B_1 of the perfect frame II is of Shanley type, i.e. at the bifurcation the local unloading develops and the load parameter increases up to limit point L close to the bifurcation point B_1. After buckling the displacement w_0 is not representative – the horizontal displacement of the beam can be treated as a representative displacement.

The equilibrium paths were computed under load parameter λ and then applying the displacement control w_0 (before buckling) or u_0 (postbuckling equilibrium).

In Fig. 7.22d the sensitivity of critical load λ_{cr} to imperfection parameter α is shown. In the case of higher frame I it is less sensitive to the imperfection than lower frame II.

Example 7. Symmetric Deformation of a Shallow Circular Arch

The analysis of large displacements of planar circular arches was used for testing the accuracy of the algorithm of the reduced basis method. Let us discuss an example of the arch considered in [115, 116] by means of the basis (7.152).

In the case of a shallow arch and symmetric load the instability is associated with the snap-through to the symmetric equilibrium form. That is why only a half of the arch shown in Fig. 7.23a was considered. The initial system was composed of 6 finite elements with the Lagrangian interpolation functions of the 5th order applied to the approximation of the generalized displacements $u(\xi)$, $w(\xi)$, $\varphi(\xi)$ – cf. [113]. This gave $N = 85$ degrees of freedom for the half of arch.

At the equilibrium path λ–w_0 in Fig. 7.23b there are shown points at which the basis composed of four vectors ($n = 4$) or five vectors ($n = 5$) was modified. This is visible even more clearly in Fig. 7.23c where the relation at the axial force at the clambed edge $N_A(\lambda)$ is given. The values $e > 0.002$ of the error (7.146) were assumed as a criterion for the basis modification. Five modifications were needed for $n = 4$ and four modifications for $n = 5$. The condition number computed at the point I is $\gamma \approx 10^5$ for $n = 4$ and $\gamma \approx 10^7$ for $n = 5$, cf. Fig. 7.23d.

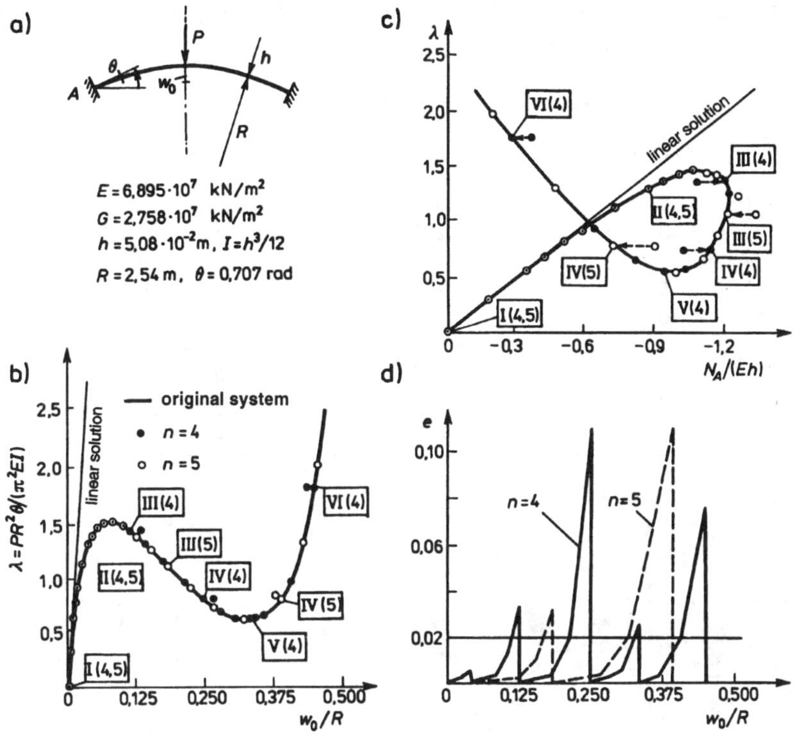

Fig. 7.23. Symmetric deformations of a shallow arch: **a**) data, **b**) deflections computed for the original FE system and for reduced basis with the basis vectors $n = 4$ and $n = 5$, respectively, **c**) relation between the axial force N_A and load parameter λ, **d**) relative error **e**) for residual forces

Example 8. Selection and Modification of the Basis Vectors for the Shallow Circular Shell

More precise discussion of the reduced basis (7.152) is done on the example of the shallow circular shell considered in Example 3 of data shown in Fig. 7.18b. The mesh 4×4 finite elements was used in [117] for a quarter of the shell. The bicubic approximation for displacements u_1, u_2 and rotations φ_1, φ_2 needs $N = 743$ DOF.

The results of computations of condition number γ of the Gram matrix (7.155) depend on parameter η in (7.152). In Table 7.5 there are given values of γ for $\eta = \lambda$ and $\eta = s$. The arc-length parameter s was also accepted as control parameter $\tau = s$ in the Crisfield constraint formula (7.121) but with $\beta = 0$ in (7.130), i.e. for the equation $\Delta \mathbf{q}_a^T \Delta \mathbf{q}_a = (\Delta\, {}^m s)^2$.

The condition number value γ was computed at the original point I of coordinates $\lambda = 0$, $w/R = 0$ in Fig. 7.24. If the condition $\gamma \approx 10^5$ is

Table 7.5. Condition number of the Gram matrix **G** of the path derivatives (7.152) with respect to parameter $\eta = \lambda$ and $\eta = s$

Number of basis vectors	Condition number γ with respect to $\eta = \lambda$	$\eta = s$
2	$1.066 \cdot 10^2$	1.0
3	$4.207 \cdot 10^2$	1.489
4	$9.741 \cdot 10^4$	$1.508 \cdot 10$
5	$7.206 \cdot 10^5$	$8.514 \cdot 10$
6	$7.595 \cdot 10^7$	$1.012 \cdot 10^3$
7	$6.929 \cdot 10^8$	$5.055 \cdot 10^3$
8	$2.876 \cdot 10^{10}$	$3.635 \cdot 10^4$
9	$1.720 \cdot 10^{12}$	$3.613 \cdot 10^5$
10	$7.234 \cdot 10^{13}$	$2.524 \cdot 10^6$

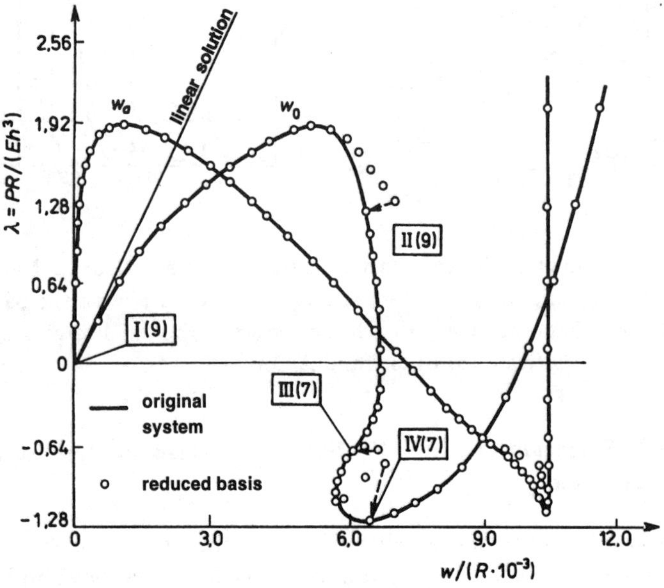

Fig. 7.24. Equilibrium paths for the circular shallow shell from Fig. 7.18b with points of the reduced basis modifications

assumed then five vectors $n = 5$ are needed for $\eta = \lambda$ and $n = 9$ for $\eta = s$, respectively.

In Fig. 7.24 the equilibrium path λ–w_0 is shown. The basis vectors were modified at the points I, ... ,IV, which correspond to the load parameter $\lambda = 0.0,\ 1.85,\ -0.765,\ -1.232$ according to condition $e < 0.01$. The number of basis vectors, shown in parentheses, was computed for $\gamma \approx 10^5$.

Example 9. Nonlinear Analysis of a Shallow Spherical Lattice Framework

The considered dome-shaped lattice framework formulated on the sphere $x_1^2 + x_2^2 + (x_3 + 7.2))^2 = 60.84$ is shown in Fig. 7.25a. The total number of DOF is $N = 111$. The computations, reported in [32], were carried out under arc-length control $\tau \equiv s$ and for three vectors of the reduced basis

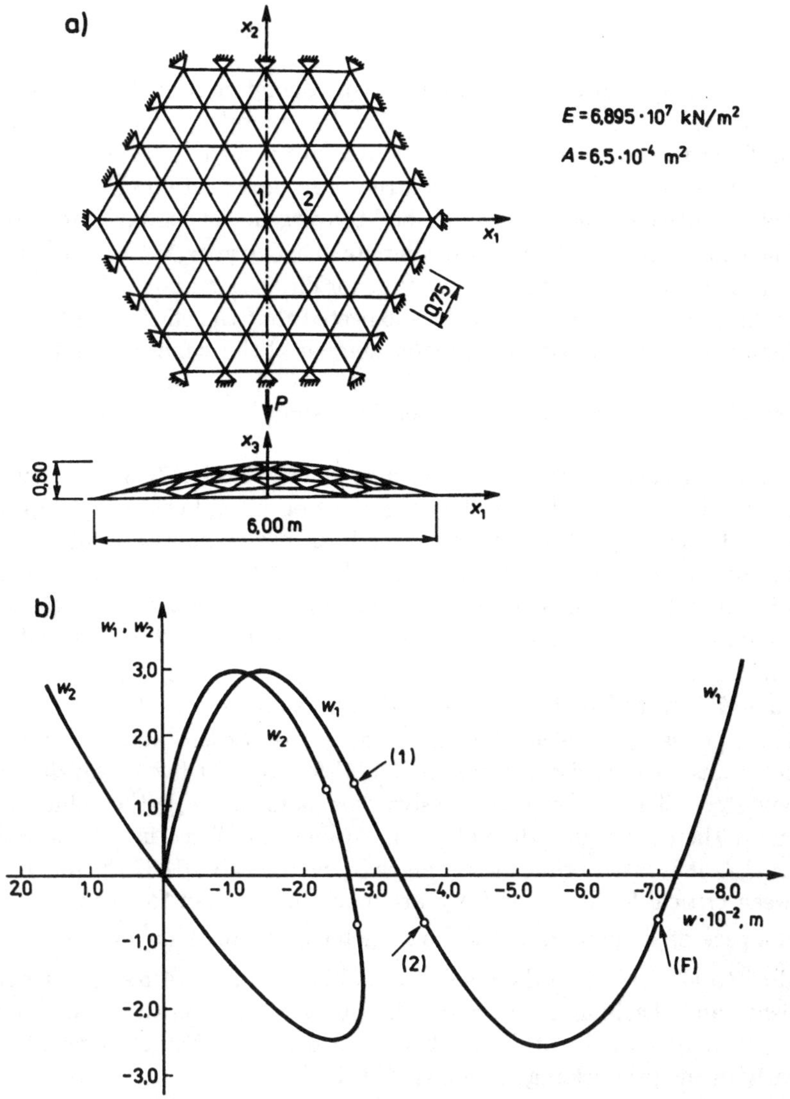

Fig. 7.25. Spherical lattice framework under concentrated load P applied to node 1: **a)** data, **b)** equilibrium paths.

Table 7.6. Computational time for the spherical lattice framework

Method	Time CPU (s)
Direct reduction	14.28
Indirect reduction	27.36
Original system analysis	26.83

(7.156). The equilibrium paths 1 and 2, corresponding to deflections w_1 and w_2, respectively, are shown in Fig. 7.25b. Two modifications were made for the indirect reduction (the limit points were overpassed by means of the first basis). In the case of direct reduction the third modification (F) was needed at the indirect reduction because of bad convergence of the mNR method.

The computational time of the computer CDC Cyber 174 is listed in Table 7.6. It is worth noticing that the time for direct reduction is half the time for indirect reduction approach. It is expected that the efficiency of indirect reduction should increase for a greater number of the structure DOFs.

Example 10. Buckling of a Two-member Frame

A symmetric, planar, two-member frame shown in Fig. 7.26a is known in literature as *William's frame.* The equations of initial buckling (7.76)b and linearized buckling (7.66) were used for both the computation of buckling loads and evaluation of the nonlinearity of prebuckling state. The assembly of 16 planar frame elements was analysed of total amount $N = 45$ DOF. From among the great number of results analysed in [177] the values of load parameter λ and parameters $\overset{\shortparallel}{y}$ and β defined in (7.175) and (7.178), are given in Table 7.7 for different values of height parameter $\kappa = H/L$.

The eigenvalue problem was analysed by the *Stodola method.* The parameter was computed according to (7.175) for the first five eigenvalues and eigenvectors. The indices α of decisive eigenpairs $(\lambda_\alpha, \mathbf{v}_\alpha)$ (the values of λ_α are more then 10^2 higher than other eigenvalues) are listed in the last row of Table 7.7. It is visible that for a shallow frame of $\kappa = 0.01882$ the differences between critical loads λ_{iI} and λ_{iL} are much bigger than for higher frames. In the case of shallow frames the parameter $\overset{\shortparallel}{\zeta} \approx 1.0$. That means that according to (7.174) the orthogonal part of displacement vector $\overset{\perp}{\mathbf{q}}$ practically vanishes and the limit point can occur (for $\kappa = 0.01882$ the limit point is at $\lambda_L = 6.50$). On the other hand, for $\kappa > 0.10$ high frames behave nearly linearly in the prebuckling state – cf. Fig. 7.26c.

The buckling modes are shown in Fig. 7.26b. In case of the shallow frame there are great differences between the initial buckling modes and linearized buckling modes. The first buckling mode computed on the basis of linearized

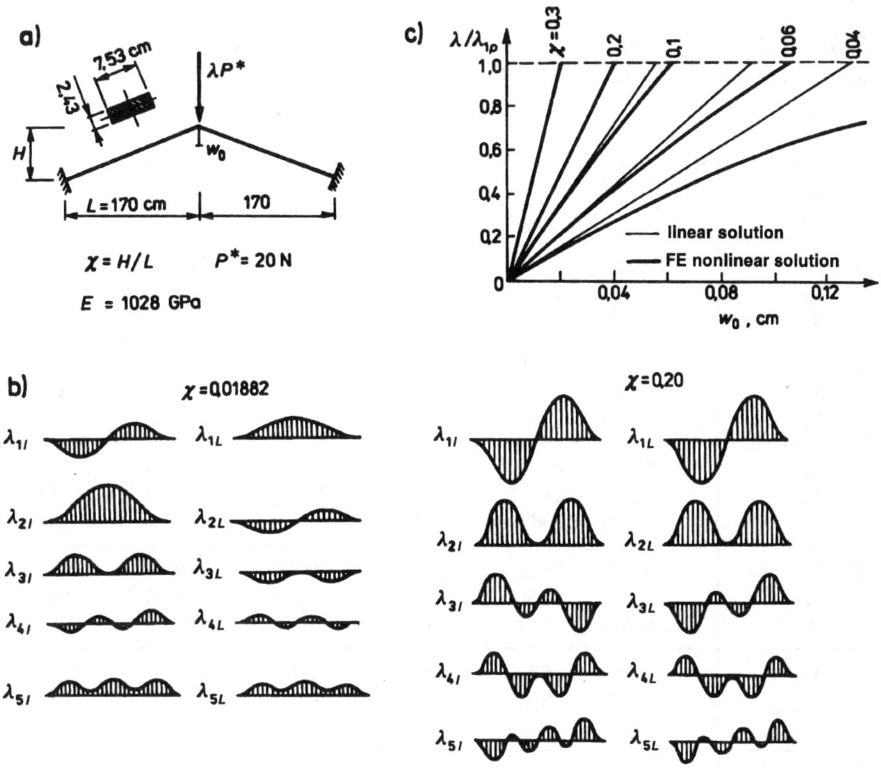

Fig. 7.26. Two-bar frame: **a)** data, **b)** buckling modes for initial buckling eigenvalues λ_{iI} and linearized buckling values λ_{iL} for the height parameters $\kappa = 0.01882,\ 0.20$

buckling equation is associated with symmetric displacement which corresponds to the snap-through phenomena. Buckling modes computed for a high frame of $\kappa = 0.20$ are nearly the same for initial buckling and linearized buckling approaches, cf. Fig. 7.26b.

Example 11. Computation of Displacements of a Portal Frame by Modal Superpositon Method

A portal frame is loaded by vertical forces $P = \lambda P$ and a horizontal force αP. In case I the coefficient $\alpha = 0.001$ implies the value of parameter $\ddot{\zeta} = 0.000016$, case II corresponds to $\alpha = 0.5$ and $\ddot{\zeta} = 0.798$.

Twelve frame elements used in computations [89] gave the mechanical system of $N = 33$ DOF. For eigenvectors $n = 4$ was applied in formula (7.181a) to compute the equilibrium path $\lambda - u_5$ according to the modal superposition method (MS). Only the antisymmetric buckling modes were active (1st and 3rd).

Table 7.7. Critical values of load parameter λ and parameters of nonlinearity $\ddot{\zeta}$ and β for a two-member frame in Fig. 7.26

		χ					
α	λ_α β_α	0.01882	0.04	0.06	0.10	0.20	0.30
1	λ_{1I}	19.180	29.09	40.77	64.99	122.20	170.8
	λ_{1L}	7.903	25.20	37.99	63.27	121.70	170.3
	β_1	142%	15.4%	7.3%	2.7%	0.68%	0.29%
2	λ_{2I}	24.81	56.91	79.75	127.1	239.1	334.1
	λ_{2L}	12.49	36.80	67.33	118.5	235.3	331.5
	β_2	99%	55%	18.4%	7.2%	1.85%	0.79%
3	λ_{3I}	37.52	86.08	120.6	192.3	362.5	505.4
	λ_{3L}	17.97	41.74	91.3	172.2	351.8	498.9
	β_3	109%	106%	32%	11.7%	3%	1.31%
4	λ_{4I}	56.75	93.89	254.4	256.5	489.4	683.6
	λ_{4L}	22.64	55.97	93.9	210.8	464.9	668.6
	β_4	150%	68%	64%	22%	5.3%	2.2%
5	λ_{5I}	86.82	167.5	241.1	384.4	724.6	1010.2
	λ_{5L}	31.23	84.8	170.3	927.2	690.9	989.0
	β_5	178%	97%	42%	17.5%	4.9%	2.1%
$\ddot{\zeta}$		0.993	0.959	0.861	0.0980	0.0035	0.0014
Numbers of active modes		2, 5	4, 5	4	4	4	4

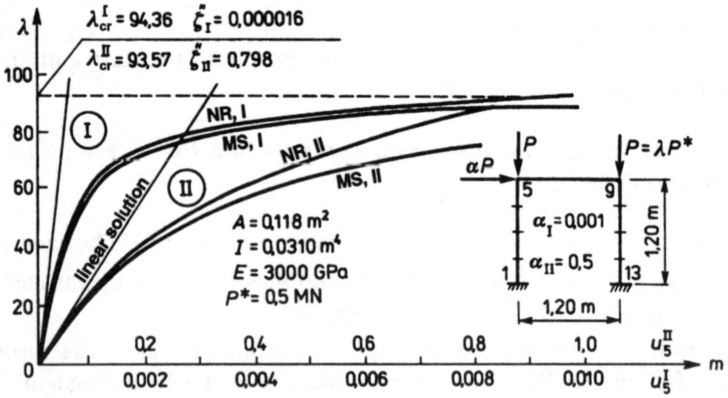

Fig. 7.27. Horizontal displacements of a portal frame for $\alpha_1 = 0.001$ and $\alpha_2 = 0.5$ computed by means of modal superposition (MS) and NR method (NR)

In case I of weak nonlinearity the equilibrium path NR,I computed by the NR method is close to the path MS,I computed by means of the modal superposition method, cf. Fig. 7.27. The differences between the paths NR,II and MS,II are bigger for case II.

Example 12. Buckling of Circular Arches and a Circular Ring

A pinned planar arch of the vertex angle 20 under external pressure p is considered, cf. Fig. 7.28a. The computations reported in [170] were carried out by means of the finite elements shown in Fig. 7.28b. The following functions were used to approximate the arch element displacements:

$$u(s) = a_1 \cos\varphi - a_2 \sin\varphi - a_3 R(1 - \cos\varphi + a_4 s + a_7 s^2 + a_8 s^3 \qquad (7.186a)$$

$$w(s) = a_1 \sin\varphi + a_2 \cos\varphi + a_3 R \cos\varphi + a_5 s^2 + a_6 s^3 \,. \qquad (7.186b)$$

In Table 7.8 the results of computations are put together for different vertex angles 2φ and thickness ratios $h/R = 0.1,\ 0,001$. Twelve finite elements were for $N = 54$ DOF of the FE assembly. In case of spatially fixed pressure

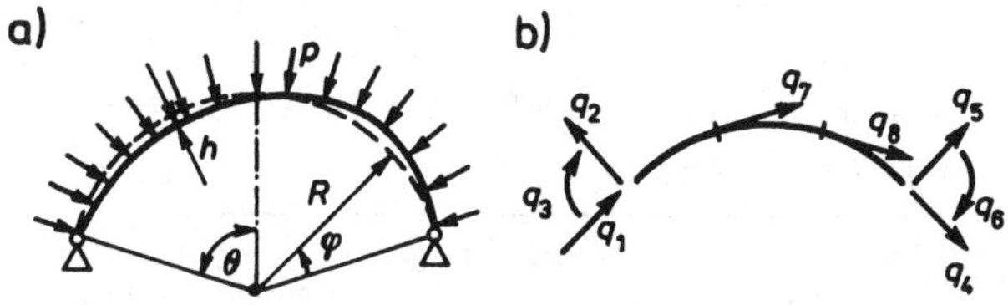

Fig. 7.28. **a**) Buckling of a circular arch under external pressure, **b**) finite element of 8 DOF

Table 7.8. Buckling loads $\lambda_{cr} = p_{cr}\, EI/R^3$ for pinned circular arches

Vertex angle 2θ	Loads: spatially fixed $h/R = 0.1$	spatially fixed $h/R = 0.001$	follower $h/R = 0.1$	follower $h/R = 0.001$
180°	3.268 (3.269)	3.307 (3.271)	2.994	3.032
160°	4.501 (4.512)	4.556 (4.514)	4.049	4.100
140°	6.174 (6.216)	6.263 (6.217)	5.577	5.654
120°	8.602 (8.726)	8.772 (8.727)	7.900	8.041
100°	12.397 (12.781)	12.820 (12.781)	11.646	11.997
60°	30.647	36.006	30.118	35.063

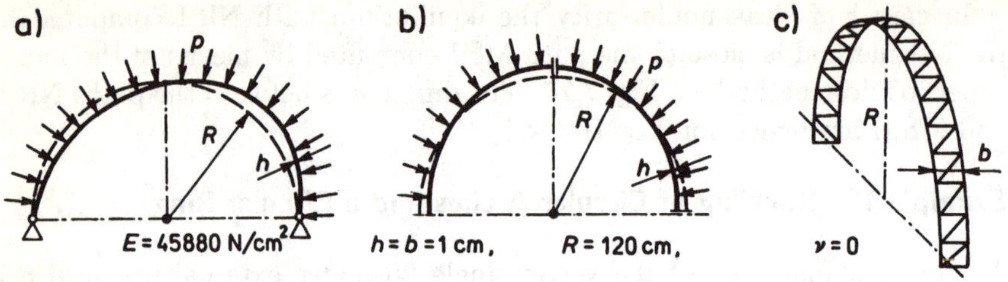

Fig. 7.29. Buckling under external pressure: **a)** semi-circular pinned arch, **b)** circular ring, **c)** division of the arch into 2D triangle elements

Table 7.9. Critical load $\lambda_{cr} = p_{cr}\, EI/R^3$ for the semi-circular pinned arch

Loads	Results according to: [49] 10 ES	[119] $N = 59$	$n = 3$
follower	3.001	2.9997	2.9997
fixed	4.002	3.9993	3.9997

Table 7.10. Critical load $\lambda_{cr} = p_{cr}\, EI/R^3$ for the circular ring (number of EFs and DOFs for a quarter of ring)

Loads	Results according to: [49] 20 ES	40 ES	[119] $N = 119$	$n = 4$	$n = 5$
follower	3.049	3.000	2.9997	6.7224	3.0037
fixed	3.327	2.327	3.2709	3.2709	3.2709

the results in brackets are taken from [47] where analytical solutions were obtained.

The results of computations carried out for the follower pressure by means of (7.67a) are compared with the results for spatially fixed pressure obtained on the basis of formula (7.67b). It is visible that the differences between buckling loads for dead (spatially fixed) and follower loads decrease for small vertex angles. In case of $2\varphi = 60°$ there is only about 3% of discrepancy. This confirms the conclusion that for shallow arches the load matrix $\mathbf{K}^{(p1)}$ can be omitted at computation of buckling loads.

In Fig. 7.29a the first buckling mode is shown for a semi- circular pinned arch. The buckling mode for a circular ring is shown in Fig. 7.29b. The results of computation by the FE method are given in Tables 7.9 and 7.10.

In case of the arch reaching the exact value of buckling load for follower pressure needs more finite elements than for spatially fixed pressure. The results of computations put together in Tables 7.9 and 7.10 are close to the exact values by means of analytical formulae: $\lambda_{cr} = 3.0$ for follower pressure, cf. [164], $\lambda_{cr} = 3.273$ computed in [19] for the arch under spatially fixed pressure and $\lambda_{cr} = 4.0$ for the ring under the same type of pressure [164].

In Tables 7.9 and 7.10 there are shown results obtained in [49] by 2D triangle finite elements, cf. Fig. 7.29c, and results of computations reported in [119], associated with the reduced basis (7.152). All the results in Table 7.10 were obtained for a quarter of the ring according to the symmetry of buckling mode shown in Fig. 7.29b.

Example 13. Buckling of Cylindrical Shell under External Pressure

In papers [49, 102, 103] a cylindrical shell of data shown in Fig. 7.30a was considered. In the case of simple supporting of both edges the buckling pressure is $\lambda_{cr} = 35.549$ and $m = 8$ circumferential halfwaves occur [164]. In [49] the buckling load for the same shell but under spatially fixed (dead) pressure was computed as 7% higher than in case of follower pressure.

In the case of free edge the term (7.45) causes nonsymmetry of the load stiffness matrix. That is why the computations were carried out on the basis of two buckling equations (7.182). The computations were done for a segment of the shell of the vertex angle $\varphi = 2\pi/m$, where: m – number of

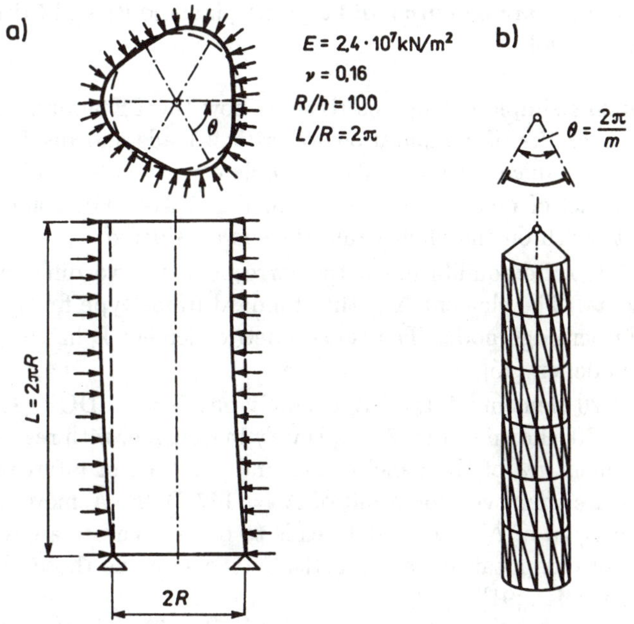

Fig. 7.30. Buckling of a circular shell: **a)** data and circumferential buckling mode, **b)** FE discretization of the shell segment.

Table 7.11. Critical load for a cylindrical shell (Fig. 7.28) under external pressure

Load	$\lambda_{cr} = p_{cr}/35549$ for θ: $\pi/2$	$\pi/3$	$\pi/4$	$\pi/5$
follower $\lambda^{(0)}$	–	0.5552	0.8942	1.4043
$\lambda^{(1)}$	0.9701	0.5542	0.8997	1.4041
spatially fixed	0.9704	0.6237	0.9534	1.4627

circumferential halfwaves, cf. Fig. 7.29b. 72 triangle finite elements, were used in the segment which gave the total number of degrees of freedom $N = 600$ DOF.

The results of computations discussed in [49, 103] are put together in Table 7.11. In case of shell with free edge the lowest buckling load occurs at $m = 6$. The differences of critical values of load parameters $\lambda^{(0)}$ and $\lambda^{(1)}$, used in (7.182) and calculated according to (7.184),correspond to $\mu = 0.0018$. Such a low value of the parameter μ indicates a small influence of the nonsymmetry of the load stiffness matrix $\mathbf{K}^{(p1)}$ on the critical buckling pressure.

The spatially fixed critical buckling pressure is about 12% higher than in case of follower pressure and about 56% lower than for a simple supported both edges.

Example 14. FE Approximation of Large Displacements and Critical Loads of a Circular Panel

The aim of this simple example is to show how the FE approximation can influence the results of computations of critical loads and displacements of deformable mechanical systems. The example corresponds to a long, elastic, cylindrical panel of cross-section shown in Fig. 7.31a. For computations a strip of unit width on the plain strain state is considered.

In Fig. 7.31b the equilibrium paths correspond to two different types of finite elements. The element A is the standard frame-type finite element of three DOFs per edge node. The isoparametric element B has 6 generalized DOFs per node, cf. [45].

In case of 20 elements A the FE assembly has $N = 57$ DOF. The equilibrium path A/20, shown in Fig. 7.31b, is very inaccurate, with respect to both the displacement w_0 of the panel center and the type of bifurcation point. The results are improved for A/50 of $N = 147$ DOF. A more precise approximation B/20 of $N = 120$ DOF leads to quite accurate approximations and results of computations are practically the same as those obtained by analytical formulae [91].

It is worth emphasizing that in stability problems and in the computation of large displacements a correct approximation of all the components of the displacement field can significantly influence the results. Such conclusions

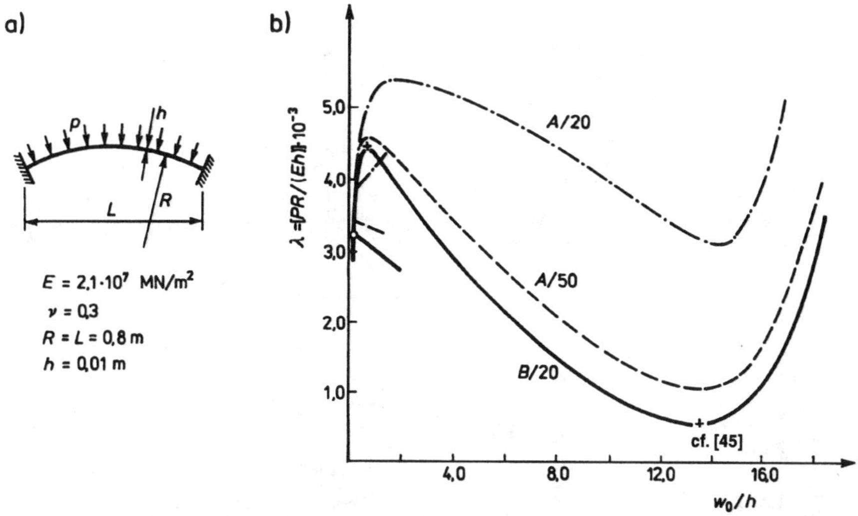

Fig. 7.31. Large displacements of a cylindrical panel: **a)** data, **b)** equilibrium paths for different FE approximations

can be related, to for instance high arches in case of omitting the terms with a_7 and a_8 in the $u(s)$ approximation (7.186a).

Example 15. Stability Boundary for a Two-member Truss

The stability boundary is discussed for a simple, two-member truss, shown in Fig. 7.32a. It is known in the literature as the *von Mises truss*. It is assumed that two independent, spatially fixed loads, P_1 and P_2 are applied to the truss.

In order to consider the local buckling of members an equivalent element with artificial spring of nonlinear characteristics was formulated in [172], cf. Fig. 7.32b. In the case of initially straight member, i.e. for $\beta_0 = 0$, the following approximate formula was derived:

$$\frac{N}{N_{\mathrm{cr}}} = \left(1 + \frac{\beta^2}{12} + \frac{\beta^4}{45}\right) \frac{\beta}{\sin\beta}, \tag{7.187}$$

where: N, N_{cr} – current and buckling axial forces, β – angle of rotation shown in Fig. 7.32b.

Due to equivalent finite elements the truss can be considered as a discrete system of $N = 4$ DOF. The displacements u_0, v_0 enable us to analyze the global stability of the frame, the rotations β_1 and β_2 make possible to trace local buckling of the members and their postbuckling displacements.

The extended set of incremental equations (7.77) was used in [172], written

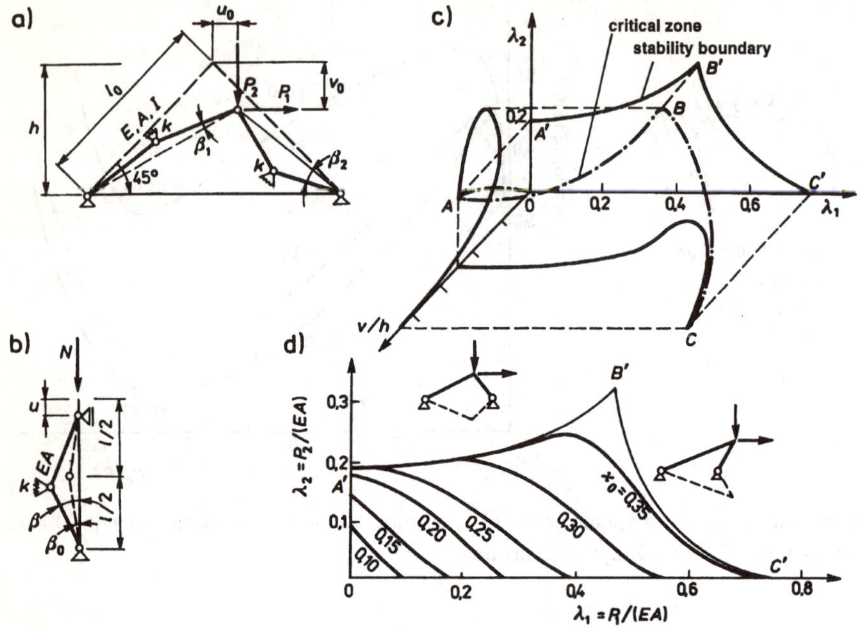

Fig. 7.32. Two-member frame under two independent loads: **a)** computational model of the frame, **b)** equivalent finite element, **c)** critical zone and critical boundary for global instability of the frame, **d)** critical boundaries for local buckling taken into account

in the following form:

$$\mathbf{K}_T(\mathbf{q})\Delta\mathbf{q} - \mathbf{Q}_1^*\Delta\lambda_1 - \mathbf{Q}_2^*\Delta\lambda_2 = \mathbf{R}(\mathbf{q}), \tag{7.188a}$$

$$\mathbf{t}^{\mathrm{T}}\Delta\mathbf{q} + t_5\Delta\lambda_1 + t_6\Delta\lambda_2 = \Delta\tau, \tag{7.188b}$$

$$\mathbf{d}^{\mathrm{T}}\Delta\mathbf{q} = -D(\mathbf{q}). \tag{7.188c}$$

Equation (7.188b) corresponds to the Riks–Wempner constraint equation (7.119). Equation (7.188c) is associated with the zero value of the stability determinant $D(\mathbf{q}+\Delta\mathbf{q}) = 0$. After the determinant is expanded into the Fourier series and only term linear with respect to $\Delta\mathbf{q}$ are conserve. Equation (7.188c) can be derived with the partial derivatives

$$d_\alpha \equiv \frac{\partial D}{\partial q_\alpha} = \frac{\partial d}{\partial K_{\beta\gamma}} K_{\beta\gamma\alpha i}, \tag{7.189}$$

treated as components of the vector $\mathbf{d}$.

As a result of the solution of the set of equations (7.188) the following vector can be computed

$$\tilde{\tilde{\mathbf{q}}} = \{\mathbf{q}, \lambda_{1\mathrm{cr}}, \lambda_{2\mathrm{cr}}\} \ . \qquad (7.190)$$

In Fig. 7.32c the stability boundary is shown as a result of projection of the critical zone on the plane (λ_1, λ_2). The stability boundary is composed of two curves which are associated with different instability modes. The first mode corresponds with the w_0 trajectory placed in-between the hinge supports. The second mode corresponds to the w_0 trajectory being out of the support hinges. The load local buckling of members cancel the singular point at the stability boundary.

It is worth emphasizing that if geometrical nonlinearities are considered then the stability boundary can be concave. It is against the Papkovich theorems [132] who proved that for linear buckling problems the stability boundary is to be convex or bounded by planes (more exactly by the hyper-planes in the load parameter space).

Example 16. Stability Boundary for a Circular Arch under Three Parameter Load

The computation of derivatives d_α is a difficult problem. According to (7.189) the components of three-dimensional matrix $K_{\alpha\beta\gamma}$ can be computed only for simple finite elements and at a small number of total degrees of freedom N. That is why in [169] an iterative approach was proposed. It depends on the computation the critical points on planes (λ_i, λ_j) going along piecewise lines of equilibrium states close to the critical state (for linear piecewise load program).

An arch shown in Fig. 7.33a is under three independent spatially fixed loads λ_1, λ_2, and λ_3. Eight finite elements ELEB (the same element was used in Example 16) approximate the arch. Data for the computations were taken from [135] where the arch was analyzed by means of an analytical approach.

In subsequent Figs. 7.33b, c the stability bounds are shown on the planes (λ_1, λ_2), and (λ_3, λ_1) for the fixed values $\lambda_3 = 0.0$, $\lambda_2 = 0.0, 0.75$ respectively. The stability surface is shown in Fig. 7.33d where the capital letter B is added at number of points corresponding to loss of stability by bifurcation.

Referring to the remark at the end of the previous Example the stability surface in Fig. 7.33d is concave with respect to the origin point $\lambda_1 = \lambda_2 = \lambda_3 = 0$ because of geometrical nonlinearities taken into account in the formulated model.

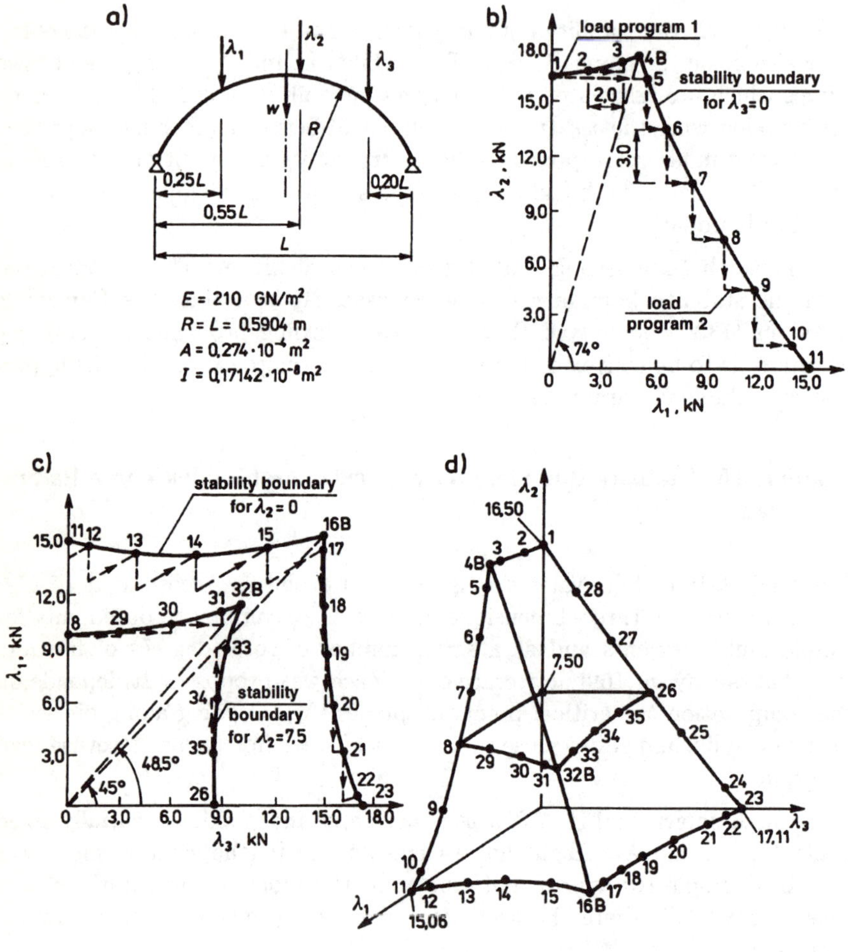

Fig. 7.33. Circular arch under three independent loads: **a)** data, **b)** stability boundary on plane (λ_1, λ_2), **c)** stability boundaries on plane (λ_3, λ_1), **d)** stability surface

7.6 Final Remarks

The limited scope of Chap. 7 made it impossible to consider more exstensively the application of FEM to the analysis of stability of structures. In the frame of static stability problems the coupling of global and local models was not discussed. Also problems of flexural-torsional buckling of thin-walled structures were omitted. The influence of imperfections on instability of various structures was considered in a very limited scope. The same concerns the accuracy of FEM approximations.

The problems discussed in Sects. 7.1–7.4 have been illustrated by numerical examples gathered in Sect. 7.5. There were selected rather simple examples in order to interpret basic ideas rather than computational details.

The references besides the quoted detailed papers include more general positions of the literature in order to offer to the potential reader a deeper insight into the FEM stability analysis of structures.

As an extension of this chapter the book [170] can be referred to where a great number of structure stability problems was discussed by means of FEM.

References

1. Abbott JP (1967) An efficient algorithm for the determination of certain bifurcation points. J Comp Appl Math 4:19–27
2. Almroth BO, Stern P, Brogan FA (1978) Automatic choice of global shape functions in structural analysis. AIAA J 16:525–528
3. Argyris JH, Chan ASL (1972) Application of finite elements in space and time. Ing Archiv 41:235–257
4. Argyris JH, Scharpf DD (1969) Finite elements in time and space. The Aeron J Royal Aeron Soc 73:1041–1944
5. Argyris JH, Symeonidis S (1981) Nonlinear finite element analysis of elastic systems under nonconservative loading – natural formulation. Part I: Quasistatic problems. Comp Meth Appl Mech Eng 26:75–123
6. Babuška I, Osborn J (1991) Eigenvalue problems. In: Ciarlet PG, Lions JL (eds) Handbook of numerical analysis, Vol II. Elsevier Science Publishers BV (North-Holland)
7. Bahgat BM, Willmert KD (1974) Finite element vibrational analysis of planar mechanisms. Mechanism and Machine Theory 11:47–71
8. Bailey CD (1975) A new look at Hamilton's principle. Found Phys 5:433–451
9. Bajer CI (1984) Non-stationary division by the space–time finite element method in vibration analysis. In: Petyt M, Wolf HF (eds) Proc Second Int Conf on Recent Advances in Structural Dynamics. ISVR, Southampton, pp 161–170
10. Bajer CI (1985) Microcomputer applications of non-rectangular space–time finite elements in vibration analysis – direct joint-by-joint procedure. In: Adey RA (ed) Engineering software IV, Vol 7. Springer Verlag, Berlin, pp 3–16
11. Bajer CI (1986) Triangular and tetrahedral space–time finite elements in vibration analysis. Int J Num Meth Eng 23:2031–2048
12. Bajer CI (1987) Notes on the stability of non-rectangular space–time finite elements. Int J Num Meth Eng 24:1721–1739
13. Baruch M, Riff R (1982) Hamilton principle, Hamilton's law, 6n correct formulations. AIAA J 20:687–692
14. Bathe K–J (1986) Finite-Elemente-Methoden. Springer-Verlag, Berlin
15. Bathe K–J, Dvorkin EN (1983) On the automatic solution of nonlinear finite element equations. Comp Struct 17:871–879
16. Bathe K–J, Ozdemir H (1976) Elastic–plastic large deformation static and dynamic analysis. Comp Struct 8:81–92
17. Bathe K–J, Ramm E, Wilson EL (1975) Finite element formulation for large deformation dynamic analysis. Int J Num Meth Eng 9:353–386

18. Bathe K–J, Wilson EL (1976) Numerical methods in finite element analysis. Prentice Hall, Englewood Cliffs
19. Batoz JL (1979) Curved finite elements and shell theories with particular reference to the buckling of a circular arch. Int J Num Meth Eng 14:1262–1267
20. Batoz JL, Dhatt G (1979) Incremental displacement algorithms for nonlinear problems. Int J Num Meth Eng 14:1262–1267
21. Bergan PG, Horrigmoe G, Krakeland B, Soreide TH (1978) Solution techniques for non-linear finite element problems. Int J Num Meth Eng 12:1677–1696
22. Besseling JF (1974) Non-linear analysis of structures by the finite element method as a supplement to a linear analysis. Comp Meth Appl Mech Eng 3:173–194
23. Bishop RED, Gladwell GML, Michaelson S (1965) The matrix analysis of vibration. University Press, Cambridge
24. Bolotin VV (1956) Dynamic stability of elastic systems (in Russian). Gosud Izd Tech–Teor Lit, Moscow
25. Borri M, Lanz M, Mategazza R (1985) Comment on "Time finite element discretization of Hamilton's law of varying action". AIAA J 23:1457–1458
26. Brush DO Almroth BO (1975) Buckling of bars, plates and shells. McGraw–Hill, New York
27. Brzeziński J, Pietrzakowski M (1979) Vibration analysis of a simple hybrid system by space–time element method (in Polish). Arch Bud Masz 26:511–526
28. Budiansky B, Hutchison JW (1979) Buckling – progress and challenge. In: Trends in solid mechanics (dedicated to the 65th birthday of WT Koiter). Sijthoff and Noordhoff, Delft, pp 165–205
29. Bufler H (1984) Pressure loaded structures under large deformations. ZAMM 64:287–295
30. Bushnell D (1985) Static collapse – a survey of methods and modes of behaviour. Finite Elem Eng Design 1:156–205
31. Chan ASL, Hsiao KM (1985) Nonlinear analysis using a reduced number of variables. Comp Meth Appl Mech Eng 52:899–913
32. Chan ASL, Lau TB (1987) Further developments of reduced basis method for geometric nonlinear analysis. Comp Meth Appl Mech Eng 62:127–144
33. Cheung YK (1976) Finite strip method in structural analysis. Pergamon Press, Oxford
34. Ciarlet PG (1978) The finite element method for elliptic problems. North-Holland, New York
35. Cichoń Cz (1984) Large displacement in-plane analysis of elastic–plastic frames. Comp Struct 19:737–745
36. Cichoń Cz (1985) Nonlinear finite element analysis of stability of bar structures (in Polish). Monograph 38, Cracow Univ of Technology, Kraków
37. Crisfield MA (1978) A combined Rayleigh–Ritz/finite element method for the non-linear analysis of stiffened plate structures. Comp Struct 8:679–689
38. Crisfield MA (1981) A fast incremental/iterative solution procedure that handles 'snap–through'. Comp Struct 13:55–62
39. Crisfield MA (1982) Solution procedure for non-linear structural problems In: Hinton E, Owen DRJ, Taylor C (eds) Recent advances in non-linear computational methods. Pineridge–Press, Swansea, pp 1–39

40. Crisfield MA (1984) Accelerating and damping the modified Newton–Raphson method. Comp Struct 18:395–407

41. Cyganecki W (1979) Criterion of choice of space–time element dimensions (in Polish). Arch Inż Ląd 25:389–397

42. Cyganecki W (1980) On importance of choice of space–time element dimensions (in Polish). Arch Inż Ląd 26:717–726

43. Demkowicz L (1994) Asymptotic convergence in finite and boundary element methods. Part I: Theoretical results, Computer and Mathematics with Applications, 27(12):69–84

44. Denis JE, Jr, More JJ (1977) Quasi–Newton methods – motivation and theory. SIAM Rev 19:46–89

45. Duong N–C, Waszczyszyn Z (1982) Stability of elastic–plastic arches and circular panels (in Polish). Eng Trans 30:247–267

46. Duvaut G, Lions JL (1976) Inequalities in mechanics and physics. Springer–Verlag, Berlin

47. Dym CL (1974) Stability theory and its applications to structural mechanics. Noordhof, Leyden

48. Endo T, Oden JT, Becker EB, Miller T (1984) A numerical analysis of contact and limit point behaviour in a class of problems of finite elastic deformation. Comp Struct 18:899–910

49. Floegl H, Mang H (1981) Zum Einfluß der Verchiebungsabhängigkeit ungleichformigen hydrostatischen Druckes auf das Ausbeulen dünner Schalen allgemeiner Form. Ing Archiv 50:15–30

50. Frey F, Cescotto S (1977) Some new aspects of the incremental total Lagrangian description in nonlinear analysis. In: Proc Int Conf Finite Elements in Nonlinear Mechanics. Geilo (Norway), Vol 1, pp 5.1–5.20

51. Fried I (1984) Orthogonal trajectory accession to the nonlinear equilibrium curve. Comp Meth Appl Mech Eng 44:283–297

52. Gajewski A, Życzkowski M (1988) Optimum structural design under stability constraints. Kluwer, Dordrecht

53. Gallagher RH (1975) Perturbation procedures in nonlinear finite element structural analysis, In: Computational mechanics. Lect Notes in Math 461. Springer, Berlin, pp 75–89

54. Geradin M, Idelsohn S, Hogge M (1981) Computational strategies for the solution of large nonlinear problems via quasi–Newton methods. Comp Struct 13:73–81

55. Gryboś R (1980) Stability of structures under impact loading (in Polish). PWN, Warszawa

56. Gurson AL (1977) Continuum theory of ductile rupture by void nucleation and growth. J Eng Math Tech 99:2–16

57. Gurtin ME (1964) Variational principles for linear initial-value problems. Quart Appl Math 22:252–256

58. Hagedorn P (1971) Die Umkehrung der Stabilitätssätze von Lagrange–Dirichlet und Routh. Arch Rat Mech Anal 42:281–316

59. Haythornthwaite R.M, Onat ET (1965) The load carrying capacity of initially flat circular steel plates under reversed loading. J Areo Sci 22:867–869

60. Herrera I, Bielak J (1974) A simplified version of Gurtin's variational principle. Arch Rat Mech Anal 53:131–149

61. Hibbitt HD (1979) Some follower forces and load stiffness. Int J Num Meth Eng 14:937–941

62. Hughes TJR, Hulbert GM (1988) Space-time finite element methods for elastodynamics: formulations and error estimates. Comp Meth Appl Mech Eng 66:339–363

63. Huseyin K (1975) Nonlinear theory of elastic stability. Noordhoff, Leyden

64. Huseyin K (1978) Vibration and stability of multiple parameter systems. Noordhoff, Alphen aan den Rijn

65. Hutchison JW (1974) Plastic buckling. In: Advances of Applied Mechanics, 14, pp 67–143

66. Idelsohn SR, Cardona A (1984) Recent advances in reduction methods in nonlinear structural dynamics, In: Proc 2nd Int Conf on Recent Advances in Structural Dynamics. Southampton

67. Idelsohn SR, Cardona A (1985) A load-dependent basis for reduced nonlinear structural dynamics. Comp Struct 20:203–210

68. Jaeger JC, Carlslaw HS (1979) Conduction of heat in soils. Oxford University Press

69. Kacprzyk Z (1981) Vibration analysis of industrial chimney under seismic load (in Polish). Arch Inż Ląd 27:507–516

70. Kacprzyk Z (1982) Space-time superelement (in Polish). Arch Inż Ląd 28:47–55

71. Kacprzyk Z (1984) On weight functions in space–time element method (in Polish). Scientific Reports, Civil Engineering 85, Warsaw University of Technology, Warsaw

72. Kacprzyk Z, Lewiński T (1983) Comparison of some numerical integration methods for the equations of motion of systems with finite number of degrees of freedom. Rozpr Inż 31:213–240

73. Kaliski S (ed) (1986) Technical mechanics, Vol 3: vibrations and waves (in Polish). Polish Scientific Publishers (PWN), Warsaw

74. Kaliski S, Dżygadło Z, Solarz L, Włodarczyk E (1966) Vibrations and waves in solid bodies (in Polish). Polish Scientific Publishers (PWN), Warsaw

75. Kączkowski Z (1975) The method of finite space–time elements in dynamics of structures. J Techn Phys 16:69–84

76. Kączkowski Z (1976) The space–time finite element method (in Polish). Arch Inż Ląd 22:365–378

77. Kączkowski Z (1979) General formulation of the stiffness matrix for the space–time finite elements. Arch Inż Ląd 25:351–357

78. Kączkowski Z (1982) On variational principles in thermoelasticity. Bull Acad Pol Sci, Sér Sci Techn 20:81–86

79. Kączkowski Z (1983) On applying non-rectangular space–time elements (in Polish). Mech Teor Stos 21:531–542

80. Kączkowski Z (1985) On applying space–time element method for heat conduction problems (in Polish). Arch Inż Ląd 31:361–373

81. Kączkowski Z (1986) Uncoupled systems of equations in the space-time element method (STEM) (in Polish). Arch Inż Ląd 32:39–50

82. Kączkowski Z (1989) Die Methode der Raum-Zeit-Elemente (MERZE) in Anwendung auf instationäre Wärmeleitungsprobleme. ZAMM 69:179–181

83. Kączkowski Z (1991) Statics of bars and bar structures. In: Życzkowski M (ed) Studies in applied mechanics, Vol 26: Strength of structural elements. PWN Polish Sci Publishers and Elsevier, Warszawa Amsterdam, pp 3–179

84. Kączkowski Z, Borkowski A (1988) Zur numerischen Stabilität des Verfahrens der Raum-Zeit-Elemente mit schrägen Rändern. ZAMM 68:387–388

85. Kączkowski Z, Witkowska Z (1978) Transfer matrix in space–time element method (in Polish). Arch Inż Ląd 24:59–66

86. Kleiber M (1986) On plastic localization and failure in plane strain and round void-containing tensile bars. Int J Plast 2:201–211

87. Kleiber M (1989) Incremental finite element modelling in non-linear solid mechanics. PWN–Ellis Horwood

88. Kleiber M (1991) Computational coupled non-associative thermoplasticity. Comp Meth Appl Mech Engng 90:943–967

89. Kleiber M, Wieczorek M (1982) Approximate mothod for nonlinear analysis of elastic frames (in Polish). Eng Transactions 30:269–281

90. Koiter WT (1945) On the stability of elastic equilibrium (in Dutch). Doctoral thesis. Paris, Delft, Amsterdam. English version by Air Force Flight Dyn Lab Tech Rep AFFDL–TR–70–25, Feb 1970

91. Kornishin MS, Isanbaeyva FS (1968) Flexible plates and panels (in Russian). Nauka, Moscow

92. Kowalczyk B (1976) Matrices and their applications (in Polish). WNT, Warsaw

93. Kröplin BH, Dinkler D (1982) A creep type strategy used for tracing the load path in elastoplastic post-buckling analysis. Comp Meth Appl Mech Eng 32:365–376

94. Kruszewski J (1969) Application of finite element method to calculations of ship structure vibrations. Eur Shipbuilding 3:28–42

95. Kruszewski J (ed) (1975) Rigid finite element method (in Polish). Arkady, Warsaw

96. Kruszewski J (ed) (1984) Rigid finite element method in structural dynamics (in Polish). Arkady, Warsaw

97. Kruszewski J, Wittbrodt E, Walczyk Z (1993) Vibrations in mechanical systems, computational approach (in Polish). WNT, Warsaw

98. Langer L (1979) Spurious damping in numerical solutions of equations of motion. Arch Inż Ląd 25:359–369

99. Lee S–L, Manuel FS, Rossow EC (1968) Large deflections and stability of elastic frames. Proc ASCE, J Eng Mech Div 94:521–547

100. Lewiński T (1984) Stability analysis of a difference scheme for the vibration equation with a finite number of degrees of freedom. Zastosowania Matematyki 18:473–486

101. Lu L–W (1965) Inelastic buckling of steel frames. Proc ASCE, J Struct Div 91:185–214

102. Mang H (1980) Symmetricability of pressure stiffness matrices for shells with load free edges. Int J Num Meth Eng 15:981–990

103. Mang HA, Gallagher RH (1983) On the unsymmetric eigenproblem for the buckling of shells under pressure loading. Trans ASME, J Appl Mech 50:95–100

104. Margues JMCC (1986) Nonlinear finite element solutions with quasi- and secant-Newton method. In: Numerical methods for nonlinear problems, 3. Pineridge–Press, Swansea, pp 1117–1147

105. Matthies H, Strang G (1979) The solution of nonlinear equations. Int J Num Meth Eng 14:1613–1626

106. Meek JL, Tan HS (1986) Instability analysis of thin plates and arbitrary shells using a faceted shell element with Loof nodes. Comp Meth Appl Mech Eng 57:148–170

107. Mote CD Jr (1971) Global-local finite elements. Int J Num Meth Eng 3:565–574

108. Nagy DA (1979) Modal representation of geometrically nonlinear behaviour by the finite element method. Comp Struct 10:683–688

109. Nagy DA, König M (1979) Geometrically nonlinear finite element behaviour using buckling mode superposition. Comp Meth Appl Mech Eng 19:447–484

110. Newmark NM (1959) A method of computation for structural dynamics. Proc ASCE, J Eng Mech Div 85:67–94

111. Nickell RE (1976) Nonlinear dynamics by mode superposition. Comp Meth Appl Mech Eng 7:107–129

112. Noor AK (1981) Recent advances in reduction methods for nonlinear problems. Comp Struct 13:31–44

113. Noor AK, Greene WH, Hartley S (1977) Nonlinear finite element analysis of curved beams. Comp Meth Appl Mech Eng 12:289–307

114. Noor AK (1981) Recent advances in reduction methods for nonlinear problems. Comp Struct 13:31–44

115. Noor AK, Peters JM (1980) Nonlinear analysis via global–local mixed finite element approach. Int J Num Meth Eng 15:1363–1380

116. Noor AK, Peters JM (1980) Reduced basis technique for nonlinear analysis of structures. AIAA J 18:455–462

117. Noor AK, Peters JM (1981) Tracing post-limit-point paths with reduced basis technique. Comp Meth Appl Mech Eng

118. Noor AK, Peters JM (1981) Bifurcation and post-buckling analysis of laminated composite plates via reduced basis technique. Comp Meth Appl Mech Eng 29:271–295

119. Noor AK, Peters JM (1983) Recent advances in reduction methods for instability analysis of structures. Comp Struct 16:67–80

120. Noor AK, Peters JM (1983) Multiple-parameter reduced basis technique for bifurcation and post-buckling analysis of composite plates. Int J Num Meth Eng 19:1782–1803

121. Noor AK, Peters JM (1983) Instability analysis in space trusses. Comp Meth Appl Mech Eng 40:199–218

122. Nowacki W (1963) Dynamics of elastic systems. Chapman and Hall/Wiley, London

123. Nowacki W (1974) Structural analysis (in Polish). Polish Scientific Publishers (PWN), Warsaw

124. Nowacki W (1975) Dynamic problems of thermo-elasticity. Noordhoff, Leyden

125. Oden JT (1969) A general theory of finite elements, II Applications. Int J Num Meth Eng 1:247–259

126. Oden JT, Demkowicz LF (1996) Applied functional analysis for science and engineering. CRC Press, Boca Raton

127. Oden JT, Reddy JN (1976) The mathematical theory of finite elements. Wiley–Interscience, New York

128. Oden JT, Reddy JN (1976) Variational methods in theoretical mechanics. Springer-Verlag, Berlin

129. Otnes RK, Enochson L (1972) Digital time series analysis. Wiley, New York

130. Padovan J, Tovichakchaikul S (1982) Self-adaptive predictor–corector algorithms for static nonlinear structural analysis. Comp Struct 15:365–377

131. Panovko JG, Gubanova II (1987) Stability and vibration of elastic systems (in Russian), ed 3. Nauka, Moscow

132. Papkovich PF (1963) Papers on ship structural mechanics (in Russian), Vol 4. Sudostroyeniye, Leningrad

133. Pelc J (1984) Nonlinear shape functions in space–time element method (in Polish). Arch Inż Ląd 30:53–63

134. Pica A, Hinton E (1980) The quasi-Newton BFGS method in the large deflection of plates. In: Numerical Methods for Nonlinear Problems, 1. Pineridge–Press, Swansea, pp 355–366

135. Plaut RH (1978) Stability of shallow arches under multiple loads. Proc ASCE, J Eng Mech Div 104:1015–1026

136. Podhorecka A (1988) Space-time element method in geometrically nonlinear problems (in Polish). Mech Teor Stos 26(4):683–699

137. Podhorecki A (1986) The viscoelastic space–time element. Computers and Structures 23(4):535–544

138. Powell G, Simons J (1981) Improved iteration strategy for non-linear structures. Int J Num Meth Eng 17:1455–1467

139. Ralston A (1965) A first course in numerical analysis. McGraw Hill, London

140. Ramm E (1981) Strategies for tracing non–linear response near limit points. In: Non-linear finite element analysis in structural mechanics. Proc Europe-US Workshop. Springer, Berlin, pp 63–86

141. Ramm E, Matzenmiller A (1987) Computational aspects of elasto–plasticity in shell analysiss. In: Owen DRJ, Hinton E, Oñate E (eds) Computational plasticity – models, software and applications, Part I. Pineridge–Press, Swansea, pp 711–734

142. Reiss R, Haug EJ (1978) Extremum principles for linear initial-value problems of mathematical physics. Int J Eng Sci 16:231–251

143. Rheinboldt WC (1981) Numerical analysis of continuation methods for nonlinear structural problems. Comp Struct 13:103–123

144. Riks E (1972) The application of Newton's method to the problem of elastic stability. Trans ASME, J Appl Mech 39:1060–1066

145. Riks E (1972) An incremental approach to the solution of snapping and buckling problems. Int J Solids Struct 15:529–551

146. Riks E (1984) Some computational aspects of the stability analysis of nonlinear structures. Comp Meth Appl Mech Eng 47:219–259

147. Riks E (1984) Bifurcation and stability – a numerical approachs. In: Innovative methods for nonlinear problems. Pineridge–Press, Swansea, pp 313–344

148. Riks E (1987) Progress in collapse analysis. Trans ASME, J Press Vessel Tech 109:33–41

149. Sabir AB, Lock AC (1972) The application of finite elements to the large deflection geometrically non-linear behaviour of cylindrical shellss. In: Variational methods in engineering. Southampton Univ Press, pp 7/66–7/74

150. Safjan A (1988) Nonlinear structural analysis via reduced basis technique. Comp Struct 29:1055–1061

151. Savin GN, Goroško OA (1962) Dynamics of cable of variable length (in Russian). Izd AN USSR, Kiev

152. Schmidt WF (1978) Adaptive step size selection for use with the continuation method. Int J Num Meth Eng 12:677–694

153. Schweizerhof KH, Ramm E (1984) Displacement dependent pressure loads in nonlinear finite element analysis. Comp Struct 18:1099–1114

154. Schweizerhof KH, Wriggers P (1986) Consistent linearization for path following methods in nonlinear FE analysis. Comp Meth Appl Mech Eng 59:261–279

155. Sewell MJ (1972) A survey of plastic bucklings. In: Leipholz HHE (ed) Stability. Study No 6, Solid Mech Div, Univ of Waterloo, pp 85–197

156. Shanley FR (1947) Inelastic column theory. J Areo Sci 14:261–267

157. Shariff P, Popov EP (1971) Nonlinear buckling analysis of sandwich arches. Proc ASCE, J Eng Mech Div 97:1397–1412

158. Shidharan S, Graves–Smith TR (1981) Postbuckling analysis with finite strips. Proc ASCE, J Eng Mech Div 107

159. Simo JC, Taylor RL (1986) A return mapping algorithm for plane stress elastoplasticity. Int J Num Meth Eng 22:649–670

160. Sokołowski M (ed) (1978) Technical mechanics, Vol 4: Elasticity (in Polish). Polish Scientific Publishers (PWN), Warsaw

161. Szabo BA, Babuška I (1991) Finite element analysis. John Wiley and Sons, New York

162. Thompson JMT (1982) Instabilities and catastrophes in science and engineering. Wiley, Chichester

163. Thompson JMT, Hunt GW (1973) A general theory of elastic stability. Wiley, London

164. Timoshenko SP, Gere JM (1961) Theory of elastic stability. McGraw–Hill, New York Toronto London

165. Tvergaard V (1987) Effect of plasticity on post-buckling behaviours. In: Buckling and Postbuckling. Lect Notes in Physics, Springer, Berlin

166. Waszczyszyn Z (ed) (1981) Current methods of the stability analysis of structures (in Polish). Ossolineum, Wrocław

167. Waszczyszyn Z (1983) Numerical problems of nonlinear stability analysis of elastic structures. Comp Struct 17:13–24

168. Waszczyszyn Z (ed) (1987) Selected problems of stability of structures (in Polish). Ossolineum, Wrocław

169. Waszczyszyn Z, Cichoń Cz (1988) Nonlinear stability analysis of structures under multiple parameter loads. Eng Comp 5:0–14

170. Waszczyszyn Z, Cichoń Cz, Radwańska M (1994) Stability of structures by finite element methods. Elsevier, Amsterdam

171. Waszczyszyn Z, Janus–Michalska M (1998) Numerical approach to the 'exact' finite element analysis of in-plane finite displacements of framed structures. Comp Struct (in press)

172. Waszczyszyn Z, Pytel E, Duong N–C (1982) Numerical analysis of nonlinear instability problems of elastic trusses under multiple–parameter loads (in Polish). Eng Trans 30:131–150

173. Wecker HJ (ed) (1978) Continuation methods. Academic Press, New York

174. Weinitschke HJ (1985) On the calculation of limit and bifurcation points in stability problems of elastic shells. Int J Solids Struct 21:79–95

175. Wempner GA (1971) Discrete approximation related to nonlinear theories of solids. Int J Solids Struct 17:1581–1599

176. Werner B, Spence A (1984) The computation of symmetry–breaking bifurcation points. SIAM J Num Anal 28:388–399

177. Wieczorek M (1986) Numerical analysis of structures sensitive to buckling (in Polish). Appendix to Biuletyn WAT 11/411, Military Techn Univ, Warszawa

178. Witkowski M (1983) On space–time in structural mechanics (in Polish). Scientific Reports, Civil Engineering 80, Warsaw University of Technology, Warsaw

179. Witkowski M (1985) Triangular space–time elements in wave propagation problems (in Polish). Eng Trans 33:549–564

180. Wittbrodt E (1983) Dynamics of systems with changing in time configuration using finite element methods (in Polish). Scientific Report, Mechanics 44, Technical University of Gdańsk

181. Wriggers P, Miehe C, Kleiber M, Simo JC (1992) On the coupled thermomechanical treatment of necking problems via finite element methods. Int J Num Meth Eng 33:869–883

182. Wriggers P, Wagner W, Stein E (1987) Algorithms for nonlinear contact constraints with application to stability problems of rods and shells. Comp Mech 2:1–6

183. Zienkiewicz OC (1977) The finite element method. McGraw-Hill, London

184. Zurmühl R, Falk S (1986) Matrizen und ihre Anwendungen, Teil 2. Numerische Methoden. Springer-Verlag, Berlin

185. Życzkowski M (1991) Stability of bars and bar structures. In: Życzkowski M (ed) Studies in applied mechanics, Vol 26: Strength of structural elements. PWN Polish Sci Publishers and Elsevier, Warszawa Amsterdam, pp 247–401

Part III

Finite Difference Method

1 Introduction

The finite difference method (FDM) is an old numerical solution method of a broad class of boundary value and initial value problems. Rapid development of computers since early sixties, resulted in reevaluation of the existing numerical methods and search for new ones. The finite element method (FEM) emerged then, and due to its numerous advantages soon took a leading position among all computer methods of discrete analysis. A general, universal approach to computer analysis of variety of boundary value problems defined over domains of arbitrarily shapes was possible for the first time. Unfortunately, fascination with the FEM, due to unquestionable successes achieved in its application to solution of various scientific and engineering problems, caused stagnation in development of other computer analysis methods including the FDM.

One must realize that the analysis carried out by means of the classical FDM, i.e., the one based on regular structured grids is difficult to automate. These difficulties emerge, first of all, in discretization of boundary conditions, especially in the case of arbitrarily shaped domains. However, results of the recent research on development of the FDM generalized for arbitrary unstructured grids (GFDM), clearly indicate potential power of this method, comparable with the power of the FEM. Various results show that the generalized FDM may not only become equally universal, versatile, and suitable to full automation as the FEM, but it is even more convenient in some areas of application. Moreover the GFDM falls into the wider class of so called meshless methods (MM). Meshless methods have been under intensive development in the recent years as a powerful alternative to the FEM. All these reasons justify the presentation of the FDM in this handbook, and its discussion as one of the leading contemporary methods of computer analysis.

The following main topics of interest will be briefly presented here:

- formulation and finite difference discretization of boundary-value problems,
- outlines of the classical FD approach,
- curvilinear finite difference (CFD) approach using irregular but structured grids, based on a mapping of subsequent local configurations of nodes,
- generalized finite difference approach (GFDM) based on: arbitrary irregular grids, density controlled mesh generation, Voronoi tessalation and Delaunay triangulation of the domain, moving weighted least squares

(MWLS) local approximation used to generation of FD formulas and post-processing,
- adaptive generalised finite difference solution approach, based on a'posteriori error estimation and multigrid analysis,
- outlines of the mathematical foundations for GFDM in elliptic problems.

The limited scope of this book excludes the possibility of many other important topics presenting the actual state of art of the method formulation like:

- GFDM on a differential manifold and its use in shell analysis,
- mixed global and local GFDM resulting from constrained optimization problems,
- generalized finite strip approach (GFSM) i.e. combined 3D analysis of bodies of revolution, or prismatic ones by the GFDM (2D) and Fourier series (1D),
- various GFDM/FEM combinations and unification of their algorithms,
- a summary of the current GFDM features, and a brief discussion on the other meshless methods (MM),

being discussed here and limits the quantity of presented sample GFDM applications. The purpose of this book, however, is to present concepts and formulation of the method rather than solutions. Therefore, only a few simple illustrative numerical examples are included.

1.1 Formulation of Boundary-value Problems for Finite Difference Analysis

The basic idea of the FDM is simple: derivatives of functions are replaced by the corresponding finite differences of their values at selected points called nodes. A discrete set of nodes forms a mesh. Due to such an approximation of a required function and its derivatives, a given boundary-value problem is converted into a system of simultaneous algebraic equations where the nodal values of that function are usually unknown. This concept, however, may be realized using various formulations of boundary-value problems outlined below.

The classical FDM emerged as an approximate discrete method of analysis of boundary-value problems given in the form of differential equations and appropriate boundary conditions (local formulation). Later on, the range of applications was broadened to problems given in the variational form. More complex formulations of boundary-value problems may be analyzed by the GFDM now. It is worth noticing that several of these formulations may be also presented as particular cases of the weighted residual method. To avoid going into details, it is assumed here, that each time the spaces being dealt with provide functions smooth sufficiently to perform all necessary differentiations.

In a considered domain $\Omega \subset \mathbb{R}^n$ with boundary $\partial\Omega$ a function $u(P)$ at each point P is sought for, using one of the following formulations

Local

$$\mathcal{L}u = f \quad \text{for } P \in \Omega \tag{1.1}$$

$$\mathcal{L}_b u = g \quad \text{for } P \in \partial\Omega \tag{1.2}$$

where $\mathcal{L}, \mathcal{L}_b$ are given differential operators, and f, g are known functions of P. In the weighted residual form the weighting factor is $w(\mathbf{x}) = \delta(\mathbf{x} - \mathbf{x}_0)$.

Global

A variety of well known variational formulations in mechanics belongs to this group. These may be posed either in the form of functional optimization or as variational principles.

In the first case functionals given in the general form are considered

$$I(u) = \frac{1}{2}\mathcal{B}(u,u) - \mathcal{L}(u) \tag{1.3}$$

In energy functionals the first term of $I(u)$ represents internal energy of the system, while the second one is the work done by external load. Unknown function $u(P)$ may denote displacements u , strains ε , stresses σ or all of them. Function u realizing extremum (minimum, stationary point) of the functional $I(u)$ is sought. Two situations may be distinguished here:

(i) unconstrained optimization problem

$$u \in V \tag{1.4}$$

if extremum is searched for in the whole space of admissible solutions V

(ii) constrained optimization problem

$$u \in V_{ad} \tag{1.5}$$

if extremum is searched for in a subspace $V_{ad} \subset V$ determined by the constraints imposed.

An alternative global boundary value problem formulation may be given in the form of variational principle (e.g. the principle of virtual work)

(i) $$\mathcal{B}(u,\delta u) = \mathcal{L}(\delta u) \quad \text{for } \delta u \in V \tag{1.6}$$

(ii) $$\mathcal{B}(u,\delta u) \geq \mathcal{L}(\delta u) \quad \text{for } \delta u \in V_{ad} \tag{1.7}$$

The global approach involves integration over the domain Ω. The weighting factor is $w(P) = 1$ if $P \in \Omega$, otherwise $w(P) = 0$.

Global / local

Boundary-value problems in which the global approach is applied to subdomains Ω_i, $i = 1, 2, \ldots, n$ belong to this group. Methods of discrete analysis using such an approach have many names like cell collocation [52] or the finite volume. Usually Ω_i is assumed as a subdomain assigned to the node P_i. It is most often represented by the so called Voronoi (or Thiessen) polygon.

In the global / local approach the weighting factor is $w(P) = 1$ if $P \in \Omega_i$, otherwise $w(P) = 0$.

Mixed global/local

The mixed approach is understood as a combination of the global and local formulations, and is given in a form of the constrained optimization problem. The functional $I(u)$ is optimized satisfying, at the same time, constraints given as differential equations $\mathcal{L}u = f$ inside the considered domain Ω, and $\mathcal{L}_b u = g$ on its boundary $\partial\Omega$.

Though this problem might be considered as a particular case of the global formulation, it is worth of separate treatment in the FD solution approach.

Combined

Various combinations of the FD with other methods, especially with the FEM [37] may be considered.

Useful FDM / FEM combinations include:

- each method working on the different part of the domain Ω
- processing done by the FEM and followed by the FDM postprocessing
- use of the FDM to generate the FE characteristics of the problem
- use of the FEM approximation to generate the FD formulas.

All boundary-value problem formulations presented above may be used in the GFD discretization of differential operators as proposed in [37, 63, 80, 83, 81]. The global approach additionally involves integration, and uses local approximation based on FD stars, resulting from the neighborhood of Gauss points rather than from nodes.

1.2 FDM Discretization

In the FDM operations $\mathcal{L}u$ and $\mathcal{L}_b u$ appearing in the local formulation (1.1) and (1.2), as well as any other differential operations performed during solution of boundary value problems are discretized. Discretization of a given differential operator $\mathcal{L}$ at a selected point P_i relies, in the simplest case, on replacement of the operator value $\mathcal{L}u_i \equiv \mathcal{L}u(P_i)$ at this point by a relevant finite difference operator $\mathrm{L}u_i$. Such FD operator may be either generated

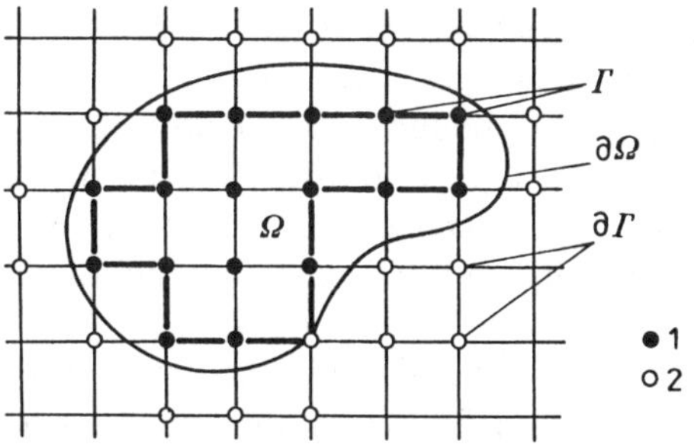

Fig. 1.1. FD mesh in domain Ω; • – internal nodes Γ, ∘ – external nodes $\partial\Gamma$

directly (classical FDM), or composed of operators discretizing particular derivatives. The complete set of all such derivatives of function u of order $1,\ldots,n$ is denoted as $\mathcal{D}u$. The value $\mathcal{D}u(P_i)$ of the operator $\mathcal{D}u$ at a point P_i is approximated, in the simplest case, by a linear combination

$$\mathcal{D}u(P_i) \approx \sum_j \alpha_{j(i)} u(P_{i+j}) \tag{1.8}$$

of function value $u_{i+j} \equiv u(P_{i+j})$ at this point, and m points $P_{i+j}, j = 1,\ldots,m$ chosen from its neighborhood, where $\alpha_{j(i)}$ are coefficients resulting from the FD approximation.

Right hand side of this relation is called the finite difference formula (or scheme) approximating the operator value $\mathcal{D}u(P_i)$ (left hand side). Points P_{i+j}, $j = 0,1,\ldots,m$ used in this formula are called nodes. These form a configuration of nodes called the FD star, where the point P_i is the central node of this star. All nodes in the considered domain constitute a mesh. Star configuration depends on the form and order of the differential operator $\mathcal{L}$ approximated by the FD formula, on the type of assumed degrees of freedom (d.o.f.) as well as on the required precision of approximation and on kind of mesh considered.

For the purpose of discretization in the whole domain Ω and on its boundary $\partial\Omega$ a set of points Γ, $P_i \in \Omega, i = 1,2,\ldots,N$ later called internal nodes, and a set of points $\partial\Gamma$, $P_i \notin \Omega$, $i = N+1, N+2,\ldots,N+M$ located on the boundary $\partial\Omega$ and in its outside neighborhood, called external nodes are selected (Fig. 1.1). These together $(\Gamma \cup \partial\Gamma)$ constitute the mesh for the boundary-value problem under consideration.

Using appropriate finite difference operators and a given formulation of the boundary-value problem, FD equations be may generated

$$\mathbf{Au} = \mathbf{f} \tag{1.9}$$

These simultaneous algebraic equations are, linear or nonlinear, depending on the type of the problem considered. In the simplest case elements of the vector $\mathbf{u}$ are unknown nodal values $u(P_i)$, $i = 1, \ldots, N$, otherwise $\mathbf{u}$ also contains all types of unknown generalized d.o.f.

In the local formulation of boundary-value problems the FD equations are (1.9) generated by collocation, inside the domain Ω, i.e. at all N internal nodes Γ for Eq. (1.1), and on the boundary for the boundary conditions (1.2). However, instead of collocation the other weighted residual methods may be applied as well.

In the global formulation the FD approximation of differential operators is used to transform a given functional into function of unknown discrete d.o.f. mentioned above. Later, FD equations (1.9) are generated by enforcing stationary conditions of that functional. In the global approach one may use a variational principle or a weak form, and taking advantage of FD operators discretize the considered problem by the Galerkin or an equivalent method.

Compared with the local approach, the global FDM displays some advantages such as:

- usually the order of difference operators required is lowered by the factor of .5;
- symmetry of the simultaneous FD equations is obtained; the reciprocity property is preserved then, i.e. when a node i belongs to the star of node j , the reverse relation holds true as well; the local formulation may yield symmetry only in the case of a regular mesh and domain, and appropriate boundary conditions;
- in the case of a complex structure it is easier to consider various strengthenings (supports), variable thicknesses, layered texture, thermal and other effects for which energy may be determined;
- the solution procedure is similar to the finite element one.

On the other hand it has to be stressed that the way leading to the FD equations is much more complex and laborous here (integration is needed). Moreover, long time experience shows [17] that the final results are quite sensitive to the way the approximation and the integration are applied.

The above considerations refer to the space discretization only. It is worth to note that the FDM is a well established and effective solution method of the time dependent problems, i.e., the initial value and initial-boundary ones. These topics, however, will not be discussed here because, while spatial discretization may be done in the same way as in boundary value problems, the time integration is beyond the scope of this chapter, and is described e.g. in [2, 91].

1.3 The Basic FDM Procedure

Prior to formulation of the FDM details, essential steps of the method are presented here. One starts from initial information on a considered boundary value problem, like formulation (differential equation(s) or functional,...), domain, boundary conditions, solution precision required etc. The basic FD solution algorithm steps are as follows:

- domain discretization
  - mesh generation: location of nodes inside the domain and on its boundary (sometimes also outside the boundary), as well as mesh type (in the case of the classical FDM) are selected
  - domain partition into subdomains assigned to particular nodes
- local approximation and problem discretization
  - selection and classification of FD stars
  - assumption of a number and kind of degrees of freedom at nodes
  - assumption of local approximation type and order
- generation of FD operators (formulas) corresponding to differential operators
  - inside the domain, in a way appropriate to formulation of the problem
  - on the boundary, relevant to boundary conditions
- generation of FDM equations inside the domain
  - local formulation – nodal collocation
  - global formulation – integration, assembling followed by the Galerkin or equivalent variational approach
- discretization of boundary conditions
- solution of simultaneous FD equations
- solution for the required final results (postprocessing)

The forthcoming chapters describe how these essential steps of the FD algorithm are realized in both classical and generalized FDM.

2 The Classical FDM

The original version of the method using regular meshes is called the classical FDM. It is broadly described in numerous monographs [8, 15, 23, 68, 69, 89, 91–97] and, therefore, it will be only briefly presented here.

Regular meshes used by the FDM may be of various types usually dependent on the domain shape, in order to best fit its boundaries. Some typical examples of such meshes are shown on Fig. 2.1, though rectangular and triangular grids are the most frequently applied. In addition, FD stars typical for the Laplace operator are marked there.

The subsequent steps of the procedure appropriate to the classical FDM will be discussed below. Simplifications resulting from mesh regularity will be underlined.

2.1 Domain Discretization

Assumption of mesh type and modulus is sufficient to define locations of all nodes as well as domain partition into subdomains assigned to these nodes. Beyond the boundary zone all these subdomains are the same.

Intrinsic problems may arise, however, in the case of discretization of curvilinear boundaries by means of regular meshes.

2.2 Selection of Stars and Generation of FD Operators

Due to mesh regularity, it is sufficient to deal with the same FD star and scheme in all internal nodes. In the classic FDM one may either assume a FD scheme at first and then obtain a FD star, or one may reverse this order. Both these situations will be considered here, as what is done first and what next depends on a chosen generation method of FD formulas.

In the traditional approach one begins with discretization of derivatives appearing in the considered differential operator.

FD formulas for derivatives of a function may be obtained by replacing these derivatives with relevant difference quotient. For a function of one variable, one of the following definitions of difference quotient approximating

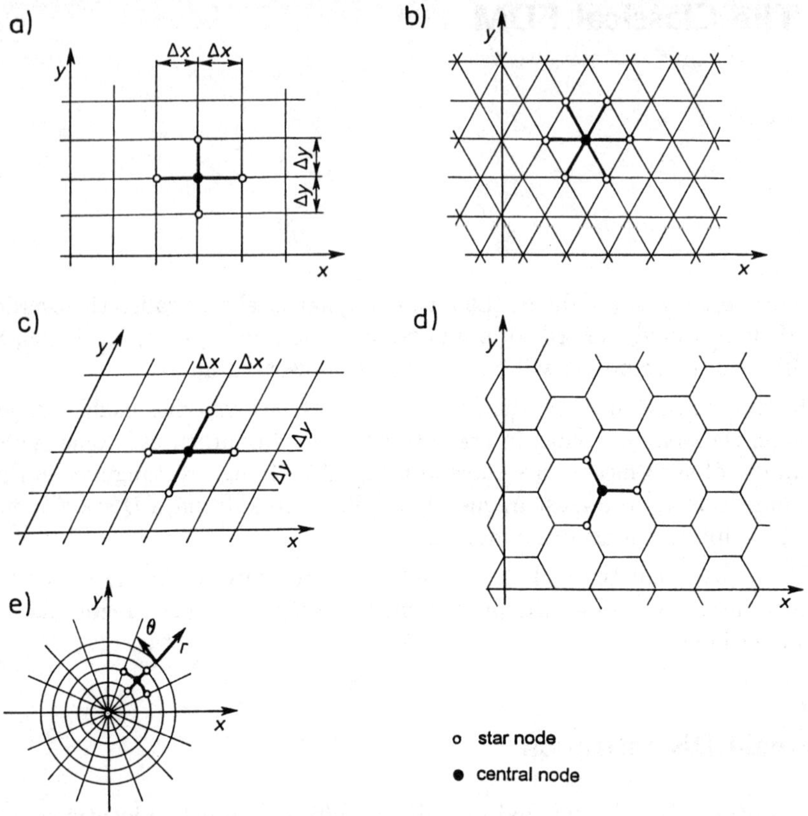

Fig. 2.1. Samples of regular 2D FD meshes: **a)** rectangular, **b)** triangular, **c)** skewed, paralellogram type, **d)** hexagonal, **e)** polar

the first derivative of function may be used

$$\left.\frac{\mathrm{d}u}{\mathrm{d}x}\right|_{x=x_i} \equiv u_i' = \begin{cases} \dfrac{u_{i+1}-u_i}{h} & + \; O(h), \\ \dfrac{u_{i+1}-u_{i-1}}{2h} & + \; O(h^2) \\ \dfrac{u_i-u_{i-1}}{h} & + \; O(h). \end{cases} \tag{2.1}$$

FD formulas for higher order derivatives may be obtained through proper composition of the formulas for the first order derivative. Thus for the second derivative one has

$$\left.\frac{\mathrm{d}^2u}{\mathrm{d}x^2}\right|_{x=x_i} \equiv u_i'' = \frac{u_{i-1}-2u_i+u_{i+1}}{h^2} + O(h)\,. \tag{2.2}$$

FD formulas are often presented in a graphic form as FD schemes. For the examples given above one has

$$\left.\frac{\mathrm{d}}{\mathrm{d}x}\right|_{x=x_i} \approx \begin{cases} \frac{1}{h}\boxed{-1}\!-\!\boxed{1} \\ \frac{1}{2h}\boxed{-1}\!-\!\boxed{0}\!-\!\boxed{1} \\ \frac{1}{h}\boxed{-1}\!-\!\boxed{1} \end{cases} \tag{2.3}$$

$$\left.\frac{\mathrm{d}^2}{\mathrm{d}x^2}\right|_{x=x_i} \approx \frac{1}{h^2}\left(\boxed{1}\!-\!\boxed{-2}\!-\!\boxed{1}\right) \tag{2.4}$$

FD scheme for an arbitrary linear differential operator $\mathcal{L} = \sum_j c_j \frac{\mathrm{d}^j}{\mathrm{d}x^j}$ may be obtained by composing FD schemes for particular derivatives. Thus, e.g., for the operator $\mathcal{L}u = u'' + au' + bu$ one has

$$\mathcal{L}u\big|_{x=x_i} \approx \frac{1}{h^2}\left(\boxed{1-\tfrac{ah}{2}}\!-\!\boxed{\text{-}2+bh^2}\!-\!\boxed{1+\tfrac{ah}{2}}\right) \tag{2.5}$$

Operators involving partial derivatives are considered in a similar way. Let us consider a regular rectangular mesh (Fig. 2.2) as an example. Using the same approach as mentioned above and denoting $u_{ij} \equiv u(x_i, y_j)$ one may find, e.g., the following formulas

$$\begin{aligned} \left.\frac{\partial u}{\partial x}\right|_{i,j} &= \frac{1}{2h}(u_{i+1,j} - u_{i-1,j}) + O(h^2) \\ \left.\frac{\partial u}{\partial y}\right|_{i,j} &= \frac{1}{2k}(u_{i,j+1} - u_{i,j-1}) + O(k^2) \\ \left.\frac{\partial^2 u}{\partial x^2}\right|_{i,j} &= \frac{1}{h^2}(u_{i-1,j} - 2u_{i,j} + u_{i+1,j}) + O(h) \\ \left.\frac{\partial^2 u}{\partial x \partial y}\right|_{i,j} &= \frac{1}{4hk}(u_{i-1,j-1} - u_{i-1,j+1} + u_{i+1,j+1} - u_{i+1,j-1}) + O(hk) \end{aligned} \tag{2.6}$$

In the above approach FD formulas derived from the difference quotient determine relevant FD stars in a unique way. It is worth noticing that for each linear differential operator $\mathcal{L}$ a relevant FD formula has the form of a linear combination

$$\mathcal{L}u\big|_i \approx \sum_{j=0}^{m} B_{j(i)} u\left(P_{i+j}\right) \tag{2.7}$$

of function values $u(P_{i+j})$ considered as the basic unknowns (and d.o.f.) of the method.

Summation is extended over all nodes $j = 0, 1, \ldots, m$ of the star, where $B_{j(i)}$ are known coefficients. Formula (2.7) is of a general character, as the operator $\mathcal{L}$ may represent a given differential equation, boundary conditions, or may appear in the functional $I(u)$.

It is worth noticing that taking into account formula (2.7) one may reverse the situation, i.e., begin by assuming a'priori a suitable star and subsequently find the appropriate FD formula. This may be accomplished in several ways:

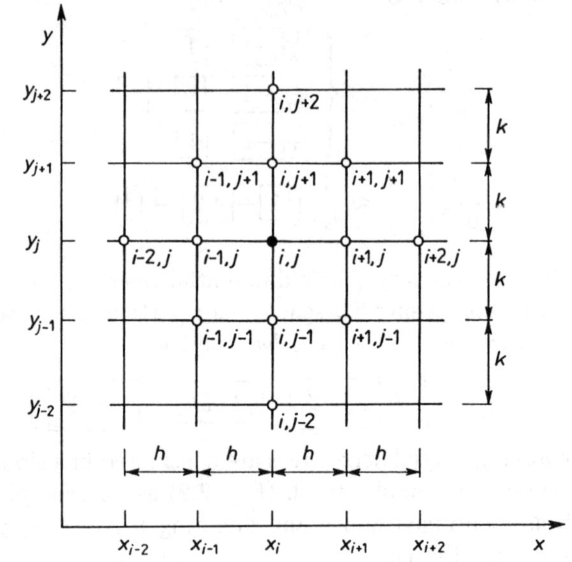

Fig. 2.2. Rectangular FD mesh

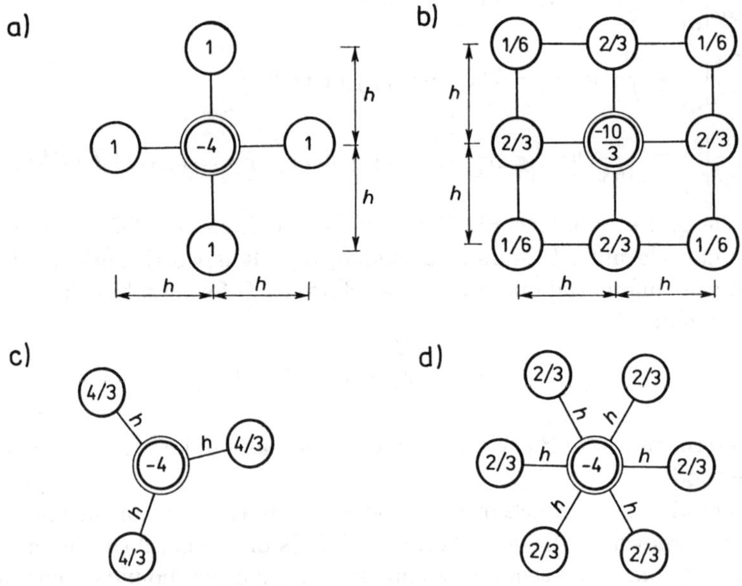

Fig. 2.3. Laplace FD operator $\nabla^2 h^2$: **a)** standard – rectangular mesh, **b)** higher order – rectangular mesh, **c)** standard – hexagonal mesh, **d)** standard – triangular mesh

(i) spanning a local approximation of function

$$u(P) \approx \sum_{j=0}^{m} \varphi_j(P) u\left(P_{i+j}\right) \tag{2.8}$$

using nodal values $u(P_{i+j})$ in the star, as well as the assumed bases functions $\varphi(P)$, and find the value of a given operator $\mathcal{L}$ at a point P_i

$$\mathcal{L}u\left(P_i\right) \approx \sum_{j=0}^{m} \mathcal{L}\varphi_j(P)\big|_i u\left(P_{i+j}\right) \tag{2.9}$$

Thus the coefficients of the difference formula approximating operator $\mathcal{L}u(P_i)$ are

$$B_{j(i)} \equiv \mathcal{L}\varphi_j(P)\big|_i \tag{2.10}$$

(ii) expressing the formula (2.7) in the form

$$\mathcal{L}u\big|_i \approx \sum_{j=0}^{m} \sum_{s=0}^{n} \alpha_{j(i)}^{(s)} u^{(s)}\left(P_{i+j}\right) \tag{2.11}$$

where, besides the nodal values of function $u(P_{i+j})$, also the nodal values of its derivatives $u^{(s)}(P_{i+j})$ were assumed as degrees of freedom. Coefficients of the FD formula $\alpha_{j(i)}^{(s)}$ are not known a'priori and have to be found. This may be accomplished in one of the two following ways:

- by expanding both sides of Eq. (2.11) into the Taylor series at the point P_i and requiring that the corresponding consecutive terms of these series up to the term of the highest possible order be equal;
- by requiring the formula (2.11) to be exact for the highest possible number of consecutive monomials (e.g. x^i in 1D, and $x^i y^j$ in 2D, etc., where $i, j = 0, 1, \dots$)

As opposed to the first way (i) of evaluation of coefficients $\alpha_{j(i)}^{(s)}$, based on differentiation of a local approximation of function $u(P)$, both variants of the approach (ii) are easily applicable to arbitrary irregular meshes. However, development into the Taylor series seems to be more advantageous, as it may also provide us with the error term of the local approximation used.

Examples of several typical FD operators generated on regular meshes are presented in Figs. 2.3–2.5. All of these use nodal values of the function only, and do not include its derivatives, i.e., $n = 0$. Except for formulas (4.3b) and (4.4b), which require a special approach involving weighting factors (cf. Sect. 4.2.4), all these formulas may be derived using any of the approaches mentioned above.

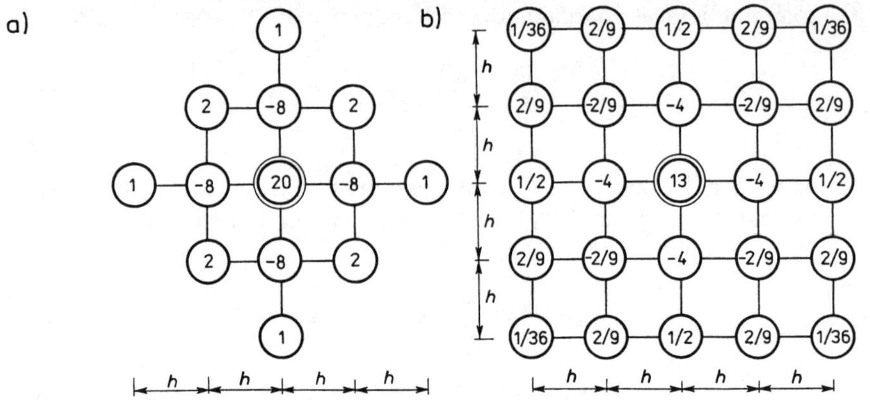

Fig. 2.4. Biharmonic FD operator $\nabla^4 h^4$ – square mesh: **a)** standard, **b)** higher order obtained by composition of ∇^2 higher order operators

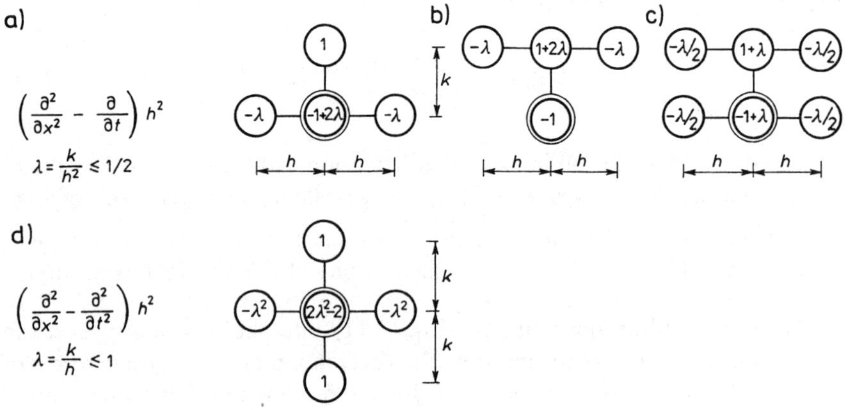

Fig. 2.5. Time dependent FD operators – rectangular mesh, parabolic operator: **a)** explicit, **b)** standard implicit, **c)** Crank–Nickolson implicit, **d)** standard hiperbolic

2.3 Generation of the FD Equations

In the local formulation FD equations are generated by collocation at all N internal nodes Γ of the domain Ω

$$\mathcal{L}u\big|_i \approx \sum_{j=0}^{m} B_{j(i)} u_{i+j} = f_i, \; i = 1, \ldots, N \tag{2.12}$$

In the global formulation after FD discretization one deals with integration at first, followed by assembling, like in the FEM.

Consider a given functional (1.3). One may denote it briefly as

$$I(u) = \int_\Omega F(u)\,\mathrm{d}\Omega \tag{2.13}$$

Integration is performed in a numerical way. In the simplest case

$$I \approx \sum_{i=0}^{N} F(u_i)\Delta\Omega_i = I(u_1, \dots, u_N) \tag{2.14}$$

where $\Delta\Omega_i$ is the surface area of a subdomain assigned to node P_i . The FD equations for unknown nodal values of function u_i are derived from stationary conditions

$$\frac{\partial I}{\partial u_i} = 0 \quad \text{for } i = 1, 2, \dots, N \tag{2.15}$$

imposed on the functional $I(u)$.

In the case of the global formulation, considered in a weak form, one may use the Galerkin discretization method

$$(\mathcal{L}u, \delta u)_\Omega = (f, \delta u)_\Omega \tag{2.16}$$

$$(\mathcal{L}_b u, \delta u)_{\partial\Omega} = (g, \delta u)_{\partial\Omega} \tag{2.17}$$

where $\delta u(x)$ is a test function and u is a trial function.

2.4 Imposition of Boundary Conditions

In the local formulation boundary conditions may be imposed either at the time when FD formulas involving boundary nodes are generated, or afterwards, to complete simultaneous FD equations derived inside the domain.

In the global formulation natural and essential boundary conditions may be distinguished. While boundary conditions of the first type are embedded in the functional directly, the essential ones have to be imposed additionally, as constraints, before minimization of the functional is performed. It is worth to distinguish two cases:

- boundary condition is imposed on an unknown function only;
- differential operator appears in the boundary condition.

In the first case discretization of the boundary condition is obvious, namely

$$u(P_i) = g_i, \quad P_i \in \Omega, \quad i = N+1, \dots, M \tag{2.18}$$

In the second case discretization

$$\mathcal{L}_b u_i \approx \sum_{j=0}^{m} B_{j(i)} u_{i+j} = g_i \quad \text{for } i = N+1, \dots, M \tag{2.19}$$

is done in a similar way as for the operator $\mathcal{L}$, described above.

Discretization of the boundary conditions becomes really complicated when nodes are not located on the domain boundary. Such a situation is typical for domains of arbitrary irregular shapes.

FD discretization of a differential operator on a boundary is usually done in one of the following ways:

- additional, so called fictitious nodes located in the neighborhood of the boundary but outside the considered domain are assumed; boundary conditions are used then to express value of unknown function u at these external nodes in terms of values of that function at internal nodes $(P_i \in \Gamma)$ of the domain Ω ;
- FD formulas are generated using only internal and boundary nodes $(P_i \in \Gamma)$; the central node P_i is located on the boundary or close to it.

In the first case one achieves better approximation because the central node is located closer to the FD star center of gravity. On the other hand, introducing external fictitious nodes one may introduce into the solution certain undesired side effects of numerical nature, like appearance of new, spurious eigenvalues and eigenvectors due to change of mass / stiffness ratio in the original problem. These false solutions may be crucial, however, in the analysis of dynamic and stability problems (e.g. change of the lowest eigenfrequency!). Thus the use of fictitious nodes, convenient in the analysis of elliptic boundary value problems, may essentially affect solutions of hyperbolic and parabolic ones.

Values of the searched function at m external nodes present additional unknowns of the method. Their presence, justified by better approximation of differential operators on a domain boundary, requires completion of the basic system of simultaneous FD equations by s additional ones.

In the local formulation these additional FD equations are obtained by collocation applied to the given differential equation at boundary nodes in the same way as it is done at internal nodes. This way both the given differential equation and boundary conditions are enforced at the boundary nodes.

In the global formulation the finite difference equations resulting from discretization of essential boundary conditions (differential type) at all boundary nodes constitute constraints of the given functional (2.13).

In the second case, i.e., when only the internal and boundary nodes are taken into account, one does not need to generate additional FD equations, and the spurious solutions mentioned above are avoided. However, the approximation quality of the boundary conditions is much worse then. In order to provide a reasonable quality of the solution, FD formulas and discretization of boundary conditions should be based on an approximation of at least the same order as the one used inside the domain. The above remarks will be illustrated with two examples of 1D discretization .

In the first example taken from the paper [5] a negative influence of fictitious nodes on the solution may be noticed. The critical force and the corresponding form of stability loss in a clamped cylindrical shell subjected to axial compression are sought. Half of that shell was analyzed. It was divided into two segments of lengths s_1 and s_2. Each of these was covered with a regular mesh of a density different in each segment. Solutions of the eigenproblem obtained for denser and denser meshes are presented in Fig. 2.6 for the first and the second critical force, as well as for the corresponding buckling forms. These solutions were found with (the first and the second critical force) and without (only the first critical force) fictitious nodes. It should be stressed that the first eigenvector found without fictitious nodes is of the same form as the second one determined with additional nodes included. Thus it may be noticed that the use of fictitious nodes located outside of the shell introduced an additional, false solution for the lowest critical force and the relevant eigenform. Moreover, this false solution proved to be very sensitive to the increase of mesh density.

The second example shows how the solution precision depends on the discretization of boundary conditions. Deflection of simply supported beam under uniformly distributed load (Fig. 2.7) is analyzed here. For the same FD discretization the problem was solved with (case A) and without (case B) fictitious nodes. In the case B the boundary conditions were discretized using three different approximation orders (2, 3, 4).

The local formulation of the problem is well known

$$w^{IV} = \frac{q}{EJ} \quad \text{for } x \in \Omega; \quad w = 0 \quad \text{and} \quad w'' = 0 \quad \text{for } x = \pm\frac{l}{2}. \tag{2.20}$$

In the discretized form one has

- inside the domain

$$w^{IV} = h^{-4}(w_{i-2} - 4w_{i-1} + 6w_i - 4w_{i+1} + w_{i+2}) + O(h) \tag{2.21}$$

  where $h = l/4$, and due to symmetry $i = 1, 2$
- on the boundary
  case A

$$w_0'' = h^{-2}(w_{-1} - 2w_0 + w_1) + O(h) \tag{2.22}$$

  case B

$$w_0'' = \begin{cases} h^{-2}(w_0 - 2w_1 + w_2) + O(h) \\ h^{-2}(2w_0 - 5w_1 + 4w_2 - w_3) + O(h^2) \\ \frac{1}{12}h^{-2}(35w_0 - 104w_1 + 114w_2 - 56w_3 + 11w_4) + O(h^3) \end{cases} \tag{2.23}$$

The final results are presented in Table 2.1.

Due to the rough FD discretization used, the influence of the way the boundary conditions are discretized on the quality of the final solution is significant. Thus for the same order of approximation the results in case A are much better than in case B_a. The precision of the solution improves

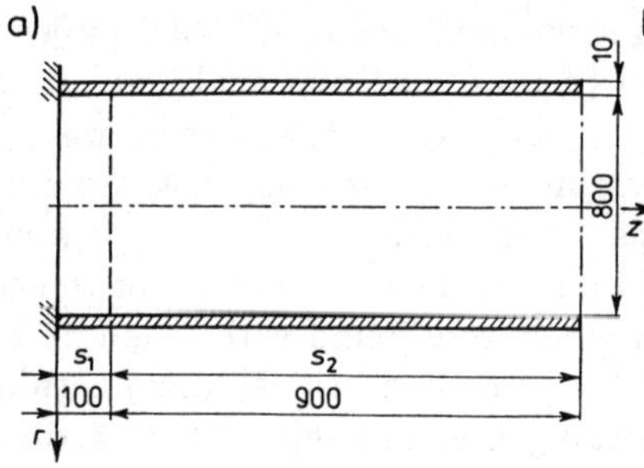

b)

number of nodes		Critical force lb/in with fictitious nodes		Critical force lb/in without fictitious nodes
s_1	s_2	1st eigenvalue	2nd eigenvalue	1st eigenvalue
26	16	7,008	16,752	16,779
26	31	8,840	16,604	16,629
26	46	11,947	16,603	16,093
26	61	13,856	15,651	15,670
26	76	14,816	15,461	15,465
51	91	15,151	15,350	15,353

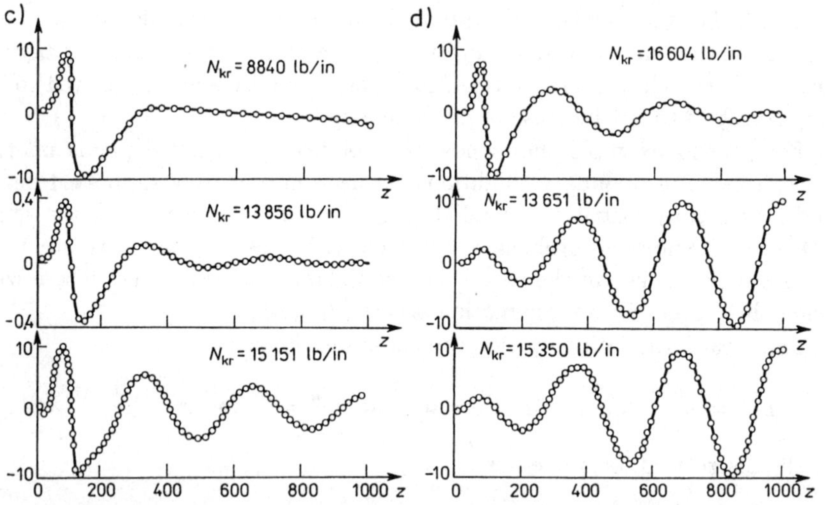

Fig. 2.6. Influence of fictitious nodes on the critical force and buckling mode of an axially compressed clamped cylindrical shell: **a)** shell, **b)** critical force, **c)** first (false) buckling mode, **d)** second (true first) buckling mode

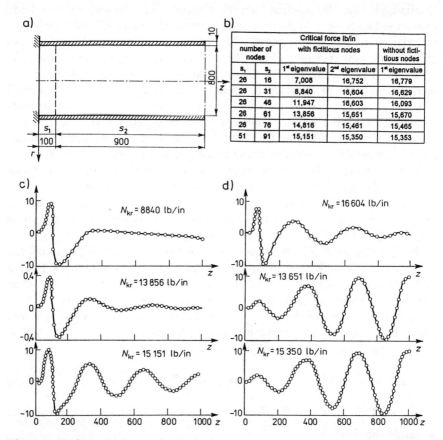

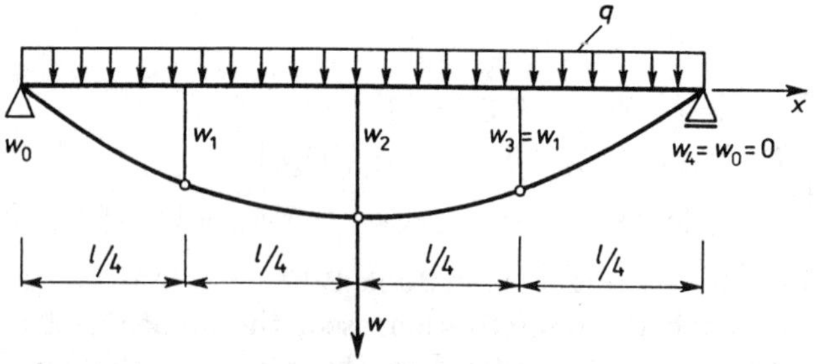

Fig. 2.7. Deflection of simply supported uniformly loaded beam

Table 2.1. Deflection of the free supported beam

Case	Order	$w_1 \times 1536 \times EJ/ql^4$	$w_2 \times 1536 \times EJ/ql^4$	Remarks
A	2	15.00	21	-
B_a	2	1.50	3	-
B_b	3	6.00	9	-
B_c	4	14.25	20	exact

essentially when the approximation order is increased (cases B_a, B_b, B_c). To obtain a result of reasonable quality without reverting to external nodes, one needs to apply the same order of approximation of the function both on the boundary and inside the domain (case B_c) rather than the same approximation order of the boundary and domain operators (case B_a).

Application of improved FD formulas to discretization of 1D boundary value problems is examined and discussed in [18].

2.5 Solution of Simultaneous FD Equations

In elliptical boundary value problems, Eqs. (2.12) and (2.19) or (2.18) (local formulation) as well as (2.15) or (2.16) and (2.17) (global formulation), lead to the set of simultaneous algebraic equations

$$\mathbf{A}\mathbf{u} = \mathbf{f} \tag{2.24}$$

with respect to unknowns $\mathbf{u} = [u_i]$, where $\mathbf{f} = [f_i]$, $i = 1, \ldots, n$, is the right hand side vector.

For linear problems the matrix $\mathbf{A}$ is symmetric in the global formulation, and usually non-symmetric in the local one (mainly due to boundary conditions). In the latter case relevant solution methods like Gaussian elimination, frontal method for non-symmetric matrices [26], projection method [30, 58] etc., including viscotic and dynamic and dynamic relaxation may be applied.

Solution of the FD equations is particularly convenient if the matrix $\mathbf{A}$ has the tridiagonal form. The "pragonka" method [95], based on a recurrent solution approach is very useful then. Having dealt with equation of the form

$$a_{i-1}x_{i-2} + b_{i-1}x_{i-1} + c_{i-1}x_i = d_{i-1} \tag{2.25}$$

with unknowns $x_1, \ldots, x_n$, one finds forwarding step by step

$$x_i = c_{i-1}^{-1}(d_{i-1} - a_{i-1}x_{i-2} - b_{i-1}x_{i-1}) \quad \text{for } i = 2, 3, \ldots, N \tag{2.26}$$

determining at last x_N. Going backward up to the first equation step by step, the required solution is finally found from the formula

$$x_{i-2} = a_{i-1}^{-1}(d_{i-1} - b_{i-1}x_{i-1} - c_{i-1}x_i) \quad \text{for } i = N-1, \ldots, 2 \tag{2.27}$$

In the case of non-linear equations, a variety of methods, like numerous variants of the Newton–Raphson approach with relaxation as well as self-correcting [59] methods may be used.

2.6 Postprocessing – Evaluation of the Required Final Results

Solution of simultaneous FD equations for the basic unknowns usually does not provide the user directly with the quantities required. For instance, if nodal displacements constitute FD unknowns, one is usually interested in strains and stresses as well, both at nodes and beyond them. In such case one needs to evaluate certain difference operators at any point required, based on already determined discrete (nodal) values of d.o.f. Approach similar to one discussed above may be applied in order to find these operators. More detailed explanation is given in the Chap. 6.

2.7 Numerical Examples

Example 1.
Analysis of deflections of a simply supported, uniformly loaded beam shown in Fig. 2.7; q – load, EJ – bending stiffness

Problem formulation

Local. Find the beam deflection $w(x)$ defined by equation

$$\frac{\mathrm{d}^2 w}{\mathrm{d}x^2} = -M(x), \qquad M(x) = \frac{ql^2}{2EJ}\left(1 - \frac{x^2}{l^2}\right) \tag{2.28}$$

for $x \in [-l, l]$, and

$$w(-l) = w(l) = 0 \tag{2.29}$$

Global. Find the beam deflection $w(x)$ using boundary conditions $w(-l) = w(l) = 0$; apply the following three formulations

(i) Principle of the total potential energy minimum

$$\min_w I(w), \; I(w) = \int_{-l}^{l} \left[\frac{1}{2}\left(\frac{\mathrm{d}w}{\mathrm{d}x}\right)^2 - M(x)w\right] \mathrm{d}x \tag{2.30}$$

$$w(-l) = w(l) = 0 \tag{2.31}$$

(ii) Galerkin approach to weak forms of equation (2.28)

$$\int_{-l}^{l} \left[\frac{\mathrm{d}^2 w}{\mathrm{d}x^2} + M(x)\right] v \,\mathrm{d}x = 0 \tag{2.32}$$

or after integration by parts and taking advantage of the boundary conditions

$$\int_{-l}^{l} \left[\frac{\mathrm{d}w}{\mathrm{d}x} \frac{\mathrm{d}v}{\mathrm{d}x} - M(x)v \right] \mathrm{d}x = 0, \tag{2.33}$$

where w is a trial function and v is a test function vanishing at the boundary $v(-l) = v(l) = 0$.

FD discretization

Advantage is taken of symmetry, i.e., $w_4 = w_0, w_3 = w_1$ and only half of the beam is considered.

Local formulation. FD collocation formula applied to equation (2.28)

$$w_{i-1} - 2w_i + w_{i+1} = M_i \frac{l^2}{4}, \quad i = 1, 2 \tag{2.34}$$

FD simultaneous equations

$$\begin{aligned} w_0 - 2w_1 + w_2 &= 3Q \\ w_1 - 2w_2 + w_1 &= 4Q \end{aligned} \tag{2.35}$$

where $Q \equiv \frac{ql^4}{32EJ}$.

FD boundary conditions

$$w_1 = w_4 = 0 \tag{2.36}$$

FD solution

$$w_1 = 5Q = \frac{20}{19} w_{1\text{EXACT}}, \qquad w_2 = 7Q = \frac{21}{20} w_{2\text{EXACT}} \tag{2.37}$$

Global formulation. Integration is an essential part of each global formulation. It is usually executed in a numerical manner. Precision of such integration, however, may significantly influence the solution quality. In the FDM one may integrate numerically around nodes or between them (like in the FEM), appropriately for the FD formulas used.

(i) Functional of the beam total potential energy is approximated by FD expressions as follows

$$\begin{aligned} I(w) \approx 2 \Bigg\{ &\frac{1}{2} \left[\left(2\frac{w_1 - w_0}{l} \right)^2 \frac{l}{2} + \left(2\frac{w_2 - w_1}{l} \right) \frac{l}{2} \right] - \\ &- (M_0 w_0 + M_1 w_1) \frac{l}{4} - (M_1 w_1 + M_2 w_2) \frac{l}{4} \Bigg\} = \\ &= \frac{2}{l} \left[w_1^2 + (w_2 - w_1)^2 - 4(\frac{3}{2} w_1 + w_2) Q \right] \end{aligned} \tag{2.38}$$

Imposing the functional stationary conditions

$$\partial I / \partial w_1 = 0 \quad \text{and} \quad \partial I / \partial w_2 = 0$$

one arrives at the same set of FD equations (2.35) as in the local formulation. Simple rectangular integration rule is applied here between nodes, with function evaluated in nodal mid-points, where the first derivative of w may be computed with the highest accuracy.

(ii) Discretizing of the week Galerkin type form leads to

$$\begin{aligned} l2\frac{l}{2}\Bigg[&\left(4\frac{-w_1-2w_0+w_1}{l^2}+M_0\right)v_0\frac{1}{2}+\\ &+\left(4\frac{w_0-2w_1+w_2}{l^2}+M_1\right)v_1+\\ &+\left(4\frac{w_1-2w_2+w_1}{l^2}+M_2\right)v_2\frac{1}{2}\Bigg]=\\ &=[4(-2w_1+w_2)+12Q]\,v_1+\\ &\qquad+[8(w_1-w_2)+16Q]\,\frac{1}{2}v_2=0 \end{aligned} \tag{2.39}$$

or

$$\begin{aligned} 2\Bigg[&2\frac{w_1-w_0}{l}2\frac{v_1-v_0}{l}\frac{l}{2}+2\frac{w_2-w_1}{l}2\frac{v_2-v_1}{l}\frac{l}{2}\\ &-\frac{M_0v_0+M_1v_1}{2}\frac{l}{2}-\frac{M_1v_1+M_2v_2}{2}\frac{l}{2}\Bigg]=\\ =&\frac{4}{l}(2w_1-w_2-3Q)v_1+\frac{4}{l}(2w_2-w_1-2Q)v_2=0 \end{aligned} \tag{2.40}$$

Taking advantage of the independence of test functions v_1 and v_2, one finally obtains from each of the above equalities the same FD equations (2.35) as before. Such an agreement is not a rule, however, and depends on the approximation and integration type used, as it will be shown in the next example. Here again the rectangular integration rule has been applied: between nodes in the second integral and around nodes in the first one, as in this case the second order derivatives are evaluated with the highest precision at nodes.

Example 2.
Stress analysis in a square prismatic bar subject to torsion

Problem formulation

Find the maximum shear stress in a prismatic bar of square cross-section (Fig. 2.8) subject to torsional moment M_s

$$M_s=\int_{-a/2}^{a/2}\int_{-a/2}^{a/2}F\,dxdy,\quad \tau_{xz}=\frac{\partial F}{\partial y},\quad \tau_{yz}=-\frac{\partial F}{\partial x}, \tag{2.41}$$

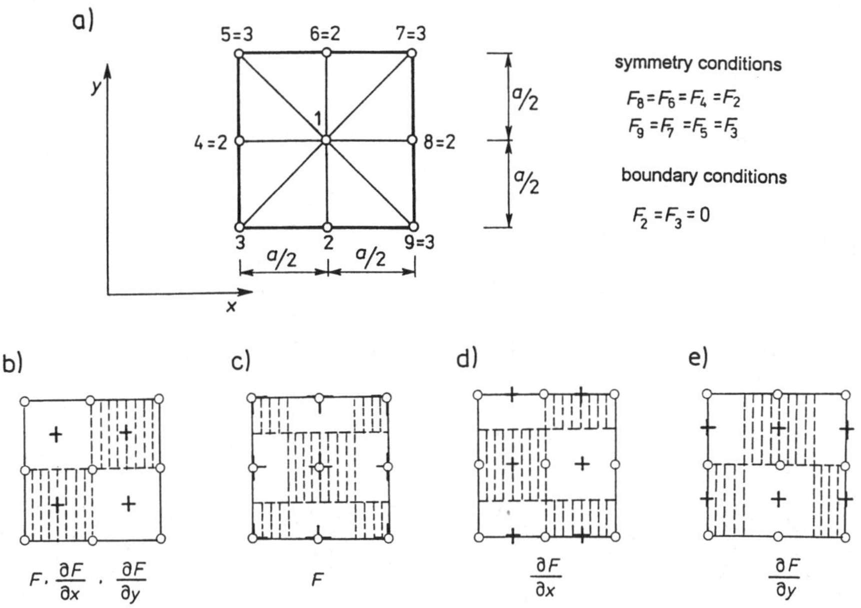

Fig. 2.8. Discretization of the prismatic bar of square cross-section subject to torsion **a)** mesh and integration domain for FD operators, F, $\frac{\partial F}{\partial x}$, $\frac{\partial F}{\partial y}$, **b)** simplest approximation, **c,d,e)** improved approximation

Local.

$$\begin{aligned} \nabla^2 F &= -2G\theta \quad \text{in} \quad \Omega \\ F &= 0 \qquad\quad \text{on} \quad \partial\Omega \end{aligned} \tag{2.42}$$

Global. Only one approach namely minimization of the functional is applied here

$$\min_F I\,, \quad I = \int_{-a/2}^{a/2}\int_{-a/2}^{a/2} \left\{ \frac{1}{2}\left[\left(\frac{\partial F}{\partial x^2}\right)^2 + \left(\frac{\partial F}{\partial y^2}\right)^2\right] - 2G\theta F \right\} dxdy$$
$$F = 0 \quad \text{on} \quad \partial\Omega$$

FD Discretization

Regular square mesh is assumed as shown in Fig. 2.8a with advantage taken of symmetry; hence $F_8 = F_6 = F_4 = F_2$, $F_9 = F_7 = F_5 = F_3$. Moreover due to the boundary conditions imposed $F_3 = F_2 = 0$.

Local formulation. For the standard finite difference Laplace operator (Fig. 2.3a) one has

$$F_2 + F_4 + F_6 + F_8 - 4F_1 = -2G\theta\left(\frac{a}{2}\right)^2 \tag{2.43}$$

while for the improved one (Fig. 2.3b)

$$\frac{1}{6}(F_3+F_5+F_7+F_9)+\frac{2}{3}(F_2+F_4+F_6+F_8)-\frac{10}{3}F_1=-2G\theta\left(\frac{a}{2}\right)^2 \tag{2.44}$$

From the first and the second equation one obtains respectively

$$F_1=\frac{1}{8}G\theta a^2=0.848F_{1\,\mathrm{EXACT}} \tag{2.45}$$

and

$$F_1=\frac{3}{20}G\theta a^2=1.018F_{1\,\mathrm{EXACT}} \tag{2.46}$$

Global formulation. Two variants differing in the way of integration are investigated here

(i) linear approximation and integration over an element (nodes 1, 8, 7, 6, – Fig. 2.8b)

$$\begin{aligned}I=4\Bigg\{\frac{1}{2}\left[\left(\frac{F_7+F_8}{2}-\frac{F_1+F_6}{2}\right)\frac{2}{a}\right]^2+\left[\left(\frac{F_6+F_7}{2}-\frac{F_1+F_8}{2}\right)\frac{2}{a}\right]^2\\ -2G\theta\frac{1}{4}(F_1+F_6+F_7+F_8)\Bigg\}\frac{a^2}{4}=F_1^2-\frac{1}{2}G\theta a^2F_1\end{aligned} \tag{2.47}$$

$$\frac{\partial I}{\partial F_1}=0\ \rightarrow\ F_1=\frac{1}{4}G\theta a^2=1.696F_{1\,\mathrm{EXACT}} \tag{2.48}$$

(ii) linear approximation and integration around the nodes where an operator value $\left(F,\frac{\partial F}{\partial x},\frac{\partial F}{\partial y}\right)$ is the most accurate (Fig. 2.8c,d,e)

$$\begin{aligned}I=\frac{1}{2}\Bigg\{&\left[2\left(2\frac{F_8-F_1}{a}\right)^2\frac{a^2}{4}+4\left(2\frac{F_7-F_8}{a}\right)^2\frac{a^2}{8}\right]\\ &+\left[2\left(2\frac{F_6-F_1}{a}\right)^2\frac{a^2}{4}+4\left(2\frac{F_7-F_8}{a}\right)^2\frac{a^2}{8}\right]\Bigg\}\\ &-2Q\theta\left(F_1\frac{a^2}{4}+4F_8\frac{a^2}{8}+4F_7\frac{a^2}{16}\right)=2F_1^2-\frac{1}{2}G\theta a^2F_1\end{aligned} \tag{2.49}$$

$$\frac{\partial I}{\partial F_1}=0\ \rightarrow\ F_1=\frac{1}{8}G\theta a^2\approx 0.848F_{1\,\mathrm{EXACT}} \tag{2.50}$$

(iii) second order approximation by means of the Simpson integration formula between nodes. Slightly better results are obtained using the same operator values as in the case (ii) and applying Simpson quadratures

$$I=\frac{8}{3}F_1^2-\frac{8}{3}G\theta a^2F_1 \tag{2.51}$$

$$\frac{\partial I}{\partial F_1}=0\ \rightarrow\ F_1=\frac{1}{6}G\theta a^2\approx 1.130F_{1\,\mathrm{EXACT}} \tag{2.52}$$

Evaluation of shearing stresses and torsional moment

Maximum shear stress. The maximum shear stress is

$$\tau_{\max} = \tau_{zy8} = \tau_{xy6} = \dots \tag{2.53}$$

Using the forward difference quotient derivative approximation at node 8 one obtains

$$\tau_{\max} \approx -2\frac{F_8 - F_1}{a} = \frac{2}{a}F_1. \tag{2.54}$$

Better result may be calculated using the second order approximation involving the central difference quotient. Let us notice first, that the above result reflects the actual value most accurately at the mid-point between points 1 and 8 rather than at the point 8 itself. Thus

$$\frac{1}{2}(\tau_{zy1} + \tau_{zy8}) \approx \frac{2}{a}F_1 \quad \rightarrow \quad \tau_{zy8} \approx \frac{4}{a}F_1 - \tau_{zy1} \tag{2.55}$$

Applying the central difference quotient at mid-point 1, one finds the shear stresses there

$$\tau_{zy1} \approx -\frac{F_8 - F_4}{a} = 0\,, \quad \tau_{zx1} \approx \frac{F_6 - F_2}{a} = 0 \tag{2.56}$$

and

$$\tau_{\max} \approx \frac{4}{a}F_1. \tag{2.57}$$

The same result is obtained using the second order FD formula for the first derivative directly at point 8

$$\tau_{\max} \approx \frac{2}{a}(-3F_4 + 4F_1 - F_8)\frac{1}{2} = \frac{4}{a}F_1. \tag{2.58}$$

Torsional moment.

(i) Use of the simplest approximation and integration around nodes

$$M_s \approx 2\left[F_1\frac{a^2}{4} + (F_2 + F_4 + F_6 + F_8)\frac{a^2}{8} + \right.$$
$$\left. +(F_3 + F_5 + F_7 + F_9)\frac{a^2}{16}\right] = \frac{1}{2}a^2F_1. \tag{2.59}$$

Hence the maximum shear stress is

$$\tau_{\max} \approx \frac{2}{a}F_1 = 4\frac{M_s}{a^3} = 0.832\tau_{\max\ \mathrm{EXACT}}. \tag{2.60}$$

(ii) Using the second order approximation and Simpson type integration between nodes

$$M_s \approx 2\,[16F_1 + 4(F_2 + F_4 + F_6 + F_8) +$$
$$+(F_3 + F_5 + F_7 + F_9)]\,\frac{a^2}{36} = \frac{8}{9}a^2F_1. \tag{2.61}$$

Hence the maximum shear stress is

$$\tau_{\max} \approx \frac{4}{a} F_1 = \frac{9}{2} \frac{M_s}{a^3} = 0.936 \tau_{\max\ \mathrm{EXACT}} . \tag{2.62}$$

It is worth noticing that in the FDM, like in the FEM, one may obtain results of different precision on the same mesh. This is because these also depend on the order and the kind of approximation used. Moreover, in the global formulation essential influence of the way and precision of numerical integration on the solution quality may be observed. Experience shows, e.g., that integration over "elements" is more appropriate for odd difference operators while around nodes for even ones.

2.8 Advantages and Disadvantages of the Classical FDM

The classical FDM has several advantages which in the past made it an effective discrete method of analysis. It is successfully applied to certain class of boundary-value (e.g., analysis of rectangular plates) and initial problems, especially in fluid mechanics [2]. The main advantages of the classical FDM are:

- easy interpretation, and application to problems defined in regularly shaped domains (rectangular, circular annular, etc.) if regular mesh of an appropriate type is used. This is due to convenient generation of the mesh, appropriate FD stars, formulas and equations. FD schemes for each internal node as well as subdomains assigned to each node are the same;
- solid mathematical basis (e.g. existence of solution, convergence, and stability). Proofs usually take advantage of the mesh regularity, and cannot be extended to the FDM generalized for arbitrary unstructured grids;
- long time practice and experience in the use of the FDM; in consequence its strong and weak points, application preferences and limitations, i.e., the knowledge of problems in which the FDM is useful or not, are well known.

On the other hand the classical FDM has serious disadvantages essentially limiting its use. Thus intrinsic difficulties arise when:

- the method is applied to domains of arbitrary irregular shapes; curvilinear boundaries are especially troublesome, because discretization of boundary conditions requires individual treatment (e.g. Fig. 2.9), and is difficult to automate;
- local increase of mesh density is needed, e.g., for the purpose of better approximation due to presence of concentrated loadings, boundary corners, notches, cracks etc.; this is also typical situation for the adaptive solution approach;
- combination of domains of different dimensionality (e.g. 1D beam, 2D plate and 3D foundation) is considered;

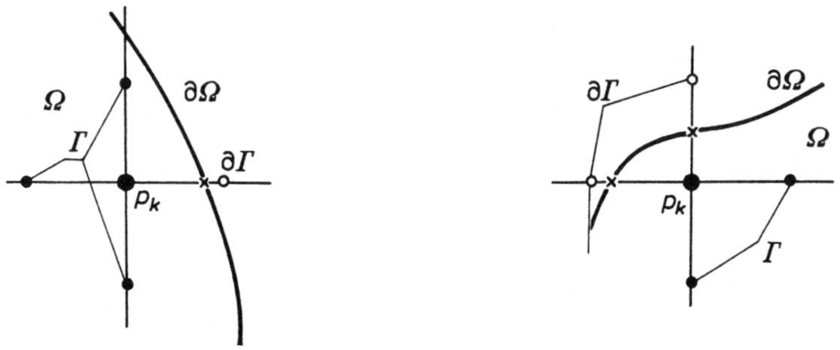

Fig. 2.9. Samples situations of curvilinear boundary and rectangular mesh

- general fully automated approach to analysis of boundary-value problems is required.

Due to these, and several other limitations, the classical FDM, as opposed to the FEM, may be treated as a tool of analysis of only selected boundary-value problem classes rather than as a general solver.

It has to be stressed that all limitations of the FDM mentioned above may be eliminated if arbitrary irregular meshes are used. Mesh irregularity results in a variety of new difficulties, but these may be overcome, and then the FDM becomes a general tool of analysis of boundary-value problems (GFDM). This method may be fully automated. Its flexibility, versatility and effective analysis capabilities may be similar or sometimes even better than those exhibited by the FEM. The next chapter outlines the formulation of such a method.

3 Curvilinear Finite Difference Method

3.1 Introduction

History of the FDM generalized for irregular grids (GFDM) is long. The R.H. Mac Neal paper [70] which appeared as early as in 1953 may be considered as the oldest publication on this subject. However, real development of the GFDM concepts, algorithms and applications started in early seventies. That research was carried out independently in several centers mainly: in England [9, 73, 99, 110] by G. Davies, B. Ford, P. Mullord, C. Snell, D.G. Vesey, M.J. Wyatt, in the USA by W.H. Frey, P.S. Jensen, R.A. Nay and S. Utku, N. Perrone [74, 17, 29, 86, 88], in Australia by P.C.M. Lau, S.K. Kwok [45, 46, 47, 50, 51], in Poland by W. Kączkowski, R. Tribiłło and J. Cendrowicz [7, 35, 44, 101, 102] (Warsaw–Białystok) as well as J. Orkisz, T. Liszka, J. Krok, M. Pazdanowski, W. Tworzydło [37, 43, 55–60, 63, 64, 82, 83, 105, 106] (Cracow). Mathematical bases were also developed [11, 15, 20, 25, 31, 32, 36].

This effort made the GFD a general, fully automated and effective tool of analysis of wide class of both boundary and initial-value problems with versatility and potential power comparable to the FEM.

Though detailed concepts of the GFDM differ from each other, two, currently the most advanced basic GFDM approaches may be distinguished.

One, called by its author Curvilinear Finite Difference (CFD) method, is based on arbitrary curvilinear structured grids. In the next section the main concepts of this method as proposed by W.H. Frey [17], P.C.M. Lau [50, 51] and S.K. Kwok [45, 46, 47] will be presented.

The other approach, which is the most general one, may use arbitrary fully irregular, unstructured grids. Such grids and corresponding primary GFDM formulation in 2D were given for the first time by P.S. Jensen [29].

In subsequent years the GFDM was significantly developed as to become fully mature, general and powerful method of discrete analysis. In early seventies the research carried out first of all in England [9, 73, 99, 110] and the USA [29, 86, 88] was reported. The most general and complete of these early GFDM versions was proposed by M.J. Wyatt et al. [110] in 1975.

In the following chapter the outlines of the currently most developed version of the GFDM will be described. It is based on concepts, algorithms and solutions obtained by the author and his numerous coworkers [33, 34, 37, 39,

40, 53, 55–64, 79–83, 90, 105, 106] including first of all research performed by T. Liszka, J. Krok, M. Pazdanowski and W. Tworzydło as well as their over twenty years experience in the development and various applications of GFD. The basic GFDM version, some of its extensions as well as chosen applications and outlines of the GFDM mathematical bases will be presented.

In what follows the name GFDM will be reserved only for the FDM generalized for arbitrary irregular, non-structured grids. The GFDM based on unstructured irregular grids falls in a broader class of discrete methods called nowadays meshless ones. In fact the GFDM is the oldest and, therefore, possibly the most developed meshless method. Following E. Oñate at al. [78] one may classify meshless methods as those in which the approximation can be constructed entirely in terms of nodes rather than elements.

Meshless methods are the subject of rapidly increasing interest nowadays. Therefore, in the last chapter of this part, the meshless methods including their classification, description of the basic concepts and review of the corresponding references will be presented very briefly.

3.2 Concept of the CFD

In the classical FDM one deals with structured, regular grids. In the Curvilinear Finite Difference (CFD) method [47] grids defined in the global coordinate system (x, y) (or x, y, z in 3D) still remain structured, though only locally. At each node a different curvilinear coordinate system (α, β) may be defined in a flexible way, based on a local configuration (called stencil, frame, star) of neighboring nodes. The only constraint is that subsequent parts (stencils, frames, stars) of such grid have to be mapped, and remapped in a non-singular way, onto a reference (basic) regular mesh, usually rectangular one. This reference mesh is used in order to define mapping $x = x(\alpha, \beta)$, $y = y(\alpha, \beta)$ and for easy generation of the FD formulas in the local coordinate system (α, β).

In the CFD method, the most important concept lies in transformation of the FD formulas derived in a local curvilinear coordinate system (α, β) to equivalent ones required in the global cartesian coordinate system (x, y) As described in [47] within every defined local coordinate system a complete polynomial surface fit of the unknown function, expressed in terms of its nodal values, is sought to approximate its true value. Partial derivatives in the local curvilinear coordinate system can, therefore, be obtained by successive differentiation of the approximate complete polynomial surface fit with the respect to the curvilinear coordinates (α, β) . These local partial derivatives are then transformed back to the global cartesian coordinate system to give the required FD approximation of the global partial derivatives. In the original CFD version [17, 50, 51] such operation is performed using the Chain Rule of Partial Differentiation (CRPD). This implies, however, that the transformation of the FD formulas of, say the n-th order partial derivatives, is independent of the FD formulas for any partial derivatives with order higher than n. S.K. Kwok found [47] that the above implication may have a significant negative influence on the numerical precision of the CFD method, especially when the local curvilinear mesh is highly irregular. Thus he proposed [47] a new approach which aims at improving the numerical accuracy of the original CFD method when applied to irregular grids that cannot be mapped conformally onto the local curvilinear coordinate system. In this Improved CFD method, the transformation matrix that correlates the local and global partial derivatives is derived from the Taylor series expansion of the unknown surface rather than use of the CRPD. This surface is fitted to the same unknown function values $u_i, i = 0, 1, \ldots, N$ that used for its approximation in the local coordinate system.

3.3 CFD Formulas for Local Partial Derivatives

In a considered domain a mesh of nodes described in the global cartesian coordinate system x, y is given. This mesh may be irregular, however, it has to allow for subsequent parts of it to be mapped on a regular reference grid. Let us consider a fixed set of nodes of this mesh that may be mapped on a rectangle (called stencil or frame) located on the reference grid (Fig. 3.1). For these nodes a local curvilinear coordinate system (α, β) mapped onto the relevant cartesian one (α, β) defined on the basic rectangular stencil is introduced. Suppose a function $u(\alpha, \beta)$ at an arbitrary point located within the stencil is approximated by the following 2nd-order complete polynomial

$$\tilde{u} = \mathbf{p}^t \mathbf{a} \tag{3.1}$$

where

$$\mathbf{p} = \{1 \ \ \alpha \ \ \beta \ \ \alpha^2 \ \ \alpha\beta \ \ \beta^2 \ \ \alpha\beta^2 \ \ \alpha^2\beta \ \ \alpha^2\beta^2\}, \quad \mathbf{a} = \{a_1, \ldots, a_9\} \tag{3.2}$$

In order to express the nine unknown coefficients $a_1, \ldots, a_9$ in terms of

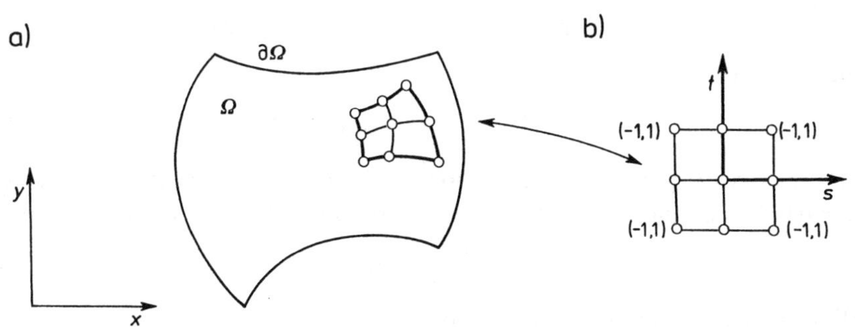

Fig. 3.1. CFD elements; **a)** real, **b)** reference

the nodal function values, the same number of nodes is required to define a local curvilinear stencil (Fig. 3.1). By applying relation (3.1) to each node of the stencil the following equation is obtained

$$\mathbf{q} = \mathbf{A}\mathbf{a} \;\rightarrow\; \mathbf{a} = \mathbf{A}^{-1}\mathbf{q} \;\rightarrow\; \text{if } \det \mathbf{A} \neq 0 \tag{3.3}$$

where

$$\mathbf{q} = \{u_1, \ldots, u_9\} \tag{3.4}$$

$$\mathbf{A} = \begin{bmatrix} 1 & \alpha_1 & \beta_1 & \ldots & \alpha_1^2\beta_1^2 \\ 1 & \alpha_2 & \beta_2 & \ldots & \alpha_2^2\beta_2^2 \\ \ldots & \ldots & & & \\ 1 & \alpha_9 & \beta_9 & \ldots & \alpha_9^2\beta_9^2 \end{bmatrix} \tag{3.5}$$

Hence, within the local curvilinear mesh $\mathbf{u}$ can be obtained from the following interpolating function

$$\tilde{u} = \mathbf{p}^t\mathbf{A}^{-1}\mathbf{q} = \tilde{\mathbf{N}}\mathbf{q}, \quad \tilde{\mathbf{N}} \equiv \mathbf{p}^t\mathbf{A}^{-1} \tag{3.6}$$

where $\tilde{\mathbf{N}}$ is the matrix of shape functions expressed in terms of the local coordinates.

The same result may be directly obtained by using Lagrange interpolation of an appropriate order

$$\tilde{u}(\alpha, \beta) = \sum_{i=1}^{I}\sum_{j=1}^{J} u_{ij} L_i^I(\alpha) L_j^J(\beta) \equiv \mathcal{F}(\mathbf{q}, \alpha, \beta) \tag{3.7}$$

Here $L_i^I(\alpha)$ and $L_j^J(\beta)$ are Lagrangian polynomials and $\mathbf{q} = \{u_{ij}\}$, $i = 1, \ldots, I, j = 1, \ldots, J$ is the vector of nodal values of unknown function u, taken from the stencil. In the general case $I \times J$ mesh rather than 3×3 one (Fig. 3.1) may be used.

Applying approach known from the isoparametric finite elements, mapping between (x, y) and (α, β) coordinates in the same way as for function $\tilde{u}$ is assumed i.e.

$$\begin{aligned} \tilde{x}(\alpha, \beta) &= \mathcal{F}(\mathbf{x}, \alpha, \beta) \qquad & \mathbf{x} &= \{x_{11}, \ldots, x_{IJ}\} \\ \tilde{y}(\alpha, \beta) &= \mathcal{F}(\mathbf{y}, \alpha, \beta) \qquad & \mathbf{y} &= \{y_{11}, \ldots, y_{IJ}\} \end{aligned} \tag{3.8}$$

or in the matrix notation

$$\tilde{x} = \mathbf{p}^t\mathbf{A}^{-1}\mathbf{x}, \qquad \tilde{y} = \mathbf{p}^t\mathbf{A}^{-1}\mathbf{y} \tag{3.9}$$

By successive partial differentiation of Eq. (3.6) or (3.7) with respect to the curvilinear coordinates α and β, the finite difference approximation

$$\mathbf{D}_l u' = \mathbf{H}\mathbf{q} \tag{3.10}$$

(FD formulas) of the partial derivatives

$$\mathbf{D}_l u = \{\; u \;\; u_{,\alpha} \;\; u_{,\beta} \;\; u_{,\alpha\alpha} \;\; u_{,\alpha\beta} \;\; u_{,\beta\beta} \;\} \tag{3.11}$$

in the local coordinate system can be obtained, where

$$\mathbf{H} \equiv \left\{ p^t \; p^t_{,\alpha} \; p^t_{,\beta} \; p^t_{,\alpha\alpha} \; p^t_{,\alpha\beta} \; p^t_{,\beta\beta} \right\}^t \mathbf{A}^{-1} \tag{3.12}$$

is a 6×9 matrix resulting from differentiation.

3.4 Transformation of the Local Partial Derivatives to the Global x, y Plane

As discussed above two approaches are applied. At first the original CFD idea using chain rule of differentiation will be described. Transformation of the first derivatives of the function u taken with respect to the local coordinates α, β to the relevant derivatives found in the global cartesian coordinates is given by the relations

$$\begin{bmatrix} u_{,x} \\ u_{,y} \end{bmatrix} = \begin{bmatrix} \alpha_{,x} & \beta_{,x} \\ \alpha_{,y} & \beta_{,y} \end{bmatrix} \begin{bmatrix} u_{,\alpha} \\ u_{,\beta} \end{bmatrix} = \frac{1}{J} \begin{bmatrix} y_{,\beta} & -y_{,\alpha} \\ -x_{,\beta} & x_{,\alpha} \end{bmatrix} \begin{bmatrix} u_{,\alpha} \\ u_{,\beta} \end{bmatrix} \tag{3.13}$$

where

$$J = x_{,\alpha} y_{,\beta} - x_{,\beta} y_{,\alpha} \tag{3.14}$$

is the Jacobian determinant. As differentiation of the global coordinates x and y given by formulas (3.8) is easy to perform in the local coordinate system, the above relations uniquely define the first order global derivatives of u, α and β .

The second order derivatives may be found in the similar way as to obtain

$$\mathbf{D}_g u = \mathbf{G} \mathbf{D}_l u \tag{3.15}$$

where

$$\mathbf{D}_g u = \left\{ u \; u_{,x} \; u_{,y} \; u_{,xx} \; u_{,xy} \; u_{,yy} \right\} \tag{3.16}$$

$$\mathbf{G} \equiv \begin{bmatrix} \alpha_{,x} & \beta_{,x} & 0 & 0 & 0 \\ \alpha_{,y} & \beta_{,y} & 0 & 0 & 0 \\ \alpha_{,xx} & \beta_{,xx} & \alpha^2_{,x} & 2\alpha_{,x}\beta_{,x} & \beta^2_{,x} \\ \alpha_{,xy} & \beta_{,xy} & \alpha_{,x}\alpha_{,y} & \alpha_{,x}\beta_{,y} + \alpha_{,y}\beta_{,x} & \beta_{,x}\beta_{,y} \\ \alpha_{,yy} & \beta_{,yy} & \alpha^2_{,y} & 2\alpha_{,y}\beta_{,y} & \beta^2_{,y} \end{bmatrix} \tag{3.17}$$

$$\alpha_{,xx} = J^{-1} \left(\alpha_{,x} y_{,\alpha\beta} + \beta_{,x} y_{,\beta\beta} \right) - J^{-2} y_{,\beta} \left(\alpha_{,x} J_{,\alpha} + \beta_{,x} J_{,\beta} \right) \tag{3.18}$$

$$\alpha_{,xy} = J^{-1} \left(\alpha_{,y} y_{,\alpha\beta} + \beta_{,y} y_{,\beta\beta} \right) - J^{-2} y_{,\beta} \left(\alpha_{,y} J_{,\alpha} + \beta_{,y} J_{,\beta} \right) \tag{3.19}$$

$$\alpha_{,yy} = -J^{-1} \left(\alpha_{,y} x_{,\alpha\beta} + \beta_{,y} x_{,\beta\beta} \right) + J^{-2} x_{,\beta} \left(\alpha_{,y} J_{,\alpha} + \beta_{,y} J_{,\beta} \right) \tag{3.20}$$

$$\beta_{,xx} = -J^{-1} \left(\alpha_{,x} y_{,\alpha\alpha} + \beta_{,x} y_{,\alpha\beta} \right) + J^{-2} y_{,\alpha} \left(\alpha_{,x} J_{,\alpha} + \beta_{,x} J_{,\beta} \right) \tag{3.21}$$

$$\beta_{,xy} = -J^{-1} \left(\alpha_{,y} y_{,\alpha\alpha} + \beta_{,y} y_{,\alpha\beta} \right) + J^{-2} y_{,\alpha} \left(\alpha_{,y} J_{,\alpha} + \beta_{,y} J_{,\beta} \right) \tag{3.22}$$

$$\beta_{,yy} = J^{-1} \left(\alpha_{,y} x_{,\alpha\alpha} + \beta_{,y} x_{,\alpha\beta} \right) - J^{-2} x_{,\alpha} \left(\alpha_{,y} J_{,\alpha} + \beta_{,y} J_{,\beta} \right) \tag{3.23}$$

$$J_{,\alpha} = x_{,\alpha\alpha} y_{,\beta} + x_{,\alpha} y_{,\alpha\beta} - x_{,\alpha\beta} y_{,\alpha} - x_{,\beta} y_{,\alpha\alpha} \tag{3.24}$$

$$J_{,\beta} = x_{,\alpha\beta} y_{,\beta} + x_{,\alpha} y_{,\beta\beta} - x_{,\beta\beta} y_{,\alpha} - x_{,\beta} y_{,\alpha\beta} \tag{3.25}$$

In the second approach used to transform FD formulas for the local partial derivatives to the global ones the Taylor series expansion of the unknown function is used with the respect to the central node u_1 . The function $u(x,y)$ at each stencil node is interpolated by the second order surface determined by the function value together with its first and second order derivatives

$$u_k = u_1 + h_k u_{1,x} + k_k u_{1,y} + \frac{1}{2} h_k^2 u_{1,xx} + h_k k_k u_{1,xy} + \frac{1}{2} k_k^2 u_{1,yy} \tag{3.26}$$

where

$$u_k = u\left(x_k, y_k\right) \ , \ h_k = x_k - x_1 \ , \ k_k = y_k - y_1 \tag{3.27}$$

By writing this equation at each node of the stencil the following equation is obtained

$$\mathbf{q} = \mathbf{P}\mathbf{D}_g u' \tag{3.28}$$

for the FD approximation $\mathbf{D}_g u'$ of the global derivatives $\mathbf{D}_g u$ where

$$\mathbf{P} = \begin{bmatrix} 1 & h_1 & k_1 & h_1^2/2 & h_1 k_1 & k_1^2/2 \\ 1 & h_2 & k_2 & k_2^2/2 & \dots & \dots \\ \vdots & & & & & \\ 1 & h_9 & k_9 & h_9^2/2 & h_9 k_9 & k_9^2/2 \end{bmatrix} \tag{3.29}$$

Now $\mathbf{q}$ in Eq. (3.28) may be substituted into Eq. (3.10) as to obtain required relation between derivatives in the local and global coordinate systems

$$\mathbf{D}_l u' = \mathbf{H}\mathbf{P}\mathbf{D}_g u' \tag{3.30}$$

By inverting this equation one gets

$$\mathbf{D}_g u' = (\mathbf{H}\mathbf{P})^{-1}\mathbf{D}_l u' \tag{3.31}$$

equivalent to Eq. (3.15) obtained from the chain rule of differentiation. Further substitution of Eq. (3.10) into Eq. (3.31) yields the final desirable FD approximation of the function u and its global derivatives

$$\mathbf{D}_g \mathbf{u}' = \mathbf{B}\mathbf{q} \, , \, \mathbf{B} \equiv (\mathbf{H}\mathbf{P})^{-1}\mathbf{H} \tag{3.32}$$

This approach involves more computational effort than the first one (inversion of 5×5 matrices). On the other hand the improved CFD method provided [47] reasonable solutions, even for those global irregular grids which cannot be mapped conformally onto the local curvilinear (α, β) coordinate system. Moreover the method is highly systematic and consequently convenient to automation.

3.5 Remarks

Following the paper [17] an example of CFD evaluation of the first and the second order derivatives of function u determined for nodal configuration (star) shown in Fig. 3.2 is presented here. The FD formulas obtained in the

	x	y
0	0,0000	0,0000
1	0,7138	0,1875
2	-0,1875	0,6563
3	-0,6250	-0,2813
4	0,0938	-0,7158
5	0,5313	1,0000
6	-0,7500	0,2813
7	-0,5625	-0,8433
8	0,8725	-0,6725

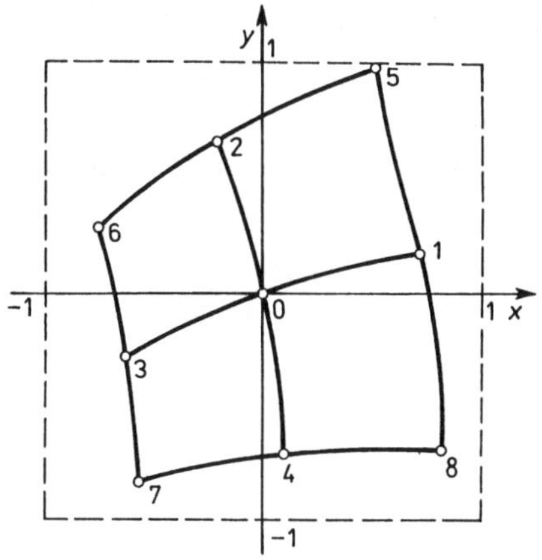

Fig. 3.2. Sample CFD element

way described above are given in equation (3.6)

$$\begin{bmatrix} u_x \\ u_y \\ u_{xx} \\ u_{yy} \\ u_{xy} \end{bmatrix} = \begin{bmatrix} 0 & 0.695 & -0.237 & -0.695 & 0.237 & 0 & 0 & 0 & 0 \\ 0 & 0.142 & 0.679 & -0.142 & -0.679 & 0 & 0 & 0 & 0 \\ -4.310 & 1.852 & 0.527 & 2.000 & -0.079 & -0.329 & 0.329 & -0.329 & 0.329 \\ -3.851 & 0.210 & 1.810 & -0.049 & 1.880 & 0.193 & -0.193 & 0.193 & -0.193 \\ 0.496 & 0.321 & -0.799 & 0.469 & -0.487 & 0.437 & -0.437 & 0.437 & -0.437 \end{bmatrix} \begin{bmatrix} u_1 \\ u_2 \\ \vdots \\ u_9 \end{bmatrix} . \tag{3.33}$$

It is worth stressing here, that the CFD concept of FD formulas generation discussed above, was applied first in the paper [108]. However, mapping from a given irregular mesh to the basic regular (triangular) one was done there at once for all nodes rather than on isoparametric element by element basis, when each time a different local curvilinear coordinate system could be used.

Many authors [17, 45, 46, 47, 50, 51, 108] contributed to formulation and development of the CFD. The most advanced version of the CFD was proposed by S.K. Kwok [47]. It includes also extension to analysis of problems defined on a differential manifold [45, 46].

The CFD approach presents a significant generalization of the classical FD method widening its potential field of applications. It constitutes a step towards the general fully automatic FD method using arbitrary irregular meshes.

On the other hand the CFD has some clear disadvantages. These result, first of all, from restrictions in mesh generation caused by the requirement of mapping of such meshes onto regular ones. This way grids used in the CFD, though flexible, still remain structured. Automatic generation of a mesh satisfying the mapping constraints and being, at the same time, appropriate to the boundary value problem considered with the domain shape and discretization of boundary conditions taken into account may be difficult.

Complications arise especially when dealing with problems like

- h-adaptive solution approach including controlled local mesh density increase (decrease);
- analysis involving unknown moving boundary (crack development, elastic-plastic boundary, contact of deformable bodies, fluid free surface, etc.)

These drawbacks may be removed when applying arbitrary irregular grids as discussed in the next chapter.

4 FD Method Generalized for Arbitrary Irregular Grids

4.1 Introduction

If compared to the classical version of the FDM at regular meshes [8, 15, 23, 68, 69, 89–97] the FEM proved to be much more successful in treatment of boundary conditions, especially at irregular domains and in local condensations of nodes which improve accuracy in the zones of rapidly growing solution gradients. Using an arbitrary irregular mesh of nodal points, however, one can overcome these difficulties simultaneously preserving the basic advantages of the FDM. On the other hand some new problems arise, mainly associated with the automatic generation of well-conditioned FD formulas.

These problems include automatic:

- mesh generation and modification, as well as topology determination, taking into account:
  - node generation and modification
  - search for nodes in the neighborhood of the considered point (node)
  - domain partition by assignment of subdomains to individual nodes
  - optimal triangulation of the mesh
- optimal star generation and classification to avoid singular or ill-conditioned FD schemes
- function approximation – choice of degrees of freedom (d.o.f.) as well as the type and order of the moving local approximation
- generation of the GFD operators
- integration of the GFD formulas (global type formulation)
- generation of the GFD equations
- discretization of boundary conditions
- solution of the GFDM equations
- evaluation of quantities required at the end of analysis
- full automation of the GFDM and its effective computer implementation.

The basic GFDM approach discussed in this chapter provides a successful solution of these problems. An extended GFDM version presenting the current development of the method involves also

- introduction of the generalized d.o.f.
- mixed global and local GFDM

- generalized Finite Strip approach – the GFDM and the Fourier analysis combination
- GFDM on a differential manifold
- GFDM/FEM combinations and unification
- higher order approximation GFDM
- a'posteriori error analysis
- adaptive GFDM approach
- multigrid solution approach.

Several of these topics will be discussed here.

The basis of the GFDM was published in the early seventies. P.S. Jensen [29] was the first to introduce fully arbitrary mesh. He considered Taylor series expansions interpolated on six-node stars in order to derive the FD formulas approximating derivatives of up to the second order. While he used that approach to the solution of boundary value problems given in the local formulation, R.A. Nay and S. Utku [74] extended it to the analysis of problems posed in the variational (energy) form. However, these very early GFDM formulations were later essentially improved and extended by many other authors including: N. Perrone, R. Kao and V. Pavlin [86, 88]; Z. Kączkowski, R. Tribiłło, M. Syczewski and J. Cendrowicz [7, 35, 101, 102]; G. Davies, B. Ford, P. Mullord, C. Snell, D.G. Vesey, M.J. Wyatt [9, 73, 99, 109, 110]. Out of these papers, representing early stage of the GFDM development, the most advanced version was given in [110]. One may notice there, e.g., the very first application of the weighted moving least squares approximation to the generation of the FD formulas as well as their integration over element subdomains; use of the generalized d.o.f., and partition of unity based interpolation of the nodal function values. Numerous researchers applied the GFDM later to solution various boundary-value problems of technical nature (cf. [10, 14, 103, 104]).

Here outlines of the most complete, general and updated GFDM approach implemented in the form worked out by the author and his coworkers [33, 34, 37–40, 43, 53, 55–64, 79–84, 90, 105, 106] will be presented. It is based on an essential development of the initial P.S. Jensen concept [29] in the course of over twenty years of independent research aimed at obtaining versatile general tool of effective analysis of boundary value problems, fully competitive with the FEM.

The research mentioned above applied to the GFDM includes: The initial general GFDM formulation [61, 62, 110], its fully mature basic version, local [63], global [60], weighted moving least squares approximation [37], the GFDM in data smoothing [33, 34], mesh generation [56, 57, 53, 81, 84, 90], mathematical basis [11, 25, 31, 32], the GFDM on differential manifold [105, 106], various FEM/GFDM combinations (including postprocessing, stress recovery) [40], a unified GFDM/FEM system [37, 41], mixed global and local GFDM formulation, the Generalized Finite Strip approach (GFDM/FEM and Fourier analysis combinations) [38], error analysis [34, 39, 79, 81, 84], the adaptive GFDM [79, 81, 39], multigrid GFDM solution approach [79, 81, 84].

The first complete presentation of such basic approach to the GFDM was done by T. Liszka [54] and later developed by W. Tworzydło [106]. Several general presentations of the GFDM were made by the author including [82, 83]. His recent presentations also include formulation of the adaptive multigrid GFDM [79, 81, 84].

Nowadays the GFDM, like the FEM, is an effective, general tool of linear and nonlinear analysis of a wide class of boundary-value problems. Some examples of less typical domains of applications are e.g.:

- reinforced pneumatic structures (differential manifold, large deformations)
- railroad rail and vehicle wheel analysis including
  - 3D elastic analysis (Generalized Finite Strip arbitrary high precision approach with error control)
  - residual stress analysis (shakedown, mixed global–local approach)
- roller straightening of railroad rails (3D highly nonlinear, transient b.v. problem)
- reservoir simulation – (adaptive approach to singularity)
- physically based enhancement of experimental measurements (data smoothing, inverse, ill-conditioned problems).

4.2 Basic GFDM Version

4.2.1 Mesh Generation and Modification, Mesh Topology Determination

First of all the notion of an "arbitrary irregular" grid should be defined. The idea of irregular grids in the FDM is not new [8, 9, 29, 70]. Initially a rectangular mesh with variable irregular spacing was used (Fig. 4.1a). Such a grid allowed for flexible fitting of nodes to a domain boundary of arbitrary shape, and local mesh condensation inside, if necessary. On the other hand, essential increase of the number of unnecessary nodes (as well as FD unknowns) is observed.

Evolution of irregular 2D grids started from the grid being partially regular in subdomains (Fig. 4.1b), then irregular but with restricted topology (Fig. 3.1) to totally irregular (Fig. 4.1c) mesh introduced by P.S. Jensen [29, 63]. The first two grid types still present structured meshes, while the last type may be classified as an unstructured one.

Similar evolution happened with respect to the FD grids on a differential manifold. Those used by Kwok when defining the CFD method [46, 47], though arbitrarily curvilinear, were structured due to mapping restrictions. Totally irregular FD grids were introduced there for the first time by

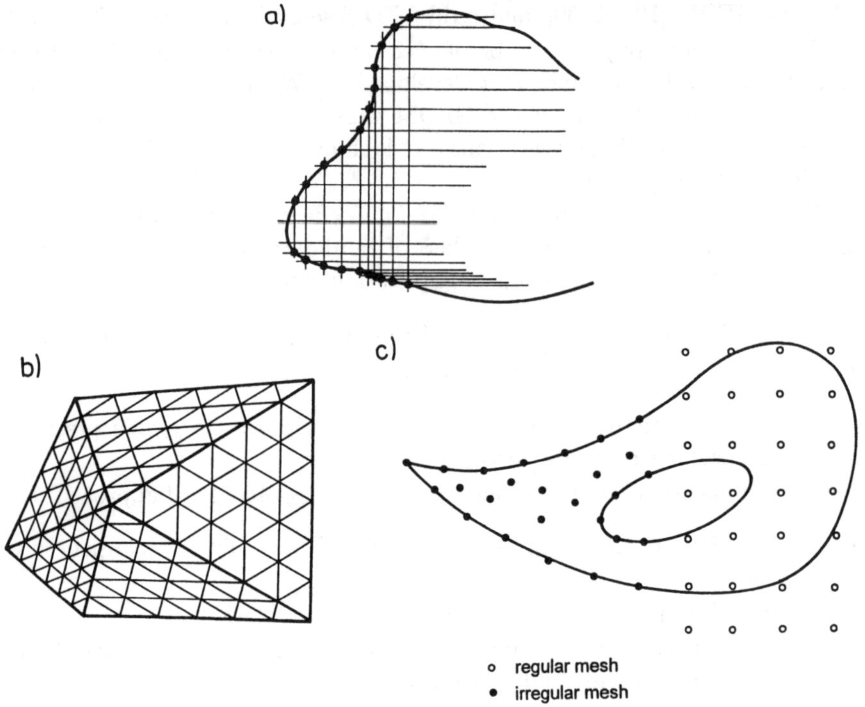

Fig. 4.1. Samples of irregular meshes: **a)** variable rectangular, **b)** regular inside irregular triangular, **c)** arbitrary irregular

W. Tworzydło, who proposed and applied the GFDM on a differential manifold [105, 106]. 3D GFDM meshes were considered by T. Liszka [54].

Discrete methods based on unstructural grids are called today the meshless ones (cf. Sect. 7.1).

In what follows "arbitrary irregular grid" refers to a mesh that has basically no restrictions. However, due to practical reasons, as well as for the purpose of generation of well-conditioned FD schemes, it may sometimes satisfy certain requirements, e.g., regularity in subdomains with guaranteed smooth transition from dense to coarse mesh zones.

Generation of an irregular mesh for the GFDM may be done in the same way as for the FEM. Thus various automatic FEM mesh generators [19] can be adapted here. On the other hand in order to recommended to take into account some special features of the FD technique and use FD oriented mesh generators is recommended.

A reasonable FD mesh generator should have the following basic features:

- require minimum user supplied amount of necessary initial data
- take full advantage of the domain shape; the domain may be single-,

double-, or multiconnected, and may be composed of primitives like circles, rectangles, triangles, etc.

- take into account smooth transition between zones (subdomains) of different mesh density as to provide well conditioned FD schemes
- assume nodes with a'priori imposed locations
- provide required and controlled precision of FD solutions
- provide easy modifications (node shift, insertion or annihilation) needed for the adaptive approach,
- in the case of the multigrid analysis subsequent meshes may contain the previous ones
- allow for moving boundaries when analyzing processes
- the ranges of stars are coextensive thus the boundary between two regions is not a single line of nodes; if the entire domain is divided into several identical regions then additional fictitious nodes should be introduced into the mesh [54] to assure proper continuity of solution (symmetric, skew symmetric,...); auxiliary nodes outside the boundary may be also introduced for discretization of boundary conditions with derivative terms
- alphanumerical and graphical input and output of the initial data, final output and intermediate modifications
- provide economical use of computer memory
- irregular mesh generation should be easily followed by domain partition into nodal subdomains, by the optimal triangulation (in 2D) and full determination of the mesh topology, if needed.

These requirements are met by the irregular FD mesh generator proposed by T. Liszka [56, 57] with further modifications done by the author [79, 81, 84], allowing for very effective adaptive GFDM approach. The basic concepts of this Liszka's mesh generator for 2D domains (extension to 3D are straightforward) will be presented below while modifications will be discussed in the next chapter.

The Liszka's generator is based on the mesh density control. Out of an arbitrary dense regular square (background) mesh with "millions and millions" of nodes required nodes are chosen ("sieved") according to a prescribed local mesh density. It is possible to generate a mesh with concentration zones in any required location, with needed density and automatic smooth transition between these zones.

Local mesh density ρ_i defined in 1D at the node P_i is

$$\rho_i = \lg_2^{-1} \frac{r_i}{r_{min}} \tag{4.1}$$

where r_i is the characteristic, locally determined mesh modulus, and r_{min} is the modulus of the most dense background mesh. Out of the background mesh ("sieved") those nodes are selected which satisfy the condition

$$p \geq \overline{\rho}^{-1}. \tag{4.2}$$

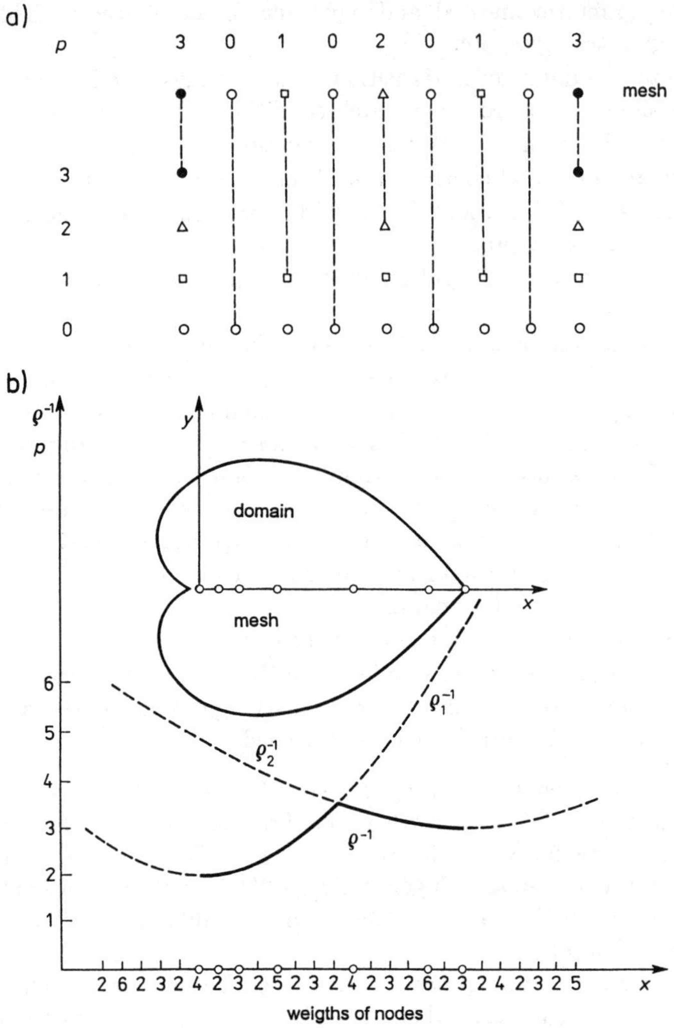

Fig. 4.2. a) 1D mesh nodal weights, **b)** 1D mesh generation (x-axis) *(continued in the next pages)*

Here

$$p = \lg_2 \frac{r_i}{r_{min}} \tag{4.3}$$

characterizes the local mesh modulus $r_i = 2^p r_{min}$ (Fig. 4.2a) while $\overline{\rho} = \overline{\rho}(x)$ is the prescribed mesh density. In the case of several (Fig. 4.2b) locally imposed mesh densities $\overline{\rho}_j(x)$, $j = 1, 2, \ldots$

$$\overline{\rho}^{-1}(x) = \inf_j \overline{\rho}_j^{-1}(x). \tag{4.4}$$

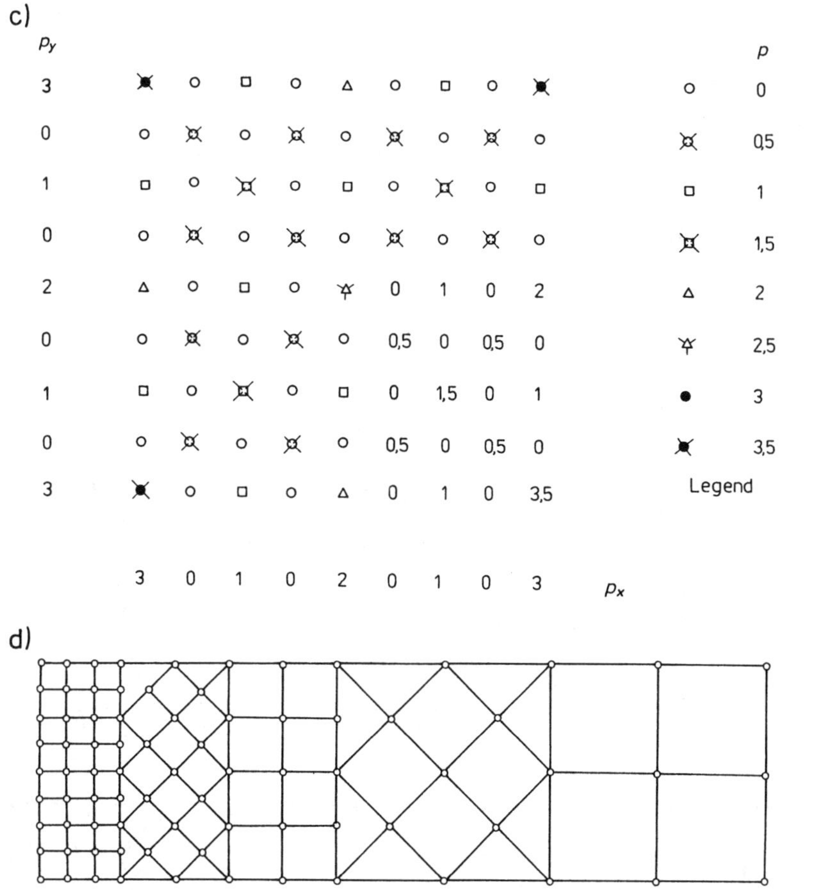

Fig. 4.2. *(continued)* **c)** p-node weight in 2D mesh, **d)** Transition zones for the regular square mesh)

The 1D definition (4.1) of the mesh density is extended to 2D domains as follows

$$\rho = \rho(x, y) = \begin{cases} \inf(\rho_x, \rho_y) & \text{if } \rho_x \neq \rho_y \\ \rho_x + 0.5 & \text{if } \rho_x = \rho_y \end{cases} \tag{4.5}$$

where ρ_x and ρ_y are 1D mesh densities found in directions of the x and y axes respectively (Fig. 4.2c).

Using the above mesh density definition and imposing appropriate 2D densities $\overline{\rho}_j(x, y)$ one may achieve:

- smooth transition between dense and coarse zones; e.g., in the case of regular square zones modulus ratio is $\sqrt{2}$ which corresponds to density change equal to 1/2 (Fig. 4.2d)

e)

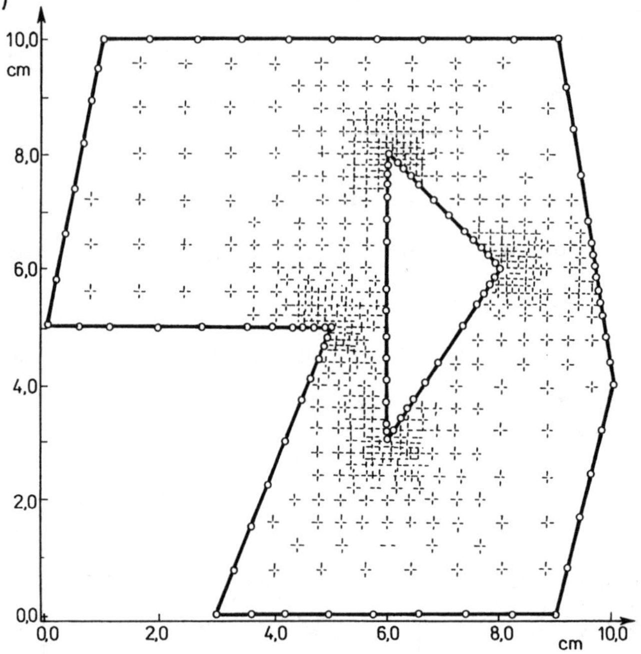

f)

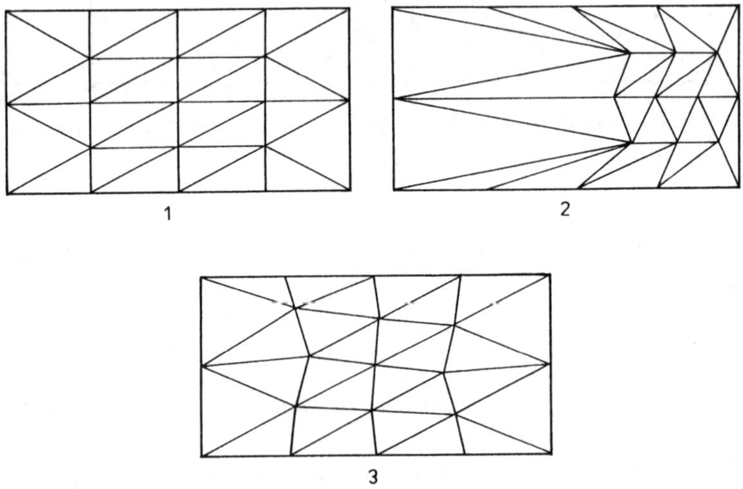

Fig. 4.2. *(continued)* **e)** Sample mesh with automatic concentration zones, **f)** Mesh smoothing examples: 1,2 – initial meshes, 3 – final mesh

- mesh generation with only minimum amount of initial data input by the user, including domain boundary definition and $\overline{\rho}_j$ density functions defined at selected points (e.g. vertices) where node concentration is required (Fig. 4.2e).

It is worth mentioning that the quality of the FD operators is sensitive to grid smoothness, thus sharp changes of mesh density should be avoided. The Liszka's mesh generator automatically meets this requirement because of built in mesh density control. Smoothness improvement may be also achieved by appropriate nodes shifts. Internal nodes having no fixed locations may be then considered. Each shifted node P_i, $i = 1, 2, \ldots$, is moved towards the gravity center of the polygon formed by all triangles having P_i as the common node. Nodes are moved one by one in an iterative process until their final locations are reached. In this way FD stars configuration is improved. On the other hand, initial regularity of the mesh, if any, is destroyed (Fig. 4.2f), as well as any concentration of nodes, because mesh homogeneity is approached. In order to avoid this effect one may keep "frozen" regular zones, as well as high density mesh areas, allowing for shifts only between "frozen" zones (transition zones).

4.2.2
Domain Partition into Nodal Subdomains – Voronoi Tessalation – Delaunay Triangulation and Mesh Topology Determination

Irregular mesh generated in a manner presented above results in a set of nodes scattered throughout the domain and its boundary. Many reasons exists for which the domain needs to be subdivided into subdomains assigned to individual nodes. These reasons include analysis of nonhomogeneous problems, use of the global approach where integration is required, determination of mesh topology and FD star selection.

In the case of regular meshes a unique domain subdivision is known a'priori. However, for irregular meshes the problem is far from being trivial. Various domain subdivisions used on the same mesh may have significant influence on FD analysis results [5]. Partition criterion should satisfy the following general requirements:

- every point of the domain Ω should belong to subdomain Ω_i of the closest node P_i
- whole domain should be covered with subdomains $\Omega = \bigcup_i \Omega_i$ and Ω_i cannot overlap
- partition of Ω should be unique
- in the case of regular grids partition should yield the well known results
- at each node P_i a strict correspondence between its FD star and the nodal subdomain Ω_i is welcome
- partition criterion should be easy to automate and be effective in calculations.

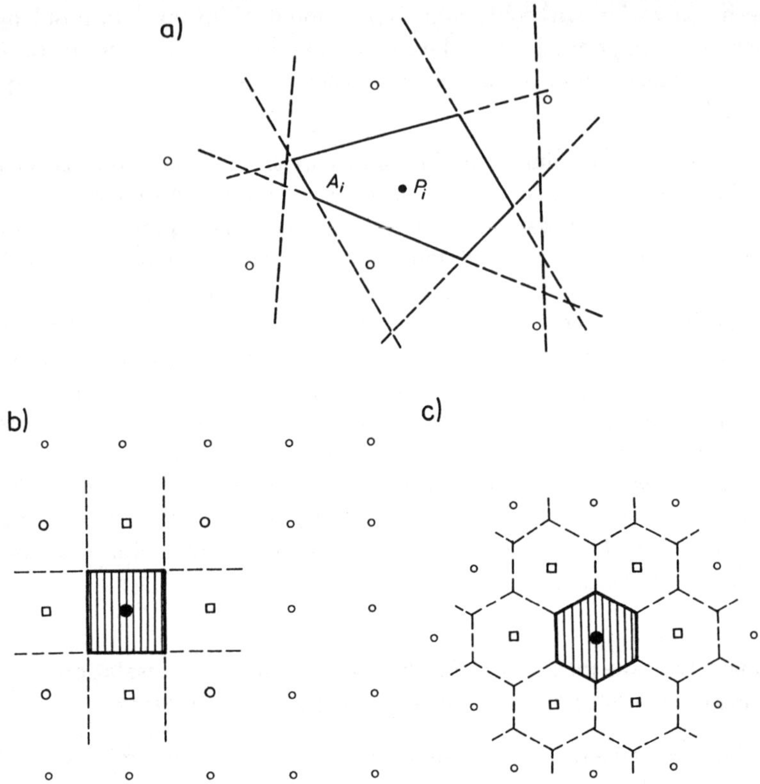

Fig. 4.3. Voronoi (Thiessen) tessalation: **a)** generation of a single polygon, **b)** Voronoi polygons for regular rectangular mesh, **c)** Voronoi polygons for regular triangular mesh *(continued in the next page)*

These requirements, partially of heuristic nature, are satisfied by the Thiessen polygons recently more often called Voronoi or Dirichlet ones [89]. Use of Voronoi (Thiessen) polygons is very popular now. However, taking advantage of them in the GFDM analysis was proposed and intensively used already in the seventies by J. Cendrowicz as well as by the author [82, 60]. Such a polygon may be obtained for a node P_i in a simple though not effective way as the internal envelope of all symmetry lines between P_i and every node in the domain (Fig. 4.3a). Examples of Thiessen polygons generated for regular (triangular, rectangular) meshes (Fig. 4.3b,c) and domain partition are shown in Fig. 4.3d,e.

Nowadays such partition process, called Voronoi tessalation, may be performed by some fast algorithms [89, 53, 90, 84] using "divide and conquer" (Fig. 4.4a) as well as "incremental insertion" (Fig. 4.4b)techniques [89]. The optimal performance provides partition using at best $N \log N$ operations.

It is useful to introduce notion of Thiessen (Voronoi) neighbors. In 2D

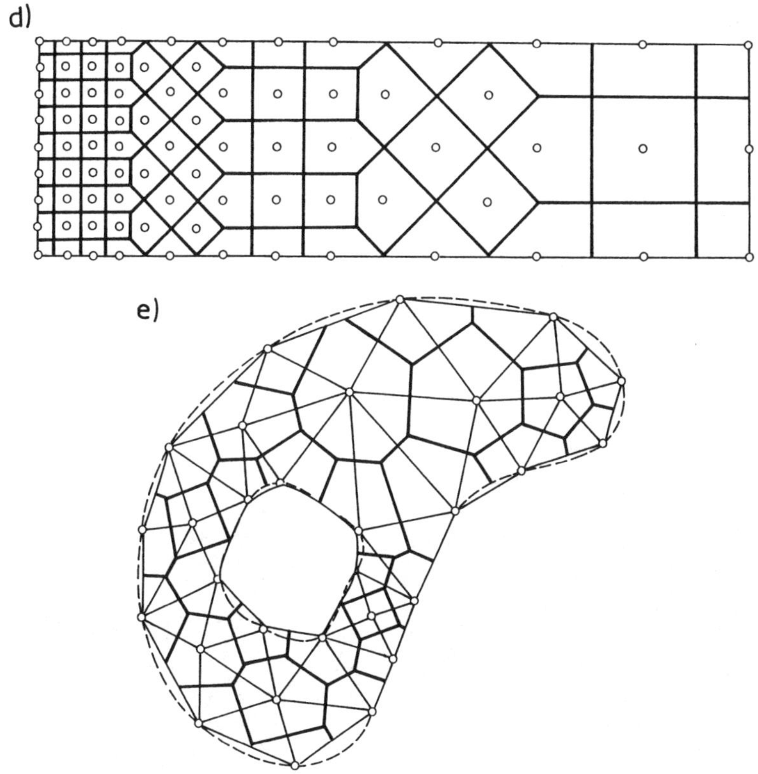

Fig. 4.3. *(continued)* **d)** Voronoi polygons for smooth transition from a course to a finer mesh, **e)** Voronoi polygons for arbitrary mesh

domains the "strong" neighbors are these nodes whose Thiessen polygons have a common side. If only one point of polygons is common the neighbor is "weak". The set of all Thiessen neighbors of a considered node P_i forms its Thiessen neighborhood.

Joining all pairs of the Thiessen neighbors yields the optimal triangular grid called Delaunay grid [89]. It may be also obtained by other, direct procedures called Delaunay triangulation. Optimality of such triangulation is understood in the following way: For given fixed locations of all nodes, each of quadrilaterals consisting of any two adjacent triangles is divided into two triangles in such a manner that their smallest angle reaches its largest possible value.

As opposed to domain partition, done by means of the Voronoi tessalation, the Delaunay triangulation may not be unique. Non-uniqueness happens if nodes of any two adjacent triangles are located on the same circle, like in the case of the rectangular mesh.

The domain topology is determined now by:

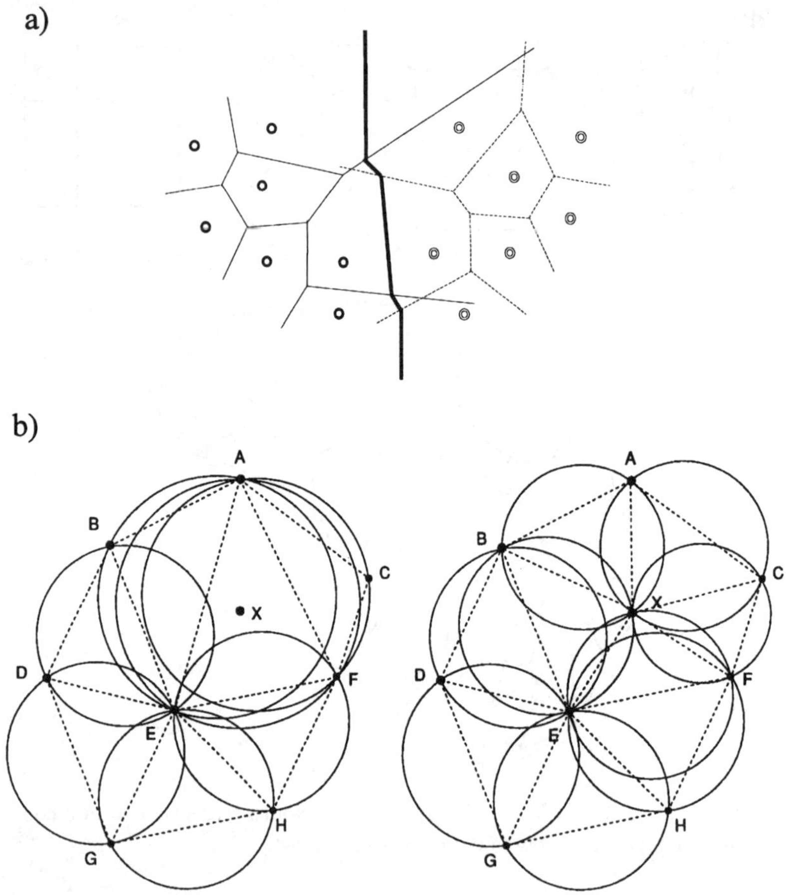

Fig. 4.4. Voronoi tessalation: **a)** divide-and-conquer method, **b)** an incremental insertion technique

- Voronoi tessalation (domain partition into nodal subdomains), and list of Voronoi neighbors assigned to each node,
- Delaunay triangulation (domain partition into triangular elements), and list of triangles involving each node.

Domain subdivision into Thiessen (Voronoi) polygons may be compared to the technique known from the FEM and used also in the CFD [47] for the generation of the FD formulas. It is performed in the following way:

- irregular mesh is mapped (if possible) part by part onto a reference regular mesh,
- partition is done on this reference domain and remapped onto the original one.

It is worth mentioning that such approach:

- cannot be applied to each arbitrary irregular mesh, but only to such one that may be mapped onto a regular grid,
- does not meet all requirements mentioned earlier, e.g., the first one.

Fig. 4.5 presents an example of an application of both techniques to the same domain and mesh.

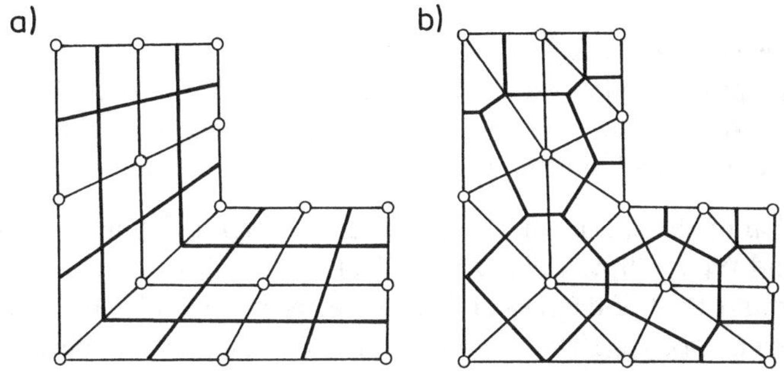

Fig. 4.5. Domain partition into subdomains: **a)** CFD concept, **b)** Voronoi (Thiessen) tessalation

4.2.3
FD Stars Selection and Classification

Several nodes used together for the generation of FD operators at a node assumed as the central one form a FD star. FD stars play similar role in the FDM as the elements in the FEM, i.e., they are used for spanning a local approximation of the searched function. The number and locations of nodes selected for each star are the decisive factors affecting precision of FD formulas approximation. It is easy to propose unique selection criteria in the case of regular meshes. If the mesh is irregular, both FD stars and formulas usually are different for each node.

Beyond boundary zones, inside the domain, all FD stars are the same in regular meshes, hence only one generation is sufficient. If a mesh is irregular, both FD stars and formulas usually differ from node to node. The most important feature of any selection criteria then is to avoid ill-conditioned and/or singular FD stars [22, 101].

Generation of a FD scheme corresponding to a given differential operator requires certain minimum number of nodes. However, if these nodes are located unfortunately, the minimum number may be insufficient to generate the FD operator (singularity), or FD schemes obtained are ill-conditioned.

Using as an example the Laplace operator ∇^2 one may notice that although at least four nodes are required (Fig. 4.6a), even six nodes may not

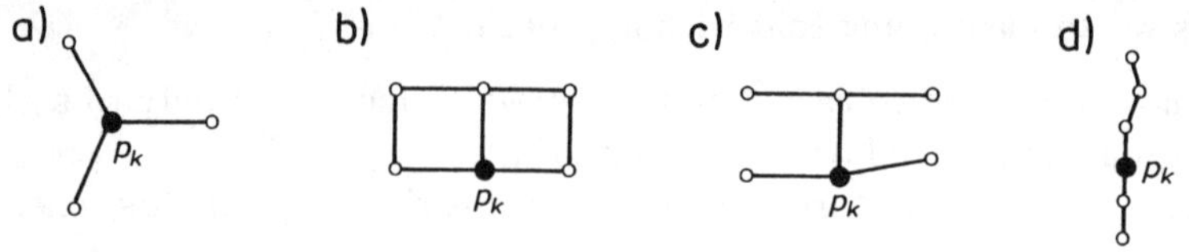

Fig. 4.6. FD stars for the harmonic operator ∇^2: **a)** well-conditioned, **b)** singular, **c)** ill-conditioned, **d)** singular

be sufficient (Fig. 4.6b,c,d).

Thus acceptable node selection criterion should take into account not only the distance of each node to the central one, but also their mutual configuration. Moreover the following problems have to be considered:

- dimension of the space where a given differential is defined, as well as its order and type (elliptic, hyperbolic, ...)
- number and type of degrees of freedom at each node of the FD star
- well known classical FD stars have to be obtained for regular meshes
- algorithm simplicity and its easy automation
- computer effectiveness (computational time, memory requirement, ...)

Moreover two situations are distinguished:

- FD formula is generated at a node (e.g. in the local FDM),
- FD formula is derived at arbitrary point outside of any node. One deals with this situation in the case of data smoothing or in the global FDM formulation.

In what follows a class of 2D problems will be considered that may require the complete set of function derivatives of the first and the second order. Nodal function values are assumed as the only d.o.f.

Let us consider at first FD operator generation at a node. A variety of star selection criteria were proposed and applied [82]. Only two of them considered the best ones will be discussed here:

(i) "Cross" criterion (Fig. 4.7a).
The domain is divided into four sectors corresponding to quadrants of the carthesian coordinate system originating at the central node. Each of its semi axes is assigned to one of these quadrants. In each sector two nodes closest to the origin are selected. If this is not possible, e.g., at the boundary, missing nodes are supplemented by not get selected nodes closest to the central one, so as to provide the total number of nine nodes in each star.

(ii) "Voronoi neighborhood" criterion (Fig. 4.7b)
Star at a node A is formed by the Thiessen (Voronoi) neighbors of that node, i.e., by its closest neighborhood.

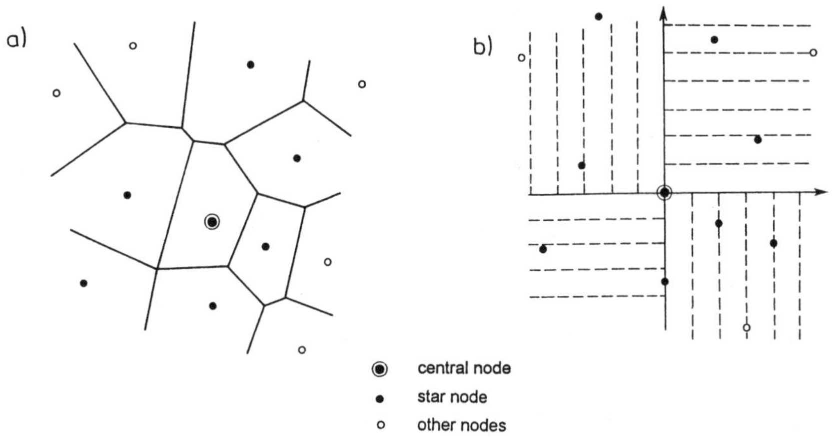

Fig. 4.7. Star selection: **a)** by Thiessen neighbors, **b)** by the cross criterion

Advantages of the first of these criteria are: its simplicity, constant number of nodes star and easy computer implementation. On the other hand the criterion output may slightly depend on the orientation of the local coordinate system. Moreover, FD star reciprocity is not guaranteed, i.e., if a node j belongs to the star of a node i the reverse does not necessarily happen.

The second criterion is more complex for practical use. The number of nodes in FD stars varies and may not suffice to generate a complete set of second order difference operators. On the other hand, as opposed to the first criterion, the second one is objective, and guarantees reciprocity. Moreover, for both rectangular and triangular regular meshes, the well known classical FD stars are obtained, while the first criterion works well only for the rectangular mesh.

Different criteria are suggested if FD formulas are derived for the central point not located at any node. Such a situation happens e.g. during postprocessing or numerical integration (global formulation) performed over elements rather than over subdomains surrounding nodes (e.g. Voronoi polygons). Consider a point P located inside a triangle spanning the nodes i, j, k . Let the closest neighborhood of these nodes be denoted by P_i, P_j and P_k (Fig. 4.8). In relation to the number of nodes required in the considered FD stars, some variants of selection criteria are possible yielding more and more nodes. Thus the FD star at point P may consist of the following nodes:

(a) minimum case:
all nodes common to the Thiessen (Voronoi) neighbors of each pair of nodes i, j, k

$$P_P^a = (P_i \cap P_j) \cup (P_j \cap P_k) \cup (P_k \cap P_i) \tag{4.6}$$

(b) intermediate case

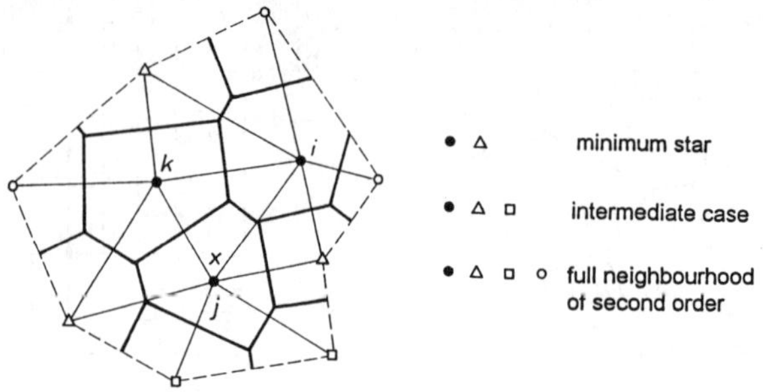

Fig. 4.8. FD star selection for a non-nodal points

nodes including the minimum case (P_P^a) and the Thiessen (Voronoi) neighbors of the node (nodes) closest to point P , e.g.

$$P_P^b = P_P^a \cup P_j \tag{4.7}$$

(c) complete case
all nodes belonging to the Thiessen (Voronoi) neighbors of any of the nodes i, j and k.

$$P_P^c = P_i \cup P_j \cup P_k \tag{4.8}$$

Together they form the so called second neighborhood of the point P.

On the domain boundary, especially in its corners (if any), the number of nodes in a star P_P^a may not be sufficient to generate all FD operators of the required second order. Star configurations P_P^b, P_P^c may be useful then. The last variant is especially worth to be applied the final postprocessing, as all nodal values are already known. The analysis is of local character then, so there is no essential change in computational time whether applying P_P^a or P_P^b stars.

It is worth stressing that practically no boundary-value problem requires use of a totally irregular mesh in the whole domain. On the contrary, in order to reduce the necessary computational effort it is convenient to use – beyond a boundary neighborhood and concentration zones – meshes regular in subdomains, joined by smooth transition zones.

Classification of FD stars is introduced, based on the notion of "equivalence class" of star configurations. It is done for the purpose of generation of FD formulas only once for each class. Automatic identification of such stars is done by means of comparison of appropriate arrays containing of the local coordinates of all nodes belonging to these stars.

4.2.4 Local Approximation Technique and FD Stars (Schemes) Generation

In the GFDM one deals with the following approximation problem:

find a function and/or its derivative(s) value(s) at a point of interest using the discrete data given at points irregularly scattered over a considered domain.

In the simplest case this data may be represented by function values only. However, values of operators, e.g., function derivatives, may also be given.

One needs to solve the above approximation problem for the purpose of derivation of the FD operators as well as for postprocessing of the final results, taking advantage of nodal values of the function and its derivatives determined earlier. One may require either interpolation or smoothing properties of the function approximating the discrete data. This depends on intended application.

The required approximation might be searched for either by a global approach or by a local one. In the first case the function is defined simultaneously over the entire domain. It is based on all points (nodes) in the domain, while a local type approximation uses only a few surrounding nodes in the neighborhood of the point of interest.

In general local approximation methods may be faster than global ones. However, usually there are difficulties if continuity requirements are to be satisfied.

In what follows a local approach will be presented and used in the GFDM. It is based on the so called moving weighted least squares approximation technique.

The first idea is due to Shepard [98]. He proposed to find the required global approximation point by point, each time fixing locally a constant determined by the weighted least squares procedure. This approach was later extended using higher order polynomial approximation.

In the GFDM such approach, based not only on the function value but also on its local expansion into the Taylor's series was proposed already in the seventies [61, 62, 63, 110]. It was used either to generate the FD formulas at nodes or to approximate function and its derivatives at any required point. Both data interpolation [55, 63] and smoothing [33] may be achieved this way. The result depends on the choice of the weighting factor.

Rigorous mathematical bases of the moving weighted least squares approximation using local polynomials were given by P. Lancaster and K. Salkauskas [78]. Due to its high quality, approximation of this kind is nowadays often used more and more. T. Belytschko et al. [3] effectively applied the local polynomial base enriched by asymptotic solution at the crack tip or by jump function.

In the case of moving weighted least squares approximation using polynomials in the local base is, of course, fully equivalent to the local expansion in the Taylor series mentioned above. However, Taylor expansion seems to be more advantageous because it additionally provides the approximation error term, and offers simple interpretation of derivatives.

The moving weighted least squares interpolation technique based on the local Taylor series is presented here following papers [37, 55, 63, 110] and applied, first of all, to generation of FD formulas in a 2D domain Ω. Generation of FD formulas (schemes) on arbitrary irregular grids is based on the concept of simultaneous derivation of FD formulas for the complete set of derivatives

$$\mathbf{D}u = \left\{ u, \frac{\partial u}{\partial x}, \frac{\partial u}{\partial y}, \ldots, \frac{\partial^n u}{\partial y^n} \right\} \tag{4.9}$$

up to a required order n, rather than for a given particular differential operator, as in the classical FDM. Though the approach is general and may be applied to differential operators of any order, as shown below, considerations will be limited here to the generation of FD schemes for the complete set of derivatives of the first and the second order only. FD formulas for higher order derivatives may be obtained either directly, in the same way, or composing the relevant lower order FD schemes.

Consider function $u(x, y)$ determined in the domain Ω. Values $u_i \equiv u(x_i, y_i)$, $i = 1, 2, \ldots, N$ of this function at points (nodes) irregularly scattered over the domain are given. Interpolation of this function at any point inside the domain Ω and on its boundary $\partial\Omega$ is required. For any sufficiently differentiable function the local approximation

$$u = \tilde{u}_0(x, y; n) + e \tag{4.10}$$

of the function $u = u(x, y)$ around a current point (x_0, y_0) is done by means of the nth order Taylor series expansion at this point. It may be expressed as:

$$\begin{aligned} \tilde{u}_0(x, y; n) &= u_0 + \sum_{l=1}^{n} \frac{1}{l!} \left(h\frac{\partial}{\partial x} + k\frac{\partial}{\partial y} \right)^l u(x, y)|_{(x_0, y_0)} = \\ &= \sum_{l=0}^{n} \sum_{i=0}^{l} \frac{1}{l!} \binom{l}{i} h^{l-i} k^i \frac{\partial^l}{\partial x^{l-i} \partial y^i} u(x, y)|_{(x_0, y_0)} \end{aligned} \tag{4.11}$$

or briefly as

$$\tilde{u}_0(x, y; n) = \mathbf{p}^t \mathbf{D}u \tag{4.12}$$

where

$$\mathbf{p}^t = \left[1, h, k, \frac{1}{2}h^2, \ldots, \frac{1}{n!}k^n \right] \tag{4.13}$$

are local interpolants,

$$\begin{aligned} &h = x - x_0,\ k = y - y_0,\ u_0 = u(x_0, y_0) \\ &e = \frac{1}{(n+1)!} \left(h\frac{\partial}{\partial x} + k\frac{\partial}{\partial y} \right)^{n+1} u(x_0 + \xi h, y_0 + \xi k)\ ,\ \ \xi \in [0, 1] \end{aligned} \tag{4.14}$$

is the truncation error term with $O(\overline{\rho})^{n+1}$, $\overline{\rho} = (h^2 + k^2)^{1/2}$ order magnitude.

The point (x_0, y_0) may represent an arbitrary point or a node in the domain Ω . In the latter case the function value u_0 is considered as given, otherwise it is unknown, and then it is evaluated together with the derivatives.

In order to determine the function and all its derivatives $\mathbf{D}u$ up to the order n at the point (x_0, y_0) , the above local Taylor approximation is spanned at (x_0, y_0) over a sufficient number of m nodes taken from its neighborhood (FD star or more). Singularity and ill-conditioning of the FD operators may be avoided and their stability improved when the considered number of nodes m exceeds the minimum number $\frac{1}{2}(n+1)(n+1)$ necessary to generate the set of all derivatives up to the order n .

One relates m element vector of nodal values

$$\mathbf{q} = \{u_1 \ldots u_m\} \tag{4.15}$$

to the corresponding values of the Taylor series

$$\mathbf{q} = \mathbf{PD}u + \mathbf{e} \tag{4.16}$$

where

$$\mathbf{P} \equiv \begin{bmatrix} 1 & h_1 & k_1 & \frac{1}{2}h_1^2 & h_1k_1 & \frac{1}{2}k_1^2 & \ldots & \frac{1}{n!}k_1^n \\ 1 & h_2 & k_2 & \frac{1}{2}h_2^2 & h_2k_2 & \frac{1}{2}k_2^2 & \ldots & \frac{1}{n!}k_2^n \\ . & . & . & . & . & . & \ldots & . \\ 1 & h_m & . & . & . & . & \ldots & \frac{1}{n!}k_m^n \end{bmatrix} \tag{4.17}$$

and the m term truncation error vector is

$$\mathbf{e} = \left\{ \pm O\left(\overline{\rho}_1^{n+1}\right), \ldots, \pm O\left(\overline{\rho}_m^{n+1}\right) \right\} \tag{4.18}$$

It is worth stressing that the Taylor series truncation errors provide valuable information on relative accuracy of the elements of $\mathbf{q}$.

The weighted least squares approximation $\mathbf{D}u'$ of derivatives $\mathbf{D}u$ is obtained solving the above set of over determined equations (4.16)

$$\mathbf{Wq} = \mathbf{WPD}u' + \mathbf{Wr} \tag{4.19}$$

Here

$$\mathbf{r} = \mathbf{q} - \mathbf{PD}u' \tag{4.20}$$

is the vector of residuals at nodes $i = 1, \ldots, m,$ and $\mathbf{W} = \lceil w_i \rfloor$ is the diagonal matrix consisting of the weighting factors w_i at these nodes. The sum of squares of the weighted residuals is minimized

$$\min_{\mathbf{D}u'} \mathbf{r}^t \mathbf{W}^2 \mathbf{r} \tag{4.21}$$

at all m nodes of the FD star considered.

Based on the necessary condition

$$\frac{\partial \left(\mathbf{r}^t \mathbf{W}^2 \mathbf{r}\right)}{\partial (\mathbf{D}u')} = 0 \tag{4.22}$$

one may now express the approximate and the exact derivatives respectively as

$$\mathbf{D}u' = \mathbf{Bq}, \quad \text{and} \quad \mathbf{D}u = \mathbf{Bq} + \mathbf{e}_1 \tag{4.23}$$

Here

$$\mathbf{B} = \begin{cases} \left(\mathbf{P}^t\mathbf{W}^2\mathbf{P}\right)^{-1}\mathbf{P}^t\mathbf{W}^2 & \text{for} \quad m > \frac{1}{2}(n+1)(n+2) \\ \mathbf{P}^{-1} & \text{for} \quad m = \frac{1}{2}(n+1)(n+2) \end{cases} \tag{4.24}$$

is the weighted generalized least squares inverse of the matrix $\mathbf{P}$, and $\mathbf{e}_1 = \mathbf{D}u - \mathbf{D}u'$ is the relevant FD approximation error.

Having obtained the derivatives $\mathbf{D}u'$ one may now present the approximation of the function u in the following form

$$u = \mathbf{p}^t\mathbf{D}u' + e_2 = \tilde{\mathbf{N}}\mathbf{q} + e_2 \tag{4.25}$$

Here

$$\tilde{\mathbf{N}} = \tilde{\mathbf{N}}(x, y; x_0, y_0.; n) \equiv \mathbf{p}^t\mathbf{B} \tag{4.26}$$

interpolates the nodal function values $\mathbf{q}$ in a neighborhood of the point (x_0, y_0), while

$$e_2 = e + e_1 \ , \quad e_1 \equiv \mathbf{p}^t\mathbf{e}_1 \tag{4.27}$$

is the compound approximation error consisting of the error e due to truncation of the Taylor series above the order n, and the FD approximation error e_1. Note that in the MWLS approximation technique [55, 63, 110, 49, 48] $\tilde{\mathbf{N}}$, providing a local type approximation at arbitrary point (x, y) in a neighbourhood of the point (x_0, y_0), is not yet a matrix of the required pseudo-shape functions. This is because the basic concept of the MWLS approach is to generate a new local approximation at each location of the moving point (x_0, y_0). Therefore, only the value of the matrix $\tilde{\mathbf{N}}(x, y; x_0, y_0; n)$ obtained when the points (x, y) and (x_0, y_0) are identified is retained each time. Thus assuming $x = x_0$, $y = y_0$ one may determine the matrix

$$\overline{\mathbf{N}}(x, y; n) \equiv \tilde{\mathbf{N}}(x, y; x, y; n) \equiv [N_i] \tag{4.28}$$

of the MWLS pseudo-shape functions $\overline{N}_i$. As $h = k = 0$, it is expressed by the first row of the matrix $\mathbf{B}$ (4.24). Its explicit analytical form may be derived only for the lowest order local approximation ($n = 0$, Sheppard) when

$$\overline{N}_i(x, y) = \frac{w_i(x, y)}{\sum_j w_j(x, y)} \quad \text{and} \quad \sum_i \overline{N}_i(x, y) = 1. \tag{4.29}$$

Otherwise it may be found, point by point, in a tedious numerical way. It is worth stressing, however, that in the GFDM solution approach determination of the function u and its derivatives at nodes, and (global approach) at Gaussian integration points only is sufficient.

Finally the MWLS approximation of the function $u(x,y)$ may be presented in the form well known in the FEM

$$u^h = \overline{\mathbf{N}}\mathbf{q} \quad \text{or} \quad u^h = \sum_j^m \overline{N}_j u_j . \tag{4.30}$$

In each location of the moving point (x,y) it involves m of its neighbor points (nodes), selected out of the whole domain Ω by means of a choice of the FD star, the weighting factor form, etc.

Due to star selection the domain of influence of a particular node, e.g. P_i is limited to its neighborhood.. Usually it is the second Voronoi neighborhood defined by the formula (4.8). Beyond the domain of influence, as well as at all its nodes $\overline{N}_i = 0$, except P_i, where $\overline{N}_i = 1$. Approximation in the whole domain Ω is of the same form (4.30) but summation is extended over all nodes $i = 1, 2, \ldots, N$.

It is worth to notice, that in the approach presented here the derivatives $\mathbf{D}u$ are approximated rather by relevant FD formulas $\mathbf{D}u'$ than by differentiation of the approximate function u^h . Consequently values of the both approximations may differ from each other. However, the values of $\mathbf{D}u'$ are not inferior to $\mathbf{D}u^h$ ones, and are obtained much faster.

The weights w_i are crucial to the approximation quality. It is commonly accepted that these are non-negative functions, and their choice should obey the rule "the longer the distance $\overline{\rho}_i$ between points (x_i, y_i) and (x_0, y_0) , the smaller the influence of the approximation error e at (x_i, y_i) on the evaluation of the derivatives $\mathbf{D}u$ at (x_0, y_0) ". Increase of the approximation error ought to be compensated by a relevant decrease of the weighting function. Thus for the purpose of interpolation

$$w_i = \overline{\rho}_i^{\,-n-1} , \quad i = 1, 2, \ldots, m \tag{4.31}$$

are proposed [110, 63] as inversely proportional to the error terms in $\mathbf{e}$ (cf. (4.16), (4.18)). Such a choice is dimensionally consistent and allows for easy change of scale.

In general terms the weighting function $w = w(\overline{\rho})$ may be classified as:

(i) singular, if $\lim_{\overline{\rho}_i \to 0} w = \infty$ (interpolation approach), and non-singular otherwise;

(ii) using the infinite support where almost everywhere $w > 0$, and with a finite support otherwise.

Singular weights, like the one assumed above, provide interpolation over the given discrete data. Such approach is convenient e.g. for generation of the FD equations resulting from the local formulation of boundary-value problems. On the other hand sometimes, e.g. for experimental (or rough numerical) data smoothing, use of non-singular weights allowing for the best approximation is more appropriate. The weighting factor

$$w_i^2 = \left(\overline{\rho}_i^2 + \frac{g^4}{\overline{\rho}_i^2 + g^2}\right)^{-n-1} \tag{4.32}$$

proposed in [33] proved to be very effective for such a purpose. Here $g \geq 0$ is a free parameter (distance) of approximation. In the extreme cases i.e. for $g = 0$ the formula (4.31) is obtained, and when $g \to 0$ the weight $w \to g^{-n-1}$ becomes independent on $\overline{\rho}$. The weighting function (4.32) provides $w'(0) = 0, w''(0) = 0, w'''(0) = 0$.

The infinite support of weight functions and a limited size of FD stars selected for approximation may result in some discontinuities of the local approximation. On the other hand this effect may be arbitrarily controlled and decreased by an appropriate choice of these functions. The weighting functions (4.31) and (4.32) already proved their numerical effectiveness in the solution process, and high quality of approximation obtained in this way.

Discontinuities would completely disappear when all discrete function values u_i in the considered domain Ω are used for the local approximation at each point. Effect of the same type may be formally obtained by application of weighting functions defined on a finite support. Such a choice of weights, however, means practically that node selection is based on the distance criterion only, disregarding their distribution. Such an approach may be acceptable to data smoothing or interpolation, but not for generation of the FD equations, where in consequence one might expect poor solution effectiveness due to excessive number of unknowns, ill-conditioness, slow convergence. On the other hand in the case of a finite support the mathematical basis of the whole approach is more sound, so existence of solution and convergence of the approach may be discussed easier. This is possibly one of the main reasons of growing popularity [49, 78, 3, 12] of various non-singular weighting functions, defined on a finite support, in spite of their clear disadvantages.

An intermediate approach is also possible. Namely use of the previously selected stars, and weighting factors defined on a finite support but mapped onto the infinite one. Application of weights (4.31) and (4.32), as well as the mapping

$$\tilde{\rho} = a\overline{\rho}\left(a^2 - \overline{\rho}^2 + \varepsilon^2\right)^{-\frac{1}{2}} \tag{4.33}$$

is suggested. Here the true internal distance $\overline{\rho} \in [0, a]$ is mapped onto the extended internal $\tilde{\rho} \in [0, \infty)$, a is the limit value of $\overline{\rho}$ where the modified weighting function

$$\tilde{w}(\tilde{\rho}) \equiv \begin{cases} w(\tilde{\rho}) & \text{for } 0 \leq \overline{\rho} \leq a \\ 0 & \text{for } a < \overline{\rho} \end{cases} \tag{4.34}$$

differs from 0; ε^2 is an arbitrary small number.

Altogether, despite certain discontinuities obtained, use of the weights (4.22) or (4.32), and nodes selected by taking advantage from the mesh topology like FD stars, neighborhood of nodes and triangles, seems to be the best option for a high quality and effective local approximation.

The following considerations are restricted basically to the development of the local approximation using the second order Taylor series. For the first and the second order derivatives

$$\mathbf{D}u \equiv \{\, u_x \;\; u_y \;\; u_{xx} \;\; u_{yy} \;\; u_{xy} \,\} \equiv \{{}^q u\}, \quad q = 1, 2, \ldots, 5 \tag{4.35}$$

the following FD approximation formulas [54] are obtained

$$\mathbf{D}u' = \mathbf{Bq}, \quad \mathbf{B} = \mathbf{\Phi}^{-1}\mathbf{C} \tag{4.36}$$

or at the node (x_i, y_i)

$${}^q u'_i = \sum_{j=0}^{m} {}^q B_{j(i)} u_{i+j}, \quad i = 1, 2, \ldots, N \tag{4.37}$$

where

$$\mathbf{\Phi} = \mathbf{\Phi}^t \equiv \begin{bmatrix} \varphi_{2,0} & \varphi_{1,1} & \frac{1}{2}\varphi_{3,0} & \frac{1}{2}\varphi_{1,2} & \varphi_{2,1} \\ & \varphi_{0,2} & \frac{1}{2}\varphi_{2,1} & \frac{1}{2}\varphi_{0,3} & \varphi_{1,2} \\ & & \frac{1}{4}\varphi_{4,0} & \frac{1}{4}\varphi_{2,2} & \frac{1}{2}\varphi_{3,1} \\ & \text{sym.} & & \frac{1}{4}\varphi_{0,4} & \frac{1}{2}\varphi_{1,3} \\ & & & & \varphi_{2,2} \end{bmatrix}, \tag{4.38}$$

$$\mathbf{C} \equiv \begin{bmatrix} h_1\rho_1^{-6} & h_2\rho_2^{-6} & h_3\rho_3^{-6} & \ldots & h_m\rho_m^{-6} \\ k_1\rho_1^{-6} & \cdot & \cdot & \cdot & \cdot \\ \frac{1}{2}h_1^2\rho_1^{-6} & \cdot & \cdot & \cdot & \cdot \\ \frac{1}{2}k_1^2\rho_1^{-6} & \cdot & \cdot & \cdot & \cdot \\ h_1k_1\rho_1^{-6} & \cdot & \cdot & \cdot & h_mk_m\rho_m^{-6} \end{bmatrix}, \tag{4.39}$$

$$\varphi_{r,s} = \sum_{j=1}^{m} h_j^r k_j^s \rho_j^{-6}. \tag{4.40}$$

The matrix $\mathbf{B}$ q-th row ${}^q B_{j(i)}$, $j = 0, 1, \ldots, m$ presents coefficients of the FD formula, generated at the node (x_i, y_i) over a m-node star for the derivative ${}^q u'_i$.

Particular form of the matrices $\mathbf{\Phi}$ and $\mathbf{C}$ depends on both the number and configuration of nodes in the FD star. In Table 4.1 FD coefficients ${}^q B_{j(0)}$ obtained for some first and the second order derivatives, as well as for the Laplace operator generated on the regular square mesh by means of: (a) the weighted least squares approach described above, and (b) the classical FDM approach are given. It is worth stressing that the FD formula for the Laplace operator, resulting from the first approach, was already known in the literature [15] as the formula of particularly high precision. This time, however, it was obtained as a by-product of the general approximation technique (MWLS).

Table 4.1. Coefficients of the FD formulas for the first and second derivatives as well as for the Laplace operator generated on regular square mesh by **a)** MWLS, **b)** the classic way

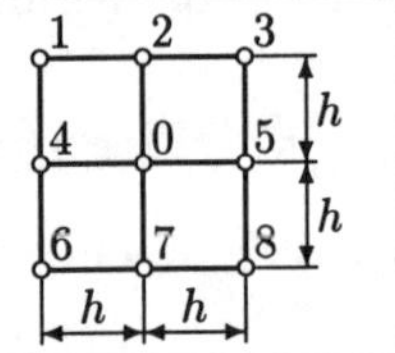		case	node								
operator	factor		0	1	2	3	4	5	6	7	8
u_x	h^{-1}	b					$-\frac{1}{2}$	$\frac{1}{2}$			
		a		$-\frac{1}{20}$		$\frac{1}{20}$	$-\frac{2}{3}$	$\frac{2}{3}$	$-\frac{1}{20}$		$\frac{1}{20}$
u_{xx}	h^{-2}	b	-2				1	1			
		a	$-\frac{5}{3}$	$\frac{1}{12}$	$-\frac{1}{6}$	$\frac{1}{12}$	$\frac{5}{6}$	$\frac{5}{6}$	$\frac{1}{12}$	$-\frac{1}{6}$	$\frac{1}{12}$
u_{xy}	h^{-2}	b		$-\frac{1}{4}$	$\frac{1}{4}$				$\frac{1}{4}$		$-\frac{1}{4}$
		a		$-\frac{1}{4}$		$\frac{1}{4}$			$\frac{1}{4}$		$-\frac{1}{4}$
$\nabla^2 u$	h^{-2}	b	-4		1		1	1		1	
		a	$-\frac{10}{3}$	$\frac{1}{6}$	$\frac{2}{3}$	$\frac{1}{6}$	$\frac{2}{3}$	$\frac{2}{3}$	$\frac{1}{6}$	$\frac{2}{3}$	$\frac{1}{6}$

Moreover, results evaluated by the MWLS approach for the same data (see Fig. 3.2) are presented for comparison with the formulas generated by means of the CFD method for first and second derivatives of the function determined at irregular nodes

$$\begin{bmatrix} u_x \\ u_y \\ u_{xx} \\ u_{yy} \\ u_{xy} \end{bmatrix} = \begin{bmatrix} 0.139 & 0.538 & -0.152 & -0.497 & 0.132 & 0.031 & -0.223 & -0.037 & 0.069 \\ -0.063 & 0.132 & 0.608 & -0.155 & -0.521 & 0.069 & 0.038 & -0.040 & -0.068 \\ -3.340 & 1.596 & -0.355 & 1.075 & -0.201 & -0.057 & 1.048 & -0.067 & 0.301 \\ -3.415 & -0.085 & 1.805 & -0.068 & 1.515 & 0.256 & -0.327 & 0.310 & 0.009 \\ 0.034 & 0.421 & -0.276 & 0.937 & -0.531 & 0.282 & -0.860 & 0.352 & -0.359 \end{bmatrix} \begin{bmatrix} u_0 \\ u_1 \\ u_2 \\ \vdots \\ u_8 \end{bmatrix} . \tag{4.41}$$

4.2.4.1
Higher Order FD Operators

FD operators of order higher than two may be obtained in several ways. The direct generation may be done in the same way as for the first and the second order derivatives. Disadvantages of such approach are: time consuming selection of appropriate, large, stars, and frequent solution of simultaneous local equations (or inversion of matrices) much larger than for terms of lower order. Use of only the first and the second order operators is, therefore, more

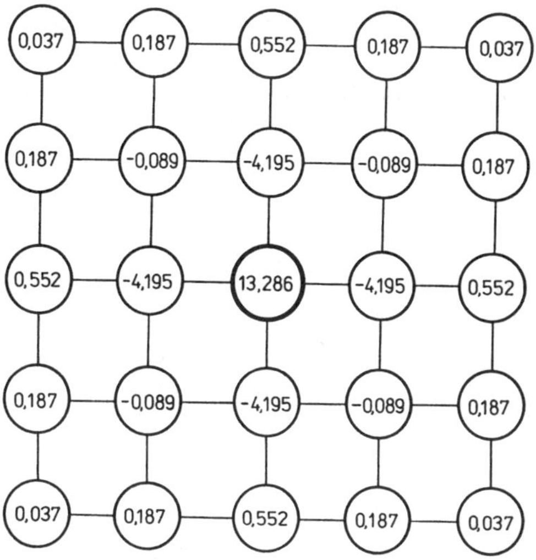

Fig. 4.9. FD biharmonic operator ∇^4 obtained directly from the MWLS method

convenient. These may be used either to lower the order of differential equations obtained, by decomposition of the given higher order ones, (the number of unknowns is increased then), or to compose the lower order operators into the higher order ones (the bandwidth is increased then). Having done e.g. the third order differential operator

$$\frac{\partial^3}{\partial x \partial^2 y} = \frac{\partial}{\partial x}\frac{\partial^2}{\partial y^2} = \frac{\partial^2}{\partial y^2}\frac{\partial}{\partial x} \tag{4.42}$$

a relevant difference operator may be composed

$${}^7B_{j(0)} = {}^1B_{j(0)}\,{}^4B_{i(j)} = {}^4B_{j(0)}\,{}^1B_{i(j)} \tag{4.43}$$

by simple multiplication of the appropriate operators of the first and the second order. An example of the FD formula for the biharmonic differential operator ∇^4 obtained directly is shown in Fig. 4.9. It may be compared with the one obtained by composition of the harmonic operators $\nabla^2\nabla^2$ (Fig. 2.4b). Each time the MWLS procedure was used for operator generation.

4.2.4.2
MWLS in Postprocessing

MWLS approximation is a powerful tool in postprocessing because it may provide us with values of a considered function, and its derivatives at any point, based on discrete data from scattered points. These results may be directly obtained using at each point of interest the approach defined in the formulas (4.16)–(4.26). Though precise, this approach is time consuming as solution of the local equations is needed at each point where approximation is required. One may consider also, therefore, a simplified but faster approach [55]. A simple and effective solution is to find approximation $\overline{u}$ at the considered point (x, y) as the combination

$$\overline{u}(x,y) = \sum_{i=1}^{m} {}_m v_i(x,y) u_i \tag{4.44}$$

of function values u_i at nodes taken from a neighborhood of this point, and the normalized weighting factors

$${}_m v_i(x,y) = \frac{w_i(x,y)}{\sum_{j=1}^{m} w_j(x,y)} \tag{4.45}$$

Here

$$w_i(x,y) = \hat{\rho}_i^{-n-1} \ , \quad \text{and} \quad \hat{\rho}_i = \left[(x-x_i)^2 + (y-y_i)^2\right]^{1/2} \tag{4.46}$$

is the distance between the current point (x, y) and a node (x_i, y_i). For $n = 0$ the Sheppard approximation is obtained. If additionally $m = 3$ linear shape functions in triangle are dealt with, well known in the FEM, hence ${}_3v_i(x, y) \equiv N_i(x, y)$.

In the general case the choice of m nodes from a neighborhood is left for the user. Various options are possible e.g. those discussed in Sect. 4.2.3. In the same way as the function u_i one may approximate the nodal values of its derivatives ${}^q u_i$, $q = 1, 2, \ldots$.

The above procedure may be extended from using only the nodal values u_i into application of their local approximation $\tilde{u}_i$ in the neighborhood of a chosen point of interest (x_0, y_0) [55]. First, at each node in the domain, the local approximations based on the Taylor series (4.11) $\tilde{u}(x, y; n)$ is spanned, taking advantage of nodal derivatives $\mathbf{D}u'$ found already. Next, these local approximations are combined the same way as above (4.44)

$$\overline{u}(x,y) = \sum_{k=1}^{m} {}_m v_i(x,y) \tilde{u}_i(x,y;n) \tag{4.47}$$

As mentioned before, one may combine local approximations of the derivatives in the same way as functions.

4.2.4.3
Generalized Degrees of Freedom

It is commonly accepted that in the FEM not only value of function but also of its derivatives, or their combinations may be considered as degrees of freedom (d.o.f.). As opposed to the classical FDM, various operators may be assumed as d.o.f. [37, 110] in the GFDM. As they are assigned to nodes only, and there are no constraints resulting from the element structure, their use in the GFDM is even more flexible than in the FEM.

Thus in the GFDM, additionally other types of degrees of freedom like various derivatives of u, or values Au of any given operator A acting on u, e.g. the Laplace operator $\nabla^2 u$ are introduced, besides the nodal values of the unknown function u. The local "best fit" for that function is to be determined using the Taylor expansion of the unknown function $u(x,y)$.

Consider as an example the FD star (Fig. 4.10) in which the vector of all d.o.f. is of the form

$$\mathbf{q} = \left\{ u_1 \quad u_2 \quad \frac{\partial u_2}{\partial x} \quad \frac{\partial u_2}{\partial y} \quad u_3 \quad \frac{\partial u_3}{\partial n} \quad \frac{\partial^2 u_1}{\partial n^2} \quad \nabla^2 u_5 \right\} \tag{4.48}$$

All these d.o.f. will be used to determine a surface $\tilde{u}(x,y;n)$ of the order n, locally strictly tangent at the central point (x_0,y_0) to the unknown function $u(x,y)$ by means of the Taylor series. The required function $\tilde{u}$ is uniquely determined by the values $\frac{\partial^{r+s}}{\partial x^r \partial y^s} u(x,y)|_{(x_0,y_0)}$ of its derivatives at the point (x_0,y_0). Denoting as before (4.9) the vector $\mathbf{D}u$ of these derivatives as well as the vector $\mathbf{p}$ of the basic local functions (4.13), and expanding into the Taylor series the unknown function u at the arbitrary point (x,y) with the respect to the point (x_0,y_0), one obtains the local approximation of that function of the same type as before (4.25).

Expanding into the Taylor series the derivatives of u in the same way

$$\frac{\partial u}{\partial x} \approx \frac{\partial \mathbf{p}^t}{\partial x} \mathbf{D}u' \,, \qquad \frac{\partial u}{\partial y} \approx \frac{\partial \mathbf{p}^t}{\partial y} \mathbf{D}u' \,, \text{ etc} \tag{4.49}$$

one forms the basic functions for the FD approximation of any operator appearing in the vector $\mathbf{q}$, such as the harmonic operator $\nabla^2 u \approx \nabla^2 \mathbf{p}^t \mathbf{D}u'$ and the normal derivative

$$\frac{\partial u}{\partial n} = \frac{\partial u}{\partial x} + m\frac{\partial v}{\partial y} = \frac{\partial \mathbf{p}^t}{\partial n} \mathbf{D}u' \tag{4.50}$$

$$\frac{\partial \mathbf{p}^t}{\partial n} = 0 \;\; 1 \;\; m \;\; lh \;\; lk + mh \;\; mk \;\; \ldots \} \tag{4.51}$$

on the boundary of the domain, defined in a local coordinate system (n,s). Finally for the star shown in Fig. 4.10 one obtains known relationship (4.16)

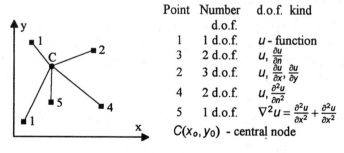

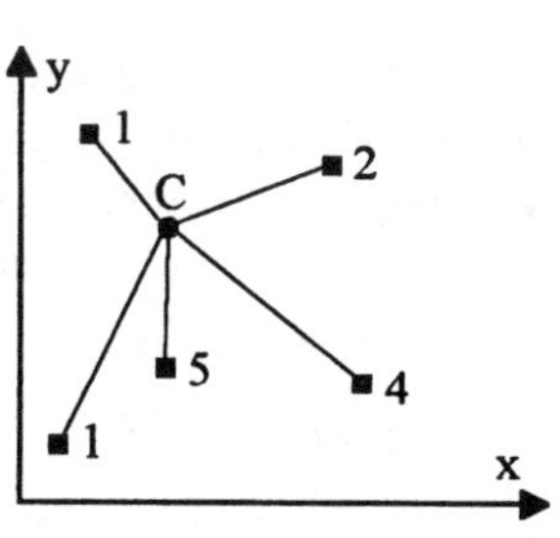

Point	Number d.o.f.	d.o.f. kind
1	1 d.o.f.	u - function
3	2 d.o.f.	$u, \frac{\partial u}{\partial n}$
2	3 d.o.f.	$u, \frac{\partial u}{\partial x}, \frac{\partial u}{\partial y}$
4	2 d.o.f.	$u, \frac{\partial^2 u}{\partial n^2}$
5	1 d.o.f.	$\nabla^2 u = \frac{\partial^2 u}{\partial x^2} + \frac{\partial^2 u}{\partial x^2}$

$C(x_0, y_0)$ - central node

Fig. 4.10. FD star; central node $C(x_0, y_0)$

and a relevant over determined system of algebraic equations for unknown $\mathbf{D}u'$ where this time

$$\mathbf{P}=\begin{bmatrix} 1 & h_1 & k_1 & \frac{1}{2}h_1^2 & h_1k_1 & \frac{1}{2}k_1^2 & \cdots & \frac{1}{n!}k_1^n \\ 1 & h_2 & k_2 & \frac{1}{2}h_2^2 & h_2k_2 & \frac{1}{2}k_2^2 & \cdots & \frac{1}{n!}k_2^n \\ 0 & 1 & 0 & h_2 & k_2 & 0 & \cdots & 0 \\ 0 & 0 & 1 & 0 & h_2 & k_2 & \cdots & \frac{1}{(n-1)!}k_2^{n-1} \\ 1 & h_3 & k_3 & \frac{1}{2}h_3^2 & h_3k_3 & \frac{1}{2}k_3^2 & \cdots & \frac{1}{n!}k_3^n \\ 0 & 1 & m & lh_3 & (lk_3+mh_3) & mk_3 & \cdots & m\frac{1}{(n-1)!}k_3^{n-1} \\ . & . & . & . & . & . & . & . \\ 0 & 0 & 0 & 1 & 0 & 1 & . & . \end{bmatrix}, \tag{4.52}$$

and the vector of error term magnitudes is

$$\mathbf{e}_1=\left\{\pm O\left(\rho_1\right)^{n+1}, \pm O\left(\rho_2\right)^{n+1}, \pm O\left(\rho_2\right)^{n}, \pm O\left(\rho_2\right)^{n}, \pm O\left(\rho_3\right)^{n+1},\right.$$
$$\left.\pm O\left(\rho_3\right)^{n}, \pm O\left(\rho_4\right)^{n+1}, \pm O\left(\rho_4\right)^{n-1}, \pm O\left(\rho_5\right)^{n-1}\right\} \tag{4.53}$$

In a similar way as before one minimizes the sum of squares of the weighted residuals and solves the problem according to the following formulas (4.21)–(4.26). This time, however, the weighting factor w_i has to be modified to account for the derivative error. Derivative of the order s has error of the $n+1-s$ order. Following the error terms of the vector $\mathbf{e}$ (4.53) the formulas (4.31) and (4.32) were extended for the case of singular and non-singular weighting factors as to obtain respectively

$$w_i=\overline{\rho}_i^{\,-n-1+s}, \quad w_i^2=\left(\overline{\rho}_i^2+\frac{g^4}{\overline{\rho}_i^2+g^2}\right)^{-n-1+s}, \quad i=1,2,\ldots,m \tag{4.54}$$

In the case of a differential operator presenting a combination of various derivatives, s is assumed as the order of the lowest derivative, including the function itself.

4.2.5 Generation of the FD Equations

Having obtained FD operators for the complete set of derivatives up to the n -th order $\mathbf{D}u'=\mathbf{B}\mathbf{q}$ at any point (node) of interest, we may easily combine them to generate a FD formula for any required differential operator. Consider e.g. a class of linear differential operators of the second order

$$\mathcal{L}=c_0+c_1\frac{\partial}{\partial x}+c_2\frac{\partial}{\partial y}+c_3\frac{\partial^2}{\partial x^2}+c_4\frac{\partial^2}{\partial x\partial y}+c_5\frac{\partial^2}{\partial y^2}=$$
$$=\mathbf{c}^t\mathbf{D}u=\sum_{q=0}^{5}c_q\,{}^q u \tag{4.55}$$

where

$$\mathbf{c} = \{c_q\}\ , \quad c_q = \text{const}\ , \quad q = 0, \ldots, 5 \tag{4.56}$$

are given constant coefficients. The Laplace operator has e.g. coefficients

$$\mathbf{c} = \{\, 0 \ \ 0 \ \ 0 \ \ 1 \ \ 0 \ \ 1 \,\} \tag{4.57}$$

A FD operator value $\mathrm{L}u$ corresponding to $\mathcal{L}u$ is then equal to

$$\mathrm{L}u \equiv \mathbf{c}^t\mathbf{D}u' = \mathbf{c}^t\mathbf{Bq} \tag{4.58}$$

In a similar way FD discretization of any differential operator resulting from boundary-value problems may be obtained.

In the local formulation (1.1), (1.2) the above approach is used for FD collocation

$$\mathrm{L}u_i = f_i\ , \quad i = 1, \ldots, N \tag{4.59}$$

where

$$\mathcal{L}u_i \approx \mathrm{L}u_i \equiv \mathbf{c}^t(\mathbf{Bq})_i = \sum_{q=0}^{5}\sum_{j=0}^{m} c_q\, {}^{q}B_{j(i)} u_{i+j} \tag{4.60}$$

of a given differential equation at all N internal nodes of the considered domain, as well as for collocation of boundary conditions as discussed in the classical FDM.

In the global (Galerkin, etc.) formulation a weak form of the problem or a variational principle is to be discretized, and integration is to be performed later. It is sufficient to discretize the differential operators at nodes or at the Gaussian points only in a way dependent on the integration type assumed. Discretization of integrands is done in the same way as mentioned above, e.g. using for the function u and its derivatives $\mathbf{D}u$ their FD approximation given by formulas (4.23), and (4.30).

4.2.5.1
Integration in the GFDM

There are three basic ways of numerical integration in the GFDM.

(i) Subdivision of the domain Ω into subdomains Ω_i, $i = 1, \ldots, N$ assigned to each node, and integration over these subdomains. This may be done by means of the Voronoi tessalation and integration over Voronoi polygons Ω_i. In the simplest case the nodal of values function F_i are multiplied by the surface areas Ω_i .

$$I \approx \sum_{i=1}^{N} F_i \Omega_i \tag{4.61}$$

(ii) Subdivision of the domain Ω into arbitrary background triangular elements with nodes located at their vertices, and integration over these triangles. The Delaunay triangulation seems to be the best choice here. Integration is then performed using the same quadratures as in the FEM.

(iii) Subdivision of the domain Ω into subdomains (triangles, squares, ...) in a way independent of nodes, and integration over these subdomains as in the FEM.

The first way follows the traditional FDM approach while the other two follow the FEM one. This is possible because the difference between the GFDM and the FEM concerns, first of all, the way and range of approximation, while the integration domain may be the same in both methods.

4.2.6 Discretization of Boundary Conditions

High quality discretization of boundary conditions (or essential boundary conditions in the global formulation of boundary-value problems) always presented one of the major difficulties in the FDM, especially when dealing with irregular (curvilinear) domains and regular meshes. Though use of arbitrary irregular grids removed various obstacles and provided much better bases for such a discretization, there are still problems to solve. As it is visible in the simple example (cf. Sect. 2.4), unknown function order and approximation quality in the boundary neighborhood, should be at least the same as inside the domain. Such requirement, however, is not satisfied, usually because of a lower order of differential operators on the boundary than inside the domain, and due to the offset of the central node, located on the boundary, from the star center of gravity where approximation is expected to be the most accurate.

Various concepts have been proposed to deal with reasonable FD discretization of boundary conditions. A brief discussion is presented below using as an example boundary conditions of the type

$$u + \beta \frac{\partial u}{\partial n} = g \quad \text{on } \partial\Omega \tag{4.62}$$

that appears in the second order elliptic problems. If $\beta = 0$, the boundary condition may be easily satisfied by $u_j = g_j,\ j = 1, 2, \ldots, J$ at boundary nodes $P_j,\ j = 1, 2, \ldots, J$. There are several options available, if $\beta \neq 0$:

(i) use of fictitious external nodes
(ii) use of internal nodes only
(iii) use of generalized d.o.f. like $\frac{\partial u}{\partial n}$ etc. in a similar way as in the FEM.

The first two options have been already presented in Sect. 2.4 when discussing the classical FDM approach. The last option, mentioned above, seems to be the most attractive though it is not certain yet if it is always effective. However, essential improvement of boundary conditions discretization quality may be expected if a higher order approximation approach is applied as proposed in Sect. 4.3.4.

4.2.7 Solution of Simultaneous FD Equations

The GFDM leads to the simultaneous FD equations of the same character as in Sect. 2.5. Consequently the same solution methods may be applied, but arbitrary mesh irregularity opens also possibility for effective application of new solution approaches. The most promising seems to be the use of the new multigrid approach, as discussed in Sect. 4.3.5 and Sect. 4.3.6.

5 Adaptive GFDM Approach

An overview of the adaptive GFDM multigrid solution approach is given in this chapter. New concepts of error analysis and mesh modification as well as prolongation, restriction and solution approach are presented.

Adaptivity is understood here mainly in the "h" sense, i.e., due to appropriate modifications of mesh density. However, a "p" type approach is also considered, using higher order GFD operators in the a'posteriori analysis integrated into the solution process.

5.1 A'posteriori Error Analysis

Any adaptive approach is based on error analysis, especially the a'posteriori one. Various criteria of the both local and global nature are considered. Criteria applied so far in the FEM [1, 112], but expressed in terms of the GFDM might be used here. A general review of such a'posteriori error criteria may be found in [1, 39]. These are useful for various discrete methods. However, for the purpose of the adaptive GFDM special FD oriented criteria are proposed as well. These provide control of both

(i) magnitude of local residuum at selected points, and
(ii) solution convergence at nodes common to the subsequent meshes used in the multigrid solution approach.

Ad (i): ***GFD residual error analysis***

The GFD operators are applied in order to determine a'posteriori residual error norms for any discrete solution obtained.

Using the locally defined boundary value problem $\mathcal{L}u = f$ in Ω, $\mathcal{L}_b u = g$ on $\partial\Omega$, one may consider the corresponding residuals $r \equiv \mathcal{L}u - f$, ${}_b r \equiv \mathcal{L}_b u - g$ and require that:

$$||r|| \leq \eta_1 ||f|| \qquad \text{and} \qquad ||{}_b r|| \leq \eta_2 ||g|| \tag{5.1}$$

where η_1, η_2 are assumed admissible error level threshold magnitudes. The above norms $||.||$ may be taken either globally over the whole domain Ω or over subdomains $\Omega_i, i = 1, 2, \ldots, N$ or locally at any point P_i of this domain or its boundary.

Introducing local GFD representations of the operators $\mathcal{L}$ and $\mathcal{L}_b$ at any point P_i, based on star nodes $i+j, j=1,2,...,m$, one may approximately evaluate the residuum r_i (or ${}_br_i$) at each considered point P_i by means of an appropriate GFD operator. The residuals r_i and ${}_br_i$ may provide a useful error indication. However, the values of residuals evaluated by the GFD approximation depend on precision of both the discrete solution examined as well as a local GFD operator applied. This last dependence is inconvenient, but may be eliminated by taking into account higher order terms in the local approximation used.

Thus consider a given n-th order differential equality $\mathcal{L}u = f$ in 2D domain Ω. After the FD discretization the value $\mathcal{L}u_i$ of that operator may be presented as follows

$$\mathcal{L}u_i = \mathrm{L}u_i + e_i' + e_i = f_i. \tag{5.2}$$

Here L is a basic FD operator of the n-th order, e_i' is the error introduced by the FD approximation to the operator value $\mathcal{L}u_i$, which exact within the $2n$-th order Taylor series expansion, while e is the error of this series due to truncation of the terms of order higher than $2n$.

In the FD solution approach evaluation of the true residual error $\| \mathcal{L}u_i - f_i \|$ for a given solution u_i may be replaced either by calculation of

$$\varepsilon_i = \| \mathrm{L}u_i + e_i' - f_i \| \tag{5.3}$$

if only the truncation error e_i is measured, or by means of the simplified formula

$$\tilde{\varepsilon}_i = \| \mathrm{L}u_i - f_i \| \tag{5.4}$$

i.e. by evaluation of the sum $e_i' + e_i$ of both errors. Obviously ε_i is a more precise tool of error analysis, however, knowledge of the error e_i' is required then. In what follows a simple and effective approach to evaluate the FD approximation error e_i is proposed.

Consider the FD operator $\mathrm{L}u_i$ as a combination of the nodal function (and/or derivatives) values, and expand it into the $2n$-th order Taylor series with the respect to the point P_i as to obtain

$$\mathrm{L}u_i = \mathcal{L}u_i + e_i'. \tag{5.5}$$

The error term due to FD approximation of derivatives is of the form

$$e_i' = \sum_{l=1}^{n} \sum_{j=0}^{n+l} \alpha_{lj} h_i^{l-j} k_i^j \frac{\partial^{n+l} u_i}{\partial x^{n+l-j} \partial y^j}. \tag{5.6}$$

Coefficients α_{lj} are derived when the Taylor series expansion is applied to the formula (5.5). As both the FD approximation of the n-th order derivatives $\mathbf{D}u'$ and nodal FD solution $u_1, \ldots, u_N$ of the problem are known, one may evaluate higher order derivatives in e_i' by the operator composition and find the errors e_i' (5.6). Such an approach may be easily implemented everywhere inside the domain Ω . The boundary region may require, however, a special treatment e.g. expansion of the higher order derivatives at boundary nodes into the Taylor series with the respect to the closest internal nodes. Fulfillment of the boundary condition as well as satisfaction of the equation given inside the domain is then also taken into account then.

The question arises where the residuals should be evaluated. One may expect that in 1D problems the largest residuals appear close to the mid points between neighboring nodes, while in 2D problems may be found close to the centroids of triangles (quadrilaterals) generated on any arbitrary irregular grids. More detailed suggestions will be proposed when discussing mesh modification and the multigrid approach.

It has to be stressed that the above local residual error criterion does not require the use of any reference solution based on a'posteriori smoothing, as it is usually the case in the other criteria. Here smoothing is built-into the FD operator generation procedure for $\mathcal{L}$ and $\mathcal{L}_b$ by means of the moving weighted least squares approximation [49, 48, 63, 98].

Ad (ii): ***Analysis of the convergence rates at nodes common to subsequent meshes***

A multigrid FD solution approach using denser and denser meshes is considered. A solution convergence rate

$$\beta_i^k \equiv \frac{||u_i^k - u_i^{k-1}||}{||u_i^k||} \leq \eta_3 \tag{5.7}$$

is examined at those nodes P_i, $i = 1, 2, \ldots$ which preserve the same locations in the subsequent meshes $\ldots, k-2, k-1, k, \ldots$; η_3 is an imposed error threshold value resulting from the required solution precision. Use of any other nodes is also possible, though less effective.

Using both error indicators r_i (and ${}_b r_i$), and β_i^k defined above, one may generate a series of more and more dense meshes, and establish an adaptive solution process. It is worth to note that both ill-conditioned problems and very slowly convergent solutions may be detected and also controlled this way.

The GFDM error indicators may be also applied to the adaptive solutions obtained by means of the global GFDM, or other discrete methods like the FEM as well.

5.2 A'posteriori Solution Smoothing

The moving weighted least squares approach [49, 48, 63, 98] is used for smoothing purposes if a reference solution is required e.g. for analysis using other than residual error norms [1, 39]. Either the direct GFD approximation procedure [40] may be applied here, or its specific version called the global–local method of the physically based approximation [33]. This last concept, devised for the purpose of smoothing the experimental data, is extended now [34] to numerical results. Advantage is taken from all information available for the problem at hand, in order to provide the "very best" possible approximation of the given data.

5.3 Adaptive Mesh Generation and Modification

The adaptive approach considered here uses a mesh refinement concept. Basically any mesh generator, providing modifications based on a'posteriori solution errors, could be used for the adaptive GFDM solution approach. However use of a special GFD oriented mesh generator seems to be more effective. It is convenient to define and use the notion of irregular mesh density in order to control mesh modifications. Advantage is taken from a simple idea of density of the regular square mesh. Such a density is in the inverse proportion to the side of square assigned to a node [53, 56, 90, 84]. In the case of irregular mesh one may use Voronoi tessalation in order to find Voronoi polygons assigned to each node. Converting each Voronoi polygon into an equivalent square of the same surface area Ω_i , one may determine the local mesh density ρ_i at a node i , the same way as if it was done for the regular mesh. Thus the relative local mesh density at a node P_i is defined like in [56]

$$\rho_i^{-1} = \rho_{max}^{-1} + \frac{1}{2}\log_2\left(\frac{k\Omega_i}{\Omega_{min}}\right) \tag{5.8}$$

Here, ρ_{max} is the largest density in the whole domain, corresponding to the Voronoi polygon with the smallest surface area Ω_{min}. For internal nodes $k = 1$, for the boundary ones $k = 2\pi/\alpha$, where α is the angle defining fraction of the polygon contained inside the domain ($\alpha = \pi$ for smooth boundary).

Mesh density $\rho(x, y)$ at an arbitrary point (x, y) of the domain Ω is obtained through the interpolation of nodal values, done on the optimal triangular Delaunay mesh by using interpolation formula (4.44) i.e. by means of the linear triangular shape functions in the simplest case.

Mesh modifications may be necessary due to requirements imposed on mesh density upon smoothness requirement and results a'posteriori error analysis.

Mesh smoothness requirement

A special care has to be taken to avoid abrupt changes in mesh density during mesh generation and modification.

The following approach is, therefore, suggested

- perform the Voronoi tessalation of the domain Ω and find the value

$$\eta_{ij} = \frac{\sqrt{\Omega_i} - \sqrt{\Omega_j}}{\rho_{ij}}, \quad \rho_{ij}^2 = (x_i - x_j)^2 + (y_i - y_j)^2$$

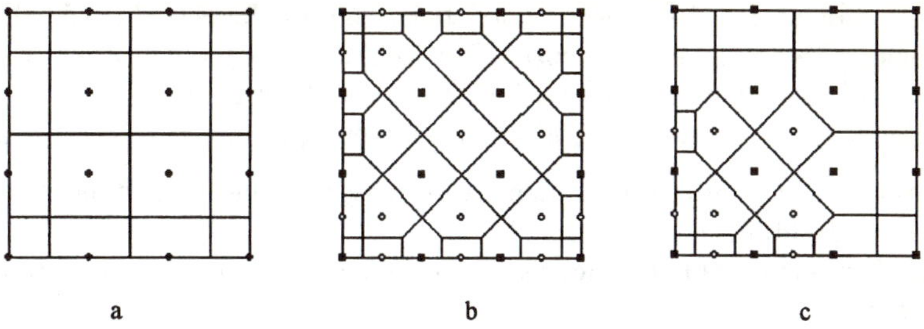

Fig. 5.1. Adaptive mesh modification; **a)** initial mesh, **b)** proposed new mesh, **c)** accepted new mesh, □ – old nodes ∘ – new nodes

characterizing the mesh density change between each pair of nodes P_i and P_j ;

- check whether $\eta_{ij} \leq \eta_{\text{admissible}}$ in the whole domain Ω
- at each node P_I for which this condition is violated find the neighbor node P_J with the maximum value $\eta_{IJ}^{max} = \max_j \eta_{ij}$;
- insert a new node at the point closest to the midpoint between each pair of nodes P_I and P_J and belonging to the one level denser mesh (Fig. 5.1b);
- repeat the whole procedure until everywhere the gradient of the mesh density change is sufficiently small i.e. until everywhere $\eta_{ij} \leq \eta_{\text{admissible}}$.

Mesh refinement strategy due to a'posteriori error analysis

The Liszka's idea of "sieving" FD mesh generator [49, 48], together with a new mesh modification strategy described below is proposed here. The modification is based on a concept of "correction filters" generation to the Liszka's sieve method. They are found based upon error indicators (residuals) defined above. The following are subsequent steps of the algorithm suggested:

(i) plan and generate the initial coarse mesh by the Liszka's method (cf. Sect. 4.2.1) (Fig. 5.1a);
(ii) perform Voronoi tessalation and Delaunay triangulation, determine mesh topology (cf. Sect. 4.2.2)
(iii) solve the boundary value problem in question, and find the potential locations of new nodes (Fig. 5.1b) using one level denser sieve (1/2 in 2D problems or 1 in 1D ones) than the one applied to the actual mesh;
(iv) examine the error indicators at potential locations of new nodes. Insert new nodes at points where admissible error norms (e.g. the average error value) are exceeded (Fig. 5.1c);
(v) unless all error norms admissible for the final solution are satisfied, return to the second step of the algorithm.

This way old nodes remain in their locations and new nodes are added. However, one may also wish to remove the old nodes sometimes, e.g. when they are totally surrounded by examined points with sufficiently low values of error indicators. Only those nodes, belonging to the actual mesh, that do not belong to one level less dense mesh ($-1/2$ in 2D problems or -1 in 1D ones), and are not prescribed as fixed ones may be removed.

Some features of the Liszka's type mesh generator are worth to be noted

- mesh of any required mesh density p_i may be generated using the criterion (4.1)–(4.5);
- simple selection of nodes belonging to one level more or one level less dense meshes;
- any given mesh may be converted into an equivalent Liszka's type;
- after Voronoi tessalation and Delaunay triangulation mesh may be used for both the GFDM and FEM purposes.

5.4 Higher Order Approximation FD Solution Approach

Besides error estimation, formulas (5.5) and (5.6) may also be used for the purpose of an iterative analysis

$$\mathrm{L}u_i^{(k)} = f_i - e_i^{\prime(k-1)}; \quad i = 1, 2, \ldots, N; \quad k = 1, 2, \ldots, \tag{5.9}$$

providing FD solution for $\mathbf{q}$ is based on higher order approximation and suffering only from the truncation error e. The error e_i' precisely indicates the difference between the value of the n-th order FD operator $\mathrm{L}u_i$, and the value $\mathcal{L}\tilde{u}_i$ found from application of the n-th order differential operator $\mathcal{L}$ to the $2n$-th order Taylor series expansion $\tilde{u}$ of the exact function $u(x, y)$.

Initially the error e_i' is not known. It may be found, however, in an iterative way starting form $e_i^{\prime(0)} = 0$. The basic FD solution is obtained then as usually. Afterwards i.e. for $k \geq 1$ the error $e_i^{\prime(k)}$ may be evaluated from formulas (5.5) and (5.6). FD approximation of derivatives of the order $n < s \leq 2n$ appearing in (5.6) is needed. They may be determined a'posteriori, by an appropriate composition of the n-th order operators $\mathbf{D}u'$ as discussed before (4.42), (4.43), and use of the FD solution already obtained in the preceding step of iteration. The following should be noted

- always the same basic FD equations (5.9) are solved, though with a modified right hand side
- quality of the final solution depends only on the order $2n$ of the Taylor series applied i.e. only on the truncation error e . On the other hand it does not depend on precision of the FD operator L used to approximation. This is because precision of each operator L is individually adjusted by an appropriate form of the error e' as to reach the exact value for the polynomials of the order $2n$.
- fast convergence of the solution process is observed.

Altogether, using the FD operator L of the lowest order allowing for approximation of a given n-th order differential operator $\mathcal{L}$ – one may obtain, in the above way a solution of the $2n$-th smoothness order. However, if the exact solution is of a lower smoothness order than $2n$, this approach may be adjusted to account e.g. for discontinuities of functions or their certain derivatives.

As a trivial, though illustrative 1D example of the approach described above, FD analysis of deflections of a simply supported uniformly loaded beam of the length $2l$ (cf. (2.28)) is presented here. The differential operator is of the second order ($n = 2$), and the 4-th (2x2) order local approximation is considered. FD discretization using only one node, located in the beam center, is assumed. Thus its displacement w_2 is the only one unknown of the problem. The FD operator $\mathrm{L}w_2$ is of the form (2.4).

Its development into the Taylor series yields the higher order approximation $\mathrm{L}w_2 = w_2'' + \frac{l^2}{12} w_2^{IV}$, and the relevant FD equation of the problem

$$\mathrm{L}w_2 = -\frac{1}{2}\frac{ql^2}{EJ} + \frac{l^2}{12} w_2^{IV} \tag{5.10}$$

with the basic solution $w_2 = \frac{1}{4}\frac{ql^4}{EJ}$. Now the error term $e_2' = \frac{1}{12}\frac{ql^2}{EJ}$ may be estimated by using twice the operator (2.4). It is applied to the function first, and to the second derivatives later. For corrected right hand side one at once obtains from there the final solution $w_2 = \frac{5}{24}\frac{ql^4}{EJ}$. This is the exact analytical value as well, because the truncation error e is zero for the fourth order local approximation used in this problem. Such a result was obtained for the most rough discretization, using only one node between the beam supports.

The above simple example indicates how much may be gained due to higher order approximation. Essential improvement of results may be expected e.g. when this technique is applied to high quality discretization of the boundary conditions.

5.5 Multigrid Approach

Following the general idea of multigrid analysis [24], proposed, developed and applied here are new concepts of prolongation and restriction and solution algorithm [81, 84]. In the multigrid approach one deals simultaneously with many meshes varying from coarse to fine, and constituting a series. Usually, though not necessarily, each finer mesh contains all nodes of the preceding more coarse one.

The solution obtained for a coarser mesh is extended to a finer one by means of the so called "prolongation" procedure. Conversely residuum evaluated on a finer mesh is reduced to a coarser mesh by the "restriction" procedure. Concepts of the proposed prolongation and restriction are described below.

5.5.1 Prolongation

The prolongation procedure consists of three essential steps

Step 1. Points at potential locations of the new nodes are searched for. These locations depend on the strategy adopted. In 2D domain the centroids of triangles taken from the optimal mesh triangulation may be considered. However, the best strategy, worked out so far, is to assume locations of new nodes obtained by the Liszka's type mesh generator after increase of the existing mesh density by 1/2 (in 2D) or 1 (in 1D).

Step 2. One examines the local residual $r_i$and ${}_b r_i$, $i \in I_{\text{examined locations}}$ in all potential locations found in step 1, and inserts new nodes wherever the conditions $||r_i|| \leq \eta_1 ||f_i||$ and $||{}_b r_i|| \leq \eta_2 ||g_i||$ are violated. The residuum r_i (and ${}_b r_i$) is approximately evaluated at each considered point by an appropriate GFD operator built only on nodes of the currently examined (old) mesh where a solution of the b.v. problem was found.

Step 3. Having assumed a new node P_i a new GFD operator $(L_{i+j}^{(m_p+1)})$) is built including this node and m_p old nodes. From the collocation requirement

$$\sum_{j=0}^{m_p} L_{i+j}^{(m_p+1)} u_{i+j} = f_i \quad i \in I^{\text{NEW}} \tag{5.11}$$

one finds the solution for the prolongation formula

$$u_i = \frac{1}{L_i^{(m_p+1)}} \left(f_i - \sum_{j=1}^{m_p} L_{i+j}^{(m_p+1)} u_{i+j} \right) \quad \to \quad u_i^{\text{NEW}} = \sum_j \alpha_{ij} u_j^{\text{OLD}} + b_i \tag{5.12}$$

It extends the solution u_{i+j}, $j = 1, \ldots, m_p$ found at the old nodes to a solution u_i required at the new node P_i. Here α_{ij} and b_i result from comparison of the formulae.

It is worth to mention here, that prolongation formula of the same type may be also built at each new node i, one by one, using the local variations $\delta u_i^{\text{NEW}} \neq 0$ and $\delta^{\text{OLD}} = 0$ in a considered variational formulation $\delta \Pi(u^{\text{OLD}}, u^{\text{NEW}}) = 0$. Such an approach may be used not only in the GFD solution procedure but also in the FEM or BEM analysis as well.

5.5.2 Restriction

Using the most fine mesh one evaluates residuals

$$r = f - Lu \tag{5.13}$$

Having found these residuals one needs to determine equivalent residuals for a subsequent coarser mesh. The following basic assumptions are proposed

- Virtual work

$$\delta W = \int_{\Omega} r \delta u \, \mathrm{d}\Omega \tag{5.14}$$

done by residuals ("residual forces") r on virtual "displacements" δu is of the same value for the both old and new meshes

- Virtual displacements of new nodes

$$\delta u_i^{\mathrm{NEW}} = \sum_j \alpha_{ij} u_j^{\mathrm{OLD}} \tag{5.15}$$

are found from the prolongation formula determined above.

Following the above assumptions one may find a GFD evaluation of the virtual work done on the new mesh

$$\begin{aligned}
\delta W &\approx \sum_l \Omega_l^{\mathrm{NEW}} r_l^{\mathrm{NEW}} \delta u_l^{\mathrm{NEW}} + \sum_k \Omega_k^{\mathrm{NEW}} r_k^{\mathrm{OLD}} \delta u_k^{\mathrm{OLD}} = \\
&= \sum_l \Omega_l^{\mathrm{NEW}} r_l^{\mathrm{NEW}} \sum_s \alpha_{ls} \delta u_s^{\mathrm{OLD}} + \sum_k \Omega_k^{\mathrm{NEW}} r_k^{\mathrm{OLD}} \delta u_k^{\mathrm{OLD}} = \\
&= \sum_k \left(\sum_l \Omega_l^{\mathrm{NEW}} r_l^{\mathrm{NEW}} \alpha_{lk} + \Omega_k^{\mathrm{NEW}} r_k^{\mathrm{OLD}} \right) \delta u_k^{\mathrm{OLD}}
\end{aligned}$$

and on the old mesh

$$\delta W = \sum_k \tilde{\Omega}_k^{\mathrm{OLD}} \tilde{r}_k^{\mathrm{OLD}} \delta u_k^{\mathrm{OLD}} \tag{5.16}$$

where $k \in I^{\mathrm{OLD}}$, $l \in I^{\mathrm{NEW}}$.

Virtual "displacements" $\delta u_i^{\mathrm{OLD}}$ are linearly independent from each other. Thus the residuals on the old mesh $\tilde{r}_i^{\mathrm{OLD}}$ equivalent in terms of the virtual work to the residuals $r_l^{\mathrm{NEW}}, r_k^{\mathrm{OLD}}$ determined on the new mesh are

$$\tilde{r}_k^{\mathrm{OLD}} = \left(\tilde{\Omega}_k^{\mathrm{OLD}} \right)^{-1} \left(\sum_l \Omega_l^{\mathrm{NEW}} r_l^{\mathrm{NEW}} \alpha_{lk} + \Omega_k^{\mathrm{NEW}} r_k^{\mathrm{OLD}} \right) \tag{5.17}$$

In the above formulas are given:

- residuals r^{NEW} and r^{OLD} for new and old nodes in the new mesh
- Voronoi polygons surface areas Ω^{NEW}, Ω^{OLD} for the new and old meshes are required
- residuals $\tilde{r}^{\mathrm{OLD}}$ on the old mesh equivalent to those on the new one.

5.5.3 Numerical Example

FD multigrid (non-adaptive) analysis of deflections of a simply-supported, uniformely loaded beam given in Sect. 2.7 is presented. Three meshes are given a'priori, namely: coarse (node numbers: 0, 4, 0; spacing $\frac{1}{2}$), intermediate (node numbers: 0, 2, 4, 2, 0; spacing $\frac{1}{4}$), and fine (node numbers: 0, 1, 2, 3, 4, 3, 2, 1, 0; spacing $\frac{1}{8}$). Advantage is taken of the beam symmetry Exact solution for *mesh 1*

$$\frac{0 - 2w_4 + 0}{(\frac{1}{2})^2} = -\frac{16}{128} \rightarrow w_4 = \frac{2}{128}$$

Prolongation from *mesh 1* to *mesh 2*

$$\frac{0 - 2w_2 + w_4}{(\frac{1}{4})^2} = -\frac{12}{128} \rightarrow \begin{bmatrix} w_2 \\ w_4 \end{bmatrix} = \frac{1}{2}\begin{bmatrix} 1 \\ 2 \end{bmatrix} w_4 + \frac{1}{128 * 32}\begin{bmatrix} 12 \\ 0 \end{bmatrix}$$

Prolongation from *mesh 2* to *mesh 3*

$$\begin{aligned} \frac{0 - 2w_1 + w_2}{(1/8)^2} &= -\frac{7}{128} \\ \frac{w_2 - 2w_3 + w_4}{(1/8)^2} &= -\frac{15}{128} \end{aligned} \Rightarrow \begin{bmatrix} w_1 \\ w_2 \\ w_3 \\ w_4 \end{bmatrix} = \frac{1}{2}\begin{bmatrix} 1 & 0 \\ 2 & 0 \\ 1 & 1 \\ 0 & 2 \end{bmatrix} \begin{bmatrix} w_2 \\ w_4 \end{bmatrix} + \frac{1}{128 * 128}\begin{bmatrix} 7 \\ 0 \\ 15 \\ 0 \end{bmatrix}$$

Defect (residuum) on *mesh 3*

$$\begin{bmatrix} r_1 \\ r_2 \\ r_3 \\ r_4 \end{bmatrix} = -\frac{1}{128}\begin{bmatrix} 7 \\ 12 \\ 15 \\ 16 \end{bmatrix} - 64 \begin{bmatrix} -2 & 1 & 0 & 0 \\ 1 & -2 & 1 & 0 \\ 0 & 1 & -2 & 1 \\ 0 & 0 & 2 & -2 \end{bmatrix} \begin{bmatrix} 47.5 \\ 88.0 \\ 115.5 \\ 128.0 \end{bmatrix} \frac{1}{128 * 64} = \frac{1}{128}\begin{bmatrix} 0 \\ 1 \\ 0 \\ 9 \end{bmatrix}$$

Restriction from *mesh 3* to *mesh 2*

"Virtual work"

$$\delta I = 2\,\frac{1}{8}(r_1 \delta w_1 + r_2 \delta w_2 + r_3 \delta w_3) + \frac{1}{8} r_4 \delta w_4 = 2\frac{1}{4}\tilde{r}_2 \delta w_2 + \frac{1}{4}\tilde{r}_4 \delta w_4$$

from prolongation $2 \rightarrow 3$

$$\delta w_1 = \frac{1}{2}\delta w_2, \qquad \delta w_3 = \frac{1}{2}(\delta w_2 + \delta w_4)$$

hence equivalent residuals on *mesh 2* are

$$\begin{bmatrix} \tilde{r}_2 \\ \tilde{r}_4 \end{bmatrix} = \frac{1}{4}\begin{bmatrix} 1 & 2 & 1 & 0 \\ 0 & 0 & 2 & 2 \end{bmatrix} \begin{bmatrix} r_1 \\ r_2 \\ r_3 \\ r_4 \end{bmatrix} = \frac{1}{128}\begin{bmatrix} 0.5 \\ 4.5 \end{bmatrix}$$

Restriction from *mesh 2* to *mesh 1*

"Virtual work"

$$\delta I = 2\,\frac{1}{4} r_2 \delta w_2 + \frac{1}{4} r_4 \delta w_4 = \frac{1}{2}\tilde{\delta}_4 \delta w_4$$

from prolongation $1 \rightarrow 2$

$$\delta w_2 = \frac{1}{2}\delta w_4 \qquad \tilde{r}_4 = \frac{1}{2}[1 \quad 1]\begin{bmatrix} r_2 \\ r_4 \end{bmatrix} = \frac{2.5}{128}$$

Exact solution for the correction term on *mesh 1*

$$v_4 = \frac{2.5}{128 * 8}$$

Prolongation from *mesh 1* to *mesh 2*

$$\begin{bmatrix} v_2 \\ v_4 \end{bmatrix} = \frac{1}{2} \begin{bmatrix} 1 \\ 2 \end{bmatrix} v_4 + \frac{1}{128 * 32} \begin{bmatrix} 0.5 \\ 0 \end{bmatrix} = \frac{1}{128 * 64} \begin{bmatrix} 11 \\ 20 \end{bmatrix}$$

Prolongation from *mesh 2* to *mesh 3*

$$\begin{bmatrix} v_1 \\ v_2 \\ v_3 \\ v_4 \end{bmatrix} = \frac{1}{2} \begin{bmatrix} 1 & 0 \\ 2 & 0 \\ 1 & 1 \\ 0 & 2 \end{bmatrix} \begin{bmatrix} v_2 \\ v_4 \end{bmatrix} + \frac{1}{128 * 128} \begin{bmatrix} 0 \\ 0 \\ 0 \\ 0 \end{bmatrix} + \frac{1}{128 * 64} \begin{bmatrix} 5.5 \\ 11 \\ 15.5 \\ 20 \end{bmatrix}$$

The final solution, exact for *mesh 3*

$$\begin{bmatrix} w_1 \\ w_2 \\ w_3 \\ w_4 \end{bmatrix} = \frac{1}{128 * 64} \begin{bmatrix} 47.5 \\ 88 \\ 115.5 \\ 128 \end{bmatrix} - \frac{1}{128 * 64} \begin{bmatrix} 5.5 \\ 11 \\ 15.5 \\ 20 \end{bmatrix} = \frac{1}{128 * 64} \begin{bmatrix} 42 \\ 77 \\ 100 \\ 108 \end{bmatrix}$$

It is worth to note that the exact solution for the fine mesh 3 was obtained after 1.5 cycles only.

5.5.4 Multigrid Adaptive Solution Procedure

Adaptivity is applied in the multigrid approach considered here. One carries the solution procedure, and at the same time one designs subsequent meshes based on the results of a'posteriori error analysis. Therefore, sufficiently precise solution is needed not only on the finest mesh but also on intermediate ones. Thus the following steps of the general solution approach are proposed:

- design and generate an initial coarse mesh and find the FD solution of the boundary-value problem considered, exact for this mesh and FD operator used
- evaluate correction terms resulting from the approximation error of the FD operator applied, and find the relevant higher order solution
- find one level denser (and one level less dense as well) mesh (± 1 for 1D and $\pm 1/2$ for 2D), and this way determine the potential locations of new nodes (or nodes which might be removed)
- in these new locations find residuals based on a considered problem formulation and the moving weighted least squares approximation locally spanned over the old nodes only; adjust these residuals for higher order terms; decide upon magnitude of these residuals where to insert new nodes, and where to remove the old ones

- find prolongation formulas from the old nodes to each new node, derived from the finite difference equations of the boundary value problem considered, written for each new node; each of these equations contains only one new node (central), all other nodes are the old ones
- use prolongation to extend the solution found for the initial mesh up to the currently finest one
- use the virtual work equivalence for old and new meshes to determine the restriction formula from the new to old nodes
- when a subsequent finest mesh is reached generate or modify the GFD stars and operators involving all new and old nodes. Find these operators adjustments for the higher order terms. Use these operators in order to find residuals. Return step by step to the preliminary coarse mesh using the restriction formulas. For this mesh find the exact solution for correction terms and use prolongation to extend them to the finest mesh. A smoothing step may be applied then followed by a next cycle of residual evaluation, restriction etc.
- control convergence of the solution using the exact results in the same nodes but for finer and finer meshes

5.6 Remarks

An adaptive multigrid approach to the meshless finite difference method generalized for arbitrary irregular grids is presented here. This is an integrated approach including original concepts of a'posteriori error analysis, solution smoothing, mesh generation and modification as well as multigrid solution procedure. The method is designed as a general and powerful tool of analysis of large and very large discrete boundary-value problems finally oriented on 3D ones.

A variety of 1D and 2D tests were done to examine convergence rate and solution effectiveness. The results obtained so far are very promising. The approach implementation includes distributed, and object-oriented programming [90] concepts. Domain decomposition may be also considered. Two modes of operation are considered, namely the basic mode and a fully distributed one. Both these modes can be used in concurrent work. The basic mode is preferred for smaller tasks while the fully distributed one is dedicated for solution of large problems.

6 On Mathematical Foundations of the FDM for Elliptic Problems[1]

The Finite Difference Method is one of the oldest methods widely used to solve ordinary and partial differential equations. It was the subject of many treatises published since 1950's, e.g. [8, 15, 23, 68, 69, 93, 94, 95, 96, 97]. Theoretical proofs of stability and consistency refer generally to the classical version of the FDM, i.e., a method based on regular meshes only. Few papers concerning mathematical foundations of the FDM with arbitrary irregular meshes appeared in recent years. Most of these deal with linear elliptic and parabolic differential equations.

This section presents some schemes of convergence proofs for the classical FDM, which are transferable to irregular meshes. Generally, however, proofs obtained for regular meshes are not applicable to irregular ones.

6.1 FDM at Regular Meshes

A FDM solution of an elliptic boundary-value problem is convergent to the exact solution, when the applied difference scheme fulfills assumptions of consistency and stability. A classical approach to convergence proofs, called truncation error technique, is shown on a simple example below.

Let us consider the Poisson's equation in a two-dimensional domain Ω, with the boundary condition of Dirichlet's type on a boundary Γ (Fig. 1.1)

$$-\Delta u = f \quad \text{in } \Omega, \tag{6.1}$$

$$u = g \quad \text{on } \Gamma. \tag{6.2}$$

In a node (x_i, y_j) the function u is expanded into the Taylor's series. Its values are evaluated at nodes $(x_{i\pm1}, y_j)$ and $(x_i, y_{j\pm1})$:

$$\begin{aligned}
u(x_{i+1}, y_j) &= u(x_i, y_j) + hu_{,x}(x_i, y_j) + 0.5h^2 u_{,xx}(x_i, y_j) + O(h^3)\\
u(x_{i-1}, y_j) &= u(x_i, y_j) - hu_{,x}(x_i, y_j) + 0.5h^2 u_{,xx}(x_i, y_j) + O(h^3)\\
u(x_i, y_{j+1}) &= u(x_i, y_j) + hu_{,y}(x_i, y_j) + 0.5h^2 u_{,yy}(x_i, y_j) + O(h^3)\\
u(x_i, y_{j-1}) &= u(x_i, y_j) - hu_{,y}(x_i, y_j) + 0.5h^2 u_{,yy}(x_i, y_j) + O(h^3)
\end{aligned} \tag{6.3}$$

where $x_{i\pm1} = x_i \pm h$, $y_{i\pm1} = y_i \pm h$.

[1] This chapter is written by A. Karafiat with the Author's cooperation

Neglecting the terms of order higher than 2, one may obtain an approximate value of the Laplace operator $\Delta_h u$ from the resulting system of equations:

$$\Delta_h u(x_i, y_j) = h^{-2}[u(x_{i+1}, y_j) + u(x_{i-1}, y_j) + u(x_i, y_{j+1}) + \\ +u(x_i, y_{j-1}) - 4u(x_i, y_j)]. \tag{6.4}$$

The difference between the exact and approximate value of the Laplace operator at (x_i, y_j) may be estimated by $O(h)$

$$|\Delta u(x_i, y_j) - \Delta_h u(x_i, y_j)| \leq Mh. \tag{6.5}$$

Direct evaluation results in a sharper inequality

$$|\Delta u(x_i, y_j) - \Delta_h u(x_i, y_j)| \leq Mh^2. \tag{6.6}$$

This is a consequence of the following completeness property: for each polynomial p of the order lower than three

$$\Delta_h p(x_i, y_j) = \Delta p(x_i, y_j) \ \ \forall \ (x_i, y_j) \in \Omega. \tag{6.7}$$

The boundary-value problem (6.1), (6.2) is then replaced by the approximate problem

$$-\Delta_h u_h(x_i, y_j) = f(x_i, y_j), \text{ for } (x_i, y_j) \in \Omega, \tag{6.8}$$

$$u_h(x_i, y_j) = g(x_i, y_j), \text{ for } (x_i, y_j) \in \Gamma, \tag{6.9}$$

and the obtained approximate solution u_h is called the second order approximation of the exact solution u, because of evaluation $O(h^2)$ in (6.6) (cf. [11]).

Here (6.8) and (6.9) form simultaneous algebraic linear equations

$$\mathbf{Au} = \mathbf{f}, \tag{6.10}$$

where $\mathbf{u} = (u_1, \ldots, u_N)$, $\mathbf{f} = (f_1, \ldots, f_N)$ and $\mathbf{A}$ is a $N \times N$ matrix which fulfills the following assumptions:

$$a_{ii} > 0\,, \quad i = 1, \ldots, N \tag{6.11}$$

$$a_{ij} \leq 0\,, \quad i, j = 1, \ldots, N\,,\ i \neq j \tag{6.12}$$

$$\sum_{j=1}^{N} a_{ij} \geq 0\,, \quad i = 1, \ldots, N \tag{6.13}$$

$$\sum_{i=1}^{N}\sum_{j=1}^{N} a_{ij} > 0, \tag{6.14}$$

which are obvious for the elliptic difference scheme (6.4).

Matrix $\mathbf{A}$ of size $N \times N$ is reducible [107] if and only if there is a permutation $(m_1, \ldots, m_N)$ of the index sequence $(1, \ldots, N)$, and there exists a number $k \in \{1, \ldots, N-1\}$, such that

$$a_{kl} = 0\,, \quad \forall k \in \{m_1, \ldots, m_k\}\,, \quad l \in \{m_{k+1}, \ldots, m_N\}. \tag{6.15}$$

In the opposite case the matrix **A** is irreducible. This is the case when the domain Ω is connected and the mesh is "connected" too.

For vectors

$$\mathbf{w} = (w_1, \ldots, w_N)\,, \qquad \mathbf{v} = (v_1, \ldots, v_N) \tag{6.16}$$

the partial ordering in $\mathbb{R}^N$ is introduced by

$$\mathbf{w} \leq \mathbf{v} \Leftrightarrow w_i \leq v_i\,, \qquad i = 1, \ldots, N. \tag{6.17}$$

Irreducibility of the matrix **A** together with conditions 6.11 -(6.14) yields the monotonicity of **A**, i.e.

$$\forall \mathbf{w}, \mathbf{v} \in \mathbb{R}^N\,, \qquad \mathbf{A}\mathbf{w} \leq \mathbf{A}\mathbf{v} \Rightarrow \mathbf{w} \leq \mathbf{v} \tag{6.18}$$

(cf. [107]). Substituting the vector $\mathbf{e} = (1, \ldots, 1)$ one obtains

$$\begin{aligned} A(Mh^2\mathbf{e} \pm \mathbf{w}) \geq \mathbf{0} \Rightarrow Mh^2\mathbf{e} \pm \mathbf{w} \geq \mathbf{0} \Rightarrow \pm\mathbf{w} \leq Mh^2\mathbf{e} \\ \Rightarrow |\mathbf{w}| \leq Mh^2|\mathbf{e}| = M_1h^2. \end{aligned} \tag{6.19}$$

This is the stability condition of the difference scheme [107].

Consistency (6.6) and stability (6.19) imply convergence of the method at the nodes

$$|(u - u_h)(x_i, y_j)| \leq Mh^2\,, \qquad \forall\,(x_i, y_j) \in \overline{\Omega}. \tag{6.20}$$

A similar proof is applied to difference solutions of parabolic equations, e.g., the heat conduction equation

$$u_{,t}(t, \mathbf{x}) - c^2\Delta u(t, \mathbf{x}) = 0\,, \qquad t \in (0, T), \mathbf{x} \in \Omega, \tag{6.21}$$

with initial

$$u(0, \mathbf{x}) = u_0(\mathbf{x})\,, \qquad \mathbf{x} \in \Omega, \tag{6.22}$$

and boundary conditions

$$u(t, \mathbf{x}) = u_1(t, \mathbf{x})\,, \qquad t \in (0, T), \mathbf{x} \in \Gamma. \tag{6.23}$$

Assumptions (6.11)–(6.14) for monotonicity of the matrix **A** cause a limitation of a time step with respect to a space step. This is the stability condition for the corresponding difference scheme [15, 32].

An analogous method is used in the estimation process for general elliptic or parabolic equations. Detailed convergence proofs are given in monographs [15, 23, 68, 69, 91–97].

J. Cea considered in [6] difference approximation as an external approximation based on piecewise constant functions. By means of Hilbert space methods he proved convergence of the FDM for equations of linear elasticity and for an elliptic equation of the fourth order. In [21] convergence of difference solutions of variational inequalities was shown, using a similar approach. Another method, typical for the finite elements was used in [95].

6.2 Finite Difference Method at Irregular Meshes

Two basic ways of generalization of the proofs mentioned above to irregular meshes will be presented here. The first one is based on the direct application of the previous approach to a mesh consisting of irregularly distributed nodes [11, 31, 32]. The proof was extended to these meshes and the convergence of the FDM was shown for elliptic equations with different boundary conditions. Second order local interpolation was assumed, based on the node and five other nodes from its neighborhood. An unknown function $u(x)$ was expanded into the Taylor series at each of these auxiliary nodes with respect to the central one. Approximate difference values of all the first and the second order derivatives of u (completeness condition (6.7)) were found at the central node from the interpolation conditions. In estimates (6.6), (6.20) $O(h^2)$ is replaced by $O(h)$, which is the highest rate of convergence which may be proved for any irregular mesh [4], although in numerical experiments the rate $O(h^2)$ is frequently observed [11, 31].

The second type proof is based on a division of the domain Ω into triangles $T_n \in \mathcal{T}_n$ and the allocation of a relevant subdomain $H(\mathbf{x}^k)$ with boundary $\Gamma(\mathbf{x}^k)$ to each node $\mathbf{x}^k$. The subdomains are cut off by bisectrices of segments joining neighboring nodes (Fig. 4.3, [25]). These subdomains are in fact Thiessen (Voronoi) polygons (cf. Sect. 4.2.2).

Equation (6.1) integrated over the domain $H(\mathbf{x}^k)$ and allocated to the node $\mathbf{x}^k$ is replaced by

$$-\int_{\Gamma(\mathbf{x}^k)} \frac{\partial u}{\partial n}\,\mathrm{d}s = \int_{H(\mathbf{x}^k)} f\,\mathrm{d}\mathbf{x} \tag{6.24}$$

using the Green's theorem.

$$\int_{H(\mathbf{x}^k)} \Delta u\,\mathrm{d}\mathbf{x} = \int_{\Gamma(\mathbf{x}^k)} \frac{\partial u}{\partial n}\,\mathrm{d}s \tag{6.25}$$

Taking advantage of the finite difference formula

$$\frac{\partial u}{\partial n}(\mathbf{x}) \approx \frac{u(\mathbf{x}^i) - u(\mathbf{x}^k)}{h_{ik}} \tag{6.26}$$

on the part b_{ik} of the boundary $\Gamma(\mathbf{x}^k)$ adjacent to the subdomain around the node $\mathbf{x}^i$, $h_{ik} = |\mathbf{x}^i - \mathbf{x}^k|$, one obtains

$$\Delta_h u(\mathbf{x}^k) = \sum_i \int_{b_{ik}} \frac{u(\mathbf{x}^i) - u(\mathbf{x}^k)}{h_{ik}}\,\mathrm{d}s, \tag{6.27}$$

$$\int_{H(\mathbf{x}^k)} f\,\mathrm{d}\mathbf{x} = f(\mathbf{x}^k) \int_{H(\mathbf{x}^k)} \mathrm{d}\mathbf{x}. \tag{6.28}$$

Hence the difference equation becomes

$$-\Delta_h u(\mathbf{x}^k) = f(\mathbf{x}^k) \int_{H(\mathbf{x}^k)} \mathrm{d}\mathbf{x}. \tag{6.29}$$

This method was initiated in 1953 by R.H. Mac Neal [70]. To prove its convergence it was shown [36] that the bilinear form

$$a_h(u_h, v_h) = -\sum_k \Delta_h u_h(\mathbf{x}^k) v_h(\mathbf{x}^k) \tag{6.30}$$

is elliptic and bounded in the Hilbert space of mesh functions

$$V = \{v : X_h \to \mathbb{R}\}, \tag{6.31}$$

where X_h is a set of nodes, with the scalar product

$$\begin{aligned}(u_h, v_h) = &\sum_{\mathbf{x}^k \in X_h} u_h(\mathbf{x}^k) v_h(\mathbf{x}^k) h^2 + \\ &+ \sum_{x^i \in T^k} [u_h(\mathbf{x}^i) - u_h(\mathbf{x}^k)][v_h(\mathbf{x}^i) - v_h(\mathbf{x}^k)]\end{aligned} \tag{6.32}$$

and norm

$$||v_h|| = (v_h, v_h)^{1/2} \tag{6.33}$$

where

$$h = \max_{i,k} h_{ik} \tag{6.34}$$

is the diameter of the circle circumscribed about the largest triangle, and $T^k = \{\mathbf{x}^i \in X_h\ ,\ \mathbf{x}^i \text{ is a neighbor of } \mathbf{x}^k\}$.

Stability and convergence of this version of the FDM are consequences of the above properties, and are satisfied for a wide class of elliptic boundary-value problems of the second order, e.g., for a plane problem of linear elasticity [25] and the Poisson equation.

The presented approach was examined in details by B. Heinrich in his book [25]. The convergence of the method in the norm 6.33 and in the norm of the space of continuous functions was proved there. The difference operators were considered as values of linear, continuous operators on special shape functions defined like those in the finite element method. The proofs for elliptic equations were extended to parabolic equations. This method is, however, applicable only to some differential operators. On the other hand, estimates better than before may be obtained.

The two methods presented here outline the most popular approaches to convergence proofs for the FDM with arbitrary irregular meshes. In spite of the efforts undertaken, the state of art in the mathematical foundations of the FDM is far behind development of algorithms, codes and applications of the method. This refers in particular to

- schemes of higher order, where the presented approaches are inadequate;

- schemes using "excessive" nodes (cf. the weighted moving minimization method), which are more stable and effective than the schemes mentioned above;
- schemes including various generalized unknowns (degrees of freedom);
- nonlinear problems;
- algorithms using various combinations of both FDM and FEM;
- problems defined on differential manifolds;
- error analysis and an adaptive approach.

In the last five years so called meshless methods appeared and were rapidly developed. The described version of the FDM, generalized for arbitrary irregular grids, is the oldest and possibly the most developed of them. It may follow e.g. some ideas elaborated by I. Babuška [71] and J.T. Oden [13] with their coworkers. Usual polynomial basis is here replaced by selected polynomials, corresponding to the differential equation, multiplied by partition of unity functions. This approach allows to keep local approximability property by polynomials, whereas conformity is enforced by regularity of partition of unity functions. This technique, fulfilling the above assumptions of the Finite Element Method, makes the rich mathematical structure of the FEM applicable in this situation. For simple boundary-value problems, approximation errors were estimated in Sobolev space norms like in the FEM. For details we refer e.g. to [13].

The presented Finite Difference Method satisfies approximability condition but conformity does not hold. A modification of this method by using regular weight functions with finite local supports should result in its new mathematical equipment.

Concluding, majority of problems of mathematical foundations of the FDM with arbitrary irregular meshes remains still open, although there is a good chance now for a significant progress.

7 Final Remarks

The finite difference method (FDM) is an old solution method of a broad class of boundary-value and initial-value problems. However, its power and scope of application were limited until the effective generalization for irregular meshes was performed (GFDM). Following the earlier studies and the recent developments like error analysis, adaptivity and multigrid solution approach, the GFDM, like the FEM, presents nowadays a general integrated solution tool displaying a variety of useful features. Some of these are briefly listed below:

(i) Various formulations. The main idea of the FD approach is to replace differential operations by difference ones. This concept may be applied to various formulations of boundary value problems. In the current state of development the GFDM may be used for solution of b.v. problems in such formulations as: local, global (given functional or variational principle), global in subdomains, mixed global and local (e.g. constrained optimization problem). Some of these formulations may be also presented as particular cases of the weighted residual ones. Combinations of the GFDM with other methods, e.g. the FEM or the BEM also may be considered. It is assumed that in each of these formulations one deals with spaces providing functions sufficiently smooth to perform the necessary differentiations,

(ii) Arbitrary irregular non structured grids may be used. Their generation may be done by enforcing a'priori given local mesh density,

(iii) Moving local approximation is applied based on the truncated Taylor series and the weighted least squares fit. Moreover enhancement functions such as local asymptote singular solutions (e.g. jumps, corner solutions, etc.) may also be included,

(iv) Consistency requirements are met, i.e., constant, linear and higher order functions are exactly reproduced if needed, due to interpolation nature of the local approximations used,

(v) FD solutions based on the higher order approximation may be effectively obtained,

(vi) Full adaptivity based on a local and global error control

(vii) High computational efficiency oriented on solution of large boundary value problems.

The basis of the method, and outlines of the adaptive multigrid approach have been presented here. However, several important topics are left beyond the scope of this book, and may be studied directly from the references cited below. The following problems enumerated below are especially worth to be noted:

- GFDM on a differential manifold; the method formulation and application to analysis of pneumatic shells are given in [105, 106],
- mixed global and local GFDM resulting from constrained optimization problems [85, 87], and applied to analysis of residual stresses as well as to experimental data smoothing,
- generalized finite strip approach (GFSM). This is a combination of the GFDM (2D) and the Fourier analysis (1D) for the purpose of 3D analysis of bodies of revolution and prismatic ones; Full error control is included [38],
- various GFDM/FEM combinations and unification in one computer system [37, 39, 41],
- effective computer implementation including distributed programming [90], object oriented programming and symbolic operations [43, 37, 42].

7.1 Meshless Methods

In the recent years more and more attention is paid to development and application of discrete approaches to analysis of boundary-value problems, based on nodes rather than on elements, and therefore using unstructured grids. The local approximation is most often based on the moving weighted least squares approach (MWLS). Generally the name meshless methods (MM) is used then, though a variety of different specific names is used as proposed by their authors (cf. review papers [3, 12]). Quite often the same, in fact, method has various names and "rediscoveries" are being made. Therefore in what follows a classification of these methods is given as proposed by the author. It is based on the main concepts of these methods rather than on their names. All these methods yield approximation of the form $u^h = \overline{\mathbf{N}}\mathbf{q}$, $\overline{\mathbf{N}} = [\overline{N}_i]$ well known from the FEM and also obtained by the GFDM (cf. (4.30)). Particular groups of these methods differ from each other by the way the "pseudo shape functions" $\overline{N}_i$ are obtained and by the specific form of these functions. However the condition $\sum_i \overline{N}_i = 1$, called partition of unity always has to be satisfied. Thus one may distinguish:

(i) *Methods based on the weighted least squares (MWLS) local approximation* The following methods may be included into this group:

- generalized finite difference (GFDM) discussed in this book, and developed since 1972
- diffuse element method (DEM) proposed in [75]. Though the concept of this method proposed in 1992 did not introduce anything new when compared with the current GFDM development [110, 60, 63, 82, 83,

49, 48], the paper [75] accomplished an important role as a trigger of wide research on the meshless methods [3, 78, 12, 13, 71]

- element free Galerkin (EFG).
  Under this name very intensive research has been recently undertaken by T. Belytschko and his numerous coworkers (cf. review paper [3]) mainly using the Galerkin formulation of boundary value problems. The classical MWLS approach is used with polynomial interpolants to derive the local pseudofunctions $\overline{N}_i$ and their derivatives. The asymptotic solutions and/or jump functions may enrich the local approximation bases when needed. Successful analysis, especially of complex fracture mechanics problems is reported
- finite point (FPM) [78, 77] and finite volume (FVM) [27, 111]
  These methods proposed by S. Idelsohn, E. Oñate, R.L. Taylor and O.C. Zienkiewicz use some basic concepts of the finite difference method. Advantage is also taken of the MWLS approximation (FPM) and Voronoi tessalation of the domain used for integration purposes (FVM) as discussed in Sect. 4.2.5.1 (cf. (4.61)). Large fluid mechanics problems were effectively analysed [78, 77].

(ii) kernel methods
These methods are based on an interpolation using the kernel estimate

$$u^h(x) = \int_\Omega w(\mathbf{x} - \mathbf{y}, h) u(\mathbf{x})\, \mathrm{d}\mathbf{y}$$

of the function $u(\mathbf{x})$ defined on a domain Ω. Assuming the weight function w (cf. pp. 393–394) as the kernel one may obtain the required pseudo shape functions $\overline{N}_i(\mathbf{x})$.
The following particular methods are considered:

- smooth particle hydrodynamic (SPH). This is the second oldest MM initiated by L.B. Lucy [67] and developed by J.J. Monaghan [72]. In the discrete form this method is not consistent i.e. the lowest order polynomials are not reproduced. This drawback is eliminated by the improved SPH version called:
- reproducing kernel particle method (RKPM) introduced by W.K. Liu with coworkers [66] and intensively developed in a series of papers [3].

(iii) partition of unity methods PUM
These methods are based on the general concept $1 = \sum_i \overline{N}_i$ called partition of unity. Once such a partition is executed the above equality may be multiplied by any function ψ as to generate new "shape functions" $\psi\overline{N}_i$ and/or perform other operations e.g. differentiation. The function ψ may represent e.g. a complete basis of monomials and/or asymptotic solutions etc. Two basic approaches were used in the PUM, differing in the way the basic shape-functions $\overline{N}_i$ are defined and introduced:

- partition of unity finite element method (PUFEM)
  This approach is due to I. Babuška and Melenk [71]. They used the classical FEM to generate shape functions N_i .
- hp-clouds method
  A. Duarte and J.T. Oden [13] proposed the use of the MWLS to generate the pseudo shape functions $\overline{N}_i$. In fact, the simplest Sheppard method is applied and $\overline{N}_i$ are defined by formulas (4.29). Moreover the h-p adaptive approach was successfully introduced. Recently also a combination of hp-clouds and GFDM has been considered [65].

(iv) particle in cell methods (PIC)
These methods are applied, first of all, to dynamic problems. The continuous body is partitioned into subdomains. Their masses are reduced to particles located in their centres of gravity. The body motion is described then as a motion of a cloud of particles following the principles of mechanics. Effective, valuable results were obtained using this method [28, 100] despite several limitations and drawbacks resulting from its simplicity.

As mentioned in [3] the meshless methods are in a state of rapid development nowadays. Although, in fact, some of them have been discovered quite a time ago (GFDM – 25 years, SHP – 20 years), it is only recently that they have captured the interests of a broader group of researchers. There are many aspects of these methods which could benefit from improvements like the techniques for treating discontinuities and other local effects (e.g. singularities). Meshless methods and wavelets are manifestations of the same basic trend towards methods with localized approximations (e.g. MWLS). Although they already form a separate group of discrete methods, some of the meshless methods still require further improvement before they can equal the prominence of the finite element method. However, the current development of the GFDM and the potential power of the PUM are already very promising. The greatest challenge of meshless methods seems to lie in development of the speed and robustness which characterize low-order finite elements. Ease of use of distributed and parallel computing as well as symbolic operations which the meshless methods display, yields a reasonable hope for breakthrough in their effective application.

References

1. Ainsworth M, Oden JT (1997) A posteriori error estimation in finite element analysis. Comp Meth Appl Mech Engng 142:1–88
2. Anderson DA, Tannenhill JC, Fletcher RH (1984) Computational fluid mechanics and heat transfer. McGraw–Hill, Washington
3. Belytschko T, Krongauz Y, Organ D, Fleming M, Krysl P (1996) Meshless methods: An overview and recent developments. Comp Meth Appl Mech Engng 139:3–47
4. Birkhoff G, Gulati S (1974) Optimal few-point discretizations of linear source problems. SIAM J Num Anal 11:700–728
5. Bushnell D, Almroth BO, Brogan F (1971) Finite difference energy method for nonlinear shell analysis. Comp Struct 3: 361–388
6. Cea J (1964) Approximation variationelle des problems aux limites. Ann Inst Fourier (Grenoble) 14:345–444
7. Cendrowicz J, Tribiłło R (1978) Variational approach to static analysis of plates of arbitrary shape (in Polish). Arch Inż Ląd 24: 411–421
8. Collatz L (1966) The numerical treatment of differential equations. Springer, Berlin, 3 ed
9. Davies G, Ford B, Mullord P, Snell O ((1973) Application of an irregular mesh finite difference approximation to the plate bucking problem. Proc of the Int Conf Variational Meth Eng, vol II, chap 11, Southampton University Press
10. Dekker K (1980) Semi-discretization methods for partial differential equation on non-rectangular grids. Int J Num Meth Eng 15:405–419
11. Demkowicz L, Karafiat A, Liszka T (1984) On some convergence results for FDM with irregular mesh. Comp Meth Appl Mech Eng 42:343–355
12. Duarte CA (1995) A review of some meshless methods to solve partial differential equations. Technical Reports, 1995-06, TICAM, Universities of Texas at Austin
13. Duarte CAM, Oden JT (1995) Hp clouds – a meshless method to solve boundary-value problems. TICAM Report 95-05, TICAM, The University of Texas at Austin, Austin
14. Dyke PP, Phelps GP (1982) The use of irregular finite difference grids for coastal sea problems. Num Meth Fluid Dyn, 259–271
15. Forsythe GE, Wasow WR (1960) Finite difference methods for partial differential equations. Wiley, New York
16. Frederick CO, Wong YC, Edge FW (1970) Two-dimensional automatic mesh generation for structural analysis. Int J Num Meth Eng 2:133–144
17. Frey WH (1977) Flexible finite-difference stencils from isoparametric finite elements. Int J Num Meth Eng 11:1653–1665

18. Gawain TH, Ball RE (1978) Improved finite difference formulas for boundary-value problems. Int J Num Meth Eng 12:1151–1160

19. George PL (1991) Automatic mesh generation. Wiley

20. Girault V (1974) Theory of a finite difference method on irregular networks. SIAM J Num Anal 11:260–282

21. Glowinski R, Lions JL, Tremolieres L (1976) Numerical analysis of variational inequalities (in French). Paris, Dunod

22. Godoy LA (1986) Ill-conditioned stars in the finite difference method for arbitrary meshes. Comp Struct 22:469–473

23. Godunov SK, Riabenskij VS (1973) Difference schemes (in Russian). Moscow, Nauka

24. Hackbusch W (1992) Multi-grid methods and applications. Springer-Verlag, Berlin–Heidelberg–New York–Tokyo

25. Heinrich B (1987) Finite difference methods on irregular networks. Berlin, Akademie-Verlag

26. Hood P (1976) Frontal solution program for unisymmetric matrices. Int J Num Meth Eng 10:379–400

27. Idelsohn S, Oñate E (1994) Finite element and finite volumes. Two good friends. Int J Numer Methods Eng 37:3323–3341

28. Jach K, Leliwa-Kopystyński J, Mroczkowski M, Świerczyński R, Wolański P (1994) Free particle modelling of hypervelocity asteroid collisions with the earth. Planet Space Sci 42(12):1123–1137

29. Jensen PS (1972) Finite difference techniques for variable grids. Comp Struct 2:17–29

30. Kaczmarz S (1937) Solution of simultaneuous linear algebraic equations (in German). Bull Intern Acad Pol Classe Sci Math Nat, (A) Sci Math 355–357

31. Karafiat A (1989) On convergence of solution for FDM with irregular mesh in Neumann problem. Comp Meth Appl Mech Eng 72:91–103

32. Karafiat A (1991) Discrete maximum principle for parabolic boundary-value problems. Ann Polon Math 53:253–265

33. Karmowski W, Orkisz J (1993) A physically based method of enhancement of experimental data – concepts, formulation andapplicationto identification of residual stresses. In: Proc IUTAM Symp on Inverse Problems in Engng Mech, Tokyo 1992; On Inverse Problems in Engineering Mechanics, Springer Verlag, pp 61–70

34. Karmowski W, Orkisz J (1997) A'posteriori error estimation based on smoothing by the global–local physically based approximation. Proc of 13th Polish Conf on Comp Meth in Mechanics, Poznań, Poland

35. Kączkowski Z, Tribiłło R (1975) A generalisation of the finite difference method (in Polish). Arch Inż Ląd 21(2):287–293

36. Kellog RB (1964) Difference equations on a mesh arising from a general triangulation. Math Comp 18:203–210

37. Krok J, Orkisz J (1990) A unified approach to the FE generalized variational FD method in nonlinear mechanics. Concept and numerical approach. In: Discretization methods in structural mechanics. IUTAM/IACM Symp, Vienna, 1989; Springer-Verlag, Berlin–Heidelberg, pp 353–362

38. Krok J, Orkisz J (1997) 3D elastic stress analysis in railroad rails and vehicle wheels by the adaptive FEM/FDM and Fourier series. Proc of the 13th Polish Conf on Comp Meth in Mechanics, Poznań, Poland, pp 661–668

39. Krok J, Orkisz J (1997) A unified approach to the adaptive FEM and FDM in nonlinear mechanics. Concepts and tests. Proc of 13th Polish Conf on Comp Meth in Mechanics, Poznań, Poland, pp 653–660

40. Krok J, Orkisz J (1986) Application of the generalized FD approach to stress evaluation in the FE solution. Proc Int Conf on Comp Mech, Tokyo, pp 31–36

41. Krok J, Orkisz J, Stanuszek M (1993) A unique FDM/FEM system of discrete analysis of boundary value problems in mechanics. Proc of 11th Polish Conf on Comp Meth in Mechanics, Kielce-Cedzyna, Poland, vol 1, pp 466–472

42. Krok J, Schaefer R, Orkisz J,Leżański P, Przybylski P (1996) Basic concepts of an open distributed system for cooperative design and structure analysis. Comp Assisted Mech and Engrg Sci 3:169–186

43. Kucwaj J, Orkisz J, Tworzydło W (1984) Application of the symbolic programming to the solution of boundary problems by the Finite Difference Methods (in Polish). Arch Inż Ląd 30(4):595–606

44. Kurowski Z (1978) Automatic generation of triangular meshes in 2D domains (in Polish). Mechanika i Komputer 1:379–389

45. Kwok SK (1985) Geometrically nonlinear analysis of general thin shells using a curvilinear finite difference (CFD) energy approach. Comp Struct 20:683–697

46. Kwok SK (1984) Numerical computations of tensor quantities on a curved surface by the curvilinear finite difference (CFD) method. Comp Struct 18:1087–1114

47. Kwok SK (1984) An improved curvilinear finite difference (CFD) method for arbitrary mesh systems. Comp Struct 18:719–731

48. Lancaster P, Salkauskas K (1990) Curve and surface fitting. Academic Press Inc

49. Lancaster P, Salkauskas K (1981) Surfaces generated by moving least-squares method. Math Comp 155(37):141–158

50. Lau PC (1979) Curvilinear finite difference method for biharmonic equation. Int J Num Meth Eng 14:791–812

51. Lau PC (1979) Curvilinear finite difference methods for tree-dimensional potential problems. J Comput Phys 32:325–344

52. Lau PC, Brebbia CA (1978) The cell collocation method in continuum mechanics, Int J Mech Sci 20

53. Leżański P, Orkisz J, Przybylski P (1997) Mesh generation for adaptive multigrid FDM and FEM analysis. Proc of 13th Polish Conf on Comp Meth in Mechanics, Poznań, Poland, pp 743–750

54. Liszka T (1977) Finite difference at arbitrary irregular meshes and advantages of its use in problems of mechanics. PhD Thesis. Cracow University of Technology, Cracow, Poland

55. Liszka T (1984) An interpolation method for an irregular net of nodes. Int J Num Meth Eng 20:1599–1612

56. Liszka T (1981) An automatic generation of irregular grids in two-dimensional analysis (in Polish). Mechanika i Komputer 4: 181–186

57. Liszka T (1979) Program of irregular mesh generation for the finite difference method (in Polish). Mechanika i Komputer 2:219–277

58. Liszka T, Orkisz J (1983) On an iterative, always convergent method of solution of simultanous linear algebraic equations (in Polish). Mechanika i Komputer 5:131–142

59. Liszka T, Orkisz J (1983) Solution of nonlinear problems of mechanics by the finite difference method at arbitrary meshes (in Polish). Mechanika i Komputer 5:117–130

60. Liszka T, Orkisz J (1980) The finite difference method for arbitrary irregular meshes – a variational approach to applied mechanics problems. 2nd Int Congress on Num Meth Eng, Paris, pp 277–288

61. Liszka T, Orkisz J (1976) Modified finite difference methods at arbitrary irregular meshes and its application in applied mechanics. Proc of the 18th Polish Conf On Mechanics of Solid, Wisła, Poland

62. Liszka T, Orkisz J (1977) Finite difference methods of arbitrary irregular meshes in non-linear problems of applied mechanics. 4th Int Conference on Structural Mechanics in Reactor Technology, San Francisco, California

63. Liszka T, Orkisz J (1980) The finite difference method at arbitrary irregular grids and its applications in applied mechanics. Comp Struct 11:83–95

64. Liszka T, Orkisz J, Tworzydło W (1981) Finite deformations of membranes by the finite difference methods (in Polish). Arch Inż Ląd 26(1):37–49

65. Liszka T, Tworzydło W (1997) An h-p Adaptive cloud method for a class of second order PDE's. Fifth Pan American Congress of Applied Mechanics – CAM, San Juan, Puerto Rico

66. Liu WK, Jun S, Zhang YF (1995) Reproducing kernel particle methods. Int J Numer Methods Eng 20:1081–1106

67. Lucy LB (1977) A numerical approach to the testing of the fission hypothesis. The Astron J 8(12):1013–1024

68. Marczuk GI (1983) Numerical analysis of mathematical physics problems (in Polish). Warsaw, PWN

69. Marczuk GI, Sajdurov VV (1979) Improved precision of the finite difference schemas (in Russian). Moscow, Nauka

70. MacNeal RH (1953) An asymetrical finite difference network. Quart Appl Math 11:295–310

71. Melenk JM, Babuška I (1996) The partition of unity finite element method: Basic theory and applications. Comp Meth Appl Mech Engrg 139:289–314

72. Monaghan JJ (1982) Why particle methods work. SIAM J Sci Stat Comput 3(4):422

73. Mullord P (1979) A general mesh finite difference method using combined nodal and element interpolation. Appl Math Modeling 3:433–440

74. Nay RA, Utku S (1973) An alternative for the finite element method. Variat Meth Eng 3:62–74

75. Nayroles B, Touzot G, Villon P (1992) Generalizing the finite element method: diffuse approximation and diffuse elements. Computational Mechanics 10:307–318

76. Oñate E, Idelsohn SR (1992) A comparison between finite element and finite volume methods in CFD. In: Hirsch C et al (eds) Computational fluid dynamics'92, Elsevier Science Publishers BV, vol 1, pp 93–100

77. Oñate E, Idelsohn S, Zienkiewicz OC, Taylor RL (1996) A finite point method in computational mechanics. Applications to convective transport and fluid flow. Int J for Num Meth in Eng 39:3839–3866

78. Oñate E, Idelsohn S, Zienkiewicz OC, Taylor RL, Sacco C (1996) A stabilized finite point method for analysis of fluid mechanics problems. Comp Meth Appl Mech Engrg 139: 315–346

79. Orkisz J (1995) Adaptive analysis of b.v. problems by the finite difference method at arbitrary irregular meshes – concept and formulation. 12th Symp on the Unifaction of Analytical Comp and Experimental Solution Methodologies, Worcester-Danvers, Mass, USA

80. Orkisz J (1995) Finite difference method (in Polish). In: Kleiber M (ed) Computer methods in solid mechanics, PWN, Warsaw

81. Orkisz J (1997) Adaptive approach to the finite difference method for arbitrary irregular grids. Interdisciplinary Symp on Advances in Comp Mech, Univ of Texas, Austin

82. Orkisz J (1979) Computer approach to the of finite difference method (in Polish). Mechanika i Komputer 2:7–69

83. Orkisz J et al (1989) Numerical analysis method. The finite difference method. In: Fatigue Design Handbook, Publ Soc Automotive Engineers, Warrendale, USA, pp 192–203

84. Orkisz J, Leżański P, Przybylski P (1997) Multigrid approach to adaptive analysis of boundary-value problems by the meshless GFDM. IUTAM/IACM Symp on Discretization Meth in Struct Mechanics II, Vienna

85. Orkisz J, Pazdanowski M (1995) Analysis of residual stresses by the generalised finite difference method. Proc of the Fourth Int Conf Computational Plasticity, Barcelona, pp 189–200

86. Pavlin V, Perrone N (1975) Finite difference energy techniques for arbitrary meshes. Comp Struct 5:45–58

87. Pazdanowski M (1994) Application of the generalized FDM to analysis of residual stresses in bodies under cyclic loading (in Polish). PhD Thesis. Cracow Univ of Technology, Cracow, Poland

88. Perrone N, Kao R (1975) A general finite difference method for arbitrary meshes. Comp Struct 5:45–58

89. Preparata FP, Shamos MI (1985) Computational geometry. An introduction. New York, Springer Verlag

90. Przybylski P (1995) The distributed algorithm of a unstructural triangular mesh generator – object-oriented approach. Proc of 12th Polish Conf on Comp Meth in Mechanics, Warszawa-Zegrze, Poland:

91. Richtmyer RD (1957) Difference methods for initial-value problems. New York, Wiley

92. Richtmyer RD, Morton KW (1967) Difference methods for boundary-value problems. New York, Wiley

93. Samarskij AA (1974) Introduction into theory of finite difference schemes (in Russian). Moscow, Nauka

94. Samarskij AA (1977) Theory of finite difference schemes (in Russian). Moscow, Nauka

95. Samarskij AA, Andreev VB (1976) Finite difference methods for elliptic equations (in Russian). Moscow, Nauka

96. Samarskij AA, Gulin AV (1973) Stability of finite difference schemes (in Russian). Moscow, Nauka

97. Samarskij AA, Popov JP (1975) Finite difference schemes of gas dynamics (in Russian). Moscow, Nauka

98. Shepard D (1968) A two dimensional interpolation function for irregularly spaced points. In: Proc ACM National Conference, pp 517–524

99. Snell C, Vesey DG, Mullord P (1981) The application of a general finite difference method to some boundary value problems. Comp Struct 13:547–552

100. Sulsky D, Chen Z, Schreyer HL (1994) A Particle methods for history-dependent materials. Comp Meth Appl Mech Engng 118: 179–196

101. Syczewski M, Tribiłło R (1981) Singularities of sets used in the mesh method. Comp Struct 14:509–511

102. Syczewski M, Tribiłło R (1984) Division of an area in the analysis of surface plates by the mesh method. Comp Struct 18:813–818

103. Thacker WC (1980) A brief review of techniques for generating irregular computational grids. Int J Num Meth Eng 15:1335–1341

104. Tseng AA, Gu SX (1989) A finite difference scheme with arbitrary mesh for solving high-order partial differential equations. Comp Struct 31:319–328

105. Tworzydło W (1987) Analysis of large deformation of membrane shells by the generalized finite difference method. Comp Struct 27:39–59

106. Tworzydło W (1989) The FDM in arbitrary curvilinear coordinates formulation. Numerical approach and application. Int J Num Meth Eng 28:261–277

107. Varga RS (1962) Matrix iterative analysis. Prentice-Hall, London

108. Wen-Hua-Chu (1971) Development of a general finite difference approximation for a general domain. Part I: Machine transformation. J Comput Phys 8:392–408

109. Wyatt MJ, Davies G, Snell C (1976) Truncation error control in generalized finite element method. J Engrg Mech Div, Proc ASCE, EM4:736–741

110. Wyatt MJ, Davies G, Snell C (1975) A New Difference Based Finite Element Method. Instn Engineers, 59(2):395–409

111. Zienkiewicz OC, Oñate E (1991) Finite elements versus finite volumes. Is there a choice? In: Wriggers P, Wagner W (eds) Non-Linear Computational Mechanics. State of the Art. Springer, Berlin

112. Zienkiewicz OC, Zhu JZ (1987) A simple error estimator and adaptive procedure for practical engineering Analysis. Int J Num Meth Eng 24:337–357

Part IV

Boundary Element Method

1 Introduction

The origin of the Boundary Element Method (BEM) can be traced back to the classical analysis of the nineteenth century. In this period a possibility of transforming a certain class of boundary problems into equivalent systems of integral equations was noted in some publications. Quite often, purely mathematical works were inspired by physical problems, in particular by problems of motion in the gravitation fields of Newton forces, and by boundary value problems of elasticity. As a result a broad branch of the theory of elliptic differential equations is called a potential theory in the classical analysis.

The interest in the potential theory increased in particular when Fredholm's fundamental works were published. Some valuable theoretical results were obtained. A possibility of constructing numerical solutions using a combination of the method of potentials with the finite difference method or some other approximate method was pointed out. A number of valuable applications of the potential theory in elasticity were presented by Kupradze and his coworkers [32]. Initially, the method of boundary integral equations, or, in general, the potential theory, met with little interest. The reason was that there was still no effective tool, which could solve large systems of algebraic equations produced by a finite-dimensional approximation of problems formulated in the form of integral equations.

The situation changed completely when fast and easily available computers appeared. The method of boundary integral equations started a fast development under the name of the Boundary Element Method (BEM), which is associated with the emphasis placed on the approximation method in the numerical solution.

The method was popularized as an effective tool of engineering calculations competing with the Finite Element Method (FEM). It turned out that BEM is in certain cases more efficient and accurate than FEM.

From among the books and monographs covering the foundations of BEM and its application to various issues of mechanics [1, 2, 3, 5] are worth mentioning. Some fundamental theoretical papers about BEM have also appeared. In these papers one can find some methods of estimating the accuracy and convergence of BEM, as well as a theorem of the equivalence of its variational formulation and the method of boundary integral equations (cf. [11]).

It must be emphasized that the Boundary Element Method should be treated as a method of finite-dimensional approximation of problems formulated in the language of boundary integral equations. The advantages of this formulation, especially when the influence of the boundary shape and boundary conditions on the solution is investigated, caused a fast BEM development. It still finds new applications both as a method of construction of numerical solutions and in theoretical research. Moreover, the method is conceptually simple and can be successfully combined with other methods (e.g. FEM). It seems that this fact and the existence of a growing number of well-tested and efficient programs justifies the interest in the method.

The essence of the method can be explained by an example. It seems that in the case of a contribution dedicated to the readers acquainted with mechanics, and especially continuum mechanics, an example which makes use of well-known concepts is the reciprocal work theorem. The theorem states a relation between the work done by the forces of system 1 on displacements caused by the forces of system 2, denoted by $\mathcal{L}_{12}$, and the work done by the forces of system 2 on displacements caused by the forces of system 1, denoted by $\mathcal{L}_{21}$.

Assuming linearized kinematics (i.e. small displacements and their gradients) it can be shown that for an arbitrary stable linearly-elastic body the following equality holds:

$$\mathcal{L}_{12} = \mathcal{L}_{21} \, . \tag{1.1}$$

Let us consider a prismatic beam (Fig. 1.1) with bending stiffness $EJ = 1$ and length $2l$, under loading $b(x)$, which is normal to the beam axis. This loading will be treated as the forces of system 2 (or state 2). The forces

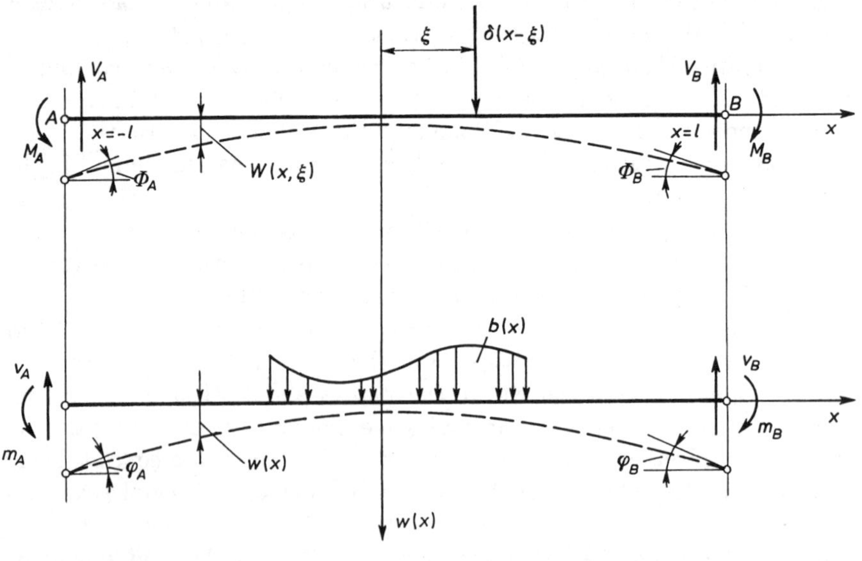

Fig. 1.1. Beam problem to illustrate the boundary element method

of system 1 (or state 1) will be represented by a concentrated force $b = 1$ applied at an arbitrary point $x = \xi$. Let us denote for simplicity

$$\Omega = \{x : x \in < -l; l >\} . \tag{1.2}$$

As is well-known, the deformed axis of the prismatic beam satisfies the equation[1]:

$$\frac{\mathrm{d}^4}{\mathrm{d}x^4}[w(x)] = \frac{b(x)}{EJ} \qquad \forall x \in \mathrm{Int}\Omega , \tag{1.3}$$

where $EJ = 1$, and the boundary conditions:

$$\begin{aligned} & w|_{x=-l} = w_A, \qquad w|_{x=l} = w_B, \\ & \left.\frac{\mathrm{d}w^2}{\mathrm{d}x^2}\right|_{x=-l} = m_A, \qquad \left.\frac{\mathrm{d}w^2}{\mathrm{d}x^2}\right|_{x=l} = m_B, \end{aligned} \tag{1.3a}$$

or

$$\begin{aligned} & \left.\frac{\mathrm{d}w}{\mathrm{d}x}\right|_{x=-l} = \varphi_A, \qquad \left.\frac{\mathrm{d}w}{\mathrm{d}x}\right|_{x=l} = \varphi_B, \\ & \left.\frac{\mathrm{d}w^3}{\mathrm{d}x^3}\right|_{x=-l} = v_A, \qquad \left.\frac{\mathrm{d}w^3}{\mathrm{d}x^3}\right|_{x=l} = v_B. \end{aligned} \tag{1.3b}$$

Denoting the solution of the equation by $W(x, \xi)$:

$$\frac{\mathrm{d}^4}{\mathrm{d}x^4}[W(x, \xi)] = \delta(x - \xi) , \qquad x, \xi \in (-\infty, \infty) , \tag{1.4}$$

where $\delta(x - \xi)$ is the so-called *Dirac delta function,* the work done by the forces of system 2 on the displacements caused by the forces of system 1 is given by the integral

$$\mathcal{L}_{21} = \int_\Omega b(\xi) W(x, \xi)\, \mathrm{d}\xi . \tag{1.5}$$

Integrating by parts we obtain:

$$\begin{aligned} & \int_\Omega b(\xi) W(x, \xi)\, \mathrm{d}\xi - v_A W_A + v_B W_B + m_A \left.\frac{\mathrm{d}W}{\mathrm{d}\xi}\right|_{\xi=-l} - m_B \left.\frac{\mathrm{d}W}{\mathrm{d}\xi}\right|_{\xi=l} = \\ & \quad = \int_\Omega \delta(x - \xi) w(\xi)\, \mathrm{d}\xi - \left.\frac{\mathrm{d}W^3}{\mathrm{d}\xi^3}\right|_{x=-l} w_A + \left.\frac{\mathrm{d}W^3}{\mathrm{d}\xi^3}\right|_{x=l} w_B \\ & \quad + \left.\frac{\mathrm{d}W^2}{\mathrm{d}\xi^2}\right|_{x=-l} \varphi_A - \left.\frac{\mathrm{d}W^2}{\mathrm{d}\xi^2}\right|_{x=l} \varphi_B . \end{aligned} \tag{1.6}$$

[1] Symbol $\forall$ in equation (1.3) denotes a general quantifier; read *"for each ... "*.

With the notation

$$\begin{aligned}\frac{\mathrm{d}W}{\mathrm{d}\xi} &= \Phi(\xi, x) - \text{rotation angle},\\ \frac{\mathrm{d}^3W}{\mathrm{d}\xi^3} &= V(\xi, x) - \text{shear force},\\ \frac{\mathrm{d}^2W}{\mathrm{d}\xi^2} &= M(\xi, x) - \text{bending moment},\end{aligned} \tag{1.7}$$

it is easy to see that we obtained the equation:

$$\begin{aligned}\mathcal{L}_{21} &= \int_\Omega b(\xi)W(x, \xi)\,\mathrm{d}\xi - v_A W_A + v_B W_B + m_A\Phi_A - m_B\Phi_B =\\ &= w - V_A w_A + V_B w_B + M_A\varphi_A - M_B\varphi_B = \mathcal{L}_{12}\,,\end{aligned} \tag{1.8}$$

which is the Betti's theorem for the prismatic beam (cf. Fig. 1.1). We notice that it is at the same time an integral formula which allows us to find the deflection of the beam under the loading $b(x)$, but under the condition that in addition to the displacement boundary conditions, viz.:

$$\begin{aligned} w|_{x=-l} &= w_A, & \left.\frac{\mathrm{d}w}{\mathrm{d}x}\right|_{x=-l} &= \varphi_A,\\ w|_{x=l} &= w_B, & \left.\frac{\mathrm{d}w}{\mathrm{d}x}\right|_{x=l} &= \varphi_B,\end{aligned} \tag{1.9}$$

we know the boundary conditions constraining the generalized section forces (so-called associated boundary conditions):

$$\begin{aligned} \left.\frac{\mathrm{d}w^2}{\mathrm{d}x^2}\right|_{x=-l} &= m_A, & \left.\frac{\mathrm{d}w^3}{\mathrm{d}x^3}\right|_{x=-l} &= v_A,\\ \left.\frac{\mathrm{d}w^2}{\mathrm{d}x^2}\right|_{x=l} &= m_B, & \left.\frac{\mathrm{d}w^3}{\mathrm{d}x^3}\right|_{x=l} &= v_B,\end{aligned} \tag{1.10}$$

and the function $W(x, \xi)$ which is a solution of Eq. (1.4) and is called the *fundamental solution.* The function can easily be found by integration of Eq. (1.4) considering

$$\frac{\mathrm{d}}{\mathrm{d}x}[H(x)] \overset{\text{def}}{=} \delta(x)\,, \tag{1.11}$$

where $H(x)$is the so-called *Heaviside function.* The differentiation above must be understood in the sense of distributions (cf. Chap. 2 of Part II). After integration we find:

$$W(x, \xi) = \frac{1}{12}(x-\xi)^3\,\mathrm{sign}(x-\xi)\,, \tag{1.12}$$

where

$$\mathrm{sign}(x-\xi) = H(x-\xi) - H(-x+\xi)\,. \tag{1.13}$$

Of course, among the eight boundary conditions only four are usually known, e.g. in the form (1.11). The missing boundary values can be found using (1.9). Substituting $x = \pm l$ we obtain two equations

$$w_{A(B)} = \int_\Omega b(\xi) W(l-\xi)\,\mathrm{d}\xi - [v_A W_A - v_B W_B - m_A \Phi_A + m_B \Phi_B]_{x=\pm l} +$$
$$+ [V_A w_A - V_B w_B - M_A \varphi_A + M_B \varphi_B]_{x=\pm l} \tag{1.14}$$

which provide the unknown displacements w_A and w_B. We arrive at the equations that can be used to determine the rotation angles $\varphi_A = \frac{\mathrm{d}w}{\mathrm{d}x}\big|_{x=-l}$ and $\varphi_B = \frac{\mathrm{d}w}{\mathrm{d}x}\big|_{x=l}$, when we perform the differentiation operation on identity (1.9) and substitute $x = \pm l$ or $x = -l + \epsilon$ and $x = (l - \epsilon)$:

$$\varphi_{A(B)} = \int_\Omega b(\xi) \left[\frac{\mathrm{d}W(x,\,\xi)}{\mathrm{d}x}\right]_{x=\pm l} \mathrm{d}\xi - v_A \left.\frac{\mathrm{d}W}{\mathrm{d}x}\right|_{x=-l} + v_B \left.\frac{\mathrm{d}W}{\mathrm{d}x}\right|_{x=l}$$
$$+ m_A \left.\frac{\mathrm{d}^2W}{\mathrm{d}x^2}\right|_{x=-l} - m_B \left.\frac{\mathrm{d}^2W}{\mathrm{d}x^2}\right|_{x=l} + w_A \left.\frac{\mathrm{d}W^4}{\mathrm{d}x^4}\right|_{x=(-l+\epsilon)} \tag{1.15}$$
$$- w_B \left.\frac{\mathrm{d}W^4}{\mathrm{d}x^4}\right|_{x=l-\epsilon} + \varphi_A \left.\frac{\mathrm{d}W^3}{\mathrm{d}x^3}\right|_{x=(-l+\epsilon)} - \varphi_B \left.\frac{\mathrm{d}W^3}{\mathrm{d}x^3}\right|_{x=l-\epsilon} .$$

Attention should be paid to the sixth and seventh component on the right-hand side of Eq. (1.15). Based on the definition of function $W(x,\,\xi)$ it is easy to realize that:

$$\left.\frac{\mathrm{d}^4W}{\mathrm{d}x^4}\right|_{\xi=l} = \frac{\mathrm{d}}{\mathrm{d}x}\left[\left.\frac{\mathrm{d}W^3}{\mathrm{d}x^3}\right|_{\xi=-l}\right] = \frac{\mathrm{d}}{\mathrm{d}x}[V_A] = \delta(x-l)$$

and for $x = l$ the expression loses sense (the local value of distribution is an undefined notion). This is the reason why $x = (-l + \epsilon)$ and $x = l - \epsilon$ was adopted.

Since the boundary of the domain Ω, which is a one-dimensional manifold, reduces to a point (i.e. a zero-dimensional manifold), equations (1.14) and (1.15) are algebraic equations (in a limit sense, i.e. when $\epsilon \to 0$). When the state equation, i.e. a counterpart of Eq. (1.3), is a partial differential equation, and the domain is a manifold in two or more dimensions, counterparts of Eqs. (1.14) and (1.15) are integral equations on boundary manifolds which have a one order lower dimension. For instance in Sect. 3.3.1 integral equations (3.122) and (3.124) are analogues of Eqs. (1.14) and (1.15), defined on a one-dimensional boundary manifold.

The whole Part IV of this volume is devoted to the construction of counterparts of formula (1.8) and so-called *boundary integral equations*, i.e. counterparts of Eqs. (1.14) and (1.15), for various types of operators encountered in continuum mechanics. It is assumed that the domain Ω in which the operators are defined, i.e. the body under consideration, is an arbitrary manifold with the dimension not larger than 3. The presented material is arranged in such a way that the reader goes from the simplest applications (i.e. problems described by self-adjoint elliptic operators) to more difficult problems related to some formal operations that allow us to generalize the method. Starting from the application of the method in the analysis of linear elastic systems and going through some nonlinear problems, the method is also shown to be effective in synthesis problems.

However, to help the readers who would like to get acquainted with a particular application of the method, the presentation is arranged in such a way that each chapter, and even each section, forms as coherent and complete unit as possible.

2 Mathematical Foundations

2.1 Formal Outline of the Method

The formal foundations of the method presented in this chapter have been formulated using a language which is slightly different from the one commonly used, but close to the one of applied mathematics.

Let A be a symmetric and positively defined operator in a Hilbert space, i.e. A is such that:

1) $A : V \to V, \qquad V$ – Hilbert space
2) $(Au,\, v) = (u,\, Av) \qquad \forall\, u,\, v \in V\,,$
3) $(Au,\, u) \geq a^2 \|u\|^2, \qquad a > 0\,.$

The closure of its domain dom $A \subset V$ in a metric induced by the scalar product:

$$[u,\, v] \overset{\text{def}}{=} (Au,\, v) \qquad \forall\, u,\, v \in \operatorname{dom} A$$

is a Hilbert space V_A.

Definition 2.1. The *generalized solution* of the equation

$$Au = f\,; \qquad f \in V\,; \qquad A : V_A \to V\,,$$

is such $u \in V_A$ that

$$[u,\, v] \equiv (f,\, v) \qquad \forall\, v \in V_A\,.$$

It can be shown that such a solution exists and is unique [35].

Let A in particular be an operator satisfying the conditions 1), 2), 3), and let δ denote the so-called Dirac delta function (or δ – *distribution*).

Definition 2.2. The element $\mathbf{u} \in V_A$ such that:

$$\mathbf{u} : \; A\mathbf{u} \equiv \delta\,,$$

is called a *fundamental* (or basic) solution. Since $\mathbf{u}$ is a generalized solution, according to the definition 2.1:

$$[\mathbf{u},\, v] \equiv [\delta,\, v] \qquad \forall\, v \in V_A\,.$$

Let $u = \mathbf{u} * f$ be a generalized function (the symbol $*$ denotes a convolution of distributions). It is a solution of the equation:

$$Au = f, \quad A : V_A \to V_A, \tag{2.1}$$

since

$$[u, v] = [\mathbf{u} * f, v] = (A(\mathbf{u} * f), v) = (A\mathbf{u} * f, v) = (\delta * f, v) = (f, v),$$

so, if the fundamental solution is known, the solution of Eq. (2.1) can be written in the form of a convolution.

Let us consider the following boundary value problem:

$$\begin{aligned} Au &= f & \forall x &\in \Omega, \\ B_{S_i}u &= g_i & \forall x &\in \partial\Omega_{B_i}, \\ C_{S_i}u &= h_i & \forall x &\in \partial\Omega_{C_i}, \end{aligned} \tag{2.2}$$

where supp $f = \Omega$,

$$\bigcup_i [\partial\Omega_{B_i} \cup \partial\Omega_{C_i}] = \partial\Omega,$$

$$\bigcap_i [\partial\Omega_{B_i} \cap \partial\Omega_{C_i}] \bigcup [\bigcap_i \partial\Omega_{B_i}] \bigcup [\bigcap_i \partial\Omega_{C_i}] = \emptyset.$$

The function

$$u_1 = \mathbf{u} * f \tag{2.3}$$

is not a solution of problem (2.2), since, in general, it does not satisfy the boundary conditions, which can be shown as follows:

$$\begin{aligned} B_{S_i}u_1 &= B_{S_i}(\mathbf{u} * f) = B_{S_i}\mathbf{u} * f \neq g_i & \forall x &\in \partial\Omega_{B_i}, \\ C_{S_i}u_1 &= C_{S_i}(\mathbf{u} * f) = C_{S_i}\mathbf{u} * f \neq h_i & \forall x &\in \partial\Omega_{C_i}. \end{aligned}$$

A solution of the considered problem could, however, be built using the fundamental solution. For this purpose we apply integration by parts to the product $[u, v]$:

$$\begin{aligned} [u, v] = \int_\Omega v\, Au \,\mathrm{d}\Omega &= \sum_{i=0}^{k} \int_{\partial\Omega} B_{S_i}uC_{S_{k-i}}v \,\mathrm{d}\Gamma - \int_\Omega \mathrm{D}u\,\mathrm{D}v\,\mathrm{d}\Omega, \\ [v, u] = \int_\Omega u\, Av \,\mathrm{d}\Omega &= \sum_{i=0}^{k} \int_{\partial\Omega} B_{S_{k-i}}vC_{S_i}u \,\mathrm{d}\Gamma - \int_\Omega \mathrm{D}v\,\mathrm{D}u\,\mathrm{d}\Omega, \end{aligned} \tag{2.4}$$

where $k = m - 1$ and $m \in \boldsymbol{N}$.

Subtracting equations (2.4) side by side we obtain a generalization of the so-called *Green formula:*

$$\int_\Omega [v\, Au - u\, Av] \,\mathrm{d}\Omega = \sum_{i=0} \int_{\partial\Omega} [B_{S_i}uC_{S_{k-i}}v - b_{S_{k-i}}vC_{S_i}u] \,\mathrm{d}\Gamma, \tag{2.5}$$

and substituting $v = \mathbf{u}$:

$$\begin{aligned} u &= \int_\Omega u\, A\, \mathbf{u}\, \mathrm{d}\Omega \;=\; \int_\Omega u\, \delta\, \mathrm{d}\Omega \\ &= \int_\Omega \mathbf{u} f\, \mathrm{d}\Omega - \sum_{i=0} [B_{S_i} u C_{S_{k-i}} \mathbf{u} \;-\; b_{S_{k-i}} \mathbf{u} C_{S_i} u]\, \mathrm{d}\Gamma\,. \end{aligned} \tag{2.6}$$

In this way the solution of (2.2) has been expressed in terms of the right-hand side of the equation and the boundary conditions. However, it is not an effective solution, since on the same boundary segment only one type of boundary conditions can be prescribed. On the segment $\partial\Omega_{B_i}$ the function g_i is given, but $C_{S_i} u$ is unknown. The unknown, *associated* boundary conditions can be determined from certain equations. Before presenting these equations we write formula (2.6) in a concise and convenient form of a convolution. For this purpose we represent the boundary operator in the following form:

$$\begin{aligned} B_{S_i} &= B_i \delta_\sigma \\ C_{S_i} &= C_i \delta_\sigma\,;\; \delta_\sigma = \delta(s-\sigma)\,;\; s, \sigma \in \partial\Omega \end{aligned} \tag{2.7a}$$

and define an operator:

$$\overline{A} \overset{\text{def}}{=} \sum_{i=0}^{k} [B_i * \delta_\sigma \ldots C_{k-i} * \delta_\sigma u \;-\; B_{k-i} * \delta_\sigma u C_i * \delta_\sigma \ldots]\, \delta_s\,. \tag{2.7b}$$

Considering the fact that

$$g * \delta_s \;=\; \int_\Omega g(x)\, \delta(x-s)\, \mathrm{d}\Omega \;=\; \int_{\partial\Omega} g(s)\, \mathrm{d}\Gamma_s\,,$$

we can write formula (2.6) in the form of a convolution:

$$u * \delta = u = \mathbf{u} * [A - \overline{A}]\, u = \begin{cases} u(x) & \forall\, x \in \Omega \\ \alpha u(x) & \forall\, x \in \partial\Omega \\ 0 & \forall\, x \notin \Omega \end{cases} \tag{2.8}$$

The value of $\alpha \in (0, 1)$ depends on the regularity of the boundary and the form of the operator, cf. Eq. (2.29) and Sect. 2.6.

The equations necessary to determine the associated boundary conditions are obtained when the operator B_i is applied to the solution (2.8):

$$\alpha B_i u \;=\; u * B_i \delta \;=\; B_i \mathbf{u} * [A - \overline{A}] u\,,\;\; i \in (0, \ldots, m-1),\;\; \alpha \in (0,1). \tag{2.9}$$

This equation is defined on the boundary $\partial\Omega$ of the domain Ω.

The outlined formalism is a complete idea of the boundary integral equations. As presented, the method consists in replacing problem (2.2) by convolution equation (2.9) called the *boundary integral equation.*

The ability to solve the latter equation decides about the usefulness and effectiveness of the method. The approximate manner of solving Eq. (2.8) is called the Boundary Element Method. The distinction between the two notions: Method of Boundary Integral Equations (MBIE) and Boundary Element Method (BEM) is essential and will be complied with in this presentation.

One of the important advantages of the MBIE can be shown on the following simple example. Assume we are interested in the function

$$w = Ku, \tag{2.10}$$

where u is a solution of problem (2.2) and K a certain linear operator. Using Eq. (2.6) we find:

$$Ku = K[\mathfrak{u} * [A - \overline{A}]u = K\mathfrak{u} * [A - \overline{A}]u = \mathfrak{w} * [A - \overline{A}]u \tag{2.11}$$

and an appropriate boundary equation

$$\alpha B_i w = B_i \mathfrak{w} * [A - \overline{A}]u; \quad \mathfrak{w} = K\mathfrak{u}; \qquad \alpha \in (0,1). \tag{2.12}$$

Equation (2.12) can be solved using the BEM, i.e. with the same accuracy as the one with which problem (2.2) can be solved. Unlike the Finite Element Method, there is no necessity of a double approximation, i.e. an approximation of the solution u and the operator K, cf. Sect. 3.1.2, Eqs. (3.41) and (3.42).

2.2 Elliptic Equations of the Second Order

The application of the MBIE to particular types of differential operators consists in the presentation of an extended form of integral equations and a fundamental solution or a method of its construction. The extended version of the boundary equations is an immediate consequence of the Green formula. It turns out that the notion of a self-adjoint operator, which is narrower from the notion of a symmetric operator, is more convenient for the formulation. Therefore, we recapitulate the following definitions:

Definition 2.3. Let U and V be Banach spaces and let U^* and V^* be respective dual spaces. Let A be a continuous linear operator $U \to V$. The operator[1] $A^* : \ \forall g \in V^* \ \exists A^* : g(A(u)) = (A^*g)(u)$ is called *adjoint* to A. The adjoint operator is therefore such that $A^* : U^* \to V^*$. It can be shown that $\|A\| = \|A^*\|$.

For a Hilbert space the adjoint operator is defined as:

$$A^* : (Au, v) = (u, A^*v), \qquad u \in \operatorname{dom} A, \qquad v \in \operatorname{dom} A^*. \tag{2.13}$$

Definition 2.4. An operator is called *self-adjoint* when $A = A^*$.

[1] Symbol $\exists$ denotes a particular quantifier; read *"there exists ... such that ... "*.

In a particular case of second order elliptic operators the following notation is used:

$$M \stackrel{\text{def}}{=} \sum_{\alpha,\beta=1}^{3} \frac{\partial}{\partial x_\alpha}\left[a_{\alpha\beta}\frac{\partial}{\partial x_\beta}\right] + \sum_{\alpha=1}^{3} e_\alpha \frac{\partial}{\partial x_\alpha} + c\,, \tag{2.14}$$

where

$$e_\alpha \stackrel{\text{def}}{=} b_\alpha - \sum_{\alpha,\beta=1}^{3} \frac{\partial a_{\alpha\beta}}{\partial x_\beta}$$

and $a_{\alpha\beta}$, b_α, c are certain $C^1(\Omega)$ functions. Here $\Omega = \overline{\Omega}$ is a closed subset of space $\boldsymbol{E}^3$ with the boundary $\partial\Omega = \Gamma$ which satisfies the Lyapunov conditions (cf. Supplement 2.6). An operator adjoint to operator M reads:

$$N \stackrel{\text{def}}{=} \sum_{\alpha,\beta=1}^{3} \frac{\partial}{\partial x_\alpha}\left[a_{\alpha\beta}\frac{\partial}{\partial x_\beta}\right] - \sum_{\alpha=1}^{3} \frac{\partial}{\partial x_\alpha}[e_\alpha] + c\,, \tag{2.15}$$

which can be verified basing on definition (2.13).

It is convenient to use the following operator definitions:

$$\begin{aligned} P &\stackrel{\text{def}}{=} \sum_{\alpha,\beta=1}^{3} \left[n_\beta a_{\alpha\beta}\frac{\partial}{\partial x_\alpha}\right] + \gamma\,, \\ Q &\stackrel{\text{def}}{=} \sum_{\alpha,\beta=1}^{3} \left[n_\beta a_{\alpha\beta}\frac{\partial}{\partial x_\alpha}\right] + (\gamma - b)\,, \end{aligned} \tag{2.16}$$

where γ is an arbitrary function and

$$b = \sum_{\alpha=1}^{3} e_\alpha n_\alpha\,,$$

with n_α denoting the coordinates of a unit vector which is outer normal to the boundary $\partial\Omega$, i.e. $n_\alpha = \cos(\mathbf{n},\, x_\alpha)$.

With the introduced notation the following identities can be derived using integration by parts (or the Green theorem):

$$\begin{aligned} \int_\Omega vMu\,\mathrm{d}\Omega &= \int_\Gamma vPu\,\mathrm{d}\Gamma\,; \qquad \Gamma = \partial\Omega\,, \\ \int_\Omega uNv\,\mathrm{d}\Omega &= \int_\Gamma uQv\,\mathrm{d}\Gamma\,, \end{aligned} \tag{2.17}$$

and the generalized Green formula:

$$\int_\Omega [vMu - uNv]\,\mathrm{d}\Omega = \int_\Gamma [vPu - uQv]\,\mathrm{d}\Gamma\,. \tag{2.18}$$

In a particular case where

$$b_\alpha = \sum_{\beta=1}^{3} \frac{\partial a_\alpha \beta}{\partial x_\beta}, \tag{2.19}$$

the operator M becomes self-adjoint (i.e. $M = N$), for which the Green formula can be written in the form:

$$\int_\Omega [vMu - uMv]\,\mathrm{d}\Omega = \int_\Gamma [vPu - uPv]\,\mathrm{d}\Gamma\,. \tag{2.20}$$

Assuming $v = \mathbf{u}$, where $\mathbf{u}$ is a fundamental solution for the operator M, i.e.

$$M\mathbf{u} = \delta\,,$$

and

$$Mu = f \quad \forall\, x \in \operatorname{Int}\Omega\,,$$

it is possible to obtain, basing on Eq. (2.18), an integral formula for the solution of the inhomogeneous equation:

$$u = \int_\Omega \mathbf{u} f\,\mathrm{d}\Omega + \int_\Gamma [uP\mathbf{u} - \mathbf{u}\,Pu]\,\mathrm{d}\Gamma\,. \tag{2.21}$$

However, for this formula we need to know both the value of the solution on the boundary of the domain Ω and the value that operator P acting on function u assumes on the boundary, so:

$$\begin{aligned} u &= k \quad \forall\, x \in \Gamma \\ Pu &= h\,. \end{aligned} \tag{2.22}$$

In a properly formulated problem either the function k is known, i.e. we have the *Dirichlet type boundary condition,* or the function h is given, i.e. the *Neumann boundary condition* is specified. The so-called *mixed boundary condition* presents the last possibility:

$$\begin{gathered} u|_{\Gamma_1} = k\,, \\ Pu|_{\Gamma_2} - h\,, \\ \Gamma_1 \cap \Gamma_2 = \emptyset \wedge \Gamma_1 \cup \Gamma_2 = \Gamma\,. \end{gathered} \tag{2.23}$$

Briefly speaking, on a certain segment of the boundary only one of the functions k or h is known. Therefore, in order to use formula (2.22) the value taken by the operator P acting on function u on boundary Γ_1, as well as the value of the function u on boundary Γ_2 must be determined.

The equations necessary to achieve this goal can be derived by performing a limit transition:

$$\operatorname{Int}\Omega \ni x \to \partial\Omega = \Gamma\,, \tag{2.24}$$

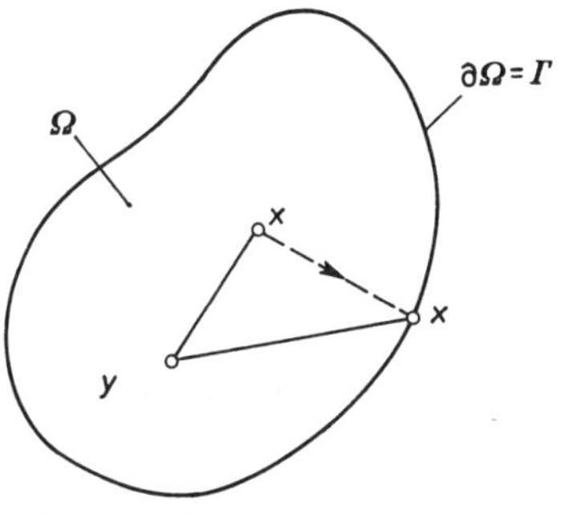

Fig. 2.1. Limit transition of point x from the inside of domain Ω to its boundary $\partial\Omega$

i.e. examining the behaviour of the integrals

$$\int_{\Gamma_1} P\mathrm{u}\,\mathrm{d}\Gamma = \int_{\Gamma}\left[\sum_{\alpha,\beta=1}^{3} n_\beta a_{\alpha\beta}\frac{\partial}{\partial x_\alpha} + \gamma\right]u\,\mathrm{d}\Gamma \quad \text{and} \quad \int_{\Gamma_2}\mathrm{u}\,\mathrm{d}\Gamma \tag{2.25}$$

while crossing the boundary Γ of domain Ω.

For the considered class of second order elliptic equations the fundamental solution has the following general form (depending on the space dimension):

$$\mathrm{u}(x,y) = \begin{cases} \dfrac{1}{(n-2)\sigma[A(y)]^{\frac{1}{2}}}\left[\sum_{\alpha,\beta=1}^{3} a_{\alpha\beta}^{-1}(y)\,(x_\alpha - y_\alpha)(x_\beta - y_\beta)\right]^{\frac{-1}{2(n-2)}}; \\ \qquad n>0 \wedge n\neq 2, \\ \dfrac{1}{2\pi\,[A(y)]^{\frac{1}{2}}}\ln\left[\sum_{\alpha,\beta=1}^{3} a_{\alpha\beta}^{-1}(y)\,(x_\alpha - y_\alpha)(x_\beta - y_\beta)\right]^{\frac{-1}{2}}; \\ \qquad n=2, \end{cases}$$

with the following notation: n – dimension of the $\boldsymbol{E}^n$ space; σ – area of a hyper-sphere in $\boldsymbol{E}^n$; $\{x_\alpha\}$, $\{y_\alpha\}$ – coordinates of a pair of points x, y; $A(y) = \det[a_{\alpha\beta}]$; $a_{\alpha\beta}^{-1}$ – reciprocals of the equation components.

The fundamental solution is a weakly singular function of the pair of points, which for all $x \neq y$ satisfies the equation:

$$\sum_{\alpha,\beta=1}^{3} a_{\alpha\beta}\frac{\partial^2 u}{\partial x_\alpha \partial x_\beta} = 0 \quad \forall\, x,\, y \in \boldsymbol{E}^n\,. \tag{2.26}$$

In the potential theory the following definitions have been accepted and are still in use:

– the function

$$u(x) = \int_{\Omega}\mathrm{u}(x,\,y)f\,(y)\,\mathrm{d}\Omega_y\,; \qquad x,\,y,\in \operatorname{Int}\Omega$$

is called a *volume potential* with density $f(y)$, where dom $f = \Omega$;

– the function

$$v(x) = \int_\Gamma \mathbf{u}(x, y)\, \mu(y)\, \mathrm{d}\Gamma_y; \qquad x \in \operatorname{Int}\Omega$$

is called a *single layer potential* with density $\mu(y)$, where $y \in \Gamma = \partial\Omega$ and $\operatorname{dom}\mu = \Gamma$ and

– the function

$$w(x) = \int_\Gamma P\mathbf{u}(x, y)\nu(y)\, \mathrm{d}\Gamma_y$$

is called a *double layer potential* with density $\nu(y)$, where $y \in \Gamma = \partial\Omega$ and $\operatorname{dom}\nu = \Gamma$ (in the classical potential theory the name is used for $P = \partial/\partial\mathbf{n}$, where $\mathbf{n}$ is the outer normal).

Owing to the properties of the fundamental solution, the defined potentials are singular integrals. The rank of singularity guarantees the existence of each of the potentials in the sense of the Cauchy principal value. A procedure of finding the principal integrals is described in detail in [21, 39] and requires an examination of the behaviour of the potentials on a hyper-sphere with a radius $\rho \to 0$. It can be shown that for $n = 3$ the single layer potential preserves continuity in the whole $\boldsymbol{E}^3$ space, while its normal derivative exhibits a jump on the boundary Γ of domain Ω:

$$\left[\frac{\partial v}{\partial \mathbf{n}}\right]^{\pm} = \pm\frac{\mu}{2a} + \int_\Gamma \frac{\partial \mathbf{u}}{\partial \mathbf{n}}\mu(y)\, \mathrm{d}\Gamma_y\,, \tag{2.27}$$

where $\mathbf{n} = \{n_\alpha\}$ is a vector normal to the boundary Γ and

$$a = \sum_{\alpha,\beta=1}^{3} a_{\alpha\beta} n_\alpha n_\beta\,. \tag{2.27a}$$

The sign $(+)$ corresponds to the inner boundary and the sign $(-)$ to the outer one. The double layer potential also exhibits a jump:

$$[w]^{\pm} = \pm\frac{\nu}{2} = \int_\Gamma P\mathbf{u}\nu(y)\, \mathrm{d}\Gamma_y\,. \tag{2.28}$$

The above formulas are valid at all points on the boundary which satisfies the Lyapunov conditions (cf. the supplement) and under the assumption that the respective densities are continuous functions.

Knowing the behaviour of the potentials while crossing the boundary of domain Ω it is possible to obtain the boundary integral equations as a result of the mentioned limit transition $\operatorname{Int}\Omega \ni x \to \Gamma$:

$$\left.\begin{matrix} \alpha k|_{\Gamma_1} \\ \alpha \mathbf{u}|_{\Gamma_2} \end{matrix}\right\} = \int_{\Gamma_1} uP\mathbf{u}\, \mathrm{d}\Gamma_y - \int_{\Gamma_2} \mathbf{u}h\, \mathrm{d}\Gamma_y - \int_{\Gamma_1} kP\mathbf{u}\, \mathrm{d}\Gamma_y - \int_{\Gamma_2} \mathbf{u}P u\, \mathrm{d}\Gamma_y + \int_\Omega \mathbf{u}f\, \mathrm{d}\Omega\,. \tag{2.29}$$

In the integrals $\alpha = 0$ for $x \in \boldsymbol{E}^n \backslash \Omega$, $\alpha \in (0, 1)$ for $x \in \Gamma = \partial\Omega$. This is a set of Fredholm integral equations of the first kind on part Γ_1 of the boundary and of the second kind on part Γ_2, which can be used to determine u on Γ_2 and Pu on Γ_1. In the case of Dirichlet boundary conditions the set is obviously reduced to a single equation of the first kind and in the case of the Neumann boundary condition also to a single equation, but of the second kind.

Assuming that once the boundary equations (2.29) are solved the unknown functions $u, \forall\, x \in \Gamma_2$ and $Pu, \forall\, x \in \Gamma_1$ are determined from the integral formula (2.21), the solution of the boundary problem:

$$\begin{aligned} M\,u &= f \qquad \forall\, x \in \operatorname{Int}\Omega\,, \\ u &= k \qquad \forall\, x \in \Gamma_1\,, \\ P\,u &= h \qquad \forall\, x \in \Gamma_2\,, \end{aligned} \tag{2.30}$$

with $\Gamma_1 \cap \Gamma_2 = \emptyset \wedge \Gamma_1 \cup \Gamma_2 = \Gamma$, can be found. The solution has the form:

$$\begin{aligned} u(x) = &\int_\Omega \mathbf{u}(x,\, y) f(y)\, \mathrm{d}\Omega_y + \int_{\Gamma_1} k(y) P\mathbf{u}(x,\, y)\, \mathrm{d}\Gamma_y \\ &+ \int_{\Gamma_2} u(y) P\mathbf{u}(x,\, y)\, \mathrm{d}\Gamma_y - \int_{\Gamma_1} Pu(y)\mathbf{u}(x,\, y)\, \mathrm{d}\Gamma_y \\ &- \int_{\Gamma_2} h(y)\mathbf{u}(x,\, y)\, \mathrm{d}\Gamma_y\,, \qquad \forall\, x \in \operatorname{Int}\Omega,\ y \in \Gamma\,. \end{aligned} \tag{2.31}$$

In particular, for $\Gamma = \Gamma_1$ we obtain a solution of the Dirichlet problem and taking $\Gamma = \Gamma_2$ we obtain a solution of the Neumann problem. In case the coefficients $a_{\alpha\beta}$ are not constant the solution of the inhomogeneous equation can be written as a convolution only if understanding of the notion of fundamental solution is modified (see [20]).

The presented method is called a *direct approach* of the MBIE. Sometimes, it is more convenient to use an indirect approach which results from the reasoning presented below. When point $x \in \Omega$, the integral formula giving a solution of the problem (2.30) reads:

$$u = \int_\Omega \mathbf{u} f\, \mathrm{d}\Omega + \int_\Gamma uP\mathbf{u}\, \mathrm{d}\Gamma - \int_\Gamma \mathbf{u}Pu\, \mathrm{d}\Gamma\,. \tag{2.32}$$

For $x \in \boldsymbol{E}^n \backslash \Omega$ Eq. (2.29) gives

$$0 = -\int_\Gamma \mathbf{u}P\hat{u}\, \mathrm{d}\Gamma + \int_\Gamma \hat{u}P\mathbf{u}\, \mathrm{d}\Gamma\,. \tag{2.33}$$

Assuming that the first solution u and another one denoted by $\hat{u}$ and called a *solution of an external problem* satisfy the following condition:

$$\forall\, x \in \Gamma\,, \qquad \hat{u} = u\,, \tag{2.34}$$

it is possible to subtract Eq. (2.33) by sides from Eq. (2.32) and write:

$$u = \int_\Omega \mathrm{u} f \, \mathrm{d}\Omega + \int_\Gamma \mathrm{u}\mu \, \mathrm{d}\Gamma \,, \tag{2.35}$$

where $\mu = Pu - P\hat{u}$.

Moreover, assuming that the solutions u and $\hat{u}$ satisfy the condition

$$\forall\, x \in \Gamma \,, \qquad Pu = P\hat{u} \,, \tag{2.36}$$

it is possible to obtain in a similar way:

$$u = \int_\Omega \mathrm{u} f \, \mathrm{d}\Omega + \int_\Gamma \nu P \mathrm{u} \, \mathrm{d}\Gamma \,, \tag{2.37}$$

where $\nu = u - \hat{u}$.

Using Eq. (2.35) and placing point x on boundary Γ we can build an equation to determine such a function μ that the boundary condition

$$\forall\, x \in \Gamma \,, \qquad u = k \,, \tag{2.38}$$

is satisfied. Consequently we have:

$$k = \int_\Omega \mathrm{u} f \, \mathrm{d}\Omega + \int_\Gamma \mathrm{u}\mu \, \mathrm{d}\Gamma \,. \tag{2.39}$$

After solving this equation, the integral formula (2.35) can be used to solve a Dirichlet problem.

If the boundary condition has the form

$$\forall\, x \in \Gamma : \qquad Pu = h \,, \tag{2.40}$$

the following formula can be found on the basis of Eq. (2.35):

$$Pu = \int_\Omega P_x \mathrm{u} f \, \mathrm{d}\Omega_y + \int_\Gamma P_x \mathrm{u}\mu \, \mathrm{d}\Gamma_y . \tag{2.41}$$

When point x is brought to boundary Γ and when the boundary condition (2.40) is used we obtain the equation

$$h = \int_\Omega P_x \mathrm{u} f \, \mathrm{d}\Omega_y + \int_\Gamma P_x \mathrm{u}\mu \, \mathrm{d}\Gamma_y - \frac{\mu}{2a} \,, \tag{2.42}$$

in which the quantity a is defined by Eq. (2.27a). Equation (2.42) can then be used to determine such a function μ, defined on the boundary Γ, that substituted into Eq. (2.35) allows the solution of a Neumann problem to be obtained.

The solution of a Dirichlet problem can also be constructed basing on formula (2.37). As a result of bringing the point x to the boundary and using the condition (2.37) the following integral equation is obtained:

$$k = \int_\Omega \mathrm{u} f \, \mathrm{d}\Omega + \int_\Gamma \nu P \mathrm{u} \, \mathrm{d}\Gamma + \frac{1}{2} \nu \,. \tag{2.43}$$

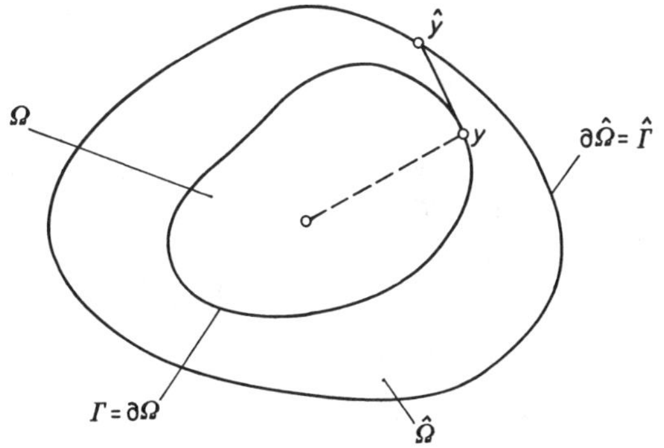

Fig. 2.2. Domain $\hat{\Omega}$ with boundary $\partial\hat{\Omega}$, which encloses subdomain Ω

The function ν determined from this equation can be substituted into Eq. (2.37), which gives the solution sought.

Formula (2.37) can also be employed in order to find the solution of a Neumann problem. Applying operator P the following identity is found

$$Pu = \int_{\Omega} f P_x \mathbf{u} \, \mathrm{d}\Omega_y + \int_{\Gamma} \nu P_x(P_y \mathbf{u}) \, \mathrm{d}\Gamma_y \tag{2.44}$$

and can be used to construct an appropriate integral equation. This idea is, however, ineffective since once the operator P is applied the kernel of the equation becomes strongly singular, which is very troublesome in applications.

In both presented variations of the MBIE, i.e. in the direct and indirect approach, the resulting integral equations are singular. The fact is found to be inconvenient in the process of finite-dimensional approximation of the equations. The troubles can be avoided using a slightly modified version of the indirect approach. Let $\hat{\Omega}$ with boundary $\partial\hat{\Omega} = \hat{\Gamma}$ be such a subdomain of $\boldsymbol{E}^n$ that

$$\Omega \subset \hat{\Omega} \wedge \partial\Omega \cap \partial\hat{\Omega} = \emptyset \,. \tag{2.45}$$

Let function μ be defined on the boundary of domain $\hat{\Omega}$, i.e. dom $\mu = \hat{\Gamma}$. Solution of the equation

$$\begin{aligned} M\,u = f\,; &\quad \forall\, x \in \operatorname{Int}\hat{\Omega}; \quad \operatorname{dom} f = \Omega \subset \hat{\Omega}, \\ u = \mu\,; &\quad \forall\, x \in \hat{\Gamma} \end{aligned}$$

is given by the integral formula:

$$u = \int_{\Omega} \mathbf{u} f \, \mathrm{d}\Omega + \int_{\Gamma} \mathbf{u} \mu \, \mathrm{d}\Gamma \,. \tag{2.46}$$

The function μ can be chosen in such a way that the condition

$$u = k\,; \quad \forall\, x \in \Gamma\,, \tag{2.47}$$

is satisfied. For this purpose the following equation can be employed:

$$k = \int_{\Omega} \mathrm{u} f \,\mathrm{d}\Omega_y + \int_{\hat{\Gamma}} \mathrm{u}\mu \,\mathrm{d}\Gamma_y ; \quad \forall\, x \in \Gamma . \tag{2.48}$$

Since $\mathrm{u} = \mathrm{u}(x, y)$ and $x \in \Gamma$, moreover $y \in \operatorname{Int}\Omega$ in the first integral and $y \in \hat{\Gamma}$ in the second one, and additionally $\Gamma \cap \hat{\Gamma} = \emptyset$, therefore Eq. (2.48) is nonsingular.

A similar procedure can be applied for the following boundary condition

$$Pu = h ; \qquad \forall\, x \in \Gamma . \tag{2.49}$$

An appropriate integral equation has then a form similar to Eq. (2.42)

$$Pu = h = \int_{\Omega} f P_x \mathrm{u} \,\mathrm{d}\Omega_y + \int_{\hat{\Gamma}} \mu P_x \mathrm{u} \,\mathrm{d}\Gamma ; \quad \forall\, x \in \Gamma \tag{2.50}$$

and can be obtained in an analogical manner. This equation is nonsingular for the same reasons as presented previously.

It can be easily verified that a function derived from formula (2.46) and Eqs. (2.48) or (2.50) are a solution of the boundary value problem of the Dirichlet and Neumann types, respectively. For this purpose we assume that the function μ defined on the boundary $\hat{\Gamma}$ is augmented with the value zero in the whole domain $\boldsymbol{E}^n \backslash \hat{\Gamma}$. Additionally, we assume that the function f is extended with zero to the domain $\boldsymbol{E}^n \backslash \Omega$. With these assumptions the solution (2.46) can be written in the form of a convolution:

$$u = \mathrm{u} * (f + \mu) ,$$

and further we obtain:

$$Mu = M\mathrm{u} * (f + \mu) = Mu * (f + \mu) = \delta * (f + \mu) = f ; \ \forall\, x \in \Omega .$$

This means that the equation is satisfied $\forall x \in \operatorname{Int}\Omega$. Moreover, the boundary condition is fulfilled, since obviously

$$u = k ; \qquad \forall\, x \in \Gamma ,$$

if μ is a solution of Eq. (2.48) and

$$Pu = Pu * (f + \mu) = h ; \qquad \forall\, x \in \Gamma ,$$

if μ is a solution of Eq. (2.50).

The procedure described above is called a *nonsingular (indirect)* MBIE. Due to the mentioned advantages it seems recommendable for practical applications.

2.3 Higher-Order Elliptic Equations

The presented idea of constructing the equations of the MBIE can be extended to the case of elliptic operators of an arbitrary order $2m$ which are an

m-fold combination of a second order operator M. The basic integral formula can be obtained applying the Green identity to the second order operator. Adopting

$$v = M\hat{v}$$

Eq. (2.20) gives:

$$\int_\Omega [M\hat{v}Mu - uM(M\hat{v})]\,\mathrm{d}\Omega = \int_\Gamma [M\hat{v}Pu - uP(M\hat{v})]\,\mathrm{d}\Gamma\,. \tag{2.51}$$

Similarly, adopting $u = M\hat{u}$ it is possible to write:

$$\int_\Omega [vM(M\hat{u}) - M\hat{u}Mv]\,\mathrm{d}\Omega = \int_\Gamma [vP(M\hat{u}) - M\hat{u}Pv]\,\mathrm{d}\Gamma\,. \tag{2.52}$$

Adding by sides and changing the notation

$$u \equiv \hat{u} \text{ and } v \equiv \hat{v}$$

we obtain:

$$\int_\Omega [vM^2u - uM^2v]\,\mathrm{d}\Omega = \int_\Gamma [vP(Mu) - MuPv + MvPu - uP(Mv)]\,\mathrm{d}\Gamma\,, \tag{2.53}$$

where $M \circ M$ is denoted by M^2.

If $v = \mathbf{u}$ is a fundamental solution, i.e.

$$M^2\mathbf{u} = \delta\,, \tag{2.54}$$

then the solution of the boundary value problem

$$\begin{aligned} M^2u = f\,; &\quad \forall\, x \in \operatorname{Int}\Omega\,, \\ u = k\,; &\quad \forall\, x \in \partial\Omega = \Gamma\,, \\ P\,u = h\,, & \end{aligned} \tag{2.55}$$

can be written as the integral formula:

$$u = \int_\Omega \mathbf{u}f\,\mathrm{d}\Omega - \int_\Gamma [\mathbf{u}P(Mu) - Mu P\mathbf{u} + M\mathbf{u}h - kP(M\mathbf{u})]\,\mathrm{d}\Gamma\,. \tag{2.56}$$

Like Eq. (2.20) for second order equations, Eq. (2.56) contains unknown functions defined on boundary Γ of domain Ω. These functions are as follows:

$$\begin{aligned} P(Mu) : &\quad \operatorname{dom} P(Mu) = \Gamma\,, \\ Mu : &\quad \operatorname{dom} Mu = \Gamma\,. \end{aligned} \tag{2.57}$$

They can be determined from appropriate boundary integral equations obtained from Eq. (2.56) as a result of the limit transition $\operatorname{Int}\Omega \ni x \to \Gamma$. The form of these equations is shown in Sect. 2.4.2 for the biharmonic operator.

Following the pattern it is possible to construct the Green formula for polyharmonic operators, which makes application of the MBIE to equations of the following type possible:

$$M^{2m}u = f\,. \tag{2.58}$$

For practical reasons the case $m > 2$ is not sufficiently interesting to quote integral formulas, which can be easily derived along the lines shown above.

2.4 Examples of MBIE Application

The aim of this section is to apply the information given above to the operators which are most frequently used while formulating boundary problems of mechanics.

2.4.1 Poisson Equation

The Poisson equation is constructed using the Laplace operator, i.e.

$$M = \Delta = \sum_{\alpha=1}^{n} \frac{\partial^2}{\partial x_\alpha^2}\,, \tag{2.59}$$

where n denotes the dimension of the $\boldsymbol{E}^n$ space. Accordingly, the equation has the form:

$$\sum_{\alpha=1}^{n} \frac{\partial^2 u(x_\alpha)}{\partial x_\alpha^2} = f(x_\alpha)\,, \tag{2.60}$$

and the fundamental solution is the function

$$\mathrm{u}(x,\,y) = \begin{cases} \frac{1}{4\pi}\cdot\frac{1}{r} & \text{for } n = 3\,, \\ \frac{1}{2\pi}\ln\left(\frac{1}{r}\right) & \text{for } n = 2\,, \end{cases} \tag{2.61}$$

in which the following notation has been used:

$$x = x(x_\alpha) \in \boldsymbol{E}^n, \qquad y = y(y_\alpha) \in \boldsymbol{E}^n,$$

$$r = \left[\sum_{\alpha=1}^{n} (x_\alpha - y_\alpha)^2\right]^{\frac{1}{2}}, \quad n = (2,\,3)\,.$$

The solution of the Poisson equation (2.60) can then be expressed as the following integral formula:

$$u = \int_\Omega \mathrm{u}(x,\,y) f(y)\,\mathrm{d}\Omega_y + \int_\Gamma \left[u(y)\frac{\partial \mathrm{u}(x,\,y)}{\partial n} - \mathrm{u}(x,\,y)\frac{\partial u(y)}{\partial n}\right]\mathrm{d}\Gamma_y\,, \tag{2.62}$$

and consequently, as a result of the limit transition Int $\Omega \ni x \to \Gamma = \partial\Omega$, the boundary integral equation is obtained:

$$\frac{1}{2}u = \int_{\Omega} \mathrm{u}(x, y) f(y)\, \mathrm{d}\Omega_y + \int_{\Gamma} \left[u(y) \frac{\partial \mathrm{u}(x, y)}{\partial n_y} - \mathrm{u}(x, y) \frac{\partial u(y)}{\partial n_y} \right] \mathrm{d}\Gamma_y\,, \tag{2.63}$$

with $x \in \Gamma$. In the case of mixed boundary conditions, i.e.

$$\begin{aligned} \forall\, x \in \Gamma_1 \qquad u(x) &= k\,, \\ \forall\, x \in \Gamma_2 \qquad \frac{\partial u(x)}{\partial n} &= h\,, \\ \Gamma_1 \cap \Gamma_2 = \emptyset \wedge \Gamma_1 \cup \Gamma_2 &= \Gamma\,, \end{aligned} \tag{2.64}$$

two boundary integral equations are obtained:

$$\begin{aligned} \frac{1}{2}k(x) &= \int_{\Omega} \mathrm{u}(x,y) f(y)\, \mathrm{d}\Omega_y \\ &\quad + \int_{\Gamma_1} \left[k(y) \frac{\partial \mathrm{u}(x,y)}{\partial n_y} - \mathrm{u}(x,y) \frac{\partial u(y)}{\partial n_y} \right] \mathrm{d}\Gamma_y \qquad \forall\, x \in \Gamma_1, \\ \frac{1}{2}u(x) &= \int_{\Omega} \mathrm{u}(x,y) f(y)\, \mathrm{d}\Omega_y \\ &\quad + \int_{\Gamma_2} \left[u(y) \frac{\partial \mathrm{u}(x,y)}{\partial n_y} - \mathrm{u}(x,y) h(y) \right] \mathrm{d}\Gamma_y \qquad \forall\, x \in \Gamma_2\,. \end{aligned} \tag{2.65}$$

They preserve their validity under the assumption that the boundary of the domain Ω satisfies the Lyapunov conditions (see the supplement). The solutions of the Dirichlet and Neumann problems are obtained for $\Gamma_1 = \emptyset$ and $\Gamma_2 = \emptyset$, respectively. Using an abbreviated notation, cf. Eq. (2.29):

$$\alpha u = \int_{\Omega} \mathrm{u} f\, \mathrm{d}\Omega + \int_{\Gamma} \left(u \frac{\partial \mathrm{u}}{\partial n} - \mathrm{u} \frac{\partial u}{\partial n} \right) \mathrm{d}\Gamma\,, \qquad \alpha = \left\{ 0, \frac{1}{2}, 1 \right\}, \tag{2.66}$$

the boundary integral equations (2.63) are obtained for $\alpha = 1/2$, while the integral formula (2.62), which gives the solution of the boundary value problem in the domain Ω, for $\alpha = 1$. The solution is extended by the function $u = 0$ to the domain $\boldsymbol{E}^n \backslash \Omega$, so $\alpha = 0$ should be assumed $\forall\, x \in [\boldsymbol{E}^n \backslash \Omega]$.

It should be emphasized that the form of Eq. (2.66) should be understood exclusively as a simplified notation. In reality, for different values of α the sense of the expressions changes considerably.

As a result of the indirect approach described previously the solution is expressed as the following integral formula:

$$u(x) = \int_{\Omega} \mathrm{u}(x, y) f(y)\, \mathrm{d}\Omega_y + \int_{\Gamma} \mathrm{u}(x, y) \mu(y)\, \mathrm{d}\Gamma_y\,, \tag{2.67}$$

with the function $\mu(y) : \; y \in \Gamma$ determined from the following boundary integral equation:

$$k(x) = \int_{\Omega} \mathrm{u}(x, y) f(y)\, \mathrm{d}\Omega_y + \int_{\Gamma} \mathrm{u}(x, y)\mu(y)\, \mathrm{d}\Gamma_y , \qquad x, y \in \Gamma \tag{2.68}$$

in the case of the Dirichlet boundary condition and the following boundary equation:

$$h(x) = \int_{\Omega} f(y) \frac{\partial \mathrm{u}(x, y)}{\partial n_x}\, \mathrm{d}\Omega_y - \frac{1}{2}\mu(x) + \int_{\Gamma} \mu(y) \frac{\partial \mathrm{u}(x, y)}{\partial n_x}\, \mathrm{d}\Gamma_y , \quad x, y \in \Gamma \tag{2.69}$$

for the Neumann boundary condition (cf. Sect. 1.2).

In both approaches the boundary integral equations have weakly singular kernels. The boundary equations with regular kernels, which are much more convenient in practical applications, can be obtained following the indirect (nonsingular) approach described in Sect. 2.2.

The solution is expressed (cf. Sect. 2.2) as an integral formula:

$$u = \int_{\hat{\Omega}} \mathrm{u}(x, y) f(y)\, \mathrm{d}\Omega_y + \int_{\hat{\Gamma}} \mathrm{u}(x, y)\mu(y)\, \mathrm{d}\hat{\Gamma}_y , \tag{2.70}$$

$$\hat{\Omega} : \Omega \subset \hat{\Omega} \wedge \partial\Omega = \Gamma \cap \hat{\Gamma} = \partial\hat{\Omega} = \emptyset; \qquad \operatorname{dom} f \subset \Omega .$$

For the Dirichlet boundary condition, the function $\mu(x) : \; x \in \Gamma$ should be determined from the equation:

$$k(x) = \int_{\hat{\Omega}} \mathrm{u}(x, y) f(y)\, \mathrm{d}\Omega_y + \int_{\hat{\Gamma}} \mathrm{u}(x, y)\mu(y)\, \mathrm{d}\hat{\Gamma}_y , \tag{2.71}$$

and in the case of the Neumann boundary condition from the formula:

$$h(x) = \int_{\hat{\Omega}} f(y) \frac{\partial \mathrm{u}(x, y)}{\partial n_x}\, \mathrm{d}\Omega_y + \int_{\hat{\Gamma}} \mu(y) \frac{\partial \mathrm{u}(x, y)}{\partial n_x}\, \mathrm{d}\hat{\Gamma}_y . \tag{2.72}$$

The kernels of the equations are regular, since in the first integrals $y \in \operatorname{dom} f \subset \operatorname{Int} \Omega$, and in the second ones $y \in \hat{\Gamma}$ while $x \in \Gamma$ and the domains Ω and $\hat{\Omega}$ are separate ($\hat{\Gamma} \cap \Gamma = \emptyset$).

2.4.2 Biharmonic Equations

In the case of a biharmonic equation the operator M is a combination of the Laplace operators:

$$M = \Delta\Delta = \sum_{\alpha=1}^{n} \frac{\partial^2}{\partial x_\alpha^2} \left[\sum_{\beta=1}^{n} \frac{\partial^2}{\partial x_\beta^2} \right] , \tag{2.73}$$

where n denotes the dimension of the $\boldsymbol{E}^n$ space. The equation has accordingly the form:

$$\sum_{\alpha=1}^{n} \frac{\partial^2}{\partial x_\alpha^2} \left[\sum_{\beta=1}^{n} \frac{\partial^2 u}{\partial x_\beta^2} \right] = f . \tag{2.74}$$

The fundamental solution of the biharmonic equation is, depending on the space dimension, one of the functions:

$$\mathbf{u} = \begin{cases} l\frac{1}{2\pi} r^2 \ln(r) & \text{for } n = 2 , \\ \frac{r}{4\pi} & \text{for } n = 3 , \end{cases} \tag{2.75}$$

where

$$r = \left[\sum_{\alpha=1} (x_\alpha - y_\alpha)^2 \right]^{1/2} .$$

Making use of the properties of the fundamental solution and Eq. (2.56) we can obtain the integral formula:

$$u(x) = \int_\Omega \mathbf{u}(x, y) f(y)\, \mathrm{d}\Omega_y + \int_\Gamma \left\{ \mathbf{u}(x, y) \frac{\partial}{\partial n_y} [\Delta u(y)] - \Delta u(y) \frac{\partial \mathbf{u}(x, y)}{\partial n_y} \right.$$
$$\left. + \Delta \mathbf{u}(x, y) \frac{\partial \mathbf{u}(y)}{\partial n_y} - u(y) \frac{\partial}{\partial n_y} [\Delta \mathbf{u}(x, y)] \right\} \mathrm{d}\Gamma_y , \tag{2.76}$$

with

$x \in \operatorname{Int} \Omega$,
$y \in \operatorname{Int} \Omega$ in the first integral,
$y \in \Gamma = \partial\Omega$ in the second integral.

From among the four functions u, $\frac{\partial u}{\partial \mathbf{n}}$, Δu, $\frac{\partial}{\partial \mathbf{n}}(\delta u)$ only two are known on the boundary, and the remaining ones must be determined from the boundary integral equations.

In the case of the following boundary conditions:

$$u = k ; \forall x \in \Gamma , \qquad \frac{\partial u}{\partial \mathbf{n}} = h , \tag{2.77}$$

the first of these equations is a result of the limit transition $\operatorname{Int} \Omega \ni x \to \Gamma = \partial\Omega$ in Eq. (2.76)

$$\frac{1}{2} k(x) = \int_\Omega \mathbf{u}(x, y) f(y)\, \mathrm{d}\Omega_y + \int_\Gamma \left\{ \mathbf{u}(x, y) \frac{\partial}{\partial n_y} [\Delta u(y)] - \Delta u(y) \frac{\partial \mathbf{u}(x, y)}{\partial n_y} \right.$$
$$\left. + \Delta \mathbf{u}(x, y) h(y) - k(y) \frac{\partial}{\partial n_y} [\Delta \mathbf{u}(x, y)] \right\} \mathrm{d}\Gamma_y , \tag{2.78}$$

where as above

$x \in \Gamma$,
$y \in \operatorname{Int}\Omega$ in the first integral,
$y \in \Gamma$ in the second integral.

The second equation can be obtained performing the operation $\frac{\partial}{\partial n_x}$, $x \in \Gamma$ on Eq. (2.76) and then placing the point x on the boundary Γ:

$$\frac{1}{2}\,h(x) = \int_\Omega \frac{\partial \mathrm{u}(x,\,y)}{\partial n_x} f(y)\,\mathrm{d}\Omega_y + \int_\Gamma \left\{ \frac{\partial \mathrm{u}(x,\,y)}{\partial n_x} \frac{\partial}{\partial n_y}[\Delta u(y)] - \Delta u(y) \frac{\partial}{\partial n_x}\left[\frac{\partial \mathrm{u}(x,\,y)}{\partial n_y}\right] + \Delta\left[\frac{\partial \mathrm{u}(x,\,y)}{\partial n_x}\right] h(y) + k(y)\frac{\partial}{\partial n_y}\left[\Delta\left(\frac{\partial \mathrm{u}(x,\,y)}{\partial n_x}\right)\right]\right\} \mathrm{d}\Gamma_y\,. \tag{2.79}$$

Equations (2.78) and (2.79) form a system of boundary integral equations, from which the unknown functions $\Delta u(x)$ and $\frac{\partial}{\partial \mathbf{n}}[\delta u(x)]$ can be determined.

In the case of a biharmonic equation the number of different forms of the possible boundary conditions is larger than for the case of the harmonic equation. A detailed discussion of all their versions and the boundary integral equations associated does not seem useful. It is enough to point out that, assuming a sufficient regularity of the integrated functions, the appropriate boundary equations can be obtained by changing the sequence of differentiation and integration.

The integral formula (2.76) and boundary equations (2.78) can be written in an abbreviated form:

$$\alpha u(x) = \int_\Omega \mathrm{u}(x,\,y) f(y)\,\mathrm{d}\Omega_y + \int_\Gamma \left\{ \mathrm{u}(x,\,y)\frac{\partial}{\partial n_y}[\Delta u(y)] - \Delta u(y)\frac{\partial \mathrm{u}(x,\,y)}{\partial n_y} + \Delta \mathrm{u}(x,\,y)\frac{\partial \mathrm{u}(y)}{\partial n_y} - u(y)\frac{\partial}{\partial n_y}[\Delta \mathrm{u}(x,\,y)] \right\} \mathrm{d}\Gamma_y\,, \tag{2.80}$$

where

$$\begin{aligned} \alpha &= 0\,; && \forall x \in \boldsymbol{E}^3 \backslash \bar{\Omega}, \\ \alpha &= \frac{1}{2}\,; && \forall x \in \partial\Omega, \\ \alpha &= 1\,; && \forall x \in \operatorname{Int}\Omega. \end{aligned}$$

The set of boundary equations is a set of equations with singular kernels. If the boundary operator is a harmonic operator or, in general, a differential operator of the second order, then the kernels are weakly singular.

In practical applications to the theory of plates the problem of a plate with a free boundary leads to a boundary condition with a differential operator of the third order and cannot be solved in the presented manner. The nonsingular indirect approach presented previously proves then especially practical.

Before quoting the equations of the latter approach adapted to the mentioned problem, it is useful to recap an obvious property of the fundamental solution.

If u_N denotes a function

$$\mathrm{u}_N = N[\mathrm{u}], \tag{2.81}$$

in which N is a linear differential operator with constant coefficients, then function

$$u_N = \mathrm{u}_N * f \tag{2.82}$$

is a solution of the equation

$$\Delta\Delta u_N = N[f], \tag{2.83}$$

since

$$\Delta\Delta u_N = \Delta\Delta[\mathrm{u}_N] * f = N[\Delta\Delta\mathrm{u}] * f = N[\delta] * f = N[f].$$

Let a domain $\hat{\Omega}$ with a boundary $\partial\hat{\Omega} = \hat{\Gamma}$ be such that

$$\Omega \subset \hat{\Omega} \wedge \partial\Omega \cap \partial\hat{\Omega} = \emptyset.$$

Moreover, let the following functions

$$q_1 \,;\, \mathrm{dom}\, q_1 = \hat{\Gamma}, \qquad q_2 \,;\, \mathrm{dom}\, q_2 = \hat{\Gamma}$$

be defined on the boundary $\partial\hat{\Omega} = \hat{\Gamma}$. Then the solution of the biharmonic equation can be expressed in the form

$$u = \int_\Omega \mathrm{u} f \,\mathrm{d}\Omega + \int_\Gamma \mathrm{u} q_1 \,\mathrm{d}\Gamma + \int_{\hat{\Gamma}} \mathrm{u}_{,n}\, q_2 \,\mathrm{d}\Gamma. \tag{2.84}$$

where $f : \mathrm{dom}\, f \subset \Omega$ and $\mathrm{u}_{,n} = \partial\mathrm{u}/\partial\mathbf{n}$.

For the following boundary conditions:

$$\begin{aligned} Nu &= k \quad \forall x \in \Gamma, \\ Pu &= h, \end{aligned} \tag{2.85}$$

the set of boundary integral equations takes the form:

$$\begin{aligned} k(x) = &\int_\Omega N_x[\mathrm{u}(x, y)] f(y) \,\mathrm{d}\Omega_y + \int_{\hat{\Gamma}} N_x[\mathrm{u}(x, y)] q_1(y) \,\mathrm{d}\Gamma_y \\ &+ \int_{\hat{\Gamma}} N_x[\mathrm{u}_{,}n(x, y)] q_2 \,\mathrm{d}\Gamma_y, \end{aligned} \tag{2.86}$$

$$\begin{aligned} h(x) = &\int_\Omega P_x[\mathrm{u}(x, y)] f(y) \,\mathrm{d}\Omega_y + \int_{\hat{\Gamma}} P_y[\mathrm{u}(x, y)] q_1(y) \,\mathrm{d}\Gamma_y \\ &+ \int_{\hat{\Gamma}} P_y[\mathrm{u}_{,}n(x, y)] q_2 \,\mathrm{d}\Gamma_y, \end{aligned} \tag{2.87}$$

with $x \in \Gamma = \partial\Omega$, $y \in \mathrm{Int}\,\Omega$ in the first integrals and $y \in \hat{\Gamma} = \partial\hat{\Omega}$ in the other integrals. This is, like before, a set of Fredholm integral equations of the first kind. However, these equations are nonsingular since $x \neq y$, so they are particularly convenient from the viewpoint of a finite-dimensional approximation. It should be added that the singular indirect approach is practically not used.

2.4.3
Equations of Linear Elasticity

The set of equations of linear elasticity was given in Chap. 2 of Part I, however, from the MBIE viewpoint a displacement formulation equivalent to the set (2.25) is more convenient. The equation governing the problem reads:

$$L\mathbf{u} + \mathbf{b} = \mathbf{0}, \tag{2.88}$$

where $L \overset{\text{def}}{=} \mu\Delta \ldots + (\mu + \lambda)\text{grad}[\text{div} \ldots]$ is the Lame operator, $\mathbf{u}$ is the displacement vector, $\mathbf{b}$ is a vector of body forces, μ and λ are Kirchhoff and Lame constants, respectively.

Defining operator P

$$P \overset{\text{def}}{=} \Sigma\mathbf{n}, \tag{2.89}$$

where

$$\Sigma \overset{\text{def}}{=} \mu[\nabla \ldots + \nabla^T \ldots] + \lambda\mathbf{I}\,\text{div} \ldots, \tag{2.89a}$$

$\nabla \ldots \overset{\text{def}}{=} \text{grad} \ldots$, $\mathbf{n}$ is an outer normal vector, it is possible to associate a stress vector, or a vector of internal surface forces (cf. Chap. 2 of Part I)

$$\mathbf{p} = P\mathbf{u} \tag{2.90}$$

with the displacement vector $\mathbf{u}$. A convenient notation introduced in Chap. 2 of Part I is the matrix notation, in which the operators defined above take the form:

$$L \overset{\text{def}}{=} \begin{bmatrix} \mu\Delta + (\mu+\lambda)\frac{\partial^2}{\partial x_1{}^2} & (\mu+\lambda)\frac{\partial^2}{\partial x_1 \partial x_2} & (\mu+\lambda)\frac{\partial^2}{\partial x_1 \partial x_3} \\ (\mu+\lambda)\frac{\partial^2}{\partial x_1 \partial x_2} & \mu\Delta + (\mu+\lambda)\frac{\partial^2}{\partial x_2{}^2} & (\mu+\lambda)\frac{\partial^2}{\partial x_2 \partial x_3} \\ (\mu+\lambda)\frac{\partial^2}{\partial x_1 \partial x_3} & (\mu+\lambda)\frac{\partial^2}{\partial x_2 \partial x_3} & \mu\Delta + (\mu+\lambda)\frac{\partial^2}{\partial x_3{}^2} \end{bmatrix}, \tag{2.91}$$

$$P = \begin{bmatrix} \begin{array}{l}(2\mu+\lambda)n_1\frac{\partial}{\partial x_1} \\ \quad + \mu\left(n_2\frac{\partial}{\partial x_2} + n_3\frac{\partial}{\partial x_3}\right)\end{array} & \lambda n_1\frac{\partial}{\partial x_2} + \mu n_2\frac{\partial}{\partial x_1} & \lambda n_1\frac{\partial}{\partial x_3} + \mu n_3\frac{\partial}{\partial x_1} \\ \lambda n_2\frac{\partial}{\partial x_1} + \mu n_1\frac{\partial}{\partial x_2} & \begin{array}{l}(2\mu+\lambda)n_2\frac{\partial}{\partial x_2} \\ \quad + \mu\left(n_1\frac{\partial}{\partial x_1} + n_3\frac{\partial}{\partial x_3}\right)\end{array} & \lambda n_2\frac{\partial}{\partial x_3} + \mu n_3\frac{\partial}{\partial x_2} \\ \lambda n_3\frac{\partial}{\partial x_1} + \mu n_1\frac{\partial}{\partial x_3} & \lambda n_3\frac{\partial}{\partial x_2} + \mu n_2\frac{\partial}{\partial x_3} & \begin{array}{l}(2\mu+\lambda)n_3\frac{\partial}{\partial x_3} \\ \quad + \mu\left(n_2\frac{\partial}{\partial x_2} + n_1\frac{\partial}{\partial x_1}\right)\end{array} \end{bmatrix} \tag{2.92}$$

Having adopted these definitions, the boundary value problem can be expressed in one of the following forms. The first is the so-called *displacement problem* of elasticity (the name is determined by the boundary condition):

$$\begin{aligned} L\mathbf{u} + \mathbf{b} = \mathbf{0}\,; &\quad \forall\, x \in \operatorname{Int}\Omega\,, \\ \mathbf{u} = \mathbf{k}\,; &\quad \forall\, x \in \partial\Omega = \Gamma\,. \end{aligned} \tag{2.93}$$

The second is the so-called *stress problem* of elasticity:

$$\begin{aligned} L\mathbf{u} + \mathbf{b} = \mathbf{0}\,; &\quad \forall\, x \in \operatorname{Int}\Omega\,, \\ P\mathbf{u} = \mathbf{h}\,; &\quad \forall\, x \in \Gamma\,. \end{aligned} \tag{2.94}$$

The third is the so-called *mixed problem* of elasticity:

$$\begin{aligned} L\mathbf{u} + \mathbf{b} = \mathbf{0}\,; &\quad \forall\, x \in \operatorname{Int}\Omega\,, \\ \mathbf{u} = \mathbf{k}\,; &\quad \forall\, x \in \Gamma_u\,, \\ P\mathbf{u} = \mathbf{h}\,; &\quad \forall\, x \in \Gamma_p\,, \end{aligned} \tag{2.95}$$

where $\Gamma_u \cup \Gamma_p = \Gamma \wedge \Gamma_u \cap \Gamma_p = \emptyset$.

The basis of the indirect formulation of the MBIE can be found in the Betti formula stating the principle of reciprocal work. It can be shown that from the formal viewpoint it is analogical to the Green formula. Starting from the rule of integration by parts:

$$\begin{aligned} \int_\Omega \mathbf{v} \frac{\partial}{\partial x_j}\left(\frac{\partial \mathbf{w}}{\partial x_m}\right) \mathrm{d}\Omega = \int_{\partial\Omega} \left(\mathbf{v}\frac{\partial \mathbf{w}}{\partial x_m} n_j + \mathbf{v}\frac{\partial \mathbf{w}}{\partial x_j} n_m\right) \mathrm{d}\Gamma \\ - \int_\Gamma \left(\frac{\partial \mathbf{v}}{\partial x_j}\frac{\partial \mathbf{w}}{\partial x_m} + \frac{\partial \mathbf{v}}{\partial x_m}\frac{\partial \mathbf{w}}{\partial x_j}\right) \mathrm{d}\Omega, \end{aligned} \tag{2.96}$$

it is possible to obtain (cf. [28]):

$$\int_\Omega \mathbf{v} L\mathbf{u}\, \mathrm{d}\Omega = \int_\Gamma (\mathbf{v}P\mathbf{u})\, \mathrm{d}\Gamma - \int_\Omega e(\mathbf{u},\, \mathbf{v})\, \mathrm{d}\Omega\,,$$

$$\int_\Omega \mathbf{u} L\mathbf{v}\, \mathrm{d}\Omega = \int_\Gamma (\mathbf{u}P\mathbf{v})\, \mathrm{d}\Gamma - \int_\Omega e(\mathbf{v},\, \mathbf{u})\, \mathrm{d}\Omega\,,$$

where

$$e(\mathbf{u},\, \mathbf{v}) \overset{\text{def}}{=} \sum_{i,m} \left[\mu \frac{\partial v_i}{\partial x_m}\frac{\partial u_i}{\partial x_m} + (\mu + \lambda)\left(\frac{\partial v_i}{\partial x_i}\frac{\partial u_m}{\partial x_m} + \frac{\partial v_i}{\partial x_m}\frac{\partial u_m}{\partial x_i}\right)\right].$$

Subtracting by sides the two equations it is easy to find the formula

$$\int_\Omega (\mathbf{v}L\mathbf{u} - \mathbf{u}L\mathbf{v})\, \mathrm{d}\Omega = \int_\Gamma (\mathbf{v}P\mathbf{u} - \mathbf{u}P\mathbf{v})\, \mathrm{d}\Gamma\,, \tag{2.97}$$

in which the reciprocal work theorem can be recognized.

Assuming that the displacement state $\mathbf{v}$ is caused by a concentrated volume force $\mathbf{b} = \mathbf{I}\,\delta(P)$ applied at point $P \in \operatorname{Int}\Omega$, it is easy to conclude that it satisfies the equation

$$LU = -\mathbf{I}\,\delta(P); \qquad \mathbf{U} = \mathbf{v}\,, \tag{2.98}$$

i.e. it is a fundamental solution for the operator L. The fundamental solution of the Lame equation is known in the elasticity theory as Green tensor (or Kelvin matrix). The element $U_{ij}(x, y)$ denotes a displacement measured along the i-th axis of the coordinate system induced at point x by a force parallel to the j-th axis, which acts at point y. The Green tensor, called further a matrix of fundamental solutions, is symmetric with respect to indices, i.e. $U_{ij} = U_{ij}$, and also with respect to the points $\mathbf{U}(x, y) = \mathbf{U}(y, x)$. The function

$$\mathbf{v} = \mathbf{U} * \mathbf{b} \tag{2.99}$$

is a solution of the Lame equation since

$$L\mathbf{v} = L\mathbf{U} * (-\mathbf{b}) = -\mathbf{b}\,. \tag{2.100}$$

Substituting the convolution (2.99) into Eq. (2.97) the basic integral formula is obtained:

$$\mathbf{u} = \int_\Omega \mathbf{U}\mathbf{b}\,\mathrm{d}\Omega - \int_\Gamma [\mathbf{u}P\mathbf{U} - \mathbf{U}P\mathbf{u}]\,\mathrm{d}\Gamma\,. \tag{2.101}$$

Once the transition $\operatorname{Int}\Omega \ni x \to \Gamma$ is performed and the properties of the fundamental solution, precisely of the function $P(\mathbf{U})$, are considered, we obtain

$$\frac{1}{2}\mathbf{u}(x) = \int_\Omega \mathbf{U}(x, y)\mathbf{b}(y)\,\mathrm{d}\Omega_y - \int_\Gamma [\mathbf{u}(y)P\mathbf{U}(x, y) - \mathbf{U}(x, y)P\mathbf{u}(y)]\,\mathrm{d}\Gamma_y\,, \tag{2.102}$$

where $x \in \Gamma$, $y \in \operatorname{Int}\Omega$ in the first integral, and $y \in \Gamma$ in the second one. This equality makes it possible to obtain the following boundary integral equation

$$\frac{1}{2}\mathbf{k}(x) = \int_\Omega \mathbf{U}(x, y)\mathbf{b}(y)\,\mathrm{d}\Omega_y - \int_\Gamma [\mathbf{k}(y)P\mathbf{U}(x, y) - \mathbf{U}(x, y)P\mathbf{u}(y)]\,\mathrm{d}\Gamma_y \tag{2.103}$$

in the case of a displacement boundary condition, and the following

$$\frac{1}{2}\mathbf{u}(x) = \int_\Omega \mathbf{U}(x, y)\mathbf{b}(y)\,\mathrm{d}\Omega_y - \int_\Gamma [\mathbf{u}(y)P\mathbf{U}(x, y) - \mathbf{U}(x, y)\mathbf{h}(y)]\,\mathrm{d}\Gamma_y \tag{2.104}$$

in the case of a stress boundary condition.

The mixed boundary conditions lead to two integral equations:

$$\frac{1}{2}\mathbf{k}(x) = \int_{\Omega} \mathbf{U}(x,y)\mathbf{b}(y)\,\mathrm{d}\Omega_y - \int_{\Gamma_u} [\mathbf{k}(y)P\mathbf{U}(x,y) - \mathbf{U}(x,y)P\mathbf{u}(y)]\,\mathrm{d}\Gamma_y, \quad \forall\, x \in \Gamma_u,$$

$$\frac{1}{2}\mathbf{u}(x) = \int_{\Omega} \mathbf{U}(x,y)\mathbf{b}(y)\,\mathrm{d}\Omega_y - \int_{\Gamma_p} [\mathbf{u}(y)P\mathbf{U}(x,y) - \mathbf{U}(x,y)\mathbf{h}(y)]\,\mathrm{d}\Gamma_y, \quad \forall\, x \in \Gamma_p. \tag{2.105}$$

Equations (2.103), (2.104) and (2.105) are singular (cf. [35]), i.e. they have kernels which are integrable in the sense of the Cauchy principal value). The same type of equations can be obtained using the singular indirect formulation. A detailed discussion of this approach does not seem useful, since on one hand it is rarely applied in practice, and on the other hand its basic formulas can be easily derived from the results quoted in Sect. 2.2 if only the vector (one-column matrix) $\mathbf{b}$ is substituted for the function f and the definition of the operator P from Eq. (2.92) is adopted.

However, the nonsingular version of the indirect approach deserves a comment. In the case of elasticity equations a physical interpretation of the formulation can be produced, similarly to the physical sense attributed to the Green formula. Functions $\mu(x); x \in \hat{\Gamma}$ can be interpreted as a certain set of forces, chosen in such a way that the boundary conditions on the contour $\Gamma = \partial\Omega$ are satisfied. A certain set of internal surface forces (tractions) on the contour Γ is associated with each set of boundary forces $\mu(x)$ prescribed on the boundary $\hat{\Gamma}$, This observation justifies the name of the set of *fictitious forces* used for the function μ and the name of *fictitious load method* used for the approach.

2.5 BEM as Finite-Dimensional Approximation of MBIE

As mentioned before, the lack of an efficient method of solving the integral equations was an obstacle in dissemination of the MBIE. In approximate methods based on a finite-dimensional approximation it is necessary to solve large sets of algebraic equations, which practically became feasible only with the use of computers.

From a completely formal viewpoint the concept of finite-dimensional approximation of the MBIE is based on an idea common to all the approximate methods. It is assumed that a given problem can be considered in a space equipped with a Schauder base [29, 49] (each Banach space containing dense and countable sets is such a space). Since the Schauder base is countable, each function can be written in the form of the series:

$$\mathbf{u} = \sum_{i=1}^{\infty} \varphi_i \mathbf{e}_i, \tag{2.106}$$

where $\mathbf{e}_i$ are the elements of a minimum complete base (Schauder base), and $\varphi_i \in \boldsymbol{R}$ are the series coefficients. The approximation consists in the substitution of a finite series for the series (2.106):

$$\tilde{\mathbf{u}} = \sum_{i=1}^{l} \varphi_i \mathbf{e}_i; \qquad l \in \boldsymbol{N} \text{ – natural number set.} \tag{2.107}$$

Since the base is composed of known functions, the problem is reduced to finding the series coefficients in Eq. (2.107), which leads to a set of algebraic equations. A concept of approximation with a set of piecewise constant functions is the simplest and then we obtain

$$\tilde{\mathbf{u}} = \sum_{i=1}^{l} \varphi_i \chi_i, \tag{2.108}$$

where χ_i is a characteristic function for i-th interval Δ_i,

$$\chi_i = \begin{cases} 0\,; & \forall x \notin \Delta_i, \\ 1\,; & \forall x \in \Delta_i, \end{cases} \tag{2.109}$$

while $\varphi_i \in \boldsymbol{R}$ are real numbers. As can be seen this leads to the known collocation method.

The physical sense of such an approximation can be shown using the properties of the solution of the Flamant problem. The displacement of a point (x_n, y_n) caused by concentrated forces applied at points $(\xi, 0)$, cf. Fig. 2.3, can be expressed in the form:

$$u_i = \sum_{j=1}^{m} u_{ij} P_j, \tag{2.110}$$

where u_{ij} is the displacement resulting from the solution of the Flamant problem. As is well-known, the solution of the Flamant problem (in two-dimensional case, and the Bousinesque problem in three-dimensional case) has all the properties of so-called *influence functions*. The matrix of fundamental solutions has an identical physical sense. It is a matrix of influence functions, constructed however for an infinite body (and not half-infinite) and for loading with a concentrated volume force (and not with a boundary force). The Kelvin matrix thus represents displacements of a point $P \in \boldsymbol{E}^n$; $n = (2, 3)$ caused by a unit concentrated force acting at a point $Q \in \boldsymbol{E}^n$. The element $U_{ij}(P, Q)$ is an i-th coordinate of the displacement of the point P induced by a j-th unit coordinate of the force vector applied at the point Q (see Fig. 2.4).

The idea of the simplest finite-dimensional approximation of the MBIE is based on the interpretation of the fundamental solution matrix given above. Let the contour $\partial\Omega$ of the domain Ω be divided with nodal points into (in general curvilinear) segments $\Delta\partial\Omega_j$; $j \in (1, \ldots, n)$. The given function $\mathbf{k}(x)$; $x \in \Gamma$, which specifies the boundary condition, can be expressed in

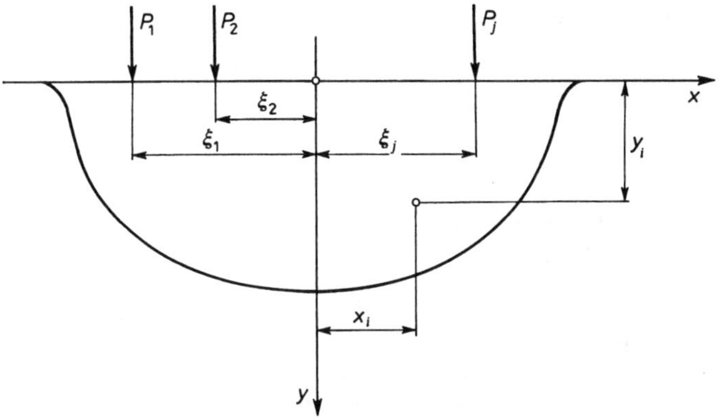

Fig. 2.3. Illustration of the Flamant problem

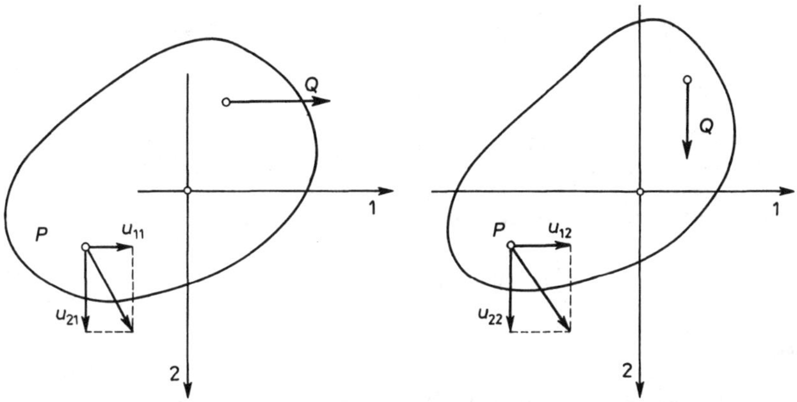

Fig. 2.4. Illustration of the components of the fundamental solution matrix

the form (2.108). Under the assumption that $\mathbf{k}_j$ is the value of the function $\mathbf{k}(x_j)$, with x_j - the centre of the segment $\Delta\partial\Omega_j$, we have:

$$\mathbf{k}(x) \approx \sum_{j=1}^{n} \chi_j \mathbf{k}_j; \qquad n \text{ – natural number,} \tag{2.111}$$

where

$$\chi_i = \begin{cases} 1\,; & \forall\, x \in \Delta\partial\Omega_j = \Delta\Gamma_j, \\ 0\,; & \forall\, x \notin \Delta\partial\Omega_j = \Delta\Gamma_j. \end{cases}$$

It is possible to express an unknown function $\mathbf{h}(x)$ in a similar way:

$$\mathbf{h}(x) = P\mathbf{u}(x) \approx \sum_{j=1}^{n} \chi_i \mathbf{h}_j, \tag{2.112}$$

with unknown coefficients χ_j.

With these assumptions the integral formula (2.101) takes the form:

$$\mathbf{u} = \int_{\Omega} \mathbf{U}(x, y)\mathbf{b}(y)\, \mathrm{d}\Omega_y + \sum_{j=1}^{n} \int_{\Delta\Gamma_j} \mathbf{k}_j(y) P\mathbf{U}(x, y)\, \mathrm{d}\Gamma_y$$
$$+ \sum_{j=1}^{n} \int_{\Delta\Gamma_j} \mathbf{U}(x, y)\mathbf{h}_j(y)\, \mathrm{d}\Gamma_y \,. \tag{2.113}$$

The first of the integrals can be calculated analytically or numerically. For convenience we denote:

$$\mathbf{u}_0 i = \mathbf{u}_0(x_i) = \int_{\Omega} \mathbf{U}(x_i, y)\mathbf{b}(y)\, \mathrm{d}\Omega_y \,; \quad x_i \in \Delta\Gamma_i \,. \tag{2.114}$$

Each of the integrals in the first sum can be written as:

$$\int_{\Delta\Gamma_j} \mathbf{k}_j(y) P\mathbf{U}(x_i, y)\, \mathrm{d}\Gamma_y = \mathbf{k}_j \mathbf{a}_{ij} \,; \quad x_i \in \Delta\Gamma_i, \tag{2.115}$$

with

$$\mathbf{a}_{ij} = \int_{\Delta\Gamma_j} P\mathbf{U}(x_i, y)\, \mathrm{d}\Gamma_y; \; x_i \in \Delta\Gamma_i \,; \quad y \in \Delta\Gamma_j \,.$$

The same refers to the integrals of the second sum, i.e.

$$\int_{\Delta\Gamma_j} \mathbf{u}(x_i, y)\mathbf{h}_j\, \mathrm{d}\Gamma_j = \mathbf{h}_j \mathbf{b}_{ij} \,, \tag{2.116}$$

where

$$\mathbf{b}_{ij} = \int_{\Delta\Gamma_j} \mathbf{U}(x_i, y)\, \mathrm{d}\Gamma_y; \; x_i \in \Delta\Gamma_i \,; \quad y \in \Delta\Gamma_j \,.$$

Finally, the expression (2.113) becomes an algebraic equation

$$\mathbf{u} = \mathbf{u}_{0i} + \sum_{j=1}^{n} \mathbf{a}_{ij}\mathbf{k}_j + \sum_{j=1}^{n} \mathbf{b}_{ij}\mathbf{h}_j, \qquad i \in (1, \ldots, n), \tag{2.117}$$

physical sense of which can easily be recognized.

The displacement of an arbitrary point $x_i \in \operatorname{Int}\Omega$ is a sum of displacements caused by the volume forces and the forces acting on the contour. The latter ones can be treated as reaction forces of the neglected part of the body, which supplements the domain and completes the space.

The approximation of the boundary integral equations has an analogical form

$$\frac{1}{2}\mathbf{k}_i = \mathbf{u}_{0i} + \sum_j \mathbf{a}_{ij}\mathbf{k}_j + \sum_j \mathbf{b}_{ij}\mathbf{h}_j \,. \tag{2.118}$$

It is a set of algebraic equations which can be used to determine:

- the unknown set of coefficients $\mathbf{h}_j$ in the case of a displacement boundary condition,
- the unknown set of coefficients $\mathbf{k}_j$ in the case of a stress boundary condition.

The presented physical interpretation of the finite-dimensional approximation has a sense only for boundary problems of the elasticity theory. However, the presented method, as one of formal possibilities, is quite universal.

It should be noticed that the elements of the coefficient matrices $\mathbf{a}_{ii}$ and $\mathbf{b}_{ii}$ are values of singular integrals, understood in the sense of the principal value. These integrals cannot be calculated numerically. There are methods of overcoming this difficulty and they will be discussed in Chap. 3. One of the simplest methods is of course the application of a non-singular version of the MBIE. In this case the set of equations can be automatically generated under the condition that the free components $\mathbf{u}_{0i}$ are calculated as a result of numerical integration, which is always possible.

To conclude, it should be emphasized that the set of equations (2.117) is non-symmetric. From the practical viewpoint, this is of course an essential inconvenience, however, an undoubted advantage of the method is a smaller (e.g. in comparison with the FEM) dimension of the problem, and the properties mentioned in Sect. 2.1. The other advantages of the method, which are important from the viewpoint of numerical methods, will be mentioned in Chap. 3.

2.6 Supplement

The rule of integration by parts and, consequently, the Green formula resulting from it, require that the boundary $\partial\Omega$ of the domain Ω be sufficiently regular. In the potential theory [21] this assumption becomes a condition that the boundary of the domain $\Omega \in \boldsymbol{E}^3$ should be a Lyapunov surface. The requirement has been quoted in the previous sections without an explanation of the term and therefore we now present the definition.

Definition 2.5. The Lyapunov surface is a surface that satisfies the following conditions:

1. There exists a unique normal vector at each point of the surface,
2. It is always possible to adopt a sufficiently small radius that a sphere with the radius and the centre at an arbitrary point x on the surface contains a part of the surface that has one and only one common point with an arbitrary straight line parallel to a normal at this point,
3. The measure of an angle between normals at points x and y does not exceed the quantity $A|x - y|$, where $|x - y|^\lambda$ is a distance between the points, A and λ are real numbers such that $0 < \lambda \leq 1$ and $A > 0$.

The above definition preserves its validity also in the case when the domain $\Omega \subset \boldsymbol{E}^2$ and its boundary is a curve. It is enough to replace the term "sphere" with "circle" in item 2. Similarly, replacing the term "sphere" with "hyper-sphere", the definition can be generalized to an n-dimensional case of a "hyper-surface".

The regularity requirements for the boundary of the domain can be reduced. In the MBIE literature it is usually assumed that the boundary is a surface composed of smooth segments. This means that with the exception of a finite number of curves the boundary is a C^1-continuous surface, and in a two-dimensional case the boundary is a C^1-curve with the exception of a finite number of points.

The segment-wise smooth surface becomes a Lyapunov surface under the additional assumption that the derivative satisfies the Hölder condition:

$$\frac{f_i'(\xi_k) - f_i'(0)}{|\xi_k|^\lambda} < A\,; \qquad \boldsymbol{R} \ni A > 0;\ i,k \in (1,2),\ 0 < \lambda \leq 1\,,$$

where ξ_k are the local coordinates of a point on the surface. This remark allows us to notice that all the quoted formulas still hold when the requirements concerning the degree of regularity of the boundary are reduced provided that one considers the possible jump of the derivative at the points where the condition 3 in the definition 2.5 is not satisfied. The value of the coefficient α in the abbreviated form of the integral formulas, e.g. in Eq. (2.29), must also be corrected. Instead of the value 1/2 the coefficient adopts a value from the interval (0, 1) depending on the angle between the left- and right-hand sided gradient at the discontinuity point [29]. In such cases, the formalism of the distribution theory turns out to be very useful [19, 20], although not indispensable.

The physical sense of the necessary modification of integral formulas is well illustrated by the Green identity in the plate theory, which contains components representing the derivative jump. Interpreting the integral formula as the Betti theorem (in analogy to the interpretation of the Green identity in the elasticity theory, shown in Sect. 2.4.3), the derivative jump has a sense of concentrated reaction forces at the plate corners (see Sect. 3.3.1).

Finally, the applicability of the method to hyperbolic equations should be referred to. The fundamental idea in this case is to perform a Laplace (or Fourier) transformation of the hyperbolic operator into an appropriate differential convolution operator (or, to be more precise, into a family of operators dependent on the transformation parameter), to which the formalism of Boundary Integral Equations can be applied. The details of such a procedure are discussed in Sect. 3.2 where the applications of the BEM to the solution of dynamic problems of elasticity are presented.

Other possibilities to extend the application of the MBIE are discussed in the examples included in subsequent chapters.

3 Boundary Element Method in Linear Theory of Elasticity

3.1 BEM in Statics of Elastic Medium

3.1.1 Boundary Integral Equations for Static Elasticity

The body considered is linear elastic and occupies a domain Ω in the Euclidean space $\boldsymbol{E}^d$ ($d = 2$ or 3) with a boundary $\partial\Omega \equiv \Gamma$. The boundary is assumed to be a Lyapunov surface. The body forces $\mathbf{b} = (b_i)$ and surface tractions $\mathbf{p} = (p_i)$ acting on the body induce a displacement field $\mathbf{u}(x) = (u_i(x))$, $x = (x_k)$ and associated strain and stress fields $\boldsymbol{\sigma}(x) = \{\sigma_{ij}(x)\}$ and $\boldsymbol{\epsilon}(x) = \{\epsilon_{ij}(x)\}$, respectively. Assuming that $\|\mathbf{u}(x)\| = \mathit{0}(x)$ and $\|\Delta\mathbf{u}(x)\| = \mathit{0}(\mathbf{u})$, which allows for a linear approximation of the deformation description (cf. Part I), the state of the body is defined by the following relations (cf. [44]):

– equilibrium equations

$$\operatorname{div}\boldsymbol{\sigma} + \mathbf{b} = \mathbf{0}\,, \qquad x \in \Omega\,, \tag{3.1}$$

– constitutive equations

$$\boldsymbol{\sigma} = \mathbf{C}\boldsymbol{\epsilon}\,, \tag{3.2}$$

where $\mathbf{C}$ is the elasticity matrix,

– geometric equations

$$\boldsymbol{\epsilon} = \frac{1}{2}\left(\nabla\mathbf{u} + \nabla^T\mathbf{u}\right), \tag{3.3}$$

supplemented by the boundary conditions

$$\mathbf{u}(x) = \mathbf{u}^0(x), \qquad x \in \Gamma_u\,, \tag{3.4}$$

$$\mathbf{p}(x) = \boldsymbol{\sigma}\mathbf{n} = \mathbf{p}^0(x), \qquad x \in \Gamma_p\,, \tag{3.5}$$

where $\Gamma_u \cup \Gamma_p = \Gamma$ and $\Gamma_u \cap \Gamma_p = \emptyset$.

In the case of a homogeneous and isotropic body the coefficients of the matrix $\mathbf{C}$ take the form:

$$C_{ijkl} = \lambda\delta_{ij}\delta_{kl} + \mu(\delta_{ik}\delta_{jl} + \delta_{il}\delta_{jk})\,, \tag{3.6}$$

where λ is the Lame constant, $\mu = G$ is the Kirchhoff constant and δ_{ij} is the Kronecker symbol. The presented set of equations can be substituted by so-called displacement equations (cf. Sect. 2.4.3). They are obtained by the substitution (3.3) → (3.2) → (3.1) and using Eq. (3.6). As described in detail in Sect. 2.4.3, these equations can be replaced by an equivalent set of integral equations:

$$\mathbf{c}(x)\mathbf{u}(x) + \int_\Gamma \mathbf{P}(x,y)\mathbf{u}(y)\,\mathrm{d}\Gamma(y) = \\ = \int_\Gamma \mathbf{U}(x,y)\mathbf{p}(y)\,\mathrm{d}\Gamma(y) + \int_\Omega \mathbf{U}(x,z)\mathbf{b}(z)\,\mathrm{d}\Omega(z)\,, \qquad (3.7)$$

in which

$$\mathbf{P}(x,y) \stackrel{\text{def}}{=} [P_y \mathbf{U}(x,y)]^T$$

and P is the stress vector operator, cf. Eq. (2.92). If a point $x \in \mathrm{Int}\Omega$ then $\mathbf{c} = \mathbf{I}$ and Eq. (3.7) is the well-known Somigliana integral formula. If, on the other hand, $x \in \Gamma$ then Eq. (3.7) becomes a boundary integral equation (cf. Sect. 2.4.3) and the coefficients $\mathbf{c}$ take the form:

$$\mathbf{c}(x) = \mathbf{I} + \lim_{\varepsilon\to 0} \int_{\Gamma_\varepsilon} \mathbf{P}(x,y)\,\mathrm{d}\Gamma(y)\,, \qquad (3.8)$$

where Γ_ε denotes a sphere with the centre at point x and a radius ε. The coefficients $\mathbf{c}$ can have the values $\mathbf{c} = \alpha\,\mathbf{I}$ with $\alpha \in (0,\,1)$, cf. [28], and the meaning of the parameter has been described in Sect. 2.6. For a boundary

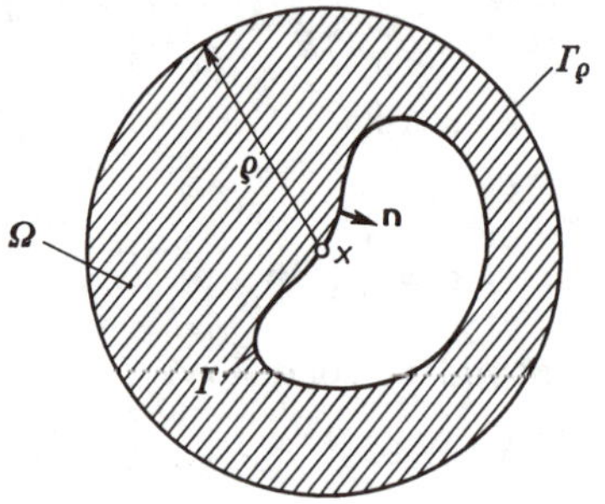

Fig. 3.1. Infinite domain

which is sufficiently smooth $\alpha = 1/2$ holds. For infinite domains $\bar{\Omega} = \boldsymbol{E}^3 \backslash (\Omega \cup \Gamma)$, i.e. for so-called external boundary value problem, Eq. (3.7) is still valid, but only if the same additional conditions are satisfied. Let Γ_ϱ be a sphere with a radius ϱ and the centre at point $x \in \Gamma$, where $\Gamma = \partial\Omega$ (Fig. 3.1).

Formula (3.7) takes now the form:

$$\mathbf{c}(x)\mathbf{u}(x) + \int_{\Gamma} \mathbf{P}(x,y)\mathbf{u}(y)\,\mathrm{d}\Gamma(y) + \int_{\Gamma_\varrho} \mathbf{P}(x,y)\mathbf{u}(y)\,\mathrm{d}\Gamma(y) = \\ = \int_{\Gamma} \mathbf{U}(x,y)\mathbf{p}(y)\,\mathrm{d}\Gamma(y) + \int_{\Gamma_\varrho} \mathbf{U}(x,y)\mathbf{p}(y)\,\mathrm{d}\Gamma(y) \\ + \int_{\Omega} \mathbf{U}(x,z)\mathbf{b}(z)\,\mathrm{d}\Omega(z)\,. \tag{3.9}$$

If the following condition

$$\lim_{\varrho\to\infty} \int_{\Gamma_\varrho} [\mathbf{P}(x,y)\mathbf{u}(y) - \mathbf{U}(x,y)\mathbf{p}(y)]\,\mathrm{d}\Gamma(y) = 0 \tag{3.10}$$

is satisfied then Eq. (3.9) takes the form (3.7). The condition (3.10) is satisfied if $\mathbf{u}(y) = O(\varrho^{-1})$ and $\mathbf{p}(y) = O(\varrho^{-2})$. It is easy to see that it guarantees the satisfaction of conditions in infinity by a solution of a well-posed equivalent boundary value problem. Beside the above direct formulation the indirect formulation presented in Sect. 2.4.3 can be employed. Considering the finite-dimensional approximation, the non-singular formulation described there is the most convenient.

3.1.2 Finite-Dimensional Approximation

The boundary integral equations, obtained from the formula (3.7) with $x \in \Gamma$, can be solved in an approximate manner discretizing the boundary Γ of the domain Ω and approximating the functions defined on the boundary (cf. [12, 31, 34, 54]).

Let us divide the boundary of an elastic body into boundary elements Γ^e, so that $\Gamma = \bigcup_{e=1}^{K} \Gamma^e$. For three-dimensional problems they are surface segments of triangular (Fig. 3.2a) or quadrilateral (Fig. 3.2b) shape. In a particular case of two-dimensional problems the boundary elements are line segments (Fig. 3.2c).

The division into elements defines a set of nodal points. Cartesian coordinates $x = (x_k)$ of each point within a boundary element Γ^e can be expressed in terms of coordinates of its nodal points $(x)_e^w = (x_k)_e^w\,, k = 1, \ldots, d-1$:

$$x(\xi) = \mathbf{M}^w(\xi)(x)_e^w\,, \qquad x \in \Gamma^e\,. \tag{3.11}$$

The matrix of interpolation functions (shape functions) $\mathbf{M}^w(\xi)$ can be expressed as follows:

$$\mathbf{M}^w(\xi) = \mathbf{I}\,M^w(\xi)\,, \tag{3.12}$$

where $\mathbf{I}$ is a square unit matrix with the dimensions $d \times d$ ($d = 3$ for three-dimensional problems, $d = 2$ for two-dimensional problems), $M^w(\xi)$, $w = 1, 2, \ldots, W_e$ is a shape function and W_e is a number of nodal points in element Γ^e.

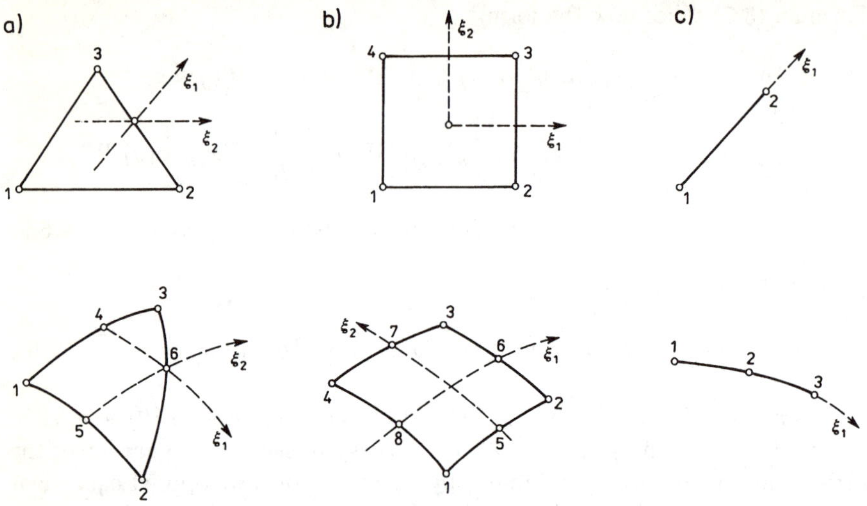

Fig. 3.2. Boundary elements: **a)** triangular, **b)** quadrilateral, **c)** boundary elements for two-dimensional problems

The shape functions $M^w(\xi)$ are defined in a local set of coordinates $(\xi) = (\xi_1, \xi_2)$ and can be described as in the finite element method (cf. Part I). For instance for a quadrilateral surface element (Fig. 3.2b) with a quadratic interpolation (number of nodes $W_e = 8$) the shape functions have the form:

$$
\begin{aligned}
M_1(\xi) &= \frac{1}{4}(1-\xi_1)(1-\xi_2)(-\xi_1-\xi_2-1)\,, \\
M_2(\xi) &= \frac{1}{4}(1+\xi_1)(1-\xi_2)(\xi_1-\xi_2-1)\,, \\
M_3(\xi) &= \frac{1}{4}(1+\xi_1)(1+\xi_2)(\xi_1+\xi_2-1)\,, \\
M_4(\xi) &= \frac{1}{4}(1-\xi_1)(1+\xi_2)(-\xi_1+\xi_2-1)\,, \\
M_5(\xi) &= \frac{1}{2}(1-\xi_1^2)(1-\xi_2)\,, \quad M_6(\xi) = \frac{1}{2}(-\xi_2^2)(1+\xi_1)\,, \\
M_7(\xi) &= \frac{1}{2}(1-\xi_1^2)(1+\xi_2)\,, \quad M_8(\xi) = \frac{1}{2}(-\xi_2^2)(1-\xi_1)\,.
\end{aligned}
$$

The shape functions for a quadratic curvilinear element in Fig. 3.2c (number of nodes $W_e = 3$) read:

$$
M_1(\xi) = \frac{1}{2}\,\xi_1(1+\xi_1)\,, \quad M_2(\xi) = 1-\xi_1^2\,, \quad M_3(\xi) = \frac{1}{2}\,\xi_1(\xi_1-1).
$$

The displacement field $\mathbf{u}(x)$ and the tractions $\mathbf{p}(x)$ are approximated in the local coordinate set $\xi = (\xi_i)$ of each element Γ^e using their nodal values

$(\mathbf{u})_e^w$ and $(\mathbf{p})_e^w$, respectively, and suitable shape functions:

$$\mathbf{u}(x(\xi)) \approx \tilde{\mathbf{u}}(x(\xi)) = \mathbf{N}^w(\xi)(\mathbf{u})_e^w, \qquad x \in \Gamma^e, \tag{3.13}$$

$$\mathbf{p}(x(\xi)) \approx \tilde{\mathbf{p}}(x(\xi)) = \bar{\mathbf{N}}^w(\xi)(\mathbf{p})_e^w, \qquad x \in \Gamma^e, \tag{3.14}$$

where $\mathbf{N}^w(\xi) = \mathbf{I}\,N_w(\xi)$ and $\bar{\mathbf{N}}^w(\xi) = \mathbf{I}\,\bar{N}_w(\xi)$ are shape function matrices, $N_w(\xi)$ and $\bar{N}_w(\xi)$ are the interpolation functions.

The functions $N_w(\xi)$ and $\bar{N}_w(\xi)$, similarly to the functions $M_w(\xi)$, are polynomials of order κ. In the case of $\kappa = 0$ piecewise constant approximation functions are used (cf. Sect. 2.5), and for $\kappa = 1$, 2 and 3 we have the approximation with a set of linear, quadratic and cubic functions, respectively. Usually, the same shape functions are used for the approximation of the displacements and boundary tractions, i.e. $N_w(\xi) = \bar{N}_w(\xi)$. The same interpolation functions are often used for approximation of both the boundary functions and the boundary itself, i.e. $\mathbf{N}^w(\xi) = \mathbf{M}^w(\xi)$, and we then have an isoparametric case.

As a result of the finite-dimensional approximation above the formula (3.7) takes the form:

$$\mathbf{c}(x)\mathbf{u}(x) + \sum_{e=1}^{K}\sum_{w=1}^{W_e}(\mathbf{u})_e^w \int_\Gamma^e \mathbf{P}[x, y(\xi)]\mathbf{N}^w(\xi)J(\xi)\,\mathrm{d}\Gamma(\xi) =$$

$$= \sum_{e=1}^{K}\sum_{w=1}^{W_e}(\mathbf{p})_e^w \int_\Gamma^e \mathbf{U}[x, y(\xi)]\bar{\mathbf{N}}^w(\xi)J(\xi)\,\mathrm{d}\Gamma(\xi) + \mathbf{B}(x), \tag{3.15}$$

where

$$\mathbf{B}(x) = \int_\Omega \mathbf{U}(x, z)\mathbf{b}(z)\,\mathrm{d}\Omega(z). \tag{3.16}$$

The Jacobian $J(\xi)$, defined by the relation

$$J(\xi) = \mathrm{d}\Gamma/\mathrm{d}\xi = \det[\mathbf{g}_1 \cdot \mathbf{g}_2], \qquad \mathbf{g}_i = [\partial\mathbf{M}^w(\xi)/\partial\xi_i](x)_e^w, \tag{3.17}$$

has been introduced because of the exchange of the global coordinate set for the local one $(\xi) = (\xi_1, \xi_2)$, with $\mathrm{d}\xi = \mathrm{d}\Gamma(\xi)$.

Introducing a global numbering of the boundary nodes x^β, $\beta = 1, 2, \ldots, W$, Eq. (3.15) can be rewritten as:

$$\mathbf{r}(x) \stackrel{\text{def}}{=} \sum_{\beta=1}^{W}[\mathbf{P}^\beta(x)\mathbf{u}_\beta - \mathbf{U}^\beta(x)\mathbf{p}_\beta] - \mathbf{B}(x), \tag{3.18}$$

where

$$\mathbf{P}^\beta(x) = \mathbf{c}(x) + \sum_e \int_{\Gamma^e} \mathbf{P}[x, y(\xi)]\mathbf{N}^\beta(\xi)J(\xi)\,\mathrm{d}\Gamma(\xi), \tag{3.18a}$$

$$\mathbf{U}^\beta(x) = \sum_e \int_{\Gamma^e} \mathbf{U}[x, y(\xi)]\bar{N^\beta}(\xi)J(\xi)\,\mathrm{d}\Gamma(\xi). \tag{3.18b}$$

The summation in Eqs. (3.18a) and (3.18b) is performed over all the boundary elements which contain node β, $\mathbf{u}_\beta$ and $\mathbf{p}_\beta$ denote the respective displacement and traction vectors at a node with coordinates $x = x^\beta$.

The residual vector $\mathbf{r}(x)$ defined by Eq. (3.18) is in general other than zero due to the approximate character of the assumed form of the displacements (3.13) and tractions (3.14). The unknown values of the nodal displacements and forces can be determined from the minimization of $\mathbf{r}(x)$ requiring that in each boundary element Γ^e, $e = 1, 2, \dots, K$ the following condition is satisfied:

$$\int_\Gamma^e \mathbf{T}^\alpha(x)\mathbf{r}(x)\,\mathrm{d}\Gamma(x) = 0\,, \tag{3.19}$$

where $\mathbf{T}^\alpha(x) = \mathbf{I}T_\alpha(x)$, $x \in \Gamma^e$, $e = 1, 2, \dots, K$ is a matrix of weight functions. In the case of the collocation the matrix of weight functions has the form:

$$\mathbf{T}^\alpha(x) = \mathbf{I}\,\delta(x - x^\alpha)\,, \tag{3.20}$$

where x^α are collocation points. It is most convenient to adopt the boundary nodes $x = x^\alpha$ as the collocation points and then the condition (3.15) reads:

$$\mathbf{r}(x^\alpha) = \mathbf{0}\,. \tag{3.21}$$

This implies that we require Eq. (3.15) to be satisfied at each boundary node $x = x^\alpha$, $\alpha = 1, 2, \dots, W$. In result we obtain the following algebraic formula:

$$\begin{aligned} \mathbf{c}_\alpha\mathbf{u}_\alpha + [\hat{\mathbf{H}}_{\alpha 1}\dots\hat{\mathbf{H}}_{\alpha\alpha}\dots\hat{\mathbf{H}}_{\alpha w}][\mathbf{u}_1\dots\mathbf{u}_\alpha\dots\mathbf{u}_w]^T &= \\ = [\mathbf{G}_{\alpha 1}\dots\mathbf{G}_{\alpha\alpha}\dots\mathbf{G}_{\alpha w}][\mathbf{p}_1\dots\mathbf{p}_\alpha\dots\mathbf{p}_w]^T + \mathbf{B}_\alpha\,, \end{aligned} \tag{3.22}$$

in which

$$\hat{\mathbf{H}}_{\alpha\beta} = \sum_e \int_\Gamma^e \mathbf{P}[x^\alpha, y(\xi)]\mathbf{N}^\beta(\xi)J(\xi)\,\mathrm{d}\Gamma(\xi)\,, \tag{3.23}$$

$$\mathbf{G}_{\alpha\beta} = \sum_e \int_\Gamma^e \mathbf{U}[x^\alpha, y(\xi)]\bar{N}^\beta(\xi)J(\xi)\,\mathrm{d}\Gamma(\xi) \tag{3.24}$$

and

$$\mathbf{c}(x^\alpha) = \mathbf{c}_\alpha, \qquad \mathbf{B}(x^\alpha) = \mathbf{B}_\alpha\,. \tag{3.25}$$

The matrices $\hat{\mathbf{H}}_{\alpha\beta}$ and $\mathbf{G}_{\alpha\beta}$ have the dimensions $d \times d$, where $d = 3$ or 2.

Eq. (3.22) written for all W boundary nodes takes the form:

$$\begin{bmatrix} \mathbf{H}_{11} & \mathbf{H}_{12} & \dots & \mathbf{H}_{1\alpha} & \dots & \mathbf{H}_{1w} \\ \mathbf{H}_{\alpha 1} & \mathbf{H}_{\alpha 2} & \dots & \mathbf{H}_{\alpha\alpha} & \dots & \mathbf{H}_{\alpha w} \\ \mathbf{H}_{w1} & \mathbf{H}_{w2} & \dots & \mathbf{H}_{w\alpha} & \dots & \mathbf{H}_{ww} \end{bmatrix} \begin{Bmatrix} \mathbf{u}_1 \\ \mathbf{u}_2 \\ \vdots \\ \mathbf{u}_\alpha \\ \vdots \\ \mathbf{u}_w \end{Bmatrix} =$$

$$= \begin{bmatrix} \mathbf{G}_{11} & \mathbf{G}_{12} & \dots & \mathbf{G}_{1\alpha} & \dots & \mathbf{G}_{1w} \\ \mathbf{G}_{\alpha 1} & \mathbf{G}_{\alpha 2} & \dots & \mathbf{G}_{\alpha\alpha} & \dots & \mathbf{G}_{\alpha w} \\ \mathbf{G}_{w1} & \mathbf{G}_{w2} & \dots & \mathbf{G}_{w\alpha} & \dots & \mathbf{G}_{ww} \end{bmatrix} \begin{Bmatrix} \mathbf{p}_1 \\ \mathbf{p}_2 \\ \vdots \\ \mathbf{p}_\alpha \\ \vdots \\ \mathbf{p}_w \end{Bmatrix} + \begin{Bmatrix} \mathbf{B}_1 \\ \mathbf{B}_2 \\ \vdots \\ \mathbf{B}_\alpha \\ \vdots \\ \mathbf{B}_w \end{Bmatrix}, \tag{3.26}$$

where

$$\begin{aligned} \mathbf{H}_{\alpha\beta} &= \hat{\mathbf{H}}_{\alpha\beta} && \text{for } \alpha \neq \beta , \\ \mathbf{H}_{\alpha\beta} &= \hat{\mathbf{H}}_{\alpha\beta} + \mathbf{c}_\alpha && \text{for } \alpha = \beta . \end{aligned} \tag{3.27}$$

Eq. (3.26) can be written in the following short form:

$$\mathbf{Hu} = \mathbf{Gp} + \mathbf{B} , \tag{3.28}$$

in which the column matrices $\mathbf{u}$ and $\mathbf{p}$ contain the nodal values of boundary displacements $\mathbf{u}_\alpha$ and tractions $\mathbf{u}_\alpha$, $\alpha = 1, 2, \dots , W$, respectively, and the column matrix $\mathbf{B}$ reflects the volume forces.

The integrals from the volume forces over Ω can be calculated numerically. For this purpose the domain Ω is also discretized to obtain a set of so-called internal cells Ω^q , $q = 1, 2, \dots , Q$. The geometry and description of the cells Ω^q, which have the form of volume segments (for $d = 3$) or surface segments (for $d = 2$), resembles finite elements. Cartesian coordinates $x = (x_k)$, $k = 1, \dots , d$ of a point within a cell Ω^q can be expressed in terms of nodal point coordinates $(x)_q^w = (x_k)_q^w$, $k = 1, \dots , d$:

$$x(\eta) = \mathbf{V}^v(\eta)(x)_q^v , \qquad x \in \Omega^q , \tag{3.29}$$

where $\mathbf{V}^v = \mathbf{I}\, V^v(\eta)$ and $V^v(\eta)$ are interpolation functions. The volume force $\mathbf{b}$ is approximated in a local coordinate set $(\eta) = (\eta_1, \eta_2, \eta_3)$ of each internal cell Ω^q by means of the nodal values $(\mathbf{b})_q^v$ and interpolation functions $\mathbf{V}^v(\eta)$:

$$\mathbf{b}\,(x(\eta)) \approx \tilde{\mathbf{b}}\,(x(\eta)) = \mathbf{V}^v(\eta)(\mathbf{b})_q^v , \qquad x \in \Omega^q . \tag{3.30}$$

Then, Eq. (3.16) becomes:

$$\mathbf{B}(x^\alpha) = \sum_{q=1}^{Q}\sum_{v=1}^{V}(\mathbf{b})_q^v \int_{\Omega^q} \mathbf{U}[x^\alpha, z(\eta)]\mathbf{V}^v(\eta)\psi(\eta)\,\mathrm{d}\Omega(\eta)\,, \tag{3.31}$$

in which the Jacobian $\psi(\eta)$ has the form:

$$\psi(\eta) = \mathrm{d}\Omega/\mathrm{d}\eta = \det[(\mathbf{a}_1 \times \mathbf{a}_2)\cdot \mathbf{a}_3]\,, \qquad \mathbf{a}_i = [\partial\mathbf{V}^v(\eta)/\partial\eta_i](x)_q^v\,, \tag{3.32}$$

with $\mathrm{d}\eta = \mathrm{d}\Omega(\eta)$.

It is worth mentioning that, for a certain class of volume forces, the integral over the domain Ω can be expressed as a boundary integral. The problem is discussed in Sect. 3.1.3.

Computation of the matrices $\mathbf{H}$ and $\mathbf{G}$ with an appropriate accuracy is an essential stage of discretization. Coefficients of the matrices depend on boundary integrals like:

$$J_{\alpha e}^w = \int_{\Gamma^e} F^w(x^\alpha, \xi)\,\mathrm{d}\xi\,, \tag{3.33}$$

where $F^w(x^\alpha, \xi) = R[x^\alpha, y(\xi)]N_w(\xi)J(\xi)$, $R = (R_{ij})$ and $R_{ij} = U_{ij}$ or $R_{ij} = P_{ij}$. In order to calculate $J_{\alpha e}^w$ it is first necessary to examine the behaviour of the expression under the integral $F^w(x^\alpha, \xi)$ within a boundary element Γ^e. Two cases can be distinguished: when $x^\alpha \notin \Gamma^e$ and when $x^\alpha \in \Gamma^e$.

In the first case the integrated expression is limited and the integral $J_{\alpha e}^w$ is computed using the Gauss quadrature:

$$J_{\alpha e}^w = \sum_{r=1}^{m_1}\sum_{s=1}^{m_2} A_r^{(m_1)} A_s^{(m_2)} F^w(x^\alpha, \xi^{(m_1)}\xi^{(m_2)})\,, \tag{3.34}$$

where $A_r^{(m_1)}$ and $A_s^{(m_2)}$ are weight coefficients of order m_1 and m_2, respectively. A proper choice of m_1 and m_2 should assure the same accuracy of computation of all matrix elements (cf. [34, 54]).

When $x^\alpha \in \Gamma^e$ the expression under the integral becomes singular. For three-dimensional problems it is a singularity of the $1/r$ type for $R_{ij} = U_{ij}$ and of the $1/r^2$ for $R_{ij} = P_{ij}$ with $r = |\mathbf{x} - \mathbf{y}|$. In this case the singular component has to be extracted from the expression under the integral and integrated analytically, while the remaining (non-singular) part of the expression is computed numerically (cf. [27]).

There is a method which allows a numerical calculation of the singular integral and the constant matrix $\mathbf{c}(x^\alpha)$ by considering a unit displacement of the body treated as rigid. We assume that $\mathbf{p} = \mathbf{0}$ and $\mathbf{b} = \mathbf{0}$ and, having applied the unit displacement $\mathbf{u} = \mathbf{I}_l$ in direction l ($l = 1$, 2 or 3), we obtain from Eq. (3.28)

$$\mathbf{H}\,\mathbf{I}_l = \mathbf{0} \tag{3.35}$$

or

$$\mathbf{c}_\alpha + \hat{\mathbf{H}}_{\alpha\alpha} + \sum_{\alpha\neq\beta} \hat{\mathbf{H}}_{\alpha\beta}) = \mathbf{0}\,. \tag{3.35a}$$

The diagonal submatrices $\mathbf{H}_{\alpha\alpha}$ can be determined from this equation as follows:

$$\mathbf{H}_{\alpha\alpha} = \mathbf{c}_\alpha + \hat{\mathbf{H}}_{\alpha\alpha} = -\sum_{\alpha\neq\beta} \hat{\mathbf{H}}_{\alpha\beta}\,. \tag{3.36}$$

For an infinite or half-infinite body the relation (3.36) reads:

$$\mathbf{H}_{\alpha\alpha} = \mathbf{I} - \sum_{\alpha\neq\beta} \mathbf{H}_{\alpha\beta}\,. \tag{3.37}$$

Considering boundary conditions, Eq. (3.28) takes the form of a set of algebraic equations:

$$\mathbf{AX} = \mathbf{WY} + \mathbf{B} \equiv \mathbf{F}\,, \tag{3.38}$$

where the column matrix $\mathbf{X}$ contains unknown nodal values of forces on the boundary Γ_u and displacements on the boundary Γ_p, the column matrix $\mathbf{Y}$ depends on the specified boundary conditions, and the square matrices $\mathbf{A}$ and $\mathbf{W}$ depend on the coefficients of matrices $\mathbf{H}$ and $\mathbf{G}$.

The matrix $\mathbf{A}$ is a full non-symmetric matrix. It is the main disadvantage of the presented collocation method. If the domain Ω is divided into subdomains Ω_s, $s = 1, 2, \ldots, S$ with respective boundaries Γ_s, then for each subdomain an equation like (3.28) can be written, which together with appropriate force and displacement compatibility conditions on Γ_s leads to a block-banded coefficient matrix $\mathbf{A}$ (see Sect. 3.1.4). This concept of filling the matrix $\mathbf{A}$ is called a subdomain method.

The described collocation method has found the largest application in computational practice (cf. [1, 2, 3, 12, 31, 34, 54]). On one hand it leads to a full and non-symmetric coefficient matrix $\mathbf{A}$, on the other hand it is simple and effective in applications, since it requires only one integration over the boundary.

A set of algebraic equations with a symmetric coefficient matrix can be obtained employing other approximations. However, the computation time necessary to determine coefficients of the latter matrix is much longer. If, for instance, the Galerkin method is used, the matrix of weight functions $\mathbf{T}^\alpha$ is equal to the matrix of shape functions $\mathbf{N}^\alpha$, $\alpha = 1, 2, \ldots, W$ and the formulas for submatrices $\mathbf{H}_{\alpha\beta}$ and $\mathbf{G}_{\alpha\beta}$ derived from Eq. (3.19) require a double integration over the boundary (cf. [45]).

The matrix equation (3.38) can be solved with one of the known methods, however, while choosing it one should remember that the coefficient matrix $\mathbf{A}$ is not a banded and symmetric matrix. Most frequently the Gauss elimination method is used to solve the set of equations (3.38).

Once the displacements and forces on the whole boundary Γ are known, the displacements inside the body are computed from Eq. (3.7) assuming $\mathbf{c} = \mathbf{I}$. Components of the stress tensor $\boldsymbol{\sigma} = (\sigma_{ij})$ are determined from a relation obtained applying the stress operator Σ to Eq. (3.7), cf. Eq. (2.89a) in Sect. 2.4.3:

$$\boldsymbol{\sigma}(x) = \int_{\Gamma} \mathbf{D}(x,y)\mathbf{P}(y)\,\mathrm{d}\Gamma(y) - \int_{\Gamma} \mathbf{S}(x,y)\mathbf{u}(y)\,\mathrm{d}\Gamma(y) + \int_{\Omega} \mathbf{D}(x,z)\mathbf{b}(y)\,\mathrm{d}\Omega(z)\,, \tag{3.39}$$

where

$$\mathbf{D}(x,y) = \Sigma_x[\mathbf{U}(x,y)], \qquad \mathbf{S}(x,y) = \Sigma_x[\mathbf{P}(x,y)], \qquad x \in \Omega,\ y \in \Gamma \tag{3.40}$$

and

$$\mathbf{D}(x,z) = \Sigma_x[\mathbf{U}(x,z)], \qquad x \in \Omega,\ z \in \Omega\,, \tag{3.41}$$

with the operator Σ_x defined by formula (2.89a).

It results from Eq. (3.39) that in order to determine the components of the stress state at an arbitrary point $x \in \Omega$ the distribution of all the displacements and tractions has to be known. Therefore the stresses are computed at the second stage after the set of algebraic equations (3.38) has been solved.

The Somigliana formula and the relation (3.39) have after discretization the following form:

$$\mathbf{u}(x) = \mathbf{G}(x)\mathbf{p} - \mathbf{H}(x)\mathbf{u} + \mathbf{B}(x), \qquad x \in \Omega, \tag{3.42}$$

$$\boldsymbol{\sigma}(x) = \mathbf{G}_D(x)\mathbf{p} - \mathbf{H}_S(x)\mathbf{u} + \mathbf{B}_D(x), \qquad x \in \Omega, \tag{3.43}$$

where the matrices $\mathbf{G}(x)$, $\mathbf{H}(x)$, $\mathbf{G}_D(x)$ and $\mathbf{H}_S(x)$ depend on boundary integrals with the respective kernels $\mathbf{U}(x,y)$, $\mathbf{P}(x,y)$, $\mathbf{D}(x,y)$ and $\mathbf{S}(x,y)$ calculated for a given point $x \in \Omega$ and for $y \in \Gamma$, while the column matrices $\mathbf{B}(x)$ and $\mathbf{B}_D(x)$ depend on integrals over the domain Ω with the respective kernels $\mathbf{U}(x,z)$ and $\mathbf{D}(x,z)$, and with the volume forces $\mathbf{b}(z)$ for $x \in \Omega$ and $z \in \Omega$. It is worth noting that while using Eqs. (3.42) and (3.43) the displacement and stress components at a given point $x \in \Omega$ are computed from a finite-dimensional approximation of appropriate analytical formulas. This assures a better accuracy than that obtained in other numerical methods which require a double approximation for stress computation (i.e. a finite-dimensional approximation of operator P acting on approximate, discrete values of displacements).

In many cases (e.g. in problems with large gradients of tractions, in sensitivity analysis and shape optimization) it is important to know the most accurate possible values of boundary tractions. One of the ways to determine the stresses at the boundary is to use Eq. (3.39) with $x \to \Gamma$. However, this approach is not convenient because of strong singularities occurring in the kernels $\mathbf{D}(x,y)$ and $\mathbf{S}(x,y)$ (e.g. in three-dimensional problems these singularities are $1/r^2$ and $1/r^3$, respectively). A simpler way to determine the tractions is to derive them from the known displacements and tractions. For this purpose, it is convenient to introduce a local Cartesian coordinate set $\bar{x}_k$, $k \in [1, 2, 3]$ with unit vectors $\bar{\mathbf{e}}_k$ at a point at which the stress state is to be found. The coordinate set is such that the axes $\bar{x}_1$ and $\bar{x}_2$ are tangential to the boundary and $\mathbf{n} = \bar{\mathbf{e}}_3$. A part of the stress state components can then be expressed in terms of tractions $\bar{p}_k$:

$$\bar{\sigma}_{k3} = \bar{p}_k, \qquad k = 1, 2, 3. \tag{3.44}$$

The other components of the stress state are calculated using the constitutive equation:

$$\begin{aligned} \bar{\sigma}_{11} &= \frac{1}{1-\nu}[\nu\bar{p}_3 + 2\mu(\bar{\epsilon}_1 1 + \nu\bar{\epsilon}_2 2)] \, , \\ \bar{\sigma}_{22} &= \frac{1}{1-\nu}[\nu\bar{p}_3 + 2\mu(\bar{\epsilon}_2 2 + \nu\bar{\epsilon}_1 1)] \, . \end{aligned} \tag{3.45}$$

The displacement distribution within a boundary element is expressed as:

$$\bar{\mathbf{u}}^e = \mathbf{Q}^T \mathbf{N} \mathbf{u}^e w \, , \tag{3.46}$$

where $\mathbf{Q} = \bar{\mathbf{e}}_\alpha \otimes \mathbf{e}_i$ is a transformation matrix and the symbol ($\otimes$) denotes the (outer) tensor product.

The strains $\bar{\epsilon}_{ij}$, determined in the local coordinate set, are obtained by differentiation of the displacements from Eq. (3.46):

$$\bar{\epsilon}_{ij} = \frac{1}{2}\left(\frac{\partial \bar{u}_j}{\partial \bar{x}_i} + \frac{\partial \bar{u}_i}{\partial \bar{x}_j}\right) \, . \tag{3.47}$$

Consequently , they depend on shape function derivatives and nodal displacement values.

3.1.3 Particular Form of Volume Forces

In order to enable integrals over the domain Ω to be numerically calculated, a discretization of the interior of the body by means of cells Ω^q , $q = 1, 2, \dots, Q$ has been proposed. However, there is a broad class of volume forces for which the integrals over the domain Ω can be transformed into boundary integrals. This is the case for conservative volume forces. They can be represented as a gradient of a certain scalar function called a volume force potential. This class encompasses the self weight and centrifugal forces emerging from a constant velocity rotation of a body around a fixed axis. A load resulting from a stationary temperature field $T(x)$ which satisfies the equation

$$\Delta T(x) = 0 \, , \tag{3.48}$$

also belongs to this class of volume forces.

The transformation of integrals over the domain Ω into integrals over the boundary Γ is performed expressing the fundamental solution for the elasticity theory U_{ij} in terms of the Galerkin tensor G_{ij}:

$$U_{ij} = \left[G_{ij,k} - \frac{1}{2(1-\nu)} G_{ik},j \right]_{,k} , \tag{3.49}$$

where the tensor G_{ij} is defined as

$$G_{ij} = (r/a)\delta_{ij} \tag{3.50}$$

for three-dimensional problems and as

$$G_{ij} = r^2 \ln(r^{-1})/a \tag{3.51}$$

for two-dimensional problems and the notation $a = 8\pi\mu$ is used.

If an elastic body with a constant mass density ϱ is placed in a stationary gravitational field g_i, then

$$b_j = \varrho g_j \tag{3.52}$$

and the volume integral expressing the volume force influence is transformed into the boundary integral:

$$\begin{aligned} B_i &= \varrho g_j \int_\Omega \left(G_{ij,k} - \frac{1}{2(1-\nu)} G_{ik,j} \right)_{,k} \mathrm{d}\Omega \\ &= \varrho g_j \int_\Gamma \left(G_{ij,k} - \frac{1}{2(1-\nu)} G_{ik,j} \right) n_k \, \mathrm{d}\Gamma . \end{aligned} \tag{3.53}$$

If a body rotates with a constant angular velocity ω around an axis containing the origin of a coordinate system, then the volume forces are defined by the equation

$$b_j = g_{ij} x_i , \tag{3.54}$$

in which $g_{ij} = \varrho(\delta_{ij}\omega_m\omega_m - \omega_i\omega_j)$. The volume force is then reduced to a boundary integral of the form:

$$B_i = g_{ij} \int_\Omega \left\{ x_j \left(G_{ik,m} - \frac{1}{2(1-\nu)} G_{im,k} \right) n_m - \frac{(1-2\nu)}{2(1-\nu)} G_{ik} n_j \right\} \mathrm{d}\Gamma . \tag{3.55}$$

For a stationary temperature field $T(x)$ the volume force influence is determined by the following boundary integral:

$$B_i = G \left(\frac{1+\nu}{1-\nu} \right) \alpha \int_\Gamma (G_{ik,kj} T - G_{ik,k} T_{,j}) n_j \, \mathrm{d}\Gamma , \tag{3.56}$$

where α is the thermal expansion coefficient. The temperature T and its derivative $T_{,i}$ on the boundary Γ is obtained after solving the boundary value problem described by the differential equation (3.48) with appropriate boundary conditions (cf. Sect. 2.4.1).

3.1.4 Bodies of Revolution

In the case of bodies of revolution the application of the BEM and the Fourier series reduces the dimension of the problem by two. The boundary integral equations are then defined only on the generator L, the rotation of which around the symmetry axis generates the surface of the body (cf. [47]).

Introducing a cylindrical coordinate system (ϱ, θ, z) and expanding displacements $\mathbf{u}(x)$ and tractions $\mathbf{p}(x)$ into a Fourier series we obtain:

$$\mathbf{m}(x) = 0.5\mathbf{m}^0(x) + \sum_{n=1}^{\infty}[{}^s\mathbf{m}^n(\hat{x})\sin n\theta_x + {}^c\mathbf{m}^n(\hat{x})\cos n\theta_x]\,, \quad \mathbf{m} = \mathbf{u}, \mathbf{p}\,. \tag{3.57}$$

The fundamental solutions can be similarly developed:

$$\mathbf{T}^{n\{{}^s_c\}}(\hat{x},\, \hat{y}) = \int_{-\pi}^{\pi} \mathbf{T}(x,\, y) \begin{Bmatrix} \sin n\theta \\ \cos n\theta \end{Bmatrix} \mathrm{d}\theta\,, \qquad \mathbf{T} = \mathbf{U},\, \mathbf{P}\,;\; \theta = \theta(x) - \theta(y)\,, \tag{3.58}$$

where the symbol ($\hat{}$) above x and y means that the points x and y are defined by coordinates ϱ and z, ${}^s\mathbf{m}^n$ and ${}^c\mathbf{m}^n$ are Fourier expansion coefficients.

The boundary integral equation in the direct version can be expressed in terms of two vector equations (for simplicity volume forces have been neglected):

$$\begin{aligned}\mathbf{c}(\hat{x})^s\mathbf{u}^n(\hat{x}) = \int_l \{&[\mathbf{U}^{nc}(\hat{x},\hat{y})^s\mathbf{p}^n(\hat{y}) - \mathbf{U}^{ns}(\hat{x},\hat{y})^c\mathbf{p}^n(\hat{y})] \\ &- [\mathbf{P}^{nc}(\hat{x},\hat{y})^s\mathbf{u}^n(\hat{y}) - \mathbf{P}^{ns}(\hat{x},\hat{y})^c\mathbf{u}^n(\hat{y})]\}\, \varrho(\hat{y})\, \mathrm{d}L(\hat{y})\,,\end{aligned} \tag{3.59}$$

$$\begin{aligned}\mathbf{c}(\hat{x})^c\mathbf{u}^n(\hat{x}) = \int_l \{&[\mathbf{U}^{nc}(\hat{x},\hat{y})^c\mathbf{p}^n(\hat{y}) + \mathbf{U}^{ns}(\hat{x},\hat{y})^s\mathbf{p}^n(\hat{y})] \\ &- [\mathbf{P}^{nc}(\hat{x},\hat{y})^c\mathbf{u}^n(\hat{y}) + \mathbf{P}^{ns}(\hat{x},\hat{y})^s\mathbf{u}^n(\hat{y})]\}\, \varrho(\hat{y})\, \mathrm{d}L(\hat{y})\,,\end{aligned} \tag{3.60}$$

$$n = 0, 1, 2, \ldots \infty\,,$$

where l is the generator of the axisymmetrical surface. It is discretized with one-dimensional boundary elements (cf. Fig. 3.2c) similarly to two-dimensional problems. Finally, we obtain the discrete form of the integral equations (3.59) and (3.60):

$$\mathbf{H}_n^n\mathbf{u}^* = \mathbf{G}_n^n\mathbf{p}^*\,, \tag{3.61}$$

where $\mathbf{H}_n$ and $\mathbf{G}_n$ are square matrices $6W \times 6W$, W is a total number of nodes on L, ${}^n\mathbf{u}^*$ and ${}^n\mathbf{p}^*$ are column matrices, which contain $6W$ values of nodal displacements ${}^s\mathbf{u}^n$, ${}^c\mathbf{u}^n$ and tractions ${}^s\mathbf{p}^n$, ${}^c\mathbf{p}^n$, respectively.

If, in addition to the axisymmetrical geometry of the body, boundary conditions are also axisymmetrical, then Eq. (3.60) with ${}^s\mathbf{p}^n$ and ${}^s\mathbf{u}^n$ equal to zero is sufficient to solve the problem, and the matrices in Eq. (3.60) reduce to the dimension $3W \times 3W$ $(n = 0)$. For a non-symmetrical boundary condition Eq. (3.61) is solved successively for each value of n.

3.1.5 Anisotropic and Heterogeneous Bodies

The elastic properties of an anisotropic material are incorporated in the tensor $\mathbf{C} = (c_{ijkl})$. Because of the symmetry of the elasticity tensor: $c_{ijkl} = c_{jikl} = c_{klij}$, there are only 21 non-zero components of the $\mathbf{C}$ tensor. It is a maximum number of functions describing the most general case of anisotropy of an elastic material. For a higher order of symmetry the number of functions is smaller. For an anisotropic body the displacement equation quoted in Sect. 2.4.3 changes. If the body is homogeneous, then components of the elasticity tensor $\mathbf{C}$ are constant and the equation becomes:

$$\frac{1}{2}\mathbf{C}\,[\operatorname{div}\operatorname{grad}\mathbf{u} + \operatorname{div}\operatorname{grad}^T\mathbf{u}] + \mathbf{b} = \mathbf{0}\,. \tag{3.62}$$

It is a type of equation discussed earlier in Sect. 2.2. Eq. (3.62) can be replaced by an equivalent integral formula which is formally identical with Eq. (3.7). A fundamental solution of this equation is a particular case of the solution quoted in Sect. 1.2 and reads (cf. [51]):

$$U_{ij}(x,y) = \frac{1}{8\pi^2|x-y|}R_{ij}(v_1,v_2)\,, \tag{3.63}$$

with

$$R_{ij}(v_1,v_2) = \oint_{|\xi|=1} [c_{ijkm}\xi_k\xi_m]^{-1}\,\mathrm{d}s(\xi)\,. \tag{3.64}$$

In a general case, for a given type of anisotropy, the integral is calculated numerically [55]. The kernel $\mathbf{P}(x,\,y)$ is defined as: $\mathbf{P}(x,\,y) = [P_{n(y)}\mathbf{U}(y,\,x)]^T$, where the components of the stress vector $P_n(\mathbf{u})$ have the form: $(P_n\mathbf{u})_i = c_{ijkl}u_{k,l}n_j$.

In the case of heterogeneous bodies the components of tensor $\mathbf{C}$ are functions and the displacement equation is written as:

$$\frac{1}{2}\operatorname{div}\{\mathbf{C}\,[\operatorname{grad}\mathbf{u} + \operatorname{grad}^T\mathbf{u}]\} + \mathbf{b} = \mathbf{0}\,. \tag{3.65}$$

It is an elliptic equation, but with coefficients which are functions. Such equations can be reformulated into an integral form (cf. Sect. 2.2), but fundamental solutions in a closed form are known only for some particular cases of heterogeneity. The formalism of boundary integral equations is in this case problematic. Nevertheless, it can be applied with success if the heterogeneity can be approximated by piecewise constant functions.

Let us divide the domain of the body Ω into subdomains Ω_m such that $\bigcup_m \Omega_m = \Omega$ and $\bigcap_m \Omega_m = \emptyset$. The heterogeneous body can be "approximated" by a set of homogeneous bodies with properties defined by constant

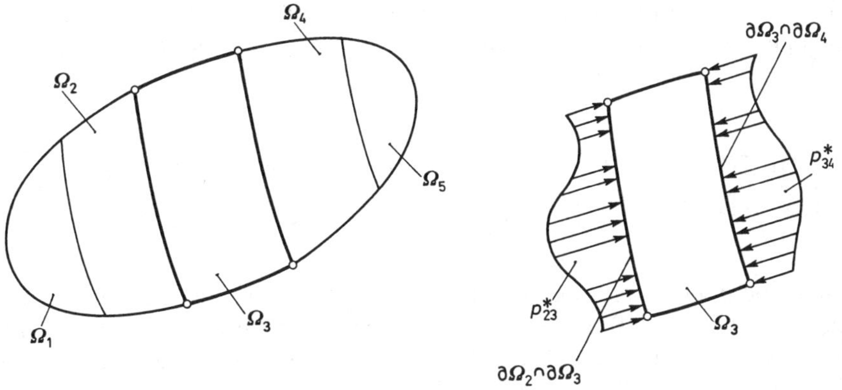

Fig. 3.3. Piecewise heterogeneous body: **a)** subdomain division, **b)** extracted subdomain Ω_3

$c_{ijkl}^{(m)}$ if we assume that

$$c_{ijkl}(x) = \sum_m c_{ijkl}^{(m)} \kappa^{(m)} \; ; \qquad \forall_{ijkl} \; , \tag{3.66}$$

where

$$\kappa^{(m)} = \begin{cases} 0 \; ; \; \forall x \notin \Omega_m \; , \\ 1 \; ; \; \forall x \in \Omega_m \; . \end{cases}$$

For each of such bodies a boundary value problem can be formulated. Reaction forces $\mathbf{p}_{m-1,m}^*$ now occur on the separation surfaces, which should be determined from displacement compatibility conditions for adjacent subdomains:

$$\mathbf{u}_{m-1}^* = \mathbf{u}_m^* \; , \qquad x \in \partial\Omega_{m-1} \, , \cap \partial\Omega_m \; , \tag{3.67}$$

with $\mathbf{p}_{m-1,m}^* = -\mathbf{p}_{m,m-1}^*$. The boundary value problem for each of the approximating bodies can be solved using the formalism of boundary integral equations. In the particular case when $M \in (1, 2)$, i.e. $\Omega = \Omega_1 \cup \Omega_2$ and $\Gamma^* = \partial\Omega_1 \cap \partial\Omega_2$, after a finite-dimensional approximation two sets of algebraic equations is obtained:

$$[\mathbf{H}_1 \mid \mathbf{H}_1^*] \begin{bmatrix} \mathbf{u}_1 \\ \mathbf{u}_1^* \end{bmatrix} = [\mathbf{G}_1 \mid \mathbf{G}_1^*] \begin{bmatrix} \mathbf{p}_1 \\ \mathbf{p}_{1,2}^* \end{bmatrix} + \mathbf{B}_1 \; , \tag{3.68a}$$

$$[\mathbf{H}_2^* \mid \mathbf{H}_2] \begin{bmatrix} \mathbf{u}_2^* \\ \mathbf{u}_2 \end{bmatrix} = [\mathbf{G}_2^* \mid \mathbf{G}_2] \begin{bmatrix} \mathbf{p}_{2,1}^* \\ \mathbf{p}_2 \end{bmatrix} + \mathbf{B}_2 \; . \tag{3.68b}$$

Considering the displacement compatibility conditions $\mathbf{u}_1^* = \mathbf{u}_2^* = \mathbf{u}^*$ and the indicated property of the reactions $\mathbf{p}_{1,2}^* = \mathbf{p}_{2,1}^* = \mathbf{p}^*$, we obtain a matrix equation valid in the whole domain Ω

$$\begin{bmatrix} \mathbf{H}_1 & \mathbf{H}_1^* & -\mathbf{G}_1^* & \mathbf{0} \\ \mathbf{0} & \mathbf{H}_2^* & \mathbf{G}_2^* & \mathbf{H}_2 \end{bmatrix} \begin{Bmatrix} \mathbf{u}_1 \\ \mathbf{u}^* \\ \mathbf{p}^* \\ \mathbf{u}_2 \end{Bmatrix} = \begin{bmatrix} \mathbf{G}_1 & \mathbf{0} \\ \mathbf{0} & \mathbf{G}_2 \end{bmatrix} \begin{Bmatrix} \mathbf{p}_1 \\ \mathbf{p}_2 \end{Bmatrix} + \begin{Bmatrix} \mathbf{B}_1 \\ \mathbf{B}_2 \end{Bmatrix} . \tag{3.69}$$

Taking boundary conditions into account Eq. (3.69) can be transformed into the form of Eq. (3.38), in which the unknown quantities on the boundary $\partial\Omega$ as well as the displacements and forces on $\Gamma^* = \partial\Omega_1 \cap \partial\Omega_2$ are contained in the column matrix $\mathbf{X}$ while $\mathbf{A}$ is a block-banded matrix.

3.2 BEM in Dynamics of Elastic Medium

3.2.1 Boundary Integral Equations for Dynamic Elasticity

Let us consider an elastic body occupying domain $\Omega \in \boldsymbol{E}^d$ ($d = 2$ or 3), subjected to external volume forces $\mathbf{b}(x, t)$, $x \in \Omega$, with the boundary conditions:

$$\mathbf{u}(x, t) = \mathbf{u}_0(x, t) ; \qquad \forall\, x \in \Gamma_u , \tag{3.70}$$

$$\mathbf{p}(x, t) = \mathbf{p}_0(x, t) ; \qquad \forall\, x \in \Gamma_p , \tag{3.71}$$

where $t \in T = [0, t_k]$ is the time, $\Gamma_u \cup \Gamma_p = \Gamma$ and $\Gamma_u \cap \Gamma_p = \emptyset$. Within the linear theory, the motion of such a body is described by the hyperbolic differential equation (cf. [44]):

$$L_d\mathbf{u}(x,t) \equiv L\mathbf{u}(x,t) - \varrho\ddot{\mathbf{u}}(x,t) = -\mathbf{b}(x,t) , \tag{3.72}$$

where the elliptic operator L is a Lame operator, see Eq. (2.88) in Sect. 2.4.3, and ϱ is mass density. The solution $\mathbf{u}(x, t)$ of Eq. (3.72) should additionally satisfy the initial conditions:

$$\mathbf{u}(x,0) = \mathbf{u}^0(x) , \quad \dot{\mathbf{u}}(x,0) = \dot{\mathbf{u}}^0(x) , \qquad \forall\, x \in \Omega . \tag{3.73}$$

The solution can be extended $\forall\, t \in (-\infty, \infty)$. We then obtain:

$$L_d\hat{\mathbf{u}}(x,t) = -\hat{\mathbf{b}}(x,t) , \ \forall\, t \in (-\infty,\infty) , \tag{3.74}$$

where $\hat{\mathbf{u}}(x,t) = H(t)\mathbf{u}(x,t)$, $H(t)$ is the Heaviside function and $\hat{\mathbf{b}}(x,t)$ is determined from the relation:

$$\hat{\mathbf{b}}(x,t) = H(t)\mathbf{b}(x,t) + \varrho[\mathbf{u}^0(x)\dot{\delta}(t) + \dot{\mathbf{u}}^0(x)\delta(t)] , \tag{3.75}$$

with $\delta(t)$ and $\dot{\delta}$ are the Dirac distribution and its derivative respectively. The theorem of reciprocal work for two different dynamic states: $\hat{\mathbf{u}}$, $\hat{\mathbf{p}}$ and $\hat{\mathbf{b}}$ at time τ, and $\hat{\mathbf{u}}'$, $\hat{\mathbf{p}}'$ and $\hat{\mathbf{b}}'$ at time $t-\tau$ has the form:

$$\int_0^t \int_\Omega [\hat{\mathbf{b}}(x,t)\hat{\mathbf{u}}'(x,t-\tau) - \hat{\mathbf{b}}'(x,t-\tau)\hat{\mathbf{u}}(x,\tau)]\, \mathrm{d}\tau \mathrm{d}\Omega$$
$$+ \int_0^t \int_\Gamma [\hat{\mathbf{p}}(x,t)\hat{\mathbf{u}}'(x,t-\tau) - \hat{\mathbf{p}}'(x,t-\tau)\hat{\mathbf{u}}(x,\tau)]\, \mathrm{d}\tau \mathrm{d}\Gamma = 0\,. \quad (3.76)$$

We assume that the fields of displacements $\hat{\mathbf{u}}' = \mathbf{U}(x,y,t)$ and forces $\hat{\mathbf{p}}' = \mathbf{P}(x,y,t)$ is caused by a transient unit concentrated force (i.e. an impulse) $\hat{\mathbf{b}}' = \delta_{1j}\delta(x-y)\delta(t)$ acting at point y of an infinite elastic space. Matrices $\mathbf{U}(x,y,t) = [U_{ij}(x,y,t)]$ and $\mathbf{P}(x,y,t) = [P_{ij}(x,y,t)]$ are fundamental solutions of the dynamic elasticity theory, which can be expressed as follows:

$$U_{ij}(x,y,t) = \frac{1}{4\pi\varrho c_2^2}\left\{\frac{\delta_{ij}}{r}\delta\left(t-\frac{r}{c_2}\right)\right.$$
$$\left. + c_2^2\left\{\frac{1}{r}\left[\left(t-\frac{r}{c_1}\right)H\left(t-\frac{r}{c_1}\right) - \left(t-\frac{r}{c_2}\right)H\left(t-\frac{r}{c_2}\right)\right]\right\}_{,ij}\right\} \quad (3.77)$$

for three-dimensional problems and

$$U_{ij}(x,y,t) = \frac{1}{2\pi\mu}\left\{\delta_{ij}f_0\left(t,\frac{r}{c_2}\right) + c_2^2\left[f_2\left(t,\frac{r}{c_1}\right) - f_2\left(t,\frac{r}{c_2}\right)\right]_{,ij}\right\}, \quad (3.78)$$

with

$$f_0(t,a) = \frac{H(t-a)}{\sqrt{t^2-a^2}}\,,$$
$$f_2(t,a) = H(t-a)\{t\ln(t+\sqrt{t^2-a^2}) - \sqrt{t^2-a^2}\}\,,$$

for two-dimensional problems (plane strain conditions). The tensor $\mathbf{P}(x,y,t)$ can in general be expressed as:

$$\mathbf{P}(x,y,t) = [P_{ij}(x,y,t)]_{d\times d} = [P_{n(y)}\mathbf{U}(x,y,t)]^T\,, \quad (3.79)$$

where P_n is the operator defined in Eq. (2.89) and $d = \dim(\boldsymbol{E}^d)$. In Eqs. (3.77) and (3.78) $c_1 = [(\lambda+2\mu)/\varrho]^{1/2}$ denotes the velocity of a dilatational wave, and $c_2 = (\mu/\varrho)^{1/2}$ denotes the velocity of a torsional wave in an elastic medium. With the above relations, the principle of reciprocal work allows us to write an integral formula for a dynamic problem of the elasticity theory[1]:

$$\mathbf{c}(x)\hat{\mathbf{u}}(x,t) + \int_{\Gamma} \mathbf{P}(x,y,t) * \hat{\mathbf{u}}(y,t)\,\mathrm{d}\Gamma(y) =$$
$$= \int_{\Gamma} \mathbf{U}(x,y,t) * \hat{\mathbf{p}}(y,t)\,\mathrm{d}\Gamma(y) + \int_{\Omega} \mathbf{U}(x,y,t) * \hat{\mathbf{b}}(y,t)\,\mathrm{d}\Omega(y)\,, \quad (3.80)$$

where the symbol (*) denotes a time convolution. The formula is valid for internal problems (i.e. for limited bodies) and for external problems (i.e. for an unlimited elastic medium).

The boundary integral equation constructed on the basis of Eq. (3.80) can be solved in the time domain, using a time step method, or in the transform domain, using the Laplace or Fourier transform.

3.2.2 Time Step Method

There are two ways of solving the integral equation (3.80) with the time step method. In the first approach each time step Δt is considered as a new problem. At the end of each step displacements and velocities inside the domain Ω must be computed and treated as pseudo-initial conditions for the next step. In the second approach integration over time is carried out always starting from the initial moment $t = 0$ and therefore the displacements and velocities do not have to be calculated for intermediate steps. The second approach is especially effective for problems with homogeneous initial conditions. Below the second approach is presented, in which, to simplify the discussion, such initial conditions and zero volume forces are adopted.

The time interval $t \in [0, t_k]$ is divided into M time steps $\Delta t = t_m - t_{m-1}\,,\ m = 1, 2, \ldots, M$. Assuming that $\hat{\mathbf{b}} = \mathbf{0}$ the integral formula (3.80) can be written for time t_m as:

$$\mathbf{c}(x)\mathbf{u}(x,t_m) + \int_{t_{m-1}}^{t_m}\int_{\Gamma} [\mathbf{P}(x,y,t_m-\tau)\mathbf{u}(y,\tau) - \mathbf{U}(x,y,t_m-\tau)\mathbf{p}(y,\tau)\,\mathrm{d}\Gamma\mathrm{d}\tau$$
$$= \sum_{r=1}^{m-1}\int_{t_{r-1}}^{t_r} [\mathbf{P}(x,y,t_m-\tau)\mathbf{u}(y,\tau) - \mathbf{U}(x,y,t_m-\tau)\mathbf{p}(y,\tau)\,\mathrm{d}\Gamma\mathrm{d}\tau,$$
$$m = 1, 2, \ldots, M. \quad (3.81)$$

It is assumed that the displacement field $\mathbf{u}(y,\tau)$ and traction field $\mathbf{p}(y,\tau)$ vary linearly within each time step Δt:

$$\begin{aligned} \mathbf{u}(y,\tau) &= N_1^m \mathbf{u}^{m-1}(y) + N_2^m \mathbf{u}^m(y)\,, \\ \mathbf{p}(y,\tau) &= N_1^m \mathbf{p}^{m-1}(y) + N_2^m \mathbf{p}^m(y)\,, \end{aligned} \quad (3.82)$$

[1] In order to simplify the notation, symbol "$\hat{()}$" will be omitted in further discussion.

where the interpolation functions have the form:

$$\begin{aligned} N_1^m &= \frac{t_m - \tau}{\Delta t} \{H[\tau - (m-1)\Delta t] - H[\tau - m\Delta t]\} \ , \\ N_2^m &= \frac{\tau - t}{\Delta t} \{H[\tau - (m-1)\Delta t] - H[\tau - m\Delta t]\} \ , \end{aligned} \tag{3.83}$$

moreover $\mathbf{u}^m(y) = \mathbf{u}(y, t_m)$, $\mathbf{p}^m(y) = \mathbf{p}(y, t_m)$ and H is the Heaviside function. The integral formula (3.81) now takes the form:

$$\begin{aligned} \mathbf{c}(x)\mathbf{u}(x,t_m) + \int_\Gamma \mathbf{P}^m(x,y)\mathbf{u}(y,t_m)\mathrm{d}\Gamma(y) &= \\ = \int_\Gamma \mathbf{U}^m(x,y)\mathbf{p}(y,t_m)\mathrm{d}\Gamma(y) + \mathbf{R}(x,t_m) \ , \end{aligned} \tag{3.84}$$

where

$$\mathbf{U}^m(x,y) = \frac{1}{\Delta t}\int_{t_{m-1}}^{t_m} (\tau - t_{m-1})\mathbf{U}(x,y,t_m - \tau)\,\mathrm{d}\tau \ , \tag{3.84a}$$

$$\mathbf{P}^m(x,y) = \frac{1}{\Delta t}\int_{t_{m-1}}^{t_m} (\tau - t_{m-1})\mathbf{P}(x,y,t_m - \tau)\,\mathrm{d}\tau \tag{3.84b}$$

and

$$\begin{aligned} \mathbf{R}(x,t_m) = \frac{1}{\Delta t}\int_\Gamma \Bigg[& \mathbf{p}(x,t_{m-1})\int_{t_{m-1}}^{t_m} (t_m - \tau)\mathbf{U}(x,y,t_m - \tau)\,\mathrm{d}\tau \\ & - \mathbf{u}(x,t_{m-1})\int_{t_{m-1}}^{t_m} (t_m - \tau)\mathbf{P}(x,y,t_m - \tau)\,\mathrm{d}\tau\Bigg]\,\mathrm{d}\Gamma(y) \\ & + \sum_{r=1}^{m-1}\int_\Gamma\int_{t_{r-1}}^{t_r} [\mathbf{U}(x,y,t_m - \tau)\mathbf{p}(y,\tau) \\ & - \mathbf{P}(x,y,t_m - \tau)\mathbf{u}(y,\tau)]\,\mathrm{d}\tau\,\mathrm{d}\Gamma(y) \ . \end{aligned} \tag{3.84c}$$

For each time $t = t_m$ Eq. (3.84) can be transformed into a set of algebraic equations as described before. For this purpose, similarly to the static case, it is necessary to discretize the boundary using boundary elements and approximate displacements and tractions. As a result the following set of algebraic equations is obtained for step m :

$$\mathbf{A}_m\mathbf{X}_m = \mathbf{W}_m\mathbf{Y}_m + \mathbf{R}_m \ , \qquad m = 1, 2, \ldots, M \ , \tag{3.85}$$

where $\mathbf{A}^m$ and $\mathbf{W}^m$ are square matrices with components dependent on boundary integrals with kernels $\mathbf{U}^m(x, y)$ and $\mathbf{P}^m(x, y)$, $\mathbf{X}^m$ is a column matrix containing unknown values of displacements and tractions at time t_m, $\mathbf{Y}^m$ is a column matrix dependent on given values of displacements and tractions and on boundary integrals with kernels $\mathbf{U}^m(x, y)$ and $\mathbf{P}^m(x, y)$, $\mathbf{R}^m$ is a column matrix with coefficients dependent on the history of the dynamic process prior to time t_m. Integration over time can be performed analytically (cf. [37]), which is especially easy for three-dimensional problems due to the form of the fundamental solution.

3.2.3 Integral Transformation Method

Using the Laplace transformation:

$$\bar{f}(x,s) = \mathcal{L}\{f(x,t)\} = \int_0^{\infty} f(x,t)e^{-st}\mathrm{d}t\,, \tag{3.86}$$

or the Fourier transformation:

$$\tilde{f}(x,\omega) = \mathcal{F}\{f(x,t)\} = \int_{-\infty}^{\infty} f(x,t)e^{-\mathrm{i}\omega t}\,\mathrm{d}t\,, \tag{3.87}$$

the hyperbolic equation (3.72) is reformulated into an elliptic equation and the time convolutions disappear in the integral formula (3.80). In the case of the Laplace transformation Eq. (3.80) becomes:

$$\mathbf{c}(x)\bar{\mathbf{u}}(x,s) + \int_{\Gamma} \bar{\mathbf{P}}(x,y,s)\bar{\mathbf{u}}(y,s)\,\mathrm{d}\Gamma(y) =$$

$$= \int_{\Gamma} \bar{\mathbf{U}}(x,y,s)\bar{\mathbf{p}}(y,s)\,\mathrm{d}\Gamma(y) + \int_{\Omega} \bar{\mathbf{U}}(x,y,s)\bar{\mathbf{b}}(y,s)\,\mathrm{d}\Omega(y)\,, \tag{3.88}$$

where $\bar{\mathbf{U}}(x,y,s) = \mathcal{L}[\mathbf{U}(x,y,t)]$ and $\bar{\mathbf{P}}(x,y,s) = \mathcal{L}[\mathbf{P}(x,y,t)]$. The integral formula (3.88) represents a family of integral expressions dependent on the transformation parameter s. The solution procedure is thus similar to the case of static problems. It consists in solving the equation for a series of the complex parameters s using a boundary element discretization of the transforms of the boundary displacements $\mathbf{u}(x,s)$ and tractions $\mathbf{p}(x,s)$ with given boundary and initial conditions. At the final stage of the procedure a numerical inversion of the obtained displacement and force transforms must be performed.

If the Fourier transformation is employed, the integral equation (3.88) has a similar form and only the transformation parameter ω has a different sense.

The dynamic description of the medium in the domain of integral transforms allows the application of the presented concept to quasi-static and dynamic problems of viscoelastic bodies (cf. [38]).

The state equation for a viscoelastic body can be written as:

$$\mathbf{P}(D)\mathbf{s}(x,t) = \mathbf{F}(D)\mathbf{e}(x,t)\,, \tag{3.89}$$

where $\mathbf{P}(D)$ and $\mathbf{F}(D)$ are linear differential operators:

$$\mathbf{P}(D) = \sum_{r=0}^{m} a_r D^r\,, \quad \mathbf{F}(D) = \sum_{r=0}^{m} b_r D^r\,, \quad D^r = \partial^r/\partial t^r\,, \tag{3.90}$$

while $\mathbf{s} = (s_{ij})$ and $\mathbf{e} = (e_{ij})$ are stress and strain deviators, respectively.

Using an elastic-viscoelastic analogy the initial boundary value problem of viscoelasticity is solved in the domain of the Laplace transforms similarly to the elasticity problem, but the Lame constants μ and λ in the fundamental solution must be replaced by the following quantities dependent on the parameter of the integral transformation s:

$$\mu_v(s) = \mathbf{F}(s)/[2\mathbf{P}(s)]\,, \qquad \lambda_v(s) = \lambda + (2/3)[\mu - \mu_v(s)]\,. \tag{3.91}$$

A disadvantage of the integral transformation method is the necessity of computations in the domain of complex variables and the numerical inversion of the transforms at the final stage of the solution. However, there is a large class of stochastic dynamics problems in which a solution in the Fourier transform domain in terms of spectral densities of stochastic space-time fields of displacements, forces and stresses gives a sufficient characteristic of the analyzed dynamic process without the necessity to return to the time domain (cf. [5]).

3.2.4 Steady-State Vibrations in Dynamic Elasticity

An analysis of vibrations under harmonic load is of great practical significance. The influence of initial conditions can then be neglected, the volume forces $\mathbf{b}(x, t)$ as well as the boundary fields of displacements $\mathbf{u}(x, t)$ and tractions $\mathbf{p}(x, t)$ can be expressed as follows:

$$\mathbf{b}(x,t) = \tilde{\mathbf{b}}(x,\omega)e^{-i\omega t}\,, \quad \mathbf{u}(x,t) = \tilde{\mathbf{u}}(x,\omega)e^{-i\omega t}\,, \quad \mathbf{p}(x,t) = \tilde{\mathbf{p}}(x,\omega)e^{-i\omega t}\,, \tag{3.92}$$

where $\tilde{\mathbf{b}}(x,\omega)$, $\tilde{\mathbf{u}}(x,\omega)$ and $\tilde{\mathbf{p}}(x,\omega)$ are complex amplitudes of the volume forces, displacements and tractions, respectively.

The initial boundary value problem is now reduced to the boundary value problem described by a differential equation with an elliptic operator L_d^ω:

$$L_d^\omega(x,\omega) \equiv L\tilde{\mathbf{u}}(x,\omega) + \varrho\,\omega^2\tilde{\mathbf{u}}(x,\omega) = -\tilde{\mathbf{b}}(x,\omega)\,, \tag{3.93}$$

together with boundary conditions defined for the displacement and traction amplitudes:

$$\tilde{\mathbf{u}}(x,\omega) = \tilde{\mathbf{u}}^0(x,\omega)\,, \quad x \in \Gamma_u\,, \quad \tilde{\mathbf{p}}(x,\omega) = \tilde{\mathbf{p}}^0(x,\omega)\,, \quad x \in \Gamma_p\,. \tag{3.94}$$

The boundary integral equation now becomes:

$$\mathbf{c}(x)\tilde{\mathbf{u}}(x,\omega) + \int_\Gamma \tilde{\mathbf{P}}(x,y,\omega)\tilde{\mathbf{u}}(y,\omega)\,\mathrm{d}\Gamma(y) =$$

$$= \int_\Gamma \tilde{\mathbf{U}}(x,y,\omega)\tilde{\mathbf{p}}(y,\omega)\,\mathrm{d}\Gamma(y) + \int_\Omega \tilde{\mathbf{U}}(x,y,\omega)\tilde{\mathbf{b}}(y,\omega)\,\mathrm{d}\Omega(y)\,, \tag{3.95}$$

where $\tilde{\mathbf{U}}(x,y,\omega)$ is a fundamental solution for the operator L_d^ω.

The latter equation can be obtained in a formal manner substituting $s = i\omega$ in Eq. (3.88) which describes non-stationary vibrations in the domain of the Laplace transforms.

Discretizing the boundary of the body using boundary elements and approximating the amplitudes $\tilde{\mathbf{u}}(x, \omega)$ and $\tilde{\mathbf{p}}(x, \omega)$ for each value of $\omega = (\omega_l)$, $l = 1, 2, \ldots$ by means of nodal values and shape functions we obtain the following algebraic form of Eq. (3.95):

$$\tilde{\mathbf{H}}(\omega)\mathbf{u}(\omega) = \tilde{\mathbf{G}}(\omega)\mathbf{p}(\omega) + \tilde{\mathbf{B}}(\omega)\,, \qquad \omega = (\omega_l)\,, \qquad l = 1, 2, \ldots\,, \tag{3.96}$$

where the square complex matrices $\tilde{\mathbf{H}}(\omega)$ and $\tilde{\mathbf{G}}(\omega)$ are computed for a sequence of values ω_l, $l = 1, 2, \ldots$.

Considering the boundary conditions (3.94) Eq. (3.96) becomes a set of algebraic equations:

$$\tilde{\mathbf{A}}(\omega)\tilde{\mathbf{X}}(\omega) = \tilde{\mathbf{Y}}(\omega)\,, \qquad \omega = (\omega_l)\,, \qquad l = 1, 2, \ldots\,. \tag{3.97}$$

The equation has a unique solution for limited bodies except for the case when ω^2 is not an eigenvalue of the operator L_d^{ω}.

In the case of homogeneous boundary conditions and zero volume forces Eq. (3.97) takes the form:

$$\tilde{\mathbf{A}}(\omega)\tilde{\mathbf{X}}(\omega) = 0\,. \tag{3.98}$$

The necessary and sufficient condition for a non-trivial solution of Eq. (3.98) to exist is as follows:

$$\det\tilde{\mathbf{A}}(\omega) = 0\,. \tag{3.99}$$

The roots of Eq. (3.99) are eigenvalues of the eigenproblem in Eq. (3.98). Since the matrix $\mathbf{A}(\omega)$ is full and non-symmetric, numerical calculation of its eigenvalues is cumbersome (cf. [43]). Knowing the eigenvalues we can compute the forms of the operator using Eq. (3.98).

3.2.5 Alternative BEM Formulation of Dynamic Theory for Elastic Medium

The main disadvantage of the presented methods of solving the dynamic problem of elasticity is the necessity to compute the coefficient matrix for each moment of time (method of time steps) or for a sequence of values of an integral transformation parameter (method of integral transformation). It results from the fact that the fundamental solution for elasto-dynamics is time dependent. An alternative approach presented below is based on the fundamental solution for the static theory of elasticity $\mathbf{U}(x, y)$ and requires only one computation of the coefficient matrix (cf. [42]). The approach enables us to generate a mass matrix for an elastic system on the basis of the discretization of the boundary of the system.

Applying the Green formula based on the fundamental solution for operator L to Eq. (3.72) we obtain:

$$\mathbf{c}(x)\mathbf{u}(x,t) + \int_\Gamma \mathbf{P}(x,y)\mathbf{u}(y,t)\,\mathrm{d}\Gamma(y) = \int_\Gamma \mathbf{U}(x,y)\mathbf{p}(y,t)\,\mathrm{d}\Gamma(y) + \int_\Omega \mathbf{U}(x,y)\mathbf{b}(y,t)\,\mathrm{d}\Omega(y) - \varrho\int_\Omega \mathbf{U}(x,y)\ddot{\mathbf{u}}(y,t)\,\mathrm{d}\Omega(y)\,. \quad (3.100)$$

The last integral on the right-hand side of Eq. (3.100) contains an unknown acceleration field $\ddot{\mathbf{u}}(y,t) = (\ddot{u}_i(y,t))$ inside the domain Ω. We assume that the field can be expressed as a sum of products of unknown time functions $\ddot{\mathbf{s}}(t) = \left(\ddot{s}_i^j(t)\right)$ and known basis functions $f(y) = \left(f^j(y)\right)$, $y \in \Omega$:

$$\ddot{u}_i(y,t) = \ddot{s}_i^j(t) f^j(y)\,, \qquad j = 1,\, 2,\, \dots\,,\, W\,. \quad (3.101)$$

The basis functions $f^j(y)$ can be selected in many ways. One of the simplest classes of basis functions is described by the relation:

$$f^j(y) = R(A_j,\, y)\,, \quad (3.102)$$

where $R(A_j,\, y)$ is a distance between boundary points A_j and point y. If the expression $\delta_{li} f^j$ is interpreted as a volume pseudo-force, then, in an infinite elastic space, the force generates a pseudo-displacement field Ψ_{li}^j which satisfies the equation:

$$L(\Psi_{li}^j) = \delta_{li} f^j\,. \quad (3.103)$$

Applying now the Green formula and considering Eq. (3.101) the following form of the inertial component in the formula (3.100) is obtained:

$$\ddot{\mathbf{s}}^j(t)\int_\Omega \mathbf{U}(x,y) f^j(y)\,\mathrm{d}\Omega(y) = \ddot{\mathbf{s}}^j(t)\{-\mathbf{c}(x)\boldsymbol{\Psi}^j(x) + \int_\Gamma \mathbf{U}(x,y)\boldsymbol{\Sigma}^j(y)\,\mathrm{d}\Gamma(y) - \int_\Gamma \mathbf{P}(x,y)\boldsymbol{\Psi}^j(y)\,\mathrm{d}\Gamma(y)\}\,, \quad (3.104)$$

where $\boldsymbol{\Psi}^j(y)$ and $\boldsymbol{\Sigma}^j(y)$ are vectors of pseudo-displacements and pseudo-stresses induced in the infinite elastic space by the volume pseudo-force $\delta_{li} f^j$. Finally, Eq. (3.100) becomes:

$$\mathbf{c}(x)\mathbf{u}(x,t) + \int_\Gamma \mathbf{P}(x,y)\mathbf{u}(y,t)\,\mathrm{d}\Gamma(y) - \int_\Gamma \mathbf{U}(x,y)\mathbf{p}(y,t)\,\mathrm{d}\Gamma(y) = \varrho\left\{-\mathbf{c}(x)\boldsymbol{\Sigma}^j(x) + \int_\Gamma \mathbf{U}(x,y)\boldsymbol{\Sigma}^j(y)\,\mathrm{d}\Gamma(y) - \int_\Gamma \mathbf{P}(x,y)\boldsymbol{\Psi}^j(y)\,\mathrm{d}\Gamma(y)\right\}\ddot{\mathbf{s}}^j(t) + \int_\Omega \mathbf{U}(x,y)\mathbf{b}(y,t)\,\mathrm{d}\Omega(y), \quad x \in A_j\,, \quad (3.105)$$

where A_j coincides with the j-th nodal point on the boundary Γ.

Approximating the fields of displacements $\mathbf{u}(y,t)$, tractions $\mathbf{p}(y,t)$ as well as the fields of pseudo-displacements $\mathbf{\Psi}^j(y)$ and pseudo-forces $\mathbf{\Sigma}^j(y)$ using shape functions and nodal values, cf. Eqs. (3.13) and (3.14), we obtain:

$$\mathbf{Hu}(t) + \mathbf{Gp}(t) - \varrho\,(\mathbf{H\Psi} - \mathbf{G\Sigma})\ddot{\mathbf{s}}(t) = \mathbf{B}(t)\,. \tag{3.106}$$

The matrices $\mathbf{H}$ and $\mathbf{G}$ are defined precisely like in the static problem, but the matrices $\mathbf{\Psi}$ and $\mathbf{\Sigma}$ contain nodal values of functions Ψ_i^j and Σ^j, respectively.

The relation (3.101) written for all nodal points $y \in A_j$, $j = 1, 2, \ldots, W$, generates the matrix equation:

$$\ddot{\mathbf{u}}(t) = \mathbf{D}\ddot{\mathbf{s}}(t)\,, \tag{3.107}$$

in which $\mathbf{D}$ is a square matrix containing the values of the function $f^j(y)$ at the nodal points. Finally, a set of ordinary differential equations is obtained, which in a matrix form reads:

$$\mathbf{M}\ddot{\mathbf{u}}(t) + \mathbf{Hu}(t) = \mathbf{Gp}(t) + \mathbf{B}(t)\,. \tag{3.108}$$

In this equation the mass matrix is defined as follows:

$$\mathbf{M} = -\varrho\,(\mathbf{H\Psi} - \mathbf{G\Sigma})\mathbf{D}^{-1}\,. \tag{3.109}$$

Considering the boundary conditions Eq. (3.108) becomes:

$$\underline{\mathbf{M}}\ddot{\mathbf{u}}_x + \underline{\mathbf{H}}\mathbf{u}_x = \underline{\mathbf{G}}\mathbf{p}_0 + \underline{\underline{\mathbf{H}}}\mathbf{u}_0 + \underline{\underline{\mathbf{M}}}\ddot{\mathbf{u}}_0\,, \tag{3.110}$$

where $\mathbf{u}_x$ is a column matrix of unknown displacements on the boundary Γ_p, $\mathbf{p}_0$ and $\mathbf{u}_0$ are column matrices of forces and displacements determined by the boundary conditions (3.70) and (3.71). The matrices $\underline{\mathbf{M}}$, $\underline{\mathbf{H}}$, $\underline{\mathbf{G}}$, $\underline{\underline{\mathbf{H}}}$ and $\underline{\underline{\mathbf{M}}}$ are defined as follows:

$$\begin{aligned} \underline{\mathbf{K}} &= \mathbf{K}_{pp} - \mathbf{G}_{pu}\mathbf{G}_{uu}^{-1}\mathbf{K}_{up}\,, \\ \underline{\underline{\mathbf{K}}} &= \mathbf{G}_{pu}\mathbf{G}_{uu}^{-1}\mathbf{K}_{uu} - \mathbf{K}_{pu}\,, \end{aligned} \tag{3.111}$$

where $\mathbf{K}$ represents the matrices $\mathbf{M}$, $\mathbf{H}$ and $\mathbf{G}$, while the indices u and p denote the parts of the boundary Γ_u and Γ_p.

The motion equation (3.110) is solved using a direct time integration method, e.g. the Newmark, Wilson or Houbolt method.

Free vibrations are a special case described by the motion equation (3.110). We assume that the system is subjected to no external excitations ($\mathbf{u}_0 = \mathbf{0}$ and $\mathbf{p}_0 = \mathbf{0}$) and that the displacements exhibit a harmonic variation in time:

$$\ddot{\mathbf{u}}_x = -\omega^2\mathbf{u}_x\,. \tag{3.112}$$

The motion equation (3.110) is then reduced to the form:

$$(\underline{\mathbf{H}} - \omega^2\underline{\mathbf{M}})\mathbf{u}_x = \mathbf{0}\,. \tag{3.113}$$

Equation (3.113) represents a generalized eigenproblem in the alternative approach to the boundary element method.

3.3 Boundary Element Method in Theory of Plates

3.3.1 Statics of Plates

Let us consider an isotropic homogeneous plate with thickness h, the middle plane of which occupies a domain $\Omega \in \boldsymbol{E}^2$ with a boundary Γ that possesses vertices A_k, $k = 1, 2, \ldots, K$ (Fig. 3.4). The points of the domain Ω and boundary Γ are denoted by $x = (x_i)$ and $y = (y_i)$, respectively ($i = 1, 2$).

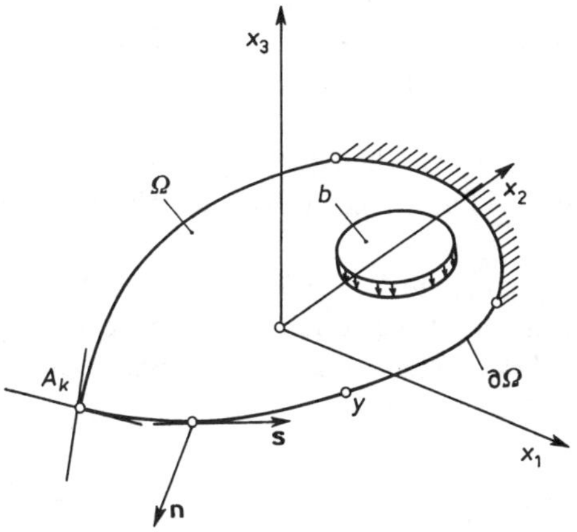

Fig. 3.4. Isotropic homogeneous plate with vertices A_k

The deflection of the plate $w(x)$, $x \in \Omega$, determined within the classical theory of thin plates, satisfies the biharmonic differential equation:

$$\nabla^4 w(x) = b(x)/D, \; x \in \Omega\,, \tag{3.114}$$

where $D = Eh^3/[12(1-\nu^2)]$ is a bending stiffness of the plate and $b(x)$ is the transversal loading of the plate.

Equation (3.114) must be supplemented by boundary conditions on Γ, which define the supports of the plate.

A boundary integral equation for the plate can be easily derived from the reciprocal work theorem which, for two mutually independent mechanical states, is expressed as follows (cf. [48]):

$$\int_\Omega (w\nabla^4 w' - w'\nabla^4 w)\,\mathrm{d}\Omega = \int_\Gamma [w'v_n - \varphi'_n m_n + m'_n\varphi_n - v'_n w]\,\mathrm{d}\Gamma + \sum_{k=1}^{K} [\,[\![m_{ns}]\!]w' - w[\![m'_{ns}]\!]\,]_{A_k}\,, \tag{3.115}$$

where

$$\{\varphi'_n, m'_n, m'_{ns}, v'_n\} = \{\hat{\Phi}_n, \hat{M}_n, \hat{M}_{ns}, \hat{V}_n\}(w') . \tag{3.116}$$

The operators $\hat{M}_n$, $\hat{M}_{ns}$, $\hat{\Phi}_n$, $\hat{V}_n$ are defined as follows

$$\hat{M}_n = -\nabla^2 + (1-\nu)\left[n_2^2\frac{\partial^2}{\partial x_1^2} + n_1^2\frac{\partial^2}{\partial x_2^2} - 2n_1n_2\frac{\partial^2}{\partial x_1 \partial x_2}\right] , \tag{3.117}$$

$$\hat{M}_{ns} = -(1-\nu)\left[n_1n_2\left(\frac{\partial^2}{\partial x_2^2} - \frac{\partial^2}{\partial x_1^2}\right) - (n_1^2 - n_2^2)\frac{\partial^2}{\partial x_1 \partial x_2}\right] , \tag{3.118}$$

$$\hat{\Phi}_n = \frac{\partial}{\partial n} ; \qquad \hat{V}_n = -\frac{\partial \nabla^2}{\partial n} + \frac{\partial \hat{M}_n}{\partial s} , \tag{3.119}$$

where $\partial/\partial s$ denotes the differentiation in the direction of a tangent $\mathbf{s}$ to the boundary of the plate and the symbol $[\![\,.\,]\!]$ denotes the jump of the quantity at the vertex.

If the boundary is smooth then the last component in Eq. (3.115) vanishes and the equation is a Green identity for the biharmonic equation, cf. Eq. (2.76).

The quantities appearing in Eq. (3.115) have the following physical interpretation: φ_n is an angle of deflection, m_n is a bending moment, m_{ns} is a twisting moment and v_n is a substitute transverse shear force.

Let us denote the fundamental solution of the biharmonic equation by $W(x, y)$, i.e.

$$\nabla^4 W(x, y) = \delta(x-y) , \qquad x, y \in \boldsymbol{E}^2 . \tag{3.120}$$

The solution has the form:

$$W(x,y) = \frac{r^2}{8\pi}\ln r , \qquad r = |x-y| . \tag{3.121}$$

The boundary integral equation can be written as (cf. Sect. 2.4.2):

$$\begin{aligned} \frac{1}{2}w(x) + \int_\Gamma [V_n(x,y)w(y) - M_n(x,y)\varphi_n(y) \\ + \Phi_n(x,y)m_n(y) - W(x,y)v_n(y)]\, \mathrm{d}\Gamma(y) \\ + \sum_{k=1}^{K} [[\![M_{ns}(x, A_k)]\!]\, w(A_k) \quad W(x, A_k)\, [\![m_{ns}(A_k)]\!]] \\ = \frac{1}{D}\int_\Omega W(x,y)b(y)\, \mathrm{d}\Omega(y) , \end{aligned} \tag{3.122}$$

where

$$\{\Phi_n, M_n, M_{ns}, V_n\} = \{\hat{\Phi}_n, \hat{M}_n, \hat{M}_{ns}, \hat{V}_n\}W . \tag{3.123}$$

At each point of the boundary Γ there are two unknown functions: the deflection w or the substitute transverse shear force v_n and the deflection angle φ_n or the moment m_n. Therefore, an additional boundary equation is necessary.

It can be obtained by differentiation of the integral equation (3.122) in the direction of the normal $\mathbf{n}$ at point x. We then have (cf. Sect. 2.4.2):

$$\begin{aligned}\frac{1}{2}\varphi_n(x) + \int_\Gamma [\underline{V}_n(x,y)w(y) - \underline{M}_n(x,y)\varphi_n(y) \\ + \underline{\Phi}_n(x,y)m_n(y) - \underline{w}(x,y)v_n(y)]\, \mathrm{d}\Gamma(y) \\ + \sum_{k=1}^{K} [\,[\![\underline{M}_{ns}(x,A_k)]\!]\, w(A_k) - \underline{W}(x,A_k)\, [\![m_{ns}(A_k)]\!]\,] \\ = \frac{1}{D}\int_\Omega \underline{W}(x,y)b(y)\, \mathrm{d}\Omega(y)\,, \quad (3.124)\end{aligned}$$

where

$$\{\underline{W}, \underline{\Phi}_n, \underline{M}_n, \underline{M}_{ns}, \underline{V}_n\} = \frac{\partial}{\partial n}\{W, \Phi_n, M_n, M_{ns}, V_n\}\,. \quad (3.125)$$

The set of integral equations (3.122) and (3.124) together with the given boundary conditions allow us to solve the boundary value problem of plate bending. The boundary conditions can have one of the following forms:

- clamped edge: $w = \varphi_n = 0\,,\ m_{ns}(A_k) = w(A_k) = 0$ on Γ,
- simply supported edge: $w = m_n = 0\,,\ w(A_k) = 0\,,\ m_{ns}(A_k) \neq 0$ on Γ,
- free edge: $v_n = m_n = 0\,,\ m_{ns}(A_k) = 0\,,\ w(A_k) \neq 0$ on Γ.

The boundary Γ is divided into boundary elements Γ^e, $e = 1, 2, \ldots, E$, which have the form of curvilinear or straight line segments. The boundary functions a, $a = w, \varphi_n, m_n, v_n$ are approximated within each element using the shape functions $\mathbf{N}(\xi)$ and the nodal values:

$$a^e(\xi) = N^w(\xi)a^{ew}\,, \quad a = w, \varphi_n, m_n, v_n\,, \quad (3.126)$$

where ξ is a local coordinate set of element Γ^e. As a result we obtain the discrete matrix form of the integral equations (3.122) and (3.124):

$$\begin{bmatrix} \mathbf{H}_{11} & \mathbf{H}_{12} \\ \mathbf{H}_{21} & \mathbf{H}_{22} \end{bmatrix} \mathbf{u} = \begin{bmatrix} \mathbf{G}_{11} & \mathbf{G}_{12} \\ \mathbf{G}_{21} & \mathbf{G}_{22} \end{bmatrix} \mathbf{p} + \mathbf{B}\,, \quad (3.127)$$

where

$$\mathbf{u} = [w_1 w_2 \ldots w_w \varphi_{n1}\varphi_{n2}\ldots\varphi_{nw}]^T\,, \quad \mathbf{p} = [m_{n1}m_{n2}\ldots m_{nw}v_{n1}v_{n2}\ldots v_{nw}]^T$$

and the matrix coefficients $\mathbf{H}_{11}$, $\mathbf{H}_{12}$, $\mathbf{H}_{21}$, $\mathbf{H}_{22}$ and $\mathbf{G}_{11}$, $\mathbf{G}_{12}$, $\mathbf{G}_{21}$, $\mathbf{G}_{22}$ depend on the boundary integrals of the fundamental solutions.

The coefficients $c_i = c$ occur only in the diagonal matrix elements $\mathbf{H}_{11}$, $\mathbf{H}_{22}$ and can be determined in the case of a non-smooth boundary e.g. enforcing a unit displacement on the plate treated as a rigid body. The matrix $\mathbf{B}$ depends on the transverse loading $b(x)$.

3.3.2 Dynamics of Plates

The initial boundary value problem of plate vibrations is described by the equation:

$$L_d w(x,t) \equiv \nabla^4 w(x,t) + \frac{\varrho h}{D}\frac{\partial^2 w(x,t)}{\partial t^2} = \frac{b(x)}{D}\,, \quad x \in \Omega\,, \tag{3.128}$$

which, in addition to the boundary conditions, should be supplemented with the initial conditions:

$$w(x,0) = w_0(x)\,, \quad \dot{w}(x,0) = \dot{w}^0(x)\,, \quad x \in \Omega\,. \tag{3.129}$$

Similarly to the dynamic problem of elasticity, the solution of Eq. (3.118) can be extended $\forall\, t \in (-\infty\,, \infty)$. We then obtain:

$$\hat{b}(x,t) = H(t)b(x,t) + \varrho h[w_0(x)\dot{\delta}(t) + \dot{w}^0(x)\delta(t)]\,. \tag{3.130}$$

The application of the reciprocal works principle for the dynamic problem leads to the following set of integral equations:

$$\begin{aligned}
&\begin{bmatrix} c(x) & 0 \\ 0 & c(x) \end{bmatrix} \begin{Bmatrix} w(x,\, t) \\ \varphi_n(x,\, t) \end{Bmatrix} \\
&+ \int_\Gamma \begin{bmatrix} V_n(x,\, y,\, t) & -M_n(x,\, y,\, t) \\ \underline{V}_n(x,\, y,\, t) & -\underline{M}_n(x,\, y,\, t) \end{bmatrix} * \begin{Bmatrix} w(y,\, t) \\ \varphi_n(y,\, t) \end{Bmatrix} \mathrm{d}\Gamma(y) \\
&+ \int_\Gamma \begin{bmatrix} \Phi_n(x,\, y,\, t) & -W(x,\, y,\, t) \\ \underline{\Phi}_n(x,\, y,\, t) & -\underline{W}(x,\, y,\, t) \end{bmatrix} * \begin{Bmatrix} m_n(y,\, t) \\ v_n(y,\, t) \end{Bmatrix} \mathrm{d}\Gamma(y) \\
&+ \sum_{k=1}^{K} \begin{bmatrix} M_{ns}(x,\, A_k,\, t) & -W(x,\, A_k,\, t) \\ \underline{M}_{ns}(x,\, A_k,\, t) & -\underline{W}(x,\, A_k,\, t) \end{bmatrix} \begin{Bmatrix} m_n(A_k,\, t) \\ v_n(A_k,\, t) \end{Bmatrix} \\
&= \frac{1}{D}\int_\Omega \begin{bmatrix} W(x,\, y,\, t) & 0 \\ 0 & \underline{W}(x,\, y,\, t) \end{bmatrix} * \begin{Bmatrix} \hat{b}(y,\, t) \\ \hat{b}(y,\, t) \end{Bmatrix} \mathrm{d}\Omega(y)\,,
\end{aligned} \tag{3.131}$$

where $W(x,\, y,\, t)$ is the fundamental solution for operator L_d,

$$L_d W(x,y,t) = \delta(x-y)\delta(t)\,, \tag{3.132}$$

while the quantities $\Phi_n(x,\, y,\, t)$, $M_n(x,\, y,\, t)$, $M_{ns}(x,\, y,\, t)$, $V_n(x,\, y,\, t)$ as well as $\underline{\Phi}_n(x,\, y,\, t)$, $\underline{M}_n(x,\, y,\, t)$, $\underline{M}_{ns}(x,\, y,\, t)$, $\underline{V}_n(x,\, y,\, t)$ are defined in Eqs. (3.123) and (3.125).

The set of equations (1.131) can be solved using the time step method or the integral transformation method, described previously. If the loading is a harmonic function,

$$b(x,t) = \tilde{b}(x,\omega)e^{-i\omega t}\,, \tag{3.133}$$

then also the deflection of the plate has a similar character,

$$w(x,t) = \tilde{w}(x,\omega)e^{-i\omega t}\,, \tag{3.134}$$

and the equation of plate vibrations reads:

$$L_d^\omega\,\tilde{w}(x,\omega) \equiv \nabla^4\tilde{w}(x,\omega) - \Lambda^4\tilde{w}(x,\omega) = \tilde{b}(x,\omega)/D\,, \tag{3.135}$$

where $\Lambda^4 = \omega^2\varrho h/D$.

The fundamental solution for operator L_d^ω satisfies the equation:

$$L_d^\omega\,\tilde{W}(x,y,\omega) = \delta(x-y) \tag{3.136}$$

and has the form:

$$\tilde{W}(x,y,\omega) = i\,[H_0^{(1)}(\Lambda r) - H_0^{(1)}(\mathrm{i}\Lambda r)]/(8\Lambda^2)\,, \tag{3.137}$$

with a Hankel function of the first kind and zero order $H_0^{(1)}$.

The integral equation for stationary vibrations can also be obtained performing the Fourier transformation of Eq. (3.131).

3.4 Numerical Examples

Example 1

The boundary element method is especially effective in the case of stress concentrations. For illustration, computation results for a symmetric straight tooth of a gear wheel [5] are quoted. Because of symmetry and according to a common practice, a single tooth subjected to a unit force and supported at the base was considered. A discretization of the contour with 118 linear boundary elements and diagrams of the computed principal stresses at both sides of the tooth are presented in Fig. 3.5a. The maximum value of the σ_1 stress is 3.621 units. The value is very close to that obtained using the conformal mapping method. For comparison FEM calculations were performed. The mesh with 396 finite elements (with the former division of the boundary preserved) is shown in Fig. 3.5b. The maximum principal stress was 3.126 units. For the same order of discretization the FEM usually gives less accurate computation results than the BEM (cf. [46]). This refers particularly to the case where stresses at the boundary are determined.

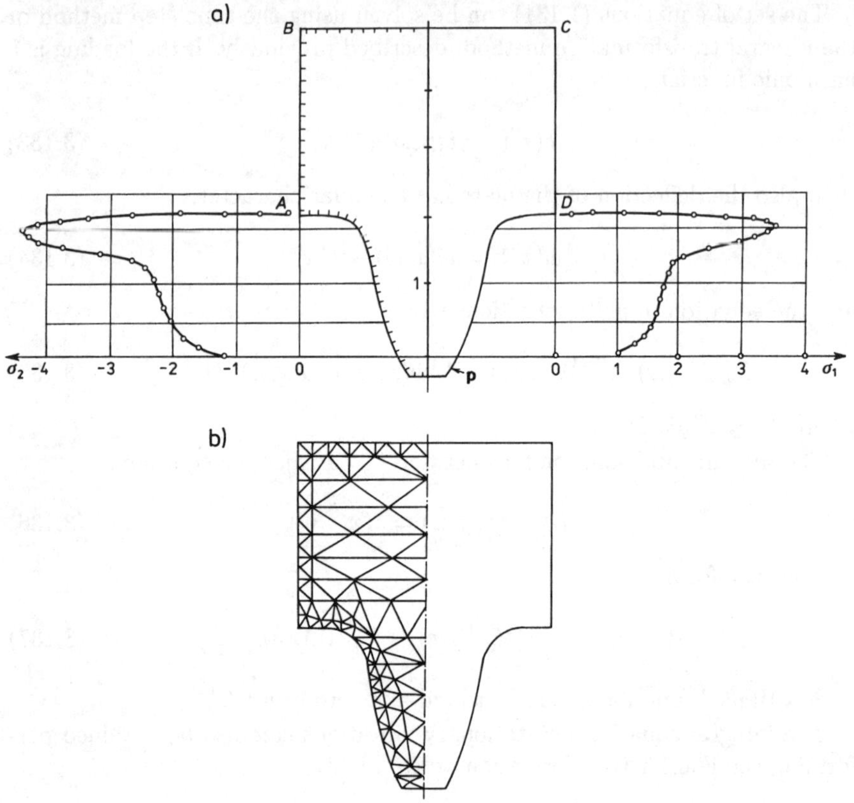

Fig. 3.5. Symmetric straight tooth of a gear wheel: **a)** discretization with boundary elements and distribution of principal stresses, **b)** discretization with finite elements

Example 2

In this example the stress state in a flange connecting steel pipes was determined [33]. Due to symmetry an angular segment of 15° was considered (Fig. 3.6a). Loading is composed of an internal pressure 4.5 N/mm^2 and a prestressing force in the screw 50 kN. To discretize the problem 57 surface boundary elements were employed (Fig. 3.6b). The same problem was solved using 64 twelve-noded isoparametric finite elements, so that the discrete model contained 501 nodes (Fig. 3.6c). The results of the computations in the form of meridional and circumferential stresses are presented in Fig. 3.6d–g. The largest stress concentration occurs at the point where the flange is connected with the conic part. According to the author of the solution [33] the example is relatively costly to analyze with the BEM, since the quotient of the surface area of the examined structure to its volume is quite large. The BEM is especially effective for compact bodies, for which the quotient is small.

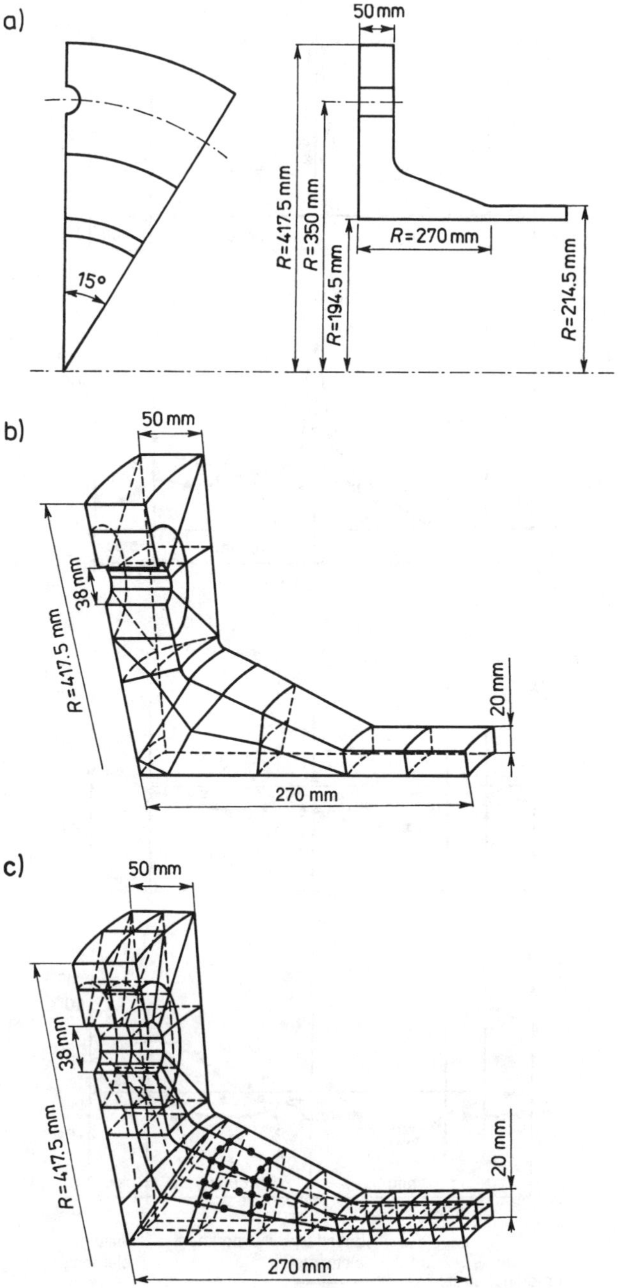

Fig. 3.6. Steel flange connecting two pipes: **a)** dimensions, **b)** discretization with boundary elements, **c)** discretization with finite elements, *(continued in the next page)*

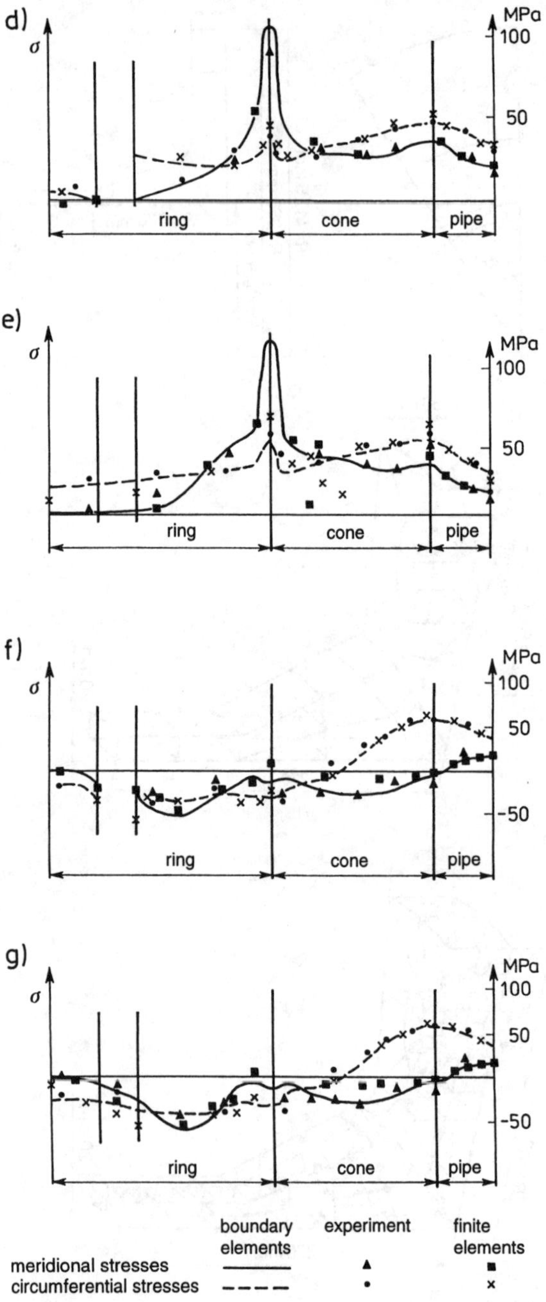

Fig. 3.6. *(continued)* Steel flange connecting two pipes: **d–g)** diagrams of meridional and circumferential stresses

Example 3

To illustrate an application of the boundary element method to dynamic problems, in the version described in Sect. 3.5.2, a problem of free vibrations of a steel arch presented in Fig. 3.7a was solved (cf. [15]). It was assumed that the structure satisfies the plane strain conditions and is made of a material

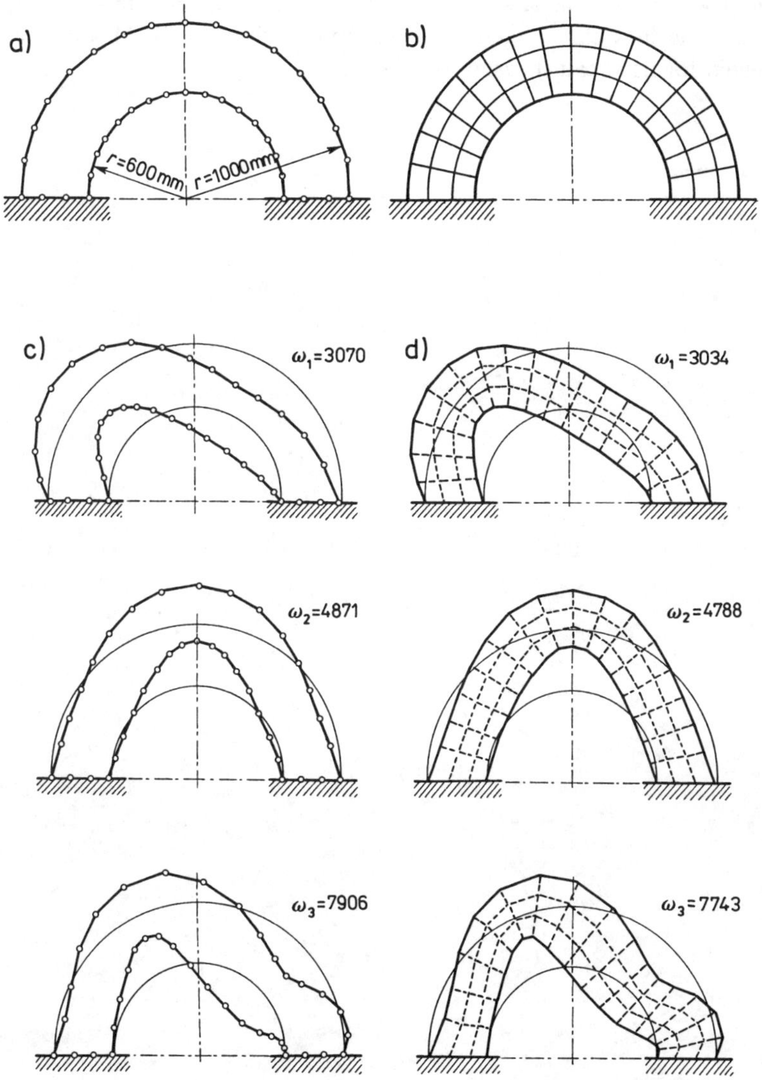

Fig. 3.7. Steel arch: **a)** discretization with boundary elements, **b)** discretization with finite elements, **c)** the first three forms of free vibrations (BEM), **d)** the first three forms of free vibrations (FEM)

with the following properties: Young's modulus $E = 0.21 \cdot 10^{12}$ Pa, Poisson's ratio $\nu = 0.3$, density $\varrho = 8000$ kg/m^3.

To discretize the structure 38 linear boundary elements were used (Fig. 3.7a). The first three eigenfrequencies were computed: $\omega_1 = 3070$ s^{-1}, $\omega_2 = 4871$ s^{-1} and $\omega_3 = 7906$ s^{-1}. At the same time, calculations by means of the finite element method were performed. In the discretization 48 four-noded elements were used (Fig. 3.7b). The frequencies obtained with this method were: $\omega_1 = 3034$ s^{-1}, $\omega_2 = 4788$ s^{-1} and $\omega_3 = 7743$ s^{-1}. Fig. 3.7c presents the first three eigen-forms obtained using the BEM and Fig. 3.7d the eigen-forms obtained using the FEM.

4 Boundary Element Method in Nonlinear Problems

In Chap. 2 of Part I a complete set of equations describing the motion and deformation of a deformable body, was derived and the nonlinear character of the process was emphasized. It seems useful to draw attention to the sources of nonlinearity.

The first source is structural or material properties, which only in special cases can be described with linear constitutive equations. This kind of nonlinearity is called a *physical nonlinearity.* Sect. 4.2 is devoted to solving the physically nonlinear problems using the BEM.

The second source of nonlinearity is the character of the relationship between displacements and an adopted deformation or strain measure (it is worth reminding that the two notions differ in nonlinear mechanics [14, 52, 56, 57]). This kind of nonlinearity, as opposed to the previous one, is called a *geometrical nonlinearity.* Such problems are considered from the BEM viewpoint in Sect. 4.1.

Another kind of nonlinearity is found in the problems with nonlinear boundary conditions. The contact problems, which are of great practical significance, belong to this class and will be discussed in Sect. 4.3.

From among the books which extensively deal with the application of the BEM in nonlinear mechanical problems the monographs [40, 50] and chapters in [1, 2, 3, 36, 41] should be referred to.

4.1 Geometrical Nonlinearities

Let us first recapitulate some facts related to geometrical nonlinearities.

Let us denote a reference configuration by C^0 and assume it to be identical with an initial configuration of the body. The position vector of a particle in this configuration is denoted by x^0. Coordinates (x_i^0) of the vector are called *material coordinates* (or Lagrange coordinates).

Let us denote the current configuration of the body by C^τ. It is a set of positions of the particles in a reference space at time $t = \tau$. Coordinates (x_i) of a position vector of a particle in the current configuration are called *space coordinates* (or Euler coordinates).

The motion of the body is a time dependent family of mappings of the initial configuration C^0 onto the current configuration C^τ; $\tau \in [0\,,\ \tau_k]$ and can be represented by a vector function

$$\mathbf{x} = \mathbf{x}(\mathbf{x}^0, \tau)\,.$$

It is assumed that the mapping $C^0 \to C^\tau$ is bijective which implies the invertibility of the motion function. Defining the displacement as a vector

$$\mathbf{u}(\mathbf{x}^0) = \mathbf{x}(\mathbf{x}^0, \tau) - \mathbf{x}^0\,,$$

we can treat it as a material vector, i.e. as a vector field the domain of which is configuration C^0, or as a spatial vector, i.e. a vector field the domain of which is the current configuration:

$$\mathbf{u}(\mathbf{x}) = \mathbf{x} - \mathbf{x}^0(\mathbf{x}, \tau)\,.$$

The same refers to other functions describing a state of the body. They can be defined in configuration C^0 and this description is called material, or they can be defined in configuration C^τ and then we deal with a spatial description. An updated description is in a limit sense $\Delta t \to 0$ a convective description. In our further consideration both descriptions will be used.

For a perfectly rigid body the distance between an arbitrary pair of particles remains constant, i.e.

$$\forall\,\tau \in [0, t]:\; |\Delta\mathbf{x}| = |\Delta\mathbf{x}^0|\,.$$

A feature that distinguishes deformable bodies is a change of the distance during motion, which is a function of coordinates of the particle, so

$$\Delta\mathbf{x}^0 = \mathbf{x}^0(\mathbf{x} + \Delta\mathbf{x}, t) - \mathbf{x}^0(\mathbf{x}, t)\,.$$

Developing $\Delta\mathbf{x}^0$ in a series it is easy to find a linear estimation:

$$\Delta\mathbf{x}^0 = \frac{\partial\mathbf{x}^0}{\partial\mathbf{x}}\Delta\mathbf{x}\,.$$

The tensor

$$\mathbf{F}^{-1} \stackrel{\text{def}}{=} \frac{\partial\mathbf{x}^0}{\partial\mathbf{x}}\,; \qquad \operatorname{dom}\mathbf{F}^{-1} = C^\tau\,,$$

defined in the current configuration, is called a *spatial deformation gradient.* It is an operator which assigns a vector $d\mathbf{x}^0$ between two particles (belonging to the initial configuration) to a vector $d\mathbf{x}$ between their current positions (cf. Chap. 2 in Part I).

The deformation gradient is a basic deformation measure, which can be used to construct other measures, such as the spatial (Cauchy) deformation tensor

$$\mathbf{c} \stackrel{\text{def}}{=} \mathbf{F}^{-T}\mathbf{F}^{-1}\,; \qquad \operatorname{dom}\mathbf{c} = C^\tau\,,$$

or the strain (Almansi–Hammel) tensor

$$\mathbf{e} \stackrel{\text{def}}{=} \frac{1}{2}(\mathbf{g} - \mathbf{c})\,; \qquad \operatorname{dom}\mathbf{e} = C^\tau\,,$$

where $\mathbf{g}$ is a metric tensor of the reference space. It is assumed that it is a Euclidean space, which can always be globally parametrized with Cartesian coordinates. This allows us to take $\mathbf{g} = \mathbf{I}$ without restricting the generality. It is easy to find a relationship between the displacements and deformation. On the basis of the displacement vector definition one can write

$$\nabla \mathbf{u} = \mathbf{I} - F^{-1}, \qquad \text{where } \nabla \stackrel{\text{def}}{=} \text{grad}\ldots,$$

and further, using the definition of the deformation tensor,

$$\mathbf{c} = \mathbf{I} - \nabla \mathbf{u} - \nabla^T \mathbf{u} + \nabla^T \mathbf{u} \nabla \mathbf{u}.$$

Similarly, from the definition of the strain tensor,

$$\mathbf{e} = \frac{1}{2}(\nabla \mathbf{u} + \nabla^T \mathbf{u} - \nabla^T \mathbf{u} \nabla \mathbf{u}).$$

It can be seen that the last two relationships are nonlinear. They have been quoted in order to avoid an impression that the nonlinear character of analogical equations in Chap. 2 of Part I results from an incremental formulation or a special choice of the reference configuration.

The process of deformation is by nature nonlinear and therefore the state equations (cf. the set of equations in Sect. 2.6 of Part I) are nonlinear irrespective of the character of constitutive equations. In engineering practice a linearized set of equations is most frequently used, which describes the deformation in an approximate manner [10]. It is worth reminding that the linearization requires an assumption that only "small deformations" are considered, i.e. so small that the product $\nabla^T \mathbf{u} \nabla \mathbf{u}$ is negligible in comparison with the displacement gradient:

$$\nabla^T \mathbf{u} \nabla \mathbf{u} = \mathit{0}(\nabla \mathbf{u}).$$

Moreover, it is assumed that the displacements are sufficiently small to identify the configurations $C^0 \approx C^\tau$ (a so-called *solidification principle* in the strength of materials). The consequence of the latter assumption is that the material and spatial description coincide. Both Piola–Kirchhoff stress tensors are identical with the Cauchy stress tensor. There is no reason, either, to distinguish the notions of deformation and strain. In other words, the initial configuration is treated as an equilibrium configuration.

The above recapitulation of assumptions of the linear approximation of the deformation description makes it possible to notice that the set of equations (3.1)–(3.6), quoted in Chap. 3 of Part I is of course not completely a linearized description. In the incremental formulation the assumption concerning the displacement gradients is accepted, but the configurations are distinguished.

The emphasis on the adopted assumptions is important, since the application of the BEM to geometrically nonlinear problems involves a combination of the method with an incremental formulation. Due to the distinction between the configurations, a displacement equation equivalent to the set of equations in Chap. 3 of Part I has the form:

$$\frac{1}{2}\mathbf{C}\,\mathrm{div}[\nabla(\Delta\mathbf{u}) + \nabla^T(\Delta\mathbf{u})] + \mathrm{div}[\nabla(\Delta\mathbf{u})\tilde{\sigma}] + \Delta\mathbf{b} = \varrho\Delta\ddot{\mathbf{u}}\,; \quad \forall\mathbf{x} \in \mathrm{Int}(C^\tau), \tag{4.1}$$

with boundary conditions:

$$\Delta\sigma\mathbf{n} = \Delta\mathbf{p}\,; \quad \forall\mathbf{x} \in \partial_p C^\tau\,,$$
$$\Delta\mathbf{u} = \Delta\mathbf{g}\,; \quad \forall\mathbf{x} \in \partial C^\tau\,,$$

where $\mathbf{C}$ is the material constant tensor of a hyper-elastic homogeneous body, $\Delta\mathbf{u}$ is a displacement increment with respect to configuration C^τ, $\Delta\mathbf{b} = \varrho\Delta\mathbf{f}$ is a volume force increment ($\varrho\Delta\mathbf{f}$ is a body force increment), $\Delta\ddot{\mathbf{u}}$ is an inertia force increment and $\tilde{\sigma}$ is the second Piola–Kirchhoff tensor. The second component in Eq. (4.1) distinguishes it from the known Lame equation. Figuratively speaking, this component takes into account the fact that configuration C^τ is an equilibrium configuration if the stress state from the previous configuration is considered. Consequently, at the first step Eq. (4.1) is a Lame equation and problems emerge in subsequent steps. Attempting to use the BEM, it would be necessary to construct a fundamental solution for the operator defined by Eq. (4.1). Since the second component varies at each step, the fundamental solution would have to be constantly corrected. To avoid this problem, the state equation can be solved iteratively at each step. Let us denote the Lame operator by L, and the following operator by L_1:

$$L_1 \stackrel{\mathrm{def}}{=} \frac{1}{2}\mathbf{C}\,\mathrm{div}[\nabla(\ldots) + \nabla^T(\ldots)] + \mathrm{div}[\nabla(\ldots)\tilde{\sigma}]\,.$$

The inertia forces have been neglected, which means that we limit our consideration to a static problem. In the iterative method, in succeeding approximations, we look for such a fictitious volume force $\Delta\hat{\mathbf{b}}$ that

$$L[\Delta\mathbf{u}] + \Delta\hat{\mathbf{b}} = L_1[\Delta\mathbf{u}] + \Delta\mathbf{b}\,. \tag{4.2}$$

Taking into account the relation

$$L_1 = L + \mathrm{div}[\nabla(\ldots)\tilde{\sigma}]\,,$$

it is easy to find the fictitious volume force for the k-th iteration:

$$\Delta\hat{\mathbf{b}}_k = \Delta\hat{\mathbf{b}}_{k-1} + \mathrm{div}[\nabla(\Delta\mathbf{u}_{k-1})]\tilde{\sigma}\,, \quad k \in (1,\ldots,n)\,,\ \Delta\mathbf{b} = \Delta\mathbf{b}_0\,. \tag{4.3}$$

Iterations can be terminated when

$$|\Delta\mathbf{b}_0 - \Delta\mathbf{b}_n| \le \varepsilon\,,$$

where ε is an assumed, sufficiently small positive number. At each step it is necessary to repeat the procedure described in Sect. 2.4.3 and Sect. 3.1 n times, and then, when the iterations have been finished, the stress increment:

$$\Delta\tilde{\sigma}_k = \int_\Gamma \{\Delta\mathbf{u}_k P[\Sigma(\mathbf{U})] - \Sigma(\mathbf{U})P(\Delta\mathbf{u}_k)\}\,\mathrm{d}\Gamma - \int_\Omega \Sigma(\mathbf{U})\Delta\hat{\mathbf{b}}_k\,\mathrm{d}\Omega \tag{4.4}$$

and the gradient of the stress increment

$$\nabla[\Delta\mathbf{u}_k] = \int_\Gamma [\Delta\mathbf{u}_k P(\nabla\mathbf{U}) - \nabla\mathbf{U}P(\Delta\mathbf{u}_k)]\,\mathrm{d}\Gamma - \int_\Omega \nabla\mathbf{U}\Delta\hat{\mathbf{b}}_k\,\mathrm{d}\Omega \tag{4.5}$$

should be computed, where $\Omega \subset \boldsymbol{E}^3$ is the volume occupied by configuration C^τ in the reference space, $\Gamma = \partial\Omega$ is the boundary of the domain Ω, $\Delta\hat{\mathbf{b}}_k$ is the fictitious force determined in the k-th iteration, P is an operator defined in Eq. (2.92) of Sect. 2.4.3, and Σ is a matrix operator $\Delta\boldsymbol{\sigma} = \Sigma(\Delta\mathbf{u})$, cf. Sect. 2.4.3, which for a Hooke's material has the form:

$$\Sigma \stackrel{\text{def}}{=} \begin{bmatrix} (\lambda+2\mu)\frac{\partial}{\partial x_1} & \lambda\frac{\partial}{\partial x_2} & \lambda\frac{\partial}{\partial x_3} \\ \lambda\frac{\partial}{\partial x_1} & (\lambda+2\mu)\frac{\partial}{\partial x_2} & \lambda\frac{\partial}{\partial x_3} \\ \lambda\frac{\partial}{\partial x_1} & \lambda\frac{\partial}{\partial x_2} & (\lambda+2\mu)\frac{\partial}{\partial x_3} \\ \mu\frac{\partial}{\partial x_2} & \mu\frac{\partial}{\partial x_1} & 0 \\ 0 & \mu\frac{\partial}{\partial x_3} & \mu\frac{\partial}{\partial x_2} \\ \mu\frac{\partial}{\partial x_3} & 0 & \mu\frac{\partial}{\partial x_1} \end{bmatrix}, \tag{4.4a}$$

On the basis of the calculated increments, the configuration must be updated:

$$\mathbf{x}_{k+1} = \mathbf{x}_k + \Delta\mathbf{u}_k\,, \tag{4.6}$$

and the stress state associated with the configuration must be computed:

$$\boldsymbol{\sigma}_{k+1} = \tilde{\boldsymbol{\sigma}}_k + \nabla(\Delta\mathbf{u}_k)\tilde{\boldsymbol{\sigma}}_k + \tilde{\boldsymbol{\sigma}}_k\nabla^T(\Delta\mathbf{u}_k) + \Delta\tilde{\boldsymbol{\sigma}}_k\,. \tag{4.7}$$

Repetition of the presented operations at each step makes it possible to solve a geometrically nonlinear problem. A basis is the incremental formulation described in Part I, which is called *updated,* i.e. where the current configuration is the reference configuration. It is thus necessary to update the division of the boundary and, consequently, the finite-dimensional approximation. This inconvenience can be avoided using the material description. It means that, at each step, the problem is mapped onto the initial configuration, which in this way becomes a reference configuration, common for all steps.

The relation between the stress increment in the current configuration and the increment of the first Piola–Kirchhoff tensor referred to the initial configuration has the form (cf. Part I):

$$\Delta\tilde{\boldsymbol{\sigma}} = \{\Delta\tilde{\boldsymbol{\sigma}}(\mathbf{F}-1)^T + \tilde{\boldsymbol{\sigma}}\nabla^T(\Delta\mathbf{u}) + \Delta\tilde{\boldsymbol{\sigma}}\nabla^T(\Delta\mathbf{u})\}J\,, \tag{4.8}$$

in which

$$J = \det\left[\frac{\partial x}{\partial x^0}\right] \approx 1 + \sum_{k=1}^{N+1} \operatorname{div}(\Delta\mathbf{u}_k)\,.$$

Linearization of Eq. (4.8) and substitution of the geometric equations into the constitutive relations, and the latter ones into the incremental motion (or equilibrium) equation gives the following displacement equations:

$$\begin{aligned}
&\frac{1}{2}\mathbf{C}\,\mathrm{Div}\left\{[\nabla^0(\Delta\mathbf{u}_N)+\nabla^{0T}(\Delta\mathbf{u}_N)]J\right\}\\
&-\frac{1}{2}\mathbf{C}\,\mathrm{Div}\left\{[\nabla^0(\Delta\mathbf{u}_N)+\nabla^{0T}(\Delta\mathbf{u})]\sum_{k=0}^{N}\nabla^0(\Delta\mathbf{u}_k)J\right\}\\
&+\mathrm{Div}[\tilde{\sigma}\nabla^0(\Delta\mathbf{u}_N)J]+\Delta\mathbf{b}^0=\mathbf{0}\,; \qquad \forall\, x^0\in C^0, \qquad (4.9)
\end{aligned}$$

where the notation

$$\mathrm{Div}\stackrel{\mathrm{def}}{=}\sum_{i=1}^{3}\frac{\partial}{\partial x_i^0}=\mathrm{tr}\nabla^0\,,\quad \nabla^0=\frac{\partial}{\partial x^0}\,,\quad \Delta\mathbf{b}^0=\Delta\mathbf{b}(x^0)$$

has been used. Equation (4.9) must be supplemented by the boundary conditions:

$$\begin{aligned}
\Delta\sigma\mathbf{n}^0&=\Delta\mathbf{p}^0\,; \qquad \forall\,\mathbf{x}^0\in\partial_p C_0\,,\\
\Delta\mathbf{u}&=\Delta\mathbf{g}\,; \qquad \forall\,\mathbf{x}^0\in\partial_u C_0\,,
\end{aligned} \qquad (4.10)$$

where $\mathbf{n}^0$ is an outer normal vector to the boundary ∂C_0.

Equation (4.9) is significantly different from the Lame equation. Obviously, a part identical to the Lame operator can be separated and solved with the iterative method. However, the expression which defines the fictitious volume force will now be incomparably more extended. Multiple numerical calculation of the fictitious force is therefore not more cumbersome than updating the boundary division. Equation (4.9) has been quoted here for completeness, since, from the BEM viewpoint, it is not more convenient than the formulation which operates on the reference configuration which is updated at each step.

4.2 Physical Nonlinearities

We consider a physically nonlinear body which occupies a domain Ω with a boundary Γ in the space $\boldsymbol{E}^d$, $d=2$ or 3. Its state is described by the set of equations:

$$\mathrm{div}\boldsymbol{\sigma}+\mathbf{b}(x)=\mathbf{0}\,,\qquad \forall\, x\in\Omega\,, \qquad (4.11a)$$

$$\boldsymbol{\epsilon}=\frac{1}{2}(\nabla\mathbf{u}+\nabla^T\mathbf{u})\,, \qquad (4.11b)$$

$$\boldsymbol{\sigma}=\mathbf{f}(\boldsymbol{\epsilon}) \qquad (4.11c)$$

and boundary conditions:

$$\mathbf{u}(x)=\mathbf{u}^0(x)\,,\qquad \forall\, x\in\Gamma\,, \qquad (4.12a)$$

$$\mathbf{p}(x)=\boldsymbol{\sigma}\mathbf{n}=\mathbf{p}^0(x)\,,\qquad \forall\, x\in\Gamma_p\,, \qquad (4.12b)$$

where $\Gamma_u \cup \Gamma_p = \Gamma$ and $\Gamma_u \cap \Gamma_p = \emptyset$. The source of nonlinearity is the constitutive relation (4.11c), and the form of Eq. (4.11b) implies that the kinematics of the body is linearized. The strain tensor $\boldsymbol{\epsilon}$ can be decomposed as

$$\boldsymbol{\epsilon} = \boldsymbol{\epsilon}^e + \boldsymbol{\epsilon}^n , \tag{4.13}$$

with an elastic strain tensor $\boldsymbol{\epsilon}^e$ and an inelastic (irreversible) strain tensor $\boldsymbol{\epsilon}^n$. The latter tensor can contain plastic strains, rheological strains etc. Decomposing the stress tensor in a similar way, the Hooke's law can be used to describe the elastic properties of the body

$$\boldsymbol{\sigma}^e = 2\mu\boldsymbol{\epsilon}^e + \lambda\,\mathbf{I}\,\mathrm{tr}\boldsymbol{\epsilon}^e , \tag{4.14}$$

and further, after substitution of Eq. (4.13)

$$\boldsymbol{\sigma}^e = 2\mu(\boldsymbol{\epsilon} - \boldsymbol{\epsilon}^n) + \lambda\,\mathbf{I}\,\mathrm{tr}(\boldsymbol{\epsilon} - \boldsymbol{\epsilon}^n) , \tag{4.15}$$

or

$$\boldsymbol{\sigma}^e = 2\mu\boldsymbol{\epsilon} + \lambda\,\mathbf{I}\,\mathrm{tr}\boldsymbol{\epsilon} - \boldsymbol{\sigma}^n , \tag{4.16}$$

where the stress tensor

$$\boldsymbol{\sigma}^n = 2\mu\boldsymbol{\epsilon}^n + \lambda\,\mathbf{I}\,\mathrm{tr}\boldsymbol{\epsilon}^n \tag{4.17}$$

can be treated as an "initial stress" field.

The inelastic body considered can thus be treated as a fictitious elastic body with a certain initial stress field $\boldsymbol{\epsilon}^n$. Using Eqs. (4.11a), (4.11b), (4.13) and (4.16) the following state equation can be found:

$$L\mathbf{u}(x) + \mathbf{b}^*(x) = \mathbf{0} , \qquad \forall\, x \in \Omega , \tag{4.18}$$

and the boundary conditions are:

$$\mathbf{u}(x) = \mathbf{u}^0 , \qquad \forall : x \in \Gamma_u , \tag{4.19a}$$

$$\mathbf{p}^*(x) = \boldsymbol{\sigma}\mathbf{n} + \mathbf{p}^n = \mathbf{p} + \mathbf{p}^n , \qquad \forall\, x \in \Gamma_p , \tag{4.19b}$$

where

$$\mathbf{p}^n = \boldsymbol{\sigma}^n\mathbf{n} , \tag{4.20}$$

$$\mathbf{b}^* = \mathbf{b} - 2\mu\,\mathrm{div}\boldsymbol{\epsilon}^n - \lambda\,\mathbf{I}\,\mathrm{div}[\mathrm{tr}\boldsymbol{\epsilon}^n] = \mathbf{b} - \mathbf{b}^* = \mathrm{div}\boldsymbol{\sigma}^n , \tag{4.21}$$

and L is the Lame operator, cf. Eq. (2.91).

If $\mathbf{b}^*$ and $\mathbf{p}^*$ are understood as fictitious fields of volume forces and tractions, respectively, then Eq. (4.18) with boundary conditions (4.19) describes a boundary value problem similar to the ones encountered in the elasticity theory (cf. Chap. 3). An integral formula for a problem formulated in such a way can be written as:

$$\mathbf{c}(x)\mathbf{u}(x) = \int_\Gamma \mathbf{U}(x,y)\mathbf{p}^*(y)\,\mathrm{d}\Gamma(y) - \int_\Gamma \mathbf{P}(x,y)\mathbf{u}(y)\,\mathrm{d}\Gamma(y) + \\ + \int_\Omega \mathbf{U}(x,y)\mathbf{b}^*(y)\,\mathrm{d}\Omega(y)\,, \tag{4.22}$$

where the fictitious volume forces $\mathbf{b}^*$ and tractions $\mathbf{p}^*$ depend on inelastic strains and are determined by Eqs. (4.21) and (4.19b).

The integral formula (4.22) enables unknown displacements and tractions to be computed provided the distribution of inelastic strains $\boldsymbol{\epsilon}^n$ (treated as "initial") is known. Since this distribution is most frequently unknown, an incremental method can be applied and the problem can be treated as quasi-static. We can then imagine that we deal with an elastic body for which the volume forces and tractions are modified according to Eqs. (4.21) and (4.19b) for a given loading increment. The correcting components for the actual volume forces $\Delta\mathbf{b}_n = \Delta\mathbf{b}_* - \Delta\mathbf{b}$ and tractions $\Delta\mathbf{p}_n = \Delta\mathbf{p}_* - \Delta\mathbf{p}$ are determined in an iterative process until such fictitious quantities $\Delta\mathbf{b}_*$ and $\Delta\mathbf{p}_*$ are found that the behaviour of the nonlinear body can be described with the set of linear equations (4.18), (4.19a) and (4.19b). The process is repeated for each step associated with an increment of external forces. In a particular case of an elastic-plastic body, it is also necessary to determine a zone in which the plastic state is reached at a certain loading step. A yield condition

$$\mathcal{F}(\boldsymbol{\sigma}\,,\,\boldsymbol{\epsilon}^p\,,\,\kappa) = 0\,,$$

verified at each step, is employed to determine the plastic zone. In this equation κ is a hardening parameter.

The approach described above is called a method of fictitious volume forces and tractions. Its inconvenience is caused by the necessity to calculate derivatives of inelastic strains with respect to space coordinates within the domain Ω, cf. Eq. (4.21). This disadvantage can be avoided by reformulating the last integral in Eq. (4.22):

$$I \equiv \int_\Omega \mathbf{b}^*\,\mathrm{d}\Omega = \int_\Omega \mathbf{U}\mathbf{b}\,\mathrm{d}\Omega - \int_\Omega \mathbf{U}\mathrm{div}\boldsymbol{\sigma}^n\,\mathrm{d}\Omega\,. \tag{4.23}$$

Applying the Gauss–Ostrogradsky theorem to the last component, we obtain:

$$I = \int_\Omega \mathbf{U}\mathbf{b}\,\mathrm{d}\Omega + \int_\Omega \mathbf{E}\boldsymbol{\sigma}^n\,\mathrm{d}\Omega - \int_\Gamma \mathbf{U}\mathbf{p}^n\,\mathrm{d}\Gamma\,, \tag{4.24}$$

where

$$\mathbf{E} = \frac{1}{2}(\nabla\mathbf{U} + \nabla^T\mathbf{U})\,. \tag{4.25}$$

Eventually, the integral formula (4.22) becomes:

$$\mathbf{cu} = \int_\Gamma \mathbf{Up}\,\mathrm{d}\Gamma - \int_\Gamma \mathbf{Pu}\,\mathrm{d}\Gamma + \int_\Omega \mathbf{Ub}\,\mathrm{d}\Omega + \int_\Omega \mathbf{E}\boldsymbol{\sigma}^n\,\mathrm{d}\Omega\,. \tag{4.26}$$

Since the following relationship between strains and inelastic stresses is valid:

$$\int_\Omega \mathbf{E}\boldsymbol{\sigma}^n\,\mathrm{d}\Omega = \int_\Omega \mathbf{T}\boldsymbol{\epsilon}^n\,\mathrm{d}\Omega\,, \tag{4.27}$$

where

$$\mathbf{T} = \Sigma(\mathbf{U})\,, \tag{4.28}$$

and Σ is a matrix differential operator defined in Eq. (4.4a), the integral formula (4.26) can also be expressed in terms of inelastic strains $\boldsymbol{\epsilon}^n$:

$$\mathbf{cu} = \int_\Gamma \mathbf{Up}\,\mathrm{d}\Gamma - \int_\Gamma \mathbf{Pu}\,\mathrm{d}\Gamma + \int_\Omega \mathbf{Ub}\,\mathrm{d}\Omega + \int_\Omega \mathbf{T}\boldsymbol{\epsilon}^n\,\mathrm{d}\Omega\,. \tag{4.29}$$

In an incremental formulation, Eqs. (4.26) and (4.29) adopt the form

$$\mathbf{cu} = \int_\Gamma \mathbf{U}\Delta\mathbf{p}\,\mathrm{d}\Gamma - \int_\Gamma \mathbf{P}\Delta\mathbf{u}\,\mathrm{d}\Gamma + \int_\Omega \mathbf{U}\Delta\mathbf{b}\,\mathrm{d}\Omega + \Delta\mathbf{C}^n\,, \tag{4.30}$$

in which the term $\Delta\mathbf{C}^n$ is defined as follows:

– for the "initial strain" approach

$$\Delta\mathbf{C}^n = \int_\Omega \mathbf{T}\Delta\boldsymbol{\epsilon}^n\,\mathrm{d}\Omega\,, \tag{4.31}$$

– for the "initial stress" approach

$$\Delta\mathbf{C}^n = \int_\Omega \mathbf{E}\Delta\boldsymbol{\sigma}^n\,\mathrm{d}\Omega\,. \tag{4.32}$$

The calculation of stresses at internal points of the domain Ω is one of the most important solution stages in the inelastic problems of mechanics of deformable bodies. Adopting the Hooke's law we obtain for the elastic part of the stress tensor

$$\boldsymbol{\sigma} = \Sigma(\mathbf{u}) - \boldsymbol{\sigma}^n\,. \tag{4.33}$$

The displacement derivatives, present in Eq. (4.33), can be calculated numerically, using the finite difference technique, or analytically, differentiating integral formula (4.26) or (4.31) for $x \in \Omega$, $\mathbf{c} = \mathbf{I}$. The main difficulty in the application of the latter approach, which allows us to avoid the "double approximation", is caused by the fact that the kernels $\mathbf{E}$ and $\mathbf{T}$ have a singularity of order $O(r^{-2})$ in three-dimensional problems. Therefore, in order to differentiate the last integral with kernel $\mathbf{T}$ with respect to x, it is necessary to split the integration domain Ω into an infinitesimal sphere Ω' with the centre at point x and the remaining volume $\Omega - \Omega'$. Differentiation of the integral over Ω' results in an expression, the contribution of which is negligible because of the weak regularity condition for $\boldsymbol{\epsilon}^n$ (cf. [50]). Differentiation of the integral over $\Omega - \Omega'$ gives, next to an integral with kernel div$\mathbf{T}$, existing in the sense of the Cauchy principal value, an additional "convective" component, defined on the sphere $\partial\Omega'$ and induced by the fact that the domain $\Omega - \Omega'$ changes together with the position of the point x (cf. [4]):

$$\frac{\partial}{\partial x_m}\int_{\Omega-\Omega'} \mathbf{T}(x,y)\boldsymbol{\epsilon}^n(y)\,\mathrm{d}\Omega(y) = \int_{\Omega-\Omega'} \operatorname{div}\mathbf{T}\boldsymbol{\epsilon}^n\,\mathrm{d}\Omega(y) - \boldsymbol{\epsilon}^n\int_{\partial\Omega'} \mathbf{T}\mathbf{n}'(\partial\Omega')\,, \tag{4.34}$$

where $\mathbf{n}' = (n'_m)$ is a normal to the surface of the sphere $\partial\Omega'$. The last integral in Eq. (4.34), defined on the surface $\partial\Omega'$, can be calculated analytically and written explicitly. The final components of the stress state inside the domain Ω are expressed in terms of the following integral formula:

$$\boldsymbol{\sigma}(x) = \int_\Gamma \mathbf{D}\mathbf{p}\,\mathrm{d}\Gamma - \int_\Gamma \mathbf{S}\,\mathrm{d}\Gamma + \int_\Gamma \mathbf{D}\mathbf{b}\,\mathrm{d}\Gamma + \mathbf{C}^n_\sigma(x)\,, \tag{4.35}$$

in which

$$\mathbf{C}^n_\sigma = \int_\Omega \mathbf{T}^\sigma\boldsymbol{\epsilon}^n\,\mathrm{d}\Omega + \mathbf{I}^\sigma\boldsymbol{\epsilon}^n(x) \tag{4.36}$$

or

$$\mathbf{C}^n_\sigma = \int_\Omega \mathbf{E}^\sigma\boldsymbol{\sigma}^n\,\mathrm{d}\Omega + \mathbf{J}^\sigma\boldsymbol{\sigma}^n(x)\,. \tag{4.37}$$

The tensors $\mathbf{D}$ and $\mathbf{S}$ are defined by Eqs. (3.40) and (3.41), while $\mathbf{T}^\sigma$, $\mathbf{E}^\sigma$, $\mathbf{I}^\sigma = (I^\sigma_{ijkl})$ and $\mathbf{J}^\sigma = (J^\sigma_{ijkl})$ are defined as:

$$\mathbf{T}^\sigma = \Sigma(\mathbf{T})\,, \qquad \mathbf{E}^\sigma = \Sigma(\mathbf{E}) \tag{4.38}$$

and

$$I^\sigma_{ijkl} = -\frac{2\mu(1-\nu)^{-1}}{\alpha(\alpha+2)}\{[(\alpha^2-2)-\nu(\alpha^2-4)]\delta_{ijk}\delta_{lj} + [1-\nu(\alpha+2)]\delta_{ij}\delta_{kl}\}\,, \tag{4.39}$$

$$J^\sigma_{ijkl} = -\frac{(1-\nu)^{-1}}{\alpha(\alpha+2)}\{[(\alpha^2-2)-\nu(\alpha^2-4)]\delta_{ik}\delta_{lj} + [1-\nu(\alpha+2)]\delta_{ij}\delta_{kl}\}\,, \tag{4.40}$$

where $\alpha = 2$ for two-dimensional and $\alpha = 3$ for three-dimensional problems. The kernels $\mathbf{T}^\sigma$ and $\mathbf{E}^\sigma$ have a singularity order $O(r^{-3})$ and therefore the integrals in Eqs. (4.36) and (4.37) require special treatment during numerical integration.

It is worth noticing that the integral expression (4.35) enables the stresses to be computed only inside the domain Ω. To compute the stresses at the boundary different relations must be used. They are obtained from the relations between strains, displacements and tractions in each boundary element.

They do not require integration, their form for elastic problems has been presented in Eqs. (3.44) and (3.45) in Chap. 3, but the expressions for stresses (3.45) must be extended to account for inelastic strains.

The integral formulas derived above describe in an analytical way the inelastic problem of mechanics. In order to solve it numerically, it is necessary to perform a finite-dimensional discretization of the formulas (4.26) or (4.29) and (4.35) following the approach presented in Sect. 3.1.2. In addition to the discretization of the boundary with boundary elements, the inside of the domain, where inelastic strains $\boldsymbol{\epsilon}^n$ or stresses $\boldsymbol{\sigma}^n$ occur, must also be discretized with internal cells Ω^q, $q = 1, 2, \ldots, Q$. The inelastic strains $\boldsymbol{\epsilon}^n$ and inelastic stresses $\boldsymbol{\sigma}^n$ are approximated within each cell similarly to the volume forces, cf. Eq. (3.30).

After discretization, Eqs. (4.29) and (4.35) take the following matrix form:

$$\mathbf{Hu} = \mathbf{Gp} + \mathbf{R}\boldsymbol{\epsilon}^n + \mathbf{B}\,, \tag{4.41}$$

$$\boldsymbol{\sigma} = \mathbf{G}_D\mathbf{p} - \mathbf{H}_S\mathbf{u} + \mathbf{M}\boldsymbol{\epsilon}^n + \mathbf{B}_D\,, \tag{4.42}$$

where matrices $\mathbf{H}, \mathbf{G}, \mathbf{G}_D, \mathbf{H}_S, \mathbf{B}$ nd $\mathbf{B}_D$ are defined like in the linear problems of elasticity (cf. Chap. 3), matrix $\mathbf{R}$ depends on the integral with kernel $\mathbf{T}$ and elements of matrix $\mathbf{M}$ depend on the integral with kernel $\mathbf{T}^\sigma$ and the term $\mathbf{I}^\sigma$. The column matrix $\boldsymbol{\epsilon}^n$ contains values of inelastic strains at nodes of the cells Ω^q.

If we apply the approach using the initial stresses, then the following equations are counterparts of Eqs. (4.41) and (4.42):

$$\mathbf{Hu} = \mathbf{Gp} + \bar{\mathbf{R}}\boldsymbol{\sigma}^n + \mathbf{B}\,, \tag{4.41a}$$

$$\boldsymbol{\sigma} = \mathbf{G}_D\mathbf{p} - \mathbf{H}_S\mathbf{u} + \bar{\mathbf{M}}\boldsymbol{\sigma}^n + \mathbf{B}_D\,, \tag{4.42a}$$

where matrices $\bar{\mathbf{R}}$ and $\bar{\mathbf{M}}$ depend on integrals with kernels $\mathbf{E}$ and $\mathbf{E}^\sigma$, respectively. Having considered the boundary conditions, Eq. (4.41) becomes:

$$\mathbf{AX} = \mathbf{F} + \mathbf{R}\boldsymbol{\epsilon}^n\,, \tag{4.43}$$

where matrices $\mathbf{A}$ and $\mathbf{F}$ are defined like in the linear elasticity problems, $\mathbf{X}$ is a column matrix of unknown displacements and forces at nodes of boundary elements. The internal stresses can be expressed as follows:

$$\boldsymbol{\sigma} = -\bar{\mathbf{A}}\mathbf{X} + \bar{\mathbf{F}} + \mathbf{M}\boldsymbol{\epsilon}^n\,, \tag{4.44}$$

where elements of the square matrix $\bar{\mathbf{A}}$ are generated from matrices $\mathbf{G}_D$ and $\mathbf{H}_S$ in such a way that they are associated with unknown boundary values included in the vector $\mathbf{X}$, and the column matrix $\bar{\mathbf{F}}$ depends on the given boundary conditions and volume forces. The solution of Eq. (4.43) with $\det\mathbf{A} \neq 0$ reads:

$$\mathbf{X} = \mathbf{K}\boldsymbol{\epsilon}^n + \mathbf{V}\,, \tag{4.45}$$

where

$$\mathbf{K} = \mathbf{A}^{-1}\mathbf{R},$$

$$\mathbf{V} = \mathbf{A}^{-1}\mathbf{F}.$$

Considering the solution (4.45), Eq. (4.44) takes the form:

$$\boldsymbol{\sigma} = \mathbf{Z}\boldsymbol{\epsilon}^n + \boldsymbol{\sigma}_E, \tag{4.46}$$

where

$$\mathbf{Z} = \mathbf{M} - \bar{\mathbf{A}}\mathbf{A}^{-1}\mathbf{R},$$

$$\boldsymbol{\sigma}_E = \bar{\mathbf{F}} - \bar{\mathbf{A}}\mathbf{A}^{-1}\mathbf{F}.$$

The square matrix $\mathbf{Z}$ transforms the inelastic strain field into an inelastic stress field and the column matrix $\boldsymbol{\sigma}^E$ contains the values of elastic stresses at internal points. Equations (4.45) and (4.46) in an incremental form become:

$$\Delta\mathbf{X} = \mathbf{K}\Delta\boldsymbol{\epsilon}^n + \Delta\mathbf{V}, \tag{4.47}$$

$$\Delta\boldsymbol{\sigma} = \mathbf{Z}\Delta\boldsymbol{\epsilon}^n + \Delta\boldsymbol{\sigma}_E. \tag{4.48}$$

Equations (4.47) and (4.48) must be completed with a constitutive equation of the medium, which is expressed in an incremental form as follows:

$$\Delta\boldsymbol{\sigma} = \mathbf{C}_\Delta \Delta\boldsymbol{\epsilon}, \tag{4.49}$$

where the matrix of material properties $\mathbf{C}_\Delta = (\mathbf{C}^\Delta_{ijkl})$ depends on the history of the loading process.

In the literature a rate description of the considered problem is often employed. In a limit case, when $\Delta t = \delta t \to 0$, dividing Eqs. (4.47)–(4.49) by Δt and tending to the limit, the following set of rate equations is obtained:

$$\dot{\mathbf{X}} = \mathbf{K}\dot{\boldsymbol{\epsilon}}^n + \dot{\mathbf{V}}, \tag{4.47a}$$

$$\dot{\boldsymbol{\sigma}} = \mathbf{Z}\dot{\boldsymbol{\epsilon}}^n + \dot{\boldsymbol{\sigma}}^E, \tag{4.48a}$$

$$\dot{\boldsymbol{\sigma}} = \mathbf{C}_\Delta \dot{\boldsymbol{\epsilon}}. \tag{4.49a}$$

Equations (4.47)–(4.49) and (4.47a)–(4.49a) form a complete set of equations for a physically nonlinear problem in the incremental description and in the rate description, respectively. In our consideration of geometrical (Sect. 4.1) and physical (Sect. 4.2) nonlinearities we have so far limited our interest to static problems and we have assumed that the two nonlinear cases occur separately. An analysis of more complex cases using the BEM is also possible, e.g. when the geometrical and physical nonlinearities occur simultaneously [41] or when we deal with nonlinear dynamic problems [7, 30].

4.3 Other Nonlinear Problems

Beside the nonlinear problems discussed previously there is a class of problems which are of great practical significance and, despite a linearized deformation description and linear constitutive equations (Hooke's material), are nonlinear. These are so-called *contact problems* with unilateral constraints. Let us recap that constraints are understood as any restrictions imposed on the motion or deformation of a body. Below we shall confine our discussion to holonomic and simple constraints, i.e. to constraints imposed on displacements of the body (and not e.g. on deformation gradient). One of the possible forms of such an equation is the function

$$f(\mathbf{u}) = 0\,, \tag{4.50}$$

and so-called displacement boundary conditions can serve as an example. If constraints are described by an equality, then they are called ***bilateral constraints.*** Another kind of constraints can be written in the form of an inequality

$$f(\mathbf{u}) \leq 0\,. \tag{4.51}$$

They are called ***unilateral constraints.*** A simply supported beam with an obstacle (Fig. 4.1) can serve as a simple example of the constraints mentioned. The constraints imposed at points A and B are obviously bilateral constraints. On the other hand, the rigid body Ω represents unilateral constraints. As long as the displacements of the beam axis satisfy the condition

$$|\mathbf{u}| < d\,; \quad \forall x \in [A, B]\,,$$

the latter constraints are inactive and the solution does not differ from the one derived in the strength of materials. However, it can happen that

$$|\mathbf{u}| - d = 0\,; \quad \forall x \in [C, D]\,,$$

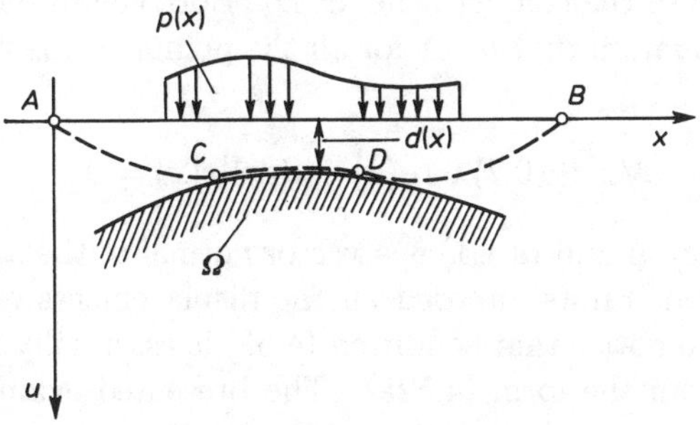

Fig. 4.1. Simply supported beam with an obstacle

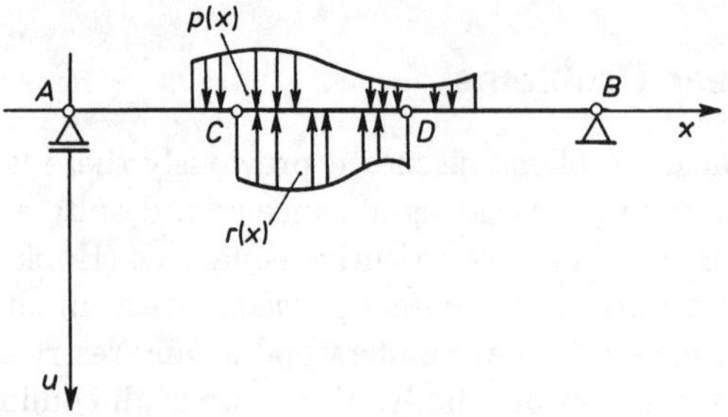

Fig. 4.2. Forces normal to the axis of an undeformed beam

which implies that in the interval $[C, D]$ the constraints are active. Obviously, basing on a postulate, the constraints can be removed and replaced by appropriate reaction forces. Limiting the discussion to perfect constraints (without friction), for which the work done by the reactions on displacements conforming with the constraints is zero, we can consider a simply supported beam loaded additionally by the reaction forces instead of the beam with an obstacle. Since the constraints are perfect, the appropriate reaction forces are normal to the obstacle (or, equivalently, to the axis of the beam). Taking into account the solidification principle, valid in the strength of materials (cf. Sect. 4.1) it is possible to assume that they are normal to the axis of the undeformed beam (Fig. 4.2). The measure of the reaction forces with respect to the normal is then unknown, but it is (so far) the only unknown, usually determined from the constraint equation. In the considered case the location of the points C and D, defining the so-called *contact zone* (i.e. the line segment along which the constraint reactions occur), is also unknown. The equations of unilateral constraints can be written as follows:

$$\forall\, x : |u(x)| < d, \qquad \mathbf{r}(x) = 0\,, \tag{4.52a}$$

$$\forall\, x : |u(x)| = d, \qquad \mathbf{r} \cdot \mathbf{n} \leq 0\,. \tag{4.52b}$$

Obviously, condition (4.52a) applies to the intervals $[A, C]$ and $[D, B]$ and condition (4.52b) to the contact zone $[C, D]$. Both conditions can be written together in an equation that holds for all the points on the axis of the beam, namely

$$\forall\, x \in (0, l) : [\mathbf{u}(x) - \mathbf{d}(x)]\mathbf{r}(x) = 0\,, \tag{4.53}$$

where $\mathbf{d}(x) = d(x) \cdot \mathbf{n}$ and $\mathbf{n}(0, 1)$ is a vector normal to the axis. A discussion of the form of constraints imposed on the displacements of the axis of the beam allows us to notice that condition (4.53) is essentially nonlinear, which clearly results from the form (4.52a). The presented example of the beam explains the observation that all the problems with unilateral constraints are nonlinear, and the source of nonlinearity lies in the boundary conditions.

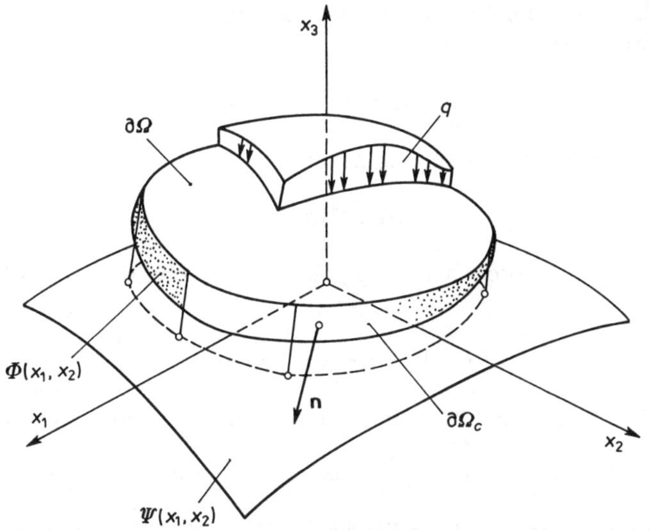

Fig. 4.3. Contact problem for a deformable body with unilateral constraints

After this introduction we shall formulate a contact problem with unilateral constraints for any deformable body (Fig. 4.3) within the framework of the linearized elasticity theory and with an assumption that only perfect, holonomic and simple constraints are considered. The following notation has been adopted:

- $\Omega \subset \boldsymbol{E}^3$ – volume occupied by the body,
- $\partial\Omega$ – boundary of the body,
- $\Psi(x_1, x_2) = x_3$ – function describing the boundary of the rigid obstacle body (target),
- $\Phi(x_1, x_2) = x_3$ – function describing the boundary $\partial\Omega$ of the body Ω,
- $\partial\Omega_c = \Gamma_c$ – unknown contact surface between the body and the obstacle.
- $\partial\Omega_p = \Gamma_p$ – part of the boundary of the body, subjected to external tractions,
- $\mathbf{n}$ – outer vector normal to the boundary $\partial\Omega$:

$$\mathbf{n} = \frac{\nabla\hat{\Phi}}{|\nabla\hat{\Phi}|}; \qquad \hat{\Phi} = \Phi(x_1, x_2) - x_3, \qquad \nabla = \frac{\partial}{\partial \mathbf{x}}, \tag{4.54}$$

$\forall\, x \in \Gamma_c : \Phi(x_1, x_2) = \Psi(x_1, x_2)$ – solidification principle,

- $\mathbf{u}(\mathbf{x})$ – displacement of the points of the body,
- $d = \Phi(x_i) - \Psi(x_i)$; $i = (1,\ 2)$ – distance between the boundary of the body and the obstacle,
- $\mathbf{d}[x_1,\ x_2,\ d]$ – vector measure of the distance between the boundary of the body and the obstacle.

With the adopted notation the following equation can be written for points in the contact zone:

$$(\mathbf{d}+\mathbf{u})\mathbf{n} = 0\,; \quad \forall\, x \in \Gamma_c\,. \tag{4.55}$$

For points which belong to Γ_p we have:

$$(\mathbf{d}+\mathbf{u})\mathbf{n} \neq 0\,. \tag{4.56}$$

Obviously, reaction forces act at all points where condition (4.55) is satisfied and hence:

$$\forall\, x \in \Gamma_c : \mathbf{n}r = \mathbf{r} \;\text{ and }\; r < 0\,, \tag{4.57}$$

while at the other boundary points the reaction is a zero vector, so:

$$\forall\, x \in \Gamma_p : \mathbf{n}r = \mathbf{r} \;\text{ and }\; r = 0\,. \tag{4.58}$$

Instead of a conjunction of conditions (4.55) and (4.57) for the contact zone and a conjunction of conditions (4.56) and (4.58) outside of it, one can, in a similar way as in the beam example, write a condition valid for all the boundary points:

$$(\mathbf{d}+\mathbf{u})\mathbf{r} = 0\,. \tag{4.59}$$

Finally, we have the following boundary value problem to solve:

$$L[\mathbf{u}] + \mathbf{b} = \mathbf{0}\,; \qquad \forall\, x \in \text{Int}\Omega\,, \tag{4.60}$$

$$\boldsymbol{\sigma}\mathbf{n} = \mathbf{p}\,; \qquad \forall\, x \in \Gamma_p\,, \tag{4.61}$$

$$\boldsymbol{\sigma}\mathbf{n} = \mathbf{r}\,; \qquad \forall\, x \in \Gamma_c\,. \tag{4.62}$$

Due to the form of the boundary conditions the problem can be classified as a stress problem of the elasticity theory. However, essential differences must be emphasized. In the problem considered neither the reaction forces nor the contact zone Γ_c are known. The reaction forces can be determined in a usual manner from the constraint equation (4.59), which, however, is now a nonlinear condition. Moreover, one should not forget about condition (4.57), which is a formal characteristic of the unilateral constraints. It is easier to discuss further particular features of the problem (4.59)–(4.62) when it is written using the formalism of the boundary integral equations. An integral formula which allows us to determine the displacements of any point (particle) of a body has the form:

$$\mathbf{u}(x) = \int_{\Omega} \mathbf{U}(x,y)\mathbf{b}(y)\,\mathrm{d}\Omega_y + \int_{\Gamma_p} \{\mathbf{u}(y)P[\mathbf{U}(x,y)] - \mathbf{p}(y)\mathbf{U}(x,y)\}\,\mathrm{d}\Gamma_y$$
$$+ \int_{\Gamma_c} \{\mathbf{u}(y)P[\mathbf{U}(x,y)] - \mathbf{r}(y)\mathbf{U}(x,y)\}\,\mathrm{d}\Gamma_y\,, \qquad \forall\, x \in \text{Int}\Omega\,. \tag{4.63}$$

An effective application of the formula requires of the following boundary integral equation to be solved:

$$\frac{1}{2}\,\mathbf{u}(x) = \int_{\Omega} \mathbf{U}(x,y)\mathbf{b}(y)\,\mathrm{d}\Omega_y + \int_{\Gamma_p} \{\mathbf{u}(y)P[\mathbf{U}(x,y)] - \mathbf{p}(y)\mathbf{U}(x,y)\}\,\mathrm{d}\Gamma_y$$
$$+ \int_{\Gamma_c} \{\mathbf{u}(y)P[\mathbf{U}(x,y)] - \mathbf{r}(y)\mathbf{U}(x,y)\}\,\mathrm{d}\Gamma\,, \quad \forall\, x \in \Gamma = \Gamma_c + \Gamma_p\,. \tag{4.64}$$

For completeness, a constraint equation in the form (4.59) and equilibrium conditions for the set of active forces and reactions have to be added:

$$\int_{\Omega} \mathbf{b}(x)\,\mathrm{d}\Omega_x + \int_{\Gamma_p} \mathbf{p}(x)\,\mathrm{d}\Gamma_x + \int_{\Gamma_c} \mathbf{r}(x)\,\mathrm{d}\Gamma = \mathbf{0}\,, \tag{4.65a}$$

$$\int_{\Omega} \mathbf{b}(x) \times (\mathbf{x} - \mathbf{x}_A)\,\mathrm{d}\Omega_x + \int_{\Gamma_p} \mathbf{p}(x) \times (\mathbf{x} - \mathbf{x}_A)\,\mathrm{d}\Gamma_x$$
$$+ \int_{\Gamma_c} \mathbf{r}(x) \times (\mathbf{x} - \mathbf{x}_A)\,\mathrm{d}\Gamma_x = \mathbf{0}\,, \quad \forall\, \mathbf{x}_A\,. \tag{4.65b}$$

In Eq. (4.65b) a vector notation has been employed for simplicity. The quoted conditions (4.65) play a double role. Firstly, they give an equation that makes it possible to determine the additional unknown, i.e. the contact zone Γ_c. Secondly, they represent conditions of existence of a solution. Without going into formal details of the formulation, some attention will be paid to the physical sense of the characteristic features of the problem considered.

Let us begin with Eqs. (4.65) treated as conditions for a solution to exist. From the viewpoint of mechanics they require the active force system and the reaction force system to be mutually opposite (or, in other words, they form together a zero or balanced system). Let us recapitulate that an arbitrary system of vectors, and in particular the active force system, can be reduced to one of the three forms:

1. The reduced system is a moment vector (i.e. its sum is a zero vector),
2. The reduced system is a resultant force (i.e. the moment calculated with respect to points on a straight line called a *central axis* is a zero vector),
3. The reduced system is a wrench (i.e. it is composed of a sum vector and a moment vector which, calculated with respect to points on the central axis, is parallel to the sum vector).

It is easy to notice that if reduction of the force system leads to case 1, then there is no solution of the problem. This results immediately from Eq. (4.57) (the reaction system can not have a zero sum if it is not a zero system). If reduction of the active force system gives system 3, then there is no solution either. It follows from the assumption of perfect constraints that the moment can not be a vector parallel to the sum vector. Consequently, a solution can exist only when the system of active forces is reduced to a resultant force. The existence is further determined by the following conditions:

1. The point of intersection of the boundary of the body $\partial\Omega$ with the central axis belongs to the boundary of the obstacle,
2. The sense of the resultant force vector is opposite to the sense of the outer normal to the obstacle at the intersection point.

The above conditions have been formulated under the assumption that the solidification principle resulting from a linear approximation of kinematics holds (see Sect. 4.1). In other words, the existence of a resultant of the active force system with a simultaneous satisfaction of conditions 1 and 2 guarantees a possibility to assure equilibrium of the body using smooth constraints. Let us further assume that the conditions for the existence of a solution are satisfied. This allows us to focus attention on the construction of an approximate solution using a finite-dimensional approximation with boundary elements, described in Sect. 2.5. Employing the notation introduced in that section, the integral formula (4.63) can be written as:

$$\mathbf{u}_i = \mathbf{u}_{0i} + \sum_{j\in I_1\cup I_2} \mathbf{a}_{ij}\hat{\mathbf{u}}_j + \sum_{j\in I_1} \mathbf{b}_{ij}\mathbf{p}_j - \sum_{j\in I_2} \mathbf{b}_{ij}\mathbf{r}_j\,, \tag{4.66}$$

where I_1 is a set of indices attributed to boundary segments $\Delta\Gamma_{p,j}$ which lie outside the contact zone, I_2 is a set of indices attributed to boundary segments $\Delta\Gamma_{c,j}$ which lie in the contact zone, $\hat{\mathbf{u}}_j$ are coefficients of boundary displacement approximation $\hat{\mathbf{u}}(x) = \sum_j \kappa_j\hat{\mathbf{u}}_j$ with a set of piecewise constant functions, $\mathbf{p}_j$ are approximation coefficients for tractions $\mathbf{p}(x) = \sum_j \kappa_j\mathbf{p}_j$ and $\mathbf{r}_j$ are, as the previous quantities, approximation coefficients for the reaction system $\mathbf{r}(x) = \sum_j \kappa_j\mathbf{r}_j$. The boundary integral equation (4.64) also takes the form of a set of algebraic equations:

$$\frac{1}{2}\hat{\mathbf{u}}_i = \hat{\mathbf{u}}_{0i} + \sum_{j} \mathbf{a}_{ij}\hat{\mathbf{u}}_j + \sum_{j\in I_1} \mathbf{b}_{ij}\mathbf{p}_j - \sum_{j\in I_2} \mathbf{b}_{ij}\mathbf{r}_j\,. \tag{4.67}$$

The constraint equation, which is used to determine the reaction forces, now becomes:

$$\begin{gathered}\sum_{j\in I_2}(\mathbf{d}_j + \mathbf{u}_j)\mathbf{r}_j = 0\,,\\ \mathbf{r}_j = \lambda_j\mathbf{n}_j\,; \qquad \lambda_j < 0\,,\end{gathered} \tag{4.68}$$

where $\mathbf{d}_j$ denotes a matrix form of an approximation of vector $\mathbf{d}$ with a set of piecewise constant functions, i.e.

$$\mathbf{d}_j\{\hat{x}_{1j}, \hat{x}_{2j}, \chi_j[\Phi(\hat{x}_{1j}, \hat{x}_{2j}) - \Psi(\hat{x}_{1j}, \hat{x}_{2j})]\}\,,$$

where κ_j is a characteristic function for segment $\Delta\Gamma_{c,j}$ and x_{1j}, x_{2j} are coordinates of the centre of the segment. Derivation of the solution begins with finding a simplest reduced system of external active forces, i.e.

– the sum vector

$$\mathbf{S} = \int_{\Omega} \mathbf{b}\, d\Omega + \int_{\Gamma_p} \mathbf{p}\, d\Gamma\,, \tag{4.69}$$

– the moment vector

$$\mathbf{M} = \int_{\Omega} \mathbf{x} \times \mathbf{b}\, d\Omega + \int_{\Gamma_p} \times \mathbf{p}\, d\Gamma\,, \tag{4.70}$$

which is for simplicity determined with respect to the origin of the coordinate system. These integrals are calculated only once, and their analytical calculation usually causes no problems. A condition for the existence of the resultant force reads:

$$\mathbf{S} \cdot \mathbf{M} = 0\,. \tag{4.71}$$

Assuming that the latter condition is satisfied, we determine the central axis:

$$\mathbf{x}_s = \frac{\mathbf{S} \times \mathbf{M}}{|\mathbf{S}|} + \beta \mathbf{S}\,, \tag{4.72}$$

which is the line of action of the resultant. The next step is to determine point C, which is an intersection point of the central axis and the boundary of the obstacle (due to linearization of kinematics the point belongs to the boundary of the body). When point C has been found, the first approximation of the contact zone should be adopted as a set of line segments $\Delta\Gamma_{c,j}$ with the obvious requirement that $C \in \Delta\Gamma_{c,j}$ (for simplicity we are discussing a two-dimensional case). Having determined normal vectors $\mathbf{n}_j$ associated with the centres of the segments $\Delta\Gamma_{c,j}$ and adopting a set of numbers $(r_j) : j \in I_2 \wedge r_j < 0$ one obtains the first approximation of the reaction force system:

$$\mathbf{r}_j = \mathbf{n}_j r_j\,.$$

The set should be selected in such a way that its resultant is a vector opposite to the resultant of the active force set, i.e.

$$\mathbf{S}_r = -\mathbf{S}\,.$$

Using Eqs. (4.68), it is possible to eliminate the number of unknowns, equal to the number of segments in the contact zone, from Eqs. (4.67). The unknowns have a physical interpretation of displacement vector coordinates of the segments $\Delta\Gamma_{c,j}$ (nodal points). Solving the set of equations (4.66), displacements of all nodal points on the boundary can be calculated as the functions of the reaction system assumed. Next, condition (4.68) in the form

$$d_{nj} + u_{nj} \begin{cases} > 0 \\ = 0\,, \\ < 0 \end{cases} \qquad \begin{aligned} d_{nj} &= d_j n_j\,, \\ u_{nj} &= u_j n_j\,. \end{aligned} \tag{4.73}$$

should be verified. If the sum, referring to points outside the contact zone, is less than zero, then the situation illustrated in Fig. 4.4 takes place, which

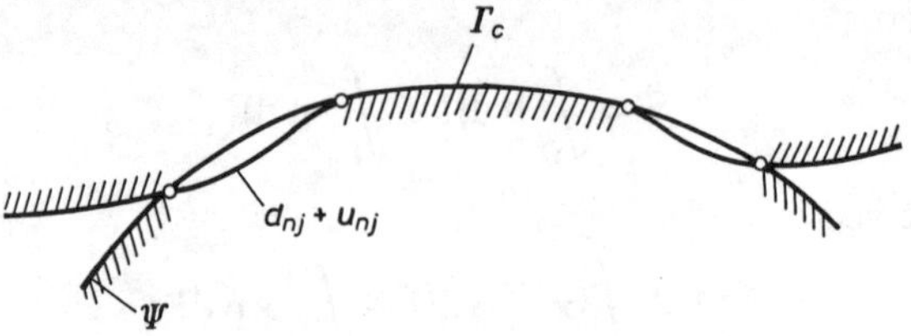

Fig. 4.4. Physical illustration of the case when the sum defined in Eq. (4.73) is smaller than zero

means that the contact zone assumed was too small. If we find a similar inequality, but for the points inside the contact zone, then reaction forces are too small. A reverse inequality for points in the contact zone implies that the assumed reaction forces are too large. With the above indications, a numerical solution of the contact problem with unilateral constraints can be constructed in subsequent approximations. In order to simplify the iterative process, it is convenient not to change the number of segments in the contact zone, but to modify their measure instead, e.g. length (in a two-dimensional case) or area (in a three-dimensional case). This should be done in such a way that the measure of the whole boundary does not change. The described concept of an approximate solution using the BEM is effective in practice, although numerical experience is not extensive. This leaves the research area open for comparisons with the FEM. The usage of adaptive techniques, especially advisable in this case, is an open issue similarly to other BEM applications.

4.4 Numerical Examples

Example 1

A strip of an aluminium sheet with a hole is subjected to tension [50]. The following material parameters have been adopted: Young's modulus $E = 7 \cdot 10^4$ MPa, plastic limit $\sigma_0 = 243$ MPa, inclination angle tangent for the tension diagram $H' = 0.032\,E$, Poisson's ratio $\nu = 0.2$. Fig. 4.5a shows the discretization of the boundary with 33 boundary elements and the division of the inside of the body with internal cells. Only a part of the domain in which we expect plastic strains is discretized with the cells. This is undoubtedly an advantage of the BEM. Fig. 4.5b presents a comparison of stresses obtained from numerical calculations and experimental measurements. The points in the plastified area for different values of the ratio $2\sigma_a/\sigma_0$ are shown in Fig. 4.5c. The plastic zones (filled with lines) found by employing a mesh of isoparametric finite elements shown in Fig. 4.5d are also drawn.

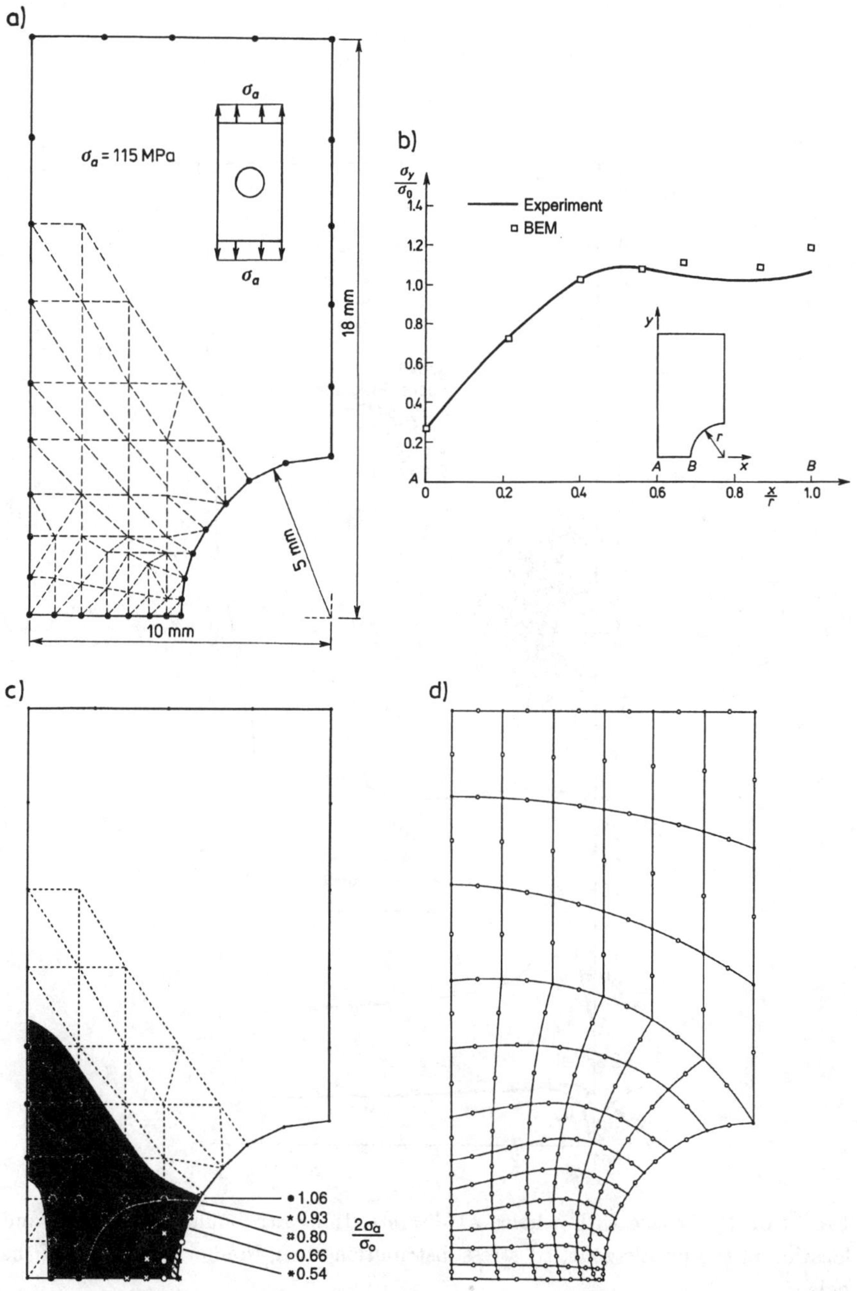

Fig. 4.5. Strip of an aluminium sheet with a hole, subjected to tension: **a)** discretization with boundary elements, **b)** comparison of stresses obtained from numerical calculations and experimental measurements, **c)** location of the plastic zone, **d)** discretization with finite elements

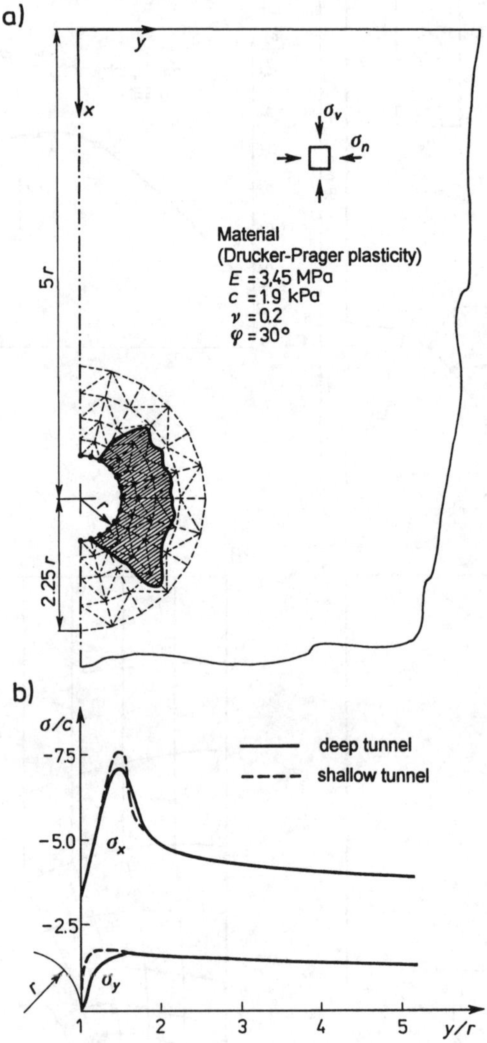

Fig. 4.6. Half-space with a hole: **a)** discretization with boundary elements and location of the plastic zone, **b)** stress distribution along the horizontal axis of the hole

Example 2

The example illustrates a particular advantage of the BEM, namely the possibility to analyze infinite and half-infinite domains. A half-space with a hole of radius $r = 40$ m satisfies plane strain conditions [50]. The material of the half-space is modelled with the Drucker–Prager plasticity. The initial stress state, varying linearly with the depth, has been assumed:

$$\sigma_v = \bar{\sigma}_v + \gamma h \text{ (vertical stresses)} ,$$
$$\sigma_h = 0.4\bar{\sigma}_v \text{ (horizontal stresses)} ,$$

where $\bar{\sigma}_v$ is a constant pressure ($\bar{\sigma}_v = 2.07$ kPa), γ is a unit weight ($\gamma = 2.46$ MN/m^3), h is a distance from the ground surface. Discretization of the boundary of the hole and of the area adjacent to the hole, as well as the final distribution of the plastic zone is presented in Fig. 4.6a. The stress distribution along a horizontal axis crossing the centre of the hole is shown in Fig. 4.6b. The stress distribution for the case of a hole in an infinite space is also plotted for comparison.

5 Boundary Element Method in Synthesis Problems

In Chaps. 3 and 4 we have presented the applications of the boundary element method to the analysis of linear and nonlinear problems for deformable systems. These problems consist in the computation of the response of a system (in the form of displacements, strains or stresses) to an external mechanical action given as boundary and initial conditions as well as volume forces. We have assumed that both the shape of the boundary and material parameters are known. The information obtained from such an analysis is, however, insufficient for the design of machine or structure parts. The process of design and construction is in its essence a synthesis problem, which can be understood as a search for geometrical and material properties of mechanical systems which satisfy criteria determined in advance. Such synthesis problems include optimization problems and sensitivity analysis.

Optimization algorithms are not commonly applied by designers in the structural synthesis. An optimal solution is usually searched for in an informal manner. However, when there is no precedent solution for a problem, the designer lacks experience or his intuition fails, the application of the formalism of the optimal control theory becomes indispensable. There are numerous examples that the existing and functioning mechanical systems have been designed basing on an optimization algorithm.

From the mechanical viewpoint, one of the most useful, but also most difficult elements of the synthesis is the problem of optimal shape design.

In Part VI a concept of so-called *adjoint systems* has been thoroughly discussed, which, combined with a suitably adapted notion of a material derivative, seems to be a very effective formal tool in shape optimization and sensitivity analysis. However, it is an analytical tool and numerical methods have to be applied in order to obtain quantitative results. One of the methods, which is especially useful and predisposed because of its features, is the BEM (cf. [5, 6]). The formalism of boundary integral equations seems to generate an analytical formulation of the optimal shape design problem, which is not less convenient than the previously mentioned concept of adjoint systems. The BEM is in this case a natural discretization method.

Another class of problems in which the method of boundary integral equations seems to be almost irreplaceable, is the optimal control of boundary conditions, in the language of mechanics – control of system constraints (cf. [22, 24]).

5.1
Boundary Element Method in Optimal Control Problems

Before discussing the applications of the method of boundary integral equations and the BEM in optimization, it is worthwhile to recapitulate a model formulation of a typical optimization problem. It is composed of three fundamental elements:

I. An equation governing a system, i.e. so-called *state equation:*

$$A(\mathbf{u}, \mathbf{v}, \mathbf{y}) = 0\,, \tag{5.1}$$

where $\mathbf{u}$ is a state vector (e.g. a displacement vector), $\mathbf{v}$ is a control vector (with parameters that can be selected), $\mathbf{y}$ is a vector of parameters that can not be modified.

II. Constraints (restrictions) imposed on the control or the state vector:

$$\begin{aligned} G_i(\mathbf{u}, \mathbf{v}) &\leq \emptyset\,, \qquad g_k(\mathbf{u}, \mathbf{v}) \leq 0\,, \\ H_j(\mathbf{v}) &\leq \emptyset\,, \qquad h_i(\mathbf{v}) \leq 0\,, \end{aligned} \tag{5.2}$$

where G is an operator defined on space $U \times V$, H is an operator defined on space U, g_k is a functional defined on space $U \times V$ and h_i is a functional defined on space U.

III. A functional called an *objective function* or a *quality functional:*

$$J[\mathbf{u}(\mathbf{v}, \mathbf{y}), \mathbf{v}]\,. \tag{5.3}$$

Distinguishing the state equation (5.1) is a matter of tradition. From a formal viewpoint, it can be treated as a restriction of type (5.2_1). The following definitions will appear convenient in the further discussion:

– a set of admissible control vectors,

$$\mathcal{V} \stackrel{\text{def}}{=} \{\mathbf{v} \in V; G_i(\mathbf{u}, \mathbf{v}) \leq \emptyset \wedge H_j(\mathbf{v}) \leq \emptyset \wedge g_k(\mathbf{u}, \mathbf{v}) \leq 0 \wedge h_i(\mathbf{v}) \leq 0\}\,, \tag{5.4}$$

– a set of achievable states,

$$\mathcal{U} \stackrel{\text{def}}{=} \{\mathbf{u} \in U : A(\mathbf{u}, \mathbf{v}, \mathbf{y}) = \emptyset \wedge \mathbf{v} \in\}\,. \tag{5.5}$$

The problem of optimization or optimal control consists in finding such a pair of vectors

$$(\tilde{\mathbf{u}}, \tilde{\mathbf{v}}) \in \mathcal{U} \times \mathcal{V}\,,$$

that

$$J(\tilde{\mathbf{u}}, \tilde{\mathbf{v}}) = \inf_{<u,v> \in \mathcal{U} \times \mathcal{V}} J\,, \tag{5.6}$$

where $\tilde{\mathbf{u}}$ is called an *optimal state* of a system (or an optimal trajectory) and $\tilde{\mathbf{v}}$ is called an *optimal control.* It should be mentioned that the optimization problem differs from an ordinary minimization, since the lower bound of the functional (i.e. the global minimum) is searched for within a closed set.

The following remark justifies the exchangeable usage of the notions *optimization problem* and *control problem.* Let us denote the required state of the system by $\tilde{\mathbf{u}}$ and adopt an objective function in the form:

$$J(\mathbf{u}, \mathbf{v}) = \|\tilde{\mathbf{u}} - \mathbf{u}(\mathbf{v}, \mathbf{y})\| \,, \tag{5.7}$$

where $\|\ldots\|$ is a norm of space u. The optimal control assures that the state of the system is the closest possible to the desirable one. The problem of optimization in mechanics is discussed in detail in Part V. In this chapter the discussion is confined to these elements of the formulation, which are directly related to the MBIE formalism. Therefore, only two of the possible types of optimal control problems are essentially considered below, i.e.

- the problem of optimal control of constraints (boundary conditions),
- the problem of boundary control (optimal shape control).

It has been assumed that the reader has a sufficient knowledge of the contents of the previous chapters in Part IV.

5.1.1 Optimal Control of Constraints

Let us write the problem of constraint control using the language of boundary integral equations, referring to the previously quoted typical formulation (5.1), (5.2) and (5.3). The state equation takes an integral form, cf. Eq. (3.7):

$$\mathbf{u}(x) = \int_{\Omega} \mathbf{U}(x,y)\mathbf{b}(y)\,\mathrm{d}\Omega(y) - \int_{\Gamma} [\mathbf{P}(x,y)\mathbf{u}(y) - \mathbf{U}(x,y)\mathbf{p}(y)]\,\mathrm{d}\Gamma(y)\,. \tag{5.8}$$

Let us recap that in the second integral $\mathbf{u}(y)$ denotes the vector of boundary displacements and $\mathbf{p}(y) = P[\mathbf{u}(y)]$ is an external loading of the boundary associated with the displacements. Depending on the type of the problem only one of the mentioned functions is known on the whole boundary or each of them is known on certain separate parts of the boundary. In further derivation all problems are for convenience transformed into a stress problem. This implies that in the case of a displacement boundary condition (in the form of a constraint equation, cf. Sect. 4.3)

$$f(\mathbf{u}) = 0 \,, \tag{5.9}$$

we introduce reactions $\mathbf{r}(y), y \in \Gamma$. They are additional unknowns determined from the constraint equation. In other words, the constraints will always be represented by the distribution of reaction forces which will further be treated as the control. As is usual for stress problems, the boundary displacements should be determined from a boundary integral equation (cf. Sect. 1.4.3). From the viewpoint of the theory of optimal processes, the equation is a restriction (of an operator type) imposed on the state vector $\mathbf{u}$.

We rewrite the state equation considering the fact that the control via the reaction forces takes place:

$$\mathbf{u}(x) = \int_{\Omega} \mathbf{U}(x,y)\mathbf{b}(y)\,\mathrm{d}\Omega(y) - \int_{\Gamma_u} [\mathbf{P}(x,y)\mathbf{u}(y) - \mathbf{U}(x,y)\mathbf{r}(y)]\,\mathrm{d}\Gamma(y)$$
$$+ \int_{\Gamma_p} [\mathbf{P}(x,y)\mathbf{u}(y) - \mathbf{U}(x,y)\mathbf{p}(y)]\,\mathrm{d}\Gamma(y)\,, \tag{5.10}$$

where Γ_p is a part of the boundary where external forces are applied, Γ_u is the remaining part of the boundary, $\partial\Omega = \Gamma = \Gamma_u \cup \Gamma_p \wedge \Gamma_u \cap \Gamma_p = \emptyset$ and $\mathbf{p}(y)$, $y \in \Gamma$ are external boundary forces. In the first place among the restrictions we have the boundary equation

$$\frac{1}{2}\mathbf{u}(x) = \int_{\Omega} \mathbf{U}(x,y)\mathbf{b}(y)\,\mathrm{d}\Omega(y) - \int_{\Gamma_u} [\mathbf{P}(x,y)\mathbf{u}(y) - \mathbf{U}(x,y)\mathbf{r}(y)]\,\mathrm{d}\Gamma(y)$$
$$+ \int_{\Gamma_p} [\mathbf{P}(x,y)\mathbf{u}(y) - \mathbf{U}(x,y)\mathbf{p}(y)]\,\mathrm{d}\Gamma(y)\,. \tag{5.11}$$

In the case of static problems the system of reaction forces has to satisfy the equilibrium equations, which represent restrictions imposed on the control:

$$\begin{aligned}\int_{\Omega} \mathbf{b}\,\mathrm{d}\Omega + \int_{\Gamma_u} \mathbf{r}\,\mathrm{d}\Gamma + \int_{\Gamma_p} \mathbf{p}\,\mathrm{d}\Gamma = 0\,,\\ \int_{\Omega} \mathbf{x}\times\mathbf{b}\,\mathrm{d}\Omega + \int_{\Gamma_u} \mathbf{x}\times\mathbf{r}\,\mathrm{d}\Gamma + \int_{\Gamma_p} \mathbf{x}\times\mathbf{p}\,\mathrm{d}\Gamma = 0\,.\end{aligned} \tag{5.12}$$

We notice that on the part of the boundary $\hat{\Gamma}_u \subset \Gamma_u$ a certain kind of constraints can be given, i.e.

$$\hat{\mathbf{u}} : f[\hat{\mathbf{u}}(y)] = 0; \qquad \forall\, y \in \hat{\Gamma}_u\,.$$

Obviously, reactions on this part of the boundary can not be arbitrary. Therefore, another restriction appears:

$$\forall\, y \in \hat{\Gamma}_u;\ \mathbf{r}(y) = P[\hat{\mathbf{u}}(y)] = \hat{\mathbf{r}}(y)\,. \tag{5.13}$$

In a particular case when the constraint equation has the form

$$\hat{\mathbf{u}}(y) = 0; \qquad \forall\, y \in \hat{\Gamma}_u\,,$$

with $\hat{\Gamma}_u \neq \emptyset \wedge \Gamma_u \backslash \hat{\Gamma}_u \neq \emptyset$, equilibrium of the system is always preserved and restrictions (5.12) can be neglected. Another possible restriction we can quote reads:

$$\int_{\Gamma} \alpha[\mathbf{r}(y)]\,\mathrm{d}\Gamma \leq c\,, \tag{5.14}$$

which can for instance represent a limit cost of constraint imposition.

Another restriction, important in practice, is local:

$$\forall\, y \in \hat{\Gamma}_u : |\mathbf{r}(y)| \leq c; \qquad c \in \boldsymbol{R}_+\,, \tag{5.15}$$

where $\boldsymbol{R}_+$ is a set of positive real numbers and $\hat{\Gamma}_u \subset \Gamma_u$, which implies that reactions on the part of the boundary, not necessarily identical with Γ_u, are not allowed to exceed a certain value (e.g. the strength of the support). We notice that restriction (5.15) eliminates all point supports. In the case of smooth constraints the following condition must be added to the set of restrictions:

$$\forall\, y \in \hat{\Gamma}_u : \hat{\mathbf{r}}(y) = \lambda\, \mathrm{grad}(\Gamma_u), \qquad \lambda \in \boldsymbol{R} \tag{5.16}$$

(also here $\hat{\Gamma}_u \subset \Gamma_u$ can be assumed). Another kind of restriction is the one imposed on stresses, e.g. in the form

$$\max_{\alpha}\{\sigma_\alpha\} \leq \sigma_0 : \alpha \in (1,2,3)\,, \tag{5.17a}$$

where σ_α denotes a principal stress. This can be written in a compact form as

$$g[\Sigma(\mathbf{u})] \leq 0\,, \tag{5.17b}$$

with the operator Σ defined in Eq. (2.89a). The set of restrictions can be extended. However, it should be remembered that adding restrictions makes an effective solution of a problem more difficult.

The last element of the formulation is the objective function. Its form results almost exclusively from practical circumstances. The reduction of deformability can be an objective of optimization and then the functional has one of the forms:

$$J(\mathbf{u}) = \int_\Omega \mathbf{u}^2\, \mathrm{d}\Omega\,, \qquad J(\mathbf{u}) = \int_\Omega |\mathbf{u}|\, \mathrm{d}\Omega\,, \tag{5.18a}$$

$$J(\mathbf{u}) = \sup_{y \in \mathrm{Int}\Omega} |\mathbf{u}(y)|\,, J(\mathbf{u}) = \mathbf{u}(y_0) : \ y_0 \in \mathrm{Int}\Omega\,. \tag{5.18b}$$

Each functional is a certain measure of the displacement field (a norm of an appropriate space). In practice, functionals (5.18a) and (5.18b) are used only in exceptional cases, since they produce many formal difficulties.

A mechanical system, in which the stress field has certain properties, e.g.

$$\begin{aligned} J(\mathbf{u}) &= |\max_{\alpha}\{\sigma_\alpha\} - \sigma_0|\,, \\ J(\mathbf{u}) &= F[\Sigma(\mathbf{u})]\,, \end{aligned} \tag{5.19}$$

can be an objective of optimization. Obviously, in this case Eqs. (5.17a) and (5.17b) must be excluded from the set of restrictions. Instead, a restriction of deformability can be added:

$$g(\mathbf{u}) \leq 0\,. \tag{5.20}$$

It should be emphasized that, owing to the formalism of the boundary integral equations, the problem of optimal control of constraints (boundary conditions) has been formulated in an explicit analytical form. This makes it possible to propose a formal optimality condition. For this purpose an appropriate Lagrange functional must be constructed (cf. [17]):

$$\Phi[\mathbf{u},\mathbf{r},\lambda,\chi,\lambda^*,\chi^*] = J(\mathbf{u},\mathbf{r}) + \sum_i \langle G_i, \lambda_i^* \rangle + \sum_j \langle H_j, \chi_j^* \rangle + \sum_k \lambda_k g_k + \sum_l \chi_l h_l \,, \tag{5.21}$$

where the symbol $\langle\ ,\ \rangle$ denotes a scalar product and $\lambda_i^*, \chi_j^*, \lambda_k, \chi_l$ are generalized Lagrange multipliers (elements of an appropriate dual space). The optimality condition can now be written as a stationarity condition for a Lagrange functional

$$\partial\Phi = 0 \,, \tag{5.22}$$

in which ∂ denotes the Gateaux derivative. Equation (5.22) must be augmented with conditions concerning the Lagrange multipliers (cf. [17]).

The procedure described, and in particular the optimality condition (5.22), allows us to construct an effective analytical solution only in very simple cases which are trivial from the practical viewpoint. It has been presented, since it is a good illustration of the effectiveness of the discussed formalism. Moreover, there is another important reason. The approach is a basis of approximate numerical solutions. A finite-dimensional boundary element approximation (cf. Sect. 2.5 and 3.1.2) is a starting point for a numerical application. The state equation becomes then an algebraic equation:

$$\mathbf{u}_i = \mathbf{u}_{0i} + \sum_{j=1}^{n} \mathbf{a}_{ij}\hat{\mathbf{u}}_j - \sum_{j\in I_1} \mathbf{b}_{ij}\mathbf{r}_j - \sum_{j\in I_2} \mathbf{b}_{ij}\mathbf{p}_j \,, \tag{5.23}$$

where I_1 is a set of indices denoting segments of the boundary Γ_u, i.e. $\Delta\Gamma_{u,j}$, I_2 is a set of indices denoting segments of the boundary Γ_p, i.e. $\Delta\Gamma_{p,j}$, and $\hat{\mathbf{u}}_j$ are displacements of nodal points on the contour. The following boundary equation has an analogical form:

$$\frac{1}{2}\hat{\mathbf{u}}_i = \hat{\mathbf{u}}_{0i} + \sum_{j=1}^{n} \mathbf{a}_{ij}\hat{\mathbf{u}}_j - \sum_{j\in I_1} \mathbf{b}_{ij}\mathbf{r}_j - \sum_{j\subset I_2} \mathbf{b}_{ij}\mathbf{p}_j \,. \tag{5.24}$$

Restrictions also become algebraic expressions after discretization. Let us write them in a general form related to the notation in (5.2):

$$\begin{aligned} G_i(\mathbf{u}_k,\mathbf{r}_i) &\le \emptyset \,, & H_j(\mathbf{r}_k) &\le \emptyset, \\ g_k(\mathbf{u}_k,\mathbf{r}_i) &\le 0 \,, & h_i(\mathbf{r}_i) &\le 0 \,. \end{aligned} \tag{5.25}$$

Obviously, the objective function is also an algebraic formula:

$$J[\mathbf{u}_k,\mathbf{r}_i] \,. \tag{5.26}$$

In this way, the finite-dimensional approximation reduces the considered optimization problem to a parametric optimization, e.g. coefficients of the approximation of a function $\mathbf{r}$ with a set of piecewise constant functions

$$\mathbf{r} = \sum_j \mathbf{r}_j \chi_j \tag{5.27}$$

are now the control. To solve the problem means to find such a set of coefficients $\{\mathbf{r}_j^*\}$ that

$$J[\mathbf{u}_k(\mathbf{r}_i^*), \mathbf{r}_n] = \min_{\{r_j\} \in R} J \,, \tag{5.28}$$

where R denotes a set of admissible coefficient values, i.e.

$$R \stackrel{\text{def}}{=} \{\mathbf{r}_j, G_l(\mathbf{u}_k, \mathbf{r}_i) \leq \emptyset \,, \quad H_j(r_k) \leq \emptyset \,, \quad g_m(\mathbf{u}_n, \mathbf{r}_i) \leq \emptyset \,, \quad h_n(\mathbf{r}_i) \leq \emptyset\} \,.$$

The set can be determined on the basis of the stationarity condition for an appropriate Lagrange functional which is analogical to the optimality condition (5.22)

$$\partial \Phi[\lambda_j, \chi_i, \mathbf{r}_k] = 0 \,. \tag{5.29}$$

In practice, direct methods of searching for a minimum of the functional (e.g. the Gauss method, the penalty method or gradient methods, described in Part V, cf. [9, 17]) are more effective.

5.1.2 Optimal Control of the Boundary

Shape optimization belongs to the most interesting problems from the viewpoint of application, but at the same time it is one of the most difficult problems from the mathematical standpoint. The discussed formalism of boundary integral equations turns out to be extremely effective also in this case. It could be argued that it is intrinsically predisposed to formulate the problem of shape optimization (cf. [25, 26]).

The state equation is, as before, an integral formula:

$$\begin{aligned} \mathbf{u}(x) = & \int_{\{\Omega\}} \mathbf{U}(x,y)\mathbf{b}(y)\,\mathrm{d}\Omega(y) - \int_{\{\Gamma_u\}} [\mathbf{P}(x,y)\mathbf{u}(y) - \mathbf{U}(x,y)\mathbf{r}(y)]\,\mathrm{d}\Gamma(y) \\ & - \int_{\{\Gamma_u\}} [\mathbf{P}(x,y)\mathbf{u}(y) - \mathbf{U}(x,y)\mathbf{p}(y)]\,\mathrm{d}\Gamma(y) \,. \end{aligned} \tag{5.30}$$

It seems that it does not differ from Eq. (5.10). In reality there is an essential difference: the domain of integration is unknown, which has been emphasized by the notation. Obviously, we shall further attempt to formulate the problem in such a way that the control appears under integrals over a known domain. It is a classical operation (cf. [18]), but as a result integro-differential equations are obtained.

We shall consider a class of so-called *star domains.* A star domain is such a domain Ω that its boundary $\partial\Omega = \Gamma$ can be mapped in a continuous manner onto a unit circle (or, more generally, a sphere). In other words, any radius drawn from a point $x_0 \in \mathrm{Int}\Omega$ has one and only one intersection point with the contour Γ. The discussion is further limited to a two-dimensional case. An extension to a three-dimensional case is a formal operation. It can be concluded from the above definition of the star domain that the equation of the boundary can be written as

$$R(\varphi) = R[1 + f(\varphi)], \qquad f(\varphi) \in C_{2\pi}\,, \tag{5.31}$$

where $\{R, \varphi\}$ denote polar coordinates. Relation (5.31), rewritten as

$$\varrho = R[1 + f(\varphi)], \qquad \psi = \varphi\,, \tag{5.32}$$

can be treated as a change of variables. This allows us to write the formula (5.30) and the boundary integral equation with integrals over a fixed domain. The function $f(\varphi)$, which defines the mapping of the contour we search for onto a circle, appears under the integrals as a result of transformations of the domain differential $\mathrm{d}\Omega$ and the arc differential $\mathrm{d}\Gamma$. Taking into account the Jacobian of the mapping (5.32):

$$J = \det|\frac{\partial(\rho.\psi)}{\partial(R,\varphi)}| = 1 + f(\varphi)\,, \tag{5.33}$$

and writing the transformation of the arc differential as

$$\mathrm{d}\Gamma = \sqrt{\varrho^2 + (\varrho')^2}\,\mathrm{d}\psi = h(\varrho,\varrho')\,\mathrm{d}\psi\,, \tag{5.34}$$

the integral formula (5.30) can be expressed as follows:

$$\begin{aligned}\mathbf{u} = &\int_{2\pi}^{0} B(P,Q)\,\mathrm{d}\psi_Q - \int_{\psi_1}^{\psi_2} [\mathbf{g}(\hat{Q})\mathbf{P}(P,\hat{Q}) - \mathbf{U}(P,\hat{Q})\mathbf{r}(\hat{Q})]h(\varrho,\varrho')]\,\mathrm{d}\psi_{\hat{Q}}\\ &- \int_{\psi_1}^{\psi_2} [\mathbf{u}(\hat{Q})\mathbf{P}(P,\hat{Q}) - \mathbf{U}(P,\hat{Q})\mathbf{q}(\hat{Q})]h(\varrho,\varrho')]\,\mathrm{d}\psi_{\hat{Q}}\,,\end{aligned} \tag{5.35}$$

where:

$$\begin{aligned}&\Gamma_u = \{\varrho,\psi : \varrho(\psi) = R[1 + f(\psi)] \wedge \psi \in (\psi_1,\psi_2)\}\,,\\ &\Gamma_p = \{\varrho,\psi : \varrho(\psi) = R[1 + f(\psi)] \wedge \psi \in (\psi_1,\psi_2)\}\,,\\ &\mathbf{g}(Q) :\ \mathbf{u}(Q) = \mathbf{g}(Q)\,; \qquad \forall Q \in \Gamma_u\,,\\ &\mathbf{q}(Q) :\ P[\mathbf{u}(Q)] = \mathbf{q}(Q)\,;\ \forall Q \in \Gamma_p\end{aligned} \tag{5.36}$$

and

$$B(\varrho,\psi) = \int_0^{\varrho(\psi)} \mathbf{U}\mathbf{b}\varrho\,\mathrm{d}\varrho\,. \tag{5.37}$$

The formula (5.37) is useful after the following boundary equation has been solved:

$$\frac{1}{2}\mathbf{u}(\hat{P}) = \int_0^{2\pi} B(\hat{P},Q)\,\mathrm{d}\psi_Q - \int_{\psi_1}^{\psi_2} [\mathbf{g}(\hat{Q})\mathbf{P}(\hat{P},\hat{Q}) - \mathbf{U}(\hat{P},\hat{Q})\mathbf{r}(\hat{Q})]h(\varrho,\varrho')\,\mathrm{d}\psi_{\hat{Q}}$$
$$- \int_{\psi_1}^{\psi_2} [\mathbf{u}(\hat{Q})\mathbf{P}(\hat{P},\hat{Q}) - \mathbf{U}(\hat{P},\hat{Q})\mathbf{q}(\hat{Q})]h(\varrho,\varrho')\,\mathrm{d}\psi_{\hat{Q}}\,, \qquad \hat{P},\hat{Q} \in \Gamma\,. \tag{5.38}$$

This integral equation is used to determine $\mathbf{u} = \mathbf{u}(\mathbf{q})$ on part Γ_p of the boundary and $\mathbf{r} = \mathbf{r}(\mathbf{g})$ on part Γ_u.

Referring to the remark in the previous section, the boundary equation (5.38) can be understood as a restriction. The following conditions, which assure correctness of solutions, must be included in the set of restrictions:

$$\forall\psi: \quad R_1 \leq \varrho(\psi) \leq R_2\,, \tag{5.39}$$

$$\forall\Omega: \quad \int_\Omega \mathrm{d}\Omega = c; c > 0\,. \tag{5.40}$$

Then, the restrictions imposed on the stress field can also be included:

$$F[\Sigma(\mathbf{u})] \leq 0\,. \tag{5.41}$$

For completeness we also add the objective function. It can have one of the forms (5.18) or (5.19), with a stipulation that if we use Eq. (5.19) then we eliminate restriction (5.41).

The solution of the above problem consists in finding such a shape $\partial\Omega^*$ which belongs to the set of admissible contours and for which the objective function reaches the lower bound.

Let us recapitulate that the contour is defined by the function $f(\psi)$ which plays the role of the control. In other words, the function is the unknown which is determined from the condition of functional minimization.

Before further consideration we focus our attention on Eq. (5.38). It is easy to notice that

$$h(\varrho,\varrho') = \{[1 + f(\psi)]^2 + [f'(\psi)]^2\}^{1/2}\,.$$

Thus it is an integro-differential equation, which makes the problem much more complicated. This difficulty can be overcome if we adopt an assumption that only linear mappings are considered, i.e.

$$f(\psi) = \sum_{k=1}^{n} (\alpha_k + \beta_k\psi)\chi_k\,, \tag{5.42}$$

where α_k, $\beta_k \in \boldsymbol{R}$, $\psi \in \Delta\psi_k$, $n < \infty$, κ_k is a characteristic function for segment $\Delta\Gamma_k$ of the boundary Γ.

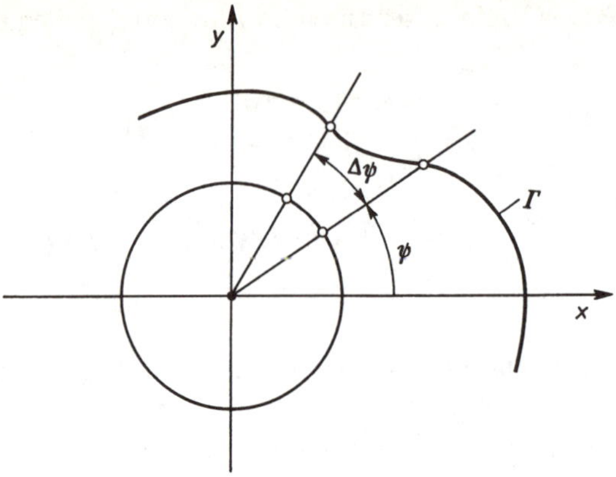

Fig. 5.1. Division of boundary Γ into segments contained by radii which determine angular domains $\Delta\psi$

The above assumption implies that the boundary curve has been divided into segments $\Delta\Gamma_k$ contained by radii which determine angular domains $\Delta\psi_k$ and each of the segments can be linearly mapped on a circular arc. This assumption limits the generality of the solution, but does not make an attempt to find a solution not restricted by Eq. (5.42) impossible. For this purpose it is enough to examine the limit for $n \to \infty$.

The approximation described above is not related to a boundary element approximation. Owing to the assumption (5.42) the state equations, the restrictions and the objective function become functions of the parameters (α_k, β_k), which now play the role of the control. Selecting a set of parameters (α_k, β_k) so that the restrictions are satisfied, we can, however, obtain an unacceptable solution, since a formal condition of contour continuity, i.e. a formal counterpart of an assumption about continuity of function $f(\psi)$), is missing among the restrictions. The condition of contour continuity is written in the form:

$$\sum_k [\alpha_k + \beta_k \psi_k] \cos\left[\sum_k \Delta\psi_k\right] = 0\,,$$
$$\sum_k [\alpha_k + \beta_k \psi_k] \sin\left[\sum_k \Delta\psi_k\right] = 0\,, \tag{5.43}$$

where $\Delta\psi_k$ is an angular area between radii k and $k+1$, while ψ_k is a measure of the angle which determines the location of segment $\Delta\Gamma_k$. Condition (5.43) should be added to the set of restrictions. The described procedure reduces the problem of shape optimization to a parametric optimization problem. A complete formulation of the latter problem can easily be obtained substituting the following identities

$$k(\varrho, \varrho') = \left\{ \left[1 + \sum_k (\alpha_k + \beta_k \psi) \right]^2 + \beta_k^2 \right\}^{1/2} ,$$

$$\varrho = R \left[1 + \sum_k (\alpha_k + \beta_k \psi) \chi_k \right]^{1/2} , \quad B(P, Q) = B(\alpha_k, \beta_k, P, Q) \tag{5.44}$$

into Eqs. (5.35) and (5.38)–(5.41) and considering a new form of the objective function:

$$J(\mathbf{u}) = J(\mathbf{u}(\alpha_k, \beta_k), \alpha_k, \beta_k) . \tag{5.45}$$

Then, the solution of the shape optimization problem consists in finding such a set of parameters (α_k^*, β_k^*) which satisfies the restrictions and minimizes the functional, i.e.

$$J[\mathbf{u}^*, \alpha_k^*, \beta_k^*] = \inf_{\{\alpha_k, \beta_k\}} J[\mathbf{u}, \alpha_k, \beta_k] .$$

The above problem can be solved analytically using the Lagrange formalism. However, it is not an effective method in general. A solution can be effectively constructed in a numerical way via a finite-dimensional approximation with boundary elements. As a result we obtain a set of equations, which differ from formulas (5.23)–(5.28) only in the fact that the coefficients of the equations (i.e. $\mathbf{a}_{ij}, \mathbf{b}_{ij}$) are now functions of the parameters (α_k^*, β_k^*).

The finite-dimensional version of the problem is a typical example of nonlinear programming and the methods mentioned in Sect. 5.1.1 can be used to solve it.

To sum up, it should be emphasized that the formalism of boundary integral equations is an effective tool in the problems of shape and constraint optimization. The formulation is not less general than other known concepts which use the notion of *boundary variation* (e.g. so-called *material derivative*). It is because eventually the class of linear boundary mappings, such as presented in Eq. (5.42), is always considered.

5.2 Boundary Element Method in Sensitivity Analysis

As mentioned before, the problems of optimization, and in particular of optimal shape control, belong to especially difficult problems. A lot of effort must usually be spent in order to obtain a solution. Moreover, in optimization one can encounter multi-valued or only locally optimal solutions. In practice, an improved *quasi-optimal* solution is often satisfactory, and in particularly difficult cases the information concerning a method of improving the solution is even valuable. In these cases the methods of sensitivity analysis, discussed in Part VI, are of particular importance, provided that there is an effective possibility of constructing a sensitivity operator. The BEM turns out to be a convenient and natural numerical technique which allows us to construct an approximate form of the sensitivity operator. Moreover, the sensitivity analysis can be treated as an initial stage of optimization, since the sensitivity operator determines the gradient of the quality functional and the restricting conditions (cf. [5, 6, 8, 16]).

5.2.1 Boundary Element Method in Sensitivity Analysis of Statically Loaded Bodies

Let us assume that the mechanical state of the body, characterized by a certain set of functionals $J_\alpha, \alpha = 1, 2, \dots, K-2$, depends on the choice of a set of design parameters $\mathbf{a} = (a_r), r = 1, 2, \dots, R$. We also assume, as in Part VI, that the functionals have the form:

$$J_\alpha = \int_\Omega \Psi_\alpha(\boldsymbol{\sigma}, \boldsymbol{\epsilon}, \mathbf{u}, \mathbf{a})\,\mathrm{d}\Omega + \int_\Gamma h_\alpha(\mathbf{u}, \mathbf{p}, \mathbf{a})\,\mathrm{d}\Gamma\,, \quad \alpha = 0, 1, 2, \dots, K-2\,, \tag{5.46}$$

where $\Omega = \Omega(\mathbf{a})$ is a domain of the body, $\Gamma = \Gamma(\mathbf{a})$ is its boundary, $\boldsymbol{\sigma} = \boldsymbol{\sigma}(\mathbf{a})$ is a stress state, $\boldsymbol{\epsilon} = \boldsymbol{\epsilon}(\mathbf{a})$ is a strain state, $\mathbf{u} = \mathbf{u}(\mathbf{a})$ are displacements, $\mathbf{p} = \mathbf{p}(\mathbf{a})$ are external tractions (which can for instance represent reactions of constraints dependent on the shape). Below, we neglect the volume forces, which means they do not belong in the selected design parameters $\mathbf{a}$. Let us examine the influence of the modification of the boundary shape on the functionals J_α.

We consider an infinitesimal variation of the shape of the body described by a continuous and differentiable vector field $\delta\mathbf{g} = (\delta g_j)$:

$$x_j^* = x_j + \delta g_j\,. \tag{5.47}$$

The transformation field $\mathbf{g} = \mathbf{g}(\mathbf{x}; \mathbf{a})$ modifies the shape of the external boundary Γ, where $\mathbf{a} = (a_r), r = 1, 2, \dots, R$ is a set of design shape parameters. The variable x is defined in the domain Ω with boundary Γ which does not undergo the transformation, while the variable x^* is defined in the transformed domain $\Omega^* = \Omega(\mathbf{a})$ with boundary $\Gamma^* = \Gamma(\mathbf{a})$. The variation of the transformation field $\mathbf{g}$ is defined as:

$$\delta g_k = \frac{\partial g_k}{\partial a_r}\delta a_r = v_k^r \delta_r\,, \tag{5.48}$$

where the quantity $v_k^r = \partial g_k / \partial a_r$ can be considered as a "velocity" of the transformation associated with the shape parameter a_r.

First variations of the functionals $J_\alpha, \alpha = 1, 2, \dots, K-2$ can be expressed as:

$$\delta J_\alpha = \frac{\mathrm{D}J_\alpha}{\mathrm{D}a_r}\delta a_r\,, \tag{5.49}$$

where $(\mathrm{D}/\mathrm{D}a_r)$ is a generalization of a material derivative, i.e.

$$\frac{\mathrm{D}\mathbf{q}}{\mathrm{D}a_r} = \frac{\partial \mathbf{q}}{\partial a_r} + v_k^r \,\mathrm{grad}\mathbf{q}\,, \qquad \mathbf{q} = \mathbf{u}, \boldsymbol{\epsilon}, \boldsymbol{\sigma}\,, \tag{5.50}$$

and a total material derivative of the tractions is

$$\frac{\mathrm{D}p_i}{\mathrm{D}a_r} = \frac{\partial \sigma_{ij}}{\partial a_r} n_j + \sigma_{ij,k} n_j v_k^r + \sigma_{ij}(n_j n_l - \delta_{jl}) n_k v_{k,l}^r\,, \tag{5.51}$$

where $\mathbf{n} = (n_j)$ is an outward unit vector normal to the boundary Γ. Defining the elements of the sensitivity tensor S_α:

$$S_{\alpha r} \stackrel{\mathrm{def}}{=} \frac{\mathrm{D}J_\alpha}{\mathrm{D}a_r}\,, \tag{5.52}$$

we can write

$$\delta J_\alpha = S_\alpha^T \delta \mathbf{a}\,, \tag{5.53}$$

so the sensitivity of the functional to a variation of parameters (a_r) is determined by the sensitivity operator S_α^T. The elements of the sensitivity operator can be expressed in an extended form (cf. Chap. 1 of Part VI):

$$\begin{aligned}\frac{\mathrm{D}J_\alpha}{\mathrm{D}a_r} = &\int_\Gamma [\,\Psi_\alpha - \boldsymbol{\sigma}\cdot\boldsymbol{\epsilon}^a + \mathbf{b}\cdot\mathbf{u}^a + (h_\alpha + \mathbf{p}\cdot\mathbf{u}^a)_{,n} - 2(h_\alpha + \mathbf{p}\cdot\mathbf{u}^a)\mathcal{K}\,]\, n_k v_k^r \,\mathrm{d}\Gamma \\ &+ \int_{\Gamma_u} \left(\frac{\partial h_\alpha}{\partial \mathbf{u}} - \mathbf{p}^a\right)\cdot\left(\frac{\mathrm{D}\mathbf{u}^0}{\mathrm{D}a_r} - \mathbf{u}^0_{,k} v_k^r\right) \mathrm{d}\Gamma_u \\ &+ \int_{\Gamma_p} \left(\frac{\partial h_\alpha}{\partial \mathbf{p}} + \mathbf{u}^a\right)\cdot\left(\frac{\mathrm{D}\mathbf{p}^0}{Da_r} - \mathbf{p}^0_{,k} v_k^r\right) \mathrm{d}\Gamma_p + \int_L [\![h_\alpha + \mathbf{p}\cdot\mathbf{u}^a]\!]\, v_\nu^r \,\mathrm{d}L\,,\end{aligned} \tag{5.54}$$

where the expression under the integral $[\![h_\alpha + \mathbf{p}\cdot\mathbf{u}^a]\!] = (h_\alpha + \mathbf{p}\cdot\mathbf{u}^a)^+ - (h_\alpha + \mathbf{p}\cdot\mathbf{u}^a)^-$ represents a jump of the quantity $(h_\alpha + \mathbf{p}\cdot\mathbf{u}^a)$ along curve L which separates parts Γ_u and Γ_p of the boundary, and $\mathcal{K}$ is an average curvature of the boundary.

For the given boundary conditions the total material derivatives $(\mathrm{D}\mathbf{u}^0/\mathrm{D}a_r)$ on Γ_u and $(\mathrm{D}\mathbf{p}^0/\mathrm{D}a_r)$ on Γ_p are known and can be expressed in terms of the transformation velocity and gradients of the given boundary conditions $\mathbf{u}^0$ and $\mathbf{p}^0$. Let us recap that

$$\mathbf{u}_\alpha^{a0} = -\frac{\partial h_\alpha(\mathbf{u},\mathbf{p})}{\partial \mathbf{p}} \qquad \text{on } \Gamma_u\,, \tag{5.55a}$$

$$\mathbf{p}_\alpha^{a0} = \frac{\partial h_\alpha(\mathbf{u},\mathbf{p})}{\partial \mathbf{u}} \qquad \text{on } \Gamma_p \tag{5.55b}$$

are the boundary conditions on boundary Γ of the adjoint system $(\mathrm{AS})_\alpha$. It is visible in Eq. (5.54) that the sensitivity operator depends only on the boundary quantities which characterize the state of the body, so-called *basic system* (BS) and adjoint system $(\mathrm{AS})_\alpha$. The sensitivity operator can thus be effectively formulated under the condition that solutions for both systems are known. The solution for the basic system (BS) is a solution of a boundary value problem of elasticity and has been described in Chap. 3 of this part. The solution for the boundary system $(AS)_\alpha$ requires the following boundary integral equation to be solved:

$$\mathbf{c}(x)\mathbf{u}^a_\alpha(x) = \int_\Gamma [\mathbf{U}(x,y)\mathbf{p}^a_\alpha(y) - \mathbf{P}(x,y)\mathbf{u}^a_\alpha(y)]\,\mathrm{d}\Gamma(y) + \mathbf{B}^a_\alpha(x)\,, \tag{5.56}$$

where

$$B^a_{\alpha j} = \int_\Omega U_{jk}[b^a_k - c_{klsr}\epsilon^{ai}_{\alpha sr,l} - \sigma^{ai}_{\alpha kl,l}]\,\mathrm{d}\Omega\,, \tag{5.57}$$

and the initial strain field ϵ^{ai}_α, the initial stress field σ^{ai}_α and the volume forces are defined as:

$$\epsilon^{ai}_\alpha = \frac{\partial\Psi_\alpha(\boldsymbol{\sigma},\boldsymbol{\epsilon},\mathbf{u})}{\partial\sigma} \quad \text{in}\ \Omega\,, \tag{5.58a}$$

$$\sigma^{ai}_\alpha = \frac{\partial\Psi_\alpha(\boldsymbol{\sigma},\boldsymbol{\epsilon},\mathbf{u})}{\partial\epsilon} \quad \text{in}\ \Omega\,, \tag{5.58b}$$

$$\mathbf{b}^a_\alpha = \frac{\partial\Psi_\alpha(\boldsymbol{\sigma},\boldsymbol{\epsilon},\mathbf{u})}{\partial\mathbf{u}} \quad \text{in}\ \Omega\,. \tag{5.58c}$$

A constitutive law for the adjoint systems reads:

$$\sigma^a_\alpha = \mathbf{C}(\epsilon^a_\alpha - \epsilon^{ai}_\alpha) - \sigma^{ai}_\alpha\,, \tag{5.59}$$

where $\mathbf{C} = (C_{ijkl})$ is defined in Eq. (3.6). A discrete version of the integral equation (5.56) can be constructed in the way presented in Chap. 3. Finally for both systems we obtain sets of algebraic equations:

$$\mathbf{AX}^s = \mathbf{WY}^s + \mathbf{B}^s\,, \qquad s = (UP)\,,\ (US)_\alpha\,, \tag{5.60}$$

where the unknown state variables of the basic and adjoint systems which are needed in sensitivity computations for the functional J_α, are contained in the column matrix $\mathbf{X}^s$. It is worth noticing that matrices $\mathbf{A}$ and $\mathbf{F}$, depending on boundary integrals, are similar for the basic and adjoint systems and therefore it is sufficient to compute them only once.

5.2.2 Boundary Element Method in Sensitivity Analysis of Dynamically Loaded Bodies

Let now a dynamic state of the body be determined by the system of functionals $J_\alpha, \alpha = 1, 2, \ldots, K-2$ (cf. Part VI):

$$J_\alpha = \int_T\int_\Omega \Psi_\alpha(\boldsymbol{\epsilon},\mathbf{u},\mathbf{a})\,\mathrm{d}\Omega\mathrm{d}t \int_T\int_\Gamma h_\alpha(\mathbf{u},\mathbf{a})\,\mathrm{d}\Gamma\mathrm{d}t\,, \tag{5.61}$$

where Ψ_α are continuous functions of strains $\boldsymbol{\epsilon} = \boldsymbol{\epsilon}(\mathbf{x},t,\mathbf{a})$ and of displacements $\mathbf{u} = \mathbf{u}(\mathbf{x},t,\mathbf{a})$ in $\Omega \times T$, while h_α are continuous functions of displacements $\mathbf{u} = \mathbf{u}(\mathbf{x},t,\mathbf{a})$ on $\Gamma \times T$, $T = [0, t_K]$, $t \in T$. Like previously, we can write

$$\delta J_\alpha = \frac{\mathrm{D}J_\alpha}{\mathrm{D}a_r}\delta a_r\,, \tag{5.62}$$

where elements $S_{\alpha r} = (\mathrm{D}J_\alpha/\mathrm{D}a_r)$ of a dynamic sensitivity operator S_α are expressed as follows (cf. Chap. 1 of Part VI):

$$\begin{aligned}\frac{\mathrm{D}J_\alpha}{\mathrm{D}a_r} = &\int_T\int_\Gamma [\,\Psi_\alpha - \boldsymbol{\sigma}\cdot\boldsymbol{\epsilon}^a + \mathbf{b}\cdot\mathbf{u}^a - \rho\ddot{\mathbf{u}}\cdot\mathbf{u}^a + (h_\alpha + \mathbf{p}\cdot\mathbf{u}^a)_{,n} \\ &\quad - 2(h_\alpha + \mathbf{p}\cdot\mathbf{u}^a)\mathcal{K}\,]\,n_k v_k^r\,\mathrm{d}\Gamma\mathrm{d}t \\ &+ \int_T\int_{\Gamma_u}\left(\frac{\partial h_\alpha}{\partial\mathbf{u}} - \mathbf{p}^a\right)\cdot\left(\frac{\mathrm{D}\mathbf{u}^0}{\mathrm{D}a_r} - \mathbf{u}^0_{,k}v_k^r\right)\mathrm{d}\Gamma_u\mathrm{d}t \\ &+ \int_T\int_{\Gamma_p}\mathbf{u}^a\cdot\left\{\frac{\mathrm{D}\mathbf{p}^0}{\mathrm{D}a_r} - \mathbf{p}^0_{,k}v_k^r\right\}\mathrm{d}\Gamma_p\mathrm{d}t + \int_T\int_L [\![h_\alpha + \mathbf{p}\cdot\mathbf{u}^a]\!] v_\nu^r\,\mathrm{d}L\mathrm{d}t\,,\end{aligned} \tag{5.63}$$

L is a curve that separates Γ_u and Γ_p, $(\dot{})$ denotes a derivative with respect to time T and $\alpha = 1, 2, \ldots, K-2$. In order to calculate the sensitivity of functionals J_α non-stationary adjoint problems are considered, which are described in the domain of time $t^a \in T^a$, $T^a = [0, t_F]$, with t^a related to t by

$$t^a = t_F - t\,. \tag{5.64}$$

Eq. (5.64) implies that the conjugate process proceeds in the opposite direction with respect to the dynamic phenomena in the basic system (BS). It can be noticed that the sensitivities of the dynamic functionals (5.63) depend only on the boundary state variables for the basic system (BS) and adjoint systems $(\mathrm{AS})_\alpha$.

The initial boundary value problem for the adjoint systems $(\mathrm{AS})_\alpha$ is determined by

- boundary conditions:

$$\mathbf{u}_\alpha^{a0}(x, t^a) = 0 \qquad \text{on}\ \Gamma_u\,, \tag{5.65}$$

$$\mathbf{p}_\alpha^{a0}(x, t^a) = \frac{\partial h_a(\mathbf{u}(x,t))}{\partial\mathbf{u}} \qquad \text{on}\ \Gamma_p\,, \tag{5.66}$$

- initial conditions:

$$\mathbf{u}_\alpha^a(x, 0) = 0 \quad \text{in}\ \Omega\,, \qquad \frac{\partial\mathbf{u}_\alpha^a(x,0)}{\partial t^a} = 0 \quad \text{in}\ \Omega \tag{5.67}$$

and a set of boundary integral equations in the domain of time $t^a \in T^a$:

$$\begin{aligned}\mathbf{c}(x)\mathbf{u}_\alpha^a(x, t^a) = &\int_\Gamma [\mathbf{U}(x, y, t^a) * \mathbf{p}_\alpha^{a0}(y, t^a) \\ &- \mathbf{P}(x, y, t^a) * \mathbf{u}_\alpha^a(y, t^a)]\,\mathrm{d}\Gamma(y) + \hat{\mathbf{B}}_\alpha^a(x, t^a)\,,\end{aligned} \tag{5.68}$$

where

$$\hat{\mathbf{B}}^a_\alpha(x,t^a) = \int_\Omega U_{jk}(x,y,t^a) * [b^a_k(y,t^a) - \sigma^{ai}_{\alpha kl,l}(y,t^a)]\,\mathrm{d}\Omega\,, \tag{5.69}$$

and the initial stresses and the volume forces are defined as:

$$\sigma^{ai}_\alpha(x,t^a) = \frac{\partial\Psi_\alpha(\epsilon(x,t),\mathbf{u}(x,t))}{\partial\epsilon} \quad \text{in}\ \Omega\,, \tag{5.70}$$

$$\mathbf{b}^a_\alpha(x,t^a) = \frac{\partial\Psi_\alpha(x,t)}{\partial\mathbf{u}} \quad \text{in}\ \Omega\,. \tag{5.71}$$

A discrete version of Eq. (5.69) can be obtained using the time step method (cf. Sect. 3.2.2) or the alternative approach (cf. Sect. 3.2.5). An analysis of the sensitivity of eigenfrequencies to shape variation is a special case. The problem of free vibrations can also be described by means of a variational equation

$$\boldsymbol{A}(\mathbf{u},\mathbf{u}) = \omega^2\boldsymbol{B}(\mathbf{u},\mathbf{u})\,, \tag{5.72}$$

where

$$\boldsymbol{A}(\mathbf{u},\mathbf{u}) = \int_{\Omega(a)} \boldsymbol{\sigma}(\mathbf{u})\cdot\boldsymbol{\epsilon}(\mathbf{u})\,\mathrm{d}\Omega\,, \qquad \boldsymbol{B}(\mathbf{u},\mathbf{u}) = \int_{\Omega(a)} \varrho\,\mathbf{u}\cdot\mathbf{u}\,\mathrm{d}\Omega\,, \tag{5.73}$$

$\mathbf{u}(x)$ is an eigenfunction associated with the eigenvalue ω^2, $\boldsymbol{\sigma}(\mathbf{u})$ and $\boldsymbol{\epsilon}(\mathbf{u})$ are strain and stress tensors, respectively. A variation of the frequency of free vibrations is given by the relation:

$$\delta\omega = \frac{\mathrm{D}\omega}{\mathrm{D}a_r}\delta a_r\,, \tag{5.74}$$

and the sensitivity $(\mathrm{D}\omega/\mathrm{D}a_r)$ can be calculated as

$$\frac{\mathrm{D}\omega}{\mathrm{D}a_r} = (2\omega)^{-1}\int_\Gamma [\boldsymbol{\sigma}(\mathbf{u})\cdot\boldsymbol{\epsilon}(\mathbf{u}) - \omega^2\varrho\mathbf{u}\cdot\mathbf{u}]n_k v^r_k\,\mathrm{d}\Gamma\,. \tag{5.75}$$

As can be seen, the relation between the variation of the boundary shape of an elastic body and the variation of the frequency depends on the eigenvalue and the eigenfunction defined on the boundary. In order to determine the frequencies and forms of the vibrations the eigenvalue problem described by Eqs. (3.83) and (3.84) or (3.95) has to be solved.

5.2.3 Discretization of Sensitivity Boundary Integrals with Boundary Elements

The analysis of sensitivity of the considered functionals $J_\alpha, \alpha = 1, 2, \ldots, K-2$ for static problems, non-stationary dynamic problems and individual vibration frequencies $J_{K-1} = \omega$ indicates that the relation between the boundary shape variations and $J_\alpha, \alpha = 1, 2, \ldots, K-1$ has the form of a boundary integral, in which the expression under the integral depends in a general case on state variables of the basic and adjoint systems. The fact is important in numerical sensitivity calculations by means of the BEM.

The dependence of each functional $J_\alpha, \alpha = 1, 2, \dots, K-1$ on the set of shape parameters $\mathbf{a} = (a_r), r = 1, 2, \dots, R$ can be described as follows:

$$\delta J_\alpha = \mathbf{S}_\alpha^T \delta\mathbf{a}\,, \tag{5.76}$$

where

$$\mathbf{S}_\alpha = [S_{\alpha 1}, S_{\alpha 2}, \dots, S_{\alpha r}, \dots, S_{\alpha R}]^T \tag{5.77}$$

is a sensitivity matrix the elements of which are defined as material derivatives of the functionals J_α with respect to a_r, namely

$$S_{\alpha r} = \frac{\mathrm{D}J_\alpha}{\mathrm{D}a_r}\,, \tag{5.78}$$

while

$$\delta\mathbf{a} = [\delta a_1, \delta a_2, \dots, \delta a_r, \dots, \delta a_R] \tag{5.79}$$

is a matrix of variations of shape parameters. The sensitivity matrix $\mathbf{S}_\alpha$ contains gradients of the functional J_α with respect to the shape parameters a_r, $r = 1, 2, \dots, R$.

In order to calculate the sensitivity matrix $\mathbf{S}_\alpha$ the transformation field v_k^r associated with the shape parameter a_r must be determined. The selection of shape parameters is an essential step in the sensitivity analysis and shape optimization. The simplest and most natural way to describe the shape of a boundary is to adopt the positions of boundary nodes as decision variables. Let the set of all boundary elements Γ^e, $e = 1, 2, \dots, E$ be expressed in the form of the following sum of sets:

$$\{\Gamma^e\} = \bigcup_{p=1}^{P} \{\Gamma_p^e\}\,, \tag{5.80}$$

where $\{\Gamma_p^e\}$ is a set of all the boundary elements that contain node p. In the case of space problems boundary elements Γ_p^e are represented by quadrilateral or triangular surface segments. In two-dimensional problems the boundary elements Γ_p^e are curvilinear segments.

The discretization of boundary Γ with boundary elements should be performed in such a way that the curve L is generated by the edges of elements Γ_p^e. In two-dimensional problems L is reduced to the points which should be identified with boundary nodes.

Among many types of boundary elements which can be applied in computations of sensitivity analysis, elements with linear shape functions assure the required accuracy without the need of long numerical calculations. An application of these elements in boundary parametrization will be discussed below in detail. The vector field of shape transformation $\mathbf{g}(\mathbf{x})$ can be expressed within each boundary element Γ_p^e as follows:

$$\mathbf{g}_p^e(\xi) = M_p^e(\xi)\mathbf{b}^p + \underline{M}_p^e(\xi)\underline{\mathbf{b}}^p + \underline{\underline{M}}_p^e(\xi)\underline{\underline{\mathbf{b}}}^p\,, \tag{5.81}$$

where

$$M_p^e(\xi) = 1 - \frac{1}{L_{1p}^e}\xi_1 - \frac{1}{L_{2p}^e}\xi_2 \,, \tag{5.82}$$

$$\underline{M}_p^e(\xi) = \frac{1}{L_{1p}^e}\xi_1 \,, \quad \underline{\underline{M}}_p^e(\xi) = \frac{1}{L_{2p}^e}\xi_2 \,, \tag{5.83}$$

$\xi = (\xi_j), j = 1, \ldots, d-1$, is a local coordinate set of a boundary element and L_{1p}^e, L_{2p}^e are side lengths of element Γ_p^e. In a two-dimensional case we have $\xi = \xi_1$. The vector shape parameters $\mathbf{b}^p, \underline{\mathbf{b}}^p$ and $\underline{\underline{\mathbf{b}}}^p$ are defined as follows:

$$\mathbf{g}_p^e(\xi) = \begin{cases} \mathbf{b}^p \text{ for } \xi_1 = \xi_2 = 0 \,, \\ \underline{\mathbf{b}}^p \text{ for } \xi_1 = L_{1p}^e \text{ and } \xi_2 = 0 \,, \\ \underline{\underline{\mathbf{b}}}^p \text{ for } \xi_1 = 0 \text{ and } \xi_2 = L_{2p}^e \,. \end{cases} \tag{5.84}$$

The vector shape parameters $\mathbf{b}^p = (b_k^p)$, $k = 1, \ldots, d$; $1, 2, \ldots, P$ are associated with the shape parameters $\mathbf{a} = (a_r), r = 1, 2, \ldots, R$. And thus, for instance, for space problems ($d = 3$) it holds that:

$$a_{3p-2} = b_1^p \,; \quad a_{3p-1} = b_2^p \,; \quad a_{3p} = b_3^p \,. \tag{5.85}$$

The number of shape parameters a_r equal to R is related to the number of nodal points P by the relation $R = mP$. The field of shape transformations v_k^r can now be expressed within each boundary element Γ_p^e using the functions $M_p^e(\xi)$. Neglecting, for simplicity, the integrals over L, the elements of the sensitivity matrix $S_{\alpha r}$ can be expressed in the following forms:

– for static problems and eigenvalue problems

$$S_{\alpha,mp-h} = \sum_{e=1}^{E_p} \int_{\Gamma_p^e} W_\alpha^e(\xi) n_{m-h} M_p^e(\xi) \, \mathrm{d}\Gamma_p^e(\xi) \,, \tag{5.86}$$

– for non-stationary problems

$$S_{\alpha,mp-h} = \sum_{e=1}^{E_p} \sum_{i=1}^{N_T} \left\{ w_i \int_{\Gamma_p^e} W_\alpha^e(\xi, t_i) n_{m-h} M_p^e(\xi) \, \mathrm{d}\Gamma_p^e(\xi) \right\} \,, \tag{5.87}$$

$$h = 2, 1 \text{ for } d = 3 \,,$$
$$p = 1, 2, \ldots, P \,; \; h = 1, 0 \text{ for } d = 2 \,,$$

where E_p denotes the number of boundary elements which contain node p, $\{t_i\}_{i=1}^{N_T}$ and $\{w_i\}_{i=1}^{N_T}$ are the positions and weights of the integration quadrature in a time interval $T = [0, t_F]$, respectively. Functions W_α^e are defined within each boundary element Γ_p^e by state variables for the basic system (BS) and adjoint systems $(\mathrm{AS})_\alpha$, $\alpha = 1, 2, \ldots, K-2$.

The presented method of boundary parametrization, which employs nodal coordinates, is the simplest. However, it has serious drawbacks. The number of shape parameters is usually quite large in this description and in the optimization process there is a tendency to generate unrealistic and discontinuous boundary shapes. Therefore to describe the boundary shape it is convenient to select non-straight functions (Bezier functions, B-splines), known from applications in computer graphics. Then, the shape parameters $\mathbf{a} = (a_r)$ can be identified with controlling points of the functions. The presented method of computing the sensitivity matrix which takes advantage of the method of adjoint systems and the numerical discretization with boundary elements is highly effective.

5.2.4 Sensitivity Analysis in Boundary Shape Optimization

Let us return to the boundary shape optimization problem considered in Sect. 5.1.2. We assume that the shape of the boundary of the body is defined by a set of parameters $\mathbf{a} = (a_r)$. In this way the optimal shape control is reduced to a parametric optimization problem.

Assuming that the objective function is represented by a certain functional $J_0(\mathbf{a}, \boldsymbol{\epsilon}, \boldsymbol{\sigma}, \mathbf{u})$, the problem can be formulated as follows: find a set of parameters $\mathbf{a}^*$ such that

$$J_0(\mathbf{a}^*, \boldsymbol{\epsilon}^*, \boldsymbol{\sigma}^*, \mathbf{u}^*) = \inf_{a \in U} J_0 \,, \tag{5.88}$$

where U denotes a set of admissible values of the shape parameters, i.e. of the control (cf. Sect. 5.1), which is determined by a set of restrictions in the form of functionals $J_\alpha, \alpha = 1, 2, \dots, K-1$, expressed in terms of stresses, strains and displacements or frequencies, and by the limitation of the cost $J_c = J_K$, so that

$$J_\alpha - c_\alpha \leq 0 \,, \qquad \alpha = 1, 2, \dots, K, \tag{5.89}$$

where $C_\alpha, \alpha = 1, 2, \dots, K$ are given constants. If the cost is proportional to its volume or weight, then we can write:

$$J_c = \int_\Omega C \, \mathrm{d}\Omega \,, \tag{5.90}$$

where C is a unit material cost. The Lagrange formalism allows us to propose an optimality condition for the choice of shape parameters. For this purpose a Lagrangian should be constructed

$$\Phi = J_0 + \sum_{\alpha=1}^{K} \lambda_\alpha (J_\alpha - c_\alpha) \,, \tag{5.91}$$

where $\lambda_\alpha, \alpha = 1, 2, \dots, K$ are Lagrange multipliers ($\lambda_\alpha \geq 0$), and a stationary point $\mathbf{a}^*$ should be found, in which the Gateaux differential becomes zero:

$$\partial\Phi = \partial J_0 + \sum_{\alpha=1}^{K} \lambda_\alpha \partial J_\alpha = 0\,. \tag{5.92}$$

It is easy to notice that the use of definition (5.52) makes it possible to write condition (5.92) in the form

$$\sum_{\alpha=0}^{K} \lambda_\alpha S_\alpha \delta\mathbf{a} = 0\,, \qquad \lambda_0 = 1\,. \tag{5.93}$$

Using the notation

$$\sum_{\alpha=0}^{K} [\lambda_\alpha S_\alpha] \overset{\text{def}}{=} S\,, \tag{5.93a}$$

we obtain

$$S\,\delta\mathbf{a} = 0\,, \tag{5.94}$$

where S is an operator of shape variation sensitivity, so the problem of shape optimization is closely related to sensitivity analysis, which has been expected. An optimal solution, defined by a set of parameters $\mathbf{a}^*$, is thus a point in the space of admissible control which is a stationary point of the functional Φ or, in other words, in which the sensitivity operator becomes zero. Taking into account that

$$S(\mathbf{a})\,\delta\mathbf{a} = 0\,; \qquad \forall\,\delta\mathbf{a} \tag{5.95}$$

it is easy to conclude that

$$S(\mathbf{a}^*) = 0\,, \qquad \mathbf{a}^* \in U\,. \tag{5.96}$$

The construction of the sensitivity operator has been discussed in Sect. 5.2.1 and 5.2.2. In the case of a cost limitation expressed by Eq. (5.90), the component $S_c = S_K$ of the sensitivity operator can be written as a boundary integral using a material derivative:

$$S_K = \int_\Gamma C n_k v_k \, \mathrm{d}\Gamma\,. \tag{5.97}$$

The optimality condition (5.94) together with Eq. (5.97) can be expressed, cf. Eqs. (5.54), (5.63) and (5.75), as

$$\int_\Gamma [W_0(\mathbf{u}, \mathbf{u}^a, \mathbf{p}, \mathbf{p}^a, \boldsymbol{\epsilon}, \boldsymbol{\epsilon}^a, \boldsymbol{\sigma}, \boldsymbol{\sigma}^a) + \lambda_1 C]\mathbf{n}\cdot\mathbf{v}\,\mathrm{d}\Gamma = 0 \tag{5.98}$$

for static and eigenvalue problems, and as

$$\int_T \left\{ \int_\Gamma \left[W_0(\mathbf{u}, \mathbf{u}^a, \mathbf{p}, \mathbf{p}^a, \boldsymbol{\epsilon}, \boldsymbol{\epsilon}^a, \boldsymbol{\sigma}, \boldsymbol{\sigma}^a; t) + \frac{\lambda_1}{t_K} C \right] \right\} \mathbf{n}\cdot\mathbf{v}\,\mathrm{d}\Gamma \mathrm{d}t = 0 \tag{5.99}$$

for non-stationary dynamic problems. The functions W_0 are now functions of state variables of the basic and adjoint systems. Taking into account the relation $\delta \mathbf{a} = \delta \mathbf{g}_n = \mathbf{n} \cdot \mathbf{v}$, an expanded version of condition (5.96) can be written as

$$W_0 = -\lambda_1 C = \text{const} \quad \text{on } \Gamma\,, \tag{5.100}$$

$$W_{0T} \equiv \int_T W_0 \mathrm{d}t = -\lambda_1 C = \text{const} \quad \text{on } \Gamma\,. \tag{5.101}$$

The above conditions can be directly applied in the process of generation of an optimal structural shape of the body via an appropriate shift of the boundary nodes, in order to make the functions W_0 (for static problems and free vibrations) or W_{0T} (for non-stationary dynamic problems) evenly distributed on the boundary Γ. The optimal solution is thus found in an iterative procedure. At each step the basic system (BS) and the adjoint system $(\mathrm{AS})_\alpha$ must be solved by means of the boundary element method.

5.3 Numerical Examples

Example 1

The problem of sensitivity analysis and optimization of an anchor screw made from aluminium ($E = 73000$ MPa, $\nu = 0.34$) is considered, cf. [5]. The quality functional J_0 has been given by Eq. (5.46) in which the function $\Psi_0 = W(\sigma)$ represents the stress potential associated with the unit volume, the function h_0 is defined on the boundary Γ_u by the relation $h_0 = -\mathbf{p} \cdot \mathbf{u}^0$. As a result the functional J_0 represents a complementary energy. For linear elastic systems the complementary energy is a measure of average compliance of a system. Minimization of the adopted functional is thus equivalent to the shape optimization with respect to a minimum of average compliance (or rather maximum of average stiffness) of the system.

The boundary of the body has been divided into 30 linear boundary elements (Fig. 5.2a). The first variation of the adopted functional has been computed, assuming that the head of the screw is modified so that the boundary nodes are shifted towards the inside of the body in the direction normal to the boundary by a value $\delta a_n = 1$ mm each node. The results of numerical calculations of the functional variation are presented in Fig. 5.2a. It is visible that a modification of the lower part of the screw has the largest influence on the compliance variation.

The problem of optimization of the screw head with respect to the minimum average compliance has been solved with restriction that the area of the domain remains constant. The iterative optimization process stopped after 13 iterations. The final shape of the geometrical form is presented in Fig. 5.2b. The distribution of equivalent stresses according to Huber's hypothesis has also been computed before and after the optimization (Fig. 5.2c). It is worth

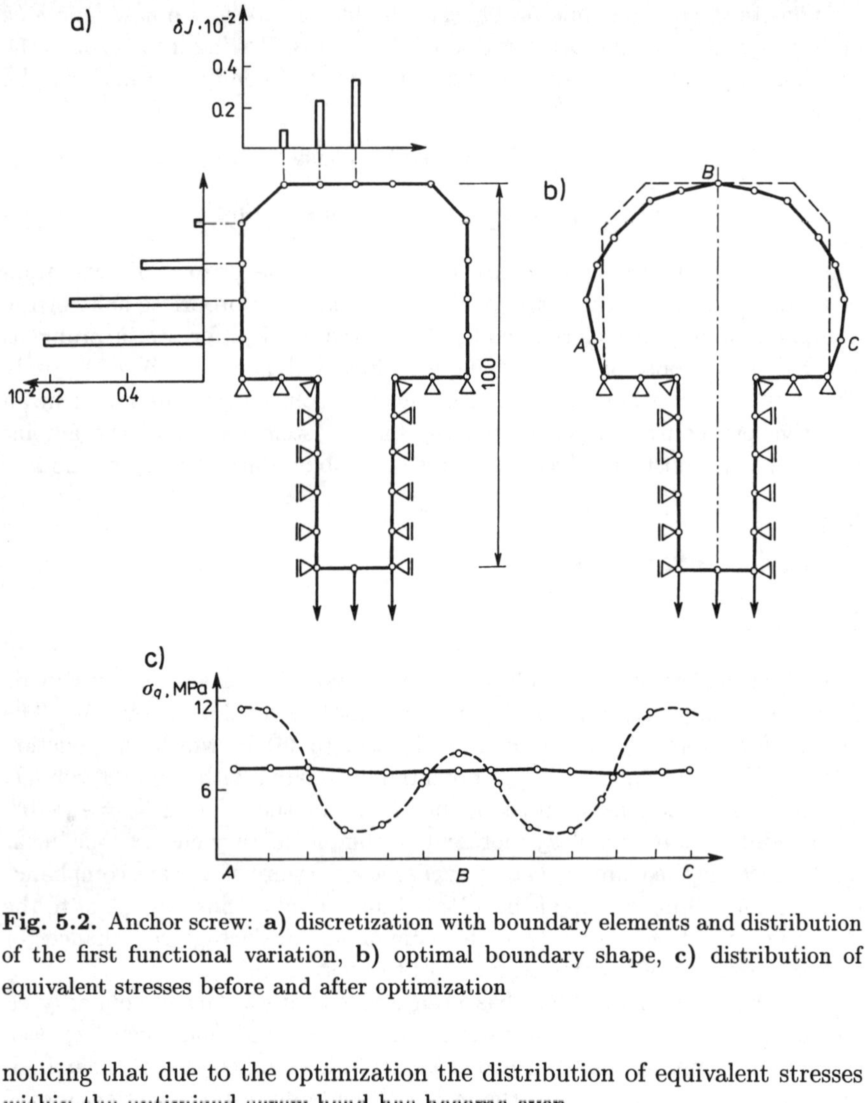

Fig. 5.2. Anchor screw: **a)** discretization with boundary elements and distribution of the first functional variation, **b)** optimal boundary shape, **c)** distribution of equivalent stresses before and after optimization

noticing that due to the optimization the distribution of equivalent stresses within the optimized screw head has become even.

Example 2

The example illustrates sensitivity analysis and optimization of a freely vibrating flat steel panel ($E = 2 \cdot 10^{12}$ Pa, $\nu = 0.3$, $\varrho = 8000$ kg/m^3), cf. [8]. The boundary of the panel has been discretized with 40 linear boundary elements (Fig. 5.3a). Employing the approach described in Sect. 3.2.5 the first three frequencies of free vibrations have been computed: $\omega_1 = 12400$ rad/s, $\omega_2 = 28080$ rad/s and $\omega_3 = 32680$ rad/s. The sensitivity of the eigenfrequencies to the modification of the lower edge of the panel has been analyzed by

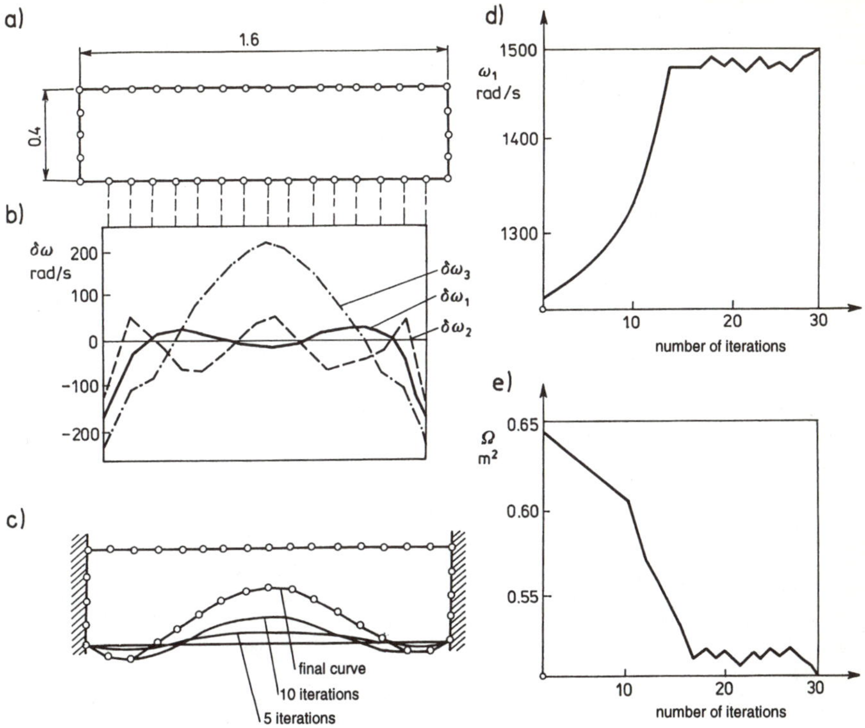

Fig. 5.3. Rectangular panel: **a)** boundary element discretization, **b)** distribution of the frequency variation of eigenvibrations, **c)** optimal shape of the lower edge, **d)** variation of the first eigenfrequency for different iteration numbers, **e)** variation of the area of the panel for different iteration numbers

shifting the nodes in the normal direction to the inside by $\delta a_n = 0.04$ m. The frequency variations are presented in Fig. 5.3b.

The first eigenfrequency maximization has been the criterion of boundary optimization, augmented with restrictions on the area and shape parameters. The optimal shape of the panel lower edge has been reached after 18 iteration steps (Fig. 5.3c). The first frequency of free vibrations increased by 20% (Fig. 5.3d) and the area decreased by 49% (Fig. 5.3e).

6 Final Remarks

The boundary element method has some characteristic features, two of which deserve to be emphasized.

The first feature is associated with the properties of the formalism of boundary integral equations. It is the so-called weak formulation of the problem, mentioned in Chap. 2. In a natural way, the solutions have a generalized character, which is not only of formal importance. It is also essential for practical reasons, since it allows an approximation of the required solution in a computationally convenient manner, without losing of formal correctness. A convenient finite-dimensional approximation is the boundary element method. In the literature the method is prevalently identified with the approximation concept, cf. [1, 2, 3]. In this context, it is also confronted and compared with the finite element method. A circumstance that is then usually neglected is the fact that in order to take a full advantage of the FEM it is necessary to formulate a problem appropriately (cf. Chap. 2 of Part II). It is not always the simplest formulation. Usually, it requires a non-conventional mathematical formalism, which is often hard to interpret in terms of mechanics. In case if the MBIE formalism, which is the basis of the BEM, the formulation has an obvious mechanical sense. Due to this, approximate solutions can be constructed without the necessity to reformulate the problem and with no anxiety as to their formal correctness.

The second feature of the BEM/MBIE is its dynamic development. It is not only a qualitative progress, related to the utilization of possibilities rooted in the formalism, but also a quantitative progress. The applications in the optimal control problems, discussed in Chap. 5, are an example of the qualitative development. The notion of quantitative progress has been used to describe all the proposals and analyses aiming at an improvement of the application potential of the method, i.e. the research oriented towards selected numerical methods, which are usually adapted to a certain class of boundary value problems.

A different research branch, also related to the quantitative development, but having a distinctly fundamental character, are all the papers related to the assessment of the convergence and accuracy of the method (cf. [11]). It seems that in particular a lot remains to be done in the domain of constructing *a posteriori* estimates. The formulation of an adaptive boundary element method depends on the solution of the problem [53]. It is one of the most up-to-date and attractive problems.

Then, the papers which employ the formalism of the MBIE in sensitivity analysis [5, 6, 8, 22, 23, 25] should be mentioned. Such a formulation has a lot of advantages, and its development is undoubtedly a part of the quantitative progress.

Having summarized the characteristic features of the method, it is time to mention its limitations and complexity from the viewpoint of applicability.

The first observation is that the method in its original shape can only be applied to linear self-adjoint operators, and even then the knowledge of the fundamental solution is a limitation. For a series of operators it is unknown, or known, but not in a closed form. In fact, there is a possibility to extend the applicability of the method to a certain class of evolution operators (cf. Sect. 3.2), and also nonlinear ones (cf. Chap. 4). However, as it often happens, a fundamental research is in this case delayed with respect to the numerical practice. The intuitive character of these extensions is definitely a weakness of the method.

A completely different inconvenience is the lack of symmetry of the set of algebraic equations obtained as a result of the finite-dimensional approximation. It is an essential disadvantage, which often balances the gain resulting from the reduction of the problem dimension. In fact, some symmetrization methods for the set of equations are known, but it is difficult to treat them as sufficiently general and effective. The problem of a symmetric formulation of the method belongs to research topics examined comprehensively.

There are still some problems posed by the calculation of singular integrals (existing in the sense of the Cauchy principal value), for which classical numerical integration methods fail. It seems that the nonsingular formulation is a radical solution of this problem, however, it has some drawbacks which have not yet been sufficiently interpreted from the formal viewpoint. Therefore, the problem can still be considered open.

The MBIE/BEM is undoubtedly a method of the future, constantly finding new interesting applications, and therefore worth attention.

Finally, it is customary to produce a list of disadvantages and advantages of the MBIE/BEM and attempt to define its place among the other known numerical methods. The authors are aware of the dangers involved. Firstly, such conclusions are intrinsically subjective and do not exactly evaluate the reality, but rather express likes and dislikes of the ones writing them. Secondly, the presented method is far from its final shape, which can make some of the assessments outdated. Therefore, the list of advantages and disadvantages below should be treated as opinions of the authors only, which will be verified by time.

Advantages of the BEM/MBIE

1. It is a general numerical method with broad applications in mechanics. The method has well established mathematical foundations (cf. Chap. 2) and is an effective numerical tool. The BEM can be applied to linear and nonlinear problems (in the latter case under the condition that the problem can be approximated as incrementally-linear).

2. The method usually requires a discretization of the boundary of a body and not its interior like in the other numerical methods, e.g. in the FEM or FDM. The reduction of the problem dimension by an order of one makes the preparation of input data simpler and leads to a set of algebraic equations which is much smaller than for the FEM or FDM.
3. For the same order of discretization the method usually gives more accurate results than the FEM, which is especially important in problems with large stress gradients (e.g. in fracture mechanics [13]). The accuracy can still be improved when special boundary elements or special integration techniques are applied.
4. The application of singular fundamental solutions or Green functions yields a possibility to consider problems in infinite or half-infinite domains without an introduction of artificial boundaries, as is done in the FEM.
5. The method makes it possible for unknown quantities inside the domain to be calculated without the necessity of its discretization and only at the points of interest.
6. It can be employed to construct stiffness matrices for very large (also infinite) homogeneous finite elements (called super-elements), which radically reduces the number of unknowns in the FEM formulation.
7. Problems which are not correctly posed (called over-determined, i.e. those in which there is too much information about the solution on the boundary, but some other information necessary for the formulation to be well-posed is missing), cumbersome to be tackled by the other methods, can successfully be solved by the MBIE, and the BEM assures a numerical effectiveness of such an approach.

Disadvantages of the BEM/MBIE

1. A full and non-symmetric coefficient matrix is obtained.
2. The method can only be applied in cases where the fundamental solution is known, and for anisotropic or heterogeneous materials such solutions are unknown or do not have a closed form.
3. The method is not effective for bodies for which one or two dimensions are small in comparison with the other dimensions. The reason is that ill-conditioned matrices may appear in such problems, especially for shells.
4. In comparison with the FEM, there is a limited library of general and professional BEM programs or packages. It is a temporary stage, since several renowned institutions are working on such packages.

Finally, it should be noticed that the BEM/MBIE and the FEM can be regarded as complementary in the sense that their advantages and disadvantages mutually compensate. Extensive research on the combination of the two methods lead to hybrid approaches which adopt the advantages of both methods and eliminate their disadvantages.

References

1. Banerjee PK, Butterfield R (1981) Boundary element methods in engineering science. McGraw–Hill, London

2. Beskos DE, (ed) (1987) Boundary element methods in mechanics. North-Holland, Amsterdam

3. Brebbia CA, Telles JCF, Wrobel LC (1984) Boundary element techniques. Springer-Verlag, Berlin

4. Bui HD (1978) Some remarks about the formulation of the three-dimensional thermo-elasto-plastic problems by integral equations. Int J Solids Structures 14:935–939

5. Burczyński T (1995) Boundary element method in mechanics (in Polish). WNT, Warsaw

6. Burczyński T (1993) Applications of BEM in sensitivity analysis and optimization. Computational Mechanics 3(1/2):29–44

7. Burczyński T, Adamczyk T (1988) Analysis of nonlinear systems using the boundary element method (in Polish). Mechanika i Komputer 7:149–164, PWN, Warsaw

8. Burczyński T, Fedeliński P (1992) Boundary elements in shape design sensitivity analysis and optimal design of vibrating structures. Engineering Analysis with boundary Elements 9(3):195–202

9. Cea J (1976) Optimization. Theory and algorithms (Polish translation). PWN, Warsaw

10. Ciarlet PG (1988) Mathematical elasticity. Vol I: Three-dimensional elasticity. North-Holland, Amsterdam

11. Costabel M, Stephan E, Wendland WL (1983) On boundary integral equations of the first kind for bi-Laplacian in the polygonal plane domain. Annali Scoula Normale Superiore – Pisa, Serie IV, Vol X, No 2

12. Cruse TA (1974) An improved boundary integral equation method for three-dimensional elastic stress analysis. Comp Struct, 5(4):741–754

13. Cruse TA (1988) Boundary element analysis in computational fracture mechanics. Kluwer Academic Publishers, Dordrecht

14. Eringen AC (1962) Nonlinear theory of continuous media. McGraw–Hill, London

15. Fedeliński P (1991) Application of the boundary element method to the structural shape optimization of vibrating mechanical systems (in Polish). Doctoral thesis, Politechnika Śląska, Gliwice

16. Fedeliński P, Burczyński T (1991) Shape optimal design of vibrating structural using boundary elements. ZAMM 71(6):726–728

17. Findeisen W, Szymanowski J, Wierzbicki A (1981) Computational methods of Optimization (in Polish). PWN, Warsaw

18. Gelfand JM, Fomin SN (1970) Variational calculus (Polish translation). PWN, Warsaw

19. Gelfand JM, Šivov G (1958) Generalized functions, Vol 2: Generalized functions and operations on the functions (in Russian). Izd Fiz–Mat Literat, Moscow

20. Gelfand JM, Šivov G (1958) Generalized functions, Vol 3: Some issues of the theory of differential equations (in Russian). Izd Fiz–Mat Literat, Moscow

21. Giunter NM (1957) Potential theory (Polish translation). PWN, Warsaw

22. Grabacki J (1985) Constraint control for a deformable body (in Polish), Chap 6 in: Control in mechanics. PTMTS

23. Grabacki J (1991) Boundary integral equations in sensitivity analysis. Appl Math Modelling 15(4):170–181

24. Grabacki J (1980) Minimum deformability design and control of constraints. In: Leipholz HHE (ed) Structural control, IUTAM. North-Holland Publ Comp & SM Publ, pp 281–295

25. Grabacki J (1988) Shape optimization and shape sensitivity analysis of elastic plates. SM Arch 13(2):103–120

26. Grabacki J, Mamoń M (1988) The method of boundary integral equations in shape optimization and sensitivity analysis (in Polish). Prace Kom Mech Stos PAN Oddz Kraków: Mechanika, Vol 13

27. Hall WS (1988) Integration methods for singular boundary element integrands. In: Brebbia CA (ed) Boundary elements X, Vol 1, Mathematical and computational aspects. Springer-Verlag, Berlin, pp 219–236

28. Hartmann F (1980) Computing the C-matrix in non-smooth boundary points. In: Brebbia CA (ed) New developments in boundary element methods. Butterworths, London, pp 367–379

29. Kołodziej W (1970) Selected aspects of mathematical analysis (in Polish). PWN, Warsaw

30. Kontoni DPN, Beskos DE (1988) Boundary element formulation for dynamic analysis of nonlinear systems. Engineering analysis 5(3):114–125

31. Kuhn G, Möhrmann W (1983) Boundary element method in elastostatics: theory and applications. Appl Math Modelling 7:97–105

32. Kupradze WD (ed) (1976) Three-dimensional problems of the mathematical theory of elasticity and thermoelasticity (in Russian). Izd Nauka, Moscow

33. Lachat JC (1975) A further development of the boundary integral technique for elasto-statics. PhD Thesis, Southampton University

34. Lachat JC, Watson JO (1976) Effective numerical treatment of boundary integral equation: a formulation for three-dimensional elasto-statics. Int J Num Meth Eng 10(5):991–1005

35. Ladyzenskaya DA, Uralceva NN (1973) Linear and Quasi-Linear elliptic equations (in Russian). Izd Nauka, Moscow

36. Maier G, Novati G, Perego U (1988) Plastic analysis by boundary elements. In: Stein E and Wendland W (eds) Finite element and boundary element Techniques from mathematical and engineering point of view, CISM. Springer-Verlag, pp 213–272

37. Manolis GD, Ahmad S, Banerjee PK (1985) Boundary element method implementation for three-dimensional transient elasto-dynamics. Chap 2 in: Banerjee PK and Watson JO (eds) Developments in boundary element methods – 4. Elsevier Applied Science Publishers, London

38. Manolis GD, Beskos DE (1981) Dynamic stress concentration studies by the boundary integrals and Laplace transform. Int J Num Meth Eng 17:573–599

39. Mikhlin SG (1962) Multidimensional singular integrals (in Russian). Izd Fizmatgiz, Moscow

40. Mukherjee S (1982) Boundary element methods in creep and fracture. Applied Science Publishers, London

41. Mukherjee S, Chandra A (1984) Boundary element formulation for large strain – large deformation problems of plasticity and viscoplasticity. Chap 2 in: Banerjee PK and Mukherjee S (eds) Developments in boundary element methods – 3. Elsevier Science Publishers, London

42. Nardini D, Brebbia CA (1985) Boundary integral formulation of mass matrices for dynamic analysis. Chap 7 in: Brebbia CA (ed) Topics in boundary element Research, Vol 2. Springer-Verlag, Berlin

43. Niwa Y, Kobayashi S, Kitahara M (1982) Determination of eigenvalues by the boundary element methods. Chap 6 in: Banerjee PK and Shaw RP (eds) Developments in boundary element methods – 2. Applied Science Publishers, London

44. Nowacki W (1970) Theory of elasticity (in Polish). PWN, Warsaw

45. Perreira P (1988) A numerical integration scheme for the Galerkin approach in boundary elements. In: Brebbia CA (ed) Boundary elements X, Vol 1, Mathematical and computational aspects. Springer-Verlag, Berlin, pp 297–311

46. Radaj D, Möhrmann W, Schilberth G (1984) Economy and convergence of notch stress analysis using boundary and finite element methods. Int J Num Meth Eng 20:565–572

47. Rizzo FJ, Shippy DJ (1986) A boundary element method for axisymmetric elastic bodies. Chap 3 in: Banerjee PK and Watson JO (eds) Developments in boundary element method – 4. Elsevier Applied Science Publishers, London

48. Stern M, Lin TI (1986) Thin elastic plates in bending. Chap 4 in: Banerjee PK and Watson JO (eds) Developments in boundary element method – 4. Elsevier Applied Science Publishers, London

49. Szwartz L (1988) A course in mathematical analysis (Polish translation). PWN, Warsaw

50. Telles JCF (1983) The boundary element method applied to inelastic problems. Lecture Notes in Engineering, Vol 1, Springer-Verlag, Berlin

51. Truesdell C, Toupin RA (1960) The classical field theories. Encyclopedia of physics, Vol III/1, Springer-Verlag, Berlin

52. Truesdell C, Noll W (1965) The Non-Linear field theories of mechanics. Encyclopedia of physics, Vol III/3, Springer-Verlag, Berlin

53. Umetani S (1987) Adaptive boundary element methods in elastostatics. CMP, Southampton

54. Watson JO (1979) Advanced implementation of the boundary element method for two- and three-dimensional elasto-statics. Chap 3 in: Banerjee PK and Butterfield R (eds) Developments in boundary element methods – 1. Applied Science Publishers, London

55. Wilson RB, Cruse TA (1978) Efficient implementation of anisotropic three-dimensional boundary-integral equation stress analysis. Int J Num Meth Eng 12:1383–1397

56. Woźniak C (1985) Mechanics of continuous media. Part IV in: Applied mechanics. Vol I: Foundations on mechanics (in Polish). PWN, Warsaw

57. Wesołowski Z (1978) Nonlinear theory of elasticity. Part II in: Applied mechanics, Vol IV: Elasticity (in Polish). PWN, Warsaw

Part V

Optimization Methods

1 Numerical Approaches to Structural Optimization

1.1 Introduction

The problems of structural analysis were discussed in previous Parts. It means that all parameters defining material, geometrical properties, topology, loading and supporting conditions were specified in advance. The displacement, strains and stress fields within a structure are derived during the analysis step. From mathematical point of view, the structural analysis results in solution of proper set of equations describing the behavior of assumed model of structure. Assuming a continuous structure model its behavior will be described by a set of differential equations with unknown variables being the displacement and/or stress fields within structure domain. On the other hand, introducing the discrete model of actual structure and using, for instance, finite element or boundary element methods, the behavior of structure is described using a set of linear or nonlinear algebraic equations in the form

$$\mathbf{K} \cdot \mathbf{q} = \mathbf{Q} \tag{1.1}$$

where $\mathbf{K}$ denotes the global stiffness matrix in FEM, $\mathbf{Q}$ is a vector of generalized nodal forces, and $\mathbf{q}$ is the unknown vector of nodal displacements. The solution of the set (1.1), formally written in the form

$$\mathbf{q} = \mathbf{K}^{-1} \cdot \mathbf{Q} \tag{1.2}$$

specifies the values of generalized displacements of a model at nodes of finite element mesh used for structure discretization. Using these values, the distribution of displacements, strains and stresses within each element of structure can be easily calculated.

The results obtained during analysis step can be unsatisfactory from the point of view of the designer when he observes, for instance, excessive stresses above allowable levels in some domains of structure or unsatisfactory use of material. As the results of such observation, the value of some parameters defining the structure should be modified and the analysis of a modified structure should be repeated. The consecutive calculations can bring the designer closer or further to the assumed objective in the form of an improved structure. The final result depends mainly on complexity of the problem and experience of the designer. Such process goes thus beyond the analysis of structure and brings the designer closer to its optimization, that is provides for such properties of structure which will finally satisfy the objective set by the designer. The search for an optimal structure can be intuitive, based mainly on earlier experience of the designer, or much more rigorous based on mathematical methods of optimization. The latter offers, in some sense, the automatic attainment of the assumed objective (or at least close to it) assuming that this objective is formulated in the mathematical form, namely the search of an extremal value of the assumed objective function or functional describing the local or global structural property.

The main goal of this chapter is to familiarize the reader with the modern and now widely applied optimization methods used in mechanical and civil engineering. We do not intend to discuss the mathematical foundations of optimization methods, or the particular algorithms. The readers interested in these topics are referred to numerous monographs devoted to mathematical methods of optimization (see, for instance, [9, 12]) and optimization in mechanical and civil engineering ([5, 6, 10, 11, 17, 18]). In what follows, it is assumed that the optimized structure is modeled by means of finite elements and the state of stress, strain and displacement fields in its actual configuration can be obtained as the result of solution of the set of equations (1.1). The actual structure configuration is identified by parameters defining its material properties, geometry and topology, assembling of its elements, supporting and loading conditions etc. Some of these parameters are prescribed in advance and are not modified during the process of search for an optimal structure. The remaining parameters modified in consecutive steps of optimization process, constitute a set of design variables or parameters. The complete specification of structural design parameters will be presented in the subsequent chapters of this book devoted to sensitivity analysis of structures.

Considering the mathematical problems of structural optimization, the *objective function* depending on design parameters has to be specified. Its value, varying together with the variation of design parameters, will enable to select the best values of the design parameters belonging to their admissible set.

The structural optimization is generally aimed at reduction of the structure cost or objective function expressed in terms of selected design variables or parameters. However, in practical engineering problems, there is a need to introduce some *constraints* on the design parameters and on some measures of structural behavior (depending on those parameters) in order to satisfy some specific purposes or to avoid the failure of a structure. These constraints can be formulated either in the form of equalities or inequalities. The typical examples of equality constraints are equations of equilibrium (or motion). The inequality constraints can be imposed on displacements at some selected points within structural domain, stresses, free frequency and so on. The design parameters themselves can also be subjected to some geometric constraints.

The optimal set of design parameters, obtained as the result of solution, has to guarantee that the equality constraints are satisfied. However, not all inequality constraints have to attain their limit values. It can happen that only some of these constraints will be satisfied as equalities. This subset will be called the set of *active constraints.*

Looking for an optimal structure, one can introduce several objective functions which should be taken simultaneously into account during optimization process. Such approach is called the *multi-criterion optimization* which will not be considered here. Another type of optimization is a *discrete optimization*, in which we deal with finite set of values of design parameters. This case occurs when the structure is designed using the standardized elements or elements from catalogues. This kind of optimization will be discussed in Chap. 3 of this Part.

The discrete model of a structure is generated in order to perform its numerical analysis. It leads, as the result, to fundamental set of equations (1.1). In order to perform the optimization process, some design parameters are selected. Their values can be modified during the optimization process. Denoting the set of these parameters by vector $\mathbf{s}$, it can be assumed that stiffness matrix and in some cases the vector of nodal forces depend on these parameters, that is $\mathbf{K} = \mathbf{K}(\mathbf{s})$ and $\mathbf{Q} = \mathbf{Q}(\mathbf{s})$. Next, the objective function, depending on state variables and design parameters, has to be selected and mechanical and geometrical constraints have to be imposed. The optimal solution of optimization problem will be equivalent to finding the minimum of objective function at which the equilibrium equations as well as constraints imposed on state variables and design parameters are satisfied. Thus, to construct the optimization model, the following has to be defined:

1. design parameters,
2. objective (or cost) function,
3. behavior and design constraints.

The specification of design parameters depends on the nature of a particular problem. In bar- or beam-type elements, the design parameters can be considered as cross-sectional dimensions, material properties, length of elements, or configuration parameters, such as locations of nodes, element orientation and so on. The detailed characterization of structural design parameters is presented in Part VI of this book.

The objective function usually represents the volume or the weight of a structural material (minimum volume or weight design). These two possible objective functions are equivalent for homogeneous structures. However, they are different for composite materials or when different materials are used for structure elements. The design for minimum of weight is mostly used in aerospace industry, where the project of airplane or rocket with the least weight constitutes the fundamental objective of optimization. In other engineering constructions, the additional elements, such as cost of manufacturing, transport and assembling and so on should be taken into account. All these cost components depend in some way on selected design parameters and then the final cost of material and fabrication can be related to these parameters

$$V_m = \sum_{i=1}^{n} l_i A_i + \sum_{w=1}^{m} V_w \tag{1.3}$$

where l_i and A_i denote the lengths and cross-sections of elements while V_w denotes material volumes of nodes. The weight of structure can be expressed in the form

$$G_m = \sum_{i=1}^{n} \gamma_i l_i A_i + \sum_{w=1}^{m} G_w \tag{1.4}$$

where γ_i is a unit weight of particular elements and G_w $(w = 1, 2, \ldots, m)$ denote the weights of nodes. The cost of material C_m can be expressed similarly introducing the unit costs of elements and nodes. Generally, the cost of structure can be expressed as the sum of material cost and manufacturing cost

$$C = C_m + C_p = \sum_{i=1}^{n} c_i l_i A_i + \sum_{w=1}^{m} c_w V_w + C_p(l_i, A_i, l_w, A_w) \tag{1.5}$$

where the cost of manufacturing can be expressed in terms of design parameters of elements l_i, A_i and nodes l_w, A_w. Taking additionally into account the cost of transportation and assemblage we should note that these costs increase nonlinearly with the dimensions of elements. Thus, we can write

$$V_m = \sum_{i=1}^{n} l_i A_i (1 + l_i^{\alpha}) + \sum_{w=1}^{m} V_w \tag{1.6}$$

where the exponent α characterizes the non-linearity of cost increase. For $\alpha = 0$ we get the cost of material only. For greater elements lengths, the optimal solution will lead to their segmentation and introduction of assembling hinges, whose cost should be also added to the total structural cost. For a continuous variation of cross-sectional area of the element the manufacturing cost may increase, so the cost function can also be dependent on the cross-sectional area gradient $\mathrm{d}A/\mathrm{d}x$.

The introduced mechanical constraints follow from the requirements of stiffness and strength or reliability of a structure. In the elastic range, the strength constraint is usually expressed in the form of upper bound imposed on some equivalent maximal stress

$$\sigma_i^m \le \sigma_{0i} \tag{1.7}$$

where σ_i^m denotes the maximal stress in i-th element and σ_{0i} is an admissible limit stress in that element. The value of stress in that element depends on design parameters, $\sigma_i^m = \sigma_i^m\{l_s, A_s\}$, and results from the solution of boundary value problem (1.1). This dependence is implicit what is indicated by using the { } brackets. Similarly, the stiffness constraint can be expressed in the form

$$u_i^m \leq u_{0i} \tag{1.8}$$

where u_i^m denotes the maximal displacement component in i-th elements and u_{0i} is an upper bound. Here, once again u_i^m is implicit function of design parameters, $u_i^m = u_i^m\{l_s, A_s\}$. Depending on assumed stress and displacement upper bounds as well as on the span of structure, one of these constraints will be active and the second one will be satisfied in the form of inequality. With increasing structural length, the displacement constraint plays a more important role in optimal design. There exists also some range of structural dimensions in which both constraints are active and optimal structure has to satisfy simultaneously both constraints in the form of equalities. Instead of local constraints imposed on each particular element of a discrete model of a structure, there can be formulated some global stress and displacement constraints specified, for instance, in the form

$$u_p = \left\{ \frac{1}{n} \sum_{i=1}^{n} \frac{1}{l_i} \int_0^{l_i} |u_i|^p \, \mathrm{d}x \right\}^{\frac{1}{p}} \leq u_0 \tag{1.9}$$

where $p > 1$. For p tending to infinity, the measure u_p tends to the maximal local displacement, and then (1.9) can be treated as the mean value of displacement within structure domain. Similar measure can be introduced for stress components.

The above stress and displacement constraints are the most used mechanical constraints in the case of static and dynamic loading. Another class of constraints can be obtained for the free structure vibration described by the equation

$$\mathbf{Ku} - \omega^2 \mathbf{Mu} = 0 \tag{1.10}$$

where ω denotes the free vibration frequency and $\mathbf{M}$ is the mass matrix. During optimization process we are looking for optimal solution satisfying constraints imposed on eigen-frequencies in the form

$$\omega_1 \geq \omega_{01} \,, \quad \omega_2 \geq \omega_{02} \,. \tag{1.11}$$

Since the eigen-frequencies can be expressed by the Rayleigh quotients

$$\omega_1^2 = \frac{\mathbf{Ku}_1 \cdot \mathbf{u}_1}{\mathbf{Mu}_1 \cdot \mathbf{u}_1} \,, \qquad \omega_2^2 = \frac{\mathbf{Ku}_2 \cdot \mathbf{u}_2}{\mathbf{Mu}_2 \cdot \mathbf{u}_2} \,, \tag{1.12}$$

where $\mathbf{u}_1$, $\mathbf{u}_2$ denote the eigen-vectors, then the constraints (1.11) can be expressed in terms of design parameters, noting that $\mathbf{K} = \mathbf{K}(\mathbf{s})$ and $\mathbf{M} = \mathbf{M}(\mathbf{s})$.

Similar constraints can be formulated for designing structures undergoing buckling, by introducing the upper limit of critical load factor

$$\lambda_c < \lambda_{0c} \,. \tag{1.13}$$

A much more complex class of design problems is obtained when two levels of loading are assumed. Let us assume that λ_s denotes the service load factor. Assuming elastic response of structure under service load, the stresses cannot exceed the elastic limit and the displacements are limited by their upper bounds. Thus, the stress and displacement constraints previously discussed are generated. However, another constraint can be added in the form of a specified safety factor with respect to limit plastic state or failure due to cracking. Such constraint can be formulated by considering the structure at the limit plastic state when plastic failure mode occurs and the elastic deformations can be neglected. Denoting the load multiplier in the limit state by $\lambda_c > \lambda_s$, we can require

$$\lambda_c \geq \eta \lambda_s \tag{1.14}$$

where η is a prescribed safety factor with respect to service load. In this kind of optimization, two material models and two stress states have to be considered, namely the elastic and limit states.

Other types of constraints are the geometrical constraints on design variables or parameters, expressed in the form of inequalities. For truss structures, for instance, the lower and upper bounds can be imposed on bar cross-sections in the form

$$A_i^- \leq A_i \leq A_i^+ \,. \tag{1.15}$$

Generally, the geometrical constraints reflect the fabrication cost, technological as well as aesthetic requirements.

1.2 Optimal Design Problem Formulation

In presenting the methods of optimal design, the finite element model of structure will be considered and the cross-sectional sizes of elements will play the role of design parameters. The shape of structure, its topology and location of element nodes are assumed to be fixed. If, for instance, the structure is modeled by bar and disk elements, then the design parameters will be cross-sectional areas of bar elements and thicknesses of disk elements.

The objective function in structural design, is in general material and manufacturing cost. This cost is always closely related to structure weight and then the minimization of weight becomes the main objective of the optimization procedure. Denoting by a_i $(i = 1, \ldots, n)$ the design variables being the bar cross-sections or disk thicknesses, the objective function, or in other words structural cost, can be expressed as a linear form

$$W = \sum_{i=1}^{n} l_i a_i \tag{1.16}$$

where l_i denotes the weight coefficients of the i-th element, that is the product of material density and bar length or disk area.

The weight minimization problem is subjected to constraints imposed on displacements and stresses. The lack of such constraints will result in structure with vanishing bar cross-sections or disk thicknesses, and then the infinite displacements or stresses may occur. Thus, at some selected m points of structure the maximal displacements are limited. These displacement constraints have the form

$$u_j = \bar{u}_j\ , \quad j = 1, \ldots, m, \tag{1.17}$$

where $\bar{u}_j$ is an upper bound at j-th point of structure subjected to specified loading conditions. The strength of structural elements must also be considered in the design requirements in the form of stress limitation. In bar type elements, the elastic strains range of materials cannot be exceeded in the case of tension and the buckling safety factor has to be preserved in compression. In two-dimensional elements, constraint on stress state has to be introduced by using the equivalent stress expressed, for instance, by the von Misses criterion. Because the stress constraints can usually be considered for different loading cases, they can be written in the form

$$\sigma_{il} \leq \bar{\sigma}_i\ , \quad i = 1, \ldots, n; \quad l = 1, \ldots, c, \tag{1.18}$$

where σ_{il} denotes the equivalent stress in i-th element for l-th loading case and $\bar{\sigma}_i$ is an allowable stress level in this element. Besides the displacement and stress constrains, other types of behavior constraints have sometimes to be taken into account. To avoid, for instance, resonance phenomena, the lowest free-frequency of a structure has to lie within some prescribed interval. To prevent the loss of stability under applied load, a lower limit can be imposed on the critical buckling load multiplier. In view of limitation of this chapter, these constraints will not be considered. The main attention will be paid to displacement and stress constraints expressed by (1.17) and (1.18).

The optimal design problem for an arbitrary structure can now be formulated as follows:

Determine

$$\min W(a) = \sum_{i=1}^{n} l_i a_i \tag{1.19}$$

subject to behavioral constraints

$$h_j(\sigma, \mathbf{u}, a) \leq 0, \quad j = 1, \ldots, m \tag{1.20}$$

and geometrical constraints

$$a_{i\ \max} \geq a_i \geq a_{i\ \min}\ . \tag{1.21}$$

The behavioral constraints (1.20), in view of (1.17) and (1.18), can be written in the form

$$h_j(a) \equiv \begin{cases} u_j(a) - \bar{u}_j \leq 0, \\ \sigma_{il} - \bar{\sigma}_i \leq 0. \end{cases} \tag{1.22}$$

In addition, according to (1.21) the design variables are subjected to side constrains. The lower and upper bounds $a_{i\ \min}$ and $a_{i\ \max}$ reflect the fabrication and aesthetic requirements as well as the conditions of proper behavior of structure during its service.

The linear objective function (1.19) can be represented by a set of constant weight planes in an n-dimensional *design space*, $a_i \in \mathbf{E}^n$. The whole domain of design space is divided into two sub-domains: the first one in which the constraints (1.20) and (1.21) are violated, and the second one called the *feasible domain* in which at every point a_i ($i = 1, \ldots, n$) the constraints (1.20) and (1.21) are fulfilled. The solution of optimization problem results in finding such point within feasible domain, for which the objective function attains its minimum.

Thus, looking for the optimal point in the design space we have to determine a trajectory leading from a given starting point to the optimal point. Finding this trajectory is generally obtained by an iterative two-step algorithm. In the analysis step, the constraints are evaluated and new search direction is determined, whereas in synthesis step the objective function is reduced and new values are assigned to design variables. The applicability of such algorithms, particularly to large structural systems, is thus largely dependent on the number of reanalyses needed to attain a sufficient decrease of value of the objective function. In the last decade, the methods of *mathematical programming* and *optimality criteria* have been employed to solve this problem. The advantages of mathematical programming methods are their generality permitting consideration of any type of problem and convergence to optimal solution (if it exists). The essential disadvantage lies in the computation time, considerably increasing with the problem size. On the other hand, more intuitive optimality criteria methods reduce the number of required reanalyses, which do not grow considerably with increase of number of design parameters. However, the main drawback of this method is the lack of generality and often unpredictable loss of convergence of the iterative procedure. In the next sections of this chapter, we shall present these two methods based on mathematical programming and on optimality design criteria.

1.3 Mathematical Programming Methods

1.3.1 Problem Statement

The *mathematical programming* (MP) problem can be formulated as follows:

Find

$$\min f(x_i) \tag{1.23}$$

subject to

$$h_j(x_i) \geq 0, \qquad j = 1, \ldots, m, \tag{1.24}$$

$$x_{i\,\max} \geq x_i \geq x_{i\,\min}\,, \quad i = 1, \ldots, n. \tag{1.25}$$

The constraints (1.25) imposed on design variables x_i are distinguished from the general set (1.24) due to their simplicity and possibility of separate treatment in most mathematical programming algorithms. In particular, there is no need to incorporate them into Lagrange function from which the necessary optimality conditions can be derived. The Lagrange function associated with the problem (1.23)–(1.25) can be written in the form

$$L(x_i, \lambda_j) = f(x_i) - \sum_{j=1}^{m} \lambda_j h_j(x_i) \tag{1.26}$$

where all multipliers λ_j are non-negative. The necessary optimality conditions follow from stationarity of (1.26) and take the form of well-known Kuhn–Tucker conditions [9]

$$\frac{\partial f}{\partial x_i} - \sum_{j=1}^{m} \lambda_j \frac{\partial h_j}{\partial x_i} = 0 \qquad \text{when} \;\; x_{i\,\min} < x_i < x_{i\,\max}\,, \tag{1.27}$$

$$\frac{\partial f}{\partial x_i} - \sum_{j=1}^{m} \lambda_j \frac{\partial h_j}{\partial x_i} > 0 \qquad \text{when} \;\; x_i = x_{i\,\min}\,, \tag{1.28}$$

$$\frac{\partial f}{\partial x_i} - \sum_{j=1}^{m} \lambda_j \frac{\partial h_j}{\partial x_i} < 0 \qquad \text{when} \;\; x_i = x_{i\,\max}\,, \tag{1.29}$$

where $\lambda_j > 0$ for $h_j = 0$ and $\lambda_j = 0$ if $h_j > 0$.

When the objective function $f(x_i)$ and all constraints $h_j(x_i)$, $j = 1, \ldots, m$, are linear function of design variables x_i, then the problem (1.23), (1.25) is called a *linear programming problem* (LP). It is the simplest case of MP for which a lot of typical iterative algorithms have been proposed. One of the most frequently used is the simplex algorithm [9]. If any of the functions f or h_j is a nonlinear function of design variables, then the problem (1.23)–(1.25) is called a *nonlinear programming problem* (NP). In this case the solution methods are much less typical. The particular case of NP is an *unconstrained minimization problem* written in the form

Determine

$$\min f(x_i) \tag{1.30}$$

for

$$x_i \in \mathbf{E}^n$$

where $\mathbf{E}^n$ is the n-dimensional space of design variables. This problem is very important because it constitutes a foundation for all methods of nonlinear programming. Another particular case is a *linearly constrained minimization problem* in which constraints (1.24) are linear functions of design variables. Solution of this type of problems can be obtained by simple adapting of unconstrained minimization algorithms. Moreover, some general optimization algorithms use the technique of transformation of original problem into a series of linearly constrained sub-problems. This feature is widely used in modern minimal weight optimization problems.

Generally speaking, the problem of MP will be called a *convex programming problem* when the objective function is convex in design space and all constraints are concave. The feasible domain is the convex set and any local solution is also a global one. Additionally, the dual solutions, which are very important in structural optimization, are valid only for convex problems. The dual solutions are used effectively in *separable programming problems.* Generally, a separable programming problem is stated as follows:

Find

$$\min f(x_i) = \sum_{i=1}^{n} f_i(x_i) \tag{1.31}$$

subjected to

$$h_j(x_i) = \sum_{i=1}^{n} h_{ji}(x_i) \geq 0, \quad j = 1, \ldots, m, \tag{1.32}$$

$$x_{i\ \max} \geq x_i \geq x_{i\ \min}\,, \tag{1.33}$$

where each function $f_i(x_i)$ and $h_{ji}(x_i)$ depends only on the single variable x_i.

As mentioned previously, only the constraints (1.24) have to be associated with Lagrange multipliers, or in other words with dual variables x_j while the constraints (1.25) imposed directly on design variables can be considered separately. Let X denote the set of all points belonging to the design space which satisfy constraints (1.25), that is

$$X = \{\mathbf{x} : x_{i\ \min} \leq x_i \leq x_{i\ \max}\,;\ i = 1, \ldots, n\} \tag{1.34}$$

and let Λ be a set of all dual points satisfying the condition of non-negativity, that is

$$\Lambda = \{\lambda : \lambda_j \geq 0;\ j = 1, \ldots, m\}\,. \tag{1.35}$$

The *primary nonlinear programming problem* will be stated as follows

Find

$$\min f(\mathbf{x}) \quad \text{for } \mathbf{x} \in X \tag{1.36}$$

subjected to

$$h_j(x) \geq 0, \quad j = 1, \ldots, m.$$

The function $f(\mathbf{x})$ will be called the *primary function.* If this function is strictly convex and all functions $h_j(\mathbf{x})$ are concave, then there exists a unique dual nonlinear programming problem given by

$$\max_{\lambda\in\Lambda}\min_{\mathbf{x}\in X} L(\mathbf{x},\lambda) \tag{1.37}$$

where $L(\mathbf{x},\lambda)$ is a Lagrange function defined by (1.26). The dual problem can be formulated in the form

Find

$$\max l(\lambda) \quad \text{for } \lambda\in\Lambda \tag{1.38}$$

subjected to

$$\lambda_j \geq 0, \quad j=1,\ldots,m$$

where

$$l(\lambda) = \min_{\mathbf{x}\in X} L(\mathbf{x},\lambda) \tag{1.39}$$

is defined as the *dual function* to primary function $f(\mathbf{x})$.

Let us note that the primary problem (1.36) involves n design variables, m general constraints and $2n$ side constraints imposed directly on design variables. On the other hand the dual problem (1.38) depends on m dual variables and involves only m constraints imposed on these variables. Since the dual problem does not involve the general constraints, then it can be solved using slightly modified algorithms for unconstrained minimization. Applying such algorithms requires the knowledge of the gradient of dual function. It can easily be derived, noting that it is defined by constraints of primary problem, namely

$$\frac{\partial l}{\partial\lambda_j} = -h_j[\mathbf{x}(\lambda)] \tag{1.40}$$

where $\mathbf{x}(\lambda)$ denotes the primary space point that minimizes function $L(\mathbf{x},\lambda)$ on the set X for given λ. Thus the dual function (1.39) can also be written as

$$l(\lambda) = f[\mathbf{x}(\lambda)] - \sum_{j=1}^{m} h_j[\mathbf{x}(\lambda)]\,. \tag{1.41}$$

Applying any numerical method for solution of the dual problem in order to calculate the value of the dual function, the values of primary constraints h_j have to be known. Due to this, the gradient (1.40) of dual function can be calculated without any additional computations. To find the value of $l(\lambda)$ it is necessary to find the point $\mathbf{x}$ minimizing Lagrange function. This requirement follows formally from relation (1.39). For certain class of problems this in an easy task. Considering the separable programming problem (1.31)–(1.33), the dual function takes the form

$$l(\lambda) = \sum_{i=1}^{n}\left\{x_{i\,\min} \leq x_i^{\min} \leq x_{i\,\max} : \left[f_i(x_i) - \sum_{j=1}^{m}\lambda_j h_{ji}(x_i)\right]\right\} \tag{1.42}$$

and the search for $x_i^{\min}$ is reduced to one-dimensional minimization with respect to each of n design variables. This one dimensional minimization is so simple in some cases that the solution with respect to x_i can be obtained in a closed form. When this case occurs then the dual function (1.42) can be expressed explicitly in terms of x_i .

1.3.2 Unconstrained Minimization

Most algorithms for solving a minimization problem are iterative. They require first of all to estimate an initial starting point $x^{(0)}$ in the design space and next a series of solutions $x^{(k)}$ constructed so that a point $x^{(k+1)}$ is a better estimation of optimal solution when compared to point $x^{(k)}$. Thus, it has to be satisfied

$$f(\mathbf{x}^{(k+1)}) < (\mathbf{x}^{(k)}) \tag{1.43}$$

The usually applied sequential methods of minimizing of $f(\mathbf{x})$ are based on search methods in the design space along some directions which are called *search directions.* A new estimate of optimal solution is obtained according to the rule

$$\mathbf{x}^{(k+1)} = \mathbf{x}^{(k)} + \alpha^{(k)}\mathbf{s}^{(k)}(\mathbf{x}^{(k)}), \tag{1.44}$$

where $\mathbf{s}^{(k)}$ denotes the search direction and $\alpha^{(k)}$ is step measure in this direction. Thus, for sufficiently small $a > 0$ there has to be

$$f(\mathbf{x}^{(k)} + \alpha^{(k)}) > f(\mathbf{x}^{(k)}) \tag{1.45}$$

Assuming differentiability of objective function, the condition (1.45) is equivalent to

$$\mathbf{s}^{(k)T}\mathbf{g}^{(k)} < 0, \tag{1.46}$$

where $\mathbf{g}^{(k)} = \nabla f(\mathbf{x}^{(k)})$ denotes the gradient of $f(\mathbf{x})$ at point $\mathbf{x}^{(k)}$.

A basic *descent algorithm* for unconstrained minimization relying on the above idea is shown in Fig. 1.1. Each of iteration steps of these algorithms involves two stages. Firstly, the search direction $\mathbf{s}^{(k)}$ is calculated at point $\mathbf{x}^{(k)}$ and secondly a step length $\alpha^{(k)}$ is evaluated, from which a new point $\mathbf{x}^{(k+1)}$ is calculated by using (1.44). The step length $\alpha^{(k)}$ can be arbitrary assumed so that condition (1.45) is satisfied or can be obtained as the result of one-dimensional minimization along search direction $\mathbf{s}^{(k)}$. Thus, $\alpha^{(k)}$ has to satisfy the following requirement

$$f(\mathbf{x}^{(k+1)}) = \min_{a} \Phi(\alpha) \tag{1.47}$$

where $\Phi(\alpha)$ denotes the objective function $f(\mathbf{x})$ along direction $\mathbf{x}^{(k)}$ treated as one parameter function of α only. Thus

$$\Phi(\alpha) = f(\mathbf{x}^{(k)} + \alpha\mathbf{s}^{(k)}) \tag{1.48}$$

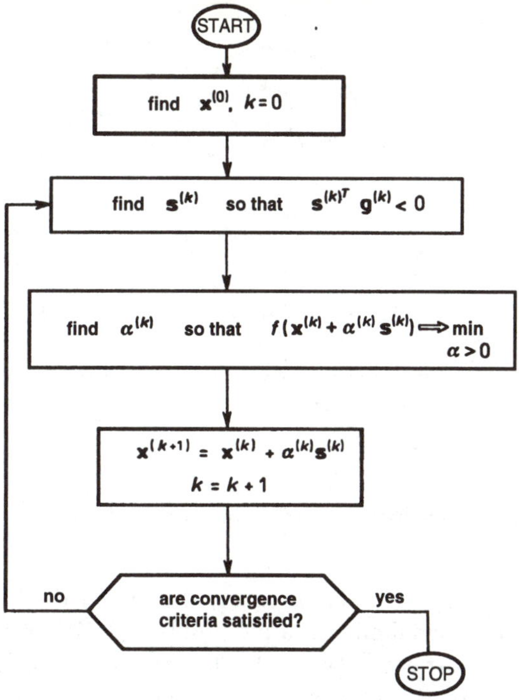

Fig. 1.1. Flow-chart of minimization along search direction

and in each iteration step of descent algorithm the following equation has to be solved

$$\Phi'(\alpha) = \frac{\mathrm{d}\Phi}{\mathrm{d}\alpha} = 0. \tag{1.49}$$

In view of (1.48), $\Phi'(\alpha)$ is expressed by

$$\Phi'(\alpha) = \sum_{i=1}^{n} \frac{\partial f}{\partial x_i}(\mathbf{x}^{(k)} + \alpha \mathbf{s}^{(k)}) \frac{\partial}{\partial \alpha}(x_i^{(k)} + \alpha s_i^{(k)}) = g(\mathbf{x}^{(k)} + \alpha \mathbf{s}^{(k)})^T \mathbf{s}^{(k)}. \tag{1.50}$$

Note, moreover, that the condition (1.44) implies the orthogonality of gradient $\mathbf{g}^{(k+1)}$ of function $f(\mathbf{x})$ of point $\mathbf{x}^{(k+1)}$ and search direction $\mathbf{s}^{(k+1)}$, namely

$$\mathbf{g}^{(k+1)} \mathbf{s}^{(k)} = 0. \tag{1.51}$$

This property is shown in Fig. 1.2. Let us note that in view of (1.51), the search direction $\mathbf{s}^{(k)}$ is tangent to the contour of function $f(\mathbf{x}) = f(\mathbf{x}^{(k+1)})$. Moreover, there exist

$$\Phi'(\alpha) < 0 \quad \text{for } \alpha < \alpha^{(k)}; \qquad \Phi'(\alpha) > 0, \quad \text{for } \alpha > \alpha^{(k)}. \tag{1.52}$$

The equalities (1.52) can be useful for estimation of the optimal step length $\alpha^{(k)}$.

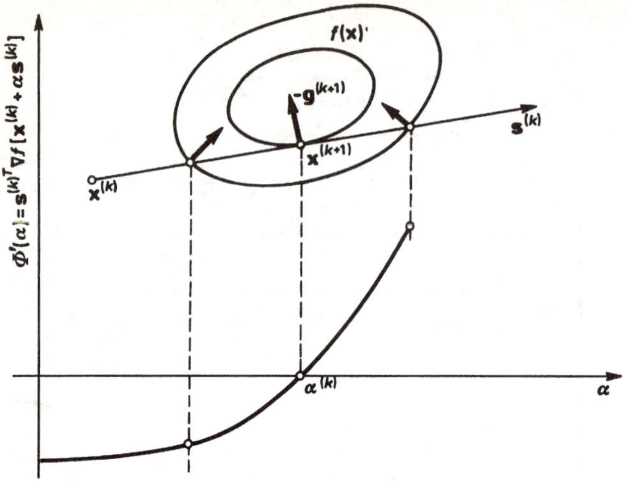

Fig. 1.2. Gradient orthogonality to search direction

One of the most important steps of any descent method constitutes the solution of one-dimensional minimization problem (1.47), or equivalently the solution of the nonlinear equation (1.49) The procedure of finding the proper step length α^* as the result of directional minimization of $\Phi(\alpha)$ is called *line search.* Most often this procedure is performed iteratively and is terminated when some assumed convergence criteria are satisfied. To find a directional minimum, a separate calculation of function value and its gradient must be performed. Thus, the effectiveness of minimum calculation affects essentially the effectiveness of the superior optimization algorithm.

Many line search techniques are based on approximation of the considered function by using the polynomial of proper degree. Depending on whether only the function values or function and its derivatives are known, one or several points along search direction will be used to determine the approximation polynomial. Applying, for instance, the Newton–Raphson method for solving the equation (1.49), only one point is used at which first and second derivatives $\Phi'(\alpha)$ and $\Phi''(\alpha)$ have to be calculated in each iteration step. However, only the information on the first derivative is available in most practical problems. In this case, two points have to be used to determine the approximation polynomial. Assuming the initial interval $[\alpha_1, \alpha_2]$ containing the minimum α^* is known, the gradual reduction of this interval leads to estimation of this minimum with satisfactory accuracy. The reduction of previous interval can be based on new estimation of minimum, according to the rule

$$\alpha_3 = \alpha_2 - \beta(\alpha_2 - \alpha_1) \quad \text{for } 0 \le \beta \le 1. \tag{1.53}$$

It is obvious from (1.53) that $\alpha_1 \le \alpha_3 \le \alpha_2$ and we can narrow the interval of the next iteration step according to sign of $\Phi'(\alpha_3)$. The value of coeffi-

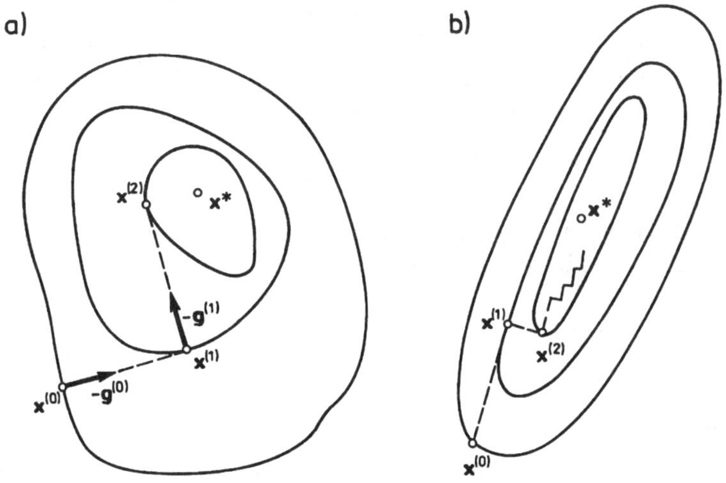

Fig. 1.3. Steepest descent method

cient β appearing in (1.53) depends on the assumed interpolation formula being used. Assuming $\beta = \frac{1}{2}$ we narrow the interval by half in each iteration step (bisection iteration). Much more effective approximation can be obtained exploring the information on values of $\Phi(\alpha)$ and its derivative $\Phi'(\alpha)$ in order to construct a polynomial of order two or three in $\Phi(\alpha)$ and next to find analytically the minimum of that polynomial. When the information on the first derivative is not available, one can use a three point pattern for approximation of $\Phi(\alpha)$ by means of quadratic polynomial, basing only on function values at these points.

There is no doubt that among the descent algorithms the steepest descent method is the fundamental technique for unconstrained optimization. Here the negative gradient vector of minimized function is taken as search direction. Although this method itself is not a very efficient, it constitutes a basis for all minimization methods using gradient information. Moreover, the convergence conditions for this method can be established and next used as reference conditions for other minimization algorithms. According to the assumed search direction $\mathbf{s}^{(k)} = -\mathbf{g}^{(k)} = -\nabla f(\mathbf{x}^{(k)})$, the condition (1.46) is always satisfied, if only $\mathbf{g}^{(k)} \neq 0$. Thus , starting at the point $\mathbf{g}^{(k)}$, we search along direction of negative gradient for a new point $\mathbf{g}^{(k+1)}$ in which the function attains its directional minimum. It is obvious from (1.51) that the successive search directions are orthogonal each to other. The illustration of how this method works is given in Fig. 1.3. It can be seen from Fig. 1.3a that the method performs very well on an objective function with nearly circular contours. However, when the contours of constant function values are very narrow and elongated (Fig. 1.3b) then convergence of the method can be very slow. It is due mainly to the errors in directional minimization which yield the errors in estimation of consecutive search direction.

The other approach to unconstrained minimization is the conjugate gradient method. Let the function $f(\mathbf{x})$ be expressed in a quadratic form, that is the minimization problem is formulated as follows

$$\text{Find} \qquad \min f(\mathbf{x}) = \frac{1}{2}\mathbf{x}^T\mathbf{A}\,\mathbf{x} - \mathbf{b}^T\mathbf{x} \tag{1.54}$$

where the matrix $\mathbf{A}$ is symmetric and positive definite. The solution of problem (1.54) can be written in the form

$$\mathbf{x}^* = \mathbf{A}^{-1}\mathbf{b}. \tag{1.55}$$

The search directions $\mathbf{s}^{(i)}$, $i = 1, \ldots, n$ will be called *mutually conjugate* with respect to the positive definite matrix $\mathbf{A}$ (or $\mathbf{A}$-orthogonal) when the following relations hold

$$\mathbf{s}^{(i)T}\mathbf{A}\mathbf{s}^{(j)} = 0 \quad \text{for } i \neq j. \tag{1.56}$$

Let us note that the conjugate directions defined by (1.56) are linearly independent. It can easily be shown that an arbitrary vector $\mathbf{v}$ can be expressed in the form

$$\mathbf{v} = \sum_{i=1}^{n} \frac{\mathbf{s}^{(i)T}\mathbf{A}\,\mathbf{v}}{\mathbf{s}^{(i)T}\mathbf{A}\mathbf{s}^{(i)}}\,\mathbf{s}^{(i)} \tag{1.57}$$

Using (1.56) and (1.57), the solution (1.55) of the problem (1.54) can be written as

$$\mathbf{x}^* = \sum_{i=1}^{n} \gamma_i \mathbf{s}^{(i)} = \sum_{i=1}^{n} \frac{\mathbf{s}^{(i)T}\mathbf{b}}{\mathbf{s}^{(i)T}\mathbf{A}\mathbf{s}^{(i)}}\,\mathbf{s}^{(i)} \tag{1.58}$$

This indicates that the minimization problem (1.54) can be solved in n iteration steps in which successive coefficients γ_i are derived without knowledge of solution $\mathbf{x}^*$. This result constitutes the foundation of conjugate directions methods by which the minimum of any quadratic function can be found at most in n iterations, starting from an arbitrary initial estimation of point $\mathbf{x}^*$. We say that such class of iterative procedures is characterized by the *second-order convergence.* Performing directional minimization along consecutive mutually conjugate directions the quadratic function (1.54) is minimized in the domain bounded by these directions. It follows from the fact that gradient vector of $f(\mathbf{x})$ at point $\mathbf{x}^{(k+1)}$ is orthogonal to all previous descent directions $\mathbf{s}^{(i)}$, namely

$$\mathbf{g}^{(k+1)T}\mathbf{s}^{(i)} = 0. \quad i = 1, \ldots, k. \tag{1.59}$$

The conjugate gradient algorithm is a particular case of conjugate direction methods, in which the set of conjugate directions is generated from $\mathbf{A}$-orthogonalization of successive gradient vectors. The steepest descent vector is selected as the first search direction and then $\mathbf{s}^{(k+1)}$ is determined as the linear combination of $-\mathbf{g}^{(k+1)}$ and $\mathbf{s}^{(k)}$. There exist many implementations of conjugate gradient method, depending on evaluation of conjugate directions. One of the most representative is the Fletcher–Reeves method [9]. Its algorithm is as follows:

Step 1: Select an arbitrary starting point $\mathbf{x}^{(0)}$;
Calculate $f(\mathbf{x}^{(0)})$, $\mathbf{g}^{(0)} = f(\mathbf{x}^{(0)})$;
Assume $\mathbf{s}^{(0)} = -\mathbf{g}^{(0)}$, $k = 0$;
Step 2: Perform line search in direction $\mathbf{s}^{(k)}$ and find optimal $\alpha^{(k)}$;
$\mathbf{x}^{(k+1)} = \mathbf{x}^{(k)} + \alpha^{(k)}\mathbf{s}^{(k)}$
Step 3: $\mathbf{g}^{(k+1)} = f(\mathbf{x}^{(k+1)})$;
$\dfrac{\mathbf{g}^{(k+1)T}\mathbf{g}^{(k+1)}}{\mathbf{g}^{(k)T}\mathbf{g}^{(k)}}$;
Step 4: Find conjugate step direction $\mathbf{s}^{(k+1)} = -\mathbf{g}^{(k+1)} + \beta^{(k)}\mathbf{s}^{(k)}$
Step 5: Set $k = k + 1$ and go to step two.

It is important to remember that the successive search directions are mutually conjugate only if directional minimization in step two is exact. Moreover, since the vectors $[\mathbf{s}^{(1)}, \ldots, \mathbf{s}^{(i)}]$ and $[\mathbf{g}^{(1)}, \ldots, \mathbf{g}^{(i)}]$ bound the same sub-domain of design space, then Eq. (1.59) is equivalent to

$$\mathbf{g}^{(k+1)} = -\mathbf{g}^{(k+1)} + \beta^{(k)}\mathbf{s}^{(k)} \tag{1.60}$$

It means that the gradient of objective function at point $\mathbf{x}^{(k+1)}$ is orthogonal to all previously calculated gradients. Note that in this algorithm only the function value and its gradient are evaluated at each iteration step. Thus, this algorithm can be applied for an arbitrary objective function and not only to quadratic function (1.54). If an initial starting point is selected close enough to the optimal point $\mathbf{x}^*$, then arbitrary function $f(\mathbf{x})$ can be approximated in the neighborhood of this point by quadratic form (1.54) with satisfactory accuracy and search directions generated by Fletcher–Reeves algorithm are nearly conjugate. It can thus be expected that the point $\mathbf{x}^{(n)}$ generated in n steps of this algorithm is closer to an optimum $\mathbf{x}^*$ than the initial point $\mathbf{x}^{(0)}$, so that the quadratic approximation (1.54) of $f(\mathbf{x})$ will be better at $\mathbf{x}^{(n)}$ than at $\mathbf{x}^{(0)}$. This suggests a strategy of restarting the algorithm after each n-th step with the initial search direction $\mathbf{s}^{(0)} = -\mathbf{g}^{(0)}$.

Another possible modification is to introduce a convergence test to stop the algorithm. Similarly, if directional minimization is not accurate enough, it can happen that $\mathbf{g}^{(k+1)T}\mathbf{s}^{(k)} < 0$ and the algorithm should be stopped in this step and restarted.

Generally, the fundamental properties of conjugate gradient methods indicate their greater effectiveness than that of the steepest descent method. It is due to the fact that these methods take into account an additional information related to the curvature of objective function. These methods are particularly useful when the number of dėsign variables is large, due to requirement of relatively small amount of computer storage, in opposition to Newton methods.

The Newton method is much more efficient in comparison with gradient methods due to its convergence of the second order. Applying this method, we put $\alpha^{(k)} = 1$ in the general iteration formula (1.44) and assume the search direction $\mathbf{s}^{(k)}$ in the form

$$\mathbf{s}^{(k)} = -[\mathbf{H}^{(k)}]^{-1}\mathbf{g}^{(k)}, \tag{1.61}$$

where $\mathbf{H}$ denotes the Hessian of objective function, that is the matrix of its second derivatives with respect to design variables. When applied to the quadratic function (1.54), Newton method generates its minimum in one step. For an arbitrary nonlinear function $f(\mathbf{x})$, the minimum is evaluated during an iterative procedure, according to the rule

$$\mathbf{x}^{(k+1)} = \mathbf{x}^{(k)} - [\mathbf{H}^{(k)}]^{-1}\mathbf{g}^{(k)}, \tag{1.62}$$

where $\mathbf{x}^{(k+1)}$ is a minimal point of the local quadratic approximation of f(x) at point $\mathbf{x}^{(k)}$ following from the expansion in a second order Taylor series in the vicinity of this point, thus

$$\tilde{f}(\mathbf{x}) = f(\mathbf{x}^{(k)}) + (\mathbf{x} - \mathbf{x}^{(k)})^T \nabla f(\mathbf{x}^{(k)}) + \frac{1}{2}(\mathbf{x} - \mathbf{x}^{(k)})^T \nabla^2 f(\mathbf{x}^{(k)})(\mathbf{x} - \mathbf{x}^{(k)}) \tag{1.63}$$

Assuming that the Hessian of objective function at point $\mathbf{x}^{(k)}$ is positive definite, the search direction calculated in Newton method is always the downhill direction, since $\mathbf{s}^{(k)T}\mathbf{g}^{(k)} = -\mathbf{s}^{(k)T}\mathbf{H}^{(k)}\mathbf{s}^{(k)} < 0$. However, since the unit step length is used, $\alpha^{(k)} = 1$, the condition of decreasing function value in successive iteration may not be always fulfilled. This results in some complications of Newton algorithm consisting in adding directional minimization of $f(\mathbf{x})$ to formula (1.62). This approach gives the optimal step-length $\alpha^{(k)}$ in each iteration step. If the Hessian $\mathbf{H}^{(k)}$ remains always positive-definite in the vicinity of $\mathbf{x}^{(k)}$ then satisfaction of (1.45) can be assured by selection of a sufficiently small step-length α in direction $\mathbf{s}^{(k)}$. On the other hand, when the Hessian is not necessarily positive definite, as in the case of non-convex functions, the classical Newton method fails. In this case, the steepest descent iteration can replace the Newton iteration, or if Newton search direction is uphill, we can move simply in the opposite direction. There exist also more sophisticated procedures modifying Hessian in such a way that the resulting modified matrix becomes positive definite. One example of such modification can be written in the form

$$\tilde{\mathbf{H}}^{(k)} = \mathbf{H}^{(k)} + \varepsilon^{(k)}\mathbf{I}, \quad \varepsilon > 0, \tag{1.64}$$

where $\mathbf{I}$ denotes the unit matrix and $\tilde{\mathbf{H}}^{(k)}$ is a modified positive definite matrix to be used as the replacement of $\mathbf{H}$ in formula (1.61) at the k-th iteration step.

The main disadvantage of Newton method is the necessity of direct calculation of the second derivatives of objective function in each iteration step, and the solution of associated linear equations (1.62). To avoid this one can replace the Hessian matrix, or more precisely the inverse of hessian by approximate matrix which is calculated from available information on the first derivatives only. This leads to the so called *quasi-Newton methods.* The idea results from expansion of the gradient of objective function in Taylor series in the vicinity of $\mathbf{x}^{(k+1)}$, which can be written in the form

$$\mathbf{g}^{(k)} - \mathbf{g}^{(k+1)} = \mathbf{H}^{(k+1)}(\mathbf{x}^{(k)} - \mathbf{x}^{(k+1)}) + \Delta, \tag{1.65}$$

where $\Delta \to 0$ for $\mathbf{x}^{(k)} \to \mathbf{s}^{(k+1)}$. Denoting by

$$\mathbf{y}^{(k)} = \mathbf{g}^{(k+1)} - \mathbf{g}^{(k)}, \qquad \mathbf{s}^{(k)} = \mathbf{x}^{(k+1)} - \mathbf{x}^{(k)} \tag{1.66}$$

and neglecting the second order term Δ, we get $\mathbf{y}^{(k)} = \mathbf{H}^{(k+1)}\mathbf{s}^{(k)}$. Assuming then $\mathbf{S}^{(k+1)}$ as an approximation of $[\mathbf{H}^{(k+1)}]^{-1}$, it can be constructed in such a way that it satisfies quasi-Newton equation resulting from expansion (1.65)

$$\mathbf{S}^{(k+1)}\mathbf{y}^{(k)} = \mathbf{s}^{(k)}, \quad k \geq 0. \tag{1.67}$$

The matrix $\mathbf{S}^{(k+1)}$ has to be computed from $\mathbf{S}^{(k)}$ by adding some correction term $\mathbf{C}^{(k)}$ depending on $\mathbf{S}^{(k)}$, $\mathbf{y}^{(k)}$ and $B^{(k)}$ only, Thus we have

$$\mathbf{S}^{(k+1)} = \mathbf{S}^{(k)} + \mathbf{C}^{(k)}. \tag{1.68}$$

The correction term $\mathbf{C}^{(k)}$ should preserve the symmetry and positive definiteness of matrix $\mathbf{S}$ in order to obtain the downhill search direction at point $\mathbf{x}^{(k)}$, which is defined as

$$\mathbf{s}^{(k)} = -\mathbf{S}^{(k)}\mathbf{g}^{(k)}. \tag{1.69}$$

The different variants of quasi-Newton approach have been constructed and they differ only in methods of calculating the correction term in Eq. (1.68). One of the most frequently approaches is the Davidon–Fletcher–Powell method, or in other words the variable metric method. It consists of calculating the conjugate directions in such a way that the matrix $\mathbf{S}^{(k)}$ in each consecutive iteration is a better approximation of matrix $\mathbf{H}^{-1}$. When the matrix $\mathbf{S}^{(k)}$ is equivalent to $\mathbf{H}^{-1}$, the procedure stops and the actual point $\mathbf{x}^{(k)}$ can be treated as the minimal point $\mathbf{x}^*$. The updating formula (1.68), which modifies the matrix $\mathbf{S}^{(k)}$ takes here the form

$$\mathbf{S}^{(k+1)} = \mathbf{S}^{(k)} + \frac{\mathbf{s}^{(k)}\mathbf{s}^{(k)T}}{\mathbf{s}^{(k)T}\mathbf{y}^{(k)}} - \frac{\mathbf{S}^{(k)}\mathbf{y}^{(k)}\mathbf{y}^{(k)T}\mathbf{S}^{(k)}}{\mathbf{y}^{(k)T}\mathbf{S}^{(k)}\mathbf{y}^{(k)}}\,. \tag{1.70}$$

It can be proved that for positive definite matrix $\mathbf{S}^{(k)}$, also $\mathbf{S}^{(k+1)}$ is positive definite when an exact line search is performed. Using this method to quadratic function, one can show the second order convergence property of the method and the fact that the search directions are conjugate. Starting from the unit matrix as an initial approximation $\mathbf{S}^{(0)}$, the Davidon–Fletcher–Powell method becomes the conjugate gradient method. Another interesting variant of quasi-Newton approach is Wolf–Broyden–Davidon method, in which modification of matrix $\mathbf{S}^{(k)}$ follows according to the rule

$$\mathbf{S}^{(k+1)} = \mathbf{S}^{(k)} + \frac{\left[\mathbf{s}^{(k)} - \mathbf{S}^{(k)}\mathbf{y}^{(k)}\right]\left[\mathbf{s}^{(k)} - \mathbf{S}^{(k)}\mathbf{y}^{(k)}\right]^T}{\mathbf{y}^{(k)T}\left[\mathbf{s}^{(k)} - \mathbf{S}^{(k)}\mathbf{y}^{(k)}\right]} . \tag{1.71}$$

Similarly as previously, the matrix $\mathbf{S}^{(k)}$ is transformed into $\mathbf{H}^{-1}$ after n steps. The advantage of this method is also no need for directional minimization to generate exactly the next point $\mathbf{x}^{(k+1)}$. This feature is particularly attractive when applied to an arbitrary objective function $f(\mathbf{x})$. We must take into account, however, the loss of positive definiteness of the matrix $\mathbf{S}^{(k)}$ during the iteration process, which may cause a 'breakdown' of the algorithm.

Comparing the class of the quasi-Newton methods with conjugate gradient methods, it should be stressed that the quasi-Newton methods exhibit a stronger stability of convergence but require larger computer storage ($n^2/2$ additional words).

Comparing all discussed methods of unconstrained optimization, one can observe that the variable metric methods are the most effective and among them particularly the Wolf–Broyden–Davidon and Fletcher–Powell–Davidon methods. For some cases the Newton method converges faster to optimal solution but its use is limited. Firstly, with increasing number of design variables the computational effort increases very fast due to necessity of calculating $n(n+1)/2$ second-order derivatives of objective function and reversing the Hessian matrix. Secondly, this method is very sensitive for proper selection of the initial starting point. The bad selection can result in the lack of convergence. The conjugate gradient method, despite the similar second order convergence property, is not as good as the Newton or quasi-Newton methods. Its advantage lies in using smaller computer storage as compared to the methods of variable metric. It is the result of a simpler algorithm of calculating search directions. Thus, it is often more useful to apply the conjugate gradient method of Wolf–Broyden–Davidon, even though the calculation speed is decreasing.

1.3.3 Linearly Constrained Minimization Problem

The problem considered in this section is formulated as follows

Find

$$\min f(\mathbf{x}) \tag{1.72}$$

subject to

$$\sum_{i=1}^{n} c_{ij} x_i \geq b_j , \qquad j = 1, \ldots, m;$$
$$x_{i\,\max} \geq x_i \geq x_{i\,\min} , \quad i = 1, \ldots, n.$$

The effective methods of solution of the problem (1.72) are the so called *projection methods.* They consist of finding the minimum of $f(\mathbf{x})$ by moving along downhill directions lying on polyhedron boundaries of the feasible domain. Therefore each point generated during minimization process is feasible and the value of the objective function decreases in each iteration step. Simultaneously, in each iteration step, an estimation of the Lagrange multiplier associated with active constraints is also generated.

Let a feasible point $\mathbf{x}^{(k)}$ be given with q constraints active, that is $\mathbf{c}_j^T\mathbf{x}^{(k)} - \mathbf{b}_j = 0$, $j = 1, \ldots, q$. The remaining constraints are passive. Let, moreover, $\mathbf{N}_q$ denote the matrix composed of the active constraints gradients

$$\mathbf{N}_q = [\mathbf{c}_1, \ldots, \mathbf{c}_q]. \tag{1.73}$$

We assume here that the side constraints imposed directly on design variables will be treated similarly to regular linear constraints, Thus, the columns of matrix $\mathbf{N}_q$ can contain vectors $\mathbf{c}_j$, being the base vectors of design space. Assuming the regularity of active constraints, then $\mathbf{N}_q$ is the rectangular matrix of dimensions $n \times q$ and rank $q < n$. Any k-th iteration of the solution procedure must obviously lead from the point $\mathbf{x}^{(k)}$ to another feasible point $\mathbf{x}^{(k+1)}$ according to the general formula (1.44). In what follows, let us omit, for the sake of simplicity, the iteration index k and denote by upper script $*$ a new point at iteration $(k+1)$. Thus, the formula (1.44) can be written in the form

$$\mathbf{x}^* = \mathbf{x} + \alpha\mathbf{s}, \tag{1.74}$$

where α is a step-length in search direction $\mathbf{s}$. When the vector $\mathbf{s}$ denotes the downhill direction, then we have $\mathbf{s}^T\mathbf{g} < 0$, where $\mathbf{g} = \nabla f(\mathbf{x})$ is as usually the gradient of objective function. In addition, let the vector $\mathbf{s}$ lie at the intersection of hyper-planes of active constraints, so that

$$\mathbf{N}_q^T\mathbf{s} = 0. \tag{1.75}$$

Due to this, all currently active constraints remain still active at point $\mathbf{x}^*$. Thus, the projection of negative gradient on constraints intersection can be selected as search direction, what can be written as

$$\mathbf{s} = -\mathbf{P}_q\mathbf{g}\,, \tag{1.76}$$

where $\mathbf{P}_q$ is a projection matrix of gradient $\mathbf{g}$. To find the form of matrix $\mathbf{P}_q$, let us note that any vector $\mathbf{v}$ can be represented as a difference between the projection of this vector on direction of constraints intersection $\mathbf{P}_q\mathbf{v}$ and the vector $\mathbf{N}_q\lambda$ orthogonal to this direction, namely

$$\mathbf{v} = \mathbf{P}_q\mathbf{v} - \mathbf{N}_q\lambda\,, \tag{1.77}$$

where $\lambda \in E^q$. Replacing $\mathbf{v}$ with negative gradient and using (1.75), we get

$$\mathbf{N}_q^T[\mathbf{P}_q\mathbf{g}] = \mathbf{N}_q^T(\mathbf{g} - \mathbf{N}_q\lambda) = 0. \tag{1.78}$$

Since the matrix $\mathbf{P}_q$ is of rank q, then Eq. (1.78) can be solved with respect to λ, yielding

$$\lambda = (\mathbf{N}_q^T \mathbf{N}_q)^{-1} \mathbf{N}_q^T \mathbf{g} . \tag{1.79}$$

Thus, Eq. (1.76) can be written in the form

$$\mathbf{s} = -\mathbf{P}_q \mathbf{g} = -\mathbf{g} + \mathbf{N}_q \lambda \tag{1.80}$$

and then the projection matrix $\mathbf{P}_q$ is expressed as

$$\mathbf{P}_q = \mathbf{I} - \mathbf{N}_q (\mathbf{N}_q^T \mathbf{N}_q)^{-1} \mathbf{N}_q^T . \tag{1.81}$$

The search direction $\mathbf{s}$ given by (1.80) will be called the *projected gradient.* It is easy to show that this direction satisfies condition (1.75). Also the condition (1.46) is satisfied since $\mathbf{s}^T \mathbf{g} = -\mathbf{s}^T \mathbf{s} < 0$, if $\mathbf{s} \neq 0$. For $\mathbf{s} = 0$, it follows from (1.80) that

$$\mathbf{g} - \mathbf{N}_q \lambda = 0 . \tag{1.82}$$

Since the matrix $\mathbf{N}_q$ is composed of gradients of active constraints, then Eq. (1.82) yields that Kuhn–Tucker conditions are satisfied for non-negative components of vector λ. Thus, the process of linearly constrained minimization can be terminated. Assume now that $\mathbf{s} = 0$ and at least one of the components of vector λ, say λ_r, is negative. In this case, we can find new feasible direction as the result of dropping out the constraint associated with the vector λ_r from the active set of constraints. It means that we assume $\mathbf{c}_r^T \mathbf{x} - \mathbf{b}_r > 0$ and project the gradient of objective function on the direction of intersection of remaining $q-1$ active constraints

$$\tilde{\mathbf{s}} = -\mathbf{P}_{q-1} \mathbf{g} \tag{1.83}$$

where $\mathbf{P}_{q-1}$ is a new projection matrix calculated from (1.81) with $\mathbf{N}_q$ replaced by $\mathbf{N}_{q-1}$. The matrix $\mathbf{N}_{q-1}$ follows form $\mathbf{N}_q$ by deleting the column $\mathbf{c}_r$. The new search direction (1.83) is then the feasible direction indicating the downhill direction of the objective function. It can easily be verified if we notice that just dropped constraint cannot be violated due to

$$\mathbf{c}_r^T \mathbf{s} = -\frac{1}{\lambda_r} \tilde{\mathbf{s}}^T \tilde{\mathbf{s}} > 0 . \tag{1.84}$$

The procedure of removing the constraint at the point which satisfies (1.82) is shown in Fig. 1.4a. It is obvious that the vector λ satisfying (1.82) can be identified with the vector of Lagrange multipliers and it can be treated as first-order approximation of these multipliers. Let us consider the calculation of step-length α in direction $\mathbf{s}$ appearing in Eq. (1.74). It can lead to addition of a new constraint to a set of actual active constraints. Iteration (1.74) has to be performed in such a way that the objective function is minimal in $\mathbf{s}$ direction, keeping in mind that a new point $\mathbf{x}^*$ has to be still a feasible point. Therefore, there must exist a maximum allowable step-length α, which can

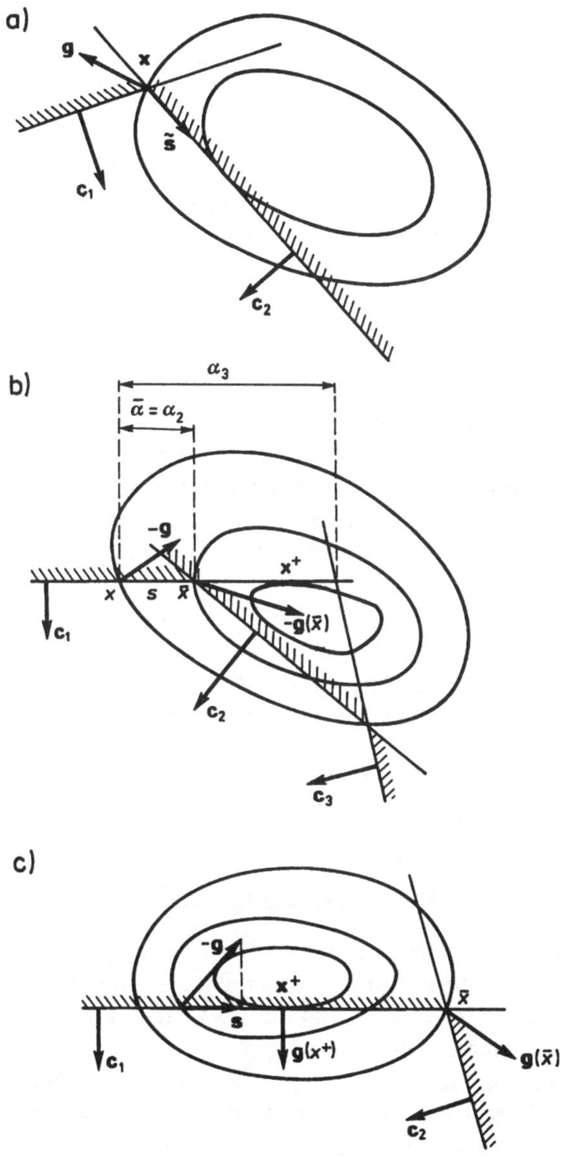

Fig. 1.4. Linearly constrained optimization

be calculated as

$$\bar{\alpha} = \min_{j=q+1,\dots,n} \{\alpha_j > 0 : \mathbf{x} + \alpha_j \mathbf{s} \text{ is feasible}\}. \tag{1.85}$$

Since

$$\alpha_j = \frac{\mathbf{c}_r^T \mathbf{x} - \mathbf{b}_j}{\mathbf{c}_r^T \mathbf{s}}, \quad j = q+1, \dots, n, \tag{1.86}$$

then obviously α_j is positive only if $\mathbf{c}_r^T \mathbf{s} < 0$. As it is shown in Fig. 1.4b, α_j represents the distance from hyper-plane of previously inactive constraints. Calculating $\bar{\alpha}$ from (1.85), we can compute a point $\bar{\mathbf{x}} = \mathbf{x} + \alpha \mathbf{s}$. If $\mathbf{s}^T \mathbf{g}(\bar{\mathbf{x}}) \leq 0$, then $\bar{\mathbf{x}}$ is a minimal point of objective function $f(\mathbf{x})$ along direction $\mathbf{s}$, since for every $\alpha > \bar{\alpha}$ at least one constraint will be violated. We can thus assume $\mathbf{x}^* = \bar{\mathbf{x}}$ and add a newly counted constraint to a set of active constraints. This means that the corresponding gradient vector of this constraint is added to matrix $\mathbf{N}_q$, and new matrix $\mathbf{N}_{q+1}$ of dimensions $n \times (q+1)$ is created. The corresponding projection matrix $\mathbf{P}_{q+1}$ is calculated again from (1.81), with $\mathbf{N}_q$ replaced by $\mathbf{N}_{q+1}$, and iteration process can be repeated at point $\mathbf{x}^*$. If, however, $\mathbf{s}^T \mathbf{g}(\bar{\mathbf{x}}) > 0$ at point $\bar{\mathbf{x}}$, then there exists $\tilde{\alpha} \in [0, \alpha]$ so that $\tilde{\mathbf{x}} = \mathbf{x} + \tilde{\alpha} \mathbf{s}$ is a minimum point of $f(\mathbf{x})$ along direction $\mathbf{s}$, see Fig. 1.4c. Thus, the value $\tilde{\alpha}$ can be found from directional minimization, see Sect. 1.3.3. The set of active constraints remains unchanged in this case. We can put $\mathbf{x}^* = \tilde{\mathbf{x}}$ and repeat the iteration process at point $\mathbf{x}^*$.

Similarly to the descent method for unconstrained minimization, the gradient projection method for linearly constrained minimization is much more inefficient due to it its slow convergence rate property. It is therefore obvious to consider the methods similar to Newton, quasi-Newton or conjugate gradients for the case of constrained minimization. One of the main problems is then to define the proper projection operator of gradient vector. If, as previously, $\mathbf{H}$ denotes the Hessian of objective function, then the projection operator can be assumed in the form

$$\tilde{\mathbf{P}}_q = \mathbf{I} - \mathbf{H}^{-1} \mathbf{N}_q \left(\mathbf{N}_q^T \mathbf{H}^{-1} \mathbf{N}_q \right)^{-1} \mathbf{N}_q^T . \tag{1.87}$$

The operator (1.87) is called an *oblique projection operator* weighted by $\mathbf{H}^{-1}$. The matrix $\tilde{\mathbf{P}}_q \mathbf{H}^{-1}$ can be considered as a projection of inverse of hessian matrix and hence the search direction $\mathbf{s} = -\tilde{\mathbf{P}}_q \mathbf{H}^{-1}$ is a search direction of Newton method along intersection of hyper-planes of active constraints. It can be shown for quadratic objective function and linear equality constraints that the optimal point is found in one iteration step, as it was the case for Newton algorithm for unconstrained minimization.

The projection method (1.87), using the second derivatives of objective function will be called *the projection algorithm of the second order.* Applying this method to optimization problem with linear constraints (1.72), the minimization is performed iteratively, similarly as for the projection algorithm of first order. The main difference between these algorithms is that now the second order estimation of Lagrange multipliers is known, and given as

$$\lambda = \left(\mathbf{N}_q^T \mathbf{H}^{-1} \mathbf{N}_q\right)^{-1} \mathbf{N}_q^T \mathbf{H}^{-1} \mathbf{g} \tag{1.88}$$

and then the search direction can be written in the form

$$\mathbf{s} = -\mathbf{H}^{-1}(\mathbf{g} - \mathbf{N}\lambda). \tag{1.89}$$

The directional minimization can be performed approximately by using step-length in this direction. This kind of 'projected' Newton method is probably the most suitable for solution of linearly constrained minimization problems assuming that the Hessian of the objective function and its inverse can be calculated in each iteration step. On the other hand, however, due to the fact that the Hessian cannot necessarily be positive definite at minimal point, the method of enforcing positive definiteness has to be used much more often than in the unconstrained case.

Another approach for solving the considered minimization problem is to use quasi Newton methods for approximation of hessian matrix and its inverse. This can be done by using the first order information from previous iteration steps. For instance, the Wolf–Broyden–Davidon formula (1.71) can be used for this task, due to its advantage that the exact directional minimization is not necessary. For this approach the Hessian inverse $\mathbf{H}^{-1}$ in Eqs. (1.88) and (1.89) is replaced by matrix $\mathbf{S}^{(k)}$ constituting an approximation of inverse of Hessian in k-th iteration. The similar strategy can be obtained by combining the Davidon–Fletcher–Powell method and gradient projection method, what results in search direction defined by

$$\mathbf{s}^{(k)} = -\mathbf{S}_q^{(k)} \mathbf{g}^{(k)} \tag{1.90}$$

as long as active constraints set remains unchanged. In fact, if initial approximation matrie of $\mathbf{H}^{-1}$ is orthogonal to gradients of constraints, that is $\mathbf{S}_q^{(0)} = \mathbf{P}_q$ as follows from (1.81), then all consecutive approximations are orthogonal if $\mathbf{S}^{(k)}$ will be modified using Davidon–Fletcher–Powell formula (1.70). As the result of that, all search directions are residing in initial subdomain of linear constraints. Obviously, when the set of active constraints is changed then formula (1.70) has to be modified. In the first order projection method, the matrix $\mathbf{S}_q$ is an approximation of matrix $\tilde{\mathbf{P}}_q \mathbf{H}^{-1}$, where $\tilde{\mathbf{P}}_q$ is defined by (1.87). If objective function is in positive definite quadratic form and if $n - q$ successive iterations are performed with unchanged set of active constraints, then starting with $\mathbf{S}_q^{(0)} = \mathbf{P}_q$, $\mathbf{S}_q^{(n-q)}$ will become equal to $\tilde{\mathbf{P}}_q \mathbf{H}^{-1}$. Additionally, search directions generated in these $n - q$ iterations are $\mathbf{H}$-conjugate.

The methods of conjugate gradients can be modified in a similar way. If, for instance, in algorithm of Fletcher–Reeves (1.60) the gradient of objective function $\mathbf{g}^{(k)}$ will be replaced by $\mathbf{P}_q \mathbf{g}^{(k)}$, then the convergence speed of gradient projection methods will be improved. For quadratic objective function and linear constraints, the optimal solution is obtained during $n - g$

iteration. For an arbitrary objective function, the modified Fletcher–Reeves method has to be initialized again when projection matrix $\mathbf{P}_q$ is changed as the result of changing the set of active constraints. The previous information is lost in this way and that fact can be considered as a disadvantage when compared with methods of approximation of the Hessian inverse.

The determination of a set of active constraints at each iteration step constitutes an important part of each projection algorithm. Conditions for increasing a set of active constraints are rather simple and they are based on determination of length step in direction to closest constraints, inactive in this moment, according to condition (1.85). It is much more difficult to determine which of active constraints should be removed from the active set in particular iteration. The simplest way is to preserve all active constraints until a minimum in sub-domain bounded in these constraints will be found. Only at this point, signs of the Lagrange multipliers are checked, what can eventually lead to removing some constraints from the active set. It can be shown that this procedure will be stopped at points of strong local minima after a finite number of basis changes. The main drawback of this strategy is the necessity of finding a minimum in each consecutive sub-domain of active constraints set. If the initial set of active constraints differs considerably from the similar set at the optimum point, then this fact influences considerably the calculation effort during process of searching the final solution.

An alternative approach consists of calculating the estimation of Lagrange multipliers estimations in each iteration step and next removing the constraint associated with the most negative Lagrange multiplier from the active set. This method corresponds to the rule used in simple algorithm of linear programming, where the estimations of Lagrange multipliers are exact. It should be noted, however, that for nonlinear objective function the estimation of Lagrange multipliers at points far from minimum in actual design sub-domain can be very inaccurate. The error is of order $0(h)$ for the first order estimation and of $0(h^2)$ for estimations of the second order, where h denotes the distance between actual and minimal points. This inaccuracy in estimation can cause the phenomenon of 'zigzagging', in which a constraint is dropped out from the active set and then reintroducing to this set in successive iterations, as long as the sign of associated multiplier will be stabilized. This phenomenon results in slowing the solution process and also the oscillations can occur. However, some effective methods of reducing the zigzagging effect exist. One of possible solutions is, for instance, assuming that if the constraint dropped out from the active set is added again, then it is retained in the active set until a stationary point is achieved. Another method of determination of the active constraints set assumes that the constraint is removed from this set only if it results in considerable reduction of objective function. The benefit associated with removing of j-th constraints can be approximately calculated as $\frac{1}{2}\tilde{\lambda}_j^2/u_j$, where $\tilde{\lambda}_j$ is a second order estimation of Lagrange multiplier and u_j is a j-th diagonal element of matrix

$(\mathbf{N}_q^T\mathbf{H}^{-1}\mathbf{N}_q)^{-1}$. Thus, the strategy of removing constraints from the active set can be as follows: find the q-th constraint from the subset of constraints for which $\tilde{\lambda}_j < 0$, that maximizes the values of $\frac{1}{2}\tilde{\lambda}_j^2/u_j$. This constraint should be removed from the active set only if the additional reduction of objective function is γ times greater than without removing it. Thus, the q-th constraint will be deleted if

$$-\mathbf{s}^T\mathbf{s} = \mathbf{s}^T\mathbf{H}\,\mathbf{s} \leq \frac{1}{\gamma}\frac{\tilde{\lambda}_q^2}{u_q} \tag{1.91}$$

where γ is the assumed positive constant. In the gradient projection method, for which Hessian of objective function is not calculated, the condition (1.91) is replaced by

$$-\mathbf{s}^T\mathbf{g} = \mathbf{s}^T\mathbf{H}\,\mathbf{s} \leq \frac{1}{\gamma}\frac{\tilde{\lambda}_q^2}{v_q} \tag{1.92}$$

where λ_q denotes the first order estimation of the proper Lagrange multiplier, and v_q is a q-th diagonal element of matrix $(\mathbf{N}_q^T\mathbf{N}_q)^{-1}$. The validity of the proposed procedure depends strongly on quality of multiplier estimation, and then it is advised to use this procedure only if

$$|\mathbf{s}^T\mathbf{g}| \leq \varepsilon \tag{1.93}$$

where ε is the assumed tolerance. The condition (1.93) guarantees sufficient accuracy of multipliers estimations since it can be satisfied only close to a stationary point, where these estimates are sufficiently exact.

1.3.4 Minimization Methods for General Nonlinear Programming Problem

The general methods of solution of problem of nonlinear programming (1.23)–(1.25) will be briefly discussed in this section. Recently used methods for solving this general problem can be classified as follows: direct method of solving the problem by the use of sequentially one-dimensional minimization along usable feasible directions; methods of linearization, in which nonlinear programming problem is replaced by a series of linear programming sub-problems; and transformation methods, in which constrained problem is converted into a sequence of unconstrained problems.

Using direct approach, the gradient projection method discussed already in Sect. 1.3.2 can be modified in order to incorporate nonlinear constraints. Modification consists of projecting the gradient of objective function on plane tangent to hyper-surface of active constraints. The nonlinear constraint can be approximated in the neighborhood of admissible boundary point $\tilde{\mathbf{x}}$ in the linear form

$$\tilde{h}_j(\mathbf{x}) = h_j(\tilde{\mathbf{x}}) + (\mathbf{x}-\tilde{\mathbf{x}})^T\nabla h_j(\tilde{\mathbf{x}}), \quad j = 1,\ldots,q. \tag{1.94}$$

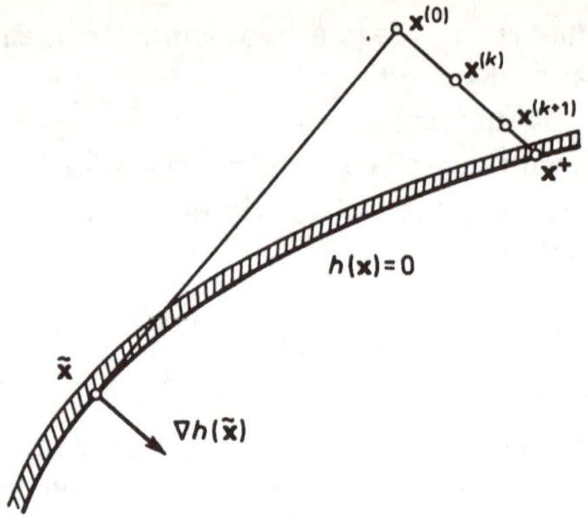

Fig. 1.5. Nonlinearly constrained optimization

Similarly as for linear constraints, search direction generated by proper projection of the objective function gradient is given by (1.76). The projection matrix $\mathbf{P}_q$ is still calculated using (1.71), where now $\mathbf{N}_q$ is composed of gradients of active constraints and has a form

$$\mathbf{N}_q = [\nabla h_j(\tilde{\mathbf{x}}), \ldots, \nabla h_q(\tilde{\mathbf{x}})]. \tag{1.95}$$

Let us note that now, in contrarily to the case of linear constraints, the components of matrix $\mathbf{N}_q$ are no longer constant and they depend on value $\tilde{\mathbf{x}}$. This leads to substantial changes in the algorithm of gradient projection method. Firstly, the matrix $\left(\mathbf{N}_q^T\mathbf{N}_q\right)^{-1}$ has to be calculated at each iteration step, since it depends on the actual value of $\tilde{\mathbf{x}}$. Secondly, a nonlinear set of equations has to be solved in *minimization phase* in order to determine maximal allowable step length in search direction. Finally, the point obtained as the result of directional minimization cannot necessary be a feasible point. Thus, a restoration phase has to be introduced, which leads back to the boundary of feasible design domain before the next downhill direction of objective function will be calculated. To do this, a linear approximation of active constraints (1.94) can be used. The comeback on the boundary of feasible domain is performed along direction which is normal to intersection of constraints at the initial point, see Fig. 1.5. According to (1.94) and (1.95), a new feasible point $\mathbf{x}^*$ such that $h_j(\mathbf{x}^*) = 0$ for $j = 1, \ldots, q$, can be iteratively generated, according to the formula

$$\mathbf{x}^{(k+1)} = \mathbf{x}^{(k)} - \mathbf{N}_q\left(\mathbf{N}_q^T\mathbf{N}_q\right)^{-1}\mathbf{h}(\mathbf{x}^{(k)}) \tag{1.96}$$

The convergence of this restoration process depends obviously on the distance of $\mathbf{x}^{(0)}$ to the constraint surface and next depends on the step length of directional minimization. Simultaneously, the condition of decreasing of objective function must be retained, that is $f(\mathbf{x}^*) < f(\tilde{\mathbf{x}})$. In the similar way, the other algorithms for linearly constrained problems can be adapted, such as conjugate gradients, Newton, or quasi-Newton methods.

Another direct method uses the concept of feasible directions. According to this concept, each typical iteration (1.74) consists of directional minimization in s direction which does not leave immediately the feasible domain. The feasibility condition at the boundary point in which q constraints are active has the form

$$\mathbf{s}^T \nabla h_j \geq 0, \quad j = 1, \ldots, q, \tag{1.97}$$

where the nonlinear constraints are strict inequalities. Any vector $\mathbf{s}$ satisfying (1.97) is contained, at least partially, in the feasible domain. Moreover, feasible direction $\mathbf{s}$ is called *usable*, if it indicates the downhill direction of the objective function. The usable direction satisfies then the condition

$$\mathbf{s}^T \nabla f \leq 0\,. \tag{1.98}$$

If the actual point is not a local minimum of the objective function, then the inequalities (1.97) and (1.98) define a cone of feasible and usable search directions. The problem of finding proper search direction can be formulated as a linear programming problem. Feasible and usable search direction, along which the maximal reduction of the objective function occurs, is defined as the solution of the following problem:

Find

$$\max \beta$$

subjected to:

$$\begin{aligned} \mathbf{s}^T \nabla f + \beta &\leq 0\,, \\ \mathbf{s}^T \nabla h_j - \theta_j \beta &\geq 0\,, \quad j = 1, \ldots, q \\ -1 \leq s_i &\leq 1\,, \end{aligned} \tag{1.99}$$

where θ_j are given positive constants. Let us note that the normalization condition imposed on vector $\mathbf{s}$ can be replaced by another bound of length of vector $\mathbf{s}$. Obviously, for $\beta_{\max} \geq 0$, the inequalities (1.97) and (1.98) are strict and selected, vector $\mathbf{s}$ is a feasible and usable direction. For $\beta_{\max} = 0$, the initial point is a point of local minimum. The coefficients θ_j, called the *push-off factors* are the measure of rejection of vector $\mathbf{s}$ from boundary of feasible domain. For small values of θ_j, the feasible direction is chosen so that $\mathbf{s}^T \nabla f + \beta = 0$. It means that the value of the objective function decreases relatively fast but minimization follows the direction almost equivalent to the boundary of feasible domain. On the other hand, for great values of θ_j, the search direction $\mathbf{s}$ tends to follow the contours of constant values of objective function. The risk of running out from feasible domain is in that case considerably reduced but the value of objective function decreases much slower. Intermediate values of θ_j (usually equal to one) are some compromise for which both the objective function and constraints decrease with similar rate in vicinity of the initial point.

Quite another problem is a selection of step length of feasible and usable direction. This selection can be performed similarly as in the gradient projection method. In the feasible direction method, the zigzagging phenomenon can also be observed, depending on the selection of the actual set of active constraints. The procedure of selection of this set, discussed earlier, can also be useful here.

One of the most natural approaches to solution of the problem of nonlinear programming is its transformation into a *sequence of lincar programming subproblems*, defined by

Find

$$\min f(\tilde{\mathbf{x}}) + (\mathbf{x} - \tilde{\mathbf{x}})^T \nabla f(\tilde{\mathbf{x}}) \tag{1.100}$$

subjected to

$$h_j(\nabla x) + (x - \tilde{x})^T \nabla h_j(\tilde{x}) \geq 0\,, \quad j = 1, \ldots, m\,,$$

where $\tilde{\mathbf{x}}$ is a point at which the objective function and constraints are linearized. The solution of problem (1.100) and finding a minimal point constitute a starting point for the next iteration. Such approach to nonlinear programming problems, falling into the category of linearization methods, seems to be natural for an engineer and it allows for easy implementation by using the standard computer procedures of linear programming. However, this very attractive approach suffers from some important limitations. First of all, the convergence to local minimum cannot be achieved if this minimum is not located at a vertex of feasible domain. In such cases, the process converges to a vertex which is not an optimal solution or it oscillates indefinitely between two vertices. In addition, the problem (1.100) can have a unique solution only if the number of constraints is greater than the number of design variables, that is $m > n$.

An interesting modification of sequential linear programming is an *approximate programming* in which the linearized problem (1.100) is supplemented by additional constraints imposed on design variables in the form

$$\tilde{x}_i - \alpha_i \leq x_i \leq \tilde{x}_i + \beta_i \tag{1.101}$$

where α_i and β_i are properly selected positive constraints, called *move limits*. After solution of the problem (1.100)–(1.101), the objective function and constraints are again linearized and move limits α_i and β_i are eventually modified.

The transformation methods of constrained minimization depend generally on modification of the objective function as the result of adding some penalty terms associated with constraint violation. The modified objective function is next subjected to unconstrained minimization. Depending on the form and way of using the penalty term, two approaches can be distinguished: *interior penalty function* called sometimes a barrier function and *external penalty function* called also simply penalty function.

In interior penalty function approach, some auxiliary function is constructed, which is a combination of objective function and constraints. This unconstrained auxiliary function has minima inside the feasible domain. The auxiliary function constructs a barrier at the boundary of feasible domain preventing the violation of the constraints. Gradually removing the effect of constraints in auxiliary function is generated by controlled changes of some parameter in the sequence of unconstrained minimization problems, whose solutions are interior points of the feasible domain, converging to the minimal point of original problem. The methods of interior penalty function are particularly interesting in structural optimization problems due to the fact that even prematurely terminating of algorithm generates always a feasible solution which corresponds usually to a better design (not necessary optimal) than the initial design. The transformation of objective function in interior penalty approach has the form

$$\Phi(\mathbf{x}, r) = f(\mathbf{x}) + rB(\mathbf{x})\,, \quad r > 0\,, \tag{1.102}$$

where r is a controlled parameter, while the barrier function $B(\mathbf{x})$ is positive inside the feasible domain and tends to infinity for points approaching the boundary of this domain. Therefore the internal penalty function cannot be applied for problems with equality constraints. The barrier functions mostly used are the logarithmic functions

$$B(\mathbf{x}) = -\sum_{j=1}^{m} \log[h_j(\mathbf{x})] \tag{1.103}$$

and the inverse functions

$$B(\mathbf{x}) = \sum_{j=1}^{m} \frac{1}{h_j(\mathbf{x})}\,. \tag{1.104}$$

It can be proved that if a sequence $\{\mathbf{x}^{(k)}\}$ is monotonic and decreasing to zero ($r^{(k)} \to 0$ for $k \to \infty$), then the solutions of a sequence of unconstrained problems

$$\min_{x} \Phi(\mathbf{x}, r^{(k)}) \tag{1.105}$$

with starting point inside feasible domain are all the interior points $\{\mathbf{x}^{(k)} = \mathbf{x}(r^{(k)})\}$ converging to optimal point $\mathbf{x}^*$ of original constrained problem. Moreover, there exists

$$\lim_{k\to\infty} \Phi(\mathbf{x}^{(k)}, r^{(k)}) = \lim_{k\to\infty} f(\mathbf{x}^{(k)}) = f(\mathbf{x}^*) \tag{1.106}$$

and a function $f(\mathbf{x}^{(k)})$ is monotonicly decreasing.

The basic steps of any algorithm of interior penalty function method are as follows:

1. Select a monotonically decreasing sequence $\{r^{(k)}\} \to 0$, for $k \to \infty$. Find an interior starting point $\mathbf{x}^{(0)}$ and set $k = 0$.

2. Assuming $\mathbf{x}^{(k)}$ as an initial point, minimize $\Phi(\mathbf{x}, r^{(k)})$ in order to find $\mathbf{x}^{(k+1)} = \mathbf{x}(r^{(k)})$.
3. Check the convergence criterion. If it is not satisfied, set $k = k + 1$ and go to step 2.

Although the interior penalty function approach is an effective tool for solving unconstrained minimization problems, it suffers from some essential disadvantages. Namely, for decreasing values of controlling parameter $r^{(k)}$ the minimization of auxiliary function $\Phi(\mathbf{x}, r^{(k)})$ becomes more and more difficult. This is due to increase of ill-conditioned nature of he Hessian matrix of Φ. Thus, the proper selection of a sequence of $\{r^{(k)}\}$ becomes very important for future success of the algorithm. On the one hand, $r^{(k)}$ should be sufficiently small so that minimal point $\mathbf{x}(r^{(k)})$ can approach the boundary of feasible domain. However, on the other hand, the parameter r should be large enough to assume effective minimization of auxiliary function $\Phi(\mathbf{x}, r^{(k)})$. To satisfy these two opposite requirements, one can assume in practice

$$r^{(k+1)} = r^{(k)}\varphi \tag{1.107}$$

where the coefficient φ can vary in an internal $(0.1 \div 0.5)$ depending on the nature of the problem and applied algorithm of unconstrained minimization.

The external penalty function approach, in contrary to internal penalty transformation, takes into account the cost of constraint violation. The auxiliary function is assumed here in the form

$$\Psi(\mathbf{x}, r) = f(\mathbf{x}) + \frac{1}{r} D(\mathbf{x}), \quad r > 0, \tag{1.108}$$

where penalty function $D(\mathbf{x}) \geq 0$ for all $\mathbf{x} \in \mathbf{E}^n$ if only $\mathbf{x}$ is an exterior point of feasible domain. The auxiliary function $\Psi(\mathbf{x}, r) \rightarrow \infty$ with increasing constraints violation. Some of the mostly used penalty functions are the quadratic functions

$$D(\mathbf{x}) = \sum_{j=1}^{m} [\min(0, h_j(\mathbf{x}))]^2 \tag{1.109}$$

and Zangwill function

$$D(\mathbf{x}) = \sum_{j=1}^{m} \min[0, h_j(\mathbf{x})] \tag{1.110}$$

Contrarily to previous penalty approach, the external penalty function transformation allows for handling the equality constrained problems. For all constraints being equality constraints, the auxiliary function can be assumed in the form

$$\Psi(\mathbf{x}, r) = f(\mathbf{x}) + \frac{1}{r} \sum_{j=1}^{m} [h_j(\mathbf{x})]^2 \tag{1.111}$$

The controlling parameter r is used effectively for increasing penalty associated with constraint violation, that is the magnitude of penalty increases as $r \to 0$. For small values of r it is obvious that the minimum point of unconstrained problem

$$\min_{x} \Psi(\mathbf{x}, r) \tag{1.112}$$

will lay in a domain when $D(\mathbf{x})$ is small. Thus, for a decreasing sequence of $\{r^{(k)}\}$, one can expect that the solution points of problem (1.112) will minimize objective function $f(\mathbf{x})$. Theoretically, for $r^{(k)} \to 0$, the solutions of sub-problems (1.130) will converge to solution of original constrained problem.

Thus, it can be observed that the behaviors of both penalty function approaches are similar. The only difference between both algorithms depends on different selection of starting point $\mathbf{x}^{(0)}$. This point has to be inside the feasible domain for interior penalty function method, whereas for the second method, we start from outside of this domain, and the sequence of points $\mathbf{x}^{(k)}$, calculated in consecutive iteration steps, approaches the optimal point $\mathbf{x}^*$ from the exterior of feasible domain.

The use of Lagrange function seems to be one of the most effective methods of constrained minimization. It is due to information of curvature of the constraints, which is incorporated in their quadratic approximation contained in the Lagrange function. Assume for simplicity that the minimization problem with only equality constraints is considered here, that is:

Find

$$\min f(\mathbf{x}) \tag{1.113}$$

subjected to

$$h_j(\mathbf{x}) = 0, \quad j = 1, \dots, q.$$

The extension of problem (1.113) for the case of inequality constraints can be made using the method of selection of active constraint set, based on information carried by Lagrange multipliers. The base for *recursive quadratic programming* methods is constituted by stationarity conditions of Lagrange function

$$L(\mathbf{x}, \lambda) = f(\mathbf{x}) - \sum_{j=1}^{q} \lambda_j h_j(\mathbf{x}) \tag{1.114}$$

which can be written in the form of nonlinear equations

$$\begin{cases} \nabla f(\mathbf{x}) - \sum_{j=1}^{q} \lambda_j h_j(\mathbf{x}) = 0 & \\ h_j(\mathbf{x}) = 0 & j = 1, \dots, q \end{cases} \tag{1.115}$$

Using the Newton method for solution of set (1.115), the better approximation $(\mathbf{x}, \lambda)$ can be obtained from an estimate $(\hat{\mathbf{x}}, \hat{\lambda})$ following from a solution of a system of linear equation written as

$$\begin{bmatrix} \mathbf{G}(\hat{\mathbf{x}},\hat{\lambda}) & -\mathbf{N}(\hat{\mathbf{x}}) \\ -\mathbf{N}^T(\hat{\mathbf{x}}) & 0 \end{bmatrix} \begin{Bmatrix} \mathbf{x}-\hat{\mathbf{x}} \\ \lambda-\hat{\lambda} \end{Bmatrix} = \begin{Bmatrix} -\nabla f(\hat{\mathbf{x}}) + \mathbf{N}(\hat{\mathbf{x}})\hat{\lambda} \\ h(\hat{\mathbf{x}}) \end{Bmatrix} \tag{1.116}$$

where $\mathbf{G}(\mathbf{x},\lambda)$ denotes the Hessian matrix of Lagrange function (1.114) and $\mathbf{N}(\mathbf{x})$ is a gradient matrix of constraints (see Eq. (1.113)). Let us note that the correction vector $\mathbf{d} = \hat{\mathbf{x}} - \mathbf{x}$ satisfying (1.116) can be also found as a minimum point of the quadratic form

$$\min_d \frac{1}{2}\mathbf{d}^T\mathbf{G}(\hat{\mathbf{x}},\hat{\lambda})\mathbf{d} + \mathbf{d}^T\nabla f(\hat{\mathbf{x}}) \tag{1.117}$$

subjected to linearized constraints

$$\mathbf{N}^T(\hat{\mathbf{x}})\mathbf{d} + \mathbf{h}(\hat{\mathbf{x}}) = 0\,. \tag{1.118}$$

The equations (1.117)–(1.118) suggest then the transformation of the original problem (1.113) into a sequence of quadratic sub-problems, in which objective functions are second order approximation of Lagrange function calculated together with active constraints, and where the constraints are replaced by their linear approximations (1.118). A further improvement of this approach can depend on replacing in each iteration step the calculation of Hessian matrix $\mathbf{G}(\hat{\mathbf{x}},\hat{\lambda})$ by its approximation as it is done in quasi-Newton methods.

The Lagrange function (1.114) plays also an important role in another method, which is called the *multiplier method.* It is well known that if there exists some λ^* for which $\mathbf{x}^*$ is the solution of unconstrained minimization problem $\min\limits_x L(\mathbf{x},\lambda^*)$ for satisfied equality constraints, then $\mathbf{x}^*$ is also the solution of original problem (1.113). Thus, the problem

Find

$$\min_x L(\mathbf{x},\lambda) \quad \text{for proper } \lambda \tag{1.119}$$

subjected to

$$h_j(\mathbf{x}) = 0, \quad j = 1,\dots,q,$$

is equivalent to the original problem (1.113) in the sense that $\mathbf{x}^*$ is a minimum point for both problems and $\mathbf{x}^*$ is an associated vector of multipliers. Let us consider now the solution of equivalent problem (1.119) by using the external penalty function method given in the form (1.111). Since the penalty term is now added to Lagrange function, then the *extended Lagrange function* for equality constrained problem is formed as

$$X(\mathbf{x},\lambda,r) = f(\mathbf{x}) - \sum_{j=1}^{q}\lambda_j h_j(\mathbf{x}) + \frac{1}{r}\sum_{j=1}^{q}[h_j(\mathbf{x})]^2\,. \tag{1.120}$$

For given values of λ and r, the multiplier method depends on applying the unconstrained minimization algorithm to the function (1.120) in order to obtain a point $\mathbf{x}(\lambda,r)$, the values of λ and r are adjusted in consecutive iteration steps so that $\mathbf{x}(\lambda, r)$ converges to $\mathbf{x}^*$ as the number of iterations increases. If in particular case, $\lambda = 0$, then (1.120) is reduced to the classical transformation (1.111) of external penalty function approach. On the other hand, for $\lambda = \lambda^*$, the minimization of $X(\mathbf{x}, \lambda^*, r)$ yields the solution of the original problem for an arbitrary value of $r > 0$. These two extreme cases imply that by updating the Lagrange multipliers in consecutive iterations so that they tend to λ^*, the convergence of the process is guaranteed without the need of assuming very small values of r. Thus, the ill-conditioning associated with penalty approaches can be avoided. The correction formula for Lagrange multipliers is usually assumed in the form

$$\lambda_j^{(k+1)} = \lambda_j^{(k)} - \frac{2}{r^{(k)}} h_j(\mathbf{x}^{(k)}) \tag{1.121}$$

where k denotes the iteration index. The problem of suitable correction of vector λ can also be considered as maximization problem of an auxiliary dual function

$$l_r(\lambda) = \min_x X(\mathbf{x}, \lambda, r) \tag{1.122}$$

by using, for instance, a steepest ascent method (a contradiction to steepest descent method) with a step length $2/r^{(k)}$ in the dual space. In fact, the first derivatives of the dual function $l_r(\lambda)$ are given by negative values of constraints (see Eq. (1.40)). Therefore, the multiplier method can be interpreted as a *primary–dual optimization method* of some kind with limited searching in dual space for optimal values of Lagrange multipliers.

1.4 Optimality Criteria Approach

Most of the used optimality criteria methods are concerned with the optimal design of statically determinate and indeterminate trusses. The procedure following from the applied optimality criteria depends mainly on the character of constraints imposed in particular optimal design problem.

The best known example of application of an optimality criteria method is the so called *fully stressed design* criterion which can be applied to the case where the only constraints in a given design problem are stress constraints in the form (1.18). According to this criterion, in the optimal structure the maximal allowable stress level is attained in each member, at least for one loading case. Generating a sequence of solutions using this criterion, the values of design variables at $(k+1)$ iteration step are calculated by means of the formula

$$a_i^{(k+1)} = a_i^{(k)} \max_{l=1,\ldots,c} \frac{\sigma_{il}^{(k)}}{\bar{\sigma}_i} \,. \tag{1.123}$$

For statically determinate structures, for which internal forces are constant and independent of design variables (i.e. bar cross-sections), the optimal solution is obtained in one iteration step, and requires only one structural analysis. In the statically indeterminate cases, for which internal forces depend on values of design parameters, the fully stressed design criterion is approximate and the relation (1.123) has to be applied recursively as long as the change in values of design parameters in two consecutive iteration steps will be sufficiently small. It is worth noticing that the concept of fully stressed design does not involve explicitly the objective function. This criterion is only dependent on the allowable stress level and leads to a vertex in the design space, which is not necessarily an optimal point. However, the formula (1.123) generally produces solution point very close to optimal design in a small number of iterations, independently of the number of design variables. This feature allows for the effective use of this intuitive approach to automated sizing of even large structures.

Consider a case of single displacement constraint. The displacement of an arbitrary node of discrete model of structure can be derived from the following relation

$$u = \mathbf{q}^T \tilde{\mathbf{f}} = \mathbf{q}^T \mathbf{K} \tilde{\mathbf{q}} \tag{1.124}$$

where $\tilde{\mathbf{f}}$ denotes the respective unit load vector and $\tilde{\mathbf{q}}$ is the corresponding nodal displacement vector. Since the global stiffness matrix $\mathbf{K}$ is a sum of element stiffness matrices $\mathbf{K}_i$, then the displacement constraint, following from (1.124), can be expressed in the form

$$h(a) = u - \bar{u} = \sum_{i=1}^{n} \frac{c_i}{a_i} - \bar{u} \leq 0 \tag{1.125}$$

where $\bar{u}$ is the specified value and c_i are defined by

$$c_i = (\mathbf{q}^T \mathbf{K} \tilde{\mathbf{q}}) a_i \,. \tag{1.126}$$

Let us note that for statically determinate structure the coefficients c_i are constant. For example, for truss members they take the form

$$c_i = \frac{f_i \tilde{f}_i l_i}{E_i} \tag{1.127}$$

where f_i and $\tilde{f}_i$ are the forces in i-th node for actual and unit loads, respectively, and l_i, E_i denote number lengths and Young's moduli of the i-th bar.

Therefore, the problem of structural weight minimization (1.19) subjected to equality constraint (1.125) is formulated explicitly in terms of the design variables a_i. The solution of this problem can be obtained from the stationary point of Lagrange function

$$L(a_i, \lambda) = \sum_{i=1}^{n} l_i a_i + \lambda \left(\sum_{i=1}^{n} \frac{c_i}{a_i} - \bar{u} \right) \tag{1.128}$$

where λ denotes the Lagrange multiplier. The stationary condition of function L provides the redesign relation

$$a_i^2 = \lambda \frac{c_i}{l_i} \tag{1.129}$$

where the multiplier λ can be obtained from the explicit equality constraint (1.125). Let us note, however, that using criterion (1.129) not all coefficients c_i expressed by (1.127) can be positive. This leads to subdivision of the design variables set into active and passive subsets. The active design parameters, given by (1.129), can now be expressed by using constraints (1.125) in the form

$$a_i = \left[\frac{1}{\bar{u} - u_0} \sum_{k=1}^{n} \sqrt{l_k c_k} \right] \sqrt{\frac{c_i}{l_i}}, \quad i = 1, \ldots, \tilde{n}, \tag{1.130}$$

while the remaining passive variables can be sized to an arbitrary small values. In expression (1.130) $\tilde{n}$ denotes the number of active variables for which $c_i > 0$ and u_0 is the contribution of passive variables to displacement u, that is

$$u_0 = \sum_{i>\tilde{n}} \frac{c_i}{a_i} . \tag{1.131}$$

Since for statically determinate structure the coefficients c_i are constant, then the optimal solution (1.130) is obtained in one iteration step. In statically indeterminate case c_i are no longer constant since they depend on design variables and the optimal solution is obtained in an iterative fashion by repeated application of (1.130) by assuming weak redistribution of internal forces in each iteration step.

The physical sense of the optimality criterion (1.129) is much more obvious when this relation is written in the form

$$\varepsilon_i = \frac{e_i}{l_i a_i} = \text{const}, \tag{1.132}$$

where

$$e_i = \mathbf{q}^T \mathbf{K} \tilde{\mathbf{q}} = \frac{c_i}{a_i} \tag{1.133}$$

denotes the mutual strain energy of two states in i-th element. Thus, this criterion states that the mutual strain energy density is the same for each active design variable. This observation does not apply obviously to the set of passive design variables which are associated with the negative mutual strain energy.

When several stress and displacement constraints are introduced simultaneously in the minimum weight design problem then the optimality criteria approach assumes that the subset of active constraints in the optimal solution is known in advance. Thus, only $\tilde{n}$ active design variables can be modified during optimization process. The remaining variables are the passive design variables whose values are determined by minimum size constraint or maximal strength of material, that is

$$a_i = \max(a_{i\ \min}\,, \tilde{a}_i)\,, \quad i > \tilde{n}, \tag{1.134}$$

where $\tilde{a}_i$ follows from material strength criterion in the form

$$\tilde{a}_i = a_i \max_{l=1,\dots,c} \left\{ \frac{\sigma_{il}}{\bar{\sigma}_i} \right\}. \tag{1.135}$$

Let us note that for statically determinate trusses, the valuc of $\tilde{a}$ determined by (1.135) is in fact the minimum value that the design variables can take without violating the stress constraint.

The satisfaction of displacement constraints at several points can be guaranteed by using the technique applied for the single displacement constraint, by introducing as many unit load factors as there are displacement constraints. Since all these constraints are assumed to be active, then they can be written in the following explicit form

$$\sum_{i=1}^{n} \frac{c_{ij}}{a_j} - \bar{u}_j = 0\,, \quad j = 1, \dots, m. \tag{1.136}$$

Introducing m Lagrange multipliers, the Lagrange function can be formed

$$L(a_i, \lambda_j) = \sum_{i=1}^{n} l_i a_i + \sum_{j=1}^{m} \lambda_j \left(\sum_{i=1}^{n} \frac{c_{ij}}{a_j} - \bar{u}_j \right). \tag{1.137}$$

The stationary conditions of (1.137) yield the following design formula

$$a_j = \sqrt{\frac{1}{l_j} \sum_{j=1}^{m} \lambda_j c_{ij}}\,, \quad i = 1, \dots, n. \tag{1.138}$$

Introducing, similarly as in (1.132), the concept of the mutual strain energy density, the optimality criterion can be written in the form

$$\sum_{j=1}^{m} \lambda_j \varepsilon_{ij} = 1\,. \tag{1.139}$$

For statically determinate structure the coefficients c_{ij} are constant and then optimality criterion is exact. The relation (1.138) specifies then optimal values of active set of design variables. In the case of statically indeterminate structures the above procedure gives only the approximation of optimal solution and then it has to be used in the iterative way.

The optimization problem is thus reduced to seeking optimal values of Lagrange multipliers in each analysis step, such that the displacement constraints (1.136) are satisfied. Since it is not possible to do analytically, as in the case of single displacement constraint, frequently some intuitive numerical procedures are proposed. One of the possible methods is to treat all displacement constraints separately by using relation (1.130), and next, to select the largest value obtained for each design variables as an approximate solution. Another approach assumes the scaling of all Lagrange multipliers proportionally to the square root of the respective displacement. The Newton–Raphson method can also be used for solving the nonlinear set of equations (1.136) and (1.138). Using all these optimality criteria techniques, we should however be aware of their major drawback associated with the proper selection of active design variables and constraints.

1.5 Concluding Remarks

The modern approaches and methods used in optimal design of structures were briefly presented in this chapter. A powerful and general approach to this design problem was achieved by replacing the original problems with a sequence of approximate explicit problems and solving them using either primal or dual algorithms. A primal approach consists in applying a mixed method. The properties of this method lay between properties of purely mathematical programming and optimality criteria methods. On the other hand, dual methods lead to generalization of rather intuitive optimality criteria methods with powerful algorithm for identifying critical constraints. Among the advantages and disadvantages of primal and dual methods the following are worth to be outlined:

1. The dual methods are most usually effective and economical from calculation cost point of view but their convergence can suffer some instability. Oppositely, the primal methods offer a better control of convergence process at the price of higher cost of calculations.
2. The primary methods can also be applied to more complex objective functions than the weight of structure or to the problem with inseparable approximation of constraints. On the other hand, the dual methods can easily be adapted to discrete optimization problems.
3. The computer implementation of structural optimization methods seems to be simpler for dual algorithms than for primal ones.

2 Applications of Linear Programming

2.1 Main Notions of Mathematical Programming

Mathematical Programming (MP) is a branch of mathematics dealing with the search of constrained minima or maxima of functions. In this section we recall only those elements of the MP-theory that are directly related to structural optimization. A full presentation of this theory including the proofs of the theorems cited in the sequel can be found in numerous books on the subject, e.g. in the monograph written by Zangwill [34].

Mathematical Programming deals with problems of the following type:

find

$$\min f(x_i)$$

subject to constraints

$$\begin{aligned} g_j(x_i) &= 0, \quad \text{for} \quad j = 1, 2, \ldots, m', \\ g_j(x_i) &\leq 0, \quad \text{for} \quad j = m'+1, m'+2, \ldots, m, \\ x_i &\geq 0, \quad \text{for} \quad i = n'+1, n'+2, \ldots, n. \end{aligned} \tag{2.1}$$

A cost function f and constraint functions g_j depend upon n variables x_i. Most results in the MP-theory are derived under assumption that these functions are continuous and convex.

Taking the minimization of the cost function and the inequality sign $\leq$ in (2.1) does not restrict generality of formulation, since the maximization problem and the constraints of type $\geq$ can be transformed into (2.1) by changing the sign of the functions f and g_j. Note that there are both equality and inequality constraints in the problem (2.1) and that there are two types of variables: free and nonnegative.

The constraints of the problem (2.1) determine in the n-dimensional space of variables a certain convex region Ω. Each point $\mathbf{x}_0 \in \Omega$ is called an admissible solution of the problem (2.1). The point $\mathbf{x}_*$ corresponding to the minimum of f over Ω is called an optimum solution or shortly a solution of (2.1). A correctly posed problem (2.1) has such a solution and that solution corresponds to finite minimum value f_* of the cost function.

An important notion of the MP-theory is duality. It appears that each problem of the type (2.1) has its dual defined as find

$$\max L(x_i, u_j)$$

subject to constraints

$$\begin{aligned} \frac{\partial L}{\partial x_i} &= 0, \quad \text{for} \quad i = 1, 2, \ldots, n', \\ \frac{\partial L}{\partial x_i} &\geq 0, \quad \text{for} \quad j = n'+1, n'+2, \ldots, n, \\ u_j &\geq 0, \quad \text{for} \quad j = m'+1, m'+2, \ldots, m. \end{aligned} \tag{2.2}$$

In this definition L is Lagrange function

$$L(x_i, u_j) = f(x_i) + \sum_{j=1}^{m} u_j g_j(x_i) \tag{2.3}$$

and u_j are Lagrange multipliers called also dual variables.

The following theorem holds for a pair of dual problems (2.1), (2.2):

Duality Theorem. *If the primal problem (2.1) has a solution* $\mathbf{x}_* \in \mathcal{R}^n$ *corresponding to finite* f_**, then there exists such a* $\mathbf{u}_* \in \mathcal{R}^m$ *that the pair* $(\mathbf{x}_*, \mathbf{u}_*)$ *is a solution of the dual problem (2.2) with* $L(\mathbf{x}_*, \mathbf{u}_*) = f_*$.

The dual pair of solutions $\mathbf{x}_*, \mathbf{u}_*$ satisfies the Kuhn–Tucker conditions of optimality:

$$\begin{aligned} & & \frac{\partial L}{\partial u_j} &= 0, \quad \text{for} \quad j = 1, 2, \ldots, m', \\ \frac{\partial L}{\partial u_j} &\leq 0, & u_j &\geq 0, \quad \text{for} \quad j = m'+1, m'+2, \ldots, m, \\ & & \frac{\partial L}{\partial x_i} &= 0, \quad \text{for} \quad i = 1, 2, \ldots, n', \\ \frac{\partial L}{\partial x_i} &\geq 0, & x_i &\geq 0, \quad \text{for} \quad i = n'+1, n'+2, \ldots, n, \\ & & \frac{\partial L}{\partial u_j} u_j &= 0, \quad \text{for} \quad j = 1, 2, \ldots, m, \\ & & \frac{\partial L}{\partial x_i} x_i &= 0, \quad \text{for} \quad i = 1, 2, \ldots, n, \end{aligned} \tag{2.4}$$

Note that the conditions $(2.4)_1$ to $(2.4)_4$ coincide with the constraints of the dual problems. On the other hand, the complementary slackness conditions $(2.4)_5$, $(2.4)_6$ express a simple rule: if the k-th constraint is satisfied as inequality by the solution of one of the problems (2.1), (2.2), then the k-th variable should be equal zero in the solution of the dual problem.

If any of functions f, g_j is non-linear, than we are dealing with the problem of Non-Linear Programming (NLP). A particular case of it is the problem of Quadratic Programming (QP). Such a problem has quadratic cost function and linear constraints. If both the cost function and the constraints are linear, then we obtain the problem of Linear Programming (LP):

$$\min\{\mathbf{c}^T\mathbf{x} \mid \mathbf{A}\mathbf{x} \geq \mathbf{b}, \ \mathbf{x} \geq \mathbf{0}\}. \tag{2.5}$$

The following matrices appear in such a problem: the matrix of cost parameters $\mathbf{c} \in \mathcal{R}^n$, the LHS-matrix of constraints $\mathbf{A} \in \mathcal{R}^{m \times n}$, the RHS-matrix of

Table 2.1. Simplex table for problem (2.5)

x_1	x_2	$\dots$	x_n		
$-A_{11}$	$-A_{12}$	$\dots$	$-A_{1n}$	b_1	≥ 0
$-A_{21}$	$-A_{22}$	$\dots$	$-A_{2n}$	b_2	≥ 0
$\vdots$	$\vdots$	$\dots$	$\vdots$	$\vdots$	$\vdots$
$-A_{m1}$	$-A_{m2}$	$\dots$	$-A_{mn}$	b_m	≥ 0
c_1	c_2	$\dots$	c_n	min	

Table 2.2. Simplex table for problem (2.6)

u_1	u_2	$\dots$	u_m		
$-A_{11}$	$-A_{21}$	$\dots$	$-A_{m1}$	c_1	≤ 0
$-A_{12}$	$-A_{22}$	$\dots$	$-A_{m2}$	c_2	≤ 0
$\vdots$	$\vdots$	$\dots$	$\vdots$	$\vdots$	$\vdots$
$-A_{1n}$	$-A_{2n}$	$\dots$	$-A_{mn}$	c_n	≤ 0
b_1	b_2	$\dots$	b_m	max	

constraints $\mathbf{b} \in \mathcal{R}^m$, and the matrix of variables $\mathbf{x} \in \mathcal{R}^n$. To simplify the notation we took into account inequality constraints and nonnegative variables only ($m' = n' = 0$).

It is easy to show that the general definition (2.2) leads to the following dual problem for (2.5):

$$\max\{\mathbf{b}^T\mathbf{u} \mid \mathbf{A}^T\mathbf{x} \leq \mathbf{c},\ \mathbf{u} \geq \mathbf{0}\}. \tag{2.6}$$

Note that for linear problems the notion of duality is symmetric: the dual of (2.6) coincides with (2.5). This is not true for the non-linear problem (2.1).

It is convenient to put LP-problems into the so-called simplex tables: Table 2.1 represents the primal problem (2.5) and Table 2.2 represents the dual one (2.6).

We close the presentation of MP-theory with remarks on the existence and the uniqueness of solutions of LP-problems. There exists a solution of LP-problem when its constraints are not contradictory, i.e. when the set Ω is not empty. If Ω is not bounded, then the constrained minimum of the cost function can be attained at $-\infty$ or its maximum at $+\infty$. In applications such situations indicate false formulation of the problem. However, finite extreme of the dual problems (2.5), (2.6) do not ensure the uniqueness of solutions $\mathbf{x}_*, \mathbf{u}_*$. It may happen that one of those solutions is degenerated, i.e. has zero elements. Then the solution of the dual problem is non-unique: many values of variables correspond to the same value of the cost function. Typical situations regarding the existence and the uniqueness of solutions for the LP-problems are shown in Fig. 2.1.

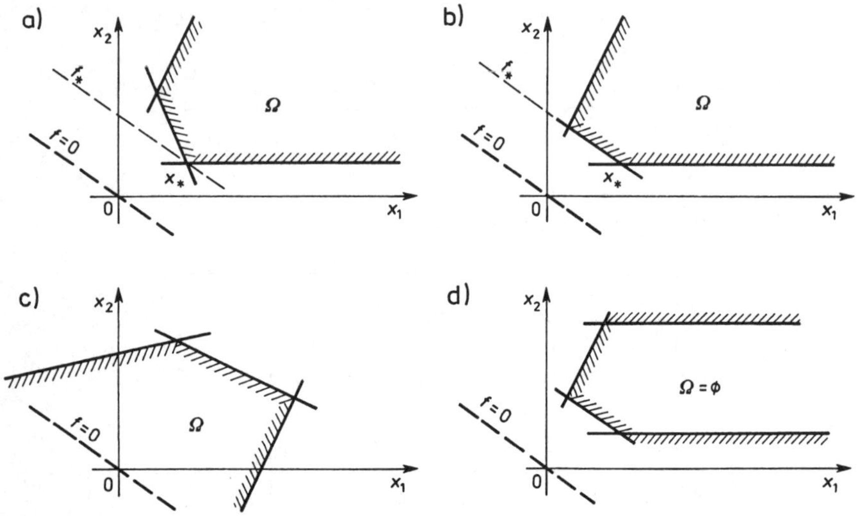

Fig. 2.1. Different possibilities for the solutions of problem (2.5): **a)** the solution $\mathbf{x}_*$ exists, it is unique and it corresponds to the finite value f_*, **b)** the solution $\mathbf{x}_*$ exists but it is non-unique, **c)** the cost function is not bounded from below ($f_* \to -\infty$), **d)** there is no solution $\mathbf{x}_*$ since constraints are contradictory

Numerous methods were proposed for solving MP-problems. A detailed description of such algorithms can be found, e.g., in the book [20]. Readers interested in the comparison of computer codes should consult the paper [29]. In general one can say that the efficiency of the MP-solvers decreases rapidly with increasing non-linearity of the problem. The limit of practical applicability encompasses the order of several tenth variables for NLP-problems, several hundred variables for QP-problems and several thousand variables for LP-problems.

The situation in Linear Programming is the most favourable. There exists general solver, called Simplex Method, and its implementations are commercially available for each computer platform. Such systems use sophisticated techniques, like sparse matrix approach, problem decomposition, memory swapping etc. Therefore large LP-problems can be solved in practically acceptable time and under commonly available computational power. It should be noted that the Simplex Method gives simultaneously solutions of both dual problems. The FORTRAN code of that method is given in the Appendix to the book [4].

2.2 Matrix Model of Structure

In this section we present matrix relations that govern the behaviour of skeletal structures made of elastic or rigid-plastic materials. We restrict ourselves to geometrically linear problems and to static loads. A detailed derivation of the governing relations for particular types of skeletal structures can be found in Part I of [35] (elastic behaviour) and in the monograph [4] (elastic-plastic and rigid-plastic behaviours). Having in mind the convenience of the reader we recall now those elements of the matrix description that will be used in the optimization problems.

According to the usual F.E.M. approach one should begin with the description of a generic finite element (a single bar) and then derive the global relations. In order to save space, we skip the local description and we present only the global description of the structure assembled from individual elements.

2.2.1 Kinematics and Equilibrium

Let the matrix $\mathbf{q} \in \mathcal{R}^n$ represent the displacements of the structure and let the matrix $\mathbf{s} \in \mathcal{R}^m$ describe its strains[1]. Assuming that displacements remain small, we have the following linear kinematic relation:

$$\mathbf{s} = \mathbf{B}\,\mathbf{q}. \tag{2.7}$$

This is an algebraic counterpart of the differential equation (2.29), Part II and it corresponds to the special case (no displacements of the supports) of the equation (5.17) in Part I of [35].

For each of the kinematic variables we assign such a static variable that the product of the both variables has the meaning of the virtual work. Hence, the external loads are represented by the matrix $\mathbf{Q} \in \mathcal{R}^n$ and the internal forces are represented by the matrix $\mathbf{S} \in \mathcal{R}^m$ such that the virtual work principle is reduced to the following equality of scalar products:

$$\mathbf{s}^T\,\mathbf{S} = \mathbf{q}^T\,\mathbf{Q} \tag{2.8}$$

It is easy to show by substituting into this equality the right hand side of (2.7) that the equation of static equilibrium should read

$$\mathbf{Q} = \mathbf{B}^T\,\mathbf{S} \tag{2.9}$$

This is an algebraic counterpart of the differential equation (2.31), Part II and a special case (no equilibrium equations for the supports) of the equation (5.28) in Part I of [35].

[1] We mean here the generalized quantities: the entries of $\mathbf{q}$ can be the rotations of nodes as well as their linear displacements, the entries of $\mathbf{s}$ can include the elongations of bars as well as the rotations of their ends.

For isostatic structures the matrix $\mathbf{B}$ is square and non-singular. This allows us to find unique internal forces by solving the matrix equation (2.9):

$$\mathbf{S} = \mathbf{B}^{-T}\,\mathbf{Q} \tag{2.10}$$

Here $\mathbf{B}^{-T}$ means an inverse of $\mathbf{B}^{T}$. Since optimization of isostatic structures is trivial, we are not going to consider such systems.

For hyperstatic (statically redundant) structures the matrix $\mathbf{B}$ is rectangular and the index $r = m - n$ determines the degree of redundancy. The solution

$$\mathbf{S} = \mathbf{B}^{-T}\,\mathbf{Q} + \mathbf{B}^{0T}\,\mathbf{Z} \tag{2.11}$$

of the equilibrium equation (2.9) is then non-unique, because it contains r free parameters grouped in the matrix $\mathbf{Z}$.

A pair of new matrices appears in (2.11): the generalized inverse $\mathbf{B}^{-} \in \mathcal{R}^{n\times m}$ and the kernel matrix $\mathbf{B}^{0} \in \mathcal{R}^{r\times m}$. These matrices fulfil the conditions

$$\mathbf{B}^{-}\,\mathbf{B} = \mathbf{I}_n,\ \ \mathbf{B}^{0}\,\mathbf{B} = \mathbf{0}, \tag{2.12}$$

where $\mathbf{I}_n$ is the identity matrix of rank n and 0 is $(r \times n)$-matrix with zero entries. The definition (2.12) is not unique: for a given matrix $\mathbf{B}$ one can find several matrices $\mathbf{B}^{-}$, $\mathbf{B}^{0}$. It does not preclude efficient calculation of such matrices on computer (compare the program *Solve* given in the Appendix of the book [4]). Physically the calculation of $\mathbf{B}^{-}$ and $\mathbf{B}^{0}$ in (2.11) corresponds to the selection of the primary system in the Direct Stiffness Method. The entries of $\mathbf{Z}$ are the redundant stresses – the unknowns of that method.

In kinematics the matrices $\mathbf{B}^{-}$, $\mathbf{B}^{0}$ have the following meaning: the relation (2.7) can be inverted, i.e. written as

$$\mathbf{q} = \mathbf{B}^{-}\mathbf{s}, \tag{2.13}$$

provided the generalized strains fulfil the compatibility condition

$$\mathbf{B}^{0}\,\mathbf{s} = \mathbf{0}. \tag{2.14}$$

Let us consider two simple skeletal systems that illustrate the introduced matrix relations: the truss shown in Fig. 2.2 and the frame depicted in Fig. 2.3. As commonly known, the kinematics of a truss is determined by the linear displacements u_i, v_i of the nodes and by the elongations ΔL^j of the bars. The former are grouped in the matrix $\mathbf{q}$, the latter – in the matrix $\mathbf{s}$. For the truss shown in Fig. 2.2 this gives

$$\mathbf{q} = \{v_1, u_2, v_2, u_4, v_4\},$$
$$\mathbf{s} = \{\Delta L^1, \Delta L^2, \dots, \Delta L^6\}.$$

Hence, $n = 5$, $m = 6$, $r = 1$, and the degree of redundancy is equal to one.

Ignoring the reactions of supports we group the rest of the nodal forces in the matrix $\mathbf{Q}$ and we assign the axial forces in the bars to the matrix $\mathbf{S}$:

$$\mathbf{Q} = \{V_1, U_2, V_2, U_4, V_4\},$$

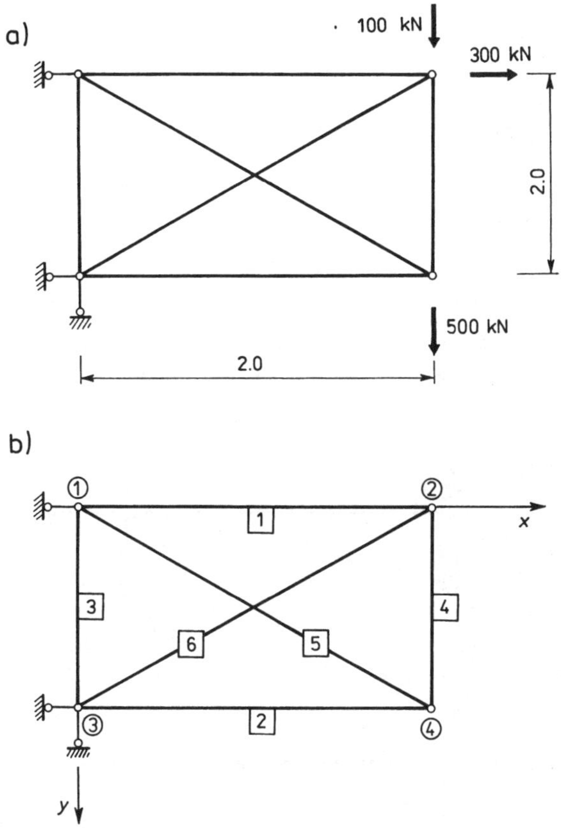

Fig. 2.2. Example of truss: **a)** dimensions and loading, **b)** computational scheme

$$\mathbf{S} = \{N^1, N^2, \ldots, N^6\}.$$

The matrix **B** can be constructed by imposing unit displacements of the nodes and examining elongations of the bars. However, it is easier to find its transposed form writing the equilibrium equations for the nodes. Such a procedure gives

$$\mathbf{B}^T = \begin{bmatrix} 0 & 0 & -1 & 0 & -0.707 & 0 \\ 1 & 0 & 0 & 0 & 0 & 0.707 \\ 0 & 0 & 0 & -1 & 0 & -0.707 \\ 0 & 1 & 0 & 0 & 0.707 & 0 \\ 0 & 0 & 0 & 1 & 0.707 & 0 \end{bmatrix}$$

For framed structures not only the linear displacements of the nodes are important but their rotations as well: a position of the i-th node in the deformed configuration is determined by 3 parameters u_i, v_i, φ_i, where φ_i denotes the angle of rotation in radians. As shown in Fig. 2.4, we must take

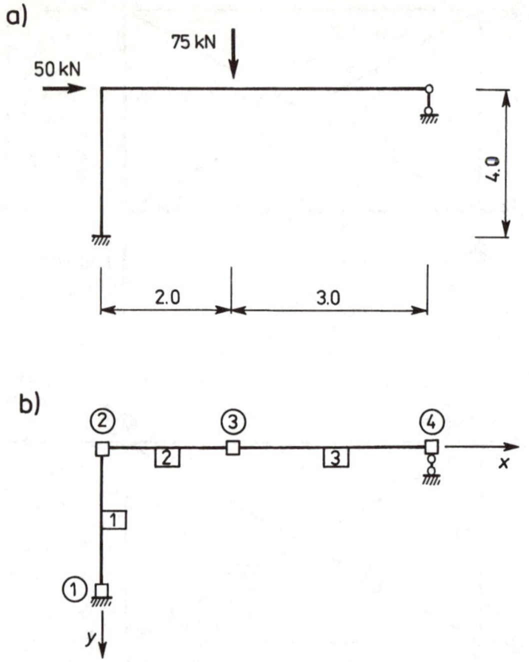

Fig. 2.3. Example of frame: **a)** dimensions and loading, **b)** computational scheme

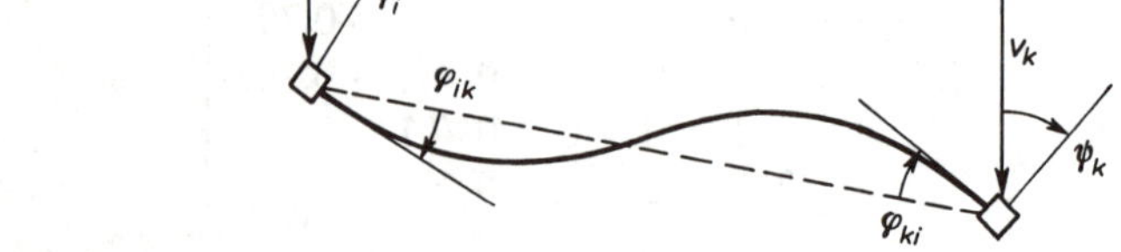

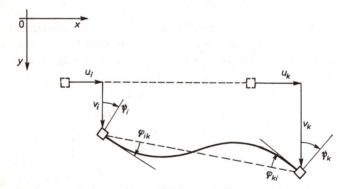

Fig. 2.4. Kinematic variables for j-th element of frame

into account for frames not only the elongation of the generic element but its bending as well.

A convenient strain measure for the bending mode is the angle of rotation of the end cross-section of the bar. Such angles should be taken relative to the chord of the bar, as shown in Fig. 2.4. Hence, the deformed state of the j-th element in bending is described by 3 parameters $\varphi_{ik}, \varphi_{ki}, \Delta L^j$.

The static counterparts of the generalized displacements u_i, v_i, φ_i are the forces U_i, V_i and the moment Φ_i acting on the i-th node. The stress state of the j-th bar is determined by the bending moments M_{ik}, M_{ki} acting at its ends. Note that the axial force N^j and the shear force T^j are treated as the internal reactions and are not incorporated into the matrix **S**. They would become the generalized stresses if we would take into account the elongation ΔL^j and if we would substitute the Timoshenko beam for the Euler beam.

The axial stiffness of bars is much higher than their bending stiffness. Therefore, the axial strains are usually neglected. This leaves 2 variables $\varphi_{ik}, \varphi_{ki}$ in the description of the deformed state of the j-th element. One should remember, however, that inextensible bars reduce the number of degrees of freedom of the nodes. For example, the node 2 of the frame depicted in Fig. 2.3 is deprived then from the possibility of vertical displacement and horizontal displacements of the nodes 2 and 3 become identical.

Hence, the frame depicted in Fig. 2.3 can be described by the following matrices of state variables:

$$\begin{aligned}
\mathbf{q} &= \{u_2, \varphi_2, v_3, \varphi_3, \varphi_4\} \\
\mathbf{Q} &= \{U_2, \Phi_2, V_3, \Phi_3, \Phi_4\} \\
\mathbf{s} &= \{\varphi_{12}, \varphi_{21}, \varphi_{23}, \varphi_{32}, \varphi_{34}, \varphi_{43}\} \\
\mathbf{S} &= \{M_{12}, M_{21}, M_{23}, M_{32}, M_{34}, M_{43}\}
\end{aligned}$$

We see that $n = 5$ and $m = 6$ which gives the degree of redundancy $r = 1$.

Similarly as it was for a truss, the matrix **B** of a frame can be constructed either kinematically or statically. The latter method means that the rows of $\mathbf{B}^T$ are found by considering equilibrium of the nodes. The kinematic approach is more convenient for frames with orthogonal bars. It follows from (2.7) that the entries of the k-th column of **B** are the generalized strains caused by the unit k-th displacement. Let us introduce a hinge into each junction element-node. Then let us suppress all degrees of freedom for the resulting mechanism, except the k-th degree of freedom. Imposing the unit value of the k-th displacement and calculating the resulting rotations of hinges, we obtain

$$\mathbf{B} = \begin{bmatrix} -0.25 & 0 & 0 & 0 & 0 \\ -0.25 & 1 & 0 & 0 & 0 \\ 0 & 1 & -0.5 & 0 & 0 \\ 0 & 0 & -0.5 & 1 & 0 \\ 0 & 0 & 0.333 & 1 & 0 \\ 0 & 0 & 0.333 & 0 & 1 \end{bmatrix}$$

Using the procedure *Solve* [4] we calculate the generalized inverse

$$\mathbf{B}^- = \begin{bmatrix} -4 & 0 & 0 & 0 & 0 & 0 \\ 0 & 0 & 1 & -0.6 & 0.6 & 0 \\ 0 & 0 & 0 & -1.2 & 1.2 & 0 \\ 0 & 0 & 0 & 0.4 & 0.6 & 0 \\ 0 & 0 & 0 & 0.4 & -0.4 & 1 \end{bmatrix}$$

and the kernel

$$\mathbf{B}^0 = \begin{bmatrix} -1 & 1 & -1 & 0.6 & -0.6 & 0 \end{bmatrix}$$

These matrices will be used in Sect. 2.4.

2.2.2
Constitutive Relations – Linear Elastic Model

The generalized strains and stresses of a structure made of linear elastic material are coupled by the relations

$$\begin{aligned} \mathbf{S} &= \mathbf{C}\,\mathbf{s}, \\ \mathbf{s} &= \mathbf{C}^{-1}\,\mathbf{S}, \end{aligned} \tag{2.15}$$

where $\mathbf{C}$ is a $m \times m$ matrix of elasticity. The equations (2.15) are the structural counterparts of the Hooke's law (2.30) considered in Part II. They are also a special case of the constitutive law (5.78) given in Part I of [35].

For trusses $\mathbf{C}$ is diagonal and it contains the axial stiffnesses of individual bars:

$$C^j = \frac{EA^j}{L^j}. \tag{2.16}$$

Here E is the Young's modulus of material, A^j is the cross-sectional area of the j-th bar and L^j denotes the length of that bar.

Given

$$E = 2.0 \cdot 10^5\ \mathrm{MPa}, \qquad A^1 = A^2 = \ldots = A^6 = 28.3\ \mathrm{cm}^2$$

the elasticity matrix of the truss depicted in Fig. 2.2 reads:

$$\mathbf{C} = \mathrm{diag}\,\{1, 1, 1, 1, 0.707, 0.707\} \cdot 28.3 \cdot 10^4\ \mathrm{kN\,m^{-1}}.$$

For frames $\mathbf{C}$ is quasi-diagonal: it is composed of the submatrices $\mathbf{C}^j$ of individual bars. Depending upon the assumed model of a bar, one obtains specific $\mathbf{C}^j$. For inextensible bars

$$\mathbf{C}^j = \begin{bmatrix} \frac{4EJ^j}{L^j} & \frac{2EJ^j}{L^j} \\ \frac{2EJ^j}{L^j} & \frac{4EJ^j}{L^j} \end{bmatrix} \tag{2.17}$$

Let the frame shown in Fig. 2.3 be made of the I-section with the dimensions:

$$B = 150\ \mathrm{mm},\ H = 250\ \mathrm{mm},\ B' = 10\ \mathrm{mm},\ H' = 20\ \mathrm{mm}.$$

Such a cross-section has the following geometric characteristics:

$$A = 84\ \mathrm{cm}^2,\ J = 9395\ \mathrm{cm}^4,\ W = 844\ \mathrm{cm}^4.$$

Assuming $E = 2 \cdot 10^5$ MPa and neglecting the axial strains, we obtain

$$\mathbf{C} = \begin{bmatrix} 1 & 0.5 & 0 & 0 & 0 & 0 \\ 0.5 & 1 & 0 & 0 & 0 & 0 \\ 0 & 0 & 2 & 1 & 0 & 0 \\ 0 & 0 & 1 & 2 & 0 & 0 \\ 0 & 0 & 0 & 0 & 1.333 & 0.667 \\ 0 & 0 & 0 & 0 & 0.667 & 1.333 \end{bmatrix} \cdot 2.11 \cdot 10^4 \mathrm{kN\,m}.$$

2.2.3 Constitutive Relations – Rigid-Plastic Model

We are going to use this model in Sect. 2.4 devoted to the optimum design of structures having a prescribed ultimate load carrying capacity. Skipping detailed derivations that can be found in books [4] and [19], we recall in the sequel the final form of the matrix relations that describe the behaviour of a structure made from rigid-perfectly-plastic material.

In order to be able to apply Linear Programming we assume that the yield surfaces of structural cross-sections are piecewise-linear. Then the constitutive relations for the assembled structure can be written as:

$$\begin{aligned} \mathbf{F} = \mathbf{N}^T\mathbf{S} - \mathbf{\Lambda} &\le \mathbf{0}, \\ \dot{\mathbf{s}} &= \mathbf{N}\dot{\lambda}, \\ \dot{\lambda} &\ge \mathbf{0}, \\ \dot{\lambda}^T\mathbf{F} &= 0. \end{aligned} \tag{2.18}$$

The first inequality determines admissible stress states: a projection of the vector $\mathbf{S}$ on the normal to the k-th yield plane can not be longer than the distance Λ_k of that plane from the origin of the reference frame $0, S_1, S_2, \ldots, S_m$ (Fig. 2.5). The distance Λ_k is called the plastic modulus and the difference F_k is called the plastic potential of the k-th yield plane. The plastic modulae and the plastic potentials are grouped, respectively, in the matrices $\mathbf{\Lambda}, \mathbf{F} \in \mathcal{R}^p$, where p is the number of yield planes. The matrix $\mathbf{N} \in \mathcal{R}^{m\times p}$ collects the unit normals to such planes. Hence, it is the gradient matrix of the yield surface.

The remaining conditions (2.18) define the associated plastic flow rule. It says that the direction of the plastic strain rate vector $\dot{\mathbf{s}}$ is perpendicular to the active yield plane and that the length of that vector is determined by the non-negative plastic multiplier $\dot{\lambda}_k$. The latter can assume a positive value only when $\mathbf{S}$ touches the k-th plane ($F_k = 0$). Such an yield plane is called active. The plastic multipliers are grouped in the matrix $\dot{\lambda} \in \mathcal{R}^p$. They are conjugate (dual) variables for the plastic modulae, since the scalar product

$$d = \dot{\mathbf{s}}^T\mathbf{S} = \dot{\lambda}^T\mathbf{\Lambda} \tag{2.19}$$

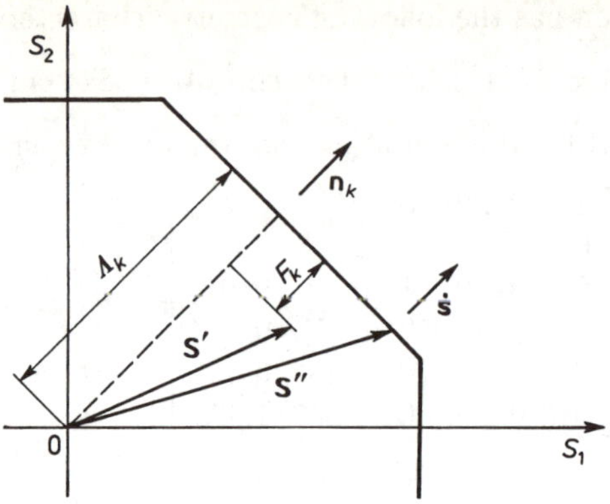

Fig. 2.5. Piecewise-linear yield curve and associated flow rule

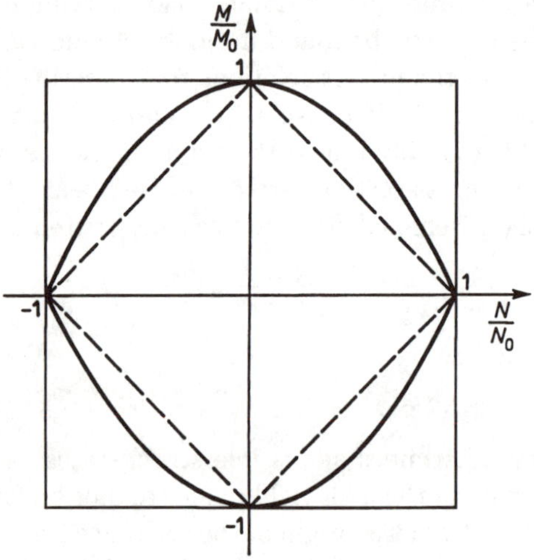

Fig. 2.6. Exact and linearized yield curves for rectangular cross-section

gives the power dissipated during plastic deformation.

Let us recall the most common yield conditions $(2.18)_1$. We restrict ourselves to the cross-sections of frames, since the rigid-perfectly-plastic model is seldom used for trusses: their compressed members can buckle prior to yielding. In principle one could begin with the most general case taking into account the influence of all generalized stresses – the bending moment M, the shear force T and the axial force N – on the yielding of the cross-section. However in the most frames the normal stresses are much more significant than the shear stresses. This allows us to treat T as an internal reaction.

The yield curve for the rectangular cross-section loaded by the bending moment M and the axial force N consists of 2 parabola shown in Fig. 2.6.

The corresponding stress admissibility condition

$$\mp\frac{M}{M_0} + \left(\frac{N}{N_0}\right)^2 \leq 1. \tag{2.20}$$

is obviously non-linear. Two important characteristics of the cross-section appear in (2.20): the yield moment

$$M_0 = \sigma_0 W_{\mathrm{pl}} \tag{2.21}$$

and the yield axial force

$$N_0 = \sigma_0 A, \tag{2.22}$$

where σ_0 is the yield stress of material and W_{pl} is the plastic modulus of the cross-section. For rectangular cross-section of the width B and the height H we have

$$W_{\mathrm{pl}} = \frac{1}{4}\,BH^2. \tag{2.23}$$

The simplest and the most frequently used piecewise-linear approximation of (2.20) is the following set of 4 inequalities:

$$\begin{aligned} M + \beta N - M_0 &\leq 0,\\ -M + \beta N - M_0 &\leq 0,\\ -M - \beta N - M_0 &\leq 0,\\ M - \beta N - M_0 &\leq 0, \end{aligned} \tag{2.24}$$

where

$$\beta = \frac{M_0}{N_0}. \tag{2.25}$$

Using (2.24) instead of (2.20) we replace the non-linear condition by an inscribed rhombus. Physically such an approximation means substituting the rectangular solid cross-section (Fig. 2.7a) by the sandwich one shown in Fig. 2.7b. The plastic modulus of the ideal sandwich cross-section is

$$W_{\mathrm{el}} = W_{\mathrm{pl}} = BH'H. \tag{2.26}$$

For I-section (Fig. 2.7c) we have

$$W_{\mathrm{pl}} \approx BH'H + \frac{1}{4}B'H^2. \tag{2.27}$$

Hence, it is reasonable to neglect the resistance of the web and to replace the I-section by the ideal sandwich one.

The conditions (2.24) must be fulfilled at both ends of a bar. Therefore, for the entire structure we must take into account $p = 8m$ inequalities. They can be put into the form $(2.18)_1$ if we take

$$\begin{aligned} \mathbf{N} &= \operatorname{diag}[\mathbf{N}^1\ \mathbf{N}^2\ \ldots \mathbf{N}^m],\\ \mathbf{F} &= \{\mathbf{F}^1\ \mathbf{F}^2\ \ldots \mathbf{F}^m\},\\ \mathbf{\Lambda} &= \{\mathbf{\Lambda}^1\ \mathbf{\Lambda}^2\ \ldots \mathbf{\Lambda}^m\}, \end{aligned} \tag{2.28}$$

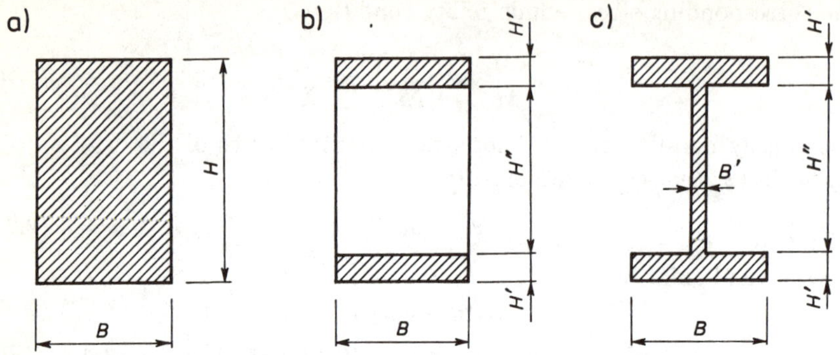

Fig. 2.7. Typical cross-sections: **a)** rectangular, **b)** ideal sandwich, **c)** rolled I-section

where the submatrices

$$
\begin{aligned}
\mathbf{N}^j &= \begin{bmatrix} 1 & -1 & -1 & 1 & 0 & 0 & 0 & 0 \\ 0 & 0 & 0 & 0 & 1 & -1 & -1 & 1 \\ \beta & \beta & -\beta & -\beta & \beta & \beta & -\beta & -\beta \end{bmatrix}, \\
\mathbf{F}^j &= \{F^I_{ik}, F^{II}_{ik}, F^{III}_{ik}, F^{IV}_{ik}, F^I_{ki}, F^{II}_{ki}, F^{III}_{ki}, F^{IV}_{ki}\} \\
\mathbf{\Lambda}^j &= \{1,1,1,1,1,1,1,1\} \cdot M_0^j
\end{aligned}
\tag{2.29}
$$

represent plastic properties of the j-th element.

If axial forces are absent (beams, grillages) or if their influence on yielding is minor as compared to the influence of bending moments, then the admissibility condition (2.24) can be replaced by 2 inequalities:

$$
\begin{aligned}
M - M_0 &\leq 0, \\
-M - M_0 &\leq 0.
\end{aligned}
\tag{2.30}
$$

The first one takes care of positive bending moments, whereas the second one constrains negative bending moments.

Imposing conditions (2.30) at each end of a bar, we obtain $2m$ inequalities of the type $(2.18)_1$. The matrices $\mathbf{N}^j, \mathbf{F}^j, \mathbf{\Lambda}^j$ have then the following internal structure:

$$
\begin{aligned}
\mathbf{N}^j &= \begin{bmatrix} 1 & -1 & 0 & 0 \\ 0 & 0 & 1 & -1 \end{bmatrix}, \\
\mathbf{F}^j &= \{F^I_{ik}, F^{II}_{ik}, F^I_{ki}, F^{II}_{ki},\}, \\
\mathbf{\Lambda}^j &= \{1,1,1,1\} \cdot M_0^j.
\end{aligned}
\tag{2.31}
$$

For example, rigid-perfectly-plastic properties of the frame depicted in

Fig. 2.3 are described by:

$$\mathbf{N} = \begin{bmatrix} 1 & -1 & 0 & 0 & 0 & 0 & 0 & 0 & 0 & 0 & 0 & 0 \\ 0 & 0 & 1 & -1 & 0 & 0 & 0 & 0 & 0 & 0 & 0 & 0 \\ 0 & 0 & 0 & 0 & 1 & -1 & 0 & 0 & 0 & 0 & 0 & 0 \\ 0 & 0 & 0 & 0 & 0 & 0 & 1 & -1 & 0 & 0 & 0 & 0 \\ 0 & 0 & 0 & 0 & 0 & 0 & 0 & 0 & 1 & -1 & 0 & 0 \\ 0 & 0 & 0 & 0 & 0 & 0 & 0 & 0 & 0 & 0 & 1 & -1 \end{bmatrix},$$

$$\mathbf{F} = \{F_{12}^I, F_{12}^{II}, F_{21}^I, F_{21}^{II}, F_{23}^I, F_{23}^{II}, F_{32}^I, F_{32}^{II}, F_{34}^I, F_{34}^{II}, F_{43}^I, F_{43}^{II}\},$$

$$\mathbf{\Lambda} = \{M_0^1, M_0^1, M_0^1, M_0^1, M_0^2, M_0^2, M_0^2, M_0^2, M_0^3, M_0^3, M_0^3, M_0^3\}.$$

2.2.4 Elastic Analysis

Relations (2.7), (2.9), (2.19) constitute a complete set of $(2m + n)$ linear algebraic equations that enable us to find the elastic response $\mathbf{q}_*, \mathbf{s}_*, \mathbf{S}_*$ of the structure to the given load $\mathbf{Q}_*$. Since such a set is too large one reduces its size choosing as the primary unknowns either displacements or redundant stresses. The first method is simpler: one takes the matrix equation of equilibrium (2.9), replaces the generalized stresses by the generalized strains following from the constitutive law (2.19) and substitutes displacements for strains according to the kinematic equation (2.7). The results is the set of n equations

$$\mathbf{K}\,\mathbf{q} = \mathbf{Q} \tag{2.32}$$

that can be solved with respect to the unknown displacements q_i. The stiffness matrix

$$\mathbf{K} = \mathbf{B}^T\,\mathbf{C}\,\mathbf{B} \tag{2.33}$$

is symmetric. In practice Eq. (2.33) is applied for the j-th element and the submatrices $\mathbf{K}^j$ are assembled into the global matrix $\mathbf{K}$. Such a procedure allows us to avoid storing large matrices $\mathbf{B}$ and $\mathbf{C}$. Note that Eq. $(1.7)_1$, Part II, remains valid for each type of structure. In the case of a bar the prescribed integration can be performed analytically since the exact shape functions (Hermitian polynomials) are known.

After the displacements $\mathbf{q}_*$ were calculated, one obtains the internal forces from the expression

$$\mathbf{S} = \mathbf{C}\,\mathbf{B}\,\mathbf{q} \tag{2.34}$$

that follows Eqs. (2.7), (2.19). Equation (2.34) is also used in practice at the element level.

Let us calculate the elastic response of the truss shown in Fig. 2.2 to the load

$$\mathbf{Q}_* = \{0,\ 300,\ 100,\ 0,\ 500\}\ \text{kN}.$$

Substituting into Eq. (2.33) the matrices **B** and **C** that we found in Sects. 2.2.1 and 2.2.2, we obtain the following stiffness matrix:

$$\mathbf{K} = \begin{bmatrix} 38.3 & 0 & 0 & -10.0 & -10.0 \\ & 38.3 & -10.0 & 0 & 0 \\ & & 38.3 & 0 & -28.3 \\ & \text{sym.} & & 38.3 & 10.0 \\ & & & & 38.3 \end{bmatrix} \cdot 10^4\,\mathrm{kN\,m^{-1}}.$$

The solution of Eq. (2.32) is

$$\mathbf{q}_* = \{0.113,\ 0.205,\ 0.484,\ -0.113,\ 0.547\}\ \mathrm{cm}.$$

Substituting this result into Eq. (2.34) we obtain the axial forces in the bars:

$$\mathbf{S}_* = \{579,\ -321,\ -321,\ 179,\ 454,\ -395\}\ \mathrm{kN}.$$

2.2.5 Ultimate Load Analysis

Let us consider a structure made from the rigid-perfectly-plastic material. The plastic resistance of that structure is defined by the given matrix of plastic modulae $\mathbf{\Lambda}_*$. Let the loading be

$$\mathbf{Q} = \mu\, \mathbf{Q}_*, \tag{2.35}$$

where μ is the load factor and $\mathbf{Q}_*$ is the given reference load. Increasing μ slowly from zero we reach finally its ultimate value μ_* that corresponds to the ultimate load carrying capacity of the structure. If $\mathbf{Q}_*$ corresponds to the service load, then μ_* indicates the safety level against the plastic collapse.

Finding μ_* is the aim of the part of the Theory of Plasticity called the Ultimate Load Analysis [28]. The advantages of this approach are the simplicity and the good agreement with experiments for such popular structural materials as steel and reinforced concrete. However, one should not forget the assumptions laying behind the Ultimate Load Analysis: the loading is quasi-static and proportional, the displacements are small, no plastic hardening occurs and the structure remains stable up to the plastic collapse.

The following two fundamental theorems govern the ultimate load factor:

Static Theorem. *μ_* is the largest of the statically admissible load factors μ_S;*

Kinematic Theorem. *μ_* is the smallest of the kinematically admissible load factors μ_K.*

The load factor is called *statically admissible* if it corresponds to the distribution of internal forces $\mathbf{S}_S$ that satisfies simultaneously the equilibrium equation (2.9) and the admissibility condition $(2.18)_1$. On the other hand, the load factor is called *kinematically admissible* if a plastic collapse mechanism $(\dot{\mathbf{s}}_K, \dot{\mathbf{q}}_K)$ can be associated with it. In such a mechanism the power generated by the load must be dissipated by the plastic strain. Using (2.19), (2.35) we can express this energy balance as

$$\mathbf{Q}_K^T \dot{\mathbf{q}}_K = \mu_K \mathbf{Q}_*^T \dot{\mathbf{q}}_K = \mathbf{S}_K^T \dot{\mathbf{s}}_k = \mathbf{\Lambda}_*^T \dot{\lambda}_K .$$

Hence,

$$\mu_K = \frac{\mathbf{\Lambda}_*^T \dot{\lambda}_K}{\mathbf{Q}_*^T \dot{\mathbf{q}}_K} \tag{2.36}$$

i.e. the kinematically admissible load factor is equal to the ratio of the dissipated power and the power generated by the reference load.

Under the assumption of piecewise-linear yield surface the Ultimate Load Analysis is reduced to the solution of the following dual pair of LP-problems [7, 4]:

$$\max\{\, \mu \mid -\mathbf{B}^T\mathbf{S} + \mu \mathbf{Q}_* = \mathbf{0},\ \mathbf{N}^T\mathbf{S} \le \mathbf{\Lambda}_* \}, \tag{2.37}$$

$$\min\{\, \mathbf{\Lambda}_*^T \dot{\lambda} \mid -\mathbf{B}\dot{\mathbf{q}} + \mathbf{N}\dot{\lambda} = \mathbf{0},\ \dot{\lambda} \ge \mathbf{0} \}. \tag{2.38}$$

The primal problem (2.37) reproduces directly the Static Theorem. The dual one (2.38) can be derived from the Kinematic Theorem. Namely, instead of minimizing (2.33) one can minimize the dissipated power keeping

Table 2.3. Simplex table for problem (2.37) – determining ultimate load factor for frame shown in Fig. 2.3

M_{12}	M_{21}	M_{23}	M_{32}	M_{34}	M_{43}	μ		
−0.25	−0.25	0	0	0	0	−50.0	0	= 0
0	1.0	1.0	0	0	0	0	0	= 0
0	0	−5.0	−5.0	0.333	0.333	−75.0	0	= 0
0	0	0	1.0	1.0	0	0	0	= 0
0	0	0	0	0	1.0	0	0	= 0
−1.0	0	0	0	0	0	0	158.0	≤ 0
1.0	0	0	0	0	0	0	158.0	≤ 0
0	−1.0	0	0	0	0	0	158.0	≤ 0
0	1.0	0	0	0	0	0	158.0	≤ 0
0	0	−1.0	0	0	0	0	158.0	≤ 0
0	0	1.0	0	0	0	0	158.0	≤ 0
0	0	0	−1.0	0	0	0	158.0	≤ 0
0	0	0	1.0	0	0	0	158.0	≤ 0
0	0	0	0	−1.0	0	0	158.0	≤ 0
0	0	0	0	1.0	0	0	158.0	≤ 0
0	0	0	0	0	−1.0	0	158.0	≤ 0
0	0	0	0	0	1.0	0	158.0	≥ 0
0	0	0	0	0	0	1.0	max	

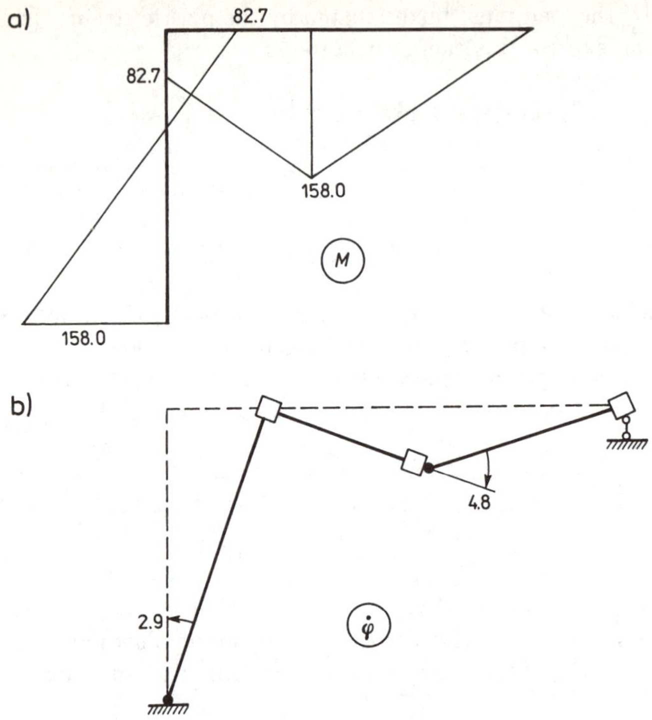

Fig. 2.8. Ultimate state of frame shown in Fig. 2.3: **a)** distribution of bending moments, **b)** collapse mechanism

the power generated by $\mathbf{Q}_*$ normalized. In addition to that one has to ensure the kinematic compatibility of the plastic strain rates and the plastic displacement rates, as well as the non-negativeness of the plastic multipliers. Thus, we arrive at the cost function and the constraints of the LP-problem (2.38).

Solving the dual pair (2.37), (2.38) we obtain the ultimate load factor μ_*, the distribution of internal forces at plastic collapse $\mathbf{S}_*$ and the collapse mechanism $(\dot{\boldsymbol{\lambda}}_*, \dot{\mathbf{q}}_*)$. Only the value μ_* is unique.

Let us take the ultimate load analysis of the frame depicted in Fig. 2.3 as an example. Assuming the constant dimensions of the cross-section given in Sect. 2.2.2 throughout the frame, neglecting the resistance of the web and taking the yield stress $\sigma_0 = 210$ MPa we obtain according to (2.21) and (2.22):

$$M_0 = 158\,\text{kN m}, \; N_0 = 1260\,\text{kN}, \; \beta = 0.125.$$

Let the reference load be (Fig. 2.3a):

$$\mathbf{Q}_* = \{50.0,\ 0,\ 75.0,\ 0,\ 0\}.$$

Simplex table 2.3 refers to the problem (2.37) formulated for the bending-only model of the considered frame. Solving that problem we obtain the ultimate load factor $\mu_* = 1.20$ and the distribution of the bending moments at plastic collapse (Fig. 2.8a). The collapse mechanism (Fig. 2.8b) follows from the solution of the dual problem. If we would take axial forces into account, then instead of 17 constraints for 7 variables we would have 32 constraints for 10 variables in the problem (2.37). The ultimate load factor would drop to 1.19. The difference is of the order of 1 % which confirms the rationality of applying the simpler model.

2.3 Optimum Design of Elastic Trusses

In the present section we consider optimum design of elastic trusses subjected to static loads. Our main aim will be the optimum topology of the truss since the optimization of cross-sectional areas under given topology is the subject of Sect. 3. We take the weight of the truss as the cost function, neglecting the costs of supports, nodal connectors, assembly work etc. Assuming material of the truss to be homogeneous (steel) we replace the minimization of weight by the minimization of volume.

2.3.1 Requirements of Civil Engineering Code

Design of a structure begins usually with an initial solution: the designer chooses the configuration of the structure and the characteristics of cross-sections taking into account given data (outer dimensions of the structure, loading, material, etc.) and his personal experience. After that the analysis of the structure is performed in order to assure that the structure fulfils all the requirements of the Civil Engineering Code.

In Poland the code appropriate for steel trusses in Civil Engineering is PN-76/B-03200, [26]. According to that code one has to ensure that none of the following ultimate states occurs under the given loading:

I. Ultimate Load Carrying Capacity,
II. Ultimate Service State.

The first ultimate state can be associated with:

a) rupture of the most strained bar;
b) local or global loss of stability;
c) conversion of a structure into mechanism;
d) failure caused by fatigue;
e) excessive elastic or plastic strain, displacement or rupture causing the change of shape of structure;
f) loss of stability of structure taken as a rigid body (horizontal displacement or falling).

The second ultimate state can be associated with:

a) excessive strains that prevent normal usage of the structure;
b) excessive displacements that do not cause the loss of stability and do not change the shape of the structure;
c) excessive vibrations;
d) damage caused by an inappropriate maintenance or insufficient resistance against corrosion, fire and other external agents.

Ultimate states of type I are checked against the norm loading, whereas the characteristic loading is taken when checking against the states of type II. A detailed presentation of the code requirements for steel structures can be found, e.g., in the book [24]. The above classification of the ultimate states was taken also from that source.

The above listing shows that in order to develop the project of a steel truss that confirms to the Civil Engineering Code one has to solve several problems falling out of the scope of Structural Analysis. Additionally such a project should be acceptable from the point of view of production and construction. Typically the process of design consists of several iterative steps. At each of them some constraints induced by the code or by technology become satisfied. Both the Optimum Elastic Design considered in this Section and the Optimum Plastic Design introduced in Sect. 2.4 should be understood as the intermediate stages in such an iterative design process.

Let us restrict ourselves to the ultimate states Ia, Ib related to the stress state of the truss and to the state IIb related to its displacements. The stresses in bars subjected to tension should not exceed the admissible stress for the material of which the truss has been made. Hence, the inequality

$$\frac{N^j}{A^j} \leq f_d. \tag{2.39}$$

should hold. For bars in compression the constraint

$$-\frac{N^j}{\varphi A^j} \leq f_d, \tag{2.40}$$

applies, where φ is the buckling coefficient dependent upon the slenderness of the bar and on the kinematic conditions at its ends. Formulae (2.39), (2.40) are valid for welded trusses. If nodes are bolted-on or riveted then one should distinguish between the brut and net areas of the cross-section.

The ultimate state IIb is excluded if the maximum displacement of a node does not exceed the admissible value:

$$\max q_i \leq q_0. \tag{2.41}$$

The code [26] defines q_0 dependent on the type of structure and its dimensions. For example, the vertical displacement of a roof truss having a span $L < 15$ m should not exceed the value $L/300$.

Structures in Civil Engineering are subjected to several loadings and the code requires that they should resist each single loading case and the most unfavourable combination of them. Therefore, the envelopes of stresses and displacements are calculated during elastic analysis. Such envelopes allow the designer to substitute the extreme values of N^i and q_i into the left hand side of the formulae (2.39), (2.40) and (2.41).

2.3.2 General Formulation of Optimum Design Problem

In order to become a well-defined optimum design problem we should specify its four components: the cost function, the constraints, the design variables and the state variables. Let us begin with the declaration what do we mean under the optimum truss: it is the truss of minimum weight satisfying the conditions (2.39), (2.40), (2.41) for a given loading.

Let the considered truss be composed of m prismatic bars. If A^j is the cross-sectional area and L^j is the length of the j-th bar then the volume of the truss is

$$f = \sum_{j=1}^{m} A^j L^j = \mathbf{A}^T \mathbf{L}, \tag{2.42}$$

where $\mathbf{A} \in \mathcal{R}^m$, $\mathbf{L} \in \mathcal{R}^m$. Thus we have defined the cost function.

Looking for optimum design of a truss we can modify the following:

a) connections between nodes,
b) co-ordinates of nodes,
c) cross-sectional areas of bars.

The topological optimum design that follows from the choice a) is not well suited for Mathematical Programming. Already in 1960's the attempts to apply Linear Programming for such problems were made [8] but the results were not satisfactory. Nowadays it is usually assumed that the connections between nodes are given and the optimization concerns the nodal co-ordinates and the cross-sectional areas. Let us consider such choices of the design variables in the sequel.

The co-ordinates of nodes related to the global reference frame $0, x, y$ can be grouped in the matrix $\mathbf{w} = \{\mathbf{w}_1, \mathbf{w}_2, \ldots, \mathbf{w}_w,\}$, where $\mathbf{w}_i = \{x_i, y_i\}$ holds for the i-th node and w is the number of nodes. Let

$$\mathbf{w} = \bar{\mathbf{w}} + \mathbf{G}^T \mathbf{X}, \tag{2.43}$$

where $\bar{\mathbf{w}} = \{\bar{\mathbf{w}}_i\}$ groups fixed co-ordinates, $\mathbf{G} = \{\mathbf{G}_i\} \in \mathcal{R}^{t \times n}$ is the given shape matrix and the matrix $\mathbf{X} \in \mathcal{R}^t$ contains non-negative design variables. Relation (2.43) is general enough: it reflects the circumstance that some nodal co-ordinates can be prescribed a priori and it covers the common case of interdependence of the co-ordinates. For example, if the co-ordinates of the i-th node are fixed then $\mathbf{w}_i = \bar{\mathbf{w}}_i$ and $\mathbf{G}_i = \mathbf{0}$. An illustration of the interdependence of nodal co-ordinates can be searching for the optimum height of a truss with parallel chords. Then all vertical co-ordinates of nodes belonging to a particular chord are represented by a single design variable.

Since the length of the j-th bar connecting the nodes i and k is equal to

$$L^j = \sqrt{(x_k - x_i)^2 + (y_k - y_i)^2}, \tag{2.44}$$

the entries of $\mathbf{L}$ are non-linear functions of the design variables when the positions of nodes are otimized.

Let us turn now to the optimization of cross-sectional areas. Similarly as for Eq. (2.43) we take

$$\mathbf{A} = \bar{\mathbf{A}} + \mathbf{G}^T\mathbf{X}, \tag{2.45}$$

where the matrix $\bar{\mathbf{A}} = \{\bar{A}^j\}$ groups the fixed areas and $\mathbf{G} = \{\mathbf{G}^j\}$ is a given shape matrix. If the cross-sectional area of the j-th bar is not subject to optimization then $\mathbf{G}^j = \mathbf{0}$. Note that the matrix $\mathbf{G} \in \mathcal{R}^{t\times m}$ allows us also to take into account the unification of the cross-sectional areas: the number of decision variables t is usually much less than the number of bars m.

Let us illustrate Eqs. (2.43), (2.45) on the truss shown in Fig. 2.2. First we assume that both co-ordinates of the nodes 1, 2, 3 and the horizontal co-ordinate of the node 4 are fixed, whereas the vertical co-ordinate of the node 4 is to be optimized. Then

$$\begin{aligned}
\mathbf{X} &= \{y_4\},\ (t = 1),\\
\bar{\mathbf{w}} &= \{0, 0, 2, 0, 0, 2, 2, 0\}\\
\mathbf{G} &= [0\,0\,0\,0\,0\,0\,0\,1].
\end{aligned}$$

Next we would like to optimize the cross-sectional areas of truss under the assumption $A^5 = A^6$. Then we obtain

$$\begin{aligned}
\mathbf{X} &= \{A^1, A^2, \ldots, A^5\},\ (t = 5),\\
\bar{\mathbf{A}} &= \{0, 0, 0, 0, 0, 0\}\\
\mathbf{G} &= \begin{bmatrix} 1 & 0 & 0 & 0 & 0 & 0\\ 0 & 1 & 0 & 0 & 0 & 0\\ 0 & 0 & 1 & 0 & 0 & 0\\ 0 & 0 & 0 & 1 & 0 & 0\\ 0 & 0 & 0 & 0 & 1 & 1 \end{bmatrix}.
\end{aligned}$$

Obviously it is not reasonable to keep such sparse matrices explicitly in the computer memory, since relations between the design variables and the nodal co-ordinates or the cross-sectional areas can be coded in other way. However, we shall keep Eqs. (2.43), (2.45) for their simplicity.

Let us proceed with the definition of the constraints of our problem. Firstly, in many practical situations the condition $\mathbf{X} \geq \mathbf{0}$ turns out to be insufficient. One introduces then an admissible range for each design variable. In matrix notation this can be written as

$$\mathbf{X}^- \leq \mathbf{X} \leq \mathbf{X}^+, \tag{2.46}$$

where $\mathbf{X}^- \geq \mathbf{0}$, $\mathbf{X}^+ \geq \mathbf{0}$ are given column matrices of dimension t. Such constraints exclude optimum solutions that are non-acceptable from practical point of view (bars having too small or too large area of the cross-section). On the other hand, one should keep in mind that a wrong choice of the admissible range can lead to contradictory constraints and the lack of solution. Inside the "box" defined by the inequalities (2.46) we allow any values of the design variables. This means that we are dealing with the *continuous optimization* in contrast to the *discrete optimization*. The latter concerns the choice of optimum values from a given finite set and will be described in Sect. 3.

The code requirements (2.39), (2.40) allow us to define for the j-th bar the ultimate axial force in tension

$$N^{+j} = f_d A^j \tag{2.47}$$

and the ultimate axial force in compression

$$N^{-j} = -\varphi f_d A^j. \tag{2.48}$$

Grouping such ultimate forces in the matrices $\mathbf{S}^- \in \mathcal{R}^m$ and $\mathbf{S}^+ \in \mathcal{R}^m$ we can express the stress constraints as

$$\mathbf{S}^- \leq \mathbf{S} \leq \mathbf{S}^+. \tag{2.49}$$

The constraint (2.41) is satisfied when the absolute value of each entry of $\mathbf{q}$ is not greater than q_0. Introducing the column matrices

$$\mathbf{q}_0^- = \{-1, -1, \ldots, -1\} \cdot q_0, \qquad \mathbf{q}_0^+ = \{1, 1, \ldots, 1\} \cdot q_0,$$

of the dimension n we can express the constraints on displacements as

$$\mathbf{q}^- \leq \mathbf{q} \leq \mathbf{q}^+. \tag{2.50}$$

Thus, optimum design of the elastic truss under constraints on stresses and displacements can be formulated as:

find

$$\min f = \mathbf{A}^T \mathbf{L},$$

under constraints

$$\begin{array}{lrcl} \text{a) on state variables} & \mathbf{q} & = & \mathbf{K}^{-1}\mathbf{q}_0, \\ & \mathbf{S} & = & \mathbf{CBq}, \\ \text{b) on stresses} & \mathbf{S}^- & \leq \mathbf{S} \leq & \mathbf{S}^+, \\ \text{c) on displacements} & \mathbf{q}^- & \leq \mathbf{q} \leq & \mathbf{q}^+, \\ \text{d) on design variables} & \mathbf{X}^- & \leq \mathbf{X} \leq & \mathbf{X}^+. \end{array} \tag{2.51}$$

The character of the cost function and of the constraints depends upon whether we are optimizing the nodal co-ordinates or the cross-sectional areas. In the former case the matrices $\mathbf{A}$ and $\mathbf{S}^+$ are fixed whereas the matrices $\mathbf{L}, \mathbf{K}, \mathbf{C}, \mathbf{B}$ and $\mathbf{S}^-$ depend upon the design variables. In the latter case the matrices $\mathbf{L}$ and $\mathbf{B}$ are fixed and $\mathbf{A}, \mathbf{K}, \mathbf{C}$ are variable.

Hence, the optimum design of elastic truss turns out to be the NLP-problem with a non-linear cost function and non-linear constraints. Except for trivial cases of trusses composed of few bars such problems can not be solved directly by means of the NLP-algorithms because not all constraints are given explicitly. On must, therefore, apply simplified procedures.

2.3.3 Linearized Problem

The simplest and yet efficient way to simplify the problem (2.51) is to linearize it. Let $(\mathbf{X}, \mathbf{S}, \mathbf{q})$ be the current solution satisfying all constraints in (2.51). Taking linear approximation in the vicinity of that point we are looking for the increment of decision variables $d\mathbf{X}$ that minimises the increment of volume

$$df = \mathbf{c}^T d\mathbf{X}, \tag{2.52}$$

where

$$\mathbf{c} = \nabla \mathbf{L}^T \mathbf{A} + \nabla \mathbf{A}\mathbf{L}. \tag{2.53}$$

The matrices of bar lengths $\mathbf{L}$ and cross-sectional areas $\mathbf{A}$ as well as their gradients

$$\nabla \mathbf{L} \in \mathcal{R}^{m \times t}, \qquad \nabla L_{ij} = \frac{\partial L_i}{\partial X_j}, \qquad \nabla \mathbf{A} \in \mathcal{R}^{m \times t}, \qquad \nabla A_{ij} = \frac{\partial A_i}{\partial X_j}$$

are known from the current solution.

The linearized stress constraints can be written as

$$\mathbf{H}^+ d\mathbf{X} \le d\mathbf{S}^+, \qquad \mathbf{H}^- d\mathbf{X} \le d\mathbf{S}^-, \tag{2.54}$$

where

$$\begin{aligned} \mathbf{H}^+ &= \nabla \mathbf{S} - \nabla \mathbf{S}^+, & \mathbf{H}^- &= \nabla \mathbf{S} - \nabla \mathbf{S}^-, \\ d\mathbf{S}^+ &= \mathbf{S}^+ - \mathbf{S}, & d\mathbf{S}^- &= \mathbf{S}^- - \mathbf{S}. \end{aligned} \tag{2.55}$$

The matrices of internal forces $\mathbf{S}, \mathbf{S}^+, \mathbf{S}^-$ and their gradients

$$\nabla \mathbf{S} \in \mathcal{R}^{m \times t}, \qquad \nabla S_{ij} = \frac{\partial S_i}{\partial X_j},$$

$$\nabla \mathbf{S}^+ \in \mathcal{R}^{m \times t}, \qquad \nabla S_{ij}^+ = \frac{\partial S_i^+}{\partial X_j}, \qquad \nabla \mathbf{S}^- \in \mathcal{R}^{m \times t}, \qquad \nabla S_{ij}^- = \frac{\partial S_i^-}{\partial X_j}.$$

are again known from the current solution.

The displacement constraints have simpler form after linearization, since $\mathbf{q}_0$ does not depend upon $\mathbf{X}$:

$$\begin{aligned} \nabla \mathbf{q}\, d\mathbf{X} &\le d\mathbf{q}^+, \\ \nabla \mathbf{q}\, d\mathbf{X} &\ge d\mathbf{q}^-, \end{aligned} \tag{2.56}$$

where

$$d\mathbf{q}^+ = \mathbf{q}^+ - \mathbf{q}, \qquad d\mathbf{q}^- = \mathbf{q}^- - \mathbf{q}. \tag{2.57}$$

Also in this case the matrices of displacements $\mathbf{q}, \mathbf{q}_0$ and their gradients

$$\nabla \mathbf{q} \in \mathcal{R}^{n \times t}, \qquad \nabla q_{ij} = \frac{\partial q_i}{\partial X_j}$$

are known.

The original problem (2.51) is highly non-linear. In order to assure the convergence of the iterative linearization procedure one has to keep changes small. This can be accomplished by imposing constraints on the increments of decision variables:

$$d\mathbf{X}^- \leq d\mathbf{X} \leq d\mathbf{X}^+. \tag{2.58}$$

When the current value of the k-th decision variable approaches its limit X_k^- or X_k^+ (compare the conditions (2.46)), then the increments $dX_k^- \leq 0$, $dX_k^+ \geq 0$ must take it into account:

$$\begin{aligned} X_k^+ - X_k < dX_k^+ &\Rightarrow dX_k^+ = X_k^+ - X_k, \\ X_k - X_k^- > dX_k^- &\Rightarrow dX_k^- = X_k - X_k^-, \end{aligned} \tag{2.59}$$

Figure 2.9 explains those conditions. Note, that the increments of decision variables can be either positive or negative.

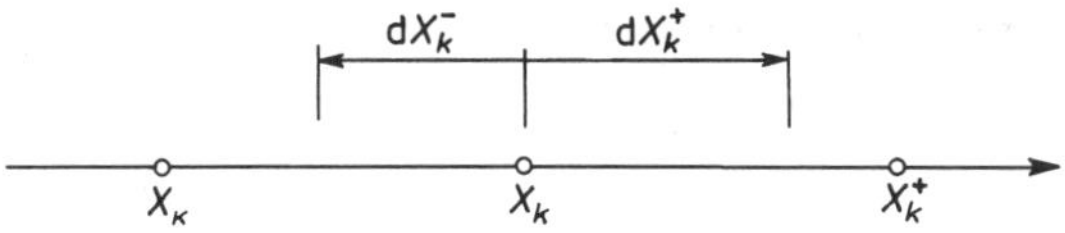

Fig. 2.9. Taking into account local and global constraints on increments of the k-th decision variable

As a result of linearization the original problem (2.51) has been replaced by the following LP-problem:

$$\begin{aligned} \min\{\mathbf{c}^T d\mathbf{X} \mid \mathbf{H}^+ d\mathbf{X} &\leq d\mathbf{S}^+, \\ \mathbf{H}^- d\mathbf{X} &\geq d\mathbf{S}^-, \\ d\mathbf{q}^- \leq \nabla \mathbf{q} d\mathbf{X} &\leq d\mathbf{q}^+, \\ d\mathbf{X}^- \leq d\mathbf{X} &\leq d\mathbf{X}^+\}. \end{aligned} \tag{2.60}$$

The internal structure of the simplex matrix for this problem is shown in Table 2.4. The increments obtained by solving (2.60) allows us to calculate the improved values of decision variables

$$\mathbf{X}' = \mathbf{X} + d\mathbf{X}_*. \tag{2.61}$$

In the next step $\mathbf{X}'$ is taken as the current solution. The iteration is stopped when the changes of the design become small enough.

Table 2.4 suggests that we have to solve the LP-problem having t unknowns and $2(m + n + t)$ constraints in each iterative step. Since many such step

Table 2.4. Structure of simplex matrix for linearized problem (2.60)

$d\mathbf{X}$		
$-\mathbf{H}^+$	$d\mathbf{S}^+$	≥ 0
$\mathbf{H}^-$	$-d\mathbf{S}^-$	≥ 0
$-\nabla\mathbf{q}$	$d\mathbf{q}^+$	≥ 0
$\nabla\mathbf{q}$	$-d\mathbf{q}^-$	≥ 0
$\mathbf{I}$	$-d\mathbf{X}^+$	≥ 0
$-\mathbf{I}$	$d\mathbf{X}^-$	≥ 0
$\mathbf{c}^T$	min	

may be necessary, it is reasonable to reduce the dimensions of (2.60). One of the possibilities is the so-called *active set strategy*: at current step only those constraints are taken into account that were satisfied as equalities within the accuracy δz in the previous step. It is usually possible to apply such value of δz that the number of active constraints remains much smaller than $2(m+n+t)$. Other way to increase the efficiency of the iterative procedure is the adaptive adjustment of the parameters dX_k^-, dX_k^+. For initial iterations one can allow larger variations of the decision variables, diminishing them in the vicinity of the optimum solution.

2.3.4 Gradient Matrices

We present now the formulae for the gradient matrices $\nabla\mathbf{q}, \nabla\mathbf{S}$ and $\nabla\mathbf{S}_0$ that appear in the linearized problem (2.60). Differentiating Eq. (2.32) with respect to the decision variable X_k under assumption $\mathbf{Q}_0 = \text{const}$ we obtain:

$$\mathbf{K}\nabla\mathbf{q}_k = \tilde{\mathbf{Q}}_k, \tag{2.62}$$

where

$$\tilde{\mathbf{Q}}_k = -\nabla\mathbf{K}_k\mathbf{q}. \tag{2.63}$$

Thus, the entries of the k-th column

$$\nabla\mathbf{q}_k = \left\{\frac{\partial\mathbf{q}}{\partial X_k}\right\}$$

of the gradient matrix $\nabla\mathbf{q}$ can be found by solving the matrix equation of the Direct Stiffness Method with the right hand side replaced by the fictitious load $\mathbf{Q}_k$. The entries of the derivative of stiffness matrix

$$\nabla\mathbf{K}_k = \left[\frac{\partial\mathbf{K}}{\partial X_k}\right]$$

depend upon that whether X_k controls a nodal co-ordinate or a cross-sectional area. In the first case both $\mathbf{B}$ and $\mathbf{C}$ are functions of X_k and differentiating Eq. (2.33) with respect to that variable we obtain

$$\nabla \mathbf{K}_k = \nabla \mathbf{B}_k^T \mathbf{C}\mathbf{B} + \mathbf{B}^T \nabla \mathbf{C}_k \mathbf{B} + \mathbf{B}^T \mathbf{C} \nabla \mathbf{B}_k. \tag{2.64}$$

In the second case we have $\mathbf{B} = \text{const}$ and

$$\nabla \mathbf{K}_k = \mathbf{B}^T \nabla \mathbf{C}_k \mathbf{B}. \tag{2.65}$$

Since the set of equations (2.32) is usually solved by Cholesky–Banachiewicz decomposition, the upper and lower triangular components of $\mathbf{K}$ are known prior to the calculation of $\nabla \mathbf{q}$. This allows to compute the columns $\nabla \mathbf{q}_k$ at low cost.

It is convenient to calculate the matrices of stress gradients at the level of the k-th bar. Differentiating Eq. (2.34) with respect to the variable X_k we obtain the k-th column of the matrix $\nabla \mathbf{S}^j$:

$$\nabla \mathbf{S}_k^j = \nabla \mathbf{C}_k^j \mathbf{B}^j \mathbf{q}^j + \mathbf{C}^j \nabla \mathbf{B}_k^j \mathbf{q}^j + \mathbf{C}^j \mathbf{B}^j \nabla \mathbf{q}_k^j. \tag{2.66}$$

In this expression we see derivatives of the matrices $\mathbf{C}^j, \mathbf{B}^j$ that depend upon the nature of the k-th decision variable, as well as derivatives of the displacements of the nodes adjacent to the j-th bar.

It is more convenient to consider separately the entries of the gradient matrices $\nabla \mathbf{L}, \nabla \mathbf{A}, \nabla \mathbf{S}^+, \nabla \mathbf{S}^-$ for both types of decision variables. Let us begin with X_k controlling the cross-sectional area. Then there follows from Eq. (2.45) that $\nabla \mathbf{L}_k = \mathbf{0}, \nabla \mathbf{A}_k = \mathbf{G}_{Ak}^T$. Differentiating Eqs. (2.47), (2.48) with respect to X_k we obtain the derivatives

$$\frac{\partial N^{+j}}{\partial X_K} = R\, G_A^{JK}, \qquad \frac{\partial N^{-j}}{\partial X_k} = -\frac{1}{m_w^j} R\, G_A^{jk}, \tag{2.67}$$

that enter the matrices $\nabla \mathbf{S}_k^+, \nabla \mathbf{S}_k^-$.

When X_k controls the nodal displacement, then $\nabla \mathbf{A}_k = \mathbf{0}$ and $\nabla \mathbf{L}_k$ must be found by differentiating Eq. (2.44). Note that $\nabla \mathbf{S}_k^+ = \mathbf{0}$, since the ultimate value of axial force in tension depends only upon the area of the cross-section. However, $\nabla \mathbf{S}_k^- \neq \mathbf{0}$, since the buckling factor m_w depends upon variable length of the bar.

Explicit formulae for the gradient matrices $\nabla \mathbf{L}, \nabla \mathbf{A}, \nabla \mathbf{q}, \nabla \mathbf{S}, \nabla \mathbf{S}^+, \nabla \mathbf{S}^-$ can be found in [25].

2.3.5 Examples

Let us consider first the truss shown in Fig. 2.2. Taking $\mathbf{X}, \mathbf{w}_0, \mathbf{A}_0, \mathbf{G}_w, \mathbf{G}_A$ from Sect. 2.3.2 we are going to find the derivative of displacements with respect to the cross-sectional area A^1, i.e. the first column of the matrix $\nabla \mathbf{q}$. In order to accomplish that we must calculate the derivative of the stiffness matrix $\mathbf{C}$ with respect to A^1. The entries of that matrix are linear functions of A^j. Since E, L^1 are given, we obtain

$$\nabla \mathbf{C}_1 = \operatorname{diag}\left\{\tfrac{E}{L^1}, 0, 0, 0, 0, 0\right\} = \operatorname{diag}\{1, 0, 0, 0, 0, 0\} \cdot 10^8\ \mathrm{kN\,m^{-3}}.$$

Then there follows from Eq. (2.65) that

$$\nabla \mathbf{K}_1 = \begin{bmatrix} 0 & 0 & 0 & 0 & 0 \\ 0 & \frac{E}{L^1} & 0 & 0 & 0 \\ 0 & 0 & 0 & 0 & 0 \\ 0 & 0 & 0 & 0 & 0 \\ 0 & 0 & 0 & 0 & 0 \end{bmatrix} = \begin{bmatrix} 0 & 0 & 0 & 0 & 0 \\ 0 & 1 & 0 & 0 & 0 \\ 0 & 0 & 0 & 0 & 0 \\ 0 & 0 & 0 & 0 & 0 \\ 0 & 0 & 0 & 0 & 0 \end{bmatrix} \cdot 10^8 \, \mathrm{kN\,m^{-3}}.$$

We found already in Sect. 2.2.3 the displacements $\mathbf{q}_*$ caused by the given load. This allows us to calculate the fictitious load

$$\tilde{\mathbf{Q}}_1 = \{0, -20.47, 0, 0, 0\} \, \mathrm{kN\,m^{-3}}.$$

using Eq. (2.63). Finally, solving the set of equations (2.62) we obtain the derivative of displacements

$$\nabla \mathbf{q}_1 = \{-0.075, -0.648, -0.436, 0.075, -0.362\} \, \mathrm{m^{-1}}.$$

This derivative informs us about the influence of the first decision variable, i.e. the cross-sectional area of the bar 1, has on the entries of the displacement matrix $\mathbf{q} = \{v_1, u_2, v_2, u_4, v_4\}$. It happens that an increase of that area diminishes the absolute values of all displacements: the positive displacements v_1, u_2, v_2, v_4 have negative derivatives and the single negative displacement u_4 has positive derivative.

Optimum design of trusses according to the model described in the present Section is implemented in the computer code *Krata* (Truss) [23]. Figure 2.10 shows the results obtained for the truss taken from Fig. 2.2. The initial design was the truss of dimensions 2.0×2.0 m made from the tubes of diameters $\phi_{ext} = 50$ mm, $\phi_{int} = 40$ mm, which gives the cross-sectional area $A = 28.3$ cm^2. The tubes were made of steel R35 having the strength $R = 220$ MPa. The volume of such a truss is $f_{\mathrm{const}} = 0.0387$ m^3.

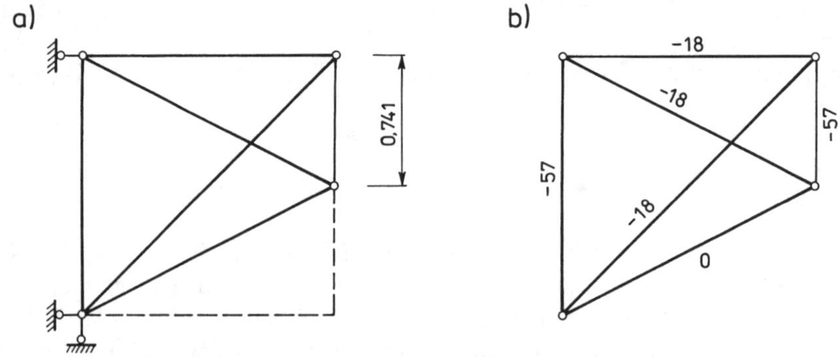

Fig. 2.10. Optimum solution for truss shown in Fig. 2.2: **a)** optimum position of node 4, **b)** changes of cross-sectional areas relative to $A_{\mathrm{const}} = 28.3$ cm^2 (in percents)

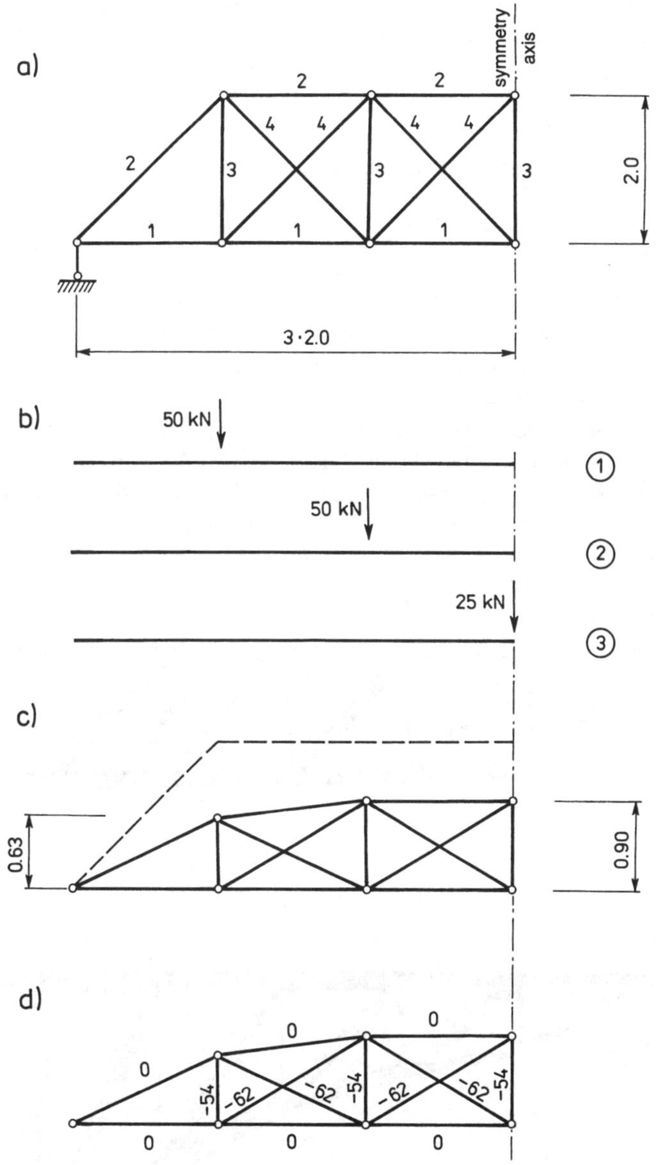

Fig. 2.11. Optimum solution for bridge truss: **a)** initial design, **b)** loading cases, **c)** optimum shape of upper chord, **d)** changes of cross-sectional areas relative to initial design

Figure 2.10a shows the result of search for the optimum position of the node 4. Taking $X = y_4$ we obtained the optimum value $y_{4*} = 0.741$ after 3 local linearizations. The volume of the optimum truss is $f'_{\text{opt}} = 0.0342$ m^3.

Keeping this configuration fixed we optimized then the areas of cross-sections. Assuming

$$\mathbf{X} = \{A^1, A^2, A^3, A^4, A^5\}, \qquad A^5 = A^6,$$

we obtained the following optimum values:

$$\mathbf{X}_* = \{23.1, 28.3, 12.1, 12.1, 23.1\}\,\text{cm}^2.$$

This time the optimum truss has the volume $f''_{\text{opt}} = 0.0262$ m^3. Hence, the gain due to the optimization of topology was about 12% and the optimization of cross-sections allowed us to spare additional 20% of the material volume. This ratio would be reversed if we had more degrees of freedom in the topological optimization.

The next example solved by the same program is the bridge truss of Fig. 2.11. The initial design is shown in Fig. 2.11a (only half of the truss is depicted due to the symmetry), together with the assignment of the bars

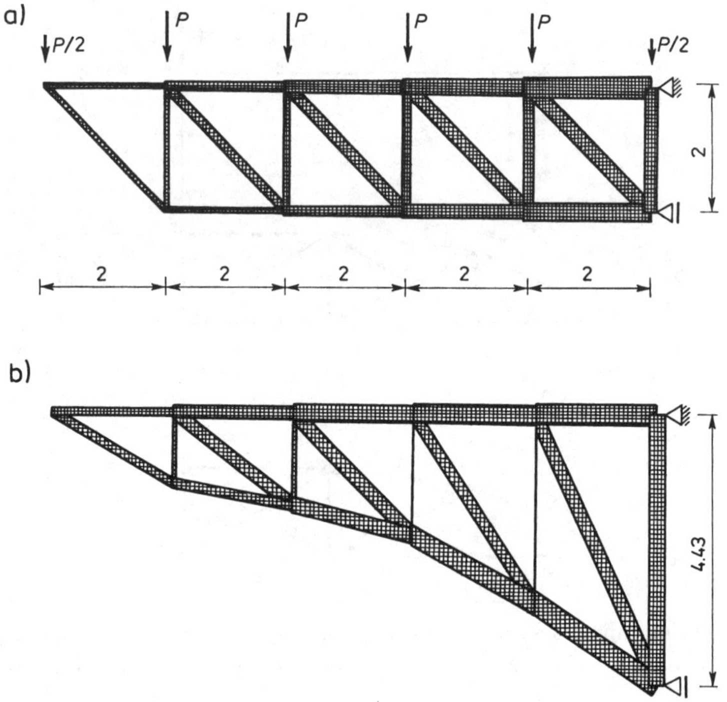

Fig. 2.12. Optimum solution for cantilever truss: **a)** initial design, **b)** optimum design

to classes. The bars belonging to each class have constant cross-sections with the following values:

$$A^I = 18.3\,\text{cm}^2, \quad A^{II} = 27.9\,\text{cm}^2, \quad A^{III} = 22.8\,\text{cm}^2, \quad A^{IV} = 27.9\,\text{cm}^2.$$

Figure 2.11b shows 3 loading cases taken into account. The dead weight of the structure has been neglected. After 5 linearization steps the solution from Fig. 2.11c was found. The optimum truss has 23% smaller volume than the initial one.

In the second stage we optimize the cross-sections, keeping the same classes and the configuration found in during the first stage. This solution also required 5 iterations. The changes of cross-sectional areas with respect to the initial design are shown in Fig. 2.11d. The result of the 2nd stage is the 24% reduction of volume. Both stress and displacement constraint were active.

One has to keep in mind when evaluating these results that our approach does not incorporate optimization of the connections between nodes. All optimum trusses turned out to be redundant. One should check, therefore, whether there exist possibly better solutions between statically determined structures.

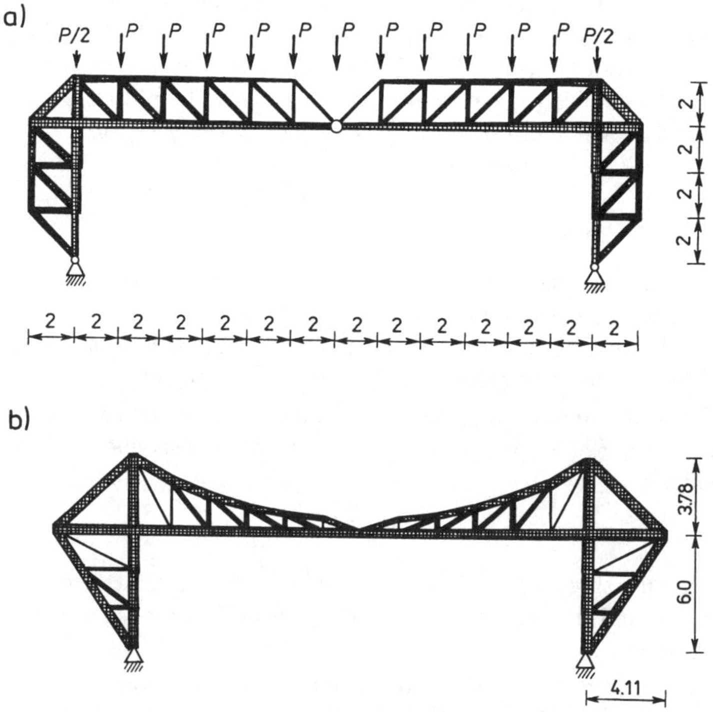

Fig. 2.13. Optimum solution for 3-hinge truss: **a)** initial design, **b)** optimum design

The last 2 examples were taken from [2]. These are the cantilever from Fig. 2.12 and the 3-hinge truss from Fig. 2.13. It can be clearly seen from those pictures that the sequential linearization leads to rational designs. However, such an optimum design requires human interaction: the designer should monitor the iteration process adjusting the range of admissible variations for the design variables and modifying the set of active constraints. Our experience in using the program *Krata* confirms the efficiency of such methodology.

2.4 Optimum Plastic Design

Dimensioning a structure based upon the constraints on maximum stresses in the elastic state has a flow that the global safety factor against overloading remains unknown. The reason for this deficiency is that such a factor depends not only on local stresses but on the global properties of the structure (such as its redundancy) as well. The desire to take into account the global safety factor together with economy considerations that suggest allowing plastic strains for structures in Civil Engineering were motivations causing the replacement of the design methodology based upon admissible stresses for the design based upon the concept of ultimate states. One of them is the development of plastic collapse mechanism (the ultimate state Ic according to the classification given in Sect. 2.3.1). In the sequel we consider optimum design of the structure having given safety margin against the plastic collapse. We begin with the question how to determine such a safety factor.

2.4.1 Ultimate Yield Factor

In Sect. 2.3 we considered the situation when plastic properties of the structure are known and the aim is to find its ultimate load. Now we want to deal with the reverse problem: what should be the yield characteristics of the structure in order to assure the given load carrying capacity.

Let the ultimate load $\mathbf{Q}_{\mathrm{ult}}$ be given which corresponds to a given safety factor μ_* in Eq. (2.35). We begin with the simplest case when the matrix of yield modulae

$$\mathbf{\Lambda} = \xi \mathbf{\Lambda}_* \tag{2.68}$$

depends upon a single parameter ξ because the reference matrix $\mathbf{\Lambda}_*$ is known. Our aim is to find the value ξ_* that corresponds to the onset of plastic collapse of the structure.

Similarly as in Sect. 2.2.5 we introduce the notions of statically admissible yield factor ξ_S and kinematically admissible yield factor ξ_K. The yield factor is statically admissible if there exists a statically admissible distribution of internal forces $\mathbf{S}_S$ satisfying everywhere the yield conditions and being in equilibrium with the load $\mathbf{Q}_{\text{ult}}$. Kinematically admissible yield factor

$$\xi_K = \frac{\mathbf{Q}_{\text{ult}}^T \dot{\mathbf{q}}_K}{\mathbf{\Lambda}_*^T \dot{\lambda}_K} \tag{2.69}$$

follows from the balance of power in the collapse mechanism $(\mathbf{q}_K, \dot{\lambda}_K)$.

The ultimate value ξ_* can be found using the following theorems:

Static theorem. *ξ_* is the smallest of all statically admissible yield factors ξ_S.*

Kinematic theorem. *ξ_* is the largest of all kinematically admissible yield factors ξ_K.*

In matrix description these theorems reduce to the following pair of dual LP-problems:

$$\min\{\xi \,|\, \mathbf{B}^T\mathbf{S} = \mathbf{Q}_{\text{ult}},\ \xi\mathbf{\Lambda}_* - \mathbf{N}^T\mathbf{S} \geq \mathbf{0}, \xi \geq 0\}, \tag{2.70}$$

$$\max\{\mathbf{Q}_{\text{ult}}^T\dot{\mathbf{q}} \,|\, \mathbf{B}\dot{\mathbf{q}} - \mathbf{N}\dot{\lambda} = \mathbf{0},\ \mathbf{\Lambda}_*^T\dot{\lambda} \leq 1, \dot{\lambda} \geq \mathbf{0}\}. \tag{2.71}$$

Contrary to the ultimate load factor which is always positive due to maximization (compare the problem (2.37)), here we must explicitly demand ξ to remain non-negative. The result $\xi_* > 0$ implies $\mathbf{\Lambda}_*^T\dot{\lambda} = 1$ for the solution of the dual problem (2.71) (compare the Kuhn–Tucker condition $(2.4)_6$). Hence, in (2.71) we maximize the external power under normalized energy dissipation.

Since we can always take sufficiently large ξ that makes internal forces induced by $\mathbf{Q}_{\text{ult}}$ statically admissible, the static problem (2.70) always has solution. The yield factor is limited from below by zero. It follows from the duality theorem that the solution $(\dot{\mathbf{q}}_*, \dot{\lambda}_*)$ exists and is unique.

Obviously taking $\mathbf{Q}_* = \mathbf{Q}_{\text{ult}}$ in (2.37), (2.38) we would get reciprocal ultimate factors, i.e. there would be $\mu_*\xi_* = 1$.

Let us calculate ξ_* for the frame shown in Fig. 2.3. We assume that the overload factor is equal to 1.20. Then

$$\mathbf{Q}_{\text{ult}} = \{60.0, 0, 90.0, 0, 0\}\ \text{kN}.$$

As a reference structure we take the frame of constant I-section dimensioned already in Sect. 2.2.5. The influence of axial forces on yielding is neglected.

Table 2.5 shows the simplex matrix of the problem (2.70). Solving it we obtain $\xi_* = 1$ and the distribution of bending moments prior to plastic collapse and the mechanism of such a collapse coincide with those depicted in Fig. 2.8. It indicates that the safety margin was taken exactly equal to the ultimate load factor.

Table 2.5. Simplex matrix for problem (2.70) – calculating ultimate yield factor for frame from Fig. 2.3

M_{12}	M_{21}	M_{23}	M_{32}	M_{34}	M_{43}	ξ		
0.25	0.25	0	0	0	0	0	66.0	$=0$
0	−1.0	−1.0	0	0	0	0	0	$=0$
0	0	5.0	5.0	−0.333	−0.333	0	90.0	$=0$
0	0	0	−1.0	−1.0	0	0	0	$=0$
0	0	0	0	0	−1.0	0	0	$=0$
1.0	0	0	0	0	0	158.0	0	≥ 0
−1.0	0	0	0	0	0	158.0	0	≥ 0
0	1.0	0	0	0	0	158.0	0	≥ 0
0	−1.0	0	0	0	0	158.0	0	≥ 0
0	0	1.0	0	0	0	158.0	0	≥ 0
0	0	−1.0	0	0	0	158.0	0	≥ 0
0	0	0	1.0	0	0	158.0	0	≥ 0
0	0	0	−1.0	0	0	158.0	0	≥ 0
0	0	0	0	1.0	0	158.0	0	≥ 0
0	0	0	0	−1.0	0	158.0	0	≥ 0
0	0	0	0	0	1.0	158.0	0	≥ 0
0	0	0	0	0	−1.0	158.0	0	≥ 0
0	0	0	0	0	0	1	min	

2.4.2
General Formulation of Optimum Design Problem

In the previous Section we assumed that plastic properties of the structure depend upon a single parameter. Now we introduce linear dependence

$$\mathbf{\Lambda} = \bar{\mathbf{\Lambda}} + \mathbf{G}^T\mathbf{X}, \tag{2.72}$$

where matrix $\bar{\mathbf{\Lambda}} \in \mathcal{R}^p$ groups given plastic modulae, $\mathbf{G} \in \mathcal{R}^{t\times p}$ is a fixed influence matrix and entries of the matrix $\mathbf{X} \in \mathcal{R}^t$ are non-negative. We assume that both the reference load $\mathbf{Q}_*$ and the safety factor μ_* are given. Hence, the ultimate loading can be found from Eq. (2.35).

We call the design $\mathbf{X}_0$ *safe* when its ultimate load factor found from the solution of the problem (2.37) satisfies the condition $\mu_0 \geq \mu_*$. There exist many safe designs having the same ultimate load carrying capacity. Our aim is to find among them the *optimum* one which corresponds to the minimum weight structure. In order to keep the cost function linear we assume additionally that the volume of the structure depends upon the decision variables in the following way:

$$f = \bar{f} + \mathbf{c}^T\mathbf{X}. \tag{2.73}$$

Here $\bar{f}$ denotes the constant component and $\mathbf{c} \in \mathcal{R}^t$ groups given non-negative multipliers.

It follows from the Static Theorem that $\mathbf{X}$ is safe when it is possible to find statically admissible distribution of internal forces. Hence, the static formulation of the plastic design reads:

$$\min\{\mathbf{c}^T\mathbf{X} \mid \mathbf{B}^T\mathbf{S} = \mathbf{Q}_{\text{ult}},\ \mathbf{G}^T\mathbf{X} - \mathbf{N}^T\mathbf{S} \geq -\bar{\mathbf{\Lambda}}, \mathbf{X} \geq \mathbf{0}\}. \tag{2.74}$$

This is correctly posed LP-problem. For any $\mathbf{Q}_{\text{ult}}, \bar{\mathbf{\Lambda}}$ it is possible to find sufficiently large $X_k > 0$ that ensures the admissibility of internal forces. Hence, there exists always a solution $(\mathbf{X}_*, \mathbf{S}_*)$ (not necessary unique) and the volume is bounded from below by zero. Note that for large $\bar{\Lambda}_k > 0$ it may happen that $\mathbf{X}_* = \mathbf{0}$. This means simply that constant components of the yield modulae are sufficient large to ensure the required load carrying capacity.

The dual problem for (2.74) reads

$$\max\{\mathbf{Q}^T_{\text{ult}}\dot{\mathbf{q}} - \bar{\mathbf{\Lambda}}^T\dot{\lambda} \mid \mathbf{B}\dot{\mathbf{q}} - \mathbf{N}\dot{\lambda} = \mathbf{0},\ \mathbf{G}\dot{\lambda} \leq \mathbf{c},\ \dot{\lambda} \geq \mathbf{0}\}. \tag{2.75}$$

The cost function here represents the external power reduced by the contribution of fixed yield modulae to the dissipated power. In the case of $\bar{\mathbf{\Lambda}} = \mathbf{0}$ simply the external power is maximized. The equality constraint ensures kinematic consistency of the displacement rates and the strain rates. The latter are expressed through the plastic multipliers according to the associated flow rule $(2.18)_2$. The inequality constraints relate the plastic multipliers, and hence indirectly the strain rates, to the coefficients of the cost function and assure non-negativeness of the plastic multipliers.

There exists a solution $(\dot{\mathbf{q}}_*, \dot{\lambda}_*)$ of (2.75) for any $\mathbf{Q}_{\text{ult}}, \bar{\mathbf{\Lambda}}$. It's uniqueness, i.e. the uniqueness of the plastic collapse mechanism for the optimum structure, is warranted only for non-degenerated[2] solutions $(\mathbf{X}_*, \mathbf{S}_*)$. In particular, there follows from the Kuhn–Tucker condition $(2.4)_6$ that $X_{k*} > 0$ implies

$$\sum_{j=1}^{p} G_{kj}\dot{\lambda}_{j*} = c_k. \tag{2.76}$$

This condition has simple physical interpretation. Let us find the power dissipated in a unit volume of the structural region controlled by the k-th decision variable. It follows from Eqs. (2.19), (2.72), (2.73) and (2.76) that

$$\frac{d_{k*}}{f_{k*}} = \frac{\dot{\lambda}^T_{k*}\mathbf{\Lambda}_k}{c_k X_{k*}} = \frac{\dot{\lambda}^T_{k*}\mathbf{g}_k X_{k*}}{c_k X_{k*}} = 1, \tag{2.77}$$

where $\dot{\lambda}_{k*}, \mathbf{\Lambda}_{k*}$ are the sub-matrices of $\dot{\lambda}_*, \mathbf{\Lambda}_*$ related to the k-th structural region and $\mathbf{g}_k$ denotes the k-th column of $\mathbf{G}$. Hence, we found that the specific dissipated power is constant (unit) for the optimum structure. In the case of $\bar{\mathbf{\Lambda}} \neq \mathbf{0}$ this conclusion is valid for the part of power generated by the

[2] A solution of LP-problem that contains no vanishing elements is called non-degenerated

variable yield modulae. This result is called the *optimality criterion* and is used for a direct search of the optimum structure (compare, e.g., [27]).

In particular, if each entry of the matrix $\mathbf{\Lambda}$ would be an independent decision variable ($\bar{\mathbf{\Lambda}} = \mathbf{0}$, $\mathbf{G} = \mathbf{I}$) then the optimality criterion would reduce to the constraints $\dot{\lambda}_{j*} = c_j$ imposed on each plastic multiplier separately. Reducing the number of decision variables we cause a linear combination of those multipliers to appear in Eq. (2.76).

The dual pair of LP-problems (2.74), (2.75) describes optimum plastic design with non-negative decision variables. Similarly as it was for elastic optimum design, such a constraint can be insufficient for engineering practice. One uses then the constrained range decision variables according to (2.46). This approach requires to replace the model (2.74), (2.75) by

$$\begin{aligned} \min\{\mathbf{c}^T\mathbf{X} \mid\ & \mathbf{B}^T\mathbf{S} = \mathbf{Q}_{\text{ult}}, \\ & \mathbf{G}^T\mathbf{X} - \mathbf{N}^T\mathbf{S} \geq -\bar{\mathbf{\Lambda}}, \\ & \mathbf{X}^- \leq \mathbf{X} \leq \mathbf{X}^+ \quad \}, \end{aligned} \tag{2.78}$$

$$\begin{aligned} \max\{\mathbf{Q}_{\text{ult}}^T\dot{\mathbf{q}} - \bar{\mathbf{\Lambda}}^T\dot{\lambda} + \mathbf{X}^{-T}\dot{\mathbf{x}}^+ - \mathbf{X}^{+T}\dot{\mathbf{x}}^+ \mid\ & \mathbf{B}\dot{\mathbf{q}} - \mathbf{N}\dot{\lambda} = \mathbf{0}, \\ & \mathbf{G}\dot{\lambda} + \dot{\mathbf{x}}^- - \dot{\mathbf{x}}^+ \leq \mathbf{c}, \\ & \dot{\lambda} \geq \mathbf{0},\ \dot{\mathbf{x}}^- \geq \mathbf{0},\ \dot{\mathbf{x}}^+ \geq \mathbf{0}\}. \end{aligned} \tag{2.79}$$

Additional inequality constraints in the static problem generated additional non-negative variables $\dot{x}_k^-, \dot{x}_k^+$ in the kinematic problem. It follows from the Kuhn–Tucker condition $(2.4)_5$ that these variables can attain positive values in the solution $(\dot{\mathbf{q}}_*, \dot{\lambda}_*, \dot{\mathbf{x}}_*^-, \dot{\mathbf{x}}_*^+)$ only when the corresponding static constraints are made active by the solution $(\mathbf{X}_*, \mathbf{S}_*)$. It means that $\dot{x}_{k*}^- > 0$ implies $X_{k*} = X_k^-$ and $\dot{x}_{k*}^+ > 0$ implies $X_{k*} = X_k^+$. In other words, the complementarity condition $\dot{x}_{k*}^-\dot{x}_{k*}^- = 0$ holds: only one of the coupled variables $\dot{x}_{k*}^-, \dot{x}_{k*}^-$ can be greater than zero.

It is important to remember that the condition of constant specific dissipated power ceases to be valid when constrained decision variables are used. The reason for that are the additional terms $\dot{x}_{k*}^-, \dot{x}_{k*}^-$ that appear on the left hand side of Eq. (2.76). They become not equal to zero when $X_{k*} = X_k^-$ or $X_{k*} = X_k^+$. Moreover, wrongly chosen constants X_k^-, X_k^+ can lead to contradictory constraints in the problem (2.78).

2.4.3 Coefficients of Volume Function

We are going now to consider the coefficients c_k that appear in the linearized form (2.73) of the volume of structure. In order to do that we must define typical cross-sections and decide which dimensions are subject to optimization.

Before that let us recall relations between the yield modulae from the stress admissibility condition $(2.18)_1$ and the yield moments of particular cross-sections. Let matrix $\mathbf{M}_0 \in \mathcal{R}^m$ group such moments. We assume that

$$\mathbf{M}_0 = \bar{\mathbf{M}}_0 + \mathbf{D}^T\mathbf{X} \tag{2.80}$$

and

$$\mathbf{\Lambda} = \mathbf{R}^T\mathbf{M}_0, \tag{2.81}$$

where $\bar{\mathbf{M}}_0 \in \mathcal{R}^m$ represents fixed yield moments and $\mathbf{D} \in \mathcal{R}^{t\times m}$, $\mathbf{R} \in \mathcal{R}^{m\times p}$ are given matrices. Since simultaneously Eq. (2.72) must hold, the influence matrices $\mathbf{D}$ and $\mathbf{G}$ should satisfy relation

$$\mathbf{G} = \mathbf{DR}. \tag{2.82}$$

For example, the matrix $\mathbf{M}_0 = \{M_0^1, M_0^2, M_0^3\}$ describes the yield moments of the frame shown in Fig. 2.3. Assuming that the cross-section of the horizontal bar is constant along its length, we take $\mathbf{X} = \{X_1, X_2\}$ and

$$\mathbf{D} = \begin{bmatrix} 1 & 0 & 0 \\ 0 & 1 & 1 \end{bmatrix}.$$

This means that X_1 controls the yield moment M_0^1 and X_2 takes care of $M_0^2 = M_0^3$. On the other hand, comparing the structure of matrix $\mathbf{\Lambda}$ given in Sect. 2.2.3 with Eq. (2.81) we see that

$$\mathbf{R} = \begin{bmatrix} 1 & 1 & 1 & 1 & 0 & 0 & 0 & 0 & 0 & 0 & 0 & 0 \\ 0 & 0 & 0 & 0 & 1 & 1 & 1 & 1 & 0 & 0 & 0 & 0 \\ 0 & 0 & 0 & 0 & 0 & 0 & 0 & 0 & 1 & 1 & 1 & 1 \end{bmatrix}.$$

Performing the multiplication of matrices according to Eq. (2.82) we obtain

$$\mathbf{G} = \begin{bmatrix} 1 & 1 & 1 & 1 & 0 & 0 & 0 & 0 & 0 & 0 & 0 & 0 \\ 0 & 0 & 0 & 0 & 1 & 1 & 1 & 1 & 1 & 1 & 1 & 1 \end{bmatrix}.$$

Let us consider now the volume of the structure. Assuming the volume to linear function of the yield modulae and taking into account Eq. (2.80) we obtain

$$f = \mathbf{a}^T\mathbf{M}_0 = \bar{f} + \mathbf{a}^T\mathbf{D}^T\mathbf{X}, \tag{2.83}$$

where $\bar{f}$ denotes the fixed part and $\mathbf{a} \in \mathcal{R}^m$ is a given matrix of coefficients. According to Eq. (2.73) the matrices $\mathbf{a}$ and $\mathbf{c}$ must fulfil the condition

$$\mathbf{c} = \mathbf{Da}. \tag{2.84}$$

This equation allows us to calculate the coefficients of the cost function in (2.74), (2.78) provided the coefficients a^j in Eq. (2.83) and the matrix $\mathbf{D}$ are known. Expressions for the coefficients a^j depend upon a particular type of the cross-section.

Let the j-th bar has rectangular cross-section shown in Fig. 2.7a. The volume of such a bar is equal to

$$f^j = B^jH^jL^j, \tag{2.85}$$

where L^j is the length. It follows from Eqs. (2.21), (2.23) that the yield moment M_0^j is linear with respect to the width B^j and quadratic with respect to the height H^j of the cross-section. In order to keep the function $f^j = f^j(M_0^j)$ linear we must assume $H^j = \text{const}$. Then

$$a^j = \frac{4}{\sigma_0^j H^j} L^j. \tag{2.86}$$

In the case of ideal sandwich cross-section (Fig. 2.7b) we take also $H^j = \text{const}$ and optimize the volume of outer layers

$$f^j = 2B^j H'^j L^j. \tag{2.87}$$

Equation (2.26) shows that the yield moment is linear with respect to the width of the outer layer B^j, as well as with respect to its thickness H'^j. Each of those dimensions can be taken as a decision variable which leads to the following results:

- for constant thickness:

$$a^j = \frac{2}{\sigma_0^j H'^j} L^j; \tag{2.88}$$

- for constant width:

$$a^j = \frac{2}{\sigma_0^j B'^j} L^j; \tag{2.89}$$

As we already mentioned speaking about the yield curves in Sect. 2.2.3, the I-section can be approximately replaced by the ideal sandwich cross-section. Relations similar to (2.88), (2.89) are obtained also for the reinforced concrete cross-section of constant height. The restricted volume of the present Section precludes us from discussing the optimum plastic design of the reinforced concrete structures. We recommend the interested reader to consult the literature on the subject, e.g. [3].

It happens often in practice that the parameters σ_0^j, H^j, B^j are the same for the entire structure. One can omit the constant factor in Eqs. (2.86), (2.88), (2.89) taking $a^j = L^j$. In particular, for the frame shown in Fig. 2.3 we would have $\mathbf{a} = \{4.0, 2.0, 3.0\}$. Then $\mathbf{c} = \{4.0, 5.0\}$ according to (2.84). Hence, in a structure made of a homogeneous material and having elements of constant dimensions of cross-sections the coefficient c_k is proportional to the sum of lengths of bars governed by the k-th decision variable. Using such coefficients in the optimality criterion (2.76) we obtain the *Foulkes rule*: the sum of rotation rates in plastic hinges developed on bars controlled by the k-th decision variable is proportional to the summary length of those bars. This rule is valid when there are no active constraints of the type (2.46).

Concluding our remarks about the dimensioning of the cross-section let us mention that the assumption $H^j = \text{const}$ is important not only from the point of view of the linearized relation volume–yield moment. This assumption assures also the constant ratio β between the yield moment and the yield axial force. This allows us to keep $\mathbf{N} = \text{const}$ in the model accounting for the influence of axial forces on yielding (compare $(2.18)_1$).

2.4.4 Formulation in Redundant Stresses

There are $t+m$ variables and $n+p$ constraints in the static formulation (2.74) of the plastic optimum design problem. It's version (2.78) has additionally $2t$ side constraints on the decision variables. It is clear, therefore, that the dimensions of the simplex matrix increase rapidly with growing number of elements and nodes of the structure. In order to assure practical applicability of the proposed approach one has to consider means of the rational usage of computer resources.

Such methods fall into 2 groups. The first one incorporates special techniques borrowed from Computer Science or Mathematics. They allow to process the LP-problem in an efficient way (sparse matrix storage schemes, special versions of the simplex algorithm, etc.). Presentation of such methods is out of the scope of the present Section. The second group is directly related to Structural Analysis. Below we introduce one of such techniques known in Elastic Analysis as the Direct Stiffness Method or Force Method.

The set of constraints in (2.74) includes n equations of equilibrium. This allows us to express n entries of $\mathbf{S}$ through $r = m - n$ redundant stresses. In the relevant Eq. (2.11) those stresses are grouped in the matrix $\mathbf{Z}$. Substituting (2.11) into the constraints of the problem (2.74) we obtain its reduced version:

$$\min\{\mathbf{c}^T\mathbf{X} \mid \mathbf{G}^T\mathbf{X} - \mathbf{N}_Z^T\mathbf{Z} \geq \mathbf{\Lambda}_Z,\ \mathbf{X} \geq \mathbf{0}\}, \tag{2.90}$$

where the matrices

$$\mathbf{N}_Z = \mathbf{B}^0\mathbf{N}, \tag{2.91}$$

$$\mathbf{\Lambda}_Z = \mathbf{N}^T\mathbf{B}^{-T}\mathbf{Q}_{\text{ult}} - \bar{\mathbf{\Lambda}} \tag{2.92}$$

are known. Problem (2.90) has $t + r$ variables and p constraints (we do not count the side constraints $X_k \geq 0$. A solution $(\mathbf{X}_*, \mathbf{Z}_*)$ of that problem gives us explicitly the optimum structure whereas the distribution of internal forces at plastic collapse can be calculated from Eq. (2.11).

The dual of (2.90) is

$$\max\{\mathbf{\Lambda}_Z^T\dot{\lambda} \mid \mathbf{N}_Z\dot{\lambda} = \mathbf{0},\ \mathbf{G}\dot{\lambda} \leq \mathbf{c},\ \dot{\lambda} \geq \mathbf{0}\}. \tag{2.93}$$

The only unknowns are now the plastic multipliers. The inequality constraints remained the same as in the non-reduced problem (2.76). The equality constraints assure now the continuity of strain rates:

$$\mathbf{N}_Z\dot{\lambda} = \mathbf{B}^0\mathbf{N}\dot{\lambda} = \mathbf{B}^0\dot{\mathbf{s}} = \mathbf{0}.$$

Deriving these formulae we used Eqs. (2.14), $(2.18)_2$, (2.91).

It is seen that the efficiency of reduction depends upon the degree of redundancy: lower r leads to bigger gain. Such a gain is attained at the cost of finding matrices $\mathbf{N}_Z$ and $\mathbf{\Lambda}_Z$.

2.4.5 Examples

Let us begin with the frame shown in Fig. 2.3. Taking the same ultimate load as used in Sect. 2.4.1, we are looking now for the optimum yield moments for the column and the beam. Hence, we assume $X_1 = M_0^1$, $X_2 = M_0^2 = M_0^3$. Table 2.6 gives the simplex matrix of problem (2.74) for this example.

The same example in the reduced version (2.90), (2.92) is represented by the simplex matrix from Table 2.7. Both versions lead to the solution shown in Fig. 2.14. Note that the sum of rotation rates in plastic hinges situated on the column is equal to 4.0 and for the beam such a sum is equal to 5.0. Both values agree with the Foulkes rule. That rule remains valid since there were no side constraints imposed on the yield moments and both optimum yield moments are greater than zero.

Comparing the optimum frame with the solution $M_0 = 158.0$ kN m we see that the column has been made less resistant to bending whereas the beam has been strengthened. The volume of the optimum design is 3% less than the volume of the frame with a constant cross-section. Small gain is caused by restricted freedom granted by 2 decision variables.

Table 2.6. Simplex matrix for problem (2.74) – optimization of frame from Fig. 2.3

M_{12}	M_{21}	M_{23}	M_{32}	M_{34}	M_{43}	M_0^1	M_0^2		
0.25	0.25	0	0	0	0	0	0	66.0	$= 0$
0	−1.0	−1.0	0	0	0	0	0	0	$= 0$
0	0	5.0	5.0	−0.333	−0.333	0	0	90.0	$= 0$
0	0	0	−1.0	−1.0	0	0	0	0	$= 0$
0	0	0	0	0	−1.0	0	0	0	$= 0$
1.0	0	0	0	0	0	−1.0	0	0	≥ 0
−1.0	0	0	0	0	0	−1.0	0	0	≥ 0
0	1.0	0	0	0	0	−1.0	0	0	≥ 0
0	−1.0	0	0	0	0	0	−1.0	0	≥ 0
0	0	1.0	0	0	0	0	−1.0	0	≥ 0
0	0	−1.0	0	0	0	0	−1.0	0	≥ 0
0	0	0	1.0	0	0	0	−1.0	0	≥ 0
0	0	0	−1.0	0	0	0	−1.0	0	≥ 0
0	0	0	0	1.0	0	0	−1.0	0	≥ 0
0	0	0	0	−1.0	0	0	−1.0	0	≥ 0
0	0	0	0	0	1.0	0	−1.0	0	≥ 0
0	0	0	0	0	−1.0	0	−1.0	0	≥ 0
0	0	0	0	0	0	4.0	5.0	min	

Table 2.7. Simplex matrix for problem (2.90) – optimization of frame from Fig. 2.3 by means of reduced formulation

Z	M_0^1	M_0^2		
−1.0	−1.0	0	−24.0	≥ 0
1.0	−1.0	0	−24.0	≥ 0
1.0	−1.0	0	0	≥ 0
−1.0	−1.0	0	0	≥ 0
1.0	0	−1.0	0	≥ 0
0.6	0	−1.0	−10.8	≥ 0
−0.6	0	−1.0	10.8	≥ 0
−0.6	0	−1.0	10.8	≥ 0
0.6	0	−1.0	−10.8	≥ 0
0	0	−1.0	0	≥ 0
0	0	−1.0	0	≥ 0
0	4.0	5.0	min	

The next example concerns optimization of the 3-bay frame shown in Fig. 2.15a. Assuming symmetry of the structure, neglecting axial forces and taking the yield moments of columns and beams as the decision variables we obtain the solution from Fig. 2.15b,c,d. Note that neither the distribution of internal forces nor the collapse mechanism is symmetric. This is due to the horizontal force present in the loading. The collapse mechanism confirms to the optimality criterion (2.76). In particular, the sum of rotation rates in the hinges developed on the outer columns is equal to 6.0, i.e. equal to the sum of lengths of those columns.

Let us finally concern the 2-storey frame of Fig. 2.16a. In the first attempt we do not impose side constraints on the yield moments and we treat the yield moment of each bar as an independent decision variable. This gives the solution shown in Fig. 2.16b. It is seen that the distribution of material in such a structure is very uneven: the lower beam is very stiff, the left hand side columns are very slender when compared to the right hand side columns and the left lower column has even zero yield moment. The last circumstance suggests that one might replace that column by a hinged strut but in general the solution is practically non-acceptable. Hence, we introduce the constraint $10.0 \leq X_k \leq 120.0$ and we restrict the diversity of cross-sections: the columns situated on one side of the frame should have the same yield moment. Figures 2.16c,d. show the relevant solution. The volume increased by 9% as compared to the previous design but now we have a solution which is technologically acceptable. The Foulkes rule is valid for all bars except the lower beam, where the upper bound on the yield moment became active. The length of that beam is 6.0 and the rate of rotation in the plastic hinge is 8.0. The difference 2.0 corresponds to the value of additional variable of type $\dot{x}_k^+$.

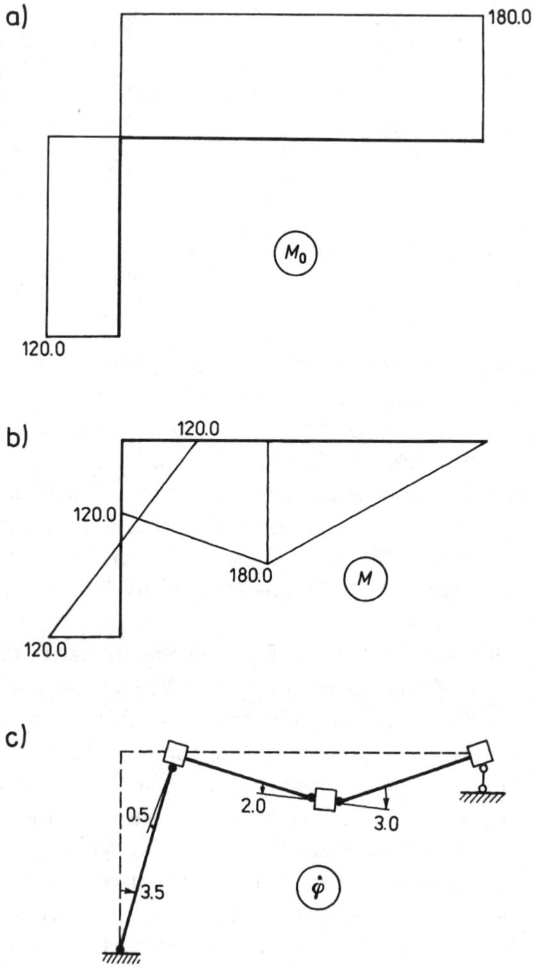

Fig. 2.14. Optimum solution for truss shown in Fig. 2.3: **a)** yield moments, **b)** bending moments, **c)** collapse mechanism

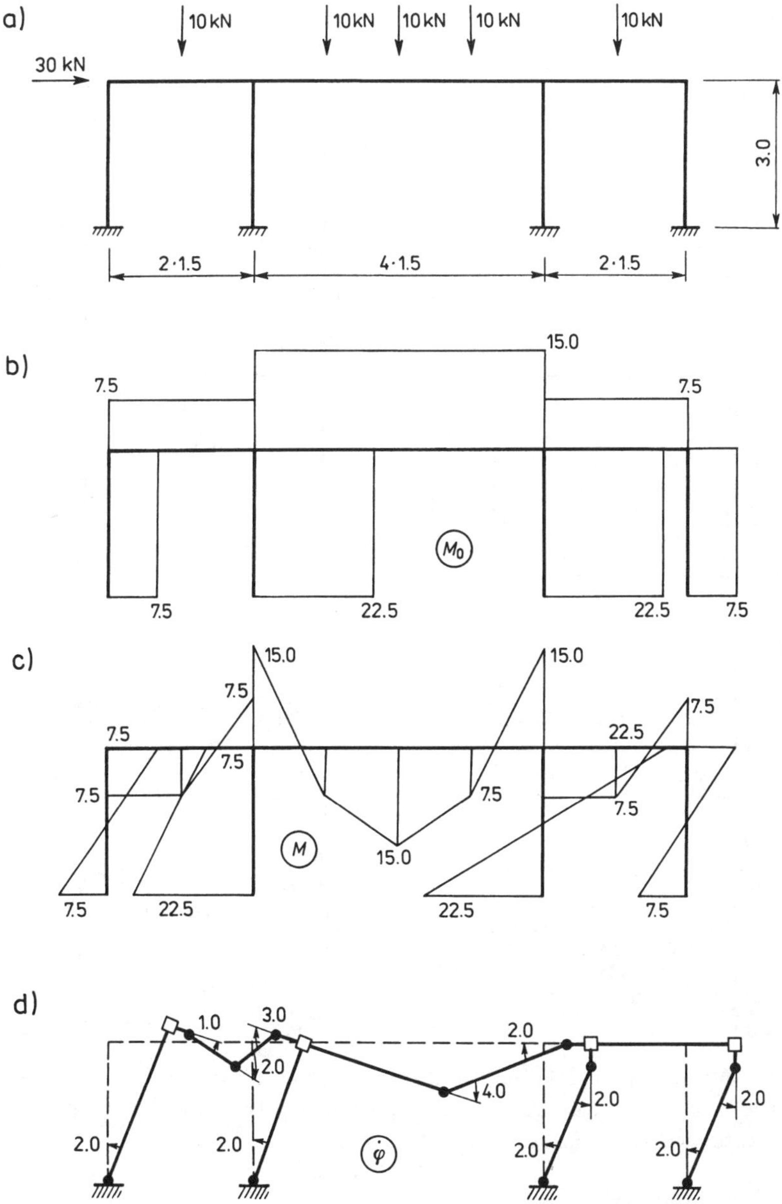

Fig. 2.15. Optimum solution for single storey frame: **a)** dimensions and loading, **b)** yield moments, **c)** bending moments, **d)** collapse mechanism

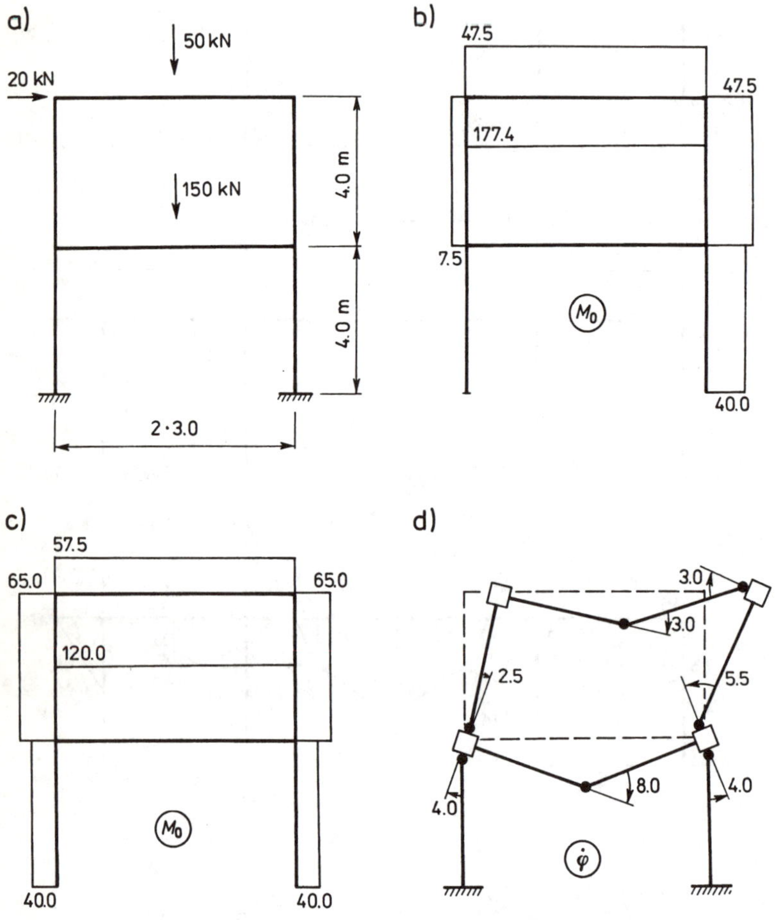

Fig. 2.16. Optimum solution for 2-storey frame: **a)** dimensions and loading, **b)** yield moments – case $X_k \geq 0$, **c)** yield moments – case $10 \leq X_k \leq 120$, **d)** collapse mechanism for case $10 \leq X_k \leq 120$

3 Applications of Nonlinear Programming

3.1 Introduction

In Chap. 1 of this Part, the basic notions have been introduced of the optimization in mechanics regarded as a mathematical programming problem. Now, the solution methods of practical optimization problems will be presented.

The first three methods to be analyzed in the sequel stem from the formulation of a structural optimization problem as a problem of nonlinear programming. The definition of nonlinear programming is given in Chap. 1. Recall that when formulating a nonlinear programming problem, the design variables are assumed to constitute subsets of the real numbers set.

There are two main approaches when dealing with nonlinear programming problems. The first one consists in solving a system of equations and inequalities which result from the necessary conditions for the existence of extremum in the considered problem. The second one is based on the gradient methods of seeking that extremum.

Our presentation will start with two methods that belong to the first approach i.e. are based on the necessary conditions. The first method (Sect. 3.3) involves linearization of constraints. The second one (Sect. 3.4) is based on the assumption that a structural optimization problem can be solved using a finite series of iterations each having a single active constraint.

Both methods will be applied to the optimization of structures with a priori given layouts. This is an important group of structures whose shapes are determined by the tasks they have to fulfil. Aerospace structures, steel building frames, large-span space roofs, etc. are typical examples of such structures.

3.2 Design of Minimum Weight Structure as a Nonlinear Programming Problem

Before formulating the problem itself, let us specify the properties of the considered structures:

- they are made of a linear elastic material,
- displacements are small,

- structural members are finite and have constant thickness (roofs) or constant cross-section area (bars).

In this section, the weight (or volume) of structure is selected as the cost function. Consequently, the cost of material may be excluded from the total cost of structure erection. Note that the total cost of structure erection comprises the cost of design, construction, assembling, etc. However, such an assumption is unacceptable if the structure layout is also subject to optimization. In that case, variations of the structure layout affect, as a rule, the cost of structure erection.

The constraints that appear in optimization problems are classified as equality and inequality constraints. The equality constraints include the equilibrium (motion) equations for the considered structure. They are typically employed in the finite element formulations (cf. Part II).

On the other hand, the inequality constraints are related to displacements, strains (stresses), eigenfrequencies, critical loads, and dimensions of cross-sections or thicknesses. In particular optimization problems only some of these constraints are imposed.

The above functions and inequalities have the following form:

$$\text{cost function} \qquad f = \mathbf{A}^{\mathrm{T}}\mathbf{l} \to \min, \tag{3.1}$$

where $\mathbf{A}^{\mathrm{T}}$ is the vector of design variables A_f having f_0 components which represent the cross-section area of a bar or the thickness of a plate (shell), $\mathbf{l}$ is the vector of length l_f (or area) of a structural member.

In many problems, the number of structural members, or the number of finite elements of a discretized structure, substantially exceeds the number of design variables. This means that the same design variable has to be assigned to a certain group of members. In such a case, the components of $\mathbf{l}$ represent not the lengths (or cross-section areas) of individual members but the sum of the lengths (or cross-section areas) of all members which were given the same value of the design variable. Then, the component l_f can be expressed as

$$l_f = \sum_{e \in \mathbf{l}_f} l_e. \tag{3.2}$$

3.2.1 Equality Constraints

In the optimization of structures subject to static loading, the equilibrium equations will be expressed in the notation that has been introduced in the chapter devoted to the finite element method, i.e.

$$\mathbf{Kq} = \mathbf{Q}_1, \mathbf{Q}_2, \ldots, \tag{3.3}$$

where $\mathbf{Q}_1, \mathbf{Q}_2, \ldots.$ denote the consecutive, mutually independent loading conditions.

If the constraints are imposed on the eigenfrequencies, or the critical loads in the incipient stability problems, the following set of homogeneous equations has to be satisfied

$$\begin{aligned} \mathbf{Ky} - \omega^2 \mathbf{My} &= 0 \qquad \text{(a)};\\ \mathbf{Kw} - p\mathbf{K}_\sigma^0 \mathbf{w} &= 0 \qquad \text{(b)} \end{aligned} \tag{3.4}$$

where $\mathbf{y}$ and $\mathbf{w}$ are the eigenvectors normalized according to

$$\begin{aligned} \mathbf{y}^{\mathrm{T}}\mathbf{My} - 1 &= 0 \qquad \text{(a)};\\ \mathbf{w}^{\mathrm{T}}\mathbf{K}_\sigma^0 \mathbf{w} - 1 &= 0 \qquad \text{(b)}. \end{aligned} \tag{3.5}$$

In Eqs. (3.5), $\mathbf{M}$ is the mass matrix, $\mathbf{K}_\sigma^0$ is the geometric matrix, or the matrix of initial stresses that are computed for a unit load and allowed to grow p-times up to the limit set by the eq. (3.4b). Note that only such problems will be considered for which $\mathbf{K}_\sigma^0$ is strictly or 'approximately' independent of the design variables A_f. These problems form a broad class in engineering practice. In the first place, statically determined structures should be mentioned for which $\mathbf{K}_\sigma^0$ and A_f are strictly independent. On the other hand, 'approximate' independence between $\mathbf{K}_\sigma^0$ and A_f may be assumed for slender masts and truss columns for which the imposed loads are mainly transmitted through the longitudinal members. A similar assumption can be made when considering stability problems of frames and plates.

3.2.2 Inequality Constraints

The constraint imposed on the displacement q_i of the i-th node has the form

$$\overset{+}{q}{}_i^0 - q_i \geq 0, \qquad q_i - \overline{q}_i^0 \geq 0 \tag{3.6}$$

where $\overset{+}{q}{}_i^0$ and $\overline{q}_i^0$ are the extreme values that can be reached by this displacement.

Similarly, the constraints imposed on the strains (or stresses) in the point $\mathbf{x}^p$ of a structural member may be written as

$$\overset{+}{\varepsilon}{}_j^0(\mathbf{x}^p) - B_{ji}(\mathbf{x}^p)q_i \geq 0, \qquad B_{ji}(\mathbf{x}^p)q_i - \overset{+}{\varepsilon}{}_j^0 \geq 0. \tag{3.7}$$

If the constraints are imposed on stresses, they have to be specified for a particular case of uni-, bi- or triaxial loading conditions. For example, for a truss member (uniaxial stress state) these constraints are as follows

$$\overset{+}{\sigma}{}_e^0 - EB_{ei}q_i \geq 0, \qquad EB_{ei}q_i - \overline{\sigma}_e^0 \geq 0. \tag{3.8}$$

In the optimization problems in which the vibration frequency ω^0 or the load multiplier p^0 must not be exceeded, these constraints become

$$\omega - \omega_0 \geq 0, \qquad p^0 - p \geq 0. \tag{3.9}$$

Since the design variables are not allowed to take just any values, for practical purposes they should be bounded from above and below. Hence,

$$\overset{+}{A}{}^{0} - A_f \geq 0, \qquad A_f - \overset{-}{A}{}^{0} \geq 0. \tag{3.10}$$

The cost function (3.1) and the constraints (3.2)–(3.10) formally define the optimization problems whose solutions will be the subject of the rest of this chapter.

3.3 Optimality Criteria Method

The most elementary problem in optimal design is a statically determined truss of minimum weight with constraints imposed only on stresses. Since the truss is statically determined, the equilibrium equations (2.3) can be solved in terms of forces which depend neither on the cross-sections nor the displacements. Having computed the forces in the truss members, their magnitudes are divided by the admissible stress $\overset{+}{\sigma}{}^{0}_{e}$ or $\overset{-}{\sigma}{}^{0}_{e}$ in order to obtain the minimum cross-sections of individual bars. By summing up the minimum volumes of bars, the minimum volume of the entire truss is obtained.

For statically undetermined structures, and constraints imposed not only on stresses, solutions of optimization problems are much more complicated. Below, the essence of the optimality criteria method is presented on the example of a truss loaded by time-independent forces. The constraints will be put on a certain number of displacements and stresses. Obviously, it is conceivable to apply this method to more complex problems. However, since emphasis is placed here on the presentation clarity, the selected example must be simple and tractable. The main idea of the method follows from the so called "fully stress design" technique known in the structural design. Recall that the fully stress design is connected with an iteration process in which the design variables and the internal forces are, alternately, assumed constant. More specifically, it means that forces S_e are first computed for the known values of A_f. Then, the cross-sections of bars are adjusted in such a way that only one bar is modified, while the changes in the remaining ones are assumed to have no effect on it. For the new design variables determined according to this scheme, the corresponding values of internal forces are found and the iteration process is repeated.

The displacement of the i-th node is given by

$$q_i = \sum_e \frac{S_e S_e^i l_e}{A_e E_e} \tag{3.11}$$

where S_e^i is the axial force in e-th rod due to the virtual load $Q_i = 1$. Summation in (3.11) is extended over all bars of the truss.

Similarly, by applying virtual unit forces to both ends of the d-th rod, the stress σ_d can be expressed as

$$\sigma_d = \frac{E_d}{l_d} \sum_e \frac{S_e S^i_e l_e}{A_e E_e} \tag{3.12}$$

Denoting

$$\frac{S_e S^i_e l_e}{E_e} = c_{ie} \tag{3.13}$$

and

$$\frac{S_e S^i_e l_e}{A_e E_e} \frac{E_d}{l_d} = d_{de}, \tag{3.14}$$

equations (3.11) and (3.12) can be rewritten as

$$q_i = \sum_e \frac{c_{ei}}{A_e}, \qquad \sigma_d = \sum_e \frac{d_{de}}{A_e}. \tag{3.15}$$

Using this notation, the constraints (3.7) and (3.9) take the following form

$$\overset{+}{q}{}^0_i - \sum_e \frac{c_{ei}}{A_e} \geq 0, \qquad \sum_e \frac{c_{ei}}{A_e} - \overset{-}{q}{}^0_i \geq 0, \tag{3.16}$$

$$\overset{+}{\sigma}{}^0_d \sum_e \frac{d_{de}}{A_e} \geq 0, \qquad \sum_e \frac{d_{de}}{A_e} - \overset{-}{\sigma}{}^0_i \geq 0. \tag{3.17}$$

In further considerations, the constraints will be imposed solely on $\overset{+}{q}{}^0_i$ and $\overset{+}{\sigma}{}^0_d$ since for the minimum constraints $\overset{-}{q}{}^0_d$ and $\overset{-}{\sigma}{}^0_d$ the analysis would be essentially the same.

It is known from the preceding chapters that the necessary condition for the cost function to reach the minimum under given constraints is the fulfillment of the so called Kuhn–Tucker conditions. Recall that these conditions follow from the Lagrangean, which in the present case reads

$$L = -f + \boldsymbol{\lambda}^{q\mathrm{T}}(\overset{+}{\mathbf{q}}{}^0 - \mathbf{q}) + \boldsymbol{\lambda}^{s\mathrm{T}}(\overset{+}{\sigma}{}^0 - \sigma), \tag{3.18}$$

where q_i and σ_d are given by (3.15).

The Kuhn–Tucker necessary conditions take the following form

$$\begin{gathered} -\frac{\partial f}{\partial A_f} - \boldsymbol{\lambda}^{q\mathrm{T}} \frac{\partial \mathbf{q}}{\partial A_f} - \boldsymbol{\lambda}^{s\mathrm{T}} \frac{\partial \sigma}{\partial A_f} = 0 \quad \text{for } f = 1, 2, ..., f_0, \\ \lambda^q_i(\overset{+}{q}{}^0_i - q_i) = 0 \quad \text{for all } q_i, \quad \text{on which constraint is imposed} \\ \lambda^s_d(\overset{+}{\sigma}{}^0 - \sigma_d) = 0 \quad \text{for all } \sigma_d, \quad \text{on which constraint is imposed} \\ \boldsymbol{\lambda}^q,\ \boldsymbol{\lambda}^s \geq 0. \end{gathered} \tag{3.19}$$

Differentiating the cost function (3.1) with respect to A_f gives

$$\frac{\partial f}{\partial A_f} = l_f. \tag{3.20}$$

When differentiating q_i with respect to A_f, an assumption is made that c_{ei} and d_{de} are independent of the design variables. This is a consequence of the fact that c_{ei} and d_{de} are connected with the internal forces which, in turn, are related to the cross-sections. These assumptions, whose origins go back to the mentioned fully stress design technique, make it possible to express the appropriate derivatives as

$$\frac{\partial q_i}{\partial A_f} = -\frac{c_{ie}}{A^2}; \quad \frac{\partial \sigma_d}{\partial A_f} = -\frac{d_{de}}{A^2}. \tag{3.21}$$

It can be thus seen that the above described approximation leads to the decomposition of q_i into the sum of terms, each of which depends only on a single design variable. In the literature this is known under the name of *separability condition.*

Using (3.19) and (3.20) and performing some algebraic manipulations, the following optimization conditions are obtained

$$\frac{1}{A_f^2 l_f}\left[\sum_{i\in I}\lambda_i^d c_{if} + \sum_{d\in D}\lambda_d^s d_{df}\right] = 1, \quad \text{for } f = 1, 2, ..., f_0, \quad \lambda_i^d, \lambda_d^s \geq 0. \tag{3.22}$$

where the summation is extended over those indices i and d for which the constraints are active. If several structural members have the same cross-section A_f, then it holds

$$c_{if} = \sum_{e\in I_f} c_{ie}, \quad d_{df} = \sum_{e\in I_f} d_{de}. \tag{3.23}$$

In (3.22), the number of unknowns is equal to the number of equations and inequalities.

3.4 Explicit Formulation of Kuhn–Tucker Necessary Conditions for Minimum Weight Structures

3.4.1 Essence of the Method

The optimality criteria method (OC) described in the preceding section is based on many simplifying assumptions. These simplifying assumptions include those ensuing from the separability conditions (3.21), and also the one stating that in a given iteration the internal forces, due to external and virtual loads alike, do not change when the design variables A_f change. A question arises whether these assumptions are really necessary. It turns out that they can be relaxed by a simple extension of the classical OC method.

To this end, the Lagrangean which contains inequality constraints, as in the OC method, should be modified by including equality constraints pertinent to the considered problem [15, 16]. The equilibrium (motion) equations are the equality constraints of primary importance. In the eigenvalue problems (vibrations, stability), equality constraints are provided by the normalization conditions for eigenvectors.

Introduction of the equality constraints has numerous advantages, out of which the following three are the most important. First, the simplifying assumptions of the OC method are avoided. Second, it is possible to express the algorithm of an optimization problem using variables typically employed in the FEM packages, such as stresses, nodal displacements, node reaction forces. It is for these reasons that the explicit form of the necessary conditions is emphasized in the heading of this subsection. Third, the approach outlined in this subsection allows for the application of an iterative optimization process in order to determine the solution sensitivity to the variations of design variables.

3.4.2 Minimum of Structure Weight with Account for Static, Stability and Free Vibrations Problems

The problem of minimum weight structure considered so far in this chapter was confined to a linear elastic material and infinitesimal strains and displacements. Now, the same problem will be solved but none of these assumptions will be made. Consider a Lagrangean that includes inequality constraints, equilibrium (motion) equations and eigenvectors normalization relations. Thus, (3.18) is extended to be

$$\begin{aligned} L = & -f + \boldsymbol{\lambda}^{e\mathrm{T}}[\mathbf{K}(\mathbf{A})\mathbf{q} - \mathbf{Q}] + \boldsymbol{\lambda}^{d\mathrm{T}}[\mathbf{K}(\mathbf{A})\mathbf{y} - \omega^2\mathbf{M}\mathbf{y}] \\ & + \boldsymbol{\lambda}^{s\mathrm{T}}[\mathbf{K}(\mathbf{A})\mathbf{w} - P_{cr}\mathbf{K}_\sigma\mathbf{w}] + \boldsymbol{\lambda}^{q\mathrm{T}}(\mathbf{q}^0 - \mathbf{q}) + \boldsymbol{\lambda}^{q\mathrm{T}}[\varepsilon^0(\mathbf{x}^p) - \mathbf{B}(\mathbf{x}^p)\mathbf{q}] \\ & + \lambda^\omega(\omega^2 - \omega_0^2) + + \lambda^p(P_{cr0} - P_{cr}) + \boldsymbol{\lambda}^{c\mathrm{T}}(\mathbf{A} - \mathbf{A}_{\min}) \\ & + \lambda_1^0(\mathbf{y}^\mathrm{T}\mathbf{M}y - 1) + \lambda_2^0(\mathbf{w}^\mathrm{T}\mathbf{K}_\sigma\mathbf{w} - 1). \end{aligned} \tag{3.24}$$

According to the Kuhn–Tucker theorem (see Chap. 1), the necessary condition for minimum structure weight is provided by the following system of equations and inequalities (some algebraic operations are explained below):

a) $\mathbf{K}\mathbf{q} - \mathbf{Q} = 0,$

b) $\mathbf{K}\boldsymbol{\lambda}^e - \boldsymbol{\lambda}^q - \mathbf{B}^\mathrm{T}\boldsymbol{\lambda}^\sigma = 0,$

c) $[\mathbf{K} - P_{\mathrm{cr}}\mathbf{K}_\sigma]\mathbf{w} = 0,$

d) $[\mathbf{K} - \omega^2\mathbf{M}]\mathbf{y} = 0,$

e) $\mathbf{y}^\mathrm{T}\mathbf{M}\mathbf{y} - 1 = 0,$

f) $\mathbf{w}^\mathrm{T}\mathbf{K}_\sigma\mathbf{w} - 1 = 0,$

g) $[\mathbf{K} - \omega^2\mathbf{M}]\boldsymbol{\lambda}^d + 2\lambda_1^0\mathbf{M}\mathbf{y} = 0,$

$$
\begin{aligned}
&\text{h)}\ \ [\mathbf{K} - P_{cr}\mathbf{K}_\sigma]\boldsymbol{\lambda}^s + 2\lambda_2^0\mathbf{K}_\sigma\mathbf{w} = 0,\\
&\text{i)}\ \ \boldsymbol{\lambda}^{\sigma\mathrm{T}}(\boldsymbol{\varepsilon}^0 - \mathbf{B}\mathbf{q}) = 0, \qquad (3.25)\\
&\text{j)}\ \ \boldsymbol{\lambda}^{q\mathrm{T}}(\mathbf{q}^0 - \mathbf{q}) = 0,\\
&\text{k)}\ \ \lambda^\omega(\omega^2 - \omega_0^2) = 0,\\
&\text{l)}\ \ \lambda^p(P_{cr} - P_{cr0}) = 0,\\
&\text{m)}\ \ -\boldsymbol{\lambda}^{d^{\mathrm{T}}}\mathbf{M}\mathbf{y} + \lambda^\omega = 0,\\
&\text{n)}\ \ \boldsymbol{\lambda}^{s^{\mathrm{T}}}\mathbf{K}_\sigma\mathbf{w} + \lambda^p = 0,\\
&\text{o)}\ \ \lambda_f^c(A_f - A_{\min}) = 0,\\
&\text{p)}\ \ -l_f A_f + \mu\boldsymbol{\lambda}^{e\mathrm{T}(e)}\mathbf{k}^{(e)}\mathbf{q}^{(e)} + \boldsymbol{\lambda}^{d\mathrm{T}(e)}[\mu\mathbf{k}^{(e)}\mathbf{y} - \omega^2 M^{(e)}\mathbf{y}]\\
&\qquad + \boldsymbol{\lambda}^{s\mathrm{T}(e)}\mu\mathbf{k}^{(e)}\mathbf{w} + \lambda_1^0\mathbf{y}^{\mathrm{T}}\mathbf{M}^{(e)}\mathbf{y} + \lambda_f^c A_f = 0 \quad \text{for } f = 1, \ldots, p,\\
&\text{r)}\ \ \boldsymbol{\lambda}^q,\ \boldsymbol{\lambda}^\sigma,\ \lambda^\omega,\ \lambda^p,\ \lambda_f^c \geq 0
\end{aligned}
$$

where $\boldsymbol{\lambda}^e$, $\boldsymbol{\lambda}^d, \boldsymbol{\lambda}^s$ are the Lagrange multipliers associated with the equilibrium equations, the governing equations for a dynamic vibration problem and the governing equations for a stability problem, respectively; λ_1^0, λ_2^0 are the multipliers related to the normalization conditions of the eigenvectors $\mathbf{y}$ and $\mathbf{w}$; $\boldsymbol{\lambda}^\sigma$, $\boldsymbol{\lambda}^q$, λ^ω, λ^p, λ_f^c are the Lagrange multipliers resulting from the constraints imposed on stresses (strains), displacements, vibration frequencies, critical force, minimum lateral dimension of a finite element, respectively.

Equations (3.25p) are the optimality conditions. They are derived differentiating the Lagrangean (3.24) with respect to the design variables A_f and then multiplying the result by A_f. Depending on the type of structure, the design variable A_f enters (3.24) with different powers. Therefore, the coefficient μ appears in (3.25p), taking the following values: 1 for trusses, 2 for sandwich structures, 3 for plates and shells, and various numbers for beams.

It is worth noticing that the product $\boldsymbol{\lambda}^{e\mathrm{T}(e)}\mathbf{k}^{(e)}\mathbf{q}^{(e)}$ in (3.25p) can be expressed as

$$\boldsymbol{\lambda}^{e\mathrm{T}(e)}\mathbf{k}^{(e)}\mathbf{q}^{(e)} = \boldsymbol{\lambda}^{e\mathrm{T}(e)}\mathbf{Q}^{(e)}. \qquad (3.26)$$

which is nothing else but the work of the reaction force vector $\mathbf{Q}^{(e)}$ in the nodes of e-th finite element on the displacements $\lambda^{e(e)}$ of these nodes. In the case of a truss, (3.26) may be recast as

$$\boldsymbol{\lambda}^{e\mathrm{T}(e)}\mathbf{Q}^{(e)} = EA_f\varepsilon^{(e)}\mathbf{B}^{(e)}, \qquad (3.27)$$

where $\varepsilon^{(e)}$ and $\mathbf{e}^{(e)}$ are defined by the familiar kinematic relationships

$$\varepsilon^{(e)} = \mathbf{B}^{(e)^{\mathrm{T}}} q^{(e)}, \quad \mathbf{e}^{(e)} = \mathbf{B}^{(e)^{\mathrm{T}}} \boldsymbol{\lambda}^{e(e)}. \qquad (3.28)$$

All the algebraic manipulations described above enable representation of equation (3.25) in terms of the quantities typically encountered in FEM codes. An additional quantity, being that of the derivative of the stiffness matrix with respect to the design variable, is not needed.

Some of the unknowns may be computed in a straightforward manner. Multiplying (3.25g) from the left by $\mathbf{y}^{\mathrm{T}}$, while (3.25h) by $\mathbf{w}^{\mathrm{T}}$, and keeping in mind that matrices $\mathbf{K}, \mathbf{M}, \mathbf{K}_\sigma$ are symmetric, it turns out that $\lambda_1^0 = \lambda_2^0 = 0$. Assuming next that normalization conditions (3.25e,f) are identical for both the Lagrange multipliers λ^e, λ^d and the vectors $\mathbf{y}$, $\mathbf{w}$, it follows from (3.25m) that $\lambda^\omega = \lambda^p = 1$.

Further manipulations of (3.25) lead to an additional equation relating the cost function, the constraints and the constraint-associated Lagrange multipliers. For this purpose, multiply (3.25a) by $\boldsymbol{\lambda}^{e\mathrm{T}}$ and from the obtained result subtract the product of (3.25b) and $\mathbf{q}^{\mathrm{T}}$. Hence,

$$\boldsymbol{\lambda}^{e\mathrm{T}}\mathbf{Q} = \mathbf{q}^{\mathrm{T}}\boldsymbol{\lambda}^q + \mathbf{q}^{\mathrm{T}}\mathbf{B}^{\mathrm{T}}\boldsymbol{\lambda}^\sigma. \tag{3.29}$$

However, since the products $q_i\lambda_i^\sigma$ are non-zero for the active constraints only, i.e. for $q_i = q_i^0$ and $\sum_i b_{ij}q_i = \varepsilon_j^0$, equation (3.29) becomes

$$\boldsymbol{\lambda}^{e\mathrm{T}}\mathbf{Q} = \sum_i \lambda_i^q q_i^0 + \sum_j \lambda_j^\sigma \varepsilon_j^0, \tag{3.30}$$

with the summation extended over the active constraints. Adding up all the equations (3.25p) and making use of the relation $\sum_i^p \mathbf{k}^{(e)}\mathbf{q}^{(e)} = \mathbf{Q}$ gives

$$-f + \mu\boldsymbol{\lambda}^{e\mathrm{T}}\mathbf{Q} + (\mu-1)\lambda^\omega\omega_0^2 + (\mu-1)\lambda^p P_{\mathrm{cr}} + \sum_f \lambda_f^c A_{\min} = 0. \tag{3.31}$$

By multiplying (3.30) by μ and subtracting (3.31) from it, the following important relation is obtained

$$f = \mu\sum_i \lambda_i^q q_i^0 + \mu\sum_f \lambda_f^\sigma \varepsilon_f^0 + \sum_f \lambda_f^c A_{\min} + (\mu-1)\lambda^\omega\omega_0^2 + \lambda^p P_{\mathrm{cr}0}. \tag{3.32}$$

Thus, the cost function is equal to the sum of products formed by the constraint limit values and the corresponding Lagrange multipliers. Due to the fact that (3.32) contains the inequality constraints exclusively, the summation is confined to the active constraints. Note that for a given active constraint the corresponding Lagrange multiplier can be directly determined from (3.32).

The term containing the $(\mu - 1)$ multiplier should be furnished with an additional comment. In this context, consider a minimum weight truss problem with constraint imposed on one of the eigenfrequencies, i.e. $\omega^2 - \omega_0^2 \geq 0$. In this case, eq.(3.32) reads

$$f = (\mu-1)\lambda^\omega\omega_0^2. \tag{3.33}$$

For the considered truss $\mu = 1$, hence its minimum weight follows to be $f = 0$. This is a direct consequence of the fact that both the stiffness matrix $\mathbf{K}$ and the mass matrix **M** (3.25d) are linearly dependent on the design variables A_f. The value of ω that results form the solution of (3.25d) depends on the ratio of the components of A_f, rather than on the individual components themselves. Clearly, **K** and **M**, thus the whole equation (3.25d), may be multiplied by the same number and the resulting value of ω will not change. If that number is selected as tending to zero, then it follows that $F \to 0$. Consequently, in real truss optimization problems, the minimum values of design variables A_f should be included in the set of constraints, i.e. $A_f - A_{\min} \geq 0$. Having done that, relation (3.33) for a truss becomes

$$f = \sum_f \lambda_f^c A_{\min}. \tag{3.34}$$

In other words, equations (3.25d) furnish the ratios of the components of **A**, while the optimality conditions (3.25p) select those that attain the minimum values.

The number of nonlinear equations (3.25a÷p) is balanced by the number of unknowns and depends on the problem considered. For example, in static problems there are $2n + p + q$ unknowns, where n is the number of degrees of freedom, p is the number of design variables, q is the number of imposed constraints. There exist many effective methods of solution for nonlinear equations systems. If, from the practical point of view, the number of unknowns is large, then the most efficient methods are the iterative ones. The iterative methods prove particularly advantageous when the equations are, as in the considered case, bilinear. For example, in (3.25a) we can choose the design variables that enter the stiffness matrix **K**. Next, the components of the displacement vector q_i may be found using a standard FEM program. The situation gets more complicated when solving equations (3.25b) since the Lagrange multipliers are not known there. If some information with regard to λ_i^q, λ_j^σ were available, equations (3.25b) could have been solved in a similar way as in (3.25a), with these multipliers playing the role of loading. Fortunately, equation (3.32) proves useful here. Namely, for a single active constraint, the Lagrange multiplier can be derived from (3.32). Solving (3.25a,b) gives q_i and λ_i^e, where the latter ones are the components of Lagrange multipliers vector associated with the equilibrium equations. Substitute now the obtained results into the optimality conditions (3.25p). These conditions will be satisfied only then if the substituted values are the solutions to the considered problem. For an arbitrary iteration, the residua D_j will emerge at the right hand sides of these equations. These, in turn, are used to obtain the design variables for the next iteration.

The concept of the above described algorithm relies on the assumption that only one constraint is active. Consider now the case with a larger number of constraints. Theoretically, each active solution is strictly satisfied, i.e. the constrained function reaches the very limit value of the constraint. This is not exactly the case in numerical analyses since the size of iteration step is finite. No matter how small the step is, it restrains the constraints from reaching their limit values. During the iteration process, the values of all active constraints differ from the theoretical values by a certain small number ε. This number depends on the size of iteration step and on the adopted numerical method. In our case, the scaling of design variables (see step 4 in the diagram below), after the displacements q_i have been found from (3.25a), makes only one constraint to reach its limit value. The other constraints remain unaffected. Consequently, every iteration step may be treated as a part of the solution of a single constraint optimization problem. Hence, the Lagrange multiplier is directly derivable form equation (3.32). It does not mean, however, that in a given multiple constraint optimization problem, the same constraint will be active in all iteration steps. It is the scaling that decides which constraint is active in a given iteration step, and whether the active constraint changes from iteration to iteration. Thus, to avoid substantial changes in the cost function, for example when switching from stress- to displacement-type constraints, it is advisable to employ dimensionless quantities in computations.

The algorithm just presented can be justified as follows. Consider a structural optimization problem with q active constraints. The weight of the structure is $\mathbf{W}_q$. Consider now the same structure but with $q-1$ constraints from the preceding problem, i.e. one of the q constraints has been removed. The new structure weight is $\mathbf{W}_{q-1} \leq \mathbf{W}_q$, which means that it does not exceed the old value. Removing successive constraints, we come to the conclusion that for a single constraint structure it follows that $\mathbf{W}_1 \leq \mathbf{W}_q$. The order of constraint removing is not arbitrary but is determined by the scaling. In other words, in a given iteration step, the scaling selects numerically the most active constraint. Therefore, within the above described algorithm, the solution of a single constraint optimization problem will never lead to a heavier structure than the one obtained from the solution of a multiple constraint problem.

3.4.3 Scheme of the Solution Algorithm

The following algorithm is concerned with minimum weight structures under static loading. Numerical examples show the algorithm at work. The algorithm consists of two parts coupled with each other. The first one relies on any FEM program of structural analysis. In the second one, an optimization procedure is carried out making use of the results furnished by the first part. In the algorithm below, the first part is marked (A), while the second one (O).

Step 1 (O). Assume $n := 0$ (n – iteration number).

Step 2 (O). Assume arbitrary vector of design variables $\mathbf{A}(0)$.

Step 3 (A). Compute displacement vector $\mathbf{q}(n)$ from (3.25a)

$$\mathbf{K}\,[\mathbf{A}(n)]\,\mathbf{q}(n) - \mathbf{Q} = 0.$$

Step 4 (O). Find maximum:

$$\max\left\{\max_i \left[\left|\frac{q_i(n)}{q_i^0}\right|\right],\ \max_j \left[\left|\frac{\varepsilon_j(n)}{\varepsilon_j^0}\right|\right]\right\} = r(n)$$

Step 5 (O). Scale design variables A_j and state variables q_i using $r(n)$

$$A_j(n) := A_j(n)r(n), \qquad q_i(n) := q_i(n)/r(n).$$

Step 6 (O). If $A_j(n) \leq A_{\min}$ then put $A_j(n) := A_{\min}$.
If $A_j(n) \geq A_{\max}$ then put $A_j(n) := A_{\max}$.

Step 7 (O). Depending on what constraint is active, compute Lagrange multipliers, either λ_i^q or λ_j^σ, from (3.32).

Step 8 (A). Compute $\boldsymbol{\lambda}^e$ from (3.25b)

$$\mathbf{K}\,[\mathbf{A}(n)]\,\boldsymbol{\lambda}^e - \boldsymbol{\lambda}^q(n) - \mathbf{B}^{\mathrm{T}}\boldsymbol{\lambda}^\sigma(n) = 0.$$

Step 9 (O). Compute residua $D_j(n)$ from (3.25p) which, according to (3.27), for a truss are

$$D_j(n) = -1 + \varepsilon_j(n)e_j(n) + \lambda_j^c(n)$$

If $|D_j(n)| \geq c_j$, then go to step 10, otherwise go to step 12 (c_j arbitrarily small number).

Step 10 (O). $n := n + 1$.

Step 11 (O). $$A_j(n) := A_j(n-1)\left[1 + \beta \frac{D_j(n-1)A_j(n-1)p}{\left\{\sum_j [D_j(n-1)A_j(n-1)]^2\right\}^{1/2}}\right]$$

(where β is the acceleration coefficient, p is the number of design variables). Go to step 3.

Step 12 (O). Stop.

3.4.4 Numerical Examples

Below two numerical examples are solved. The presented results were obtained using both the algorithm just described and, for comparison purposes, the one contained in Chap. 4. The algorithm of Chap. 4 is concerned with a discrete solution where the design variables are selected from a catalogue.

Example 1

Determine the minimum weight of a ten-bar truss shown in Fig. 3.1. The following constraints are imposed on the nodal displacements in vertical and horizontal directions: $q_i = 2$, the stresses in all bars: $\sigma^0 = 2.5 \cdot 10^4$ lb/in^2,

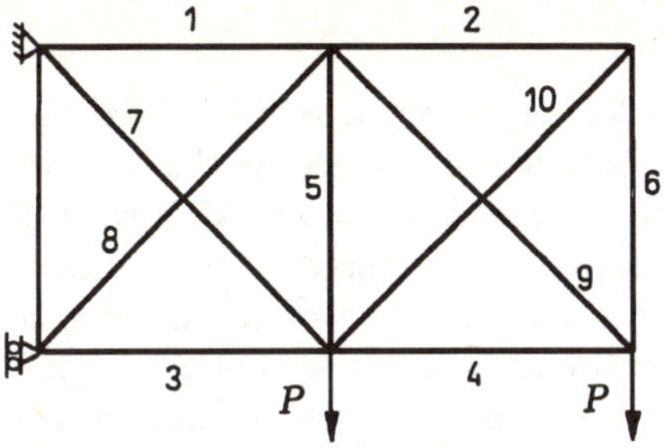

$$
\begin{aligned}
E &= 10^7 \text{ lb/in}^2\\
\rho &= 0.1 \text{ lb/in}^3\\
A_{\min} &= 0.1 \text{ in}^3\\
\sigma^0 &= 2.5 \cdot 10^4 \text{ lb/in}^2\\
q^0 &= 2.0 \text{ in}\\
P &= 10^5 \text{ lb}
\end{aligned}
$$

Fig. 3.1. Ten-bar minimum weight truss.

Table 3.1. Optimal cross-sections A_f.

Bar	Continuous solution (in^2)	Discrete solution (in^2)	Catalogue of discrete solution (in^2)	
1	30.29	36.0		
2	0.10	0.5	0.1;	0.5
3	23.59	27.0		
4	15.41	19.0	1.0;	2.0
5	0.10	0.5		
6	0.10	2.0	4.0;	7.0
7	7.56	7.0		
8	20.72	19.0	12.0;	19.0
9	21.14	19.0		
10	0.10	0.1	27.0;	36.0
Total weight (lb)	5026.19	5356.18		

the cross-sections of bars viewed as the design variables: $A_{\min} = 0.1 \text{ in}^2$. The dimensions of the structure and the material constants are specified in Fig. 3.1. Judging from their numerical values, it is clear that this example has no reference to a real structure. Nevertheless, it is often used in the relevant literature as a test problem. Its testing utility consists in a very sensitive final solution that can be applied to verify usefulness of the adopted computational methods. The results obtained within the present method confirm its practical applicability. In Table 3.1, the cross-sections of individual bars and the total volume of the minimum weight truss are juxtaposed. Throughout the example, the Anglo-Saxon unit system is used in order to enable the comparison with the results obtained by means of other methods.

Example 2

Consider the minimum weight problem of a space truss (Fig. 3.2a) being the roof of a sport hall. The structure is uniformly loaded by the wind pressure of 2.5 KPa. The constraints of the magnitude $\overset{+}{\sigma}_0 = 210$ MPa when tensile, and $\bar{\sigma}_0 = 168.5$ MPa when compressive, were imposed on the stresses acting in the bars. Moreover, the vertical displacements of the node located in the crossing point of the structure symmetry axes were bounded from above and below by $q_0 = \pm 10$ cm. For structures with large number of members, the optimization procedure involving equal number of design variables and structural members is practically unattainable. Therefore, the linking groups

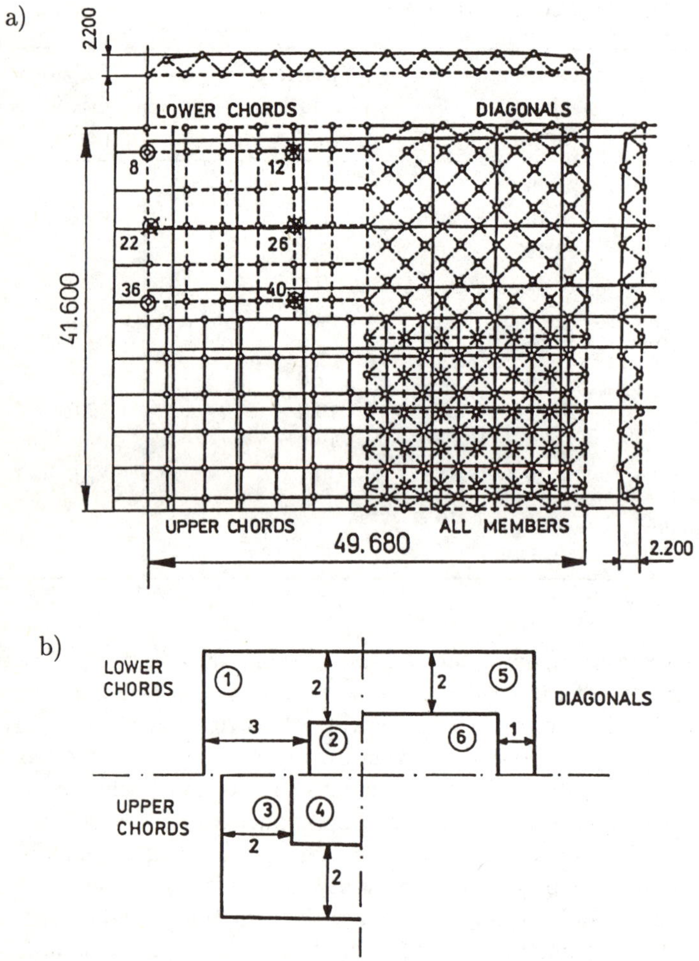

Fig. 3.2. Minimum weight space roof.

are formed consisting of a number of members to which the same design variable is assigned. In the considered case, all members were divided into six groups as indicated in Fig. 3.2b.

In view of the double symmetry, only a quarter of the structure consisting of 273 members and having 273 degrees of freedom was chosen for the further analysis. Additionally, it was assumed that the minimum cross-sectional area of a bar is limited by $A_{\min} = 10.79\ \text{cm}^2$. Table 3.2 shows the computed cross-sections for the six groups of bars.

Table 3.2. Optimum cross-sections A_j for six groups of members of space roof.

Group of bars	Catalogue of available elements (cm^2)	Continuous solution (cm^2)	Discrete solution (cm^2)
1	10.79; 21.17; 30.01; 44.16;	23.36	30.01
2		10.79	10.79
3	12.89; 21.17; 30.01; 44.16;	34.38	44.16
4		34.38	44.16
5	11.20; 15.52	12.08	15.52
6		12.08	15.52
	Total volume (m^3)	8.04	10.26

3.4.5 Solution Sensitivity to Variations of Design Variables

In the preceding subsection, the problem of minimum weight structure with multiple constraints has been solved. According to the adopted algorithm, the scaling selects numerically one of the constraints as fully active. Sometimes, a designer might need information on the extent of violation of a fully active constraint if all, or a part, of design variables are slightly altered. Such a situation may happen if the elements are picked up from the catalogues of available prefabricated elements. Usually, those elements are chosen whose properties are close to the ones obtained from the continuous solution.

The foundations of structural sensitivity analysis have been presented in Part VI. Now, the sensitivity of a fully active constraint will be examined using the method and the algorithm of the preceding subsection. Two cases will be considered. The first one deals with a statically loaded structure. The second one is the minimum weight problem with constraint imposed on the eigenfrequency of a vibrating structure.

Consider first the static problem. The constraints are denoted as follows

$$\begin{aligned} \mathbf{g}_e &= \mathbf{K}\mathbf{q} - \mathbf{Q} = 0, && \text{(a);} \\ \mathbf{h}_q &= \mathbf{q}^0 - \mathbf{q} \geq 0, && \text{(b);} \\ \mathbf{h}_\sigma &= \sigma_0 - \mathbf{B}^\mathrm{T}\mathbf{q} \geq 0, && \text{(c);} \\ \mathbf{h}_c &= \mathbf{A} - \mathbf{A}_{\min} \geq 0 && \text{(d).} \end{aligned} \tag{3.35}$$

Assuming that the following functions and their first derivatives are continuous, the variations of the above listed constraints take the form

$$\begin{aligned} \delta\mathbf{g}_e &= \frac{\partial \mathbf{g}_e}{\partial \mathbf{A}}\delta\mathbf{A} + \frac{\partial \mathbf{g}_e}{\partial \mathbf{q}}\delta\mathbf{q} = 0, && \text{(a);} \\ \delta\mathbf{h}_q &= \frac{\partial \mathbf{h}_q}{\partial \mathbf{q}}\delta\mathbf{q}, && \text{(b);} \\ \delta\mathbf{h}_\sigma &= \frac{\partial \mathbf{h}_\sigma}{\partial \mathbf{q}}\delta\mathbf{q}, && \text{(c);} \\ \delta\mathbf{h}_c &= \frac{\partial \mathbf{h}_c}{\partial \mathbf{q}}\delta\mathbf{A}, && \text{(d).} \end{aligned} \tag{3.36}$$

In the adopted notation, the vectorial forms of derivatives of the Lagrangean with respect to **A** and **u** read

$$\begin{aligned} \left(\frac{\partial \mathbf{L}}{\partial \mathbf{A}}\right)^\mathrm{T} &= -\mathbf{1}^\mathrm{T} + \boldsymbol{\lambda}^{e\mathrm{T}}\frac{\partial \mathbf{g}_e}{\partial \mathbf{A}} + \boldsymbol{\lambda}^{c\mathrm{T}}\frac{\partial \mathbf{h}_c}{\partial \mathbf{A}} = 0, && \text{(a);} \\ \left(\frac{\partial \mathbf{L}}{\partial \mathbf{q}}\right)^\mathrm{T} &= -\boldsymbol{\lambda}^{e\mathrm{T}}\frac{\partial \mathbf{g}_e}{\partial \mathbf{q}} + \boldsymbol{\lambda}^{q\mathrm{T}}\frac{\partial \mathbf{h}_q}{\partial \mathbf{q}} + \boldsymbol{\lambda}^{\sigma\mathrm{T}}\frac{\partial \mathbf{h}_\sigma}{\partial \mathbf{q}} = 0 && \text{(b).} \end{aligned} \tag{3.37}$$

Multiplying (3.37a) by $\delta\mathbf{A}$, (3.37b) by $\delta\mathbf{q}$, (3.36a) by $-\boldsymbol{\lambda}^{e\mathrm{T}}$ and adding up the resulting equations, the following linear combination is obtained

$$-\mathbf{1}^\mathrm{T}\delta\mathbf{A} + \boldsymbol{\lambda}^{q\mathrm{T}}\frac{\partial \mathbf{h}_q}{\partial \mathbf{q}}\delta\mathbf{q} + \boldsymbol{\lambda}^{\sigma\mathrm{T}}\frac{\partial \mathbf{h}_\sigma}{\partial \mathbf{q}}\delta\mathbf{q} + \boldsymbol{\lambda}^{c\mathrm{T}}\frac{\partial \mathbf{h}_c}{\partial \mathbf{A}} = 0. \tag{3.38}$$

Noting that the algorithm from the preceding subsection is a series of iterations with a single active constraint, the Lagrange multiplier may be found from (3.32) for that constraint.

If active constraints are imposed on the displacement q_i, or the stress σ_j, or the minimum cross-section area A_j, respectively, the variations of these constraints are

$$\begin{aligned} \delta q_i &= -\frac{\partial f}{f}q_{0i}, && \text{(a);} \\ \delta \sigma_j &= -\frac{\partial f}{f}\sigma_{0i}, && \text{(b);} \\ \delta A_j &= -\frac{\partial f}{f}A_{\min}, && \text{(c).} \end{aligned} \tag{3.39}$$

For example, reducing each design variable by 5%, it is found from (3.39a) that the change of q_{0i} equals $0.05q_{0i}$. This is in tune with the linear character of the equilibrium equations relative A_j. However, the following two restrictions should be kept in mind when using equations (3.39). Firstly, the variations of design variables must be small, e.g. less than a few percent of their original values. Secondly, equations (3.39) are valid exclusively for the sensitivity of active constraints. When expressed in terms of the present method, it means that every considered problem may have a single active constraint only.

A reasoning similar to that of the considered static problem may be extended to a dynamic problem with constraints imposed on the frequency of free vibrations. Let the constraints be denoted as follows

$$\begin{aligned} \mathbf{g}_d &= [\mathbf{K} - \omega_1^2 \mathbf{M}]\mathbf{y} = 0, && \text{(a);} \\ \mathbf{g}_0 &= \mathbf{y}^{\mathrm{T}} \mathbf{M} \mathbf{y} - 1 = 0, && \text{(b);} \\ h_\omega &= \omega_1^2 - \omega_0 \geq 0, && \text{(c);} \\ \mathbf{h}_c &= \mathbf{A} - \mathbf{A}_{\min} \geq 0, && \text{(d);} \end{aligned} \tag{3.40}$$

with the individual symbols as in Sect. 3.2.

The derivatives of the Lagrangean with respect to the independent variables $\mathbf{A}$, $\mathbf{y}$, ω_1^2 take the form

$$\begin{aligned} \left(\frac{\partial L}{\partial \mathbf{A}}\right)^{\mathrm{T}} &= -\mathbf{1}^{\mathrm{T}} + \boldsymbol{\lambda}^{d\mathrm{T}} \frac{\partial \mathbf{g}_d}{\partial \mathbf{A}} + \lambda^0 \frac{\partial \mathbf{g}_0}{\partial \mathbf{A}} + \boldsymbol{\lambda}^{cT} \frac{\partial \mathbf{h}_c}{\partial \mathbf{A}} = 0, && \text{(a);} \\ \left(\frac{\partial L}{\partial \mathbf{y}}\right)^{\mathrm{T}} &= \boldsymbol{\lambda}^{d\mathrm{T}} \frac{\partial \mathbf{g}_d}{\partial \mathbf{y}} + \lambda^0 \frac{\partial \mathbf{g}_0}{\partial \mathbf{y}} = 0,; && \text{(b);} \\ \frac{\partial L}{\partial \omega_1^2} &= \boldsymbol{\lambda}^{d\mathrm{T}} \frac{\partial \mathbf{g}_d}{\partial \omega_1^2} + \lambda^\omega \frac{\partial \mathbf{h}_\omega}{\partial \omega_1^2} = 0 && \text{(c).} \end{aligned} \tag{3.41}$$

Analogously as in the static case, multiply (3.41a) by $\delta\mathbf{A}$, (3.41b) by $\delta\mathbf{y}$, (3.41c) by $\delta\omega^2$, while δg_d, which is given by the formula

$$\delta \mathbf{g}_d = \frac{\partial \mathbf{g}_d}{\partial \mathbf{A}} \delta \mathbf{A} + \frac{\partial \mathbf{g}_d}{\partial \mathbf{q}} \delta \mathbf{y} + \frac{\partial \mathbf{g}_d}{\partial \omega_1^2} \delta \omega^2 = 0 \tag{3.42}$$

by $-\boldsymbol{\lambda}^{d^{\mathrm{T}}}$. Adding up the obtained equations gives

$$\delta f + \lambda^\omega \frac{\partial h_\omega}{\partial \omega_1^2} \delta \omega_1^2 + \boldsymbol{\lambda}^{cT} \frac{\partial \mathbf{h}_c}{\partial \mathbf{A}} \delta \mathbf{A} = 0. \tag{3.43}$$

Assuming as before that only one constraint is active in a given problem and making use of (3.32) to determine the value of Lagrange multiplier, we get

$$\delta \omega_1^2 = \frac{\delta f}{f} (\mu - 1) \omega_0^2, \quad \text{(a);} \qquad \delta A_j = \frac{\delta f}{f} A_{\min}, \quad \text{(b).} \tag{3.44}$$

Note that for a truss $(\mu - 1)$ in (3.44a) equals zero. Recall from the preceding sections of this chapter that for a truss, the free vibrations frequency cannot be taken as an active constraint. Consequently, relation (3.44a) ceases to be valid for trusses since the sensitivity considered here is restricted to active constraints.

4 Discrete Programming in Structural Optimization

4.1 Introductory Remarks

The preceding chapter has dealt with the structural optimization based on the assumption that the design variables are selected from the real numbers set. Consequently, it was possible to make use of the theorems, including the Kuhn–Tucker necessary conditions, that are based on the continuity of the considered functions. In the sequel, such an approach will be called the *continuous optimization.*

Practical engineering design proves to be more complex. Designers do not always have on their disposal arbitrary lists of elements from which design variables can be selected. Quite often, catalogues of prefabricated elements, thus finite sets of values for design variables, have to be used. In such cases, the methods of continuous optimization are not applicable. A rigorous formulation of an optimization problem with a finite number of available parameters (to be called the *discrete optimization* in the sequel) can be reduced to the formulation addressed in Chap. 3 (equations 3.1÷3.10) with an additional constraint imposed on the design variable A_f, namely

$$A_f \in [A_1^f, ..., A_k^f]_f, \quad A_{j<f>}^f \in R, \qquad f = 1, ..., f_0, \quad j_f = 1, ..., k_f \qquad (4.1)$$

where f_0 is the number of structural members, or groups of members, having the same properties, k_f is the number of items in the catalogue from which the f-th structural element is picked.

The structural optimization problem defined by equations (3.1)÷(3.10) and (4.1) is a discrete programming problem. The simplest way of solving such problems it would be to check a finite number of structures with all admissible combinations of elements from the available catalogues. In doing so, those combinations which violate the imposed mechanical constraints (displacements, stresses, vibration frequencies, etc.) are disregarded. Of the remaining admissible combinations, the one associated with the lightest structure is selected as the solution of the discrete programming problem (minimum volume structure). Although conceptually simple, this method is unacceptable from the practical point of view since the number of structures to be analyzed is too large.

To illustrate this point, consider a structure for which $f_0 = 10$. Assume further that all 10 design variables are selected from the same catalogue containing $k = 10$ elements. Anticipating future developments, it would be necessary to analyze $n = k^{f_0} = 10^{10}$ structures. This analysis comprises the input of data, solution of the state equations and checking of the constraints. For $f_0 = 10$ and $k = 10$, the considered structure would have to be solved 10^{10} times. Even if the individual solution took only 1 second, the whole analysis would last more than 300 years! Therefore, the following question sounds legitimate: does it make any sense to seek the solution of discrete optimization if the continuous optimization solution for the same structure takes only a few minutes on a PC? It is conceivable, though, to construct a discrete solution in such a way that the values of design variables obtained from the continuous solution be replaced by the nearest larger values picked from a catalogue. A brief word of caution is necessary with respect to this reasoning. There are many instances showing that the design variables selected according to this scheme do not lead to the structure of minimum weight in the sense of discrete optimization. To support this statement, consider the following discrete programming problem:

Find

$$\max\ x_1 - 3x_2 + 3x_3$$

taking into account the following constraints:

$$2x_1 + x_2 - x_3 \leq 4,$$

$$4x_1 - 3x_2 \leq 2,$$

$$-3x_1 + 2x_2 + x_3 \leq 3.$$

Assuming that $x_1,\ x_2,\ x_3$ are real numbers, their optimum values are found to be

$$x_1 = \frac{1}{2}, \quad x_2 = 0, \quad x_3 = 4\frac{1}{2}.$$

If additional assumption is made that $x_1,\ x_2,\ x_3$ be integers, thus discrete, the corresponding optimum solution reads $x_1 = 2$, $x_2 = 0$, $x_3 = 5$.

It seems now quite logical to ask the following two questions: Is there any way of making use of the continuous solution when solving a discrete problem? Knowing a priori that some combinations of design variables will not be the solutions to the considered problem, is it possible to control the method in such a way as to exclude those combinations from the numerical procedure? The answer to both question is yes.

Consider first the continuous solution in the context of a minimum weight structure, i.e. the answer to the first question. Such a solution is the lower limit to all discrete solutions which satisfy the imposed constraints. This stems from the fact that any discrete solution leads to a greater structure weight than the corresponding continuous solution does. Consequently, the continuous solution trims all those combinations of the design variables which lead to the structure weights that are below the lightest one obtained from the continuous solution. This is a very useful piece of information which, as it becomes clear later, makes the derivation of discrete solution much easier.

The answer to the second question is the subject of the next section.

It should be noted that the controlled enumeration method, to be presented shortly, is not the only method of solving discrete programming problems. The origins of the controlled enumeration method go back to the sixties [13, 33]. These papers, however, were not further used when analyzing large scale structures. Also, the segment method [32] is worth mentioning but its range of applicability is as yet rather limited. There exists another method which is based on the dual formulation of the original problem. In this case, the Lagrange multipliers constitute a simplification of the solution to the original problem. For more details concerning this method, which is applicable to practical problems, the reader is referred to original papers [30, 31]. For the sake of completeness, the controlled enumeration methods based on the Greenberg algorithm [13] should also be mentioned here.

4.2 Graph Representing Finite Number of All Possible Structure Volumes

Consider a structure with f_0 members, or with f_0 linking groups, associated with the same design variables A_f. This structure is to be erected using the members from a list of available elements, i.e.

$$A_j \in [A_1^f, ..., A_{k_f}^f], \tag{4.2}$$

where $A_{j_f}^f < A_{j_f+1}^f$, which means that the elements in the catalogue are arranged in increasing order.

Consistent with the notation of Chap. 3, the cost function, being in our case the material volume, is given by

$$f = \mathbf{A}^{\mathrm{T}}\mathbf{l}. \tag{4.3}$$

To depict all possible values of f, generate a tree graph [21, 22]. Select a root and assign a pair of indices $(0,0)$ to it, Fig. 4.1. The consecutive vertices of the graph are set up in the following manner: the vertex with the indices (f,i) is the father of the vertices with the indices $(f+1,k)$, where $k = 1, ..., k_{f+1}$. The vertex whose first index is f_0 has no sons. If indices of a son are (f,i) then the volume associated with that edge is $l_f A_{i_f}^f$.

The vertices of the graph can be arranged in layers numbered from 1 to f_0. In the first layer, the vertices represent all possible volume combinations of the first structural member. In the second layer, the vertices represent all possible combinations of the sum of volumes of the first two structural members. Finally, the vertices in the f_0-th layer represent all possible combinations of the volume sums for the whole structure. There are

$$n = k_1 \cdot k_2 \cdot ... \cdot k_{f_0}. \tag{4.4}$$

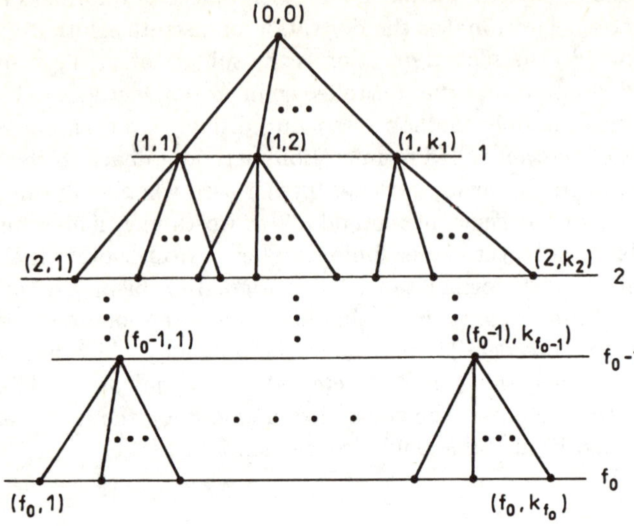

Fig. 4.1. Tree graph with vertices representing volumes of members.

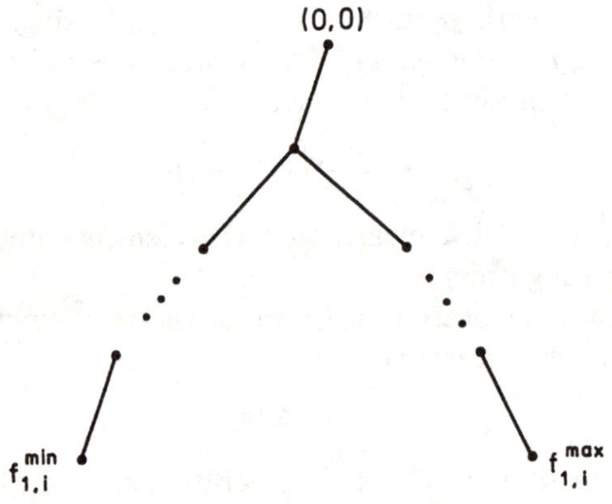

Fig. 4.2. Maximum and minimum values of cost function in graph and subgraphs.

of such combinations. If all the structural members are selected from the same catalogue $k_1 = k_2 = k_3 = ... = k_{f_0} = k$, the number of volume combinations becomes

$$n = k^{f_0}. \tag{4.5}$$

The graph with the root $(0,0)$ and the subgraphs with the roots (f,i) exhibit a certain property which is important when the minimum of discrete solution is sought. Namely, the sum of values assigned to the paths of the least catalogue values A_1^f, gives the vertex for which the volume of the considered structural members is minimum. Similarly, the sum of values assigned to the paths of the biggest catalogue values A^f determines the vertex for which the volume of the involved members is maximum. These situations are depicted in Fig. 4.2. As a consequence, in the graph based on the $A^f_{i_f} < A^f_{i_f+1}$ order, the edge vertices of a given layer represent the minimum and maximum volumes, respectively.

4.3 Minimum Weight Structure Made of Catalogued Members: Concept of an Algorithm

In this section, an algorithm will be outlined for determination of a minimum volume structure consisting of catalogued members. It is assumed that the minimum volume is a priori known from the continuous solution. In this context, to find a solution means to get such combination of design variables which leads to the least volume difference when compared with the continuous solution, and does not violate any of the imposed mechanical constraints.

Step 1. Find the continuous solution for the minimum volume structure (Chap. 3). Denote the obtained cost function by f_c. This value is the lower limit for all structures designed using the elements from a given catalogue and satisfying the imposed constraints.

Step 2. Consider all subgraphs with the roots $(1, i)$ for $i = 1, 2, ..., k_1$, i.e. having the vertices located in the first layer. The extreme values of the structure volume, represented by the vertices in the f_0 layer, are

$$f^{\min}_{1,i} = l_1 A^1_{i_1} + \sum_{f=2}^{f_0} A^f_1, \quad \text{(a)};$$
$$f^{\max}_{1,i} = l_1 A^1_{i_1} + \sum_{f=2}^{f_0} A^f_{k_f}, \quad \text{(b)}. \qquad (4.6)$$

These values and the volume f_i obtained from the continuous solution are related as follows

$$f_c < f_{1,i_{\min}}, \quad \text{(a)};$$
$$f_{1,i_{\min}} \leq f_c \leq f_{1,i_{\max}}, \quad \text{(b)}; \qquad (4.7)$$
$$f_{1,i_{\max}} < f_c, \quad \text{(c)}.$$

If inequality (4.7a) holds, the combination of the catalogue elements corresponding to $f_{1,i\min}$ and satisfying the constraints is the solution, provided that it does not belong to the subgraph $(1, i-1)$. In particular, for $i = 1$ the optimum solution is the set of the smallest catalogue elements for all structural members.

If inequality (4.7c) holds, the subgraph with vertex $(1,i)$ contains no solution of the considered discrete problem. Note that when $k = 10$ and $f_0 = 10$, then 10^9 combinations will be excluded form the further analysis. This is due to the fact that the subgraph with root $(1,i)$ has 10^9 vertices in the layer f_0.

If (4.7.b) holds, it is necessary to check for the solution in this subgraph, according to Step 3.

Step 3. Consider the subgraphs $(2,j)$ of the subgraphs $(1,i)$, for which (4.7b) is satisfied.

The extreme values of structure volume, depicted by the vertices (f_0,i), are

$$\begin{aligned} f_{2,i}^{\min} &= l_1 A_{i_1}^1 + l_2 A_{j_2}^2 + \textstyle\sum_{f=3}^{f_0} A_1^f, \quad \text{(a)};\\ f_{2,i}^{\max} &= l_1 A_{i_1}^1 + l_2 A_{j_2}^2 + \textstyle\sum_{f=3}^{f_0} A_{k_f}^f, \quad \text{(b)}. \end{aligned} \tag{4.8}$$

Proceeding further as in Step 2, find the subgraphs with roots $(2,j)$ for which

$$f_{2,i_{\min}} \leq f_c \leq f_{2,i_{\max}}. \tag{4.9}$$

Step 4,5,..., $f_0 - 1$**.** In the next steps, follow the procedures of Steps 2 and 3 as long as the layer $f_0 - 1$ is reached.

Step f_0 In the layer f_0, browse the vertices which are sons of the vertices $(f_0 - 1, i)$ obeying the following inequalities

$$f_{f_0-1,2,i_{\min}} \leq f_c \leq f_{f_0-1,2,i_{\max}}. \tag{4.10}$$

Store the vertices of the individual subgraphs $(f_0, -1, i)$ for which

$$f_{f_0,i} - f_c = \min. \tag{4.11}$$

Step $f_0 + 1$ For each combination check whether the imposed constraints are satisfied. The first structure for which these constraints are satisfied is the solution of the considered discrete optimization problem. If none of the combinations thus determined satisfies all the constraints, return to Step 2 replacing f_c from the continuous solution by the discrete value of f_c associated with that combination.

It may happen that several structures, being different combinations of the catalogued elements, have the same volume. In such a case, all these combinations have to be tested for violation of the imposed constraints. In practical applications, the number of combinations leading to the same structural volume may be large. This substantially increases the time needed to select the solution. In particular, this situation is encountered when two or more structural members (or group of members) have equal values of l_f (length of a bar, area of a finite element) and, additionally, each of the design variables is selected from the same catalogue.

4.4
Numerical Examples

Two numerical examples of the structural optimization (as in Chap. 3) were considered in [1, 14]. The results are juxtaposed in Tables 3.1 and 3.2 in order to facilitate comparison of the two solutions.

References

1. Bauer J, Gutkowski W, Iwanow Z (1981) A discrete model for lattice structures optimization. Engng Optim 5:121–128
2. Bletzinger KU, Kimmich S (1985) Strukturoptimierung. Technical Report Konzepte SFB 230, Heft 7, Universität Stuttgart, Stuttgart
3. Borkowski A (1977) Optimization of slab reinforcement by linear programming. Comp Meth Appl Mech Eng 12:1–17
4. Borkowski A (1988) Analysis of skeletal structural systems in the elastic and elastic-plastic range. Developments in Civil Engng, 20. Elsevier, Amsterdam
5. Brandt AM (1977) Kryteria i metody optymalizacji konstrukcji. Polish Scientific Publishers (PWN), Warsaw
6. Brandt AM (1978) Podstawy optymalizacji elementów konstrukcji budowlanych. Polish Scientific Publishers (PWN), Warsaw
7. Chyras A (1983) Mathematical models for the analysis and optimization of elastoplastic structures. Ellis Horwood Ltd., Chichester
8. Dorn WS, Gomory RE, Greenberg HJ (1964) Automatic design of optimal structures. Journal Mechanique 3:2–51
9. Findeisen A, Szymanowski J, Wierzbicki A (1980) Metody obliczeniowe optymalizacji. Polish Scientific Publishers (PWN), Warsaw
10. Fox RL (1971) Optimization methods for engineering design. Addison–Wesley, New York
11. Gallagher RM, Zienkiewicz OC (eds) (1973) Optimum structural design: theory and applications. Wiley & Sons, London
12. Gill PE, Murray W (1974) Numerical methods for constrained optimization. Academic Press, London
13. Greenberg DE, Lee WH (1986) Optimal synthesis of frameworks under elastic and plastic performance constraints using discrete sections. J Struct Mech 14:401–420
14. Gutkowski W, Bauer J, Iwanow Z (1986) Minimum weight design of space frames from a catalogue. In: Heki K (ed) Proc IASS Symp on Shells, Membranes and Space Frames, Osaka, Japan, 1986, volume 3. Elsevier, Amsterdam, pp 229–236
15. Gutkowski W, Bauer J, Iwanow Z (1990) Explicit formulation of Kuhn-Tucker necessary conditions in structural optimization. Int J Comp Struct 37:753–758
16. Gutkowski W, Bauer J, Iwanow Z, Kupść J (1987) Computer aided design of space trusses. In: Proc IASS Coll on Space Structures and Sport Buildings, Beijing, China, Oct 1987. Science Press, Beijing, pp 234–237
17. Haftka RT, Gürdal Z, Kamat MP (1990) Elements of structural optimization. Kluwer, Dordrecht

18. Haug EJ, Arora JS (1979) Applied optimal design. Mechanical and structural systems. Wiley & Sons, London

19. Horne MR, Morris LJ (1981) Plastic design of low-rise frames. The MIT Press, Cambridge, MA

20. Hörnlein HREM (1993) Structural optimization - a survey. In: Hörnlein HREM, Schittkowski K (eds) Software Systems for Structural Optimization. Birkhäuser–Verlag, Stuttgart, pp 1–32

21. Iwanow Z (1981) The enumeration method according to the increasing value of the objective function in the optimization of bar structures. Bull Acad Pol Sci, Ser Sci Tech XXIX:481–486

22. Iwanow Z (1990) An algorithm for finding an ordered sequence of values of a discrete linear function. Control and Cybernetics 19:129–154

23. Jóźwiak S, Borkowski A (May 1989) Intelligent computer aided design of trusses. In: Proc IX Polish Conference on Computer Methods in Mechanics, Rytro. Cracow University of Technology, pp 345–362

24. Łubiński M, Czernecki J, Giżejowski M, Iwanowska B, Kleśta L, Patorski J (1980) Designing elements of steel structures. Technical Report (in Polish), Warsaw University of Technology, Warsaw

25. Pedersen P (1972) On the optimal layout of the multipurpose trusses. Int J Comp Struct 2:695–712

26. Polish Committee of Normalization, Warsaw. PN-76/B-03200: Steel structures. Static analysis and design (in Polish), (1977)

27. Save M, Prager W (eds) (1990) Optimality criteria, volume 1 of Structural optimization. Plenum Press, New York

28. Sawczuk A (1982) Introduction to mechanics of plastic structures (in Polish). Polish Scientific Publishers (PWN), Warsaw

29. Schittkowski K (1993) Mathematical optimization: an introduction. In: Hörnlein HREM, Schittkowski K (eds) Software systems for structural optimization. Birkhäuser–Verlag, Stuttgart, pp 33–42

30. Schmit LA, Fleury C (1980) Discrete-continuous variable structural synthesis using dual methods. AIAA J 18:1515–1524

31. Sempulveda A, Cassis JH (1986) An efficient algorithm for the optimum design of trusses with discrete variables. Int J Num Meth Engng 23:1111–1130

32. Templeman AB, Yates DF (1983) A segmental method for the discrete optimum design of structures. Engng Optim 6:145–155

33. Toakley AR (1968) Optimum design using available sections. Proc ASCE, J Struct Div 94:1219–1241

34. Zangwill WI (1969) Nonlinear Programming: a Unified Approach. Prentice–Hall, Englewood Cliffs

35. Życzkowski M (ed) (1988) Technical mechanics, volume 9: Strength of structural elements (in Polish). Polish Scientific Publishers (PWN), Warsaw

Part VI

Methods of Sensitivity Analysis

1 Introduction

In classical problems of structural mechanics it is usually assumed that geometric form of a structure, support conditions, loading, and material parameters are specified. Then the analysis problem is concerned with determination of state fields of displacements, strains and stresses, so that equilibrium, geometric compatibility, and boundary conditions are satisfied. Most structure mechanics textbooks consider this class of analysis problems. Now, the question arises how these state fields vary when the structural parameters undergo modification. In particular, one could study the variation of scalar functionals of the state fields or variation of their extremal values. This type of information is needed when the design or existing structure are to be modified. A similar question arises when the effect of imperfections and material parameter scatter is considered and their effect on variation of stress or displacement is studied.

A closely related class of problems occurs when the structure suffers from damage and the stiffness parameters are affected by damage growth. The variation of state fields and their functionals is needed in order to assess the damage effect, or to identify the damage by measuring static or dynamic displacements and strains with their variation during service life.

The sensitivity analysis is concerned with this class of problems. Considering the structure modification, a small variation of structure parameters can be considered, and the variational methods applied to study the evolution of state fields from the initial equilibrium state. When a finite parameter variation occurs, the sensitivity can be analysed by solving additional boundary-value problems.

The sensitivity analysis is directly related to the optimal design methods where the design parameters or functions are to be specified in order to minimize the objective function (structure cost or material volume) subject to the constraints imposed on state fields and design parameters. Specifying the sensitivity gradients of objective function and constraints, the optimality conditions can be stated and used in generating the optimal solutions analytically or numerically. Alternatively, the gradient optimization techniques can be applied in reaching the optimal solution through iterative steps of analysis and redesign.

A separate class of problems for which the sensitivity analysis is of major importance is related to structure parameter identification. This class of *inverse problems* is gaining its importance especially for more advanced nonlinear and inelastic models accounting for plastic deformations and damage. The experimentally specified displacement or strain fields at specified points are used in order to determine the structure material or configuration parameters and also loading parameters. The optimization procedure is now aimed at minimizing the distance norm between actual and theoretical state fields predicted by the assumed structure model.

The aim of this Part is to introduce the reader to the methods of sensitivity analysis by discussing both variational formulation and numerical aspects. It turns out that in many cases the analytical expressions can be obtained for sensitivity gradients or variations and they can easily be incorporated in the process of structure redesign or modification. It is hoped that the sensitivity analysis will become a permanent element of structural mechanics and used as an effective tool in structural optimization, identification, or redesign.

2 Classification of Structural Design Variables and Parameters

In the structure analysis problems the *state fields*, that is displacement, strain, and stress fields $\mathbf{u}(\mathbf{x})$, $\boldsymbol{\varepsilon}(\mathbf{x})$ and $\boldsymbol{\sigma}(\mathbf{x})$ are to be determined from the boundary-value problem for specified structural parameters or fields. However, these parameters may vary and have different character needing classification. When the elasticity or compliance matrices $\mathbf{C}(\mathbf{x})$ and $\mathbf{D}(\mathbf{x})$ vary within the structure domain, the respective state field variations can be determined from a properly formulated new boundary-value problem associated with material parameter variation. For surface or beam structures (plates, shells, frames, etc.) the cross-sectional stiffness moduli depend on both material stiffness and cross-sectional dimensions. For instance, for a rectangular beam the flexural stiffness equals $S_b = EI = Ebh^3/12$ where E denotes the Young modulus and $I = bh^3/12$ is the cross-sectional moment of inertia (b, h are beam width and depth, respectively). Thus both cross-sectional dimensions and material elasticities affect the flexural or extensional stiffness. We can therefore put material or cross-sectional dimensional variables within the same class becouse their variation does not affect boundary or loading conditions.

A separate class of structure variables is associated with its shape or configuration, cf. [10,12,32]. Consider, for instance, a structure presented schematically in Fig. 2.1. The structure shape is specified by its external boundaries S_T, S_0 and S_u where S_T denotes the loaded boundary portion, S_0 is the traction-free boundary portion and S_u denotes the supported boundary with specified displacement components. However, the shape variables are also constituted by interfaces S_c, S_r and S_d where S_c denotes *composite interface* separating materials of different stiffness moduli, S_r is the *reinforcing interface* inducing traction discontinuity and S_d is the *relaxation interface* characterized by displacement discontinuity and relaxing stresses induced by initial strain fields or boundary displacements.

Variation of structural shape induces in general the variation of boundary and continuity conditions on modified surfaces. For beam or truss structures, configuration variables correspond to joint positions whose variation affects both equilibrium and compatibility conditions.

Topological structure variables correspond to type of microstructure within the material or to type of connectivity between structural members within a discrete structure. For instance, comparing a voided or inclusion stiffened material with a homogeneous material, we can state that these materials have different topologies as they differ by a number of free boundaries or

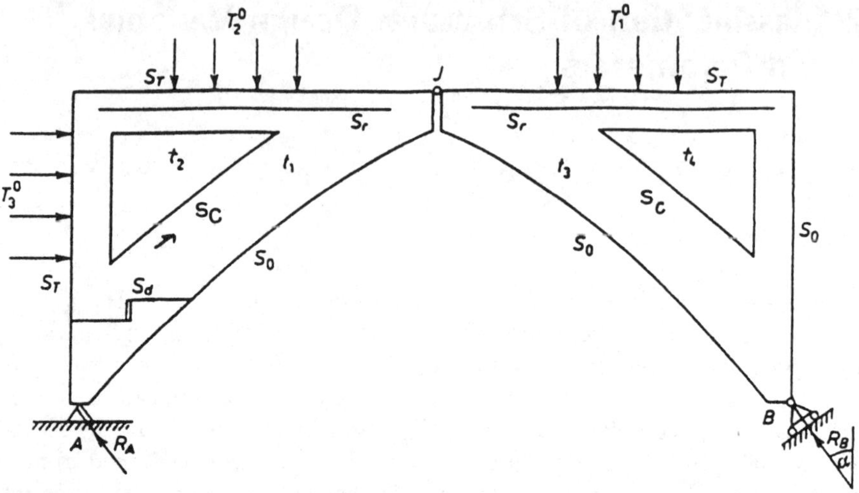

Fig. 2.1. Scheme of structure and design variables: dimensional: t_1, t_2, t_3, t_4; shape: S_0, S_T, S_u, S_i; support: R_A, R_B; connection: K_J

interfaces. Similarly, for a truss structure, number of joints and types of connections between joints specify the structure topology.

Support and loading conditions constitute a separate class of design variables or parameters specifying boundary conditions. For instance, for point supports, their position η, orientation φ and reaction $\mathbf{R}$ can be modified. Support variation may also be induced by shape variation (for instance, a support along direction normal to the boundary surface). Similarly, for external loading, its position, orientation and surface distribution can be modified, cf. [36, 39]. An additional class of design variables can be generated by initial strain distortion fields $\varepsilon^d(\mathbf{x})$ or initial boundary displacements $\mathbf{u}^d(\mathbf{x})$ inducing self-equilibrated stress fields within the structure. The distortion fields may affect the stresses and displacements due to loading and their optimal distribution constitutes an important engineering problem cf. [18].

Let us summarize our discussion by listing state and design variables:

1. Analysis: *state fields* $\mathbf{u}$, ε, $\boldsymbol{\sigma}$,
2. Redesign, optimization, identification: *design variables*
   (a) material or cross-sectional variables: $\mathbf{C}$, $\mathbf{D}$, h, b,
   (b) shape or configuration variables: S_T, S_0, S_u, S_c, S_r, S_d,
   (c) support and loading variables: η, φ, $\mathbf{R}$, $\lambda\mathbf{T}^0$,
   (d) distortion variables ε^d, $\mathbf{u}^d$,
   (e) joint or connection parameters (stiffness and strength): J_s, J_d,
   (f) topological parameters (number of joints or connections): n, m. (positive integers).

Consider now the analysis functional depending on both state and design variables

$$\Pi = \Pi(\mathbf{u},\, \varepsilon,\, \boldsymbol{\sigma};\, \mathbf{C},\, S_T,\, S_0,\, S_u,\, S_c,\, S_r,\, S_d,\\ \eta,\, \varphi,\, \mathbf{R},\, \lambda\mathbf{T}^0,\, \varepsilon^d,\, \mathbf{u}^d,\, J_s,\, J_d,\, n,\, m) \qquad (2.1)$$

In classical analysis problems the design variables are specified and state fields may vary. In particular, the functional in Eq. (1) may correspond to potential or complementary energy of the structure. The stationarity condition of the functional with respect to the displacement field

$$\delta\Pi_u(\mathbf{u}) = 0, \qquad (2.2)$$

or with respect to the stress field

$$\delta\Pi_\sigma(\boldsymbol{\sigma}) = 0, \qquad (2.3)$$

provide equilibrium and geometric compatibility conditions within a class of kinematically admissible displacement fields or statically admissible stress fields and thus constitute solutions to boundary-value problems.

A more general Hellinger–Reissner functional $\Pi^m(\mathbf{u},\, \varepsilon,\, \boldsymbol{\sigma})$ specifies hybrid (or mixed formulation of boundary-value problems for free fields $\mathbf{u}$, ε and $\boldsymbol{\sigma}$, so that

$$\delta\Pi_u^m(\mathbf{u},\, \varepsilon,\, \boldsymbol{\sigma}) = 0\,, \qquad \delta\Pi_\varepsilon^m(\mathbf{u},\, \varepsilon,\, \boldsymbol{\sigma}) = 0\,, \qquad \delta\Pi_\sigma^m(\mathbf{u},\, \varepsilon,\, \boldsymbol{\sigma}) = 0\,, \quad (2.4)$$

where $\delta\Pi_u^m$, $\delta\Pi_\varepsilon^m$, $\delta\Pi_\sigma^m$, are the functional variations with respect to $\mathbf{u}$, ε and $\boldsymbol{\sigma}$.

Assume now that the design variables undergo variations for fixed boundary conditions. The state fields will then vary and the functional (2.1) or any other response functional

$$G = G\,(\mathbf{u},\, \varepsilon,\, \boldsymbol{\sigma};\, \mathbf{C},\, S_T\,,\, S_0\,,\, S_u\,,\, S_c\,,\, S_r\,,\, S_d\,,\, \ldots\ldots) \qquad (2.5)$$

will also vary due to design variation. In order to specify the variation of G explicitly in terms of design variable variation, that is

$$\delta G = \mathbf{S}\cdot\delta\mathbf{s} \qquad (2.6)$$

where $\delta\mathbf{s}$ denotes collectively the variation of the design vector $\mathbf{s}$ and $\mathbf{S}$ is the conjugate sensitivity vector, we have to determine or eliminate variations of state fields occurring in (2.5). This can be achieved by solving a set of additional boundary-value problems. Alternatively, one may specify $\mathbf{S}$ in terms of primary and adjoint state fields determined from the solution of an adjoint problem. The dot in (2.6) denotes the scalar product and integral over the structure domain in the case of integral functionals. The details of sensitivity analysis will be discussed is subsequent sections.

3 Fundamental Variational Theorems

Let us now briefly discuss the variational formulation of the analysis problem within the assumptions of small strain theory and linear or non-linear elastic model. This exposition, though concerned with classical theorems, is aimed at laying the foundation for sensitivity analysis. Consider a three-dimensional structure loaded at its boundary portion δS_T and supported on the boundary portion δS_u . The actual stress, strain, and displacement states at equilibrium are denoted by $\boldsymbol{\sigma}$, $\boldsymbol{\varepsilon}$ and $\mathbf{u}$. In a Cartesian reference system the fundamental equations are expressed as follows.

- Equilibrium equations:

$$\sigma_{ij,i} + f_i = 0 \tag{3.1}$$

- Strain–displacement relations:

$$\varepsilon_{ij} = \frac{1}{2}(u_{i,j} + u_{j,i}) \tag{3.2}$$

- Constitutive relations:

$$\sigma_{ij} = C_{ijkl}\varepsilon_{kl} \tag{3.3}$$

- Boundary conditions:

$$\sigma_{ij}n_i = T_j^0 \qquad \text{on} \quad S_T \tag{3.4}$$

$$u_i = u_i^0 \qquad \text{on} \quad S_u \tag{3.5}$$

where comma denotes differentiation with respect to the space coordinate x_i , σ_{ij} is the Cauchy stress tensor for which the index i specifies the normal orientation to the physical plane and index j specifies the orientation within the plane. Obviously in view of stress and strain tensor symmetry there is $\sigma_{ij} = \sigma_{ji}$, $\varepsilon_{ij} = \varepsilon_{ji}$, or in matrix notation $\boldsymbol{\sigma} = \boldsymbol{\sigma}^T$, $\boldsymbol{\varepsilon} = \boldsymbol{\varepsilon}^T$. The body force vector is denoted by $\mathbf{f}$. Using matrix notation, the set of Eqs. (3.1)–(3.5) can be expressed as follows

$$\begin{gathered} \boldsymbol{\nabla}^T\sigma + \mathbf{f}^T = 0, \qquad \varepsilon = \frac{1}{2}(\boldsymbol{\nabla}\mathbf{u}^T + \mathbf{u}\boldsymbol{\nabla}^T) = \mathbf{B}\,\mathbf{u} \qquad \text{within } V \\ \mathbf{n}\sigma = \boldsymbol{\sigma}^T\mathbf{n} = \mathbf{T}^0 \quad \text{on } S_T\,, \qquad \mathbf{u} = \mathbf{u}^0 \quad \text{on } S_u\,, \end{gathered} \tag{3.6}$$

where $\boldsymbol{\nabla}$ is the column gradient vector.

3.1 Virtual Work Equation

Consider a kinematically admissible displacement field $\mathbf{u}^k(\mathbf{x})$, continuous within the structure domain, Ω_s, $\mathbf{u}^k(\mathbf{x}) \in C^0$, and satisfying the boundary conditions (3.5) on S_u. Similarly, the statically admissible stress field $\boldsymbol{\sigma}^s(\mathbf{x})$ satisfies the equilibrium equations (3.1) within Ω_s and boundary conditions (3.4) on S_T. In view of (3.1) and (3.2), the following equality can be derived for those two fields

$$\int \sigma_{ij}^s \varepsilon_{ij}^k \, \mathrm{d}V = \int \sigma_{ij}^s u_{i,j}^k \, \mathrm{d}V = \int \sigma_{ij}^s n_i u_j^k \, \mathrm{d}S + \int f_i u_i^k \, \mathrm{d}V \tag{3.7}$$

where $\mathrm{d}V$ and $\mathrm{d}S$ denote the volume and surface elements. This *virtual work equation* can also be written in the form

$$\int \sigma_{ij}^s \varepsilon_{ij}^k \, \mathrm{d}V = \int T_i^0 u_i^k \, \mathrm{d}S_T + \int T_i u_i^0 \, \mathrm{d}S_u + \int f_i u_i^k \, \mathrm{d}V \tag{3.8}$$

or in vector notion

$$\int \boldsymbol{\sigma}^s \cdot \boldsymbol{\varepsilon}^k \, \mathrm{d}V = \int \mathbf{T}^0 \cdot \mathbf{u}^k \, \mathrm{d}S_T + \int \mathbf{T} \cdot \mathbf{u}^0 \, \mathrm{d}S_u + \int \mathbf{f} \cdot \mathbf{u}^k \, \mathrm{d}V \tag{3.9}$$

where dot between two symbols denotes the scalar product $\boldsymbol{\sigma} \cdot \boldsymbol{\varepsilon} = \boldsymbol{\sigma}^T \cdot \boldsymbol{\varepsilon} = \sigma_{ij}\varepsilon_{ij}$. In deriving the virtual work equation, the Green–Gauss theorem was used

$$\int \frac{\partial \phi}{\partial x_i} \, \mathrm{d}V = \int \phi n_i \, \mathrm{d}S \tag{3.10}$$

where ϕ is a differentiable scalar function. For the vector function $\mathbf{A}$ we have

$$\int \frac{\partial A_i}{\partial x_j} \, \mathrm{d}V = \int A_i n_j \, \mathrm{d}S \, . \tag{3.11}$$

In particular, there is

$$\int \operatorname{div} \mathbf{A} \, \mathrm{d}V = \int \frac{\partial A_i}{\partial x_i} \, \mathrm{d}V = \int A_i n_i \, \mathrm{d}S = \int \mathbf{A} \cdot \mathbf{n} \, \mathrm{d}S \, . \tag{3.12}$$

The virtual work equation can be regarded as the weak (integral) expression of equilibrium or geometric compatibility conditions. Assume that (3.8) occurs for any kinematically admissible field $\mathbf{u}^k$, that is continuous field satisfying the boundary condition $\mathbf{u}^k = \mathbf{u}^0$ on S_u. Since

$$\sigma_{ij}^s \varepsilon_{ij}^k = \sigma_{ij}^s u_{j,i}^k = (\sigma_{ij}^s u_j^k)_{,i} - \sigma_{ij,i}^s u_j^k \, , \tag{3.13}$$

Eq. (3.8) can be presented in the form

$$\int (\sigma_{ij,i}^s + f_j) u_j^k \, \mathrm{d}V + \int (T_j^0 - \sigma_{ij}^s n_i) u_j^k \, \mathrm{d}S_T = 0 \, . \tag{3.14}$$

This equation expresses in a weak form the equilibrium conditions for a structure. In fact, requiring (3.14) to occur for each field $\mathbf{u}^k$, then, according to the fundamental lemma of the variational calculus, there is

$$\sigma^s_{ij,i} + f_j \quad \text{within } V, \qquad \sigma^s_{ij} n_i = T^0_j \quad \text{on } S_T\,. \tag{3.15}$$

Assume now that (3.8) occurs for each statically admissible field $\boldsymbol{\sigma}^s(\mathbf{x})$, that is satisfying Eq. (3.15). Since there is

$$\begin{aligned} \int T^0_j u^k_j \,\mathrm{d}S_T + \int f_j u^k_j \,\mathrm{d}V &= \int \sigma^s_{ij} n_i u^k_j \,\mathrm{d}S_T - \int \sigma^s_{ij,i} u^k_j \,\mathrm{d}V \\ &= \int \sigma^s_{ij} u^k_{j,i} \,\mathrm{d}V - \int T^0_j u^k_j \,\mathrm{d}S_u \end{aligned} \tag{3.16}$$

then from (3.8) it follows that

$$\int \sigma^s_{ij}(\varepsilon^k_{ij} - u^k_{i,j})\,\mathrm{d}V - \int T^s_j(u^0_j - u^k_j)\,\mathrm{d}S_u = 0\,. \tag{3.17}$$

The equation expresses in a weak form the kinematic compatibility of the strain field. In fact if (3.17) is to occur for each statically admissible field $\boldsymbol{\sigma}^s$, then

$$\varepsilon^k_{ij} = \frac{1}{2}(u^k_{i,j} + u^k_{j,i}) \quad \text{within } V, \qquad u^k_j = u^0_j \quad \text{on } S_u\,. \tag{3.18}$$

3.2 Theorems of Potential and Complementary Energies

Consider a linear elastic material for which the specific strain and stress energies are

$$U(\varepsilon) = \frac{1}{2} C_{ijkl}\, \varepsilon_{ij}\, \varepsilon_{kl} > 0\,, \qquad W(\sigma) = \frac{1}{2} D_{ijkl}\, \sigma_{ij}\, \sigma_{kl} > 0\,, \tag{3.19}$$

where $\mathbf{C} = \mathbf{D}^{-1}$ is the elasticity matrix and $\mathbf{D}$ is the elastic compliance matrix. The functions $U(\varepsilon)$ and $W(\boldsymbol{\sigma})$ are the elastic potentials, so that

$$\boldsymbol{\sigma} = \frac{\partial U(\varepsilon)}{\partial \varepsilon} = \mathbf{C}\varepsilon\,, \qquad \varepsilon = \frac{\partial W(\boldsymbol{\sigma})}{\partial \boldsymbol{\sigma}} = \mathbf{D}\boldsymbol{\sigma}\,. \tag{3.20}$$

The potential energy of the structure is

$$\Pi(\mathbf{u}) = \int U(\varepsilon)\,\mathrm{d}V - \int \mathbf{T}^0 \cdot \mathbf{u}\,\mathrm{d}S_T - \int \mathbf{f}\cdot\mathbf{u}\,\mathrm{d}V \tag{3.21}$$

where the first term represents the elastic energy of the structure the second term represents the potential energy of external forces, and the last term corresponds to the work of body forces. For the actual state fields $\mathbf{u}$, ε, $\boldsymbol{\sigma}$, the potential energy can be specified by the work of surface tractions and body forces. In fact, in view of the virtual work equation (3.8) written for actual fields $\mathbf{u}$, ε, $\boldsymbol{\sigma}$, we have

$$\Pi(\mathbf{u}) = -\frac{1}{2}\int \mathbf{T}^0 \cdot \mathbf{u}\,\mathrm{d}S_T + \frac{1}{2}\int \mathbf{T}\cdot\mathbf{u}^0\,\mathrm{d}S_u - \int \mathbf{f}\cdot\mathbf{u}\,\mathrm{d}V\,. \tag{3.22}$$

Assume now that the structure is rigidly supported on S_u: $\mathbf{u}^0 = \mathbf{0}$, and the body forces vanish, $\mathbf{f} = \mathbf{0}$. Then from (3.22) it follows that

$$\Pi(\mathbf{u}) = -\frac{1}{2}\int \mathbf{T}^0 \cdot \mathbf{u}\, \mathrm{d}S_T < 0\,. \tag{3.23}$$

On the order hand, when only displacements are imposed on S_u, that is $\mathbf{T}^0 = \mathbf{0}$, then

$$\Pi(\mathbf{u}) = -\frac{1}{2}\int \mathbf{T} \cdot \mathbf{u}^0\, \mathrm{d}S_u > 0\,. \tag{3.24}$$

It is seen that the potential energy is negative in the case of specified loading on S_T and positive in the case of specified displacements on S_u.

The variation of $\Pi(\mathbf{u})$ induced by the variation of the displacement and stress fields from the equilibrium state equals

$$\begin{aligned}\delta\Pi(\mathbf{u}) &= \int \frac{\partial U(\boldsymbol{\varepsilon})}{\partial \boldsymbol{\varepsilon}} \cdot \delta\boldsymbol{\varepsilon}\, \mathrm{d}V - \int \mathbf{T}^0 \cdot \delta\mathbf{u}\, \mathrm{d}S_T - \int \mathbf{f} \cdot \delta\mathbf{u}\, \mathrm{d}V \\ &= \int \boldsymbol{\sigma} \cdot \delta\boldsymbol{\varepsilon}\, \mathrm{d}V - \int \mathbf{T}^0 \cdot \delta\mathbf{u}\, \mathrm{d}S_T - \int \mathbf{f} \cdot \delta\mathbf{u}\, \mathrm{d}V = 0\,. \end{aligned} \tag{3.25}$$

The variation of $\Pi(\mathbf{u})$ thus vanishes in view of the virtual work equation (3.8). In fact, setting $\mathbf{u}^k = \mathbf{u}$, $\boldsymbol{\varepsilon}^k = \delta\boldsymbol{\varepsilon}$, $\delta\mathbf{u} = \mathbf{0}$ on S_u, Eq. (3.8) is equivalent to (3.25). In order to demonstrate that the stationary state of $\Pi(\mathbf{u})$ corresponds to minimum, consider the difference between $\Pi(\mathbf{u}^k)$ for any arbitrary kinematically admissible field $\mathbf{u}^k$ and $\Pi(\mathbf{u})$ for the actual equilibrium state. We have

$$\begin{aligned}\Pi(\mathbf{u}^k) - \Pi(\mathbf{u}) &= \int [U(\boldsymbol{\varepsilon}^k) - U(\boldsymbol{\varepsilon})]\, \mathrm{d}V - \int \mathbf{T}^0 \cdot (\mathbf{u}^k - \mathbf{u})\, \mathrm{d}S_T - \int \mathbf{f} \cdot (\mathbf{u}^k - \mathbf{u})\, \mathrm{d}V \\ &= \int [U(\boldsymbol{\varepsilon}^k) - U(\boldsymbol{\varepsilon}) - \boldsymbol{\sigma} \cdot (\boldsymbol{\varepsilon}^k - \boldsymbol{\varepsilon})]\, \mathrm{d}V \\ &= \frac{1}{2}\int (\boldsymbol{\varepsilon}^k - \boldsymbol{\varepsilon}) \cdot \mathbf{C}\, (\boldsymbol{\varepsilon}^k - \boldsymbol{\varepsilon})\, \mathrm{d}V \geq 0\,. \end{aligned} \tag{3.26}$$

It is seen that $\Pi(\mathbf{u}^k) \geq \Pi(\mathbf{u})$ and the potential energy reaches its minimum at the equilibrium state.

Consider now the complementary energy

$$\bar{\Pi}(\boldsymbol{\sigma}) = \int W(\boldsymbol{\sigma})\, \mathrm{d}V - \int \mathbf{T} \cdot \mathbf{u}^0\, \mathrm{d}S_u\,. \tag{3.27}$$

When the structure is rigidly supported ($\mathbf{u}^0 = \mathbf{0}$ on S_u) we have

$$\bar{\Pi}(\boldsymbol{\sigma}) = \int W(\boldsymbol{\sigma})\, \mathrm{d}V = \frac{1}{2}\int \boldsymbol{\sigma} \cdot \mathbf{D}\boldsymbol{\sigma}\, \mathrm{d}V = \int \mathbf{T}^0 \cdot \mathbf{u}\, \mathrm{d}S_T > 0 \tag{3.28}$$

and when the structure is subjected to displacements on S_u ($\mathbf{T}^0 = \mathbf{0}$ on S_T), we can write

$$\bar{\Pi}(\boldsymbol{\sigma}) = -\int W(\boldsymbol{\sigma})\, \mathrm{d}V = -\frac{1}{2}\int \mathbf{T} \cdot \mathbf{u}^0\, \mathrm{d}S_u < 0\,. \tag{3.29}$$

Generally, there is

$$\bar{\Pi}(\boldsymbol{\sigma}) + \Pi(\mathbf{u}) = 0\,. \tag{3.30}$$

Consider now the variation of the complementary energy due to statically admissible variation $\delta\boldsymbol{\sigma}$, satisfying equilibrium equations and boundary condition $\mathbf{n}\cdot\delta\boldsymbol{\sigma} = 0$ on S_T. We have

$$\begin{aligned}\delta\bar{\Pi}(\boldsymbol{\sigma}) &= \int \frac{\partial W(\boldsymbol{\sigma})}{\partial \boldsymbol{\sigma}} \cdot \delta\boldsymbol{\sigma}\,\mathrm{d}V - \int \delta\mathbf{T}\cdot\mathbf{u}^0\,\mathrm{d}S_u \\ &= \int \boldsymbol{\varepsilon}\cdot\delta\boldsymbol{\sigma}\,\mathrm{d}V - \int \delta\mathbf{T}\cdot\mathbf{u}^0\,\mathrm{d}S_u = 0\end{aligned} \tag{3.31}$$

and the variation vanishes in view of the virtual work equation (3.8). In fact, setting $\delta\boldsymbol{\sigma} = \boldsymbol{\sigma}^s$, $\mathbf{n}\cdot\delta\boldsymbol{\sigma}^T = 0$ on S_T and $\boldsymbol{\nabla}^T\delta\boldsymbol{\sigma} = 0$ within V, we obtain (3.31). To demonstrate that the complementary energy reaches a minimum at the stationary state, we consider the energy difference, thus

$$\begin{aligned}\bar{\Pi}(\boldsymbol{\sigma}^s) - \bar{\Pi}(\boldsymbol{\sigma}) &= \int [W(\boldsymbol{\sigma}^s) - W(\boldsymbol{\sigma})]\,\mathrm{d}V - \int (\mathbf{n}\cdot\boldsymbol{\sigma}^s - \mathbf{T})\cdot\mathbf{u}^0\,\mathrm{d}S_u \\ &= \int [W(\boldsymbol{\sigma}^s) - W(\boldsymbol{\sigma}) - \boldsymbol{\varepsilon}\cdot(\boldsymbol{\sigma}^s - \boldsymbol{\sigma})]\,\mathrm{d}V \\ &= \frac{1}{2}\int (\boldsymbol{\sigma}^s - \boldsymbol{\sigma})\cdot\mathbf{D}(\boldsymbol{\sigma}^s - \boldsymbol{\sigma})\,\mathrm{d}V > 0\end{aligned} \tag{3.32}$$

and the complementary energy reaches a minimum for the actual stress field.

These theorems can easily be extended for nonlinear elastic materials with monotonic stress–strain characteristic in the uniaxial case, Fig. 3.1b. The specific strain and stress energies now are

$$U(\boldsymbol{\varepsilon}) = \int_0^{\varepsilon} \boldsymbol{\sigma}(\boldsymbol{\varepsilon})\cdot\mathrm{d}\boldsymbol{\varepsilon}\,, \qquad W(\boldsymbol{\sigma}) = \int_0^{\sigma} \boldsymbol{\varepsilon}(\boldsymbol{\sigma})\cdot\mathrm{d}\boldsymbol{\sigma}\,, \tag{3.33}$$

and the potential relations occur

$$\boldsymbol{\varepsilon} = \frac{\partial W(\boldsymbol{\sigma})}{\partial \boldsymbol{\sigma}}\,, \qquad \boldsymbol{\sigma} = \frac{\partial U(\boldsymbol{\varepsilon})}{\partial \boldsymbol{\varepsilon}}\,, \qquad U(\boldsymbol{\varepsilon}) + W(\boldsymbol{\sigma}) = \boldsymbol{\sigma}\cdot\boldsymbol{\varepsilon}\,. \tag{3.34}$$

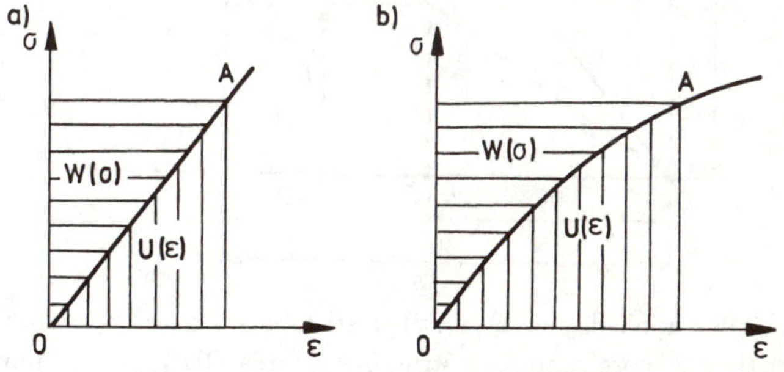

Fig. 3.1. Strain and stress energies: **a)** for a linear elastic material, **b)** for a non-linear material

In what follows, the convex elastic potentials will be considered for which the following convexity inequalities occur

$$\begin{aligned} U(\varepsilon_2) - U(\varepsilon_1) - \frac{\partial U}{\partial \varepsilon_1} \cdot (\varepsilon_2 - \varepsilon_1) &= \\ &= U(\varepsilon_2) - U(\varepsilon_1) - \sigma_1 \cdot (\varepsilon_2 - \varepsilon_1) > 0\,, \\ W(\sigma_2) - W(\sigma_1) - \frac{\partial W}{\partial \sigma_1} \cdot (\sigma_2 - \sigma_1) &= \\ &= W(\sigma_2) - W(\sigma_1) - \varepsilon_1 \cdot (\sigma_2 - \sigma_1) > 0\,. \end{aligned} \tag{3.35}$$

These inequalities can be also be expressed as integrals along paths connecting $\boldsymbol{\sigma}_1, \varepsilon_1$ and $\boldsymbol{\sigma}_2, \varepsilon_2$, thus

$$\int_1^2 (\sigma - \sigma_1) \cdot \mathrm{d}\varepsilon > 0\,, \qquad \int_1^2 (\varepsilon - \varepsilon_1) \cdot \mathrm{d}\sigma > 0\,. \tag{3.36}$$

It follows from (3.35) and (3.26), (3.32), that the convexity inequalities assure the minima of potential and complementary energies.

The functionals $\Pi(\mathbf{u})$ and $\bar{\Pi}(\boldsymbol{\sigma})$ can be used as global measures of structure stiffness and compliance. In fact, for a rigidly supported structure ($\mathbf{u}^0 = \mathbf{0}$ on S_u) the complementary energy $\bar{\Pi}(\boldsymbol{\sigma})$ is expressed by (3.28) as the half of work of surface tractions and is positive. It can therefore be assumed that the complementary energy is a measure of *global compliance* of a structure rigidly supported on S_u. For icreasing displacements of the loaded boundary S_T the

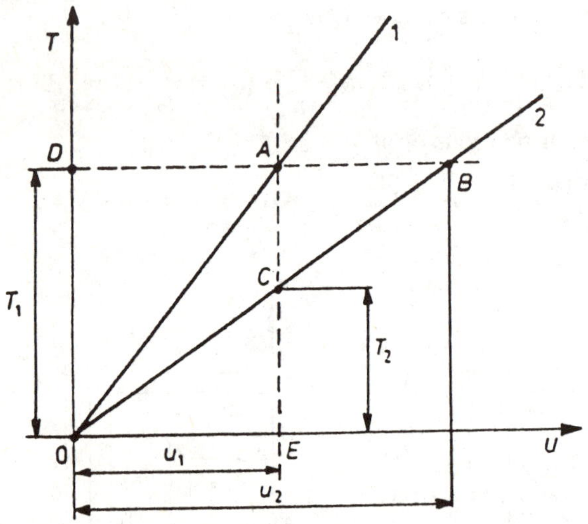

Fig. 3.2. Under applied load T, a stiffer structure 1 stores less elastic energy (area OAD) than a more compliant structure 2 (area ODB). Under imposed displacement u_1, the stored energy of structure 2 (area OEC) is less than that of structure 1 (area OEA)

complementary energy increases. On the other hand, since $\Pi(\mathbf{u}) = -\bar{\Pi}(\boldsymbol{\sigma})$, the potential energy is negative and decreases for a more compliant structure. The potential energy can therefore be assumed as a measure of *global stiffness* of the loaded structure. When displacement imposed structure is considered, the potential energy is positive and increases for increasing stiffness but the complementary energy is negative and decreases for a stiffer structure. Fig. 3.2 illustrates the dependence of the elastic energy on the stiffness modulus. For a stress controlled process the elastic energy increases for a more compliant response and for a constant strain process, the energy increases with the stiffness modulus.

3.3 Mixed (Hybrid) Variational Principles

The principles of potential and complementary energies were formulated by using kinematically admissible displacement or statically admissible stress fields $\mathbf{u}^k(\mathbf{x})$ and $\boldsymbol{\sigma}^s(\mathbf{x})$. However, the stationarity of more general functionals can also be proved for two or three field functionals. Assume first that $\boldsymbol{\sigma}(\mathbf{x})$ and $\mathbf{u}(\mathbf{x})$ are two independent (free) fields and consider the functional

$$\begin{aligned}\Pi_1(\boldsymbol{\sigma}, \mathbf{u}) = & \int W(\sigma_{ij})\,\mathrm{d}V - \int T_j u_j^0\,\mathrm{d}S_u \\ & + \int u_j(\sigma_{ij,i} + f_i)\,\mathrm{d}V - \int (T_j - T_j^0) u_j\,\mathrm{d}S_T\,. \end{aligned} \tag{3.37}$$

The present functional can be regarded as the complementary energy functional. However, the stress field is not statically admissible, hence, the equilibrium and boundary conditions on S_T are introduced as constraints on $\boldsymbol{\sigma}(\mathbf{x})$, with $\mathbf{u}(\mathbf{x})$ being the Lagrange multiplier. It is easy to demonstrate the stationarity of this functional. In fact, we have

$$\begin{aligned}\delta\Pi_1(\boldsymbol{\sigma}, \mathbf{u}) = & \int \varepsilon_{ij}\delta\sigma_{ij}\,\mathrm{d}V + \int u_j\delta\sigma_{ij}n_i\,\mathrm{d}S - \int \delta\sigma_{ij}u_{i,j}\,\mathrm{d}V \\ & - \int \delta T_j u_j^0\,\mathrm{d}S_u + \int \delta u_j(\sigma_{ij,i} + f_i)\,\mathrm{d}V - \int (T_j - T_j^0)\delta u_j\,\mathrm{d}S_T \\ = & \int \delta\sigma_{ij}(\varepsilon_{ij} - u_{(i,j)})\,\mathrm{d}V + \int \delta T_j(u_j - u_j^0)\,\mathrm{d}S_u \\ & + \int \delta u_j(\sigma_{ij,i} + f_i)\,\mathrm{d}V - \int (T_j - T_j^0)\delta u_j\,\mathrm{d}S_T\,. \end{aligned} \tag{3.38}$$

Requiring $\delta\Pi_1 = 0$, the following equations are obtained

$$\begin{aligned} \varepsilon_{ij} = \frac{1}{2}(u_{i,j} + u_{j,i}) = u_{(i,j)}\,, \qquad & u_i = u_i^0 \qquad \text{on } S_u\,, \\ \sigma_{ij,i} + f_i = 0\,, \qquad & T_i = T_i^0 \qquad \text{on } S_T\,. \end{aligned} \tag{3.39}$$

Thus the stationarity conditions $\delta\Pi_1 = 0$ generates equilibrium and compatibility conditions for two free fields $\mathbf{u}(\mathbf{x})$ and $\boldsymbol{\sigma}(\mathbf{x})$ occuring in (3.37).

The two field functionals are called *Hellinger–Reissner functionals* as they were formulated by Hellinger (in 1913) and Reissner (in 1950), cf. [49]. An alternative form of (3.37) is

$$\Pi_2(\boldsymbol{\sigma}, \mathbf{u}) = \int [\sigma_{ij} u_{i,j} - W(\sigma_{ij}) - f_i u_i] \, \mathrm{d}V - \int T_i (u_i - u_i^0) \, \mathrm{d}S_u - \int T_i^0 u_i \, \mathrm{d}S_T . \tag{3.40}$$

This functional can be regarded as the potential energy functional with the boundary condition constraint on S_u.

More general three-field functionals can be constructed by assuming $\boldsymbol{\sigma}$, $\boldsymbol{\varepsilon}$ and $\mathbf{u}$ as free fields. We have the Washizu functional [49]

$$\begin{aligned} \Pi_3(\boldsymbol{\sigma}, \boldsymbol{\varepsilon}, \mathbf{u}) = & \int [U(\boldsymbol{\varepsilon}) - \mathbf{f} \cdot \mathbf{u} - \sigma_{ij} (\varepsilon_{ij} - u_{(i,j)})] \, \mathrm{d}V \\ & - \int \mathbf{T} \cdot (\mathbf{u} - \mathbf{u}^0) \, \mathrm{d}S_u - \int \mathbf{T}^0 \cdot \mathbf{u} \, \mathrm{d}S_T . \end{aligned} \tag{3.41}$$

The stationarity condition implies field equations for three fields, thus

$$\begin{aligned} \sigma_{ij} &= \frac{\partial U}{\partial \varepsilon_{ij}} , \\ \sigma_{ij,i} + f_i &= 0 , \\ \varepsilon_{ij} &= u_{(i,j)} , \\ u_i &= u_i^0 \qquad \text{on } S_u , \\ \sigma_{ij} n_i &= T_j^0 \qquad \text{on } S_T . \end{aligned} \tag{3.42}$$

An alternative form of (3.41) is

$$\begin{aligned} \Pi_4(\boldsymbol{\sigma}, \boldsymbol{\varepsilon}, \mathbf{u}) = & \int [\boldsymbol{\sigma} \cdot \boldsymbol{\varepsilon} - U(\boldsymbol{\varepsilon}) + (\sigma_{ij,i} + f_j) u_j] \, \mathrm{d}V \\ & - \int \mathbf{T} \cdot \mathbf{u}^0 \, \mathrm{d}S_u - \int (\mathbf{T} - \mathbf{T}^0) \cdot \mathbf{u} \, \mathrm{d}S_T . \end{aligned} \tag{3.43}$$

These functionals can be applied to determine state fields and also to analyse their sensitivity with respect to design variables.

4 Sensitivity Analysis for Varying Material Parameters

Let us consider now the case when the loaded elastic body undergoes variation of its state fields due to variation of the elasticity or compliance matrices. Assume the matrices $\mathbf{C}$ and $\mathbf{D}$ to depend on the material functions $s_k(\mathbf{x})$ $(k = 1, 2, \ldots, M)$, so the Hookes law takes the form

$$\boldsymbol{\sigma} = \mathbf{C}(s_k)\boldsymbol{\varepsilon}\,, \qquad \boldsymbol{\varepsilon} = \mathbf{D}(s_k)\boldsymbol{\sigma}\,. \tag{4.1}$$

More generally, one can assume the specific strain and stress energies to depend on s_k, $U = U(\boldsymbol{\varepsilon}, s_k)$, $W = W(\boldsymbol{\sigma}, s_k)$, and the constitutive relations can be generated for both linear and non-linear materials (Fig. 4.1). We have

$$\boldsymbol{\sigma} = \frac{\partial U}{\partial \boldsymbol{\varepsilon}} = \boldsymbol{\sigma}(\boldsymbol{\varepsilon}, s_k)\,, \qquad \boldsymbol{\varepsilon} = \frac{\partial W}{\partial \boldsymbol{\sigma}} = \boldsymbol{\varepsilon}(\boldsymbol{\sigma}, s_k)\,, \tag{4.2}$$

and the incremental relations corresponding to small variations δs_k are

$$\begin{aligned} \delta\boldsymbol{\sigma} &= \frac{\partial \boldsymbol{\sigma}}{\partial \boldsymbol{\varepsilon}}\,\delta\boldsymbol{\varepsilon} + \frac{\partial \boldsymbol{\sigma}}{\partial s_k}\,\delta s_k = \frac{\partial^2 U}{\partial \boldsymbol{\varepsilon}\partial \boldsymbol{\varepsilon}}\,\delta\boldsymbol{\varepsilon} + \frac{\partial^2 U}{\partial \boldsymbol{\varepsilon}\partial s_k}\,\delta s_k = \delta\boldsymbol{\sigma}' + \delta\boldsymbol{\sigma}'' \\ \delta\boldsymbol{\varepsilon} &= \frac{\partial \boldsymbol{\varepsilon}}{\partial \boldsymbol{\sigma}}\,\delta\boldsymbol{\sigma} + \frac{\partial \boldsymbol{\varepsilon}}{\partial s_k}\,\delta s_k = \frac{\partial^2 W}{\partial \boldsymbol{\sigma}\partial \boldsymbol{\sigma}}\,\delta\boldsymbol{\sigma} + \frac{\partial^2 W}{\partial \boldsymbol{\sigma}\partial s_k}\,\delta s_k = \delta\boldsymbol{\varepsilon}' + \delta\boldsymbol{\varepsilon}'' \end{aligned} \tag{4.3}$$

For the linear elastic materials specified by (4.1), we have accordingly

$$\begin{aligned} \delta\boldsymbol{\sigma} &= \mathbf{C}\,\delta\boldsymbol{\varepsilon} + \delta\mathbf{C}\,\boldsymbol{\varepsilon} = \mathbf{C}\,\delta\boldsymbol{\varepsilon} + \frac{\partial \mathbf{C}}{\partial s_k}\,\delta s_k\boldsymbol{\varepsilon} = \delta\boldsymbol{\sigma}' + \delta\boldsymbol{\sigma}'' \\ \delta\boldsymbol{\varepsilon} &= \mathbf{D}\,\delta\boldsymbol{\sigma} + \delta\mathbf{D}\,\boldsymbol{\sigma} = \mathbf{D}\,\delta\boldsymbol{\sigma} + \frac{\partial \mathbf{D}}{\partial s_k}\,\delta s_k\boldsymbol{\sigma} = \delta\boldsymbol{\varepsilon}' + \delta\boldsymbol{\varepsilon}'' \end{aligned} \tag{4.4}$$

The terms $\delta\boldsymbol{\sigma}', \delta\boldsymbol{\sigma}''$ and $\delta\boldsymbol{\varepsilon}', \delta\boldsymbol{\varepsilon}''$ are shown in Fig. 4.1a for a linear material. Here $\delta\boldsymbol{\sigma}', \delta\boldsymbol{\varepsilon}'$ are the stress and strain variations, the terms $\delta\boldsymbol{\sigma}'', \delta\boldsymbol{\varepsilon}''$ are induced by variation of elasticity or compliance matrices at constant strain or stress. These terms play the role of *initial* stresses or strains generating stress redistribution within the structure.

For a non-linear material, Eqs. (4.3) are identical to (4.4) provided the *tangent stiffness* and *compliance* matrices are introduced

$$\mathbf{C}^t = \frac{\partial^2 U}{\partial \boldsymbol{\varepsilon}\partial \boldsymbol{\varepsilon}}\,, \qquad \mathbf{D}^t = \frac{\partial^2 W}{\partial \boldsymbol{\sigma}\partial \boldsymbol{\sigma}}\,. \tag{4.5}$$

Figure 4.1b. presents the terms $\delta\boldsymbol{\sigma}', \delta\boldsymbol{\sigma}''$ and $\delta\boldsymbol{\varepsilon}', \delta\boldsymbol{\varepsilon}''$ for a non-linear elastic material.

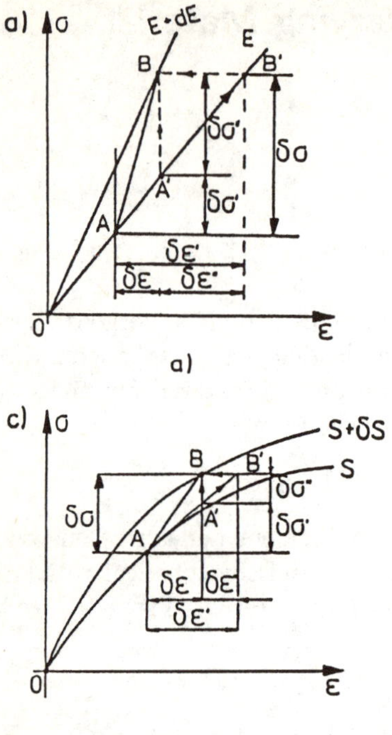

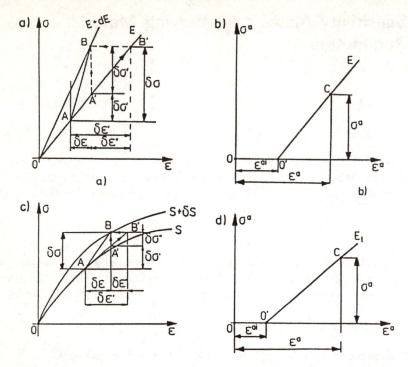

Fig. 4.1. **a)** Variation of the stiffness modulus E and the decompositions $\delta\sigma = \delta\sigma' + \delta\sigma''$, $\delta\varepsilon = \delta\varepsilon' + \delta\varepsilon''$ for a linear material, **b)** the adjoint material response, **c)** Variation of structural parameter for a non-linear material, **d)** the adjoint material model

The equilibrium and geometric compatibility conditions are expressed in a weak form by virtual work and virtual stress equations, thus

$$
\begin{aligned}
&\int \boldsymbol{\sigma} \cdot \delta\boldsymbol{\varepsilon}\, \mathrm{d}V = \int \mathbf{T}^0 \cdot \delta\mathbf{u}\, \mathrm{d}S_T + \int \mathbf{f} \cdot \delta\mathbf{u}\, \mathrm{d}V\,, \\
&\int \delta\boldsymbol{\sigma} \cdot \boldsymbol{\varepsilon}\, \mathrm{d}V = \int \delta\mathbf{T}^0 \cdot \mathbf{u}^0\, \mathrm{d}S_u + \int \delta\mathbf{f} \cdot \mathbf{u}\, \mathrm{d}V\,,
\end{aligned}
\tag{4.6}
$$

where $\mathbf{T}^0$ and $\mathbf{u}^0$ are the specified tractions and displacements on the boundary portions S_T and S_u. Assume that in the general case the body forces depend on s_k, thus $\mathbf{f} = \mathbf{f}(s_k)$. This case may occur, for instance when the self-weight forces in disks or plates of varying thickness are accounted for.

Equations (4.3) and (4.4) can be used in formulating the incremental boundary-value problems for determination of variations $\delta\boldsymbol{\sigma}$, $\delta\boldsymbol{\varepsilon}$ and $\delta\mathbf{u}$ induced by the variation of stiffness or compliance moduli with fixed loading and support conditions. Let us regard $\delta\boldsymbol{\sigma}'' = \delta\mathbf{C}\boldsymbol{\varepsilon}$ as the *initial stress field* introduced into the structure. As the total stress variation $\delta\boldsymbol{\sigma}$ must be self-equilibrated, the following set of equations is obtained for $\delta\boldsymbol{\sigma}$, $\delta\boldsymbol{\varepsilon}$, $\delta\mathbf{u}$, namely

$$\begin{gathered}\delta\sigma'_{ij,i} + \delta\sigma''_{ij,i} = 0\,, \qquad \delta\varepsilon_{ij} = \frac{1}{2}(\delta u_{i,j} + \delta u_{j,i})\,,\\ \delta\sigma'_{ij} = C_{ijkl}\delta\varepsilon_{kl}\,,\\ \delta\sigma'_{ij}n_i = -\delta\sigma''_{ij}n_i \quad \text{on } S_T\,, \qquad \delta u_i = 0 \quad \text{on } S_u\,.\end{gathered} \tag{4.7}$$

This set of equations specifies the boundary value problem for which the field of body forces $\delta f_j = \delta\sigma''_{ij,i}$ and boundary tractions on S_T, $\delta T^0_j = -\delta\sigma''_{ij}n_i$ are generated by the initial stress field $\delta\boldsymbol{\sigma}'' = \delta\mathbf{C}\,\boldsymbol{\varepsilon}$. Alternatively, we can also depart from the second equation (4.4) and regard the field $\delta\boldsymbol{\varepsilon}'' = \delta\mathbf{D}\,\boldsymbol{\sigma}$ as the *initial strain field* introduced into the structure of the elasticity matrix $\mathbf{C}$. The following set of equations is used to determine the fields $\delta\boldsymbol{\sigma}$, $\delta\boldsymbol{\varepsilon}$ and $\delta\mathbf{u}$

$$\begin{gathered}\delta\sigma_{ij,i} = 0\,, \qquad \delta\varepsilon_{ij} = \frac{1}{2}(\delta u_{i,j} + \delta u_{j,i})\,,\\ \delta\sigma_{ij} = C_{ijkl}(\delta\varepsilon_{kl} - \delta\varepsilon''_{kl})\,,\\ \delta\sigma_{ij}n_i = 0 \quad \text{on } S_T\,, \qquad \delta u_i = 0 \quad \text{on } S_u\,.\end{gathered} \tag{4.8}$$

Obviously, the incremental problems (4.7) and (4.8) are equivalent. When elasticity matrix depends on n parameters, then for the variation of each parameter, the solutions of type (4.8) should be provided in order to calculate the variations $\delta\boldsymbol{\sigma}$, $\delta\boldsymbol{\varepsilon}$ and $\delta\mathbf{u}$ for any combination of parameter variation. Denoting these parameters by $p_1, p_2, \ldots, p_n$, we can write

$$\begin{aligned}\delta\boldsymbol{\sigma} &= \boldsymbol{\sigma}^{(1)}\delta p_1 + \boldsymbol{\sigma}^{(2)}\delta p_2 + \cdots + \boldsymbol{\sigma}^{(n)}\delta p_n\,,\\ \delta\boldsymbol{\varepsilon} &= \boldsymbol{\varepsilon}^{(1)}\delta p_1 + \boldsymbol{\varepsilon}^{(2)}\delta p_2 + \cdots + \boldsymbol{\varepsilon}^{(n)}\delta p_n\,,\end{aligned} \tag{4.9}$$

where $\boldsymbol{\sigma}^{(1)}, \boldsymbol{\sigma}^{(2)}, \ldots$ and $\boldsymbol{\varepsilon}^{(1)}, \boldsymbol{\varepsilon}^{(2)}, \ldots$ are the sensitivity derivatives with respect to the parameters $p_1, p_2, \ldots, p_n$. The specification of all sensitivity derivatives requires the solution of the primary problem and of n problems associated with variation of constitutive parameters. The initial stresses and strains associated with parameter variation are

$$\boldsymbol{\sigma}^{(k)''} = \frac{\partial\mathbf{C}}{\partial p_k}\,\boldsymbol{\varepsilon}\,, \qquad \boldsymbol{\varepsilon}^{(k)''} = \frac{\partial\mathbf{D}}{\partial p_k}\,\boldsymbol{\sigma}\,, \qquad k = 1, 2, \ldots, n. \tag{4.10}$$

The present method will be called the *direct sensitivity method.* It requires $n+1$ solutions of boundary-value problems for n design variables.

In cases when we are interested in particular global or local properties of the structure, it is more convenient to apply the *adjoint structure method* requiring the solution of only one additional (adjoint) problem associated with the considered functional of state fields. Consider a general integral functional of the form

$$G = \int \psi(\boldsymbol{\sigma}, \mathbf{u}, s_k)\,\mathrm{d}V + \int h(\mathbf{u}, \mathbf{T})\,\mathrm{d}S \tag{4.11}$$

where the integrand functions are continuous and differentiable with respect to their arguments. Our aim is to express the variation of G explicitly in terms of variation of the design vector $s_k(\mathbf{x})$. As for any statically admissible field $\boldsymbol{\sigma}^s(\mathbf{x})$ and for any kinematically admissible field $\mathbf{u}^k(\mathbf{x})$, the virtual work equation holds

$$\int \boldsymbol{\sigma}^s \cdot \boldsymbol{\varepsilon}^k \,\mathrm{d}V - \int \mathbf{f}^s \cdot \mathbf{u}^k \,\mathrm{d}V - \int \mathbf{T}^s \cdot \mathbf{u}^k \,\mathrm{d}S = 0 \tag{4.12}$$

this equation will be used as the *constraint equation* which must be satisfied by variations of stress, strain and displacement fields. Let us identify the static fields $\boldsymbol{\sigma}^s$, $\mathbf{T}^s$ and $\mathbf{f}^s$ with the actual fields within the structure. The kinematic fields $\mathbf{u}^k$ and $\boldsymbol{\varepsilon}^k$ will be ascribed to an *adjoint structure* of the same shape, material properties but with different boundary and support conditions, admitting also existence of the initial strain field. Denote the adjoint state fields by $\mathbf{u}^a$, $\boldsymbol{\varepsilon}^a$, $\boldsymbol{\sigma}^a$ and the initial strain field by $\boldsymbol{\varepsilon}^{ai}$. The functional (4.11) augmented by the virtual work equation (4.12) now takes the form

$$\begin{aligned}\bar{G} = &\int \psi(\boldsymbol{\sigma}, \mathbf{u}, s_k)\,\mathrm{d}V + \int h(\mathbf{u}, \mathbf{T})\,\mathrm{d}S \\ &- \int \boldsymbol{\sigma} \cdot \boldsymbol{\varepsilon}^a \,\mathrm{d}V + \int \mathbf{f} \cdot \mathbf{u}^a \,\mathrm{d}V + \int \mathbf{T} \cdot \mathbf{u}^a \,\mathrm{d}S\,.\end{aligned} \tag{4.13}$$

The Lagrange multiplier was assumed for simplicity as 1, since its value does not affect the sensitivity expression. Considering the variation of (4.13) and requiring stationarity of G with respect to $\boldsymbol{\sigma}$, $\mathbf{u}$ and $\mathbf{u}^a$, the equilibrium equations and boundary conditions for both primary and adjoint structures are obtained. The variation with respect to s_k will express the functional sensitivity. Thus, we have

$$\begin{aligned}\delta\bar{G} = &\int \left(\frac{\partial\psi}{\partial\boldsymbol{\sigma}} \cdot \delta\boldsymbol{\sigma} + \frac{\partial\psi}{\partial\mathbf{u}} \cdot \delta\mathbf{u} + \frac{\partial\psi}{\partial s_k}\,\delta s_k\right)\mathrm{d}V \\ &+ \int \left(\frac{\partial h}{\partial\mathbf{u}} \cdot \delta\mathbf{u} + \frac{\partial h}{\partial\mathbf{T}} \cdot \delta\mathbf{T}\right)\mathrm{d}S \\ &- \int (\boldsymbol{\sigma} \cdot \delta\boldsymbol{\varepsilon}^a + \delta\boldsymbol{\sigma} \cdot \boldsymbol{\varepsilon}^a)\,\mathrm{d}V + \int (\delta\mathbf{f} \cdot \mathbf{u}^a + \mathbf{f} \cdot \delta\mathbf{u}^a)\,\mathrm{d}V \\ &+ \int (\delta\mathbf{T} \cdot \mathbf{u}^a + \mathbf{T} \cdot \delta\mathbf{u}^a)\,\mathrm{d}S\,.\end{aligned} \tag{4.14}$$

Assume the stress $\boldsymbol{\sigma}^a$ in the adjoint structure to satisfy the constitutive equations

$$\boldsymbol{\sigma}^a = \mathbf{C}(\boldsymbol{\varepsilon}^a - \boldsymbol{\varepsilon}^{ai}) \tag{4.15}$$

where $\boldsymbol{\varepsilon}^{ai}$ will be specified later. Noting that $\delta\mathbf{u} = \delta\mathbf{u}^0 = \mathbf{0}$ on S_u and

$\delta \mathbf{T} = \delta \mathbf{T}^0 = \mathbf{0}$ on S_T, we have

$$\begin{aligned}\delta \bar{G} = & \int \left(\frac{\partial \psi}{\partial s_k} \delta s_k + \delta \mathbf{f} \cdot \mathbf{u}^a \right) \mathrm{d}V - \left[\int (\boldsymbol{\sigma} \cdot \delta \boldsymbol{\varepsilon}^a - \mathbf{f} \cdot \delta \mathbf{u}^a) \, \mathrm{d}V - \int \mathbf{T} \cdot \delta \mathbf{u}^a \, \mathrm{d}S \right] \\ & - \left\{ \int \left[(\boldsymbol{\varepsilon}^a - \boldsymbol{\varepsilon}^{ai}) \cdot \delta \boldsymbol{\sigma} - \mathbf{f}^a \cdot \delta \mathbf{u} \right] \, \mathrm{d}V - \int \mathbf{T}^{a0} \cdot \delta \mathbf{u} \, \mathrm{d}S_T \right\} \\ & - \left\{ \int \left[\left(\boldsymbol{\varepsilon}^{ai} - \frac{\partial \psi}{\partial \boldsymbol{\sigma}} \right) \cdot \delta \boldsymbol{\sigma} + \left(\mathbf{f}^a - \frac{\partial \psi}{\partial \mathbf{u}} \right) \cdot \delta \mathbf{u} \right] \mathrm{d}V \right. \\ & \left. + \int \left(\mathbf{T}^{a0} - \frac{\partial h}{\partial \mathbf{u}} \right) \cdot \delta \mathbf{u} \, \mathrm{d}S_T - \int \left(\mathbf{u}^{a0} - \frac{\partial h}{\partial \mathbf{T}} \right) \cdot \delta \mathbf{T} \, \mathrm{d}S_u \right\} \end{aligned} \tag{4.16}$$

The equilibrium conditions of the primary structure are expressed by the virtual work principle

$$\int \boldsymbol{\sigma} \cdot \delta \boldsymbol{\varepsilon}^a \, \mathrm{d}V - \int \mathbf{f} \cdot \delta \mathbf{u}^a \, \mathrm{d}V - \int \mathbf{T} \cdot \delta \mathbf{u}^a \, \mathrm{d}S = 0 \tag{4.17}$$

and the expression in square brackets of (4.16) vanishes. The term in the first parentheses can be transformed to

$$\begin{aligned} & \left\{ \int \left[(\boldsymbol{\varepsilon}^a - \boldsymbol{\varepsilon}^{ai}) \cdot \delta \boldsymbol{\sigma} - \mathbf{f}^a \cdot \delta \mathbf{u} \right] \, \mathrm{d}V - \int \mathbf{T}^a \cdot \delta \mathbf{u} \, \mathrm{d}S \right\} \\ & \quad = \left\{ \left[\int (\boldsymbol{\sigma}^a \cdot \delta \boldsymbol{\varepsilon} - \mathbf{f}^a \cdot \delta \mathbf{u}) \, \mathrm{d}V - \int \mathbf{T}^a \cdot \delta \mathbf{u} \, \mathrm{d}S_T \right] \right. \\ & \quad \left. + \int (\boldsymbol{\varepsilon}^a - \boldsymbol{\varepsilon}^{ai}) \cdot \frac{\partial \mathbf{C}}{\partial s_k} \, \boldsymbol{\varepsilon} \delta s_k \mathrm{d}V \right\}. \end{aligned} \tag{4.18}$$

Since the equilibrium conditions of the adjoint structure are expressed by the virtual work equation

$$\int \boldsymbol{\sigma}^a \cdot \delta \boldsymbol{\varepsilon}^a \, \mathrm{d}V - \int \mathbf{f}^a \cdot \delta \mathbf{u}^a \, \mathrm{d}V - \int \mathbf{T} \cdot \delta \mathbf{u} \, \mathrm{d}S_T = 0 \tag{4.19}$$

the expression in the square brackets of (4.18) vanishes. The expression in the second parenthesis of (4.16) vanishes when

$$\begin{aligned} & \boldsymbol{\varepsilon}^{ai} = \frac{\partial \psi}{\partial \boldsymbol{\sigma}}, \qquad \mathbf{f}^a = \frac{\partial \psi}{\partial \mathbf{u}}, \qquad \text{within } V, \\ & \mathbf{T}^{a0} = \mathbf{n} \boldsymbol{\sigma}^a = \frac{\partial h}{\partial \mathbf{u}} \quad \text{on } S_T, \qquad \mathbf{u}^{a0} = -\frac{\partial h}{\partial \mathbf{T}} \quad \text{on } S_u. \end{aligned} \tag{4.20}$$

Note that Eqs. (4.20) specify the initial strain field, body force field, and boundary conditions for the adjoint structure. If these conditions are satisfied, the variation of G is expressed as follows

$$\delta G = \delta \bar{G} = \int \left[\frac{\partial \psi}{\partial s_k} + \frac{\partial \mathbf{f}}{\partial s_k} \cdot \mathbf{u}^a - (\boldsymbol{\varepsilon}^a - \boldsymbol{\varepsilon}^{ai}) \cdot \frac{\partial \mathbf{C}}{\partial s_k} \boldsymbol{\varepsilon} \right] \delta s_k \, \mathrm{d}V = \int S_k \delta s_k \, \mathrm{d}V \tag{4.21}$$

that is in terms of variations of s_k and the state fields of both primary and adjoint structures. It is seen that the variation of G is expressed as an integral over structure domain and the integrand $S_k(\boldsymbol{\sigma}, \mathbf{u}, \boldsymbol{\sigma}^a, \mathbf{u}^a)$ is specified once the state fields of both structures have been determined. Since

$$\mathbf{CD} = \mathbf{I}, \qquad \frac{\partial \mathbf{C}}{\partial s_k}\mathbf{D} = -\mathbf{C}\frac{\partial \mathbf{D}}{\partial s_k}, \tag{4.22}$$

the sensitivity expression (4.21) can be presented as follows

$$\delta \bar{G} = \int \left[\frac{\partial \psi}{\partial s_k} + \frac{\partial \mathbf{f}}{\partial s_k} \cdot \mathbf{u}^a - \boldsymbol{\sigma}^a \cdot \frac{\partial \mathbf{D}}{\partial s_k} \boldsymbol{\sigma} \right] \delta s_k \, \mathrm{d}V = \int S_k \delta s_k \, \mathrm{d}V \tag{4.23}$$

When the design variables depend on a set of parameters p_l , $l = 1, 2, \ldots, n$), so that

$$s_k = s_k(\mathbf{x}, p_l), \qquad \delta s_k = \frac{\partial s_k}{\partial p_l} \delta p_l \tag{4.24}$$

then Eq. (4.21) takes the form

$$\delta \bar{G} = \left\{ \int \left[\frac{\partial \psi}{\partial p_l} + \frac{\partial \mathbf{f}}{\partial p_l} \cdot \mathbf{u}^a - (\boldsymbol{\varepsilon}^a - \boldsymbol{\varepsilon}^{ai}) \cdot \frac{\partial \mathbf{C}}{\partial s_k} \boldsymbol{\varepsilon} \right] \mathrm{d}V \right\} \delta p_l = S^{(l)} \, \delta p_l \tag{4.25}$$

where $S^{(l)}$ are the sensitivity derivatives with respect to the parameters p_l , thus

$$S^{(l)} = \frac{\partial G}{\partial p_l}, \qquad \delta G = S^{(l)} \, \delta p_l \tag{4.26}$$

For physically non-linear structures, the same expressions for sensitivity of G are obtained provided the matrices $\mathbf{C}$ and $\mathbf{D}$ are replaced by the tangent matrices $\mathbf{C}^t$ and $\mathbf{D}^t$.

So far, we have presented both direct and adjoint structure approaches for generating sensitivity variations and derivatives. The number of adjoint structures equals the number of functionals whose sensitivity is to be determined. In the case of n parameters and m functionals, the direct method requires $n + 1$ solutions and the adjoint method requires $m + 1$ solutions. Thus the economy of each method depends on the values of n and m. and on the convenience in generating respective solutions. Let us emphasize that the incremental problem (4.8) for the direct method and the adjoint problem (4.19), (4.20) are linear.

Consider now the variation of global structural compliance measured by the complementary energy $\bar{\Pi}(\boldsymbol{\sigma}, s_k)$. Let us recall that for a linear structure with traction boundary-value problem the complementary energy is proportional to the work of surface tractions. The variation of $\bar{\Pi}$ equals

$$\begin{aligned} \delta \bar{\Pi}_s &= \int \frac{\partial W}{\partial \boldsymbol{\sigma}} \cdot \delta \boldsymbol{\sigma} \, \mathrm{d}V - \int \delta \mathbf{T} \cdot \mathbf{u}^0 \, \mathrm{d}S_u + \int \frac{\partial W}{\partial s_k} \delta s_k \, \mathrm{d}V \\ &= \left[\int \boldsymbol{\varepsilon} \cdot \delta \boldsymbol{\sigma} \, \mathrm{d}V - \int \delta \mathbf{T}^0 \cdot \mathbf{u} \, \mathrm{d}S_u \right] + \frac{1}{2} \int \boldsymbol{\sigma} \cdot \frac{\partial \mathbf{D}}{\partial s_k} \boldsymbol{\sigma} \delta s_k \, \mathrm{d}V \, . \end{aligned} \tag{4.27}$$

In view of the virtual work equation, the term in square brackets vanishes and the variation of $\bar{\Pi}$ is expressed as follows

$$\delta\bar{\Pi}_s = \int \frac{\partial W}{\partial s_k}\,\delta s_k\,\mathrm{d}V = \frac{1}{2}\int \boldsymbol{\sigma}\cdot\frac{\partial \mathbf{D}}{\partial s_k}\,\boldsymbol{\sigma}\,\delta s_k\,\mathrm{d}V = \frac{1}{2}\int \sigma_{ij}\,\frac{\partial D_l}{\partial s_k}\,\sigma_{lm}\,\delta s_k\,\mathrm{d}V\,. \tag{4.28}$$

Similarly, the variation of the potential energy $\Pi(\mathbf{u}, s_k)$ equals

$$\begin{aligned}\delta\Pi_s &= \left[\int \frac{\partial U}{\partial \boldsymbol{\varepsilon}}\cdot\delta\boldsymbol{\varepsilon}\,\mathrm{d}V - \int \delta\mathbf{T}\cdot\mathbf{u}^0\,\mathrm{d}S_T\right] + \int \frac{\partial U}{\partial s_k}\,\delta s_k\,\mathrm{d}V \\ &= \int \frac{\partial U}{\partial s_k}\,\delta s_k\,\mathrm{d}V = \frac{1}{2}\int \boldsymbol{\varepsilon}\cdot\frac{\partial \mathbf{C}}{\partial s_k}\,\boldsymbol{\varepsilon}\,\delta s_k\,\mathrm{d}V\,.\end{aligned} \tag{4.29}$$

Because of the stationary property of Π with respect to $\mathbf{u}$, the terms within square brackets vanish. It is easy to demonstrate that $\delta\bar{\Pi}_s = -\delta\Pi_s$.

The second variations of the potential and complementary energies can easily be derived by using the stationarity properties of $\Pi(\mathbf{u}, s_k)$ and $\bar{\Pi}(\boldsymbol{\sigma}, s_k)$ with respect to state fields $\boldsymbol{\sigma}$ and $\mathbf{u}$. For two equilibrium states $(\mathbf{u}, s_k)$ and $(\mathbf{u}+\delta\mathbf{u}_s\,,\ s_k+\delta s_k)$, we have

$$\delta\Pi_u(\mathbf{u}, s_k) = 0, \qquad \delta\Pi_u(\mathbf{u}+\delta\mathbf{u}_s\,,\ s_k+\delta s_k) = 0 \tag{4.30}$$

and

$$\delta^2\Pi_{u,s} = \int \delta\boldsymbol{\varepsilon}_s\cdot\frac{\partial^2 U}{\partial\boldsymbol{\varepsilon}\partial\boldsymbol{\varepsilon}}\,\delta\boldsymbol{\varepsilon}_s\,\mathrm{d}V + \int \delta\boldsymbol{\varepsilon}_s\cdot\frac{\partial^2 U}{\partial\boldsymbol{\varepsilon}\partial s_k}\,\delta s_k\,\mathrm{d}V = 0. \tag{4.31}$$

Here $\delta\mathbf{u}_s$, $\delta\boldsymbol{\varepsilon}_s$ denote variations of state fields due to variation of design variables s_k . Similarly, for the complementary energy we have

$$\delta^2\bar{\Pi}_{\sigma,s} = \int \delta\boldsymbol{\sigma}_s\cdot\frac{\partial^2 W}{\partial\boldsymbol{\sigma}\partial\boldsymbol{\sigma}}\,\delta\boldsymbol{\sigma}_s\,\mathrm{d}V + \int \delta\boldsymbol{\sigma}_s\cdot\frac{\partial^2 W}{\partial\boldsymbol{\sigma}\partial s_k}\,\delta s_k\,\mathrm{d}V = 0. \tag{4.32}$$

In view of Eqs. (4.31) and (4.32) the second variations of Π and $\bar{\Pi}$ are expressed as follows

$$\begin{aligned}\delta^2\Pi_s &= -\int \delta\boldsymbol{\varepsilon}_s\cdot\frac{\partial^2 U}{\partial\boldsymbol{\varepsilon}\partial\boldsymbol{\varepsilon}}\,\delta\boldsymbol{\varepsilon}_s\,\mathrm{d}V + \int \frac{\partial^2 U}{\partial s_k\partial s_l}\,\delta s_k\delta s_l\,\mathrm{d}V = 0, \\ \delta^2\bar{\Pi}_s &= -\int \delta\boldsymbol{\sigma}_s\cdot\frac{\partial^2 W}{\partial\boldsymbol{\sigma}\partial\boldsymbol{\sigma}}\,\delta\boldsymbol{\sigma}_s\,\mathrm{d}V + \int \frac{\partial^2 W}{\partial s_k\partial s_l}\,\delta s_k\delta s_l\,\mathrm{d}V = 0.\end{aligned} \tag{4.33}$$

It is seen that the first variations of state fields are required in order to determine the second order sensitivities. These variations can be determined by applying the direct method of sensitivity analysis, see Eqs. (4.8).

The second variation of the general state functional (4.11) can be derived from Eq. (4.21) or (4.23). We have

$$\begin{aligned}\delta^2 G_s = \int \Big(& \frac{\partial^2\psi}{\partial s_k \partial s_l}\,\delta s_k \delta s_l + \frac{\partial^2\psi}{\partial \mathbf{u} \partial s_k}\,\delta \mathbf{u} \delta s_k + \frac{\partial^2\psi}{\partial \boldsymbol{\sigma} \partial s_k}\,\delta \boldsymbol{\sigma} \delta s_k \\ & + \mathbf{u}^a \cdot \frac{\partial^2 \mathbf{f}}{\partial s_k \partial s_l}\,\delta s_k \delta s_l + \frac{\partial \mathbf{f}}{\partial s_k} \cdot \boldsymbol{\sigma}\,\delta s_k \\ & + \boldsymbol{\sigma}^a \cdot \frac{\partial^2 \mathbf{D}}{\partial s_k \partial s_l}\,\boldsymbol{\sigma}\,\delta s_k \delta s_l + \boldsymbol{\sigma}^a \cdot \frac{\partial \mathbf{D}}{\partial s_k}\,\delta \boldsymbol{\sigma} \delta s_k \Big)\,\mathrm{d}V\,. \end{aligned} \tag{4.34}$$

In view of virtual work equations for stress and strain increments of primary and adjoint structures there is

$$\begin{aligned} \int \delta\boldsymbol{\sigma}^a \cdot \delta\boldsymbol{\varepsilon}\,\mathrm{d}V &= \int \delta\mathbf{f}^a \cdot \delta\mathbf{u}\,\mathrm{d}V + \int \delta\mathbf{T}^{a0} \cdot \mathbf{u}\,\mathrm{d}S_T\,, \\ \int \delta\boldsymbol{\sigma} \cdot \delta\boldsymbol{\varepsilon}^a\,\mathrm{d}V &= \int \delta\mathbf{f} \cdot \delta\mathbf{u}^a\,\mathrm{d}V + \int \delta\mathbf{T} \cdot \mathbf{u}^{a0}\,\mathrm{d}S_u\,, \end{aligned} \tag{4.35}$$

where

$$\begin{aligned} \delta\mathbf{f}^a &= \frac{\partial^2\psi}{\partial\mathbf{u}\partial\mathbf{u}}\,\delta\mathbf{u} + \frac{\partial^2\psi}{\partial\mathbf{u}\partial s_k}\,\delta s_k\,, \\ \delta\mathbf{T}^{a0} &= \frac{\partial^2 h}{\partial\mathbf{u}\partial\mathbf{u}}\,\delta\mathbf{u}\,, \\ \delta\mathbf{u}^{a0} &= -\frac{\partial^2 h}{\partial\mathbf{T}\partial\mathbf{T}}\,\delta\mathbf{T}\,, \\ \delta\boldsymbol{\varepsilon}^a &= \frac{\partial^2\psi}{\partial\boldsymbol{\sigma}\partial\boldsymbol{\sigma}}\,\delta\boldsymbol{\sigma} + \frac{\partial^2\psi}{\partial\boldsymbol{\sigma}\partial s_k}\,\delta s_k\,, \end{aligned} \tag{4.36}$$

and the expression for the second variation takes the form

$$\begin{aligned}\delta^2 G_s = \int \Big[& -\delta\boldsymbol{\sigma} \cdot \frac{\partial^2\psi}{\partial\boldsymbol{\sigma}\partial\boldsymbol{\sigma}}\,\delta\boldsymbol{\sigma} - 2\delta\boldsymbol{\sigma} \cdot \mathbf{D}\,\delta\boldsymbol{\sigma} + \delta\mathbf{u} \cdot \frac{\partial^2\psi}{\partial\mathbf{u}\partial\mathbf{u}}\,\delta\mathbf{u} \\ & + 2\frac{\partial^2\psi}{\partial\mathbf{u}\partial s_k}\,\delta\mathbf{u}\delta s_k + \Big(\frac{\partial^2\psi}{\partial s_k \partial s_l} + \boldsymbol{\sigma}^a\,\frac{\partial^2\psi}{\partial s_k \partial s_l}\,\boldsymbol{\sigma} \Big)\,\delta s_k \delta s_l \Big]\,\mathrm{d}V \\ & + \int \delta\mathbf{u} \cdot \frac{\partial^2\psi}{\partial\mathbf{u}\partial\mathbf{u}}\,\delta\mathbf{u}\,\mathrm{d}S_T - \int \delta\mathbf{T} \cdot \frac{\partial^2 h}{\partial\mathbf{T}\partial\mathbf{T}}\,\delta\mathbf{T}\,\mathrm{d}S_u \end{aligned} \tag{4.37}$$

where the dependence of the body force field on the design variable was neglected.

Consider now some functionals that are essential for assessment of structural behaviour. For instance, if we need to specify the mean or local stress level, let us consider the functional

$$G_1(\boldsymbol{\sigma}) = \left[\frac{1}{\Omega} \int \phi^p(\boldsymbol{\sigma})\,\mathrm{d}\Omega \right]^{\frac{1}{p}} \tag{4.38}$$

where $\phi(\boldsymbol{\sigma})$ is a homogeneous function of stress of degree one (for instance, the effective stressi associated with the Huber–Mises yield condition, $\phi(\boldsymbol{\sigma}) =$

$\sigma_e = (3J_2)^{1/2}$, where J_2 is the second invariant of stress deviator), and p is a positive real number. When $p \to \infty$ then $G_1 \to \sup \phi(\boldsymbol{\sigma})$ and Eq. (4.38) represents the *maximal effective local stress* within the considered domain Ω of the structure. For finite values of p, the functional (4.38) represents the averaged value of the effective stress. It is therefore a convenient scalar measure of stress that enables continuous transition from mean to maximal local values. The sensitivity of this functional is specified by Eq. (4.21) and the adjoint structure is subjected to initial strain, so that

$$\begin{gathered}\boldsymbol{\varepsilon}^{ai} = \frac{\phi^{p-1}}{\left[V\int \phi^p(\boldsymbol{\sigma})\,\mathrm{d}V\right]^{\frac{p-1}{p}}}\,\frac{\partial \phi}{\partial \boldsymbol{\sigma}} \qquad \text{within } V, \\ \mathbf{T}^{a0} = \mathbf{0} \quad \text{on } S_T\,, \qquad \mathbf{u}^{a0} = \mathbf{0} \quad \text{on } S_u\,.\end{gathered} \tag{4.39}$$

An alternative scalar measure of stress level can be assumed as the excess effective stress over a specified level σ_0. The respective functional then has the form

$$G_2(\boldsymbol{\sigma}) = \frac{1}{n+1}\int \langle \phi(\boldsymbol{\sigma}) - \sigma_0 \rangle^{n+1}\,\mathrm{d}V \tag{4.40}$$

where $\langle x \rangle = x$ for $x > 0$ and $\langle x \rangle = 0$ for $x < 0$. Thus, the effective stress below σ_0 does not contribute to (4.40) and $G_2(\boldsymbol{\sigma})$ can be regarded as the "penalty" functional. The initial strain of the adjoint structure then is

$$\boldsymbol{\varepsilon}^{ai} = \langle \phi(\boldsymbol{\sigma}) - \sigma_0 \rangle^n \,\frac{\partial \phi}{\partial \boldsymbol{\sigma}} \qquad \text{within } V. \tag{4.41}$$

Similar measures can be introduced for displacements or strains.

Local values of displacements or stress can also be expressed by bilinear functionals. In order to determine the displacement u_q at $\mathbf{x} = \mathbf{x}_0$ along the q-direction, let us apply in the adjoint structure the unit force $f_q^a = 1$ along this direction. Denoting by $\boldsymbol{\sigma}^a$, $\boldsymbol{\varepsilon}^a$ and $\mathbf{u}^a$ the state fields of the adjoint structure due to force f_q^a, we obtained

$$f_q^a u_q = \int \boldsymbol{\sigma}^a \cdot \boldsymbol{\varepsilon}\,\mathrm{d}V = \int \boldsymbol{\sigma}^a \cdot \boldsymbol{\varepsilon}^a\,\mathrm{d}V \tag{4.42}$$

where $\mathbf{u}^{a0} = \mathbf{0}$ on S_u. The variation of u_q can be obtained from (4.42). We have

$$\begin{aligned} f_q^a \delta u_q &= \int \delta\boldsymbol{\sigma}^a \cdot \boldsymbol{\varepsilon}\,\mathrm{d}V + \int \boldsymbol{\sigma}^a \cdot \delta\boldsymbol{\varepsilon}^a\,\mathrm{d}V = \int \boldsymbol{\sigma}^a \cdot \delta\boldsymbol{\varepsilon}\,\mathrm{d}V \\ &= \int \boldsymbol{\sigma}^a \cdot (\mathbf{D}\delta\boldsymbol{\sigma} + \delta\mathbf{D}\,\boldsymbol{\sigma})\,\mathrm{d}V = \int \boldsymbol{\sigma}^a \cdot \delta\mathbf{D}\,\boldsymbol{\sigma}\,\mathrm{d}V \\ &= \int \boldsymbol{\sigma}^a \cdot \frac{\partial \mathbf{D}}{\partial s_k}\,\boldsymbol{\sigma}\,\delta s_k\,\mathrm{d}V = -\int \boldsymbol{\varepsilon}^a \cdot \frac{\partial \mathbf{C}}{\partial s_k}\,\boldsymbol{\varepsilon}\,\delta s_k\,\mathrm{d}V\,. \end{aligned} \tag{4.43}$$

It is seen that the expression identical to (4.21) is obtained. Formally, we set $h = 0$ in Eq. (4.11) and the integrand of first terms is

$$\psi(\mathbf{u}) = \delta(\mathbf{x} - \mathbf{x}_0)\,\delta_{qk} u_k(\mathbf{x}) \tag{4.44}$$

where $\delta(\mathbf{x}-\mathbf{x}_0)$ is the Dirac function and δ_{qk} denotes the Kronecker symbol. The displacement component now is

$$u_q(\mathbf{x}_0) = \int \psi(\mathbf{u})\,\mathrm{d}V = \int \delta(\mathbf{x}-\mathbf{x}_0)\,\delta_{qk} u_k(\mathbf{x})\,\mathrm{d}V \tag{4.45}$$

and the unit force at $\mathbf{x}_0$ is obtained from (4.20), thus

$$f_k^a = \frac{\partial \psi}{\partial u_k} = \delta(\mathbf{x}-\mathbf{x}_0)\,\delta_{qk}\,, \quad \mathbf{x} \in V\,. \tag{4.46}$$

The augmented functional $\bar{G}$ now has the form

$$\bar{G} = u_q(\mathbf{x}_0) - \int \boldsymbol{\sigma}\cdot\boldsymbol{\varepsilon}^a\,\mathrm{d}V + \int \mathbf{T}\cdot\mathbf{u}^a\,\mathrm{d}S = \int \boldsymbol{\sigma}^a\cdot\boldsymbol{\varepsilon}\,\mathrm{d}V\,. \tag{4.47}$$

In order to specify the stress component (for instance, acting normally to the plane of unit normal vector $\mathbf{n}_q$), introduce in an adjoint system a unit edge dislocation $\Delta u_{qq} = 1$ corresponding to the displacement discontinuity along the normal vector. Denote by $\boldsymbol{\sigma}^a$, $\boldsymbol{\varepsilon}^a$ and $\mathbf{u}^a$ the state fields due to dislocation at $\mathbf{x} = \mathbf{x}_0$ with the boundary conditions $\mathbf{T}^{a0} = \mathbf{0}$ on S_T and $\mathbf{u}^{a0} = \mathbf{0}$ on S_u. By the virtual work equation, we have

$$\sigma_{qq}\Delta u_{qq} = \int \mathbf{T}^0\cdot\mathbf{u}^a\,\mathrm{d}S_T - \int \boldsymbol{\sigma}\cdot\boldsymbol{\varepsilon}^a\,\mathrm{d}V = \int \mathbf{T}^0\cdot\mathbf{u}^a\,\mathrm{d}S_T \tag{4.48}$$

and the variation of σ_{qq} is expressed by Eq. (4.43).

5 Structural Shape Variation

The shape of a structure is specified by its external boundary S, formed in general by a set of regular surfaces intersecting along edges. For arch or shell structures, the line or surface shape specify the structure shape. For bar structures (trusses, frames), the shape is specified by node positions and connections between nodes. When interfaces exist, their shapes can also constitute the design variables.

5.1 Shape Description

Shape generation and visualization constitutes a separate research area of geometric modeling, cf. Mortenson [31]. In a parametric description, the cartesian coordinates of a line or a surface are specified in terms of parametric variables (s, t). For instance, $x_i = x_i(s, d_k)$ for a line and $x_i = x_i(s, t, d_k)$ for a surface $(i = 1, 2, 3)$. Such representations depend on a set of parameters d_k $(k = 1, 2, \ldots, m)$ and the proper shape algorithm generates coordinate positions and their variations associated with parameter variation.

Consider first simple examples of shape variation shown in Fig. 5.1. A straigth segment AB of Fig. 5.1a is infinitesimally transformed to a segment A′B′. This transformation can be achieved by normal transformations of points A and B equal to $\dot{\varphi}(\mathrm{A})$, $\dot{\varphi}(\mathrm{B})$ and tangential transformations $\dot{\varphi}_t(\mathrm{A})$, $\dot{\varphi}_t(\mathrm{B})$, (dots over the symbol denote variations or velocities). Similarly, a curvilinear segment AB of Fig. 5.1b undergoes transformation to A′B′, which can be achieved by specifying normal transformation on AB and two tangential components at A and B. A more convenient way is to specify the transformation velocity vector $\dot{\varphi}(s)$ along the whole segment so that both normal and tangential components of $\dot{\varphi}$ are prescribed on the segment. The transformation vector can be specified as a function of material points in the reference configuration or as a function of space points in the actual configuration. Figure 5.1c presents a closed polygonal domain bounded by straight segments AB–BC–CD–DA. Now, the transformation is specified in terms of parallel translation of polygon sides, thus

$$\dot{\varphi}_n = \left[\dot{\varphi}_n^{(1)}, \dot{\varphi}_n^{(2)}, \dot{\varphi}_n^{(3)}, \dot{\varphi}_n^{(4)}\right]^T .$$

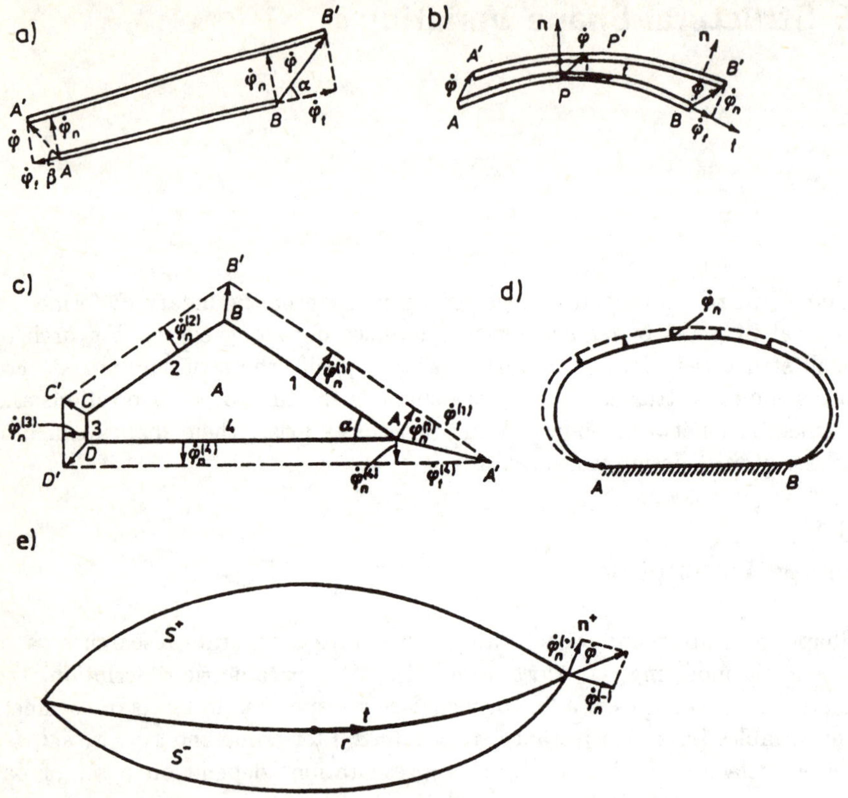

Fig. 5.1. Shape transformation : **a)** linear segment OB, **b)** arc AB, **c)** polygonal domain, **e)** domain bounded by a regular surface, **e)** domain bounded by two regular surfaces

Note that sides undergo also tangential transformation

$$\dot{\varphi}_t = \left[\dot{\varphi}_t^{(1)}, \dot{\varphi}_t^{(2)}, \dot{\varphi}_t^{(3)}, \dot{\varphi}_t^{(4)}\right]^T .$$

However, the tangential transformation velocities are not independent but are related to normal transformation velocities so that, in general, we have

$$\dot{\varphi}_t = \mathbf{T}\,\dot{\varphi}_n\,. \tag{5.1}$$

For instance, at the corner A, we have

$$\begin{bmatrix} \dot{\varphi}_t^{(1)} \\ \dot{\varphi}_t^{(2)} \end{bmatrix} = \begin{bmatrix} \cot\alpha & 1/\cos\alpha \\ \sin\alpha & \cot\alpha \end{bmatrix} \begin{bmatrix} \dot{\varphi}_n^{(1)} \\ \dot{\varphi}_n^{(2)} \end{bmatrix}. \tag{5.2}$$

Similar relations occur at other corners. Generally, in the case of a closed domain the normal transformation velocity vector specifies infinitesimal domain transformation. In Fig. 5.1d, the domain is bounded by a regular surface. In

Fig. 5.1e, two curves Γ specify the closed domain. The normal transformation rates $[\dot{\varphi}_n^+ , \dot{\varphi}_n^-]$ on S^+ and S^- are sufficient for specification of domain variation because the tangential components $\dot{\varphi}_t^+ , \dot{\varphi}_t^-$ on the edge Γ are obtained from Eq. (5.1). Let us note that the topology of the domain may vary in the course of shape transformation: in fact, some sides of a polygon, Fig. 5.1c, may disappear due to translation of its sides.

So far, we specified the rule of infinitesimal evolution of boundary or internal surfaces by defining the transformation variation (or velocity) field $\dot{\varphi}(\mathbf{x})$. This field is continued analytically, into the interior and exterior of structure domain. Assume that there is an admissible domain for structure design Ω_a such that structure domain Ω_s lies within Ω_a. Along Ω_a we have $\dot{\varphi}_n = \mathbf{0}$, and the gradients of this field can be determined inside the structure domain and on its boundary, so that

$$\dot{\varphi} = \dot{\varphi}(\mathbf{x}, t) , \qquad \mathbf{x} \in \Omega_a . \tag{5.3}$$

Obviously, this continuation of the transformation velocity field need not be unique. The present description is analogous to the Eulerian approach in describing fluid flow. In fact, the transformation velocity (5.3) can be compared to the flow velocity field $\mathbf{v}(\mathbf{x}, t)$ inducing variation of external boundaries. The usual treatments of shape transformation and sensitivity analysis were mostly based on the Eulerian approach and the material derivative concept, cf. Haug et al. [23] Dems and Mróz [12], Sokolowski and Zolesio [45], Baniczuk [2], Dems and Haftka [10]. The application of Lagrangian description was presented by Haber [20], Phelan et al. [43], Cardoso and Arora [6]. For discretized systems, the sensitivity of solution with respect to shape variation was usually generated by considering finite element node variation and associated variation of the stiffness matrix. The so called semi-analytical, analytical and direct methods were developed in numerical treatment of sensitivity analysis and their accuracy was studied, cf. Zienkiewicz and Campbell [50], Barthelemy et al. [3], Cheng and Lin [7], Pedersen et al. [41]. Dems and Mróz [14] studied the analytically derived incremental sensitivity equations and tested their accuracy.

The finite shape transformation process can be described as follows. Consider a body in the *reference configuration* C_0 and denote the material point position by $\mathbf{X}$ with respect to the assumed reference frame. The shape transformation process is conceived as a mapping of a material point $\mathbf{X}$ onto a point $\mathbf{X}^t$, so that

$$\mathbf{X}^t = \mathbf{X}^t(\mathbf{X}, t) \qquad \text{or} \qquad \mathbf{X}^t = \mathbf{X} + \varphi(\mathbf{X}, t) \tag{5.4}$$

where $\varphi(\mathbf{X}, t)$ is the transformation vector field specified over the domain in configuration C_0. If the loading is now applied to a body transformed to configuration C_t, the deformation process occurs and the material point $\mathbf{X}$ (at $\mathbf{X}^t$ after transformation) is now mapped onto $\mathbf{x}$ in the actual configuration C, cf. Fig. 5.2, so that

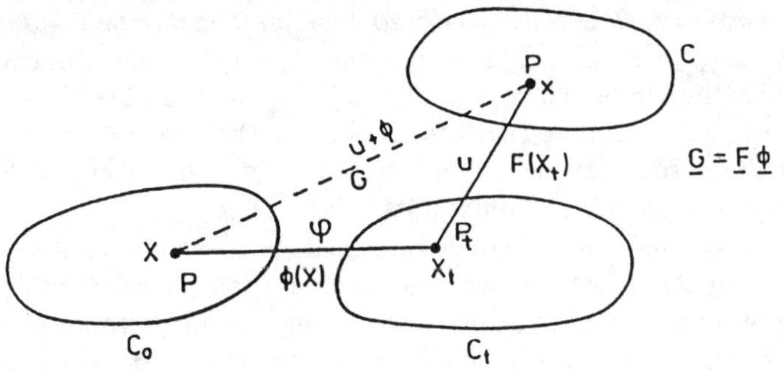

Fig. 5.2. Transformation and deformation: transition $C_0 \to C_t \to C$

$$\mathbf{x} = \mathbf{x}(\mathbf{X}^t, t) = \mathbf{x}[\mathbf{X}^t(\mathbf{X}, t), t] = \mathbf{X}^t + \mathbf{u} = \mathbf{X} + \boldsymbol{\varphi} + \mathbf{u} \tag{5.5}$$

where $\mathbf{u}(\mathbf{X}^t, t)$ denotes the displacement field specified over the configuration C_t. The transformation and deformation gradients now are

$$\begin{aligned} \boldsymbol{\phi} &= \frac{\partial \mathbf{X}^t}{\partial \mathbf{X}} = \mathbf{1} + \frac{\partial \boldsymbol{\varphi}}{\partial \mathbf{X}}, \qquad & \phi_{ij} &= \frac{\partial X_i^t}{\partial X_j} = \delta_{ij} + \frac{\partial \varphi_i}{\partial X_j}, \\ \mathbf{F} &= \frac{\partial \mathbf{x}^t}{\partial \mathbf{X}^t} = \mathbf{1} + \frac{\partial \mathbf{u}}{\partial \mathbf{X}^t}, \qquad & F_{kl} &= \frac{\partial x_k^t}{\partial X_l^t} = \delta_{kl} + \frac{\partial u_k}{\partial X_l^t}, \end{aligned} \tag{5.6}$$

where $\mathbf{1}$ denotes the unit second order tensor, or Kronecker delta, i.e. $(\mathbf{1})_{ij} = \delta_{ij}$.

The total gradient with respect to configuration C_0 can now be presented in a product form as follows

$$\mathbf{G} = \frac{\partial \mathbf{x}}{\partial \mathbf{X}} = \frac{\partial \mathbf{x}}{\partial \mathbf{X}^t}\frac{\partial \mathbf{X}^t}{\partial \mathbf{X}} \qquad \text{or} \qquad \mathbf{G} = \mathbf{F}\boldsymbol{\phi} \tag{5.7}$$

and its rate equals

$$\dot{\mathbf{G}} = \dot{\mathbf{F}}\boldsymbol{\phi} + \mathbf{F}\dot{\boldsymbol{\phi}} \qquad \text{or} \qquad \dot{\mathbf{F}} = \dot{\mathbf{G}}\boldsymbol{\phi}^{-1} - \mathbf{G}\boldsymbol{\phi}^{-1}\dot{\boldsymbol{\phi}}\boldsymbol{\phi}^{-1} \tag{5.8}$$

The deformation velocity gradient equals

$$\mathbf{L}^d = \frac{\partial \dot{\mathbf{u}}}{\partial \mathbf{X}^t} = \frac{\partial \mathbf{u}}{\partial \mathbf{X}^t}\frac{\partial \mathbf{X}^t}{\partial \mathbf{X}} = \dot{\mathbf{F}}\mathbf{F}^{-1} \tag{5.9}$$

and similarly the transformation velocity gradient is expressed as follows

$$\mathbf{L}^t = \frac{\partial \dot{\mathbf{X}}^t}{\partial \mathbf{X}^t} = \frac{\partial \dot{\mathbf{X}}^t}{\partial \mathbf{X}}\frac{\partial \mathbf{X}}{\partial \mathbf{X}^t} = \dot{\boldsymbol{\phi}}\boldsymbol{\phi}^{-1}. \tag{5.10}$$

The total velocity gradient incorporating both transformation and deformation rates can be calculated from Eqs. (5.9) and (5.10) as follows

$$\mathbf{L} = \frac{\partial \dot{\mathbf{x}}}{\partial \mathbf{x}} = \frac{\partial \dot{\mathbf{x}}}{\partial \mathbf{X}}\frac{\partial \mathbf{X}}{\partial \mathbf{x}} = \dot{\mathbf{G}}\mathbf{G}^{-1} = \dot{\mathbf{F}}\mathbf{F}^{-1} + \mathbf{F}\dot{\boldsymbol{\phi}}\boldsymbol{\phi}^{-1}\mathbf{F}^{-1} = \mathbf{L}^d + \mathbf{F}\mathbf{L}^t\mathbf{F}^{-1}. \tag{5.11}$$

The description of the transformation and deformation can now be conducted in several ways. First, we can use the material configuration C_t as a reference configuration, thus assuming

$$\mathbf{F} = \mathbf{F}(\mathbf{X}^t, t), \qquad \phi = \phi(\mathbf{X}^t, t), \qquad \dot{\varphi} = \dot{\varphi}(\mathbf{X}^t, t). \tag{5.12}$$

This is a *mixed description* since it is material (or Lagrangian) with respect to the deformation and spatial (or Eulerian) with respect to the transformation.

Alternatively, one may use the fixed reference configuration C_0 and assume

$$\mathbf{F} = \mathbf{F}(\mathbf{X}, t), \qquad \phi = \phi(\mathbf{X}, t), \qquad \dot{\varphi} = \dot{\varphi}(\mathbf{X}, t). \tag{5.13}$$

This is a *total Lagrangian description* as both transformation and deformation are referred to a fixed reference configuration C_0.

Finally, one could consider a total Eulerian (or spatial) description by assuming

$$\mathbf{F} = \mathbf{F}(\mathbf{x}, t), \qquad \phi = \phi(\mathbf{x}, t), \qquad \dot{\varphi} = \dot{\varphi}(\mathbf{x}, t). \tag{5.14}$$

and use as a reference the actual configuration C.

Consider first the mixed description referred to the transformed configuration $\mathbf{X}_t(t)$. Introduce the Green strain $\mathbf{E}$, and the second Piola–Kirchhoff stress $\mathbf{S}$ referred to this configuration, related by the constitutive equation thus

$$\mathbf{E} = \frac{1}{2}(\mathbf{F}^T\mathbf{F} - \mathbf{1}), \qquad \mathbf{S} = (\det \mathbf{F})\mathbf{F}^{-1}\boldsymbol{\sigma}\mathbf{F}^{-T}, \qquad \mathbf{S} = \frac{\partial U(\mathbf{E})}{\partial \mathbf{E}}, \tag{5.15}$$

where $\boldsymbol{\sigma}$ denotes the Cauchy stress in the actual configuration C. The incremental or rate description follows from Eq. (5.15). We have

$$\begin{aligned}
\dot{\mathbf{E}} &= \frac{1}{2}(\dot{\mathbf{F}}^T\mathbf{F} + \mathbf{F}^T\dot{\mathbf{F}}),\\
\dot{\mathbf{S}} &= (\det \mathbf{F})\mathbf{F}^{-1}(\dot{\boldsymbol{\sigma}} - \mathbf{L}^d\boldsymbol{\sigma} - \boldsymbol{\sigma}\mathbf{L}^{dT} + \boldsymbol{\sigma}\,\mathrm{tr}\,\mathbf{L}^d)\mathbf{F}^{-T},\\
\dot{\mathbf{S}} &= \frac{\partial^2 U}{\partial\mathbf{E}\partial\mathbf{E}}\dot{\mathbf{E}} = \mathbf{C}^t\dot{\mathbf{E}} \qquad \text{or} \qquad \dot{S}_{ij} = C^t_{ijkl}\dot{E}_{kl},
\end{aligned} \tag{5.16}$$

where C^t_{ijkl} denotes the tangent stiffness matrix. Assuming C_t as the reference configuration, we have $\mathbf{F} = \mathbf{F}\left(\mathbf{X}^t(t), t\right)$, $\mathbf{E} = \mathbf{E}\left(\mathbf{X}^t(t), t\right)$. If an infinitesimal transformation process occurs from C_t, then $\mathbf{X}^{t*} = \mathbf{X}^t + \dot{\varphi}$, and

$$\begin{aligned}
\dot{\mathbf{F}} &= \frac{\partial\mathbf{F}}{\partial t} + \frac{\partial\mathbf{F}}{\partial\mathbf{X}^t}\cdot\dot{\varphi} = \dot{\bar{\mathbf{F}}} + \mathbf{F}_{,i}\,\dot{\varphi}_i,\\
\dot{\mathbf{E}} &= \frac{\partial\mathbf{E}}{\partial t} + \frac{\partial\mathbf{E}}{\partial\mathbf{X}^t}\cdot\dot{\varphi} = \dot{\bar{\mathbf{E}}} + \mathbf{E}_{,i}\,\dot{\varphi}_i,
\end{aligned} \tag{5.17}$$

where a comma in the subscript indicates the partial derivative with respect to Cartesian reference coordinate x_i. Similar relations occur for other fields. Here $\dot{\bar{\mathbf{F}}}$ and $\dot{\bar{\mathbf{E}}}$ denote local rates or variations with element position fixed. Equations (5.17) are the material derivatives or variations associated with the transformation process. As we remarked earlier, the transformation process is now treated in the Eulerian description, and all field equations with boundary conditions are satisfied within variable domain C_t.

For the total Lagrangian description, both strain and stress should be referred to the fixed configuration C_0. In view of (5.7) the Green strain will be expressed as follows

$$\mathbf{E} = \frac{1}{2}(\mathbf{F}^T\mathbf{F} - \mathbf{1}) = \frac{1}{2}(\boldsymbol{\phi}^{-T}\mathbf{G}^T\mathbf{G}\boldsymbol{\phi}^{-1} - \mathbf{1}). \tag{5.18}$$

Similarly, the Kirchhoff stress mapped onto C_0 is

$$\mathbf{S}^0 = (\det\mathbf{G})\mathbf{G}^{-1}\boldsymbol{\sigma}\mathbf{G}^{-T} = (\det\boldsymbol{\phi})\boldsymbol{\phi}^{-1}\mathbf{S}\boldsymbol{\phi}^{-T} \tag{5.19}$$

The rates of $\mathbf{E}$ and $\mathbf{S}^0$ are

$$\begin{aligned} \dot{\mathbf{E}} &= \frac{1}{2}\left[\dot{\boldsymbol{\phi}}^{-T}\mathbf{G}^T\mathbf{G}\boldsymbol{\phi}^{-1} + \boldsymbol{\phi}^{-T}(\dot{\mathbf{G}}^T\mathbf{G} + \mathbf{G}^T\dot{\mathbf{G}})\boldsymbol{\phi}^{-1} + \boldsymbol{\phi}^{-T}\mathbf{G}^T\mathbf{G}\dot{\boldsymbol{\phi}}^{-1}\right] \\ \dot{\mathbf{S}}^0 &= (\det\boldsymbol{\phi})\boldsymbol{\phi}^{-1}(\dot{\mathbf{S}} - \mathbf{S}\mathbf{L}^t - \mathbf{S}\mathbf{L}^{tT} + \mathbf{S}\,\mathrm{tr}\,\mathbf{L}^t)\boldsymbol{\phi}^{-T}, \end{aligned} \tag{5.20}$$

where

$$\dot{\boldsymbol{\phi}}^{-1} = -\boldsymbol{\phi}^{-1}\dot{\boldsymbol{\phi}}\boldsymbol{\phi}^{-1}, \qquad \dot{\boldsymbol{\phi}}^{-T} = -\boldsymbol{\phi}^{-T}\dot{\boldsymbol{\phi}}^T\boldsymbol{\phi}^{-T}. \tag{5.21}$$

Substituting the constitutive equation (5.16) into (5.20), we can relate $\dot{\mathbf{S}}^0$ and $\dot{\mathbf{E}}$, namely

$$\dot{\mathbf{S}}^0 = (\det\boldsymbol{\phi})\boldsymbol{\phi}^{-1}(\mathbf{C}\dot{\mathbf{E}} - \mathbf{L}^t\mathbf{S} - \mathbf{S}\mathbf{L}^{tT} + \mathbf{S}\,\mathrm{tr}\,\mathbf{L}^t)\boldsymbol{\phi}^{-T}. \tag{5.22}$$

It is seen that the stress rate is now affected by the transformation gradient $\boldsymbol{\phi}$ and transformation velocity gradient $\mathbf{L}^t$. When the configurations C_0 and C_t coincide, then $\mathbf{X} = \mathbf{X}^t$, $\boldsymbol{\phi} = \mathbf{1}$, $\det\boldsymbol{\phi} = 1$, and (5.22) takes the form

$$\dot{\mathbf{S}}^0 + \mathbf{L}^t\mathbf{S} + \mathbf{S}\mathbf{L}^{tT} - \mathbf{S}\,\mathrm{tr}\,\mathbf{L}^t = \mathbf{C}\dot{\mathbf{E}}. \tag{5.23}$$

In order to write the incremental equilibrium equations in the total Lagrangian formulation, let us specify the first Piola–Kirchhoff stress tensor referred to the configuration C_0

$$\mathbf{P}^0 = \boldsymbol{\phi}\mathbf{S}^0 = (\det\boldsymbol{\phi})\mathbf{S}\boldsymbol{\phi}^{-T} \tag{5.24}$$

and its rate

$$\dot{\mathbf{P}}^0 = \dot{\boldsymbol{\phi}}\mathbf{S}^0 + \boldsymbol{\phi}\dot{\mathbf{S}}^0 = \boldsymbol{\phi}(\dot{\mathbf{S}}^0 + \boldsymbol{\phi}^{-1}\mathbf{L}^t\mathbf{S}\boldsymbol{\phi}^{-T}) = (\det\boldsymbol{\phi})(\dot{\mathbf{S}} - \mathbf{S}\mathbf{L}^{tT} + \mathbf{S}\,\mathrm{tr}\,\mathbf{L}^t)\boldsymbol{\phi}^{-T}. \tag{5.25}$$

The incremental equilibrium equations can now be expressed as follows

$$\begin{aligned} \mathrm{Div}\,\dot{\mathbf{P}}^0 + \rho_0\dot{\mathbf{f}}^0 &= \mathbf{0} \\ \dot{\mathbf{P}}^0\mathbf{N}^0 = \dot{\mathbf{T}}^0(\mathbf{x}) &\quad \text{on } S_T^0 \end{aligned} \tag{5.26}$$

where $\dot{\mathbf{T}}^0(\mathbf{X})$ is calculated for the specified field of external surface tractions, and $\mathbf{N}^0$ is the unit normal vector to S_T^0 in the reference configuration.

Passing now to the transformed configuration C_t as the reference configuration, we set $\boldsymbol{\phi} = \mathbf{1}$ in (5.20), (5.22) and (5.25), to obtain

$$\begin{aligned} \mathbf{E} &= \frac{1}{2}(\mathbf{F}^T\mathbf{F} - \mathbf{1}), \qquad \mathbf{P}^0 = \mathbf{S}^0 = \mathbf{S}, \\ \dot{\mathbf{S}}^0 &= \mathbf{C}\dot{\mathbf{E}} - \dot{\mathbf{S}}^i, \qquad \dot{\mathbf{S}}^i = \mathbf{L}^t\mathbf{S} + \mathbf{S}\mathbf{L}^{tT} + \mathbf{S}\,\mathrm{tr}\,\mathbf{L}^t \end{aligned} \tag{5.27}$$

and in the index notation

$$\begin{aligned} E_{ij} &= \frac{1}{2}(u_{i,j} + u_{j,i} + u_{k,i}u_{k,j}) \\ \dot{S}_{ij} &= C_{ijkl}\dot{E}_{kl} - \dot{S}^i_{ij}\,, \\ \dot{S}^i_{ij} &= \dot{\varphi}_{i,k}S_{kj} + S_{ik}\dot{\varphi}_{j,k} - S_{ij}\dot{\varphi}_{k,k} \\ \dot{E}_{ij} &= \frac{1}{2}[\dot{u}_{i,j} + \dot{u}_{j,i} + \dot{u}_{k,i}u_{k,j} + u_{k,i}\dot{u}_{k,j} - (\dot{\varphi}_{k,i}u_{j,k} + u_{i,k}\dot{\varphi}_{k,j}) \\ &\quad - (\dot{\varphi}_{k,i}u_{m,k}u_{m,j} + u_{m,j}u_{m,k}\dot{\varphi}_{k,i})] \end{aligned} \tag{5.28}$$

and the equilibrium equations

$$\begin{aligned} \dot{P}^0_{ij,i} + \rho\dot{f}^0_j &= \dot{P}^0_{ij,i} + (\dot{\varphi}_{i,k}S_{kj})_{,j} + \rho\dot{f}^0_j \\ &= (C_{ijkl}\dot{E}_{kl})_{,i} - (S_{ik}\dot{\varphi}_{j,k})_{,i} + (S_{ij}\dot{\varphi}_{k,k})_{,i} + \rho\dot{f}^0_j = 0 \end{aligned} \tag{5.29}$$

with the boundary conditions

$$\dot{S}^0_{ij}N^0_i + \dot{\varphi}_{i,k}S_{kj}N^0_i = \dot{T}^0_j(\mathbf{x}) \qquad \text{on } S^0_T\,. \tag{5.30}$$

The transition to small strain theory is obtained by neglecting displacement terms in (5.28), so that

$$\begin{aligned} E_{ij} &= \varepsilon_{ij} = \frac{1}{2}(u_{i,j} + u_{j,i}) & \boldsymbol{\sigma} &= \mathbf{S} = \mathbf{S}^0, \qquad \mathbf{n} = \mathbf{N}^0 = \mathbf{N} \\ \dot{\sigma}^0_{ij} &= \dot{S}^0_{ij} = C_{ijkl}\dot{\varepsilon}_{kl} - \dot{S}^i_{ij}\,, & \dot{S}^i_{ij} &= \dot{\varphi}_{i,k}\sigma_{kj} + \sigma_{ik}\dot{\varphi}_{j,k} - \sigma_{ij}\dot{\varphi}_{k,k} \\ \dot{E}_{ij} &= \dot{\varepsilon}_{ij} = \frac{1}{2}(\dot{u}_{i,j} + \dot{u}_{j,i}) - \dot{\varepsilon}^i_{ij} & \dot{\varepsilon}^i_{ij} &= (\dot{\varphi}_{k,i}u_{j,k} + u_{i,k}\dot{\varphi}_{k,j})\,. \end{aligned} \tag{5.31}$$

It is seen that the transformation rate field affects incremental equilibriun equations (5.29), boundary conditions (5.30), constitutive relations, and strain rate relations to rate of displacement gradients (5.31). In fact, the initial strain and stress fields $\dot{\varepsilon}^i_{ij}$ and $\dot{S}^i_{ij}$ occur in these relations, and the additional "force" field $(\dot{\varphi}_{i,k}\sigma_{kj})$ occurs in equilibrium equations.

Setting $\mathbf{F} = \boldsymbol{\phi} = \mathbf{1}$, the configurations C_0, C_t and C will coincide and from (5.16) and (5.20) it follows that

$$\begin{aligned} \dot{\mathbf{S}}^0 &= \dot{\boldsymbol{\sigma}} - \mathbf{L}\boldsymbol{\sigma} - \boldsymbol{\sigma}\mathbf{L}^T + \boldsymbol{\sigma}\,\mathrm{tr}\,\mathbf{L}, \qquad \mathbf{L} = \mathbf{L}^d + \mathbf{L}^t \\ \dot{\mathbf{S}} &= \dot{\boldsymbol{\sigma}} - (\mathbf{L} - \mathbf{L}^t)\boldsymbol{\sigma} - \boldsymbol{\sigma}(\mathbf{L} - \mathbf{L}^t)^T + \boldsymbol{\sigma}\,\mathrm{tr}(\mathbf{L} - \mathbf{L}^t) = \mathbf{C}\dot{\mathbf{E}}\,. \end{aligned} \tag{5.32}$$

Thus the constitutive equations for the Cauchy stress rate expressed in terms of the convective rate expressed in terms of the deformation velocity gradient $\mathbf{L}^d = \mathbf{L} - \mathbf{L}^t$ or difference between the total and transformation velocity gradients.

We have discussed the total Lagrangian description and by setting $\phi = \mathbf{1}$, we obtain the description referred to the configuration C_t. The mixed description provides the same resulting equations, though the method of treatment is different since then $\mathbf{X}^t$ are treated as space coordinates. We shall now discuss the derivation of sensitivity of the functional G in the case of shape variation, assuming small strain theory, that is setting $\mathbf{F} = \mathbf{1}$, $\mathbf{x} = \mathbf{X}^t$, and using the mixed description.

Assume the differentiable transformation velocity field $\dot{\varphi} = \dot{\varphi}(\mathbf{X}^t, t)$ to be specified within the whole admissible domain Ω_a, so a typical structural point undergoes transformation $\mathbf{X}^{*t} = \mathbf{X}^t + \dot{\varphi} dt = \mathbf{X}^t + \delta\varphi(\mathbf{X}^t)$. Considering small strain theory and setting $\mathbf{x} = \mathbf{X}^t$, the variations of displacement, strain and stress fields are

$$\delta\mathbf{u} = \delta\bar{\mathbf{u}} + \mathbf{u}_{,k}\delta\varphi_k\,, \qquad \delta\varepsilon = \delta\bar{\varepsilon} + \varepsilon_{,k}\delta\varphi_k\,, \qquad \delta\sigma = \delta\bar{\sigma} + \sigma_{,k}\delta\varphi_k\,, \tag{5.33}$$

where $\delta\bar{\mathbf{u}}$, $\delta\bar{\varepsilon}$ and $\delta\bar{\sigma}$ denote local variations at fixed material points and the state fields are assumed to be referred to the configuration C_t. Thus, $\mathbf{u} = \mathbf{u}(\mathbf{x}, t)$, $\varepsilon = \varepsilon(\mathbf{x}, t)$, $\sigma = \sigma(\mathbf{x}, t)$ and $\mathbf{x} = \mathbf{X}^t$. The time derivatives are expressed in a manner similar to the material derivatives in the Eulerian description, namely

$$\dot{\mathbf{u}} = \dot{\bar{\mathbf{u}}} + \mathbf{u}_{,k}\dot{\varphi}_k\,, \qquad \dot{\varepsilon} = \dot{\bar{\varepsilon}} + \varepsilon_{,k}\dot{\varphi}_k\,, \qquad \dot{\sigma} = \dot{\bar{\sigma}} + \sigma_{,k}\dot{\varphi}_k\,, \tag{5.34}$$

5.2 Variation of Volume and Surface Integrals

As the structure domain is varied, the volume, surface elements, and the unit vector normal to the boundary surface vary, thus

$$\begin{aligned} \delta(\mathrm{d}V) &= \delta\varphi_{kk}\,\mathrm{d}V, \\ \delta(\mathrm{d}S) &= (\delta_{kl} - n_k n_l)\,\delta\varphi_{k,l}\,\mathrm{d}S, \\ \delta n_j &= (n_j n_l - \delta_{jl}) n_k\,\delta\varphi_{k,l}\,, \end{aligned} \tag{5.35}$$

or in vector notation the respective rates are

$$(\mathrm{d}V)^{\cdot} = \mathrm{tr}\,\mathbf{L}^t\,\mathrm{d}V, \qquad (\mathrm{d}S)^{\cdot} = (\mathbf{1} - \mathbf{n}\otimes\mathbf{n})\cdot\mathbf{L}^t\,\mathrm{d}S, \qquad \dot{\mathbf{n}} = -\mathbf{L}^{tT}\mathbf{n} + (\mathbf{n}\cdot\mathbf{L}^t\mathbf{n})\mathbf{n}\,. \tag{5.36}$$

These formulae follow from transformation relations between deformed and undeformed configurations, cf. Dems and Mróz [12]. The surface traction vector is transformed according to the formula

$$\begin{aligned} \delta T_j &= \delta(n_i\sigma_{ij}) = \delta n_i\,\sigma_{ij} + n_i\,\delta\sigma_{ij} \\ &= n_i\,\delta\bar{\sigma}_{ij} + \sigma_{ij}(n_j n_l - \delta_{jl}) n_k\,\delta\varphi_{k,l} + n_i\sigma_{ij,k}\,\dot{\varphi}_k\,. \end{aligned} \tag{5.37}$$

The volume integral is transformed as follows

$$\begin{aligned} I_V &= \int f \, \mathrm{d}V \\ \delta I_V &= \int \delta f \, \mathrm{d}V + \int f \, \delta(\mathrm{d}V) = \int (\delta f + f \delta\varphi_{k,k}) \, \mathrm{d}V \\ &= \int [\delta \bar{f} + (f\delta\varphi_k)_{,k}] \, \mathrm{d}V = \int \delta \bar{f} \, \mathrm{d}V + \int f \delta\varphi_n \, \mathrm{d}S \end{aligned} \tag{5.38}$$

where $\delta\varphi_n = \delta\varphi_k n_k$ denotes the normal component of the transformation vector $\delta\boldsymbol{\varphi}$.

The surface integral specified over the regular surface S with its boundary Γ

$$I_S = \int f \, \mathrm{d}S \tag{5.39}$$

is transformed according to the following formula

$$\begin{aligned} \delta I_S &= \int \delta f \, \mathrm{d}S + \int f \, \delta(\mathrm{d}S) = \int [\delta f + f(\delta_{kl} - n_k n_l)\delta\varphi_{k,l}] \, \mathrm{d}S \\ &= \int [\delta f + f(\delta\varphi_{\alpha,\alpha} - 2\kappa_m \delta\varphi_n)] \, \mathrm{d}S \end{aligned} \tag{5.40}$$

where $\delta\varphi_{\alpha,\alpha}$ $(\alpha = 1, 2)$ are the tangential components of $\delta\boldsymbol{\varphi}$ referred to a curvilinear reference system within the surface and κ_m is the mean curvature of the surface. The last expression can be transformed as follows

$$\begin{aligned} \delta I_S &= \int (\delta f + f_{,\alpha}\delta\varphi_\alpha - 2f\kappa_m\delta\varphi_n) \, \mathrm{d}S + \oint f\delta\varphi_\mu \, \mathrm{d}l \\ &= \int (\delta_n f - 2f\kappa_m \delta\varphi_n) \, \mathrm{d}S + \oint f\delta\varphi_\mu \, \mathrm{d}l \end{aligned} \tag{5.41}$$

where we applied the Green–Gauss theorem for a continuous and differentiable vector $\mathbf{v}$ specified on a regular surface boundary by a piecewise smooth closed curve Γ, that is

$$\int_S v_{\alpha,\alpha} \, \mathrm{d}S = \int_\Gamma v_\alpha \mu_\alpha \, \mathrm{d}l = \int_\Gamma v_\mu \, \mathrm{d}l \tag{5.42}$$

where l is the arc length and μ denotes the unit normal vector to Γ and tangential to S, pointing toward the exterior of S. The symbol $\delta_n f$ denotes the variation along the direction normal to S, thus

$$\delta_n f = \delta \bar{f} + f_{,n} \, \delta\varphi_n \tag{5.43}$$

and $\delta\varphi_n = \delta\varphi_\alpha \mu_\alpha = \delta\varphi_i \mu_i$ is the component of the transformation vector $\delta\boldsymbol{\varphi}$ in the direction of μ. In order to calculate the variation of a surface integral over a piecewise regular surface, Eq. (5.41) can be directly applied over each regular surface, with subsequent addition of consecutive variations.

In particular, the variation of the surface integral over a *closed* piecewise regular surface S takes the form

$$\delta I_S = \int (\delta_n f - 2f\kappa_m \delta\varphi_n)\,\mathrm{d}S + \sum_i \int (f^+ \delta\varphi_\mu^+ + f^- \delta\varphi_\mu^-)\,\mathrm{d}l_i \tag{5.44}$$

where the sum of line integrals is taken over all edges of the surface S and "+" and "−" signs refer to quantities evaluated on the two regular sections intersecting along the edge Γ. Note that the pair $(\delta\varphi_\mu^+, \delta\varphi_\mu^-)$ of tangential transformations at the edge is uniquely specified by the pair $(\delta\varphi_n^+, \delta\varphi_n^-)$ of normal surface variations on the edge.

5.3 Sensitivity Analysis with Shape Transformation

Consider a body of regular boundary undergoing a regular transformation, such that $\delta\boldsymbol{\varphi}(\mathbf{x}, t)$ is smooth, differentiable and possesses a continuous space derivative on the boundary. On the other hand, when $\delta\boldsymbol{\varphi}(\mathbf{x}, t)$ is discontinuous or its normal component $\delta\varphi_n$ has a discontinuous space derivative on the boundary, the *singular* shape transformation occurs. For instance, when a sharp wedge notch or a crack is formed on the boundary, the singular transformation takes place. The state fields $\boldsymbol{\sigma}$ and $\boldsymbol{\varepsilon}$ then have singularities at notch root or crack front and the formulae derived for regular transformation cannot be directly applied for singular cases. There may also be a singular boundary undergoing regular transformation.

Consider now the simplest case of regular boundary and regular transformation. Considering the augmented functional $\bar{G}$, its variation associated with shape transformation is

$$\begin{aligned}\delta\bar{G} = &\int \delta\psi\,\mathrm{d}V + \int \psi\,(\delta\mathrm{d}V) + \int \delta h\,\mathrm{d}S + \int h\,(\delta\mathrm{d}S)\\ &- \int \delta(\boldsymbol{\sigma}\cdot\boldsymbol{\varepsilon}^e)\,\mathrm{d}V - \int \boldsymbol{\sigma}\cdot\boldsymbol{\varepsilon}^e\,(\delta\mathrm{d}V)\\ &+ \int \delta(\mathbf{f}\cdot\mathbf{u}^a)\,\mathrm{d}V + \int \mathbf{f}\cdot\mathbf{u}^a\,(\delta\mathrm{d}V)\\ &+ \int \delta(\mathbf{T}\cdot\mathbf{u}^a)\,\mathrm{d}S + \int \mathbf{T}\cdot\mathbf{u}^a\,(\delta\mathrm{d}S)\,.\end{aligned} \tag{5.45}$$

In view of Eqs. (5.35) and (5.41), we can further transform (5.45) as follows

$$\begin{aligned}\delta\bar{G} = &\int \left(\frac{\partial\psi}{\partial\boldsymbol{\sigma}}\cdot\delta\bar{\boldsymbol{\sigma}} + \frac{\partial\psi}{\partial\mathbf{u}}\cdot\delta\bar{\mathbf{u}}\right)\mathrm{d}V + \int \psi n_k\,\delta\varphi_k\,\mathrm{d}S\\ &+ \int \left[\frac{\partial h}{\partial\mathbf{u}}\cdot\delta\mathbf{u} + \frac{\partial h}{\partial\mathbf{T}}\cdot\delta\mathbf{T} + h(\delta_{kl} - n_k n_l)\,\delta\varphi_{k,l}\right]\mathrm{d}S\\ &- \int (\delta\bar{\boldsymbol{\sigma}}\cdot\boldsymbol{\varepsilon}^a + \boldsymbol{\sigma}\cdot\delta\bar{\boldsymbol{\varepsilon}}^a - \delta\bar{\mathbf{f}}\cdot\mathbf{u}^a + \mathbf{f}\cdot\delta\bar{\mathbf{u}}^a)\,\mathrm{d}V\end{aligned}$$

$$
\begin{aligned}
&- \int (\boldsymbol{\sigma}\cdot\boldsymbol{\varepsilon}^a - \mathbf{f}\cdot\mathbf{u}^a)\delta\varphi_n \,\mathrm{d}S \\
&+ \int [\delta\mathbf{T}\cdot\mathbf{u}^a + \mathbf{T}\cdot\delta\mathbf{u}^a + \mathbf{T}\cdot\mathbf{u}^a(\delta_{kl} - n_k n_l)\,\delta\varphi_{k,l}]\,\mathrm{d}S\,. \qquad (5.46)
\end{aligned}
$$

Note that the local variations $\delta\bar{\boldsymbol{\sigma}}$ and $\delta\bar{\mathbf{u}}$ occur in volume integrals and that the total variations $\delta\boldsymbol{\sigma}$ and $\delta\mathbf{u}$ occur in surface integrals. Consider a linearly elastic material for which the constitutive relations are

$$
\delta\bar{\boldsymbol{\sigma}} = \mathbf{C}\delta\bar{\boldsymbol{\varepsilon}} \qquad (5.47)
$$

that is, relating local variations by Hooke's law. Note that

$$
\int (\boldsymbol{\varepsilon}^a - \boldsymbol{\varepsilon}^{ai})\cdot\delta\bar{\boldsymbol{\sigma}}\,\mathrm{d}V = \int (\boldsymbol{\varepsilon}^a - \boldsymbol{\varepsilon}^{ai})\cdot\mathbf{C}\delta\bar{\boldsymbol{\varepsilon}}\,\mathrm{d}V = \int \boldsymbol{\sigma}^a\cdot\delta\bar{\boldsymbol{\varepsilon}}\,\mathrm{d}V \qquad (5.48)
$$

since $\boldsymbol{\sigma}^{ai} = \mathbf{C}(\boldsymbol{\varepsilon}^a - \boldsymbol{\varepsilon}^{ai})$. Substituting Eq. (5.48) into Eq. (5.46) and using the definition of the adjoint state, we finally obtain, cf. Dems and Mróz [12]

$$
\begin{aligned}
\delta\bar{G} = &\int [\psi - \boldsymbol{\sigma}\cdot\boldsymbol{\varepsilon}^a + \mathbf{f}\cdot\mathbf{u}^a + (h + \mathbf{T}\cdot\mathbf{u}^a)_{,n} - (h + \mathbf{T}\cdot\mathbf{u}^a)\,2\kappa_m]\,\delta\varphi_n\,\mathrm{d}S \\
&+ \int \left(\frac{\partial h}{\partial\mathbf{T}} + \mathbf{u}^a\right)\cdot(\delta\mathbf{T}^0 - \mathbf{T}^0_{,k}\delta\varphi_k)\,\mathrm{d}S_T \\
&+ \int \left(\frac{\partial h}{\partial\mathbf{u}} - \mathbf{T}^a\right)\cdot(\delta\mathbf{u}^0 - \mathbf{u}^0_{,k}\delta\varphi_k)\,\mathrm{d}S_u \\
&+ \int \delta\bar{\mathbf{f}}\cdot\mathbf{u}^a\,\mathrm{d}V + \sum_r \int [h + \mathbf{T}\cdot\mathbf{u}^a]\delta\varphi_\mu\,\mathrm{d}l_r \qquad (5.49)
\end{aligned}
$$

where [] denotes the discontinuity of the enclosed quantity along the edge of intersection of two regular surfaces. Note that the boundary conditions on S_T and S_u are specified not only on the boundary but in the whole admissible domain $\mathbf{T}^0 = \mathbf{T}^0(\mathbf{x},\mathbf{n})$, $\mathbf{u}^0 = \mathbf{u}^0(\mathbf{x},\mathbf{n})$, so their variations and gradients are known.

The sensitivity of G is now expressed totally as the surface integral in terms of primary and adjoint state fields. Note that the first term of (5.49) can be transformed into a volume integral by using the Green–Gauss theorem.

When the potential energy variation is considered, the formula (5.49) provides for the case of variation of S_T

$$
\begin{aligned}
\delta\Pi(\mathbf{u},\mathbf{f},\mathbf{T}^0,\boldsymbol{\varphi}) = &\int [U - \mathbf{f}\cdot\mathbf{u} - (\mathbf{T}^0\cdot\mathbf{u})_{,n} + 2\mathbf{T}^0\cdot\mathbf{u}\kappa]\,\delta\varphi_n\,\mathrm{d}V \\
&- \int (\delta\mathbf{T}^0 - \mathbf{T}^0_{,k}\delta\varphi_k)\cdot\mathbf{u}\,\mathrm{d}S_T\,. \qquad (5.50)
\end{aligned}
$$

In the case of translation of the boundary by the vector $\delta\mathbf{a} = \delta\boldsymbol{\varphi}$ and fixed surface tractions $\mathbf{T}^0$, we have for $\mathbf{f} = \mathbf{0}$

$$
\delta\Pi = \delta a_k \int (U\delta_{kj} - \sigma_{ij}u_{i,k})n_j\,\mathrm{d}S = A_k(\mathbf{u})\,\delta a_k \qquad (5.51)
$$

and for the general functional we have

$$\delta G = \delta a_k \int [(\psi - \boldsymbol{\sigma} \cdot \boldsymbol{\varepsilon}^a)\delta_{kj} + \sigma_{ij} u^a_{i,k} + \sigma^a_{ij} u_{i,k}] n_j \, \mathrm{d}S = S_k(\mathbf{u}, \mathbf{u}^a) \, \delta a_k \, . \quad (5.52)$$

Consider now the case of an interface S_c in the interior of a structure, separating two materials of different stiffness matrices. The displacement and surface traction vectors are continuous on S_c, thus we have

$$\begin{aligned} [\mathbf{u}] &= \mathbf{0} & [\mathbf{T}] &= \mathbf{n}[\sigma] = \mathbf{0} \qquad \text{on } S_c \\ [\mathbf{u}_{,k}] &= [\mathbf{u}_{,n}] n_k = \mathbf{a} n_k & [\mathbf{T}_{,k}] &= [\mathbf{T}_{,n}] n_k \end{aligned} \quad (5.53)$$

where $\mathbf{a}$ denotes the interface discontinuity vector. The continuity condition of $\mathbf{u}$ and $\mathbf{T}$ on S_c subjected to variation is expressed as follows

$$\begin{aligned} [\delta \mathbf{u}] &= [\delta \bar{\mathbf{u}}] + [\mathbf{u}_{,n}] \, \delta\varphi_n = \mathbf{0} \\ [\delta \mathbf{T}] &= \delta \mathbf{n}[\sigma] + \mathbf{n}[\delta \bar{\sigma}] + \mathbf{n}[\sigma_{,k}] \, \delta\varphi_k = \mathbf{0}. \end{aligned} \quad (5.54)$$

From (5.54) it follows that

$$\begin{aligned} [\delta \bar{\mathbf{u}}] &= -[\mathbf{u}_{,n}] \, \delta\varphi_n = -\mathbf{a} \, \delta\varphi_n \\ [\delta \bar{\mathbf{T}}_i] &= n_i [\delta \bar{\sigma}_{ij}] = -[\sigma_{ij}] n_k \, \delta\varphi_{k,j} - [\sigma_{ij,k}] n_i \, \delta\varphi_k \, . \end{aligned} \quad (5.55)$$

Setting $\delta \bar{\mathbf{f}} = \mathbf{0}$ within V and $h = 0$ on S, we obtain the sensitivity variation from the general formula

$$\begin{aligned} \delta \bar{G} = & \int \{ [\psi] - [\boldsymbol{\sigma} \cdot \boldsymbol{\varepsilon}^a] + [\mathbf{f}] \cdot \mathbf{u}^a + \mathbf{T} \cdot [\mathbf{u}^a_{,n}] + \mathbf{T}^a \cdot [\mathbf{u}_{,n}] \} \, \delta\varphi_n \, \mathrm{d}S_c \\ & + \int [\mathbf{T} \cdot \mathbf{u}^a] \delta\varphi_\mu \, \mathrm{d}l \end{aligned} \quad (5.56)$$

where the last integral is specified along the line Γ at the intersection of the interface with the boundary surface, and $[\mathbf{u}^a_{,n}] = \mathbf{a}^a$ is the adjoint discontinuity vector. Neglecting the last term of (5.56) and setting $[\mathbf{f}] = \mathbf{0}$, we obtain

$$\begin{aligned} \delta \bar{G} &= \int \{ [\psi] - [\boldsymbol{\sigma} \cdot \boldsymbol{\varepsilon}^a] + \mathbf{T} \cdot \mathbf{a}^a + \mathbf{T}^a \cdot \mathbf{a} \} \, \delta\varphi_n \, \mathrm{d}S_c \\ &= \int [P_{jk}] n_j n_k \delta\varphi_n \, \mathrm{d}S = \int H(\mathbf{u}, \mathbf{u}^a) \, \delta\varphi_n \, \mathrm{d}S_c \end{aligned} \quad (5.57)$$

where P_{jk} is the mutual energy momentum tensor

$$P_{jk} = (\psi - \boldsymbol{\sigma} \cdot \boldsymbol{\varepsilon}^a) \, \delta_{jk} + \sigma_{ij} u^a_{j,k} + \sigma^a_{ij} u_{j,k} \, , \quad (5.58)$$

$H(\mathbf{u}, \mathbf{u}^a)$ is the generalized sensitivity force at the interface. For the potential energy there is

$$\delta \Pi = \int H^p \delta\varphi_n \, \mathrm{d}S_c \quad (5.59)$$

where

$$H^p(\mathbf{u}) = [U] - \mathbf{T} \cdot \mathbf{a} = \mathbf{n} \cdot \mathbf{P}^p \mathbf{n} \qquad P^p_{jk} = U\delta_{jk} - \sigma_{ij} u_{i,k} \tag{5.60}$$

and P^p_{jk} is the energy momentum tensor introduced by Eshelby [17].

The interface S_c in plates or disks may also represent the thickness discontinuity. Such interfaces and their optimization were treated by Dems and Mróz [15] . The sensitivity and optimal design of stiffened disks and plates by curvilinear stiffening and relaxing interfaces with strong displacement discontinuity were considered by the same authors [13].

6 Sensitivity Analysis for Beam Structures

6.1 Fundamental Equations

The general sensitivity analysis of the two preceding sections can now be applied to beams subjected to flexure and extension. Assume the Navier hypothesis and specify the displacement field

$$u(x,y,z) = -z\,\frac{\mathrm{d}w}{\mathrm{d}x} = -zw', \qquad v = 0, \qquad w(x,y,z) = w(x) \tag{6.1}$$

where the x-axis coincides with the beam axis and z-axis is normal to the x-axis within the flexure plane. The prime denotes differentiation with respect to axial position x.

Denoting by M and Q the bending moment and the shear force in the beam cross-section, the equilibrium equations are

$$Q = M', \qquad Q' = -p, \tag{6.2}$$

where $p(x)$ is the transverse beam loading. Denoting the deflection slope by $\psi(x) = w'$ and curvature $\kappa(x) = -w''$, the constitutive relation is

$$M = EI\kappa = -EIw'' \tag{6.3}$$

and the equilibrium equation takes the form

$$M'' = -p \qquad \text{or} \qquad (EIw'')'' = p \tag{6.4}$$

where $I = \int z^2\,\mathrm{d}A$ denotes the moment of inertia of the cross-section. When the axial force N is present, Eq. (6.4) is modified, namely

$$(EIw'')'' - Nw'' = p\,. \tag{6.5}$$

The boundary conditions depend on a kind of support and are expressed in terms of deflection, its slope, bending moment, and shear force. For the case of elastic support, the relation between static and geometric quantities is provided. Generally, the boundary conditions can be classified as follows

Geometric conditions	Static conditions	Elastic support
$w = w^0$	$Q = Q^0$	$Q \pm k_1 W = \alpha$
$\psi = w' = \psi^0$	$M = M^0$	$M \pm k_2 w' = \beta$

where w^0, ψ^0, Q^0 and M^0 denote the values at the beam ends, and k_1, k_2 are the stiffnesses of end supports. The values of α and β are equal to zero when there are no initial deflections or rotations. Thus, for a supported beam there is $w = w^0$, $M = M^0$ at the end, for a constrained beam there is $W = W^0$, $\psi = \psi^0$, and for the statically supported beam there is $M = M^0$, $Q = Q^0$.

The virtual work equation can be written in the form

$$\int_0^l pw \, \mathrm{d}x + [Qw - M\psi]_0^l = \int_0^l M\kappa \, \mathrm{d}x \tag{6.6}$$

where $[\cdot]_0^l$ denotes the difference of values of the enclosed function for $x = l$ and $x = 0$. Writing (6.6) for two states: static field p_1, M_1, Q_1, and kinematically admissible field w_2, we have

$$\int_0^l p_1 w_2 \, \mathrm{d}x + Q_{l1} w_{l2} - Q_{01} w_{02} - M_{l1} w'_{l2} + M_{01} w'_{02} = \int_0^l M_1 \kappa_2 \, \mathrm{d}x \, . \tag{6.7}$$

Similarly, for the static field p_2, M_2, Q_2, and the kinematically admissible field, w_1, the virtual work equation is

$$\int_0^l p_2 w_1 \, \mathrm{d}x + Q_{l2} w_{l1} - Q_{02} w_{01} - M_{l2} w'_{l1} + M_{02} w'_{01} = \int_0^l M_2 \kappa_1 \, \mathrm{d}x \, . \tag{6.8}$$

From (6.7) and (6.8) it follows that the work reciprocal equation occurs

$$\begin{aligned} &\int_0^l p_1 w_2 \, \mathrm{d}x + Q_{l1} w_{l2} - Q_{01} w_{02} - M_{l1} w'_{l2} + M_{01} w'_{02} \\ &\quad = \int_0^l p_2 w_1 \, \mathrm{d}x + Q_{l2} w_{l1} - Q_{02} w_{01} - M_{l2} w'_{l1} + M_{02} w'_{01} \, . \end{aligned} \tag{6.9}$$

The effect of axial force can be accounted for by introducing the equivalent lateral loading

$$P_N = N w'' . \tag{6.10}$$

Equation (6.7) then takes the form

$$\begin{aligned} &\int_0^l (p_1 + N w'') w_2 \, \mathrm{d}x + Q_{l1} w_{l2} - Q_{01} w_{02} - M_{l1} w'_{l2} + M_{01} w'_{02} \\ &\quad = \int_0^l M_1 \kappa_2 \, \mathrm{d}x = \int_0^l EI w''_2 w''_1 \, \mathrm{d}x \, . \end{aligned} \tag{6.11}$$

As the following integral equality occurs

$$\int_0^l N w'' w_2 \, \mathrm{d}x = - \int_0^l N w' w'_2 \, \mathrm{d}x - [N w'_1 w_2]_0^l \tag{6.12}$$

Eq. (6.11) can be transformed as follows

$$\begin{aligned} &\int_0^l p_1 w_2 \, \mathrm{d}x + [Q_{l1} + N w'_{l1}] w_{l2} - M_{l1} w'_{l2} - [Q_{01} + N w'_{l1}] w_{02} + M_{01} w'_{02} \\ &\quad = \int_0^l (EI w''_1 w''_2 + N w'_1 w'_2) \, \mathrm{d}x \end{aligned} \tag{6.13}$$

where $M_l = M(l)$, $Q_l = Q(l)$, $M_0 = M(0)$ and $Q_0 = Q(0)$.

Consider now the potential energy of the beam clamped at the end $x = 0$ and simply supported at the end $x = l$. We have

$$\Pi(w) = \int_0^l \frac{1}{2} EI\kappa^2 \, dx + \int_0^l \frac{1}{2} N(w')^2 \, dx - \int_0^l pw \, dx + M^0(l)w'(l). \quad (6.14)$$

In (6.14) the elastic energy due to beam extension by axial force N was neglected. It is easy to demonstrate that the stationarity condition of the potential energy corresponds to the equilibrium condition of the beam. In fact, from (6.13) and (6.14) it follows that

$$\delta\Pi(w) = \int_0^l M\delta\kappa \, dx + \int_0^l Nw'\delta w' \, dx - \int_0^l p\delta w \, dx + M^0(l)\delta w'(l) = 0. \quad (6.15)$$

Integrating by parts two first terms of (6.15), and using Eqs. (6.2)–(6.5), we obtain

$$\delta\Pi(w) = -\int_0^l [M'' + (Nw')' + p]' \, \delta w \, dx + (M^0 - M)\delta w'|_0^l + (Q - Nw')\delta w|_0^l \, . \quad (6.16)$$

Requiring all three terms of (6.16) to vanish for arbitrary variations δw, we obtain the equilibrium equation (6.5), support condition at $x = 0$: $\delta w(0) = \delta w'(0) = 0$, and the boundary condition at $x = l$: $M(l) = M^0$, $\delta w(l) = 0$.

The strong minimum of the potential energy at the equilibrium state for $N > 0$ can easily be demonstrated. In fact, for any kinematically admissible field $w^k(x)$ we have

$$\begin{aligned}\Pi(w^k) - \Pi(w) = &\int_0^l [U(\kappa^k) - U(\kappa)] \, dx + \int_0^l \frac{1}{2} N[(w^{k\prime})^2 - (w')^2] \, dx \\ &- \int_0^l p(w^k - w) \, dx + M^0(l)[w^{k\prime}(l) - w'(l)]. \end{aligned} \quad (6.17)$$

In view of the virtual work equation for the field $w^k(x) - w(x)$, Eq. (6.17) can be presented in the form

$$\begin{aligned}\Pi(w^k) - \Pi(w) = &\int_0^l [U(\kappa^k) - U(\kappa) - M(\kappa^k - \kappa)] \, dx \\ &+ \int_0^l \frac{1}{2} N(w^{k\prime} - w')^2 \, dx \end{aligned} \quad (6.18)$$

where $U(\kappa) = \frac{1}{2}M\kappa = \frac{1}{2}EI\kappa^2$ is the specific bending energy of the beam. As the specific energy is a convex function, the integrand of (6.18) is positive-definite. Similarly, for $N > 0$ (tension force) the second term is also positive, hence $\Pi(w^k) > \Pi(w)$. Neglecting the axial force ($N = 0$), the complementary energy of the beam is

$$\bar{\Pi}(M) = \int_0^l \frac{M^2}{2EI} \, dx - w^0(l)Q(l) - \psi^0(l)M(l) \, . \quad (6.19)$$

Requiring the variation of $\bar{\Pi}(M)$ to vanish,

$$\delta\bar{\Pi}(M) = \int_0^l \kappa\delta M \,\mathrm{d}x - w^0(l)\delta Q(l) - \psi^0(l)\delta M(l) = 0 \tag{6.20}$$

we obtain the conditions of geometric compatibility

$$\kappa = -w'', \qquad w(l) = w^0(l), \qquad \psi^0 = \psi^0(l). \tag{6.21}$$

6.2 Sensitivity of Potential and Complementary Energies

Assume, as previously, that the complementary energy is used a measure of global structure compliance and the potential energy is a measure of global stiffness. Let us investigate the variations of these measures when the structural parameters, such as cross section area, beam length, support position, hinge location, etc. undergo variation.

Consider a beam built-in at the end $x = 0$ and supported at the end $x = l$, so that the deflection and the slope are specified, thus $w(l) = w^0(l)$, $\psi^0 = \psi^0(l)$. The complementary energy (6.19) is now not only a function of M but also of design variables: cross sectional stiffness $S = EI$, length l, support position s, etc. We can therefore write

$$\bar{\Pi} = \bar{\Pi}(M, S, l, s, \ldots). \tag{6.22}$$

In order to analyse the sensitivity of $\bar{\Pi}$ with respect to cross-sectional stiffness, let us express the variation of $\bar{\Pi}$

$$\delta\bar{\Pi} = \delta\bar{\Pi}_M + \delta\bar{\Pi}_s \tag{6.23}$$

where $\delta\bar{\Pi}_M$ is given by (6.20). As for the actual solution there is $\delta\bar{\Pi}_M = 0$, the total variation is

$$\delta\bar{\Pi} = \delta\bar{\Pi}_s = -\int_0^l \frac{M^2\delta S}{2S^2}\,\mathrm{d}x - \int_0^l \frac{M^2}{2(EI)^2}\delta(EI)\,\mathrm{d}x \tag{6.24}$$

The sensitivity variation thus equals the integral of the derivative of the specific stress energy with respect to cross-sectional stiffness $S = EI$. Similarly, considering the potential energy $\Pi(M, S, l, s, \ldots)$, we obtain

$$\delta\bar{\Pi} = \delta\bar{\Pi}_u = \delta\bar{\Pi}_s \tag{6.25}$$

where

$$\delta\Pi_u = \int_0^l M\delta\kappa\,\mathrm{d}x - \int_0^l p\delta w\,\mathrm{d}x = 0. \tag{6.26}$$

Since $\delta\Pi_u = 0$ in the equilibrium state, we have

$$\delta\Pi = \delta\Pi_s = -\delta\bar{\Pi}_s \tag{6.27}$$

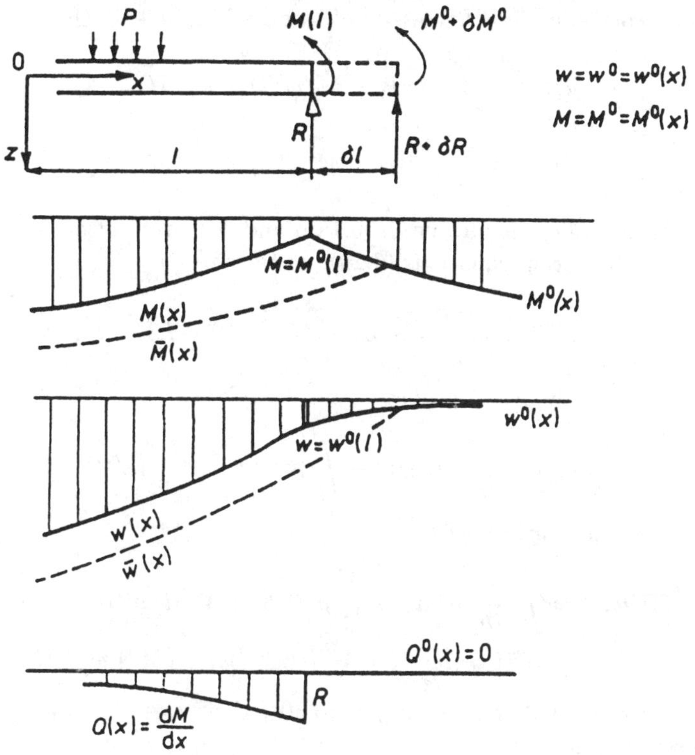

Fig. 6.1. Variation of beam length: continuation of fields of bending moment $M^0 = M^0(x)$, deflection $w^0 = w^0(x)$, and shear force $Q^0 = Q^0(x)$

Consider now the beam length variation. Referring to Fig. 6.1, assume that at $x = l$ the bending moment $M^0(l)$ and the deflection $w^0(l)$ are specified. The boundary values are continued in the vicinity of the beam end, so that field of boundary data $w^0(x)$ and $M^0(x)$ is specified within the interval of admissible length variation. The actual state fields $w(x)$ and $M(x)$ are specified within the beam length $0 \le x \le l$ and can be continued analytically beyond the actual length l. Obviously, for $x = l$, there is $M(l) = M^0(l)$, $w(l) = w^0(l)$ that is there is a continuity of state and boundary data fields.

Let the beam be transformed by moving its end $x = l$ through the infinitesimal distance δl. The transformation field $\delta\varphi(x)$ can be constructed within the beam span by setting $\delta x^* = x + \delta\varphi(x)$, where $\delta\varphi(x)$ is a continuous function satisfying the condition $\delta\varphi = 0$ at $x = 0$ and $\delta\varphi = \delta l$ for $x = l$. The respective variations δM and δw at $x = l$ are

$$\delta M = \delta\bar{M} + M'\delta l, \qquad \delta w = \delta\bar{w} + w'\delta l, \qquad \delta w' = \delta\bar{w}' + w''\delta l, \qquad (6.28)$$

where $\delta\bar{M}$, $\delta\bar{w}$ and $\delta\bar{w}'$ are variations at fixed material points. Let us write the virtual work equations for beams of lengths l and $l + \delta l$, thus

$$\begin{aligned} \int_0^l M\kappa \,\mathrm{d}x &= \int_0^l pw \,\mathrm{d}x - M^0 w'(l) - Rw^0(l) \\ \int_0^l M\bar{\kappa} \,\mathrm{d}x &= \int_0^l p\bar{w} \,\mathrm{d}x - M^0 \bar{w}'(l) - R\bar{w}^0(l) \end{aligned} \tag{6.29}$$

where $\bar{w}(x)$ and $\bar{\kappa}(x)$ denote the deflection and curvature after beam transformation. Subtracting equations (6.29), we obtain

$$\int_0^l M\delta\bar{\kappa} \,\mathrm{d}x = \int_0^l p\delta\bar{w} \,\mathrm{d}x - M^0 \delta\bar{w}'(l) - R\delta\bar{w}^0(l) \,. \tag{6.30}$$

The potential energy of the beam

$$\Pi(w, l) = \int_0^l U(\kappa) \,\mathrm{d}x - \int_0^l pw \,\mathrm{d}x + M^0(l) w'(l) \tag{6.31}$$

now undergoes the variation

$$\begin{aligned} \delta\Pi(w, l) = {} & \int_0^l \frac{\partial U}{\partial \kappa} \delta\bar{\kappa} \,\mathrm{d}x - \int_0^l p\delta\bar{w} \,\mathrm{d}x + M^0(l)\delta\bar{w}'(l) \\ & + (U - pw)\delta l + M^0(l) w''(l)\delta l + (M^0)' w'(l)\delta l . \end{aligned} \tag{6.32}$$

In view of the virtual work equation (6.30), the expression (6.32) can be presented as follows

$$\delta\Pi = -R\,\delta\bar{w}(l) + \left[U - pw + M^0(l) w''(l) + (M^0)' w'(l)\right] \delta l \,. \tag{6.33}$$

Note that $w^0(l+\delta l) = w(l) + \delta\bar{w} + w'(l)\delta l$. As the fields of state and boundary data are continuous at $x^* = l + \delta l$, the following equalities hold

$$\begin{aligned} \delta\bar{w}(l) &= \left\{[w^0(l)]' - w'(l)\right\} \delta l \,, \\ \delta\bar{M}(l) &= \left\{[M^0(l)]' - M'(l)\right\} \delta l \,. \end{aligned} \tag{6.34}$$

It is assumed that $Q^0(x) = 0$ for $x > l$, hence $R = Q^0 - Q = -Q = -M'$, where R is the support reaction at the beam end. In view of these relations, the expression (6.33) is finally transformed as follows

$$\begin{aligned} \delta\Pi(w, l) &= \left\{-R[(w^0)' - w'] - U - pw\right\} \delta l \\ &= \left\{M'[(w^0)' - w'] - U - pw\right\} \delta l \,. \end{aligned} \tag{6.35}$$

When the beam end is simply supported, Fig. 6.2, we obtain

$$M^0(l) = 0, \qquad Q^0(l) = 0, \qquad R = -Q = -M', \qquad w^0 = (w^0)' = 0 \tag{6.36}$$

and the sensitivity variation (6.35) is

$$\delta\Pi(w, l) = \left[-Rw' - \frac{1}{2} M\kappa - pw\right] \delta l. \tag{6.37}$$

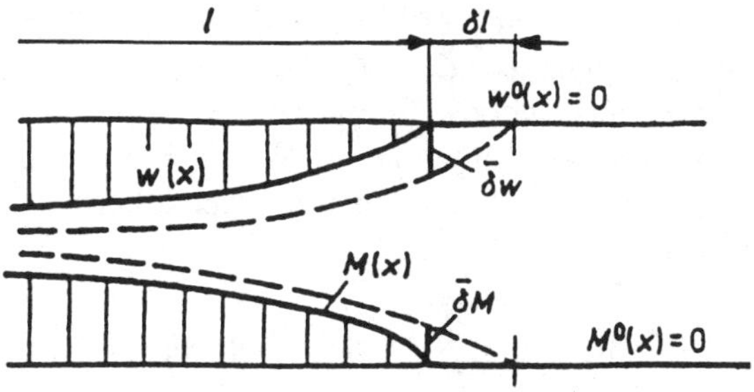

Fig. 6.2. Continuation of deflection of bending moment fields along the beam

For a built-in end, the following relations occur at $x = l$

$$w^0 = (w^0)' = 0, \qquad w = w' = 0, \qquad \delta\bar{w} = -w'\delta l, \qquad \delta\bar{w}' = -w''\delta l, \tag{6.38}$$

and

$$\delta\Pi(w,l) = -[U + pw]\delta l = -\left[\frac{1}{2}M\kappa + pw\right]\delta l\,. \tag{6.39}$$

If there is no loading at the beam end, $p(l) = 0$, then from (6.39) it follows that $\delta\Pi = -U\delta l < 0$. Thus, the global beam stiffness decreases for increasing beam length.

Similar formulae can be derived by considering the complementary energy

$$\bar{\Pi}(M,l) = \int_0^l \frac{M^2}{2EI}\,\mathrm{d}x + Rw^0(l) = \int_0^l W(M)\,\mathrm{d}x + Rw^0(l) \tag{6.40}$$

and its variation

$$\delta\bar{\Pi}(M,l) = \int_0^l \kappa\,\delta\bar{M}\,\mathrm{d}x + W(M)\delta l + \delta Rw^0(l) + R(w^0)'\delta l\,. \tag{6.41}$$

Writing the virtual work equation for local stress variation

$$\int_0^l \kappa\,\delta\bar{M}\,\mathrm{d}x = -\delta\bar{M}w'(l) - \delta\bar{R}w^0(l) \tag{6.42}$$

and using the relations

$$\begin{aligned} \delta\bar{M} &= \left[(M^0)' - M'\right]\delta l, \\ \delta\bar{R} = -\delta\bar{M}' &= \left[M'' - (M^0)''\right]\delta l = \left[p + (M^0)''\right]\delta l \end{aligned} \tag{6.43}$$

we obtain

$$\delta\bar{\Pi}(M,l) = \left\{R\left[(w^0)' - w'\right] + W(M) + pw\right\}\delta l\,. \tag{6.44}$$

It is seen that $\Pi + \bar{\Pi} = 0$, $\delta\Pi + \delta\bar{\Pi} = 0$.

The present analysis can be applied for the case of variation of other design parameters. Consider, for instance, the variation of a rigid support position from $x = s$ to $x^* = s + \delta s$. Denoting the shear forces at both sides of support

$x = s^+$ and $x = s^-$ by Q^+ and Q^-, where $Q^+ = (M^+)'$ and $Q^- = (M^-)'$ the support reaction equals $R = Q^+ - Q^-$. The shear forces and the bending moment derivatives thus undergo jumps at $x = s$, but the deflection is a continuous and differentiable function $w(s^+) = w(s^-) = 0$, $w(s^+)' = w(s^-)'$ for $x = s$. For a simply supported or built-in beam at both ends, the potential energy and its variation are

$$\Pi(w, s) = \int_0^l U(\kappa)\,\mathrm{d}x - \int_0^l pw\,\mathrm{d}x \tag{6.45}$$

and

$$\delta\Pi(w, s) = \int_0^l M\,\delta\bar{\kappa}\,\mathrm{d}x - \int_0^l p\,\delta\bar{w}\,\mathrm{d}x = -R\,\delta\bar{w} = Rw'\,\delta s \tag{6.46}$$

since $\delta w(s+\delta s) = \delta\bar{w}(s) + w'(s)\delta s = 0$ and $\delta\bar{w}(s) = -w'(s)\delta s$. Similarly, the variation of the complementary energy equals

$$\bar{\Pi} = \int_0^l W(M)\,\mathrm{d}x, \qquad \delta\bar{\Pi} = -\int_0^l \kappa\,\delta\bar{M}\,\mathrm{d}x = -Rw'\,\delta s. \tag{6.47}$$

In the case of a segmented beam with segments of varying thickness, the design parameters can be constituted by segment positions $l_1, l_2, \ldots, l_m$. Consider a typical segment interface at $x = l$ and its variation $x^* = l + \delta l$. The beam stiffness on both sides of interface is $S^+ = (EI)^+$ and $S^- = (EI)^-$. Since $M^+ = M^-$ for $x = l^+$ and $x = l^-$, then the beam curvature and the specific energy suffer discontinuity at $x = l$, thus

$$\begin{aligned} \kappa^+ &= \frac{M}{(EI)^+}, \quad & U^+ &= \frac{1}{2}(EI)^+(\kappa^+)^2 = \frac{M^2}{2(EI)^+}, \\ \kappa^- &= \frac{M}{(EI)^-}, \quad & U^- &= \frac{1}{2}(EI)^-(\kappa^-)^2 = \frac{M^2}{2(EI)^-}, \end{aligned} \tag{6.48}$$

The variation of the potential energy now equals

$$\delta\Pi(w, l) = [U^-(l) - U^+(l) - M(\kappa^+ - \kappa^-)]\,\delta l = \left[\frac{M^2}{2(EI)^+} - \frac{M^2}{2(EI)^-}\right]\delta l\,. \tag{6.49}$$

This formula can be extended for the case of simultaneous variation of several segment lengths and positions. If a rigid support is placed at segment connection and moves with the interface, the formulae (6.47) and (6.49) can be combined to provide

$$\delta\Pi(w, l) = [U^-(l) - U^+(l) + Rw'(l)]\,\delta l\,. \tag{6.50}$$

When a hinge exists at $x = r$ in the beam, a finite rotation $\theta = (w')^- - (w')^+$ occurs at the hinge cross-section. For an ideal hinge the bending moment vanishes, $M(r) = 0$. When the hinge is translated through the distance δr,

Table 6.1. Formulae for sensitivities of potential and complementary energies for beam

	Sensitivity of potential and complementary energies	Beam transformation
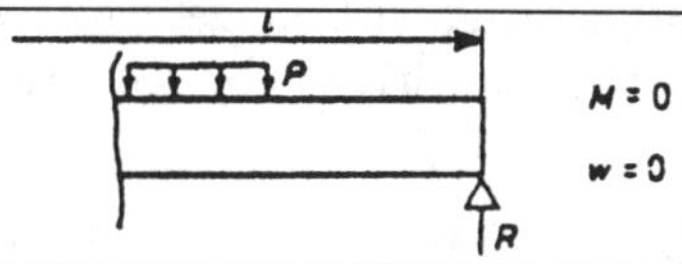	$\delta\Pi^p = Rw'\delta l$ $\delta\Pi^c = -\delta\Pi^p$	translation of simply supported end
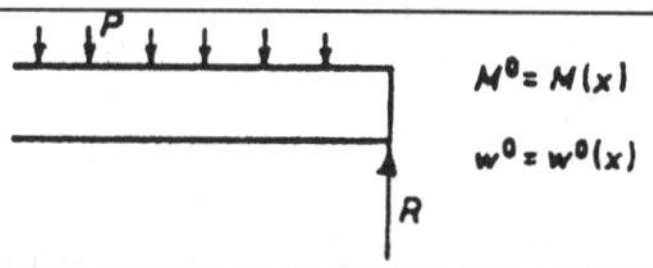	$\delta\Pi^p = \{-R\left[(w^0)' - w'\right] + U(\kappa) + pw\}_{x=l} \times \delta l$	translation of loaded and supported end on curvilinear support
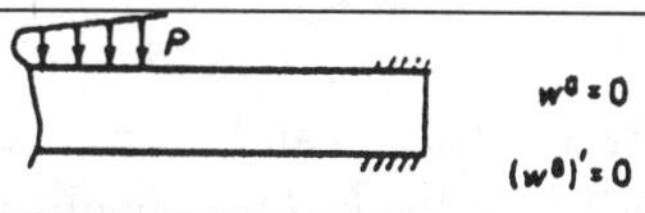	$\delta\Pi^p = -U(\kappa)\,\delta l < 0$ $\delta\Pi^c = W(M)\,\delta l > 0$	translation of built-in end
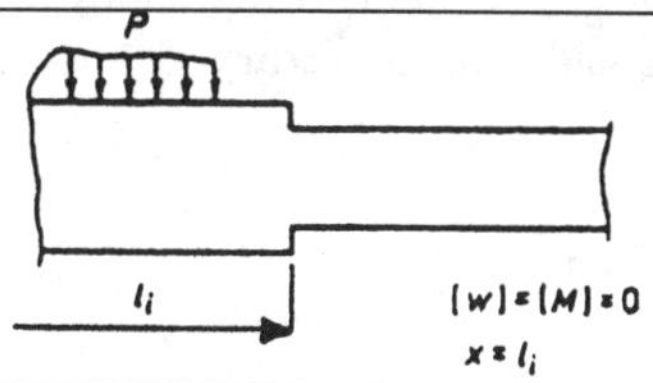	$\delta\Pi^p = (U^+ - U^-)\delta l_i = \left[\frac{M^2}{2(EI)^+} - \frac{M^2}{2(EI)^-}\right]_{x=l} \delta l_i$	translation of segment interface
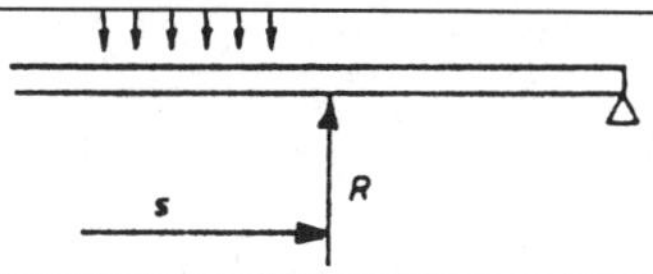	$\delta\Pi^p = Rw'\delta s$ $\delta\Pi^c = -Rw'\delta s$	translation of rigid support
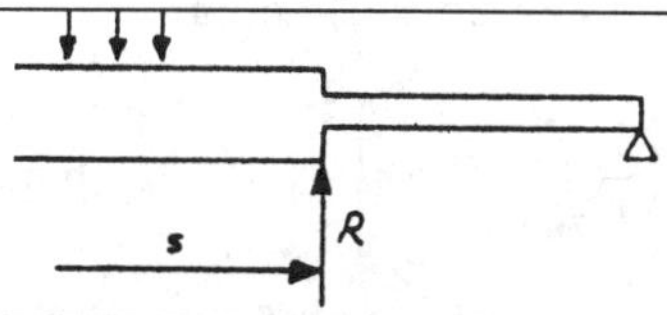	$\delta\Pi^p = [U^+ - U^- + Rw']_{x=s}\,\delta s$	translation of rigid support and segment interface
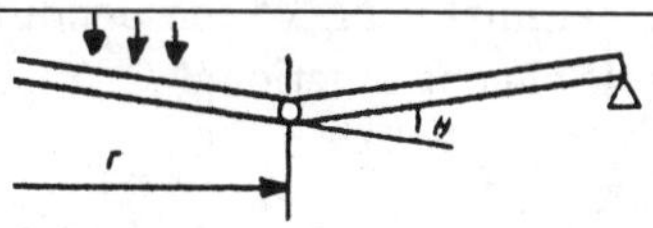	$\delta\Pi^p = (Q\theta)_{x=r}\,\delta r$	hinge translation
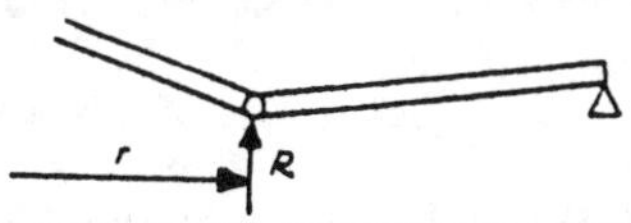	$\delta\Pi^p = [Q^-(w')^- - Q^+(w')^+]_{x=r}\,\delta r$	hinge and rigid support translation

the variation of the potential energy equals

$$\delta\Pi(w,r) = \int_0^l M\,\delta\bar{\kappa}\,\mathrm{d}x - \int_0^l p\,\delta\bar{w}\,\mathrm{d}x = -M'\theta\,\delta r = Q\theta\,\delta r\,. \tag{6.51}$$

In fact, the bending moment vanishes, $M(r) = 0$, and we have $M(r+\delta r) \approx M'\delta r$, so the term on the right-hand side of (6.51) expresses the work of the moment $M(r+\delta r)$ on the finite rotation.

If the hinge is supported by a rigid support moving with hinge, we have $R = Q^+(r) - Q^-(r)$, $\theta = w'(r)^+ - w'(r)^-$. Writing the virtual work equation for both beam portions, we obtain

$$\delta\Pi(w,r) = [Q^-(w')^- - Q^+(w')^+]\,\delta r \tag{6.52}$$

The results obtained are summarized in Table 6.1

6.3 Variation of Arbitrary Functionals

Similarly as in the general case, the sensitivity variations can be expressed explicitly by introducing the adjoint system for which the loading conditions are specified by the sensitivity functional. Consider an arbitrary functional of the bending moment field

$$G(M) = \int_0^l \phi(M,S)\,\mathrm{d}x \tag{6.53}$$

or the displacement functional

$$H(u) = \int_0^l \psi(u,S)\,\mathrm{d}x \tag{6.54}$$

where $S = EI$. In particular, the following form can be considered

$$G(M) = \left[\frac{1}{l}\int_0^l \phi^p(M)\,\mathrm{d}x\right]^{\frac{1}{p}} = \left[\frac{1}{l}\int_0^l M^p\,\mathrm{d}x\right]^{\frac{1}{p}}. \tag{6.55}$$

For $p \to \infty$ the functional $G(M)$ tends to the maximal value of the bending moment, $G(M) \to \max|M|$. Similarly, the displacement functional

$$H(u) = \left[\int_0^l u^p\,\mathrm{d}x\right]^{\frac{1}{p}}. \tag{6.56}$$

tends to the maximum of deflection for $p \to \infty$.

Consider now the adjoint beam of the same support conditions but with induced initial curvature distortions

$$\kappa^i(x) = \frac{\partial\phi(M)}{\partial M}\,. \tag{6.57}$$

This field will induce the deflection $w^i(x)$ which in general will not satisfy boundary conditions. In a statically indeterminate structure there will be self-equilibrated moment, curvature, and deflection fields $M^r(x)$, $\kappa^r(x)$ and $w^r(x)$, so that the adjoint deflection field

$$w^a(x) = w^i(x) + w^r(x) \tag{6.58}$$

will satisfy the support conditions. Denote by $M^a(x)$, $\kappa^a(x)$ and $w^a(x)$ the state fields of the adjoint structure, so that

$$\kappa^a = \kappa^i + \kappa^r, \qquad M^a = M^r = S(\kappa^a - \kappa^i) = S\kappa^r. \tag{6.59}$$

Let us analyze the variation of functional (6.53) due to the internal support variation. We have

$$\delta G = \int_0^l \frac{\partial \phi}{\partial M}\,\delta M\,\mathrm{d}x = \int_0^l \kappa^i\,\delta M\,\mathrm{d}x = \int_0^l (\kappa^a - \kappa^r)\,\delta M\,\mathrm{d}x = \int_0^l \kappa^a\,\delta M\,\mathrm{d}x \tag{6.60}$$

since

$$\int_0^l \kappa^r\,\delta M\,\mathrm{d}x = \int_0^l M^r\,\delta\kappa\,\mathrm{d}x = 0 \tag{6.61}$$

and the field $M^r(x)$ is self-equilibrated. Considering the support position at $x = s$ and $x = s + \delta s$, we can write the virtual work equation

$$\delta G = \int_0^l \kappa^a\,\delta M\,\mathrm{d}x = -w^a\delta R - w\delta R^a - [R(w^a)' + R^a w']\,\delta s\,. \tag{6.62}$$

The first term on the right-hand side corresponds to variation of the support reaction R; the second corresponds to variation of the support reaction R^a in the adjoint beam, the third term corresponds to support translation. Setting $R^a = 0$, the optimal position of support is specified by the conditions

$$w^a = (w^a)' = 0 \qquad \text{at } x = s \tag{6.63}$$

requiring vanishing deflection and slope in the adjoint structure. Some optimal support problems were treated in [36] for beams and arches.

Similarly, considering the functional (6.54), let us apply the lateral loading

$$p^a = \frac{\partial \psi(u)}{\partial u} \tag{6.64}$$

in the adjoint structure. The variation of $H(u)$ due to rigid support translation is expressed as follows

$$\begin{aligned}\delta H &= \int_0^l \frac{\partial H}{\partial u}\,\delta u\,\mathrm{d}x = \int_0^l p^a \delta u\,\mathrm{d}x = \int_0^l M^a\,\delta\kappa\,\mathrm{d}x = \int_0^l \kappa^a\,\delta M\,\mathrm{d}x \\ &= -(w^a\delta R + w\delta R^a) - [R(w^a)' + R^a w']\,\delta s\end{aligned} \tag{6.65}$$

that is by the same formula as that given by Eq. (6.62). In particular, when the local deflection variation at A is considered, the adjoint structure is loaded

Table 6.2. Sensitivity variations for arbitrary stress and deflection functionals

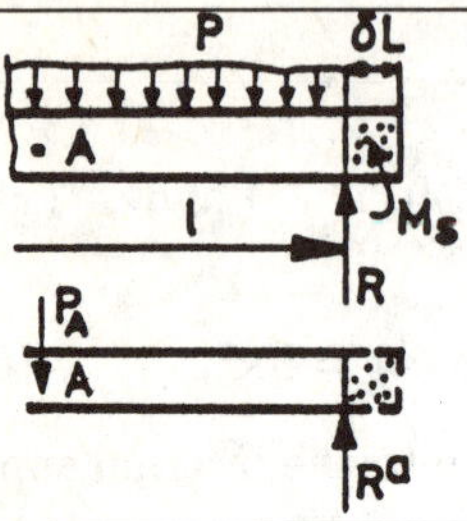	$P^a\delta w_A = [-R^a w' - R(w^a)' + pw^a + M\kappa^a]_{x=l}\,\delta l = H_l\delta l$	translation of simply supported boundary
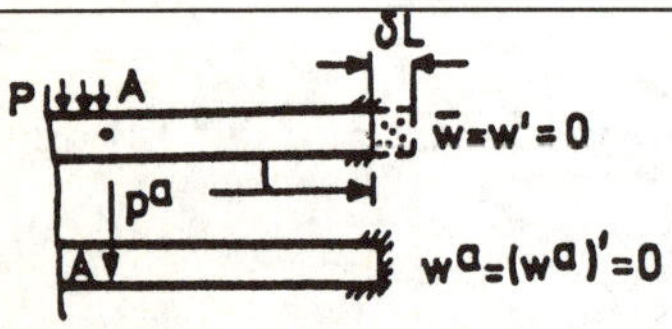	$P^a\delta w_A = 2U^m(l)\delta l = M\kappa^a\,\delta l = H_l\delta l$	translation of built-in boundary
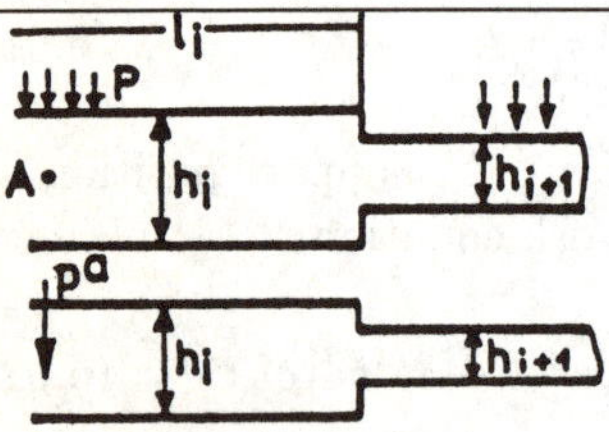	$P^a\delta w_A = 2[(U^m)^+ - (U^m)^-]\,\delta l_i = [M^a\kappa^+ - M^a\kappa^-]_{x=l_i}\,\delta l_i$	translation of interface S_i between two segments of different stiffness
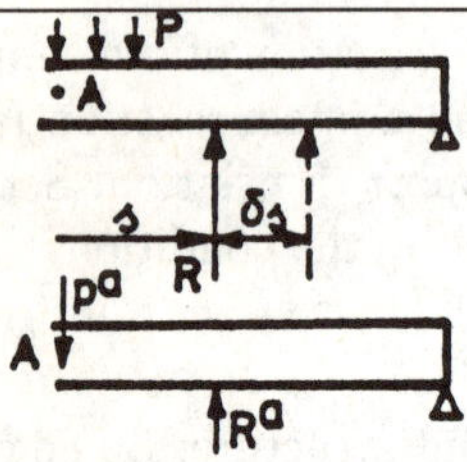	$P^a\delta w_A = -[R^a w' + R(w^a)']_{x=s}\,\delta s$	translation of internal rigid support
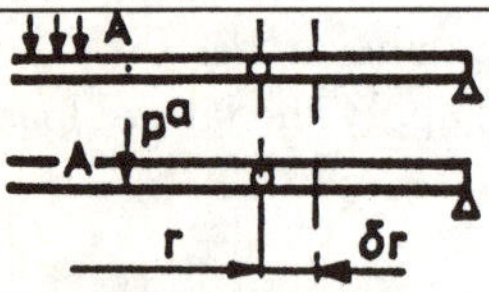	$P^a\delta w_A = (Q^a\theta + Q\theta^a)_{x=r}\,\delta r$ $\theta = (w')^- - (w')^+$ $\theta^a = (w^{a\prime})^- - (w^{a\prime})^+$	translation of hinge
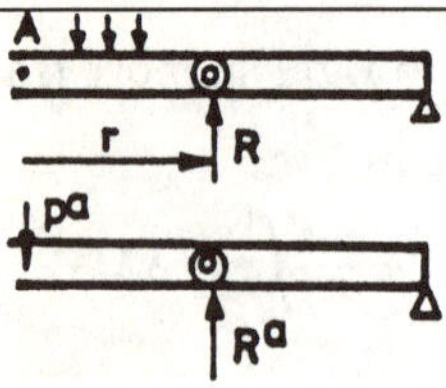	$P^a\delta w_A = [(Q^a)^-(w')^- + Q^-(w^{a\prime})^- - (Q^a)^+(w')^+ - Q^+(w^{a\prime})^+]_{x=r}\,\delta r$	translation of supported hinge

by a concentrated force P^a at A. We then have

$$p^a(\mathrm{A}) = \int_0^l M^a \kappa \,\mathrm{d}x - R^a w(s) \tag{6.66}$$

and

$$\begin{aligned} P^a \delta w(\mathrm{A}) &= \int_0^l M^a \,\delta\kappa \,\mathrm{d}x - \delta R^a w(s) - R^a w'(s)\delta s \\ &= -(w^a \delta R + w \delta R^a) - [R(w^a)' + R^a w']\,\delta s\,. \end{aligned} \tag{6.67}$$

Table 6.2 presents the sensitivity expressions for various design variations in the beam.

7 Sensitivity Analysis in Static and Dynamic Thermo-elasticity

The present chapter is devoted to problems of sensitivity analysis of structures dynamically loaded and subjected to temperature field. Consider the thermally isotropic and physically non-linear body B occupying the domain of volume V with a regular boundary S, Fig. 7.1. On the boundary portion S_q the heat flux q^0 is specified and on the portion S_ϑ the temperature ϑ^0 is prescribed. The surface tractions $\mathbf{T}^0$ are specified on the boundary portion S_T, and displacements $\mathbf{u}^0$ on the boundary portion S_u. In addition, the body force field $\mathbf{f}(\mathbf{x})$ and the heat source field $Q(\mathbf{x})$ are prescribed within the body domain.

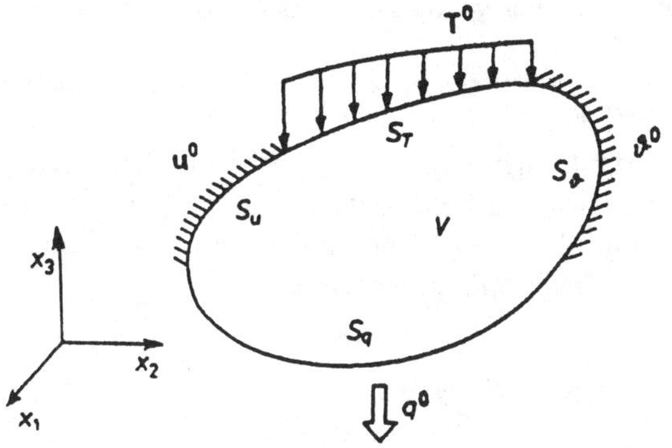

Fig. 7.1. Primary body under thermomechanic loading

Assume, similarly as previously, that both thermal and mechanical material properties depend on a set of material parameters a_k, the variation of body shape is specified by the transformation field $\varphi(\mathbf{x}, b_p)$ defined within V and its exterior, with dependence on shape parameters b_p. The temperature, displacement, strain and stress fields are denoted by ϑ, $\mathbf{u}$, ε and σ.

7.1 Static Sensitivity Analysis

Assuming that temperature, displacement, strain and stress fields are steady, the thermo-mechanical state is specified by heat conduction and equilibrium equations. The heat conduction equation has the form

$$\lambda\vartheta_{,ii} + Q = 0 \tag{7.1}$$

where $\lambda(a_k)$ denotes the heat conductivity coefficient. The following boundary conditions are specified on S_q and S_ϑ , namely

$$-\lambda\vartheta_{,i}n_i = q^0 \quad \text{on } S_q\,, \qquad \vartheta = \vartheta^0 \quad \text{on } S_\vartheta\,. \tag{7.2}$$

The stress and strain fields satisfy the equilibrium and geometric compatibility conditions

$$\sigma_{ij,i} + f_i = 0\,, \qquad \varepsilon_{ij} = \frac{1}{2}(u_{i,j} + u_{j,i}) \tag{7.3}$$

and the boundary conditions are

$$\sigma_{ij}n_i = T_j^0 \quad \text{on } S_T\,, \qquad u_j = u_j^0 \quad \text{on } S_u\,. \tag{7.4}$$

The nonlinear stress–strain relation is assumed in a form

$$\sigma_{ij} = S_{ij}(\varepsilon_{lm}, \vartheta, a_k). \tag{7.5}$$

It is also assumed that the heat source and body force fields Q and $\mathbf{f}$ can depend on the set of material design variables a_k .

Consider now an arbitrary functional specified in the form

$$G = \int_{V(b_p)} \Psi(\boldsymbol{\sigma}, \boldsymbol{\varepsilon}, \mathbf{u}, \vartheta, a_k)\,\mathrm{d}V + \int_{S(b_p)} h(\mathbf{T}, \mathbf{u}, q, \vartheta)\,\mathrm{d}S \tag{7.6}$$

and determine its first variation with respect to design variables a_k and b_p . The sensitivity analysis will be discussed separately for material and shape parameter variation. Both, direct and adjoint state methods will be considered in deriving the functional sensitivities.

7.1.1 Material Parameter Sensitivity

Assuming that only the material (or cross-sectional parameters) can vary, the first variation of the functional (7.6) can be written as follows

$$\delta G = \Bigg[\int (\Psi_{,\boldsymbol{\sigma}} \cdot \boldsymbol{\sigma}_{,a_k} + \Psi_{,\boldsymbol{\varepsilon}} \cdot \boldsymbol{\varepsilon}_{,a_k} + \Psi_{,\mathbf{u}} \cdot \mathbf{u}_{,a_k} + \Psi_{,\vartheta}\vartheta_{,a_k} + \Psi_{,a_k})\,\mathrm{d}V$$
$$+ \int (h_{,\mathbf{T}} \cdot \mathbf{T}_{,a_k} + h_{,\mathbf{u}} \cdot \mathbf{u}_{,a_k} + h_{,q}q_{,a_k} + h_{,\vartheta}\vartheta_{,a_k})\,\mathrm{d}S\Bigg]\,\delta a_k\,. \tag{7.7}$$

In order to specify (7.7) by the direct method, the state field sensitivities with respect to material parameters should be determined. These sensitivities can be obtained by solving K additional boundary-value problems. The governing equations for these problems are obtained by differentiating the state equations and boundary conditions with respect to material parameters. Denote by ϑ^k, $\mathbf{q}^k$ and $\mathbf{u}^k$, $\boldsymbol{\varepsilon}^k$, $\boldsymbol{\sigma}^k$, $\mathbf{T}^k$ the sensitivities $\vartheta_{,a_k}$, $\mathbf{q}_{,a_k}$ and $\mathbf{u}_{,a_k}$, $\boldsymbol{\varepsilon}_{,a_k}$, $\boldsymbol{\sigma}_{,a_k}$, $\mathbf{T}_{,a_k}$. Differentiating the heat conduction equation (7.1) and the boundary conditions (7.2) with respect to the parameter a_k, the following set of thermal state sensitivity equations is obtained

$$
\begin{aligned}
\lambda \vartheta^k_{,ii} + Q^k &= 0 && \text{in } V, \\
-\lambda \vartheta^k_{,i} n_i = q^{0k} &= -\frac{1}{\lambda}\lambda_{,a_k} q^0 && \text{on } S_q, \\
\vartheta^k &= 0 && \text{on } S_\vartheta .
\end{aligned} \tag{7.8}
$$

where the heat source field is

$$
Q^k = Q_{,a_k} - \frac{1}{\lambda}\lambda_{,a_k} Q \quad \text{in } V. \tag{7.9}
$$

Next, differentiating equilibrium equations and boundary conditions (7.3) and (7.4), the k-th mechanical state sensitivity equations are

$$
\begin{aligned}
&\sigma^k_{ij,i} + f^k_j = 0, \qquad \varepsilon^k_{ij} = \frac{1}{2}(u^k_{i,j} + u^k_{j,i}) \qquad \text{in } V, \\
&T^{0k}_j = 0 \quad \text{on } S_T, \qquad u^{0k}_j = 0 \quad \text{on } S_u,
\end{aligned} \tag{7.10}
$$

where the body force field is

$$
f^k_j = f_{j,a_k} . \tag{7.11}
$$

The constitutive equation of the sensitivity problem is obtained by differentiating (7.5), thus

$$
\sigma^k_{ij,i} = S_{ij,\varepsilon_{lm}}\, \varepsilon^k_{lm} + S_{ij,\vartheta}\, \vartheta^k + S_{ij,a_k} . \tag{7.12}
$$

Let us note that the relations (7.12) are linear and the last two terms on the right hand side can be regarded as the specified field of initial stresses. Generating solutions of the boundary-value problems (7.8)–(7.9) and (7.10)–(7.12) for all design parameters, the state sensitivities are determined. The functional sensitivity is then specified by (7.7). Thus, for K design parameters, the direct method requires the solutions of $K+1$ problems of heat conduction equations and $K+1$ problems of elasticity in order to specify the thermal and mechanical state sensitivities and the sensitivity of functional G. This sensitivity takes the form

$$
\begin{aligned}
\frac{\mathrm{d}G}{\mathrm{d}a_k} &= \int (\Psi_{,\sigma} \cdot \sigma^k + \Psi_{,\varepsilon} \cdot \varepsilon^k + \Psi_{,\mathbf{u}} \cdot \mathbf{u}^k + \Psi_{,\vartheta}\vartheta^k + \Psi_{,a_k})\,\mathrm{d}V \\
&+ \int h_{,\mathbf{T}} \cdot \mathbf{T}^k \,\mathrm{d}S_u + \int h_{,\mathbf{u}} \cdot \mathbf{u}^k \,\mathrm{d}S_T \\
&+ \int h_{,\vartheta}\vartheta^k \,\mathrm{d}S_q + \int h_{,q}\left(\frac{1}{\lambda}\lambda_{,a_k} q + q^k\right)\mathrm{d}S_\vartheta .
\end{aligned} \tag{7.13}
$$

The second method of sensitivity analysis, based on adjoint states requires the solution of adjoint problems of heat conduction and elasticity, thus providing the fields ϑ^a and $\mathbf{u}^a$, ε^a, σ^a. The adjoint elasticity problem is defined by the equations

$$\sigma^a_{ij,i} + f^a_j = 0, \qquad \varepsilon^a_{ij} = \frac{1}{2}(u^a_{i,j} + u^a_{j,i}) \qquad \text{in } V, \tag{7.14}$$

and the boundary conditions

$$\sigma^a_{ij} n_i = T^{a0}_j = h_{,u} \quad \text{on } S_T, \qquad u^a_j = u^{a0}_j = -h_{,T_j} \quad \text{on } S_u. \tag{7.15}$$

The adjoint body is subjected to the action of initial strains, stresses, and body forces

$$\varepsilon^{ai}_{ij} = \Psi_{,\sigma_{ij}}, \qquad \sigma^{ai}_{ij} = \Psi_{,\varepsilon_{ij}}, \qquad f^a_j = \Psi_{,u_j} \qquad \text{in } V, \tag{7.16}$$

so the constitutive equations are

$$\sigma^a_{ij} = S_{ij,\varepsilon_{lm}}(\varepsilon^a_{lm} - \varepsilon^{ai}_{lm}) - \sigma^{ai}_{ij}. \tag{7.17}$$

The adjoint problem of heat conduction is specified by the equation

$$\lambda \vartheta^a_{,ii} + Q^a = 0 \qquad \text{in } V \tag{7.18}$$

where the heat source field is expressed as follows

$$Q^a = \Psi_{,\vartheta} - (\varepsilon^a_{ij} - \Psi_{,\sigma_{ij}}) S_{ij,\vartheta} \tag{7.19}$$

and the boundary conditions are

$$\vartheta^a = \vartheta^{a0} = h_{,q} \quad \text{on } S_\vartheta, \qquad q^{a0} = -\lambda \vartheta^a_{,i} n_i = -h_{,\vartheta} \quad \text{on } S_q. \tag{7.20}$$

In view of (7.12), (7.15)–(7.16) and (7.19)–(7.20), Eq. (7.7) can be transformed as follows

$$\begin{aligned} \delta G = \Big\{ \int \big[(\mathbf{S}^T_{,\varepsilon} \cdot \boldsymbol{\varepsilon}^{ai} + \boldsymbol{\sigma}^{ai}) \cdot \boldsymbol{\varepsilon}_{,a_k} + \boldsymbol{\varepsilon}^{ai} \cdot \mathbf{S}_{,a_k} + \mathbf{f}^a \cdot \mathbf{u}_{,a_k} \\ + Q^a \vartheta_{,a_k} + \boldsymbol{\varepsilon}^a \mathbf{S}_{,\vartheta} \vartheta_{,a_k} + \Psi_{,a_k} \big] \, dV \\ - \int \mathbf{u}^{a0} \cdot \mathbf{T}_{,a_k} \, dS_u + \int \mathbf{T}^{a0} \cdot \mathbf{u}_{,a_k} \, dS_T \\ + \int \vartheta^{a0} q_{,a_k} \, dS_\vartheta - \int q^{a0} \vartheta_{,a_k}) \, dS_q \Big\} \, \delta a_k . \end{aligned} \tag{7.21}$$

provided $\mathbf{T}_{,a_k}$ vanishes on S_T, $\mathbf{u}_{,a_k}$ vanishes on S_u, and similarly $q_{,a_k}$ and $\vartheta_{,a_k}$ vanish on S_q and S_ϑ. Transforming the first and third integral of (7.21), with the use of (7.14) and (7.18), next transforming the last integral with the use of (7.12), the following expression for the sensitivity of G is obtained

$$\frac{dG}{da_k} = \int [\mathbf{u}^a \cdot \mathbf{f}_{,a} - (\boldsymbol{\varepsilon}^a - \Psi_{,\boldsymbol{\sigma}}) \cdot \mathbf{S}_{,a} - \lambda_{,a_k} \vartheta^a_{,i} \vartheta_{,i} + \vartheta^a Q_{,a} + \Psi_{,a_k}] \, dV . \tag{7.22}$$

This sensitivity is expressed in terms of integrand derivatives of Eq. (7.6) and state fields of both primary and adjoint problems of heat conduction and elasticity. Thus dG/da_k is specified in terms of two solutions of elasticity problems and two solutions of heat conduction problems.

Comparing direct and adjoint methods of sensitivity analysis, it is easy to note that for K material design parameters and m functionals for which sensitivity derivatives are to be determined, the direct method requires $2(K+1)$ solutions and the adjoint state method requires $2(m+1)$ solutions. Thus, the selection of proper method depends on the ratio m/K and the convenience of generating the respective solutions.

The methods of sensitivity analysis discussed in this section can be used to generate both sensitivity of local functions and global functionals. Consider the local constraint in a general form

$$g(\mathbf{x}_0) = g(\boldsymbol{\sigma}, \boldsymbol{\varepsilon}, \mathbf{u}, \vartheta, a_k) \qquad \text{for } \mathbf{x}_0 \in V \tag{7.23}$$

or

$$g(\mathbf{x}_0) = g(\mathbf{T}, \mathbf{u}, q, \vartheta) \qquad \text{for } \mathbf{x}_0 \in S. \tag{7.24}$$

The first variation of (7.23) can be expressed as follows

$$\delta g = (g_{,\boldsymbol{\sigma}} \cdot \boldsymbol{\sigma}_{,a_k} + g_{,\boldsymbol{\varepsilon}} \cdot \boldsymbol{\varepsilon}_{,a_k} + g_{,\mathbf{u}} \cdot \mathbf{u}_{,a_k} + g_{,\vartheta}\vartheta_{,a_k} + g_{,a_k})\delta a_k \tag{7.25}$$

Using the direct method, the state sensitivities are calculated and Eq. (7.25) is presented in the form

$$\delta g = (g_{,\boldsymbol{\sigma}} \cdot \boldsymbol{\sigma}^k + g_{,\boldsymbol{\varepsilon}} \cdot \boldsymbol{\varepsilon}^k + g_{,\mathbf{u}} \cdot \mathbf{u}^k + g_{,\vartheta}\vartheta^k + g_{,a_k})|_{\mathbf{x}=\mathbf{x}_0}\, \delta a_k \tag{7.26}$$

When the number of constraints (7.23) or (7.24) is small, the adjoint method is more effective. Replacing, for instance, (7.23) by its equivalent global form

$$G = \int g(\boldsymbol{\sigma}, \boldsymbol{\varepsilon}, \mathbf{u}, \vartheta, a_k)\, \delta(\mathbf{x} - \mathbf{x}_0)\, \mathrm{d}V \tag{7.27}$$

where $\delta(\mathbf{x} - \mathbf{x}_0)$ is the Dirac function, it is easy to see that

$$\Psi = g\delta(\mathbf{x} - \mathbf{x}_0) \quad \text{in } V, \qquad h = 0 \quad \text{on } S, \tag{7.28}$$

and the first variation of g follows from Eq. (7.22), namely

$$\delta g(\mathbf{x}_0) = \Big[(g_{,a_k} + g_{,\boldsymbol{\sigma}} \cdot \mathbf{S}_{,a_k})|_{\mathbf{x}=\mathbf{x}_0} + \int (\mathbf{u}^a \cdot \mathbf{f}_{,a_k} - \boldsymbol{\varepsilon}^a \cdot \mathbf{S}_{,a_k} - \lambda_{,a_k}\vartheta^a_{,i}\vartheta_{,i} + \vartheta^a Q_{,a_k})\, \mathrm{d}V \Big]\, \delta a_k\,. \tag{7.29}$$

Let us note that the adjoint fields are singular at the point $\mathbf{x} = \mathbf{x}_0$.

7.1.2 Sensitivity Analysis for Shape Variation

Consider now the case of shape variation of the body B specified by the transformation field $\boldsymbol{\varphi}(\mathbf{x}, b_p)$, so that $\mathbf{x}^t = \mathbf{x} + \boldsymbol{\varphi}$ and the set of shape parameters b_p $(p = 1, 2, \ldots, P)$ specifies the actual shape. Assume now that the material parameters are fixed during the transformation process.

Let us now specify the sensitivity of the functional (7.6) due to infinitesimal shape variation associated with the shape parameter variation. Introduce the shape transformation velocity field v_k^p $(k = 1,2,3)$ associated with the parameter b_p, namely

$$v_k^p = \frac{\partial \phi_k}{\partial b_p} \,. \tag{7.30}$$

Using (7.30), the material derivatives of volume and surface elements, and of the unit normal vector are

$$\begin{aligned} \frac{\mathrm{D}(\mathrm{d}V)}{\mathrm{D}b_p} &= v_{k,k}^p \,\mathrm{d}V \,, \\ \frac{\mathrm{D}(\mathrm{d}S)}{\mathrm{D}b_p} &= (\delta_{kl} - n_k n_l) v_{k,l}^p \,, \\ \frac{\mathrm{D}n_i}{\mathrm{D}b_p} &= (n_i n_l - \delta_{il}) n_k v_{k,l}^p \,, \end{aligned} \tag{7.31}$$

where δ_{kl} denotes the Kronecker symbol. If the integrands of (7.6) are continuous and differentiable ,the first variation of G can be expressed as follows

$$\delta G = \frac{\mathrm{D}G}{\mathrm{D}b_p}\,\delta b_p = \left\{ \int \left[\frac{\mathrm{D}\Psi}{\mathrm{D}b_p}\,\mathrm{d}V + \Psi\,\frac{\mathrm{D}(\mathrm{d}V)}{\mathrm{D}b_p} \right] + \int \left[\frac{\mathrm{D}h}{\mathrm{D}b_p}\,\mathrm{d}S + h\,\frac{\mathrm{D}(\mathrm{d}S)}{\mathrm{D}b_p} \right] \right\} \delta b_p \tag{7.32}$$

where $\mathrm{D}G/\mathrm{D}b_p$ denotes the sensitivity of the functional G with respect to the shape parameter b_p. Accounting for (7.31), Eq. (7.32) can be rewritten in the form

$$\begin{aligned} \delta G = \Bigg\{ \int \Bigg(& \Psi_{,\boldsymbol{\sigma}} \cdot \frac{\mathrm{D}\boldsymbol{\sigma}}{\mathrm{D}b_p} + \Psi_{,\boldsymbol{\varepsilon}} \cdot \frac{\mathrm{D}\boldsymbol{\varepsilon}}{\mathrm{D}b_p} + \Psi_{,\mathbf{u}} \cdot \frac{\mathrm{D}\mathbf{u}}{\mathrm{D}b_p} + \Psi_{,\vartheta}\,\frac{\mathrm{D}\vartheta}{\mathrm{D}b_p} + \Psi v_{k,k}^p \Bigg) \mathrm{d}V \\ & + \int \Bigg[h_{,\mathbf{T}} \cdot \frac{\mathrm{D}\mathbf{T}}{\mathrm{D}b_p} + h_{,\mathbf{u}} \cdot \frac{\mathrm{D}\mathbf{u}}{\mathrm{D}b_p} + h_{,q}\,\frac{\mathrm{D}q}{\mathrm{D}b_p} + h_{,\vartheta}\,\frac{\mathrm{D}\vartheta}{\mathrm{D}b_p} \\ & \qquad + h\,(\delta_{kl} - n_k n_l) v_{k,l}^p \Bigg] \mathrm{d}S \Bigg\} \delta b_p \end{aligned} \tag{7.33}$$

When the direct method is used, the sensitivities of state fields occurring in (7.33) are to be determined. The material derivatives of state equations (7.1)–(7.5) with respect to b_p are to be stated and solved. Denoting by $(\cdot)^p$ the sensitivities $\mathrm{D}(\cdot)/\mathrm{D}b_p$, from (7.1) and (7.2) the following thermal sensitivity equations are obtained

$$\begin{aligned} \lambda \vartheta^p_{,ii} + Q^p &= 0 && \text{in } V, \\ -\lambda \vartheta^p_{,i} n_i = q^{0p} &= \frac{\mathrm{D}q^0}{\mathrm{D}b_p} - q^0 n_k n_i v_{k,i}^p \\ &\quad - \lambda(\vartheta_{,k} n_i + \vartheta_{,i} n_k) v_{k,i}^p && \text{on } S_q \,, \\ \vartheta^p = \vartheta^{0p} &= \frac{\mathrm{D}\vartheta^0}{\mathrm{D}b_p} && \text{on } S_\vartheta \,. \end{aligned} \tag{7.34}$$

where the heat source field for the problem associated with variation of b_p is expressed in the form

$$Q^p = \frac{DQ}{Db_p} - \lambda(2\vartheta_{,ki}v^p_{k,i} + \vartheta_{,k}v^p_{k,ii}) \,. \tag{7.35}$$

Similarly, differentiating (7.3)–(7.5), the sensitivity equations for mechanical fields are

$$\begin{gathered} \sigma^p_{ij,i} + f^p_j = 0, \qquad \varepsilon^p_{ij} = \frac{1}{2}(u^p_{i,j} + u^p_{j,i}) \qquad \text{in } V, \\ u^p_j = u^{0p}_j = \frac{Du^0_j}{Db_p} \qquad \text{on } S_u \,, \\ \sigma^p_{ij}n_i = T^{0p}_j = \frac{DT^0_j}{Db_p} - \sigma_{ij}(n_i n_l - \delta il)n_k v^p_{k,l} \qquad \text{on } S_\vartheta \,, \\ \sigma^p_{ij} = S_{ij,\varepsilon_{lm}}(\varepsilon^p_{lm} - \varepsilon^{ip}_{lm}) + S_{ij,\vartheta}\vartheta^p \,, \end{gathered} \tag{7.36}$$

where the fields of body forces and initial strains are specified by the relations

$$f^p_j = \frac{Df_j}{Db_p} - \sigma_{ij,k}v^p_{k,i} \,, \qquad \varepsilon^{ip}_{ij} = \frac{1}{2}(u_{i,k}v^p_{k,j} + u_{j,k}v^p_{k,i}) \,. \tag{7.37}$$

The elastic sensitivity equations are linear, though the primary problem may be non-linear in view of physical non-linearity. The elasticity matrix of the sensitivity problem equals the tangential matrix of the primary problem. The last term on the right hand side of fourth equation (7.36) can be regarded as the initial stress field. The solutions of boundary value problems (7.36) and (7.37) provide the state field sensitivities, so that

$$\frac{D\sigma}{Db_p} = \sigma^p, \qquad \frac{D\varepsilon}{Db_p} = \varepsilon^p - \varepsilon^{ip}, \qquad \frac{D\mathbf{u}}{Db_p} = \mathbf{u}^p, \qquad \frac{D\mathbf{T}}{Db_p} = \mathbf{T}^p. \tag{7.38}$$

Solving equations (7.34)–(7.37) for all values of p, the sensitivity of the functional G is specified by the equation (7.33) in the form

$$\begin{aligned} \frac{DG}{Db_p} = &\int \left[\Psi_{,\sigma}\cdot\sigma^p + \Psi_{,\varepsilon}\cdot(\varepsilon^p - \varepsilon^{ip}) + \Psi_{,\mathbf{u}}\cdot\mathbf{u}^p + \Psi_{,\vartheta}\,\vartheta^p + \Psi v^p_{k,k}\right] dV \\ &+ \int h\,(\delta_{kl} - n_k n_l)v^p_{k,l}\,dS + \int \left(h_{,\mathbf{T}}\cdot\frac{D\mathbf{T}^0}{Db_p} + h_{,\mathbf{u}}\cdot\mathbf{u}^p\right) dS_T \\ &+ \int \left\{h_{T_i}\left[T^p_i + \sigma_{ij}(n_j n_l - \delta_{jl})n_k v^p_{k,l}\right] + h_{,\mathbf{u}}\cdot\frac{D\mathbf{u}^0}{Db_p}\right\} dS_u \\ &+ \int \left(h_{,q}\cdot\frac{Dq^0}{Db_p} + h_{,\vartheta}\,\vartheta^p\right) dS_q \\ &+ \int \left\{h_{,q}\left[q^p + q n_k n_l v^p_{k,l} + \lambda(\vartheta_{,k}n_l + \vartheta_{,l}n_k)v^p_{k,l}\right] + h_{,\vartheta}\frac{D\vartheta^0}{Db_p}\right\} dS_\vartheta \,. \end{aligned} \tag{7.39}$$

Thus, for P shape parameters, the sensitivity vector for functional G is specified by the direct method by solving $P+1$ boundary-value problems for the steady heat condition and $P+1$ problems for elastic fields.

Let us now discuss the adjoint state method. Using the formulae

$$\frac{\mathrm{D}f}{\mathrm{D}b_p} = f_{,b_p} + f_{,k}v^p_{k,i}\,, \qquad \frac{\mathrm{D}f_{,i}}{\mathrm{D}b_p} = (f_{,b_p})_{,i} + f_{,ki}v^p_{k,i}\,, \tag{7.40}$$

valid for any continuous and differentiable function $\mathbf{f}$, Eq. (7.33) can be transformed as follows

$$\begin{aligned}\delta G = \Bigg\{ &\int \left(\Psi_{,\sigma}\cdot\sigma_{,b_p} + \Psi_{,\varepsilon}\cdot\varepsilon_{,b_p} + \Psi_{,\mathbf{u}}\cdot\mathbf{u}_{,b_p} + \Psi_{,\vartheta}\,\vartheta_{,b_p}\right)\mathrm{d}V \\ &+ \int \Big\{ \Psi n_k v^p_k + h_{T_i}\left[T^p_i + \sigma_{ij,k}n_j v^p_k + \sigma_{ij}(n_j n_l - \delta_{jl})n_k v^p_{k,l}\right] \\ &+ h_{,\mathbf{u}}\cdot\left(\mathbf{u}_{,b_p} + \mathbf{u}_{,k}v^p_k\right) \\ &+ h_{,q}\left[q_{,b_p} - \lambda\vartheta_{,ik}n_i v^p_k - \lambda\vartheta_{,j}(n_j n_l - \delta_{jl})n_k v^p_{k,l}\right] \\ &+ h_{,\vartheta}\left(\vartheta_{,b_p} + \vartheta_{,k}v^p_k\right) + h\left(\delta_{kl} - n_k n_l\right)v^p_{k,l}\Big\}\,\mathrm{d}S\Bigg\}\,\delta b_p\,. \end{aligned} \tag{7.41}$$

In order to eliminate in (7.41) the local derivatives of thermal and elastic fields with respect to b_p, the adjoint problems (7.14)–(7.17) for elastic fields and (7.18)–(7.20) for thermal fields are to be solved. Using the relation

$$\sigma_{ij,b_p} = S_{ij,\varepsilon_{lm}}\,\varepsilon_{lm,b_p} + S_{ij,\vartheta}\,\vartheta_{,b_p} \tag{7.42}$$

and equations (7.14), (7.16)–(7.19), the first integral on the right side of (7.41), denoted by $\mathcal{A}$, can be presented in the form

$$\begin{aligned}\mathcal{A} &= \int \left[\left(\mathbf{S}^T_{,\varepsilon}\cdot\varepsilon^{ai} + \sigma^{ai}\right)\cdot\mathbf{S}_{,b_p} + \mathbf{f}^a\cdot\mathbf{u}_{,b_p} + Q^a\,\vartheta_{,b_p} + \varepsilon^a\cdot\mathbf{S}_{,\vartheta}\vartheta_{,b_p}\right]\mathrm{d}V \\ &= \int \left[\varepsilon^a\cdot\sigma_{,b_p} - \sigma^a\cdot\varepsilon_{,b_p} + \mathbf{f}^a\cdot\mathbf{u}_{,b_p} - \lambda\vartheta^a_{,ii}\vartheta_{,b_p}\right]\mathrm{d}V \\ &= \int \mathbf{u}^a\cdot\mathbf{f}_{,b_p}\,\mathrm{d}V + \int \left(\mathbf{u}^a\cdot\mathbf{T}_{,b_p} - \mathbf{T}^a\cdot\mathbf{u}_{,b_p}\right)\mathrm{d}S \\ &\quad + \int \lambda\vartheta^a_{,ii}\vartheta_{,b_p}\,\mathrm{d}V + \int \left(q^a\vartheta_{,b_p} - \vartheta^a q_{,b_p}\right)\mathrm{d}S\,. \end{aligned} \tag{7.43}$$

Assume for simplicity that $\mathbf{f}_{,b_p} = \mathbf{0}$ and $Q_{,b_p} = 0$, that is body forces and heat sources of the primary problem do not depend explicitly on shape parameters. Using now (7.43) in (7.41) and accounting for the boundary conditions (7.14) and (7.20) of the adjoint body, the shape sensitivity of the functional G is expressed in the form

$$\begin{aligned}\frac{\mathrm{D}G}{\mathrm{D}b_p} = &\int [\Psi - \boldsymbol{\sigma}\cdot\boldsymbol{\varepsilon}^a - \mathbf{f}\cdot\mathbf{u}^a - \lambda Q\vartheta^a + (h + \mathbf{T}\cdot\mathbf{u}^a - q\vartheta^a)_{,n} \\ &- (h + \mathbf{T}\cdot\mathbf{u}^a - q\vartheta^a)H]\, v_n^p \,\mathrm{d}S + \int \langle (h + \mathbf{T}\cdot\mathbf{u}^a - q\vartheta^a) v_v^p \rangle \,\mathrm{d}\Gamma \\ &+ \int (h_{,\mathbf{T}} + \mathbf{u}^a)\cdot \left(\frac{\mathrm{D}\mathbf{T}^0}{\mathrm{D}b_p} - \mathbf{T}_{,k} v_k^p\right) \mathrm{d}S_T \\ &+ \int (h_{,\mathbf{u}} - \mathbf{T}^a)\cdot \left(\frac{\mathrm{D}\mathbf{u}^0}{\mathrm{D}b_p} - \mathbf{u}_{,k} v_k^p\right) \mathrm{d}S_u \\ &+ \int (h_{,q} - \vartheta^a) \left(\frac{\mathrm{D}q^0}{\mathrm{D}b_p} - q_{,k} v_k^p\right) \mathrm{d}S_q \\ &+ \int (h_{,\vartheta} + q^a) \left(\frac{\mathrm{D}\vartheta^a}{\mathrm{D}b_p} - \vartheta_{,k} v_k^p\right) \mathrm{d}S_\vartheta \end{aligned} \tag{7.44}$$

where H denotes the principal curvature of the boundary surface S, $\langle\,\rangle$ denotes the discontinuity and v_n^p is the normal component of the transformation velocity field on S. Assuming further that S is composed of regular portions with Γ denoting the edge of intersection of two neighbouring portions S^+ and S^-, v_v^p being the transformation velocity component lying within the plane tangential to S^+ or S^- and normal to Γ, the second right hand integral of (7.44) specifies the sensitivity contribution along the edge of the surface S. Further, we note that the material derivatives $\mathrm{D}\mathbf{T}^0/\mathrm{D}b_p$, $\mathrm{D}\mathbf{u}^0/\mathrm{D}b_p$, $\mathrm{D}q^0/\mathrm{D}b_p$ and $\mathrm{D}\vartheta^0/\mathrm{D}b_p$ on the respective boundary portions are known and can be explicitly specified from the boundary conditions on respective portions of S.

Equations (7.39) and (7.44) specify the sensitivity of an arbitrary global functional G determined by the direct and adjoint methods. These two approaches can also be used in determination of local state sensitivities, similarly as it was demonstrated in Sect. 7.1.1.

7.2 Transient and Dynamic Sensitivity Analysis

Consider now the problem of sensitivity analysis for the case when the fields of temperature, displacement, strain and stress vary in time. Introducing, as previously, the set of parameters a_k and the set b_p specifying the body shape, we shall consider the sensitivity of an arbitrary time-dependent functional

$$\begin{aligned} G = &\int_{V(b_p)} \Psi[\mathbf{u}(\mathbf{x},t,a_k), \boldsymbol{\varepsilon}(\mathbf{x},t,a_k), \vartheta(\mathbf{x},t,a_k)]\,\mathrm{d}V \\ &+ \int_{S(b_p)} h[\mathbf{u}(\mathbf{x},t,a_k), \vartheta(\mathbf{x},t,a_k)]\,\mathrm{d}S\,. \end{aligned} \tag{7.45}$$

We shall also consider the functional F specified over body domain V and a finite time interval $\langle 0, t_f \rangle$ defined as follows

$$F = \int_0^{t_f} G \, \mathrm{d}t = \int_0^{t_f} \left[\int \Psi \, \mathrm{d}V + \int h \, \mathrm{d}S \right] \mathrm{d}t \tag{7.46}$$

where t_f denotes the terminal instant of the process.

The thermomechanic response is governed by the heat conduction equation

$$\lambda \vartheta_{,ii} + Q = c\dot{\vartheta} \tag{7.47}$$

with the boundary and initial conditions

$$\begin{aligned} \vartheta(\mathbf{x}, t) &= \vartheta^0(\mathbf{x}, t) && \text{on } S_\vartheta\,, \\ -\lambda \vartheta_{,i}(\mathbf{x}, t) n_i &= q^0(\mathbf{x}, t) && \text{on } S_q\,, \\ \vartheta(\mathbf{x}, 0) &= \vartheta_0(\mathbf{x}) && \text{in } V\,, \end{aligned} \tag{7.48}$$

and the equation of motion

$$\sigma_{ij,i} + f_j = \rho \ddot{u}_j \tag{7.49}$$

with the initial and boundary conditions

$$\begin{aligned} &\sigma_{ij}(\mathbf{x}, t) n_i = q^0(\mathbf{x}, t) && \text{on } S_T\,, \\ &u_j(\mathbf{x}, t) = u_j^0(\mathbf{x}, t) && \text{on } S_u\,, \\ &u_j(\mathbf{x}, 0) = u_{j0}(\mathbf{x}), \quad \dot{u}_j(\mathbf{x}, 0) = v_{j0}(\mathbf{x}) && \text{in } V\,, \end{aligned} \tag{7.50}$$

where $c(a_k)$ and $\rho(a_k)$ denote the specific heat and material density. The dot over the symbol denotes time differentiation. The non-linear constitutive equation is specified by (7.5).The sensitivity analysis of functionals (7.45) and (7.46) will be considered for the case of time dependent response.

7.2.1 Sensitivity Analysis with Respect to Material Parameters

Using the direct method, the variation of the functional (7.45) is expressed by (7.7), where the terms $\Psi_{,\boldsymbol{\sigma}}$, $h_{,\mathbf{T}}$ and $h_{,q}$ are deleted. In order to generate the set of equations for the sensitivities $\vartheta_{,a_k}$, $\mathbf{u}_{,a_k}$ and $\varepsilon_{,a_k}$, Eqs. (7.47)–(7.50) are differentiated with respect to a_k. A set of additional K equations of dynamic thermo-elasticity is obtained and their solutions provide the state sensitivities. Assuming the thermomechanic boundary and initial conditions to be independent of the material parameters and using the notion introduced in this chapter, the k-th problem of thermal state sensitivity is specified by the boundary conditions (7.8) and the homogeneous initial conditions $\vartheta^k(\mathbf{x}, 0)$, where ϑ^k denotes the sensitivity derivative $\vartheta_{,a_k}$. The heat source field for this problem follows from differentiation of (7.47) and is expressed as follows

$$Q^k = \lambda \left[\left(\frac{Q}{\lambda} \right)_{,a_k} - \left(\frac{c}{\lambda} \right)_{,a_k} \dot{\vartheta} \right]. \tag{7.51}$$

Similarly, differentiation of (7.49)–(7.50) provides the form of k-th dynamic problem of elasticity, satisfying homogeneous boundary and initial conditions with the body force field

$$f_j^k = f_{j,a_k} - \rho_{,a_k} \ddot{u}_j \,. \tag{7.52}$$

The constitutive equation for the sensitivity problem is provided by (7.12). Determining state sensitivities, next the sensitivity of the functional G is determined by the direct method in the form

$$\begin{aligned} \frac{\mathrm{d}G}{\mathrm{d}a_k} &= \int \left(\Psi_{,\mathbf{u}} \cdot \mathbf{u}^k + \Psi_{,\boldsymbol{\varepsilon}} \cdot \boldsymbol{\varepsilon}^k + \Psi_{,\vartheta}\, \vartheta^k + \Psi_{,a_k} \right) \mathrm{d}V \\ &\quad + \int h_{,\mathbf{u}} \cdot \mathbf{u}^k \, \mathrm{d}S_T + \int h_{,\vartheta}\, \vartheta^k \, \mathrm{d}S_q \,. \end{aligned} \tag{7.53}$$

This method requires $K+1$ solutions of non-stationary heat conduction equations and $K+1$ solutions of elastodynamics. The functional sensitivity can be calculated for any instant t from the interval $\langle 0, t_f \rangle$.

Consider next the sensitivity of functional F defined by Eq. (7.46) in a finite time interval. As δF is expressed in the form

$$\delta F = \int_0^{t_f} \delta G \, \mathrm{d}t \tag{7.54}$$

then in view of (7.53) there is

$$\begin{aligned} \frac{\mathrm{d}F}{\mathrm{d}a_k} &= \int_0^{t_f} \left[\int \left(\Psi_{,\mathbf{u}} \cdot \mathbf{u}^k + \Psi_{,\boldsymbol{\varepsilon}} \cdot \boldsymbol{\varepsilon}^k + \Psi_{,\vartheta}\, \vartheta^k + \Psi_{,a_k} \right) \mathrm{d}V \right. \\ &\quad \left. + \int h_{,\mathbf{u}} \cdot \mathbf{u}^k \, \mathrm{d}S_T + \int h_{,\vartheta}\, \vartheta^k \, \mathrm{d}S_q \right] \mathrm{d}t \,. \end{aligned} \tag{7.55}$$

Let us now determine the sensitivities of functionals G and F using the adjoint method. Consider the adjoint problems of elastodynamics and transient heat conduction. The elastodynamic problem in the time interval $0 \le \tau \le t_f$ is constituted by the equations of motion, boundary and initial conditions in the form

$$\begin{aligned} \sigma_{ij,i}^a + f_j^a &= \rho \frac{\partial^2 u_j^a}{\partial \tau^2} && \text{in } V \,, \\ \sigma_{ij}^a n_i = T_j^{a0}(\mathbf{x}, \tau) &= h_{,u_j}(\mathbf{x}, \tau) && \text{on } S_T \,, \\ u_j^a(\mathbf{x}, \tau) &= 0 && \text{on } S_u \,, \\ u_j^a(\mathbf{x}, 0) = 0, \quad \frac{\partial u_j^a(\mathbf{x}, 0)}{\partial \tau} &= 0 && \text{in } V \,, \end{aligned} \tag{7.56}$$

and the constitutive equation

$$\sigma_{ij}^a = S_{ij,\varepsilon_{lm}} (\varepsilon_{lm}^a - \varepsilon_{lm}^{ai}) - \sigma_{ij}^{ai} \tag{7.57}$$

where the fields of body force and initial stress are

$$f_j^a(\mathbf{x}, \tau) = \Psi_{,u_j}(\mathbf{x}, t), \qquad \sigma_{ij}^{ai}(\mathbf{x}, \tau) = \Psi_{,\varepsilon_{ij}}(\mathbf{x}, t), \tag{7.58}$$

The adjoint equations of the transient heat conduction problem takes the form

$$\begin{aligned} \lambda\vartheta^a_{,ii} + Q^a &= c\,\frac{\partial Q^a}{\partial \tau} && \text{in } V\,, \\ \vartheta^a(\mathbf{x},\tau) &= 0 && \text{on } S_\vartheta\,, \\ -\lambda\vartheta^a_{,i}n_i &= q^{a0}(\mathbf{x},\tau) = h_{,\vartheta}(\mathbf{x},\tau) && \text{on } S_q\,, \\ \vartheta^a(\mathbf{x},0) &= 0 && \text{in } V\,, \end{aligned} \tag{7.59}$$

where the heat source field is

$$Q^a(\mathbf{x},\tau) = \Psi_{,\vartheta}(\mathbf{x},\tau) - \varepsilon^{ai}_{ij}(\mathbf{x},\tau)S_{ij,\vartheta}\,. \tag{7.60}$$

The transformation rule between the times of primary and adjoint processes is

$$\tau = t_f - t\,. \tag{7.61}$$

This means that when the primary process proceeds forward, the adjoint process proceeds backward and the initial conditions for the adjoint process are specified at $t = t_f$ that is at the terminal instant of the primary process. In view of (7.61) the right hand terms of motion equation (7.56) and the heat conduction equation (7.59) are respectively replaced by $-\rho\ddot{u}^a_j$ and $-c\dot{\vartheta}$. In view of (7.12) and (7.56)–(7.60) the first variation of (7.45) can be presented as follows

$$\begin{aligned} \delta G = \bigg[\int \big(&\mathbf{f}^a\cdot\mathbf{u}_{,a_k} - \boldsymbol{\sigma}^a\cdot\boldsymbol{\varepsilon}_{,a_k} + \boldsymbol{\sigma}_{,a_k}\cdot\boldsymbol{\varepsilon}^a + Q^a\vartheta_{,a_k} - \mathbf{S}_{,a_k}\cdot\boldsymbol{\varepsilon}^a + \Psi_{,a_k}\big)\,\mathrm{d}V \\ &+ \int \mathbf{T}^{a0}\cdot\mathbf{u}_{,a_k}\,\mathrm{d}S_T - \int q^{a0}\vartheta_{,a_k}\mathrm{d}S_q\bigg]\,\delta a_k\,. \end{aligned} \tag{7.62}$$

Hence, using the reciprocity relation, the sensitivity of G can be presented in the form

$$\begin{aligned} \frac{\mathrm{d}G}{\mathrm{d}a_k} = \int \Big(&\mathbf{f}^a\cdot\mathbf{u}_{,a_k} - \mathbf{S}_{,a_k}\cdot\boldsymbol{\varepsilon}^a - \rho_{,a_k}\dot{\mathbf{u}}\cdot\mathbf{u}^a \\ &- \lambda_{,a_k}\vartheta^a_{,i}\vartheta_{,i} + \vartheta^a Q_{,a_k} - c_{,a_k}\dot{\vartheta}\vartheta^a + \Psi_{,a_k}\Big)\,\mathrm{d}V \\ &+ \frac{\mathrm{d}}{\mathrm{d}t}\int\Big[\rho\,(\dot{\mathbf{u}}^a\cdot\mathbf{u}_{,a_k} - \mathbf{u}^a\cdot\dot{\mathbf{u}}_{,a_k}) - c\vartheta^a\dot{\vartheta}_{,a_k}\Big]\,\mathrm{d}V \end{aligned} \tag{7.63}$$

where the values of Ψ and of primary state fields are calculated at the instant t, whereas the values of adjoint fields are calculated at the instant $\tau = t_f - t$.

In order to specify the sensitivity of functional F for the finite time interval $\langle 0, t_f\rangle$, Eq. (7.63) is substituted into (7.54), so the sensitivity of F equals

$$\begin{aligned} \frac{\mathrm{d}F}{\mathrm{d}a_k} = \int_0^{t_f}\bigg[\int\Big(&\mathbf{f}^a\cdot\mathbf{u}_{,a_k} - \mathbf{S}_{,a_k}\cdot\boldsymbol{\varepsilon}^a - \rho_{,a_k}\dot{\mathbf{u}}\cdot\mathbf{u}^a \\ &- \lambda_{,a_k}\vartheta^a_{,i}\vartheta_{,i} + \vartheta^a Q_{,a_k} - c_{,a_k}\dot{\vartheta}\vartheta^a + \Psi_{,a_k}\Big)\,\mathrm{d}V\bigg]\,\mathrm{d}t. \end{aligned} \tag{7.64}$$

Let us note that the last integral on the right hand side of (7.63) vanishes after time integration with proper account of initial conditions for both primary and adjoint problem equations.

7.2.2 Shape Sensitivity Analysis for Transient Thermomechanical Problems

Assume now that the body may undergo the shape transformation described in Sect. 7.1.2 and determine the sensitivities of functionals (7.45) and (7.46) with respect to shape parameters b_p. As the optimal control problems are not discussed here, it can be assumed that shape parameters and transformation velocity fields associated with these parameters are time independent.

Using the direct method, the first variation of functional (7.45) can again be expressed by (7.33) with vanishing terms $\Psi_{,\sigma}$, $h_{,\mathbf{T}}$ and $h_{,q}$. To specify the sensitivities $\mathrm{D}\mathbf{u}/\mathrm{D}b_p$, $\mathrm{D}\varepsilon/\mathrm{D}b_p$ and $\mathrm{D}\vartheta/\mathrm{D}b_p$, consider first the material derivatives of Eqs. (7.47)–(7.50) with respect to b_p and generate the sensitivity equations whose solutions provide the state sensitivities. Differentiating equations (7.47) and (7.48) with respect to b_p, we obtain P additional transient heat equations. Each p-th problem satisfies equations (7.34) and (7.35) where in (7.34) on the right hand side the term $c\dot{\vartheta}^p$ occurs. The initial conditions for thermal problems are

$$\vartheta^p(\mathbf{x},0) = \vartheta_{0,k} v_k^p \qquad \text{in } V \tag{7.65}$$

where it is assumed that $\vartheta_{0,b_p} = 0$, so the initial conditions are not dependent on the shape parameters b_p.

Similarly, differentiating Eqs. (7.5) and (7.49)–(7.50), we obtain a set of P additional elastodynamic sensitivity equations specified by (7.36) and (7.37), where on the right-hand side the term $\rho\ddot{u}^p$ occurs. The set (7.36)–(7.37) is associated with the initial conditions

$$\mathbf{u}^p(\mathbf{x},0) = \mathbf{u}_{0,k} v_k^p, \qquad \dot{\mathbf{u}}^p(\mathbf{x},0) = \mathbf{v}_{0,k} v_k^p, \qquad \text{in } V \tag{7.66}$$

where it is assumed, as previously, that the initial conditions are not affected by the variation of shape parameters b_p.

The state sensitivities of the primary system are again expressed by (7.38) and the sensitivity of functional G can be presented in the form

$$\frac{\mathrm{D}G}{\mathrm{D}b_p} = \int \left[\Psi_{,\mathbf{u}} \cdot \mathbf{u}^p + \Psi_{,\varepsilon} \cdot (\varepsilon^p - \varepsilon^{ip}) + \Psi_{,\vartheta}\, \vartheta^p + \Psi v_{k,k}^p\right] \mathrm{d}V + \int \left[h_{,\mathbf{u}} \cdot \mathbf{u}^p + h_{,\vartheta}\, \vartheta^p + h\,(\delta_{kl} - n_k n_l) v_{k,l}^p\right] \mathrm{d}S \tag{7.67}$$

and the sensitivity $\mathrm{D}F/\mathrm{D}b_p$ can be determined from (7.54) using the relation (7.67) specifying $\mathrm{D}G/\mathrm{D}b_p$. Thus, the sensitivities of the functionals G and F are obtained by solving primary problems of non-stationary heat conduction and elastodynamics and additional P state sensitivity problems.

The adjoint method in determining the sensitivities of G and F can be applied in a similar way. Introduce two additional problems of elastodynamics and heat condition defined over the time interval $0 \leq \tau \leq t_f$, where τ is defined by (7.61). The adjoint elastodynamic problem is specified by Eqs. (7.56)–(7.58) and the adjoint heat conduction problem is defined by (7.59)–(7.60). Using (7.40), the variation δG can be presented in a form similar to (7.41). The sensitivities of G can now be expressed in terms of primary and adjoint fields, thus

$$
\begin{aligned}
\frac{\mathrm{D}G}{\mathrm{D}b_p} = & \int \Big[\Psi - \boldsymbol{\sigma} \cdot \boldsymbol{\varepsilon}^a - \mathbf{f} \cdot \mathbf{u}^a - \rho \dot{\mathbf{u}} \cdot \mathbf{u}^a - \lambda \vartheta^a_{,i} \vartheta_{,i} + Q \vartheta^a - c \dot{\vartheta} \vartheta^a \\
& + (h + \mathbf{T} \cdot \mathbf{u}^a - q \vartheta^a)_{,n} - (h + \mathbf{T} \cdot \mathbf{u}^a - q \vartheta^a) H \, v^p_n \Big] \, \mathrm{d}S \\
& + \oint \langle (h + \mathbf{T} \cdot \mathbf{u}^a - q \vartheta^a) v^p_n \rangle \, \mathrm{d}\Gamma + \int \mathbf{u}^a \cdot \left(\frac{\mathrm{D}\mathbf{T}^0}{\mathrm{D}b_p} - \mathbf{T}_{,k} v^p_k \right) \mathrm{d}S_T \\
& - \int (\mathbf{T}^a - h_{,\mathbf{u}}) \cdot \left(\frac{\mathrm{D}\mathbf{u}^0}{\mathrm{D}b_p} - \mathbf{u}_{,k} v^p_k \right) \mathrm{d}S_u \\
& - \int \vartheta^a \left(\frac{\mathrm{D}q^0}{\mathrm{D}b_p} - q_{,k} v^p_k \right) \mathrm{d}S_q + \int (h_{,\vartheta} + q^a) \left(\frac{\mathrm{D}\vartheta^a}{\mathrm{D}b_p} - \vartheta_{,k} v^p_k \right) \mathrm{d}S_\vartheta \\
& + \frac{\mathrm{d}}{\mathrm{d}t} \int \left[\rho \left(\dot{\mathbf{u}}^a \cdot \mathbf{u}_{,b_p} - \mathbf{u}^a \cdot \dot{\mathbf{u}}_{,b_p} \right) - c \vartheta^a \dot{\vartheta}_{,b_p} \right] \mathrm{d}V
\end{aligned}
\tag{7.68}
$$

Similarly as previously, the primary solutions are determined at the instant t and the adjoint solutions at the instant $\tau = t_f - t$. The sensitivity $\mathrm{D}F/\mathrm{D}b_p$ can be calculated from (7.54) after substituting (7.68). The last integral on the right hand side of (7.68) vanishes in view of initial conditions of primary and adjoint problems.

7.3 Examples

Two simple examples will be presented in order to illustrate both direct and adjoint methods of determining the sensitivity of arbitrary functionals with respect to material and shape parameters.

Example 1: Uniform bar under temperature gradient

Consider a uniform bar of length l and constant cross section, rigidly supported between two end walls, Fig. 7.2. The heat flux q^0 is specified at its left end and constant temperature ϑ^0 is specified at the right end. The bar material is physically non-linear and its constitutive equation has the form

$$\varepsilon = \alpha \vartheta + t + a \vartheta^0 t^2 \tag{7.69}$$

where $t = \frac{\sigma}{E}$ denotes the normalized stress, a is the material parameter and α is the coefficient of thermal expansion. The heat conductance coefficient is

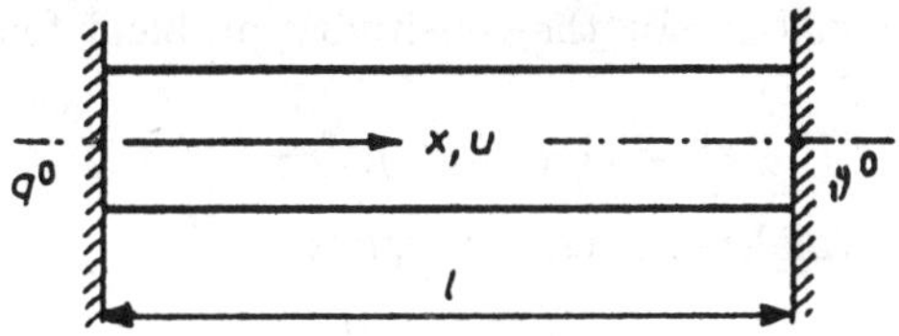

Fig. 7.2. Bar dimensions and support

denoted by λ.

Let us determine the sensitivity of mean bar displacement u_m with respect to variation of α and λ. The mean displacement is expressed as follows

$$G = u_m = \frac{1}{l_0}\int_0^l u\mathrm{d}x\,. \tag{7.70}$$

Comparing (7.70) and (7.6) it is easy to see that $\Psi = \frac{u}{l}$ and $h = 0$. The solution of the primary thermo-elastic problem

$$\begin{aligned} &\lambda\vartheta_{,xx} = 0 && \text{for } 0 \le x \le l,\\ &-\lambda\vartheta_{,x}(0) = q^0, \quad \vartheta(l) = \vartheta^0, &&\\ &t_{,x} = 0, \quad u_{,x} = \varepsilon && \text{for } 0 \le x \le l,\\ &u(0) = 0, \quad u(l) = 0, && \end{aligned} \tag{7.71}$$

provides the temperature and displacement distribution

$$\vartheta = \vartheta^0 + \frac{q^0}{\lambda}(l - x), \qquad u = \frac{q^0\alpha}{2\lambda}\,x(l - x)\,. \tag{7.72}$$

Substituting (7.72) to (7.70), after integration, the explicit dependence of the functional G on α and λ can be established. Differentiating G with respect to α and λ, the sensitivities are expressed analytically. However, to demonstrate the applicability of direct and adjoint methods, these sensitivities will now be determined by following the method described in the previous section.

In order to determine the sensitivities of ϑ and u with respect to the thermal expansion coefficient by the direct method, we shall first generate the sensitivity equations associated with (7.8)–(7.11). Equations (7.8) and (7.9) are now reduced to the form

$$\begin{aligned} &\lambda\vartheta^\alpha_{,xx} = 0 && \text{for } 0 \le x \le l,\\ &q^{\alpha 0}(0) = 0, \quad q^{\alpha 0}(l) = 0, && \end{aligned} \tag{7.73}$$

hence $\vartheta^\alpha(x) = \vartheta_{,\alpha} = 0$. The elastic sensitivity problem associated with (7.10) and (7.11) now takes the form

$$\begin{aligned} &t^\alpha_{,x} = 0, \quad u^\alpha_{,x} = \varepsilon && \text{for } 0 \le x \le l,\\ &u^{\alpha 0}(0) = 0, \quad u^{\alpha 0}(l) = 0. && \end{aligned} \tag{7.74}$$

The constitutive equation for the sensitivity problem follows from (7.69), namely

$$\varepsilon^\alpha = (1 + 2a\vartheta^0 t)(t^\alpha + t^{\alpha i}) \tag{7.75}$$

where the initial normalized stress $t^{\alpha i}$ equals

$$t^{\alpha i} = \frac{\sigma^{\alpha i}}{E} = -\frac{1}{E} S_{,\alpha} \,. \tag{7.76}$$

As the stress-strain relation is not explicitly given, $\sigma = S(\varepsilon, \sigma, \lambda)$, then in order to determine $S_{,\alpha}$ we differentiate (7.69)) and the relation $t = \frac{\sigma}{E}$ with respect to α. Noting that in view of (7.75) there is

$$S_{,\varepsilon} = \frac{E}{1 + 2a\vartheta^0 t} \tag{7.77}$$

we obtain

$$S_{,\alpha} = -\vartheta S_{,\varepsilon} = -\frac{\vartheta E}{1 + 2a\vartheta^0 t} \tag{7.78}$$

and the initial stress within the bar equals

$$t^{\alpha i} = \frac{\vartheta}{1 + 2a\vartheta^0 t} \,. \tag{7.79}$$

Solving equations (7.74)–(7.75) with account for (7.79) provides the displacement sensitivity in the form

$$u^\alpha(x) = u_{,\alpha} = \frac{q^0}{2\lambda} x(l - x) \,. \tag{7.80}$$

Thus, in view of (7.13), the sensitivity of functional (7.70) with respect to α is expressed as follows

$$\frac{\mathrm{d}G}{\mathrm{d}\alpha} = \frac{1}{2} \int_0^l u^\alpha \, \mathrm{d}x = \frac{1}{12} \frac{q^0 l^2}{\lambda} \,. \tag{7.81}$$

Similarly, the sensitivity of temperature and displacement fields with respect to heat conductance coefficient λ can be determined. The sensitivity equation associated with heat conduction is

$$\begin{aligned} &\lambda \vartheta^\lambda_{,xx} = 0 \qquad \text{for } 0 \le x \le l, \\ &q^{\lambda 0}(0) = -\frac{1}{\lambda} q^0, \qquad q^{\lambda 0}(l) = 0, \end{aligned} \tag{7.82}$$

and the temperature sensitivity field is expressed as follows

$$\vartheta^\lambda(x) = \vartheta_{,\lambda} = -\frac{q^0}{\lambda^2}(l - x) \,. \tag{7.83}$$

The sensitivity equation of elastic response takes the form

$$\begin{aligned} &t^\lambda_{,x} = 0, \qquad u^\lambda_{,x} = \varepsilon^\lambda \qquad \text{for } 0 \le x \le l, \\ &u^{\lambda 0}(0) = 0, \qquad u^{\lambda 0}(l) = 0, \\ &\varepsilon^\lambda = (1 + 2a\vartheta^0 t)(t^\lambda + t^{\lambda i}) \end{aligned} \tag{7.84}$$

where now the initial stress is specified as follows

$$t^{\lambda i} = -\frac{1}{E}\, S_{,\vartheta}\vartheta^{\lambda} = -\alpha S_{,\varepsilon}\vartheta^{\lambda} = -\frac{\alpha\vartheta^{\lambda}}{1+2a\vartheta^0 t}\,. \tag{7.85}$$

Solving (7.84), we obtain

$$u^{\lambda}(x) = u_{,\lambda} = -\frac{q^0\alpha}{2\lambda^2}\, x(l-x) \tag{7.86}$$

and in view of (7.13) there is

$$\frac{\mathrm{d}G}{\mathrm{d}\lambda} = \frac{1}{l}\int_0^l u^{\lambda}\,\mathrm{d}x = -\frac{q^0\alpha}{12\lambda^2}\, l^2. \tag{7.87}$$

Now, we shall determine the sensitivity of G by applying the adjoint state method. In view of (7.14)–(7.16) the adjoint problem is specified by the equations

$$\begin{aligned} &t^a_{,x} + \frac{1}{El} = 0, \qquad u^a_{,x} = \varepsilon^a \qquad \text{for } 0 \le x \le l,\\ &u^{a0}(0) = 0, \qquad u^{a0}(l) = 0,\\ &\varepsilon^a = (1+2a\vartheta^0 t)t^a \end{aligned} \tag{7.88}$$

so the adjoint strain field is expressed as follows

$$\varepsilon^a = \frac{1+2a\vartheta^0 t}{E}\left(\frac{1}{2} - \frac{x}{l}\right) \tag{7.89}$$

The adjoint problem of heat condition associated with (7.18)–(7.20) now has the form

$$\begin{aligned} &\lambda\vartheta^a_{,xx} + Q^a = 0 \qquad \text{for } 0 \le x \le l,\\ &q^{a0}(0) = 0, \qquad q^{a0}(l) = 0, \end{aligned} \tag{7.90}$$

where the adjoint heat source field is

$$Q^a = -\varepsilon^a S_{,\vartheta} = x\left(\frac{1}{2} - \frac{x}{l}\right). \tag{7.91}$$

Solving (7.90), the adjoint temperature distribution is

$$\vartheta^a = \frac{\alpha}{\lambda}\left(\frac{l^2}{12} - \frac{x^2}{4} + \frac{x^3}{6l}\right). \tag{7.92}$$

The first variation of the functional (7.70) generally specified by (7.22), can now be expressed in the form

$$\delta G = \int_0^l -\varepsilon^a S_{,\alpha}\,\mathrm{d}x\,\delta\alpha + \int_0^l -\vartheta^a_{,x}\vartheta_{,x}\,\mathrm{d}x\,\delta\lambda\,. \tag{7.93}$$

Substituting (7.71), (7.78), (7.84) and (7.92) into equation (7.93) and integrating, the sensitivities $\mathrm{d}G/\mathrm{d}\alpha$ and $\mathrm{d}G/\mathrm{d}\lambda$ are obtained in forms identical to those derived by the direct method.

Example 2: Axisymmetric disc non-uniformly heated

Consider now the axisymmetric disc of internal and external radii b_1 and b_2, non-uniformly heated, Fig. 7.3. The disc material is isotropic and linearly elastic with constant material parameters E, ν and α, λ. The temperature and stress conditions on disc boundaries are

$$\vartheta(b_1) = \vartheta_1\,, \qquad \vartheta(b_2) = \vartheta_2\,, \qquad \sigma_r(b_1) = 0, \qquad \sigma_r(b_2) = 0. \tag{7.94}$$

The heat conditions and equilibrium equations together with strain–displacement relations take the form

$$\begin{gathered} \vartheta_{,rr} + \frac{1}{r}\vartheta_{,r} = 0, \qquad \sigma_{r,r} + \frac{\sigma_r - \sigma_t}{r} = 0, \qquad \text{for } b_1 \le r \le b_2\,, \\ \varepsilon_r = u_{,r}\,, \qquad \varepsilon_t = \frac{u}{r}\,, \end{gathered} \tag{7.95}$$

where $u(r)$ is the radial displacement field. The solutions of (7.95) with account for (7.94) provide temperature, displacement, stress and strain distributions.

Consider now the complementary disc energy expressed as follows

$$G = \Pi_\sigma = \int_{b_1}^{b_2} W(\sigma_r, \sigma_t, \vartheta) r \, \mathrm{d}r \tag{7.96}$$

where the specific stress energy equals

$$W = \frac{1}{2E}(\sigma_r^2 - 2\nu\sigma_r\sigma_t + \sigma_t^2) + \alpha\vartheta(\sigma_r + \sigma_t). \tag{7.97}$$

Assume now the radii b_1 and b_2 to be the design variables that may undergo variation. We shall specify the sensitivity of $G = \Pi_\sigma$ associated with this variation. The transformation of disc domain can be described by the transformation velocity fields v^1 and v^2 associated with the variables b_1 and b_2,

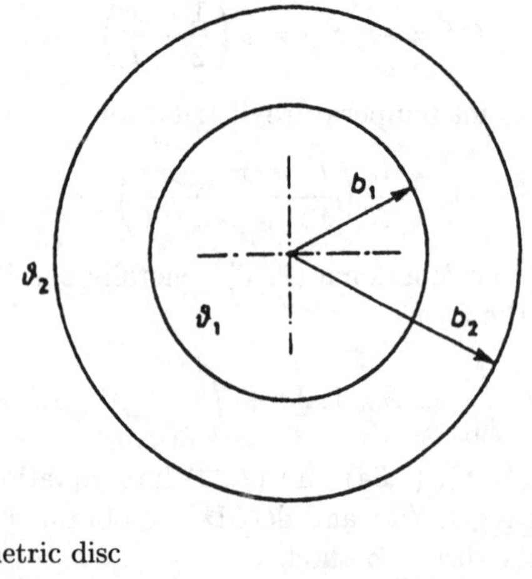

Fig. 7.3. Axisymmetric disc

so that

$$v^1 = \frac{b_2 - r}{b_2 - b_1}\,, \qquad v^2 = \frac{r - b_1}{b_2 - b_1}\,. \tag{7.98}$$

Comparing equations (7.96) and (7.6) it is easy to see that $\Psi = W$ within the disc domain and $h = 0$ on its boundaries. Using the direct method in determining the sensitivity of Π_σ , in view of (7.39) and (7.98) we have

$$\delta\Pi_\sigma = \left\{ \int \left[\varepsilon_r \sigma_r^p + \varepsilon_t \sigma_t^p + \alpha(\sigma_r + \sigma_t)\vartheta^p + W\left(v^p_{,r} + \frac{v^p}{r} \right) \right] r\, \mathrm{d}r \right\} \delta b_p \tag{7.99}$$

where ϑ^p, σ_r^p and σ_t^p are temperature and stress sensitivities with respect to b_p. The sensitivity ϑ^p can be obtained as a solution of the following heat conduction problem

$$\begin{aligned} &\vartheta^p_{,rr} + \frac{1}{r}\vartheta^p_{,r} - \vartheta_{,r}\frac{v^p}{r^2} = 0 \qquad && \text{for } b_1 \le r \le b_2\,, \\ &\vartheta^p(b_1) = 0, \qquad \vartheta^p(b_2) = 0, && p = 1, 2. \end{aligned} \tag{7.100}$$

The sensitivity equilibrium equation and boundary conditions providing the stress sensitivities σ_r^p and σ_t^p take the form

$$\begin{aligned} &\sigma^p_{r,r} + \frac{\sigma_r^p - \sigma_t^p}{r} + \sigma_{r,r} v^p_{,r} + (\sigma_r - \sigma_t)\frac{v^p}{r^2} = 0, \qquad \text{for } b_1 \le r \le b_2\,, \\ &\sigma_r^p(b_1) = 0, \qquad \sigma_r^p(b_2) = 0. \end{aligned} \tag{7.101}$$

The constitutive equation (7.5) now takes the explicit form

$$\begin{aligned} \sigma_r^p &= \frac{E}{1-\nu^2}\left[\varepsilon_r^p + \nu\varepsilon_t^p - \varepsilon_r v^p_{,r} - \nu\varepsilon_t \frac{v^p}{r^2} - (1+\nu)\alpha\vartheta^p \right], \\ \sigma_t^p &= \frac{E}{1-\nu^2}\left[\varepsilon_t^p + \nu\varepsilon_r^p - \nu\varepsilon_r v^p_{,r} - \varepsilon_t \frac{v^p}{r^2} - (1+\nu)\alpha\vartheta^p \right]. \end{aligned} \tag{7.102}$$

Solving equations (7.94)–(7.95) and (7.100)–(7.101), next substituting the results into (7.99), after integration the following sensitivity of Π_σ is obtained

$$\delta\Pi_\sigma = \frac{\mathrm{D}\Pi_\sigma}{\mathrm{D}b_1}\delta b_1 + \frac{\mathrm{D}\Pi_\sigma}{\mathrm{D}b_2}\delta b_2 \tag{7.103}$$

where the sensitivities of Π_σ with respect to b_1 and b_2 are

$$\begin{aligned} \frac{\mathrm{D}\Pi_\sigma}{\mathrm{D}b_1} &= \frac{1}{2}E\alpha^2(\vartheta_1 - \vartheta_2)^2 b_1 \left[\left(\frac{1}{2\ln(\frac{b_2}{b_1})} - \frac{b_2^2}{b_2^2 - b_1^2} \right) \right. \\ &\quad \left. + \frac{1}{2\ln(\frac{b_2}{b_1})}\left(\frac{b_2^2}{b_2^2 - b_1^2} - \frac{b_2^2 - b_1^2}{4b_1^2[\ln(\frac{b_2}{b_1})]^2} \right) \right], \\ \frac{\mathrm{D}\Pi_\sigma}{\mathrm{D}b_2} &= \frac{1}{2}E\alpha^2(\vartheta_1 - \vartheta_2)^2 b_2 \left[\left(\frac{1}{2\ln(\frac{b_2}{b_1})} - \frac{b_1^2}{b_2^2 - b_1^2} \right) \right. \\ &\quad \left. + \frac{1}{2\ln(\frac{b_2}{b_1})}\left(\frac{b_1^2}{b_2^2 - b_1^2} - \frac{b_2^2 - b_1^2}{4b_2^2[\ln(\frac{b_2}{b_1})]^2} \right) \right]. \end{aligned} \tag{7.104}$$

The sensitivity of Π_σ can also be obtained by using the adjoint system approach. As the functional considered represents the complementary energy, then in view of (7.16), the adjoint disc is subjected only to the initial strain field $\boldsymbol{\varepsilon}^{ai} = W_{,\boldsymbol{\sigma}} = \boldsymbol{\varepsilon}$ and the adjoint states are $\boldsymbol{\sigma}^a = \mathbf{0}$, $\boldsymbol{\varepsilon}^a = \boldsymbol{\varepsilon}$ and $\mathbf{u}^a = \mathbf{u}$. Further, the adjoint heat source field Q^a is, in view of (7.19), expressed as follows

$$Q^a = W_{,\vartheta} - (\boldsymbol{\varepsilon}^a - W_{,\boldsymbol{\sigma}}) \cdot \mathbf{S}_{,\vartheta} = \alpha(\sigma_r + \sigma_t) \tag{7.105}$$

with vanishing adjoint temperature at both boundaries. Using (7.44), the first variation of Π_σ calculated by the adjoint method is

$$\begin{aligned} \delta\Pi_\sigma = {} & b_1 \left[\frac{1}{2E}\sigma_t^2(b_1) - \lambda\vartheta_{,r}(b_1)\vartheta^a_{,r}(b_1) \right] \delta b_1 \\ & + b_2 \left[-\frac{1}{2E}\sigma_t^2(b_2) - \lambda\vartheta_{,r}(b_2)\vartheta^a_{,r}(b_2) \right] \delta b_2 \, . \end{aligned} \tag{7.106}$$

Solving the primary and adjoint problems and substituting the state fields into (7.106), the sensitivity of Π_σ with respect to shape variables b_1 and b_2 is obtained. The final result is identical to that obtained by the direct method, cf. (7.104).

8 Numerical Aspects of Sensitivity Analysis

Numerical analysis of sensitivity can be carried out using either direct or adjoint state approaches. In the first approach the discrete structure model is usually used to describe the structure response under specified loading or deformation history. The sensitivity analysis is next performed for the discrete model by differentiating the system equations with respect to design variables. This approach will be called *discretized* or *algebraic*. The other approach, called *analytical*, is based on analytically derived sensitivity expressions of an arbitrary functional using continuum formulations and variational framework. The respective sensitivities were presented in Chaps. 4 and 5 as the integrals over body domain or its boundary and expressed in terms of primary and adjoint states. The discretized forms are next generated in the final stage of calculations using the discrete model developed for structure analysis. The analytical approach provides exact sensitivity expressions and can be coupled with any method of analysis such as finite element or boundary equation methods. In the following, the finite element discretization will be considered and the sensitivity analysis will be carried out numerically using this technique.

To simplify our discussion, assume the response functional in the form

$$G = \int_{V(b_l)} \Psi(\mathbf{u}, \boldsymbol{\varepsilon}, a_k) \, \mathrm{d}V \tag{8.1}$$

and its sensitivity will be considered. Here a_k denotes the set of material and b_l the set of shape parameters, specifying the shape of free boundary S_0. It is assumed further that loaded and supported boundary portions S_T and S_u are fixed. The material is assumed to behave linearly and satisfy the constitutive equation

$$\boldsymbol{\sigma} = \mathbf{S}(a_k) \cdot \boldsymbol{\varepsilon} \,. \tag{8.2}$$

The exact expression of the sensitivity of (8.1) can be written as follows

$$\delta G = \delta_a G + \delta_b G = \frac{\mathrm{d}G}{\mathrm{d}a_k} \, \delta a_k + \frac{\mathrm{d}G}{\mathrm{d}b_l} \, \delta b_l \tag{8.3}$$

where the material parameter sensitivity $\mathrm{d}G/\mathrm{d}a_k$ and the shape sensitivity $\mathrm{d}G/\mathrm{d}b_l$ are expressed by the formulae (see Chaps. 4 and 5)

$$\frac{\mathrm{d}G}{\mathrm{d}a_k} = \int_V (\Psi_{,a_k} + \mathbf{f}_{,a_k} \cdot \mathbf{u}^a - \boldsymbol{\varepsilon}^a \cdot \mathbf{S}_{,a_k} \cdot \boldsymbol{\varepsilon}) \, \mathrm{d}V, \qquad k = 1, 2, \ldots, K \tag{8.4}$$

and

$$\frac{dG}{db_l} = \int_{S_0} (\Psi + \mathbf{f} \cdot \mathbf{u}^a - \boldsymbol{\sigma} \cdot \boldsymbol{\varepsilon}^a) \frac{\partial \phi_n}{\partial b_l} \, dS_0 \,, \qquad l = 1, 2, \ldots, L. \tag{8.5}$$

The adjoint boundary value problem is specified for a structure subjected to body forces $\mathbf{f}^a$ and initial stress $\boldsymbol{\sigma}^{ai}$, so that

$$\mathbf{f}^a = \Psi_{,\mathbf{u}} \,, \qquad \boldsymbol{\sigma}^{ai} = \Psi_{,\boldsymbol{\varepsilon}} \tag{8.6}$$

with the homogeneous boundary conditions $\mathbf{u}^{a0} = \mathbf{0}$ on S_u and $\mathbf{T}^{a0} = \mathbf{0}$ on S_T.

Assume the structure to be discretized by a set of E finite elements connected at W nodes. Denoting by $\mathbf{v}^e$ the vector of nodal displacements of the element e, the displacement, strain and stress within the element are

$$\mathbf{u} = \mathbf{N}(b_l)\mathbf{v}^e, \qquad \boldsymbol{\varepsilon} = \mathbf{B}(b_l)\mathbf{v}^e, \qquad \boldsymbol{\sigma} = \mathbf{S}(a_k)\mathbf{B}(b_l)\mathbf{v}^e - \mathbf{S} \cdot \boldsymbol{\varepsilon}^i - \boldsymbol{\sigma}^i, \tag{8.7}$$

where $\mathbf{N}$ is the interpolation matrix depending on position and nodal coordinates, $\mathbf{B}$ specifies the relation between the local strain and the nodal displacement vector. It is evident that the matrices $\mathbf{N}$ and $\mathbf{B}$ depend on the set of shape parameters b_l and the matrix $\mathbf{S}$ depends on the set of material parameters a_k. The stiffness matrix for the element e is expressed as follows

$$\mathbf{k}^e = \int_{V^e(b_l)} \mathbf{B}^T(b_l)\mathbf{S}(a_k)\mathbf{B}(b_l) \, dV^e \tag{8.8}$$

and the vectors of nodal forces associated with body forces, initial strains and stresses inside the element are respectively

$$\begin{aligned} \mathbf{F}^e_f &= \int_{V^e(b_l)} \mathbf{N}^T(b_l)\mathbf{f} \, dV^e, \\ \mathbf{F}^e_\varepsilon &= \int_{V^e(b_l)} \mathbf{B}^T(b_l)\mathbf{S}(a_k)\boldsymbol{\varepsilon}^i \, dV^e, \\ \mathbf{F}^e_\sigma &= \int_{V^e(b_l)} \mathbf{B}^T(b_l)\boldsymbol{\sigma}^i \, dV^e. \end{aligned} \tag{8.9}$$

The surface tractions acting on the boundary S_T are replaced by the concentrated forces acting at the boundary nodes of finite elements according to the relation

$$\mathbf{F}^e_T = \int_{S^e(b_l)} \mathbf{N}^T(b_l)\mathbf{T}^0 \, dS^e, \tag{8.10}$$

The global stiffness matrix $\mathbf{K}$ and the generalized nodal force vector $\mathbf{F}$ obtained as a sum of contributions of all elements, constitute the set of algebraic equations describing the structure response

$$\mathbf{K}(a_k, b_l)\mathbf{v} = \mathbf{F} \tag{8.11}$$

where $\mathbf{v}$ denotes the generalized nodal displacement vector specifying the displacements of all nodes.

The solution of (8.11) provides the node displacement vector $\mathbf{v}$ and the state fields $\mathbf{u}$, $\boldsymbol{\varepsilon}$, $\boldsymbol{\sigma}$ specified by (8.7). The expressions (8.1), (8.4) and (8.5) defining the functional G and its sensitivities are also discretized by the finite element discretization. Thus, the functional (8.1) now takes the form

$$G = \int \Psi \, \mathrm{d}V = \sum_{e=1}^{E} \int_{V^e(b_l)} \Psi[\mathbf{N}(b_l), \mathbf{B}(b_l), \mathbf{v}^e, a_k] \, \mathrm{d}V^e \tag{8.12}$$

so its value is calculated for given a_k and b_l as a sum of integrals for all elements of the discrete model. In order to determine the sensitivity of (8.12) expressed by (8.4) and (8.5), besides the primary state fields $\mathbf{u}$, $\boldsymbol{\varepsilon}$, $\boldsymbol{\sigma}$, also the adjoint fields $\mathbf{u}^a$, $\boldsymbol{\varepsilon}^a$, $\boldsymbol{\sigma}^a$ are to be calculated. Assuming the same discretization for the adjoint structure, body forces and initial stress in the adjoint structure can be represented by the equivalent nodal forces, thus

$$\begin{aligned}
\mathbf{F}_f^{ae} &= \int \mathbf{N}^T \mathbf{f}^a \, \mathrm{d}V^e = \int \mathbf{N}^T \Psi_{,\mathbf{u}} \, \mathrm{d}V^e \\
&= \int_{V^e(b_l)} \mathbf{N}^T(b_l) \boldsymbol{\Psi}_f[\mathbf{N}(b_l), \mathbf{B}(b_l), \mathbf{v}^e, a_k] \, \mathrm{d}V^e, \\
\mathbf{F}_\sigma^{ae} &= \int \mathbf{N}^T \boldsymbol{\sigma}^{ai} \, \mathrm{d}V^e = \int \mathbf{N}^T \Psi_{,\boldsymbol{\varepsilon}} \, \mathrm{d}V^e \\
&= \int_{V^e(b_l)} \mathbf{B}^T(b_l) \boldsymbol{\Psi}_\sigma[\mathbf{N}(b_l), \mathbf{B}(b_l), \mathbf{v}^e, a_k] \, \mathrm{d}V^e,
\end{aligned} \tag{8.13}$$

where $\boldsymbol{\Psi}_f = \Psi_{,\mathbf{u}}$ and $\boldsymbol{\Psi}_\sigma = \Psi_{,\boldsymbol{\varepsilon}}$. Thus, knowing the solution for a discrete model of the primary structure we are able to specify nodal forces of the adjoint structure. As the discretization of both structures is the same, the element stiffness matrices are the same and defined by (8.8). Performing summation for all elements of the adjoint structure, the system of equations is obtained in the form

$$\mathbf{K}(a_k, b_l)\mathbf{v}^a = \mathbf{F}^a \tag{8.14}$$

where $\mathbf{v}^a$ denotes the global displacement vector of the adjoint structure and $\mathbf{F}^a$ is the generalized nodal force vector. Let us note that the solution of the adjoint system (8.14) is much easier since the matrix $\mathbf{K}$ was already generated and transformed in solving the primary structure equations (8.11). Knowing the nodal displacement vector $\mathbf{v}^a$, the displacement and strain fields at any paint of the element e are

$$\mathbf{u}^a = \mathbf{N}(b_l)\mathbf{v}^{ae}, \qquad \boldsymbol{\varepsilon}^a = \mathbf{B}(b_l)\mathbf{v}^{ae}. \tag{8.15}$$

Now, using the relations (8.7), (8.8), (8.9) and (8.15) in (8.4), the discretized form of sensitivity of functional (8.12) with respect to material parameters

a_k can be written in the form

$$\frac{\mathrm{d}G}{\mathrm{d}a_k} = \sum_{e=1}^{E} \left[\int_{V^e} \Psi_{,a_k} \,\mathrm{d}V^e + \overset{T}{\mathbf{v}}{}^{ae} \mathbf{F}^e_{f,a_k} - \overset{T}{\mathbf{v}}{}^{ae} \mathbf{k}^e_{,a_k} \mathbf{v}^e \right], \qquad k = 1, 2, \ldots, K, \tag{8.16}$$

where the derivatives of element nodal forces associated with body forces **F** and of stiffness matrices $\mathbf{k}^e$ are

$$\mathbf{F}^e_{f,a_k} = \int \mathbf{N}^T \mathbf{f}_{,a_k} \,\mathrm{d}V^e, \qquad \mathbf{k}^e_{,a_k} = \int \mathbf{B}^T \mathbf{S}_{,a_k} \mathbf{B} \,\mathrm{d}V^e. \tag{8.17}$$

We shall now derive the discrete form of sensitivity of functional G with respect to variation of the free boundary S_0, dependent on the set of shape parameters b_l. Assume the boundary S_0 to be represented by a set P of boundaries of finite elements. The transformation function modifying the element boundary p belonging to S_0 is given in a form

$$\varphi_n = \varphi^p_l(\alpha, \beta) b_l, \qquad \frac{\partial \varphi_n}{\partial b_l} = \varphi^p_l(\alpha, \beta), \tag{8.18}$$

where $\varphi^p_l(\alpha, \beta)$ ($p = 1, 2, \ldots, P$, $l = 1, 2, \ldots, L$) is a set of given shape functions depending on the parameters α, β specified on the boundary surface. Using (8.7)–(8.9) and (8.18) in (8.5), the discretized form of sensitivity of G with respect to variation of the boundary S_0 is expressed in the form

$$\frac{\mathrm{d}G}{\mathrm{d}b_l} = \sum_{p=1}^{P} \left\{ \int_{S^p_0} \Psi \,\mathrm{d}S^p_0 + \overset{T}{\mathbf{v}}{}^{ap} \left[\int_{S^p_0} (\mathbf{N}^T \mathbf{f} - \mathbf{B}^T \mathbf{S} \mathbf{B}) \varphi^p_l \,\mathrm{d}S^p_0 \right] \mathbf{v}^p \right\}, \qquad l = 1, 2, \ldots, L. \tag{8.19}$$

This expression specifies the sensitivity in the form of sum of boundary integrals for all elements whose surface constitute the free boundary S_0.

The calculation of surface integral (8.19) may pose some problems of accuracy as the calculation of stresses on element boundaries may contain a considerable approximation error. To improve accuracy, the iterative method of Loubignac [8] can be applied. This method requires the iterative correction of nodal displacements $\mathbf{v}$ and $\mathbf{v}^a$ in order to satisfy equilibrium conditions of nodal forces obtained from calculated stress $\boldsymbol{\sigma}$ and $\boldsymbol{\sigma}^a$ and applied tractions. This method can improve accuracy of calculation of boundary stresses at the expense of several iterative solutions for both primary and adjoint structures.

The second method of improving the accuracy of sensitivity calculation is to replace the surface integral (8.19) by a volume integral. Referring to (8.5), and noting that $\varphi_n = \varphi_i n_i$, the application of Gauss theorem provides the equivalent expression to (8.5) in a form of volume integral

$$\frac{\mathrm{d}G}{\mathrm{d}b_l} = \int_V \left[(\Psi_{,\mathbf{u}} \cdot \mathbf{u}_{,i} + \Psi_{,\boldsymbol{\varepsilon}} \cdot \boldsymbol{\varepsilon}_{,i} + \mathbf{f}_{,i} \cdot \mathbf{u}^a + \mathbf{f} \cdot \mathbf{u}^a_{,i} - \boldsymbol{\sigma}_{,i} \cdot \boldsymbol{\varepsilon}^a - \boldsymbol{\sigma}^a \cdot \boldsymbol{\varepsilon}_{,i}) \frac{\partial \phi_i}{\partial b_l} + (\Psi + \mathbf{f} \cdot \mathbf{u}^a - \boldsymbol{\sigma} \cdot \boldsymbol{\varepsilon}^a) \frac{\partial \phi_{i,i}}{\partial b_l} \right] \mathrm{d}V. \tag{8.20}$$

It is now assumed that the transformation functions $\varphi_i(x_j, b_l)$, $i, j = 1, 2, 3$, are specified over the whole structure domain V. Using (8.7)–(8.9) and (8.15) in (8.20), the discrete form of the sensitivity of G expressed as volume integral (8.20) can be obtained. Assuming, as previously, that the transformation function modifying shape of the element e can be presented in the form

$$\varphi_i(x_j, b_l) = \varphi^e_{il}(x_j) b_l\,, \qquad \frac{\partial \varphi_i}{\partial b_l} = \varphi^e_{il}(x_j)\,, \tag{8.21}$$

the discrete form of the sensitivity expression (8.20) is

$$\frac{\mathrm{d}G}{\mathrm{d}b_l} = \sum_{e=1}^{E} \left[\int_{V^e} \Psi \varphi^e_{il,i}\, \mathrm{d}V^e + \overset{T}{\mathbf{v}}{}^{e} \mathbf{H} + \overset{T}{\mathbf{v}}{}^{ae} \mathbf{H}^a - \overset{T}{\mathbf{v}}{}^{e} \mathbf{M} \mathbf{v}^{ae} \right] \tag{8.22}$$

where the matrices $\mathbf{H}$, $\mathbf{H}^a$ and $\mathbf{M}$ are defined as follows

$$\begin{aligned} \mathbf{H} &= \int_{V^e} (\mathbf{N}^T_{,i} \boldsymbol{\Psi}_f + \mathbf{B}^T_{,i} \boldsymbol{\Psi}_\sigma) \varphi^e_{il}\, \mathrm{d}V^e, \\ \mathbf{H}^a &= \int_{V^e} \left[(\mathbf{N}^T \mathbf{f}_{,i} + \mathbf{N}^T_{,i} \mathbf{f}) \varphi^e_{il} + \mathbf{N}^T \mathbf{f} \varphi^e_{il,i} \right] \mathrm{d}V^e, \\ \mathbf{M} &= \int_{V^e} \left[(\mathbf{B}^T_{,i} \mathbf{D} \mathbf{B} + \mathbf{B}^T \mathbf{D} \mathbf{B}_{,i}) \varphi^e_{il} + \mathbf{B}^T \mathbf{D} \mathbf{B} \varphi^e_{il,i} \right] \mathrm{d}V^e, \end{aligned} \tag{8.23}$$

The application of (8.22) requires integration over all elements of a discretized structure. On the other hand, the use of (8.19) requires the integration over element surfaces constituting the free boundary S_0. However, the accuracy of volume integration is higher as the stress and strains are evaluated at Gauss points of all elements.

References

1. Allaire G, Kohn RV (1993) Optimal design for minimum weight and compliance in plane stress using extremal microstructures. Eur J Mech A/Solids 12:839–878
2. Banichuk NV (1983) Problems and methods of optimal structures design. Plenum Press, New York
3. Barthelemy B, Chan CT, Haftka RT (1988) Sensitivity approximation of static structures response. Finite Elements in Analysis and Design 4:249–265
4. Bendsoe M, Kikuchi N (1988) Generating optimal topologies in structural design using a homogenization method. Comp Meth Mech Eng 71:197–224
5. Berthold FJ, Stein E (1994) Optimierung von Strukturen aus isotropen hyperelastischen Materialen bei grossen Deformation. IBNM - Brecht 94/1. Univ Hannover, Rep Inst Baumechanik und Num Mech
6. Cardoso JB, Arora JS (1988) Variational method for design sensitivity analysis in non-linear structural mechanics. AIAA J 26:595–603
7. Cheng G, Lin Y (1987) A new computational scheme for sensitivity analysis. Eng Opt 12:219–235
8. Cook RD (1982) Loubignac's iterative method in finite element elastostatics. Int J Num Meth Eng 18:67–75
9. Dems K, Mróz Z (1987) Variational approach to sensitivity analysis in thermoelasticity. J Thermal Stresses 10:283–306
10. Dems K, Haftka RT (1989) Two approaches to sensitivity analysis for shape variation of structures. Mech Struct Machines 16:501–522
11. Dems K, Mróz Z (1983) Variational Approach by means of adjoint systems, Part I: Variation of material parameters within fixed domain. Int J Solids Struct 19:677–692
12. Dems K, Mróz Z. Variational approach by means of adjoint systems, Part II: Structure shape variation. Int J Solids Struct 20:(1984) 527–552
13. Dems K, Mróz Z (1991) Shape sensitivity analysis and optimal design of disks and plates with strong discontinuities of kinematic fields. Int J Solids Struct 29:437–463
14. Dems K, Mróz Z (1993) On the shape sensitivity approaches in numerical analysis of structures. Struct Optim 6:86–93
15. Dems K, Mróz Z, Szeląg D (1989) Optimal design of rib-stiffeners in disks and plates. Int J Solids Struct 25(9):973–998
16. Ding Y (1986) Shape optimization of structures: a literature survey. Comp Struct 24:985–1004
17. Eshelby JD (1956) The Continuum theory of lattice defects. In: Seitz F et al. (eds) Solid state Physics, 3, 76. Academic Press

18. Garstecki A, Mróz Z (1987) Optimal design of elastic structures subjected to loads and initial distortions. J Struct Mech 15:47–68
19. Gunther W (1962) Über reinige Randintegral der Elastomechanik. Abh Braunschweig Wiss Geselschaft 14:54–63
20. Haber RB (1987) Application of the Eulerian–Lagrangian optimization. In: Mota-Soares CA (ed) Computer aided optimal design: Structural and mechanical systems. Springer Verlag, pp 573–587
21. Haftka RT, Grandhi RV (1986) Structural shape optimization — a survey. Comp Meth Appl Eng 57:91–106
22. Haftka RT, Gurdal Z, Kamat MP (1990) Elements of structural optimization. Kluwer Acad Publishers
23. Haug HJ, Choi KK, Komkov V (1986) Design sensitivity analysis of structural systems. Academic Press
24. Hsieh CC, Arora JS (1984) Structural design sensitivity analysis with general boundary conditions: static problem. Int J Num Meth In Eng 20:1661–1670
25. Jog CS, Haber RB, Bendsoe MP (1994) Topology design with optimized, self-adaptive materials. Int J Num Meth Eng 37:1323–1350
26. Kamat MP (1993) Structural optimization – status and promise. AIAA Progress in Aeronautic and Astronomics, Series 150. Am Inst Aeron Astron, Washington, DC
27. Kleiber M, Hien TD (1992) The stochastic finite element method. Wiley & Sons
28. Knowles JK, Sternberg E (1972) On a class of conservation laws in linearized finite elastostatics. Arch Rat Mech Anal 44:187–210
29. Lee EH (1969) Elastic-plastic deformations at finite strains. J Appl Mech 36:1–6
30. Moore R (1979) Methods and applications of Interval analysis SIAM Publ, Philadelphia
31. Mortenson M (1985) Geometric modelling. Wiley & Sons
32. Mróz Z (1986) Variational approach to shape sensitivity analysis and optimal design. In: Bennet JA, Botkin ME (eds) Optimium shape. Plenum Press, pp 79–111
33. Mróz Z (1987) Sensitivity analysis and optimal design with account for varying shape and support conditions. In: Mota-Soares CA (ed) Computer aided optimal design: structural and mechanical systems, Ser F. Springer Verlag, pp 407–438
34. Mróz Z, Dems K (1993) Discrete and continuous reinforcement of materials and structures. In: Pedersen P (ed) Proc IUTAM Symp Optimal Design with Advanced Materials. Elsevier Sc Publ, pp 383–405
35. Mróz Z, Haftka RT (1994) Design sensitivity of nonlinear structures in regular and critical states. Int J Solids Struct 31:2071–2098
36. Mróz Z, Lekszycki T (1982) Optimal support reaction in elastic frame structures. Comp Struct 14:179–185
37. Mróz Z, Kamat MP, Plaut RH (1985) Sensitivity analysis and optimal design of non-linear beams and plates. J Struct Mech 13:245–266
38. Mróz Z, Piekarski J (1994) First and second order design sensitivity at bifurcation point. In: Herskovits J (ed) Advances in structural optimization. Kluwer Acad Publ

39. Mróz Z, Rozvany GI (1975) Optimal design of structures with variable support conditions. J Opt Theory Appl 15:85–101

40. Neumaier A (1990) Interval methods for systems of Equations. Cambrige University Press

41. Pedersen P, Cheng G, Rasmussen J (1989) On accuracy problems for semi-analytical sensitivity analysis. Mech Struct Machines 6:113–128

42. Petryk H, Mróz Z (1986) Time derivatives of integrals and functionals defined on varying volume and surface domains. Arch Mech 38:697–724

43. Phelan DG, Vidal C, Haber RD (1991) An adjoint variable method for sensitivity analysis of non linear elastic systems. Int J Num Methods in Engineering 31:1649–1667

44. Rozvany GIN (1989) Structural design via optimality criteria. Kluwer Acad Publ

45. Sokolowski J, Zolesio JP (1991) Introduction to shape optimization. Springer Verlag

46. Sutter TR, Camarda CJ (1988) Comparison of several methods for the calculation of vibration mode-shape derivatives. AIAA Journal 26:1506–1511

47. Szeląg D, Mróz Z (1978) Optimal design of elastic beams with unspecified support conditions. Z Angew Math Mech 58:501–510

48. Szefer G, Mróz Z, Demkowicz L (1987) Variational approach to sensitivity analysis in non-linear elasticity. Arch Mech 39:247–259

49. Washizu K (1975) Variational Methods in Elasticity and Plasticity. Pergamon Press

50. Zienkiewicz OC, Campbell JS (1973) Shape optimization and sequential linear programming. In: Optimum struct design, theory and applications. Wiley & Sons, pp 109–126

Index

Printing: Mercedesdruck, Berlin
Binding: Buchbinderei Lüderitz & Bauer, Berlin